VOLUME FIVE

STRATIGRAPHY & TIMESCALES

Carbon Isotope Stratigraphy

VOLUME FIVE

STRATIGRAPHY & TIMESCALES

Carbon Isotope Stratigraphy

Edited by

MICHAEL MONTENARI

Department of Earth Sciences and Geography
Keele University
Newcastle
United Kingdom

ELSEVIER

ACADEMIC PRESS
An imprint of Elsevier

Academic Press is an imprint of Elsevier
50 Hampshire Street, 5th Floor, Cambridge, MA 02139, United States
525 B Street, Suite 1650, San Diego, CA 92101, United States
The Boulevard, Langford Lane, Kidlington, Oxford OX5 1GB, United Kingdom
125 London Wall, London, EC2Y 5AS, United Kingdom

First edition 2020

ISBN: 978-0-12-820991-2
ISSN: 2468-5178

For information on all Academic Press publications
visit our website at https://www.elsevier.com/books-and-journals

Publisher: Zoe Kruze
Acquisitions Editor: Jason Mitchell
Editorial Project Manager: Chris Hockaday
Production Project Manager: Denny Mansingh
Cover Designer: Alan Studholme

Typeset by SPi Global, India

Contents

Contributors

Mònica Aguilera
Joint Research Unit CTFC—AGROTECNIO; Department of Crop and Forest Sciences, University of Lleida, Lleida, Spain

Per Ahlberg
Department of Geology, Lund University, Lund, Sweden

Moujahed Al-Husseini
Gulf PetroLink-GeoArabia, Manama, Bahrain

J. Javier Álvaro
Instituto de Geociencias (CSIC-UCM), Madrid, Spain

José Luis Araus
Integrative Crop Ecophysiology Group, Plant Physiology Section, Faculty of Biology, University of Barcelona, Barcelona; AGROTECNIO (Center for Research in Agrotechnology), Lleida, Spain

Pedro Barba
Departamento de Geología, Facultad de Ciencias, Universidad de Salamanca, Salamanca, Spain

José Antônio Barbosa
NEG–LABISE, Department of Geology, Federal University of Pernambuco, Recife, Brazil

Stig M. Bergström
School of Earth Sciences, The Ohio State University, Columbus, OH, United States

Jiubin Chen
Institute of Surface-Earth System Science, Tianjin University, Tianjin, China

Luis Drude de Lacerda
LABOMAR, Institute of Marine Sciences, Federal University of Ceará, Fortaleza, Brazil

Mats E. Eriksson
Department of Geology, Lund University, Lund, Sweden

Valderez Pinto Ferreira
NEG–LABISE, Department of Geology, Federal University of Pernambuco, Recife, Brazil

Juan Pedro Ferrio
Aragon Agency for Research and Development (ARAID); Unidad de Recursos Forestales, Centro de Investigación y Tecnología Agroalimentaria de Aragón (CITA), Zaragoza, Spain

Claudio Gaucher
Facultad de Ciencias, Universidad de La República, Montevideo, Uruguay

Darren R. Gröcke
Stable Isotope Biogeochemistry Laboratory (SIBL), Department of Earth Sciences, Durham University, Durham, United Kingdom

Denis A. Gruzdev
N.P. Yushkin Institute of Geology Komi SC UrB RAS, Syktyvkar, Russia

Cristoph Korte
Department of Geosciences and Natural Resource Management, University of Copenhagen, ØsterVoldgade10,1350 and Nordic Center for Earth Evolution (NordCEE), Copenhagen, Denmark

Chao Li
State Key Laboratory of Palaeobiology and Stratigraphy, Nanjing Institute of Geology and Palaeontology and Center for Excellence in Life and Paleoenvironment, Chinese Academy of Sciences, Nanjing; University of Chinese Academy of Sciences, Beijing, China

Spencer G. Lucas
New Mexico Museum of Natural History & Science, Albuquerque, NM, United States

Michael Montenari
Geography, Geology and the Environment, Keele University, Newcastle under Lyme, United Kingdom

Manoj Kumar Pandit
Department of Geology, University of Rajasthan, Jaipur, India

Natan Silva Pereira
Department of Biology, State University of Bahia, Paulo Afonso, Brazil

Agnieszka Pisarzowska
Institute of Earth Sciences, University of Silesia in Katowice, Sosnowiec, Poland

Artem N. Plotitsyn
N.P. Yushkin Institute of Geology Komi SC UrB RAS, Syktyvkar, Russia

Grzegorz Racki
Institute of Earth Sciences, University of Silesia in Katowice, Sosnowiec, Poland

Ethan J. Richardson
Geography, Geology and the Environment, Keele University, Newcastle under Lyme; 35 Thoresby Road, York, United Kingdom

Paulo Ricardo Riedel
NEG–LABISE, Department of Geology, Federal University of Pernambuco, Recife, Brazil

Wolfgang Ruebsam
Department of Organic and Isotope Geochemistry, University of Kiel, Kiel, Germany

Alcides Nóbrega Sial
NEG–LABISE, Department of Geology, Federal University of Pernambuco, Recife, Brazil

Juan Carlos Silva-Tamayo
Faculty of Sciences, Antonio Nariño University, Bogota, Colombia

Irina V. Smoleva
N.P. Yushkin Institute of Geology Komi SC UrB RAS, Syktyvkar, Russia

Jorge Spangenberg
Institute of Earth Surface Dynamics (IDYST), University of Lausanne, Lausanne, Switzerland

Lawrence H. Tanner
Department of Biological Sciences, Le Moyne College, Syracuse, NY, United States

Nicolas Thibault
Department of Geosciences and Natural Resource Managment, University of Copenhagen, Copenhagen, Copenhagen

José M. Ugidos
Departamento de Geología, Facultad de Ciencias, Universidad de Salamanca, Salamanca, Spain

María Isabel Valladares
Departamento de Geología, Facultad de Ciencias, Universidad de Salamanca, Salamanca, Spain

Jordi Voltas
Joint Research Unit CTFC—AGROTECNIO; Department of Crop and Forest Sciences, University of Lleida, Lleida, Spain

Yuandong Zhang
State Key Laboratory of Palaeobiology and Stratigraphy, Nanjing Institute of Geology and Palaeontology and Center for Excellence in Life and Paleoenvironment, Chinese Academy of Sciences, Nanjing; University of Chinese Academy of Sciences, Beijing, China

Andrey V. Zhuravlev
N.P. Yushkin Institute of Geology Komi SC UrB RAS, Syktyvkar, Russia

Preface

Over the past decades the use of carbon isotope stratigraphy has become an important tool to provide insights into dramatic environmental changes and the biological response to external and internal forcing. The present book aims to provide an update on the progress achieved in the methodologies applied and the carbon isotope stratigraphic record for a variety of stratigraphic intervals.

The first chapter by *Darren R. Gröcke* on "Carbon isotope stratigraphy—principles and applications" is highlighting that "even after all this time our understanding of the modern and ancient carbon cycle is still incomplete. Stable carbon isotopes are an extremely useful tool from which to monitor and trace the carbon cycle. In this manner, carbon isotope stratigraphy can be developed from many organic and inorganic proxies in geologic time. Carbon isotope stratigraphy is a very powerful tool that works on timescales of <10 kyr, and therefore, be used to correlate and constrain sedimentary sequences on the same timescale. Although the technique has been applied across many time intervals from many carbon proxies there are still a lot of improvements and information that can be extracted. This review on carbon isotope stratigraphy will identify, through my somewhat biased viewpoint, what requires consideration and improvement to make the technique even more powerful. These include proxy selection, stratigraphic resolution, uncertainty and reproducibility, and ultimately stratigraphic correlation. Overall, the future for carbon isotope stratigraphy is very exciting." This chapter is followed by the contribution of *Moujahed Al-Husseini and Wolfgang Ruebsam* on "Interpreting Phanerozoic $\delta^{13}C$ patterns as periodic glacio-eustatic sequences" which outlines that "the deterministic Orbital Scale of glacio-eustasy is calibrated with 405-kyr time-rock units named stratons. It predicts polar ice volumes (global sea level) varied throughout the Phanerozoic according to a periodic template that repeats every 14.58 myr. The scale represents the interval between 1.586 and 541.046 Ma with 37 orbitons, each lasting 14.58 myr and consisting of three dozons (4.86 myr, 12 stratons) and 36 stratons. In times older than c. 8.0 Ma the predicted ages of glacial maxima closely correlate to those of positive $\delta^{13}C$ plateaus as estimated by radiometric and other dating techniques. This study illustrates the application of the scale in nine examples from the early Cambrian, late Ordovician and early Silurian, Carboniferous and Cretaceous. Each example compares the predicted glacio-eustatic signal to $\delta^{13}C$ patterns (plateaus, valleys, rising, and falling limbs) and cites supporting stratigraphic evidence for the interpretation (e.g., transgression, regression, erosional unconformity, sequence boundary, cold snap, flooding interval). The examples also review empirical age estimates of stage boundaries and where inconsistent they are calibrated by the Orbital Scale. The results indicate several stage boundaries are defined in major $\delta^{13}C$ plateaus that pass to glacio-eustatic unconformities and hiatuses. The 4.86-myr dozon stands out in the examples as the easiest-to-recognize stratigraphic unit because it is bounded by two significant $\delta^{13}C$ plateaus or excursions that pass to unconformities. Dating $\delta^{13}C$ plateaus according to the Orbital Scale may therefore provide key anchors that better constrain the geologic time scale." In the third chapter by *Juan Pedro Ferrio, Mònica Aguilera, Jordi Voltas, and José Luis Araus* on "Stable carbon isotopes in archaeological plant remains" it is emphasized that "archaeological plant remains are a valuable source of information about the social and environmental conditions experienced by past societies. Full exploitation of the informative potential of plant remains involves the use of analytical techniques. In this context, carbon isotope composition of archaebotanical remains may give clues on the environmental conditions, agricultural management, food web, and dietary preferences, among others. This chapter provides a comprehensive view of the potential applications in the archaeology of analyzing the stable carbon isotope signature of plant remains. After a broad explanation of the physiological and environmental factors determining the carbon isotope signature in plants, most relevant methodological issues are discussed. Among others, the chapter offers a critical review of sampling and pretreatment protocols, uncertainties associated with atmospheric changes, and the potential alterations that plant remains may suffer prior or after deposition in the soil. Exemplified by selected case studies, methodological and conceptual clues are provided for the use of carbon isotopes in archaeology, and the still existing challenges highlighted." The following contribution by *José M. Ugidos, Pedro Barba, and María I. Valladares* provides a welcome "Review of the Upper Ediacaran-Lower Cambrian Detrital Series in Central and North Iberia: NE Africa as possible Source Area" in which it is stressed that "this work offers a review of the Upper Ediacaran and Lower Cambrian succession in the Central and North Iberian Massif. The aim is to simplify the terminology and stratigraphic descriptions to ensure that the treatment of all available geochemical data is more easily understood. In general terms, the stratigraphic succession comprises of three depositional sequences. Depositional sequence 1, of Ediacaran age, is incomplete and only appears in the Cantabrian and West Asturian

Leonese zones. Depositional sequence 2, of Late Ediacaran age, is found in the Central Iberian Zone and corresponds to a sedimentary wedge having the Cantabrian and West Asturian Leonese zones as a secondary area of origin. Depositional sequence 3, of Early Cambrian age, appears in the three zones of study. Samples from all the three sequences have been selected for analytical purposes. The samples were collected within stratigraphic series previously studied but also randomly. In total 240 analyses of major and trace elements and 36 analyses of Nd-isotopes have been considered in the current work. Geochemical data for the Upper Ediacaran detrital rocks of the Central and North Iberia provides evidence of a remarkable geochemical uniformity (trace elements, Nd-isotopes) despite the distances separating the sampling areas. The geochemical uniformity shown by the Upper Ediacaran siliciclastic sediments from the three zones is probably the most important result. So, it is proposed that the extensive volumes of first cycle detrital materials in the Arabian-Nubian Shield mixed with older crustal materials from the Saharan Metacraton and that these mixtures evolved and homogenized in a continental setting across NE Africa prior to final sedimentation in a marine basin. The Lower Cambrian succession corresponds to decreased relative proportions of detritus derived from the Arabian-Nubian Shield." The fifth chapter by *J. Javier Álvaro* on "Calibrating on $\delta^{13}C$ and $\delta^{18}O$ chemostratigraphic correlations across Cambrian strata of SW Europe and Morocco, West Gondwana" highlights that "the Cambrian remains one of the least constrained periods from the International Geological Timescale. Several global boundaries of series and stages are still awaiting formal definition and ratification. Particularly in West Gondwana, a major obstacle to the development of a common regional Cambrian timescale and its global correlation is the endemic character of its shelly fauna (mainly trilobites, archaeocyaths, echinoderms, and brachiopods) due to the rare exposure of fossiliferous slope-to-basinal strata bearing cosmopolitan taxa. Over the past two decades, there has been a growing interest in the application of isotopic chemostratigraphic methods to regional studies, in an effort to solve and calibrate the timescale resolution in the Cambrian Atlas-Ossa Morena rifting axes and their lateral flanks preserved in Morocco, the Iberian Peninsula, and South France. A critical reappraisal of the resulting chemostratigraphic methods and results applied to the recognition and regional correlation of the Ediacaran–Cambrian (BACE event), Cambrian Series 2–Miaolingian (ROECE), and Miaolingian–Furongian (SPICE) boundaries is presented here. The reliability of isotopic data is mainly controlled by (i) the preferential analysis of bulk rock in carbonates and of shales with total organic contents (TOC) lesser than 1 wt%; (ii) the unequal record of diagenetic (e.g., early vs late diagenetic dolomitization) and metamorphic processes related to Variscan tectonothermal events; and (iii) dating uncertainty of chemostratigraphic fluctuations by chronostratigraphic and radiometric constraints." This chapter is followed by the contribution of *Stig M. Bergström, Mats E. Eriksson, and Per Ahlberg* on "Ordovician $\delta^{13}C$ chemostratigraphy: A global review of major excursions and their ties to graptolite and conodont biostratigraphy" in which the authors explain that "recent studies on $\delta^{13}C$ chemostratigraphy have produced a wealth of new information of major importance for clarification of local and global stratigraphic relationships. Currently, approximately 15 named positive and negative, isotopic excursions have been identified in the Ordovician System. Some of these are significant and geographically widespread, whereas others are poorly developed and regional, thus hindering unambiguous usage. In order to correctly identify the named Ordovician excursions, it is of basic importance to tie each of these as precisely as possible to a biostratigraphic framework. Because isotopic age data useful for precise excursion dating are rarely available in the Ordovician, biostratigraphic information from geographically widespread fossil groups such a graptolites and conodonts are of critical importance for correct identification of isotopic excursions. This study is a global review of the relations between Ordovician $\delta^{13}C$ excursions and graptolite and conodont biostratigraphic units with special emphasis on Baltoscandia, North America, and Argentina, three large regions that have been centers of Ordovician chemostratigraphic research. In this report we review for each excursion its proper nomenclatural designation, relations to biostratigraphic units, its known geographic range, and key study localities." This is followed by the contribution of *Li Chao and Zhang Yuandong* on "Ordovician carbon isotope stratigraphy of China: a synthesis" in which is highlighted that "in this paper we review the published carbon isotope records ($\delta^{13}C_{carb}$) of the Ordovician System from Chinese comprising terranes, including mainly South China, North China, Tarim, and Xizang (Tibet). All the $\delta^{13}C_{carb}$ values from a number of sections in each terrane and published by different authors are calibrated to integrated graptolite and conodont biostratigraphic zones, and are correlated with each other. The $\delta^{13}C_{carb}$ curves for each block or terrane are interpreted by referring to the specific geological backgrounds and are discussed with their implications. The composite curve of $\delta^{13}C_{carb}$ from Tarim overall agrees well with the global composite curve, but with a significantly prolonged GICE and absence of data for late Katian and Hirnantian. In South China, where the published $\delta^{13}C_{carb}$ values of different facies are reviewed separately, the results show that the composite curves of three facies are generally correlative, especially well for the Tremadocian and early Floian. The MDICE excursions are not convincingly confirmed in the Middle Yangtze Platform and the Jiangnan Slope, suggesting the likely presence of potential hiatuses within a succession of uniform facies, or the very records are missed due to inadequate sampling, a common case for condense sections. The GICE excursions have been steadily recognized in all the facies, but showed a significant prolongation than that in the United States. The HICE can be steadily identified in the Upper and Middle Yangtze Platform, but distinguished by their overall low values of the $\delta^{13}C_{carb}$ values, ca. 5‰ lower than Baltoscandia and Anticosti Island, suggesting the likeliness of missing peaks records due to tectonic disturbances." This chapter is followed by the contribution of *Agnieszka Pisarzowska and Grzegorz Racki* on "Comparative carbon isotope chemostratigraphy of major Late Devonian biotic crises." Within this contribution it is highlighted that "changes in the climate-sensitive carbon cycle during the major and second-order Late Devonian global events were more complex than usually expected, i.e., they were not characterized only by positive $\delta^{13}C_{carb}$ and $\delta^{13}C_{org}$ excursions up to 5‰. In fact, the carbon (and oxygen) isotope signatures in lime mudstones are less susceptible to diagenetic bias and therefore several negative trends and spikes in different facies of three distant continents are interpreted as primary biogeochemical signals. In addition to significant intra- and inter-regional variability resulting from diverse local carbon cycling factors, the initial phase of both Kellwasser Crisis events [Frasnian-Famennian (F-F) mass extinction] and end-Devonian Hangenberg Crisis confirms the predicted causal relationship between large-scale volcanic degassing (mainly in large igneous provinces?) and the response of global carbon cycle in ^{13}C depleted signals. Potential volcanic signals were confirmed in more than 40% of the 73 F-F sites analyzed for C isotope stratigraphy, as summarized in the global database. Lowered $\delta^{13}C$ values characterize also other less known Late Devonian events, exemplified particularly by well-documented Frasnian Middlesex/punctata Event. Thus, consistently documented worldwide, paired negative $\delta^{13}C$ and $\delta^{18}O$ excursions, even only in carbonate signatures, can be cautiously regarded as a primary volcano-climatic signal and additional proxy of the volcanic trigger. In addition, the repeated negative $\delta^{13}C$ spike near the F-F boundary correlates with Hg anomalies considered to be the signature of the arc magmatism acme (and alkaline/carbonatite/kimberlite paroxysm?). However, both the carbon cycle response and time relationships between disturbances in C and Hg cycles during mass extinctions were certainly complex and required high-resolution studies in the most stratigraphically extended series and advanced quantitative modeling." The following chapter by *Andrey V. Zhuravlev, Artem N. Plotitsyn, Denis A. Gruzdev, and Irina V. Smoleva* on "Carbon isotope stratigraphy of the Tournaisian (Lower Mississippian) successions of NE Europe" gives new insights into "Biostratigraphically constrained $\delta^{13}C_{carb}$ variations in the Tournaisian successions of NE Europe (Komi Republic, Russia) are considered. Eight reference sections, characterizing platforms, slopes, and shelf depression facies, comprise the basis of this study. Of 412 studied samples of bulk carbonates, 268 samples passed all of the screening tests and were used to analyze the carbonate carbon isotope record. The established regional $\delta^{13}C_{carb}$ data allowed the recognition of the distinct, terminal Famennian and mid-Tournaisian carbon isotope events and two minor positive shifts in the sulcata Zone and Upper duplicata - lowermost quadruplicata interval. The age of the isotope shifts is underpinned by conodont biostratigraphy. These isotope events and shifts are traced through the facies gradient from the shallow-water carbonate platform to deep-water shelf depression. Because the regional $\delta^{13}C_{carb}$ variations match those known in different regions (Europe, South China, and North America), the recognized C-isotope shifts can be considered to be reliable isotope-stratigraphic markers for both regional and world-wide correlation." This chapter is followed by the contribution of *Lawrence H. Tanner and Spencer G. Lucas* on "The isotopic composition of pedogenic carbonate as a record of pCO_2: Insights from the Late Paleozoic of tropical western Pangea" in which it is highlighted that "the use of the carbon isotope composition ($\delta^{13}C$) of pedogenic calcite as a proxy for pCO_2 was established over three decades ago and remains a primary means of investigating climate history. Over time, an increased understanding of the systematics of pedogenic carbonate precipitation has improved the constraints on interpreting paleoatmospheric composition from the isotopic composition of calcrete. We review the development of the use of pedogenic $\delta^{13}C$ as a proxy for pCO_2 over geologic time, with focus on the late Paleozoic. We illustrate the discussion with specific examples from the late Carboniferous through early Permian icehouse climate interval and present new as well as previously published data from strata of Virgilian (Late Pennsylvanian) through Wolfcampian (early Permian) age collected in northern and south-central New Mexico, which was located in tropical western Pangea during the late Paleozoic. Our success in fitting pCO_2 values we generate through new analyses to previously published paleo-pCO_2 curves strengthens the rationale for use of these data as proxies for paleoclimate and suggests the potential for their use as a correlation tool." This followed by the contribution of *Alcides Nóbrega Sial, Jiubin Chen, Cristoph Korte, Manoj Kumar Pandit, Jorge Spangenberg, Juan Carlos Silva-Tamayo, Luis Drude de Lacerda, Valderez Pinto Ferreira, José Antônio Barbosa, Claudio Gaucher, Natan Silva Pereira, and Paulo Ricardo Riedel* on "Globally enhanced Hg concentration and Hg and C isotopes in Permian–Triassic Boundary successions: possible linkage to volcanism" in which the authors present "C and Hg (Hg/TOC) chemostratigraphy data on identified Permian–Triassic successions across the globe with an aim to develop a better understanding on the 'mass extinction' event during this time interval. Besides, it would also contribute to the global database on C and Hg stratigraphy and reinforce the application of the latter as a reliable proxy in global correlation, especially in cases of sedimentary sequences deposited coevally with LIP volcanism. In this endeavor, the $\delta^{13}C$ chemostratigraphic curves from thousands of kilometers apart from nine classical marine sedimentary sections from various paleolatitudinal zones as well as Hg/TOC stratigraphic pathways of eight of these sections, were examined in a transect across the Pangea supercontinent (approx. 252 Ma). A strong negative δ^{13}Ccarb shift at the Late Permian Mass Extinction (LPME), in Meishan D, Guryul Ravine and several other (Permian Triassic Boundary) PTB sections worldwide, is probably due to voluminous CO_2 degassing during volcanism in the Siberian Traps Large Igneous Province (STLIP). The C-isotope shift at the LPME is much stronger than in the PTB or Early Triassic Mass Extinction (ETME). The Hg/TOC peaks are observed at LPME and ETME horizons in the Meishan D section, whereas the peaks were recorded at LPME and PTB horizons in Hovea-3, Ursula Creek, Idrijca, and Rizvanuša sections. The Rizvanuša section displays a single peak at the ETME horizon, Zal and Abadeh successions at the LPME and ETME horizons, while the Misci section shows Hg/TOC enrichment at the LPME horizon. The largest Hg/TOC peaks at the LPME, PTB, and ETME can be associated with the beginning of Stage 2 (extrusive hiatus) of STLIP. The origin of a Hg/TOC spike in the LPME–PTB interval in seven of these sections is unclear as well as the Hg/TOC spike below the LPME at the Misci and Zal sections. The latter may, possibly be, linked to the STLIP (Stage 1) or some local volcanism. In the δ^{202}Hg (MDF) vs Δ^{201}Hg (MIF) cross plot, the majority of samples from the extinction interval plot within the 'volcanic-emission box' and a few samples, in the 'sediment/soil/peat box.' Hg-isotope signatures underline mixing of volcanic and terrestrial Hg, generating four trends with negligible variation in Δ^{201}Hg.

Lesser terrigenous-Hg influx was noticed in sections closer to the STLIP (Rizvanuša, Idrijca and Seres/Misci, all in Europe), recording close to zero Δ^{201}Hg (MIF) values. Marked influence is observed in sections far away from the STLIP (Meishan, Ursula Creek and Hovea-3) that exhibit negative a Δ^{201}Hg (MIF). The Zal and Abadeh sections, at intermediate distance from the STLIP, exhibit the highest, positive Δ^{201}Hg values (Abadeh) and the lowest, negative Δ^{201}Hg values (Zal). The Δ^{199}Hg (MIF) vs Hg (n.ng- 1) cross plot testifies that volcanic Hg has been contaminated by terrigenously sourced- Hg influx, inducing the two major curved trends. The negative C-isotope excursions and Hg/TOC enrichments in these sections are nearly coeval, substantiating a synchronism between the P–T transition biotic crises (LPME and ETME) and the start of the stage 2 of the Siberian Traps (sill-intrusions)." This contribution is followed by the chapter of *Wolfgang Ruebsam, Nicolas Thibault, and Moujahed Al-Husseini* on "Early Toarcian glacio-eustatic unconformities and chemostratigraphic black holes" in which it is highlighted that "during Late Pliensbachian and Early Toarcian times (Early Jurassic Epoch, c. 185 to 180 Ma) major environmental instabilities and perturbations affected the evolution of the carbon cycle, climate, sea level, and biota. The temporal and causal relationships between these global-scale phenomena however remain poorly understood. This study starts by refining the accuracy of the ages of Lower Toarcian ammonite zones and subzones, and carbon-isotope excursions defined in the carbon-13 isotope (δ^{13}C) reference curve. Based on an extensive review of stratigraphic data our study identified and calibrated by numerical age ammonite-dated transgressive-regressive (T-R) sequences, sequence boundaries (SB), maximum flooding intervals (MFI), and climate cycles. These calibrations placed in a common temporal framework the δ^{13}C reference curve, T-R sequences, climate cycles, and the Karoo-Ferrar Large Igneous Province (K-F LIP). The temporal framework revealed four intervals that are either fragmented, or completely missing in numerous Upper Pliensbachian and Lower Toarcian δ^{13}C records (Stratigraphic Black Holes: SBH). The identified SBHs closely correlated to erosional unconformities or sequence boundaries and global cooling. Unconformities occur in the spinatum zone (SBH 1), in the middle tenuicostatum zone (SBH 2), and in the middle serpentinum zone (SBH 3 and 4). The older three SBHs further contain global sequence boundaries SB JPl8, JTo1, and JTo2. The sea falls associated with the unconformities have amplitudes of many tens to possibly 75 m and exceed by more than an order of magnitude falls associated with thermo-eustasy or aquifer-eustasy. Sea level falls are therefore attributed to build-ups of polar ice caps and their ages and those of subsequent highstands were found to be consistent with the predictions of the Orbital Scale of glacioeustasy. We conclude tuning of insolation caused by eccentricity cycles with periods of c. 0.1, 0.405, and c. 2.4 myr drove sea-level fluctuations and also paced changes in the global carbon cycle and climate. Greenhouse gas emissions associated with volcanism in the Karoo-Ferrar Large Igneous Province affected the global carbon cycle and thereby amplified climate and sea level changes." The final chapter by *Ethan J. Richardson and Michael Montenari* on "Assessing shale gas reservoir potential using multi- scaled SEM pore network characterizations and quantifications: The Ciñera-Matallana pull-apart basin, NW Spain" is highlighting that "in recent years, unconventional hydrocarbon reservoirs (UHRs), specifically shale gas plays, have become a vast source of economically viable hydrocarbons, in response to better drilling and stimulation techniques. To further improve recovery, more detailed pore network modeling needs to be conducted by means of detailed pore network characterizations, based upon predefined pore size and pore type classification schemes. Through the quantification of the nano/microporous properties of the pore network, a better understanding on how hydrocarbon storage and migration phenomena operate within UHRs and how well they might respond to stimulation can be developed. Kasimovian-Gzhelian (Late Carboniferous) lacustrine sediments from the Upper Pastora Formation of the Ciñera-Matallana coalfield, NW Spain, were investigated to identify the potential for shale gas, by conducting a detailed pore network analysis using SEM-based porosity data, alongside traditional methods of formation evaluation. Here we show that the Upper Pastora Formation demonstrates shale gas potential and that the nanoporous network is a complex multifaceted and multiscaled system composed of various pore types and sizes, each with specific dynamic properties and varying contributions toward total porosity. The present contribution provides a full pore network characterization and quantification, based on pore type and size, followed by a quantification of the porosity percentage held within each pore type and size. This work will provide an insight into the need for pore network characterizations alongside pore property quantifications with regard to computational fluid flow models and as such the need to carry out this work in conjunction with more traditional methods of shale gas exploration."

This book was coordinated and written amidst the Covid-19 pandemic of 2020. I thank all the contributors for their tremendous efforts and all the hard work under sometimes extreme conditions. I also want to thank my magnificent team at Elsevier. Your continuous support and dedication was fantastic. I thank you all!

Michael Montenari
Keele University
United Kingdom

CHAPTER ONE

Carbon isotope stratigraphy: Principles and applications

Darren R. Gröcke*

Stable Isotope Biogeochemistry Laboratory (SIBL), Department of Earth Sciences, Durham University, Durham, United Kingdom

*Corresponding author: e-mail address: d.r.grocke@durham.ac.uk

Contents

Abstract

Even after all this time, our understanding of the modern and ancient carbon cycle is still incomplete. Stable carbon isotopes are an extremely useful tool from which to monitor and trace the carbon cycle. In this manner, carbon isotope stratigraphy can be developed from many organic and inorganic proxies in geologic time. Carbon isotope stratigraphy is a very powerful tool that works on timescales of <10 kyr, and therefore, can be used to correlate and constrain sedimentary sequences on Milankovitch timescales. Although the technique has been applied across many geological time intervals from many carbon proxies there are still a lot of improvements and information that can be extracted. This review on carbon isotope stratigraphy will identify, through my somewhat biased viewpoint, what requires consideration and improvement to make the technique even more powerful. These include proxy selection, stratigraphic resolution, uncertainty and reproducibility, and ultimately stratigraphic correlation. Overall, the future for carbon isotope stratigraphy is very exciting.

Stratigraphy & Timescales, Volume 5
ISSN 2468-5178
https://doi.org/10.1016/bs.sats.2020.08.002

1. Introduction

In an Anthropocene world, humans are in the process of altering the carbon cycle in a way that is "currently" unprecedented in geologic time (e.g., Hall et al., 2019; Steffen et al., 2020; Woodard et al., 2019). Current predictions suggest that the carbon cycle may reach a tipping point (Lenton et al., 2008; Schimel et al., 2015) after which humans will no longer be able to sustain current habitability—a runaway greenhouse event. Many approaches have been employed to understand the geological record of the carbon cycle, for example, stomatal index of fossil plants (e.g., Barclay and Wing, 2016; Hu et al., 2019; Retallack and Conde, 2020; Royer, 2001; Steinthorsdottir et al., 2013; Steinthorsdottir et al., 2019), marine alkenone biomarkers (e.g., Freeman and Hayes, 1992; Jasper and Hayes, 1990; Pagani, 2002; Pagani et al., 1999a, 1999b; Riebesell et al., 2000; Zhang et al., 2013, 2019), carbon isotopes in foraminifera (e.g., Berger, 1979; Durazzi, 1981; Harper et al., 2020; Huber et al., 2011; Kennett et al., 2000; MacLeod and Huber, 1996; Schmittner et al., 2017; Spero, 1992) and carbon cycle models (e.g., Berner, 1992, 1998; Caves et al., 2016; Isson et al., 2020; Krissansen-Totton et al., 2018; Kump et al., 2000; Li and Elderfield, 2013; Wallmann, 2001) to name just a few.

Carbon isotope ratios ($\delta^{13}C$) produced from a multitude of organic and inorganic materials have been one of the most extensively employed proxies to investigate changes in the ancient carbon cycle (e.g., Brand, 2004; Canfield et al., 2020; Cerling, 1992; Craig, 1953; Degens, 1969; Fox and Koch, 2004; Freeman and Hayes, 1992; Gröcke, 2002; Holser, 1997; Landergren, 1954; Nier and Gulbransen, 1939; Oehlert and Swart, 2014; Schidlowski, 1987; Tappert et al., 2013; Tipple et al., 2010; Zachos et al., 2001). Although $\delta^{13}C$ can be employed for stratigraphic purposes, it is by no means the simplest, most precise or accurate method to adopt. Since the 1950s researchers have used carbon isotope ratios to investigate modern and ancient environments, ecosystems and climates. Some 70+ years later and we are still learning and discovering new aspects about how carbon isotope signatures exchange, modify and incorporate into abiogenic and biogenic systems. As an example, the modern carbon cycle is still not fully understood in terms of residence time, reservoir sizes and exchange rates (e.g., Falkowski et al., 2000; Houghton et al., 2018; Isson et al., 2020).

1.1 The beginnings of isotope stratigraphy

For excellent resources on the history of the stable isotope mass spectrometer, I refer the reader to Jennings (2012) and Sharp (2017). Some other key papers about mass spectrometry include, Nier (1947), Inghram (1948), Hunter et al. (1949), McKinney et al. (1950), Matthews and Hayes (1978), De Laeter et al. (1992), De Laeter and Kurz (2006), Ireland (2013), Münzenberg (2013) and Brand et al. (2015). For those specifically interested in the history of the MAT (Mess- und Analysen-Technik) mass spectrometer, I refer the reader to Brunneé (1997).

A simple search in Google Scholar™ can provide a qualitative assessment on the history of "isotope stratigraphy," "stable isotope stratigraphy" and "carbon isotope stratigraphy" in the scientific literature (Fig. 1).

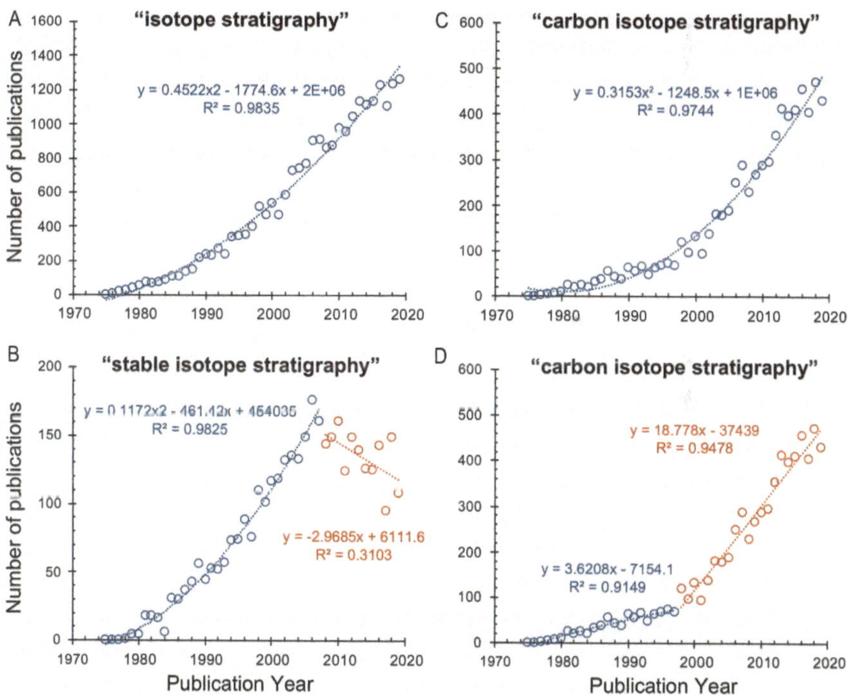

Fig. 1 Google Scholar™ records of the number of publications against publication year for search terms, (A) "isotope stratigraphy," (B) "stable isotope stratigraphy," and (C and D) "carbon isotope stratigraphy." Graphs (C) and (D) are the same except that pre- and post-1997 was separated by two distinct regression lines. Note, the decrease in the term, "stable isotope stratigraphy" since 2007.

Each record within the search was not checked to determine its relevance to the discipline, but it is clear that the term "isotope stratigraphy" has increased exponentially since the mid-1970s. The concept of investigating stratigraphic changes in stable isotopes and using them for correlation was first used by Emiliani (1955), but it was not until a decade later when the term "isotope stratigraphy" was specifically used by Gonfianti (1965). Even then it was not until the late 1970s that the term "isotope stratigraphy" was used more frequently and incorporated into scientific literature. On reflection, Emiliani (1955) developed the concept, but it was Gonfianti (1965) who coined the term "isotope stratigraphy."

Fig. 1B ("stable isotope stratigraphy") clearly indicates a shift in terminology from being more generic prior to 2007 to one that is more specific after that date (e.g., "carbon isotope stratigraphy," "nitrogen isotope stratigraphy"), and hence the decline in the use of the term. The incorporation of "carbon isotope stratigraphy" has risen exponentially since the 1970s (Fig. 1C) with a distinct increase in the term from 1997 (Fig. 1D). So, what happened in 1997 that caused the sudden increase in publications using this term? Apart from the obvious (e.g., I started my DPhil at Oxford, the first book of Harry Potter was published, or Steve Jobs returning to Apple), it would be great if one could pin-point a seminal paper that led to this increase. Using citations is not the most precise indicator of influence, but there was one seminal paper published in *Geology* in 1997 that caused a "blast" for re-interpreting negative $\delta^{13}C$ excursions in the geological record (Dickens et al., 1997). Although originally proposed two years before, first by Matsumoto (1995) and a month later by Dickens et al. (1995), the *Geology* paper reached a greater audience in Earth Sciences. The negative $\delta^{13}C$ excursion presented by Dickens et al. (1997) was the Palaeocene–Eocene Thermal Maximum (PETM, aka LPTM, IETM, EETM), and based on those initial results and many more since (e.g., Baczynski et al., 2019; Bains et al., 1999; Cui et al., 2011; Gutjahr et al., 2017; Harper et al., 2020; Kent et al., 2003; McInerney and Wing, 2011; Zachos et al., 2007, 2010; Zeebe et al., 2016) has allowed many Earth Scientist's to brand their research as concerning more relevant Recent and Anthropocene climate change (e.g., global warming and greenhouse gases): the *past is the key to the present* (*addendum*: minus humans!).

The purpose of this review is not to focus on the interpretations of $\delta^{13}C$ excursions in the geologic record, but to provide a background on carbon isotope stratigraphy in marine and terrestrial systems and how this technique can be applied more synergistically in the future. Unfortunately (or not!), a significant amount of the review will be partly based on my personal view of how carbon isotope stratigraphy is used, how the techniques employed could be improved and standardized, and how future research approaches are required in order to advance our understanding of the ancient carbon cycle. Apologies in advance to all those authors not cited, but there are only so many hours in a day. Apologies also for focussing on specific time intervals like the Jurassic, Cretaceous and Cenozoic, but researchers investigating other time periods will understand the same issues, concerns and/or examples provided. A review on the use of carbon isotopes in palaeoceanography was recently published by Mackensen and Schmiedl (2019), although it was more focused on the interpretation of the records and not how they were generated; other reviews on "isotope stratigraphy" can also be found in Tiwari et al. (2015) and Sial et al. (2015).

1.2 Definition of isotope stratigraphy

Isotope stratigraphy can be broadly defined as the variation of a stable isotope signature preserved in the sedimentary record through a period of time. It is a sub-discipline of chemostratigraphy that looks at the chemical signature of sediments through time; most chemostratigraphic studies, however, use stratigraphic changes in trace element and rare earth element abundances to define and correlate sedimentary packages (e.g., Hines et al., 2019; LaGrange et al., 2020; Mackey and Stewart, 2019). With respect to the definition of isotope stratigraphy, time can be defined as anything greater than seasonal, and therefore include research in many other disciplines, such as limnology, glaciology, archeology and pedology. Isotope stratigraphy also relies on similar principles in Darwin's theory of evolution. That is, in order for isotope stratigraphy to work and be informative about the history of biogeochemical cycles, it must rely on the fact that the stable isotope ratio (δX, where X is any isotope system of interest, e.g., ^{13}C or ^{15}N or ^{34}S) will evolve/change through time, and those changes are dictated by climate/environmental changes in that global cycle.

1.3 What makes a stable isotope ideal for stratigraphy?

There are many stable isotope systems out there that can be used for isotope stratigraphy, but many are more appropriate than others, and this will depend on the research question/s being explored. For a review of isotopes in the periodic table the reader should use the IUPAC Technical Report by Holden et al. (2018) and the excellent on-line resource maintained by the

Commission on Isotopic Abundances and Atomic Weights (https://www.ciaaw.org). Many key factors can make an isotopic system ideal for employment in isotope stratigraphy. These include: (1) a short residence-time (R_t); (2) it is globally synchronous; (3) presence in a wide variety of sedimentary environments; (4) ease (cost, time) with which to generate high-resolution records; and (5) resistance to alteration/diagenesis. These factors will be discussed throughout this review, but let's start with the basic principles of isotope stratigraphy.

2. Principles of isotope stratigraphy

As with any discipline a number of principles need to be laid out in order to generate unequivocal research results and interpretations. Unfortunately, in terms of isotope stratigraphy there has been no such documentation—at least in written form. Although the definition and structure of principles below may not be all-inclusive, it will provide a basis from which future research datasets can be more accurately and directly compared to one another; especially, in terms of reconstructing Earth's carbon cycle through geological time. However, research projects and the subsequent generation of datasets are primarily controlled by two major factors, time and money. If a research group had ample time (person hours) and no financial limits the carbon isotope curve that could be generated would be near perfect and there would be no need to write this review: that, unfortunately, as many groups and individuals are aware, is not the case. Therefore, this review provides a series of guidelines that hopefully (given time and finance) are realistic to generate appropriate and excellent carbon isotope stratigraphies. This review will provide both good and poor examples (in my opinion!) of carbon isotope stratigraphy and on how it can be improved, where possible.

2.1 Proxy selection

A proxy is defined here as the geological sample (material) that is used to generate the isotopic ratio and signature of the environment/climate/biology at the time of formation. The choice of a particular proxy will depend on a number of factors, such as the presence, preservation and type of isotopic analysis to be performed. For example, one would not attempt to generate a carbonate $\delta^{13}C$ curve from a marine setting where the calcite compensation depth (CCD) is highly variable through time and impinges onto the site. A study of this kind would generate a $\delta^{13}C$ curve with stratigraphic gaps (i.e., when the CCD rose to shallower depths) and be an incomplete record. Likewise, you would not generate a terrestrial $\delta^{13}C$ record from open ocean settings where the presence of terrestrial organic matter is limited. The above examples do not just apply to outcrop, but to core material also. For example, you would not produce a detailed stratigraphic record of belemnite $\delta^{13}C$ from a core, as it will be highly dependent on whether the core has cut through, or contains, a belemnite. For example, in one of my own studies (Gröcke et al., 2003), a belemnite isotope curve was generated for the Late Jurassic. The stratigraphic resolution was satisfactory for the Volga Basin (Russia), whereas that generated from New Zealand was very poor ($n=17$ over 4km of stratigraphy). More recently, Ullmann et al. (2016) produced 1423 stable isotope analyses of belemnites, bivalves and brachiopods from the Late Permian to Late Jurassic. Even they acknowledge a lack of resolution and state, "the Early and Middle Jurassic are poorly constrained and data density does not allow for sound interpretations of the trends" (Ullmann et al., 2016, p. 68). Of course, in these two studies the sites (e.g., palaeolatitude) were fundamental to the project and the belemnites were collected and analyzed where possible.

Table 1 lists some key proxies that have been used for constructing stratigraphic $\delta^{13}C$ curves in marine and terrestrial sediments. These proxies have different stratigraphic ranges, and some recent references have been provided that illustrate their application in carbon isotope stratigraphy. There are other proxies that have not been listed either due to their limited nature in the geological record and/or limited application to date in carbon isotope stratigraphy, such as, peat sediments, ostracods, animal collagen, corals, coccolithophores, and trilobites.

2.2 Stratigraphic resolution

As with many aspects of scientific research, the fewer data you have the less reliable your conclusions are. Therefore, in order to generate a more reliable dataset the study requires more samples to be collected and analyzed. There are many publications out there that discuss the minimum number of samples required to generate statistical significance in a population (e.g., Pearson and Grove, 2013; Phillips and Gregg, 2001). But, in terms of isotope stratigraphy this does not always apply since we are limited by what is preserved in the sedimentary rocks and not all samples will be absolutely dated in order to enable a direct comparison. The resolution of the record is dependent on time and money, but another major factor is the actual stratigraphic section

Table 1 Common proxies used for generating $\delta^{13}C$ curves in geologic time.

Marine	$\delta^{13}C_{org}$	$\delta^{13}C_{carb}$	**Age range**	**Some recent example papers**
Bulk sediment	X	X	Archaen–present	Daines et al. (2017), Galeotti et al. (2019), and Oborny et al. (2020)
Biomarkers	X		Archaen–present	French et al. (2014), Elling et al. (2019), and Lee et al. (2019)
Foraminifera		X	Late Ordovician–present	Giorgioni et al. (2019), Hermoso et al. (2020), and Sosdian et al. (2020)
Brachiopods		X	Early Cambrian–present	Mii et al. (2012), Korte et al. (2017), and Ullmann et al. (2020)
Belemnites		X	Early Jurassic–Late Cretaceous	Ullmann et al. (2016), Arabas et al. (2017), and Hesselbo et al. (2020)
Terrestrial	$\delta^{13}C_{org}$	$\delta^{13}C_{carb}$	**Age range**	
Bulk sediment	X	X	Archaen–present	Sciscio et al. (2016), Richey et al. (2018), and Liu et al. (2020)
Plant material	X		mid-Ordovician–present	Nordt et al. (2017), Graham et al. (2019), and Wan et al. (2019)
Coal	X	X ?	early Carboniferous–present	Holdgate et al. (2009), and Guo et al. (2020)
Amber	X		mid-Carboniferous–present	Tappert et al. (2013), and Dal Corso et al. (2017)
Biomarkers	X		Archaen–present	Ghosh et al. (2017), Naafs et al. (2019), and Neupane et al. (2019)
Palaeosols	X	X	Archaen–present	Baczynski et al. (2016), and van der Meulen et al. (2020)
Collagen	X		Early Jurassic–present	Stevens and Hedges (2004), and Richards et al. (2017)
Tooth enamel		X	Early Devonian–present	Du et al. (2019), Hynek et al. (2019), and Uno et al. (2011)
Ice cores	*	*	Late Pliocene–present	Schmitt et al. (2012), Eggleston et al. (2016), and Bock et al. (2017)

* refers to $\delta^{13}C$ from ice cores, but in these cases CO_2 and/or CH_4 gases are analyzed. The ? in coal $\delta^{13}C_{carb}$ indicates it is possible when there is carbonate present in coal seams, but this has not been adequately demonstrated to date.

that you are targeting to research. The stratigraphy may be condensed or expanded, and there may be obvious stratigraphic gaps (e.g., unconformities); all of which will determine the sampling strategy. But constraining yourself to a specific sedimentary type may also bias or limit our knowledge; for example, recently Caruthers et al. (2018) were able to produce a Silurian $\delta^{13}C$ curve from organic matter within evaporite deposits of the Michigan Basin. In addition, one needs to consider if the carbon isotope dataset (or other geochemical analyses) will be subsequently used to determine cyclicity in the stratigraphy. An equal sample resolution approach may not always catch the full suite of cycles. For example, the first suite of data will need to be evaluated, after which additional sampling and analyses will need to be obtained and included in the spectral analysis.

Even when I started my postgraduate studies at Oxford the sample resolution used in carbon isotope stratigraphy was relatively coarse, even on the order of a meter scale. For example, Jenkyns and Clayton (1997) sampled Toarcian (Early Jurassic) shales from England and Wales at a sample resolution ranging from every 10 cm (Hawsker Bottoms) up to 10 m (Mochras Borehole). Comparing carbon isotope records with this range of sampling resolution is fraught with danger (see Fig. 2), especially when a single $\delta^{13}C$ value may just be that or it could suggest a significant change in the isotopic curve. So, when you start to make comparisons between high-resolution Toarcian datasets such as those in Kemp et al. (2005) and Them et al. (2017) with a sample resolution of ~2 cm down to ~2 mm to that from Bodin et al. (2016) where sample resolution is on the order of 1–10 m the stratigraphic correlation will have many possibly solutions and therefore, errors (see Figs. 2 and 3). Fortunately, the Early Jurassic Toarcian oceanic anoxic event (T-OAE) has a fairly distinctive carbon isotope shift and so distinguishing it at low resolution is feasible. However, if you are wanting to explore driving mechanisms of the carbon isotope excursion you need to sample at a resolution greater to or equal to the R_t of CO_2 in the carbon cycle; on the order of 1–10 kyr in the marine reservoir (Ciais et al., 2013; Hansell et al., 2009). The R_t of CO_2 is dependent on many factors, such as, climate, ocean circulation, mid-oceanic ridge spreading rates and the hydrological cycle. Of course, without accurate chronology in the stratigraphy causal mechanisms will often be impossible to determine or even postulate. However, recent research is starting to generate carbon isotope records through sequences that can be absolutely dated (e.g., volcanic ash beds) (e.g., Hall et al., 2004; Schoene et al., 2010; Schoene et al., 2019; Yager et al., 2017) and this will enable more accurate correlations and interpretations of the carbon isotope cycle to be made. However, not every site contains such lithologies for absolute dating.

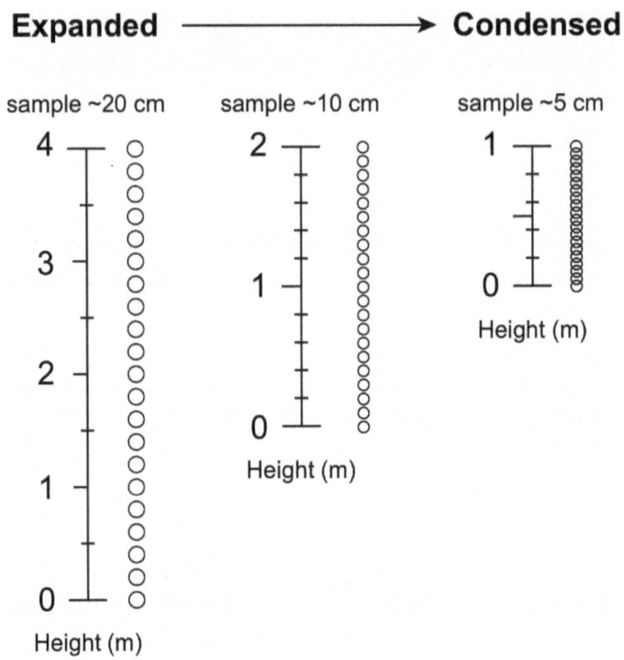

Fig. 2 An idealized sampling strategy for stratigraphic sections that are expanded to condensed with known stratigraphic markers (e.g., unconformities, biostratigraphy). Therefore, when an expanded section is sampled every 20 cm for $\delta^{13}C$, then a section half that thickness should be sampled by half (~10 cm) and subsequently, a condensed section of only 1 m should be halved again (~5 cm).

Another approach in order to ascertain time for correlation was portrayed by Gröcke et al. (1999), where each ammonite subzone was arbitrarily assigned a duration. The carbon isotope curve generated in each ammonite subzone was condensed or expanded so that time (or in the case of this plot, thickness: see fig. 2 in Gröcke et al., 1999) was the same—this approach artificially tried to account for changes in sedimentation rate and time. Although far from perfect (especially since ammonite evolution is not equal in duration), this is a much better method to employ compared to those generating a 3-pt (or more) moving average through the data to generate a "smoother" curve: I will discuss issues related to this method later. The geological record and processes making that record have already attempted to smooth the record.

A lot of researchers sample a stratigraphic section at relatively coarse resolution first, construct the carbon isotope stratigraphy and then focus on key intervals. This is a great way in which to investigate a site, but often there are

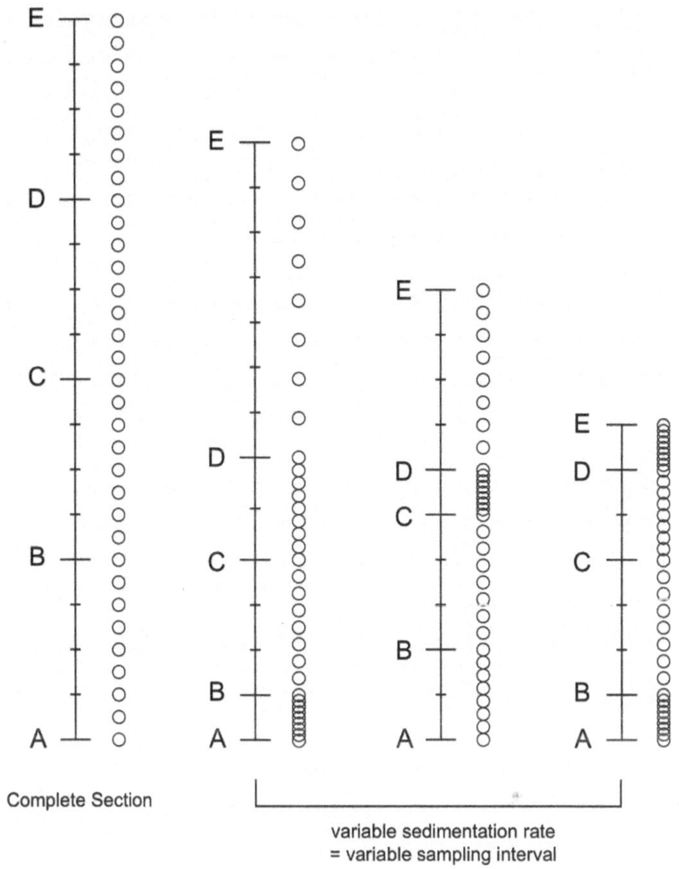

Fig. 3 An idealized sampling strategy for stratigraphic sections with time-constrained intervals A–B, B–C, C–D and D–E. A complete section is depicted on the far left. In the three other sections the thickness of the stratigraphic intervals are either expanded or condensed or the same as in the complete section. These sections require the sampling strategy to vary depending on whether the interval is condensed (higher resolution sampling) or expanded (lower resolution sampling): thus, variable sedimentation rate = variable sampling interval. Of course, sampling can remain equal throughout the stratigraphic section, however, when it comes to interpreting the $\delta^{13}C$ excursions the author needs to take into account stratigraphic time of that lithological unit (e.g., laminated shale *vs* thick-bedded sandstone).

difficulties associated with this. For example, the stratigraphic section may be in a remote area that requires significant time and finances to access (e.g., helicopter time, ground support and/or protection), or the site may change from one year to the next (e.g., slumping, landslides, snowpack).

Many researchers tackle this issue by identifying the stratigraphic interval of interest prior to fieldwork, then target and sample that interval at much higher resolution compared to the rest of the section. This may produce a record that has excellent correlation certainty at that stratigraphic interval but the remainder of the section, sampled at a much lower resolution, may be more problematic. Also, we must contend with Sod's law (aka, Murphy's law) in that the selected stratigraphic interval of interest may be slightly off or incorrect and so one might miss sampling that part in detail.

This kind of sampling approach has developed as a result of changes in research-funding priorities and scholarly evaluations. Research projects have become more focused on specific time intervals (e.g., carbon isotope excursions, extinctions) without fully appreciating and understanding the pre-excursion and post-excursion record (e.g., biogeochemical cycle background): this is not only true for traditional stable isotope stratigraphy, but also with the new, non-traditional isotopes. How do we know that the excursion of interest is actually unique? What environmental pre-cursors were occurring in the environment that could influence the excursion of interest? A potential example of this happened with respect to projects on the PETM carbon isotope excursion, where for years research only focused on that brief time interval, and then in 2005 a similar event ~2 million years younger was found, referred to as the Elmo Event (Lourens et al., 2005). Again, a similar example has been reported for the Early Jurassic T-OAE excursion, where another, rapid, negative $\delta^{13}C$ excursion was identified lower in the record at the Pliensbachian/Toarcian boundary (e.g., Bodin et al., 2016; Littler et al., 2010; Schnyder et al., 2017). Many more examples of other isotopic excursions will become apparent when higher resolution studies are performed, and other undifferentiated stratigraphic horizons are investigated (e.g., not just black shales or extinction/radiation horizons).

Unfortunately, this is the nature of the research-funding and publication beast, but as researchers we can try in the future to create a more synergetic approach when generating a stratigraphic carbon isotope curve for a site/basin/region. Fig. 2 schematically illustrates how the sample resolution strategy should differ between sections with different sedimentation rates and therefore, time. For example, if a stratigraphic interval is ~4 m in height and samples are collected every 20 cm, then a more condensed section (e.g., 2 m) should sample at half that resolution (~10 cm) and an even more condensed section would be half that (e.g., ~5 cm) (see Fig. 2). Assuming each stratigraphic section is the same duration, then the carbon isotope curves generated would be directly comparable. However, we know that this is far from the real world when sampling sediments in the field and strategies would have to be developed for each section in each basin.

Fig. 3 schematically shows an idealized complete section with equal stratigraphic units in thickness and time (i.e., constrained through geochronology and/or biostratigraphy). Sedimentary units within the basin will vary in thickness according to sedimentation rate and therefore, the sampling frequency should be adjusted accordingly assuming that depicted in Fig. 2. Three other sections are shown in Fig. 3 with varying unit thickness. It illustrates how sampling frequency will change according to sedimentary units being expanded (lower resolution sampling) or condensed (higher resolution sampling). Of course, the issue with this approach comes down to the fact that we have an understanding of time (e.g., duration). Even without this knowledge regarding precise time/duration, this type of sampling strategy should produce more reliable stratigraphic $\delta^{13}C$ curves for the purpose of correlation.

2.3 Stratigraphic uncertainty

Stratigraphic uncertainty can derive itself from variations in sample resolution (previously discussed) or through a lack of understanding of the proxy itself. Various proxies will record specific environmental conditions and a single sample will not provide all the information. For example, charcoal fragments may contain a single growth ring (i.e., annual season) or mutliple growth rings (i.e., years). Similarly, brachiopods and bivalves will have sclerochronological features that need to be considered (e.g., Gröcke and Gillikin 2008). Therefore, with respect to the carbon isotope curve generated, it may reflect a single moment in time (e.g., leaf cuticle, amber) or integrate a series of years or decades into a single sample (e.g., tree trunk, laminated sediment). One way to potentially combat this issue could be through analyzing multiple sub-samples from a single layer or taking a larger sample size to homogenise. For example, you could combine multiple leaf cuticle samples from a single sediment sample or horizon to produce a single isotopic result or analyze each cuticle fragment separately to ascertain the natural isotopic variation of the horizon and hence, the carbon ecosystem. Fig. 4 illustrates these two sampling strategies and the resultant carbon isotope curve of the sample set (this does not include instrumental and standard error produced when generating $\delta^{13}C$ values). A single sample records a $\delta^{13}C$ value, but it does not record natural variation in the system. This kind of consideration should not only apply just to terrestrial fragments

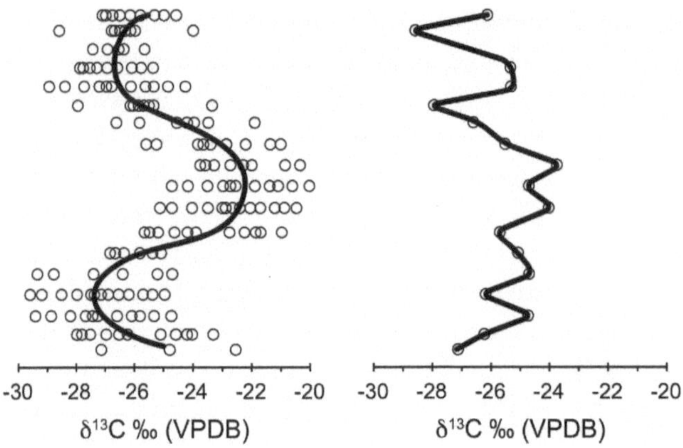

Fig. 4 (Left) An idealized $\delta^{13}C$ curve generated from multiple plant cuticle specimens from each horizon. This carbon isotope data will show the "natural" spread of the system for each horizon, followed by the running average curve (bold thick line) for the complete dataset. The bold thick line will be influenced by the number of samples analyzed within each horizon; hence, a minimum number of samples should be obtained from each horizon for $\delta^{13}C$ analysis (e.g., >5 samples). (Right) A random selection of samples from the dataset on the left generates a $\delta^{13}C$ curve that is very different and would generate an incorrect correlation.

(e.g., cuticle) but other carbon proxies as well, for example, single foraminifera analysis (e.g., Fourel et al., 2016; Fraass and Lowery, 2017; Killingley et al., 1981) and palaeosol nodules (e.g., Fox and Koch, 2004; Behrensmeyer et al., 2007).

Although one of my own pieces of work (Gröcke et al., 1999) sampled wood fragments where possible in a thick stratigraphic section from the Isle of Wight, UK, there are other studies that missed the opportunity to generate a more robust, complete $\delta^{13}C$ curve. An example of this was performed by Arens and Jahren (2000), who produced a sparse cuticle record across the Cretaceous/Tertiary (K/T, K/Pg) stratigraphic boundary from Montana, USA. Although they report an isotopic excursion just above the K/T boundary (on the order of 1.5–2‰), the $\delta^{13}C$ excursion is only recorded in a single stratigraphic sample/horizon. This type of result would often be referred to as a "flier." There is also a $\delta^{13}C$ excursion of equal magnitude in the Hauso Flat section some 0.39 Myrs later which is not considered (Arens and Jahren, 2000). Although Arens and Jahren (2000) show $\delta^{13}C$ correspondence between cuticle and bulk sediment $\delta^{13}C$ ($n=68$) to suggest that bulk sediment $\delta^{13}C$ is dominated by plant organic matter, there are many cases where the $\delta^{13}C$ difference between cuticle and bulk sediment is on the order of 1‰ (>50% of the K/T $\delta^{13}C$ excursion). Arens and Jahren (2000) only actually provide 19 results in their dataset with coupled bulk sediment and cuticle $\delta^{13}C$. What is not clear in Arens and Jahren (2000) is whether the cuticle isotopic results contain multiple cuticle specimens or a single cuticle specimen from each horizon. This kind of information is very important to report when constructing a terrestrial $\delta^{13}C$ record for correlation purposes, and for developing a greater understanding of the terrestrial carbon ecosystem.

Another issue when analysing cuticle in the geologic record is that $\delta^{13}C$ is highly variable in this proxy. For example, bulk leaf $\delta^{13}C$ variation within a modern *Ginkgo biloba* leaf is illustrated in Fig. 5. It clearly shows a distinct pattern with more ^{13}C-depleted values around the margin and photosynthetic dominant parts of the leaf compared to the more robust stem area.

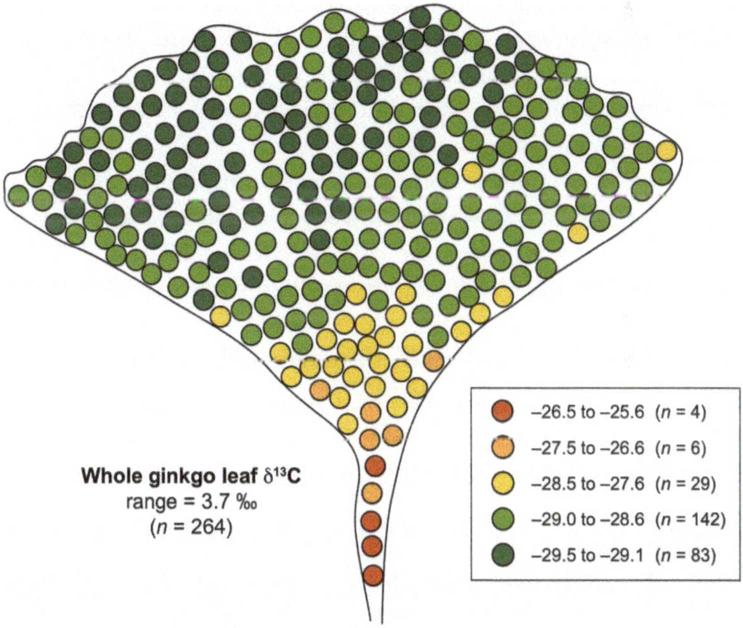

Fig. 5 Whole leaf $\delta^{13}C$ variation in a modern *Ginkgo biloba*. This outdoor ginkgo leaf collected at McMaster University was sub-sampled 264 times for $\delta^{13}C$ analysis. The $\delta^{13}C$ values in the leaf range from −25.7‰ to −29.4‰ (~3.7‰). Note, the $\delta^{13}C$ variation with more depleted values in the outer parts of the leaf compared to less depleted values at the base and stem of the leaf. This pattern has been observed in multiple *Ginkgo biloba* leaves from Canada and the United Kingdom, but this leaf was the first and with the most extreme sampling strategy.

This type of isotopic pattern has been reproduced in many ginkgo leaves from Canada and the United Kingdom (Gröcke, unpub. data): but not at this ridiculous resolution! The chance of randomly sampling a piece of cuticle with the ^{13}C-depleted values (e.g., base of leaf and stem) is only 15%, but the chance of those pieces surviving are probably higher due to their thickness and more robust nature. The $\delta^{13}C$ variation within a single leaf (*sensu* Fig. 5, 3.7‰) is greater than the cuticle K/T boundary $\delta^{13}C$ excursion reported in Arens and Jahren (2000) and many other terrestrial isotope excursions in the geologic record.

Grein et al. (2010) analyzed fossil evergreen leaves from the Eocene Messel fossil site and reported large $\delta^{13}C$ variation (up to 3‰) in different leaf species within the assemblage. In addition, they showed similar trends in the fossil leaves with more negative $\delta^{13}C$ values in the tip of the leaves *versus* the base: differences ranging from 2.2‰ to 3.0‰ in the fossil leaves and between 1.2‰ and 3.2‰ in extant leaves. A more recent study by Kipp et al. (2019) also report more negative $\delta^{13}C$ values in the length of a cycad leaf compared to the structural/stem samples. Although at much lower resolution these data also indicate that differential isotopic values in leaves relate to compositional differences or isotopic discrimination and carbon transportation during the growing season (*sensu* Leavitt and Long, 1986; Schleser, 1992; Martinelli et al., 1998; McDowell et al., 2011). To explore this further, Cretaceous leaf fossils were collected from Yankee Hill, Nebraska, and analyzed for $\delta^{13}C$ as a pilot study (see Fig. 6) to assess $\delta^{13}C$ variation in angiosperm leaves. Each leaf fragment recorded high $\delta^{13}C$ intra-variability (between 0.8‰ and 1.2‰), and a total population ($n=4$ leaves) $\delta^{13}C$ variation between −23.6‰ and −26.2‰ (range = 2.6‰). Although the leaves were not complete it was still possible to see a similar trend in $\delta^{13}C$ from the leaf margin towards the base as depicted in the modern gingko leaf (see Fig. 5).

Results of this kind have significant implications when extracting and analyzing cuticle from sedimentary rocks. For example, if the cuticle is controlled by preservation potential then the sediment may be biased toward thicker cuticle or stem material. In addition, depending on the travel path of the cuticle (e.g., river) more robust material will survive transportation, which again will provide a biased record of $\delta^{13}C$ values for that deposit. This could equally be exacerbated in environments that have low or high oxygen concentrations (e.g., shallow-marine anoxia). Unfortunately for organic geochemists, this kind of internal leaf variation from tip to base also translates itself into *n*-alkane $\delta^{13}C$ values of up to 4‰ (i.e., Gao et al., 2015).

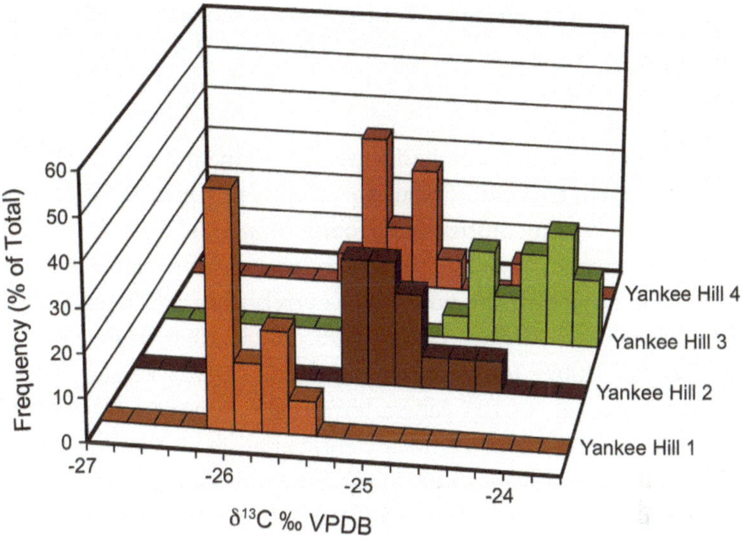

Fig. 6 Four fragmentary leaf samples of non-taxonomic use were collected from Yankee Hill, Nebraska. Fossil plants were obtained during the original excavation, with the majority of taxonomic samples reported in Wang et al. (2011). Each leaf fragment was from the plant, *Eoplatanus serrata* (Wang et al., 2011). Subsamples of each leaf were carefully removed and washed of any sediment using de-ionized water and subsequently analyzed for $\delta^{13}C$. Each fragment recorded high leaf intra-variability $\delta^{13}C$ (between 0.8‰ and 1.2‰), and a total population range from −23.6‰ to −26.2‰ (range = 2.6‰). The samples were not decalcified as there is no evidence of carbonate in the sediment.

Interestingly though, Gao et al. (2015) do not find *n*-alkane isotopic variation in *Ginkgo biloba* leaves: unpublished data in 2005 by Gröcke, Wagner and Pedentchouk also indicate no $\delta^{13}C$ variation in bulk *n*-alkanes in modern *Ginkgo biloba* leaves.

Although there is intra-leaf and inter-leaf $\delta^{13}C$ variation in fossil and extant leaves, there is still the potential to generate valuable palaeoenvironmental and palaeoclimatic information from leaves. Nguyen Tu et al. (1999) produced an exquisite dataset where they analyzed different fossil leaves from a single sedimentary deposit in order to reconstruct the spatial distribution of plants relative to the proximity of a lagoon (i.e., seawater). The $\delta^{13}C$ signature was more ^{13}C-enriched in plants that lived near the lagoon (due to environmental stress; see Gröcke, 1998) whereas those with "normal" plant $\delta^{13}C$ values lived further away from the influence of saline water. Nguyen Tu et al. (1999, 2002) suggest that the conifer, *Frenelopsis*, lived closer and more densely in an environment that was affected by saline water and/or sea spray. Further research by Aucour et al. (2008) indicate that this habitat niche for *Frenelopsis* was common in the mid-Cretaceous continental basin of south-west Europe (e.g., France, Spain).

Unfortunately, the research performed by Nguyen Tu et al. (1999) and Aucour et al. (2008) are isolated examples of how powerful the $\delta^{13}C$ of fossil leaves can be used to reconstruct palaeoenvironments. Most studies, however, focus on using leaf, wood or bulk terrestrial organic matter to construct carbon isotope curves for the purpose of correlation (e.g., Baczynski et al., 2016; Gröcke, 1997; Gröcke et al., 1999, 2005, 2006; Hasegawa, 1997; Hesselbo et al., 2007; Ruebsam et al., 2020; Suarez et al., 2017; Wu et al., 2020; Yans et al., 2010). Albeit, terrestrial isotope research is growing, our understanding of modern (and ancient) variation is often overlooked or over-simplified. For example, I distinctly remember being asked when I first presented the ginkgo results (see Fig. 5) at a conference; "*What amount of carbon isotope error [variation] would I place on individual samples in a stratigraphic record?*," to which I responded "about 10‰." The audience looked astonished, but as with all carbon isotope systems there is inherent variability due to environment, climate and/or biology. Although a single sample may be part of a 10‰ $\delta^{13}C$ terrestrial ecosystem, a stratigraphic collection of samples at high density should still produce a robust carbon isotope curve through time (see Fig. 4). The $\delta^{13}C$ curve will ultimately be governed by changes in the global carbon cycle. A researcher needs to take into account this type of $\delta^{13}C$ variation in all depositional and environmental settings. For example, many studies have even shown high $\delta^{13}C$ variability in the marine environment when individual foraminifera are analyzed (e.g., Metcalfe et al., 2015; Stap et al., 2010; Thomas et al., 2002).

Fig. 7 provides a schematic example of three terrestrial carbon isotope curves generated for the same time interval. Time is based on the first appearance datum and last appearance datum of fossil "x" (e.g., FAD x to LAD x). Within each one of these sections the sampling frequency is altered based on the number of horizons sampled to generate the $\delta^{13}C$ curve. The number of horizons sampled is different and show clearly that the greater number of horizons analyzed the more confidence you have in the generated $\delta^{13}C$ curve. Less horizon samples produce poor, indeterminant correlations. The same example is depicted in Fig. 8. However, in this scenario each section has been sampled from 13 horizons although the number of samples collected and analyzed for $\delta^{13}C$ is variable. In this case the $\delta^{13}C$ curve generated using fewer individual samples (i.e., Curve A) produces a very noisy

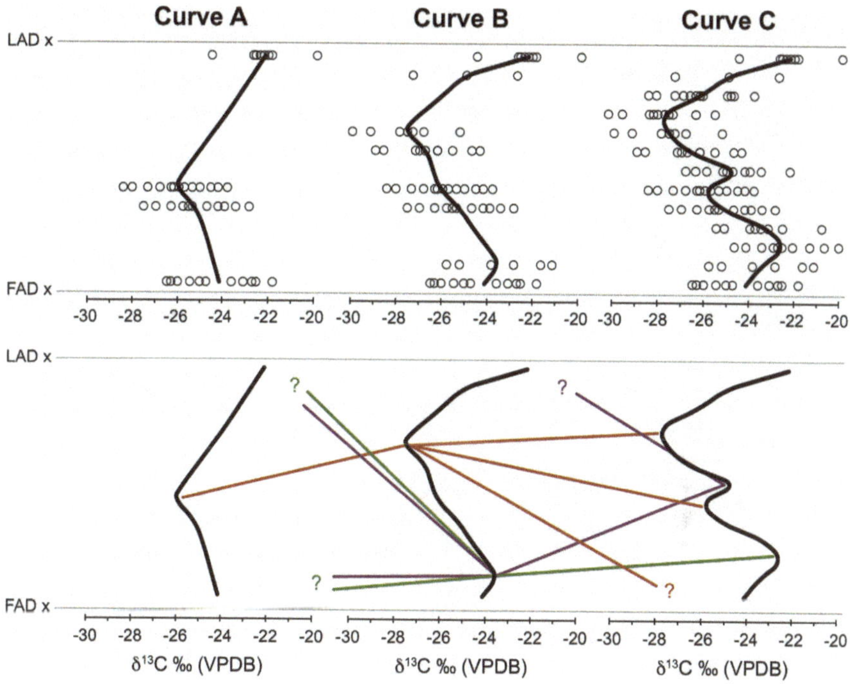

Fig. 7 A schematic example of three terrestrial carbon isotope curves generated for the same time interval based on the first appearance datum and last appearance datum of fossil "x" (e.g., FAD x to LAD x). Note, a similar result may be produced with sedimentary $\delta^{13}C$ that has variable marine *versus* terrestrial organic matter contents (for example, see the $\delta^{13}C$ scatter in Fig. 2 from Uličný et al., 2014). Each section has different sampling frequencies of multiple analyses. The top portion of the figure shows the data and resultant $\delta^{13}C$ curves for Curve A which is based on only four horizons, Curve B on eight horizons and Curve C on 13 horizons. The lower portion of the figure illustrates the potential correlations that can be made between Curve A, B and C with the latter curve being the most accurate. Only major negative and positive $\delta^{13}C$ excursions are correlated (or inferred = ?). Curve A in this case is the least correlatable section with Curves B and C.

$\delta^{13}C$ curve, whereas Curve B and C produce more robust curves. As before, this theoretical example shows that the greater the number of individual samples analyzed for $\delta^{13}C$ per horizon produces a more robust, thorough $\delta^{13}C$ curve for the purpose of correlation. In conjunction with producing a correlation curve, the $\delta^{13}C$ data is providing information about the palaeo-carbon ecosystem; a very valuable source of information for future investigations and environmental reconstructions.

Not only is there carbon isotope variation in terrestrial and marine proxies such as foraminifera, but bulk marine sediment can also produce $\delta^{13}C$

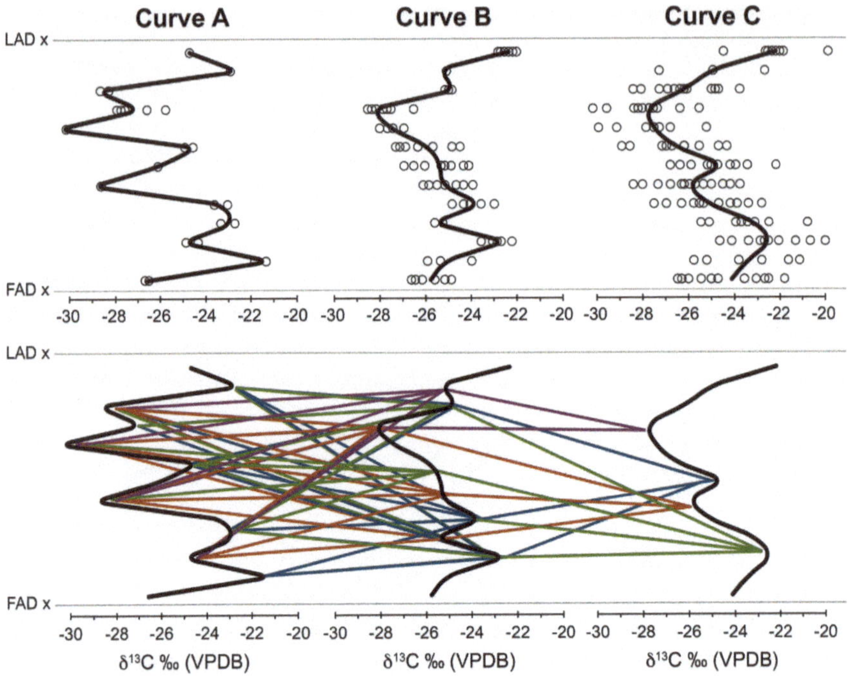

Fig. 8 A schematic example of three terrestrial carbon isotope curves generated for the same time interval based on the first appearance datum and last appearance datum of fossil "x" (e.g., FAD x to LAD x). Each section has different sampling frequencies from single to multiple $\delta^{13}C$ analyses in each horizon. The top portion of the figure shows the data and resultant $\delta^{13}C$ curve for Curve A, B and C that all have 13 horizons studied. The potential to obtain increased uncertainty in correlating is represented in Curve A due to the analysis of only one or a few samples per horizon. The lower portion of the figure shows the potential correlations that can be made between Curve A, B and C with the latter curve being the most accurate. As indicated by the correlation lines Curve A is the least correlatable section to Curve C.

variability within a specific stratigraphic horizon. Toarcian sediment samples from Port Mulgrave have been analyzed for organic $\delta^{13}C$ along a single bed over ~5 m of outcrop and produced a standard deviation of ±0.7‰ (Gröcke, 2013 unpublished data). Further work on individual 20 cm-diameter blocks from Port Mulgrave recorded $\delta^{13}C$ variation within single laminated beds from 0.1‰ to 0.3‰ (Dalby & Gröcke 2014, unpublished data). While this variation seems to be relatively minor, this example is only from a single bed of shale, in a single location and a specific time period. It is known that modern marine sediment is highly variable in $\delta^{13}C$ both spatially and temporally (e.g., Bosman et al., 2020; Gearing et al., 1984; Shultz and Calder, 1976; Zhou et al., 2018). Although geological processes relating to marine sedimentation may dilute and/or time-average this degree of variation (see Schobben and van de Schootbrugge, 2019), understanding the spatial variability of carbon isotope stratigraphy within a single basin should be constructed where possible. Creating a greater appreciation of $\delta^{13}C$ variation will help to construct a better understanding of basin dynamics; for example, determining the spatial distribution of organic matter (terrestrial *versus* marine), sedimentation packages of erosion and deposition (e.g., clinoforms) and in generating a composite, more complete $\delta^{13}C$ curve for correlation purposes.

2.4 Stratigraphic reproducibility

Terrestrial and marine environmental systems inherently show carbon isotope variability. Understanding basin $\delta^{13}C$ variability is extremely important as this will influence the isotopic framework from which correlations are made. Of course, the sedimentological and stratigraphic framework of the basin will define regions to generate the most complete section. However, there will be circumstances where this may not be the case due to cryptic surfaces that are invisible to the naked eye and/or difficult to interpret. For example, Hesselbo et al. (2000) used sedimentary laminations to estimate the onset and duration of the T-OAE $\delta^{13}C$ excursion. The laminations have since been shown, through the use of thin-section analysis by Trabucho-Alexandre (2014) to be beds bound by erosional surfaces and hence, the duration stated in Hesselbo et al. (2000) should only be considered as a "minimum." Therefore, even in a supposedly "complete" shale sedimentary sequence there is the potential to be gaps in the record, as highlighted by Trabucho-Alexandre (2014). So, abrupt/rapid negative $\delta^{13}C$ excursions in fact may not be reflecting changes in the carbon cycle (e.g., clathrate release, global wildfires), but in fact representative of stratigraphic time gaps. Trabucho-Alexandre (2014) and Them et al. (2017) have shown this illustratively: a similar depiction can be seen in Fig. 9 where a cyclic, negative or positive global $\delta^{13}C$ excursion is recorded. The theoretical study site (Fig. 9) has unconformities (cryptic surfaces) producing a resultant $\delta^{13}C$ curve that would generate incorrect correlations.

If we consider Fig. 9, imagine what the $\delta^{13}C$ stratigraphies would be like in a shallow-marine or terrestrial sedimentary sequence. This is where sequence stratigraphy coupled with carbon isotope stratigraphy could really play a part in future research. A stratigraphic basin is made up of many sedimentary sequences that form packages of clinoforms. A single stratigraphic

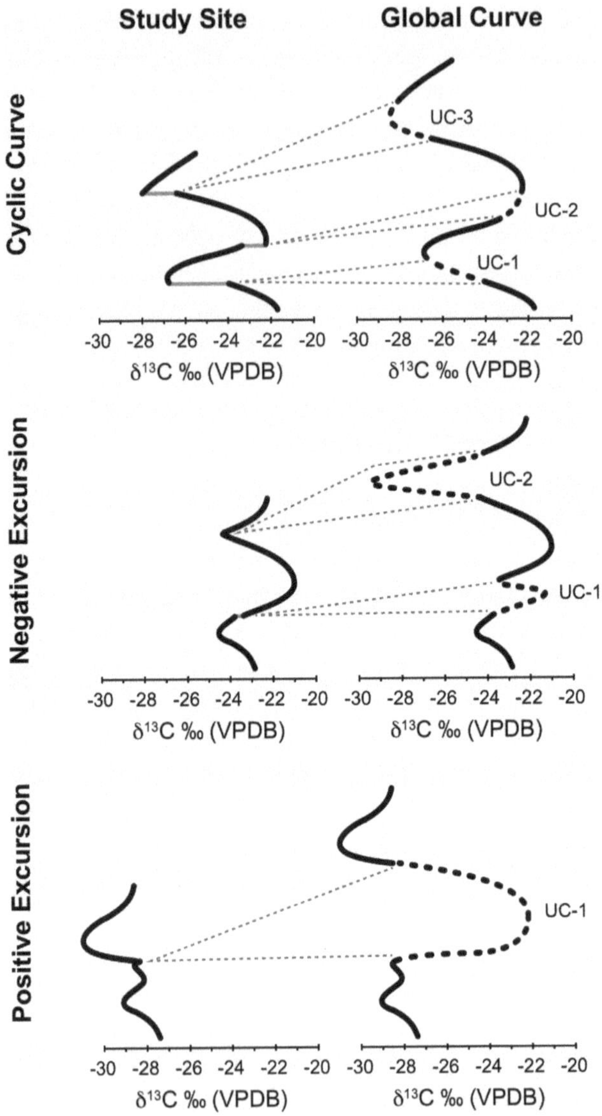

Fig. 9 Schematic $\delta^{13}C$ curves from a study site *versus* a global curve. The study site contains cryptic horizons (i.e., unconformities, UC) that result in parts of the global $\delta^{13}C$ to be missing. This produces a very different study site $\delta^{13}C$ curve and hence, the resultant $\delta^{13}C$ correlation would be erroneous. The cyclic $\delta^{13}C$ curve would produce a similar curve in the study site, but the abruptness and cause/s of the positive and negative $\delta^{13}C$ excursions would be potentially mis-interpreted. In the middle example, the negative $\delta^{13}C$ excursion is completely missing; other examples can be found in Trabucho-Alexandre (2014) and Them et al. (2017). In the bottom example, the positive $\delta^{13}C$ is completely missing and the study site would indicate only a negative $\delta^{13}C$ excursion. Many more examples of this could be produced, but the key point is to be aware of gaps in the $\delta^{13}C$ record through sedimentological investigations.

section being investigated will likely hold only one part of the basin's history (see Fig. 10). If one was to sample the deep basin record the resultant $\delta^{13}C$ curve would miss the shallower stratigraphic sequence/s, and *vice versa* if you sampled the shallower sequence. As illustrated in Fig. 10, each of the three cores sampled will produce very different isotopic curves since the complete time sequence (a to u) is not represented in any of the cores. In scenarios like this each clinoform would need sampling and the resultant curves combined to produce a composite curve (see Fig. 11). Once again, this is where sampling strategy and sample resolution will play a pivotal role in generating robust $\delta^{13}C$ curves (see Figs. 7 and 8). More investigations are emerging that have considered sequence stratigraphy when generating

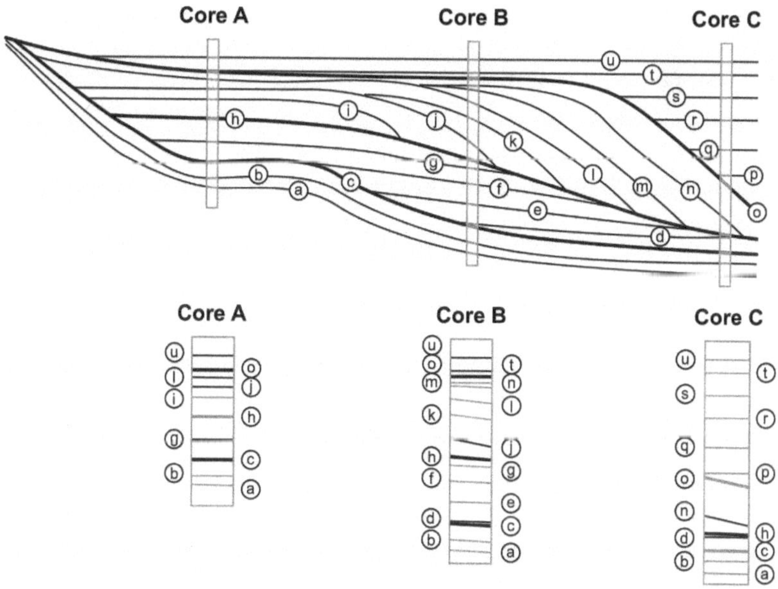

Fig. 10 A schematic sequence stratigraphic framework for a systems tract model. In this example, three cores within the sequence have been taken: thick black lines in the core represent time-stratigraphic gaps. Each core does not extract all the stratigraphic information from this depositional sequence. Thus, generating a $\delta^{13}C$ curve through these cores would need to be combined in order to produce a composite $\delta^{13}C$ curve for the basin. Even in this example some clinoforms would only have a portion their units sampled, and hence, alternative sampling strategies would need to be incorporated (see Figs. 2 and 3). A more strategic approach is illustrated in Fig. 11. *Modified from Vail, P.R., 1987. Seismic stratigraphy interpretation using sequence stratigraphy, Part 1: seismic stratigraphy interpretation procedure. In: Bally, A.W. (Ed.). Atlas of Seismic Stratigraphy. Am. Assoc. Pet. Geol. Studies in Geology 27, 1–10.*

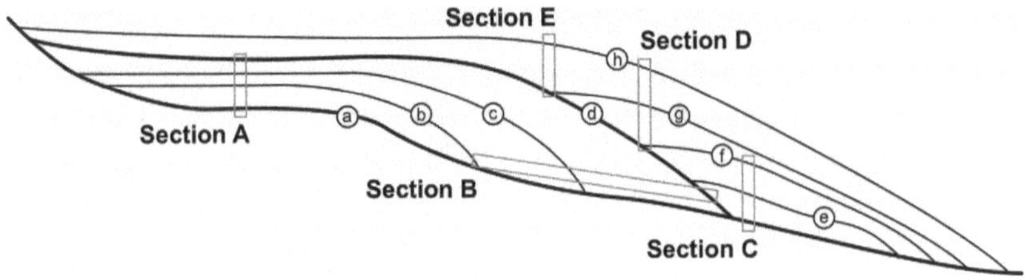

Fig. 11 A schematic sequence stratigraphic framework for a systems tract model sampled specifically for generating a composite $\delta^{13}C$ curve for the basin. In this scenario, at least five stratigraphic sections would require sampling to produce a composite $\delta^{13}C$ record. This kind of approach applies to both marine and terrestrial stratigraphic sequences—although it will be more complex and more of a challenge within terrestrial sedimentary basins.

carbon isotope curves (e.g., Ferreri et al., 1997; Henning et al., 1999; Stephens and Sumner, 2003; Gröcke and Joeckel 2008; Ludvigson et al., 2015; LeRoy and Gill, 2018). Fig. 11 shows a simplified systems tract model for a basin with corresponding timelines. In order to generate a composite $\delta^{13}C$ record from this stratigraphic basin would require a multi-site sampling strategy. However, every basin will have its own complexity and detangling this complexity for fieldwork will be required if a more complete, composite $\delta^{13}C$ curve is to be generated. Basin analysis and reconstructing the carbon cycle between shallow and deep marine environmental settings will provide valuable information on ecosystem structure, carbon pump efficiency, redox conditions and sea level variation and responses, for example.

Understanding carbon dynamics in a single basin is difficult (see the challenges reported in Suarez et al., 2017), and therefore, when comparing different basins, the challenge will be even more difficult. Biostratigraphic challenges still persist in both marine and terrestrial realms due to issues related to the fossils themselves (e.g., synchroneity, taxonomic affinities, stratigraphic ranges, habitat) and fossilization processes (e.g., preservation potential, dissolution/diagenesis). Knowing the biological marker horizon between two basins (while considering all the issues mentioned above!) a semi-quantitative correlation can be ascertained. Of course, stratigraphic sections that can generate absolute ages in the stratigraphy would help also. Although assumptions will need to be made with respect to sedimentation rates, stratigraphy, biostratigraphy and depositional environment, a multi-discipline approach in generating a $\delta^{13}C$ curve in geological time is advised.

3. Stratigraphic correlation

Now that that has all been considered (phew!), we get to the primary basis of carbon isotope stratigraphy—correlation. There are many studies being published recently that show correlations which seem to have come from the spiritual world. Correlations need to be robust and either rely solely on a robust $\delta^{13}C$ curve and/or through a variety of other geochronological, stratigraphic and biostratigraphic information to support the data. A single data point in the marine or terrestrial record should ***not*** (ever!) be used to correlate one basin to another. Absolute $\delta^{13}C$ values also should never be used—there is no reason why this should even be the case considering our understanding of the modern carbon cycle.

The first approach to generating a robust correlation is to use the broad, general trends and patterns that are produced in the $\delta^{13}C$ curve/s. One should start by tracing the $\delta^{13}C$ trends; for example, if the overall trend in $\delta^{13}C$ is increasing or decreasing. If there are no independent markers for correlation, then the major peaks or troughs should be matched up. Additionally, when correlating two stratigraphic sections, first look at the entire $\delta^{13}C$ curve, noting the overall trends that could be used. The trend *inflection point* is where the correlation line is drawn; not at the start or end of an excursion. But what defines a "trend inflection point"? It is certainly not a single data point. There is no rule of thumb here, although a series of isotopic data that is negative or positive would define an excursion. How different do the $\delta^{13}C$ values need to be to define an excursion? This will depend on the stratigraphic interval being investigated. Some time intervals have small, low magnitude excursions, whereas others will record major changes. Rather than define a $\delta^{13}C$ excursion with a precise number, it certainly needs to be outside instrumental error and sample reproducibility. Hence, an understanding of stratigraphic reproducibility is required. For example, the Toarcian sediments discussed previously should have an error bar of at least $\pm 0.7‰$ and so a $\delta^{13}C$ excursion must be greater than that and be supported by many samples and not a single data point. Kemp et al. (2005) produced a high-resolution $\delta^{13}C$ through the T-OAE from Yorkshire, and using the stratigraphic reproducibility of $\pm 0.7‰$, there are many apparent $\delta^{13}C$ shifts in this curve that would not be classified as an excursion: for example, the changes between excursion A and B in Fig. 1 of Kemp et al. (2005). So, in summary: ignore single data points unless you fill in the stratigraphy with more high-resolution data and show that it was not a flier; define the stratigraphic reproducibility of your section in each dominant lithology type; identify $\delta^{13}C$ excursions that are significant (above stratigraphic reproducibility) for correlation.

Once the major trends have been correlated, then the finer detail of the $\delta^{13}C$ curve can be constructed (e.g., short-lived negative and positive excursions) taking into consideration what has been discussed. Often though, this is where the over-enthusiastic stratigrapher will try and correlate every peak and trough, which could be feasible if you are correlating apples with apples based on sample resolution, etc., as previously discussed. But if you are comparing apples with oranges (high resolution *versus* low resolution $\delta^{13}C$ curves), then a more simplified (honest) correlation of the major $\delta^{13}C$ inflection points should only be correlated. Some examples where the above has not been adequately considered and applied can be found in Schnyder et al. (2017), Bodin et al. (2016), Hines et al. (2019) and Bastos et al. (2020): note, there are many more examples out there! Unfortunately, a lot of this comes down to experience, but also being aware of all the issues, discussed previously, that are associated with stratigraphy, sedimentology and the carbon proxy itself. No one is perfect and mistakes or assumptions would have been made in the past, but hopefully with increased awareness new carbon isotope stratigraphic records will be more robust.

Another *faux pas* that is becoming more prevalent in the literature pertains to correlating carbonate and organic matter $\delta^{13}C$ curves. Although in principle this would seem like an obvious methodological approach; since, the dominant control on $\delta^{13}C$ through geologic time is the isotopic composition of CO_2. However, it should always be considered that the carbonate and organic systems are controlled by other factors as well (e.g., alkalinity, growth rates, biosynthesis). Early work by Rau et al. (1989, 1991) and others (e.g., Hayes, 1993; Hayes et al., 1999; Kump and Arthur, 1999; Popp et al., 1989; Popp et al., 1998) have shown that pCO_2 has a significant control on the $\delta^{13}C$ value of organic matter. Differences between carbonate and organic matter carbon isotope curves can be expressed through $\Delta^{13}C‰$ (VPDB); $\Delta^{13}C‰ = \delta^{13}C_{carb} - \delta^{13}C_{org}$. Therefore, during time intervals when there are major changes in pCO_2, this will have an effect on the $\delta^{13}C$ composition of carbonate and organic matter and hence, correlations. Of course, there are many time intervals where the dominant $\delta^{13}C$ curve generated have used bulk carbonate and so one is left with no other option (e.g., Gröcke et al., 1999; Gröcke et al., 2005, 2006; Hasegawa, 1997; Jarvis et al., 2011, 2015; Jenkyns and Clayton, 1997; Tsikos et al., 2004;

Voigt et al., 2007; Weissert, 1989). As more $\delta^{13}C$ curves are generated this will become less of an issue, and then more palaeoenvironmental information can be extracted from the data. But in the meantime, carbon isotope stratigraphy should primarily rely on correlating like for like data; that is, carbonate $\delta^{13}C$ with carbonate $\delta^{13}C$, and organic $\delta^{13}C$ with organic $\delta^{13}C$. Many studies have generated and compared carbonate and organic matter $\delta^{13}C$ to discern fluctuations in carbon isotope fractionation and hence, a proxy for pCO_2 concentrations (e.g., Bodin et al., 2016; Caruthers et al., 2011; Edwards and Saltzman, 2016; Freeman and Hayes, 1992; Gröcke et al., 2005; Hasegawa, 2003; Jarvis et al., 2011; Kump et al., 1999; Pancost et al., 1999; Ruebsam et al., 2020; Sageman et al., 2006; Sarkar et al., 2003; Scopelliti et al., 2008; Thibault et al., 2018; Tsikos et al., 2004; van Hengstum and Gröcke, 2008; Voigt et al., 2007; Wu et al., 2020). In addition, there has been recent questions regarding the reliability of bulk carbonate $\delta^{13}C$ values in the Cenozoic (e.g., Jones et al., 2020; Metzger and Fike, 2013; Oehlert et al., 2012; Swart, 2008, 2015; Swart and Eberli, 2005). Conversely, Proterozoic researchers tend to favour $\delta^{13}C$ in carbonate over organic matter due to issues relating to diagenesis/catagenesis (e.g., Des Marais, 2001; Schidlowski, 2001). But it should be noted that isotope stratigraphy in Archaean and Proterozoic rocks is very different in terms of stratigraphy and isotopic excursions. For example, outcrops are often limited or not present and the isotopic changes are more significant compared to the Phanerozoic where sampling intensity is much higher, excursions are much smaller, biological evolution is more diverse and there are more basins, stratigraphic sections and environments to sample. So, what discoveries could we made if a Phanerozoic isotope stratigraphy approach was applied to Archaean and Proterozoic rocks... ?

4. Conclusions and future directions

By no means have all aspects of carbon isotope stratigraphy been covered in this review. The discipline is growing year by year. Even compared to two decades ago we have come a long way, but there is a lot more we can do to understand the ancient carbon cycle. Carbon isotope stratigraphy will become a generic tool for understanding stratigraphy and for improving global correlations—eventually, it will be the basis from which everything else is compared too. Carbon isotope curves will become more precise and robust through more detailed stratigraphic sampling and comparisons. Bed by bed data will provide unequivocal information on the carbon cycle on even shorter time scales in the geologic record. Data of this kind will be used to reconstruct comprehensive information on ecosystem carbon, and how it modified itself during more extreme climatic periods (e.g., greenhouse *versus* icehouse climate modes). Additional (or new) proxies previously underutilized in conjunction with carbon isotope stratigraphy will be developed further, paving the way to new, more all-inclusive data on the Earth climate system. Coupled $\delta^{13}C$ stratigraphies with other geochemical and bio-physical proxies will also be established to generate superior palaeoenvironmental and palaeoclimatic data that can be used in climate modeling. The future of carbon isotope research and carbon isotope stratigraphy is very promising—I hope that you will join the movement.

Acknowledgements

Where do I start? First, I must thank David McKirdy for telling me to go off during my undergraduate degree and learn whether stable isotopes can be used to reconstruct ancient diets. Second, to Michael Bird, who really showed me how carbon isotopes could be used to understand the ancient carbon cycle, and that not everything in the laboratory goes according to plan. Finally, to Prof Mark Pagani, a guiding light and breath of fresh air in academic circles: rest in peace dear friend. And then, where would I be without my stratigraphic buddies who listen to me all the time, no matter how ridiculous or senseless it sounds—thanks guys—Andrew Caruthers, Benjamin Gill, Theodore Them II and João Trabucho-Alexandre (alphabetical order guys!). I have crossed paths with many other highly respected academics in my life who have had an impact on my career and research, but a few have entered my thoughts on many occassions while I was writing this review; Larry Frakes, Jim Zachos, Mike Arthur, Isabel Montañez, Stephen Hesselbo, Hugh Jenkyns and Lisa Pratt.

References

Arabas, A., Schlögl, J., Meister, C., 2017. Early Jurassic carbon and oxygen isotope records and seawater temperature variations: insights from marine carbonate and belemnite rostra (Pieniny Klippen Belt, Carpathians). Palaeogeogr. Palaeoclimatol. Palaeoecol. 485, 119–135.

Arens, N.C., Jahren, A.H., 2000. Carbon isotope excursion in atmospheric CO_2 at the Cretaceous-tertiary boundary: evidence from terrestrial sediments. Palaios 15, 314–322.

Aucour, A.-M., Gomez, B., Sheppard, S.M.F., Thévenard, F., 2008. $\delta^{13}C$ and stomatal number variability in the Cretaceous conifer *Frenelopsis*. Palaeogeogr. Palaeoclimatol. Palaeoecol. 257, 462–473.

Baczynski, A.A., McInerney, F.A., Freeman, K.H., Wing, S.L., The Bighorn Basin Coring Project (BBCP) Science Team, 2019. Carbon isotope record of trace *n*-alkanes in a continental PETM section recovered by the Bighorn Basin Coring Project (BBCP). Paleoceanogr. Paleoclimatol. 34, 853–865.

Baczynski, A.A., McInerney, F.A., Wing, S.L., Kraus, M.J., Morse, P.E., Bloch, J.I., Chung, A.H., Freeman, K.H., 2016. Distortion of carbon isotope excursion in bulk soil organic matter during the Paleocene-Eocene thermal maximum. Geol. Soc. Am. Bull. 128, 1352–1366.

Bains, S., Corfield, R.M., Norris, R.D., 1999. Mechanisms of climate warming at the end of the paleocene. Science 285, 724–727.

Barclay, R.S., Wing, S.L., 2016. Improving the *Ginkgo* CO_2 barometer: implications for the early Cenozoic atmosphere. Earth Planet. Sci. Lett. 439, 158–171.

Bastos, L.P.H., Pereira, E., da Costa Cavalcante, D., Alferes, C.L.F., de Menezes, C.J., Rodrigues, R., 2020. Expression of Early Cretaceous global anoxic events in Northeastern Brazilian basins. Cret. Res. 110, 104390.

Behrensmeyer, A.K., Quade, J., Cerling, T.E., Kappelman, J., Khan, I.A., Copeland, P., Roe, L., Hicks, J., Stubblefield, P., Willis, B.J., Latorre, C., 2007. The structure and rate of late Miocene expansion of C_4 plants: evidence from lateral variation in stable isotopes in paleosols of the Siwalik Group, northern Pakistan. Geol. Soc. Am. Bull. 119, 1486–1505.

Berger, W.H., 1979. Stable isotopes in foraminifera. In: Lipps, J., Berger, W.H., Buzas, M.A., Douglas, R.G., Ross, C.A. (Eds.), Foraminiferal Ecology and Paleoecology. Society of Economic Paleontologists and Mineralogists, pp. 156–198. SEPM Short Course No. 6.

Berner, R.A., 1992. Weathering, plants, and the long-term carbon cycle. Geochim. Cosmochim. Acta 56, 3225–3231.

Berner, R.A., 1998. The carbon cycle and CO_2 over Phanerozoic time: the role of land plants. Phil. Trans. R. Soc. Lond. B 353, 75–82.

Bock, M., Schmitt, J., Beck, J., Seth, B., Chappellaz, J., Fischer, H., 2017. Glacial/interglacial wetland, biomass burning, and geologic methane emissions constrained by dual stable isotopic CH_4 ice core records. Proc. Natl. Acad. Sci. U. S. A. 114, E5778–E5786.

Bodin, S., Krencker, F.-N., Kothe, T., Hoffmann, R., Mattioli, E., Heimhofer, U., Kabiri, L., 2016. Perturbation of the carbon cycle during the late Pliensbachian e early Toarcian: new insight from high-resolution carbon isotope records in Morocco. J. African Earth Sci. 116, 89–104.

Bosman, S.H., Schwing, P.T., Larson, R.A., Wildermann, N.E., Brooks, G.R., Romero, I.C., Sanchez-Cabeza, J.-A., Ruiz-Fernández, A.C., Machain-Castillo, M.L., Gracia, A., Escobar-Briones, E., Murawski, S.A., Hollander, D.J., Chanton, J.P., 2020. The southern Gulf of Mexico: a baseline radiocarbon isoscape of surface sediments and isotopic excursions at depth. PLoS One 15, e0231678.

Brand, U., 2004. Carbon, oxygen and strontium isotopes in Paleozoic carbonate components: an evaluation of original seawater-chemistry proxies. Chem. Geol. 204, 23–44.

Brand, W.A., Douthitt, C.B., Fourel, F., Maia, R., Rodrigues, C., Maguas, C., Prohaska, T., 2015. Gas source isotope ratio mass spectrometry (IRMS). In: Prohaska, T., Irrgeher, J., Zitek, A., Jakubowski, N. (Eds.), New Developments in Mass Spectrometry No. 3, Sector Field Mass Spectrometry for Elemental and Isotopic Analysis. Royal Society of Chemistry, pp. 500–549.

Brunneé, C., 1997. 50 Years of MAT in Bremen. Rapid Commun. Mass Spectrom. 11, 694–707.

Canfield, D.E., Knoll, A.H., Poulton, S.W., Narbonne, G.M., Dunning, G.R., 2020. Carbon isotopes in clastic rocks and the Neoproterozoic carbon cycle. Am. J. Sci. 320, 97–124.

Caruthers, A.H., Gröcke, D.R., Kaczmarek, S.E., Rine, M.J., Kuglitsch, J., Harrisson III, W.B., 2018. Utility of organic carbon isotope data from the Salina Group halite (Michigan Basin): A new tool for stratigraphic correlation and paleoclimate proxy resource. Geol. Soc. Am. Bull. 130, 1782–1790.

Caruthers, A.H., Gröcke, D.R., Smith, P.L., 2011. The significance of an Early Jurassic (Toarcian) carbon-isotope excursion in Haida Gwaii (Queen Charlotte Islands), British Columbia, Canada. Earth Planet. Sci. Lett. 307, 19–26.

Caves, J.K., Jost, A.B., Lau, K.V., Maher, K., 2016. Cenozoic carbon cycle imbalances and a variable weathering feedback. Earth Planet. Sci. Lett. 450, 152–163.

Cerling, T.E., 1992. Use of carbon isotopes in paleosols as an indicator of the $p(CO_2)$ of the paleoatmosphere. Global Biogeochem. Cycles 6, 307–314.

Ciais, P., Sabine, C., Bala, G., Bopp, L., Brovkin, V., Canadell, J., Chhabra, A., DeFries, R., Galloway, J., Heimann, M., Jones, C., Le Quéré, C., Myneni, R.B., Piao, S., Thornton, P., 2013. Carbon and Other Biogeochemical Cycles. In: Stocker, T.F., Qin, D., Plattner, G.-K., Tignor, M., Allen, S.K., Boschung, J., Nauels, A., Xia, Y., Bex, V., Midgley, P.M. (Eds.), Climate Change 2013: The Physical Science Basis. Contribution of Working Group I to the Fifth Assessment Report of the Intergovernmental Panel on Climate Change. Cambridge University Press, Cambridge, UK, pp. 465–570.

Craig, H., 1953. The geochemistry of the stable carbon isotopes. Geochim. Cosmochim. Acta 3, 53–92.

Cui, Y., Kump, L.R., Ridgwell, A.J., Charles, A.J., Junium, C.K., Diefendorf, A.F., Freeman, K.H., Urban, N.M., Harding, I.C., 2011. Slow release of fossil carbon during the Palaeocene–Eocene thermal maximum. Nat. Geosci. 4, 481–485.

Daines, S.J., Mills, B.J.W., Lenton, T.M., 2017. Atmospheric oxygen regulation at low Proterozoic levels by incomplete oxidative weathering of sedimentary organic carbon. Nat. Commun. 8, 14379.

Dal Corso, J., Schmidt, A.R., Seyfullah, L.J., Preto, N., Ragazzi, E., Jenkyns, H.C., Delclós, X., Néraudeau, D., Roghi, G., 2017. Evaluating the use of amber in palaeoatmospheric reconstructions: the carbon-isotope variability of modern and Cretaceous conifer resins. Geochim. Cosmochim. Acta 199, 351–369.

De Laeter, J., Kurz, M.D., 2006. Alfred Nier and the sector field mass spectrometer. J. Mass Spectrom. 41, 847–854.

De Laeter, J.R., De Bievre, P., Peiser, H.S., 1992. Isotope mass spectrometry in metrology. Mass Spectrom. Rev. 11, 193–245.

Degens, E.T., 1969. Biogeochemistry of stable carbon isotopes. In: Eglinton, G., Murphy, M.T.J. (Eds.), Organic Geochemistry. Springer, pp. 304–329.

Des Marais, D.J., 2001. Isotopic evolution of the biogeochemical carbon cycle during the Precambrian. Rev. Mineral. Geochem. 43, 555–578.

Dickens, G.R., Castillo, M.M., Walker, J.C.G., 1997. A blast of gas in the latest Paleocene: simulating first-order effects of massive dissociation of oceanic methane hydrate. Geology 25, 259–262.

Dickens, G.R., O'Neil, J.R., Rea, D.K., Owen, R.M., 1995. Dissociation of oceanic methane hydrate as a cause of the carbon isotope excursion at the end of the Paleocene. Paleoceanography 10, 965–971.

Du, A., Robinson, J.R., Rowan, J., Lazagabaster, I.A., Behrensmeyer, A.K., 2019. Stable carbon isotopes from paleosol carbonate and herbivore enamel document differing paleovegetation signals in the eastern African Plio-Pleistocene. Rev. Palaeobot. Palynol. 261, 41–52.

Durazzi, J.T., 1981. Stable-isotope studies of planktonic foraminifera in North Atlantic core tops. Palaeogeogr. Palaeoclimatol. Palaeoecol. 33, 157–172.

Edwards, C.T., Saltzman, M.R., 2016. Paired carbon isotopic analysis of Ordovician bulk carbonate ($\delta^{13}C_{carb}$) and organic matter ($\delta^{13}C_{org}$) spanning the great Ordovician biodiversification event. Palaeogeogr. Palaeoclimatol. Palaeoecol. 458, 102–117.

Eggleston, S., Schmitt, J., Bereiter, B., Schneider, R., Fischer, H., 2016. Evolution of the stable carbon isotope composition of atmospheric CO_2 over the last glacial cycle. Paleoceanography 31, 434–452.

Elling, F.J., Gottschalk, J., Doeana, K.D., Kusch, S., Hurley, S.J., Pearson, A., 2019. Archaeal lipid biomarker constraints on the Paleocene-Eocene carbon isotope excursion. Nat. Commun. 10, 4519.

Emiliani, C., 1955. Pleistocene temperatures. J. Geol. 63, 538–578.

Falkowski, P., Scholes, R.J., Boyle, E., Canadell, J., Canfield, D., Elser, J., Gruber, N., Hibbard, K., Högberg, P., Linder, S., Mackenzie, F.T., Moore III, B., Pedersen, T.,

Rosenthal, Y., Seitzinger, S., Smetacek, V., Steffen, W., 2000. The global carbon cycle: a test of our knowledge of Earth as a system. Science 290, 291–296.

Ferreri, V., Weissert, H., D'Argenio, B., Buonocunto, F.P., 1997. Carbon isotope stratigraphy: a tool for basin to carbonate platform correlation. Terra Nova 9, 57–61.

Fourel, F., Martineau, F., Tóth, E.E., Görög, A., Escarguel, G., Lécuyer, C., 2016. Carbon and oxygen isotope variability among foraminifera and ostracod carbonated shells. Annales Universitatis Mariae Curie-Sklodowska, sectio AAA–Physica 70, 133–156.

Fox, D.L., Koch, P.L., 2004. Carbon and oxygen isotopic variability in Neogene paleosol carbonates: constraints on the evolution of the C_4-grasslands of the Great Plains, USA. Palaeogeogr. Palaeoclimatol. Palaeoecol. 207, 305–329.

Fraass, A.J., Lowery, C.M., 2017. Defining uncertainty and error in planktic foraminiferal oxygen isotope measurements. Paleoceanography 32, 104–122.

Freeman, K.H., Hayes, J.M., 1992. Fractionation of carbon isotopes by phytoplankton and estimates of ancient CO_2 levels. Global Biogeochem. Cycles 6, 185–198.

French, K.L., Sepúlveda, J., Trabucho-Alexandre, J., Gröcke, D.R., Summons, R.E., 2014. Organic geochemistry of the early Toarcian oceanic anoxic event in Hawsker Bottoms, Yorkshire, England. Earth Planet. Sci. Lett. 390, 116–127.

Galeotti, S., Sprovieri, M., Rio, D., Moretti, M., Francescone, F., Sabatino, N., Fornaciari, E., Giusberti, L., Lanci, L., 2019. Stratigraphy of early to middle Eocene hyperthermals from Possagno (Southern Alps, Italy) and comparison with global carbon isotope records. Palaeogeogr. Palaeoclimatol. Palaeoecol. 527, 39–52.

Gao, L., Guimond, J., Thomas, E., Huang, Y., 2015. Major trends in leaf wax abundance, δ^2H and $\delta^{13}C$ values along leaf venation in five species of C_3 plants: physiological and geochemical implications. Org. Geochem. 78, 144–152.

Gearing, J.N., Gearing, P.J., Rudnick, D.T., Requuo, A.G., Hutchins, M.J., 1984. Isotopic variability of organic carbon in a phytoplankton-based, temperate estuary. Geochim. Cosmochim. Acta 48, 1089–1098.

Ghosh, S., Sanyal, P., Kumar, R., 2017. Evolution of C_4 plants and controlling factors: insight from *n*-alkane isotopic values of NW Indian Siwalik paleosols. Org. Geochem. 110, 110–121.

Giorgioni, M., Jovane, L., Rego, E.S., Rodelli, D., Frontalini, F., Coccioni, R., Catanzariti, R., Özcan, E., 2019. Carbon cycle instability and orbital forcing during the Middle Eocene Climatic optimum. Sci. Rep. 9, 9357.

Gonfianti, R., 1965. Some results on oxygen isotope stratigraphy in the deep drilling at King Baudouin station, Antarctica. J. Geophys. Res. 70, 1815–1819.

Graham, H.V., Herrera, F., Jaramillo, C., Wing, S.L., Freeman, K.H., 2019. Canopy structure in Late Cretaceous and Paleocene forests as reconstructed from carbon isotope analyses of fossil leaves. Geology 47, 977–981.

Grein, M., Roth-Nebelsick, A., Wilde, V., 2010. Carbon isotope composition of middle Eocene leaves from the Messel Pit, Germany. Palaeodiversity 3, 1–7.

Gröcke, D.R., 1997. Carbon-isotope stratigraphy of terrestrial plant fragments in the Early Cretaceous from south-eastern Australia. In: Wolberg, D.L., Stump, E., Rosenberg, G.D. (Eds.), Dinofest^{TM} International: Proceedings of a Symposium Held at Arizona State University. The Academy of Natural Sciences, Philadelphia, USA, pp. 457–461.

Gröcke, D.R., 1998. Carbon-isotope analyses of fossil plants as a chemostratigraphic and palaeoenvironmental tool. Lethaia 31, 1–13.

Gröcke, D.R., 2002. The carbon isotope composition of ancient CO_2 based on higher-plant organic matter. Philos. Trans. Roy. Soc. Series A 360, 633–658.

Gröcke, D.R., Gillikin, D.P., 2008. Advances in mollusc sclerochronology and sclerochemistry: tools for understanding climate and environment. Geo-Mar. Lett. 28, 265–268.

Gröcke, D.R., Hesselbo, S.P., Jenkyns, H.C., 1999. Carbon-isotope composition of Lower Cretaceous fossil wood: ocean-atmosphere chemistry and relation to sea-level change. Geology 27, 155–158.

Gröcke, D.R., Joeckel, R.M., 2008. A stratigraphic test of the terrestrial carbon-isotope record of the latest Albian OAE from the Dakota Formation. Nebraska. Kansas Geological Surv, 24–30. Open-File Report 2008-2.

Gröcke, D.R., Ludvigson, G.A., Witzke, B.L., Robinson, S.L., Joeckel, R.M., Ulnar, D.F., Ravn, R.L., 2006. Recognizing the Albian–Cenomanian (OAE1d) sequence boundary using plant carbon isotopes: Dakota Formation, Western Interior Basin, USA. Geology 34, 193–196.

Gröcke, D.R., Price, G.D., Robinson, S.A., Baraboshkin, E., Mutterlose, J., Ruffell, A.H., 2005. The Upper Valanginian (Early Cretaceous) positive carbon-isotope event recorded in terrestrial plants. Earth Planet. Sci. Lett. 240, 495–509.

Gröcke, D.R., Price, G.D., Ruffell, A.H., Mutterlose, J., Baraboshkin, E., 2003. Isotopic evidence for Late Jurassic–Early Cretaceous climate change. Palaeogeogr. Palaeoclimatol. Palaeoecol. 202, 97–118.

Guo, B., Eley, Y., Hilton, J., Zou, M., Wang, Q., 2020. Carbon-isotope, petrological and floral record in coals: Implication for Bajocian (Middle Jurassic) climate change. Int. J. Coal Geol. 220, 103417.

Gutjahr, M., Ridgwell, A., Sexton, P.F., Anagnostou, E., Pearson, P.N., Pälike, H., Norris, R.D., Thomas, E., Foster, G.L., 2017. Very large release of mostly volcanic carbon during the Palaeocene–Eocene Thermal Maximum. Nature 548, 573–577.

Hall, A., Cox, P., Huntingford, C., Klein, S., 2019. Progressing emergent constraints on future climate change. Nat. Clim. Change 9, 269–278.

Hall, R.L., McNicoll, V., Gröcke, D.R., Craig, J., Johnston, K., 2004. Integrated stratigraphy in the lower and middle Fernie Formation in Alberta and British Columbia, western Canada. Riv. Ital. di Paleontol. e Stratigr. 110, 61–68.

Hansell, D.A., Carlson, C.A., Repeta, D.J., Schlitzer, R., 2009. Dissolved organic matter in the ocean: A controversy stimulates new insights. Oceanography 22, 202–211.

Harper, D.T., Hönisch, B., Zeebe, R., Shaffer, G., Haynes, L., Thomas, E., Zachos, J.C., 2020. The magnitude of surface ocean acidification and carbon release during Eocene Thermal Maximum 2 (ETM-2) and the Paleocene-Eocene Thermal Maximum (PETM). Paleoceanogr. Paleoclimatol. 35, e2019PA003699.

Hasegawa, T., 1997. Cenomanian–Turonian carbon-isotope events recorded in terrestrial organic matter from Northern Japan. Palaeogeogr. Palaeoclimatol. Palaeoecol. 130, 251–273.

Hasegawa, T., 2003. Cretaceous terrestrial paleoenvironments of northeastern Asia suggested from carbon isotope stratigraphy: increased atmospheric pCO_2-induced climate. J. Asian Earth Sci. 21, 849–859.

Hayes, J.M., 1993. Factors controlling ^{13}C contents of sedimentary organic compounds: principles and evidence. Mar. Geol. 113, 111–125.

Hayes, J.M., Strauss, H., Kaufman, A.J., 1999. The abundance of ^{13}C in marine organic matter and isotopic fractionation in the global biogeochemical cycle of carbon during the past 800 Ma. Chem. Geol. 161, 103–125.

Henning, S., Weissert, H., Bulot, L., 1999. C-isotope stratigraphy, a calibration tool between ammonite- and magnetostratigraphy: the Valanginian-Hauterivian transition. Geol. Carpath. 50, 91–96.

Hermoso, M., Godbillot, C., Minoletti, F., 2020. Enhancing our palaeoceanographic toolbox using paired foraminiferal and coccolith c alcite measurements from pelagic sequences. Front. Earth Sci. 8, 38.

Hesselbo, S.P., Gröcke, D.R., Jenkyns, H.C., Bjerrum, C., Farrimond, P., Morgans Bell, H., Green, O., 2000. Massive dissociation of gas hydrate during a Jurassic oceanic anoxic event. Nature 406, 392–395.

Hesselbo, S.P., Jenkyns, H.C., Duarte, L.V., Oliveira, L.C.V., 2007. Carbon-isotope record of the Early Jurassic (Toarcian) oceanic anoxic event from fossil wood and marine carbonate (Lusitanian Basin, Portugal). Earth Planet. Sci. Lett. 253, 455–470.

Hesselbo, S.P., Korte, C., Ullmann, C.V., Ebbesen, A.L., 2020. Carbon and oxygen isotope records from the southern Eurasian Seaway following the Triassic-Jurassic boundary: parallel long-term enhanced carbon burial and seawater warming. Earth Sci. Rev. 203, 103131.

Hines, B.R., Gazley, M.F., Collins, K.S., Bland, K.J., Crampton, J.S., Ventura, G.D., 2019. Chemostratigraphic resolution of widespread reducing conditions in the southwest Pacific Ocean during the Late Paleocene. Chem. Geol. 504, 236–252.

Holden, N.E., Coplen, T.B., Böhlke, J.K., Tarbox, L.V., Benefield, J., de Laeter, J.R., Mahaffy, P.G., O'Connor, G., Roth, E., Tepper, D.H., Walczyk, T., Wieser, M.E., Yoneda, S., 2018. IUPAC periodic table of the elements and isotopes (IPTEI) for the education community (IUPAC Technical Report). Pure Appl. Chem. 90, 1833–2092.

Holdgate, G.R., McGowran, B., Fromhold, T., Wagstaff, B.E., Gallagher, S.J., Wallace, M.W., Sluiter, I.R.K., Whitelaw, M., 2009. Eocene–Miocene carbon-isotope and floral record from brown coal seams in the Gippsland Basin of southeast Australia. Global Planet. Change 65, 89–103.

Holser, W.T., 1997. Geochemical events documented in inorganic carbon isotopes. Palaeogeogr. Palaeoclimatol. Palaeoecol. 132, 173–182.

Houghton, R.A., Baccini, A., Walker, W.S., 2018. Where is the residual terrestrial carbon sink? Global Change Biol. 24, 3277–3279.

Hu, J.-J., Xing, Y.-W., Su, T., Huang, Y.-J., Zhou, Z.-K., 2019. Stomatal frequency of *Quercus glauca* from three material sources shows the same inverse response to atmospheric pCO_2. Ann. Bot. 123, 1147–1158.

Huber, B.T., MacLeod, K.G., Gröcke, D.R., Kucera, M., 2011. Paleotemperature and paleosalinity inferences and chemostratigraphy across the Aptian/Albian boundary in the subtropical North Atlantic. Paleoceanography 26, PA4221.

Hunter, J.A., Stacy, R.W., Hitchcock, F.A., 1949. A mass spectrometer for continuous gas analysis. Rev. Sci. Instrum. 20, 333–336.

Hynek, S.A., Passey, B.H., Prado, J.L., Brown, F.H., Cerling, T.E., Quade, J., 2019. Small mammal carbon isotope ecology across the Miocene–Pliocene boundary, northwestern Argentina. Earth Planet. Sci. Lett. 321–322, 177–188.

Inghram, M.G., 1948. Modern mass spectroscopy. Adv. Electr. Electron Phys. 1, 219–268.

Ireland, T.R., 2013. Recent developments in isotope-ratio mass spectrometry for geochemistry and cosmochemistry. Rev. Sci. Instrum. 84, 011101.

Isson, T.T., Planavsky, N.J., Coogan, L.A., Stewart, E.M., Ague, J.J., Bolton, E.W., Zhang, S., McKenzie, N.R., Kump, L.R., 2020. Evolution of the global carbon cycle and climate regulation on Earth. Global Biogeochem. Cycles 34, e2018GB006061.

Jarvis, I., Lignum, J.S., Gröcke, D.R., Jenkyns, H.C., Pearce, M.A., 2011. Black shale deposition, atmospheric CO_2 drawdown and cooling during the Cenomanian–Turonian Oceanic anoxic event. Paleoceanography 26, PA3201.

Jasper, J.P., Hayes, J.M., 1990. A carbon isotope record of CO_2 levels during the late Quaternary. Nature 347, 462–464.

Jenkyns, H.C., Clayton, C.J., 1997. Lower Jurassic epicontinental carbonates and mudstones from England and Wales: chemostratigraphic signals and the early Toarcian anoxic event. Sedimentology 44, 687–706.

Jennings, K.R., 2012. A History of European Mass Spectrometry. IM Publications, Chichester, UK.

Jones, D.S., Brothers, R.W., Ahm, A.-S.C., Slater, N., Higgins, J.A., Fike, D.A., 2020. Sea level, carbonate mineralogy, and early diagenesis controlled $\delta^{13}C$ records in Upper Ordovician carbonates. Geology 48, 194–199.

Kemp, D.B., Coe, A.L., Cohen, A.S., Schwark, L., 2005. Astronomical pacing of methane release in the Early Jurassic period. Nature 437, 396–399.

Kennett, J.P., Cannariato, K.G., Hendy, I.L., Behl, R., 2000. Carbon isotopic evidence for methane hydrate instability during quaternary interstadials. Science 288, 128–133.

Kent, D.V., Cramer, B.S., Lanci, L., Wang, D., Wright, J.D., Van der Voo, R., 2003. A case for a comet impact trigger for the Paleocene/Eocene thermal maximum and carbon isotope excursion. Earth Planet. Sci. Lett. 211, 13–26.

Killingley, J.S., Johnson, R.F., Berger, W.H., 1981. Oxygen and carbon isotopes of individual shells of planktonic foraminifera from Ontong–Java Plateau, equatorial Pacific. Palaeogeogr. Palaeoclimatol. Palaeoecol. 33, 193–204.

Kipp, M.A., Stüeken, E.E., Gehringer, M.M., Sterelny, K., Scott, J.K., Forster, P.I., Strömberg, C.A.E., Buick, R., 2019. Exploring cycad foliage as an archive of the isotopic composition of atmospheric nitrogen. Geobiology 18, 152–166.

Korte, C., Thibault, N., Ullmann, C.V., Clémence, M.-E., Mette, W., Olsen, T.K., Rizzi, M., Ruhl, M., 2017. Brachiopod biogeochemistry and isotope stratigraphy from the Rhaetian Eiberg section in Austria: potentials and limitations. N. Jb. Geol. Paläont. Abh. 284, 117–138.

Krissansen-Totton, J., Arney, G.N., Catling, D.C., 2018. Constraining the climate and ocean pH of the early Earth with a geological carbon cycle model. Proc. Natl. Acad. Sci. U. S. A. 115, 4105–4110.

Kump, L.R., Arthur, M.A., 1999. Interpreting carbon-isotope excursions: carbonates and organic matter. Chem. Geol. 161, 181–198.

Kump, L.R., Arthur, M.A., Patzkowsky, M.E., Gibbs, M.T., Pinkus, D.S., Sheehan, P.M., 1999. A weathering hypothesis for glaciation at high atmospheric pCO_2 during the Late Ordovician. Palaeogeogr. Palaeoclimatol. Palaeoecol. 152, 173–187.

Kump, L.R., Brantley, S.L., Arthur, M.A., 2000. Chemical weathering, atmospheric CO_2, and climate. Annu. Rev. Earth Planet. Sci. 28, 611–667.

LaGrange, M.T., Konhauser, K.O., Catuneanu, O., Harris, B.S., Playter, T.L., Gingras, M.K., 2020. Sequence stratigraphy in organic-rich marine mudstone successions using chemostratigraphic datasets. Earth Sci. Rev. 203, 103137.

Landergren, S., 1954. On the relative abundance of the stable carbon isotopes in marine sediments. Deep-Sea Res. 1, 98–120.

Leavitt, S.W., Long, A., 1986. Stable-carbon isotope variability in tree foliage and wood. Ecology 67, 1002–1010.

Lee, C., Love, G.D., Hopkins, M.J., Kröger, B., Franeck, F., Finnegan, S., 2019. Lipid biomarker and stable isotopic profiles through Early-Middle Ordovician carbonates from Spitsbergen, Norway. Org. Geochem. 131, 5–18.

Lenton, T.M., Held, H., Kriegler, E., Hall, J.W., Lucht, W., Rahmstorf, S., Schellnhuber, H.J., 2008. Tipping elements in the Earth's climate system. Proc. Natl. Acad. Sci. U. S. A. 105, 1786–1793.

LeRoy, M.A., Gill, B.C., 2018. Evidence for the development of local anoxia during the Cambrian SPICE event in eastern North America. Geobiology 17, 381–400.

Li, G., Elderfield, H., 2013. Evolution of carbon cycle over the past 100 million years. Geochim. Cosmochim. Acta 103, 11–25.

Littler, K., Hesselbo, S.P., Jenkyns, H.C., 2010. A carbon-isotope perturbation at the Pliensbachian–Toarcian boundary: evidence from the Lias Group, NE England. Geol. Mag. 147, 181–192.

Liu, M., Sun, P., Them II, T.R., Li, Y., Sun, S., Gao, X., Huang, X., Tang, Y., 2020. Organic geochemistry of a lacustrine shale across the Toarcian Oceanic Anoxic Event (Early Jurassic) from NE China. Global Planet. Change 191, 103214.

Lourens, L.J., Sluijs, A., Kroon, D., Zachos, J.C., Thomas, E., Röhl, U., Bowles, J., Raffi, I., 2005. Astronomical pacing of late Palaeocene to early Eocene global warming events. Nature 435, 1083–1087.

Ludvigson, G.A., Joeckel, R.M., Murphy, L.R., Stockli, D.F., González, L.A., Suarez, C.A., Kirkland, J.I., Al-Suwaidi, A., 2015. The emerging terrestrial record of Aptian-Albian global change. Cret. Res. 56, 1–24.

Mackensen, A., Schmiedl, G., 2019. Stable carbon isotopes in paleoceanography: atmosphere, oceans, and sediments. Earth Sci. Rev. 197, 102893.

Mackey, J.E., Stewart, B.W., 2019. Evidence of SPICE-related anoxia on the Laurentian passive margin: paired $\delta^{13}C$ and trace element chemostratigraphy of the upper Conasauga Group, Central Appalachian Basin. Palaeogeogr. Palaeoclimatol. Palaeoecol. 528, 160–174.

MacLeod, K.G., Huber, B.T., 1996. Reorganization of deep ocean circulation accompanying a Late Cretaceous extinction event. Nature 380, 422–425.

Martinelli, L.A., Almeida, S., Brown, I.F., Moreira, M.Z., Victoria, R.L., Sternberg, L.S.L., Ferreira, C.A.C., Thomas, W.W., 1998. Stable carbon isotope ratio of tree leaves, boles and fine litter in a tropical forest in Rondônia, Brazil. Oecologia 114, 170–179.

Matsumoto, R., 1995. Causes of the $\delta^{13}C$ anomalies of carbonates and a new paradigm 'Gas-Hydrate Hypothesis'. Jour. Geol. Soc. Japan 11, 902–924.

Matthews, D.E., Hayes, J.M., 1978. Isotope-ratio-monitoring gas chromatography-mass spectrometry. Anal. Chem. 50, 1465–1473.

McDowell, N.G., Bond, B.J., Dickman, L.T., Ryan, M.G., Whitehead, D., 2011. Relationships between tree height and carbon isotope discrimination. In: Meinzer, F.C., et al. (Eds.), Size- and Age-Related Changes in Tree Structure and Function. Tree Physiology, vol. 4, pp. 255–286.

McInerney, F.A., Wing, S.L., 2011. The Paleocene-Eocene thermal maximum: a perturbation of carbon cycle, climate, and biosphere with implications for the future. Annu. Rev. Earth Planet. Sci. 39, 489–516.

McKinney, C.R., McCrea, J.M., Epstein, S., Allen, H.A., Urey, H.C., 1950. Improvements in mass spectrometers for the measurement of small differences in isotope abundance ratios. Rev. Sci. Instrum. 21, 724–730.

Metcalfe, B., Feldmeijer, W., de Vringer-Picon, M., Brummer, G.-J.A., Peeters, F.J.C., Ganssen, G.M., 2015. Late Pleistocene glacial–interglacial shell-size–isotope variability in planktonic foraminifera as a function of local hydrography. Biogeosciences 12, 4781–4807.

Metzger, J.G., Fike, D.A., 2013. Techniques for assessing spatial heterogeneity of carbonate $\delta^{13}C$ values: implications for craton-wide isotope gradients. Sedimentology 60, 1405–1431.

Mii, H.-S., Shi, G.R., Cheng, C.-J., Chen, Y.-Y., 2012. Permian Gondwanaland paleoenvironment inferred from carbon and oxygen isotope records of brachiopod fossils from Sydney Basin, southeast Australia. Chem. Geol. 291, 87–103.

Münzenberg, G., 2013. Development of mass spectrometers from Thomson and Aston to present. Int. J. Mass Spec. 349–350, 9–18.

Naafs, B.D.A., Inglis, G.N., Blewett, J., McClymont, E.L., Lauretano, V., Xie, S., Evershed, R.P., Pancost, R.D., 2019. The potential of biomarker proxies to trace climate, vegetation, and biogeochemical processes in peat: a review. Global Planet. Change 179, 57–79.

Neupane, P.C., Gani, M.R., Gani, N.D., Huang, Y., 2019. Neogene vegetation shift in the Nepalese Siwalik, Himalayas: a compound-specific isotopic study of lipid biomarkers. The Dep. Rec. 6, 192–202.

Nguyen Tu, T.T., Kvacek, J., Ulicny, D., Bocherens, H., Mariotti, A., Broutin, J., 2002. Isotope reconstruction of plant palaeoecology. Case study of Cenomanian floras from Bohemia. Palaeogeogr. Palaeoclimatol. Palaeoecol. 183, 43–70.

Nguyen Tu, T.T., Bocherens, H., Mariotti, A., Baudin, F., Pons, D., Broutin, J., Derenne, S., Largeau, C., 1999. Ecological distribution of Cenomanian terrestrial plants based on $^{13}C/^{12}C$ ratios. Palaeogeogr. Palaeoclimatol. Palaeoecol. 145, 79–93.

Nier, A.O., 1947. A mass spectrometer for isotope and gas analysis. Rev. Sci. Instrum. 18, 398–411.
Nier, A.O., Gulbransen, E.A., 1939. Variations in the relative abundance of the carbon isotopes. J. Am. Chem. Soc. 61, 697–698.
Nordt, L., Tubbs, J., Dworkin, S., 2017. Stable carbon isotope record of terrestrial organic materials for the last 450 Ma yr. Earth Sci. Rev. 159, 103–117.
Oborny, S.C., Cramer, B.D., Brett, C.E., Bancroft, A.M., 2020. Integrated silurian conodont and carbonate carbon isotope stratigraphy of the east-central Appalachian Basin. Palaeogeogr. Palaeoclimatol. Palaeoecol. 554, 109815.
Oehlert, A.M., Lamb-Wozniak, K.A., Devlin, Q.B., Mackenzie, G.J., Reijmer, J.J.G., Swart, P.K., 2012. The stable carbon isotopic composition of organic material in platform derived sediments: implications for reconstructing the global carbon cycle. Sedimentology 59, 319–335.
Oehlert, A.M., Swart, P.K., 2014. Interpreting carbonate and organic carbon isotope covariance in the sedimentary record. Nat. Commun. 5, 1–7.
Pagani, M., 2002. The alkenone–CO_2 proxy and ancient atmospheric carbon dioxide. Phil. Trans. Roy. Soc. Ser. A 360, 609–632.
Pagani, M., Arthur, M.A., Freeman, K.H., 1999a. Miocene evolution of atmospheric carbon dioxide. Paleoceanography 14, 273–292.
Pagani, M., Freeman, K.H., Arthur, M.A., 1999b. Late miocene atmospheric CO_2 concentrations and the expansion of C_4 grasses. Science 285, 876–879.
Pancost, R.D., Freeman, K.H., Patzkowsky, M.E., 1999. Organic-matter source variation and the expression of a late Middle Ordovician carbon isotope excursion. Geology 27, 1015–1018.
Pearson, J., Grove, M., 2013. Counting sheep: sample size and statistical inference in stable isotope analysis and palaeodietary reconstruction. World Archaeol. 45, 373–387.
Phillips, D.L., Gregg, J.W., 2001. Uncertainty in source partitioning using stable isotopes. Oecologia 127, 171–179.
Popp, B.N., Laws, E.A., Bidigare, R.R., Dore, J.E., Hanson, K.L., Wakeham, S.G., 1998. Effect of phytoplankton cell geometry on carbon isotopic fractionation. Geochim. Cosmochim. Acta 62, 69–77.
Popp, B.N., Takigiku, R., Hayes, J.M., Louda, J.W., Baker, E.W., 1989. The post-Paleozoic chronology and mechanism of ^{13}C depletion in primary organic matter. Am. J. Sci. 289, 436–454.
Rau, G.H., Froelich, P.N., Takahashi, T., Des Marais, D.J., 1991. Does sedimentary organic $\delta^{13}C$ record variations in Quaternary ocean [CO_2(aq)]? Paleoceanography 6, 335–347.
Rau, G.H., Takahashi, T., Des Marais, D.J., 1989. Latitudinal variations in plankton $\delta^{13}C$: implications for CO_2 and productivity in past oceans. Nature 341, 516–518.
Retallack, G.J., Conde, G.D., 2020. Deep time perspective on rising atmospheric CO_2. Global Planet. Change 189, 103177.
Richards, M.P., Pellegrini, M., Niven, L., Nehlich, O., Dibble, H., Turq, A., McPherron, S.J.P., 2017. Temporal variations in *Equus* tooth isotope values (C,N,O) from the Middle Paleolithic site of Combe Grenal, France (ca. 150,000 to 50,000 BP). J. Archaeol. Sci. Rep. 14, 189–198.
Richey, J.D., Upchurch, G.R., Montañez, I.P., Lomax, B.H., Suarez, M.B., Crout, N.M.J., Joeckel, R.M., Ludvigson, G.A., Smith, J.J., 2018. Changes in CO_2 during ocean anoxic event 1d indicate similarities to other carbon cycle perturbations. Earth Planet. Sci. Lett. 491, 172–182.
Riebesell, U., Revill, A.T., Holdsworth, D.G., Volkman, J.K., 2000. The effects of varying CO_2 concentration on lipid composition and carbon isotope fractionation in *Emiliania huxleyi*. Geochim. Cosmochim. Acta 64, 4179–4192.

Royer, D.L., 2001. Stomatal density and stomatal index as indicators of paleoatmospheric CO_2 concentration. Rev. Palaeobot. Palynol. 114, 1–28.
Ruebsam, W., Reolid, M., Schwark, L., 2020. $\delta^{13}C$ of terrestrial vegetation record stoarcian CO_2 and climate gradients. Sci. Rep. 10, 117.
Sageman, B.B., Meyers, S.R., Arthur, M.A., 2006. Orbital time scale and new C-isotope record for Cenomanian-Turonian boundary stratotype. Geology 34, 125–128.
Sarkar, A., Sarangi, S., Bhattacharya, S.K., Ray, A.K., 2003. Carbon isotopes across the Eocene-Oligocene boundary sequence of Kutch, western India: implications to oceanic productivity and pCO_2 change. Geophys. Res. Lett. 30, 1588.
Schidlowski, M., 1987. Application of stable carbon isotopes to early biochemical evolution on Earth. Annu. Rev. Earth Planet. Sci. 15, 47–72.
Schidlowski, M., 2001. Carbon isotopes as biogeochemical recorders of life over 3.8 Ga of Earth history: evolution of a concept. Precambrian Res. 106, 117–134.
Schimel, D., Pavlick, R., Fisher, J.B., Asner, G.P., Saatchi, S., Townsend, P., Miller, C., Frakenberg, C., Hibbard, K., Cox, P., 2015. Observing terrestrial ecosystems and the carbon cycle from space. Global Change Biol. 21, 1762–1776.
Schleser, G.H., 1992. $\delta^{13}C$ pattern in a forest tree as an indicator of carbon transfer in trees. Ecology 73, 1922–1925.
Schmitt, J., Schneider, R., Elsig, J., Leuenberger, D., Lourantou, A., Chappellaz, J., Köhler, P., Joos, F., Stocker, T.F., Leuenberger, M., Fischer, H., 2012. Carbon isotope constraints on the deglacial CO_2 rise from ice cores. Science 336, 711–714.
Schmittner, A., Bostock, H.C., Cartapanis, O., Curry, W.B., Filipsson, H.L., Galbraith, E.D., Gottschalk, J., Herguera, J.C., Hoogakker, B., Jaccard, S.L., Lisiecki, L.E., Lund, D.C., Martínez-Méndez, G., Lynch-Stieglitz, J., Mackensen, A., Michel, E., Mix, A.C., Oppo, D.W., Peterson, C.D., Repschläger, J., Sikes, E.L., Spero, H.J., Waelbroeck, C., 2017. Calibration of the carbon isotope composition ($\delta^{13}C$) of benthic foraminifera. Paleoceanography 32, 512–530.
Schnyder, J., Pons, D., Yans, J., Tramoy, R., Abdulanova, S., 2017. Integrated stratigraphy of a continental Pliensbachian–Toarcian Boundary (Lower Jurassic) section at Taskomirsay, Leontiev Graben, southwest Kazakhstan. In: Brunet, M.-F., McCann, T., Sobel, E.R. (Eds.), Geological Evolution of Central Asian Basins and the Western Tien Shan Range, pp. 337–356. Geol. Soc. London Spec. Pub. 427.
Schnyder, J., Pons, D., Yans, J., Tramoy, R., Abdulanova, S., 2017. Integrated stratigraphy of a continental Pliensbachian–Toarcian Boundary (Lower Jurassic) section at Taskomirsay, Leontiev Graben, southwest Kazakhstan. Geol. Soc. London Spec. Pub. 427, 337–356.
Schobben, M., van de Schootbrugge, B., 2019. Increased stability in carbon isotope records reflects emerging complexity of the biosphere. Front. Earth Sci. 7, 87.
Schoene, B., Eddy, M.P., Samperton, K.M., Keller, C.B., Keller, G., Adatte, T., Khadri, S.F.R., 2019. U-Pb constraints on pulsed eruption of the Deccan Traps across the end-Cretaceous mass extinction. Science 363, 862–866.
Schoene, B., Guex, J., Bartolini, A., Schaltegger, U., Blackburn, T.J., 2010. Correlating the end-Triassic mass extinction and flood basalt volcanism at the 100 ka level. Geology 38, 387–390.
Sciscio, L., Tsikos, H., Roberts, D.L., Scott, L., van Breugel, Y., Sinninghe Damste, J.S., Schouten, S., Gröcke, D.R., 2016. Miocene climate and vegetation changes in the Cape Peninsula, South Africa: evidence from biogeochemistry and palynology. Palaeogeogr. Palaeoclimatol. Palaeoecol. 445, 124–137.
Scopelliti, G., Bellanca, A., Erba, E., Jenkyns, H.C., Neri, R., Tamagnini, P., Luciani, V., Masetti, D., 2008. Cenomanian-Turonian carbonate and organic-carbon isotope records, biostratigraphy and provenance of a key section in NE Sicily, Italy: palaeoceanographic and palaeogeographic implications. Palaeogeogr. Palaeoclimatol. Palaeoecol. 265, 59–77.

Sharp, Z., 2017. Principles of Stable Isotope Geochemistry, second ed. https://doi.org/10.25844/h9q1-0p82.
Shultz, D.J., Calder, J.A., 1976. Organic carbon $^{13}C/^{12}C$ variations in estuarine sediments. Geochim. Cosmochim. Acta 40, 381–385.
Sial, A.N., Gaucher, C., Ferreira, V.P., Pereira, N.S., Cezario, W.S., Chiglino, L., Lima, H.M., 2015. Isotope and elemental chemostratigraphy. In: Ramkumar, M. (Ed.), Chemostratigraphy: Concepts, Techniques, and Applications. Elsevier, pp. 23–64.
Sosdian, S.M., Babila, T.L., Greenop, R., Foster, G.L., Lear, C.H., 2020. Ocean Carbon Storage across the middle Miocene: a new interpretation for the Monterey Event. Nat. Commun. 11, 134.
Spero, H.J., 1992. Do planktic foraminifera accurately record shifts in the carbon isotopic composition of seawater ΣCO_2? Marine Micropal. 19, 275–285.
Stap, L., Lourens, L., van Dijk, A., Schouten, S., Thomas, E., 2010. Coherent pattern and timing of the carbon isotope excursion and warming during Eocene Thermal Maximum 2 as recorded in planktic and benthic foraminifera. Geochem. Geophys. Geosyst. 11, Q11011.
Steffen, W., Richardson, K., Rockström, J., Schellnhuber, H.J., Dube, O.P., Dutreuil, S., Lenton, T.M., Lubchenco, J., 2020. The emergence and evolution of Earth System Science. Nat. Rev. Earth Environ. 1, 54–63.
Steinthorsdottir, M., Vajda, V., Pole, M., Holdgate, G., 2019. Moderate levels of Eocene pCO_2 indicated by Southern Hemisphere fossil plant stomata. Geology 47, 914–918.
Steinthorsdottir, M., Wohlfarth, B., Kylander, M.E., Blaauw, M., Reimer, P.J., 2013. Stomatal proxy record of CO_2 concentrations from the last termination suggests an important role for CO_2 at climate change transitions. Quat. Sci. Rev. 68, 43–58.
Stephens, N.P., Sumner, D.Y., 2003. Late Devonian carbon isotope stratigraphy and sea level fluctuations, Canning Basin, Western Australia. Palaeogeogr. Palaeoclimatol. Palaeoecol. 191, 203–219.
Stevens, R.E., Hedges, R.E.M., 2004. Carbon and nitrogen stable isotope analysis of northwest European horse bone and tooth collagen, 40,000 BP–present: palaeoclimatic interpretations. Quat. Sci. Rev. 23, 977–991.
Suarez, M.B., Suarez, C.A., Al-Suwaidi, A.H., Hatzell, G., Kirkland, J.I., Salazar-Verdin, J., Ludvigson, G.A., Joeckel, R.M., 2017. Terrestrial carbon isotope chemostratigraphy in the yellow cat member of the Cedar Mountain Formation: complications and pitfalls. In: Zeigler, K.E., Parker, W.G. (Eds.), Terrestrial Depositional Systems: Deciphering Complexities Through Multiple Stratigraphic Methods. Elsevier, 336, p. 303.
Swart, P.K., 2008. Global synchronous changes in the carbon isotopic composition of carbonate sediments unrelated to changes in the global carbon cycle. Proc. Natl. Acad. Sci. U. S. A. 105, 13741–13745.
Swart, P.K., 2015. The geochemistry of carbonate diagenesis: the past, present and future. Sedimentology 62, 1233–1304.
Swart, P.K., Eberli, G., 2005. The nature of the $\delta^{13}C$ of periplatform sediments: implications for stratigraphy and the global carbon cycle. Sed. Geol. 175, 115–129.
Tappert, R., McKellar, R.C., Wolfe, A.P., Tappert, M.C., Ortega-Blanco, J., Muehlenbachs, K., 2013. Stable carbon isotopes of C3 plant resins and ambers record changes in atmospheric oxygen since the Triassic. Geochim. Cosmochim. Acta 121, 240–262.
Them II, T.R., Gill, B.C., Caruthers, A.H., Gröcke, D.R., Tulsky, E.T.T., Martindale, R.C., Poulton, T.P., Smith, P.L., 2017. High-resolution carbon isotope records of the Toarcian Oceanic Anoxic Event (Early Jurassic) from North America and implications for the global drivers of the Toarcian carbon cycle. Earth Planet. Sci. Lett. 459, 118–126.

Thibault, N., Ruhl, M., Ullmann, C.V., Korte, C., Kemp, D.B., Gröcke, D.R., Hesselbo, S.P., 2018. The wider context of the Lower Jurassic Toarcian Oceanic Anoxic Event in Yorkshire coastal outcrops, UK. Proc. Geol. Assoc. 129, 372–391.

Thomas, D.J., Zachos, J.C., Bralower, T.J., Thomas, E., Boharty, S., 2002. Warming the fuel for the fire: evidence for the thermal dissociation of methane hydrate during the Paleocene-Eocene thermal maximum. Geology 30, 1067–1070.

Tipple, B.J., Meyers, S.R., Pagani, M., 2010. Carbon isotope ratio of Cenozoic CO_2: a comparative evaluation of available geochemical proxies. Paleoceanography 25, PA3202.

Tiwari, M., Singh, A.K., Sinha, D.K., 2015. Stable isotopes: tools for understanding past climatic conditions and their applications in chemostratigraphy. In: Ramkumar, M. (Ed.), Chemostratigraphy: Concepts, Techniques, and Applications. Elsevier, pp. 65–92.

Trabucho-Alexandre, J., 2014. More gaps than shale: erosion of mud and its effect on preserved geochemical and palaeobiological signals. In: Smith, D.G., Bailey, R.J., Burgess, P.M., Fraser, A.J. (Eds.), Strata and Time: Probing the Gaps in Our Understanding, pp. 251–270. Geol. Soc. London Spec. Pub. 404.

Tsikos, H., Jenkyns, H.C., Walsworth-Bell, B., Petrizzo, M.R., Forster, A., Kolonic, S., Erba, E., Premoli-Silva, I.P., Baas, M., Wagner, T., Sinninghe Damsté, J.S., 2004. Carbon-isotope stratigraphy recorded by the Cenomanian—Turonian Oceanic Anoxic Event: correlation and implications based on three key localities. J. Geol. Soc. London 161, 711–719.

Uličný, D., Jarvis, I., Gröcke, D.R., Čech, S., Laurin, J., Olde, K., Trabucho-Alexandre, J., Švábenická, L., Pedentchouk, N., 2014. A high-resolution carbon-isotope record of the Turonian stage correlated to a siliciclastic basin fill: Implications for mid-Cretaceous sea-level change. Palaeogeogr. Palaeoclimatol. Palaeoecol. 405, 42–58.

Ullmann, C.V., Boyle, R., Duarte, L.V., Hesselbo, S.P., Kasemann, S.A., Klein, T., Lenton, T.M., Piazza, V., Aberhan, M., 2020. Warm afterglow from the Toarcian Oceanic Anoxic Event drives the success of deep-adapted brachiopods. Sci. Rep. 10, 6549.

Ullmann, C.V., Campbell, H.J., Frei, R., Korte, C., 2016. Oxygen and carbon isotope and Sr/Ca signatures of high-latitude Permian to Jurassic calcite fossils from New Zealand and New Caledonia. Gondw. Res. 38, 60–73.

Uno, K.T., Cerling, T.E., Harris, J.M., Kunimatsu, Y., Leakey, M.G., Nakatsukasa, M., Nakaya, H., 2011. Late Miocene to Pliocene carbon isotope record of differential diet change among East African herbivores. Proc. Natl. Acad. Sci. U. S. A. 108, 6509–6514.

van der Meulen, B., Gingerich, P.D., Lourens, L.J., Meijer, N., van Broekhuizen, S., van Ginneken, S., Abels, H.A., 2020. Carbon isotope and mammal recovery from extreme greenhouse warming at the Paleocene–Eocene boundary in astronomically-calibrated fluvial strata, Bighorn Basin, Wyoming, USA. Earth Planet. Sci. Lett. 534, 116044.

van Hengstum, P., Gröcke, D.R., 2008. Stable isotope record of the Eifelian–Givetian boundary Kačák–*otomari* Event (Middle Devonian) from Hungry Hollow, Ontario, Canada. Can. J. Earth Sci. 45, 353–366.

Voigt, S., Aurag, A., Leis, F., Kaplan, U., 2007. Late Cenomanian to Middle Turonian high-resolution carbon isotope stratigraphy: new data from the Münsterland Cretaceous Basin, Germany. Earth Planet. Sci. Lett. 253, 196–210.

Wallmann, K., 2001. Controls on the Cretaceous and Cenozoic evolution of seawater composition, atmospheric CO_2 and climate. Geochim. Cosmochim. Acta 65, 3005–3025.

Wan, Z., Algeo, T.J., Gensel, P.G., Scheckler, S.E., Stein, W.E., Cressler III, W.L., Berry, C.M., Xu, H., Rowe, H.D., Sauer, P.E., 2019. Environmental influences on the stable carbon isotopic composition of Devonian and Early Carboniferous land plants. Palaeogeogr. Palaeoclimatol. Palaeoecol. 531, 109100.

Wang, H., Dilcher, D.L., Schwarzwalder, R.N., Kvaček, J., 2011. Vegetative and reproductive morphology of an extinct Early Cretaceous member of platanaceae from the Braun's Ranch Locality, Kansas, U.S.A. Int. J. Plant Sci 172, 139–157.

Weissert, H., 1989. C-isotope stratigraphy, a monitor of paleoenvironmental change: a case study from the Early Cretaceous. Surv. Geophys. 10, 1–61.
Woodard, D.L., Davis, S.J., Randerson, J.T., 2019. Economic carbon cycle feedbacks may offset additional warming from natural feedbacks. Proc. Natl. Acad. Sci. U. S. A. 116, 759–764.
Wu, Y., Tong, J., Algeo, T.J., Chu, D., Cui, Y., Song, H., Shu, W., Du, Y., 2020. Organic carbon isotopes in terrestrial Permian-Triassic boundary sections of North China: implications for global carbon cycle perturbations. Geol. Soc. Bull. 132, 1106–1118.
Yager, J.A., West, A.J., Corsetti, F.A., Berelson, W.M., Rollins, N.E., Rosas, S., Bottjer, D.J., 2017. Duration of and decoupling between carbon isotope excursions during the end-Triassic mass extinction and Central Atlantic Magmatic Province emplacement. Earth Planet. Sci. Lett. 473, 227–236.
Yans, J., Gerards, T., Gerrienne, P., Spagna, P., Dejax, J., Schnyder, J., Storme, J.-Y., Keppens, E., 2010. Carbon-isotope analysis of fossil wood and dispersed organic matter from the terrestrial Wealden facies of Hautrage (Mons Basin, Belgium). Palaeogeogr. Palaeoclimatol. Palaeoecol. 291, 85–105.
Zachos, J., Pagani, M., Sloan, L., Thomas, E., Billups, K., 2001. Trends, rhythms, and aberrations in global climate 65 Ma to present. Science 292, 686–693.
Zachos, J.C., Boharty, S.M., John, C.M., McCarren, H., Kelly, D.C., Nielsen, T., 2007. The Palaeocene–Eocene carbon isotope excursion: constraints from individual shell planktonic foraminifer records. Phil. Trans. Roy. Soc. Ser. A 365, 1829–1842.
Zachos, J.C., McCarren, H., Murphy, B., Röhl, U., Westerhold, T., 2010. Tempo and scale of late Paleocene and early Eocene carbon isotope cycles: implications for the origin of hyperthermals. Earth Planet. Sci. Lett. 299, 242–249.
Zeebe, R.E., Ridgwell, A., Zachos, J.C., 2016. Anthropogenic carbon release rate unprecedented during the past 66 million years. Nat. Geosci. 9, 325–329.
Zhang, Y.G., Pagani, M., Liu, Z., Boharty, S.M., DeConto, R., 2013. A 40-million-year history of atmospheric CO_2. Phil. Trans. Roy. Soc. Ser. A 371, 20130096.
Zhang, Y.G., Pearson, A., Benthien, A., Dong, L., Huybers, P., Liu, X., Pagani, M., 2019. Refining the alkenone-pCO_2 method I: lessons from the Quaternary glacial cycles. Geochim. Cosmochim. Acta 260, 177–191.
Zhou, F., Gao, X., Yuan, H., Song, J., Chen, J., 2018. The distribution and seasonal variations of sedimentary organic matter in the East China Sea shelf. Mar. Pollut. Bull. 129, 163–171.

Further reading

Diefendorf, A.F., Mueller, K.E., Wing, S.L., Koch, P.L., Freeman, K.H., 2010. Global patterns in leaf 13C discrimination and implications for studies of past and future climate. Proc. Natl. Acad. Sci. U. S. A. 107, 5738–5743.
Jarvis, I., Trabucho-Alexandre, J., Gröcke, D.R., Uličný, D., Laurin, J., 2015. Intercontinental correlation of organic carbon and carbonate stable-isotope records: evidence of climate and sea-level change during the Turonian (Cretaceous). Dep. Rec. 1, 53–90.
Hare, V., Loftus, E., Jeffrey, A., Ramsey, C.B., 2018. Atmospheric CO2 effect on stable carbon isotope composition of terrestrial fossil archives. Nature Commun. 9, 252.
Magozzi, S., Yool, A., Vander Zanden, H.B., Wunder, M.B., Trueman, C.N., 2017. Using ocean models to predict spatial and temporal variation in marine carbon isotopes. Ecosphere 8, e01763.
Saltzman, M.R., Thomas, E., 2012. Carbon Isotope Stratigraphy. In: Gradstein, F.M., Ogg, J.G., Schmitz, M., Ogg, G. (Eds.), The Geologic Time Scale 2012. Elsevier B.V., pp. 207–232.
Vail, P.R., 1987. Seismic stratigraphy interpretation using sequence stratigraphy, Part 1: seismic stratigraphy interpretation procedure. In: Bally, A.W. (Ed.), Atlas of Seismic Stratigraphy. Am. Assoc. Pet. Geol. Studies in Geology 27, 1–10.

CHAPTER TWO

Interpreting Phanerozoic $\delta^{13}C$ patterns as periodic glacio-eustatic sequences

Moujahed Al-Husseini[a,*] and Wolfgang Ruebsam[b]
[a]Gulf PetroLink-GeoArabia, Manama, Bahrain
[b]Department of Organic and Isotope Geochemistry, University of Kiel, Kiel, Germany
*Corresponding author: e-mail address: moujaheda@gmail.com

Contents

Stratigraphy & Timescales, Volume 5
ISSN 2468-5178
https://doi.org/10.1016/bs.sats.2020.08.005

Abstract

The deterministic *Orbital Scale* of glacio-eustasy is calibrated with 405-kyr time-rock units named *stratons*. It predicts polar ice volumes (global sea level) varied throughout the Phanerozoic according to a periodic template that repeats every 14.58 myr. The scale represents the interval between 1.586 and 541.046 Ma with 37 *orbitons*, each lasting 14.58 myr and consisting of three dozons (4.86 myr, 12 stratons) and 36 stratons. In times older than c. 8.0 Ma the predicted ages of glacial maxima closely correlate to those of positive $\delta^{13}C$ plateaus as estimated by radiometric and other dating techniques. This study illustrates the application of the scale in nine examples from the early Cambrian, late Ordovician and early Silurian, Carboniferous and Cretaceous. Each example compares the predicted glacio-eustatic signal to $\delta^{13}C$ patterns (plateaus, valleys, rising and falling limbs) and cites supporting stratigraphic evidence for the interpretation (e.g., transgression, regression, erosional unconformity, sequence boundary, cold snap, flooding interval).

The examples also review empirical age estimates of stage boundaries and where inconsistent they are calibrated by the Orbital Scale. The results indicate several stage boundaries are defined in major $\delta^{13}C$ plateaus that pass to glacio-eustatic unconformities and hiatuses. The 4.86-myr dozon stands out in the examples as the easiest-to-recognize stratigraphic unit because it is bounded by two significant $\delta^{13}C$ plateaus or excursions that pass to unconformities. Dating $\delta^{13}C$ plateaus according to the Orbital Scale may therefore provide key anchors that better constrain the geologic time scale.

1. Introduction

Carbon-13 isotope ($\delta^{13}C$) records typically consist of patterns that can be used to support biostratigraphic correlations of depositional sequences at regional and global scales (e.g., Berger and Vincent, 1981; Salzman and Thomas, 2012; Scholle and Arthur, 1980). The most common secular patterns are positive plateaus and negative valleys connected by rising and falling limbs (increasing and decreasing trends, respectively), often modulated by high-frequency orbital cycles (e.g., Ruebsam and Al-Husseini, 2020; Fig. 1). Based on a survey of $\delta^{13}C$ records in several Phanerozoic intervals, Al-Husseini (2018) noticed that the estimated ages of many globally correlative plateaus approximately coincide with the ages of major sea-level lowstands predicted by the *Orbital Scale* of glacio-eustasy (Matthews and Al-Husseini, 2010). Moreover, $\delta^{13}C$ falling and rising limbs mirror-reflect sea-level rises and falls, respectively, and $\delta^{13}C$ valleys reflect eustatic cycles (Fig. 1). The present study illustrates nine examples of how the Orbital Scale was used to calibrate and interpret $\delta^{13}C$ patterns as glacio-eustatic proxies.

The first example uses the Orbital Scale to calibrate the late Ordovician Katian to early Silurian Aeronian stages with the Ordovician/Silurian

Orbiton 1, Orbital Scale				
Orbiton / Dozon	405-kyr Straton Age @ Base (Ma)			Predicted Stratigraphic Observation
	0A-4	1	0.371	
Orbiton Zero	0A-3	2	0.776	
	0A-2	3	1.181	Highstand
SB Zero	0A-1	4	1.586	Transgression
Dozon 1C (4.86 myr)	1C-12	5	1.991	Major Lowstand
	1C-11	6	2.396	
	1C-10	7	2.801	Minor Highstand
	1C-9	8	3.206	Minor Lowstand
	1C-8	9	3.611	Minor Highstand
	1C-7	10	4.016	
	1C-6	11	4.421	Minor Lowstand
	1C-5	12	4.826	
	1C-4	13	5.231	
	1C-3	14	5.636	
	1C-2	15	6.041	Highstand
SB 1C	1C-1	16	6.446	Transgression
Dozon 1B (4.86 myr); Orbiton 1 (14.58 myr)	1B-12	17	6.851	Major Lowstand
	1B-11	18	7.256	
	1B-10	19	7.661	Minor Highstand
	1B-9	20	8.066	Minor Lowstand
	1B-8	21	8.471	Minor Highstand
	1B-7	22	8.876	
	1B-6	23	9.281	Minor Lowstand
	1B-5	24	9.686	
	1B-4	25	10.091	
	1B-3	26	10.496	
	1B-2	27	10.901	Highstand
SB 1B	1B-1	28	11.306	Transgression
Dozon 1A (4.86 myr)	1A-12	29	11.711	Major Lowstand
	1A-11	30	12.116	
	1A-10	31	12.521	Minor Highstand
	1A-9	32	12.926	Minor Lowstand
	1A-8	33	13.331	Minor Highstand
	1A-7	34	13.736	
	1A-6	35	14.141	Minor Lowstand
	1A-5	36	14.546	
	1A-4	37	14.951	
	1A-3	38	15.356	
	1A-2	39	15.761	Highstand
SB1	1A-1	40	16.166	Transgression
Orbiton 2	2C-12	41	16.571	Major Lowstand
	2C-11	42	16.976	

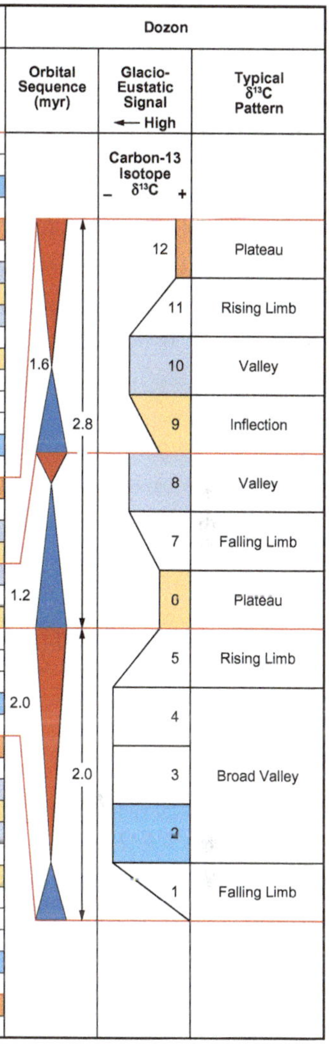

Fig. 1 Orbital scale of glacio-eustasy (Matthews and Al-Husseini, 2010) consists of stratons that tracked the long-eccentricity 0.405-myr signal (Fig. 2). Stratons form periodic orbitons (14.58 myr), dozons (4.86 myr) and third-order sequence consisting of multiples of 0.405 myr (typically, 2.0 and 2.8 myr). Stratons are numbered as consecutive integers as well as according to their positions in dozons A to C and Orbitons 1–37, and dated at their start/base. The scale predicts the ages of lowstand stratons containing major or minor unconformities, and highstands containing maximum flooding surfaces or intervals (MFS, MFI). Sequence boundaries (SB) correlate to the starts of post-glacial transgressions. In times older than c. 8.0 million years, carbon-13 isotope ($\delta^{13}C$) positive excursions (plateaus) correlate to sea level lows and negative excursions (valleys) to sea level highs. Decreasing (falling limb) and increasing (rising limb) $\delta^{13}C$ trends correlate to sea level rises and falls, respectively.

Boundary estimated at 443.4 Ma (443.8 ± 1.5 Ma in GTS 2020/01, http://www.stratigraphy.org). It interprets the glacio-eustatic signature of the latest Ordovician Gondwana glaciation from $\delta^{13}C$ records (e.g., Brenchley et al., 2003; Kaljo et al., 2001) and introduces a latest Ordovician c. 400–800 kyr hiatus in key sections that distorts time estimates and complicates chemo-biostratigraphic correlations—a *stratigraphic black hole* (SBH, Ruebsam and Al-Husseini, 2020). Example 2 supports the relationship proposed by Saltzman et al. (2004) between early Paibian SPICE (Steptoan Positive Isotope Carbon Excursion) and a sea level lowstand. It uses the Orbital Scale to date base Paibian Stage at c. 498.1 Ma (c. 497.0 Ma in GTS 2020/01), and highlights the conflicting positions and ages chosen for two early Paibian sequence boundaries (SB).

The third example calibrates and interprets the $\delta^{13}C$ record in Siberia's Ediacaran-Cambrian transition (Kouchinsky et al., 2007) and concludes its glacio-eustatic signature manifests the stable long-eccentricity 405-kyr orbital cycle in early Cambrian. The fourth and fifth examples anchor glaciations and their positive $\delta^{13}C$ excursions at c. 312.6 Ma in the Carboniferous, and near the Devonian/Carboniferous Boundary at c. 358.25 Ma (358.9 ± 0.4 Ma in GTS 2020/01). These two anchors represent a modest start at mooring a very small fraction of the extensive volume of Late Paleozoic stratigraphic data according to the Orbital Scale.

The final four examples from the Cretaceous (Late Aptian, Cenomanian to Santonian) show that $\delta^{13}C$ plateaus pass laterally into erosional unconformities/hiatuses (SBH), some of which correlate to coeval glaciations or so-called "cold snaps" and "ephemeral glaciations." These examples seek to contradict the widely held view that the Cretaceous and other greenhouse times were supposedly free of polar ice caps for extended periods and to reverse the interpretation of Cretaceous $\delta^{13}C$ plateaus as sea level highs instead of lows.

The examples cite new calibrations for several stage boundaries to an accuracy of about ±100 kyr and compare them to those determined by radiometric and other dating techniques. Most of these boundaries however are defined in unconformity-prone intervals where key age-diagnostic biozones are not always complete or represented, and radiometric dating is often inconsistent. Similar issues occur at many other Phanerozoic stage boundaries including those defined by Global Boundary Stratotype Section and Point (GSSP). These and other issues relating to regional and international chrono- and sequence-stratigraphic schemes suggest these frameworks are essentially open-ended interpretations of contested or incomplete "big data" compilations. Whether these compilations will eventually converge on the deterministic model of the Orbital Scale will require many more examples to be verified.

2. Orbital Scale of glacio-eustasy

Matthews and Frohlich (2002) developed a model that explains how variations in the eccentricity of the Earth's orbit around the Sun tune glacio-eustatic sequences with periods of c. 400 kyr, c. 2.0 and 2.8 myr. To emphasize the importance of the c. 400-kyr sequence as the fundamental building block of Phanerozoic glacio-eustatic sequences Matthews and Al-Husseini (2010) named it the *Straton*, and formulated the deterministic Orbital Scale in terms of this time-rock unit (Fig. 1). An important prediction of the scale is that 12 stratons group into a 4.86-myr *Dozon* and 3 dozons group into a 14.58-myr *Orbiton*. Dozons are bounded by significant glacio-eustatic lowstands and orbitons by even more prominent ones (Fig. 1).

The Orbital Scale represents glacio-eustatic sequences between 1.586 and 541.046 Ma with 37 orbitons (numbered 1–37 from youngest to oldest) and predicts their sequence-stratigraphic architecture as a repeating template that highlights which stratons are sea-level highs and lows, i.e., respectively, $\delta^{13}C$ valleys and plateaus (Fig. 1). From oldest to youngest in each orbiton, dozons are referred to as A to C and their stratons as A-1 to A-12, B-1 to B-12 and C-1 to C-12 (Fig. 1). Dozons are typically subdivided into third-order sequences (or composite or high-frequency sequences, HFS) consisting of several stratons (typically 5 and 7 stratons) and are interpreted at outcrop or from wells and seismic images (e.g., Catuneanu et al., 2009; Haq et al., 1987; Hardenbol et al., 1998).

3. Astronomical Time Scale (ATS)

In 2012, the International Commission on Stratigraphy (ICS, www.stratigraphy.org) adopted in their Geological Time Scale GTS 2012 (Gradstein et al., 2012) the calibration of the Astronomical Time Scale (ATS) for the Cenozoic stages spanning the past 66.0 myr (Hinnov and Hilgen, 2012). ATS uses cyclostratigraphy to estimate the duration of biostratigraphic intervals using the long-eccentricity 405-kyr orbital cycle as a chronometer (Laskar et al., 2004; Hinnov, 2018; abbreviated *E*-cycle, Fig. 2), and calibrates the interval's absolute age with radiometric dates (for ATS methodology, see Hinnov and Ogg, 2007; Huang, 2018).

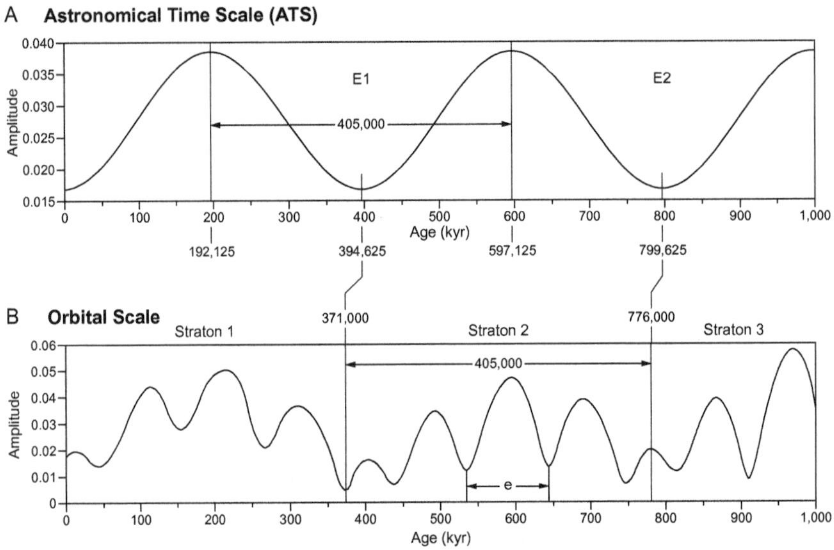

Fig. 2 (A) The Astronomical Time Scale (ATS) subdivides time according to the 405-kyr sinusoidal component of the Earth's eccentric orbit (E Cycle) and numbers them between the maxima of the sine wave. (B) The Orbital Scale defines time-rock units named Stratons between the minima of the eccentricity curve and estimates their duration as 405 kyr with Straton 1 starting at 371,000 years before present. Four c. 100-kyr short-eccentricity cycles (e) typically form a straton. The two scales differ by 23,625 years (Hinnov, 2018). Glacio-eustatic falls and stratigraphic breaks occur at eccentricity minima.

Stratons track the same 405-kyr sinusoidal chronometer of ATS; however, whereas stratons are numbered as consecutive integers between the minima of the sine wave, *E*-cycles are numbered between its maxima (Fig. 2). Importantly, although the two scales are essentially synchronized by the same 405-kyr sine wave, their approaches are completely different and their calibrations are independent. ATS is focused on deciphering the orbital signal in the narrow Milankovitch bandwidth (c. 10s to 405 kyr) but its accuracy depends on radiometric dating. In contrast, the Orbital Scale uses pattern recognition to date eccentricity-driven glacio-eustatic cycles and $\delta^{13}C$ records in the bandwidth of 0.405–4.86 myr. It uses a concise alphanumeric code to identify the $\delta^{13}C$ and eustatic signatures of specific stratons (Fig. 1). Combining the results of the two scales is therefore complimentary and provides absolute age estimates for eustatic cycles, $\delta^{13}C$ records, biozones and magnetic chrons across a broader range of periods.

Huang (2018) synthesized and recalibrated the ATS for the Mesozoic stages spanning c. 66.0–252.0 Ma (hereafter referred to as ATS 2018) but confirming this new calibration and extending ATS into the Paleozoic (and to older times) remains a challenging work-in-progress. As in many other studies, the analyses by Huang (2018) did not calibrate secular $\delta^{13}C$ patterns nor consider their significance as eustatic or climate proxies. In the following nine examples the estimated ATS 2008 and GTS 2020/01 ages of several stage boundaries are compared to those obtained with the Orbital Scale of glacio-eustasy.

4. Chemo-stratigraphic isotopes as glacio-eustatic proxies

Oxygen-18 isotopes ($\delta^{18}O$) patterns are widely used to reconstruct sea-surface temperatures and glacio-eustatic cycles (e.g. Miller et al., 2020). In the Oligocene to present-day, the minima of eccentricity cycles and sea-surface temperature (SST determined from $\delta^{18}O$ and other proxies), on one hand, correlate with the maximum ice volume hosted on Antarctica (de Boer et al., 2014; Holbourn et al., 2007, 2013; Pälike et al., 2006).

In contrast to $\delta^{18}O$ patterns, $\delta^{13}C$ patterns are recognized as much more robust correlative tools and to contain a rich orbital signal, but their interpretation as eustatic and specifically glacio-eustatic proxies is less clear (Ruebsam et al., 2020, this volume). In the present study the interpretation of $\delta^{13}C$ patterns as glacio-eustatic proxies as depicted in Fig. 1 is assumed to apply for times older than the Late Miocene Carbon Shift at c. 8.0 Ma (LMCS).

In pre-LMCS times the $\delta^{18}O$ and $\delta^{13}C$ records correlate implying they are both glacio-eustaic proxies (Fig. 1); however, they anti-correlate in post-LMCS times (Holbourn et al., 2007, 2013, 2018). The reason for the reversal across the LMCS is not well understood, but nonetheless the relationships between $\delta^{13}C$ records and glacio-eustasy appear to be valid in pre-LMCS times. Indeed $\delta^{13}C$ records carry a globally correlative orbital signal and are therefore more likely to represent orbital-forcing of climate and glacio-eustasy.

5. Part 1: Paleozoic glacio-eustasy

Haq and Schutter (2008) reconstructed the history of Paleozoic global sea level in terms of 172 short-term eustatic cycles with amplitudes ranging from tens to c. 125 m. Based on GTS 2004 (Gradstein et al., 2004) they estimated third-order cycles have durations between 0.5 and 3.0 myr and fourth-order c. 400 kyr. They reported glaciation is attributed to 28% of Paleozoic time (and suspected for another 10%) but that the glacial record is too fragmentary to conclude all sea level cycles are glacio-eustatic.

The studies of Phanerozoic eustasy by B.U. Haq and coauthors (e.g., Haq and Al-Qahtani, 2005; Haq and Schutter, 2008) do not specifically relate orbital-forcing of insolation to short-term eustatic cycles, nor do they interpret $\delta^{13}C$ patterns as glacio-eustatic proxies. The first example seeks to confirm these relationships are manifested in the late Ordovician and early Silurian interval, which includes the Hirnantian glaciation of Gondwana. It uses the Orbital Scale to predict the glacio-eustatic signature for the Katian to Aeronian interval (453.0 ± 0.7 to 438.5 ± 1.1 Ma in GTS 2020/01), which encompasses Orbiton 31 (Fig. 3). The other four Paleozoic examples extend the insights gained from the first example to other stratigraphic intervals.

6. Example 1: Mid-Paleozoic ice age

The Katian GSSP is defined at the base of the *Plectodina tenuis* conodont zone of North America's Midcontinent, which coincides with the base of the global positive Guttenberg (GICE) $\delta^{13}C$ excursion (Goldman et al., 2007; Fig. 4). In the Orbital Scale, GICE is correlated to lowstand Straton 32C-12 and top GICE to SB 31 predicted at 453.566 Ma (1.586 + 31 × 14.58 Ma). This correlation calibrates base Katian at 453.971 Ma, compared to 453.0 ± 0.7 Ma in GTS 2020/01 (Figs. 3 and 4).

SB 30 is 14.58 myr younger than SB 31 and caps Orbiton 31 at 438.986 Ma (1.586 + 30 × 14.58), slightly older than top Aeronian (438.5 ± 1.1 Ma, Fig. 3). The greatest positive $\delta^{13}C$ excursion in the early Silurian occurs in the late Aeronian lower *S. sedgewickii* graptolite zone and is interpreted as the greatest sea level lowstand in the early Silurian Llandovery Stage and a glaciation (Loydell, 1998; Fig. 5, Melchin and Holmden, 2006; also named Sandvika Excursion, Aldridge et al., 1993). In the Orbital Scale the Sandvika Excursion is correlated to lowstand Straton 31C-12 and SB 30 is positioned at top Sandvika. The anti-correlation between local sea level and $\delta^{13}C$ patterns in the Canadian Arctic is particularly evident in the early Rhuddanian and Aeronian, but less so in the *cycphus* graptolite zone (Fig. 5; Melchin and Holmden, 2006). Moreover the orbital calibrations of all three stage boundaries are well within the ±1.1 to ±1.5 myr uncertainty of GTS 2020/01.

Orbiton 31, Orbital Scale							Geological Time Scale GTS 2020		
Orbiton	Dozon	Sequence	405-kyr Straton Age @ Base (Ma)			Stratigraphic Marker	International and Baltic Stages	Epoch	Period
30	30A		30A-2	1,083	438.581	*guerchi*	Tetychian		
			30A-1	1,084	438.986	Upper *sedgwickii*	438.5 ± 1.1		
	SB 30								
Orbiton 31 (14.58 myr)	Dozon 31C (4.86 myr)	2.8 myr	31C-12	1,085	439.391	Lower *sedgwickii*, Sandvika	Aeronian	Llandovery	Silurian
			31C-11	1,086	439.796	*convolutus*			
			31C-10	1,087	440.201	*convolutus*			
			31C-9	1,088	440.606	*curtus*			
			31C-8	1,089	441.011	*curtus*			
			31C-7	1,090	441.416	*cyphus*	440.8 ± 1.2		
			31C-6	1,091	441.821	*cyphus*	Rhuddanian		
		2.0 myr	31C-5	1,092	442.226	*atavus-acinaces*			
			31C-4	1,093	442.631				
			31C-3	1,094	443.036	*ascensus-acuminatus*			
			31C-2	1,095	443.441		Juuru		
	SB 31C		31C-1	1,096	443.846	Upper HICE *persculptus*	Porkuni; Himantian	443.8 ± 1.5; Late	Ordovician
	Dozon 31B (4.86 myr)	2.8 myr	31B-12	1,097	444.251				
			31B-11	1,098	444.656	*extraordinarius*			
			31B-10	1,099	445.061	Minor Highstand	445.2 ± 1.4; Pirgu; Katian		
			31B-9	1,100	445.466	Paroveja			
			31B-8	1,101	445.871	Minor Highstand			
			31B-7	1,102	446.276				
			31B-6	1,103	446.681				
		2.0 myr	31B-5	1,104	447.086	Elkhorn	Vormsi		
			31B-4	1,105	447.491				
			31B-3	1,106	447.896	White Water			
			31B-2	1,107	448.301	Highstand			
			31B-1	1,108	448.706		Nabala		
	SB 31B								
	Dozon 31A (4.86 myr)	2.8 myr	31A-12	1,109	449.111	Waynesville, WICE			
			31A-11	1,110	449.516				
			31A-10	1,111	449.921	Minor Highstand			
			31A-9	1,112	450.326	Fairview			
			31A-8	1,113	450.731	Minor Highstand			
			31A-7	1,114	451.136				
			31A-6	1,115	451.541	KOPE			
		2.0 myr	31A-5	1,116	451.946		Rakvere		
			31A-4	1,117	452.351		Oandu; Sandbian		
			31A-3	1,118	452.756				
			31A-2	1,119	453.161	Highstand			
			31A-1	1,120	453.566				
	SB 31								
32	32A		32C-12	1,121	453.971	Guttenberg, GICE	453.0 ± 0.7		
			32C-11	1,122	454.376		Kaila		

Lesser Lowstand
Major Lowstand
Lesser Highstand
Major Highstand

Fig. 3 Periodic Orbital Scale of glacio-eustasy showing deterministic calibration of key bio- and chemo-stratigraphic markers in Katian to Aeronian Orbiton 31. *Modified after Al-Husseini, M.I., 2018. Arabian orbital sequences. In: Montenari, M. (Ed.), Stratigraphy and Timescales, 3, 219–264.*

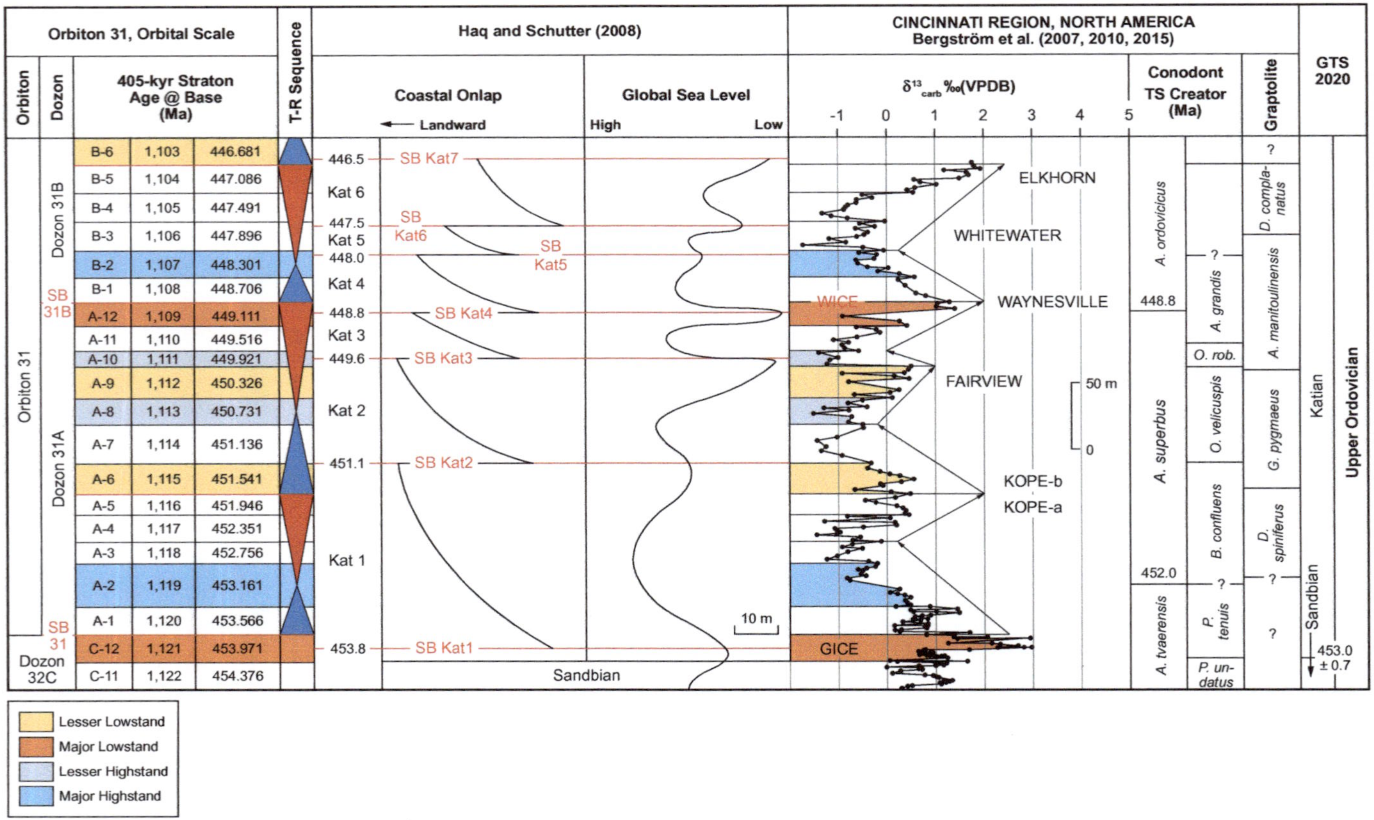

Fig. 4 Katian global sea level and coastal-onlap curves (Haq and Schutter, 2008; Snedden and Liu, 2010), and $\delta^{13}C$ record in Cincinnati Section (Bergström et al., 2010) calibrated according to GTS 2020/01 and the Orbital Scale. Age of base Katian Stage and correlative base GICE (Guttenberg) is estimated at c. 453.97 Ma at base lowstand Straton 32C-12. Calibration of base *A. ordovicicus* conodont zone at 448.8 Ma (TS Creator, Ogg and Gradstein, 2020). See Table 1 and text for discussion.

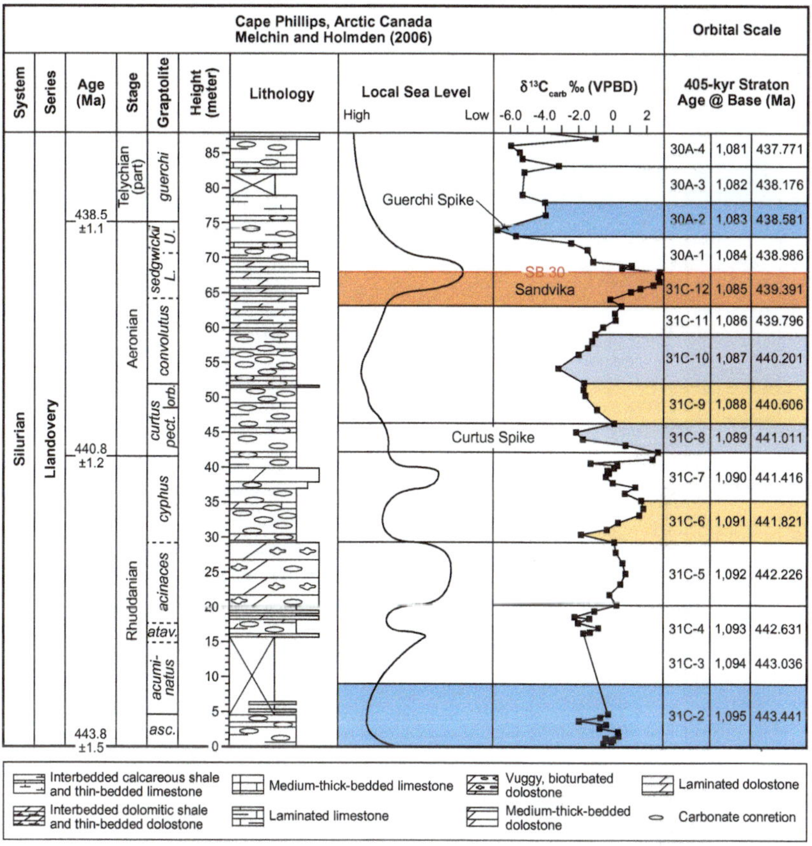

Fig. 5 Orbital calibration of Rhuddanian to lower Telychian $\delta^{13}C$ curve, local sea level and graptolite zones in the Canadian Arctic (Melchin and Holmden, 2006). Positive Sandvika Excursion is correlated to the lowest glacio-eustatic lowstand in the Silurian (Loydell, 1998) and Straton 31C-12.

Two more prominent positive excursions, Waynesville (WICE, Fig. 4, Bergström et al., 2010, 2015) and Upper HICE (Hirnantian Carbon Isotope Excursion, Brenchley et al., 2003), are correlated to lowstand stratons 31A-12 and 31B-12 (Figs. 6–9). The estimated ages of the four excursions (GICE, WICE, HICE and Sandvika) in GTS 2020/01 closely coincide with those of major lowstand stratons that bound the three 4.86-myr dozons of Orbiton 31 (Figs. 3–9; Al-Husseini, 2018, and references therein). In addition, to these four excursions several Katian positive excursions (KOPE, Fairview, Elkhorn, Bergström et al., 2010, 2015) are calibrated in Figs. 3 and 4.

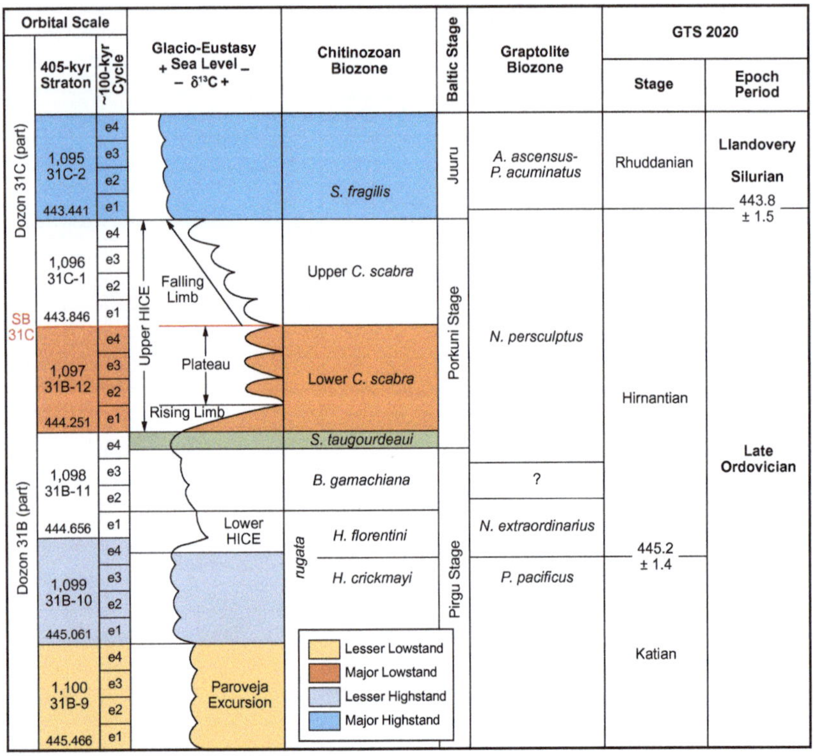

Fig. 6 Schematic interpretation of glacio-eustasy and $\delta^{13}C$ record for the Ordovician-Silurian Transition. The main Hirnantian glaciation is represented by the Upper Hirnantian carbon-isotope excursion (HICE) and the *C. scabra* chitinozoan zone. The Baltic Porkuni Stage is interpreted as a c. 900 kyr interval (c. 444.35–443.44 Ma) within the c. 1.3-myr Hirnantian Stage (c. 444.70–443.44 Ma).

6.1 Katian eustatic cycles

Bergström et al. (2010, 2015) measured $\delta^{13}C$ records in North America and the Baltic region spanning the interval from just below GICE to an upper Katian unconformity capping the Elkhorn Excursion (Figs. 3 and 4). The high-resolution $\delta^{13}C$ record in the Cincinnati region is particularly revealing—it resembles steps in a staircase between floors (Fig. 4). The 19 steps are interpreted as stratons 1122–1104 (32C-11 to 31B-5) and the alternating directions of the staircase (falling and rising limbs) as sea-level rises and falls, respectively. This interpretation can be evaluated by comparing the $\delta^{13}C$ patterns to the global Katian sequences and eustatic cycles (Haq and Schutter, 2008). The comparison however requires revising the ages of the sequences because the calibration of the Katian Stage used by Haq and

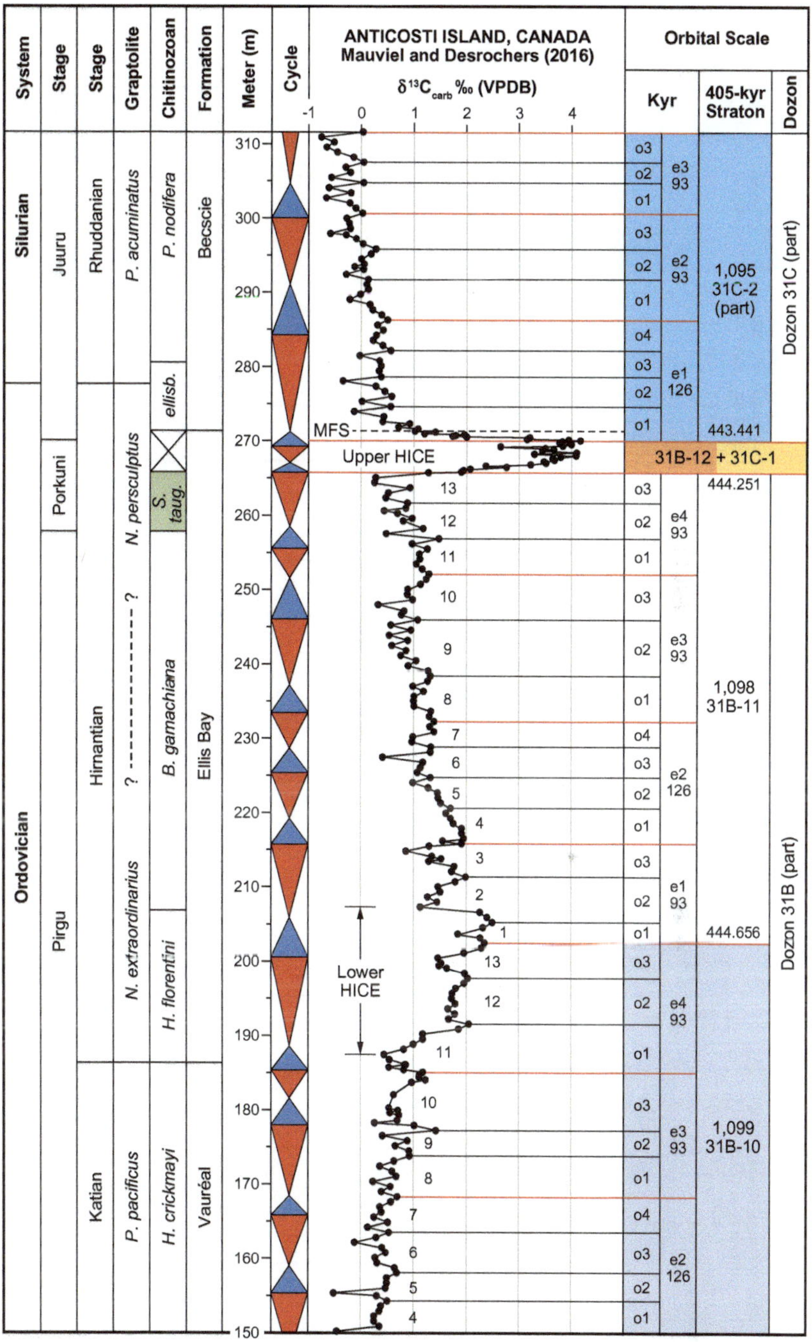

Fig. 7 High-resolution $\delta^{13}C$ record and cycles of the Ordovician-Silurian Transition in Anticosti Island (Quebec, Canada; Mauviel, 2016; Mauviel and Desrochers, 2016). The orbital calibration of the present study is constructed with c. 31-kyr obliquity "o" cycles and short-eccentrcity "e" cycles (93 and 126 kyr). Glacio-eustatic lowstand Stratons 31B-12 and 31A-1 are represented in a 4 m-thick condensed Upper HICE between an unconformity and a maximum flooding surface (MFS) at base Baltic Juuru Stage.

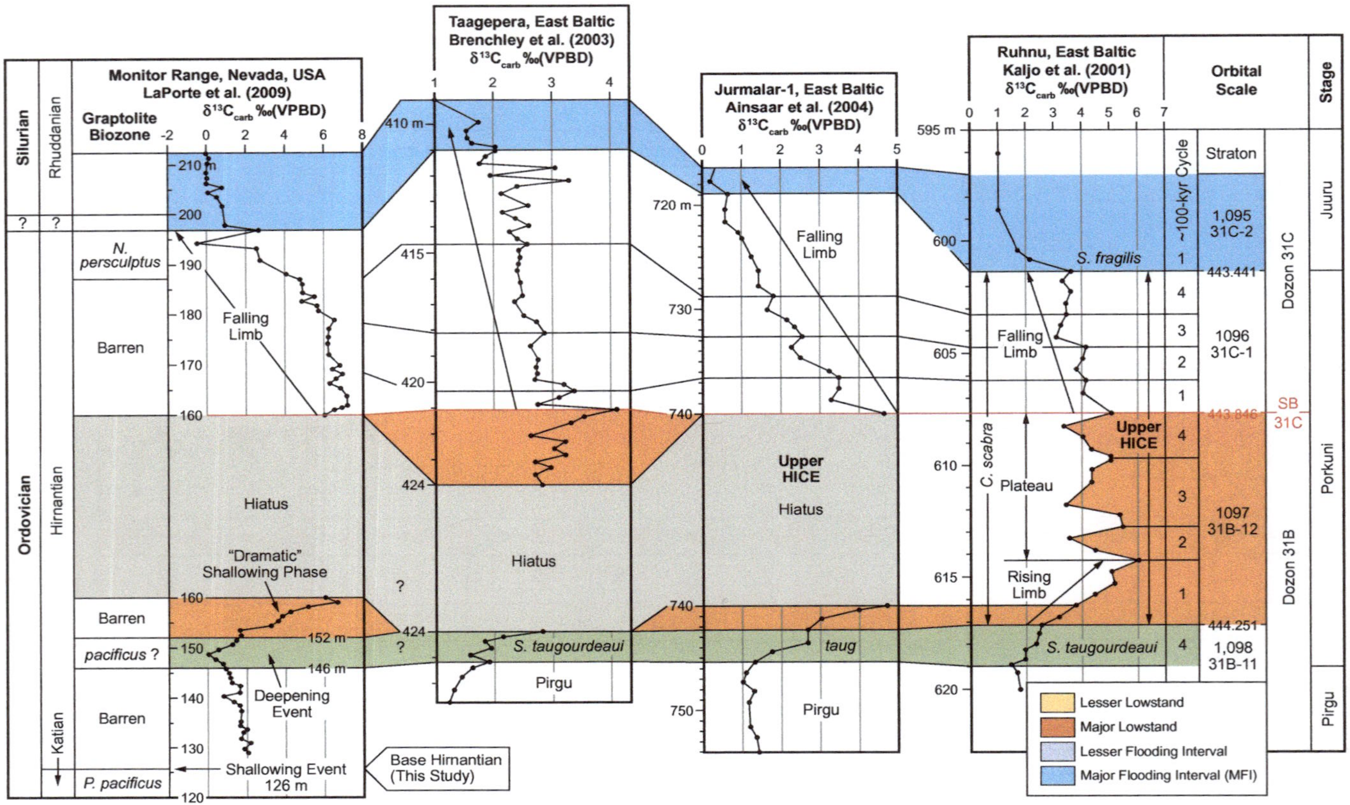

Fig. 8 Correlation of $\delta^{13}C$ records spanning the Ordovician/Silurian Boundary in East Baltic and North America. Rising and falling limbs are interpreted as sea-level fall and rise, respectively. The plateau in the Ruhnu Core apparently represents the most complete $\delta^{13}C$ record of Upper HICE and is interpreted as the glacio-eustatic signature of the main Hirnantian Glaciation.

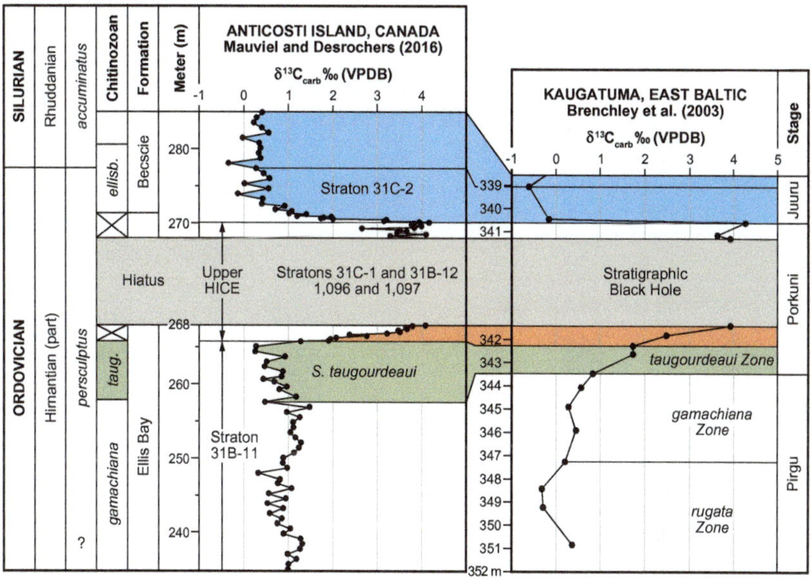

Fig. 9 Correlation of $\delta^{13}C$ records between Anticosti Island (Mauviel and Desrochers, 2016) and East Baltic Kautgatuma (Brenchley et al., 2003) showing similarity of abrupt rising and falling limbs into the Upper HICE stratigraphic black hole (SBH, hiatus) representing the main Hirnantian Glaciation.

Schutter (2008) has been revised from between 455.8 ± 1.6 and 445.6 ± 1.5 Ma in GTS 2004 (Gradstein et al., 2004), to between 453.0 ± 0.7 and 445.2 ± 1.4 Ma in GTS 2020/01, such that the stage's duration is reduced by c. 76% from 10.2 to 7.8 myr.

Haq and Schutter (2008) interpreted seven Katian sequences, dated and named Kat1 to Kat7 in Snedden and Liu (2010). In Table 1 their durations are reduced by 76% and their revised ages are plotted relative to base Katian (base GICE) as calibrated by the Orbital Scale (Fig. 4). The comparisons confirm eustatic cycles correlate to $\delta^{13}C$ valleys; for example, Cycle Kat1 correlates to the $\delta^{13}C$ valley between GICE (lowstand Straton 31C-12) and positive excursion KOPE-b (lowstand Straton 30A-6). Similar correlations between eustatic cycles and $\delta^{13}C$ valleys are evident including at the level of an individual straton: Sequence Kat5 with duration c. 0.54 myr correlates to the major negative Whitewater Excursion and Straton 30B-3.

An important four-way correlation occurs between WICE, SB Kat4 (revised to c. 448.8 Ma), the boundary between the *A. superbus* and *A. ordovicicus* conodont zones (448.8 Ma, Ogg and Gradstein, 2020), and dozon-level SB 31B (448.7 Ma). This surface closely correlates to a karst

Table 1 Katian sequences from Haq and Schutter (2008) and Snedden and Liu (2010).

Stage boundary and cycle	GTS 2008 Age at base	Age at top	Duration	GTS 2020/01 duration	Age above base GICE (B. Katian)	Revised age
Hirnantian	445.6				7.80	
Kat7	4460	445.7	0.3	0.23	7.49	446.5
Kat6	447.3	446.3	1.3	0.99	6.50	447.5
Kat5	448.0	447.3	0.7	0.54	5.96	448.0
Kat4	449.0	448.0	1.0	0.76	5.20	448.8
Kat3	450.0	449.0	1.0	0.76	4.44	449.6
Kat2	452.0	450.0	2.0	1.53	2.91	451.1
Kat1	455.5	452.0	3.5	2.68	0.23	453.8
Katian	455.8		0.3	0.23	0.0	454.0

horizon near the *superbus/ordovicicus* boundary in the Caledonian foreland, which extends across Sweden and into the eastern Baltic area (Calner et al., 2010).

6.2 Ordovician/Silurian Boundary

The stratigraphic sections, $\delta^{18}O$ and $\delta^{13}C$ records straddling the Ordovician/Silurian Boundary are riddled by unconformities and difficult to interpret and correlate regionally and globally (e.g., Brenchley et al., 2003; Ghienne et al., 2014; Holmden et al., 2013; LaPorte et al., 2009). Based on a review of several sections the glacio-eustatic signature of the Ordovician-Silurian transition is interpreted from a composite $\delta^{13}C$ curve calibrated by chitinozoan zones (Fig. 6). The interval spanning *H. crickmayi* to *S. taugourdeaui* zones containing Lower HICE is sketched from Anticosti Island (Mauviel, 2016; Mauviel and Desrochers, 2016) and calibrated with Stratons 1099 and 1098 (31B-10 and 31B-11, Figs. 6 and 7). Mauviel (2016) interpreted high-frequency T-R sequences shown in Fig. 7 and suggested they may be tuned by the 100-kyr short-eccentricity e-cycles.

During the Late Ordovician Sandbian Stage obliquity cycles had durations of c. 31 or 35 kyr and short eccentricity 95 or 128 kyr (Fang et al., 2016). Assuming the c. 5 m-thick high-frequency cycles (labeled as "o") represent 31-kyr obliquity cycles then a 405-kyr straton would consist of 13 o-cycles. This approximation suggests the interval between the maximum value in Lower HICE at c. 202 m and the base of Upper HICE at c. 266 m constitutes a straton (Fig. 7). In this calibration 3 and 4 obliquity cycles would tune with short-eccentricity e-cycles with periods of c. 93 and c. 126 kyr, respectively, and consistent with the results of Fang et al. (2016). These approximations suggest the T-R sequences of Mauviel (2016) are apparently groupings of obliquity cycles rather than short-eccentricity e-cyclers.

The interval spanning *S. taugourdeaui*, *C. scabra* and *S. fragilis* zones containing Upper HICE is sketched from the Rhunu Core (Brenchley et al., 2003; Kaljo et al., 2001) and calibrated with Stratons 1098–1095 (31B-11 to 31C-2, Figs. 6 and 8). In the $\delta^{13}C$ record of the Rhunu Core Upper HICE starts with an abrupt rising limb above the *S. taugourdeaui* zone corresponding to a regression and glacio-eustatic fall (Figs. 6 and 8, Brenchley et al., 2003). It is followed by three negative $\delta^{13}C$ cycles in the Upper HICE plateau interpreted in the present study as the maximum glaciation containing three interglacial cycles tuned by the c. 100-kyr e-cycle (Fig. 2). The Upper HICE falling limb consists of four negative steps interpreted as mainly transgressive sequences and the sea-level recovery tuned by four c. 100-kyr e-cycles. The end of the glaciation is marked by the abrupt $\delta^{13}C$ decrease interpreted as the start of the earliest Silurian flooding interval and highstand.

The Hirnantian Stage is defined by the *N. extraordinarius* and *N. persculptus* graptolite zones (Fig. 6), and its lower GSSP is defined at the boundary between the Katian *P. pacificus* and Hirnantian *N. extraordinarius* zones in the Wangjiawan North Section in China (Chen et al., 2006). In the GSSP section the chitinozoan zones are not documented rendering correlations to sections in the Baltic region uncertain (Chen et al., 2006). Moreover, in the GSSP section the $\delta^{13}C_{org}$ curve only represents Lower HICE in the *N. extraordinarius* zone but Upper HICE in the *N. persculptus* zone is not evident (Chen et al., 2006, their fig. 11) and apparently passes into the Upper HICE unconformity.

In the present study the Upper HICE unconformity is also interpreted in the Monitor Range Section (Nevada, USA, Fig. 8). In this section Finney et al. (1999) assigned the interval between 146 and 152 m to the Katian *P. pacificus* zone whereas Brenchley et al. (2003, their fig. 12) suggested this interval may correlate to the Hirnantian *S. taugourdeaui* zone in the Rhunu Core. Finney et al. (1999) interpreted the interval between 146 and 152 m as a "deepening event" and the overlying interval between 152 and 160 m as a "dramatic shallowing phase," which is considered in the present study the glacio-eustatic fall and $\delta^{13}C$ rising limb of Upper HICE.

Straton 1097 (31B-12) is interpreted as a hiatus (SBH) in the Monitor Range Section (Fig. 8). The gentle falling limb in the barren interval between 160 and 197 m is interpreted as the sea-level rise and transgression in Straton 31C-1 (not Upper HICE plateau as tentatively shown in Brenchley et al., 2003, their fig. 12). Similar interpretations are presented in Fig. 8 for the Baltic Jurmalar-1 Well (Ainsaar et al., 2004) and Taagepera Section (Brenchley et al., 2003).

The interpretations of Upper HICE in the Anticosti Island Section (Mauviel, 2016; Mauviel and Desrochers, 2016) and Baltic Kaugatuma Section (Brenchley et al., 2003) illustrate what appears to be a more long-lived SBH (Fig. 9). In these localities the plateau (glacio-eustatic lowstand) and falling limb (sea level recovery, transgression) are absent in the $\delta^{13}C$ records and the hiatus represents about 800 kyr (Stratons 1097 and 1,096).

The interpretation of Hirnantian $\delta^{13}C$ patterns as glacio-eustatic proxies in the context of the Orbital Scale appears to clarify the relationships between graptolite and chitinozoan zones in far-apart sections and to assist in recognizing unconformities and associated SBHs. The resulting interpretation implies the latest Hirnantian Gondwana ice sheets advanced abruptly in c. 100 kyr starting at c. 444.25 Ma, consisted of seven c. 100 kyr interglacial cycles, and ended with the main basal Silurian flooding at c. 443.44 Ma (compared to 443.8 ± 1.5 Ma in GTS 2020/01; Fig. 6). This scenario assumes the Rhunu $\delta^{13}C$ record does not contain a significant SBH, especially between the *S. taugourdeaui* and *C. scabra* zones.

7. Example 2: Late Cambrian spice and ice

Saltzman et al. (1998) named a prominent late Cambrian $\delta^{13}C$ positive excursion in the North American Steptoan Stage as SPICE (Steptoan Positive Isotope Carbon Excursion; Fig. 10). Saltzman et al. (2004) interpreted SPICE as a eustatic lowstand in the midcontinent of North America and positioned the base of the SAUK III transgression above an unconformity (SB SAUK II/III) near the base of the falling limb (Runkel et al., 1998). Their interpretation has been questioned by other authors who considered the SAUK III transgression started at the base of the rising limb (e.g., Schiffbauer et al., 2017; Woods et al., 2011).

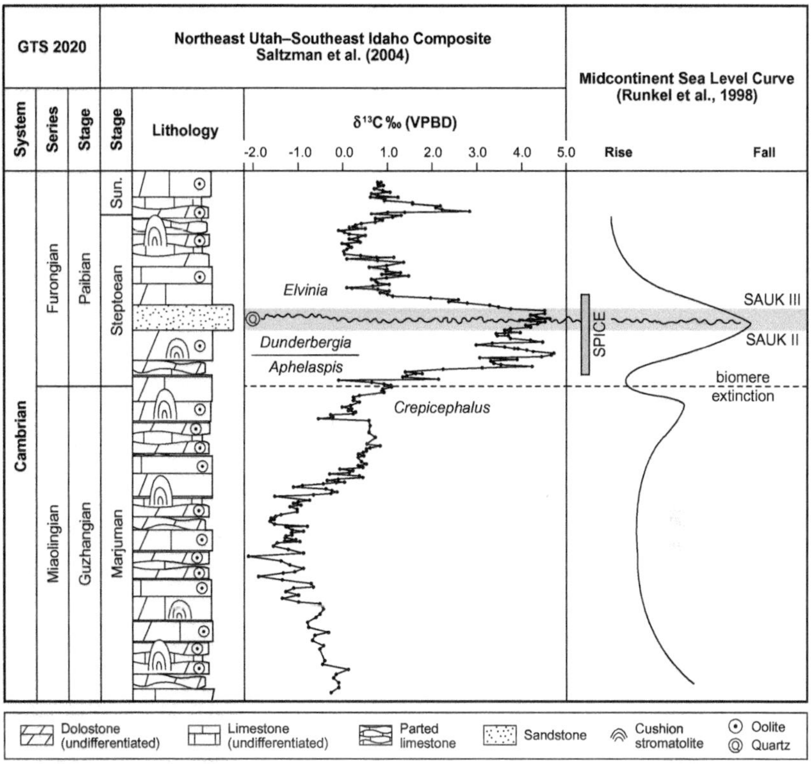

Fig. 10 Carbonate $\delta^{13}C$ composite curve from Smithfield Canyon, northern Utah compared to relative sea level curve in Upper Mississippi Valley (Saltzman et al., 2004).

In the Smithfield Canyon Section (Utah, USA, Fig. 11, Saltzman et al., 2004) SPICE consists of a plateau (approximately *Dunderbergia* trilobite zone) bounded by rising and falling limbs. In the Rheinhart Core (Iowa, USA, Fig. 11) the plateau is absent and the *Dunderbergia* zone is much reduced in thickness, but the rising and falling limbs are essentially similar to those in the Smithfield Canyon Section. The chemo-biostratigraphic correlation implies the upper part of the rising limb and the plateau in the Smithfield Canyon Section pass to an unconformity/hiatus (SBH) in the Rheinhart Core.

In the present study SPICE is interpreted as the signature of a 1.2-myr glaciation that started at 498.116 Ma (base/start Straton 35C-11) at base Paibian Stage of the Furongian Series (c. 497.0 Ma in GTS 2020/01). The $\delta^{13}C$ plateau in the Smithfield Canyon Section is correlated to lowstand Straton 35C-12, and interpreted to pass to an unconformity representing a c.

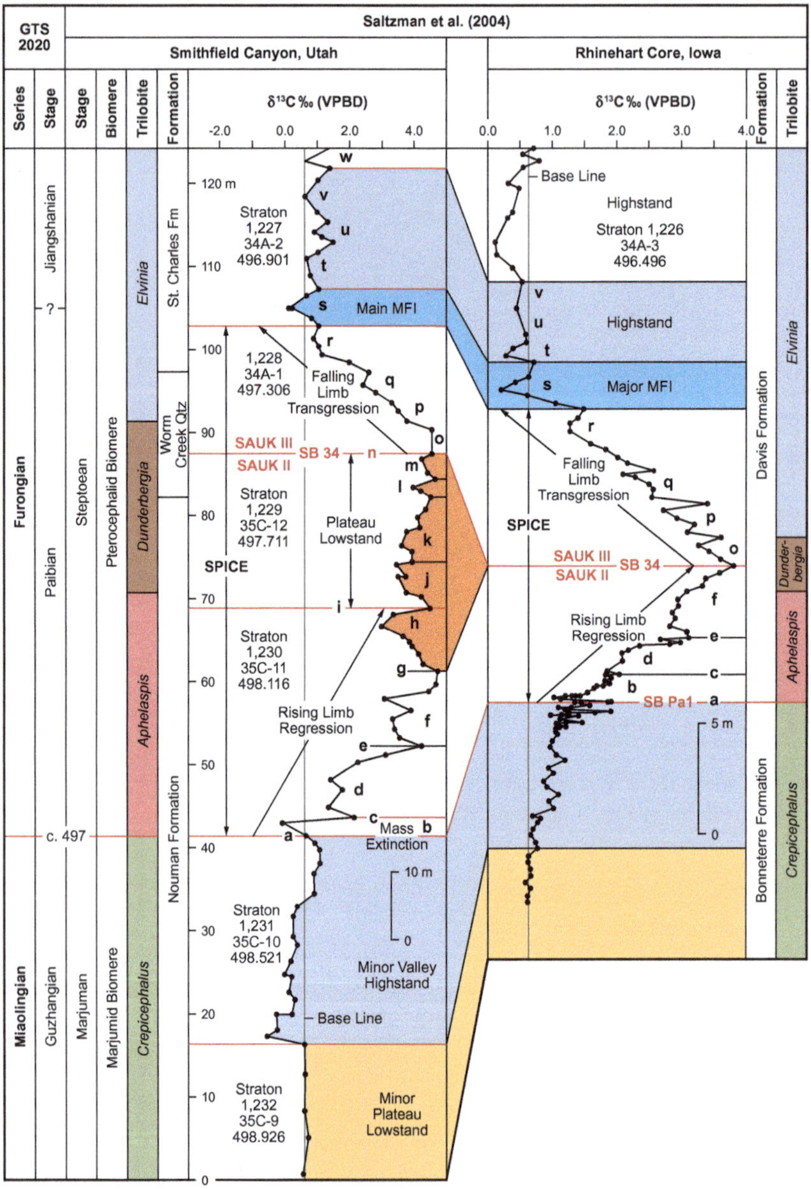

Fig. 11 SPICE (Steptoan Positive Isotope Carbon Excursion, Saltzman et al., 2004) is interpreted as a 1.2-myr glaciation. The glacio-eustatic lowstand in $\delta^{13}C$ plateau in Smithfield Canyon passes to an unconformity in the Rheinhart Core. SB Sauk II/III marks the start of transgression at SB 34, which ends with the maximum flooding interval in lower Straton 34A-2.

500-kyr SBH in the Rheinhart Core. SB SAUK II/III at the base of the falling limb corresponds to the start of the SAUK III transgression as proposed by Saltzman et al. (2004) and is correlated to the start of sea-level rise at orbiton-level SB 34 at 497.306 Ma ($1.586 + 34 \times 14.58$). In contrast, Haq and Schutter (2008, see also Snedden and Liu, 2010) positioned SB Pa1 at base Paibian Stage, which corresponds to the base of the rising limb.

SPICE has a symmetrical trapezoidal shape in the Smithfield Canyon Section and is interpreted in terms of c. 400-kyr plateau bounded by c. 400-kyr rising and falling limbs (Fig. 11). In contrast the rising limb of Upper HICE in the Rhunu Core has an estimated duration of just c. 100 kyr (Figs. 6 and 8). Alternatively, Upper HICE in the Rhunu Core may contain an unconformity between the *S. taugourdeaui* and *C. scabra* zones that hides the rising limb in an SBH.

The interpretations of sea level corresponding to SPICE and Upper HICE share a common feature. In the former case sea level is interpreted to rise in latest Ghuzhangian before the main early Paibian fall with the biomere extinction marking the start of the regression (Runkel et al., 1998; Saltzman et al., 2004; Fig. 10). A somewhat similar eustatic pattern occurs in the Monitor Range (Fig. 8) where Finney et al. (1999) interpret a "deepening event" (presumably the *S. taugourdeaui* zone) immediately preceded the "dramatic shallowing phase" (presumably the *C. scabra* zone in Upper HICE). These precursory eustatic rises followed by extinction events and glaciations seems analogous to crossing the event horizon of stratigraphic black holes (SBH).

8. Example 3: Orbiton 37 and the Ediacaran/Cambrian Boundary

The age of orbiton-level SB 37 is predicted at 541.046 Ma ($1.586 + 37 \times 14.58$) and coincides precisely with the radiometric date of 541.0 ± 0.13 Ma obtained from an ash bed below the interpreted position of the Ediacaran/Cambrian Boundary in Oman (Amthor et al., 2003; Bowring et al., 2007; GTS 2020/01). However, based on high-precision radiometric dating and detailed biostratigraphy in the Swartpunt Section in Namibia Linnemann et al. (2019) proposed revising the age of the Ediacaran/Cambrian Boundary to between 538 and 536 Ma. To fully appreciate the issues relating to the definition and dating of the Cambrian GSSP and its correlation to other sections the interested reader is referred to chapter 6.2 of Wang et al. (2020) and Geyer (2019).

Example 3 is mainly focused on recognizing the glacio-eustatic signature of Orbiton 37 from the $\delta^{13}C$ record measured in the Sukharikha Section (Russia, Fig. 12, Kouchinsky et al., 2007). An outcome of this example is a possible resolution for the conflicting ages and positions for the Ediacaran/Cambrian Boundary.

The Sukharikha Section spans Siberia's upper 'Precambrian,' Nemakit-Daldynian Stage (N-D) and Tommotion to Botomian stages, and the Sukharikha Formation is characterized by positive $\delta^{13}C$ Excursions 1p to 7p bounding negative $\delta^{13}C$ Cycles 1n to 6n (Fig. 12, Kouchinsky et al., 2007). Maloof et al. (2010) compiled $\delta^{13}C$ data from several realms and calibrated them by assuming a uniform sediment accumulation rate between the presumed Ediacaran/Cambrian Boundary at 542 Ma and the N-D/Tommotian Boundary estimated by radiometric data at c. 525 Ma (Fig. 12). They identified Sukharikha Excursions 1p to 7p but their choice for Excursions 2p and 6p (shown in black with question mark in the composite dataset in Fig. 12) are revised in the present study (shown in red). The correlations suggest Excursion 7p is an unconformity in the Sukharikha Section in uppermost N-D Stage (Fig. 12).

In Fig. 13, the Sukharikha $\delta^{13}C$ patterns are interpreted as glacio-eustatic proxies and dated using the template in Fig. 2 recalibrated for Orbiton 37 between 541.046 and 526.466 Ma. Fig. 13 also shows the global coastal-onlap sequences (Haq and Schutter, 2008; Snedden and Liu, 2010), in which the estimated ages of SBs are reduced by 1.0 myr in order to tie the presumed Ediacaran/Cambrian (=Ediacaran/N-D) Boundary at 541.0 (instead of 542.0 Ma in Haq and Schutter, 2008; Maloof et al., 2010). Now the estimated ages of SBs For1, For2 and For5 correlate almost precisely to the orbital calibrations of Excursions 1p, 2p and 4p. In contrast, SBs For3 and For6 correlate closely to the orbital calibration at the start of rising limbs near the bases of the eleventh stratons. The latter correlations imply Haq and Schutter (2008) chose the starts of regressions for SBs and included the regressive deposits as lowstands in the overlying sequences.

Uppermost Straton 1301 (37C-12) is predicted as a major glacio-eustatic lowstand below orbiton-level SB 36 at 526.466 Ma ($1.586 + 36 \times 14.58$). The lowstand is interpreted as an erosional unconformity and hiatus (SBH) that spans 1.2 myr in Stratons 37C-10 to 37C-12. The age of the hiatus between c. 527.7 and 526.5 Ma is consistent with that of SB 1Cam2 estimated at 527.0 Ma (Haq and Schutter, 2008; Fig. 13). The correlation is further supported by a radiometric date of 526.5 ± 1.1 Ma obtained from an ash bed in Bed 9 above an unconformity capping the N-D Stage

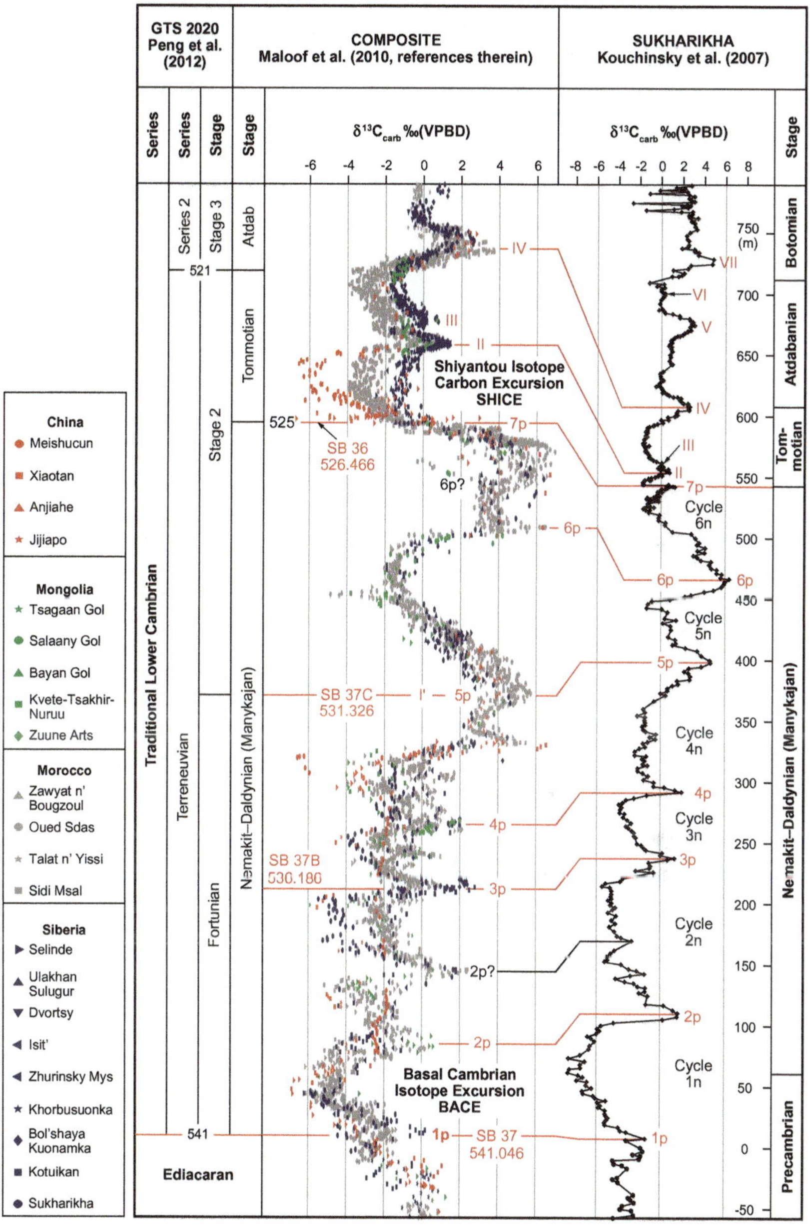

Fig. 12 Carbon-13 ($\delta^{13}C$) isotope values from China, Mongolia, Morocco and Siberia calibrated with six U-Pb dates (Maloof et al., 2010, and references therein) correlated with Sukharikha profile (Kouchinsky et al., 2007). The boundary between Fortunian Stage and Stage 2 at Excursion 5p (also I′) after Kouchinsky et al. (2007).

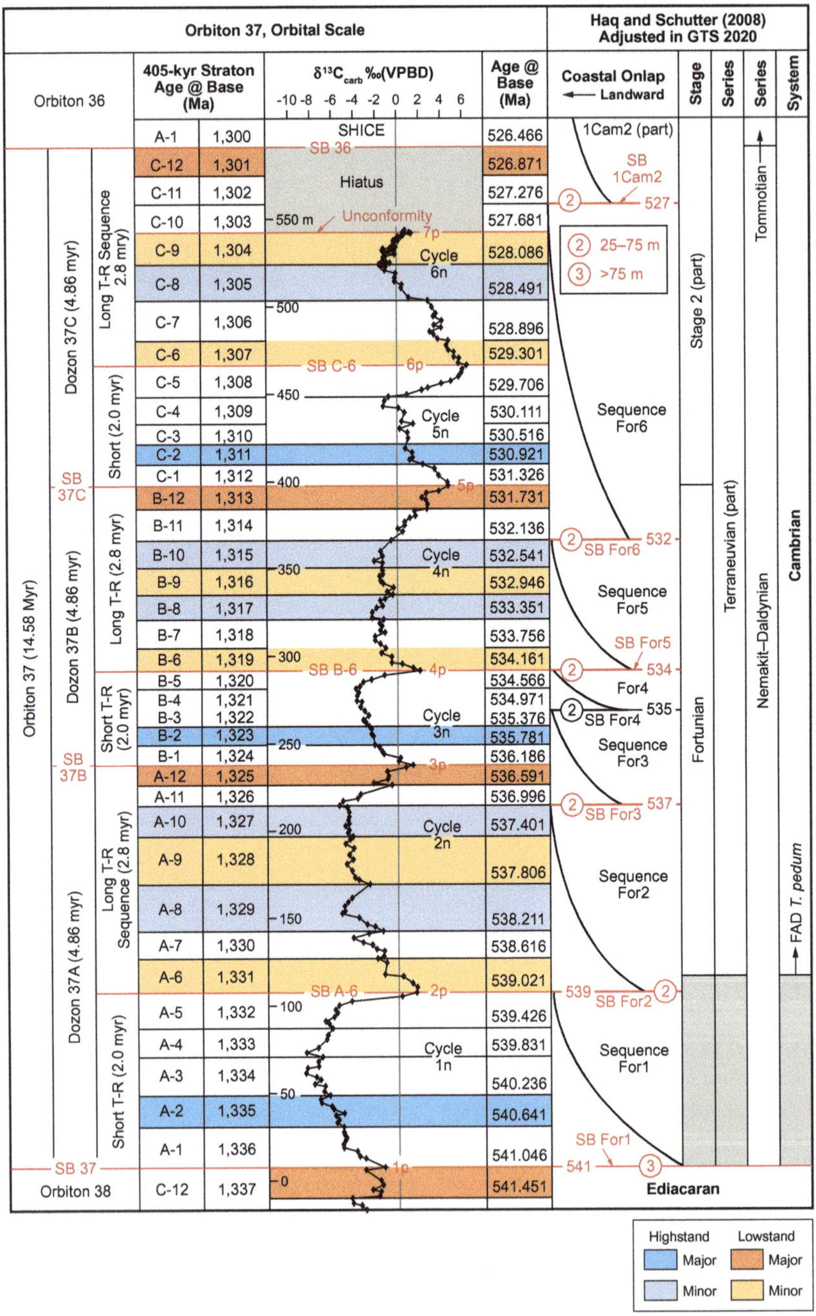

Fig. 13 Sukharikha $\delta^{13}C$ profile calibrated with Orbiton 37 compared to the global coastal-onlap sequences (Haq and Schutter, 2008). The ages of the global sequences boundaries are reduced by 1.0 myr to tie Excursion 1p at base Cycle 1n at 541.0 Ma. FAD *T. pedum* (base Cambrian) after Linnemann et al. (2019).

(top of the *W. crosbyi* assemblage zone) in the Meishucun Quarry Section in South China (Compston et al., 2008).

Two correlations between SBs and positive $\delta^{13}C$ excursions do not occur (Fig. 13): (1) SB For4 (535.0 Ma) in Cycle 3n is probably a fourth-order SB. (2) Excursion 6p is not recognized as an SB in Sequence For6 suggesting this c. 5-myr sequence is too long as a third-order sequence and may contain a sequence boundary. Importantly, Haq and Schutter (2008) did *not* use $\delta^{13}C$ patterns to interpret or date their SBs and therefore the close numerical age correlations to orbital estimates imply the Orbital Scale template applies in early Cambrian.

8.1 Ediacaran/Cambrian Boundary and the Great Unconformity

Wang et al. (2020) presented the results of a study from the deep-water Pingyin Section in South China that may resolve the conflicting age of the Ediacaran/Cambrian Boundary (Fig. 14). The $\delta^{13}C_{org}$ record in this section shows a rising limb ending in a bentonite layer dated at 541.48 ± 0.46 Ma in the plateau followed by a sharp falling limb. The overlying condensed section with uniform $\delta^{13}C_{org}$ values is dated at a second bentonite at 536.40 ± 0.47 Ma. The 541.48 ± 0.46 Ma age of the plateau implies it occurs near or within Straton 38C-12 at 541.451 Ma (Fig. 13). This result is consistent with the correlation of SB 37 to the base of the falling limb at 541.046 Ma at Excursion 1p (base Cycle 1n and so-called BAsal Cambrian Carbon Isotope Excursion, BACE, Zhu et al., 2006; Figs. 12–14).

In Namibia's Swartpunt Section two conglomeratic sequences fill 30 m-deep valleys (Shahkarami et al., 2020, and references therein), and occur above a regional unconformity that is bracketed by ash beds dated 538.99 ± 0.21 and 538.58 ± 0.19 Ma (Linnemann et al., 2019). The two sequences and underlying unconformity occur below the basal Cambrian *Treptichnus pedum* assemblage zone that defines the Cambrian GSSP (Geyer, 2019; Linnemann et al., 2019), and correlate almost precisely to lowstand Straton 37A-6 between 539.021 and 538.616 Ma (Fig. 13). The $\delta^{13}C$ values in the Swartpunt Section are almost uniform at +1.0‰ and do not record Cycle 1n (BACE, Linnemann et al., 2019) suggesting they are probably measured in the interval spanning Excursion 2p (Figs. 12 and 13).

In summary, Example 3 correlates Excursion 1p at base Cycle 1n (BACE) to SB 37 at 541.0 Ma in late Ediacaran. Cycle 1n and the depositional sequence between 541.0 and 539.0 Ma (Sequence For1) correlate to the oldest predicted 2.0-myr post-glacial sea level cycle of Orbiton

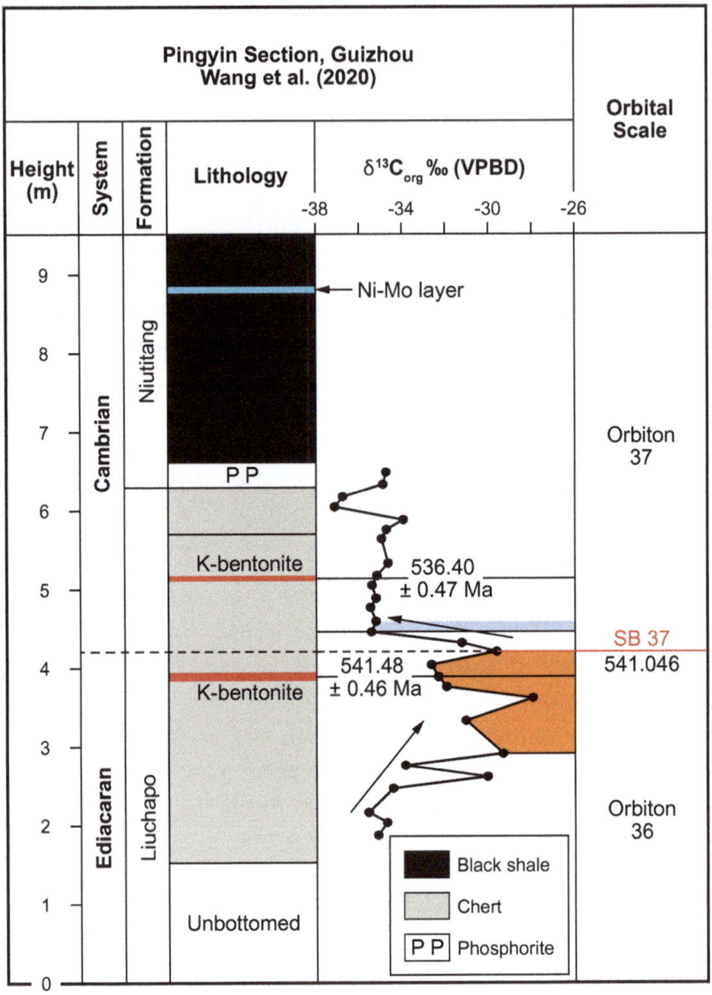

Fig. 14 The $\delta^{13}C_{org}$ record in the Pingyin Section in South China contains a plateau dated 541.48 ± 0.46 Ma bounded by rising and falling limbs (Wang et al., 2020). SB 37 predicted at 541.046 Ma is positioned at the base of the falling limb (see also Figs. 12 and 13).

37 (Stratons 37A-1 to 37A-5). Lowstand Straton 38C-12 may represent the "Great Unconformity," which has been interpreted as a glacio-eustatic fall (e.g., Germs and Gaucher, 2012; Keller et al., 2019). Excursion 2p is correlated to lowstand Straton 37A-6 and occurs just below the FAD of *T. pedum* zone marking the biostratigraphic definition of the Cambrian GSSP at c. 538.8 Ma.

9. Example 4: Moscovian glaciation

Correlations of glaciogenic deposits and positive $\delta^{13}C$ excursions are rarely considered in the literature because they are both generally poorly dated and typically occur in far-apart realms. Example 4 shows how the Orbital Scale highlights the correlation of an Early Moscovian glaciation in Middle Pennsylvanian (Carboniferous) in Argentina to a positive $\delta^{13}C$ excursion in northern Spain. In the Argentinian Paganzo Basin Gulbranson et al. (2015, their fig. 2) documented drop stones within shallow-marine clastics dated c. 313.0 Ma in an ash bed in the Huaco locality.

In the Sierra de Cuera platform in Spain's Cantabrian Mountains, Immenhauser et al. (2003, fig. 15) documented correlative $\delta^{18}O$ and $\delta^{13}C$ positive excursions of c. 4.0‰ VPBD in the Kashirian Stage and estimated the age of the "Red Marker" at c. 309.0 Ma in GTS 1989 (Harland et al., 1990). By linearly recalibrating the approximate age of the Red Marker from the Kashirian in GTS 1989 to Moscovian in GTS 2020/01, c. 309.0 Ma shifts to c. 313.0 Ma near base Straton 22B-12 at 313.03 Ma and below SB 22C at 312.626 Ma (Table 2). Sections 1 and 2 in Fig. 15 are situated on the platform interior and calibrated with Stratons 22B-7 to 22C-3. Stratons 22B-8 to 22B-10 are each approximately 12 m thick implying a sediment accumulation rate of c. 30 m/myr.

Based on a detailed study of lithofacies and stratigraphic architecture in Spain's Sierra de Cuera platform Della Porta et al. (2002) concluded the depositional sequences in the platform interior were controlled by 100–240 kyr, c. 40 m sea-level cycles and rapid tectonic subsidence (70–140 m/myr). The amplitude and periodicity of the sea-level cycles are consistent with glacio-eustatic cycles tuned in part by 100-kyr short eccentricity apparently

Table 2 Approximate age of Red Marker.

Surface	GTS 1989 (Harland et al., 1990)	GTS 2020/01
Top Moscovian	303.0	307.0 ± 0.1
Top Kashirskian	307.1	311.1[a]
Red Marker	309 (Immenhauser et al., 2003)	313[a]
Base Kashirskian	309.2	313.2[a]
Base Moscovian	311.3	315.2 ± 0.2

[a]Linear recalibration.

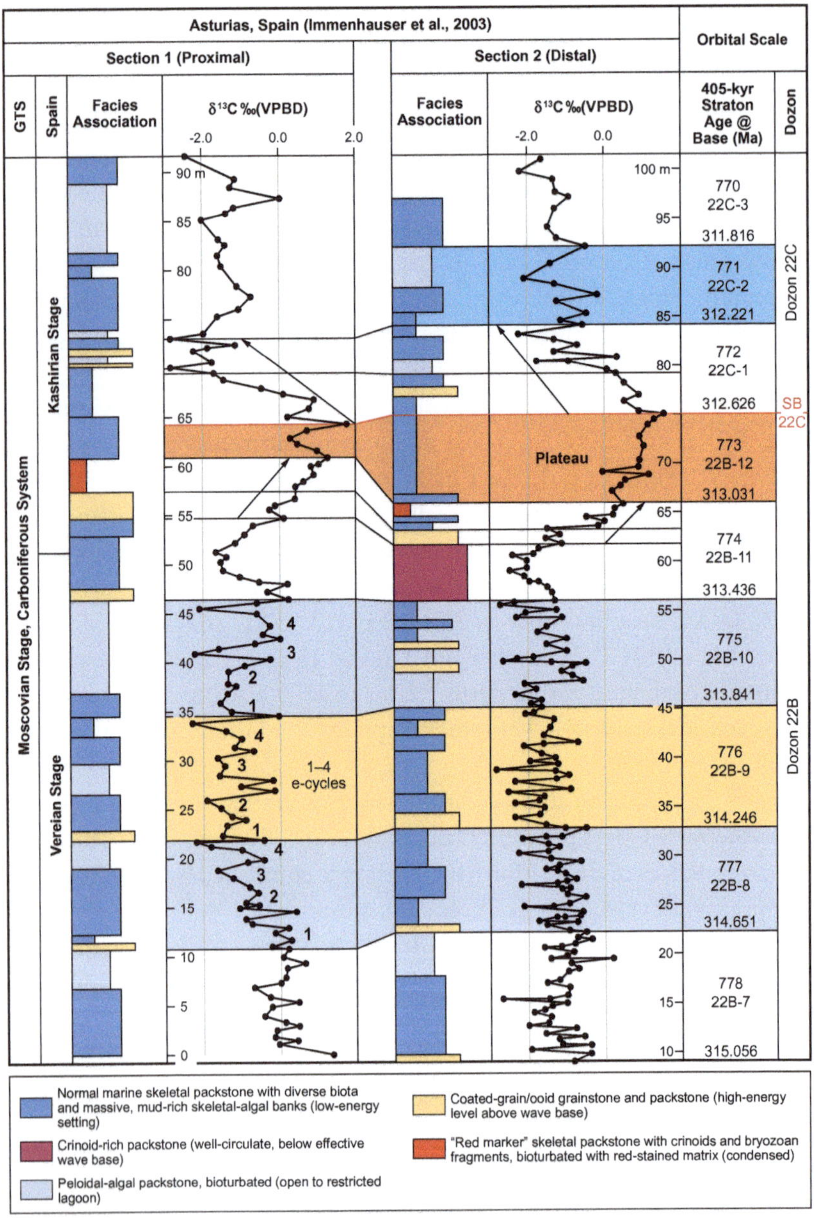

Fig. 15 Facies associations and $\delta^{13}C$ records of Sections 1 and 2, situated 1000 and 1640 m from shelf break, respectively, on the Asturius paleoshelf, northern Spain (Immenhauser et al., 2003). In the Paganzo Basin in Argentina drop stones dated c. 313 Ma approximately correlate to the $\delta^{13}C$ plateau interpreted as a glacio-eustatic lowstand.

manifested by the four negative e-cycles 1–4 in each of Stratons 22B-8 to 22B-10 (Fig. 15).

In Section 1 a sharp regression is evident by the lithofacies change at 55 m from the low-energy carbonates deposited below wave base to the overlying ooid-grainstone and packstone bed (c. 55–57 m) deposited above wave base. In Section 2 the correlative ooid-grainstone/packstone bed occurs between 62 and 63 m. Above the grainstone/packstone bed, the Red Marker is a bioturbated condensed skeletal packstone interpreted as a flooding interval and the facies association in the rising limb indicates normal marine conditions (Immenhauser et al., 2003). These interpretations suggest the rising limb is a relative sea-level highstand, leading Immenhauser et al. (2003) to interpret the geochemical record as a superposition of a temperature fall and local epeiric effects (i.e., sea level rise).

In the more proximal Section 1 above the Red Marker the $\delta^{13}C$ record consists of a single negative cycle in the plateau, followed by the falling limb and culminating in a triplet of negative $\delta^{13}C$ spikes. In the more distal Section 2 above the Red Marker the $\delta^{13}C$ record consists of a gentle rising limb within a plateau followed by the falling limb and triplet of negative $\delta^{13}C$ spikes.

The evolution of relative sea level across the positive excursion apparently contradicts the interpretations proposed in this study—i.e., the rising limb is a rising sea level not of a falling one (Fig. 1). The contradiction may be due to the excursion occurring in a rapidly subsiding basin (70–140 m/myr, Della Porta et al., 2002), which substantially exceeds the sediment accumulation rate of c. 30 m/myr. In such a setting the platform would have been too deep to accumulate deposits associated with a global glacio-eustatic lowstand.

The proposed scenario is consistent with the conclusion of Della Porta et al. (2002): "*Sierra de Cuera strata do not resemble other Pennsylvanian cycles driven by high-frequency, high-amplitude glacio-eustasy, … Tectonic subsidence might have been responsible for the lack of evidence of subaerial exposure surfaces in the examined Moscovian strata.*" This example indicates interpreting relative sea level from lithofacies and sequence architecture is appropriate at local basin scales; however, the signature of glacio-eustasy is more reliably inferred from geochemical proxies.

10. Example 5: Anchoring the Devonian/Carboniferous Boundary

The interval below the Devonian/Carboniferous Boundary is somewhat exceptional because it is associated with dated glaciations and $\delta^{13}C$

records and therefore provides a key anchor in the Orbital Scale. Establishing this and other anchors is an important first step for calibrating the Devonian and Carboniferous, which are very rich in $\delta^{13}C$ excursions (Buggisch and Joachimski, 2006). The calibration of this boundary is also chosen as an example to illustrate that glaciations do not just bound dozons, but also occur between third-order sequences (Fig. 2).

During the Late Devonian in latest Famennian times two glaciations occurred in Gondwana (Brezinski et al., 2010; Caputo et al., 2008; Isaacson et al., 2008; Kaiser et al., 2015; Lakin et al., 2016, and references therein). In South America the glaciations are dated in the *Retispora lepidophyta–Indotriradites explanata* (LE) and in the *Retispora lepidophyta–Verrucosisporites nitidus* (LN) spore zones, which correlate to the *Siphonodella praesulcata* conodont zone (Figs. 16 and 17).

The positive $\delta^{13}C$ excursion of the LN glaciation is evident in the outcrop of the Griotte Formation in La Serre (Montagne Noire, France, Brand et al., 2004); however the interval corresponding to the older LE glaciation is not sampled (shown with a question mark in Fig. 17). The Devonian–Carboniferous transition in La Serre is characterized by Cycles I to III, and the GSSP of the Famennian/Tournaisian (=Devonian/Carboniferous Boundary) is defined at the base of Bed 89 (Fig. 16, after Bless et al., 1992; Brand et al., 2004; Flajs and Feist, 1988).

In many European regions the Hangenberg Black Shale (HBS) occurs near the base of the Middle *S. praesulcata* zone (Bless et al., 1992; see distribution of HBS and equivalents in table 1 and fig. 6 of Kaiser et al., 2015). The HBS in Sauerland (Germany) is correlated to Beds 73–74 in La Serre (Fig. 16, Bless et al., 1992), and in the Holy Cross Mountains (Poland) the HBS is dated between 358.97 ± 0.11 Ma and 358.89 ± 0.20 Ma in three ash beds (Myrow et al., 2013). These dates imply the HBS and correlative Beds 73–74 were deposited in Straton 25B-6 of Dozon 25B.

Straton 25B-6 is predicted as a minor glacio-eustatic lowstand and its lower boundary at 359.2 Ma is correlated to the sharp sequence boundary at base Cycle I between Beds 68 and 69 representing the onset of the LE Glaciation (Fig. 16). The LE Glaciation is described by Kaiser et al. (2015) as "*a first and short-term glacial advance just before the onset of the HBS equivalents elsewhere* (Streel et al., 2000)," and it apparently occurred in the oldest 100-kyr e-cycle of Straton 25B-6. Cycle I is correlated to Straton 25B-6 and Beds 72–80 are interpreted as the partial melting of the LE ice sheet/sheets. The eustatic rise in La Serre Beds 72–73 (equivalent to the HBS) is interpreted as the marine flooding interval (HBS MFI) of Cycle

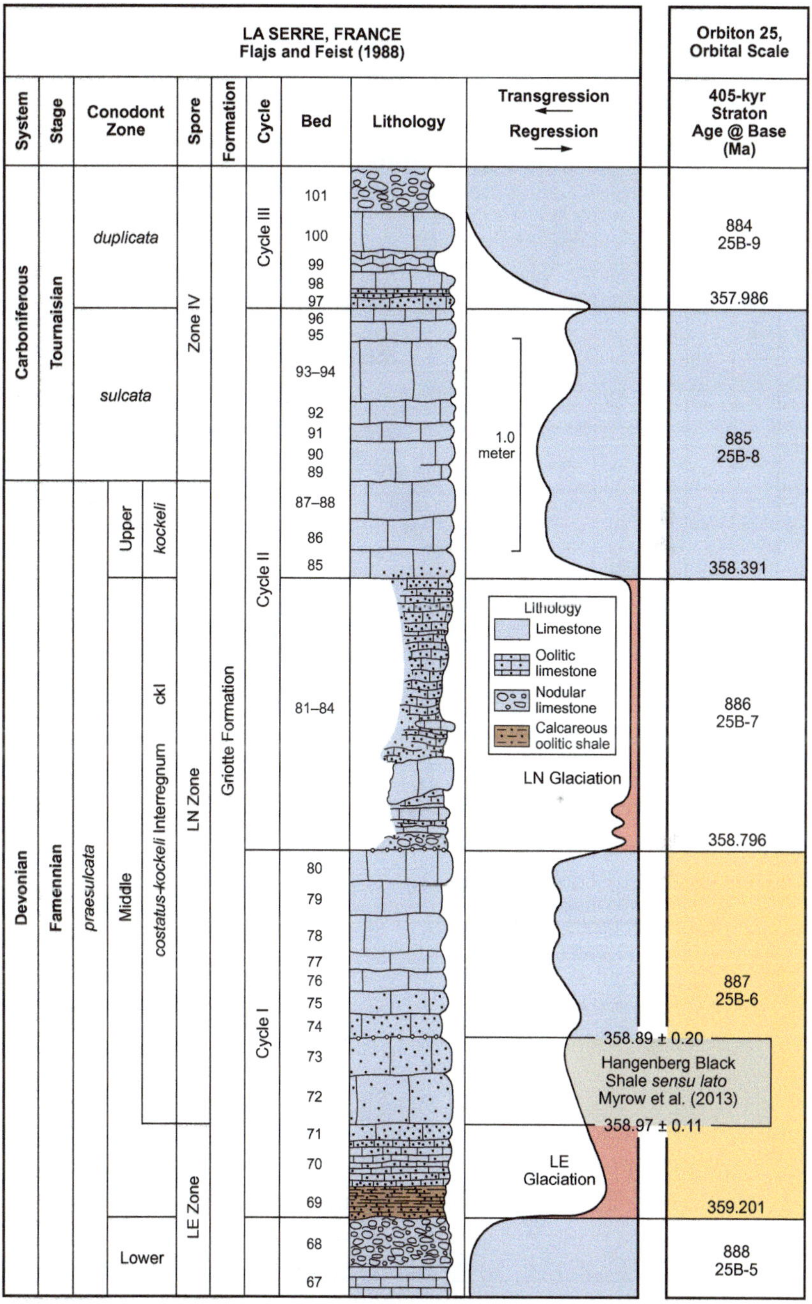

Fig. 16 GSSP of Devonian/Carboniferous boundary in La Serre (after Flajs and Feist, 1988), calibrated with the Orbital Scale, based on the U-Pb dating of the Hangenberg Black Shale *sensu lato* (HBS, Bless et al., 1992; Myrow et al., 2013). Glaciations (pink) occurred in the LE and LN spore zones (Caputo et al., 2008; Streel et al., 2000), in lowermost Straton 25B-6, and in Straton 25B-7.

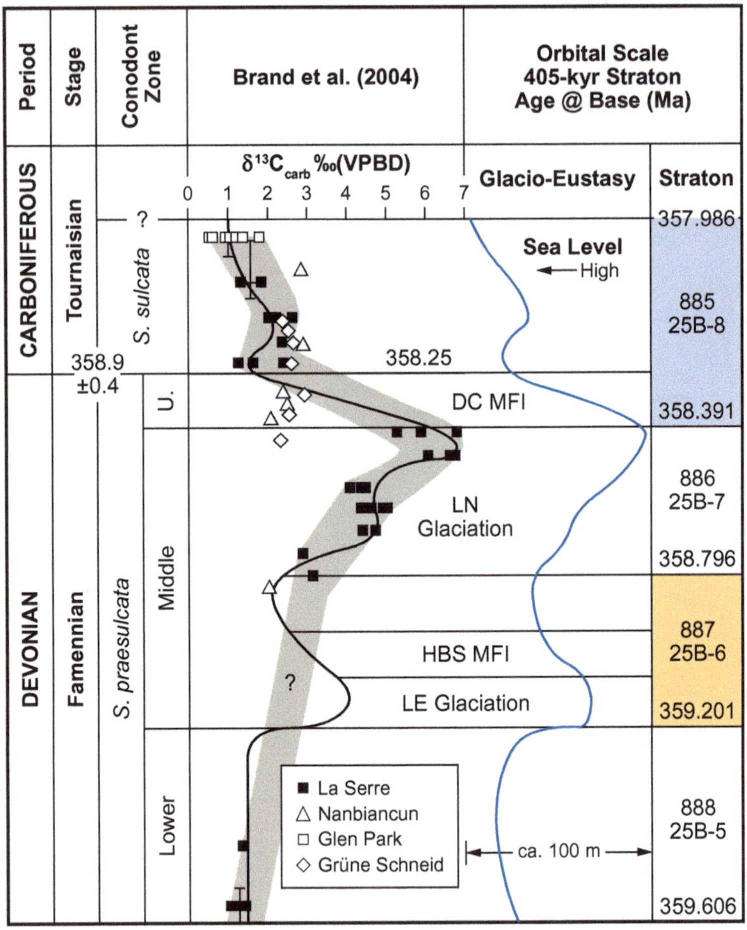

Fig. 17 Carbon-13 isotope data ($\delta^{13}C$) compiled and interpolated by gray curve (Brand et al., 2004) interpreted as a proxy for glacio-eustasy and calibrated with the Orbital Scale. The glaciation in the LE spore zone (LE Glaciation) is likely characterized by a positive $\delta^{13}C$ excursion indicated by a question mark.

1 and Beds 75–80 as a highstand, ending with a second abrupt drop in sea level at the top of Cycle I.

Cycle II (Beds 81–96) consists of two intervals of comparable thickness to Cycle I. The lower interval (Beds 81–84) represents the lowest relative sea level (c. 100 m, Kaiser et al., 2015) and maximum advance of the ice sheet/sheets of the *LN Glaciation* in Straton 25B-7. The upper interval (Beds 85–96) is a highstand and correlated to Straton 25B-8 containing a marine flooding (DC MFI) below the Devonian/Carboniferous Boundary at the top of Bed 88. The boundary would therefore be younger than base Straton 25B-8 and have an age of c. 358.25 Ma, nearly 0.65 myr younger than the average estimate of 358.9 ± 0.40 Ma in GTS 2020/01.

11. Part 2: Cretaceous glacio-eustasy

The Cretaceous is characterized by numerous globally correlative positive and negative $\delta^{13}C$ excursions; some have local names and most are related to regional oceanic anoxic events (OAE) with various alphanumeric abbreviations (e.g., OAE2, 2a, II). However, whereas Paleozoic positive excursions are correlated to sea level falls, Cretaceous ones are incorrectly correlated with the opposite sense to sea level rises (e.g., Jarvis et al., 2006, their fig. 15, and references therein). In several cases, based on proxy data, the positive excursions are associated with so-called "cold snaps" (e.g., Herrle et al., 2015; Rodríguez-López et al., 2016) or "ephemeral glaciations" in Antarctica (e.g., Miller et al., 2005a, b).

Approximately 60 fourth- and third-order global Cretaceous sequences (Haq, 2014) are associated with eustatic fluctuations of several 10s of meters (e.g., Miller et al., 2005a, b; Ray et al., 2019; Sahagian et al., 1996). Many authors however hesitate or exclude glaciations as the driving mechanism for these fluctuations because the Cretaceous is interpreted as an ice-free greenhouse period. The debate over whether the Earth was ice-free during greenhouse times (early Cenozoic, Mesozoic and several Paleozoic intervals) is reviewed in great detail in several studies (e.g., Davies et al., 2020; Miller et al., 2020; Pagani et al., 2014; Price, 1999; Ray et al., 2019) and is not repeated here.

In the following examples, the orbital template in Fig. 2 is used to predict the ages of several Cretaceous glacio-eustatic lowstands and to determine if they correlate to positive $\delta^{13}C$ excursions and unconformities (Fig. 18). In several cases the excursions coincide with stage boundaries thus allowing comparisons of their estimated numerical ages according to various GTS vintages, ATS 2018 and the Orbital Scale. Cenomanian to Santonian Examples 6–8 adopt the names of positive and negative $\delta^{13}C$ events from England (for detailed bio-, litho-, and chemostratigraphy, and nomenclature see Jarvis et al., 2006; Fig. 18). Jarvis et al. (2006) selected representative intervals from these sections to construct the English Chalk Reference Section and dated key $\delta^{13}C$ events using the Geologic Time Scale GTS 2004 (Ogg et al., 2004).

Orbiton	Dozon	Sequence	Orbital Scale 405-kyr Straton Age @ Base (Ma)			δ¹³C Event		ATS Huang (2018) (GTS 2020)
						Positive	Negative	
Orbiton 6 (part)	6B		6A-2	207	83.801			Campanian
			6A-1	208	84.206	SCBE-c	SCBE-d	
	6B	2.8 myr	6A-12	209	84.611	SCBE-a	SCBE-b	84.2 (83.6 ± 0.2)
	Dozon 6A (4.86 myr)		6A-11	210	85.016	Hawks Brow	Foreness	
			6A-10	211	85.421	Horseshoe Bay	Buckle	Santonian
			6A-9	212	85.826	ℓ1	Haven Brow	
			6A-8	213	86.231	Michel Dean	Bedwell	
			6A-7	214	86.636	Unnamed	k2	86.2 (86.3 ± 0.5)
			6A-6	215	87.041	Kingsdown	k1	
		2.0 myr	6A-5	216	87.446	Unnamed	j2	
			6A-4	217	87.851	White Fall	j1	
			6A-3	218	88.256	East Cliff	Unnamed	Coniacian
			6A-2	219	88.661	Light Point	Unnamed	
			6A-1	220	89.066	Beeding	Unnamed	
Orbiton 7 (part)	SB 6	2.8 myr	7C-12	221	89.471		Hiatus	
	Dozon 7C (4.86 myr)		7C-11	222	89.876		Unnamed	
			7C-10	223	90.281	Hitch Wood	Navigation	89.8 (89.8 ± 0.3)
			7C-9	224	90.686	h3	h4	
			7C-8	225	91.091	Caburn	Bridgewick	
			7C-7	226	91.496	Unnamed	Upper Southerham	
			7C-6	227	91.901	Pewsey	Lower Southerham	
		2.0 myr	7C-5	228	92.306	f1	Glynde	Turonian
			7C-4	229	92.711	Round Down	Lower Woolgari	
			7C-3	230	93.116	d1	d2	
			7C-2	231	93.521	Holywell	Lulworth	
			7C-1	232	93.926	OAE2-c	Unnamed	
	7C	2.8 myr	7B-12	233	94.331	OAE2-b	Unnamed	93.9 (93.9)
	Dozon 7B (4.86 myr)		7B-11	234	94.736	Sub-Plenus	Unnamed	
			7B-10	235	95.141	b2	Monument	
			7B-9	236	95.546	b1	Unnamed	
			7B-8	237	95.951	a2	MCE II	
			7B-7	238	96.356	MCE a1	Unnamed	
			7B-6	239	96.761	MCE I-a	P/B Break	Cenomanian
		2.0 myr	7B-5	240	97.166	Unnamed	Unnamed	
			7B-4	241	97.571	Unnamed	Mid-dixoni	
			7B-3	242	97.976	LCE III	Unnamed	
			7B-2	243	98.381	LCE I	LCE II	
			7B-1	244	98.786	ACBE-c	Unnamed	
	7B		7A-12	245	99.191	ACBE-a	ACBE-b	100.5 (100.5)
	7A		7A-11	246	99.596			Albian

Fig. 18 Uppermost Albian to lowermost Campanian positive and negative $\delta^{13}C$ events defined in English sections (Jarvis et al., 2006) calibrated with Orbital Scale of glacio-eustasy (see Figs. 19–26). Boundary Events (BE) occur at Albian/Cenomanian ACBE, Santonian/Campanian SCBE, and Cenomanian/Turonian also named Oceanic Anoxic Event OAE2. Early Coniacian hiatus occurs at the Beeding Unconformity (see Figs. 23 and 24).

Many of the named $\delta^{13}C$ events in England (Jarvis et al., 2006) have been recognized in other localities including continental Europe (e.g. Stoll and Schrag, 2000; Tsikos et al., 2004; Uličný et al., 2014), Japan (Hasegawa, 1997; Takashima et al., 2010), North America (Joo and Sageman, 2014, and references therein) and Tanzania (Wendler and Wendler, 2016), resulting in a global chemo-biostratigraphic framework. The final Cretaceous example presents the calibration of Late Aptian Dozon 8A, which contains a long-lived global glacio-eustatic fall of c. 30–50 m (Herrle et al., 2015; Maurer et al., 2013; van Buchem et al., 2010).

12. Example 6: Cenomanian Dozon 7B (98.786–93.926 Ma)

Lower boundary SB 7B of Dozon 7B is predicted as the start of a post-glacial transgression at 98.786 Ma (Fig. 18). In the Speeton Section the Albian/Cenomanian Boundary Event (ACBE, OAE1d, "Breistroffer") consists of a $\delta^{13}C$ plateau with the biostratigraphic stage boundary positioned between positive peaks ACBE-b and ACBE-c (Mitchell, 1995, in Jarvis et al., 2006; Figs. 18 and 19). Age estimates of the boundary vary by several million years: 97.2 Ma (Scott et al., 2009), 98.9 Ma (Hardenbol et al., 1998; Voigt and Wiese, 2000), 99.6 Ma (Ogg et al., 2004), and 100.5 Ma in ATS 2018 and GTS 2020/01. The middle estimate of 98.9 Ma is comparable to the age of SB 7B picked at ACBE-c.

In the Trunch, Dover and Kent Sections the Lower Cenomanian LCE I Event lies directly on the Albian and the interval between ACBE-a and LCE I in the Speeton Section passes to an unconformity (Fig. 19, Jarvis et al., 2006). According to Gale (1995) in some English sections the unconformity corresponds to a hiatus of up to 1.5 myr. In the Speeton Section ACBE is just c. 2 m thick and yet its $\delta^{13}C$ signature, including the ACBE-a, b and c positive spikes, can be correlated to the Gubbio Section in Italy and the Mont Risou Section in southern France (Jarvis et al., 2006, their fig. 4). In the Gubbio and Mont Risou Sections the ACBE interval is 10 m and 100 m thick, respectively. In these very thick sections the ACBE positive excursion has an amplitude of c. 0.6‰ VPBD and its shape is a symmetrical trapezoid consisting of a $\delta^{13}C$ rising limb, plateau and falling limb.

Koch and Brenner (2009) recognized the hiatus spanning the Albian/Cenomanian Boundary from the lithologic and palynologic analyses of strata along the eastern margin of the Western Interior Seaway (Nebraska, USA).

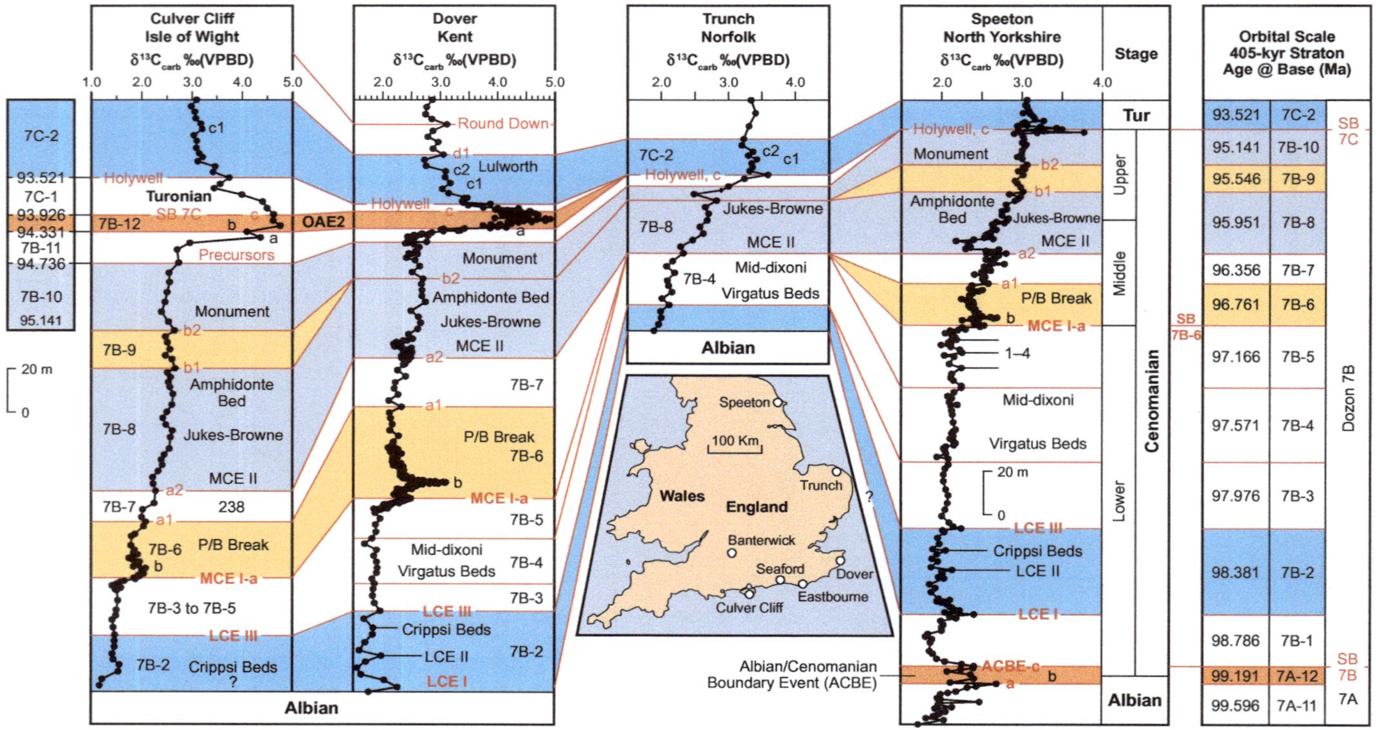

Fig. 19 Uppermost Albian, Cenomanian and lowermost Turonian Carbon-13 isotope events ($\delta^{13}C$) correlated with biostratigraphy in southern England (Jarvis et al., 2006). Dozon 7B consists of Stratons 7B-1 to 7B-12, and is correlated to the interval from ACBE-c in the Speeton Section (SB 7B) to OAE2-c in the Culver Cliff Section (SB 7C). See Fig. 1 for color scheme and explanations.

They mapped large-scale fluvial incisions (13–28 m deep), which they attributed to rapid glacio-eustatic falls.

Upper boundary SB 7C of Dozon 7B is predicted at 93.926 Ma and coincides with the 93.9 Ma estimate for the Cenomanian/Turonian Boundary in ATS 2018 and GTS 2020/01; it falls near the estimates of 94.07 ± 0.16 or 94.10 ± 0.13 Ma (Eldrett et al., 2015) but is younger than the 94.3 Ma estimate of Boulila et al. (2020; Figs. 19 and 20). One of the most prominent Mesozoic positive $\delta^{13}C$ excursions contains this stage boundary and is referred to as the Cenomanian/Turonian Boundary Event (CTBE), or Oceanic Anoxic Event 2 (OAE2, Schlanger et al., 1987; Jarvis et al., 2006; also OAE2a, OAE-2, OAE II, "Bonarelli"). The high-resolution $\delta^{13}C$ curve in Eastbourne (UK, Fig. 20) is dated by ammonite and planktonic foraminifera zones and has been correlated to several sections in the UK, Europe and N. America (e.g., Eldrett et al., 2015; Jarvis et al., 2006).

OAE2 consists of a rising limb, plateau and falling limb, which are calibrated with Stratons 7B-11, 7B-12 and 7C-1, respectively (Fig. 20). The model ages and durations for the rising limb and plateau are based on pattern recognition and comparable to within ±0.1 myr to the empirical estimates of Eldrett et al. (2015) but c. 200 kyr younger than other estimates for the Holywell Event (Fig. 20). Noteworthy is the duration of the plateau between Peaks B and C is estimated by Eldrett et al. (2015) almost precisely as 400 kyr and matches the absolute age of Straton 7B-12 within the 100 kyr uncertainty of the empirical calibrations.

The rising limb in the Eastbourne Section corresponds to the Plenus Marl and their bases coincide with a strongly burrowed erosion surface and hiatus (Jarvis et al., 2006, 2011; Fig. 20). Based on proxy data Jarvis et al. (2011) interpreted the "Plenus Cold Event" in approximately the *M. gesliniaum* ammonite zone and the "Early Turonian Thermal Maximum" in the *F. catinus* ammonite zone. The coincidence of the sub-Plenus unconformity and base of the rising limb therefore mark the start of cooling seas consistent with an abrupt glacio-eustatic fall. These sea-surface temperature (SST) patterns imply the rising limb (Straton 7B-11) and falling limb (Straton 7C-1) correspond to decreasing and increasing SSTs at mid-latitudes, respectively, consistent with polar glaciations.

Jarvis et al. (2006, their fig. 15) depicted a smoothed English Chalk Reference $\delta^{13}C$ curve and compared it with eustatic curves from India and NW Europe (Gale et al., 2002). The comparison implies OAE2 correlates to a c. 40 m flooding pulse peaking at the stage boundary. This result is opposite to the interpretation in the present study and may be explained by

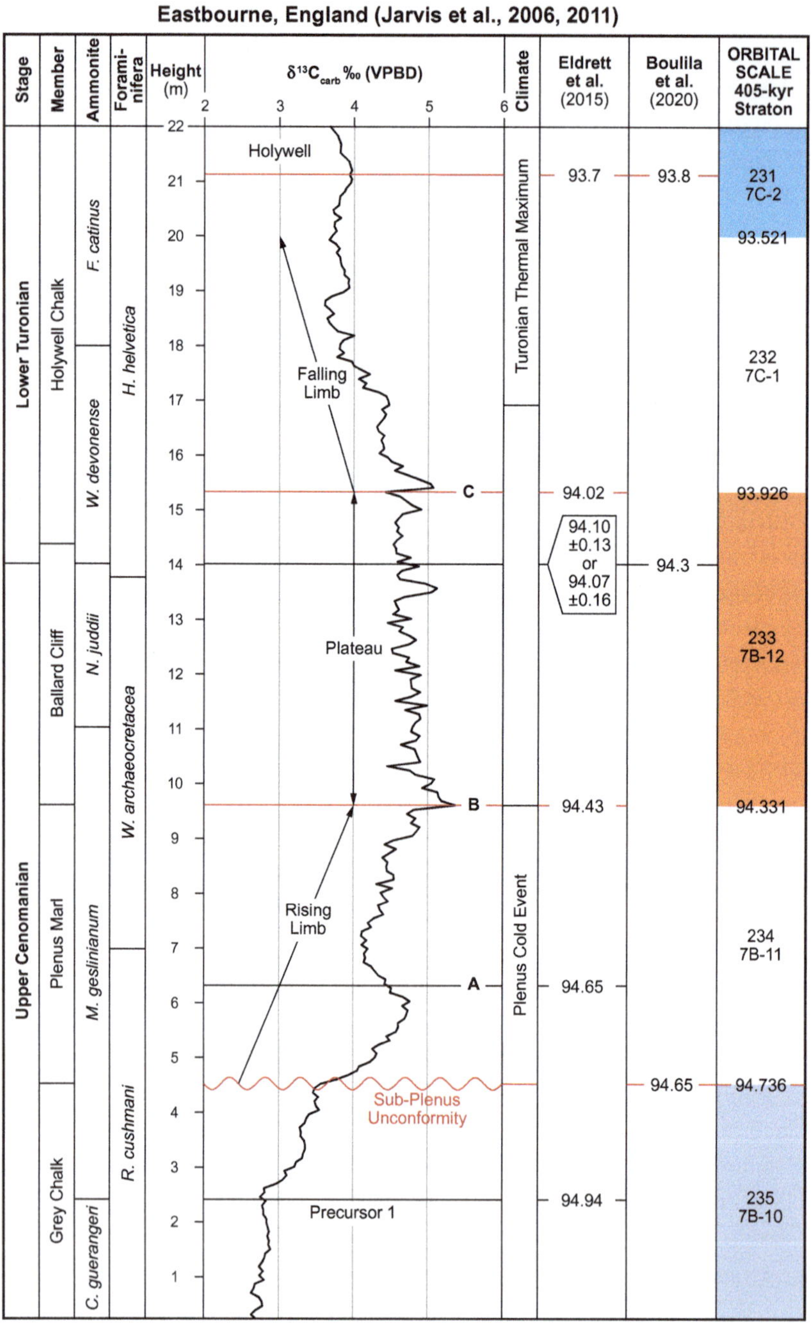

Fig. 20 Oceanic Anoxic Event OAE2 (also Cenomanian/Turonian Boundary Event, OAE-2a) in Eastbourne, UK (Jarvis et al., 2006) calibrated by several authors and the Orbital Scale. The calibrations of the stage boundary, $\delta^{13}C$ rising limb and plateau (Eldrett et al., 2015) carry uncertainties of c. 100 kyr and are matched by those of the Orbital Scale. Sea-surface temperatures (SST, climate) indicate cooling and warming (Jarvis et al., 2016) correspond to rising and falling $\delta^{13}C$ limbs and a glacio-eustatic lowstand in the plateau.

examining the traverse in Fig. 19. In the Culver Cliff Section OAE2 is a prominent plateau (Straton 7B-12) that passes to a $\delta^{13}C$ falling limb (Straton 7C-1) capped by the Holywell Event (in lower Straton 7C-2). The Holywell Event occurs in the Early Turonian Thermal Event and is interpreted in the present study as the maximum flooding surface (Holywell MFS, Figs. 18–20). By tracing the correlations of Jarvis et al. (2006; Fig. 19) it is evident the interval containing the entire OAE2 (i.e., from base rising limb to Holywell Event) becomes condensed at Dover and passes laterally to an unconformity in the Trunch and Speeton Sections.

In the Speeton Section the Cenomanian Monument Event is correlated to Straton 7B-10 and OAE2 is a thin positive $\delta^{13}C$ spike representing a c. 1.2-myr unconformity/hiatus overlain by the Turonian Holywell Event (Fig. 19). This stratigraphic geometry suggest the c. 40 m sea-level rise interpreted in India and NW Europe (Gale et al., 2002) apparently represents the Early Turonian Holywell MFS, not OAE2.

The *lower 2.0-myr sequence* of Dozon 7B is predicted between SB 7B (98.786 Ma, see above) and SB 7B-6 (base Straton 7B-6 at 96.761 Ma). SB 7B-6 is correlated to the base of the positive shift at the Mid-Cenomanian MCE I-a Event corresponding to a major unconformity in the Trunch Section (Jarvis et al., 2006; Figs. 18 and 19). The lower sequence is represented by relatively uniform $\delta^{13}C$ values with a regional flooding and sea level high between Lower Cenomanian Events LCE I and LCE III (Straton 7B-2).

The *upper 2.8-myr sequence* is interpreted between Events MCE I-a and OAE2-c (Figs. 18 and 19). It includes two minor valleys containing the Jukes Browne Event (also MCE II) and Monument Event, interpreted as sea level highs in Stratons 7B-8 and 7B-10. The valleys bound the interval between Events b1 and b2 (lowstand Straton 7B-9), which passes to an unconformity in the Trunch and Dover Sections.

Example 6 indicates the boundaries of the Cenomanian Stage are defined in condensed lowstand deposits that pass to unconformities. The upper boundary is empirically dated at c. 93.9 Ma in ATS 2018 and GTS 2020/01, consistent with the Orbital Scale, and interpreted as a glaciation. The age of its lower boundary is estimated at c. 98.8 Ma compared to 100.5 in ATS 2018 and GTS 2020/01 and in the present study ACBE is also interpreted as a glaciation.

12.1 Cenomanian glacio-eustasy in Arabia

In Oman, the uppermost Albian, Cenomanian and lower Turonian stages are represented by members A to G of the Natih Formation

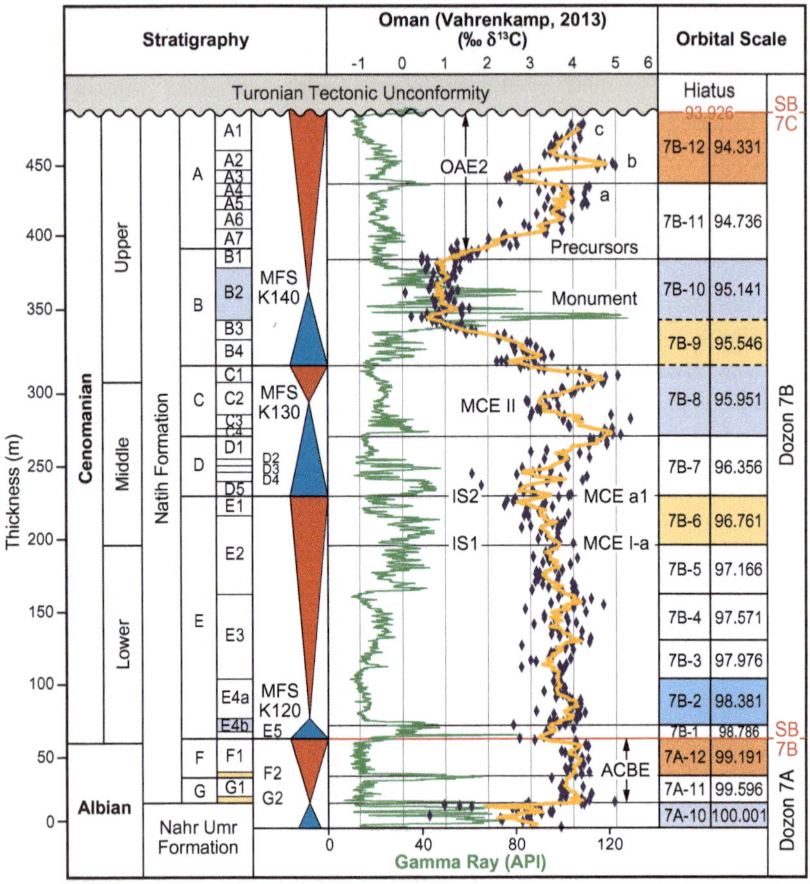

Fig. 21 Oman's Natih Members G to A and Arabian 3rd-order sequences K120 to K140 dated by biostratigraphy (van Buchem et al., 2011) and correlated by Vahrenkamp (2013) to Natih subsurface units and $\delta^{13}C$ record. Correlation to $\delta^{13}C$ markers in southern England (Jarvis et al., 2006, Fig. 19) by Vahrenkamp (2013) is modified in this study in the interval between positive excursions OAE2 (Oceanic Anoxic Event 2) and ACBE (Albian Cenomanian Boundary Event).

(Forbes et al., 2010, Fig. 21). Vahrenkamp (2013) measured $\delta^{13}C$ data across the formation and correlated Albian Natih Members G and F to ACBE (Fig. 21). The lower boundary of the overlying Natih E Member is a bored and iron-crusted hardground (van Buchem et al., 2011) and is correlated to SB 7B (Al-Husseini, 2018). Natih Member E is interpreted as Arabian Plate Sequence K120, and consists of six fourth-order sequences with MFS K120 positioned in the second sequence from the base (Homewood et al., 2008, Fig. 22). The sixth fourth-order sequence is

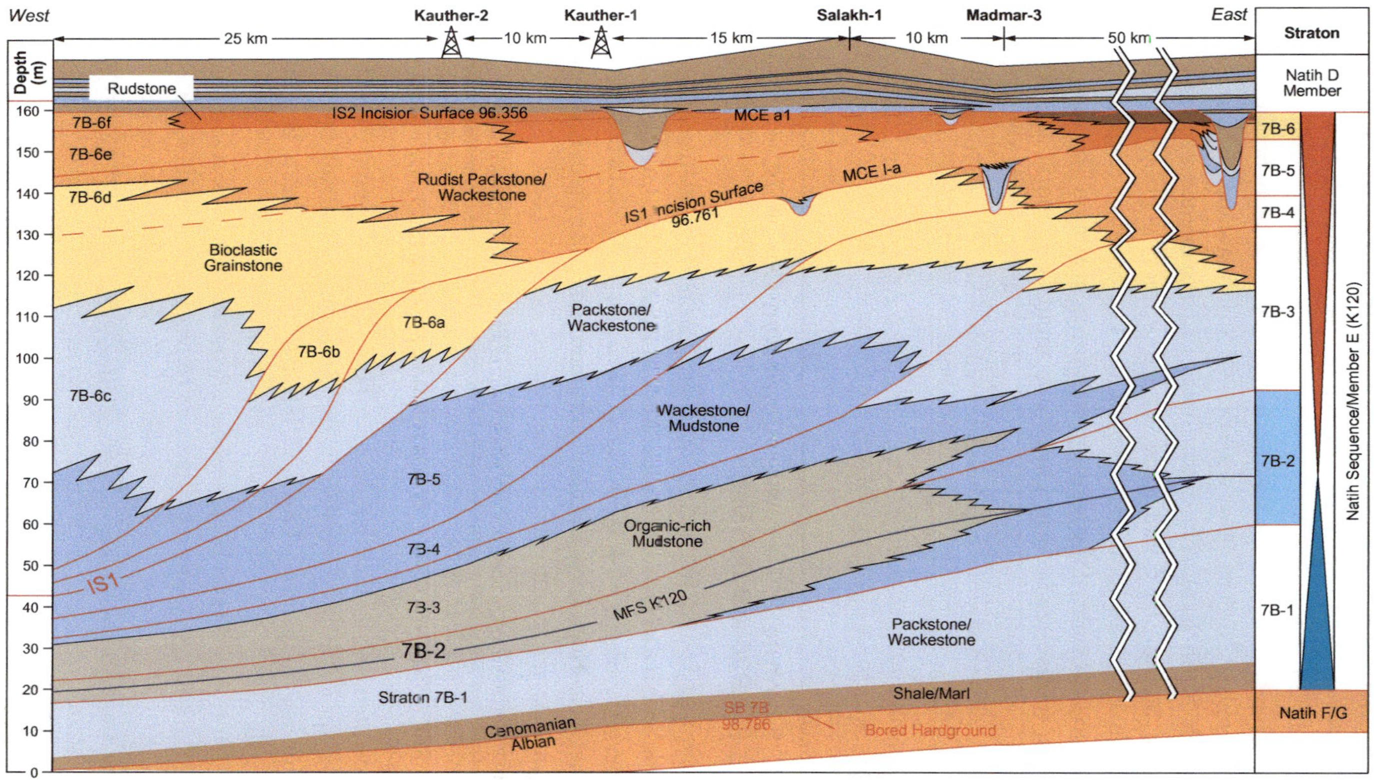

Fig. 22 Natih Member E (Sequence K120, Fig. 21) is interpreted in a 110-km traverse in North Oman as six fourth-order sequences (Homewood et al., 2008), here correlated to Stratons 7B-1 to 7B-6 with Lower Cenomanian MFS K120 in MFI Straton 7B-2. Incised channels IS1 and IS2 bound lowstand Straton 7B-6 and highlight glacio-eustatic falls of 20 and 30 m, respectively.

bounded by Incision Surfaces IS1 and IS2 characterized by erosional channels with depths of 20 and 30 m, respectively (Homewood et al., 2008; Grélaud et al., 2010; Fig. 22).

In the orbital calibration the six fourth-order Natih E sequences are correlated to Stratons 7B-1 to 7B-6 with the MFS in highstand Straton 7B-2 (organic-rich Natih E4b). The sixth fourth-order sequence is correlated to lowstand Straton 7B-6 and its lower IS1 surface to positive Middle Cenomanian Event MCE I-a at base Middle Cenomanian (Figs. 21 and 22). Based on the height of the high-dip clinoforms in Straton 7B-6 and faunal content, relative sea level in Oman was about 60 m and fluctuated by 20–30 m (Droste, 2010; van Buchem et al., 2011, and references therein). Antarctica's present-day ice cap holds the equivalent of c. 60 m of global sea level (Miller et al., 2020) suggesting the 20–30 m Cenomanian fluctuations represent short-lived buildups of ice caps on Antarctica holding the equivalent of ⅓ to ½ of its present-day ice volume.

13. Example 7: Turonian–Coniacian Dozon 7C (93.926–89.066 Ma)

Lower boundary SB 7C (93.926 Ma) of Dozon 7C is discussed above. *Upper boundary SB 6* of Dozon 7C separates Orbitons 7 and 6 and is predicted at 89.066 Ma ($1.586 + 6 \times 14.58$) in Early Coniacian (Fig. 18). In the Trunch Section negative Navigation Event of Turonian age is interpreted as Straton 7C-10, and the rising limb above it as Straton 7C-11. The $\delta^{13}C$ plateau predicted in Straton 7C-12 does not occur in any English Sections implying it passes into an unconformity marked by the Coniacian Beeding Event (Figs. 18 and 23). In the Culver Cliff Section the Beeding Unconformity merges with the Light Point Event and Stratons 7C-11, 7C-12 and 6A-1 are represented by a 1.2-myr hiatus. The missing Straton 7C-12 in England is more clearly expressed in the Bch-1 Well but not accounted for in the age calibration shown at left in Fig. 24 (Czech Republic, Uličný et al., 2014).

Haq (2014; Figs. 23 and 24) interpreted Sequence KTu5 as straddling the Turonian/Coniacian Boundary dated c. 89.8 Ma in ATS 2018, GTS 2012 and GTS 2020/01. Latest Turonian SB KTu5 occurs almost immediately below the stage boundary at 89.9 Ma (Haq, 2014) and is here correlated to base Straton 7C-11 at 89.876 Ma. Haq (2014) estimated earliest Coniacian SB KCo1 at 88.7 Ma implying it correlates to base Straton 6A-2 at 88.681 Ma and the Light Point Event (Figs. 23 and 24). These

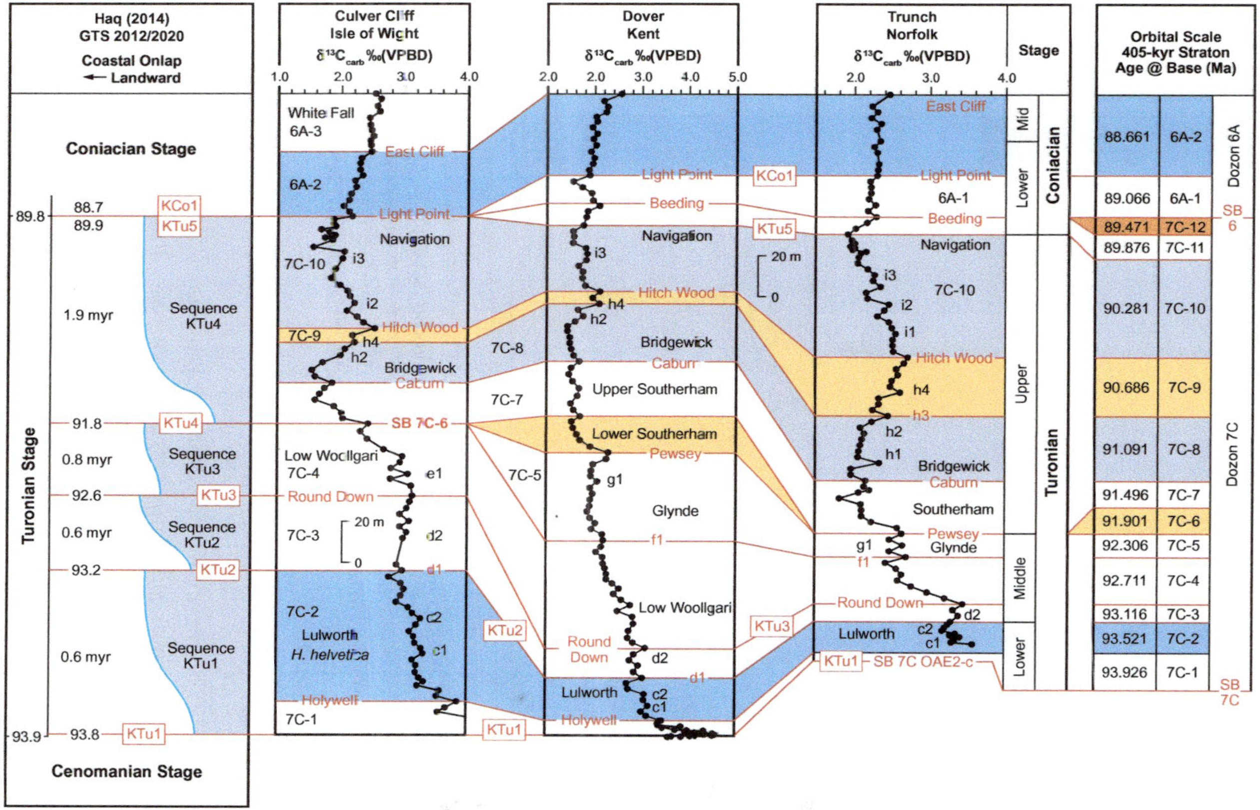

Fig. 23 Turonian–Lower Coniacian $\delta^{13}C$ events correlated with biostratigraphy in southern England (Jarvis et al., 2006). Stratons 7C-1 to 7C-11 of Dozon 7C are correlated to the interval from OAE2-c in the Dover Section to the Beeding Unconformity (Figs. 18 and 24). Straton C-12 is represented by an Early Coniacian hiatus. SBs Tu1 to Co1 and corresponding coastal-onlap sequences of Haq (2014) are correlated in this study to the $\delta^{13}C$ records.

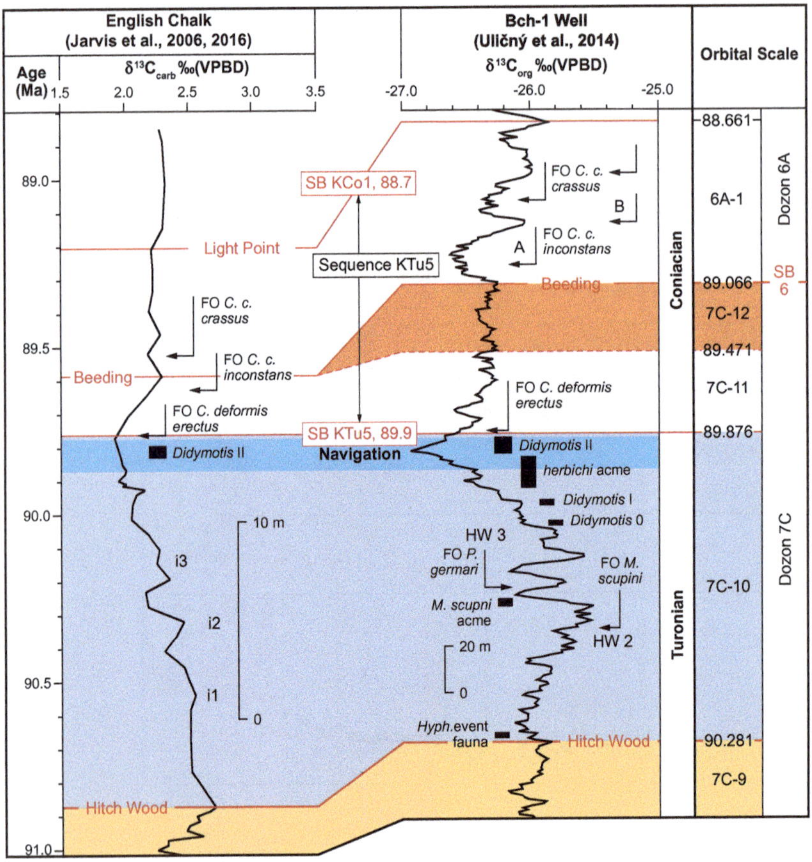

Fig. 24 Chemo-biostratigraphic correlation of English Chalk Reference Section (Jarvis et al., 2006) and Bch-1 Well (Czech Republic), with Lower Coniacian FO C. c. inconstans shown in the negative excursion "A" (Uličný et al., 2014) or in overlying positive excursion "B" (Jarvis et al., 2016). The Early Coniacian hiatus in England occurs in the Beeding Event (Beeding Unconformity), and is correlated to lowstand Straton 7C-12 (Fig. 23). SBs Tu5 and Co1 after Haq (2014).

correlations imply Sequence KTu5 consists of lowstand Stratons 7C-11 and C-12 (hiatus in England) and transgressive Straton 7A-1. In the Culver Cliff Section SB KTu5 and SB KCo1 merge into the Beeding Unconformity and Sequence KTu5 is completely missing in a 1.2-myr hiatus/SBH (Fig. 23).

The *lower 2.0-myr sequence* is interpreted between SB 7C (93.926 Ma, see above) and SB 7C-6 (base Straton 7C-6) at 91.901 Ma, Fig. 23. In the Culver Cliff Section the Middle Turonian Glynde Event is missing and the Pewsey Event is an unconformity above the *Collignoniceras woollgari* zone (Jarvis et al., 2006). Haq and Huber (2016) interpreted SB KTu4 as the greatest Turonian eustatic fall which in North European basins occurs in the upper part of the *C. woollgari* zone. They estimated its age at c. 91.8 Ma implying it correlates to SB 7C-6 (91.9 Ma) and the Pewsey Event, as well as the equivalent Mid Turonian Event at 91.95 Ma (Eldrett et al., 2015).

Haq and Huber (2016) however were reluctant to attribute the fall associated with SB KTu4 to a glaciation because MacLeod et al. (2013) concluded the climate was too warm to host Turonian ice sheets. However, the interval studied by MacLeod et al. (2013) is limited to the Early Turonian *H. helvetica* zone, dated between 93.5 and 93.0 Ma (Ogg and Gradstein, 2020) in the negative Lulworth Event (shown in the Culver Cliff Section in Fig. 23). This biozone is correlated to the major highstand in Straton 7C-2 and is interpreted as the complete melt out of Antarctica's ice cap during a very warm period.

Bornemann et al. (2008) considered the Turonian as one of the warmest periods in the Phanerozoic, and yet concluded the Pewsey Event represents a ca. 200-kyr glaciation. They also considered Antarctica the likely host of the ice sheet and its size to be equivalent to about 30 m of global sea level. Jarvis et al. (2016) reviewed the interpretation of climate in Turonian basins in England, Germany and Spain (Voigt et al., 2004; Voigt and Wiese, 2000; Wiese and Voigt, 2002) and the sequence stratigraphy in the Bohemian Basin (Uličný et al., 2014). They concluded based on $\delta^{18}O$ data a marine cooling trend started above the Glynde Event (Pewsey Event) and represents an initial period of medium-term shallowing and an influx of cool-water fauna in northern Europe (Jarvis et al., 2016, their Fig. 7), consistent with a frozen Arctic Sea.

The *upper 2.8-myr sequence* is interpreted between the Pewsey and Beeding Events with a hiatus in the Beeding Unconformity. Two minor valleys containing the Bridgewick and Navigation Events are correlated to highstand Stratons 7C-8 and 7C-10 (Fig. 23). In the Trunch Section these events bound the interval between Event h3 and the Hitch Wood Event, which is correlated to lowstand Straton 7C-9. This interval becomes thinner in the Dover and Culver Cliff Sections with the h3 and h4 Events merging into an unconformity. The Hitch Wood Event is also interpreted as a "cooling event" (Jarvis et al., 2016) and in the present study it is interpreted as a polar glaciation.

14. Example 8: Coniacian–Santonian Dozon 6A (89.066–84.206 Ma)

Lower boundary SB 6 of Dozon 6A is correlated to the Beeding Event in Early Coniacian at 89.066 Ma and is discussed above (Figs. 18, 23–25). *Upper boundary SB 6B* is predicted as a post-glacial transgression that started at 84.206 Ma in Early Campanian, precisely the 84.2 Ma estimate for the Santonian/Campanian Boundary in ATS 2018 (Fig. 18) but 1.4 myr younger than the 83.6 ± 0.2 Ma estimate in GTS 2020/01. Jarvis et al. (2006) abbreviated the name of the positive $\delta^{13}C$ event that straddles the Santonian/Campanian Boundary as SCBE (Fig. 26). Base SCBE occurs in the uppermost Santonian at the top of the *Marsupites* zone in the Seaford, Dover and Trunch Sections; top SCBE lies in basal Campanian below the first occurrence of *O. pilula* in all three sections (Jarvis et al., 2006) and is correlated to SB 6B.

The thickness of the SCBE interval is 11 and 7 m in the Trunch and Seaford Sections, respectively, but is less than 2 m in the Culver Cliff and Dover Sections (Jarvis et al., 2006; Fig. 26). Jarvis et al. (2006) reported that in the Culver Cliff Section the sharp positive $\delta^{13}C$ step at the stage boundary corresponds to a glauconitized hardground and based on the absence of the Foreness Event below and faunal records, a hiatus must be equivalent to the upper half of the Santonian *Marsupites* zone and the basal Campanian *O. pilula* zone. In the present study the SCBE is correlated to lowstand Straton 6A-12 with SB 6B at SCBE-c.

The *lower 2.0-myr sequence* is interpreted between the Beeding and Kingsdown Events (Fig. 25). The latter event represents a turning point from rising to falling limbs in the Culver Cliff and Dover Sections suggesting the end of a regression and start of a transgression. The persistent thick interval between the Light Point and East Cliff Events is interpreted as major highstand Straton 6A-2.

The *Upper 2.8-myr sequence* is subdivided into a 1.6-myr sequence between the Kingsdown and Horseshoe Bay Events and a 1.2-myr sequence between the latter event and SCBE-c (Fig. 26). Jarvis et al. (2006) indicated that in the Dover Section the Horseshoe Bay Event may be unrepresented due to substantial condensation or a major hiatus, consistent with this event representing an unconformity in lowstand Straton 6A-9. The Coniacian/Santonian Boundary is defined at the Michel Dean Event and correlated to base Straton 6A-8 at 86.231 Ma consistent with 86.2 Ma in ATS 2018

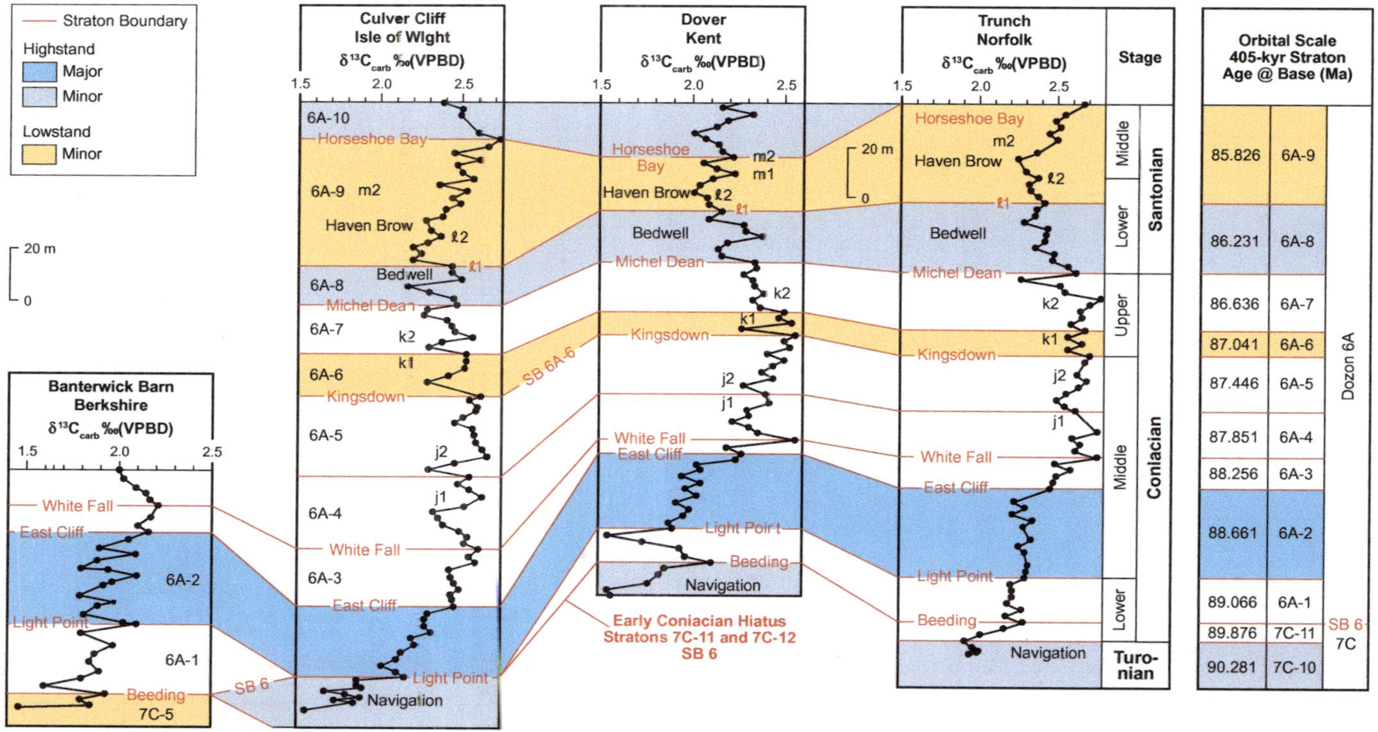

Fig. 25 Uppermost Turonian to Middle Santonian $\delta^{13}C$ events correlated with biostratigraphy in southern England (Jarvis et al., 2006) showing Stratons 7C-11 to lower 6A-9. In the Culver Cliff Section the Early Coniacian hiatus (Figs. 18, 23, and 24) extends to the Light Point Event and represents Stratons 7C-11, 7C-12, and 6A-1.

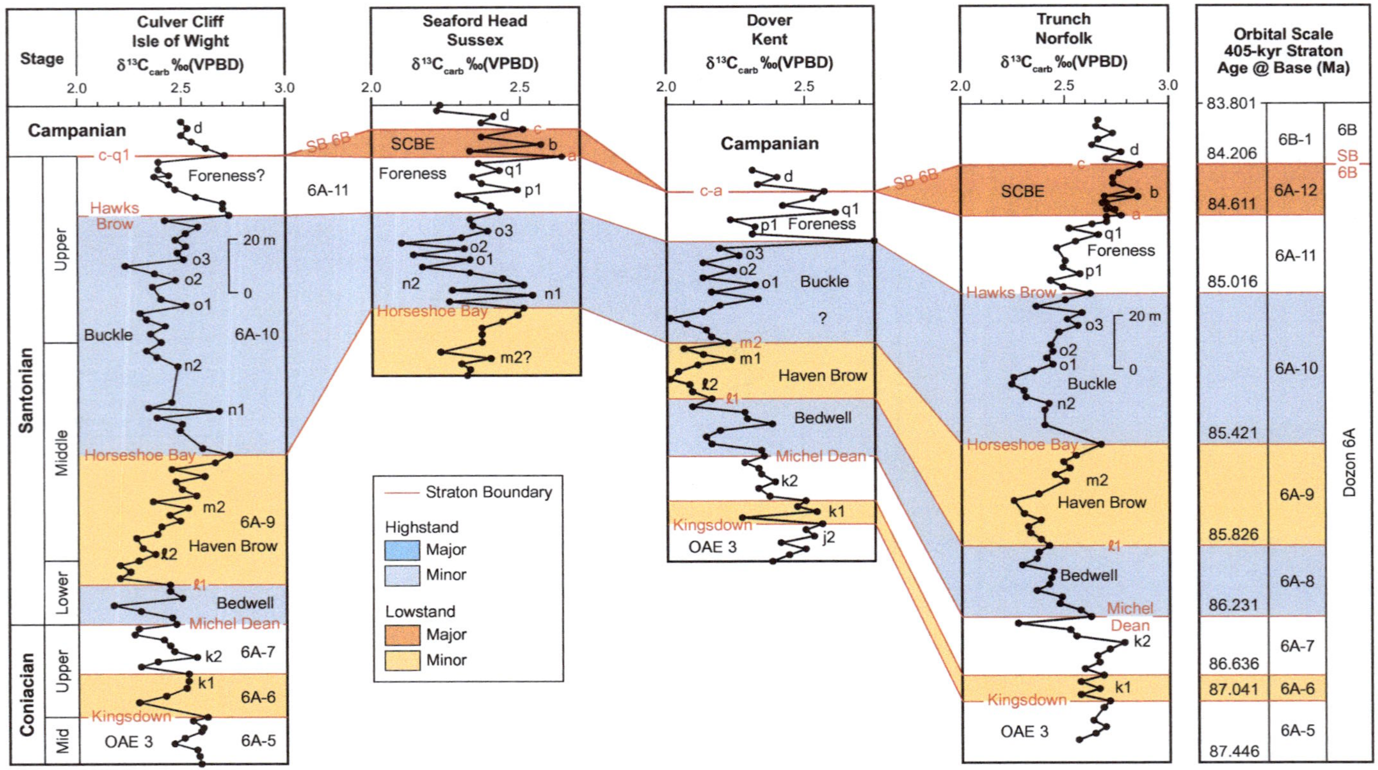

Fig. 26 Middle Coniacian–Santonian $\delta^{13}C$ events correlated with biostratigraphy in southern England (Jarvis et al., 2006) showing Stratons 6A-5 to lower 6B-1. The older stratons of Dozon 6A are represented in Fig. 25. The Santonian/Coniacian Boundary Event (SCBE) is correlated to the glacio-eustatic lowstand and unconformity in Straton 6A-12.

and 86.3 ± 0.5 Ma in GTS 2020/01. This event occurs in the middle of a falling limb and appears to mark the start of a minor highstand in Straton 6A-8.

15. Example 9: Aptian Dozon 8A (118.226–13.366 Ma)

Lower boundary SB 8 of Dozon 8A separates Orbitons 9 and 8 at 118.226 Ma (1.586 + 8 × 14.58). Identifying its stratigraphic position using empirical age calibrations is ambiguous because base Aptian estimates differ by about 5 myr: at one extreme c. 121.0 Ma (e.g., Gradstein et al., 1994; He et al., 2008; Midtkandal et al., 2016), and at the other extreme c. 126.0 Ma (Huang et al., 2010; Ogg and Hinnov, 2012; ATS 2018 and GTS 2020/01). Another difficulty in finding its stratigraphic position arises because of the great number of bio- and chemostratigraphic schemes used to represent Aptian sequences and $\delta^{13}C$ records. Nonetheless anchoring the position of SB 8 is important for resolving the ages of several global events including OAE1a (e.g., Kuhnt et al., 1998; Weissert and Breheret, 1991; also "Selli," "Goguel" black shales) and the late Aptian and early Albian long-lived glaciation (Maurer et al., 2013; van Buchem et al., 2010).

In the composite $\delta^{13}C$ curve of France's Vocontian Basin a major plateau is recorded in uppermost Lower Aptian with its top coinciding with the boundary between the *L. cabri* and *G. ferrociensis* foraminifera zones (i.e., Lower/Upper Aptian Substage Boundary, Fig. 27, Herrle et al., 2004). Leckie et al. (2002, and references therein) estimated this substage boundary at 118.2 Ma assuming base Aptian at c. 121.0 Ma. Placing SB 8 (118.226 Ma) at this boundary is not only consistent with this age estimate, but also anchors SB 8 at the turning point between the plateau (lowstand Straton 9C-12) and base of the falling limb interpreted as transgressive Straton 8A-1 (Fig. 27).

Upper boundary SB 8B of Dozon 8A is predicted at 113.366 Ma but its stratigraphic position is also difficult to identify because the Aptian upper boundary is not defined by a GSSP. Ogg and Hinnov (2012) defined a "working" Aptian upper boundary as the first occurrence of the *Leymeriella schrammeni* ammonite and dated it at c. 113.0 Ma based on a tuff dated 113.1 ± 0.3 Ma situated 85 cm below this boundary in Germany's Lower Saxony Basin (Selby et al., 2009; Fig. 27). In the present study SB 8B is positioned at the base of the $\delta^{13}C$ falling limb at c. 225 m in the Vocontian Basin.

The $\delta^{13}C$ signature of the *lower 2.0-myr sequence* is interpreted from base falling limb at 35 m in the Vocontian Basin (SB 8 at base transgressive Straton 8A-1),

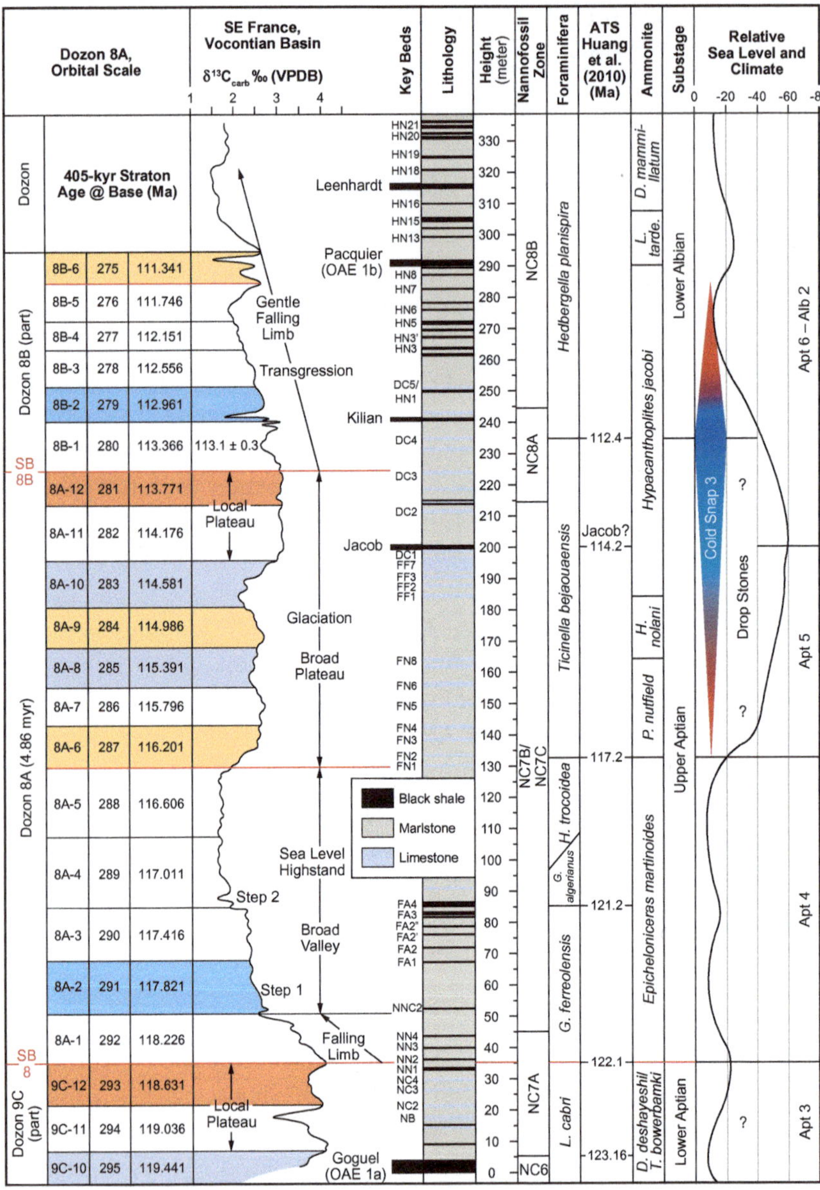

Fig. 27 Upper Aptian and Lower Albian $\delta^{13}C$ composite curve in the Vocontian Basin (Herrle et al., 2004) calibrated with Orbital Scale and ATS (Huang et al., 2010; planktonic foraminifera zones in Pioboco Core, Central Italy, with base Aptian at 125.5 Ma). Radiometric date of 113.1 ± 0.3 Ma (Selby et al., 2009) occurs near working definition of base Albian Boundary. Sea level and glaciation calibrated in Arabia according to biozones (Maurer et al., 2013), Cold Snap 3 according to $\delta^{13}C$ curve (Herrle et al., 2015, see Fig. 28) and drop stones in *T. bejaouaensis* zone (Spain, Rodríguez-López et al., 2016).

passing to the broad valley containing two negative steps (Stratons 8A-2 to 8A-5, Fig. 27). This transgressive sequence closely correlates to the *E. martinoides* ammonite zone. The upper SB of the 2.0-myr sequence is positioned near the base of the abrupt positive shift at 125 m at 116.201 Ma (base Straton 8A-6).

The *upper 2.8-myr sequence* is characterized by greater $\delta^{13}C$ values relative to the lower 2.0-myr sequence (Fig. 27). It contains two minor negative $\delta^{13}C$ valleys that are correlated to highstand Stratons 8A-8 and 8A-10 and a minor positive event at 170 m correlated to lowstand Straton 8A-9. Lowstand stratons 8A-11 and 8A-12 are interpreted in the local $\delta^{13}C$ plateau containing the Jacob Event in Straton 8A-11.

15.1 Cretaceous glacio-eustasy

In Arabia's intra-platform Bab Basin, situated in the United Arab Emirates and Oman, T-R Sequences and associated variations in the eustatic level are interpreted from outcrops, 3D seismic images and well control (Maurer et al., 2013; van Buchem et al., 2010; Fig. 27). The evidence indicates sea level fell abruptly between Aptian Sequences Apt 4 and Apt 5 by c. 30 m just above the *E. martinoides* ammonite zone. It kept falling to c. 50 m in latest Aptian Sequence Apt 5 and returned gently in Sequences Apt 6 to Albian Alb 2 to the pre-fall level in early Albian *D. mammillatum* ammonite zone. This eustatic interpretation is similarly depicted by Sahagian et al. (1996) for the Siberian Platform, and mirror-reflected in the Vocontian $\delta^{13}C$ record by the late Aptian broad plateau and early Albian falling limb.

Rodríguez-López et al. (2016) reviewed the proxy evidence for the late Aptian and early Albian "Cold Snap 3" in the northern Atlantic, and reported on the discovery in the Oliete Sub-basin (Iberian Peninsula) of an abraded boulder and two cobble-sized quartzite drop stones. They interpreted these glaciogenic deposits to have been transported by icebergs to the subtropical Atlantic Ocean situated at paleo-latitude of c. 25–30°N. The deposits are encased in black shales and fine-grained sandstones of the Escucha Formation belonging to the *Ticinella bejaouaensis* biozone (Rodríguez-López et al., 2016, Fig. 27).

Icebergs in the northern Atlantic Ocean imply an ice cap covered the late Aptian Arctic Sea; however it would have mainly consisted of floating ice and could not have caused sea level to fall by 30–50 m. The conclusion is that a continental ice cap developed on Antarctica's landmass and extended over southerly landmasses (e.g., Australia, Alley et al., 2019). Finding glaciogenic evidence for Antarctica's Mesozoic ice cap is improbable because it mainly consists of offshore buried and undated drop stones or multiple unconformities at the base of several kilometers of ice.

This final example addresses fundamental issues in the debate over whether the Earth was ice-free during extended periods of the Cretaceous or for that matter in any other Phanerozoic (and older) interval. Cold Snap 3, the glaciogenic deposits in the *T. bejaouaensis* biozone, the 30–50 m eustatic lowstand in Arabia's Bab Basin and the late Aptian broad plateau and early Albian falling limb provide indisputable evidence for a major glaciation (Fig. 27). Clearly the probability of finding glaciogenic deposits (e.g., drop stones, Rodríguez-López et al., 2016) and dating them is extremely low to negligible, and should not be a necessary requirement for proving glaciations.

16. Discussion

16.1 Finding the orbital signal in big-data compilations

Fig. 28 illustrates the Barremian to Turonian composite $\delta^{13}C$ curve of Herrle et al. (2015) and sequence-eustatic compilation of Haq (2014, and references therein) as calibrated by these authors according to GTS 2012 (Gradstein et al., 2012, unchanged in GTS 2020/01). Also shown are several $\delta^{13}C$ anchors calibrated by the Orbital Scale in the above examples. Clearly finding the predicted glacio-eustatic signal in these big-data compilations is not possible.

A more detailed examination of some intervals however reveals how the Orbital Scale could assist in calibrating and correlating $\delta^{13}C$ and eustatic curves. For example, Cold Snap 3 is clearly expressed in the composite $\delta^{13}C$ record as a late Aptian plateau and early Albian falling limb that correspond to a 30–50 m long-lived lowstand (Maurer et al., 2013; Sahagian et al., 1996; van Buchem et al., 2010, see also Ray et al., 2019, their fig. 1). The lowstand is bound by SB KAp4 and SB KAl1, but it is unclear what evidence was used to interpret the four eustatic cycles with amplitudes of 50–70 m in this interval (Haq, 2014).

The age of the late Aptian and early Albian glaciation is not random. The correlations imply it started in lowstand Straton 8A-6 at 116.2 Ma and its ice volume continued to increase throughout the late Aptian upper 2.8-myr sequence between 116.2 and 113.3 Ma. It then started to slowly melt in the early Albian lower 2.0-myr transgressive sequence between 113.3 and 111.3 Ma. It is not clear why this particular glaciation remained "stuck" for such a long time.

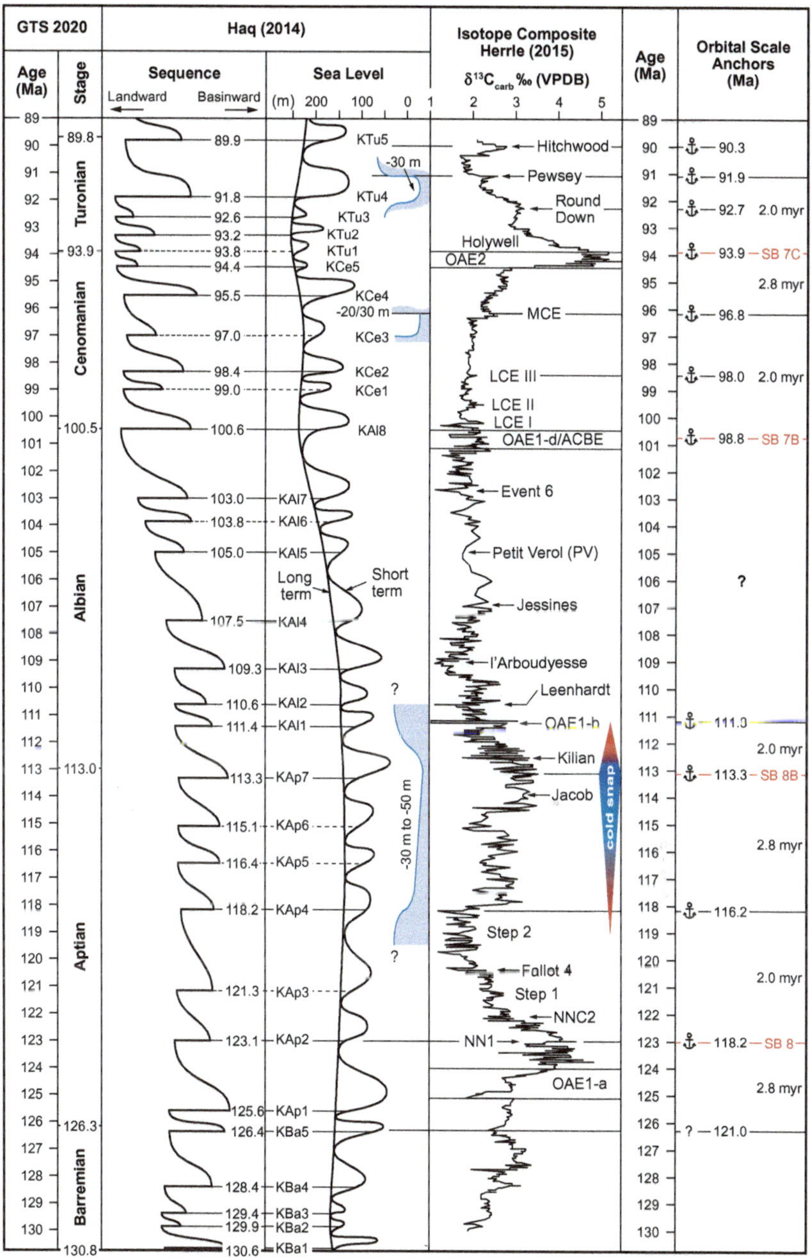

Fig. 28 Key $\delta^{13}C$ and glacio-eustatic events anchored in the Orbital Scale.

Another orbital anchor at base Straton 7C-6 (91.9 Ma) ties the lowstand of Sequence KTu4 (91.8 Ma) to the Pewsey Event (91.9 Ma) and the eustatic fall of c. 30 m (Bornemann et al., 2008; Figs. 23 and 28). Two more anchors bound Straton 7B-6 and tie Natih incised surfaces IS1 (20 m, 96.8 Ma) to MCE I-a, and IS2 (30 m, 96.4 Ma) to MCE a1 (Figs. 21 and 22). This straton may correlate to the lowstand of Sequence KCe3 (97.0 Ma, Fig. 28). Unlike the late Aptian and early Albian 30–50 m long-lived glacio-eustatic lowstand, these 20–30 m falls were short-lived. Similar and more detailed analyses may anchor other $\delta^{13}C$ events and eustatic cycles, as explained in the nine examples.

16.2 Absolute age calibrations

In the present study several global $\delta^{13}C$ anchors are calibrated near stage boundaries and therefore their age estimates according to the Orbital Scale can be compared to empirical estimates. In the case of the Ediacaran/Cambrian Boundary the comparison is qualified because the GSSP definition is biostratigraphic (base *T. pedus* zone) and apparently inconsistent with the base of BACE. A somewhat similar issue arises for the base of the Hirnantian Stage wherein the GSSP section does not apparently record Upper HICE in the *persculptus* zone. In other cases the Orbital Scale estimates are considered within the range of empirical uncertainty (Table 3).

16.3 Grand orbital cycles

Several studies have used spectral analyses to identify "grand orbital cycles" in chemo-stratigraphic records calibrated with empirical time scales (Table 4). The intervals in these studies overlap those in some of the examples and identify spectral peaks that approximately correspond to multiples of 0.4 myr, ranging from c. 1.2 myr (3 stratons) to c. 4.8 myr (12 stratons).

Boulila (2019, their fig. 2) analyzed the power spectra recorded in stacked benthic foraminifera $\delta^{18}O$ records for the Cenozoic through middle Cretaceous (Present to 115.0 Ma). This interval includes the Santonian to latest Aptian discussed in Examples 6–9 (84.0–118.2 Ma). The results of two spectral analyses identified spectral peaks corresponding to 1.2 or 1.3 myr (c. 3 stratons), consistently in both analyses 1.6 myr (4 stratons), a triplet of 2.0, 2.3 and 2.6 myr (5, c. 6 and c. 7 stratons) or a triplet of 2.3, 2.5 and 2.8 (c. 6 and 7 stratons), and 4.6 or 4.7 myr (c. 12 stratons).

Table 3 Anchors.

Cretaceous

Base	GTS 2020/01	ATS 2018	Orbital Scale	Anchor
Campanian	83.6 ± 0.2	84.2	84.206, SB 6B	Top SCBE
Santonian	86.3 ± 0.5	86.2	86.231, SB 6A-8	Michel Dean
Coniacian	89.8 ± 0.5	89.8	89.876, SB 7C-11	Top Navigation
Turonian	93.9	93.9	93.927, SB 7C-1	OAE2-c
Cenomanian	100.5	100.5	98.786, SB 7B	ACBE-c
Albian	113	113	c. 113.1, SB 8B	Top *T. bejaouaensis*
Upper Aptian	NA	NA	118.226, SB 8	Top *L. cabri*

Paleozoic

Base	GTS 2020/01	Orbital Scale		Anchor
Telychian	438.5 ± 1.1	438.581	SB 30A-2	Guerchi Spike
Aeronian	440.8 ± 1.2	441.011	SB 31C-8	Curtus Spike
Rhuddanian	443.8 ± 1.5	443.441	SB 31C-2	Top Upper HICE
Hirnantian	445.2 ± 1.4	444.656	SB 31B-11	c. Base Lower HICE
Katian	453.0 ± 0.7	453.971	SB 32C-12	Base GICE
Paibian	497.0	498.111	SB 35C-11	Base SPICE
Cambrian (Base *T. tepus*)	536–538	c. 538.8	Intra-Straton 37A-6	Middle Excursion 2p
Cambrian (GTS 2020/01)	541.0 ± 1.0	541.046	SB 37A	Excursion 1p

Table 4 Periods of grand cycles (million year).

Interval	Time series	Periods	References
0.0–115.0	$\delta^{18}O$	1.2, 1.6, 2.0, 2.3, 2.6, 4.6	Boulila (2019)
0.0–115.0	$\delta^{18}O$	1.3, 1.6, 2.3, 2.5, 2.8, 4.7	Boulila (2019)
78.0–101.0	$\delta^{13}C$	2.4, 3.4, 4.7	Sprovieri et al. (2013)
425.0–448.0	$\delta^{13}C$	4.2 range 3.7–4.8	Sproson (2020)

Sprovieri et al. (2013) conducted spectral analyses on upper Albian to lower Campanian $\delta^{13}C$ record of the Bottaccione Section (Gubbio, Italy) tuned by the 405-kyr *E*-cycle. This c. 23-myr interval, between c. 78 and 101 Ma, includes the intervals discussed in Examples 6–8. They identified $\delta^{13}C$ cycles with periods of c. 2.4, 3.4, 4.7 and 8.0 myr, and speculated the 4.7-myr periodicity represents a "homologue" implying it stands out as the dominant grand cycle.

In the Middle Paleozoic, Sproson (2020) conducted spectral analysis of $\delta^{13}C_{carb}$ and conodont apatite oxygen isotope ($\delta^{18}O$) records compiled for the Katian Stage to early Devonian mid-Lohkovian Stage (c. 425.0–448.0 Ma). In this interval, which includes Example 1, they documented a prominent cyclicity of c. 4.2 myr and grand cycles ranging between 3.7 and 4.8 myr.

The identification of cycles in $\delta^{18}O$ and $\delta^{13}C$ records with periods that are approximately multiples of c. 0.4 myr and include c. 2.0 and 2.8 myr supports the model proposed by Matthews and Frohlich (2002, fig. 1). The c. 1.2 and 1.6 myr cycles may be the two components of the 2.8 myr cycle, and the 2.0–2.6 myr cyclicity may represent more randomized tuning within the sixth straton of dozons. The detection of cycles in the 4.7–4.8 myr range suggests dozons are evident in spectral analyses but somewhat shorter than predicted at 4.86 myr. The shorter empirical estimate (e.g., 4.2–4.6 myr) may be due to hiatuses occurring at the boundaries of the 4.86-myr dozons and the approximate dating of stage boundaries.

17. Conclusions

The deterministic Phanerozoic Orbital Scale of glacio-eustasy predicts the ages of c. 405-kyr highstand and lowstand stratons within a repeating 4.86-myr dozon. This periodic glacio-eustatic signal however is generally difficult to recognize in big-data compilations of global eustatic cycles interpreted from local depositional sequences and calibrated with various empirical time scales. In contrast, carbon-13 ($\delta^{13}C$) records manifest secular patterns that are generally easy to recognize, mirror-reflect eustatic cycles and contain high-frequency cycles that are tuned by the 405-kyr long-eccentricity *E*-cycle. They generally contain globally correlative $\delta^{13}C$ plateaus that can be anchored to the predicted ages of key glacio-eustatic lowstands in the Orbital Scale.

The nine examples illustrate how unconformities, sea level highstands and lowstands, $\delta^{13}C$ plateaus and valleys were used as anchors in the

Orbital Scale. In several cases they offered a new criterion for resolving conflicting ages of stage boundaries. The application of the Orbital Scale to numerically calibrate the glacio-eustatic signal and its $\delta^{13}C$ patterns in any Phanerozoic interval is simple arithmetic but understanding and reconciling the conflicting interpretations of extensive bio- and sequence-stratigraphic datasets is a steep learning curve. This situation suggests establishing frequent $\delta^{13}C$ anchors, particularly between dozons, can greatly assist in calibrating third-order sequences as multiples of 405-kyr stratons, and fourth- and fifth-order sequences as stratons and 100 kyr cycles.

The Orbital Scale bypasses the debate over whether greenhouse times were ice-free—a proposition that has been extensively discussed in the literature. Instead this study shows examples of how $\delta^{13}C$ plateaus pass to unconformities and stratigraphic black holes (SBH) some of which are associated with cooling oceans and cold snaps. In the final Example 9 the late Aptian and early Albian is characterized by a broad $\delta^{13}C$ plateau and falling limb, a 30–50 m eustatic lowstand in Arabia, Cold Snap 3 in the Atlantic Ocean and a few glaciogenic deposits. This glaciation was first interpreted a decade ago in *GeoArabia Special Publication 4* (van Buchem et al., 2010) and published in *Terra Nova* by Maurer et al. (2013) after a major revision resulting from a somewhat skeptical peer review.

In the reply to *Terra Nova's* Editor, F. Maurer wrote "*The idea of glaciations in the Cretaceous works against the common notion of greenhouse climate in the scientific community. With our contribution we like to spur the discussion on this topic, and hope that the editors find it provocative and robust enough to accept it for publication in Terra Nova.*" We hope the present contribution will spur further interest by readers to examine the role of orbital-forcing of glacio-eustasy in other geological intervals.

Acknowledgments

The authors thank Florian Maurer for reviewing the original draft of the manuscript and making many suggestions that have improved the interpretations in some examples. In general he found the close ties between the deterministic and radiometric estimates for Paleozoic stage boundaries stage boundary quite remarkable. Michael Montenari is thanked for inviting this contribution. Arnold Egdane is thanked for designing the figures and proof-reading the manuscript.

References

Ainsaar, L., Meidla, T., Tinn, O., 2004. Middle and Upper Ordovician stable isotope stratigraphy across the facies belts in the East Baltic. In: Hints, O., Ainsaar, L. (Eds.), WOGOGOB-2004 Conference Materials. Tartu University Press, Tartu, pp. 11–12.

Aldridge, R.J., Jeppsson, L., Doring, K.J., 1993. Early Silurian oceanic episodes and events. J. Geol. Soc. Lond. 150, 501–513.
Al-Husseini, M.I., 2018. Arabian orbital sequences. In: Montenari, M. (Ed.), Stratigraphy and Timescales. vol. 3. Elsevier, pp. 219–264.
Alley, N.F., Hore, S.B., Frakes, L.A., 2019. Glaciations at high-latitude Southern Australia during the Early Cretaceous. Aust. J. Earth Sci. https://doi.org/10.1080/08120099.2019.1590457.
Amthor, J.E., Grotzinger, J.P., Schröder, S., Bowring, S.A., Ramezani, J., Martin, M.W., Matter, A., 2003. Extinction of Cloudina and Namacalathus at the Precambrian-Cambrian boundary in Oman. Geology 31, 431–434.
Berger, W.H., Vincent, E., 1981. Chemostratigraphy and biostratigraphic correlation: exercises in systemic stratigraphy. Oceanol. Acta 0399-1784 (Special Issue), 115–127.
Bergström, S.M., Young, S.A., Schmitz, B., 2010. Katian (Upper Ordovician) $\delta^{13}C$ chemostratigraphy and sequence stratigraphy in the United States and Baltoscandia: a regional comparison. Palaeogeogr. Palaeoclimatol. Palaeoecol. 296, 217–234.
Bergström, S.M., Saltzman, M.R., Leslie, S.A., Ferretti, A., Young, S.A., 2015. Transatlantic application of the Baltic Middle and Upper Ordovician carbon isotope zonation. Int. J. Earth Sci. 64 (1), 8–12.
Bless, M.J.M., Becker, R.T., Higgs, K., Paproth, E., Streel, M., 1992. Eustatic cycles around the Devonian-Carboniferous boundary and the sedimentary and fossil record in Sauerland (Federal Republic of Germany). Ann. Soc. Geol. Belg. 115, 689–702.
Bornemann, A., Norris, R.D., Friedrich, O., Beckmann, B., Schouten, S., Sinninghe Damsté, S.J., Vogel, J., Hofmann, P., Wagner, T., 2008. Isotopic evidence for glaciation during the cretaceous supergreenhouse. Science 319, 189–192.
Boulila, S., 2019. Coupling between Grand cycles and Events in Earth's climate during the past 115 million years. Sci. Rep. 9, 327.
Boulila, S., Charbonnier, G., Spangenberg, J., Gardin, S., Bruno Galbrun, B., Briard, J., Le Callonnec, L., 2020. Unraveling short-and long-term carbon cycle variations during the Oceanic Anoxic Event 2 from the Paris Basin Chalk. Global Planet. Change 186, 103–126.
Bowring, S.A., Grotzinger, J.P., Condon, D.J., Ramezani, J., Newall, M.J., Allen, P.A., 2007. Geochronologic constraints on the chronostratigraphic framework of the Neoproterozoic Huqf Supergroup, Sultanate of Oman. Am. J. Sci. 307, 1097–1145.
Brand, U., Legrand-Blain, M., Streel, M., 2004. Biochemostratigraphy of the Devonian-Carboniferous boundary global stratotype section and point, Griotte Formation, La Serre, Montagne Noire, France. Palaeogeogr. Palaeoclimatol. Palaeoecol. 205, 337–357.
Brenchley, P.J., Carden, G.A., Hints, L., Kaljo, D., Marshall, J.D., Martma, T., Meldha, T., N€olvak, J., 2003. High-resolution stable isotope stratigraphy of Upper Ordovician sequences: constraints on the timing of bioevents and environmental changes associated with mass extinctions and glaciation. GSA Bull. 115, 89–104.
Brezinski, D.K., Cecil, C.B., Skema, V.W., 2010. Late Devonian glacigenic and associated facies from the central Appalachian Basin, eastern United States. Geol. Soc. Am. Bull. 122, 265–281.
Buggisch, W., Joachimski, M.M., 2006. Carbon isotope stratigraphy of the Devonian of Central and Southern Europe. Palaeogeogr. Palaeoclimatol. Palaeoecol. 240, 68–88.
Calner, M., Lehnert, O., Nolvak, J., 2010. Palaeokarst evidence for widespread regression and subaerial exposure in the middle Katian (Upper Ordovician) of Baltoscandia: significance for global climate. Palaeogeogr. Palaeoclimatol. Palaeoecol. 296, 236–247.
Caputo, M.V., Melo, J.H.G., Streel, M., Isbell, J.L., 2008. Late Devonian and Early Carboniferous glacial records of South America. In: Fielding, C.R., Frank, T.D., Isbell, J.L. (Eds.), Resolving the Late Paleozoic Ice Age in Time and Space. vol. 441. Geological Society of America Special Paper, pp. 161–173.

Catuneanu, O., Abreu, V., Bhattacharya, J.P., Blum, M.D., Dalrymple, R.W., Eriksson, P.G., Fielding, C.R., Fisher, W.L., Galloway, W.E., Gibling, M.R., Giles, K.A., Holbrook, J.M., Jordan, R., Kendall, C.G.S.C., Macurda, B., Martinsen, O.J., Miall, A.D., Neal, J.E., Nummedal, D., Pomar, L., Posamentier, H.W., Pratt, B.R., Sarg, J.F., Shanley, K.W., Steel, R.J., Strasser, A., Tucker, M.E., Winker, C., 2009. Towards the standardization of sequence stratigraphy. Earth Sci. Rev. 92 (1–2), 1–33.

Chen, X., Rong, J., Fan, J., Zhan, R., Mitchell, C.E., Harper, D.A.T., Melchin, M.J., Peng, P., Finney, S.C., Wang, X., 2006. The Global Boundary Stratotype Section and Point (GSSP) for the base of the Hirnantian stage (the uppermost of the ordovician system). Episodes 29 (3), 183–196.

Compston, W., Zhang, Z., Cooper, J.A., Ma, G., Jenkins, R.J.F., 2008. Further SHRIMP geochronology on the early Cambrian of South China. Am. J. Sci. 308 (4), 399–420.

Davies, A., Greselle, B., Hunter, S.J., Baines, G., Robson, C., Haywood, A.M., Ray, D.C., Simmons, M.D., van Buchem, F.S.P., 2020. Assessing the impact of aquifer-eustasy on short-term Cretaceous sea-level. Cretac. Res. 112, 104445. https://doi.org/10.1016/j.cretres.2020.104445.

de Boer, B., Lourens, L.J., van de Wal, R.S.W., 2014. Persistent 400,000-year variability of Antarctic ice volume and carbon cycle is revealed throughout the Plio-Pleistocene. Nat. Commun. 5. 8 p.

Della Porta, G., Kenter, J.A.M., Immenhauser, A., Bahamonde, J.R., 2002. Lithofacies character and architecture across a Pennsylvanian inner-platform transect (Sierra de Cuera, Asturias, Spain). J. Sediment. Res. 72, 898–916.

Droste, H.J., 2010. High-resolution seismic stratigraphy of the Shu'aiba and Natih formations in the Sultanate of Oman: implications for Cretaceous epeiric carbonate platform systems. In: van Buchem, F.S.P., Gerdes, K.D., Esteban, M. (Eds.), Mesozoic and Cenozoic Carbonate Systems of the Mediterranean and the Middle East—Stratigraphic and Diagenetic Reference Models. vol. 329. Geological Society, London, Special Publication, pp. 145–162.

Eldrett, J.S., Ma, C., Bergman, S.C., Lutz, B., Gregory, F.J., Dodsworth, P., Phipps, M., Hardas, P., Minisini, D., Ozkan, A., Ramezani, J., Bowring, S.A., Kamo, S.L., Ferguson, K., Macaulay, C., Kelly, A.E., 2015. An astronomically calibrated stratigraphy of the Cenomanian, Turonian and earliest Coniacian from the Cretaceous Western Interior Seaway, USA: implications for global chronostratigraphy. Cretac. Res. 56, 316–344.

Fang, Q., Wu, H., Hinnov, L.A., Wang, X., Yang, T., Li, H., Zhang, S., 2016. A record of astronomically forced climate change in a late Ordovician (Sandbian) deep marine sequence, Ordos Basin, North China. Sediment. Geol. 341, 163–174.

Finney, S.C., Berry, W.B.N., Cooper, J.D., Ripperdan, R.L., Sweet, W.C., Jacobson, S.R., Soufiane, A., Acab, A., Noble, P.J., 1999. Late Ordovician mass extinction: a new perspective from stratigraphic sections in Central Nevada. Geology 27, 215–218.

Flajs, G., Feist, R., 1988. Index conodonts, trilobites and environment of the Devonian-Carboniferous boundary beds at La Serre (Montagne Noire, France). In: Flajs, G., Feist, R., Ziegler, W. (Eds.), Devonian-Carboniferous Boundary, Results of Recent Studies. 100. Courier Forschungsinstitut Senckenberg, pp. 53–107.

Forbes, G.A., Jansen, H.S.M., Schreurs, J., 2010. Lexicon of Oman Subsurface Stratigraphy: Reference Guide to the Stratigraphy of Oman's Hydrocarbon Basins. GeoArabia Special Publication 5, Gulf PetroLink, Bahrain. 371 p.

Gale, A.S., 1995. Cyclostratigraphy and correlation of the Cenomanian stage in Western Europe. In: House, M.R., Gale, A.S. (Eds.), Orbital Forcing Timescales and Cyclostratigraphy. vol. 85. Geological Society of London, Special Publication, pp. 177–197.

Gale, A.S., Hardenbol, J., Hathaway, B., Kennedy, W.J., Young, J.R., Phansalkar, V., 2002. Global correlation of Cenomanian (Upper Cretaceous) sequences: evidence for Milankovitch control on sea level. Geology 30, 291–294.

Germs, G.J.B., Gaucher, C., 2012. Nature and extent of a late Ediacaran (ca. 547 Ma) glacigenic erosion surface in southern Africa. S. Afr. J. Geol. 115, 91–102.

Geyer, G., 2019. A comprehensive Cambrian correlation chart. Episodes 42 (4), 1–12.

Ghienne, J.-F., Desrochers, A., Vandenbroucke, T.R.A., Achab, A., Asseln, E., Dabard, M.-P., Farley, C., Loi, A., Paris, F., Wickson, S., Veizer, J., 2014. A Cenozoic-style scenario for the end-Ordovician glaciation. Nat. Commun. 5, 4485. https://doi.org/10.1038/ncomms5485.

Goldman, D., Leslie, S.A., Nõlvak, J., Young, S.A., Bergström, M., Huff, W.D., 2007. The global Stratotype section and Point (GSSP) for the base of the Katian stage of the Upper Ordovician Series at black Knob ridge, southeastern Oklahoma, USA. Episodes 30 (4), 258–270.

Gradstein, F.M., Agterberg, F.P., Ogg, J.G., Hardenbol, J., Veen, P., Thierry, J., Huang, Z., 1994. A Mesozoic time scale. J. Geophys. Res. Solid Earth 99 (B12), 24051–24074.

Gradstein, F.M., Ogg, J.G., Smith, A., 2004. A Geologic Time Scale. Cambridge University Press. 589 p.

Gradstein, F.M., Ogg, J.G., Schmitz, M.D., Ogg, G.M. (Eds.), 2012. The Geologic Time Scale 2012. Elsevier, Oxford, UK. 1176 p.

Grélaud, C., Razin, P., Homewood, P., 2010. Channelized systems in an inner carbonate platform setting: differentiation between incisions and tidal channels (Natih Formation, Late Cretaceous, Oman). In: van Buchem, F.S.P., Gerdes, K.D., Esteban, M. (Eds.), Mesozoic and Cenozoic Carbonate Systems of the Mediterranean and the Middle East e Stratigraphic and Diagenetic Reference Models. vol. 329. Geological Society, London, Special Publications, pp. 163–186.

Gulbranson, E.L., Montañez, I.P., Tabor, N.J., Limarino, C.O., 2015. Late Pennsylvanian aridification on the southwestern margin of Gondwana (Paganzo Basin, NW Argentina): a regional expression of a global climate perturbation. Palaeogeogr. Palaeoclimatol. Palaeoecol. 417, 220–235.

Haq, B.U., 2014. Cretaceous eustasy revisited. Global Planet. Change 113, 44–58.

Haq, B.U., Al-Qahtani, A.M., 2005. Phanerozoic cycles of sea-level change on the Arabian Platform. GeoArabia 10 (2), 127–160.

Haq, B.U., Huber, B.T., 2016. Anatomy of a eustatic event during the Turonian (Late Cretaceous) hot greenhouse climate. Sci. China Earth Sci. 60, 20–29. https://doi.org/10.1007/s11430-016-0166-y.

Haq, B.U., Schutter, S.R., 2008. A chronology of Paleozoic Sea-level changes. Science 322, 64–68.

Haq, B.U., Hardenbol, J., Vail, P.R., 1987. Chronology of fluctuating sea levels since the Triassic. Science 235, 1156–1167.

Hardenbol, J., Thierry, J., Farley, M.B., de Graciansky, P.-C., Vail, P.R., 1998. Mesozoic and Cenozoic sequence chronostratigraphic framework of European basins. In: de Graciansky, P.-C., Hardenbol, J., Jacquin, T., Vail, P.R. (Eds.), Mesozoic and Cenozoic Sequence Stratigraphy of European Basins. vol. 60. Society of Economic Paleontologists and Mineralogists, Special Publication, pp. 3–13.

Harland, W.B., Armstrong, R.L., Cox, A.V., Craig, L.E., Smith, A.G., Smith, D.G., 1990. A Geologic Time Scale 1989. Cambridge University Press, New York, p. 265.

Hasegawa, T., 1997. Cenomanian-Turonian carbon isotope events recorded in terrestrial organic matter from northern Japan. Palaeogeogr. Palaeoclimatol. Palaeoecol. 130, 251–273.

He, H.Y., Pan, Y.X., Tauxe, L., Qin, H.F., Zhu, R.X., 2008. Toward age determination of the M0r (Barremian/Aptian boundary) of the Early Cretaceous. Phys. Earth Planet. In. 169 (1), 41–48.

Herrle, J.O., Kößler, P., Friedrich, O., Erlenkeuser, H., Hemleben, C., 2004. High-resolution carbon isotope stratigraphy of the Aptian to Lower Albian: a tool for reconstructing paleoceanographic changes and paleobiological evolution. Earth Planet. Sci. Lett. 218, 149–161.

Herrle, J.O., Schröder-Adams, C.J., Davis, W., Pugh, A.T., Galloway, J.M., Fath, J., 2015. Mid-Cretaceous High Arctic stratigraphy, climate, and Oceanic Anoxic Events. Geology 43 (5), 403–406.

Hinnov, L.A., 2018. Cyclostratigraphy and astrochronology in 2018. In: Montenari, M. (Ed.), Stratigraphy and Timescales. vol. 3. Elsevier Inc, pp. 1–80.

Hinnov, L.A., Hilgen, F.J., 2012. Cyclostratigraphy and astrochronology. In: Gradstein, F., Ogg, J., Ogg, G., Smith, D. (Eds.), The Geologic Time Scale 2012. Elsevier, pp. 63–83.

Hinnov, L.A., Ogg, J.G., 2007. Cyclostratigraphy and the astronomical time scale. Stratigraphy 4, 239–251.

Holbourn, A., Kuhnt, W., Schulz, M., Flores, J., Andersen, N., 2007. Orbitally-paced climate evolution during the middle Miocene "Monterey" carbon-isotope excursion. Earth Planet. Sci. Lett. 261, 534–550.

Holbourn, A.E., Kuhnt, W., Clemens, S., Prell, W., Andersen, N., 2013. Middle to late Miocene stepwise climate cooling: evidence from a high-resolution deep water isotope curve spanning 8 million years. Paleoceanography 28, 688–699.

Holbourn, A.E., Kuhnt, W., Clemens, S.C., Kochhann, K.G.D., Jöhnck, J., Lübbers, J., Andersen, N., 2018. Late Miocene climate cooling and intensification of southeast Asian winter monsoon. Nat. Commun. 9, 1584. https://doi.org/10.1038/s41467-018-03950-1.

Holmden, C., Mitchell, C.E., Della LaPorta, F., Patterson, W.P., Melchin, M.J., Finney, S.C., 2013. Nd isotope records of late Ordovician Sea-level change—implications for glaciation frequency and global stratigraphic correlation. Palaeogeogr. Palaeoclimatol. Palaeoecol. 386, 131–144.

Homewood, P., Razin, P., Grélaud, C., Droste, H.J., Vahrenkamp, V., Mettraux, M., Mattner, J., 2008. Outcrop sedimentology of the Natih Formation, northern Oman: a field guide to selected outcrops in the Adam Foothills and Al Jabal al Akhdar areas. GeoArabia 13 (3), 39–120.

Huang, C., 2018. Astronomical time scale for the Mesozoic. In: Montenari, M. (Ed.), Stratigraphy and Timescales. vol. 3. Elsevier Inc, pp. 81–150.

Huang, C., Hinnov, L.A., Fischer, A.G., Grippo, A., Herbert, T., 2010. Astronomical tuning of the Aptian Stage from Italian reference sections. Geology 38, 899–902.

Immenhauser, A., Della Porta, G., Kenter, J.A.M., Bahamonde, J.R., 2003. An alternative model for positive shifts in shallow-marine carbonate $\delta^{13}C$ and $\delta^{18}O$. Sedimentology 50, 953–959.

Isaacson, P., Díaz Martínez, E., Grader, G., Kalvoda, J., Babek, O., Devuyst, F.X., 2008. Late Devonian–earliest Mississippian glaciation in Gondwanaland and its biogeographic consequences. Palaeogeogr. Palaeoclimatol. Palaeoecol. 268, 126–142.

Jarvis, I., Gale, A.S., Jenkyns, H.C., Pearce, M.A., 2006. Secular variation in Late Cretaceous carbon isotopes: a new $\delta^{13}C$ carbonate reference curve for the Cenomanian–Campanian (99.6–70.6 Ma). Geol. Mag. 143, 561–608.

Jarvis, I., Lignum, J.S., Gröcke, D.R., Jenkyns, H.C., Pearce, M.A., 2011. Black shale deposition, atmospheric CO2 drawdown, and cooling during the Cenomanian–Turonian Oceanic Anoxic Event. Paleoceanography 26. https://doi.org/10.1029/2010PA002081.

Jarvis, I., Trabucho-Alexandre, J., Grocke, D.R., Uličný, D., Laurin, J., 2016. Intercontinental correlation of organic carbon and carbonate stable isotope records: evidence of climate and sea- level change during the Turonian (Cretaceous). Depos. Rec. 1, 53–90. Published by John Wiley & Sons Ltd.

Joo, Y.J., Sageman, B.S., 2014. Cenomanian to Campanian carbon isotope chemostratigraphy from the Western Interior Basin, U.S.A. J. Sediment. Res. 84, 529–542.
Kaiser, S.I., Aretz, M., Becker, R.E., 2015. The global Hangenberg Crisis (Devonian–Carboniferous transition): review of a first-order mass extinction. In: Becker, R.T., Konigshof, P., Brett, C.E. (Eds.), Devonian Climate, Sea Level and Evolutionary Events. vol. 423. Geological Society, London, Special Publication, pp. 387–437.
Kaljo, D., Hints, L., Martma, T., Nolvak, J., 2001. Carbon isotope stratigraphy in the latest Ordovician of Estonia. Chem. Geol. 175, 49–59.
Keller, C.B., Husson, J.M., Mitchell, R.N., Bottke, W.F., Gernon, T.M., Boehnke, P., Bell, E.A., Swanson-Hysell, N.L., Peters, S.E., 2019. Neoproterozoic glacial origin of the Great Unconformity. PNAS 116 (4), 1136–1145.
Koch, J.T., Brenner, R.L., 2009. Evidence for glacioeustatic control of large, rapid sea-level fluctuations during the Albian-Cenomanian: Dakota formation, eastern margin of western interior seaway, USA. Cretac. Res. 30 (2), 411–423.
Kouchinsky, A., Bengston, S., Pavlov, V., Runnegar, B., Tossander, P., Young, E., Ziegler, K., 2007. Carbon isotope stratigraphy of the Precambrian-Cambrian Sukharikha River section, northwester Siberian platform. Geol. Mag. 144, 609–618.
Kuhnt, W., Moullade, M., Masse, J.-P., Erlenkeuser, M., 1998. Carbon isotope stratigraphy of the Lower Aptian historical stratotype at Cassis-La Bédoule (SE France). Geol. Medit. XXV (3–4), 63–79. Marseille.
Lakin, J.A., Marshall, J.E.A., Troth, I., Harding, I.C., 2016. Greenhouse to icehouse: a biostratigraphic review of latest Devonian-Mississippian glaciations and their global effects. In: Becker, R.T., Konigshof, P., Brett, C.E. (Eds.), Devonian Climate, Sea Level and Evolutionary Events. vol. 423. Geological Society, London, Special Publication, pp. 439–464.
LaPorte, D.F., Holmden, C., Patterson, W.P., Loxton, J.D., Melchin, M.J., Mitchell, C.E., Finney, S.C., Sheets, H.D., 2009. Local and global perspectives on carbon and nitrogen cycling during the Hirnantian glaciation. Palaeogeogr. Palaeoclimatol. Palaeoecol. 276, 182–195.
Laskar, J., Robutal, P., Joutel, F., Gastineau, M., Correia, A., Levrard, B., 2004. A long term numerical solution for the insolation quantities of the Earth. Astron. Astrophys. 428 (1), 261–286.
Leckie, R.M., Bralower, T.J., Cashman, R., 2002. Oceanic anoxic events and plankton evolution: biotic response to tectonic forcing during the mid-Cretaceous. Paleoceanography 17 (3). 29 p, Wiley Online Library.
Linnemann, U., Ovtcharova, M., Schaltegger, U., Gärtner, A., Hautmann, M., Geyer, G., Vickers-Rich, P., Rich, T., Plessen, B., Hofmann, J., Zieger, J., Krause, R., Kriesfeld, L., Smith, J., 2019. New high-resolution age data from the Ediacaran–Cambrian boundary indicate rapid, ecologically driven onset of the Cambrian explosion. Terra Nova 31, 49–58.
Loydell, D.K., 1998. Early Silurian Sea-level changes. Geol. Mag. 135, 447–471.
MacLeod, K.G., Huber, B.T., Berrocoso, A.J., Wender, I., 2013. A stable and hot Turonian without glacial a δO18O excursions in indicated by exquisitely preserved Tanzanian foraminifera. Geology 41 (10), 1083–1086.
Maloof, A.C., Porter, S.M., More, J.L., Dudás, F.Ö., Bowring, S.A., Higgins, J.A., Fike, D.A., Eddy, M.P., 2010. The earliest Cambrian record of animals and ocean geochemical change. Geol. Soc. Am. Bull. 122, 1731–1774.
Matthews, R.K., Al-Husseini, M.I., 2010. Orbital-forcing glacio-eustasy: a sequence-stratigraphic time scale. GeoArabia 15 (3), 155–167.

Matthews, R.K., Frohlich, C., 2002. Maximum flooding surface and sequence boundaries: comparisons between observation and orbital forcing in the Cretaceous and Jurassic (65-190 Ma). GeoArabia 7 (3), 503–538.
Maurer, F., van Buchem, F.S.P., Eberli, G.P., Pierson, B.J., Raven, M.J., Larsen, P.-H., Al- Husseini, M.I., Vincent, B., 2013. Late Aptian long-lived glacio-eustatic lowstand recorded on the Arabian plate. Terra Nova 25, 87–94.
Mauviel, A., 2016. Integrated high-resolution chemostratigraphic and cyclostratigraphic analysis of the paleotropical carbonates spanning the Ordovician-Silurian boundary at the west end of Anticosti Island, eastern Canada. MSc thesis, University of Ottawa. 85 p.
Mauviel, A., Desrochers, A., 2016. A high resolution, continuous $\delta^{13}C$ record spanning the O/S boundary on Anticosti Island, eastern Canada. Can. J. Earth Sci. 53 (8), 795–801.
Melchin, M.J., Holmden, C., 2006. Carbon isotope chemostratigraphy of the Llandovery in Arctic Canada: implications for global correlation and sea-level change. GFF 128, 173–180.
Midtkandal, I., Svensen, H.H., Planke, S., Corfu, F., Polteau, S., Torsvik, T.H., Faleide, J.I., Grundvåg, S.-A., Selnes, H., Kürschner, W., Olaussen, S., 2016. The Aptian (Early Cretaceous) oceanic anoxic event (OAE1a) in Svalbard, Barents Sea, and the absolute age of the Barremian-Aptian boundary. Palaeogeogr. Palaeoclimatol. Palaeoecol. 463, 126–135.
Miller, K.G., Kominz, M.A., Browning, J.V., Wright, J.D., Mountain, G.S., Katz, M.E., Sugarman, P.J., Cramer, B.S., Christie-Blick, N., Pekar, S.F., 2005a. The Phanerozoic record of global sea-level change. Science 310, 1293–1298.
Miller, K.G., Wright, J.D., Browning, J.V., 2005b. Visions of ice sheets in a greenhouse world. In: Paytan, A., De La Rocha, C. (Eds.), Ocean Chemistry Throughout the Phanerozoic. vol. 217. Marine Geology Special Issue, pp. 215–231.
Miller, K.G., Browning, J.V., Schmelz, W.J., Kopp, R.E., Mountain, G.S., Wright, J.D., 2020. Cenozoic sea-level and cryospheric evolution from deep-sea geochemical and continental margin records. Sci. Adv. 6 (20). https://doi.org/10.1126/sciadv.aaz1346. 15 p.
Mitchell, S.F., 1995. Lithostratigraphy and biostratigraphy of the Hunstanton Formation (Red Chalk, Cretaceous) succession at Speeton, North Yorkshire, England. Proc. Yorks. Geol. Soc. 50, 285–305.
Myrow, P.M., Ramezani, J., Hanson, A.E., Bowring, S.A., Racki, G., Rakocinski, M., 2013. High-precision U-Pb age and duration of the latest Devonian (Famennian) Hangenberg event, and its implications. Terra Nova 26, 222–229.
Ogg, J.G., Gradstein, F.M., 2020. Geologic TimeScale Foundation. TSCreator. https://engineering.purdue.edu/.
Ogg, J.G., Hinnov, L.A., 2012. The cretaceous period. In: Gradstein, F.M., Ogg, J.G., Schmitz, M.D., Ogg, G.M. (Eds.), The Geological Time Scale. Elsevier, pp. 793–853.
Ogg, J.G., Agterberg, F.P., Gradstein, F.M., 2004. The cretaceous period. In: Gradstein, F.M., Ogg, J.G., Smith, A.G. (Eds.), A Geologic Time Scale 2004. Cambridge University Press, pp. 344–383.
Pagani, M., Huber, M., Sageman, B., 2014. Greenhouse climates. In: Treatise on Geochemistry, second ed. vol. 6. Elsevier, pp. 281–304.
Pälike, H., Norris, R.D., Herrle, J.O., Wilson, P.A., Coxall, H.K., Lear, C.H., Shackleton, N.J., Tripati, A.K., Wade, B.S., 2006. The heartbeat of the oligocene climate system. Science 314, 1894–1898.
Price, G.D., 1999. The evidence and implications of polar ice during the Mesozoic. Earth Sci. Rev. 48, 183–210.
Ray, D.C., van Buchem, F.S.P., Baines, G., Davies, A., Gréselle, B., Simmons, M.D., Robson, C., 2019. The magnitude and cause of short-term eustatic Cretaceous sea-level change: a synthesis. Earth Sci. Rev. 197. 20 p.

Rodríguez-López, J.P., Liesa, C.L., Pardo, G., Meléndez, N., Soria, A.R., Ian Skilling, I., 2016. Glacial dropstones in the western Tethys during the late Aptian–early Albian cold snap: palaeoclimate and palaeogeographic implications for the mid-Cretaceous. Palaeogeogr. Palaeoclimatol. Palaeoecol. 452, 11–27.

Ruebsam, W., Al-Husseini, M.I., 2020. Calibrating the Early Toarcian (Early Jurassic) with stratigraphic black holes (SBH). Gondw. Res. 82, 317–336.

Ruebsam, W., Thibault, N., Al-Husseini, M., 2020. Early Toarcian glacio-eustatic unconformities and chemostratigraphic black holes. In: Montenari, M. (Ed.), Stratigraphy & Timescales. vol. 5. Elsevier, pp. 629–676.

Runkel, A.C., McKay, R.M., Palmer, R., 1998. Origin of a classic cratonic sheet sandstone: stratigraphy across the Sauk II–Sauk III boundary in the Upper Mississippi Valley. Geol. Soc. Am. Bull. 110, 188–210.

Sahagian, D., Pinous, O., Olferiev, A., Zakharov, V., 1996. Eustatic curve for the Middle Jurassic-Cretaceous based on Russian Platform and Siberian stratigraphy: zonal resolution. AAPG Bull. 80, 1433–1458.

Saltzman, M.R., Runnegar, B., Lohmann, K.C., 1998. Carbon isotope stratigraphy of Upper Cambrian (Steptoean Stage) sequences of the eastern Great Basin: record of a global oceanographic event. Geol. Soc. Am. Bull. 110, 285–297.

Saltzman, M.R., Runkel, A.C., Cowan, C.A., Runnegar, B., Stewart, M.C., Palmer, A.R., 2004. The upper Cambrian SPICE ($\delta^{13}C$) event and the Sauk II-Sauk III regression: new evidence from Laurentian basins in Utah, Iowa and Newfoundland. J. Sediment. Res. 74, 366–377.

Salzman, M.R., Thomas, E., 2012. Chapter 11—carbon isotope stratigraphy. In: Gradstein, F.M., Ogg, J., Schmitz, M., Ogg, G. (Eds.), The Geological Timescale 2012. Elsevier.

Schiffbauer, J.D., Huntley, J.W., Fike, D.A., Jeffrey, M.J., Gregg, J.M., Shelton, K.J., 2017. Decoupling biogeochemical records, extinction, and environmental change during the Cambrian SPICE event. Sci. Adv. 3 (3), 7 p. https://doi.org/10.1126/sciadv.1602158.

Schlanger, S.O., Arthur, M.A., Jenkyns, H.C., Scholle, P.A., 1987. The Cenomanian–Turonian Oceanic Anoxic event, I. Stratigraphy and distribution of organic carbon-rich beds and the marine $\delta^{13}C$ excursion. In: Brooks, J., Fleet, A.J. (Eds.), Marine and Petroleum Source Rocks. 26. Geological Society of London, Special Publication, pp. 371–399.

Scholle, P.A., Arthur, M.A., 1980. Carbon isotope fluctuations in Cretaceous pelagic limestones: Potential stratigraphic and petroleum exploration tool. Am. Assoc. Pet. Geol. Bull. 64, 67–87.

Scott, R.W., Oboh-Ikuenobe, F., Benson, D.G., Holbrook, J.M., 2009. Numerical age calibration of the Albian/Cenomanian boundary. Micropaleontology 6 (1), 17–33.

Selby, D., Mutterlose, J., Condon, D.J., 2009. U–Pb and Re–Os geochronology of the Aptian/Albian and Cenomanian/Turonian stage boundaries: implications for timescale calibration, osmium isotope seawater composition and Re–Os systematics in organic-rich sediments. Chem. Geol. 265, 394–409.

Shahkarami, S., Buatois, L.A., Mángano, M.G., Hagadorn, J.W., Almond, J., 2020. The Ediacaran–Cambrian boundary: evaluating stratigraphic completeness. Precambrian Res. 345. https://doi.org/10.1016/j.precamres.2020.105721.

Snedden, J.W., Liu, C., 2010. A compiliation of Phanerozoic sea level change, coastal onlaps, and recommended sequence designations. AAPG Search Discov., 40594.

Sproson, A.D., 2020. Pacing of the latest Ordovician and Silurian carbon cycle by a ~ 4.5 Myr orbital cycle. Palaeogeogr. Palaeoclimatol. Palaeoecol. 540. https://doi.org/10.1016/j.palaeo.2019.109543.

Sprovieri, M., Sabatino, N., Pelosi, N., Batenburg, S.J., Coccioni, R., Iavarone, M., Mazzola, S., 2013. Late Cretaceous orbitally-paced carbon isotope stratigraphy from the Bottaccione Gorge (Italy). Palaeogeogr. Palaeoclimatol. Palaeoecol. 379–380, 81–94.

Stoll, H.M., Schrag, D.P., 2000. High-resolution stable isotope records from the Upper Cretaceous rocks of Italy and Spain: glacial episodes in a greenhouse planet? Geol. Soc. Am. Bull. 112, 308–319.

Streel, M., Caputo, M.V., Loboziak, S., Melo, J.H.G., 2000. Late Frasnian-Famennian climates based on palynomorph analyses and the question of the Late Devonian glaciations. Earth Sci. Rev. 52, 121–173.

Takashima, R., Nishi, H., Yamanaka, T., Hayashi, K., Waseda, A., Obuse, A., Tomosugi, T., Deguchi, N., Mochizuki, S., 2010. High-resolution terrestrial carbon isotope and planktic foraminiferal records of the Upper Cenomanian to the Lower Campanian in the Northwest Pacific. Earth Planet. Sci. Lett. 289, 570–582.

Tsikos, H., Jenkyns, H.C., Walsworth-Bell, B., Petrizzo, M.R., Forster, A., Kolonic, S., Erba, E., Premoli-Silva, I.P., Baas, M., Wagner, T., Damsté, J.S., 2004. Carbon-isotope stratigraphy recorded by the Cenomanian–Turonian Oceanic Anoxic Event: correlation and implications based on three key localities. J. Geol. Soc. Lond. 161, 711–719.

Uličný, D., Jarvis, I., Gröcke, D.R., Čech, S., Laurin, J., Olde, K., Trabucho-Alexandre, J., Švábenická, L., Pedentchouk, N., 2014. A high-resolution carbon-isotope record of the Turonian stage correlated to a siliciclastic basin fill: implications for mid-Cretaceous sea-level change. Palaeogeogr. Palaeoclimatol. Palaeoecol. 405, 42–58.

Vahrenkamp, V.C., 2013. Carbon-isotope signatures of Albian to Cenomanian (Cretaceous) shelf carbonates of the Natih Formation, Sultanate of Oman. GeoArabia 18 (3), 65–82.

van Buchem, F.S.P., Al-Husseini, M.I., Maurer, F., Droste, H.J., Yose, L.A., 2010. Sequence-stratigraphic synthesis of the Barremian—Aptian of the eastern Arabian Plate and implications for the petroleum habitat. 4 GeoArabia Special Publication, pp. 9 48.

van Buchem, F.S.P., Simmons, M.D., Droste, H.J., Davies, R.B., 2011. Late Aptian to Turonian stratigraphy of the eastern Arabian Plate: depositional sequences and lithostratigraphic nomenclature. Pet. Geosci. 17, 211–222.

Voigt, S., Wiese, F., 2000. Evidence for Late Cretaceous (Late Turonian) climate cooling from oxygen-isotope variations and palaeobiogeographic changes in Western and Central Europe. J. Geol. Soc. London 157, 737–743.

Voigt, S., Gale, A.S., Flögel, S., 2004. Midlatitude shelf seas in the Cenomanian-Turonian greenhouse world: temperature evolution and North Atlantic circulation. Paleoceanography 19. https://doi.org/10.1029/2004PA001015.

Wang, W., Zhou, M., Chu, Z., Xu, J., Li, C., Luo, T., Guo, J., 2020. Constraints on the Ediacaran-Cambrian boundary in deep-water realm in South China: evidence from zircon CA-ID-TIMS U-Pb ages from the topmost Liuchapo Formation. Sci. China Earth Sci. 63, 1176–1187. https://doi.org/10.1007/s11430-019-9590-0.

Weissert, H., Breheret, J.G., 1991. A carbonate-isotope record from Aptian-Albian sediments of the Vocontian Trough (SE France). Bull. Soc. Geol. Fr. 162 (6), 1133–1140.

Wendler, J.E., Wendler, I., 2016. What drove sea-level fluctuations during the mid-Cretaceous greenhouse climate? Palaeogeogr. Palaeoclimatol. Palaeoecol. 441, 412–419.

Wiese, F., Voigt, S., 2002. Late Turonian (Cretaceous) climate cooling in Europe: faunal response and possible causes. Geobios 35, 65–77.

Woods, M.A., Wilby, P.R., Leng, M.J., Rushton, A.W.A., Williams, M., 2011. The Furongian (late Cambrian) Steptoean positive carbon isotope excursion (SPICE) in Avalonia. J. Geol. Soc. London 168, 851–862.

Zhu, M.Y., Babcock, L.E., Peng, S.C., 2006. Advances in Cambrian stratigraphy and paleontology: integrating correlation techniques, paleobiology, taphonomy and paleoenvironmental reconstruction. Palaeoworld 15, 217222.

CHAPTER THREE

Stable carbon isotopes in archaeological plant remains

Juan Pedro Ferrio[a,b,*], Mònica Aguilera[c,d], Jordi Voltas[c,d], and José Luis Araus[e,f]

[a]Aragon Agency for Research and Development (ARAID), Zaragoza, Spain
[b]Unidad de Recursos Forestales, Centro de Investigación y Tecnología Agroalimentaria de Aragón (CITA), Zaragoza, Spain
[c]Joint Research Unit CTFC—AGROTECNIO, Lleida, Spain
[d]Department of Crop and Forest Sciences, University of Lleida, Lleida, Spain
[e]Integrative Crop Ecophysiology Group, Plant Physiology Section, Faculty of Biology, University of Barcelona, Barcelona, Spain
[f]AGROTECNIO (Center for Research in Agrotechnology), Lleida, Spain
*Corresponding author: e-mail address: jpferrio@cita-aragon.es

Contents

Abstract

Archaeological plant remains are a valuable source of information about the social and environmental conditions experienced by past societies. Full exploitation of the informative potential of plant remains involves the use of analytical techniques.

Stratigraphy & Timescales, Volume 5
ISSN 2468-5178
https://doi.org/10.1016/bs.sats.2020.08.008

In this context, carbon isotope composition of archaeobotanical remains may give clues on the environmental conditions, agricultural management, food web, and dietary preferences, among others. This chapter provides a comprehensive view of the potential applications in the archaeology of analyzing the stable carbon isotope signature of plant remains. After a broad explanation of the physiological and environmental factors determining the carbon isotope signature in plants, most relevant methodological issues are discussed. Among others, the chapter offers a critical review of sampling and pre-treatment protocols, uncertainties associated with atmospheric changes and the potential alterations that plant remains may suffer prior or after deposition in the soil. Exemplified by selected case studies, methodological and conceptual clues are provided for the use of carbon isotopes in archaeology, and the still existing challenges highlighted.

1. General introduction

In this chapter, we will present the basic principles for the application of stable carbon isotope analysis in archaeological plant remains for the reconstruction of palaeoenvironments and ancient management practices. We will consider archaeological plant remains in a broad sense, covering what is generally defined as archaeobotanical remains (seeds, grains, fruits and other plant parts), as well as wood remains (generally charcoal, but also desiccated or waterlogged wood). Since we focus here on the techniques and materials used, we will not restrict our review to studies based on purely archaeological contexts, but also those assessing plant remains recovered from natural sediments.

1.1 What makes stable isotopes so useful for palaeoreconstruction?

Isotopes are chemical forms of the same element, which share the same number of protons (what defines the element), but have different numbers of neutrons (for a general introduction about stable isotopes, see e.g. Sharp, 2007). They can be divided into unstable (radioactive) and stable (non-radioactive). Unstable isotopes disintegrate spontaneously to form other isotopes (or elements, e.g. ^{14}C becomes ^{14}N when it decays), while emitting different forms of radiation. Conversely, stable isotopes do not decay over a geologic time scale. Stable isotopes, however, are anything but constant: due to their different mass and free energy, the proportion among different isotopes of the same element in different compartments is affected by multiple physical and chemical processes. This is what makes stable isotopes particularly interesting, since they can be used as tracers for these processes.

Consequently, stable isotopes were already proposed as physiological and environmental tracers during the pioneer works performed in the first half of the 20th century (Craig, 1954; Nier, 1940; Urey, 1932).

1.2 Carbon isotopes in plant matter as physiological and environmental tracers

One of the best examples of the potential of stable isotopes as biological tracers is the case of carbon, the main constituent of organic matter. The most abundant isotopes of carbon in the biosphere are the stable forms ^{13}C and ^{12}C, which occur in a molar ratio of ca. 1:99. The variation of the heavier (and less abundant) isotope is generally expressed in differential notation, as per mil (‰) change relative to an international standard. For carbon, carbon isotope composition ($\delta^{13}C$) refers to the relative abundance of ^{13}C, as compared to the reference material (currently the Vienna Pee-Dee Belemnite). Since carbon is involved in almost any biological reaction, $\delta^{13}C$ can be used as a record of a myriad of physiological processes, reflecting the environmental conditions in which organic matter was produced. In particular, the first studies on the carbon isotope composition in plants revealed that they had lower (i.e. more negative) $\delta^{13}C$ values than inorganic compounds, such as carbonates or the atmospheric CO_2 (see Fig. 1 for an overview of the variation among the most relevant organic and inorganic pools). This phenomenon is known as carbon isotope discrimination ($\Delta^{13}C$), as it reflects the discrimination against ^{13}C occurring in plants with regard the ^{13}C of the surrounding air:

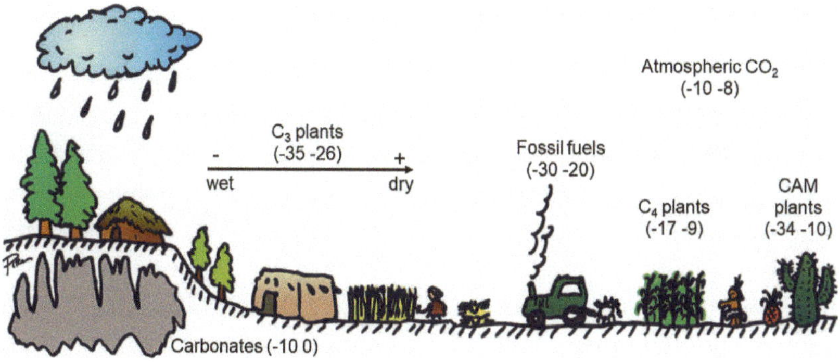

Fig. 1 Typical range of carbon isotopic composition ($\delta^{13}C$, in permil units) in different compartments of the biosphere and most relevant inorganic carbon pools.

$$\Delta^{13}C(‰) = \frac{\delta^{13}C_{air} - \delta^{13}C_{plant}}{1 + \delta^{13}C_{plant}} \quad (1)$$

Craig (1954) intuitively suggested that discrimination against the heavier ^{13}C took place in the leaves, and could be regulated by environmental conditions. Later on, Park and Epstein (1961) proposed a mechanistic model for $\Delta^{13}C$, accounting for the diffusion of CO_2, photosynthesis and secondary metabolism. These key steps, with a more detailed formulation, constitute the basis for the currently accepted models explaining $\Delta^{13}C$ in plants (Farquhar, 1983; Farquhar et al., 1982) (see Section 2 for further details).

1.3 Stable isotopes in archaeological plant remains: a historical perspective

Archaeological plant remains (either in charred, waterlogged or desiccated form) are a subproduct of plant tissues, retaining most of their original carbon. Consequently, plant remains should preserve some kind of physiological and environmental signal in their $\delta^{13}C$. A physiological effect on the isotopic composition of archaeobotanical remains was first reported by Bender (1968), who investigated the unexpectedly "young" ^{14}C dating of corn remains. This effect was attributed to the lower isotope discrimination associated with the C_4 photosynthetic pathway (see Section 2), which had been described 2 years before (Hatch and Slack, 1966). The first comprehensive studies on carbon stable isotopes in archaeological plant remains, however, were performed by DeNiro and coworkers in the 1980s (DeNiro and Epstein, 1981; DeNiro and Hastorf, 1985; Marino and Deniro, 1987). In addition to the analysis of a wide range of plant remains, they also performed a series of experiments to assess diagenetic effects, and demonstrated that the original isotopic signal of plants was generally preserved (see Section 3 for details).

In the meantime, the knowledge about $\delta^{13}C$ variation associated with plant metabolic pathways and environmental responses was consolidated (Farquhar, 1983; Farquhar et al., 1982, 1989; O'Leary, 1981). In particular, extensive research was performed on the physiological response of winter cereals (wheat and barley), aimed at the application of $\delta^{13}C$ as an integrative trait of water-use efficiency (or the ratio between dry matter production per unit water use) for plant breeding (Araus et al., 1993; Condon et al., 1987; Farquhar and Richards, 1984). These studies provided the theoretical framework for the application of $\delta^{13}C$ in archaeological grains to infer the water status of ancient cereal crops in the Western Mediterranean

(Araus et al., 1997a, b; Araus and Buxó, 1993) and the Near East (Araus et al., 1999a). These works stimulated the interest on the application of stable isotopes among the archaeobotanical community (see e.g. Fiorentino et al., 2014 and references therein).

In the last decades, the range of applications, as well as its geographic coverage, has grown steadily (Fig. 2), but we can still identify several knowledge

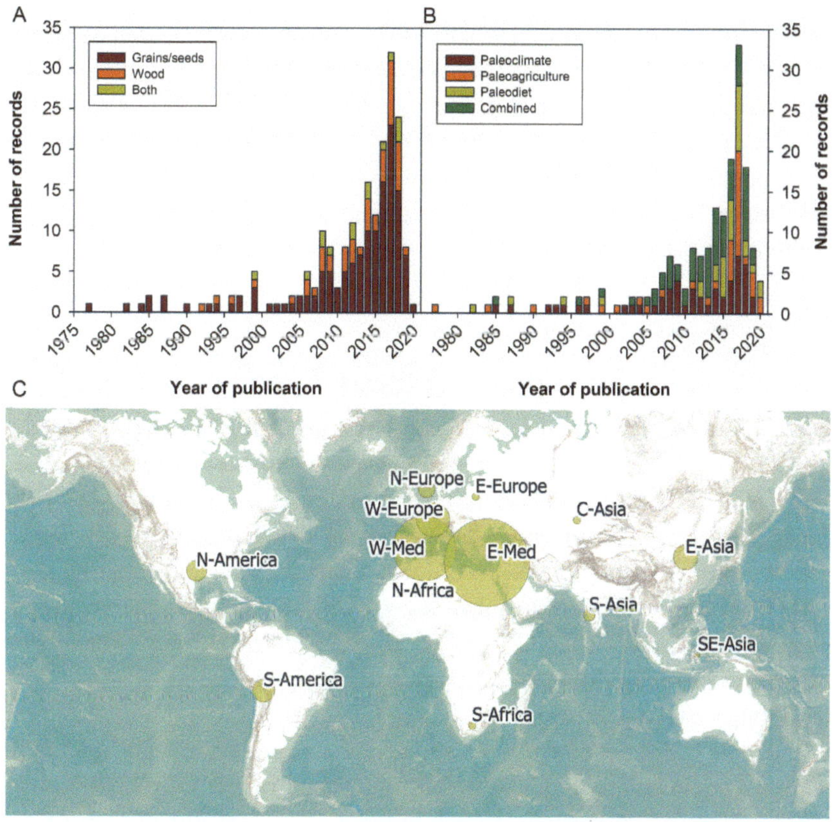

Fig. 2 Evolution over time and geographic coverage of published papers using carbon isotopes in archaeological plant remains. Upper panels display the number of papers published each year according to (A) the type of material analyzed and (B) the main aim of the study. The lower panel (C) shows the spatial coverage of the publications, with circle size proportional to the number of publications within each region. *Figures are based on a compilation of 180 publications, combining a literature search on different databases and our own archives, after excluding reviews not presenting new data. Map developed using QGIS 3.6.1-Noosa (https://qgis.org/es/site/), using the ESRI "World_Terrain_Base" layer as background (http://goto.arcgisonline.com/maps/World_Terrain_Base; sources: Esri, USGS, NOAA).*

gaps. For example, despite their widespread availability in archaeological contexts, the number of studies on charcoal is still limited (Fig. 2A), most likely due to the uncertainties regarding age and carbonization effects. It is also noticeable the progressive shift in the focus, from palaeoclimate to the study of ancient crop management, and the recent expansion in the use of reference plant material in palaeodietary studies (Fig. 2B). Regarding the region of study, the abundance of studies focusing on the Mediterranean (mainly focused on the origin and expansion of Near East agriculture) contrasts with other regions, which still remain largely underexplored (Fig. 2C). Notably, despite pioneer studies performed in the United States, Perú and South Africa (e.g. DeNiro and Hastorf, 1985; February and van der Merwe, 1992; Vogel and Van Der Merwe, 1977), articles focusing on the Americas and Africa are comparatively scarce. Conversely, Eastern Asia and Eastern Europe appear as rapidly growing areas of study, with most studies published in the last 5 years.

2. Theoretical principles

2.1 Key steps in photosynthesis: C_3 and C_4 pathways

Plants are able to fix atmospheric CO_2 through photosynthesis, and this process involves a key enzyme: Ribulose 1,5-bisphosphate carboxylase-oxigenase (RuBisCO). This enzyme can be found in the inner part of the leaves (mesophyll), inside small organelles (chloroplasts), where the actual carbon fixation (carboxylation) takes place. The enzyme converts one molecule of Ribulose 1,5-bisphosphate (1 × 5C sugar) into two molecules of 3-phosphoglycerate (2 × 3C=6C), with the addition of one carbon from CO_2. This enzyme, however, can also use O_2 to remove one carbon from the Ribulose 1,5-bisphosphate, releasing CO_2. Since CO_2 and O_2 compete for the same site in the enzyme, the efficiency of the enzyme is highly dependent on the availability of CO_2. Terrestrial plants are generally covered by a dense epidermis and a cuticle, which prevent water loss, but also limit CO_2 diffusion. As a consequence, CO_2 mainly enters the leaf through small, regulable pores in the epidermis, the stomata (Fig. 3). To overcome the limitation of RubisCO by CO_2, some plants have developed mechanisms to raise the internal CO_2 concentration through the involvement of an additional enzyme, phosphoenolpyruvate carboxylase (PEPC), also located in the mesophyll. PEPC adds CO_2 (in the form of HCO_3^-) to phosphoenolpyruvate (3C), to form a 4C acid (oxalacetate). Through different 4C intermediate compounds, this extra carbon is transported to the cells of the bundle sheath (the outer tissue of leaf veins), where it is released as CO_2,

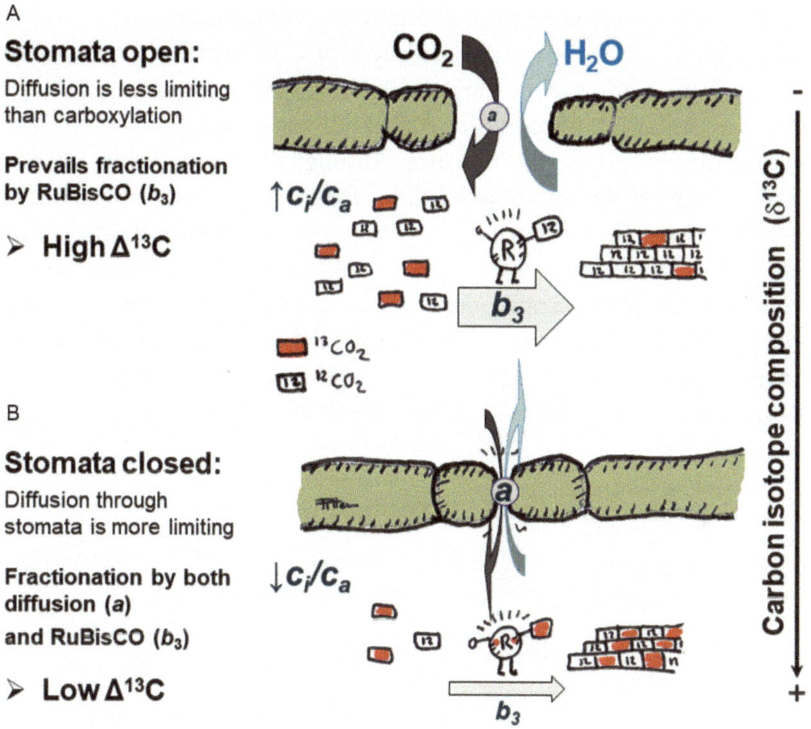

Fig. 3 Simplified scheme of the relationship between stomatal response and carbon isotope discrimination ($\Delta^{13}C$) in C_3 plants. (A) With high stomatal conductance, the CO_2 diffuses easily through the stomata, and the ratio between intercellular and air CO_2 concentration (c_i/c_a) increases, thus prevailing the fractionation by RuBisCO, which causes higher $\Delta^{13}C$. (B) When the stomata are closed, e.g. in response to drought stress, diffusion of CO_2 through the stomata becomes more limiting, and fractionation is intermediate between that caused by RubisCO (b_3) and diffusion through the stomata (a), leading to lower $\Delta^{13}C$. As reference, the arrow on the right shows changes in carbon isotope composition ($\delta^{13}C$), which are opposed in trend to $\Delta^{13}C$.

to be refixed by the RubisCO. In these plants, the bundle sheath is relatively air-tight, and therefore the internal CO_2 concentration available for RubisCO is much higher than compared with the plants which fix directly CO_2 through Rubisco in the mesophyll cells, therefore increasing the efficiency of CO_2 fixation, particularly under high temperatures. Plants that fix directly CO_2 through Rubisco are called C_3 plants, since the first compound fixed is a 3C molecule, whereas the ones involving PEPC, where the primary compound is a 4C molecule, are known as C_4 plants. All trees and most herbaceous plants from temperate regions are C_3, whereas C_4 herbaceous plants are mostly present in warmer tropical and subtropical climates.

As already advanced by Park and Epstein (1961) and modeled in detail by Farquhar et al. (1982), the diffusion of CO_2 through the stomata (and inside the leaf) and the carboxylation by RubisCO are the key processes determining carbon isotope discrimination ($\Delta^{13}C$) in C_3 plants, whereas in C_4 plants, fractionation through PEPC is also involved (Farquhar, 1983). RubisCO and PEPC display very distinct fractionation effects (ca. 29‰ and −5.7‰, respectively), which partly explains the contrasting $\Delta^{13}C$ values found in C_3 and C_4 species (see Fig. 1). Although this goes beyond the aim of this chapter, it should be noted that further fractionation occurs associated with respiration, and during the export and reconversion of primary sugars to different substances, what is broadly referred as post-photosynthetic fractionation (Badeck et al., 2005).

2.2 Carbon isotope discrimination in C_3 plants

According to the simplified model by Farquhar et al. (1982), $\Delta^{13}C$ in C_3 plants can be described as follows:

$$\Delta^{13}C(‰) = a + (b_3 - a)\frac{c_i}{c_a} \tag{2}$$

where c_a and c_i are the concentration (or partial pressure) of CO_2 in the air and inside the leaf, respectively, while a and b_3 stand for the fractionation factors due to diffusion through the air (4.4‰) and RuBisCO (30‰), respectively. As depicted in Fig. 3, this model implies that the $\Delta^{13}C$ of C_3 plants depends on the balance between these two fractionation steps. Due to CO_2 uptake by photosynthesis, c_i is lower than c_a, but when the stomata are fully open (Fig. 3A), the CO_2 diffuses easily through the stomata, and thus c_i approaches c_a, leading to higher c_i/c_a. Under such conditions, photosynthesis is less limited by CO_2 availability, and $\Delta^{13}C$ gets closer to the large fractionation caused by RuBisCO (b_3). On the contrary, when the stomata are closed (Fig. 3B), diffusion of CO_2 through the stomata is limited, and c_i becomes much smaller than c_a, leading to low c_i/c_a. Looking at Eq. (2), this means that the second term, i.e. the one including b_3, is reduced. Under such conditions, the (small) fractionation caused by diffusion (a) becomes comparatively more relevant than the (large) fractionation by RuBisCO, and plant $\Delta^{13}C$ is lower. Since stomata are sensitive to a number of environmental stressors, and particularly water shortage (Farquhar et al., 1989), in C_3 plants the $\Delta^{13}C$ of plant tissues is often positively correlated with water availability during the time they were formed (see e.g. Araus et al., 1999a, 2003; Ferrio et al., 2003).

2.3 Carbon isotope discrimination in C_4 and CAM plants

Farquhar (1983) proposed a similar simplified model for C_4 plants:

$$\Delta^{13}C(‰) = a + (b_4 + \Phi b_3 - a)\frac{c_i}{c_a} \quad (3)$$

where a, b_3, c_a and c_i stand for the same parameters as in Eq. (2), b_4 is the fractionation due to PEPC (including the dissolution of CO_2, −5.7‰) and Φ represents the fraction of CO_2 that leaks from the bundle sheath into the mesophyll. The leakiness factor Φ has a value of ca. 0.2–0.3 for most species under regular ambient conditions, but is species-specific, and has been shown to respond to environmental conditions, ranging from 0.04 to 0.70 (Henderson et al., 1992; Kromdijk et al., 2014), which complicates the interpretation of $\Delta^{13}C$ (Ellsworth and Cousins, 2016). Consequently, the term in Eq. (3) that is multiplied by c_i/c_a can be either positive or negative, and C_4 plants often (but not always) display opposite trends in $\Delta^{13}C$ in response to changes in stomatal conductance and water stress (Dercon et al., 2006; Williams et al., 2001). In any case, compared with C_3 plants, the environmental response of C_4 plants as registered in $\Delta^{13}C$ is much smaller, since the response associated with RuBisCO limitation by CO_2 is minimized.

According to the photosynthetic pathway, we can find a third group of plants showing Crassulacean acid metabolism (CAM). CAM plants are particularly adapted to arid environments because they can keep the stomata closed during the day, when the evaporative demand is higher. CAM plants capture CO_2 during the night with a mechanism similar to C_4 plants, also involving PEPC. The carbon fixed during the night is released in the form of CO_2 during the day, with the stomata closed, and is then refixed by RuBisCO. CAM plants generally show intermediate values between C_3 and C_4 plants (O'Leary, 1981; Vogel, 1993). Many CAM plants are facultative, and can behave as a typical C_3 or C_4 plants (i.e. opening stomata during the day), if conditions are favorable. This change is reflected in their $\delta^{13}C$, thus having the potential to preserve a environmental signal (Tankersley et al., 2016). Besides, environmental responses of $\delta^{13}C$ have been reported for some obligate CAM plants, such as pineapple (Medina et al., 1994). However, most CAM species are rarely preserved in archaeological contexts, and records of their $\delta^{13}C$ values measured in archaeobotanical material are very limited (DeNiro and Hastorf, 1985; Tankersley et al., 2016).

3. Methodological issues

In its early times, archaeology focused more on the collection of valuable items than on the reconstruction of past life conditions, and therefore plant remains did not attract particular attention. This changed with the emergence of ethnobotany, a concept that was first defined by Harshberger (1896), as the discipline aimed at the study of the ancient use of plants. More specifically, palaeoethnobotany has been later defined as the study of the interaction between ancient human groups and their surrounding vegetation (Helbaek, 1959), with a particular focus on plant remains from archaeological contexts (Renfrew, 1973) and, as a natural evolution, as the discipline of the study of the development of agriculture (Van Zeist and Casparie, 1984; Zohary and Hopf, 2000). With the development of palaeoethnobotany, and the broader concept of archaeobotany, the interest on plant macroremains increased, and more systematic sampling methods were implemented (Buxó, 1997; Greig, 1989). Although it is now possible to analyze $\delta^{13}C$ in pollen samples (Bell et al., 2017), we will focus here on plant macroremains, which are the most commonly used remains for isotope analyses.

3.1 Sample collection and preservation

Plant macroremains can be retrieved either dispersed through the sediment or within particular archaeological structures (e.g. hearths, storage pits) (see e.g. Fig. 4). The particular context where these remains are found is important for the interpretation of isotopic signals, since it can provide clues on whether they were naturally occurring on the site, or were purposely collected from elsewhere. The form of preservation is also relevant for isotope analyses, since their values may be altered due to diagenetic processes (see subsection below). In temperate regions, raw plant remains get degraded, and thus the usual form of preservation is carbonization. The most evident case is wood charcoal, which could be the result of unintended fires or the use of wood as fuel (e.g. in hearths). Grains, seeds and other plant material can also be burnt during food preparation, or resulting from the use of animal dung as fuel (Buxó, 1997; Van Zeist and Casparie, 1984). Carbonization takes place when plant materials are combusted under restricted oxygen supply, and due to the increase in C:N ratio and the formation of recalcitrant Maillard compounds, they become less prone to decomposition (DeNiro and Hastorf, 1985; Styring et al., 2013). Although chemically more stable, severely carbonized material becomes too brittle to be well preserved, and

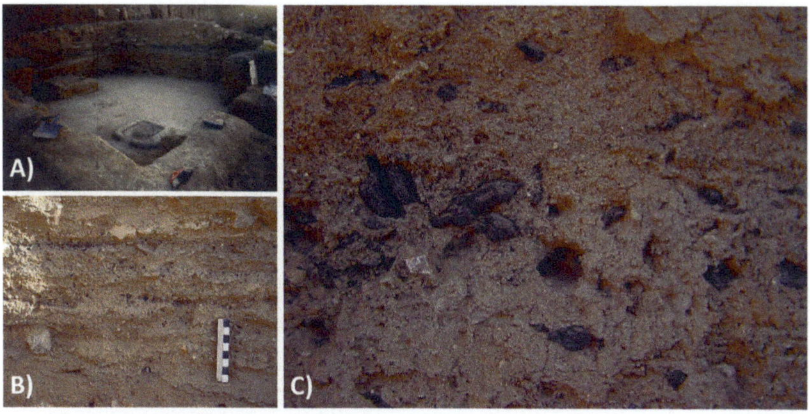

Fig. 4 A typical context where plant macroremains can be found. (A) Excavated house from the Pre-Pottery Neolithic B in Tell Halula (Middle Euphrates Valley, Syria), showing a rectangular hearth in the center. (B and C) Plant remains embedded in the sediment in Tell Halula. *Sources: J.P. Ferrio; J.L. Araus.*

deformation of plant remains (e.g. "pop-corn" effect in the case of seeds) may make their identification more difficult (Ferrio et al., 2004; Wright, 2003). Plant remains can be also preserved as mineralized tissues, although in this case the original organic carbon is lost. In arid environments, like ancient Egypt or coastal Peru, desiccated plant material can be preserved without carbonization during millennia (DeNiro and Hastorf, 1985). In wetter climates, this process may occur in dry caves, protected from rainfall. In the other extreme, plant remains can be preserved in saturated state in bogs, swamps or lake sediments. Water-logged wood, for example, can be very frequent and well preserved in lakeshore sites, where it could be used as construction material (Aguilera et al., 2011).

Large samples can be directly recovered during excavation, but smaller fragments and seeds require specialized retrieval from the sediment. The sediment can be screened dry, but most commonly wet screening or flotation methods are used (Fig. 5). Flotation techniques take advantage of the aptitude of plant remains to float in water, which allows for easy separation from the heavier mineral sediment. Generally, flotation systems include a flotation tank, where the sediment is placed over a mesh, and gently covered with water from beneath, in a way that less dense material gets out from the tank through a spout. Floated material is then further processed through different mesh sizes. In parallel, dense materials (e.g. bone fragments) are then recovered from the bottom of the tank. Screening of sediment, and subsequent identification of plant material, is a very time-consuming process, therefore

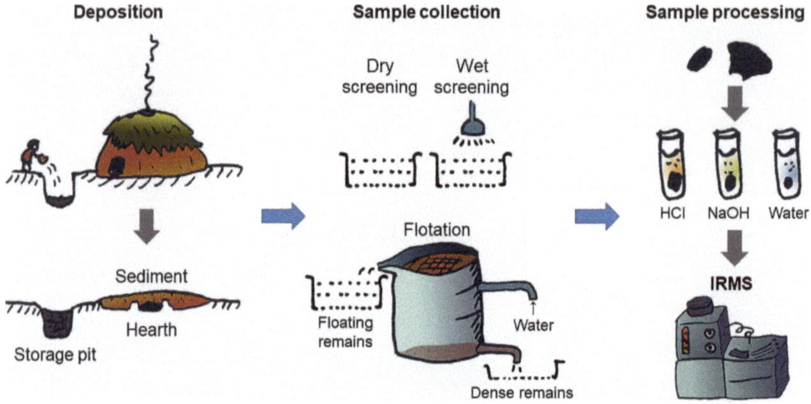

Fig. 5 Scheme summarizing the key steps involved in the analysis of stable isotopes in archaeological plant remains. Samples may be deposited in different environments, either in specific archaeological structures, or dispersed in the sediment. To recover dispersed samples from the sediment, dry or wet screening are used, the latter including the flotation method (see text for details). Recovered samples may be pre-treated to remove carbonates and humic acids, soaking with strong acid/base, although this also involves a substantial sample loss, as represented in the drawing. Once treated, samples are thoroughly rinsed with water to remove chemical agents, ground to fine power and encapsulated for isotope analysis with isotope ratio mass spectrometry (IRMS).

it cannot be performed in the whole sediment. Instead, different subsampling procedures are generally implemented, with sample size determined by the abundance of plant remains and the particular aims of the excavation (Alonso et al., 2000; Buxó, 1997).

Besides archaeological considerations, the number of samples analyzed should also be enough to avoid biased results. Heaton et al. (2009) evaluated the $\delta^{13}C$ of multiple grains from storage jars, and compared their variability with that found among wheat grains growing in a present-day agricultural field. They estimated that, for a sample size of five grains, differences smaller than 0.7‰ between contexts might be not detectable. Similarly, Riehl et al. (2014) examined the intra-field $\delta^{13}C$ variability in barley grains, and concluded that a sample size of 6–10 individual grains should be analyzed to be representative of an archaeological context. In the case of charcoal, one additional uncertainty is the existence of the "canopy effect," which causes lower $\delta^{13}C$ values for trees growing in closed canopies (Bonafini et al., 2013), and is also responsible for within-tree variations, e.g. with tree height (Hubbard et al., 2002; Voltas et al., 2013). On top of that, a "juvenile effect," also causing more depleted $\delta^{13}C$ values, is generally recognized in tree-ring studies (Leavitt, 2010; Marshall and Monserud, 1996) and confirmed in archaeological samples (Philippsen et al., 2019). For species with very sensitive stomatal response, e.g. pines, the amplitude of environmental effects on $\delta^{13}C$ is large enough to overcome intra-canopy variability (Ferrio et al., 2003, 2006b). However, for less responsive species, such as oaks, juvenile effects and other site-related influences may preclude conclusive results (Aguilera et al., 2011; Araus et al., 2014; Ferrio et al., 2003). In this regard, it is advisable to use anatomical indicators such as tree-ring curvatures for the interpretation of $\delta^{13}C$ in charcoal, to infer potential age and position-related effects of the samples (Paradis et al., 2013).

3.2 Dating of samples

An accurate dating of samples is crucial not only for the interpretation of results in the proper chronological context, but also to account for past $\delta^{13}C_{air}$ changes (see below). In the timescale covered by archaeology, radiocarbon dating is the most suitable dating technique (Nowak et al., 2017; Taylor and Aitken, 1997). ^{14}C is formed in the atmosphere as a consequence of the impact of cosmic ray neutrons on ^{14}N and oxidized to produce $^{14}CO_2$, which can be fixed by plants, as described in previous sections. Once the plant dies no further $^{14}CO_2$ is uptaken, and the remaining ^{14}C in plant tissues starts to decay over time. First studies by Arnold and Libby (1949) established the half-life of ^{14}C (i.e. the time required to reach 50% of the original ^{14}C content) in 5568 ± 30 years, although this was later corrected to the currently accepted value of 5730 ± 40 years (Godwin, 1962). The methods for ^{14}C determination have rapidly evolved since then. The solid carbon method developed by Libby's lab was already replaced by gas counting and liquid scintillation counting (LSC) during the 1950s. These two radiometric methods are based on the direct counting of beta-decay events (conversion of ^{14}C to ^{14}N), and have been widely used in radiocarbon laboratories, although they are limited by the need for large sample sizes, typically about 100 mg for accurate dating. In the late 1970s, a new method, Accelerator Mass Spectrometry (^{14}C-AMS) was developed, based on direct counting of ^{14}C atoms, instead of beta-decay events. This is currently the preferred method for archaeological material, since it allows dating tiny amounts (<0.5 mg) of valuable ancient samples (Fewlass et al., 2018; McIntyre et al., 2017; Nowak et al., 2017). Regardless of the method used, radiocarbon dating must be calibrated to account for secular changes in atmospheric ^{14}C composition, leading to variable uncertainties over time (Nowak et al., 2017; Reimer et al., 2013).

Due to its elevated cost, ^{14}C dating is normally restricted to a few selected specimens within each stratigraphic layer or archaeological structure, and is often combined with stratigraphic or cultural approaches. The actual age of the plant material analyzed is of high relevance for the purpose of environmental reconstruction, but it is also important to place it in a particular cultural context. For this reason, archaeologists normally prefer short-aged material such as grains or fruits for dating, since wood fragments may be significantly older than the sediment in which they are embedded (Nowak et al., 2017). However, one additional risk of dating based on small plant remains is that these remains can be displaced from their original sediment layers, and thus may appear outside their depositional context, leading to dating errors (Caracuta et al., 2014). Besides, physiological fractionation affects ^{14}C in a similar way to ^{13}C and, accordingly, ^{13}C can be also determined alongside with radiocarbon dating. Since physiological responses are often species-specific, a proper identification of dated material is particularly recommended (Bender, 1968; Nowak et al., 2017; Taylor and Aitken, 1997).

3.3 Sample preparation and isotope analysis

Archaeological plant samples are buried in the sediment, and thus often present contamination from different soil substances. Typical soil contaminants are carbonates in basic soils, and humic and fulvic acids in acidic soils. The early work from DeNiro and Hastorf (1985) extensively studied isotopic effects of soil contaminants on archaeological plant remains. After testing different alternatives, the authors proposed an acid-base-acid chemical wash-down in order to remove most common soil contaminants, as follows. First, carbonates are removed with a strong acid treatment by soaking the samples in HCl 6 M for 24 h (if large carbonate crusts are visible, it may require longer time). For the removal of humic substances, samples are then rinsed with distilled water until achieving a neutral pH, soaked in 1 M NaOH for additional 24 h, rinsed again and soaked in 6 M HCl for 10 min, then rinsed. If the aim is to analyze nitrogen isotopes, additional treatments are recommended in order to remove exchangeable ammonium and other nitrogen forms bound to silicates (see DeNiro and Hastorf, 1985 for details). Further studies have confirmed that this protocol generally reduces measurement uncertainty without significant alteration of the original $\delta^{13}C$ signature of charred material, but harsh treatments could affect $\delta^{15}N$ under certain conditions (Styring et al., 2013; Vaiglova et al., 2014). Likewise, intensive cleaning may imply a substantial sample loss. Brinkkemper et al. (2018) reported an average loss of ca. 50% with the acid-base-acid treatment, which is in agreement with our own experience. Therefore, unless strong evidence of recalcitrant soil contamination is observed, it is advisable to apply a more conservative protocol, e.g. using whole fragments instead of ground material, and limited to the acid treatment (Aguilera et al., 2008; Ferrio et al., 2006a). In this regard, Brinkkemper et al. (2018) confirmed that, for a wide range of preservation environments, a simple acid treatment, or even water cleaning, could provide comparable $\delta^{13}C$ values, while significantly reducing sample loss.

For isotope analysis, samples are reduced to a fine powder, encapsulated in tin capsules and then combusted. The combustion products are measured with an elemental analyzer (a gas chromatograph, GC) coupled to an isotope ratio mass spectrometer (IRMS) through a gas chromatography column (see Sharp, 2007). The grinding process is important to ensure a good homogenization of the sample, as well as to facilitate the full combustion of the material, since incomplete combustion could lead to fractionation of the resulting CO_2. What is actually measured by the IRMS is the ratio between isotopologues (i.e. molecules with different isotope composition) of the gases resulting from combustion: for carbon, typically $^{13}CO_2$ and $^{12}CO_2$ (molecular masses of 45 and 44). Due to the similar measurement requirements, it is possible to obtain the isotopic composition of nitrogen ($\delta^{15}N$) together with $\delta^{13}C$. The minimum amount of sample required depends on the equipment used and the elemental composition of the sample. Plant organic matter generally has about 40–45% of carbon, but this increases with carbonization—most often ranging between 50% and 60% in charred plant remains (see e.g. Aguilera et al., 2008; Ferrio et al., 2006b). Generally 0.3–0.5 mg is enough for charred plant remains if only $\delta^{13}C$ is measured (see e.g. Vignola et al., 2018), but for measurements including $\delta^{15}N$ this amount should be increased to at least 1 mg.

One alternative to the measurement of single samples to standard GC-IRMS is the use of microsampling techniques such as microdissection or laser ablation, which have been successfully applied in tree rings (Rinne et al., 2015; Schollaen et al., 2014). This is particularly promising for charcoal samples, offering the potential to assess intraannual variation, provided that individual tree rings can be identified (Baton et al., 2017). Besides IRMS, $\delta^{13}C$ can be also determined by AMS simultaneously with ^{14}C (Drake et al., 2012; Fiorentino et al., 2008; Fiorentino et al., 2010). This is because ^{13}C is measured together with ^{14}C during AMS dating, as ^{13}C is used to correct for isotope fractionation effects on ^{14}C. Besides being a rather expensive alternative for determining $\delta^{13}C$, AMS is not optimized for the measurement of $^{13}C/^{12}C$ ratios, and requires additional sample preparation, which has prevented a routinary use of this approach. Instead, independent IRMS and AMS measurements of subsamples (Caracuta et al., 2016; Chen et al., 2018; Vignola et al., 2017, 2018) or, more recently, the possibility to perform coupled IRMS and AMS measurements (McIntyre et al., 2017) allow for combining precise dating with accurate $\delta^{13}C$ determination.

3.4 Uncertainties associated with changes in atmospheric CO_2

The carbon present in plants is primarily derived from atmospheric CO_2. Current mean $\delta^{13}C$ of atmospheric CO_2 ($\delta^{13}C_{air}$) is about −8.5‰, although this value is steadily decreasing due to the contribution of (isotopically-depleted) fossil fuel emissions (Keeling et al., 1979). Nowadays, $\delta^{13}C_{air}$ shows substantial spatial variation, partly due to the uneven distribution of anthropogenic CO_2 sources. However, we cannot discard comparable spatial variability in the past, since this variability is also determined by ocean and terrestrial carbon cycles (Graven et al., 2017; Keeling et al., 2017). $\delta^{13}C_{air}$ also shows consistent seasonal cycles, with minima during winter and maxima during summer, as a result of seasonal changes in vegetation activity. The magnitude of this variation is particularly strong in the Northern hemisphere (ca. 0.7–0.8‰), due to the greater proportion of emerged lands and, therefore, larger amounts of vegetation in medium to high latitudes, which are most affected by seasonal cycles. This variation may introduce additional uncertainty related to, for example, the comparison of plant material formed during different periods (e.g. winter vs. summer crops).

In the archaeological context it is necessary to account for the aforementioned changes over time, particularly the large $^{13}CO_2$ depletion observed since the onset of the industrial revolution (Fig. 6A). Fortunately, the CO_2 trapped in small air bubbles within ice cores have allowed to reconstruct past changes in $\delta^{13}C_{air}$ over millennia (Bender et al., 1997; Francey et al., 1999; Indermühle et al., 1999; Leuenberger et al., 1992). Antarctic ice cores are the most suitable reference values of $\delta^{13}C_{air}$, since they provide very long records and are less prone to seasonal variations than those from the northern hemisphere (Bender et al., 1997). Ice cores provide a good estimate of long-term trends at a global scale but their accurate dating is challenging, while they cannot account for local variations. In this regard, the use of local references has been proposed based on C_4 plants (Chen et al., 2018; Marino et al., 1992; Toolin and Eastoe, 1993), under the assumption that their

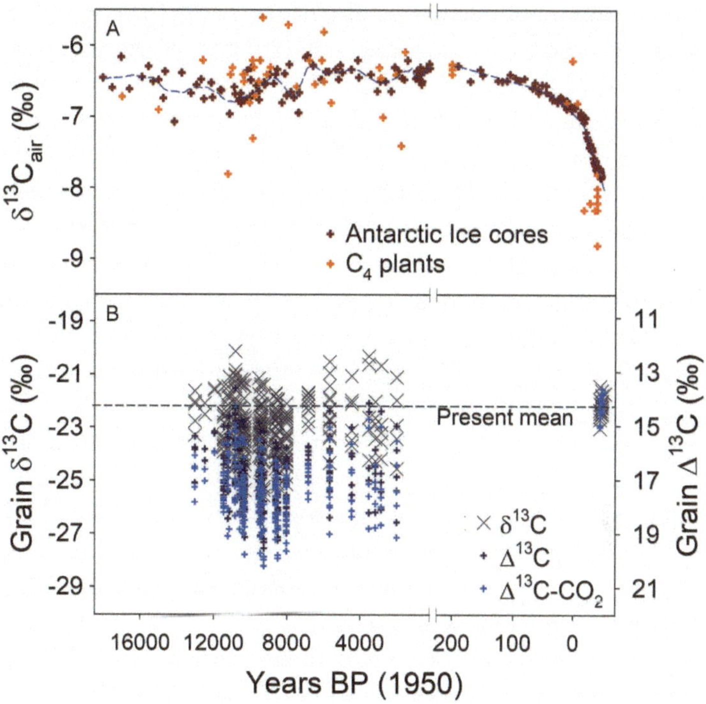

Fig. 6 (A) Long-term variation in the carbon isotope composition of atmospheric CO_2 ($\delta^{13}C_{air}$), according to different proxies: antarctic ice cores (Francey et al., 1999; Indermühle et al., 1999; Leuenberger et al., 1992) and C_4 plants (Marino et al., 1992; Toolin and Eastoe, 1993). Dashed line shows a smoothed LOESS curve used for the calculation of carbon isotope discrimination ($\Delta^{13}C$) in archaeological material, as reported in Ferrio et al. (2005). (B) Example of the use of smoothed air $\delta^{13}C$ values for the calculation of $\Delta^{13}C$ in archaeological cereal grains (data from Araus et al., 2014), and the correction for lower CO_2 concentration in the past ($\Delta^{13}C$-CO_2), as proposed by Mora-González et al. (2019).

$\Delta^{13}C$ is more or less constant, regardless of growing conditions. In higher latitudes, where C_4 plant remains are rare, long-term records available from peat mosses may be a good alternative (White et al., 1994). However, these records hold some uncertainty due to physiological imprints on $\delta^{13}C$, since both C_4 plants and mosses react to some extent to different environmental factors, such as light (Buchmann et al., 1996), temperature (Yasuda et al., 2015) or water availability (Rice and Giles, 1996; Williams et al., 2001). Considering the various spatial and dating uncertainties associated with past $\delta^{13}C_{air}$ estimates, variations smaller than ca. 0.5‰ in the archaeological record should be interpreted with caution.

Although simple subtraction or alternative corrections to $\delta^{13}C$ values have been proposed, e.g. for tree-ring records in northern latitudes (McCarroll et al., 2009), calculating $\Delta^{13}C$ (according to Eq. 1) is physiologically more meaningful. An example of the need to account for changes in $\delta^{13}C_{air}$ is shown in Fig. 6B. The $\delta^{13}C$ of cereal grains from different archaeological sites in the Near East (Araus et al., 2014) showed values comparable to those of current wheat and barley grown in the region. Without any correction, it could be mistakenly interpreted that growing conditions in the past were similar to present time. However, since current values of $\delta^{13}C_{air}$ are lower than the typical values during the Holocene, the resulting $\Delta^{13}C$ was on average ca. 2‰ higher in the past, thus indicating more favorable growing conditions (Fig. 6B).

Besides changes in $\delta^{13}C_{air}$, different studies have shown a dynamic response of $\Delta^{13}C$ to changes in atmospheric CO_2 concentration (c_a in Eq. 2), although the number of studies focused on subambient conditions (as those prevailing during the Holocene) is still scarce. For example, Polley et al. (1993) studied the response of wheat in a range of c_a from 200 to 350 ppm, and found that the ratio c_i/c_a was positively correlated with c_a. Considering Eq. (2), we would then expect, for the same environmental conditions, comparatively higher $\Delta^{13}C$ values at present (>400 ppm) than during the rest of the Holocene, prior to the Industrial Revolution (ca. 270 ppm). In other words, using present $\Delta^{13}C$ values to interpret ancient growing conditions, e.g. water availability, past $\Delta^{13}C$ would be lower than expected, thus leading to an underestimation of water availability to plants. A recent study by Hare et al. (2018) have evidenced that past values of plant $\delta^{13}C$ were affected by changes in c_a since the last deglaciation, even after accounting for changes in $\delta^{13}C_{air}$ and other environmental proxies. The effect was apparently stronger in gymnosperms than in angiosperms, showing a depletion over the entire range (ca. 180–400 ppm) of ca. −1.4‰ and ca. −0.5‰, respectively. In this context, Mora-González et al. (2019) proposed the application of the models by Polley et al. (1993) to correct present $\Delta^{13}C$ values by −0.61‰, assuming ca. 100 ppm of difference, in order to make them comparable with archaeological material. As an exercise to better understand the potential implications of this approach, we applied these corrections to data from Araus et al. (2014) by taking records from Taylor Dome, Antarctica (Indermühle et al., 1999) as reference for past CO_2 and a present (1994–2003) value of 370 ppm. To ease interpretation, we adjusted past values to the present, rather than the opposite, i.e. $\Delta^{13}C\text{–}CO_2 = \Delta^{13}C + 0.0061 \times (370\text{-}c_a \text{ past})$ (Fig. 6B). Considering these changes, the differences between present and past values would be exacerbated, further supporting a wetter environment in the past.

3.5 Effects of carbonization on $\delta^{13}C$

Although carbonization allows the preservation of archaeological plant remains, the complex chemical reactions involved during charring are likely to affect their $\delta^{13}C$. Pioneer studies from DeNiro and coworkers (DeNiro and Hastorf, 1985; Marino and Deniro, 1987) analyzed the effect of different temperature treatments on a variety of seeds and plant parts, concluding that changes in $\delta^{13}C$ were negligible. Further studies from our group (Aguilera et al., 2008; Araus and Buxó, 1993; Araus et al., 1997b; Ferrio et al., 2007) have shown that $\delta^{13}C$ of wheat and barley grains do not change significantly with carbonization, over a range of temperatures and experimental conditions. Similarly, other experiments have reported no significant effects on grain $\delta^{13}C$ of crops such as millets (Fraser et al., 2013a; Yang et al., 2011), einkorn (Styring et al., 2013), emmer or lentil (Fraser et al., 2013a). Conversely, Poole et al. (2002) found erratic changes in $\delta^{13}C$ of pea, whereas Filipović et al. (2018) reported a consistent isotopic enrichment in cherry fruit stones. As pointed out by Ferrio et al. (2007), and later by Nitsch et al. (2015), changes in $\delta^{13}C$ with carbonization are only meaningful under moderate charring conditions allowing the preservation of remains in archaeological contexts. Under harsher conditions, plant material becomes too brittle to be preserved during deposition, or too distorted to be identified. Considering multiple cereal and legume species, and within the range preventing strong distortion, Nitsch et al. (2015) concluded that a significant offset of +0.11‰ for $\delta^{13}C$ occurs as a consequence of carbonization. If existing, this offset is virtually negligible in archaeobotanical studies: it falls within the analytical error of IRMS, and is about 5–10 times smaller than the uncertainty associated with either $\delta^{13}C_{air}$ variability or within-field heterogeneity. In conclusion, and at least for most common cereal grains and pulses, there is no clear evidence that carbonization causes systematic changes in $\delta^{13}C$, although it might constitute an additional source of uncontrolled variability. Besides carbonization, other diagenetic processes associated with burial conditions might also affect the isotopic signature. DeNiro and Hastorf (1985), for example, exposed different kinds of modern plant material (either untreated or carbonized) to microbial decomposition during 4 months. They found, in some cases, a substantial depletion in $\delta^{13}C$ of untreated plants (ca. −5‰), but not clear effects on the $\delta^{13}C$ of charred material. Similarly, Fraser et al. (2013a) and

Styring et al. (2013) did not find consistent post-burial changes in the $\delta^{13}C$ of modern charred grains over 24 months. Overall, these studies confirm that carbonization prevents further alteration of the $\delta^{13}C$ due to microbial degradation. Due to the larger time scale of diagenesis we cannot discard other long-term process, but these are most likely associated with the inclusion of different minerals, which could be removed with the standard preparation methods described above.

In the case of charcoal, charring experiments have resulted in contrasting results. As reviewed in Resco et al. (2011), pine species showed a consistent depletion in $\delta^{13}C$ with increasing charring temperature, to the point that they could be corrected by accounting for concomitant changes in carbon content (Ferrio et al., 2006b). Conversely, broadleaves showed more erratic trends. By exposing several oak and *Pistacia* species to a range of temperatures varying between 300 and 500°C, Ferrio et al. (2007) found no systematic changes with carbonization, in agreement with Hall et al. (2008), but contrasting with the ca. 2‰ depletion reported by Turney et al. (2006) for *Quercus robur*. Interestingly, Philippsen et al. (2019) observed consistent differences between archaeological samples of wood and charcoal, but showing opposite trends to those found in carbonization experiments (on average, ca. 3‰ lower $\delta^{13}C$ in wood than in charcoal). However, it is likely that this offset was mainly caused by the larger diagenetic alteration of non-charred wood, rather than by the effect of carbonization (Aguilera et al., 2011; Poole et al., 2004). Overall, although there is enough indirect evidence to conclude that an environmental signal is consistently preserved in $\delta^{13}C$ of wood charcoal (Aguilera et al., 2012; Baton et al., 2017; February and van der Merwe, 1992; Ferrio et al., 2006b; Masi et al., 2013; Vernet et al., 1996; Vignola et al., 2018), species-specific assessments of carbonization effects are needed to discard any systematic bias.

4. Applications

4.1 Palaeoclimate reconstruction

The reconstruction of past climates based on $\delta^{13}C$ has been traditionally performed on a variety of fossil and subfossil plant tissues, from tree-ring chronologies to long-term records in peat mosses (Becker et al., 1991; Trimborn et al., 1995; White et al., 1994). Compared with high-resolution and (more or less) continuous records such as ice cores, lake sediments or tree-ring chronologies, archaeological plant remains have the disadvantage that they can only deliver "snapshots" of the past. However, they provide local information for areas where other palaeoenvironmental records are not available. February and van der Merwe (1992) were the first to show that the $\delta^{13}C$ of charcoal from archaeological sites of South Africa, covering a time span of 4000 years, could be used as environmental proxy. By comparing the $\delta^{13}C$ variation of extant wood along a precipitation gradient, they interpreted the isotopic variation in charcoal as evidence of changes in past precipitation. Similarly, Vernet et al. (1996) reported substantial fluctuations in the $\delta^{13}C$ of wood charcoal from Southern France, which were in agreement with other palaeoenvironmental proxies. Using modern wood as a reference for calibration, Ferrio et al. (2006b) developed a quantitative model for the estimation of changes in annual precipitation (Fig. 7). The model took advantage of the strong $\Delta^{13}C$ response to precipitation of Aleppo pine (Ferrio et al., 2003), but also accounted for changes in $\delta^{13}C_{air}$ and the carbonization effects described above. With this information, the authors could reconstruct past precipitation changes over c. 4000 years in NE Spain, showing in general wetter conditions in the past, but a dry event coinciding with the cold iron-age, and maximum precipitation, but high variability, during the Medieval Warm Period. These results are in contrast with the dry conditions reported by Drake et al. (2012) in South-Western Iberia during the Medieval Warm Period based on the analysis of miscellaneous charcoal data. Spatial variability of climate could explain these contrasting patterns. In this regard, Aguilera et al. (2009) showed contrasting responses in autumn precipitation (estimated from oak charcoal $\Delta^{13}C$) during the Iron Age Cold Epoch along a latitudinal gradient in the Eastern Iberian Peninsula. They reported drier than present conditions in the South, but wetter conditions in the North. Interestingly, the wetter autumn conditions (based on Holm oak $\Delta^{13}C$) during the Iron Age Cold Epoch contrasted with the apparently drier conditions suggested by the $\Delta^{13}C$ of Aleppo pine (Ferrio et al., 2006b). The contrasting physiological responses of oaks and pines may provide an explanation for these results. Aleppo pine is a rather thermophilous species, which mostly relies on precipitation pulses during the growing season, whereas Holm oak is better adapted to continental climates, and takes advantage of deep water reservoirs, thus being more sensitive to autumn precipitation (Del Castillo et al., 2013). Hence, a multispecies approach can provide insights into past climate seasonality. In this context, Aguilera et al. (2012) took advantage of the differential response of Aleppo pine, Holm oak and cereal grains to seasonal precipitation for reconstructing past changes in precipitation seasonality. As a leap forward in the application of multispecies approaches, Jones (2018) combined the

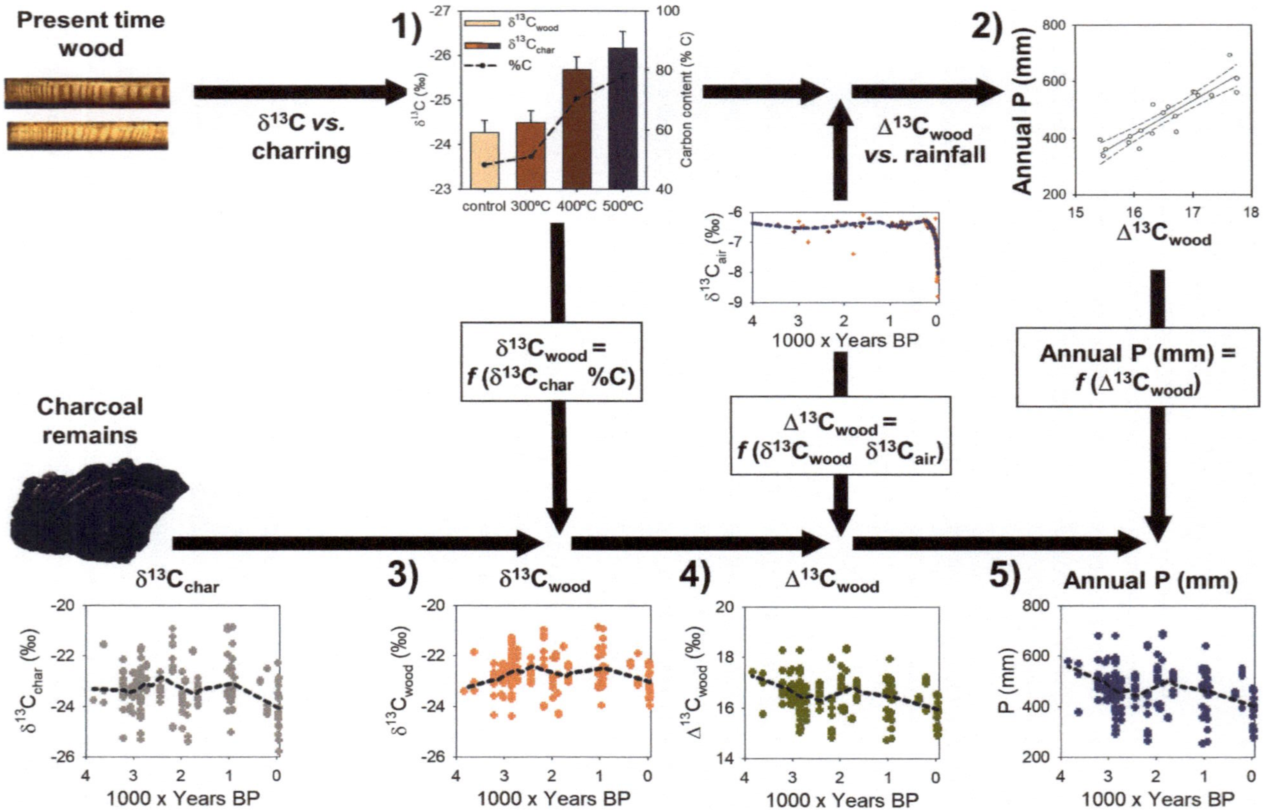

Fig. 7 Main steps in the estimation of past precipitation from the carbon isotope composition of wood charcoal ($\delta^{13}C_{char}$), according to Ferrio et al. (2006b). Extant wood is used (1) to quantify isotopic fractionation during carbonization and (eventually) correct it, based on the associated change in carbon content (%C) and (2) to model the relationship between carbon isotope discrimination ($\Delta^{13}C_{wood}$) and climate. The $\delta^{13}C_{char}$ is then (3) corrected for the effect of carbonization, estimating the original value ($\delta^{13}C_{wood}$), and (4) converted to $\Delta^{13}C_{wood}$ by accounting for past changes in $\delta^{13}C$ of atmospheric CO_2 ($\delta^{13}C_{air}$). Finally, (5) changes in mean annual precipitation are estimated from $\Delta^{13}C_{wood}$.

analysis of $\delta^{13}C$ in barley grains with the analysis of oxygen isotope composition ($\delta^{18}O$) in faunal remains in order to reconstruct changes in monsoonal activity and climate in the Indus Valley. By comparing early and late-wood $\delta^{13}C$ in a Neolithic settlement of the Jura range (France), Baton et al. (2017) observed that the oldest phases were colder and wetter, but also showed a stronger seasonality, in agreement with other regional palaeoenvironmental evidences. In this regard, the combination of intraannual tree-ring isotopic measurements and multispecies approaches might provide an unique overview of past climates.

4.2 Ancient crop management

One of the most interesting applications of stable isotopes in archaeobotanical remains ($\delta^{13}C$ in particular) is the reconstruction of growing conditions of crops, such as the potential deployment of irrigation practices and, in case of rain-fed cultivation, the trends in precipitation through time. These techniques have attracted particular attention aiming to elucidate crop status at the onset of agriculture (Araus et al., 2014; Ferrio et al., 2011; Fiorentino et al., 2014; Riehl, 2008; Wallace et al., 2013). Agriculture emerged in different areas of the world at the beginning of the Holocene, which is considered the most stable warm period in the last 400,000 years. The onset of agriculture took place in a context of rapid climate changes, from the dry-cold Younger Dryas to the warmer (and presumably wetter) early Holocene (Ferrio et al., 2011; Willcox et al., 2009). However, what pushed hunter-gatherers to settle down and begin cultivation, and, most intriguing, what made agriculture so successful as to be rapidly promoted in different areas of the world are still under debate (Balter, 2007; Ferrio et al., 2011; Fuller, 2007; Willcox et al., 2009).

4.2.1 Estimation of crop water availability

Araus and Buxó (1993) proposed for the first time the analysis of $\delta^{13}C$ in archaeological cereal grains as a way to reconstruct their water status. Subsequent works (Araus et al., 1999a, 1999b; Ferrio et al., 2005) further refined these models to attain a quantitative estimation of the water inputs received during grain filling (the period in which the grains are formed, i.e. late spring). The general validity of the quantitative models proposed by Araus and coworkers, however, has been recently questioned (Flohr et al., 2019; Wallace et al., 2013). In particular, Wallace et al. (2013) raised the need to consider temperature regime, and thus water demand, for estimation purposes and proposed a more conservative, qualitative inference.

On the other hand, Flohr et al. (2019) showed a comparatively weak response to different water regimes of $\Delta^{13}C$, particularly for wheat. Flohr et al. (2019) based their conclusions on the crop response to a tightly-regulated weekly drip-irrigation regime, where a different percentage of the crop optimum requirement was applied (Allen et al., 1998). Although this experimental set-up is hardly comparable with current (and traditional) agronomic practices—where irrigation is less frequent but more abundant—this study evidences that the quantification of water inputs based on $\Delta^{13}C$ is not straightforward, and may fail under certain conditions.

Another potential limitations of these models is that modern cultivars may show different environmental response, being less conservative and having shorter growing cycles than ancient crops (Araus, 2002). Studies comparing landraces and traditional cultivars with modern (>1970) cultivars have reported consistently higher $\Delta^{13}C$ values in the latter, between 0.3 and 1.2‰ on average, depending on the geographic origin and growing conditions (Araus et al., 2007, 2013; Royo et al., 2008). Another general trend in these studies is that modern material would show higher $\Delta^{13}C$ under more favorable conditions. Although this deserves further investigation, these studies suggest that applying modern reference data to past records may underestimate water availability, particularly during wet events. This would reinforce previous reports of wetter conditions in the past than in the present, but might have exacerbated the magnitude of arid periods. In this regard, quantitative models (e.g. Araus et al., 1999a) and proposed thresholds for qualitative assessments (e.g. Wallace et al., 2013) may need revision to account for the differential response of old genotypes.

4.2.2 Identification of water management practices

Bearing in mind these limitations, the use of $\delta^{13}C$ as a crop-based proxy offers the possibility to compare crop growing conditions for different species or under different climates for the inference of past agronomic practices. For example, Araus et al. (1997a) compared the water status of cereals (durum wheat, hulled and naked barley) and faba bean in SE Spain, and found evidence of independent cultivation of faba bean, but no clear differences in water availability among cereals. Similarly, Heaton et al. (2009) found consistent $\delta^{13}C$ covariation between einkorn and durum wheat grains recovered from storage jars in Greece, suggesting that these two cereals were cultivated under comparable conditions. Plotting the $\Delta^{13}C$ of barley and wheat over a range of sites from Spain, Ferrio et al. (2005) identified a number of outliers from the regression line, where the $\Delta^{13}C$ of wheat was higher than expected. They interpreted this finding as evidence of preferential management of wheat (either through supplemental irrigation, or by planting in better soils) over barley. Similarly, Wallace et al. (2015) comparing a range of sites from Eastern Mediterranean and Southwest Asia, and Masi et al. (2014) in Arslantepe (Eastern Anatolia, Turkey), also observed evidences of preferential management of wheat. In a study on the transition between Early to Middle Neolithic in Lesser Poland, Mueller-Bieniek et al. (2019) showed both spatial and temporal differences in the growing environment of cereal crops. On the one hand, they interpreted the higher $\Delta^{13}C$ of the plots in southern Poland during Early Neolithic as evidence of cultivation on wetter soils, surrounded by a denser vegetation landscape, compared to northern sites. On the other hand, they found a decrease in $\Delta^{13}C$ in the south during Middle Neolithic, suggesting an expansion to areas with more limited water availability (e.g. more elevated areas), in agreement with other archaeological evidences.

Alternatively, Ferrio et al. (2005) proposed the use of $\Delta^{13}C$ of wood charcoal as a reference value for the climate context, assuming that divergences with the $\Delta^{13}C$ of crops would reveal ancient irrigation practices. Mora-González et al. (2016), for example, compared the $\Delta^{13}C$ of wheat and cereal grains with those of pine and olive charcoal in the site of Terlinques (SE Iberian Peninsula). In this site, the $\Delta^{13}C$ values found in cereal grains were relatively high, and the comparison with wood charcoal, particularly after accounting for changes in WUEi in response to CO_2 (see Section 3), further supported a preferential management of the cereal crops. Masi et al. (2014) and Vignola et al. (2017) also compared the $\Delta^{13}C$ from oak charcoal with that of cereals, as an additional evidence in support of supplementary irrigation in Arslantepe (Turkey). Although this approach is indeed promising, the comparison between trees and annual crops faces some limitations, and requires a careful selection of reference species. For example, the divergence between wood and cereal $\Delta^{13}C$ could be also due to variations in precipitation seasonality, which has a species-specific response (Aguilera et al., 2012). On the other hand, the direct comparison of cereal grains and wood charcoal, even from the same archaeological context, could be affected by age divergences (Aguilera et al., 2012; Araus et al., 2014).

4.2.3 Estimation of crop productivity

Taking advantage of the tight association between water availability and grain yield of winter cereals, Araus et al. (1999b, 2001) proposed a model to estimate past cereal yields based on $\Delta^{13}C$, but accounting for changes in harvest index of recent breeds and CO_2 concentration. In agreement with the high water availability inferred for the Neolithic sites studied (Araus et al., 1997b, 1999a), estimated yields were reasonably high as compared with modern rain-fed agriculture in the area. Aguilera et al. (2008) further refined the model by accounting for changes in grain weight, estimated from charred grain dimensions (Ferrio et al., 2004, 2006a). Combining $\Delta^{13}C$-derived yield estimates with complementary evidences such as changes in weed assemblages or grain N content, Aguilera et al. (2008) showed evidence of progressive land degradation over the *ca.* 1500 years of continuous occupation of the site of Los Castillejos, in SE Spain, probably associated with agriculture intensification.

4.2.4 Use of multiproxy approaches

For the interpretation of ancient agricultural management, the combination of carbon isotopes with alternative proxies appears particularly promising. Riehl et al. (2009, 2014) linked $\Delta^{13}C$ with pedogenic carbonate records and archaeoclimatological models to interpret the interaction between climate fluctuations during the Holocene and the development of agronomic practices in the Near East. They confirmed a milder effect of climate fluctuations in the coast than in continental sites, but found unexpectedly high $\Delta^{13}C$ in some inland sites during the 4200 BP dry event, supporting the eventual onset of irrigation practices. Although this falls out of the main scope of the study, the analysis of $\delta^{15}N$, often combined with $\delta^{13}C$, has been particularly useful for the identification of ancient fertilization practices, such as manuring (Bogaard et al., 2007; Mueller-Bieniek et al., 2019; Styring et al., 2019; Vignola et al., 2017), and potential changes in soil fertility over different time scales (Aguilera et al., 2008; Araus et al., 2014).

Ferrio et al. (2012) reports another example of the combination of multiple proxies in the interpretation of $\delta^{13}C$ changes for the case of the Neolithic site of Tell Halula (Middle Euphrates, Syria) (Fig. 8). The study assessed the evolution of $\Delta^{13}C$-derived grain yield together with changes in the proportion among different crops, the distribution and diversity of weed flora, and the estimated population of the settlement. During the transition from the middle Pre-Pottery Neolithic (M-PPNB, c. 9800 BP) to the early Pottery Neolithic (E-PN, c. 8500 BP), estimated yields of wheat decreased about 20%, reaching values comparable with the low yields currently found in the area (Fig. 8A). This finding was coupled with an apparent transition toward a more extensive farming and a reduced proportion of legumes in favor of cereals, i.e. also approaching the current situation (Fig. 8B).

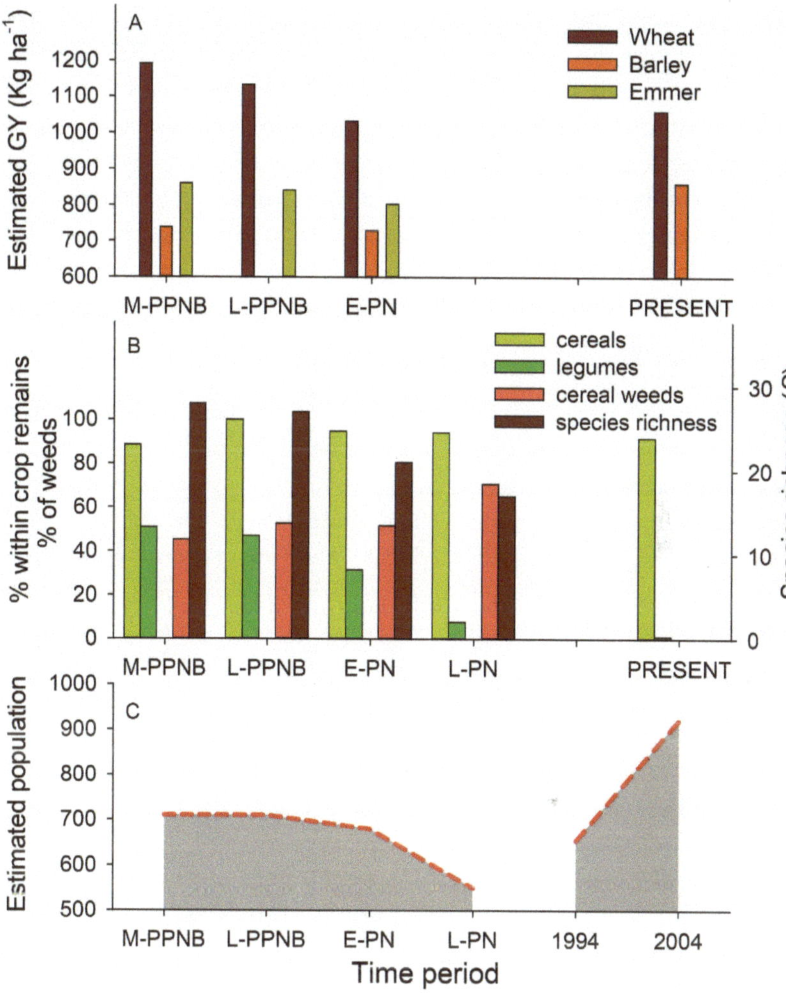

Fig. 8 Example of multiproxy application in the site of Tell Halula (Middle Euphrates Valley, Syria). (A) estimated cereal yields, based on the association between carbon isotope discrimination ($\Delta^{13}C$) and yield, and accounting for changes in harvest index and grain size. (B) changes in the ubiquity (percentage of sediment samples where a given taxon is present) of cereals and legumes (within all samples including crops), and of specialized cereal weeds (within all samples including weeds), together with the species richness (number of species) found in weed assemblages. Notice that each sediment sample may hold several taxa, therefore percentages are not complementary. (C) changes in estimated population of the site. As a reference, present-values in the study area are shown as a reference. Present population refers to the nearby village of Halula. *Data from Ferrio, J.P., Arab, G., Buxó, R., Guerrero, E., Molist, M., Voltas, J., Araus, J.L., 2012. Agricultural expansion and settlement economy in Tell Halula (Mid-Euphrates valley): a diachronic study from early neolithic to present. J. Arid Environ. 86, 104–112. https://doi.org/10.1016/j.jaridenv.2011.09.011.*

In agreement with this change in land use, the proportion of cereal weeds increased, whereas the diversity of wild flora decreased (Fig. 8B). Finally, a progressive reduction in settlement size is noticeable along with several indicators of increasing inequality (e.g. poorer domestic buildings, but larger collective works) (Fig. 8C). Under such scenario, ancient farmers might have faced an increasing demand for arable land to compensate for yield reductions, which may have contributed to the abandonment of the site at the end of the L-PN.

4.3 Reference for palaeodietary studies

Initial approaches to reconstruct palaeodiet of animals and humans were generally based on the association found between diet and the isotope composition of extant samples (Deniro and Epstein, 1978), or focused on the C_3–C_4 transition (e.g. Van Der Merwe and Vogel, 1978). Deniro and Epstein (1978), however, already highlighted that "*The limits of accuracy of this method will generally restrict its application to situations in which the diet is derived from sources with relatively large differences in their* $\delta^{13}C$ *values, such as terrestrial vs aquatic organisms or* C_3 *vs* C_4 *plants.*" In this regard, an original aim of the early works on archaeobotanical remains was to obtain local and coetaneous reference values for palaeodietary reconstruction (DeNiro and Hastorf, 1985). Nevertheless, the inclusion of plant remains as a local, direct measurement of the isotopic background has been largely obviated, despite the widespread presence of archaeobotanical remains in archaeological sites. In fact, and for the purpose of interpretating the isotope values of faunal remains, the need for a narrower referencing based on the analysis of plant remains has not been recognized until recently (Aguilera et al., 2016; Fraser et al., 2013b; Lightfoot and Stevens, 2012; Tornero et al., 2018; Treasure et al., 2016). Even for the identification of a transition from C_3 to C_4 crops (e.g. the spread of maize cultivation to different cultures in the Americas, or the expansion of millets in Eurasia), a comprehensive compilation of reference values is strongly recommended, considering alternative C_4 or CAM species consumed, and the variability in $\delta^{13}C$ among C_3 and C_4 plants (Lightfoot et al., 2018). Fortunately, the simultaneous investigation of isotopic values of human, faunal and plant remains is now becoming a common practice in archaeological research (Aguilera et al., 2018; Fernández-Crespo et al., 2019; Knipper et al., 2020; Mnich et al., 2020; Pate et al., 2016; Treasure et al., 2019). Combining isotope values from plant and animal remains is not only a promising approach for palaeodietary analyses, but has been proposed as a conceptual framework to refine palaeoclimatic inferences (Jones, 2018) and to reconstruct past husbandry practices, such as transhumans (Tornero et al., 2018).

5. Concluding remarks

For the last 20 years, the application of stable carbon isotopes to archaeobotanical remains has witnessed an increasing interest among the archaeological community. This is a direct consequence of the widening ability of this approach to infer past environmental and human-environment interactions through the entire Holocene, as registered in the archaeobotanical record. Alone or coupled with other proxies, based either on the natural abundance of stable isotopes (e.g. $^{18}O/^{16}O$, $^{15}N/^{14}N$ or $^{87}Sr/^{86}Sr$) or on alternative sources of information (e.g. flora and fauna assemblages, pollen records or pedogenic processes), the use of stable carbon isotopes will continue informing on spatially-explicit phenomena in relation to how the fates of humans and their surrounding physical environment are intertwined over their long-term coexistence. As this previously unattended source of information is complementary to approaches that can provide an absolute dating but are by far less abundant (dendrochronology) or even totally absent (written records), their combination should also boost our knowledge of past vegetation dynamics, as influenced by human activity. Even so, unanswered questions are still hindering the full exploitation of carbon isotope signature from plant remains in archaeology. We believe the interaction between plant physiologists, ecologists and archaeobotanists will be crucial for this endeavor.

Acknowledgments

J.P.F. was supported by Reference Group H09_20R (Gobierno de Aragón, Spain) and the project PID2019-106701RR-I00/AEI/10.13039/501100011033. M.A. and J.V. were supported by the project ArqueoAgro (INDEST—Universitat de Lleida). J.L.A. acknowledges the support of Catalan Institution for Research and Advanced Studies (ICREA, Generalitat de Catalunya, Spain), through the ICREA Academia Program.

References

Aguilera, M., Araus, J.L., Voltas, J., Rodríguez-Ariza, M.O., Molina, F., Rovira, N., Buxó, R., Ferrio, J.P., 2008. Stable carbon and nitrogen isotopes and quality traits of fossil cereal grains provide clues on sustainability at the beginnings of Mediterranean agriculture. Rapid Commun. Mass Spectrom. 22, 1653–1663. https://doi.org/10.1002/rcm.3501.

Aguilera, M., Espinar, C., Ferrio, J.P., Pérez, G., Voltas, J., 2009. A map of autumn precipitation for the third millennium BP in the Eastern Iberian Peninsula from charcoal carbon isotopes. J. Geochem. Explor. 102, 157–165. https://doi.org/10.1016/j.gexplo.2008.11.019.

Aguilera, M., Ferrio, J.P., Araus, J.L., Tarrús, J., Voltas, J., 2011. Climate at the onset of western Mediterranean agriculture expansion: evidence from stable isotopes of sub-fossil oak tree rings in Spain. Palaeogeogr. Palaeoclimatol. Palaeoecol. 299, 541–551. https://doi.org/10.1016/j.palaeo.2010.11.026.

Aguilera, M., Ferrio, J.P., Pérez, G., Araus, J.L., Voltas, J., 2012. Holocene changes in precipitation seasonality in the western Mediterranean Basin: a multi-species approach using $\delta^{13}C$ of archaeobotanical remains. J. Quat. Sci. 27, 192–202. https://doi.org/10.1002/jqs.1533.

Aguilera, M., Balasse, M., Lepetz, S., Zech-Matterne, V., 2016. Amender les sols cultivés avec des fumiers et déchets domestiques: une origine multimillénaire pour un enjeu majeur du développement durable. Regards croisés quand les Sci. archéologiques rencontrent l'innovation 51.

Aguilera, M., Zech-Matterne, V., Lepetz, S., Balasse, M., 2018. Crop fertility conditions in North-Eastern Gaul during the La Tène and Roman periods: a combined stable isotope analysis of archaeobotanical and archaeozoological remains. Environ. Archaeol. 23, 323–337. https://doi.org/10.1080/14614103.2017.1291563.

Allen, R.G., Pereira, L.S., Raes, D., Smith, M., 1998. Crop evapotranspiration: Guidelines for computing crop requirements. Irrig. Drain. Pap. No. 56, FAO. https://doi.org/10.1016/j.eja.2010.12.001.

Alonso, N., Tresserras, J.J., Rodríguez-Ariza, M.O., Rovira, N., 2000. Muestreo arqueobotánico de yacimientos al aire libre y en medio seco. In: Buxó, R., Piqué, R. (Eds.), La Recogida de Muestras En Arqueobotánica: Objetivos y Propuestas Metodológicas. Museu d'Arqueologia de Catalunya, Barcelona, Spain, pp. 31–48.

Araus, J.L., 2002. Plant breeding and drought in C3 cereals: what should we breed for? Ann. Bot. 89, 925–940. https://doi.org/10.1093/aob/mcf049.

Araus, J.L., Buxó, R., 1993. Changes in carbon isotope discrimination in grain cereals from the north-western Mediterranean basin during the past seven millennia. Aust. J. Plant Physiol. 20, 117–128.

Araus, J.L., Reynolds, M.P., Acevedo, E., 1993. Leaf posture, grain yield, growth, leaf structure, and carbon isotope discrimination in wheat. Crop. Sci. 33, 1273–1279.

Araus, J.L., Febrero, A., Buxo, R., Camalich, M.D., Martin, D., Molina, F., Rodriguez-Ariza, M.O., Romagosa, I., 1997a. Changes in carbon isotope discrimination in grain cereals from different regions of the western Mediterranean Basin during the past seven millennia. Palaeoenvironmental evidence of a differential change in aridity during the late Holocene. Glob. Chang. Biol. 3, 107–118. https://doi.org/10.1046/j.1365-2486.1997.00056.x.

Araus, J.L., Febrero, A., Buxó, R., Rodríguez-Ariza, M.O., Molina, F., Camalich, M.D., Martín, D., Voltas, J., et al., 1997. Identification of ancient irrigation practices based on the carbon isotope discrimination of plant seeds: a case study from the south-east Iberian Peninsula. J. Archaeol. Sci. 24, 729–740. https://doi.org/10.1006/jasc.1997.0154.

Araus, J.L., Febrero, A., Catala, M., Molist, M., Voltas, J., Romagosa, I., 1999a. Crop water availability in early agriculture: evidence from carbon isotope discrimination of seeds from a tenth millennium BP site on the Euphrates. Glob. Chang. Biol. 5, 201–212. https://doi.org/10.1046/j.1365-2486.1999.00213.x.

Araus, J.L., Slafer, G.A., Romagosa, I., Araus, J.L., 1999. Durum wheat and barley yields in antiquity estimated from ^{13}C discrimination of archaeological grains: a case study from the Western Mediterranean Basin. Aust. J. Plant Physiol. 26, 345–352. https://doi.org/10.1071/PP98141.

Araus, J.L., Slafer, G.A., Romagosa, I., Molist, M., 2001. FOCUS: estimated wheat yields during the emergence of agriculture based on the carbon isotope discrimination of grains: evidence from a 10th millennium BP site on the Euphrates. J. Archaeol. Sci. 28, 341–350. https://doi.org/10.1006/jasc.2000.0569.

Araus, J.L., Villegas, D., Aparicio, N., García del Moral, L.F., El Hani, S., Rharrabti, Y., Ferrio, J.P., Royo, C., 2003. Environmental factors determining carbon isotope discrimination and yield in durum wheat under Mediterranean conditions. Crop. Sci. 43, 170–180.

Araus, J.L., Ferrio, J.P., Buxó, R., Voltas, J., 2007. The historical perspective of dryland agriculture: lessons learned from 10 000 years of wheat cultivation. J. Exp. Bot. 58, 131–145. https://doi.org/10.1093/jxb/erl133.

Araus, J.L., Cabrera-Bosquet, L., Serret, M.D., Bort, J., Nieto-Taladriz, M.T., 2013. Comparative performance of $\delta^{13}C$, $\delta^{18}O$ and $\delta^{15}N$ for phenotyping durum wheat adaptation to a dryland environment. Funct. Plant Biol. 40, 595–608. https://doi.org/10.1071/FP12254.

Araus, J.L., Ferrio, J.P., Voltas, J., Aguilera, M., Buxó, R., 2014. Agronomic conditions and crop evolution in ancient near east agriculture. Nat. Commun. 5, 3953. https://doi.org/10.1038/ncomms4953.

Arnold, J.R., Libby, W.F., 1949. Age determinations by radiocarbon content: checks with samples of known age. Science 110, 678–680. https://doi.org/10.1126/science.110.2869.678.

Badeck, F.W., Tcherkez, G., Nogués, S., Piel, C., Ghashghaie, J., 2005. Post-photosynthetic fractionation of stable carbon isotopes between plant organs -a widespread phenomenon. Rapid Commun. Mass Spectrom. 19, 1381–1391. https://doi.org/10.1002/rcm.1912.

Balter, M., 2007. Seeking agriculture's ancient roots. Science 316, 1830–1835. https://doi.org/10.1126/science.316.5833.1830.

Baton, F., Nguyen Tu, T.T., Derenne, S., Delorme, A., Delarue, F., Dufraisse, A., 2017. Tree-ring $\delta^{13}C$ of archeological charcoals as indicator of past climatic seasonality. A case study from the Neolithic settlements of Lake Chalain (Jura, France). Quat. Int. 457, 50–59.

Becker, B., Kromer, B., Trimborn, P., 1991. A stable-isotope tree-ring timescale of the Late Glacial/Holocene boundary. Nature 353, 647–649. https://doi.org/10.1038/353647a0.

Bell, B.A., Fletcher, W.J., Ryan, P., Grant, H., Ilmen, R., 2017. Stable carbon isotope analysis of *Cedrus atlantica* pollen as an indicator of moisture availability. Rev. Palaeobot. Palynol. 244, 128–139. https://doi.org/10.1016/j.revpalbo.2017.04.008.

Bender, M.M., 1968. Mass spectrometric studies of carbon 13 variations in corn and other grasses. Radiocarbon 10, 468–472. https://doi.org/10.1017/s0033822200011103.

Bender, M., Sowers, T., Brook, E., 1997. Gases in ice cores. Proc. Natl. Acad. Sci. U. S. A. 94, 8343–8349.

Bogaard, A., Heaton, T.H.E.H.E., Poulton, P., Merbach, I., 2007. The impact of manuring on nitrogen isotope ratios in cereals: archaeological implications for reconstruction. J. Archaeol. Sci. 34, 335–343. https://doi.org/10.1016/j.jas.2006.04.009.

Bonafini, M., Pellegrini, M., Ditchfield, P., Pollard, A.M., 2013. Investigation of the "canopy effect" in the isotope ecology of temperate woodlands. J. Archaeol. Sci. 40, 3926–3935. https://doi.org/10.1016/j.jas.2013.03.028.

Brinkkemper, O., Braadbaart, F., van Os, B., van Hoesel, A., van Brussel, A.A.N., Fernandes, R., 2018. Effectiveness of different pre-treatments in recovering pre-burial isotopic ratios of charred plants. Rapid Commun. Mass Spectrom. 32, 251–261. https://doi.org/10.1002/rcm.8033.

Buchmann, N., Brooks, J.R., Rapp, K.D., Ehleringer, J.R., 1996. Carbon isotope composition of C4 grasses is influenced by light and water supply. Plant Cell Environ. 9, 392–402.

Buxó, R., 1997. Arqueología de las Plantas: La Explotación Económica de las Semillas y los Frutos en el Marco Mediterráneo de la Península Ibérica. Critica, Barcelona, Spain.

Caracuta, V., Fiorentino, G., Turchiano, M., Volpe, G., 2014. Dating historical contexts: issues, plant material, and methods to date the late Roman Site of Faragola, Apulia (SE Italy). Radiocarbon 56, 679–690. https://doi.org/10.2458/56.17463.

Caracuta, V., Weinstein-Evron, M., Yeshurun, R., Kaufman, D., Tsatskin, A., Boaretto, E., 2016. Charred wood remains in the natufian sequence of el-Wad terrace (Israel): new insights into the climatic, environmental and cultural changes at the end of the Pleistocene. Quat. Sci. Rev. 131, 20–32. https://doi.org/10.1016/j.quascirev.2015.10.034.

Chen, X.X., Yu, S.Y., Underhill, A.P., Fang, H., 2018. Radiocarbon dating and stable carbon isotopic analyses of Neolithic and Bronze Age staple crops in the lower Yellow River area and their paleodietary implications. Geoarchaeology 33, 307–313. https://doi.org/10.1002/gea.21659.

Condon, A.G., Richards, R.A., Farquhar, G.D., 1987. Carbon isotope discrimination is positively correlated with grain yield and dry matter production in field-grown wheat. Crop. Sci. 27, 996. https://doi.org/10.2135/cropsci1987.0011183X002700050035x.

Craig, H., 1954. Carbon-13 in plants and the relationship between carbon-13 and carbon-14 variations in nature. J. Geol. 62, 115–149.

Del Castillo, J., Aguilera, M., Voltas, J., Ferrio, J.P., 2013. Isoscapes of tree-ring carbon-13 perform like meteorological networks in predicting regional precipitation patterns. Eur. J. Vasc. Endovasc. Surg. 118, 352–360. https://doi.org/10.1002/jgrg.20036.

Deniro, M., Epstein, S., 1978. Influence of diet on the distribution of carbon isotopes in animals. Geochim. Cosmochim. Acta 42, 495–506. https://doi.org/10.1016/0016-7037(78)90199-0.

DeNiro, M.J., Epstein, S., 1981. Influence of diet on the distribution of nitrogen isotopes in animals. Geochim. Cosmochim. Acta 45, 341–351.

DeNiro, M.J., Hastorf, C.A., 1985. Alteration of $^{15}N^{14}N$ and $^{13}C^{12}C$ ratios of plant matter during the initial stages of diagenesis: studies utilizing archaeological specimens from Peru. Geochim. Cosmochim. Acta 49, 97–115. https://doi.org/10.1016/0016-7037(85)90194-2.

Dercon, G., Clyman, E., Diels, J., Merckx, R., Deckers, J., 2006. Differential ^{13}C isotopic discrimination in maize at varying water stress and at low to high nitrogen availability. Plant and Soil 282, 313–326.

Drake, B.L., Hanson, D.T., Boone, J.L., 2012. The use of radiocarbon-derived $\Delta^{13}C$ as a paleoclimate indicator: applications in the Lower Alentejo of Portugal. J. Archaeol. Sci. 39, 2888–2896. https://doi.org/10.1016/j.jas.2012.04.027.

Ellsworth, P.Z., Cousins, A.B., 2016. Carbon isotopes and water use efficiency in C_4 plants. Curr. Opin. Plant Biol. 31, 155–161. https://doi.org/10.1016/j.pbi.2016.04.006.

Farquhar, G., 1983. On the nature of carbon isotope discrimination in C_4 species. Aust. J. Plant Physiol. 10, 205. https://doi.org/10.1071/PP9830205.

Farquhar, G.D., Richards, R.A., 1984. Isotopic composition of plant carbon correlates with water use efficiency of wheat genotypes. Aust. J. Plant Physiol. 11, 539–552.

Farquhar, G.D., O'Leary, M.H., Berry, J.A., 1982. On the relationship between carbon isotope discrimination and the intercellular carbon dioxide concentration in leaves. Aust. J. Plant Physiol. 9, 121–137. https://doi.org/10.1071/PP9820121.

Farquhar, G.D., Hubick, K.T., Condon, A.G., Richards, R.A., 1989. Carbon isotope fractionation and plant water-use efficiency. In: Rundel, P.W., Ehleringer, J.R., Nagy, K.A. (Eds.), Stable Isotopes in Ecological Research. Ecological Studies (Analysis and Synthesis). vol. 68. Springer, New York, NY, pp. 21–40. https://doi.org/10.1007/978-1-4612-3498-2_2.

February, E., van der Merwe, N.J., 1992. Stable carbon isotope ratios of wood charcoal during the past 4000 years: anthropogenic and climatic influences. S. Afr. J. Sci. 88, 291–292.

Fernández-Crespo, T., Ordoño, J., Bogaard, A., Llanos, A., Schulting, R., 2019. A snapshot of subsistence in Iron Age Iberia: the case of La Hoya village. J. Archaeol. Sci. Rep. 28, 102037. https://doi.org/10.1016/j.jasrep.2019.102037.

Ferrio, J.P., Florit, A., Vega, A., Serrano, L., Voltas, J., 2003. $\Delta^{13}C$ and tree-ring width reflect different drought responses in *Quercus ilex* and *Pinus halepensis*. Oecologia 137, 512–518. https://doi.org/10.1007/s00442-003-1372-7.

Ferrio, J.P., Alonso, N., Voltas, J., Araus, J.L., 2004. Estimating grain weight in archaeological cereal crops: a quantitative approach for comparison with current conditions. J. Archaeol. Sci. 31, 1635–1642. https://doi.org/10.1016/j.jas.2004.04.006.

Ferrio, J.P., Araus, J.L., Buxó, R., Voltas, J., Bort, J., 2005. Water management practices and climate in ancient agriculture: inferences from the stable isotope composition of archaeobotanical remains. Veg. Hist. Archaeobotany 14, 510–517. https://doi.org/10.1007/s00334-005-0062-2.

Ferrio, J.P., Alonso, N., Voltas, J., Araus, J.L., 2006a. Grain weight changes over time in ancient cereal crops: potential roles of climate and genetic improvement. J. Cereal Sci. 44, 323–332. https://doi.org/10.1016/j.jcs.2006.07.013.

Ferrio, J.P., Alonso, N., López, J.B., Araus, J.L., Voltas, J., 2006b. Carbon isotope composition of fossil charcoal reveals aridity changes in the NW Mediterranean Basin. Glob. Chang. Biol. 12, 1253–1266. https://doi.org/10.1111/j.1365-2486.2006.01170.x.

Ferrio, J.P., Voltas, J., Alonso, N., Araus, J.L., 2007. Reconstruction of climate and crop conditions in the past based on the carbon isotope signature of archaeobotanical remains. In: Dawson, T.E., Siegwolf, R. (Eds.), Terrestrial Ecology. Elsevier Academic Press, New York, pp. 319–332. https://doi.org/10.1016/S1936-7961(07)01020-2.

Ferrio, J.P., Voltas, J., Araus, J.L., 2011. Global change and the origins of agriculture. In: Araus, J.L., Slafer, G.A. (Eds.), Crop Stress Management and Global Climate Change. CAB International, Wallingford, pp. 1–14. https://doi.org/10.1079/9781845936808.0001.

Ferrio, J.P., Arab, G., Buxó, R., Guerrero, E., Molist, M., Voltas, J., Araus, J.L., 2012. Agricultural expansion and settlement economy in Tell Halula (Mid-Euphrates valley): a diachronic study from early neolithic to present. J. Arid Environ. 86, 104–112. https://doi.org/10.1016/j.jaridenv.2011.09.011.

Fewlass, H., Talamo, S., Tuna, T., Fagault, Y., Kromer, B., Hoffmann, H., Pangrazzi, C., Hublin, J.J., Bard, E., 2018. Size matters: radiocarbon dates of <200 μg ancient collagen samples with AixMICADAS and its gas ion source. Radiocarbon 60, 425–439. https://doi.org/10.1017/RDC.2017.98.

Filipović, D., Gašić, U., Stevanović, N., Dabić Zagorac, D., Fotirić Akšić, M., Natić, M., 2018. Carbon stable isotope composition of modern and archaeological Cornelian cherry fruit stones: a pilot study. Isotopes Environ. Health Stud. 54, 337–351. https://doi.org/10.1080/10256016.2017.1392516.

Fiorentino, G., Caracuta, V., Calcagnile, L., D'Elia, M., Matthiae, P., Mavelli, F., Quarta, G., 2008. Third millennium B.C. climate change in Syria highlighted by carbon stable isotope analysis of ^{14}C-AMS dated plant remains from Ebla. Palaeogeogr. Palaeoclimatol. Palaeoecol. 266, 51–58. https://doi.org/10.1016/j.palaeo.2008.03.034.

Fiorentino, G., Caracuta, V., Volpe, G., Turchiano, M., Quarta, G., D'Elia, M., Calcagnile, L., 2010. The first millennium AD climate fluctuations in the Tavoliere Plain (Apulia, Italy): new preliminary data from the ^{14}C AMS-dated plant remains from the archaeological site of Faragola. Nucl. Instrum. Meth. B 268, 1084–1087. https://doi.org/10.1016/j.nimb.2009.10.104.

Fiorentino, G., Ferrio, J.P., Bogaard, A., Araus, J.L., Riehl, S., 2014. Stable isotopes in archaeobotanical research. Veg. Hist. Archaeobotany 24, 215–227. https://doi.org/10.1007/s00334-014-0492-9.

Flohr, P., Jenkins, E., Williams, H.R.S., Jamjoum, K., Nuimat, S., Müldner, G., 2019. What can crop stable isotopes ever do for us? An experimental perspective on using cereal carbon stable isotope values for reconstructing water availability in semi-arid and arid environments. Veg. Hist. Archaeobotany 28, 497–512. https://doi.org/10.1007/s00334-018-0708-5.

Francey, R.J., Allison, C.E., Etheridge, D.M., Trudinger, C.M., Enting, I.G., Leuenberger, M., Langenfelds, R.L., Michel, E., Steele, L.P., 1999. A 1000-year high precision record of $\delta^{13}C$ in atmospheric CO_2. Tellus B Chem. Phys. Meteorol. 51, 170–193. https://doi.org/10.3402/tellusb.v51i2.16269.

Fraser, R.A., Bogaard, A., Charles, M., Styring, A.K., Wallace, M., Jones, G., Ditchfield, P., Heaton, T.H.E., 2013a. Assessing natural variation and the effects of charring, burial and pre-treatment on the stable carbon and nitrogen isotope values of archaeobotanical cereals and pulses. J. Archaeol. Sci. 40, 4754–4766. https://doi.org/10.1016/j.jas.2013.01.032.

Fraser, R.A., Bogaard, A., Schäfer, M., Arbogast, R., Heaton, T.H.E., 2013b. Integrating botanical, faunal and human stable carbon and nitrogen isotope values to reconstruct land use and palaeodiet at LBK Vaihingen an der Enz, Baden-Württemberg. World Archaeol. 45, 492–517. https://doi.org/10.1080/00438243.2013.820649.

Fuller, D.Q., 2007. Contrasting patterns in crop domestication and domestication rates: recent archaeobotanical insights from the old world. Ann. Bot. 100, mcm048. https://doi.org/10.1093/aob/mcm048.

Godwin, H., 1962. Half-life of radiocarbon. Nature 195, 984. https://doi.org/10.1038/195984a0.

Graven, H., Allison, C.E., Etheridge, D.M., Hammer, S., Keeling, R.F., Levin, I., Meijer, H.A.J., Rubino, M., Tans, P.P., Trudinger, C.M., Vaughn, B.H., White, J.W.C., 2017. Compiled records of carbon isotopes in atmospheric CO_2 for historical simulations in CMIP6. Geosci. Model Dev. 10, 4405–4417. https://doi.org/10.5194/gmd-10-4405-2017.

Greig, J., 1989. Handbooks for Archaeologist 4: Archaeobotany. European Science Foundation, Strasbourg.

Hall, G., Woodborne, S., Scholes, M., 2008. Stable carbon isotope ratios from archaeological charcoal as palaeoenvironmental indicators. Chem. Geol. 247, 384–400.

Hare, V.J., Loftus, E., Jeffrey, A., Ramsey, C.B., 2018. Atmospheric CO_2 effect on stable carbon isotope composition of terrestrial fossil archives. Nat. Commun. 9, 252. https://doi.org/10.1038/s41467-017-02691-x.

Harshberger, J.W., 1896. The purposes of ethno-botany. Bot. Gaz. 21, 146–154. https://doi.org/10.1086/327316.

Hatch, M.D., Slack, C.R., 1966. Photosynthesis by sugar-cane leaves. A new carboxylation reaction and the pathway of sugar formation. Biochem. J. 101, 103–111. https://doi.org/10.1042/bj1010103.

Heaton, T.H.E., Jones, G., Halstead, P., Tsipropoulos, T., 2009. Variations in the $^{13}C/^{12}C$ ratios of modern wheat grain, and implications for interpreting data from Bronze Age Assiros Toumba, Greece. J. Archaeol. Sci. 36, 2224–2233. https://doi.org/10.1016/j.jas.2009.06.007.

Helbaek, H., 1959. Domestication of food plants in the old world. Science 130, 365–372. https://doi.org/10.1126/science.130.3372.365.

Henderson, S.A., Von Caemmerer, S., Farquhar, G.D., 1992. Short-term measurements of carbon isotope discrimination in several C_4 species. Aust. J. Plant Physiol. 19, 263–285.

Hubbard, R.M., Bond, B.J., Senock, R.S., Ryan, M.G., 2002. Effects of branch height on leaf gas exchange, branch hydraulic conductance and branch sap flux in open-grown ponderosa pine. Tree Physiol. 22, 575–581.

Indermühle, A., Stocker, T.F., Joos, F., Fischer, H., Smith, H.J., Wahlen, M., Deck, B., Mastroianni, D., Tschumi, J., Blunier, T., Meyer, R., Stauffer, B., 1999. Holocene carbon-cycle dynamics based on CO_2 trapped in ice at Taylor Dome, Antarctica. Nature 398, 121–126. https://doi.org/10.1038/18158.

Jones, P.J., 2018. Climate Change, Water Stress and Agriculture in the Indus Civilisation, 3000-1500 BC. University of Cambridge.

Keeling, C.D., Mook, W.G., Trans, P.P., 1979. Recent trends in the $^{13}C/^{12}C$ ratio of atmospheric carbon dioxide. Nature 277, 121–123.

Keeling, R.F., Graven, H.D., Welp, L.R., Resplandy, L., Bi, J., Piper, S.C., Sun, Y., Bollenbacher, A., Meijer, H.A.J., 2017. Atmospheric evidence for a global secular increase in carbon isotopic discrimination of land photosynthesis. Proc. Natl. Acad. Sci. U. S. A. 114, 10361–10366. https://doi.org/10.1073/pnas.1619240114.

Knipper, C., Rihuete-Herrada, C., Voltas, J., Held, P., Lull, V., Micó, R., Risch, R., Alt, K.W., 2020. Reconstructing Bronze Age diets and farming strategies at the early Bronze Age sites of La Bastida and Gatas (southeast Iberia) using stable isotope analysis. PLoS One 15, e0229398. https://doi.org/10.1371/journal.pone.0229398.

Kromdijk, J., Ubierna, N., Cousins, A.B., Griffiths, H., 2014. Bundle-sheath leakiness in C_4 photosynthesis: a careful balancing act between CO_2 concentration and assimilation. J. Exp. Bot. 65, 3443–3457. https://doi.org/10.1093/jxb/eru157.

Leavitt, S.W., 2010. Tree-ring C-H-O isotope variability and sampling. Sci. Total Environ. 408, 5244–5253. https://doi.org/10.1016/j.scitotenv.2010.07.057.

Leuenberger, M., Siegenthaler, U., Langway, C., 1992. Carbon isotope composition of atmospheric CO_2 during the last ice age from an Antarctic ice core. Nature 357, 488–490. https://doi.org/10.1038/357488a0.

Lightfoot, E., Stevens, R.E., 2012. Stable isotope investigations of charred barley (*Hordeum vulgare*) and wheat (*Triticum spelta*) grains from Danebury Hillfort: implications for palaeodietary reconstructions. J. Archaeol. Sci. 39, 656–662. https://doi.org/10.1016/j.jas.2011.10.026.

Lightfoot, E., Liu, X., Jones, P.J., 2018. A world of C_4 pathways: on the use of $\delta^{13}C$ values to identify the consumption of C_4 plants in the archaeological record. In: Lightfoot, E., Liu, X., Fuller, D.Q. (Eds.), Far From the Hearth: Essays in Honour of Martin K Jones. McDonald Institute for Archaeological Research, pp. 165–176.

Marino, B.D., Deniro, M.J., 1987. Isotopic analysis of archaeobotanicals to reconstruct past climates: effects of activities associated with food preparation on carbon, hydrogen and oxygen isotope ratios of plant cellulose. J. Archaeol. Sci. 14, 537–548. https://doi.org/10.1016/0305-4403(87)90037-9.

Marino, B.D., McElroy, M.B., Salawitch, R.J., Spaulding, W.G., 1992. Glacial-to-interglacial variations in the carbon isotopic composition of atmospheric CO_2. Nature 357, 461–466. https://doi.org/10.1038/357461a0.

Marshall, J.D., Monserud, R.A., 1996. Homeostatic gas-exchange parameters inferred from $^{13}C/^{12}C$ in tree rings of conifers. Oecologia 105, 13–21. https://doi.org/10.1007/BF00328786.

Masi, A., Sadori, L., Baneschi, I., Siani, A.M., Zanchetta, G., 2013. Stable isotope analysis of archaeological oak charcoal from eastern Anatolia as a marker of mid-Holocene climate change. Plant Biol. 15, 83–92. https://doi.org/10.1111/j.1438-8677.2012.00669.x.

Masi, A., Sadori, L., Balossi Restelli, F., Baneschi, I., Zanchetta, G., 2014. Stable carbon isotope analysis as a crop management indicator at Arslantepe (Malatya, Turkey) during the Late Chalcolithic and Early Bronze Age. Veg. Hist. Archaeobotany 23, 751–760. https://doi.org/10.1007/s00334-013-0421-3.

McCarroll, D., Gagen, M.H., Loader, N.J., Robertson, I., Anchukaitis, K.J., Los, S., Young, G.H.F., Jalkanen, R., Kirchhefer, A., Waterhouse, J.S., 2009. Correction of tree ring stable carbon isotope chronologies for changes in the carbon dioxide content of the atmosphere. Geochim. Cosmochim. Acta 73, 1539–1547. https://doi.org/10.1016/j.gca.2008.11.041.

McIntyre, C.P., Wacker, L., Haghipour, N., Blattmann, T.M., Fahrni, S., Usman, M., Eglinton, T.I., Synal, H.A., 2017. Online ^{13}C and ^{14}C gas measurements by EA-IRMS-AMS at ETH Zürich. Radiocarbon 59, 893–903. https://doi.org/10.1017/RDC.2016.68.

Medina, E., Ziegler, H., Luttge, U., Trimborn, P., Francisco, M., 1994. Light conditions during growth as revealed by $\delta^{13}C$ values of leaves of primitive cultivars of *Ananas comosus*, an obligate CAM species. Funct. Ecol. 8, 298–305. https://doi.org/10.2307/2389822.

Mnich, B., Mueller-Bieniek, A., Nowak, M., Wilczyński, J., Pospuła, S., Szostek, K., 2020. Terrestrial diet in prehistoric human groups from southern Poland based on human, faunal and botanical stable isotope evidence. J. Archaeol. Sci. Rep. 32, 102382. https://doi.org/10.1016/j.jasrep.2020.102382.

Mora-González, A., Delgado-Huertas, A., Granados-Torres, A., Contreras Cortés, F., Jover Maestre, F.J., López Padilla, J.A., 2016. The isotopic footprint of irrigation in the western Mediterranean basin during the Bronze Age: the settlement of Terlinques, southeast Iberian Peninsula. Veg. Hist. Archaeobotany 25, 459–468. https://doi.org/10.1007/s00334-016-0560-4.

Mora-González, A., Fernandes, R., Contreras Cortés, F., Granados-Torres, A., Alarcón García, E., Delgado-Huertas, A., 2019. Reporting atmospheric CO_2 pressure corrected results of stable carbon isotope analyses of cereals remains from the archaeological site of Peñalosa (SE Iberian Peninsula): agricultural and social implications. Archaeol. Anthropol. Sci. 11, 1995–2005. https://doi.org/10.1007/s12520-018-0650-6.

Mueller-Bieniek, A., Nowak, M., Styring, A., Lityńska-Zając, M., Moskal-del Hoyo, M., Sojka, A., Paszko, B., Tunia, K., Bogaard, A., 2019. Spatial and temporal patterns in Neolithic and Bronze Age agriculture in Poland based on the stable carbon and nitrogen isotopic composition of cereal grains. J. Archaeol. Sci. Rep. 27, 101993. https://doi.org/10.1016/j.jasrep.2019.101993.

Nier, A.O., 1940. A mass spectrometer for routine isotope abondance measurements. Rev. Sci. Instrum. 11, 212–216.

Nitsch, E.K., Charles, M., Bogaard, A., 2015. Calculating a statistically robust $\delta^{13}C$ and $\delta^{15}N$ offset for charred cereal and pulse seeds. Sci. Technol. Archaeol. Res. 1, 1–8. https://doi.org/10.1179/2054892315Y.0000000001.

Nowak, M., del Hoyo, M.M., Mueller-Bieniek, A., Lityńska-Zajac, M., Kotynia, K., 2017. Benefits and weaknesses of radiocarbon dating of plant material as reflected by Neolithic archaeological sites from Poland, Slovakia and Hungary. Geochronometria 44, 188–201. https://doi.org/10.1515/geochr-2015-0066.

O'Leary, M.H., 1981. Carbon isotope fractionation in plants. Phytochemistry 20, 553–567. https://doi.org/10.1016/0031-9422(81)85134-5.

Paradis, S., Dufraisse, A., Allée, P., 2013. Radius of curvature measurements and wood diameter: a comparison of different image analysis techniques. In: Damblon, F. (Ed.), Proceedings of the Fourth International Meeting of Anthracology, Brussels, 8–13 September 2008, Royal Belgian Institute of Natural Sciences. Archaeopress, Oxford, UK, pp. 173–182.

Park, R., Epstein, S., 1961. Carbon isotope fractionation during photosynthesis. Geochim. Cosmochim. Acta 21, 110–126.

Pate, F.D., Henneberg, R.J., Henneberg, M., 2016. Stable carbon and nitrogen isotope evidence for dietary variability at Ancient Pompeii, Italy. Mediter. Archaeol. Archaeom. 16, 127–133. https://doi.org/10.5281/zenodo.35526.

Philippsen, B., Olsen, J., Sørensen, S.A., Måge, B., 2019. $\delta^{13}C$ values of wood and charcoal reveal broad isotopic ranges at the base of the food web. Radiocarbon 61, 2003–2017. https://doi.org/10.1017/RDC.2019.138.

Polley, H.W., Johnson, H.B., Mayeux, H.S., Malone, S.R., 1993. Physiology and growth of wheat across a subambient carbon-dioxide gradient. Ann. Bot. 71, 347–356. https://doi.org/10.1006/anbo.1993.1044.

Poole, I., Braadbaart, F., Boon, J.J., Van Bergen, P.P.F., 2002. Stable carbon isotope changes during artificial charring of propagules. Org. Geochem. 33, 1675–1681. https://doi.org/10.1016/S0146-6380(02)00173-0.

Poole, I., Van Bergen, P.F., Kool, J., Schouten, S., Cantrill, D.J., 2004. Molecular isotopic heterogeneity of fossil organic matter: implications for $\delta^{13}C$ biomass and $\delta^{13}C$ palaeoatmosphere proxies. Org. Geochem. 35, 1261–1274. https://doi.org/10.1016/j.orggeochem.2004.05.014.

Reimer, P.J., Bard, E., Bayliss, A., Beck, J.W., Blackwell, P.G., Ramsey, C.B., Buck, C.E., Cheng, H., Edwards, R.L., Friedrich, M., Grootes, P.M., Guilderson, T.P., Haflidason, H., Hajdas, I., Hatté, C., Heaton, T.J., Hoffmann, D.L., Hogg, A.G., Hughen, K.A., Kaiser, K.F., Kromer, B., Manning, S.W., Niu, M., Reimer, R.W., Richards, D.A., Scott, E.M., Southon, J.R., Staff, R.A., Turney, C.S.M., van der Plicht, J., 2013. IntCal13 and Marine13 radiocarbon age calibration curves 0–50,000 years cal BP. Radiocarbon 55, 1869–1887. https://doi.org/10.2458/azu_js_rc.55.16947.

Renfrew, J.M., 1973. Paleoethnobotany. The Prehistoric Food Plants of the Near Est and Europe. Columbia University Press, New York.

Resco, V., Ferrio, J.P., Carreira, J.A., Calvo, L., Casals, P., Ferrero-Serrano, Á., Marcos, E., Moreno, J.M., Ramírez, D.A., Sebastià, M.T., Valladares, F., Williams, D.G., 2011. The stable isotope ecology of terrestrial plant succession. Plant Ecol. Divers. 4, 117–130. https://doi.org/10.1080/17550874.2011.576708.

Rice, S.K., Giles, L., 1996. The influence of water content and leaf anatomy on carbon isotope discrimination and photosynthesis in Sphagnum. Plant Cell Environ. 19, 118–124.

Riehl, S., 2008. Climate and agriculture in the ancient Near East: a synthesis of the archaeobotanical and stable carbon isotope evidence. Veg. Hist. Archaeobotany 17, 43–51. https://doi.org/10.1007/s00334-008-0156-8.

Riehl, S., Pustovoytov, K.E., Hotchkiss, S., Bryson, R.A., 2009. Local holocene environmental indicators in Upper Mesopotamia: pedogenic carbonate record vs. archaeobotanical data and archaeoclimatological models. Quat. Int. 209, 154–162. https://doi.org/10.1016/j.quaint.2008.06.012.

Riehl, S., Pustovoytov, K.E., Weippert, H., Klett, S., Hole, F., 2014. Drought stress variability in ancient Near Eastern agricultural systems evidenced by $\delta^{13}C$ in barley grain. Proc. Natl. Acad. Sci. 111, 12348–12353. https://doi.org/10.1073/pnas.1409516111.

Rinne, K.T., Saurer, M., Kirdyanov, A.V., Loader, N.J., Bryukhanova, M.V., Werner, R.A., Siegwolf, R.T.W., 2015. The relationship between needle sugar carbon isotope ratios and tree rings of larch in Siberia. Tree Physiol. 35, 1192–1205. https://doi.org/10.1093/treephys/tpv096.

Royo, C., Martos, V., Ramdani, A., Villegas, D., Rharrabti, Y., García Del Moral, L.F., 2008. Changes in yield and carbon isotope discrimination of Italian and Spanish durum wheat during the 20th century. Agron. J. 100, 352–360. https://doi.org/10.2134/agronj2007.0060.

Schollaen, K., Heinrich, I., Helle, G., 2014. UV-laser-based microscopic dissection of tree rings—a novel sampling tool for $\delta^{13}C$ and $\delta^{18}O$ studies. New Phytol. 201, 1045–1055. https://doi.org/10.1111/nph.12587.

Sharp, Z.D., 2007. Principles of Stable Isotopes Geochemistry. Pearson Prentice Hall, Upper Saddle River, USA.

Styring, A.K., Manning, H., Fraser, R.A., Wallace, M., Jones, G., Charles, M., Heaton, T.H.E., Bogaard, A., Evershed, R.P., 2013. The effect of charring and burial on the biochemical composition of cereal grains: investigating the integrity of archaeological plant material. J. Archaeol. Sci. 40, 4767–4779. https://doi.org/10.1016/j.jas.2013.03.024.

Styring, A.K., Höhn, A., Linseele, V., Neumann, K., 2019. Direct evidence for agricultural intensification during the first two millennia AD in northeast Burkina Faso. J. Archaeol. Sci. 108, 104976. https://doi.org/10.1016/j.jas.2019.104976.

Tankersley, K.B., Conover, D.G., Lentz, D.L., 2016. Stable carbon isotope values ($\delta^{13}C$) of purslane (*Portulaca oleracea*) and their archaeological significance. J. Archaeol. Sci. Rep. 7, 189–194.

Taylor, R.E., Aitken, M.J., 1997. Chronometric dating in archaeology. In: Advances in Archaeological and Museum Science. vol. 2. Oxford University Press, Oxford.

Toolin, L.J., Eastoe, C.J., 1993. Late pleistocene-recent atmospheric $\delta^{13}C$ record in C_4 grasses. Radiocarbon 35, 263–269.

Tornero, C., Aguilera, M., Ferrio, J.P., Arcusa, H., Moreno-García, M., Garcia-Reig, S., Rojo-Guerra, M., 2018. Vertical sheep mobility along the altitudinal gradient through stable isotope analyses in tooth molar bioapatite, meteoric water and pastures: a reference from the Ebro valley to the Central Pyrenees. Quat. Int. 484, 94–106. https://doi.org/10.1016/j.quaint.2016.11.042.

Treasure, E.R., Church, M.J., Gröcke, D.R., 2016. The influence of manuring on stable isotopes ($\delta^{13}C$ and $\delta^{15}N$) in Celtic bean (*Vicia faba* L.): archaeobotanical and palaeodietary implications. Archaeol. Anthropol. Sci. 8, 555–562. https://doi.org/10.1007/s12520-015-0243-6.

Treasure, E.R., Gröcke, D.R., Caseldine, A.E., Church, M.J., 2019. Neolithic farming and wild plant exploitation in Western Britain: archaeobotanical and crop stable isotope evidence from Wales (c. 4000–2200 cal bc). Proc. Prehist. Soc. 85, 193–222. https://doi.org/10.1017/ppr.2019.12.

Trimborn, P., Becker, B., Kromer, B., Lipp, J., 1995. Stable isotopes in tree rings: a palaeoclimatic tool for studying climatic change. In: Frenzel, B., Stauffer, B., Weiss, M.M. (Eds.), Problems of Stable Isotopes in Tree-Rings, Lake Sediments and Peat-Bogs as Climatic Evidence for the Holocene, Palaeoclimate Research. European Science Foundation, Strasbourg, pp. 163–170.

Turney, C.S.M., Wheeler, D., Chivas, A.R., 2006. Carbon isotope fractionation in wood during carbonation. Geochim. Cosmochim. Acta 70, 960–964. https://doi.org/10.1016/j.gca.2005.10.031.

Urey, H.C., 1932. A hydrogen isotope of mass 2. Phys. Rev. 39, 164–165. https://doi.org/10.1016/S0016-0032(32)90450-5.

Vaiglova, P., Snoeck, C., Nitsch, E., Bogaard, A., Lee-Thorp, J., 2014. Impact of contamination and pre-treatment on stable carbon and nitrogen isotopic composition of charred plant remains. Rapid Commun. Mass Spectrom. 28, 2497–2510. https://doi.org/10.1002/rcm.7044.

Van Der Merwe, N.J., Vogel, J.C., 1978. 13C content of human collagen as a measure of prehistoric diet in woodland North America. Nature 276, 815–816. https://doi.org/10.1038/276815a0.

Van Zeist, W., Casparie, W.A., 1984. Plants and ancient man: studies in palaeoethnobotany. In: Van Zeist, W., Casparie, W.A. (Eds.), Plants and Ancient Man: Studies in Palaeoethnobotany. Balkema, Rotterdam.

Vernet, J.L., Pachiaudi, C., Bazile, F., Durand, A., Fabre, L., Heinz, C., Solari, M.E., Thiebault, S., 1996. Le $\delta^{13}C$ de charbons de bois préhistoriques et historiques méditerranéens, de 35000 BP à l'actuel. Premiers résultats. Comptes Rendus l'Academie des Sci. série II. Sci. la Terre des planètes 323, 319–324.

Vignola, C., Masi, A., Balossi Restelli, F., Frangipane, M., Marzaioli, F., Passariello, I., Stellato, L., Terrasi, F., Sadori, L., 2017. $\delta^{13}C$ and $\delta^{15}N$ from ^{14}C-AMS dated cereal grains reveal agricultural practices during 4300–2000 BC at Arslantepe (Turkey). Rev. Palaeobot. Palynol. 247, 164–174. https://doi.org/10.1016/j.revpalbo.2017.09.001.

Vignola, C., Masi, A., Balossi Restelli, F., Frangipane, M., Marzaioli, F., Passariello, I., Rubino, M., Terrasi, F., Sadori, L., 2018. $\delta^{13}C$ values in archaeological ^{14}C-AMS dated charcoals: assessing mid-Holocene climate fluctuations and human response from a high-resolution isotope record (Arslantepe, Turkey). Rapid Commun. Mass Spectrom. 32, 1149–1162. https://doi.org/10.1002/rcm.8137.

Vogel, J.C., 1993. Variability of carbon isotope fractionation during photosynthesis. In: Ehleringer, J.R., Hall, A.E., Farquhar, G.D. (Eds.), Stable Isotopes and Plant Carbon—Water Relations. Academic Press, Inc., San Diego, pp. 29–38.

Vogel, J.C., Van Der Merwe, N.J., 1977. Isotopic evidence for early maize cultivation in New York state. Am. Antiq. 42, 238–242. https://doi.org/10.2307/278984.

Voltas, J., Camarero, J.J., Carulla, D., Aguilera, M., Ortiz, A., Ferrio, J.P., 2013. A retrospective, dual-isotope approach reveals individual predispositions to winter-drought induced tree dieback in the southernmost distribution limit of Scots pine. Plant Cell Environ. 36, 1435–1448. https://doi.org/10.1111/pce.12072.

Wallace, M., Jones, G., Charles, M., Fraser, R., Halstead, P., Heaton, T.H.E., Bogaard, A., 2013. Stable carbon isotope analysis as a direct means of inferring crop water status and water management practices. World Archaeol. 45, 388–409. https://doi.org/10.1080/00438243.2013.821671.

Wallace, M.P., Jones, G., Charles, M., Fraser, R., Heaton, T.H.E., Bogaard, A., 2015. Stable carbon isotope evidence for neolithic and bronze age crop water management in the eastern Mediterranean and southwest Asia. PLoS One 10, 1–19. https://doi.org/10.1371/journal.pone.0127085.

White, J.W.C., Ciais, P., Figge, R.A., Kenny, R., Markgraf, V., 1994. A high resolution record of atmospheric CO_2 content from carbon isotopes in peat. Nature 367, 153–156.

Willcox, G., Buxo, R., Herveux, L., 2009. Late Pleistocene and early Holocene climate and the beginnings of cultivation in northern Syria. The Holocene 19, 151–158. https://doi.org/10.1177/0959683608098961.

Williams, D.G., Gempko, V., Fravolini, A., Leavitt, S.W., Wall, G.W., Kimball, B.A., Pinter, P.J., LaMorte, R., Ottman, M., 2001. Carbon isotope discrimination by *Sorghum bicolor* under CO_2 enrichment and drought. New Phytol. 150, 285–293. https://doi.org/10.1046/j.1469-8137.2001.00093.x.

Wright, P., 2003. Preservation or destruction of plant remains by carbonization? J. Archaeol. Sci. 30, 577–583. https://doi.org/10.1016/S0305-4403(02)00203-0.

Yang, Q., Li, X., Liu, W., Zhou, X., Zhao, K., Sun, N., 2011. Carbon isotope fractionation during low temperature carbonization of foxtail and common millets. Org. Geochem. 42, 713–719. https://doi.org/10.1016/j.orggeochem.2011.06.012.

Yasuda, S., Ooki, S., Naraoka, H., Akagi, T., 2015. Carbon isotopes in *Sphagnum* from Kyushu, Japan, and their relationship with local climate. Geochem. J. 49, 495–502. https://doi.org/10.2343/geochemj.2.0375.

Zohary, D., Hopf, M., 2000. Domestication of Plants in the Old World, third ed. Oxford University Press, New York, Oxford.

CHAPTER FOUR

Review of the Upper Ediacaran-Lower Cambrian detrital series in Central and North Iberia: NE Africa as possible source area

José M. Ugidos*, Pedro Barba, and María Isabel Valladares
Departamento de Geología, Facultad de Ciencias, Universidad de Salamanca, Salamanca, Spain
*Corresponding author: e-mail address: jugidos@usal.es

Contents

Abstract

This work offers a review of the Upper Ediacaran and Lower Cambrian succession in the Central and North Iberian Massif. The aim is to simplify the terminology and stratigraphic descriptions to ensure that the treatment of all available geochemical data is more easily understood.

Stratigraphy & Timescales, Volume 5
ISSN 2468-5178
https://doi.org/10.1016/bs.sats.2020.08.004

In general terms, the stratigraphic succession comprises of three depositional sequences. Depositional sequence 1, of Ediacaran age, is incomplete and only appears in the Cantabrian and West Asturian Leonese zones. Depositional sequence 2, of Late Ediacaran age, is found in the Central Iberian zone and corresponds to a sedimentary wedge having the Cantabrian and West Asturian Leonese zones as a secondary area of origin. Depositional sequence 3, of Early Cambrian age, appears in the three zones of study. Samples from all the three sequences have been selected for analytical purposes. The samples were collected within stratigraphic series previously studied but also randomly. In total 240 analyses of major and trace elements and 36 analyses of Nd-isotopes have been considered in the current work.

Geochemical data for the Upper Ediacaran detrital rocks of the Central and North Iberia provides evidence of a remarkable geochemical uniformity (trace elements, Nd-isotopes) despite the distances separating the sampling areas.

The geochemical uniformity shown by the Upper Ediacaran siliciclastic sediments from the three zones is probably the most important result. So, it is proposed that the extensive volumes of first cycle detrital materials in the Arabian-Nubian Shield mixed with older crustal materials from the Saharan Metacraton and that these mixtures evolved and homogenized in a continental setting across NE Africa prior to final sedimentation in a marine basin. The Lower Cambrian succession corresponds to decreased relative proportions of detritus derived from the Arabian-Nubian Shield.

1. Introduction

The interpretation of the geological setting and provenance of the Upper Ediacaran-Lower Cambrian siliciclastic record in the Central Iberian zone (CIZ), West Asturian Leonese zone (WALZ) and Cantabrian zone (CZ) (Fig. 1) ranges from those proposing magmatically active margin settings mainly based on U-Pb zircon ages and tectonic discrimination diagrams (e.g., Fernández-Suárez et al., 1998; Pastor-Galán et al., 2013; Pereira et al., 2012; Rubio-Ordóñez et al., 2015) to those favoring a passive margin setting, mainly from the geochemical results of detrital rocks (Bauluz et al., 2000; Beetsma, 1995; Tassinari et al., 1996), as well as stratigraphic, petrological and geochemical data from the siliciclastic series (Ugidos et al., 1997a,b, 2003, 2010, 2016; Valladares et al., 1993, 2000, 2002a). The poor appraisal of stratigraphic studies and the misuse of geochemical data are probable causes for discrepant interpretations. Thus, the main purposes of this work are to review previous studies to examine the use of geochemical data on siliciclastic rocks and to propose an interpretation including the stratigraphic record and source areas of the siliciclastic series in the zones of study. The geochemical homogeneity of the Upper Ediacaran shales of the three zones has been established over extensive areas of tens of thousand of km^2 and consistently supports the

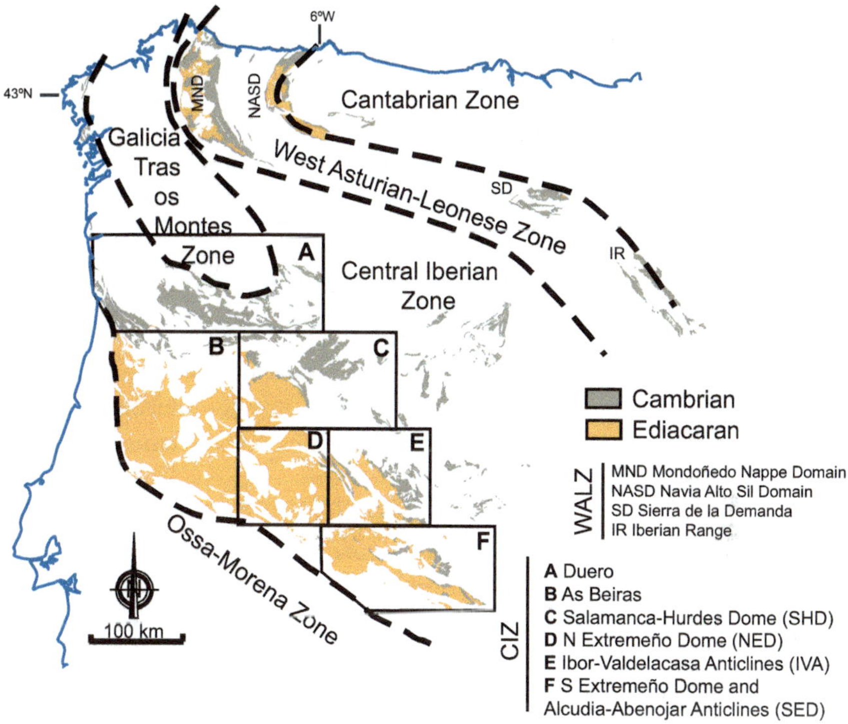

Fig. 1 Zoning of the Iberian Massif and simplified geographical distribution of Ediacaran and Cambrian detrital rocks. Based on the Mapa Geológico de España y Portugal 1:1.000.000 (Rodríguez-Fernández and Oliveira, 2015). For maps to a scale suitable for SD and IR Precambrian outcrops see Ramírez Merino et al. (1990), and Álvaro and Blanc-Vallerón (2002, references therein), respectively.

notion that the Upper Ediacaran detrital series is mainly derived from recycled homogenized detritus. The Lower Cambrian series is not as homogeneous as the Ediacaran series. However, distinctive geochemical features separate the Upper Ediacaran from the Lower Cambrian detrital series in the CIZ, while in the other two zones, this geochemical change lies within the Lower Cambrian series.

The results in this study support the notion that the Saharan Metacraton and Arabian-Nubian Shield were the source areas. It is suggested that the evolution of geological events leading up to the final sedimentation of the Upper Ediacaran and Lower Cambrian series was more complex and distinct from what is normally accepted.

In some cases, published stratigraphic terms such as Formation and Group are written incorrectly. According with norms of the International Commission on Stratigraphy the first letters of the names of formal stratigraphic units are capitalized. Informal terms are not capitalized. However, other terms (e.g., Precambrian) used by cited authors are respected in this study if there are no more specific new articles.

2. Geochemical data and the study of siliciclastic series

Three main assumptions underlie geochemical studies on siliciclastic rocks: (a) fine-grained rocks best reflect the geochemical features of source areas (Condie, 1991, 1993; Jahn and Condie, 1995; Johnsson, 1993); (b) elements such as Al, Ti, Zr, Hf, Nb, Th, rare earth elements (REE), Y, Sc, and Cr are quantitatively transported to sedimentary basins (e.g., Condie, 1991; McCulloch and Wasserburg, 1978; McLennan, 1989; McLennan and Hemming, 1992; McLennan and Taylor, 1991; Miller et al., 1986); and (c) geochemical features of sediments reflect those of magmatic rocks in the respective geotectonic settings. Consequently, abundances and ratios of these elements are potentially useful to define provenance and tectonic settings of detrital rocks and tectonic discrimination diagrams were proposed on the basis of major (Bhatia, 1983; Roser and Korsch, 1988) and trace element data for the detrital rocks (Bhatia and Crook, 1986).

However, many examples of redistribution of Th, transition metals, REE and Y (Cullers, 1988; Nesbitt and Wilson, 1992) have been mentioned as a consequence of adsorption by clays (Cullers et al., 1979; Elderfield et al., 1990), differential dissolution of minerals (Öhlander et al., 2000), mineral fractionation (Condie et al., 1995; Nesbitt and Markovicks, 1997; Nesbitt et al., 1980), generation of secondary minerals (Banfield and Eggleton, 1989; Braun et al., 1993; Cotten et al., 1995), and precipitation of Fe-oxyhydroxides from natural waters (Bau, 1999). Major chemical changes in detrital sediments are also possible during diagenesis, and many authors (e.g., Condie et al., 1995; Milodowski and Zalasiewicz, 1991; Ohr et al., 1991; Rasmussen et al., 1998) have reported evidence of REE redistribution, Sm/Nd fractionation and neodymium isotope resetting on a regional level (Bock et al., 1998; Bouch et al., 1995; Lev et al., 1999, references therein).

Many authors are been skeptical with regard to the use of tectonic discrimination diagrams because since experience has shown that sediment misclassification is common (Armstron-Altrin and Verma, 2005;

Armstrong-Altrin, 2015; Floyd et al., 1991; Haughton, 1988; Mack, 1984; McCann, 1991; Pe-Piper et al., 2008; Ryan and Williams, 2007; Velvel, 1985; Winchester and Max, 1989). The potential transport of sediments across tectonic boundaries constrains the applicability of diagrams presenting tectonic settings and sedimentary compositions (McLennan, 1989). In addition, inappropriate assumptions regarding the use of discrimination diagrams greatly reduce their applicability (McLennan et al., 1990; Rollinson, 1993). According to Pe-Piper et al. (2008): "Thus it is unlikely that a globally applicable set of elemental discriminants can identify terrigenous sediment sources." Possibly the most common inappropriate assumption is an a priori implicit uncritical acceptance of the first cycle nature of detrital sediments. Detrital materials could possibly reflect geochemical features of coeval volcanic rocks contributions, but depending on the relative proportions of different contributors to siliciclastic materials during recycling and specific processes in sedimentary environments, buffering of volcanic contributions may happen.

Therefore, the use of geochemical data to study provenance and geological settings of siliciclastic rocks requires knowledge as to whether or not the rocks are made up of recycled sediments. Due to the closure effect of analyses, the leaching of major soluble elements during weathering can result in a relative increased abundance of oxides such as Al_2O_3, TiO_2 and P_2O_5, corresponding to an increased abundance of clays and distinct heavy minerals. Hydrodynamic conditions can fractionate minerals (e.g., quartz from clays) and modify their relative abundances and element ratios in the siliciclastic detritus (Condie, 1993; McLennan, 1989; McLennan and Taylor, 1991; Morton and Hallsworth, 1999). Heavy minerals in detrital rocks may be inherited from igneous rocks, older detrital ones or mixtures of both, and may also result from weathering, diagenesis and hydrothermal processes along with one or more sedimentary cycles. Thus, a detrital rock may have different source rocks, and different populations of a mineral are expected to be characterized by varying sizes, shapes, densities and trace element contents, depending on the pressure (P)-temperature (T)-chemical (X) conditions in which they are formed. So, each population of a mineral may have a distinct physical behavior and may fractionate with respect to the others during hydrodynamic processes. A relative increase of the abundance of resistant minerals, such as Ti- and P-bearing minerals (e.g., rutile and apatite, respectively), would result in an increase of not only TiO_2 and P_2O_5 whole-rock content, but also in the respective housed trace elements, such as Nb, Ta, REE, Y and Th. The same may occur to secondary phyllosilicates or oxides bearing trace elements such as Rb, Cs, REE and Sc. A specific example is the diadochy of Sc and Fe(III) which facilitates Sc and Fe association in Fe-bearing secondary clays (Birgul, 1981) and Fe-oxyhydroxides, as well as those cemented into different-sized detrital particles (Brown et al., 2003). Fractionation of clays, metal oxide minerals and detrital particles would also cause Sc fractionation. Therefore, major and trace element abundances and element ratios of recycled detrital materials are expected to be different from those of the corresponding first cycle detritus and are not reliable for the characterization of geotectonic settings. The use of Nd isotope data without examining the possible redistribution of rare earth elements should be avoided since these processes may seriously affect $^{147}Sm/^{144}Nd$ ratios and values that differ significantly from those in the 0.100–0.130 range should be used with caution (Bock et al., 1994; Jahn and Condie, 1995; Zhao et al., 1992). Data on REE abundances and REE patterns are necessary in order to discard or identify fractionation of these elements in sediments (McDaniel et al., 1994) prior to interpreting Nd-isotope results.

3. Volcanic material and detrital series

The presence of volcanic rocks in a sedimentary succession does not necessarily imply a coeval magmatically active margin setting, given that tholeiitic rocks or those with calc-alkaline affinities are also found in post-orogenic extensional continental settings. Excessive emphasis on the local presence of volcaniclastic rocks interbedded with sedimentary strata and the failure to consider the quantitative importance of their extension relative to that of the related detrital series lacking volcaniclasts, and without geochemical data of these series may lead to flawed interpretations. Furthermore, there are no discussions on whether volcanic clasts are pyroclasts or epiclasts or how they have been distinguished after episodes of folding, alteration, and metamorphism of the sedimentary record. Also, it is necessary to determine if amphibolites are *ortho-* or *para-*rocks (Leake, 1964), especially if they are associated with beds of marls or siliciclastic-carbonated rocks, before assuming that they came from basalts or andesites.

Sedimentary series related to magmatically active margin settings are expected to have an abundant coeval igneous contribution, volcanic lithic fragments and petrological and geochemical variability of siliciclastic rocks. Also, relatively low values of La/Sc, La/Cr, Th/Cr and Th/Sc, and high values of Eu/Eu* ratios are found in fine-grained siliciclastic rocks due to contribution of mafic rocks, while geochemical and petrological homogeneity of siliciclastic rocks reflects recycling, and sands and muds of constant composition are expected to result from steady-state weathering conditions over long periods (Condie and Wronkiewciz, 1990; Condie et al., 1995; Cox and Lowe, 1995; Cox et al., 1995; Crichton and Condie, 1993; Eriksson et al., 1992; Feng et al., 1993; Garrels, 1988; McLennan, 1989; McLennan and Hemming, 1992; McLennan and Taylor, 1991; McLennan et al., 1993, 2006; Veizer and Jansen, 1985). Therefore, the distinction between first cycle and recycled siliciclastic rocks should not be very complicated, based on petrological, major and trace element data. Evidence of sedimentary homogeneity in extensive areas should be sufficient to discard magmatically active margin settings that are coeval with sedimentation.

4. Sampling and analytical techniques

For this study, random shales and sandstones sample collection began in the early 1990s, following field work campaigns, and was carried out using 1:50,000 scale sheet geological maps published by the Instituto Geológico y Minero de España (IGME) and continued combining stratigraphic and chemical data. Each sample is a composite (4–5 kg) of the corresponding outcropping material. It should be noted that the separation of the Upper Ediacaran and Lower Cambrian detrital rocks based on their geochemical features is consistent with the geological information in these maps. The only discrepancy was found in the Duero region (CIZ), where all of the collected samples revealed Lower Cambrian chemical features. Improved knowledge of the stratigraphy of the siliciclastic series of the CIZ led to certain changes in the previous sample adscription to stratigraphic units but no significant geochemical variations resulted. The analysis of samples from known stratigraphic units is presented in Tables (see Appendix) that are separated from those corresponding to samples randomly collected from other stratigraphic series. Samples of the Upper Ediacaran and/or Lower Cambrian detrital rocks of the CIZ have been collected in the following areas (Fig. 1): (A) Duero, (B) As Beiras, (C) Salamanca-Hurdes Dome (SHD), (D) North Extremeño Dome (NED), (E) Ibores-Valdelacasa Anticlines (IVA) and (F) South Extremeño Dome and Alcudia-Abenojar-Anticlines (SED).

In the SHD, 12 units (I–XII) were defined (Valladares et al., 2000). Quality of outcrops in other areas hindered a detailed study and sampling of two units (2 and 3) was only possible in the IVA. Since it is not clear if these units correspond precisely to units II and III of the SHD, Arabic numerals are used, to avoid confusion.

Samples of the corresponding WALZ detrital rocks have been collected in the Mondoñedo Nappe Domain (MND), and in the Navia and Alto Sil Domain (NASD). The scarcity of outcrops limited the number of samples and the stratigraphic study of the siliciclastic successions in the WALZ. Sampling of the Lower Cambrian shales in this zone was carried out using the Upper Ediacaran-Lower Cambrian unconformity and a carbonate bed some meters above this unconformity as stratigraphic references. The shales collected between the unconformity and the carbonate bed are called "underlying shales" and those above this bed are called "overlying shales" (Ugidos et al., 2016).

Chemical analyses were performed with ICP-AES (major elements) and ICP-MS (trace elements) at the Service d'Analyse des Roches et des Mineraux of the CRPG-CNRS (Nancy, France). The procedures for sample processing are described in Carignan et al. (2001). Groups of samples were sent for analysis over distinct years and in some samples, certain elements were not analyzed. Scandium was analyzed with ICP-OES at the Service d'Analyse des Roches et des Mineraux of the CRPG-CNRS and also at the Department of Earth and Environmental Sciences of the University of St Andrews (ICPMS Laboratory). Differences in Sc content between the duplicate samples analyzed at both laboratories are $<2\%$. Nd isotope analyses were performed at two laboratories. At the Scottish Universities Environmental Research Centre (East Kilbride, UK) Nd was separated using techniques such as those described by Barbero et al. (1995). $^{143}Nd/^{144}Nd$ ratios were measured using a VG Sector 54–30 mass spectrometer in multi-dynamic mode and mass fractionation was corrected for using an exponential law and $^{146}Nd/^{144}Nd=0.7219$. Analyses of the internal laboratory standard (JM) afforded $^{143}Nd/^{144}Nd=0.511502\pm4$ (1σ, $n=9$) which, based on a previous intercalibration, is equivalent to 0.511852 for the La Jolla standard. Measurements of Nd isotope values were also carried out at the SGIker-Geochronology and Isotopic Geochemistry facility of the University of the Basque Country (Spain). The chemical procedures for sample preparation are described in Pin and Santos Zalduegui (1997). The $^{147}Sm/^{144}Nd$ values are precise to $\pm0.2\%$ at the 95% confidence level. $^{143}Nd/^{144}Nd$ ratios were measured with TIMS using a Finnigan Mat-262 instrument in static multi-collection mode, and correcting for mass fractionation by normalization to $^{146}Nd/^{144}Nd=0.7219$ (Thirlwall, 1991; Wasserburg et al., 1981). The La Jolla isotopic standard measured under the same conditions afforded $^{143}Nd/^{144}Nd = 0.511857$. For calculations, the Ediacaran-Cambrian boundary was assumed at 541 ± 1.0 Ma (Cohen et al., 2013).

5. Summary of the published results in the three zones and new results

The following sections provide a review of the most significant characteristics already published for each of the zones.

5.1 Cantabrian zone (CZ)

In the Ediacaran siliciclastic succession of the CZ, typically referred to as Narcea group (Comte, 1959), Pérez Estaún (1978) differentiated 1500–1700 m of alternating graywackes and shales deposited by turbidity currents. Barba and Fernández (1991) and, more recently, Ugidos et al. (2016) studied 2700 m of the Ediacaran succession, consisting mainly of sandstones and shales, revealing sedimentological features characteristic of deep submarine fans (Fig. 2). The age of the Narcea group is constrained by the presence of acritarchs *Palaeogomphosphaeria caurensis* and *Sphaerocongregus variabilis* (Martín Parra, 1989; Palacios and Vidal, 1992) both of which are indicative of a Late Ediacaran age (Vidal et al., 1994). The presence of *Treptichnus* and *Rusophycus* (Jensen et al., 2010) above the angular unconformity indicates that the Herrería group (Fig. 2) sedimentation began at approximately the same time as the Late Corduban Series of the West Gondwana chronostratigraphic framework (Geyer and Landing, 2004). Thus, it was approximately at 529 Ma, i.e., during the early Stage 2 (Terreneuvian) of the International Chronostratigraphic Chart (Cohen et al., 2013). The hiatus between the Narcea group and the Herrería group is about 13 Ma (Ugidos et al., 2016).

Below the angular unconformity separating the Upper Ediacaran and Lower Cambrian detrital series, the Ediacaran rocks reveal weathering and crosscutting rubefaction (Gutiérrez-Alonso et al., 2004; van der Bosch, 1969). Calculations of the geochemical changes resulting from these processes demonstrate a loss of MgO, CaO, Na_2O, P_2O_5, Be and Sr of the altered Ediacaran shales and a gain of K_2O, Rb, and Cs in these rocks. For example, the mean values of the most important changes correspond to a loss of MgO (approximately 30%), CaO (>70%) and Na_2O (>95%), and an increase of K_2O and Rb of approximately 50% and Cs >150%, respectively. Consistently, mineral semi-quantitative analyses (XRD) revealed that the unaltered Ediacaran shales changed mean contents of albite (13%), chlorite

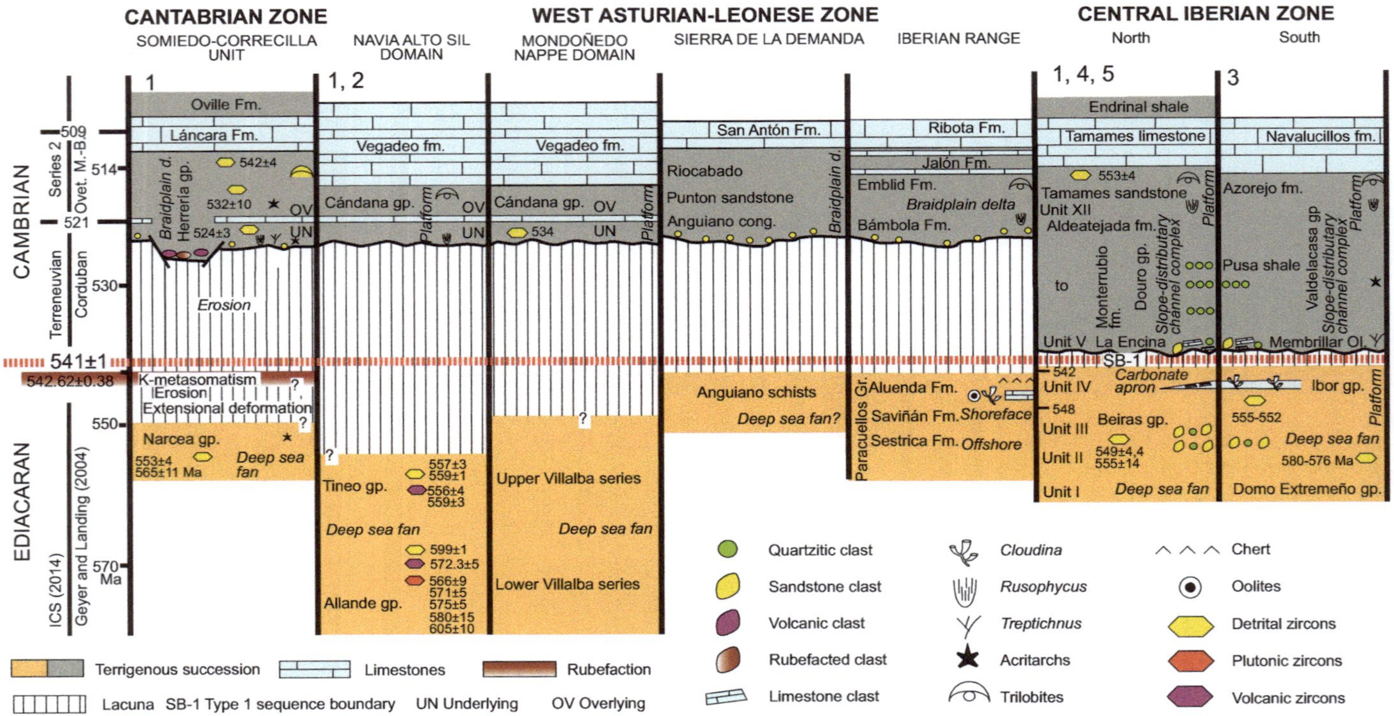

Fig. 2 Simplified correlation chart of the lithoestratigraphic units of the Late Ediacaran-Early Cambrian ages of the CZ, WALZ and CIZ. Zircon data: 1 Fernández-Suárez et al. (2014). 2 Fernández-Suárez et al. (1998); Rubio-Ordóñez et al. (2015). 3 Fernández-Suárez et al. (2000). 4 Talavera et al. (2015). 5 Gutiérrez-Alonso et al. (2003). 6 Pereira et al. (2012).

(30%) and illite (35%) to an absence of albite, a lower chlorite content (15%) and an increased abundance of illite (60%) in the rubefacted shales. The abundances of other major elements lack significant changes and also most of the other trace elements.

The weathering and K-metasomatism of the Ediacaran rocks are probably related to the Late Ediacaran sea level-fall (Knoll et al., 2004, references therein) and the intense worldwide weathering (Parnell et al., 2014) processes (the so-called sub-Cambrian weathering) and distensive tectonics resulting in high-relief areas that also favored the erosion of basement materials.

About 1000 m of stratigraphical logs were studied in the Lower Cambrian (Rubio-Ordóñez et al., 2004; Ugidos et al., 2016). The basal Lower Cambrian overlies the angular unconformity and consists of a conglomerate showing angular clasts of completely rubefacted shales that correspond to a ravinement surface and well-rounded volcanic clasts displaying a rubefaction coating. Above this conglomerate sedimented in a continental setting, probably a fan delta (Rubio-Ordóñez et al., 2004), the sequence is clearly transgressive and culminates in a carbonate level measuring some 30 m in thickness, corresponding to the maximum flooding surface. Between the angular unconformity and the carbonate level, the shales lack albite, chlorite, and K-feldspar, and they reveal a remarkable illite content (80–82%). Also, element ratios such as Ti/Zr, Zr/Sc, Zr/Nb, (La/Yb)N, Rb/Th, Rb/Zr and relatively high εNd(541) values (-3.4 to -2.1) are akin to those of the rubefacted Ediacaran shales, suggesting the contribution of Ediacaran detritus to the Lower Cambrian succession. However, the shales above the carbonate level lack albite but contain chlorite (0–22%), illite (57–76%) and K-feldspar (4–12%), and present higher Ti/Zr, La/Sc, (La/Yb)N, and Eu/Eu* ratios, but lower Zr/Sc, Zr/Nb, Rb/Th ratios and lower values (-5.8 to -4.9) of εNd(541) as compared to the Ediacaran and basal Cambrian shales. The shales below and above the carbonate level are referred to as the underlying and overlying shales, respectively. The unaltered Ediacaran shales, the equivalent rubefacted shales and the Lower Cambrian shales define the same trend in ternary diagram chlorite-quartz-illite (Ugidos et al., 2016).

In the Narcea group, 80% of Neoproterozoic zircons yield positive εHf(i) and for all zircons older than 2 Ga the U-Pb ages yield negative εHf(i) values of -0.3 to -15.7 (Naidoo et al., 2018). Another major change was detected by Zimmermann et al. (2015) who found that approximately 80% of all detrital zircons from the lower part of the Lower Cambrian Herrería group are of Neoproterozoic age, with Mesoproterozoic (only one grain), Paleoproterozoic (<6%) and Archean (<3.5%) ages are minor. On the contrary, in the upper part of this group, 38.4% of the zircons are Neoproterozoic in age, one grain is Mesoproterozoic in age and nearly 45% are Paleoproterozoic and 12.8% are Archean in age. Similar results were found by Fernández-Suárez et al. (2014) for detrital zircons from the upper part of the Herrería group: 37% Neoproterozoic, one Mesoproterozoic grain, 43% Paleoproterozoic, and 18% Archean.

5.2 West Asturian Leonese zone (WALZ)

In this zone, the Ediacaran detrital series outcrop in four regions/domains: Navia and Alto Sil Domain (NASD), Mondoñedo Nappe Domain (MND), Sierra de la Demanda (SD) and Iberian Range (IR).

5.2.1 Navia and Alto Sil Domain (NASD)

The boundary between the WALZ and CZ is the La Espina thrust (Pérez Estaún et al., 1990), along which a minimum shortening of ca. 75 km (62%) occurred during the Variscan orogeny (Gutiérrez-Alonso, 1996).

The turbiditic nature of the Ediacaran detrital rocks (Narcea group) is generally accepted in the NASD, but the considerable rock deformation makes it difficult to reach definitive conclusions. Two units have been distinguished, although the contact is unclear: the lower (Allande group) is a sandy and pelitic siliciclastic unit and the upper (Tineo group) is also siliciclastic but with more abundant pelites (Gutiérrez-Alonso et al., 2003; Rubio-Ordóñez et al., 2015).

In the Allande group there are granitic rocks with intrusive ages ranging between 590 and 566 Ma (Fernández-Suárez et al., 1998, 2014). This group also reveals flows from basalt to dacite having varying thickness, up to 3 m interbedded in siliciclastic and pyroclastic rocks, mainly crystal-rich tuffs. A dacitic tuff yielded an age of 572 ± 5 Ma. In the Tineo group there are two volcanic complexes developed in submarine environments having 1 km thick sequences of mainly ignimbritic tuffs intruded by dacitic to basaltic dykes and sills. Both complexes reveal major to intense hydrothermal alteration and the interpretation of most geochemical diagrams is "practically impossible" (Rubio-Ordóñez et al., 2015). A rhyolite from the Cudillero complex revealed an age of 559 ± 4 Ma (Gutiérrez-Alonso et al., 2004), and an ignimbritic rhyolite was dated at 557 ± 3 Ma (Rubio-Ordóñez et al., 2015). Thus, about 13 Ma separate the volcanic activities recorded in the Allande and Tineo groups. Extensions of these volcanic rocks are less than 1 and 10 km^2, respectively, within an area of about 1400 km^2 of exposed

Ediacaran detrital rocks. In these two groups, 82% and 87% of the detrital zircons are Neoproterozoic in age; Mesoproterozoic (6% and 4%, respectively), Paleoproterozoic (6% and 8%, respectively), and Archean (4% and 3%, respectively) ages are minor (Fernández-Suárez et al., 2014). Detrital zircons from two samples of the Narcea group yielded mainly Neoproterozoic (73%) ages and minor Mesoproterozoic (5%), Paleoproterozoic (11%) and Archean (11%) ages (Naidoo et al., 2018).

The Lower Cambrian succession (Cándana group, Lotze, 1961) lays unconformably upon Ediacaran strata (Fernández-Suárez et al., 2014), although the extent and nature of this unconformity have yet to be fully characterized (Díaz García, 2006; Gutiérrez-Alonso, 1996). A sample from the lower-middle (assumed here to be underlying) Cambrian (Cándana group) yielded 79% of Neoproterozoic zircons, an absence of Mesoproterozoic grains, 12% Paleoproterozoic zircons and one Archean grain (Henderson et al., 2016).

5.2.2 Mondoñedo Nappe Domain (MND)

The Ediacaran series, called the Villalba series in this domain (Martínez-Catalán, 1985), consists mainly of pelites. Martínez-Catalán (1985) distinguished a lower and an upper series forming a 2600-m-thick, fining- and thinning-upward sequence of graywackes, siltstones and shales (Fig. 2). The Villalba series is strongly deformed, making detailed stratigraphic or sedimentological studies impossible. Thus, the nature of the contact of this unit with the overlying Cambrian series (Cándana group) is controversial due to the intense tectonic deformation related to the Variscan orogeny. A concordant contact was proposed (Dozy, 1983), although other authors consider it to be discordant (Martínez-Catalán, 1985; Matte, 1968; Pérez Estaún, 1978).

5.2.3 Sierra de la Demanda (SD)

The Precambrian succession in the SD, named Anguiano schists (Colchen, 1974), consists of mudstones and siltstones with intercalated fine-grained sandstones in beds having a few centimeters of thickness. Maximum observable thickness is approximately 150 m (Ramírez Merino et al., 1990). According to Colchen (1974), the contact between the Anguiano Schists and the base of Cambrian is tectonic, while Ábalos (2001) interprets it as an angular unconformity. The basal Cambrian unit is the Anguiano conglomerate formation (Colchen, 1974), consisting mainly of 300 m of conglomerates with clasts of quartz and quartzite, and an intercalation of

35 m thick sandstones located 170 m above the base of this formation (Fig. 2). No fossils have been reported, but this unit is lithologically similar to the Bámbola Fm (see below) of the Iberian Range (Liñán et al., 2002).

5.2.4 Iberian range (IR)

The Precambrian succession in the IR, known as Paracuellos slates, consists mainly of shales and sandstones, scarce chert (Liñán and Tejero, 1988) and oolitic dolostones (Álvaro and Blanc-Valleron, 2002). Liñán and Tejero (1988) distinguished four formations (bottom to top): Sestrica, Saviñán, Frasno and Aluenda Formations, referred to as the Paracuellos Group, having a thickness ranging between 450 and 650 m (Fig. 2). More recently, Álvaro and Blanc-Valleron (2002) referred to the Frasno Formation (4 m thick) as a Bed (the smallest formal lithostratigraphic unit), and they identified in the same stratigraphic position, but in a southeastern area, another bed with a similar thickness (4 m), but consisting of carbonates, the so-called Codos Bed. The lower 100 m of the Sestrica Formation do not outcrop and were studied on drill cores. They consist of laminated black shales with intercalations of very fine- to medium-grained sandstones or laterally discontinuous siltstones beds. The uppermost 60 m of the Sestrica Formation, the 200–300 m of Saviñán Formation and the 90–200 m of Aluenda Formation comprise of alternations of laminated shales and fine- to medium-grained, graded sandstones which, in the Saviñán Formation, are arranged in thickening- and coarsening-upward sequences. Between the Saviñán and Aluenda Formations is the Frasno Bed, which is a diagenetically induced bedded chert with intercalated shales, while the Codos Bed is a dolomitized ooidal and bioclastic packstone.

The presence of the ichnospecie *Torrowangea* aff. *rosei* in the Saviñán (Liñán and Tejero, 1988) and Aluenda Formations (Álvaro et al., 2008) and the fossil *Cloudina*-like shelly fossils in de Codos Bed (Álvaro and Blanc-Valleron, 2002) let to dating the Paracuellos Group as Late Ediacaran. The sedimentation of the 100 m basal Sestrica Formations began in a platform of low energy and anoxic conditions, from offshore to hemipelagic, evolving upward (rest of the Sestrica and Saviñán Formations) to a nearshore environment affected by waves and storms, with development of sandy shoals that laterally evolve to carbonated ooidal-bioclastic shoals (Codos Bed) and ended as a siliciclastic shoreface to an offshore platform (Álvaro and Blanc-Valleron, 2002).

The Cambrian succession in the IR is named Bámbola Formation (Fig. 2). The contact between the Paracuellos Group and Bámbola Formation is controversial. This contact has been interpreted as a thrust, an angular unconformity, a concordant contact, a cartographical unconformity, a paraconformity or an erosive disconformity, depending on the area studied (Álvaro and Blanc-Valleron, 2002; Liñán et al., 2004, references therein). Álvaro et al. (2008) proposed an erosive unconformity at the lowermost conglomerate and breccia channels of the Bámbola Formation.

The Bámbola Formation consists of 300–450 m of white quartzarenites and conglomerates with quartzite pebbles, having thin levels of shales interbedded in the upper part (Liñán et al., 2002). In these shales, some ichnofossils have been found, including *Monomorphichnus lineatus, Phycodes* and *Rusophycus,* thereby suggesting a Corduban age (a north-western peri-Gondwanan stage equivalent to the Nemakit-Daldynian-Tommotian stages of the Siberian Platform; Liñan and Gámed-Vintaned, 1993) for this formation (Liñán et al., 2002). The Bámbola Formation deposited in alluvial wedges and fluvial environments with development of alluvial plain sequences in the lower part which, with the gradual sea-level rise during the Corduban, let the evolution to a platform dominated by storms and waves (Álvaro et al., 2008).

No direct evidence has been found for rubefaction of the Ediacaran series, but the Lower Cambrian shales of the IR show an increase in the mean illite content (from 39% to 59%) and K_2O (4.0–5.6%), and a decrease of the mean content of chlorite (15–5%) and feldspar (16–5%), CaO (1.0–0.6%) and Na_2O (1.9–0.4%) as compared to those of the Upper Ediacaran (table 1 ofBauluz et al., 2000). These mineralogical and geochemical features have been also found in equivalent rocks of the CZ (Ugidos et al., 2016) and the same geochemical parameters that discriminate the Ediacaran from Cambrian shales in the CIZ also separate the equivalent shales in the IR (Valladares et al., 2002a). It is inferred that the IR and CZ shared the same source area and the same settings and processes related to the worldwide weathering during sub-Cambrian times.

5.3 Central Iberian zone (CIZ)

Several comments should be made before considering this zone. The name "schist-greywacke complex" was proposed (Carrington da Costa, 1950) in reference to a very extensive region in the CIZ consisting mainly of detrital rocks of uncertain pre-Ordovician age, partly metamorphosed, and intruded by different granite bodies. For many years, this indefinite term was useful, while the geology of this region was poorly known. However, the current

1:50,000 scale sheet geological maps published since the 80s separate the Ediacaran from the Lower Cambrian detrital series and many articles dealing with different fields of geology facilitate the use of a more precise terminology. Furthermore, graywackes have been not found in the areas studied (tens of thousands km^2) and schists occupy minor extensions. Shales and fine to medium-grained lithic arenites are the dominant detrital rock types. Nowadays, the name "schist-greywacke complex" is not practical and causes confusion in the geological literature of Iberia. Therefore, it is recommended to abandon this name.

Another controversial name is Alcudian, proposed by some authors for a succession outcropping in the SED, consisting of a lower series of schists and greywakes of at least 7000 m thickness (Lower Alcudian) and an upper series of microconglomerates, sandstones and limestones of about 2000 m thickness (Upper Alcudian) (Bouyx, 1970; Ovtracht and Tamain, 1970), later referred to as the Alcudian complex (Crespo and Rey, 1971). However, these authors described local outcrops and offered imprecise thickness calculations of both units. Unfortunately, the names Lower and Upper Alcudian were later used in a chronostratigraphic sense, but authors should recognize that there are no published stratigraphic series, descriptions of lithological and sedimentological characteristics or more precise thickness, information necessary in order to define Lower and Upper Alcudian as lithostratigraphic units. However, the presence of microfossils and trace fossil was used to assign an Early Cambrian age to the Upper Alcudian (Pieren and García-Hidalgo, 1999; Simón, 2017). Currently, there are some geologists that consider the Upper Alcudian succession to be of terminal Ediacaran age while others suggest an Early Cambrian age. This may be due in part to imprecise mapping and unclear extension of the Alcudian region. Therefore, it is recommended that the names Lower and Upper Alcudian should also be abandoned, until more precise data are available.

In the CIZ, there are extensive areas occupying tens of thousands of km^2 in which the Upper Ediacaran and Lower Cambrian siliciclastic series have been not affected by intrusive bodies or significant metamorphism. However, the absence of stratigraphic, sedimentological and structural studies on a general scale, and the lithological monotony of the successions and the scarcity of trace fossils and body fossils has let to the proliferation of names for the distinct lithostratigraphic units, age attributions with insufficient geochronological data and even to remarkable differences of the proposed thickness, between 4500 and 22,000 m, for the Upper Ediacaran-Lower Cambrian succession (Álvarez Nava et al., 1988; Bouyx, 1970; Diez Balda, 1986; Diez Balda et al., 1990; Ortega and González Lodeiro, 1986; Rodriguez Alonso, 1985;

San José et al., 1990; Santamaría, 1995; Valladares et al., 2000; Vidal et al., 1994).

In this zone, most studies have been performed in the SHD, IVA, NED and SED. In the IVA and SED, three lithostratigraphic groups consisting of Upper Ediacaran and pre-trilobites Lower Cambrian detrital rocks have been differentiated (Fig. 2): (a) Domo Extremeño group consisting of shales, sandstones, siltstones and conglomerates up to 9000 m thickness, deposited in deep-sea fan environments (Álvarez Nava et al., 1988); (b) Ibor group consisting of siliciclastic-carbonate succession of up to 1100 m thick, deposited in settings corresponding to three different types of platforms, from homoclinal ramps with oolitic shoals to platforms with oolitic and stromatolitic shoals (Álvarez Nava et al., 1988; Calvet and Salas, 1988); (c) Valdelacasa group consisting of a megabreccia up to 165 m thick with boulders and blocks up to 6 m in size, some having *Cloudina* within, and some of which are carbonate blocks with signs of subaerial exposure (paleokarst). Upwards, there are up to 4000 m of sandstones, siltstones, shales, and conglomerates, consisting mainly of quartz clasts. Frequently, the shales are black and there are intercalated phosphates (Álvarez Nava et al., 1988). At the top, this group has two formations: the lower, up to 650 m thick, consists of sandstones with some shales (Azorejo formation, Fig. 2), and the upper, up to 150 m thick, is carbonated (Navalucillos formation, Fig. 2). This group reveals a shallowing upward sedimentary sequence from the basal megabreccia to a sedimentary environment that is always covered by shallow water (Zamarreño et al., 1976). For this work, some 1600 m of stratigraphical logs of partial sections have been studied in IVA and NED.

The presence of different trace fossils, such as *Torrowangea rosei, Helminthoidichnites* and *Gordia,* led to the dating of the Domo Extremeño group as Late Ediacaran. The in situ presence of skeletal fossils of different species of *Cloudina* in the Ibor group (Cortijo et al., 2010, 2015; Vidal et al., 1994; Zhuravlev et al., 2012) in the IVA and SED indicates a terminal Ediacaran age (549 ± 1 and 542 ± 1 Ma, Grotzinger et al., 1995; Amthor et al., 2003). The presence of *Cloudina* in boulders and blocks of the megabreccia at the base of the Valdelacasa group caused some authors (Jensen et al., 2007; Vidal et al., 1994) to consider that this megabreccia is also of terminal Ediacaran age and to name this group Río Huso group (Nivel de Fuentes and Membrillar Olistostrome). However, Vidal et al. (1994) state that "the delicate nature of the shell and its sedimentological context indicate that *Cloudina* preferred lower energy environments on carbonate platforms protected by some form of wave barrier, such as ooid shoals or stromatolitic reefs," which are the environments where *Cloudina* appears in the Ibor group.

The presence of trace fossils, such as *Monomorphichnus lineatus* and *Treptichnus pedum*, above the olistostromic level permitted the dating of the lower part of the Valdelacasa group as Early Corduban Series of the West Gondwana chronostratigraphic framework (Geyer and Landing, 2004), equivalent to the Fortunian Stage of the International Chronostratigraphic Chart (Cohen et al., 2013), while the presence of *Rusophycus, Psammichnites* and *Dactyloidites* (Jensen and Palacios, 2016) in the middle part and at the beginning of the upper part allowed for the dating of these parts of the Valdelacasa group as Late Corduban Series.

In the SHD, more than 21,600 m of stratigraphical logs from partial sections directly measured, meter to meter, led to the establishment of four (I–IV, Upper Ediacaran) and eight (V–XII, Lower Cambrian) lithostratigraphic units, joining features suitable to formations, based on detailed stratigraphical, sedimentological and geochemical studies (Ugidos et al., 2003, 2010; Valladares et al., 1998, 2000, 2006). The units I–III were correlated with the Domo Extremeño group, the unit IV with the Ibor group and the units V–XII with Valdelacasa group (Valladares et al., 2000). In the SHD, unit V (La Encina Olistostrome, Fig. 2) is equivalent to the olistostromic level of the base of the Valdelacasa group and, as is the case in this group, some blocks and boulders contain *Cloudina*.

The upper Ediacaran-lower Cambrian sedimentary succession in the SHD and IVA is well known, as is the correlation between the units of the two domains. In both areas, there is a cartographic unit up to 3000 m thick (López-Díaz, 1994). However, the stratigraphical position of this cartographic unit is currently unclear and its meaning in the evolution of the sedimentary basin remains unknown. The unit consists of levels of slumped shales with thicknesses ranging from metric to decametric sizes. Within the shales, there are metric blocks of sandstones, and boulders of sandstones and black or gray shales that correspond to slumps through a slope dominated by native/intrabasinal clasts. Intercalated between different levels, there is an alternation of sandstones with a fine to very fine grain in beds with centimeter thickness and shales in Tb-e incomplete Bouma sequences. This stratigraphic unit corresponds to a tectono-sedimentary event related to movement of fractures.

In the SHD and IVA, the Upper Ediacaran series outcrop S-SW of the CIZ while the Lower Cambrian series outcrop at N-NE of this zone. The grain size of the siliciclastic Upper Ediacaran series decreases S-SW from conglomerates in the north of SHD, to sandstones in IVA and south of SHD, while shales dominate to the north and south of SED.

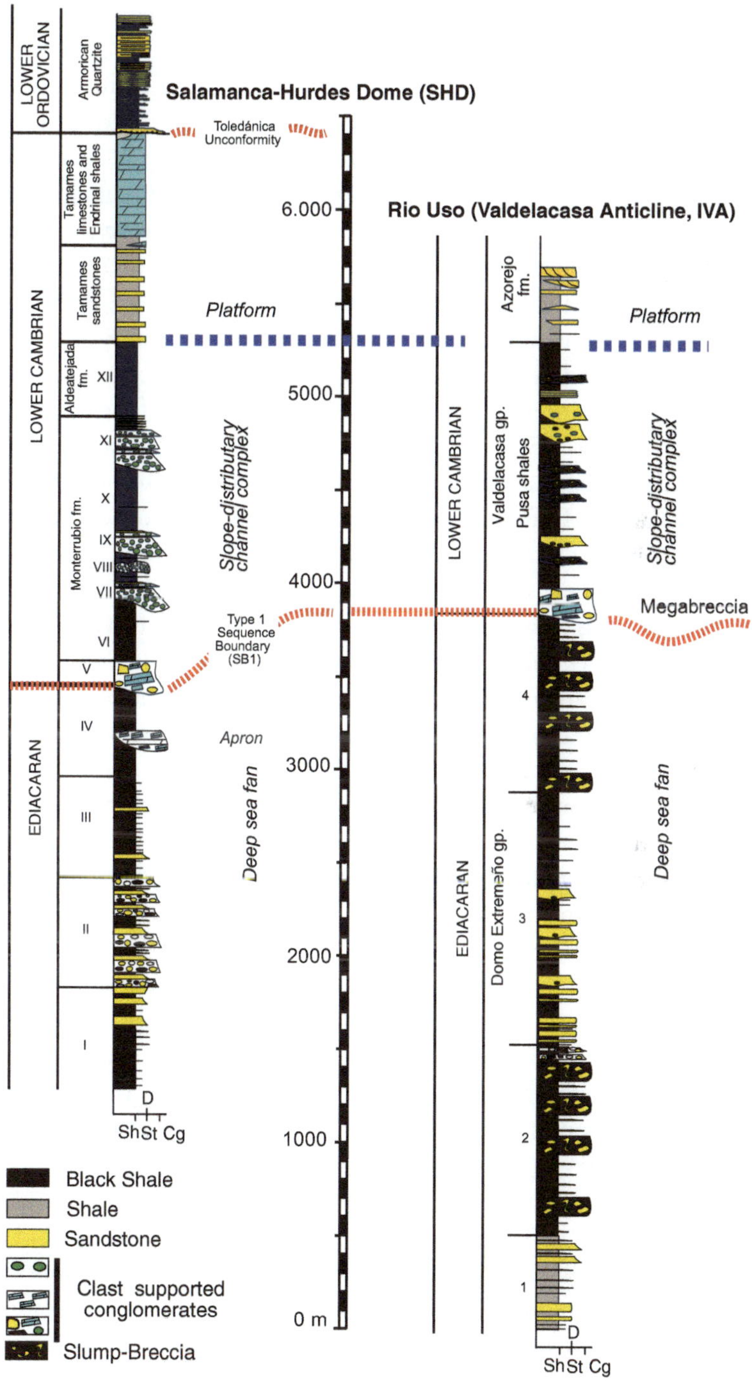

Fig. 3 Type series of the Upper Ediacaran-Lower Cambrian sedimentary succession in the SHD and IVA.

In the SHD, two depositional sequences were recognized (Figs. 2–4). They are separated by a type-1 sequence boundary, which "is interpreted to form when the rate of eustatic fall exceeds the rate of basin subsidence at the depositional-shoreline break, producing a relative fall in sea level at that position" (Van Wagoner et al., 1988), and is characterized by a strongly erosional surface that most likely corresponds to a major unconformity (Ugidos et al., 2016; Valladares et al., 2000, 2006) related to general world-wide weathering processes (Parnell et al., 2014, references therein) favored by a eustatic sea level fall at sub-Cambrian times (Knoll et al., 2004, references therein), that in the Iberian Massif received the name of Córdoba regression (Liñan and Gámed-Vintaned, 1993).

The lower depositional sequence, which is incomplete at the bottom, has a thickness of up to 2250 m in the SHD and up to 4000 m in the IVA. It contains mainly shales and sandstones (lithic arenites-lithic graywackes), with intercalations of polymictic conglomerates, thick-bedded sandstones, shales, black shales with phosphate nodules and alternations of limestone breccias, and stratified siliciclastic-carbonate rocks deposited mainly from gravity flows, such as turbidity currents and debris flows and, to a lesser extent, from submarine slides, dilute suspensions and tractive flows, in slope, base-of-slope and deep-sea fan environments (Figs. 3 and 4). The sandstones are fine-grained and sorting ranges from moderately well-sorted to well-sorted. Both the shales and the sandstones have a similar mineralogy of muscovite-illite, quartz and chlorite as the dominant minerals, together with minor albite (less than 7%). This depositional sequence is of Late Ediacaran age and exhibits lowstand system tract during the sedimentation of units I–III and Domo Extremeño group, and transgressive and highstand system tracts during the sedimentation of unit IV and Ibor group (Valladares et al., 2000).

Differences exist between SHD and IVA during transgressive and highstand system tracts. In the SHD, the unit IV contains mainly black shales, limestones breccias and stratified sandstone-limestone couples, with ooliths and bioclasts in the breccia clasts (Rodriguez Alonso, 1985), as well as clasts with wave ripples, most likely derived from a mixed rimmed shelf, although not preserved, which was located to the northeast of the CIZ, as indicated by the paleocurrents directions and deposited from debris flows and high-density turbidity currents in a mixed slope apron (Valladares, 1995; Valladares et al., 2000). However, in the IVA, the Ibor group consists of sandstones, shales, dolomites and limestones deposited in distinct platforms types, similar to as occurred in the IR. Therefore, in terminal Ediacaran the platform was at present east and northeastern.

The upper depositional sequence has a thickness of up to 2620 m and begins at the base of unit V, continuing with VI–XII units, and finishing with Tamames sandstones and Tamames limestones formations that correlated in lithology and age to the Azorejo and Navalucillos formations of the IVA. It is a shallowing upward sedimentary sequence that at the base has a megabreccia (unit V, La Encina Olistostrome, Fig. 2) of predominantly carbonated blocks with *Cloudina*. Overlying these blocks are siliciclastic blocks derived from the units IV and III (Valladares et al., 2006). This Olistostrome correlates with the Membrillar Olistostrome in IVA. In the rest of the depositional sequence (units VI–XII) black shales (units VI, VIII, X, XII), mainly deposited by settling and slides in a slope environment, alternate with sandstones and conglomerates (units VII, IX, XI) mainly deposited by debris flows and turbidite currents in distributary channels of inner deep-sea fans (Valladares et al., 2000). The units VI–XI correlate with Monterrubio formation and the unit XII with the Aldeatejada formation (Fig. 2) of Diez Balda (1980). The unit V represents the lowstand systems tract, while the units VI–XII represent an alternation of periods of poor circulation and periods with active turbiditic systems, probably having a transgressive tendency, while the highstand system tract corresponds to the two Tamames formations with siliciclastic shoreface and carbonate tidal flat and reefs of archeocyatha (Corrales and Valladares, 1980; Corrales et al., 1974; Valladares et al., 2000).

Waggoner (1999) studied the Ediacaran biota and grouped them into three major clusters from older to younger: the Avalon, White Sea and Nama Assemblages, finding cloudiniids in the Nama assemblage. It seems that there is quite a consensus that *Cloudina* has an equatorial distribution in shallow marine carbonate settings (Waggoner, 1999, 2003; Wood, 2011) favored by the high saturation of carbonate ions in surface seawater at low latitudes (Wood, 2011) and with a preference for *Cloudina* associated with thrombolite reefs (Grotzinger et al., 2003), and reefs, including Ediacaran ones, as sites of rapid and widespread *syn*-sedimentary cement formation (Wood, 2011).

During the sedimentation of unit IV and Ibor group, a stage of extensional tectonics began and resulted in high-relief areas and halfgrabens. According to paleocurrents, the Upper Ediacaran detrital material deposited in the halfgrabens came from the northeast (Valladares, 1995). It also started a worldwide sea level fall at the terminal Ediacaran (Knoll et al., 2004) and worldwide sub-Cambrian weathering (542.62 ± 0.38 Ma, Parnell et al., 2014). At the beginning of the Cambrian, a new stage of extensional tectonics also caused halfgrabens and high-relief areas, but paleocurrents indicate that detrital material came from the southwest (Ugidos et al., 2016; Valladares, 1995; Valladares et al., 2002a,b). The carbonate rocks with *Cloudina* that were on the margin of the platform remained under subaerial conditions and cemented before settling as olistostromic masses at the base of the slope through avalanches, rockfalls and slides, resulting from the destruction of the margin of platform (Valladares et al., 2000). Consequently, the olistostromes outcropping in the IVA (Nivel de Fuentes, Membrillar Olistostrome), and the outcropping in the SHD (La Encina Olistostrome) correspond to the first Cambrian sedimentary deposits. Thus, the Valdelacasa group and unit V are of Early Cambrian age.

Systematic studies on the geochemistry of the Upper Ediacaran and Cambrian detrital rocks began in the CIZ, where a combination of stratigraphic and geochemical data of 15 shales established possible criteria to separate the Upper Ediacaran from the Lower Cambrian detrital series (Valladares et al., 1993). Beetsma (1995) also published, among many other data on detrital and igneous rocks, 12 chemical analyses and 7 Nd-isotope data of metapelites that were considered to be of Late Proterozoic/Cambrian (LPC) age. Later, new stratigraphic and geochemical results revealed that chemical and Nd-isotope features of LPC metapelites coincide with those of the Upper Ediacaran shales in other areas of the CIZ (Ugidos et al., 1997a,b, 2003; Valladares et al., 2000, 2002a). Here, we, therefore, consider the LPC sedimentary rocks studied by Beetsma (1995) to be of Upper Ediacaran age.

Data from all of the zones revealed a remarkable geochemical homogeneity of the Upper Ediacaran shales and a standard of comparison was proposed, based on 100 analyses of shales from different zones: the Neoproterozoic Iberian Average Shale, NIBAS (Ugidos et al., 2010). Villaseca et al. (2014) questioned this homogeneity. These authors divided the metasedimentary rocks of the CIZ into the northern and southern Central Iberian terranes (N-CIZ and S-CIZ, respectively) both considered to be of Neoproterozoic age. This distinction is based mainly on the different geochemical features of the N-CIZ and S-CIZ detrital series. However, according to the 1:50,000 scale sheet geological maps and memoirs published by the IGME, the Upper Ediacaran and Lower Cambrian detrital series are present in both terranes. The difference is that the outcropping extension of the Lower Cambrian detrital series dominates in the N-CIZ while in the S-CIZ the Ediacaran outcrops are more extensive. Furthermore, Villaseca et al. (2014, p. 133) mention that geochronological

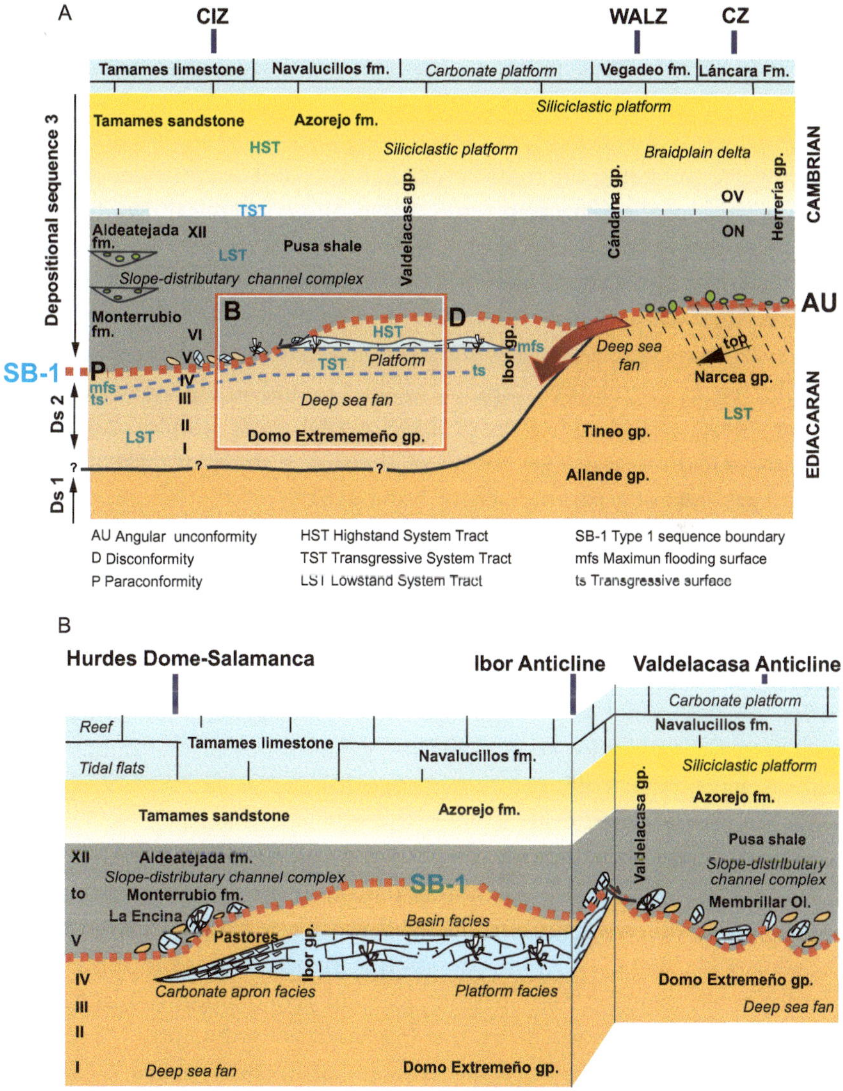

Fig. 4 (A) Framework of the relationships between the stratigraphic units of the studied domains. (B) Detail of the relationships between the Upper Ediacaran-Lower Cambrian limit in the SHD and IVA.

data for the augen orthogneisses are "… in the range of 546-464 Ma … supporting the Neoproterozoic age of most of these metasedimentary rocks." However, the intrusive ages (490–462 Ma) of orthogneisses of the

Guadarrama region (Castiñeiras et al., 2008; Navidad and Castiñeiras, 2011; Talavera et al., 2013) do not demonstrate a Neoproterozoic age of the intruded detrital series and the only Neoproterozoic age (546 ± 3 Ma, Zeck et al., 2004) corresponds to an orthogneiss located approximately 200 km to the SW of the Guadarrama region. Villaseca et al. (2014) consider the Lower Cambrian series to be Neoproterozoic and deny the geochemical homogeneity of the Ediacaran series, based on geochemical data for slates (6 samples), schists (2), paragneiss (1), hornfels (2), migmatites (4) and restitic granulites (5). One basic condition of geochemical studies of siliciclastic rocks is that samples from each group must have the same grain-size. However, many samples used by Villaseca et al. (2014) are high-grade metamorphic rocks of unknown original grain-size and stratigraphic position. Orejana et al. (2015) use seven of these detrital rocks samples and accept the supposed Ediacaran ages for all of them in a study of detrital zircon U-Pb ages, insisting on the lack of homogeneity of the Neoproterozoic rocks: "This lack of homogeneity, which contradicts previous ideas (e.g., Ugidos et al., 2010; Valladares et al., 2002a,b) …." However, these authors have never said that the detrital rocks of the CIZ shales are geochemically homogeneous, but rather, they described geochemical and isotopic differences between Ediacaran (units I–IV) and Lower Cambrian (units V–XII) shales in this zone and emphasized the homogeneity of (only) the Ediacaran rocks (Ugidos et al., 1997a,b; Valladares et al., 2002a,b). Some authors have repeatedly mentioned that Ugidos et al., and Valladares et al., propose magmatically active margins in published articles. The last example is from Rojo-Pérez et al. (2019): "…siliciclastic series laid down during the Ediacaran-Early Cambrian transition deposited in a magmatic arc setting in different positions along the African margin of Gondwana (Ugidos et al., 2003…)." This quotation is incorrect, since in this article, Ugidos et al., mention that "petrological and geochemical results suggest a stable continental source for the Precambrian sediments," "recycled orogen as source area," "West Africa craton provenance," "North Gondwana passive continental margin," and "a homogeneous and recycled source area." It seems clear that these authors do not favor the interpretation mentioned by Rojo-Pérez et al. (2019). Misquotations are clearly a serious problem.

Nd-isotope results of the Upper Ediacaran detrital rocks show εNd(541) values exceeding −4.1, except for two samples (−4.2 and −4.6) out of 61, while those of the Lower Cambrian ones have values of less than −4.1, with the exception of two samples (−3.9 and −4.0) out of 29 (see Tables 9–11).

The presence of volcanic and volcaniclastic rocks in the Upper Ediacaran series of the CIZ is a long-standing geological topic, frequently mentioned in favor of a continental, magmatically active margin setting coeval with sedimentation but unfortunately very rarely studied in detail. Possibly, the most important attempt was made by Rodríguez Alonso et al. (2004), but this study should be carefully revised. These authors proposed a magmatically active margin setting during Neoproterozoic-Early Cambrian times in the CIZ, based on the petrological and geochemical results of local volcaniclastic and massive volcanic rocks (metabasalts and metaandesites) associated with the Upper Ediacaran carbonate beds. According to the authors' descriptions these rocks were affected by secondary processes: "probable hydrothermal alteration"; "do not contain primary mineral assemblages"; show "mobility of cerium and europium ..., possibly resulting from seawater interaction"; and a metabasalt/metaandesite "shows enrichment in HREE from Ho, which is accompanied by high Zr, Hf and Th values. These may be due to a mechanical concentration of minerals." As for the volcaniclastic rocks, Rodríguez Alonso et al. (2004) say that "It is also very common to find some sequences ... clearly indicating the rapid deposition and efficient hydraulic sorting of particles during settling through the water column." The εNd results, calculated at 620 Ma (i.e., over 70 Ma older than the Late Ediacaran-Early Cambrian boundary), are −4.85 (a massive volcanic rock) and +0.24 and −1.78 (two volcaniclastic rocks). Thus, the supposed massive volcanic rock reveals a more crustal isotopic feature than the other two rocks consisting of carbonate, volcaniclasts and other detrital components. These results, along with the mentioned element changes, including REE mobility and a probable resetting of the Sm-Nd and Rb-Sr systems, make the proposed interpretation unreliable. No other specific outcrops of volcanic rocks have been reported in the Ediarcaran successions of the CIZ.

The siliciclastic rocks of the Upper Ediacaran of the CIZ are characterized by having content of Neoproterozoic zircons >71% and Paleoproterozoic zircons <18% (Fernández-Suárez et al., 2014; Pereira et al., 2012; Teixeira et al., 2011). Results from a chorite phyllite (Douro group) from Portugal reveal approximately 79% Neoproterozoic, 9% Mesoproterozoic, 9% Paleoproterozoic and 2% Archean aged zircons, respectively (Teixeira et al., 2011). However, in a metagreywacke (Beiras group), percentages are 44% Neoproterozoic, 16% Mesoproterozoic, 28% Paleoproterozoic and 12% Archean, respectively. In the Tamames formation, detrital zircons from the OD3 sample (Fernández-Suárez et al., 2014) reveal 85% Neoproterozoic, 5% Mesoproterozoic, 9% Paleoproterozoic and 3% Archean ages, which is not significantly different from other Ediacaran samples, according to a Kolmogorov-Smirnov test. However, εNd(t) of this sample is significantly more negative (Fernández-Suárez et al., 2014) than εNd(t) of the Lower Cambrian rocks of the CIZ, WALZ and CZ (<-4.1, see below). The zircons of the southern CIZ metasediments have more abundant εHf(t) positive values than negative, while in the northern CIZ, zircons with negative values are dominant (Orejana et al., 2015).

6. Geochemistry

The Chemical Index of Alteration [CIA = (Al_2O_3/Al_2O_3 + CaO* + Na_2O + K_2O)*100] of detrital rocks potentially reflects the intensity of the weathering of igneous source rocks (Nesbitt and Young, 1982). CaO* is the amount of CaO in silicate minerals after discounting that corresponding to apatite. However, in this work P_2O_5 was not considered to correspond to apatite for CIA calculation, due to the negative values of CaO* in many samples. It is inferred that the samples have other Ca-free phosphates. Also, in the A-CN-K diagram, the detritus derived from weathered igneous rocks would define a hypothetical trend parallel to A-CN join until cutting the AK join (Fedo et al., 1995; Nesbitt and Young, 1982). However, this trend may deviate from the parallel trend toward the K apex and reach the AK join at a lower position. This deviation may be due to K-metasomatism during diagenesis (Fedo et al., 1995). In the CIZ, it has been suggested that deviation trends for Upper Ediacaran and Lower Cambrian detrital rocks reflect the accumulation of detrital illite along the recycling of sediments (Ugidos et al., 2003). Whatever may be the cause of illite enrichment in rocks, this mineral may accumulate during the recycling of sediments and the resulting rocks will have lower CIA values than those corresponding to simple weathering. Thus, some interpretations may be erroneous if assuming lower chemical maturities of rocks, which, in fact, resulted from mineral fractionation when no other information is taken into account.

This study considers a total of 240 (including 106 new) analyses of Upper Ediacaran shales and fine- to medium-grained sandstones and of Lower Cambrian equivalent rocks of the CIZ and WALZ. Some previous analyses were repeated and more complete information is now offered. In the running text, there are simplified tables with the most significant features of these rocks to facilitate comparisons. These tables include data from North Portugal (Beetsma, 1995) and the CZ (Ugidos et al., 2016). Some of the tables in the running text present Al_2O_3/TiO_2 and Zr/Al_2O_3 ratios as Al/Ti and Zr/Al. Also, the molar proportions of Al_2O_3 and K_2O are presented as K/Al. Compiled (summary) Nd-isotope data (96) and new data (36) appear in a simplified and complete table, respectively. Apart from tables in the text, other Tables (A1–A16) in the Appendix show the entire analyses of all groups of shales and sandstones from stratigraphic units and randomly collected from different zones and areas within.

6.1 Ediacaran shales and sandstones

The geochemical characterization of the Upper Ediacaran detrital rocks of the CIZ is based on 55 samples of shales and 10 of sandstones collected at random, plus 51 samples of shales and 24 of sandstones collected in studied stratigraphic units in SHD and IVA. Total Ediacaran samples from the WALZ are 18 shales and 6 sandstones.

Table 1 shows mean values of selected chemical data and the corresponding standard deviations (1σ) of the Ediacaran shales from the CZ, WALZ and CIZ. All of the parameters reveal relatively low

Table 1 Synthetic table of major element abundances (wt.%) and element ratios of the Ediacaran shales. Mean values and st. dev. (1σ).

Areas/units	SiO_2	TiO_2	Al_2O_3	CIA	Al/Ti	K/Al
CZ (20)[a]	59.8(2.1)	0.85(0.1)	18.3(0.8)	68.1(2.1)	21.8(2.3)	0.223
Tineo gp.(6)	56.7(2.4)	0.91(0.1)	19.2(1.4)	70.7(3.3)	21.2(1.0)	0.207
Allande gp. (8)	58.9(1.7)	0.89(0.1)	18.8(0.8)	69.0(1.6)	21.3(1.2)	0.199
MND (4)	59.6(2.1)	1.05(0.1)	19.8(0.5)	76.6(4.0)	18.9(1.0)	0.205
SHD (6)	60.1(2.9)	0.94(0.1)	18.6(1.6)	70.1(1.2)	19.9(1.4)	0.194
NED (17)	60.4(1.6)	0.89(0.1)	18.9(0.9)	72.8(1.8)	21.3(1.0)	0.194
IVA (20)	59.0(1.7)	0.94(0.1)	19.1(1.2)	70.5(4.1)	20.4(1.3)	0.200
SED (12)	59.0(1.5)	0.94(0.1)	19.0(0.8)	70.3(3.0)	20.0(1.5)	0.190
As Beiras (11)[b]	60.6(3.9)	0.94(0.1)	20.5(2.1)	75.1(2.6)	21.9(1.0)	0.216
UNIT IV (7)	60.0(2.3)	0.91(0.0)	18.7(1.1)	69.9(3.0)	20.6(1.0)	0.202
UNIT III (8)	59.7(1.4)	0.97(0.1)	19.5(1.0)	72.6(1.7)	20.1(0.7)	0.198
UNIT II (15)	60.0(2.1)	0.99(0.1)	19.4(1.3)	74.4(1.8)	19.7(0.9)	0.204
UNIT I (8)	59.7(1.9)	1.03(0.0)	19.9(0.7)	74.7(2.7)	19.4(0.8)	0.205
UNIT 3 (11)	59.1(2.1)	1.00(0.1)	19.4(1.4)	71.6(2.0)	19.4(1.1)	0.201
UNIT 2 (2)	59.9(0.1)	0.95(0.0)	18.3(0.1)	70.2(1.3)	19.4(0.0)	0.201

Continued

Table 1 Synthetic table of major element abundances (wt.%) and element ratios of the Ediacaran shales. Mean values and st. dev. (1σ).—cont'd

Zr/Al	Ti/Nb	Ti/Zr	Zr/Nb	Nb/Hf	Th/U	(La/Yb)N
12.5(1.9)	430(10.0)	25.8(2.7)	16.8(1.5)	2.26(0.2)	3.65(0.4)	7.20(0.8)
10.6(0.8)	430(22.0)	26.9(1.3)	15.9(0.2)	2.39(0.1)	3.67(0.3)	7.41(0.8)
10.6(0.3)	421(21.5)	26.6(1.5)	15.8(0.6)	2.26(0.1)	3.60(0.3)	8.23(0.8)
10.8(0.8)	411(21.8)	25.8(1.6)	16.0(0.8)	2.25(0.1)	3.51(0.2)	8.21(0.7)
11.5(1.5)	430(15.4)	26.8(3.6)	16.2(1.6)	2.34(0.3)	3.16(0.3)	7.11(1.2)
10.7(0.8)	427(18.6)	26.7(1.4)	16.0(1.0)	2.37(0.2)	3.20(0.2)	7.47(0.8)
11.4(1.0)	437(48.1)	26.3(1.8)	16.6(1.5)	2.26(0.2)	3.41(0.3)	7.72(1.2)
11.0(1.0)	446(46.7)	27.4(2.4)	16.3(1.1)	2.30(0.2)	3.48(0.2)	7.41(0.8)
10.4(0.6)	400(18.1)	26.5(1.5)	15.2(0.8)	2.47(0.2)	3.57(0.6)	7.10(1.1)
11.2(1.6)	420(9.50)	26.4(2.5)	16.0(1.4)	2.36(0.1)	3.29(0.1)	7.35(0.9)
11.5(7.8)	437(25.0)	26.2(2.8)	16.8(1.2)	2.23(0.2)	3.40(0.9)	7.22(1.3)
12.0(0.8)	431(28.4)	25.5(1.7)	17.0(1.3)	2.13(0.2)	3.61(0.3)	7.30(1.3)
12.6(1.0)	434(15.6)	24.6(1.5)	17.7(0.9)	2.10(0.1)	3.68(0.3)	7.50(1.0)
11.8(1.1)	434(16.0)	26.1(1.5)	16.7(1.2)	2.27(0.2)	3.61(0.4)	7.52(0.5)
11.2(0.0)	440(2.10)	27.7(0.1)	15.9(0.0)	2.34(0.0)	3.43(0.0)	7.71(0.1)

[a]Ugidos et al. (2016).
[b]Beetsma (1995).
st. dev. < 0.05 compiled as 0.0. K/Al, all < 0.05. In parenthesis: number of analyses. In green: CZ and WALZ. In blue and brown: CIZ (samples at random and from units, respectively). In red: units from IVA.
REE normalized (N) to chondrite values after Pourmand et al. (2012).

standard deviations but the Ti/Nb ratios are a remarkable exception (see Section 7). Table 1 reveals the remarkable homogeneity of all of the rock groups, despite the distances between sampling areas, as well as the absence of significant vertical changes through the units I–IV and 2–3 (Fig. 3 and 4). Element ratios (Table 1) are the same regardless of the number of analyses of each group of rocks or whether or not the element ratios imply the least soluble elements (e.g., Al_2O_3/TiO_2, Zr/Al_2O_3, Nb/Hf) or these and other more mobile ones (e.g., K_2O/Al_2O_3, Th/U). The restricted ranges of Nb/Hf (2.23–2.36) and Ti/Nb (411–437) mean ratios are especially remarkable.

The Upper Ediacaran sandstones of the CZ and WALZ (Table 2) have lower abundances of SiO_2 and lower CIA values that the CIZ sandstones,

Table 2 Synthetic table of major element abundances (wt.%) and element ratios of the Ediacaran and Lower Cambrian sandstones. Mean values and st. dev. (1σ).

Areas/units	SiO_2	TiO_2	Al_2O_3	CIA	Al/Ti	K/Al
CZ (13)[a]	68.5(1.6)	0.73(0.1)	13.9(0.6)	58.3(2.3)	19.4(2.1)	0.169
NASD (6)	65.6(2.2)	0.97(0.1)	15.0(0.9)	57.6(3.2)	15.7(2.2)	0.153
SHD (3)	72.2(4.4)	0.71(0.2)	12.9(2.4)	67.8(4.1)	18.5(2.0)	0.170
IVA (6)	69.5(3.6)	0.70(0.1)	13.5(1.9)	61.9(2.4)	19.3(1.3)	0.155
As Beiras (6)[b]	76.8(3.4)	0.66(0.1)	12.1(2.2)	79.2(5.5)	18.4(1.1)	0.164
UNIT IV (3)	69.5(3.2)	0.72(0.0)	13.8(1.4)	60.9(2.8)	19.2(1.0)	0.171
UNIT III (3)	74.0(6.6)	0.66(0.2)	11.6(2.9)	69.3(5.2)	17.5(1.5)	0.149
UNIT II (9)	73.8(3.5)	0.63(0.1)	11.7(2.3)	70.5(4.1)	18.8(0.8)	0.162
UNIT I (4)	73.0(3.0)	0.66(0.1)	11.8(2.0)	65.4(3.2)	18.0(0.5)	0.150
UNIT 3 (5)	72.3(0.9)	0.67(0.0)	12.04(0.4)	63.9(1.4)	18.1(0.6)	0.182
SHD (4)	75.5(3.0)	0.57(0.1)	10.7(0.8)	62.6(13)	19.0(2.0)	0.135
Zr/Al	**Ti/Nb**	**Ti/Zr**	**Zr/Nb**	**Nb/Hf**	**Th/U**	**(La/Yb)N**
18.5(3.4)	440(25.2)	17.2(2.6)	26.02(3.6)	1.53(0.2)	3.8(0.3)	9.1(0.8)
21.7(10)	474(18.1)	19.8(5.3)	25.69(8.3)	1.62(0.4)	3.7(0.3)	8.4(0.6)
20.1(4.6)	438(7.0)	16.7(3.4)	26.83(4.7)	1.46(0.2)	3.7(0.1)	8.0(0.8)
18.2(5.1)	446(30.3)	18.0(3.7)	25.73(5.6)	1.62(0.4)	3.9(0.7)	8.7(0.8)
20.3(3.8)	400(25.3)	16.4(2.1)	24.6(2.16)	1.38(0.1)	3.4(0.4)	7.4(1.5)
17.6(3.3)	424(20.9)	18.1(2.6)	23.85(4.0)	1.69(0.3)	3.7(0.3)	8.5(0.6)
23.1(7.6)	420(85.8)	16.4(6.8)	27.08(5.1)	1.52(0.3)	3.3(0.6)	6.4(1.3)
20.7(2.8)	407(21.4)	15.8(2.7)	26.36(3.7)	1.45(0.2)	3.8(0.3)	8.1(0.7)
22.3(3.2)	420(13.1)	15.2(2.3)	27.97(3.4)	1.38(0.2)	3.5(0.3)	7.4(0.8)
24.5(2.9)	409(27.0)	13.7(1.3)	30.00(2.4)	1.29(0.1)	3.5(0.1)	8.5(0.5)
25.7(6.1)	40(77.0)	12.9(3.3)	31.7(5.2)	1.23(0.2)	4.2(1.2)	10.3(1.9)

[a]Ugidos et al. (2016).
[b]Beetsma (1995).
st. dev. < 0.05 compiled as 0.0. K/Al all < 0.05. In parenthesis: number of analyses. In blue: CZ and NASD. In brown: CIZ. In red: Cambrian sandstones.
REE normalized (N) to chondrite values after Pourmand et al. (2012).

but element ratios of the sandstones not involving major elements lack significant differences. Table 2 includes the mean Lower Cambrian sandstone (four samples).

Upper Ediacaran shales and sandstones have relatively low values of K_2O/Al_2O_3 molar ratios (<0.223 and <0.182, respectively). Values lower than 0.30 suggest a scarcity or absence of K-feldspar and that illite is the dominant K-bearing mineral (Cox et al., 1995; Cullers and Podkovyrov, 2000). They also have similar Ti/Nb, Th/U and (La/Yb)N ratios, but different Zr/Al_2O_3, Ti/Zr, Zr/Nb and Nb/Hf ratios (Tables 1 and 2). These differences are not related to zircon fractionation ("the zircon effect," Carpentier et al., 2009; Patchett et al., 1984) since Zr abundances of these rocks tend to overlap (79%) in the CIZ (46% in the CZ, 33% in the WALZ) and given the absence of covariations in diagrams such as SiO_2-Zr and SiO_2/Al_2O_3-Zr (Fig. 5A and B). Also, Ti/Nb ratios of shales and sandstones of the CIZ reaching 100% (except in one sample) overlap suggesting homogenized proportions of REE and Ti(Nb)-bearing minerals in the detrital rocks of the CIZ, and, to a lesser extent, in the other zones (92% in the CZ, 33% in the WALZ).

The REE content of the CIZ Ediacaran shales is higher than that of the corresponding sandstones, but their REE patterns are the same, although these sandstones reveal a slightly positive Ce anomaly (Fig. 6A). Many samples of the shales and sandstones show irregular HREE patterns, due mainly to the relative increased abundances of Ho and Yb (Fig. 6B). However, these irregularities vary and some mean values are buffered, as in the case of the REE pattern of the SHD samples of shales and partially

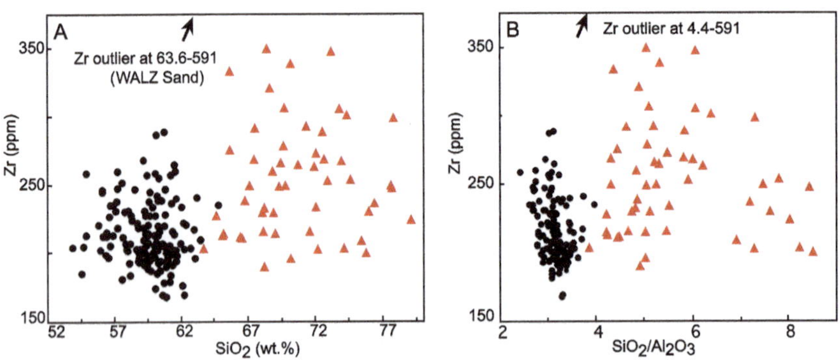

Fig. 5 Ediacaran shales (filled circles) and sandstones (red triangles) of the three zones mainly overlap in Zr content and lack covariations with SiO_2 (A) and SiO_2/Al_2O_3 (B), suggesting an absence of the "zircon effect" (see text for details).

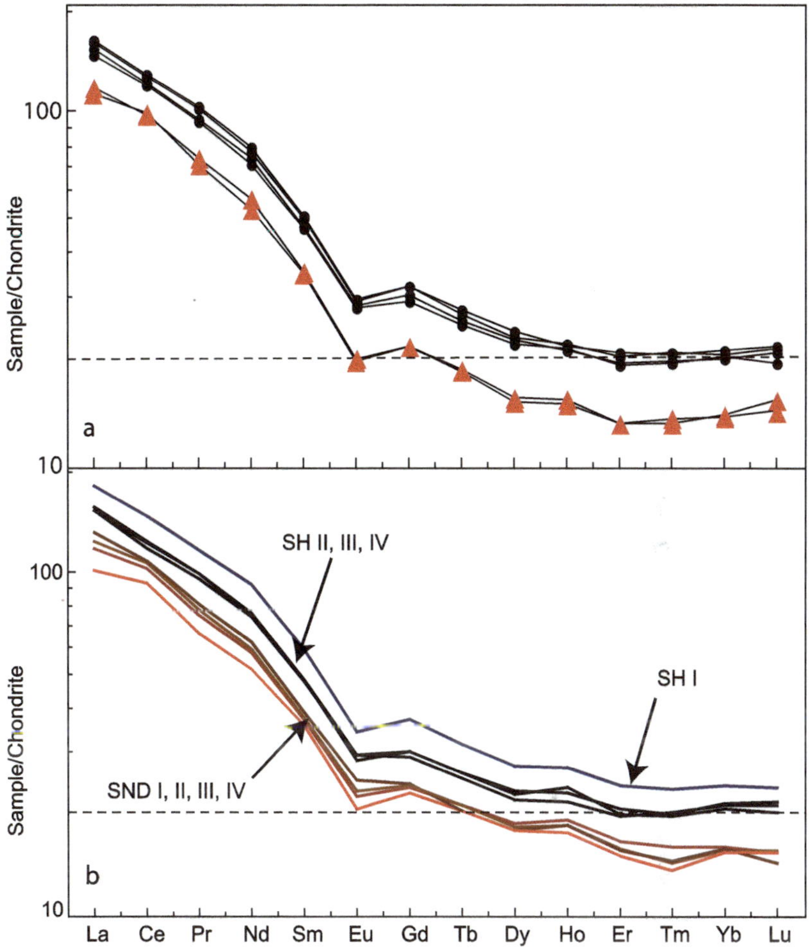

Fig. 6 Chondrite normalized REE patterns. (A) Ediacaran shales (filled circles) of the SHD (6), NED (17), IVA (20), SED (12) and two Ediacaran sandstones (red triangles) from the SHD (3) and IVA (7). (B) Ediacaran shales from units I (8), II (15), III (8), IV (7) and sandstones from units I (4), II (9), III (3) and IV (3). Many samples show irregular REE patterns. In all figures the REE patterns correspond to mean values of the number of samples indicated in parentheses. Chondrite values after Pourmand et al. (2012).

in the sandstones (Fig. 6A). In any case, the abundance of samples having irregular REE patterns decreases or the quantitative importance of HREE irregularities is minor and becomes weak or absent in domains south of SHD. The REE patterns of the Ediacaran equivalent rocks in the CZ and WALZ are parallel and similar to those of the CIZ rocks but lack irregularities (Fig. 7).

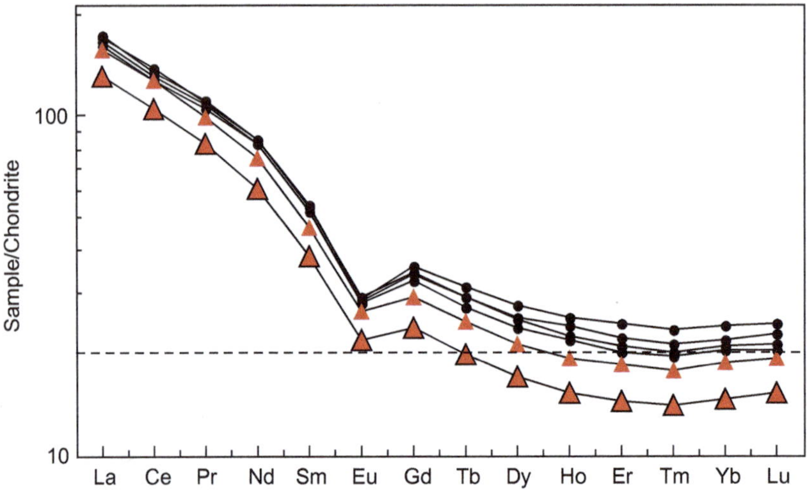

Fig. 7 Chondrite normalized REE patterns. Ediacaran shales (filled circles) from the CZ (20), WALZ (Allande and Tineo groups: 8 and 6 samples, respectively) and MND (4). Ediacaran sandstones (major red triangles) from the CZ (13) and WALZ (6, Allande group, minor red triangles). Geochemical data of the CZ are given in Ugidos et al. (2016). Chondrite values after Pourmand et al. (2012).

The overall geochemical features of the Upper Ediacaran detrital rocks of the three zones strongly support the notion that shales and sandstones are derived from the same source, given the absence of significant differences. However, diagrams such as SiO_2 vs. TiO_2 and Al_2O_3 vs. K_2O (Fig. 8A and B) reveal the same trend for the CIZ shales and sandstones, but not for the equivalent rocks of the other two zones (Fig. 8C and D). This may be due to differences in intensity of weathering and recycling, as suggested by the different CIA values of these rocks (Tables 1 and 2). Tables 3–6 present some geochemical parameters of shales and sandstones of the studied stratigraphic series in the CIZ and CZ. The values of element ratios throughout the series lack significant differences, despite the vertical separation between samples and also despite the distance between sampling areas. The CZ series has lower CIA values but element ratios not significantly different from those of the CIZ. Most likely, the shales and sandstones of the CIZ were deposited in a distal position while those of the CZ were deposited in more proximal positions and are the direct source area of the former, as shown in Fig. 4.

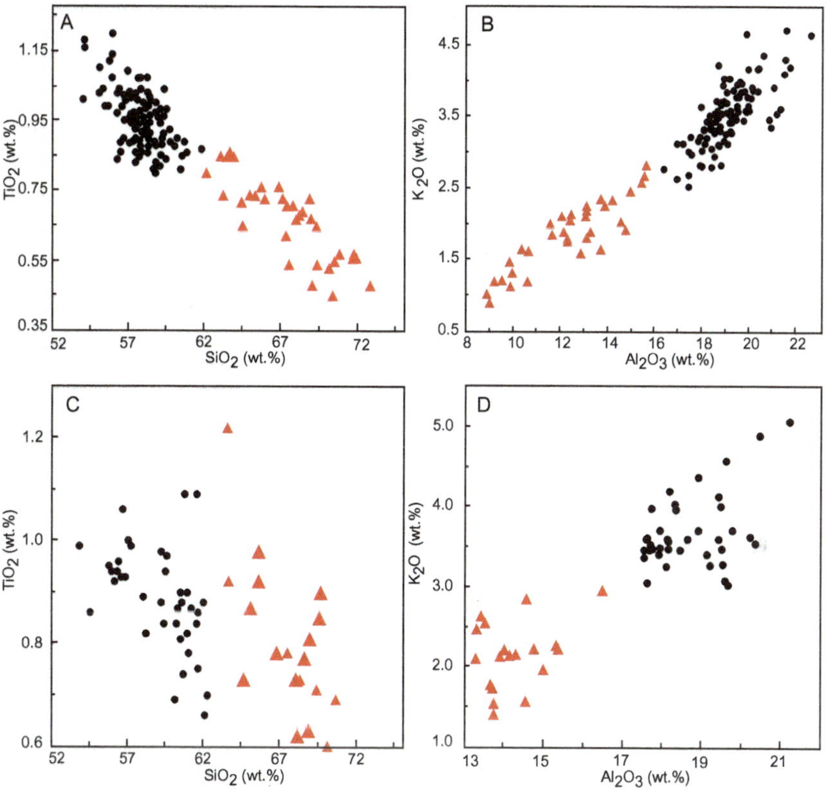

Fig. 8 (A, B) Covariations between SiO_2-TiO_2 and Al_2O_3-K_2O in Upper Ediacaran shales (filled circles) and sandstones (red triangles) of the CIZ. (C, D) Absence of covariations in the equivalent rocks of the CZ and WALZ.

6.2 Lower Cambrian Shales and Sandstones

Sampling of the Lower Cambrian shales of the CIZ has been performed in three different domains (Duero, SHD and IVA) but clear evidence to separate them into underlying and overlying ones has yet to be found. Available rocks are 32 Lower Cambrian shales and 4 sandstones collected at random and 28 shales from the studied stratigraphic units VI–XII, and 13 shales from the WALZ. Table 7 presents the mean values and standard deviation of selected chemical data for all of the groups of Lower Cambrian shales of the CIZ, WALZ and CZ. Each group of the Lower Cambrian shales of the CIZ shows similar values for the mean and median abundances of elements and element

Table 3 Chemical data of sandstones and shales of a partial series of the SHD.

Sample/unit	m	SiO_2 (wt%)	Al/Ti	K/Al	CIA	Zr/Al	Ti/Nb	Zr/Nb	Nb/Hf
SIA-39 III	446	57.48	20.70	0.217	73.83	10.71	436	16.13	2.32
SIA-38 III	332	61.22	19.43	0.203	73.95	13.50	403	17.31	2.07
SIA-32 II	282	75.82	19.76	0.125	59.89	22.50	409	30.35	1.20
SIA-31 II	267	61.30	20.04	0.204	74.28	12.58	406	17.09	2.02
SIA-37 II	250	60.07	18.98	0.198	74.69	13.25	405	16.97	2.22
SIA-36 II	190	60.21	19.57	0.198	74.64	12.12	410	16.24	2.21
SIA-30 II	135	74.18	18.73	0.107	65.38	22.69	393	27.87	1.36
SIA-35 II	133	74.19	21.96	0.210	76.31	10.81	398	15.77	2.25
SIA-29 I	78	74.20	20.44	0.202	77.30	11.97	419	17.11	2.10
SIA-34 I	38	74.21	19.84	0.196	75.74	12.45	423	17.44	2.11
SIA-27 I	-2	74.22	19.90	0.195	74.87	12.81	418	17.77	2.06
SIA-28 I	-29	74.23	17.69	0.137	64.64	26.60	412	32.32	1.25

In red: sandstones. In black: shales. m: height of sampling points within series.

Table 4 Chemical data of sandstones and shales of a partial series of the IVA.

Sample/unit	m	SiO_2 (wt.%)	Al/Ti	K/Al	CIA	Zr/Al	Ti/Nb	Zr/Nb	Nb/Hf
SJA-23 3	805	72.52	17.48	0.179	65.35	23.3	450	30.53	1.27
SJA-22 3	724	58.74	19.24	0.222	70.96	11.3	436	15.78	2.34
SJA-21 3	711	59.57	21.75	0.203	69.35	10.2	407	15.13	2.38
SJA-20 3	444	71.96	18.67	0.187	61.76	22.3	403	28.62	1.35
SJA-19 3	320	72.68	18.64	0.185	64.56	21.6	414	27.86	1.40
SJA-18 3	120	73.25	17.49	0.189	63.37	28.8	403	33.89	1.11
SJA-17 3	35	74.37	18.03	0.172	64.24	25.9	375	29.18	1.31
SJA-8 3	15	61.66	18.28	0.195	72.37	13.39	410	16.7	2.29

In red: sandstones. In black: shales. m: height of sampling points within series.

ratios, although these rocks are not as homogeneous as those of the Upper Ediacaran. The Appendix provides tables for the complete analyses of the Lower Cambrian shales that were previously considered by authors, new analyses and data, tables of samples for the stratigraphic units and tables of the underlying and overlying Lower Cambrian shales of the CZ and WALZ.

Table 5 Chemical data of sandstones and shales of a partial series of La Magdalena. CZ.

Sample	m	SiO_2 (wt.%)	Al/Ti	K/Al	CIA	Zr/Al	Ti/Nb	Zr/Nb	Nb/Hf
LR-5	381	61.6	20.6	0.212	67.3	11.4	416	16.3	2.27
LR-1	320	60.2	21.6	0.213	68.9	11.1	385	15.4	2.37
LR-6	135	68.6	18.2	0.171	56.1	22.9	448	31.2	1.33
LR-7	104	61.1	20.2	0.209	65.7	11.8	404	16.0	2.31

In red: sandstones. In black: shales. m: height of sampling points within series.

Table 6 Chemical data of sandstones and shales of a partial series of Salce. CZ.

Sample	m	SiO_2 (wt.%)	Al/Ti	K/Al	CIA	Zr/Al	Ti/Nb	Zr/Nb	Nb/Hf
BL-12	867	70.69	19.29	0.201	57.41	19.9	429	27.46	1.44
BL-11	714	61.94	19.93	0.214	66.42	11.8	419	16.43	2.24
BL-10	510	68.88	21.32	0.212	58.44	17.1	458	27.91	1.43
BL-9	317	59.19	19.31	0.251	67.13	11.1	417	14.89	2.41
BL-7	45	62.22	25.14	0.221	66.42	9.60	404	16.25	2.17

In red: sandstones. In black: shales. m: height in meters of sampling points within series.

No clear geochemical evolution is apparent from unit VI to unit XII (Table 7). However, some element ratios such as (La/Yb)N of the units XI and XII are higher that in the other units. Also, the value of εNd(541) (see below) of unit XII is the lowest found. This suggests an upward decreased contribution of juvenile materials to the detrital series.

The shales of units VI and X have relatively high contents of Mo (Mo range/mean: 1.44–16.6/7.53 ppm) and (Mo range/mean: 0.31–7.91/3.66 ppm), respectively (Table A13 in Appendix section). These two units deposited in slope environments and mostly consist of dark-gray to black shales but not all units with black shales have relatively high Mo abundances. All the other Lower Cambrian samples have Mo contents less than 0.55 ppm except one sample of unit XI (1.9 ppm). The contrasting Mo mean contents of units IV and VI are 0.89 and 7.53, respectively (data in Tables A3 and A13 in Appendix section). They may help to separate the Upper Ediacaran from the Lower Cambrian series showing black shales. However, at present, the data are insufficient to discuss the meaning of Mo abundances.

Table 7 presents three groups of Lower Cambrian rocks: the underlying and overlying shales of the CZ and WALZ, and the Lower Cambrian shales

Table 7 Mean and st. dev. (1σ) of selected major elements (wt.%) and chemical parameters of the Lower Cambrian shales collected at random and in known stratigraphic units.

Areas/units	SiO_2	TiO_2	Al_2O_3	CIA	Al/Ti	K/Al
CZ OV[a] (3)	55.9(1.3)	0.88(0.1)	21.1(0.7)	68.8(2.7)	24.3(2.0)	0.431(0.1)
NASDOV (3)	54.9(2.1)	0.90(0.1)	21.6(1.4)	75.2(1.4)	24.0(0.9)	0.294(0.0)
MND OV (1)	62.10	1.00	17.32	64.94	17.32	0.279
CZ UN[a] (5)	60.9(2.8)	0.86(0.1)	21.0(2.6)	72.8(0.6)	24.5(2.9)	0.370(0.1)
NASD UN (3)	60.3(0.8)	0.76(0.0)	19.7(1.8)	72.5(3.4)	26.0(2.5)	0.282(0.1)
MND UN (6)	58.7(2.7)	1.07(0.2)	21.4(2.6)	76.6(5.6)	20.2(1.9)	0.222(0.0)
Duero (15)	58.3(1.6)	0.90(0.1)	20.3(0.8)	71.3(3.2)	23.9(2.2)	0.239(0.0)
SHD (5)	59.8(1.2)	0.80(0.0)	19.0(0.3)	70.8(1.6)	24.4(1.1)	0.203(0.0)
IVA (12)	59.3(2.7)	0.85(0.1)	19.4(1.9)	72.3(3.2)	23.0(2.1)	0.216(0.0)
UNITXII (6)	59.9(1.5)	0.81(0.1)	19.2(1.2)	71.4(4.2)	23.5(0.6)	0.215(0.0)
UNIT XI (3)	59.7(1.3)	0.81(0.1)	19.7(0.7)	76.1(3.5)	24.4(2.5)	0.246(0.0)
UNIT X (3)	58.0(1.7)	0.70(0.1)	19.4(1.2)	74.2(4.8)	28.0(2.1)	0.221(0.0)
UNIT IX (3)	58.0(1.0)	0.84(0.1)	20.5(0.7)	75.0(4.7)	24.6(2.5)	0.207(0.0)
UNIT VIII (3)	60.1(2.5)	0.72(0.1)	18.6(1.7)	69.8(3.3)	25.8(0.6)	0.210(0.0)
UNIT VII (4)	58.8(1.6)	0.87(0.1)	21.5(1.1)	79.0(3.1)	24.8(1.9)	0.224(0.0)
UNIT VI (6)	58.5(1.4)	0.78(0.1)	18.4(2.0)	74.8(4.6)	23.7(1.4)	0.191(0.0)

Zr/Al	Ti/Nb	Ti/Zr	Zr/Nb	Nb/Hf	Th/U	(La/Yb)N
8.08(0.7)	381(19.0)	30.8(1.2)	12.4(1.0)	2.95(0.2)	3.67(0.3)	9.26(1.0)
7.97(0.3)	356(63.2)	31.4(2.3)	11.3(1.2)	3.20(0.3)	3.67(1.2)	10.6(1.6)
25.9	329	13.4	24.6	3.41	5.30	9.27
10.3(1.9)	384(20.6)	24.3(2.8)	16.0(2.4)	2.28(0.3)	3.88(0.3)	7.33(0.8)
8.85(1.1)	378(16.3)	26.5(4.0)	14.5(2.1)	2.47(0.3)	3.59(0.7)	7.89(1.0)
11.1(2.4)	410(28.9)	27.6(3.7)	15.1(2.5)	2.45(0.3)	3.63(0.3)	7.82(1.8)
7.52(1.7)	336(21.0)	31.9(5.3)	9.78(1.2)	3.63(0.6)	5.63(1.3)	12.2(1.7)
7.73(0.6)	340(29.1)	31.6(1.9)	10.8(1.5)	3.48(0.4)	4.80(1.5)	10.9(2.3)
9.22(1.0)	355(21.6)	28.0(2.3)	12.8(1.3)	2.99(0.3)	4.69(0.6)	9.79(1.2)
9.08(1.3)	340(9.9)	28.5(3.6)	12.1(1.6)	3.14(0.3)	4.84(0.9)	10.2(1.3)

Table 7 Mean and st. dev. (1σ) of selected major elements (wt.%) and chemical parameters of the Lower Cambrian shales collected at random and in known stratigraphic units.—cont'd

Zr/Al	Ti/Nb	Ti/Zr	Zr/Nb	Nb/Hf	Th/U	(La/Yb)N
9.02(1.9)	348(8.4)	27.9(3.6)	12.6(1.3)	3.10(0.4)	4.11(0.6)	9.29(1.1)
7.67(0.7)	362(6.9)	28.1(1.5)	12.9(0.9)	2.89(0.2)	3.20(0.4)	6.90(1.1)
8.85(1.0)	337(17.3)	27.7(0.9)	12.2(1.0)	3.00(0.3)	3.95(0.1)	7.38(0.6)
8.26(1.1)	363(34.3)	28.5(4.2)	12.8(1.1)	2.87(0.2)	3.51(0.5)	8.35(0.4)

[a]Ugidos et al. (2016).
st. dev. <0.05 compiled as 0.0. K/Al, all < 0.05. In parenthesis: number of analyses. OV and UN: overlying and underlying, respectively. In red: WALZ. In blue: CZ. In brown: CIZ.
REE normalized (N) to chondrite values after Pourmand et al. (2012).

of the CIZ. Only one sample of the overlying MND is available and shows features that differ from those of the other overlying shales. Comparing data from Table 7 and Table 1 reveals that the underlying shales have element ratios such as Zr/Al_2O_3, Ti/Nb, Ti/Zr, Zr/Nb, Nb/Hf, Th/U and (La/Yb)N similar to those of the Upper Ediacaran. The underlying CZ shales are characterized by relatively low contents of Fe_2O_3, MgO, CaO, Na_2O and Sr, but they have the highest K_2O, Rb and Cs contents, the lowest Rb/Cs and the highest Rb/Th ratios (Table 8). These geochemical features, together with petrological and isotopic evidence imply that these rocks, are derived from K-metasomatized and weathered Upper Ediacaran rocks (Ugidos et al., 2016). The underlying shales of the CZ and WALZ show similar mean and median values of chemical parameters (Table 8) but varying abundances of K_2O, Rb and Cs. The K_2O/Al_2O_3 and Rb/Th ratios decrease, and Sr and Rb/Cs increase from left to right in Table 8, i.e., from CZ to WALZ in the map. Thus, the contribution of detritus derived from Ediacaran K-metasomatized rocks to the underlying Lower Cambrian detrital setting decrease westwards.

Comparison between the NASD and CZ overlying shales and the Lower Cambrian shales of the CIZ reveals a certain geochemical heterogeneity that is most likely related to the different SiO_2 abundances, varying CIA values and possible presence of K-feldspar, as suggested by the relatively high K_2O/Al_2O_3 molar ratios of the CZ overlying shales (Table 8).

Only four samples of Lower Cambrian sandstones are available and have been included in Table 2. These samples were collected in the SHD. Comparison with the corresponding shales provided insignificant results and revealed very different chemical features. These sandstones show remarkable differences with the Ediacaran ones (Table 2).

Table 8 Mean (a) and median (b) element abundances of major (wt.%) and trace (ppm) elements, and element ratios of Cambrian shales.

	1a	1b	2a	2b	3a	3b	4a	4b
n samples	**5**	**5**	**3**	**3**	**6**	**6**	**3**	**3**
Fe_2O_3	3.91	3.72	6.58	7.24	7.40	7.44	5.39	5.98
MgO	1.54	1.41	2.32	2.52	2.04	1.90	2.32	2.50
CaO	bd	bd	0.17	0.17	0.18	0.07	0.18	0.18
Na_2O	bd	bd	0.97	0.28	0.85	0.38	0.13	bd
K2O	7.15	7.34	5.14	5.51	4.38	4.26	8.42	9.00
Rb	243	243	177	179	172	166	184	193
Cs	24.2	23.7	10.6	9.51	10.2	10.2	11.6	14.6
Sr	32.3	36.9	52.2	50.5	94.6	82.7	58.4	55.0
K/Al	0.370	0.372	0.282	0.306	0.222	0.217	0.431	0.463
Rb/Cs	10.6	10.2	17.7	18.9	18.0	17.5	18.7	13.3
Rb/Th	17.0	16.9	14.2	12.5	11.9	11.7	12.1	12.7
Th/Sc	0.68	0.70	0.65	0.65	0.67	0.67	0.70	0.68
Cr/Th	6.65	6.44	7.08	7.08	7.95	8.38	6.92	7.08

	5a	5b	6	7a	7b	8a	8b	9a	9b
n samples	**6**	**6**	**1**	**12**	**12**	**8**	**8**	**12**	**12**
Fe_2O_3	6.98	6.97	6.97	7.05	7.15	7.36	7.38	7.16	7.18
MgO	3.81	3.61	2.71	2.54	2.45	3.36	3.09	2.70	2.82
CaO	0.13	0.17	0.31	0.60	0.50	0.45	0.51	0.37	0.36
Na_2O	0.31	0.39	2.40	1.35	1.46	1.67	1.69	1.58	1.84
K2O	5.86	5.59	4.46	4.63	4.46	3.66	3.67	3.86	3.62
Rb	171	168	156	197	198	152	148	157	154
Cs	4.39	3.50	6.20	10.2	8.0	7.64	7.60	7.79	7.28
Sr	52.2	50.5	82.3	122	110	121	132	99.2	87.5
K/Al	0.294	0.294	0.279	0.247	0.248	0.206	0.201	0.216	0.200
Rb/Cs	44.0	42.7	25.1	25.1	26.3	20.5	20.7	20.6	20.7
Rb/Th	10.6	10.1	7.65	11.0	11.0	10.3	10.3	10.8	11.1
Th/Sc	0.75	0.75	1.23	0.96	1.02	0.72	0.72	0.80	0.82
Cr/Th	6.87	7.03	4.69	5.86	5.96	6.97	7.14	6.97	6.78

1: underlying, CZ. 2: underlying, NASD. 3: underlying, MND. 4: overlying, CZ.
5, 6: overlying NASD and MND, respectively. 7: Duero. 8: SHD. 9: IVA.

The REE patterns of the Lower Cambrian shales and sandstones of the CIZ are parallel and reveal negative and positive Ce/Ce* anomalies, respectively (Fig. 9A). Some samples from units VI–XII (SHD) have the same REE pattern irregularities (Fig. 9B) found in the Ediacaran shales sampled in the same domain. However, the REE patterns of the Lower Cambrian shales (underlying and overlying) of the CZ and WALZ lack these irregularities (Fig. 10). These features are discussed in Section 7.

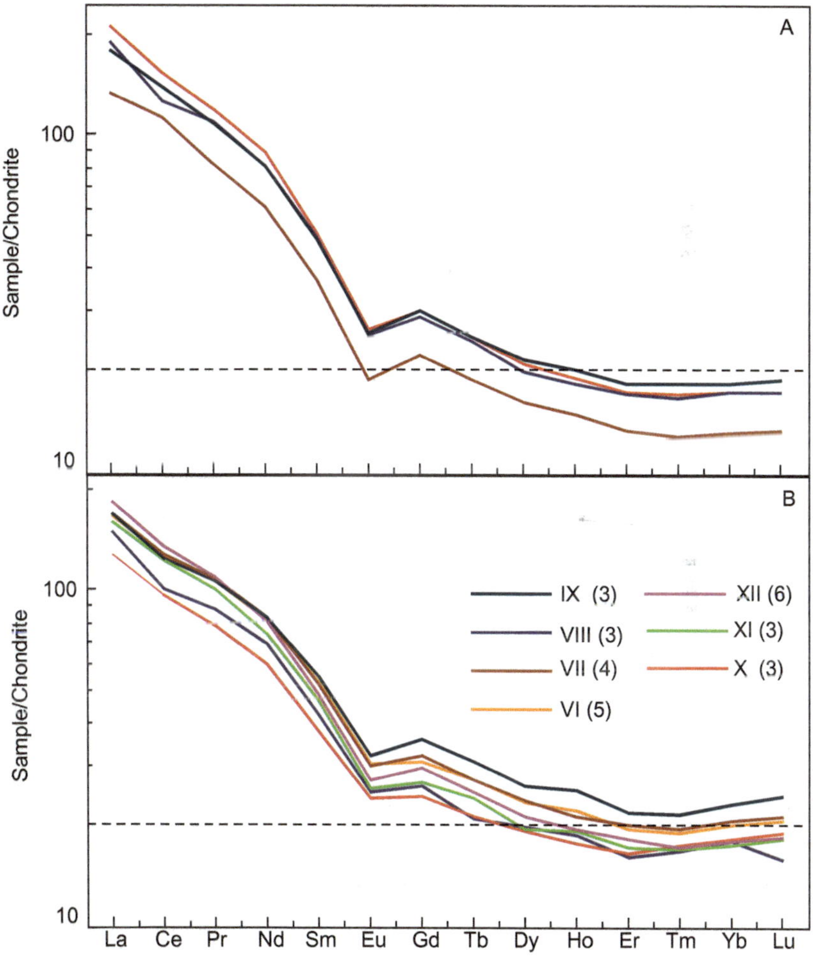

Fig. 9 Chondrite normalized REE patterns. (A) Lower Cambrian shales (three upper patterns) collected at random in Duero (13), SHD (9) and IVA (12), and sandstones (lower pattern) of the SHD (4). (B) Lower Cambrian shales from Units VI–XII. CIZ. Chondrite values after Pourmand et al. (2012).

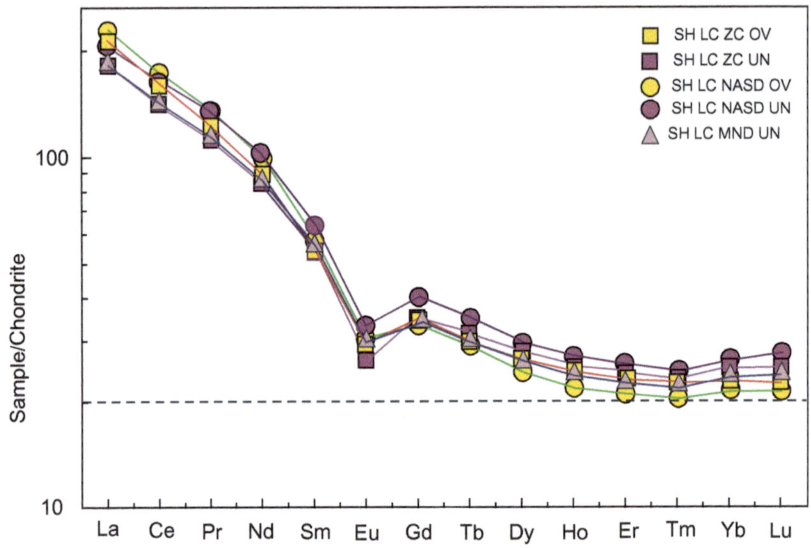

Fig. 10 Chondrite normalized REE patterns. Overlying shales (three samples each group) from CZ and NASD, yellow squares and circles, respectively. Underlying shales from CZ (5), MND (6) and NASD (3), purple squares, circles and triangles, respectively. The most significant aspects are the greater abundance of LREE in the overlying shales and the grater abundances of Yb and Lu in the underlying shales. Chondrite values after Pourmand et al. (2012).

6.3 Nd-isotope data

The Nd-isotope data for the Upper Ediacaran and Lower Cambrian detrital rocks published by different authors (Table 9) as well as new data (Tables 10 and 11) continue to indicate that there is a juvenile contribution present in the Upper Ediacaran detrital rocks which decreased significantly during Early Cambrian times in all zones and that −4.1 may be used as a general value of εNd(541), contributing to separate Upper Ediacaran from Lower Cambrian detritus. However, two Ediacaran shales (SAR-8, SAR-3, Table 10) in the MND series and the underlying Lower Cambrian shales (SAR-7, OEN-3, Table 11) have lower εNd(541) values than the corresponding rocks from the NASD and CZ. As in the CZ (Ugidos et al., 2016), the absence of covariation in diagrams Th/Sc vs. εNd(541) and εNd(541) vs. *f*Sm/Nd (Fig. 11A and B) suggest the absence of widespread coeval juvenile contribution to the Upper Ediacaran and Lower Cambrian series.

Table 9 Compilation of εNd isotope data of the Ediacaran and Cambrian shales and sandstones of the Central and North Iberia.

Authors	Zone	Lower Cambrian	Upper Ediacaran
Beetsma (1995)	CIZ		−3.1 to −2.1 (6)
			−3.1 to −1.2 (7)
Nägler et al. (1995)[a]	CIZ	−7.4 and −7.3	−3.5 and −3
	CZ	−7.2	
Tassinari et al. (1996)	CIZ		−3.8 to −1.8 (5)
López-Guijarro et al. (2008)	CIZ	−4.0	−4.2 and −2.9
	CIZ		−4.6 and −4.2
Ugidos et al. (2009)	CIZ	−7.5 to −4.2 (8)	−3.6 to −2.4 (9)
Fernández-Suárez et al. (2014)	CIZ	−6.3	−3.0
	WALZ		−3.3 to −3.0 (3)
	CZ	−10.9 (snd)	
Fuenlabrada et al. (2016)	CIZ	−5.1 to −3.9 (12)	−-3.2 to −1.6 (12)
Villaseca et al. (2014)	CIZ		−3.1 to −0.8 (5)
Ugidos et al. (2016)	CZ	−5.8 to −4.9 (3) OV	−3.6 to −1.8 (9)
	CZ	−3.4 to −2.1 (4) UN	−3.6 to −3.1 (3) **R**

[a]Some samples not compiled because of their unclear stratigraphic positions.
All data calculated at 541 Ma. **R**: rubefacted K-metasomatized shales. In green and blue: overlying and underlying shales, respectively. In red: sandstones. In parenthesis: number of samples.

Table 10 Sm-Nd isotope data of the Upper Ediacaran shales of the Central and North Iberia.

Sample	Sm	Nd	$^{143}Nd/^{144}Nd$	$^{147}Sm/^{144}Nd$	2σ	εNd(t)	$f_{Sm/Nd}$	TDM
TIN-4	9.76	49.0	0.512218	0.1205	5	−2.9	−0.39	1.47
TIN-5	8.95	46.1	0.512169	0.1174	6	−3.7	−0.40	1.50
VIL-2	8.00	40.4	0.512252	0.1198	5	−2.2	−0.39	1.41
PS-1	7.08	35.0	0.542289	0.1224	4	−1.7	−0.38	1.38
LR-2	7.90	38.9	0.512198	0.1229	5	−3.5	−0.37	1.54
RIE-1	6.89	31.5	0.512203	0.1323	5	−4.0	−0.33	1.70
GED-4	7.89	41.2	0.512165	0.1158	5	−3.6	−0.41	1.48
GED-6	8.98	45.4	0.512196	0.1196	5	−3.3	−0.39	1.49
GED-8	8.60	43.3	0.512176	0.1201	5	−3.7	−0.39	1.53

Continued

Table 10 Sm-Nd isotope data of the Upper Ediacaran shales of the Central and North Iberia.—cont'd

Sample	Sm	Nd	$^{143}Nd/^{144}Nd$	$^{147}Sm/^{144}Nd$	2σ	εNd(t)	$f_{Sm/Nd}$	TDM
GED-9	8.59	43.1	0.512187	0.1205	6	−3.5	−0.39	1.52
VIL-3	8.07	40.9	0.512191	0.1195	5	−3.4	−0.39	1.50
RIE-2	6.89	31.5	0.512163	0.1201	7	−4.0	−0.39	1.55
SAR-3	6.62	33.4	0.512155	0.1199	7	−4.1	−0.39	1.56
SAR-8	7.29	36.8	0.512131	0.1198	6	−4.6	−0.39	1.60
CORA-1	7.90	37.8	0.512291	0.1264	17	−1.9	−0.36	1.44
SVAL-4	6.77	31.7	0.512258	0.1292	6	−2.7	−0.34	1.55
CAS-2	6.55	33.2	0.512206	0.1193	8	−3.1	−0.39	1.47
SNB-1	8.31	40.7	0.512217	0.1237	9	−3.2	−0.37	1.52
MT-10	7.15	35.6	0.512259	0.1216	6	−2.2	−0.38	1.42
MT-11	7.08	34.7	0.512312	0.1234	6	−1.3	−0.37	1.36
SJA-4	9.58	43.6	0.512262	0.1328	6	−2.9	−0.32	1.60
SJA-8	8.46	41.1	0.512246	0.1245	9	−2.7	−0.37	1.49

In blue: Tineo group. In red: Allande group. In green: Villalba series (MND). In brown: CIZ.

Table 11 Sm-Nd isotope data of the Lower Cambrian shales of the Central and North Iberia.

Sample	Sm	Nd	$^{143}Nd/^{144}Nd$	$^{147}Sm/^{144}Nd$	2σ	εNd(t)	$f_{Sm/Nd}$	TDM
CN-3	10.3	52.4	0.512074	0.1189	7	5.6	−0.40	1.67
BOA-1	10.0	55.5	0.512016	0.1094	5	−6.1	−0.44	1.60
GED-3	9.03	52.2	0.511768	0.1046	5	−10.6	−0.47	1.87
TIN-2	8.29	41.9	0.512240	0.1197	6	−2.4	−0.39	1.42
TIN-3	7.11	35.5	0.512183	0.1212	5	−3.7	−0.38	1.53
VIL-5	6.31	33.6	0.512136	0.1136	7	−4.1	−0.42	1.49
OEN-2	6.54	36.4	0.511924	0.1087	6	−7.9	−0.45	1.72
SAR-7	7.81	42.2	0.512044	0.1119	6	−5.7	−0.43	1.60
SAR-9	9.02	51.7	0.512081	0.1055	5	−4.6	−0.46	1.46
OEN-3	9.55	47.4	0.511706	0.1219	8	−13.0	−0.38	2.30
FRE-2	7.88	42.9	0.512011	0.1111	17	−6.3	−0.44	1.64
FRE-5	7.80	43.4	0.511911	0.1087	7	−8.1	−0.45	1.74
MT-6	7.06	35.1	0.512036	0.1217	6	−6.6	−0.38	1.78
MT-16	6.95	37.2	0.512052	0.1130	6	−5.7	−0.43	1.61

In brown and green: overlying and underlying shales, respectively, NASD. In red and pale blue: overlying and underlying shales, respectively, MND. In blue: CIZ shales.

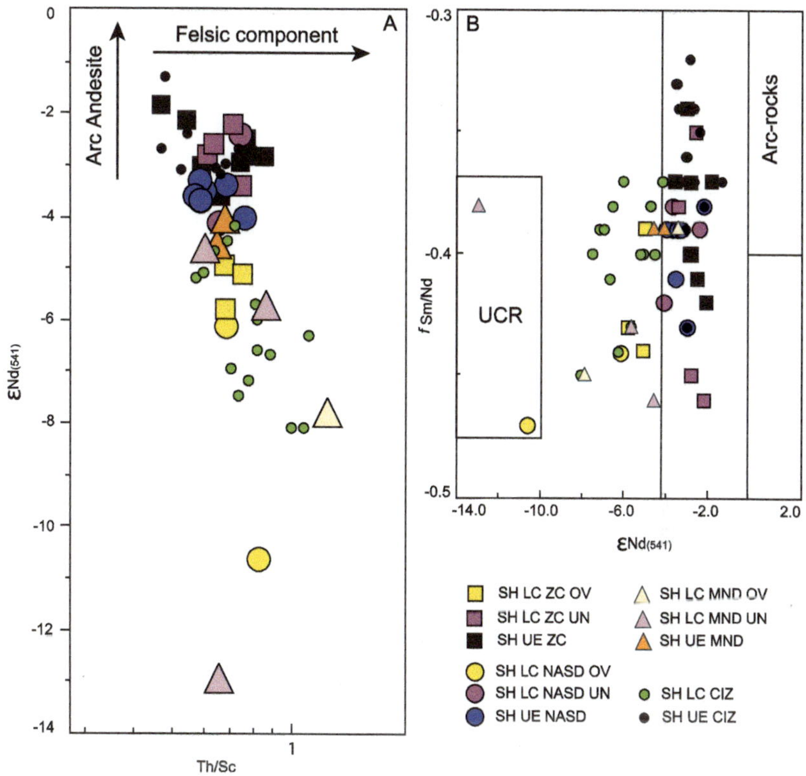

Fig. 11 (A) Th/Sc vs. εNd(t) diagram. (B) εNd(t) vs. *f*Sm/Nd diagram, with (t) refering to the age of the Ediacaran-Lower Cambrian boundary, i.e., 541 Ma (Cohen et al., 2013). Ediacaran (UE) and Lower Cambrian (LC) shales. OV and UN, overlying and underlying LC shales, respectively. Neither diagram favors significant coeval volcanic contributions to the UE and LC detrital rocks. The value of about −4.1 of εNd(541) separates the two groups of rocks. UCR, upper crustal rocks (Hu and Gao, 2008). *Panel (A) after McLennan, S.M., Hemming, S., McDaniel, D.K., Hanson, G.N., 1993. Geochemical approaches to sedimentation, provenance, and tectonics. In: Johnsson, M.J., Basu, A. (Eds.), Processes Controlling the Composition of Clastic Sediments. vol. 284. The Geological Society of America. Special Paper, pp. 21–40. Panel (B) simplified after Bock, B., McLennan, S.M., Hanson, G.N., 1994. Rare earth elements reditribution and its effects on the neodymium isotope system in the Austin Glen member of the Normenskill formation, New York, USA, Geochim. Cosmochim. Acta 58, 5245–5253.*

7. Comparison of geochemical features of the Upper Ediacaran and Lower Cambrian shales

In earlier studies by the authors, sampling carried out near the Upper Ediacaran-Lower Cambrian boundary in the CIZ was avoided due to uncertainties about mapping of the contact. Diagrams such as Al_2O_3/TiO_2 vs. Ti/Nb, TiO_2 vs. Zr/Nb and Ti/Nb vs. Rb/Zr separated the Upper Ediacaran from the Lower Cambrian shales in two fields and only 4 of 43 samples plot close to field boundaries (Valladares et al., 2000, 2002a,b). In the present study, more samples are used and some were deliberately collected close to the mapped Ediacaran-Cambrian contacts. The results are as follows: 6 out of 106 Ediacaran shale samples have Ti/Nb ratios lower than 395 (the 6 were randomly collected in IVA and SED) and 6 out of 60 Lower Cambriam samples have Ti/Nb ratios higher than 395 (4 samples collected at random and two samples from units VI and VII). It should be emphasized that these two units are the lowermost of the Lower Cambrian.

The best parameters characterizing the Upper Ediacaran and Lower Cambrian shales are presented in Tables 1 and 7. These tables show the mean and standard deviations, and median values of chemical data of groups of Upper Ediacaran and Lower Cambrian shales collected at random and in known stratigraphic units for comparative purposes. According to the results shown in Table 1, the remarkable aspects include the relatively low values of standard deviations for the chemical parameters and the absence of significant differences in the mean values of the Ediacaran shales element ratios, regardless of whether the samples are from known stratigraphic units or were collected at random in different areas of the three zones. It is not absolutely certain that 100% of the samples considered correspond to the Upper Ediacaran or Lower Cambrian, based on geochemical data, since gradual transitions of Upper Ediacaran and Lower Cambrian detrital series cannot be discarded and also due to unavoidable mapping uncertainties. However, the combination of stratigraphic and geochemical data in this work is an excellent tool and it is possible that minor sampling errors will not significantly affect the mean and median parameters used as characterization criteria or the validity of fields defined by samples in diagrams. It would be surprising if 100% of the samples plotted perfectly in separate fields.

The TiO_2 vs. Nb diagram (Fig. 12A) defines two trends with different slopes for the Ediacaran and Lower Cambrian shales of the CIZ that share relatively wide range to TiO_2 abundances. However, the Nb content of most Ediacaran samples is lower than the Lower Cambrian ones for the same TiO_2 content and therefore, the Ti/Nb ratios of the Ediacaran shales are higher (Tables 1 and 7). The two trends suggest two different sources with different populations of Ti-bearing minerals that partially mixed. A consequence of this mixing is that samples of shales with similar TiO_2 contents may have different Nb abundances and, therefore, the Ti/Nb ratios have

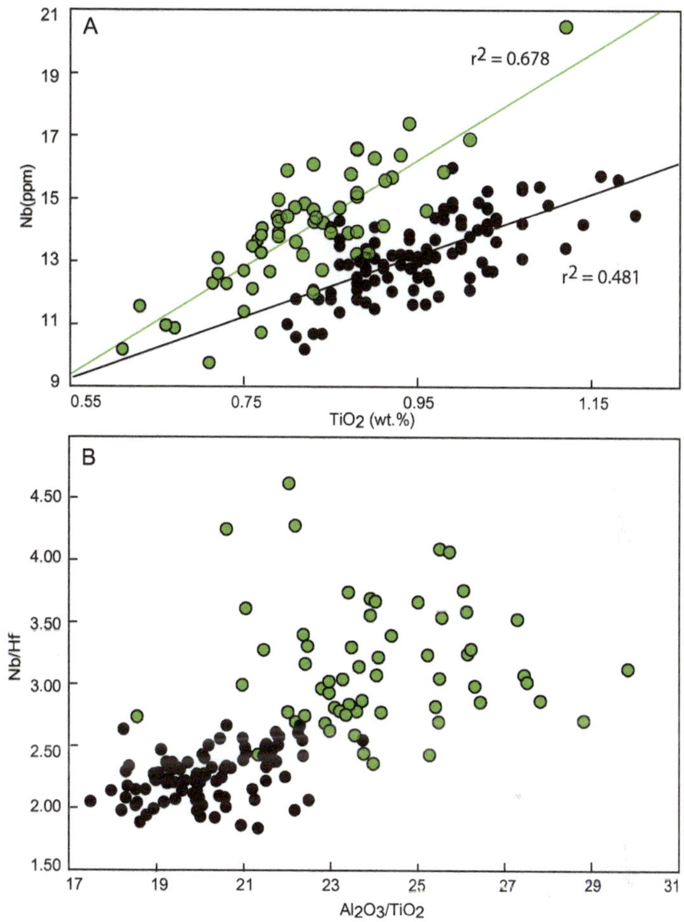

Fig. 12 Variation trends of TiO_2 vs. Nb (A) and Al_2O_3/TiO_2 vs. Nb/Hf (B) in the Ediacaran (black circles) and Lower Cambrian (green circles) shales of the CIZ.

relatively high standard deviations (see Section 6.1). Obviously, it is assumed that Ti-bearing minerals are the only Nb carriers. The same diagram for the corresponding rocks of the CZ and WALZ (Fig. 13A) reveals similar trends for the Ediacaran and underlying shales but an absence of trends for the overlying ones. Note that Ediacaran and underlying shales also share a relatively wide range of TiO_2 and Nb contents. An example of a new diagram separating the Ediacaran from the Lower Cambrian shales is Al_2O_3/TiO_2 vs. Nb/Hf for the CIZ (Fig. 12B) and the Ediacaran from the overlying shales for the CZ and WALZ (Fig. 13B).

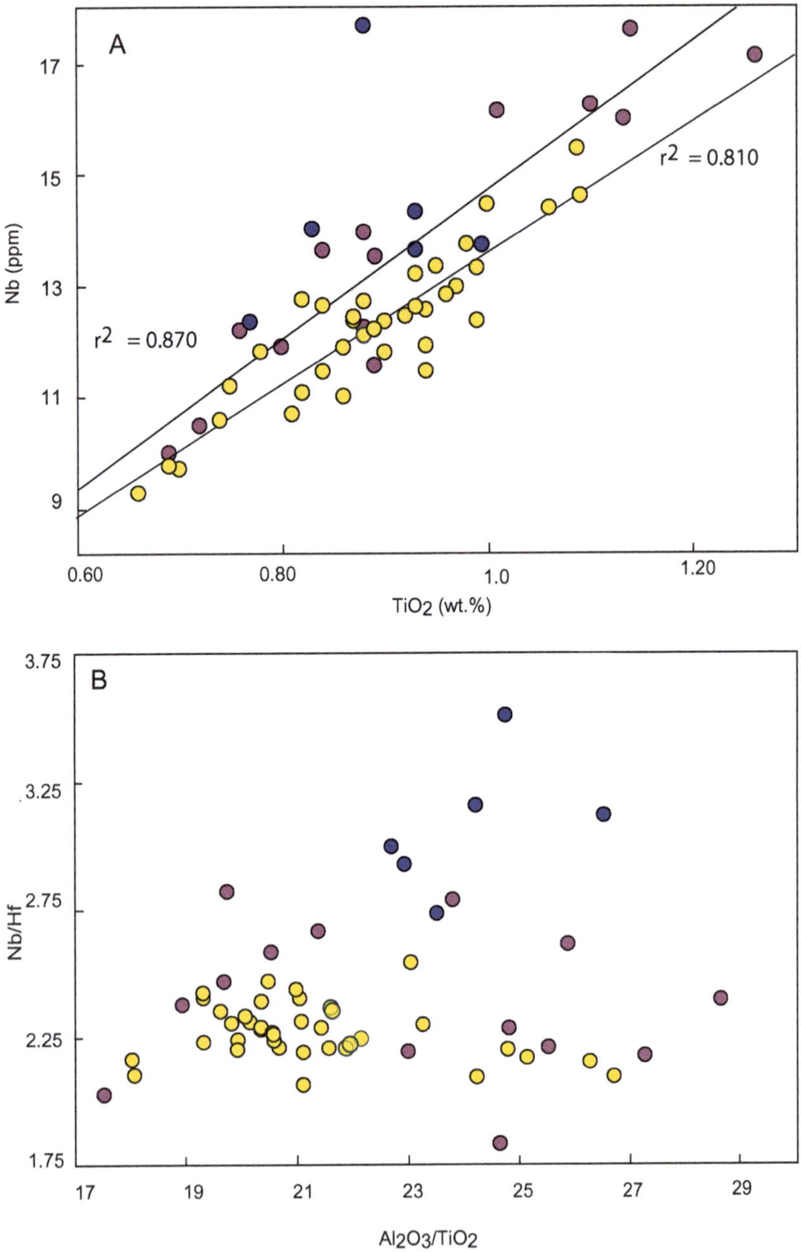

Fig. 13 Variation trends of TiO_2 vs. Nb (A) and Al_2O_3/TiO_2 vs. Nb/Hf (B) in the Ediacaran (yellow circles), and underlying and overlying (purple and blue circles, respectively) Lower Cambrian shales of the CZ, NASD and MND.

The REE patterns and Nd-isotope data also separate the Upper Ediacaran from the Lower Cambrian shales. The WALZ shales and sandstones have similar values of εNd(541). The underlying shales have also similar εNd(541) values as the Ediacaran shales. Thus, the underlying shales of these zones are dominated by the contribution of detritus from the Ediacaran rocks. However, the corresponding shales of the MND show different element abundances and CIA values, although they share certain element ratios (Table 1). These differences, along with the relatively low εNd(541) values of the underlying MND shales, suggest a higher contribution of crustal material as compared to the other underlying shales.

It is noteworthy that the REE patterns of the Ediacaran and Lower Cambrian rocks of the CIZ show irregularities while the equivalent rocks of the CZ and WALZ lack this feature. These characteristics, together with data such as the lower CIA of the CZ and WALZ Ediacaran rocks as compared to that of equivalent rocks of the CIZ, SiO_2 vs. TiO_2 and Al_2O_3 vs. K_2O diagrams strongly support the notion that Ediacaran rocks of the CZ and WALZ, were weathered and eroded, and the resulting detrital material was ultimately deposited in a marine setting at the slope and base-of-slope environments of the CIZ. Thus, irregular HREE patterns are a possible cumulative effect of the recycling and sorting of zircon as the main HREE-bearing mineral.

Also based on Nd-isotope data, separation of the Upper Ediacaran from the Lower Cambrian rocks is possible. The overlying and underlying shales consistently reveal the same chemical and isotopic features as the Lower Cambrian and Upper Ediacaran ones, respectively. Therefore, the composition of the underlying shales is dominated by the contribution of detritus derived from the Ediacaran rocks.

Zircons and Ti-bearing minerals such as rutile are typical heavy minerals found in most rock types and therefore it is expected that detrital rocks will have heterogeneous proportions of these minerals, reflecting the varying proportions of different rocks in the source areas. The abundance of heavy minerals in detrital rocks and the abundance of hosted elements are mainly dependent on the nature of the source rocks and sorting over sedimentary cycles. It is difficult to accept that the extension of the total outcropping and the thickness of the detrital Upper Ediacaran sediments having uniform chemical parameters, such as those presented, resulted from only one sedimentary cycle. Recycling of the same material with minor new contributions over time appears to be the most plausible process.

8. Zircon ages from the Upper Ediacaran and Lower Cambrian series

Results from detrital zircon geochronology must be considered in the light of studies evaluating the potential bias of zircon age patterns due to many possible processes and circumstances, including erosion rates in the source region and subsequent hydrodynamic processes (Amidon et al., 2005; Malusà et al., 2013, 2016; Moecher and Samson, 2006; Yang et al., 2012), the presence of dominant contribution of zircons derived from the youngest magmatic or metamorphic event (Cawood et al., 2014) or the recycling and mixing of detritus from different sources (Andersen et al., 2016). Also, the use of the youngest detrital zircon to provide a maximum age of deposition has been questioned and "...should never be used to correlate between different drillholes, profiles, sedimentary basins or tectonic terranes" or "...to be used with confidence, the data must be combined with a broad range of other information about the rocks from which the zircons are separated, as well as a full understanding of the prior evolution of potential source terranes" (Andersen, 2005, 2013). The application of hydraulic sorting principles and a quantitative approach to mineral fertility has greatly improved geological interpretations in modern and ancient sedimentary successions (Malusà et al., 2016).

This study reveals that Zr contents of the Ediacaran shales and related sandstones of the CIZ mainly overlap, as commented in Section 6.1. However, the similar Zr abundances in rocks of different grain size do not imply that these rocks have similar populations of zircons but rather, that there is a balanced accumulation of zircons according to grain sizes.

Stratigraphic or paleontological criteria to separate Beiras and Douro groups in Portugal have yet to be well-defined and, in general terms, they are agreed to be of Ediacaran and Cambrian ages, respectively. Accordingly to Teixeira et al. (2011), reliable biostratigraphical data are lacking from either of these metasedimentary units. A metagreywacke of the Beiras group has 44% Neoproterozoic and 28% Paleoproterozoic zircons and a chlorite phyllite of Douro group has 78% and 9%, respectively (Teixeira et al., 2011). However, the zircon percentages of the Beiras sample are akin to those of the overlying Lower Cambrian shales of the CZ and WALZ and those of the Douro sample are similar to those of Ediacaran rocks of the CIZ. Other greywacke of Beiras have 83% Neoproterozoic and 13% Paleoproterozoic zircons (Talavera et al., 2012). According to these later authors, this sample (labeled CTP-62) and five others (labeled CTC-67, Ctt-68, CTP-64, CTS-58, and CTS-55) were collected as part of the Lower Unit (Ediacaran) of the "schist and greywake complex," a confusing term, as mentioned above (Section 5.3). The CCT-68 sample has 72% Neoproterozoic and 12% Paleoproterozoic zircons, i.e., similar results to those of the other Ediacaran samples. However, the CTC-67 (metaconglomerate) sample has 63% and 24% Neoproterozoic and Paleoproterozoic zircons, respectively. These are uncommon results of the Ediacaran sandstones and shales that may be related to the poor sorting of conglomerates which can favor the presence of heterogenous sizes of zircons. May be related also to mixing of detritus from different sources, and may reflect a sorting effect on zircon abundances (see authors appearing above in this section). The other three samples were also collected as part of the Lower Unit, but the abundances of Cambro-Ordovician ages are 45%, 22% and 55% and, therefore, the ages of the corresponding sampled detrital rocks are younger than initially considered and the new results accepted (Talavera et al., 2012). However, the problem is, sampling is often carried out on outcrops where stratigraphy is not well-known and moreover, some interpretations are poorly constrained and risky.

In the WALZ, the Allande group lacks zircons in the range of 600–550 Ma, which are present in the Tineo group (Fernández-Suárez et al., 2014). A Kolmogorov-Smirnov test (excluding zircon ages in the range of 600–550 Ma) indicates that U-Pb age distributions in samples of these two groups are not significantly different and are also similar to an Ediacaran sample in the CIZ. Thus, this points to the same source for the Ediacaran rocks (Fernández-Suárez et al., 2014). In the Tineo group, the maximum age of sedimentation is 553 Ma (Fernández-Suárez et al., 2014) but in a level some meters below the sample of this age, older volcanic rocks are found (559 ± 4 Ma, Gutiérrez-Alonso et al., 2004; 557 ± 3 Ma, Rubio-Ordóñez et al., 2015). This example is not particularly significant, but it does reveal potential sources of errors if the ages of detrital series are based in a unique type of data. More important examples are found in the Herrería and Cándana groups, and that of the Tamames formation. Also, the U-Pb zircon age distribution of a sample from the Lower Cambrian Tamames formation (CIZ) does not differ from those of the Ediacaran sample, although the εNd(t) of this sample is similar to those of the Lower Cambrian rocks (Fernández-Suárez et al., 2014). As far as the authors are aware, no the other Lower Cambrian samples have been reported to present these features.

The geochemical differences between the Upper Ediacaran and Lower Cambrian rocks are clear, but it is likely that detrital zircon U-Pb age data from the three zones is insufficient to propose conclusions based on the relative abundance of zircon age populations. Present results, however, indicate that overlying Lower Cambrian detrital rocks are characterized by a relatively low percentage of Neoproterozoic zircons (37%, Fernández-Suárez et al., 2014; 38%, Zimmermann et al., 2015) and a higher percentage of Paleoproterozoic ones (43%, Fernández-Suárez et al., 2014; 52%, Zimmermann et al., 2015) while the underlying Lower Cambrian detrital rocks have a higher percentage of Neoproterozoic zircon and a lower percentage of Paleoproterozoic ones (79% and 12%, respectively, Henderson et al., 2016; 89% and 7%, Zimmermann et al., 2015). These results reinforce the idea that Ediacaran detritus contributed to the underlying Lower Cambrian detrital sediments of the CZ and WALZ. The geochemical data also support two different sources for the overlying and underlying Lower Cambrian rocks and the contribution of Ediacaran detritus to the underlying Lower Cambrian strata (Ugidos et al., 2016, and this study). The significant difference in zircon age populations in the same formation was noted by Zimmermann et al. (2015) who emphasized the distinct potential interpretations if only one sample was collected and also that "… much more care has to be taken when using detrital zircon U-Pb age geochronology to reveal maximum depositional ages, explain the geological evolution of cratonic blocks, or for correlation purposes." A sample of the upper Cambrian Oville Formation (CZ) reveals 73% Neoproterozoic, 9% Mesoproterozoic, 9% Paleoproterozoic and 5% Archean zircon ages (Henderson et al., 2016), a common result of the Ediacaran siliciclastic rocks.

Despite the varying proportions of the zircon ages, this indicates mixtures of different rocks and not significant changes in the chemical and Nd-isotope characteristics of the Ediacaran detrital rocks in all of the zones. Weathering of original compositions may easily result in mixed uniform compositions during recycling since weathering products would be similar. However, zircons remain practically unaltered and their distributions during hydrodynamic processes are dependent on their sizes. Thus, each original rock may provide the corresponding zircons of varying sizes and ages, but the new detrital rocks will selectively accumulate zircons according to grain size and specific hydrodynamics. Protracted cannibalistic recycling can join zircons from rocks of different ages in remarkable volumes of homogenized detrital rocks. Evolution of weathering, sea level changes and tectonics may lead to exposure to subaerial conditions in more extensive areas of the same source rocks, change the relative proportions of the rock types or expose new ones. In addition to zircon other heavy minerals may also reflect the same history.

The combination of recycling, sea level changes, sedimentary settings and the mixing of previous detrital material derived from the Ediacaran and Cambrian respective sources may result in varying combinations of zircon age spectra and geochemical features before final sedimentation. Interpretations may be quite different if only one sampling point is at the top or bottom of a succession. Thus, interpretations based only on detrital zircon geochronology and in the absence of other geological, stratigraphic and geochemical data may be erroneous and correlations may be unreliable.

The accumulation of zircon in the studied rocks is most likely related to recycling and sorting. Thus, varying proportions in vertical series of each group of U-Pb ages do not depend precisely on the age of the source rocks (except in the case of coeval volcanics) but may result from the time of exposure of the source rocks to the subaerial conditions and the abundance of each source rocks, as well as tectonic highs and lows.

Recently, Stephan et al. (2019) applied statistically provenance analysis to numerous U-Pb age data to identify provenance end-member populations and sedimentary mixing processes. They established three distinct provinces. The data set was filtered taking into account criteria such as: the samples exclusively used were deposited before 400 Ma and zircon age spectra only included ages of over 700 Ma. One of these provinces is the East African-Arabian Zircon Province, which includes the CIZ, WALZ and CZ.

9. Discussion of results and inferences

1. All of the groups of Ediacaran shales and sandstones from the three zones share similar features such as Al_2O_3/TiO_2, Zr/Al_2O_3, Ti/Nb and Nb/Hf ratios, REE patterns, and relatively low K_2O/Al_2O_3 molar ratios and εNd(541) values regardless of whether the samples were collected at random, at different heights of known stratigraphic series, in detrital beds separated by several Ma (e.g., the Ediacaran shales of the Allande and Tineo groups) or by distances within sampling areas. Thus, it is inferred that the Ediacaran shales and sandstones derived from the same sources. The thickness (2250–4000 m) and extension of these siliciclastic deposits (more than 60,000 km^2) strongly suggest an extensive source area that should have relatively abundant rocks of juvenile composition, and zircon populations dominated by Neoproterozoic ages, mainly Ediacaran and Cryogenian. The similar geochemical features of the Ediacaran sediments of the Narcea group and those of the CIZ, except CIA and the HREE pattern irregularities in the CIZ strongly support that the former were recycled and deposited in slope and base-of-slope environments of

the CIZ (see Fig. 4). Therefore, the repetition of geochemical features of siliciclastic rocks does not serve as proof of a direct contribution from the same original source. Weathering and recycling of a detrital succession may result in stratigraphically younger series having the same geochemical features as the older ones.

2. Probably, the most favorable setting to homogenize detrital material is a continental one, especially a passive margin setting. According to Johnsson (1993): (a) the chemical weathering of sediments in alluvial sequences is more intense than in tectonically active settings since temporary storage is longer in the former, while tectonics and rapid uplift facilitate a continuous supply of unweathered bedrocks to the surface; (b) fluvial sediments previously deposited are reincorporated into river bedload during meander growth, channel migration, and avulsion. Thus, it is possible that the homogeneity of sediments in the zones considered in this work developed in continental settings where the recycling of detrital material is possible without new contributions, i.e., in a closed system, until erosion affects the basement rocks. The only significant changes would result from the sorting that occurs during transport and sedimentation. However, this does not exclude the possibility that incoming tributaries in a fluvial system may carry sediments with distinct provenance and history (as compared to the principal source) (Johnsson, 1993). Also, sorting may cause major changes in the local concentration of dense minerals (Garzanti, 1986).
3. The combined geochemical features of the Lower Cambrian shales show significant differences as compared to those of the Ediacaran period. The former rocks present lower Ti/Nb, Zr/Al_2O_3, Zr/Nb and higher Al_2O_3/TiO_2, Ti/Zr, La/YbN and Nb/Hf ratios and lower $\varepsilon Nd(541)$ indicating a contribution of detrital material to Lower Cambrian series from a second source that is distinct from that of the Upper Ediacaran rocks or a relative increase of this source composition already present in the Upper Ediacaran sediments in lower proportions. In the CIZ, the Upper Ediacaran and Lower Cambrian successions are separated by an erosional boundary corresponding to a type-1 sequence boundary (see Section 5.3). The temporal coincidence of the worldwide eustatic sea-level fall (Knoll et al., 2004) and worldwide weathering (Parnell et al., 2014) favored a change in the proportions of sources contributing to the sediments coincident with the age of the Ediacaran-Lower Cambrian boundary (Sections 5.1, 5.2.4 and 5.3) and also favored the

erosion of Ediacaran and mixed sediments derived from the two sources as well as the erosion or lack of sedimentation of the lowermost Cambrian sediments. As a consequence, different geochemical parameters separate the Ediacaran from the Lower Cambrian sediments.

4. The local abundance of volcanic rocks in the Allande and Tineo groups separated by 13 Ma suggests independent events probably related to extensional tectonics during different ages rather than magmatically active margin settings. Remarkably, despite the presence of basaltic and andesitic rocks in the Allande and Tineo groups, the siliciclastic detrital series hosting the volcanic rocks exhibit the same chemical features and ranges of εNd(541) as those of the other series lacking volcanic rocks. Thus, volcanic contribution to the detrital series had to be relatively scarce and local, as suggested by the minor abundances of volcanic rocks as compared to those of the detrital successions. Also, acceptance of a magmatically active margin setting at Late Ediacaran times is not consistent with the relatively high $^{87}Sr/^{86}Sr$ ratio (0.7085) of the Ediacaran carbonates (unit IV) in the CIZ (Valladares et al., 2006) or with the change from relatively high and lower εNd(541) values shown by the Upper Ediacaran and Lower Cambrian siliciclastic series (>-4 and <-4, respectively). Volcanism may be related to Late Ediacaran extensional events (Valladares et al., 2002b) as has been reported in the Eastern Pyrenees (Padel et al., 2018). Also, zircons with negative εHf(t) are more abundant in the N-CIZ metasediments than in those of the S-CIZ (Orejana et al., 2015).

5. Aside from the previously mentioned geochemical differences, the Upper Ediacaran and Lower Cambrian detrital rocks reveal different abundances of detrital zircons, although few samples have yet to be studied. Neoproterozoic zircons are dominant (>75%) in the Upper Ediacaran detrital rocks, while the Lower Cambrian series have less than 38% (see Sections 5.1 and 5.2) or less than 44% if the Douro sample (Teixeira et al., 2011) is Lower Cambrian in age (see Section 8). However, it would be incorrect to accept these data as merely criteria to separate the Neoproterozoic from the Lower Cambrian series. As geochemical data and zircon ages strongly suggest, the underlying Lower Cambrian detrital series of the CZ and WALZ also received some contribution from the Upper Ediacaran sediments. Therefore, more than just one type of data should be considered to ensure the depositional age of detrital series. Special caution is needed before determining a depositional age.

10. NE Africa as possible source area

The results of Stephan et al. (2019) based on numerous U-Pb age data identified the so-called East African-Arabian Zircon Province as the source area for the CIZ, WALZ and CZ detrital rocks. Thus, in this section, the aim is to examine the extent to which other data, apart from those of the Zircon Province, are consistent with the features of the three Iberia zones. Obviously, an exhaustive bibliographic compilation is not possible here, but rather, we offer a synopsis of some of the most significant aspects.

The Arabian-Nubian Shield (ANS) covers more than $6 \times 10^6 km^2$ (Hargrove et al., 2006) stretching from NE Africa across the Arabian Peninsula, consisting mainly of juvenile rocks related to a major Cryogenian stage (about 870–635 Ma) of island arc volcano-sedimentary sequences. A second stage (630–580 Ma) of continental collision (Avigad and Gvirtzman, 2009) characterized by the abundance of calc-alkaline granite (630–580 Ma) intrusions (Bentor, 1985; Johnson et al., 2011; Morag et al., 2011, 2012; Moussa et al., 2008; Stern, 2002) is followed by a post-collision stage (580–540 Ma), which led to the stabilized ANS (Bentor, 1985; Moghazi, 2003). Abundant calc-alkaline and alkaline intrusions characterize the two later stages and by 620–580 Ma, numerous extensional terrestrial and marine basins developed, ranging from those containing only volcanic rocks to those containing only immature detrital sediments (Abd El-Rahman et al., 2010; Abu Sharib et al., 2019; Johnson, 2003; Moghazi et al., 2012; Morag et al., 2012; Samuel et al., 2011). In the North Eastern Desert, in Egypt, the Upper Precambrian Hammamat Group, derived from varying types of volcanic and plutonic rocks in different proportions, was deposited in a major river system of continental-scale proportions (Shalaby et al., 2006; Wilde and Youssef, 2002) and the immature features of these sediments reflect the rapid uplift erosion and deposition in alluvial fans and braided streams (Holail and Moghazi, 1998; Moghazi, 1999). In the central Eastern Desert, in Egypt, the 840–720 Ma group of zircon ages of the Um Had conglomerate have exclusively positive εHf(t) values, while those corresponding to 680–625 Ma have both positive and negative values. Abd El-Rahman et al. (2019) propose that the former group of zircons was derived from the rocks of the island arc stage of the ANS and the other group resulted from the contribution of older crustal materials. In the Zenifim Formation of the northern ANS, a Late

Ediacaran immature arkose was derived from the juvenile ANS rocks and shows a remarkable abundance of zircon grains in the 600–700 Ma age range, all of which have positive εHf(t) values (Avigad et al., 2015). To summarize, the ANS is characterized by an abundance of juvenile material, scarcity or absence of zircon xenocrysts in igneous rocks (Stern et al., 2010), immature detrital sediments, zircon that is less than 900 Ma and relatively high εNd(t) and positive εHf(t) values of volcanic and granitic rocks and derived sediments.

The quantitative importance and extension of detrital rocks long before the onset of Cambrian sedimentation is remarkable (Avigad and Gvirtzman, 2009; Kolodner et al., 2006). Erosion products of the ANS are also stored in the peripheral Upper Neoproterozoic sequences, across NE Africa and Arabia (Avigad et al., 2015). In both areas, the dominant depositional settings are fluvial braided-stream and alluvial fan deposits (Avigad et al., 2015; Eliwa et al., 2010; Holail and Moghazi, 1998; Khalaf, 2010; Moghazi, 1999; Wilde and Youssef, 2002).

At the end of the Ediacaran post-collisional stage (580–540 Ma) an intense weathering and erosion (Parnell et al., 2014) of the Neoproterozoic basement took place across the entire ANS and pre-dated the early Cambrian sedimentation (Johnson et al., 2011; Moghazi et al., 2012; Sandler et al., 2012) and a pervasive extensional regime affected the ANS until at least 545 Ma (Johnson et al., 2011).

To the west of the ANS is the Saharan Metacraton (SM), which extends over more than $5 \times 10^6\,km^2$ and is dominated by gneisses, metasediments and high-grade metamorphic rocks intruded by granitoids aged from 750 to 550 Ma (Abdelsalam et al., 2002). The boundary between the ANS and SM is the N-S trending Keraf-Kabus-Sekerr Suture (Abdelsalam and Stern, 1996). To the east of this suture, other areas such as a part of Sudan (Küster et al., 2008), Eritrea, and northern and western Ethiopia present similar geological and geochemical features as those of the ANS (Grenne et al., 2003; Stern, 2002; Woldemichael et al., 2010). However, in the western areas of the suture, the Precambrian basement of the SM consists mainly of older crust that was extensively reworked during the Neoproterozoic time (Abdelsalam et al., 2002; Teklay et al., 1998; Yeshanew et al., 2017). The intrusive rocks are also characterized by the presence of Proterozoic and Archean zircons, negative εHf(t) and negative εNd(t) (Harms et al., 1990; Küster et al., 2008; Stern et al., 2010; Yeshanew et al., 2017). According to these authors, and with minor exceptions, these features are in sharp contrast with those of the ANS. The joining of the ANS and SM at the end of Ediacaran resulted in an area of over $15 \times 10^6\,km^2$ of common geological history (Bentor, 1985).

It is difficult, therefore, to avoid making comparisons between Iberia and NE Africa at the end of the Ediacaran. Both regions reveal a record of sub-Cambrian weathering and erosion, contribution of Ediacaran material to those of the Lower Cambrian, source areas and general detrital sedimentary settings consistent with the features of the sediments in Iberia and consistent also with the geochemical change of the Upper Ediacaran sediments into the Lower Cambrian ones. Mixtures of detrital components from the two major sources, the ANS and SM, could supply older zircons to the resulting material and at the same time, would reduce the relatively high εNd(t) and the relative abundance of positive εHf(t) values in the ANS. Homogenization of these mixtures would take place via cannibalistic recycling in extensive continental areas, resulting from the ANS and SM joining before the final sedimentation in a marine setting.

11. Paleogeography

In the CZ, WALZ and CIZ of the Iberian Massif, the oldest sedimentary rocks, of the Late Ediacarian age and those of the Early Cambrian age constitute a complex shallowing-upward sequence, deposited in basins located in the NE African realm. The three zones of study are arranged in parallel areas, after reconstruction of the double orocline of the Variscan orogen of Iberia to its originally linear shape (Martínez-Catalán, 2011; Shaw et al., 2012), placing the CZ in a position that is more proximal to the continent and with the CIZ in a more distal position (Fig. 14).

In the CZ and WALZ, between the Late Ediacaran and Early Cambrian, an angular unconformity of tectonic origin resulted in a rise of the basement and a continentalization that may have led to these two zones being the direct source areas of the CIZ Ediacaran sedimentary wedge. This sedimentary wedge corresponds to the depositional sequence 2 in which the basal stratigraphic discontinuity is not observed, but the top corresponds to a type 1 discontinuity related to a eustatic sea level fall. In the upper part of this depositional sequence, the presence of carbonate levels formed by *Cloudina* reefs, provides an age of between 551 and 541 Ma to the Upper

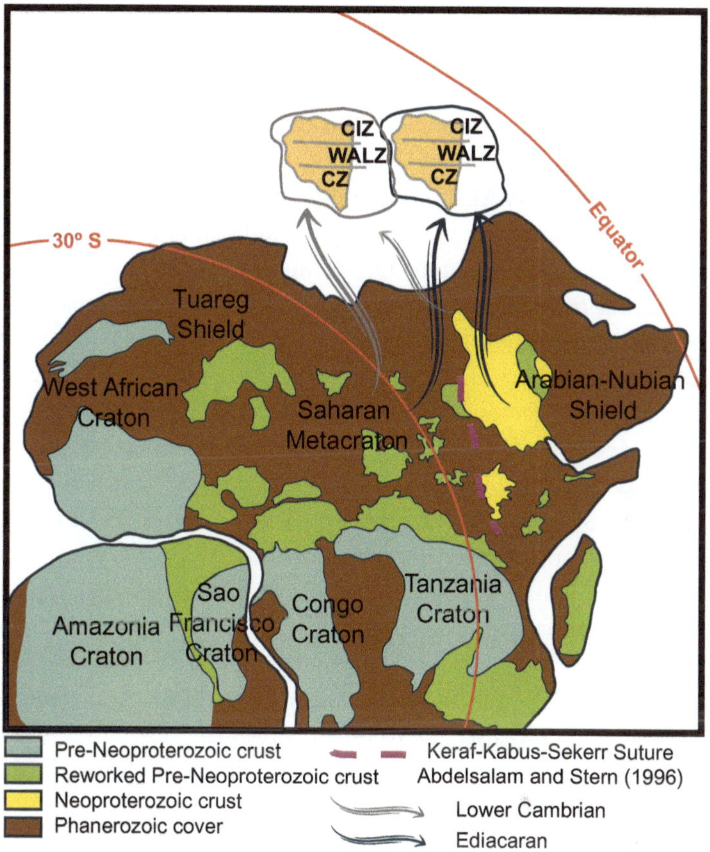

Fig. 14 Paleogeographic context of the Iberian Massif during the Late Ediacaran and Early Cambrian. Black and gray arrows indicate the dominant sources of Late Ediacaran and Early Cambrian sediments, respectively. Equator and 30° S positions at 550 Ma after Torsvik et al. (1996).

Ediacaran rocks (Amthor et al., 2003; Grotzinger et al., 1995; Narbonne et al., 2012). This age is close to that of sub-Cambrian worldwide weathering (542.62 ± 0.38 Ma, Parnell et al., 2014) that possibly caused the K-metasomatism of the Ediacaran succession in the CZ (see Section 5.1). These circumstances imply a little contribution of terrigenous sediment to the platform, and the paleogeographical situation at this time, between the equator and 30°S (see Section 5.3 and Fig. 14) at ca. 550 Ma.

The Lower Cambrian deposits define the depositional sequence 3 where the stratigraphic discontinuity of the top is located within the carbonate level (Láncara and Vegadeo limestone Formations) in the CZ and WALZ, while in the CIZ, this discontinuity is above the carbonate levels such as Tamames and Navalucillos limestones.

In the CZ and WALZ, overlying the discontinuity, there are ravinement conglomerates or clusters of alluvial fan sediments filling the structural paleo-valleys. The rest of the depositional sequence 2 are deposits ranging from braidplain deltas to carbonate platforms, with the presence of a transgressive level close to the base (521 Ma). In the CIZ, near the top of the Aldeatejada formation, a carbonate level has been recognized in allodapic facies.

In the CIZ, the sea level fall and tectonic pulses led the collapse of the platform facies and the formation of olistostromal deposits characterized by the presence of limestone blocks with karstification signs and *Cloudina*. Overlying these deposits, there are slope-distributary channel complex facies and platform facies ranging from siliciclastic to carbonate.

The stratigraphic gap between the Upper Ediacaran and Lower Cambrian deposits has a greater temporal range in the CZ and WALZ than that of the CIZ, due to the erosion of the Ediacaran deposits in the CZ and WALZ and also due to diachrony at the start of the sedimentation of the Lower Cambrian, which is earlier in the CIZ than in the CZ and WALZ.

Acknowledgments

Dr. Michael Montenari is thanked for inviting this contribution. Comments by Michael Montenari and an anonymous reviewer greatly improved the manuscript. The authors began studies of the detrital rocks of the CIZ during early 1990s and later in the CZ and WALZ. Financial support was provided by the following projects: Project No. SC1-CT90-0397 (EU Science Program). Projects PB91-0321 DGICYT, PB96-1283 DGES, BTE2002-04241-CO2-02 CGL2007-60035BTE and CGL2010-18905BTE (Ministerio de Ciencia e Innovación, Spain). Project CGL2016-78560-P (Ministerio de Economía, Industria y Competitividad, Spain).

Appendix

Table A1 Analyses of Upper Ediacaran shales. SHD. CIZ.

Sample	CORA-5	HR-42	HR-43	HR-44	HR-50	HR-52
SiO_2	54.85	62.21	60.22	63.18	60.88	59.35
TiO_2	1.18	0.89	0.94	0.89	0.80	0.95
Al_2O_3	21.50	17.57	18.76	16.97	18.00	18.58
Fe_2O_3	8.30	7.05	7.34	7.25	7.42	7.34
MgO	2.86	2.42	2.48	2.50	2.86	0.95
MnO	0.04	0.04	0.05	0.04	0.08	0.06
CaO	0.32	0.57	0.57	0.28	0.39	0.56
Na_2O	2.02	2.34	2.06	1.97	2.16	2.07
K_2O	4.10	2.97	3.30	3.12	3.03	3.44
P_2O_5	0.20	0.19	0.27	0.22	0.21	0.21
LOI	4.55	3.51	3.75	3.33	4.03	3.73
Total	99.92	99.76	99.74	99.75	99.86	98.82
Be	3.44	2.27	2.74	2.01	1.93	2.73
Rb	147	118	126	113	110	134
Sr	93.6	83.2	87.1	55.7	75.4	107
Cs	8.89	na	na	na	4.36	7.37
Ba	649	504	557	512	558	607
La	50.1	37.6	29.5	39.0	31.3	26.9
Ce	102	79.4	62.9	78.9	69.9	57.7
Pr	12.1	9.55	7.49	9.27	8.55	6.90
Nd	46.4	37.4	28.0	35.9	32.6	26.6
Sm	9.75	7.32	5.95	7.52	7.10	5.78
Eu	1.96	1.54	1.37	1.72	1.43	1.26
Gd	8.18	5.93	4.78	6.56	6.14	5.33
Tb	1.22	0.92	0.72	1.00	0.95	0.92
Dy	6.90	5.56	4.35	5.64	5.42	5.72
Ho	1.44	1.25	0.96	1.22	1.10	1.18

Continued

Table A1 Analyses of Upper Ediacaran shales. SHD. CIZ.—cont'd

Sample	CORA-5	HR-42	HR-43	HR-44	HR-50	HR-52
Er	3.91	3.02	2.64	3.12	3.17	3.30
Tm	0.62	0.46	0.45	0.51	0.52	0.51
Yb	4.27	3.22	3.17	3.21	3.27	3.49
Lu	0.57	0.46	0.44	0.47	0.51	0.54
Th	18.3	11.7	11.2	11.1	10.6	12.8
U	6.13	na	na	na	3.59	3.63
Sc	na	18.0	19.3	18.0	na	19.7
V	176	119	123	110	143	136
Cr	130	115	117	120	113	121
Co	17	12	11	13	12	24
Ni	64	42	38	54	33	61
Cu	12	30	40	11	55	39
Zn	165	112	110	116	114	132
Ga	31.1	23.5	25.0	22.1	26.3	26.8
Ge	1.28	na	na	na	1.91	1.88
As	26.1	na	na	na	4.48	6.10
Y	41.7	32.5	26.3	33.8	31.0	34.2
Zr	214	203	206	241	190	215
Nb	15.6	12.7	13.2	13.1	11.0	13.1
Mo	0.30	na	na	na	0.44	0.85
Sn	3.36	na	na	na	3.13	3.60
Sb	2.51	na	na	na	3.91	1.79
Hf	5.94	na	na	na	5.33	5.67
Ta	1.46	na	na	na	0.98	1.17
Pb	6.50	na	na	na	20.4	21.1
Bi	0.06	na	na	na	0.34	0.34

na: not analyzed; bd: below detection limit. New data in blue.

Data after Ugidos, J.M., Valladares, M.I., Barba, P., Ellam, R.M., 2003. The Upper Neoproterozoic-Lower Cambrian of the Central Iberian Zone, Spain: chemical and isotopic (Sm-Nd) evidence that the sedimentary succession records an inverted stratigraphy of its source. Geochim. Cosmochim. Acta 67, 2615–2629, Erratum: 2004, Geochim. Cosmochim. Acta 68, 1953; Ugidos, J.M., Sánchez-Santos, J.M., Barba, P., Valladares, M.I., 2010. Upper Neoproterozoic series in the Central Iberian, Cantabrian and West Asturian Leonese zones (Spain): geochemical data and statistical results as evidence for a shared homogenised source area. Precamb. Res. 178, 51–58.

Table A2 Analyses of Upper Ediacaran shales. NED. CIZ

Sample	ALC-1	ALC-2	MEM-1	MEM-2	MEM-3	SVAL-1	SVAL-2	SVAL-3	SVAL-4	ARL-1	ARL-2
SiO_2	63.07	62.37	57.60	59.63	59.16	58.65	59.28	62.69	61.33	59.20	60.99
TiO_2	0.85	0.88	0.97	0.88	0.93	0.89	1.01	0.90	0.90	0.90	0.88
Al_2O_3	17.42	18.12	20.36	19.08	19.16	19.81	20.11	18.10	18.33	19.57	18.53
Fe_2O_3	6.88	6.67	7.74	8.17	7.48	7.31	6.90	6.79	7.08	7.57	7.38
MgO	2.23	2.25	2.59	2.61	2.61	2.30	2.41	2.34	2.41	2.67	2.38
MnO	0.03	0.03	0.03	0.04	bd	bd	bd	0.03	bd	0.06	bd
CaO	0.31	0.27	0.19	0.20	0.22	0.14	0.21	0.17	0.19	0.19	0.20
Na_2O	2.46	2.10	1.28	2.00	1.39	1.58	1.78	1.71	1.17	1.91	1.93
K_2O	2.69	3.13	3.86	3.13	3.42	3.86	3.76	3.31	3.44	3.80	2.95
P_2O_5	0.18	0.17	0.17	0.16	0.19	0.15	0.19	0.16	0.18	0.16	0.18
LOI	3.73	3.86	5.07	4.06	5.29	5.16	4.21	3.66	4.83	3.85	4.44
Total	99.85	99.85	99.86	99.86	99.85	99.85	99.86	99.86	99.86	99.88	99.88
Be	2.32	2.37	2.79	2.14	2.67	2.46	2.56	2.14	2.64	2.82	2.12
Rb	92	104	134	111	123	139	128	112	115	147	105
Sr	80.9	66.9	57.7	62.9	52.6	48.6	56.3	61.1	40.2	61.0	58.0
Cs	4.42	5.22	6.35	5.46	5.57	5.79	5.58	5.90	4.35	6.82	4.96
Ba	501	549	649	586	658	660	632	588	530	685	605
La	20.6	33.5	41.9	38.0	40.1	41.1	40.2	34.1	32.6	40.6	41.9
Ce	42.1	70.2	86.1	77.9	82.7	82.5	85.3	70.1	67.5	82.8	84.4
Pr	5.41	8.07	10.2	8.98	9.47	9.68	10.1	8.44	8.20	10.0	10.2
Nd	20.7	31.7	41.0	36.1	38.7	38.6	40.8	33.0	31.7	39.1	40.3
Sm	4.50	6.52	8.41	7.22	7.77	8.15	8.50	6.56	6.77	8.13	8.24
Eu	1.07	1.42	1.79	1.66	1.70	1.72	1.80	1.44	1.50	1.81	1.79
Gd	4.75	6.21	7.11	7.04	6.71	6.91	7.20	5.77	6.02	6.38	7.11
Tb	0.82	0.91	1.13	0.99	1.02	1.05	1.10	0.85	0.89	0.95	1.07
Dy	5.33	5.61	6.69	6.23	6.22	5.61	6.18	5.40	5.33	5.71	6.33
Ho	1.15	1.13	1.28	1.23	1.25	1.18	1.27	1.07	1.06	1.13	1.26
Er	2.99	3.02	3.48	3.43	3.47	3.23	3.56	2.90	3.07	3.28	3.45
Tm	0.50	0.48	0.57	0.57	0.54	0.50	0.55	0.50	0.48	0.53	0.60
Yb	3.39	3.29	3.78	3.55	3.63	3.41	3.83	3.25	3.09	3.57	3.71

Continued

Table A2 Analyses of Upper Ediacaran shales. NED. CIZ—cont'd

Sample	ALC-1	ALC-2	MEM-1	MEM-2	MEM-3	SVAL-1	SVAL-2	SVAL-3	SVAL-4	ARL-1	ARL-2
Lu	0.53	0.54	0.62	0.57	0.59	0.55	0.60	0.49	0.51	0.58	0.52
Th	11.2	11.0	12.8	11.9	11.4	11.6	11.8	10.3	9.86	11.7	11.6
U	3.51	3.14	3.84	3.75	3.64	3.67	3.72	3.42	3.27	3.72	3.62
Sc	19.3	19.2	22.4	22.0	21.8	20.7	na	19.5	20.6	21.8	na
V	139	143	171	161	165	162	163	145	154	164	163
Cr	111	114	130	129	129	124	131	116	116	126	127
Co	14	12	14	12	13	13	24	11	14	20	12
Ni	48	46	52	57	60	62	61	40	51	58	56
Cu	26	30	34	30	41	53	35	32	41	59	43
Zn	122	121	123	120	122	130	119	101	114	143	128
Ga	22.9	24.7	28.8	27.7	27.5	28.8	28.5	24.2	24.1	29.3	27.9
Ge	1.75	1.69	1.94	1.96	2.10	1.83	1.96	1.93	1.86	2.03	1.92
As	14.1	30.5	16.1	8.61	5.81	10.6	14.7	10.9	5.18	12.1	5.21
Y	34.1	32.6	38.4	37.5	37.6	36.8	39.0	32.0	31.9	34.6	37.0
Zr	196	221	212	195	219	203	231	204	191	204	195
Nb	11.8	12.1	13.7	12.5	12.9	13.4	13.8	12.4	11.5	13.6	12.3
Mo	0.52	0.59	0.53	0.44	0.51	0.61	0.90	0.54	0.60	0.32	0.48
Sn	2.51	2.83	3.21	3.01	3.41	3.33	3.64	2.92	2.72	3.58	3.50
Sb	bd	bd	0.24	0.40	0.80	1.75	1.28	1.14	0.97	0.72	1.32
Hf	5.25	6.03	5.74	5.24	5.52	5.14	6.12	5.48	5.19	5.21	5.04
Ta	1.08	1.11	1.18	1.11	1.07	1.12	1.19	1.07	0.97	1.09	1.04
Pb	14.3	16.6	15.0	13.6	10.2	18.0	28.7	16.2	8.67	18.0	11.4
Bi	0.16	0.21	0.26	0.28	0.30	0.36	0.40	0.26	0.28	0.36	0.26

Sample	VAL-1	STAL-1	CAS-1	CAS-2	CAS-3	CAS-4
SiO_2	60.66	58.13	61.24	61.21	60.86	60.61
TiO_2	0.81	0.90	0.88	0.85	0.83	0.89
Al_2O_3	19.23	20.12	18.00	18.35	18.51	19.15
Fe_2O_3	6.91	7.64	7.38	6.71	7.35	6.79
MgO	2.40	2.57	2.60	2.25	2.51	2.46
MnO	0.05	0.03	bd	bd	bd	0.03

Table A2 Analyses of Upper Ediacaran shales. NED. CIZ—cont'd

Sample	VAL-1	STAL-1	CAS-1	CAS-2	CAS-3	CAS-4
CaO	0.23	bd	0.25	0.20	0.19	0.10
Na_2O	2.39	1.38	2.12	2.04	2.22	2.05
K_2O	3.50	3.91	2.81	3.36	3.27	3.53
P_2O_5	0.17	0.15	0.17	0.15	0.15	0.16
LOI	3.52	4.93	4.37	4.73	3.93	4.08
Total	99.87	99.85	99.82	99.85	99.82	99.85
Be	2.25	2.71	1.95	1.94	2.23	3.04
Rb	126	139	93	113	114	127
Sr	54.0	44.0	81.3	75.4	65.0	73.0
Cs	7.00	6.05	3.28	8.27	8.06	17.09
Ba	620	652	483	611	548	608
La	32.8	40.5	34.5	34.9	38.4	44.3
Ce	65.9	82.0	71.4	72.9	79.7	87.8
Pr	8.15	10.3	8.21	8.32	9.14	10.4
Nd	31.1	38.8	32.3	33.2	35.5	40.8
Sm	6.46	8.26	6.60	6.55	7.44	8.74
Eu	1.43	1.68	1.34	1.38	1.61	1.83
Gd	5.47	6.84	5.48	5.45	6.26	7.51
Tb	0.86	1.03	0.83	0.84	0.99	1.12
Dy	5.09	6.21	5.25	5.25	5.76	6.64
Ho	1.02	1.23	0.98	1.04	1.14	1.39
Er	2.82	3.31	2.86	2.98	3.26	3.81
Tm	0.47	0.55	0.46	0.46	0.48	0.63
Yb	2.95	3.60	3.02	2.99	3.38	4.06
Lu	0.46	0.52	0.46	0.44	0.51	0.58
Th	10.5	13.2	9.63	10.1	11.9	12.4
U	3.59	4.06	2.87	3.23	3.81	3.97
Sc	19.9	22.2	19.9	19.0	na	na
V	147	165	145	145	161	160

Continued

Table A2 Analyses of Upper Ediacaran shales. NED. CIZ—cont'd

Sample	VAL-1	STAL-1	CAS-1	CAS-2	CAS-3	CAS-4
Cr	109	123	119	114	123	126
Co	13	11	12	14	10	5
Ni	46	62	44	45	46	28
Cu	36	46	30	37	41	35
Zn	123	129	119	129	128	131
Ga	26.5	28.7	24.8	25.2	26.3	27.3
Ge	1.88	1.88	1.72	2.04	1.93	2.07
As	20.1	7.33	3.88	14.8	11.7	20.5
Y	29.8	35.5	30.9	32.3	33.2	41.1
Zr	178	193	184	199	168	207
Nb	11.8	13.1	11.8	12.0	12.1	12.8
Mo	0.46	0.44	0.49	0.46	0.67	0.47
Sn	3.32	3.58	2.75	3.06	2.94	3.87
Sb	6.36	1.73	0.81	0.92	5.21	0.34
Hf	4.63	5.14	4.61	4.93	4.54	5.77
Ta	1.00	1.13	0.97	0.97	1.02	1.13
Pb	16.5	7.54	7.49	12.8	26.0	16.1
Bi	0.36	0.44	0.23	0.22	0.53	0.30

na: not analyzed; bd: below detection limit. New data in blue.

Data after Ugidos, J.M., Valladares, M.I., Barba, P., Ellam, R.M., 2003. The Upper Neoproterozoic-Lower Cambrian of the Central Iberian Zone, Spain: chemical and isotopic (Sm-Nd) evidence that the sedimentary succession records an inverted stratigraphy of its source. Geochim. Cosmochim. Acta 67, 2615–2629, Erratum: 2004, Geochim. Cosmochim. Acta 68, 1953; Ugidos, J.M., Sánchez-Santos, J.M., Barba, P., Valladares, M.I., 2010. Upper Neoproterozoic series in the Central Iberian, Cantabrian and West Asturian Leonese zones (Spain): geochemical data and statistical results as evidence for a shared homogenised source area. Precamb. Res. 178, 51–58.

Table A3 Analyses of Upper Ediacaran shales. Units I–IV. SHD. CIZ

Sample	SIA-19	SIA-24	SIA-27	SIA-29	SIA-34	SSA-12	CPA-1	CPA-2B	SIA-7	SIA-12	SIA-15	SIA-21	SIA-22	SIA-31	SIA-35	SIA-36	SIA-37	CPA-3	CPA-4
Unit	I	I	I	I	I	I	I	I	II	II	II	II	II	II	II	II	II	II	II
SiO_2	60.11	61.59	60.37	57.57	60.79	56.07	60.08	60.79	59.67	62.01	59.64	58.48	57.17	61.30	56.68	60.21	60.07	59.84	60.77
TiO_2	1.07	1.04	1.00	1.03	1.00	1.10	0.99	0.97	0.98	0.92	1.04	1.09	1.07	0.94	0.99	1.02	1.03	1.00	1.00
Al_2O_3	19.91	18.90	19.90	21.05	19.84	20.86	19.45	19.27	19.65	18.33	19.51	19.96	20.37	18.84	21.74	19.96	19.55	19.41	19.22
Fe_2O_3	5.85	6.03	6.69	7.39	7.15	8.36	7.41	7.00	7.10	6.59	6.91	7.14	7.91	7.38	7.41	6.81	7.58	7.48	6.94
MgO	2.07	2.06	2.09	2.53	2.25	2.35	2.47	2.44	2.21	2.45	2.54	2.50	2.70	2.28	2.46	2.54	2.45	2.48	2.28
MnO	0.02	0.04	0.03	0.02	0.04	0.07	0.04	0.03	0.03	0.02	0.02	0.04	0.04	0.05	0.02	0.02	0.05	0.04	0.04
CaO	0.12	0.37	0.14	0.18	bd	0.26	0.20	0.22	0.21	bd	0.19	0.12	bd	0.19	bd	0.24	0.20	0.15	0.56
Na_2O	1.37	1.88	1.55	0.98	1.44	0.87	1.55	1.66	1.29	1.46	1.19	1.54	1.12	1.42	1.30	1.45	1.45	1.43	1.85
K_2O	4.66	3.94	3.57	3.91	3.59	3.46	3.47	3.42	3.96	3.52	3.95	4.17	4.17	3.54	4.20	3.65	3.57	3.5	3.53
P_2O_5	0.23	0.18	0.15	0.19	0.19	0.31	0.18	0.2	0.19	0.17	0.26	0.25	0.26	0.13	0.19	0.19	0.23	0.19	0.21
LOI	4.27	3.67	4.19	4.81	3.45	5.47	4.07	3.9	4.41	4.26	4.47	4.34	4.93	3.63	4.16	3.71	3.56	4.38	3.51
Total	99.68	99.70	99.68	99.66	99.79	99.68	99.91	99.9	99.70	99.73	99.72	99.63	99.74	99.70	99.18	99.80	99.74	99.9	99.91
Be	2.68	1.86	2.71	3.42	2.81	2.73	3.07	2.17	2.75	3.51	3.45	2.54	2.74	2.65	3.06	2.78	2.28	2.71	2.78
Rb	142	114	137	144	133	123	133	131	147	134	133	136	134	130	166	143	145	133	136
Sr	43.7	57.3	96.4	92.4	100	139	83.3	73.9	61.9	66.0	56.7	56.8	43.2	79.5	93.0	97.4	87.5	61.5	96.7
Cs	3.75	3.63	5.88	6.60	5.02	8.97	9.30	9.00	5.30	7.94	5.42	4.00	4.48	5.79	6.36	5.40	5.39	8.77	9.26
Ba	846	728	657	815	744	669	563	578	664	632	783	824	804	653	867	790	705	598	572

Continued

Table A3 Analyses of Upper Ediacaran shales. Units I–IV. SHD. CIZ—cont'd

Sample	SIA-19	SIA-24	SIA-27	SIA-29	SIA-34	SSA-12	CPA-1	CPA-2B	SIA-7	SIA-12	SIA-15	SIA-21	SIA-22	SIA-31	SIA-35	SIA-36	SIA-37	CPA-3	CPA-4
Unit	I	I	I	I	I	I	I	I	II	II	II	II	II	II	II	II	II	II	II
La	54.0	38.2	44.8	47.4	44.8	51.7	33.6	39.1	42.5	39.3	46.4	46.5	33.1	42.1	45.7	43.1	42.8	28.7	29.2
Ce	111	81.8	94.5	96.0	90.6	112	69.3	79.9	84.3	82.7	98.5	96.6	69.2	85.5	86.0	86.9	85.4	60.5	61.4
Pr	13.5	9.52	11.0	11.6	11.1	13.7	8.38	9.82	10.5	9.83	11.6	12.0	8.93	10.0	10.8	10.8	10.6	7.53	7.28
Nd	55.9	38.5	44.1	45.9	44.2	56.3	31.9	39.4	39.5	39.3	46.0	47.4	36.6	39.5	40.5	41.8	39.3	28.3	28.8
Sm	11.5	8.32	8.88	9.23	8.63	12.0	6.82	8.43	7.91	7.98	9.42	10.0	7.56	7.35	8.21	8.29	8.38	5.90	6.10
Eu	2.13	1.62	1.82	1.98	1.83	2.48	1.48	1.86	1.66	1.61	2.12	1.93	1.61	1.55	1.67	1.68	1.78	1.38	1.32
Gd	9.30	6.90	7.41	8.41	6.68	10.2	5.97	7.25	6.78	5.94	7.81	8.25	6.51	6.06	6.47	6.68	6.60	5.23	5.04
Tb	1.34	1.12	1.18	1.29	1.00	1.51	0.91	1.16	1.07	0.95	1.23	1.29	1.03	0.90	0.94	1.02	1.00	0.88	0.79
Dy	8.13	6.51	6.76	7.58	5.91	8.62	5.45	7.01	6.40	5.43	7.88	7.57	6.36	5.60	5.60	5.86	6.11	5.60	5.37
Ho	1.84	1.41	1.47	1.66	1.19	1.84	1.15	1.42	1.36	1.27	1.71	1.71	1.50	1.20	1.09	1.13	1.12	1.15	0.99
Er	4.71	3.70	3.86	4.53	3.14	4.55	3.23	4.06	3.44	3.24	4.31	4.23	3.90	3.20	3.01	3.31	3.23	3.21	3.02
Tm	0.70	0.58	0.58	0.68	0.53	0.72	0.49	0.58	0.48	0.48	0.64	0.65	0.61	0.47	0.49	0.51	0.54	0.50	0.44
Yb	4.64	3.99	3.75	4.17	3.40	4.54	3.59	4.29	3.75	3.36	4.18	4.29	3.84	3.38	3.21	3.40	3.71	3.63	3.37
Lu	0.66	0.58	0.59	0.66	0.52	0.63	0.50	0.65	0.52	0.55	0.64	0.64	0.57	0.54	0.56	0.54	0.58	0.53	0.51
Th	14.5	12.7	12.8	13.8	11.7	13.3	15.0	13.8	12.8	13.1	14.0	14.3	14.2	12.3	13.0	12.7	13.0	14.5	15.1
U	4.30	3.24	3.63	3.71	3.39	4.05	3.77	3.30	3.63	3.89	3.87	4.27	4.19	3.42	3.52	3.51	3.80	3.53	3.56
Sc	21.4	20.8	na	19.3	15.9	22.5	20.0	19.4	19.5	18.0	19.2	20.1	22.0	31.0	na	na	19.8	20.2	20.1
V	146	130	126	136	126	146	133	133	122	136	141	152	150	112	118	123	125	131	129

Cr	129	113	114	119	117	133	113	110	110	113	120	130	128	103	107	107	108	112	108
Co	17	18	35	14	11	14	11	10	20	13	15	12	15	18	10	17	15	28	18
Ni	33	35	24	30	38	22	34	27	48	25	27	27	28	48	32	24	36	37	48
Cu	19	21	28	24	25	15	27	16	43	44	22	40	26	31	20	19	27	30	39
Zn	92	90	101	77	102	102	98	94	119	90	95	108	112	114	80	73	112	106	107
Ga	28.9	24.7	28.1	29.7	26.0	28.3	25.6	25.6	26.8	25.7	27.5	28.6	29.2	25.0	29.4	26.5	27.0	26.3	27.4
Ge	na	1.49	1.62	1.62	1.37	1.50	1.40	1.30	1.66	na	na	na	na	1.69	1.31	1.25	1.27	1.30	1.20
As	23.4	17.0	12.1	9.18	3.98	14.6	9.55	5.92	6.30	16.4	1.94	10.4	3.77	11.3	0.39	11.2	9.75	5.26	8.46
Y	45.9	38.2	39.6	45.7	31.6	45.7	33.6	40.2	34.8	29.3	39.0	39.8	37.8	30.1	28.1	33.2	33.2	32.4	29.3
Zr	287	260	255	252	247	245	228	229	228	234	245	247	238	237	235	242	259	234	220
Nb	15.3	13.6	14.4	14.7	14.2	14.8	13.3	13.2	14.7	13.2	14.4	15.4	15.4	13.9	14.9	14.9	15.3	13.5	13.4
Mo	0.39	0.57	0.26	0.21	0.15	0.47	0.20	0.20	0.49	0.38	0.24	0.22	0.18	0.38	0.30	0.36	0.51	0.28	0.37
Sn	4.94	3.82	3.66	3.69	3.77	3.39	2.60	2.80	3.83	4.63	4.83	4.27	4.44	3.52	4.14	3.66	3.84	2.90	2.70
Sb	3.27	3.44	0.78	0.38	1.15	21.8	2.24	0.77	0.71	10.3	1.20	1.00	4.49	1.71	0.46	0.57	1.33	1.11	0.82
Hf	8.12	6.90	6.96	7.03	6.70	6.64	5.98	5.93	6.05	6.68	7.42	7.08	6.75	6.87	6.62	6.75	6.86	6.13	5.66
Ta	1.42	1.27	1.40	1.35	1.33	1.31	1.18	1.21	1.37	1.33	1.42	1.42	1.43	1.23	1.34	1.31	1.33	1.22	1.21
Pb	15.0	15.5	14.3	7.66	12.6	16.7	19.0	21.0	18.8	19.4	15.8	18.2	21.5	21.7	6.21	4.89	15.3	17.0	22.0
Bi	0.11	0.18	0.16	0.21	0.22	0.11	0.16	0.20	0.37	0.14	0.19	0.22	0.42	0.29	0.10	0.11	0.25	0.17	0.26

Continued

Table A3 Analyses of Upper Ediacaran shales. Units I–IV. SHD. CIZ—cont'd

Sample	CPA-5	SSA-9	CORA-3	CORA-4	SIA-3	SIA-5	SIA-6	SIA-25	SIA-38	SIA-39	HR-51	CORA-1	HR-40	HR-41	HR-45	SIA-10	SIA-13	FG-4	APL-18
Unit	II	II	II	II	III	III	III	III	III	III	III	III	IV	IV	IV	IV	IV	IV	IV
SiO_2	63.00	56.89	63.47	59.98	60.28	59.50	62.16	59.32	61.22	57.78	58.75	58.75	64.78	60.22	59.75	60.02	58.8	57.87	58.27
TiO_2	0.81	1.12	0.86	0.92	0.96	0.97	0.89	0.93	0.98	1.04	0.96	1.03	0.87	0.91	0.91	0.94	0.89	0.89	0.95
Al_2O_3	16.96	21.20	17.2	18.92	18.80	19.49	18.10	20.00	19.04	21.53	19.19	19.5	16.39	18.67	19.47	19.05	19.26	19.16	18.99
Fe_2O_3	7.72	7.98	6.9	7.57	7.47	6.95	7.26	7.36	7.02	7.32	7.62	7.57	6.45	7.29	6.77	6.35	7.99	8.59	7.28
MgO	2.95	2.60	2.47	2.62	2.74	2.46	2.28	2.51	2.24	2.23	2.76	2.97	2.17	2.59	2.67	2.41	2.83	2.57	2.79
MnO	0.05	0.06	0.02	0.02	0.05	0.04	0.06	0.04	0.05	0.05	0.06	0.02	0.04	0.05	0.05	bd	0.06	0.05	0.04
CaO	0.28	bd	0.13	0.06	bd	0.16	0.12	0.20	0.27	0.21	0.57	0.18	0.48	0.40	0.20	0.31	0.21	0.72	0.86
Na_2O	1.80	1.13	1.86	1.60	2.58	1.80	2.04	1.56	1.43	1.57	2.15	2.02	2.24	2.21	2.06	1.53	1.82	1.57	3.31
K_2O	2.64	3.55	3.13	3.72	2.83	3.63	3.33	3.58	3.56	4.30	3.55	3.65	2.77	3.49	3.77	3.85	3.27	4.04	3.26
P_2O_5	0.17	0.18	0.16	0.16	0.13	0.17	0.14	0.14	0.20	0.13	0.23	0.17	0.22	0.20	0.18	0.26	0.16	0.18	0.26
LOI	3.50	5.09	3.69	4.32	3.94	4.57	3.41	4.08	3.73	3.76	4.04	4.05	3.34	3.71	3.92	5.00	4.52	4.09	3.79
Total	99.88	99.86	99.89	99.89	99.78	99.74	99.79	99.72	99.74	99.92	99.88	99.91	99.75	99.74	99.75	99.73	99.81	99.73	99.8
Be	1.63	2.40	2.75	3.15	2.63	2.03	2.10	2.64	2.78	2.96	2.25	3.29	1.78	2.07	2.28	2.75	3.10	2.50	2.79
Rb	100	122	114	135	112	138	129	135	139	164	123	126	99.0	129	151	144	124	136	117
Sr	72.5	78.5	68.1	59.1	101	78.8	86.7	83.7	85.8	95.8	104	76.5	85.2	88.4	69.1	83.1	83.5	113	148
Cs	5.32	7.67	7.14	8.40	4.41	5.47	5.33	8.59	5.41	8.51	6.72	7.48	na	na	na	6.40	2.75	4.58	5.87
Ba	426	642	562	619	612	643	586	603	679	745	683	633	531	586	699	699	564	849	832
La	24.6	35.1	34.4	36.7	42.5	25.2	35.2	39.1	40.3	46.1	29.5	40.4	31.3	38.2	27.6	38.3	39.5	42.3	41.8

Ce	51.8	75.0	68.6	74.3	89.1	55.7	76.5	78.7	83.3	94.6	59.3	82.1	52.1	75.9	62.4	81.5	78.2	82.8	84.5
Pr	6.19	8.89	7.95	8.68	10.6	6.92	9.24	9.79	10.3	11.5	8.17	9.70	7.42	9.25	7.61	9.96	9.44	10.67	9.72
Nd	22.6	35.6	29.4	32.1	40.7	27.2	35.4	36.8	39.3	43.5	32.9	37.8	28.3	36.0	28.4	38.7	38.8	42.5	39.8
Sm	5.40	7.84	6.45	6.96	8.14	5.65	6.70	7.96	8.32	8.58	7.24	7.90	5.93	7.49	6.19	8.00	7.66	8.96	8.00
Eu	1.13	1.68	1.26	1.35	1.86	1.31	1.55	1.63	1.66	1.85	1.64	1.51	1.36	1.71	1.48	1.75	1.67	1.79	1.75
Gd	4.60	7.14	5.56	5.61	6.04	4.41	5.48	6.19	7.16	6.19	6.31	6.72	5.20	6.15	4.88	6.42	6.35	7.51	7.09
Tb	0.76	1.08	0.81	0.85	0.90	0.68	0.85	1.02	1.11	0.98	0.98	1.01	0.82	0.92	0.83	1.05	0.98	1.18	1.07
Dy	4.68	7.13	4.56	4.82	5.06	4.34	5.01	5.83	6.40	6.31	5.90	5.76	4.73	5.47	5.00	6.30	5.77	6.50	6.50
Ho	0.98	1.59	0.99	1.10	1.11	0.98	1.12	1.21	1.20	1.29	1.30	1.31	1.13	1.23	1.14	1.40	1.33	1.41	1.42
Er	2.88	4.06	2.68	3.05	2.96	2.72	2.82	3.28	3.42	3.73	3.46	3.38	3.07	2.79	2.72	3.59	3.33	3.59	3.82
Tm	0.40	0.60	0.43	0.47	0.46	0.48	0.43	0.49	0.58	0.59	0.60	0.52	0.48	0.45	0.46	0.5	0.49	0.53	0.61
Yb	2.9	4.18	3.19	3.39	3.52	3.38	3.45	3.28	3.73	3.80	3.67	3.51	3.23	3.1	3.24	3.74	3.38	3.81	3.65
Lu	0.48	0.60	0.42	0.48	0.51	0.47	0.48	0.51	0.62	0.58	0.54	0.54	0.48	0.45	0.47	0.52	0.53	0.54	0.57
Th	12.1	11.9	12.0	13.9	12.4	11.2	11.6	11.2	13.0	17.0	12.7	15.0	10.5	11.0	11.4	9.90	12.1	11.6	12.5
U	3.05	3.45	3.41	4.21	3.08	3.34	3.15	3.61	3.68	4.02	3.70	3.76	na	na	na	3.13	3.55	3.45	3.86
Sc	17.5	20.7	18.6	21.0	18.2	19.5	18.2	19.5	na	18.9	19.2	22.8	18.1	20.2	19.4	19.5	20.4	21.2	na
V	109	140	121	149	129	138	128	128	123	129	152	157	113	136	142	130	136	139	128
Cr	95	120	101	118	123	117	105	105	109	116	119	128	124	114	117	107	115	109	103
Co	18	16	12	9	20	15	21	19	15	32	11	14	12	15	9	18	18	25	33
Ni	42	30	40	42	33	13	47	48	18	51	42	25	39	49	20	23	52	53	47

Continued

Table A3 Analyses of Upper Ediacaran shales. Units I–IV. SHD. CIZ—cont'd

Sample	CPA-5	SSA-9	CORA-3	CORA-4	SIA-3	SIA-5	SIA-6	SIA-25	SIA-38	SIA-39	HR-51	CORA-1	HR-40	HR-41	HR-45	SIA-10	SIA-13	FG-4	APL-18
Unit	II	II	II	II	III	III	III	III	III	III	III	III	IV	IV	IV	IV	IV	IV	IV
Cu	26	24	18	24	11	15	34	29	27	29	36	27	32	39	16	51	36	53	32
Zn	130	99	82	108	116	98	115	124	97	114	132	103	124	114	100	91	127	101	114
Ga	22.2	26.7	22.9	25.5	26.6	27.3	25.2	27.2	25.9	29.5	26.9	26.6	22.0	25.0	27.4	26.0	26.8	27.8	26.8
Ge	1.30	1.44	1.10	1.20	2.33	1.72	1.73	1.62	1.16	1.22	2.00	1.20	na	na	na	1.97	1.67	1.86	1.45
As	3.37	8.02	6.31	7.68	30.0	5.12	5.74	0.80	12.1	24.2	20.9	5.64	na	na	na	1.31	1.26	1.78	34.5
Y	26.5	39.7	27.3	29.8	30.1	26.0	27.9	33.6	35.6	35.3	35.9	32.7	29.5	31.5	29.3	37.7	33.8	37.6	38.7
Zr	217	235	210	197	239	221	222	209	257	231	219	194	235	185	187	213	201	203	228
Nb	10.6	13.4	11.4	12.1	13.4	13.7	11.7	13.2	14.6	14.3	13.1	12.7	12.9	12.9	12.9	13.7	12.7	12.7	13.2
Mo	0.51	0.20	0.43	0.6	0.57	0.45	0.57	0.63	0.44	0.39	0.66	0.37	nd	nd	nd	1.28	1.15	1.16	1.16
Sn	1.80	3.52	2.60	2.60	3.38	3.97	3.58	3.71	3.60	3.55	3.60	2.90	na	na	na	3.96	3.19	4.21	3.38
Sb	2.43	2.30	3.06	3.43	16.2	1.56	4.65	8.16	0.92	0.40	0.55	0.86	na	na	na	1.22	1.74	4.07	0.89
Hf	5.70	6.75	5.91	5.79	6.26	5.70	6.09	5.29	7.05	6.16	5.90	5.60	na	na	na	5.89	5.35	5.05	5.90
Ta	0.99	1.23	1.06	1.14	1.22	1.23	1.11	1.20	1.26	1.35	1.19	1.20	nd	nd	nd	1.28	1.15	1.16	1.16
Pb	21.0	17.0	17.7	27.8	12.9	17.1	15.1	16.3	19.3	13.2	20.0	19.0	na	na	na	10.6	15.8	20.8	11.2
Bi	0.21	0.24	0.25	0.34	0.09	0.31	0.32	0.29	0.34	0.19	0.31	0.38	na	na	na	0.40	0.29	0.53	0.17

na: not analyzed; bd: below detection limit. New data in blue.

Data after Ugidos, J.M., Valladares, M.I., Barba, P., Ellam, R.M., 2003. The Upper Neoproterozoic-Lower Cambrian of the Central Iberian Zone, Spain: chemical and isotopic (Sm-Nd) evidence that the sedimentary succession records an inverted stratigraphy of its source. Geochim. Cosmochim. Acta 67, 2615–2629, Erratum: 2004, Geochim. Cosmochim. Acta 68, 1953; Ugidos, J.M., Sánchez-Santos, J.-M., Barba, P., Valladares, M.I., 2010. Upper Neoproterozoic series in the Central Iberian, Cantabrian and West Asturian Leonese zones (Spain): geochemical data and statistical results as evidence for a shared homogenised source area. Precamb. Res. 178, 51–58.

Table A4 Analyses of Upper Ediacaran shales. Units 2–3. IVA. CIZ

Sample	MST-5	MST-6	SJA-4	SJA-5	SJA-6	SJA-8	SJA-9	SJA-14 B	SJA-16	SJA-21	SJA-22	MT-4	MT-14
Unit	2	2	3	3	3	3	3	3	3	3	3	3	3
SiO_2	60.00	59.80	54.83	59.49	59.48	61.66	58.72	56.10	61.49	59.57	58.74	60.52	59.58
TiO_2	0.95	0.94	1.16	1.07	0.89	0.98	1.03	1.03	0.91	0.88	1.02	0.96	1.07
Al_2O_3	18.38	18.22	22.65	19.79	18.67	17.91	19.67	20.61	17.45	19.21	19.61	18.40	19.19
Fe_2O_3	7.55	7.46	7.28	7.68	7.95	7.39	7.72	7.72	7.22	6.66	7.39	7.81	7.57
MgO	2.94	0.94	2.57	2.52	2.73	2.51	2.63	2.70	2.73	2.48	2.88	2.60	2.68
MnO	0.04	0.04	0.02	0.04	0.04	0.02	0.03	0.03	0.04	0.04	0.06	0.03	0.04
CaO	0.31	0.36	0.33	0.26	0.43	0.30	0.33	0.34	0.38	0.36	0.47	0.33	0.34
Na_2O	2.01	2.26	1.50	1.72	2.52	1.70	1.47	2.18	2.39	2.39	1.72	1.80	1.57
K_2O	3.33	3.44	4.65	3.45	3.11	3.22	3.84	4.36	3.01	3.59	4.00	3.06	3.30
P_2O_5	0.24	0.23	0.17	0.16	0.12	0.17	0.18	0.27	0.26	0.17	0.19	0.25	0.22
LOI	4.06	3.59	4.76	3.72	3.94	4.03	4.28	3.95	3.51	3.95	4.40	4.02	4.24
Total	99.81	99.07	99.92	99.90	99.88	99.89	99.90	99.29	99.39	99.31	100.5	99.78	99.80
Be	2.68	2.75	3.87	3.05	2.72	2.49	2.60	2.99	2.46	2.91	3.06	3.12	2.15
Rb	125	124	173	127	120	123	144	160	117	133	140	110	107
Sr	77.0	86.2	94.2	89.3	104	83.2	77.4	102	97.1	92.8	72.2	82.0	4.07
Cs	6.54	6.81	13.1	11.0	8.40	6.83	8.70	8.84	6.80	6.69	7.21	6.28	3.95
Ba	585	613	829	645	597	622	703	844	567	765	835	531	612
La	38.5	39.0	47.0	39.5	34.0	41.5	38.5	44.4	38.8	43.1	44.5	42.7	40.0
Ce	79.3	81.4	96.8	84.8	69.9	87.4	79.0	89.5	79.6	87.5	90.2	91.6	85.7
Pr	9.53	9.64	11.6	10.4	8.83	10.3	9.69	11.0	9.78	10.7	11.3	10.3	10.1
Nd	37.3	37.5	43.6	40.4	32.7	41.1	37.6	42.4	38.2	40.7	43.0	43.1	39.8
Sm	7.74	7.51	9.58	8.42	7.13	8.46	7.86	8.52	7.85	7.66	8.56	9.75	8.59
Eu	1.67	1.58	2.08	1.71	1.41	1.70	1.67	1.73	1.67	1.56	1.73	2.14	1.82
Gd	6.42	6.32	7.97	6.76	5.19	6.63	6.11	7.11	6.70	6.14	6.77	9.24	6.90
Tb	0.97	1.01	1.23	1.03	0.83	1.01	0.95	1.11	1.03	0.98	1.03	1.67	1.14
Dy	5.67	5.88	7.59	6.34	5.10	5.81	5.63	6.62	6.11	5.92	6.14	10.1	6.19
Ho	1.11	1.17	1.56	1.32	1.04	1.15	1.18	1.30	1.20	1.20	1.22	2.10	1.38
Er	3.20	3.33	4.06	3.81	2.83	3.30	3.25	3.77	3.39	3.52	3.53	4.96	3.47
Tm	0.50	0.51	0.69	0.61	0.49	0.55	0.56	0.58	0.52	0.54	0.54	0.73	0.53

Continued

Table A4 Analyses of Upper Ediacaran shales. Units 2–3. IVA. CIZ—cont'd

Sample	MST-5	MST-6	SJA-4	SJA-5	SJA-6	SJA-8	SJA-9	SJA-14 B	SJA-16	SJA-21	SJA-22	MT-4	MT-14
Unit	2	2	3	3	3	3	3	3	3	3	3	3	3
Yb	3.39	3.51	4.56	3.82	3.18	3.56	3.63	4.03	3.46	3.74	3.64	4.23	3.59
Lu	0.53	0.54	0.67	0.56	0.41	0.47	0.56	0.63	0.54	0.58	0.56	0.66	0.58
Th	11.4	11.8	16.2	14.8	13.0	14.4	13.1	13.3	11.1	12.4	12.4	11.1	11.0
U	3.35	3.42	4.45	4.02	3.16	3.44	3.37	4.00	3.31	3.53	3.60	3.75	2.90
Sc	20.5	20.4	24.9	na	na	19.4	19.5	na	18.7	17.9	20.1	na	na
V	141	141	166	147	138	130	146	159	137	149	144	131	13.0
Cr	116	115	138	128	119	113	120	127	112	116	124	108	117
Co	22	22	7	24	23	13	9	23	16	26	23	20	11.9
Ni	55	54	39	53	56	40	35	62	46	53	54	53	45.0
Cu	38	43	33	21	38	31	26	38	44	18	30	35	34.0
Zn	108	125	110	122	125	114	106	130	129	124	122	103	124
Ga	26.5	25.1	31.9	25.7	25.0	25.1	26.4	29.0	25.2	28.8	28.1	24.0	26.0
Ge	1.80	1.67	1.42	1.32	1.26	1.18	1.30	1.96	1.94	1.89	1.87	1.61	1.58
As	13.3	16.4	31.3	26.2	27.9	15.6	11.2	20.0	21.1	24.2	14.4	17.7	7.21
Y	32.4	34.4	44.6	37.0	28.5	35.6	34.7	37.1	34.7	34.9	34.7	55.0	36.0
Zr	206	203	259	258	191	240	222	246	214	197	221	239	238
Nb	13.0	12.8	15.8	14.2	12.3	14.3	14.0	14.2	12.8	13.0	14.0	12.6	13.1
Mo	0.57	0.59	0.59	0.44	0.44	0.38	0.27	0.66	0.44	0.56	0.62	0.34	0.23
Sn	4.10	3.35	4.41	3.78	3.36	3.36	3.41	4.73	3.94	4.46	4.59	3.32	3.75
Sb	1.72	1.50	5.09	1.16	3.19	1.85	0.78	0.93	1.44	2.36	17.9	1.00	0.97
Hf	5.63	5.40	6.81	7.05	4.78	6.25	5.68	6.83	5.74	5.47	6.00	6.16	6.13
Ta	1.22	1.16	1.36	1.23	1.05	1.23	1.15	1.33	1.16	1.19	1.26	1.14	1.22
Pb	8.75	12.9	6.74	7.90	26.8	11.7	11.5	8.76	14.3	3.13	25.7	16.9	18.0
Bi	0.38	0.25	0.08	0.11	0.48	0.22	0.23	0.23	0.26	bd	0.47	0.27	0.27

na: not analyzed; bd: below detection limit. New data in blue.

Data after Ugidos, J.M., Sánchez-Santos, J.M., Barba, P., Valladares, M.I., 2010. Upper Neoproterozoic series in the Central Iberian, Cantabrian and West Asturian Leonese zones (Spain): geochemical data and statistical results as evidence for a shared homogenised source area. Precamb. Res. 178, 51–58.

Table A5 Analyses of Upper Ediacaran shales. IVA. CIZ

Sample	SJA-10	SJA-11	SJA-12	SJA-13	MST-1	LO-1	LO-2	LO-3	LO-4	LO-5	CTB-1	MT-1
SiO_2	59.55	60.76	56.71	58.01	59.35	59.15	60.30	60.11	61.12	58.83	57.86	54.66
TiO_2	0.97	0.89	0.99	0.99	0.97	0.86	0.90	0.86	0.82	0.94	0.86	1.01
Al_2O_3	19.32	17.97	20.40	21.58	18.78	18.51	18.93	18.55	17.43	18.59	18.77	19.71
Fe_2O_3	7.47	7.46	7.76	6.74	7.72	7.74	6.93	7.42	7.72	7.43	8.46	9.20
MgO	2.57	2.92	2.76	2.25	2.89	3.15	2.63	3.06	2.76	3.21	3.37	3.15
MnO	0.05	0.06	bd	bd	0.03	0.07	0.03	0.04	0.07	0.07	0.09	0.02
CaO	0.22	0.52	0.20	0.15	0.39	0.67	0.36	0.33	0.72	0.59	0.69	0.26
Na_2O	2.03	2.30	0.98	1.19	2.09	2.56	2.31	1.98	2.77	2.59	2.55	1.56
K_2O	3.87	3.64	4.19	4.72	3.53	3.54	4.04	3.52	2.53	3.72	3.37	3.97
P_2O_5	0.22	0.23	0.19	0.19	0.27	0.22	0.22	0.24	0.32	0.25	0.21	0.19
LOI	3.61	3.15	5.69	4.05	3.86	3.43	3.63	3.77	3.54	4.03	3.65	6.48
Total	99.88	99.90	99.88	99.88	99.88	99.90	100.3	99.88	99.80	100.3	99.88	99.78
Be	2.77	2.75	3.87	4.00	2.62	3.38	3.56	2.79	2.01	3.24	2.82	2.49
Rb	143	128	151	165	128	130	140	127	80.0	132	127	121
Sr	75.2	87.3	55.5	66.7	91.9	111	103	85.2	137	104	116	54.2
Cs	7.02	7.18	10.07	9.31	7.09	5.30	8.63	8.10	5.77	6.21	5.91	4.49
Ba	655	827	796	817	716	746	767	685	446	734	665	589

Continued

Table A5 Analyses of Upper Ediacaran shales. IVA. CIZ—cont'd

Sample	SJA-10	SJA-11	SJA-12	SJA-13	MST-1	LO-1	LO-2	LO-3	LO-4	LO-5	CTB-1	MT-1
La	42.1	39.1	45.1	39.5	43.8	39.4	46.7	43.2	35.2	40.0	46.0	34.9
Ce	88.1	77.5	92.8	80.2	91.5	82.6	93.5	87.3	73.7	82.6	92.8	71.5
Pr	10.6	9.4	11.7	10.6	10.9	10.2	11.4	10.6	9.08	10.0	11.3	8.73
Nd	41.2	35.3	45.7	39.7	43.6	40.4	45.1	40.6	38.6	38.3	43.6	34.9
Sm	8.13	7.27	8.37	8.24	9.12	8.45	8.94	8.23	8.96	7.85	9.05	7.54
Eu	1.80	1.47	1.79	1.63	1.94	1.61	1.83	1.67	1.95	1.52	1.60	1.70
Gd	7.45	6.33	6.67	6.33	8.19	6.98	7.84	6.76	9.46	6.75	7.64	6.39
Tb	1.08	0.89	1.05	1.03	1.12	1.09	1.12	1.03	1.37	1.05	1.14	1.04
Dy	6.19	5.57	6.24	5.80	6.86	5.99	6.83	5.90	7.68	6.14	7.08	6.44
Ho	1.25	1.07	1.17	1.15	1.36	1.24	1.40	1.18	1.38	1.18	1.42	1.44
Er	3.54	3.08	3.48	3.26	3.77	3.21	3.80	3.33	3.78	3.32	3.83	3.62
Tm	0.58	0.51	0.56	0.58	0.59	0.48	0.57	0.54	0.62	0.51	0.59	0.59
Yb	3.53	3.25	3.07	3.94	3.75	3.36	3.55	3.31	3.75	3.35	3.54	3.79
Lu	0.57	0.51	0.62	0.61	0.60	0.56	0.58	0.53	0.58	0.50	0.58	0.58
Th	13.1	13.5	15.2	15.1	13.6	14.4	14.8	14.2	10.9	12.2	14.7	10.9
U	3.84	4.08	4.15	4.26	3.87	4.13	4.09	3.66	3.74	3.50	3.99	3.43
Sc	20.8	19.9	22.5	15.3	21.6	na	19.0	na	20.6	20.4	22.2	25.5
V	155	149	147	138	151	125	137	124	154	122	135	189

Cr	122	111	116	114	118	99	112	103	113	99	105	133
Co	24	20	13	11	10	27	10	4	20	20	22	8
Ni	81	55	40	38	43	59	53	45	55	49	56	62
Cu	46	51	29	13	35	49	37	27	36	39	40	57
Zn	163	147	102	95	121	140	113	110	123	114	147	117
Ga	27.6	25.9	30.2	30.7	27.9	27.9	28.4	26.6	22.9	26.6	28.3	28.3
Ge	1.91	1.83	2.13	2.02	1.84	1.94	2.17	2.35	1.84	1.89	1.79	1.75
As	15.2	8.67	27.5	0.52	8.34	19.9	36.5	9.42	18.0	14.4	23.0	14.0
Y	35.5	32.2	34.3	30.8	39.2	34.8	40.4	34.6	41.8	31.8	38.9	38.5
Zr	215	206	206	231	228	200	220	208	179	200	207	205
Nb	13.1	13.1	14.7	16.0	14.4	12.9	14.1	13.5	10.2	12.8	14.3	12.5
Mo	0.85	1.46	0.39	bd	0.19	1.09	0.21	bd	0.86	bd	0.60	0.66
Sn	3.47	3.70	4.39	4.49	3.61	3.44	4.08	3.61	2.68	3.31	3.48	3.81
Sb	4.65	3.00	4.61	0.94	2.33	0.78	4.05	1.42	1.25	1.05	1.85	2.05
Hf	5.78	5.24	5.53	6.38	6.10	5.53	5.67	5.60	4.94	5.41	5.55	5.4
Ta	1.19	1.19	1.30	1.43	1.24	1.18	1.23	1.17	0.94	1.14	1.23	1.08
Pb	16.1	24.7	3.78	4.02	14.3	31.9	22.1	7.33	20.3	8.70	27.7	10.1
Bi	0.44	0.70	0.91	0.06	0.27	0.52	0.45	0.35	0.41	0.23	0.50	0.48

Continued

Table A5 Analyses of Upper Ediacaran shales. IVA. CIZ—cont'd

Sample	MT-8	MT-9	MT-1O	MT-11	CIB-1	CIB-2	CIB-7	ATR-1
SiO_2	57.19	57.25	60.31	58.31	59.17	61.67	59.94	60.26
TiO_2	1.14	1.20	0.95	1.01	0.83	0.84	0.92	0.88
Al_2O_3	21.36	20.95	18.92	18.73	18.41	17.92	18.81	18.66
Fe_2O_3	8.30	8.59	6.98	8.07	8.37	7.48	6.59	7.39
MgO	2.26	2.62	2.50	3.02	3.02	2.72	2.64	2.74
MnO	0.06	0.06	0.02	0.07	bd	0.03	0.03	0.04
CaO	0.14	0.17	0.27	0.47	0.42	0.19	0.34	0.27
Na_2O	0.97	0.79	2.92	2.20	2.67	2.78	3.07	2.29
K_2O	3.62	3.36	3.16	3.09	2.80	2.82	3.71	3.31
P_2O_5	0.12	0.11	0.18	0.21	0.28	0.18	0.21	0.22
LOI	4.53	4.59	3.50	3.92	3.87	3.63	3.43	3.80
Total	99.69	99.69	99.71	99.10	99.87	100.3	99.68	99.86
Be	3.15	2.57	2.27	2.81	2.48	1.79	2.77	2.21
Rb	122	111	105	100	96	90	126	116
Sr	174	84.4	75.0	123	105	99.0	91.5	82.0
Cs	5.59	6.36	3.64	5.50	7.44	6.55	8.10	5.41
Ba	656	630	595	564	505	443	775	627
La	39.0	41.3	35.8	34.2	33.5	31.4	29.0	37.1

Ce	78.3	83.0	75.4	69.8	69.4	66.0	60.8	76.6
Pr	9.61	10.3	9.17	9.09	8.83	8.24	7.41	9.24
Nd	37.5	41.1	35.6	34.9	36.0	32.0	28.9	35.6
Sm	7.27	8.23	7.15	7.40	8.19	6.86	5.83	7.77
Eu	1.72	1.82	1.50	1.63	1.91	1.40	1.14	1.69
Gd	6.26	6.89	5.94	6.09	7.51	5.36	4.79	6.40
Tb	0.96	1.06	0.97	1.01	1.11	0.87	0.76	0.98
Dy	5.60	6.37	5.58	6.04	6.47	5.45	4.62	5.89
Ho	1.27	1.46	1.28	1.28	1.33	1.15	0.96	1.21
Er	3.22	3.59	3.21	3.28	3.65	3.10	2.92	3.46
Tm	0.54	0.59	0.51	0.52	0.58	0.48	0.46	0.57
Yb	3.54	3.83	3.27	3.27	3.83	3.35	3.22	3.39
Lu	0.52	0.56	0.51	0.49	0.57	0.52	0.51	0.54
Th	11.0	11.7	10.9	9.74	11.5	10.5	12.8	12.0
U	3.21	3.40	3.32	2.92	3.84	3.61	3.57	3.99
Sc	na	22.5	19.7	22.5	na	20.2	21.6	20.5
V	147	143	130	124	165	140	126	147
Cr	131	127	109	98	121	107	98	117
Co	19	14	18	18	19	15	13	19

Continued

Table A5 Analyses of Upper Ediacaran shales. IVA. CIZ—cont'd

Sample	MT-8	MT-9	MT-1O	MT-11	CIB-1	CIB-2	CIB-7	ATR-1
Ni	53	51	44	48	59	39	33	87
Cu	37	34	32	35	43	35	26	47
Zn	115	163	99	104	134	112	97	160
Ga	27.0	27.6	25.0	25.0	26.6	23.4	25.9	25.7
Ge	1.66	1.91	1.31	1.51	2.08	1.88	1.75	2.00
As	7.65	15.2	16.6	11.0	16.9	16.1	19.9	17.7
Y	33.7	35.5	35.0	34.3	38.1	31.3	27.7	35.0
Zr	248	260	216	206	191	198	201	213
Nb	14.2	14.5	12.5	12.1	10.7	10.7	12.5	12.3
Mo	0.21	0.14	0.58	0.07	1.39	1.00	bd	0.50
Sn	3.75	3.48	3.03	3.44	3.08	2.83	3.07	3.26
Sb	0.86	1.04	6.83	1.43	2.14	1.11	3.98	3.03
Hf	6.61	7.08	5.85	5.93	5.40	5.83	5.40	5.72
Ta	1.27	1.28	1.12	1.10	1.00	0.98	1.13	1.08
Pb	11.8	11.1	9.69	17.8	18.7	20.3	13.1	21.4
Bi	0.31	0.14	0.19	0.29	0.44	0.37	0.33	0.38

na: not analyzed; bd: below detection limit. New data in blue.

Data after Ugidos, J.M., Sánchez-Santos, J.M., Barba, P., Valladares, M.I., 2010. Upper Neoproterozoic series in the Central Iberian, Cantabrian and West Asturian Leonese zones (Spain): geochemical data and statistical results as evidence for a shared homogenised source area. Precamb. Res. 178, 51–58.

Table A6 Analyses of Upper Ediacaran shales. SED. CIZ

Sample	SNB-1	VDE-1	ABJ-1	ABJ-2	NVP-1	NVP-4	NVP-5	CB-1	CB-2	VS-2	VS-4	PA-1
SiO_2	59.46	59.16	59.38	57.58	58.77	59.47	61.11	56.38	57.06	58.18	60.03	61.33
TiO_2	0.86	1.03	0.85	0.84	0.89	0.94	0.96	1.04	0.97	1.02	0.95	0.96
Al_2O_3	18.63	20.17	18.53	18.70	19.05	18.38	17.53	19.94	19.56	18.79	18.21	17.82
Fe_2O_3	7.22	7.79	7.28	7.25	7.64	7.69	7.42	8.32	8.01	8.07	7.81	7.58
MgO	3.07	2.35	3.06	3.29	2.91	2.92	2.96	3.17	3.19	3.14	3.07	3.01
MnO	0.07	0.07	0.07	0.03	0.03	0.03	0.06	0.04	0.04	0.04	0.06	0.05
CaO	0.36	0.16	0.45	0.79	0.41	0.44	0.37	0.34	0.34	0.36	0.33	0.30
Na_2O	2.57	0.56	2.90	0.78	2.35	2.11	2.42	2.34	2.01	1.86	2.44	2.32
K_2O	3.58	3.87	3.31	4.22	3.69	3.40	3.22	3.78	3.77	3.67	3.20	3.17
P_2O_5	0.24	0.14	0.28	0.20	0.21	0.27	0.21	0.20	0.22	0.22	0.22	0.19
LOI	3.81	4.58	3.76	6.06	4.23	4.31	3.89	4.71	4.90	4.69	4.18	3.77
Total	99.87	99.87	99.87	99.75	100.2	99.97	100.1	100.2	100.1	100.0	100.5	100.5
Be	2.37	3.04	2.94	3.37	3.21	2.43	2.30	2.84	2.69	2.67	2.65	2.45
Rb	132	129	129	145	130	115	104	126	127	126	114	111
Sr	93.0	79.9	99.0	35.5	75.4	66.2	68.5	63.8	56.9	72.3	90.7	70.1
Cs	5.69	7.80	6.87	15.7	8.48	5.49	4.01	4.79	6.14	7.23	6.48	6.36
Ba	614	503	513	670	550	484	531	754	594	622	545	554

Continued

Table A6 Analyses of Upper Ediacaran shales. SED. CIZ—cont'd

Sample	SNB-1	VDE-1	ABJ-1	ABJ-2	NVP-1	NVP-4	NVP-5	CB-1	CB-2	VS-2	VS-4	PA-1
La	43.1	41.9	44.8	37.2	44.6	33.9	35.5	39.2	35.7	38.4	35.2	35.1
Ce	88.9	82.7	90.4	71.9	88.9	69.9	74.5	81.4	74.2	77.3	71.1	73.9
Pr	10.8	10.7	11.0	9.39	11.4	8.49	8.92	10.1	9.09	9.80	8.98	8.97
Nd	40.6	40.5	42.8	35.0	42.9	33.2	34.5	39.8	35.1	38.4	34.9	34.7
Sm	8.31	8.41	8.59	6.96	8.73	6.92	7.36	8.43	7.55	8.00	7.28	7.21
Eu	1.68	1.84	1.79	1.38	1.63	1.45	1.58	1.74	1.65	1.73	1.58	1.56
Gd	6.95	7.49	7.28	5.70	7.21	5.89	6.35	7.73	6.62	7.01	6.27	6.22
Tb	1.10	1.15	1.05	0.89	1.10	0.90	1.02	1.24	1.08	1.10	0.98	0.98
Dy	6.13	6.83	6.04	5.31	6.38	5.38	6.07	7.55	6.44	6.47	5.76	5.90
Ho	1.17	1.32	1.19	1.03	1.25	1.06	1.17	1.50	1.30	1.28	1.14	1.17
Er	3.28	3.79	3.50	2.97	3.56	3.11	3.31	4.23	3.73	3.65	3.29	3.37
Tm	0.55	0.58	0.56	0.47	0.54	0.47	0.51	0.65	0.57	0.55	0.50	0.51
Yb	3.58	3.81	3.60	3.12	3.63	3.19	3.37	4.35	3.84	3.76	3.42	3.51
Lu	0.52	0.59	0.60	0.47	0.56	0.50	0.52	0.65	0.60	0.58	0.53	0.55
Th	13.6	13.2	13.5	11.8	14.4	10.4	10.5	11.9	11.2	12.0	10.9	10.8
U	3.71	3.71	4.26	3.05	3.99	3.22	2.87	3.49	3.56	3.39	3.09	3.14
Sc	20.4	20.2	20.6	20.1	na	20.9	20.3	24.6	24.2	22.8	20.7	20.7
V	129	125	128	126	132	139	123	157	173	151	141	136
Cr	104	116	95	95	102	114	106	126	131	123	119	117

Co	12	14	17	12	20	10	12	15	7	11	12	11
Ni	43	52	49	42	54	53	48	59	50	57	51	45
Cu	24	37	37	44	43	40	38	49	36	48	38	33
Zn	250	130	110	108	111	117	108	122	194	132	127	121
Ga	27.3	26.7	27.4	25.9	27.9	24.2	22.3	28.0	27.2	26.8	25.5	24.5
Ge	1.73	1.96	2.09	1.73	1.89	1.64	1.63	1.82	1.86	1.86	1.81	1.79
As	6.23	3.37	8.70	12.9	11.3	11.6	11.5	5.29	12.5	11.5	9.02	8.95
Y	35.0	39.2	35.0	30.5	36.6	31.2	34.0	43.6	37.8	37.2	33.0	33.7
Zr	213	269	206	181	196	188	205	212	194	207	201	216
Nb	13.7	14.7	14.1	11.8	13.1	11.6	11.7	12.7	11.9	13.0	12.1	12.4
Mo	0.37	bd	0.18	bd	0.42	bd	bd	bd	bd	bd	bd	bd
Sn	3.66	3.64	4.19	3.33	3.90	2.75	2.74	3.45	3.09	3.08	2.96	2.98
Sb	0.77	0.35	1.36	1.55	1.92	1.77	0.87	2.55	28.55	6.65	0.74	0.98
Hf	5.55	6.88	5.47	4.87	5.41	5.01	5.60	5.78	5.23	5.56	5.43	5.78
Ta	1.17	1.28	1.22	1.08	1.18	1.04	1.06	1.12	1.05	1.16	1.08	1.09
Pb	72.5	3.42	13.4	13.7	10.7	12.8	12.9	7.82	3.95	14.0	18.6	12.9
Bi	0.08	0.21	0.20	0.48	0.54	0.30	0.36	0.33	bd	0.31	0.39	0.39

na: not analyzed; bd: below detection limit. New data in blue.

Data after Ugidos, J.M., Sánchez-Santos, J.M., Barba, P., Valladares, M.I., 2010. Upper Neoproterozoic series in the Central Iberian, Cantabrian and West Asturian Leonese zones (Spain): geochemical data and statistical results as evidence for a shared homogenised source area. Precamb. Res. 178, 51–58.

Table A7 Analyses of Upper Ediacaran sandstones. SHD. CIZ.

Sample	FG-2	FG-2B*	FG-3
SiO_2	75.49	73.81	67.14
TiO_2	0.53	0.73	0.86
Al_2O_3	10.91	12.16	15.58
Fe_2O_3	4.51	5.29	6.37
MgO	1.52	1.41	1.79
MnO	0.05	0.04	0.05
CaO	0.58	0.49	0.33
Na_2O	2.09	1.71	1.58
K_2O	1.59	1.89	2.67
P_2O_5	0.11	0.09	0.10
LOI	2.46	2.12	3.26
Total	99.84	99.74	99.73
Be	1.52	1.46	1.79
Rb	58.8	69.9	97.1
Sr	95.5	78.4	67.1
Cs	4.02	5.36	4.73
Ba	290	373	463
La	21.7	31.5	29.6
Ce	44.7	74.1	68.5
Pr	5.00	7.92	7.41
Nd	19.9	28.8	28.9
Sm	4.61	6.07	5.63
Eu	0.96	1.33	1.27
Gd	4.02	5.03	4.60
Tb	0.59	0.81	0.69
Dy	3.28	4.61	3.89
Ho	0.68	0.90	0.90
Er	1.93	2.46	2.28

Table A7 Analyses of Upper Ediacaran sandstones. SHD. CIZ.—cont'd

Sample	FG-2	FG-2B*	FG-3
Tm	0.29	0.39	0.35
Yb	2.06	2.44	2.62
Lu	0.34	0.41	0.43
Th	7.96	9.34	10.8
U	2.12	2.63	2.86
Sc	na	na	na
V	65	73	93
Cr	68	79	94
Co	27	27	40
Ni	27	29	45
Cu	16	22	33
Zn	81	67	87
Ga	13.0	16.7	21.0
Ge	1.00	1.32	1.50
As	4.97	11.5	14.2
Y	19.4	26.6	22.8
Zr	209	306	250
Nb	7.19	10.2	11.7
Mo	0.23	0.22	0.29
Sn	1.35	2.22	2.89
Sb	1.02	0.97	0.45
Hf	5.47	7.60	6.73
Ta	0.80	1.00	1.22
Pb	16.4	16.0	15.7
Bi	0.13	0.22	0.22

na: not analyzed; bd: below detection limit. New data in blue. FG-2B*: earlier SIA-26.

Data after Ugidos, J.M., Valladares, M.I., Barba, P., Ellam, R.M., 2003. The Upper Neoproterozoic-Lower Cambrian of the Central Iberian zone, Spain: chemical and isotopic (Sm-Nd) evidence that t-he sedimentary succession records an inverted stratigraphy of its source. Geochim. Cosmochim. Acta 67, 2615–2629. Erratum: 2004, Geochim. Cosmochim. Acta 68, pp. 1953.

Table A8 Analyses of Upper Ediacaran sandstones. Unit 3 and at random. IVA. CIZ.

Sample	SJA-17	SJA-18	SJA-19	SJA-20	SJA-23	SJA-15	SJA-3	MT-2	MT-7	MT-12	MT-13	MST-7
Unit	3	3	3	3	3	ar	ar	ar	ar	ar	ar	ar
SiO_2	74.37	73.25	72.68	71.96	72.52	72.15	75.98	68.13	68.20	69.28	65.21	70.15
TiO_2	0.65	0.69	0.67	0.62	0.71	0.54	0.55	0.72	0.65	0.74	0.80	0.73
Al_2O_3	11.65	12.09	12.47	11.58	12.43	9.91	9.96	14.59	13.87	13.74	15.45	13.14
Fe_2O_3	4.03	4.17	4.43	4.30	4.64	4.42	4.11	5.63	4.64	5.07	6.21	4.91
MgO	1.63	1.47	1.44	1.96	1.61	1.69	1.52	2.09	1.91	2.19	2.63	1.70
MnO	0.05	0.04	0.05	0.07	0.03	0.12	0.03	0.02	0.04	0.03	0.05	**0.03**
CaO	0.54	0.66	0.38	1.10	0.50	2.50	0.83	0.38	1.26	0.63	0.73	0.57
Na_2O	2.12	2.13	2.33	1.82	2.10	2.43	1.95	3.09	3.18	3.70	2.93	2.84
K_2O	1.85	2.11	2.13	1.99	2.05	1.13	1.31	2.02	2.26	1.64	2.58	1.80
P_2O_5	0.13	0.13	0.13	0.13	0.14	0.16	0.13	0.13	0.14	0.17	0.18	0.17
LOI	2.91	2.85	2.80	3.57	2.87	3.88	2.75	3.00	3.65	2.62	3.08	2.74
Total	99.91	99.58	99.51	99.10	99.60	98.95	99.12	99.80	99.80	99.81	99.85	98.78
Be	1.26	1.47	1.76	1.93	1.32	1.39	1.24	1.74	1.77	1.64	2.04	1.67
Rb	66.8	75.8	74.6	72.2	72.5	42.9	49.1	68.1	79.6	58.1	99.2	67.7
Sr	96.8	123	120	96.3	107	279	68.9	81.0	202	172	110	120
Cs	3.00	3.62	3.53	3.56	3.84	2.45	4.16	2.48	4.59	2.83	4.87	3.56
Ba	369	421	424	414	411	224	236	299	477	382	383	331
La	33.2	34.1	30.7	30.2	31.4	26.0	24.9	26.9	28.2	33.1	32.6	**32.6**
Ce	69.7	73.1	67.4	65.9	68.7	55.3	55.3	56.1	60.3	67.9	69.1	**69.1**
Pr	7.97	8.20	7.51	7.39	7.76	6.44	6.17	6.55	6.83	7.95	7.93	**7.94**
Nd	30.3	31.5	28.7	27.9	29.8	25.1	23.9	25.9	26.3	31.1	31.6	**30.5**
Sm	5.79	6.13	5.74	5.49	5.83	5.19	4.77	5.23	5.29	6.01	6.54	**5.87**
Eu	1.26	1.32	1.23	1.25	1.28	1.20	0.95	1.17	1.18	1.32	1.36	**1.19**
Gd	4.81	5.10	4.88	4.67	4.92	4.53	3.89	4.48	4.26	4.95	5.21	**4.72**
Tb	0.73	0.78	0.75	0.70	0.74	0.70	0.57	0.69	0.66	0.77	0.83	**0.70**
Dy	4.30	4.62	4.46	4.13	4.46	4.05	3.23	4.28	3.72	4.23	4.70	**4.12**
Ho	0.84	0.92	0.86	0.81	0.89	0.78	0.66	0.94	0.79	0.90	1.08	**0.80**
Er	2.38	2.68	2.42	2.23	2.58	2.23	1.79	2.36	2.06	2.18	2.55	**2.35**
Tm	0.37	0.41	0.38	0.35	0.40	0.34	0.30	0.39	0.31	0.36	0.41	**0.36**
Yb	2.48	2.80	2.57	2.37	2.69	2.25	1.97	2.46	2.04	2.36	2.76	**2.56**
Lu	0.39	0.45	0.40	0.37	0.43	0.35	0.29	0.40	0.30	0.40	0.45	**0.40**

Table A8 Analyses of Upper Ediacaran sandstones. Unit 3 and at random. IVA. CIZ.—cont'd

Sample	SJA-17	SJA-18	SJA-19	SJA-20	SJA-23	SJA-15	SJA-3	MT-2	MT-7	MT-12	MT-13	MST-7
Unit	**3**	**3**	**3**	**3**	**3**	**ar**	**ar**	**ar**	**ar**	**ar**	**ar**	**ar**
Th	9.70	10.4	9.14	8.45	8.95	6.50	9.72	7.63	8.03	9.53	9.43	9.68
U	2.62	2.97	2.63	2.40	2.68	1.98	2.44	2.55	2.12	1.91	2.21	2.80
Sc	na	na	na	na	na	na	na	na	na	na	na	na
V	67	71	73	69	84	69	64	98	71	78	95	89
Cr	75	81	78	76	84	60	59	79	72	79	88	**92**
Co	26	25	24	30	28	34	37	20	34	25	23	**33**
Ni	28	29	33	29	34	28	22	34	26	27	37	**38**
Cu	18	20	16	19	18	17	15	27	18	17	26	**19**
Zn	76	74	77	76	78	44	44	72	53	74	96	**74**
Ga	15.2	15.6	16.4	15.4	16.7	12.9	12.1	16.9	16.6	16.5	19.9	**17.5**
Ge	1.39	1.51	1.58	1.50	1.59	1.29	1.03	1.40	1.24	1.30	1.34	**1.53**
As	11.7	10.9	8.87	9.65	9.90	13.5	10.2	7.70	12.6	8.52	18.8	**8.43**
Y	23.9	25.6	25.0	23.2	25.3	23.6	18.0	25.0	22.0	24.0	28.0	**23.4**
Zr	301	348	269	263	289	203	230	216	190	249	215	339
Nb	10.3	10.3	9.67	9.20	9.48	8.00	7.88	8.58	9.15	9.85	11.1	9.82
Mo	bd	bd	0.37	bd	0.38	bd	0.21	0.39	0.17	0.08	0.15	**0.57**
Sn	2.61	2.67	2.71	2.52	2.66	1.85	1.98	2.27	2.03	1.88	2.62	**2.07**
Sb	2.84	1.93	2.19	0.97	1.09	0.62	7.31	0.73	1.26	0.45	0.83	**0.71**
Hf	7.91	9.23	6.89	6.82	7.48	5.12	5.77	5.63	4.83	6.07	5.46	8.54
Ta	1.08	1.04	0.95	0.95	0.96	0.84	0.97	0.83	1.00	1.01	1.02	1.00
Pb	19.6	19.2	13.0	20.0	13.1	13.7	12.7	6.04	10.9	15.8	18.4	**18.0**
Bi	0.13	0.13	0.13	0.14	0.11	0.15	0.17	0.16	0.14	0.14	0.16	**0.14**

na: not analyzed; bd: below detection limit; ar: collected at random. 3: possible equivalence to unit III of the Salamanca Hurdes area (see text). New data in blue.

Data after Ugidos, J.M., Armenteros, I., Barba, P., Valladares, M.I., Colmenero, J.R., 1997b. Geochemistry and petrology of recycled orogen-derived sediments: a case study from Upper Precambrian siliciclastic rocks of the Central Iberian zone, Iberian Massif, Spain. Precamb. Res. 84, 163–180.

Table A9 Analyses of Upper Ediacaran sandstones. Units I–IV. SHD. CIZ

Sample	SIA-18	SIA-20	SIA-26	SIA-28	SIA-1	SIA-2	SIA-9	SIA-11	SIA-14	SIA-16	SIA-17	SIA-30	SIA-32	SIA-4	CR-6	CPA-2	SIA-23	APL-17	APL-21
Unit	I	I	I	I	II	II	II	II	II	II	II	II	II	III	III	III	IV	IV	IV
SiO_2	71.31	76.4	69.77	74.64	79.12	73.99	71.62	72.05	72.08	67.48	77.61	74.18	75.82	77.82	66.41	77.7	72.97	66.6	68.82
TiO_2	0.76	0.57	0.76	0.54	0.48	0.67	0.73	0.71	0.71	0.85	0.56	0.48	0.45	0.56	0.85	0.57	0.68	0.74	0.74
Al_2O_3	13.72	10.62	13.27	9.55	9.84	12.33	13.13	13.13	13.08	15.67	10.38	8.99	8.89	10.66	14.78	9.20	12.33	14.96	14.21
Fe_2O_3	4.98	4.33	6.78	5.20	3.77	4.75	5.33	5.01	5.06	5.58	4.12	6.28	4.71	3.78	6.74	5.34	5.06	5.51	5.16
MgO	1.56	1.57	1.96	2.43	1.26	1.63	1.76	1.86	1.68	1.86	1.42	2.77	2.3	1.17	2.08	1.81	1.62	2.00	1.87
MnO	0.03	0.02	0.04	0.05	bd	bd	0.02	bd	bd	0.03	0.02	0.10	0.10	0.02	0.03	0.05	0.04	0.04	0.03
CaO	0.5	0.79	0.63	1.09	bd	bd	0.17	0.05	0.06	0.09	0.06	1.09	1.36	bd	0.77	0.61	0.76	0.99	1.05
Na_2O	2.04	2.39	1.71	1.17	1.48	1.93	1.45	1.6	1.68	1.89	1.61	1.10	1.44	1.70	3.00	0.49	2.22	3.23	3.48
K_2O	2.35	1.19	1.89	1.21	1.46	1.75	2.26	2.19	2.10	2.83	1.64	0.89	1.02	1.62	1.91	1.20	1.79	2.46	2.33
P_2O_5	0.15	0.12	0.30	0.08	0.05	0.10	0.15	0.14	0.13	0.17	0.10	0.15	0.08	0.06	0.21	0.17	0.11	0.22	0.20
LOI	2.33	1.79	2.67	3.85	2.35	2.64	3.16	3.03	3.17	3.30	2.24	3.77	3.78	2.42	3.02	2.67	2.19	3.07	1.88
Total	99.73	99.79	99.78	99.81	99.82	99.8	99.78	99.78	99.76	99.75	99.76	99.80	99.95	99.81	99.8	99.81	99.77	99.82	99.77
Be	2.60	2.00	1.60	1.00	1.00	1.40	1.70	2.50	2.30	3.00	2.10	0.90	0.80	1.30	1.70	1.50	1.50	2.14	1.61
Rb	84.7	54.3	68.6	46.8	58.1	68.6	89	85.2	43.7	101	56.3	33.6	35.6	64.4	53.6	45.2	67.7	95.5	80.3
Sr	87.3	129	67.3	41.7	59.9	63.8	63.7	58.9	43.8	76.6	48.3	35.6	35.6	70.9	71.4	64.1	109	166	153
Cs	2.50	3.60	3.50	2.00	2.30	2.70	4.00	5.00	3.00	5.10	2.30	1.50	1.50	2.70	2.80	3.40	3.10	4.59	3.72
Ba	479	211	333	236	307	383	412	402	434	584	334	199	180	322	435	223	341	541	508
La	30.8	25.5	31.7	28.0	28.5	30.1	31.0	32.8	35.2	35.6	30.4	26.5	23.9	20.3	24.6	30.2	30.0	32.7	33.2

Ce	67.3	55.1	70.9	63.4	62.5	65.8	69.1	73.1	77.9	78.2	63.8	59.2	54.9	45.8	54.4	75.1	67.3	68.6	69.5
Pr	7.64	6.40	8.06	6.60	7.14	7.75	7.55	7.89	8.92	8.80	7.78	6.05	5.43	5.30	6.62	7.28	7.07	8.05	8.09
Nd	29.6	25.0	32.0	25.9	27.0	29.6	29.7	30.5	34.1	34.5	30.4	24.0	20.4	19.1	28.3	28.8	27.4	31.2	32.0
Sm	5.72	4.95	7.34	5.10	5.58	6.04	5.81	5.91	7.04	7.14	5.79	5.26	4.29	4.03	6.56	6.06	5.88	6.23	6.06
Eu	1.23	1.04	1.53	1.14	1.14	1.27	1.39	1.28	1.50	1.62	1.33	1.10	0.94	0.91	1.32	1.19	1.23	1.52	1.38
Gd	4.29	3.83	6.78	4.89	4.70	4.88	5.06	4.87	5.71	6.00	5.04	5.32	3.50	3.47	5.50	5.33	4.53	5.16	5.44
Tb	0.65	0.57	1.21	0.73	0.73	0.81	0.80	0.77	0.94	0.99	0.76	0.77	0.52	0.61	0.90	0.77	0.72	0.79	0.77
Dy	4.06	3.73	7.07	4.33	3.93	4.55	4.45	4.42	5.73	6.09	4.76	4.78	3.02	3.67	5.41	4.61	4.09	4.98	4.57
Ho	0.91	0.87	1.46	0.97	0.81	0.97	0.96	1.00	1.30	1.39	1.01	1.02	0.67	0.79	1.20	1.04	0.84	1.01	1.04
Er	2.52	2.24	3.69	2.48	2.07	2.84	2.40	2.50	3.20	3.55	2.70	2.51	1.75	2.07	2.92	2.69	2.31	2.55	2.54
Tm	0.39	0.37	0.52	0.38	0.31	0.37	0.34	0.36	0.44	0.50	0.39	0.37	0.26	0.32	0.39	0.42	0.34	0.37	0.35
Yb	2.63	2.38	3.39	2.35	2.22	2.70	2.59	2.52	3.12	3.53	2.66	2.36	1.78	2.41	2.95	2.59	2.29	2.83	2.59
Lu	0.40	0.35	0.46	0.35	0.34	0.42	0.33	0.41	0.48	0.51	0.44	0.36	0.27	0.34	0.40	0.34	0.37	0.44	0.36
Th	11.0	8.76	9.84	8.02	7.45	8.4	8.47	10.1	10.2	12.8	9.08	7.16	6.71	7.87	8.88	6.96	8.80	9.70	8.2
U	3.12	2.74	2.87	2.06	1.94	2.2	2.43	2.94	2.69	3.64	2.25	1.87	1.58	2.3	2.33	2.59	2.36	2.43	2.46
Sc	na	na	na	na	na	na	na	na	na	na	na	na	na	na	na	na	na	na	na
V	92	75	77	57	58	85	82	93	91	109	67	53	45	63	93	58	77	89	86
Cr	93	71	80	61	58	80	78	83	81	98	65	56	47	67	82	67	75	83	92
Co	29	49	27	43	36	22	25	38	34	21	47	41	37	30	30	30	31	24	31
Ni	40	28	40	27	6	22	33	27	26	23	42	34	22	6.4	45	16	30	33	29

Continued

Table A9 Analyses of Upper Ediacaran sandstones. Units I–IV. SHD. CIZ—cont'd

Sample	SIA-18	SIA-20	SIA-26	SIA-28	SIA-1	SIA-2	SIA-9	SIA-11	SIA-14	SIA-16	SIA-17	SIA-30	SIA-32	SIA-4	CR-6	CPA-2	SIA-23	APL-17	APL-21
Unit	**I**	**I**	**I**	**I**	**II**	**II**	**II**	**II**	**II**	**II**	**II**	**II**	**II**	**III**	**III**	**III**	**IV**	**IV**	**IV**
Cu	20	7	41	9	8	16	28	9	21	28	9	17	6.9	6	32	15	31	21	18
Zn	81	69	94	68	55	83	97	69	75	80	90	75	70	60	93	71	66	84	67
Ga	19.1	14.6	16.4	12.7	13.4	17.0	17.4	18.0	17.4	21.5	13.5	12.0	10.3	14.5	12.2	19.1	20.0	19.0	15.3
Ge	na	na	1.45	1.22	1.72	1.70	1.66	na	na	na	na	1.34	1.24	1.46	1.17	1.65	1.45	1.48	1.40
As	12.8	9.27	0.64	1.75	4.96	8.20	6.70	9.46	3.07	6.26	8.30	4.17	2.45	3.42	17.9	7.26	19.4	5.70	12.6
Y	21.9	21.6	41.1	26.4	22.2	26.0	24.8	24.0	28.8	31.6	22.8	30.9	18.1	21.0	30.6	28.4	23.5	26.4	26.8
Zr	293	237	250	254	224	268	216	273	234	269	250	204	200	299	212	248	253	211	260
Nb	10.8	8.35	10.4	7.86	8.05	9.60	10.8	10.2	10.2	11.8	8.12	7.32	6.59	9.44	8.84	9.85	11.0	10.50	9.16
Mo	0.50	0.40	0.20	0.10	0.20	0.30	0.30	0.50	0.30	0.20	0.20	0.10	0.20	0.40	1.10	bd	0.40	0.20	0.32
Sn	3.33	1.78	1.88	1.41	1.99	2.30	2.42	3.74	2.93	3.60	2.29	1.19	0.96	2.29	1.62	2.78	3.09	2.49	2.23
Sb	7.10	0.97	0.54	0.54	1.15	2.60	1.10	14.5	0.64	0.98	1.13	1.41	1.13	0.99	9.51	1.09	1.15	0.17	0.69
Hf	8.25	6.55	6.15	6.31	5.45	6.50	5.76	7.18	6.51	7.75	7.12	5.39	5.47	7.08	5.46	6.26	6.19	5.53	6.62
Ta	1.15	1.12	1.01	0.97	0.95	0.98	1.05	1.24	1.26	1.30	1.23	0.87	0.82	1.04	0.96	0.95	1.06	1.09	0.95
Pb	23.5	15.1	25.9	15.1	18.6	17.0	31.5	23.9	21.3	17.6	12.8	8.94	23.7	14.0	36.0	17.8	11.9	11.6	17.1
Bi	0.26	0.09	0.44	0.29	0.27	0.18	0.31	0.42	0.25	0.22	0.16	0.21	0.18	0.14	0.40	0.26	0.16	0.21	0.21

na: not analyzed; bd: below detection limit. New data in blue.

Data after Ugidos, J.M., Valladares, M.I., Barba, P., Ellam, R.M., 2003. The Upper Neoproterozoic-Lower Cambrian of the Central Iberian Zone, Spain: chemical and isotopic (Sm-Nd) evidence that the sedimentary succession records an inverted stratigraphy of its source. Geochim. Cosmochim. Acta 67, 2615–2629, Erratum: 2004, Geochim. Cosmochim. Acta 68, 1953; Ugidos, J.M., Sánchez-Santos, J.M., Barba, P., Valladares, M.I., 2010. Upper Neoproterozoic series in the Central Iberian, Cantabrian and West Asturian Leonese zones (Spain): geochemical data and statistical results as evidence for a shared homogenised source area. Precamb. Res. 178, 51–58.

Table A10 Analyses of Lower Cambrian shales. Duero. CIZ.

Sample	VCA-1	VCA-3	FRE-1	FRE-2	FRE-3	FRE-4	FRE-5	FRE-6	FRE-7	FRE-8	FRE 10	FRE-11	FRE-12	FRE-13	FRE-14
SiO_2	57.56	57.82	60.10	61.01	59.74	59.30	61.32	58.03	59.67	55.88	56.91	57.60	57.43	56.72	57.38
TiO_2	0.79	0.80	0.72	0.83	0.80	0.83	0.88	0.94	0.93	1.01	0.90	0.91	0.84	0.83	0.82
Al_2O_3	20.58	20.40	20.03	19.32	20.58	19.55	18.45	21.03	19.96	20.80	19.83	20.23	21.49	21.76	20.45
Fe_2O_3	7.97	7.76	7.17	6.96	6.56	7.56	6.49	7.22	6.04	7.66	7.01	6.68	7.28	7.51	7.99
MgO	3.09	3.01	2.63	2.29	2.37	2.79	2.33	2.31	2.57	3.05	2.52	2.68	2.30	2.31	3.08
MnO	0.08	0.08	0.05	0.06	0.07	0.07	0.07	0.06	0.06	0.06	0.07	0.06	0.07	0.07	0.09
CaO	0.65	0.60	0.19	1.00	0.64	0.34	1.69	0.63	0.29	0.60	0.99	0.29	0.27	0.21	0.41
Na_2O	1.45	1.33	1.46	1.46	1.12	1.69	2.22	1.51	0.79	0.81	1.46	1.10	1.53	1.26	1.48
K_2O	3.78	3.93	4.06	4.36	4.88	3.67	4.47	4.12	6.01	4.82	4.25	4.74	4.34	5.02	4.44
P_2O_5	0.10	0.12	0.09	0.18	0.09	0.17	0.14	0.17	0.15	0.14	0.15	0.14	0.13	0.15	0.14
LOI	3.85	4.05	3.46	2.42	3.00	3.90	1.83	3.87	3.62	4.44	4.51	4.87	4.15	4.17	3.64
Total	99.90	99.90	99.87	99.89	99.85	99.89	99.89	99.89	100.1	99.28	98.59	99.29	99.84	100.0	99.91
Be	3.70	3.74	3.95	3.77	3.65	3.35	4.00	3.47	4.17	4.09	3.82	4.51	3.75	4.42	3.67
Rb	156	162	140	197	214	145	189	173	237	201	198	235	202	196	177
Sr	166	125	108	180	142	58.0	260	140	60.3	66.7	105	82.1	112	76.8	131
Cs	8.34	9.56	3.72	8.14	9.63	5.09	6.26	5.63	40.1	7.45	7.39	8.14	7.85	9.20	8.58
Ba	642	661	824	829	928	648	898	758	1171	908	813	776	652	755	945
La	38.7	51.6	45.4	56.0	59.7	40.3	54.1	56.6	55.4	51.4	56.2	48.4	54.0	48.6	37.1
Ce	68.2	66.6	86.3	110	116	51.4	105	93.0	114	73.4	108	92.1	94.1	97.3	74.7

Continued

Table A10 Analyses of Lower Cambrian shales. Duero. CIZ.—cont'd

Sample	VCA-1	VCA-3	FRE-1	FRE-2	FRE-3	FRE-4	FRE-5	FRE-6	FRE-7	FRE-8	FRE 10	FRE-11	FRE-12	FRE-13	FRE-14
Pr	9.64	11.7	10.5	11.7	12.7	6.38	11.8	12.7	13.3	12.5	12.2	10.8	12.1	11.0	8.52
Nd	34.9	43.7	40.5	42.9	46.3	24.8	43.4	47.2	48.3	45.7	43.7	42.0	48.2	44.0	34.3
Sm	6.73	8.41	7.40	7.88	8.41	5.20	7.80	8.58	9.15	8.50	7.74	7.52	8.72	8.31	6.58
Eu	1.41	1.66	1.75	1.68	1.85	1.08	1.54	1.52	1.71	1.65	1.58	1.57	1.70	1.62	1.32
Gd	5.38	6.77	6.28	6.55	6.92	4.28	5.92	6.44	7.04	6.50	6.09	5.97	6.86	6.73	5.50
Tb	0.79	1.03	0.94	0.97	0.95	0.74	0.94	0.99	1.10	1.00	0.90	0.90	1.01	0.99	0.84
Dy	4.52	5.55	5.35	5.55	5.48	4.94	5.33	5.54	6.35	5.74	4.97	5.07	5.69	5.60	4.80
Ho	0.88	1.05	1.04	1.02	0.96	0.97	1.11	1.15	1.21	1.08	0.95	0.95	1.07	1.06	0.93
Er	2.59	2.90	2.70	2.70	2.71	2.92	2.88	3.25	3.50	3.14	2.71	2.71	2.95	2.92	2.63
Tm	0.44	0.44	0.47	0.44	0.44	0.45	0.44	0.50	0.52	0.47	0.40	0.41	0.45	0.45	0.40
Yb	2.78	2.94	2.83	2.71	2.75	3.16	2.87	3.26	3.49	3.18	2.62	2.78	3.02	3.00	2.68
Lu	0.42	0.43	0.46	0.43	0.40	0.47	0.41	0.44	0.53	0.49	0.40	0.42	0.46	0.46	0.41
Th	18.3	17.0	13.9	18.7	19.6	12.6	18.0	20.2	21.5	23.9	17.2	18.0	16.3	16.4	14.2
U	2.83	3.13	4.27	3.36	3.21	4.71	3.01	3.76	3.76	3.48	2.65	2.16	3.43	3.59	3.11
Sc	na	na	17.9	17.0	19.3	18.9	16.8	18.8	18.9	22.1	na	20.1	20.0	19.6	19.7
V	114	122	116	101	108	129	98	123	105	120	117	107	107	107	107
Cr	108	108	94	91	100	110	94	103	112	124	106	107	106	105	106
Co	20	22	29	23	19	15	28	20	23	20	19	13	16	18	19
Ni	54	53	46	40	41	38	36	43	39	42	40	41	49	47	50

Cu	34	40	35	17	12	30	10	17	9	21	20	21	27	32	27
Zn	115	120	116	112	112	126	94	105	110	126	104	109	123	133	132
Ga	28.5	29.8	26.9	26.9	28.6	26.0	26.9	29.7	30.9	32.7	29.1	29.2	30.0	30.6	28.3
Ge	1.43	1.48	1.52	2.13	2.40	1.69	2.11	1.98	2.35	2.57	2.32	2.15	2.14	2.15	1.97
As	5.02	1.48	2.64	1.93	3.82	26.8	3.28	81.1	3.95	23.5	11.5	21.1	10.4	25.4	15.5
Y	24.6	31.8	30.0	29.0	29.0	28.3	31.3	31.2	35.9	31.1	28.2	27.7	31.1	31.2	27.5
Zr	142	144	156	185	135	172	199	193	185	132	131	121	134	132	119
Nb	15.0	14.4	13.1	16.1	15.9	12.0	16.6	17.4	16.4	16.9	16.3	15.6	14.2	14.4	13.2
Mo	0.26	0.22	bd	bd	bd	bd	bd	bd	bd	0.34	0.43	bd	bd	bd	bd
Sn	4.19	4.50	4.20	4.61	5.19	4.16	3.74	4.63	6.75	6.50	5.42	4.32	4.57	4.35	4.02
Sb	3.99	3.53	0.11	bd	0.23	4.64	0.11	0.35	0.10	0.28	1.81	1.21	0.45	0.28	0.28
Hf	1.33	1.24	4.57	5.30	3.91	1.22	5.55	5.12	5.01	3.98	3.53	3.65	4.03	4.02	3.61
Ta	28.8	38.0	1.39	1.58	1.56	15.4	1.54	1.65	1.62	1.67	1.42	1.40	1.28	1.33	1.20
Pb	2.32	2.69	23.2	20.6	19.1	0.30	25.5	24.5	11.6	13.7	20.2	14.8	24.5	24.2	19.8
Bi	0.56	0.66	0.35	0.34	0.34	0.24	0.29	0.35	0.41	0.27	0.20	0.34	0.52	0.55	0.30

na: not analyzed; bd: below detection limit. New data.

Table A11 Analyses of Lower Cambrian shales. SHD. CIZ.

Sample	**MIR-22**	**GUB-24**	**GUB-25**	**GUB-26**	**GUB-27**
SiO_2	59.34	60.36	59.91	59.08	60.32
TiO_2	0.82	0.77	0.71	0.81	0.79
Al_2O_3	19.19	18.80	18.67	19.34	18.96
Fe_2O_3	6.98	7.24	7.04	7.97	7.38
MgO	6.28	2.98	3.39	3.13	3.10
MnO	2.93	0.10	0.18	0.10	0.10
CaO	0.08	0.53	0.33	0.22	0.51
Na_2O	0.85	2.02	1.30	1.91	1.94
K_2O	2.17	3.48	4.24	3.37	3.37
P_2O_5	3.33	0.16	0.18	0.15	0.15
LOI	0.14	3.58	3.84	3.89	3.84
Total	99.00	100.0	99.79	99.96	100.5
Be	3.75	3.24	3.54	3.88	3.74
Rb	143	148	165	146	144
Sr	187	135	55.1	132	143
Cs	9.87	7.13	7.96	6.49	6.77
Ba	567	575	1129	541	590
La	47.1	62.2	57.6	45.6	47.9
Ce	87.8	87.4	103	86.2	88.6
Pr	10.7	13.1	12.5	10.3	10.9
Nd	39.2	49.5	46.9	38.2	40.1
Sm	7.58	9.28	9.00	7.33	7.66
Eu	1.57	1.82	1.69	1.49	1.54
Gd	6.01	7.75	7.17	5.88	6.03
Tb	0.96	1.18	1.02	0.89	0.92
Dy	5.43	6.86	5.61	5.15	5.19
Ho	1.04	1.28	1.04	0.96	0.97
Er	2.79	3.55	2.93	2.72	2.76

Table A11 Analyses of Lower Cambrian shales. SHD. CIZ.—cont'd

Sample	MIR-22	GUB-24	GUB-25	GUB-26	GUB-27
Tm	0.41	0.51	0.44	0.41	0.41
Yb	2.72	3.46	2.96	2.79	2.76
Lu	0.42	0.52	0.45	0.42	0.42
Th	17.9	13.6	12.4	14.3	14.0
U	3.16	2.81	2.92	2.35	2.54
Sc	18.9	na	na	18.3	na
V	112	104	117	105	103
Cr	103	102	96	102	102
Co	20	19	26	20	22
Ni	50	47	99	54	51
Cu	35	24	48	26	30
Zn	111	109	407	121	129
Ga	28.1	27.7	28.6	29.1	28.0
Ge	1.41	1.99	2.05	2.00	2.05
As	6.91	2.32	26.0	1.67	bd
Y	29.3	40.9	31.6	28.8	29.0
Zr	147	149	140	148	147
Nb	14.9	14.1	12.3	14.7	14.4
Mo	0.15	0.47	bd	bd	bd
Sn	3.56	4.67	4.73	4.36	4.64
Hf	3.97	4.15	3.79	3.99	4.05
Ta	1.33	1.24	1.12	1.31	1.33
Pb	35.5	13.9	18.5	6.70	8.82
Sb	0.26	1.16	3.83	0.75	0.58
Bi	2.50	bd	0.31	bd	bd

na: not analyzed; bd: below detection limit. New data.

Table A12 Analyses of Lower Cambrian shales. IVA. CIZ.

Sample	MTA-5	MTA-6	MTA-15	MTA-16	MT-17	SJA-1	SJA-2	SJA-24	PTA-4	ESR-1	TOL-1	NA-2
SiO_2	58.23	62.19	58.37	61.10	57.23	59.59	58.51	59.55	57.67	60.44	54.15	64.66
TiO_2	0.88	0.79	0.85	0.77	0.98	0.84	0.76	0.79	0.75	0.79	1.12	0.88
Al_2O_3	19.71	17.71	19.91	17.87	21.58	19.29	20.00	18.14	19.67	19.00	23.57	16.31
Fe_2O_3	7.28	7.11	7.93	6.80	6.82	7.60	7.51	6.41	7.23	7.13	7.89	6.24
MgO	2.87	2.72	2.88	2.60	2.36	2.91	3.02	3.37	3.20	2.63	1.10	2.71
MnO	0.06	bd	0.03	0.05	0.07	0.05	0.07	0.03	0.09	0.03	bd	0.07
CaO	0.40	0.25	0.24	0.66	0.46	0.29	0.37	0.35	0.29	0.25	0.40	0.53
Na_2O	1.85	2.40	1.94	1.91	1.72	1.94	1.43	0.05	1.43	1.83	0.20	2.31
K_2O	3.56	2.77	3.54	3.31	4.37	3.40	3.69	4.92	3.62	3.72	5.75	3.68
P_2O_5	0.17	0.16	0.17	0.18	0.14	0.15	0.10	0.19	0.16	0.20	0.34	0.10
LOI	4.19	3.68	3.85	3.92	4.17	3.82	4.44	5.46	4.70	3.83	5.40	2.53
Total	99.74	99.79	99.71	99.17	99.90	99.88	99.90	99.25	98.80	99.85	99.92	100.0
Be	3.04	2.66	3.27	2.97	4.15	3.35	4.36	3.29	3.41	3.19	3.72	2.977
Rb	140	115	154	154	186	137	154	164	146	146	226	164
Sr	87.7	99.8	72.9	87.1	130	74.0	73.8	42.5	59.1	117	184	163
Cs	6.78	5.68	7.26	6.81	11.4	7.58	9.63	6.48	7.22	7.66	7.29	9.69
Ba	586	502	597	555	701	545	563	854	104	605	1074	824
La	42.7	40.3	42.2	42.6	45.6	39.3	36.2	44.4	39.2	45.4	71.0	41.6
Ce	85.7	83.0	83.4	85.6	94.5	76.7	70.8	88.7	77.1	89.3	138	81.8
Pr	10.4	9.70	9.70	9.49	11.1	9.82	8.93	10.1	9.18	10.4	16.3	9.60
Nd	40.0	35.1	39.1	37.2	42.9	36.2	33.6	37.8	34.5	40.3	60.0	35.2
Sm	8.13	7.06	7.76	6.95	8.71	6.78	6.52	7.31	6.88	7.77	10.7	6.60
Eu	1.62	1.43	1.59	1.41	1.70	1.44	1.24	1.49	1.39	1.43	2.41	1.21
Gd	6.73	5.72	6.34	5.92	6.90	5.65	5.19	5.90	5.86	5.99	9.67	5.14
Tb	1.03	0.86	1.00	0.91	1.02	0.87	0.79	0.90	0.89	0.93	1.34	0.80
Dy	6.04	5.10	5.90	5.20	6.27	5.01	4.56	5.24	5.17	5.22	7.97	4.59
Ho	1.33	1.04	1.26	1.05	1.23	1.02	0.91	1.01	1.01	1.08	1.54	0.89
Er	3.34	2.59	3.21	2.70	3.53	2.91	2.72	2.84	2.86	2.84	4.09	2.55
Tm	0.53	0.40	0.53	0.45	0.62	0.45	0.43	0.42	0.43	0.43	0.60	0.39
Yb	3.39	2.62	3.23	2.78	3.89	3.04	2.85	2.93	2.93	2.97	4.05	2.58
Lu	0.52	0.39	0.50	0.44	0.61	0.48	0.42	0.45	0.45	0.46	0.61	0.40
Th	12.6	13.1	12.6	13.3	16.2	14.5	13.2	14.2	13.1	15.4	21.2	15.9
U	3.16	2.51	2.82	2.95	3.36	2.88	2.97	2.50	3.10	3.09	6.09	2.97

Table A12 Analyses of Lower Cambrian shales. IVA. CIZ.—cont'd

Sample	MTA-5	MTA-6	MTA-15	MTA-16	MT-17	SJA-1	SJA-2	SJA-24	PTA-4	ESR-1	TOL-1	NA-2
Sc	20.7	15.9	20.1	16.3	na	19.5	19.3	16.8	18.5	17.3	21.1	15.6
V	121	93	120	102	132	124	120	102	120	104	139	94
Cr	108	83	108	91	108	106	102	95	103	98	111	87
Co	16	16	15	18	16	19	21	20	10	17	23	21
Ni	49	41	50	41	39	51	47	47	46	50	46	36
Cu	32	25	26	34	16	25	30	22	40	32	34	20
Zn	107	88	107	102	109	105	119	85	144	103	130	100
Ga	28.0	25.0	29.0	26.0	31.4	26.5	27.4	26.3	29.1	27.6	33.0	23.5
Ge	1.64	1.55	1.69	1.69	1.27	1.21	1.04	1.61	1.94	2.02	1.63	2.50
As	14.8	4.73	8.31	8.63	25.7	3.28	12.8	bd	24.5	4.05	10.2	bd
Y	35.8	27.8	35.4	30.2	38.3	31.3	29.2	30.1	30.7	30.0	42.9	25.5
Zr	195	171	187	191	225	177	160	170	139	166	211	205
Nb	14.0	13.9	13.9	13.8	15.9	12.7	12.1	13.8	12.7	14.3	20.5	15.2
Mo	0.14	0.04	0.04	0.05	0.22	0.06	0.20	bd	bd	bd	0.83	bd
Sn	4.07	3.55	4.17	3.90	5.60	3.90	4.21	3.57	4.23	3.70	4.45	2.98
Sb	1.54	3.67	5.81	1.44	4.19	1.04	1.38	1.36	0.45	2.37	1.38	0.44
Hf	5.09	4.40	4.91	4.97	5.72	4.35	4.06	4.57	3.87	4.65	5.68	5.56
Ta	1.23	1.23	1.21	1.23	1.39	1.09	1.03	1.25	1.16	1.28	1.90	1.44
Pb	14.7	10.5	8.85	21.6	12.7	5.15	5.01	3.13	19.8	18.9	33.0	14.0
Bi	0.37	0.16	0.10	0.36	0.21	0.10	0.31	0.15	0.54	0.51	0.32	0.16

na: not analyzed; bd: below detection limit. New data.

Table A13 Analyses of Lower Cambrian shales. Units VI–XII. SHD. CIZ.

Sample	CR-1	FG-5	SSA-3	SSA-4	SSA-19	SSA-25	SSA-24	SSA-28	MC-25	MC-27	CR-5	FG-1	SSA-26
Unit	**VI**	**VI**	**VI**	**VI**	**VI**	**VI**	**VII**	**VII**	**VII**	**VII**	**VIII**	**VIII**	**VIII**
SiO_2	57.58	60.09	56.61	56.83	62.6	57.45	57.73	57.23	60.17	60.2	58.71	58.57	63.01
TiO_2	0.77	0.71	0.88	0.91	0.66	0.72	0.76	0.96	0.89	0.87	0.75	0.78	0.63
Al_2O_3	18.6	17.94	20.13	20.20	14.8	18.16	20.91	23.02	21.23	20.64	19.11	19.89	16.66
Fe_2O_3	6.21	5.74	9.09	8.20	6.12	6.32	7.00	5.42	5.26	6.77	7.27	7.25	6.88
MgO	2.51	2.52	2.58	2.40	3.70	2.49	2.59	1.94	2.14	1.6	3.30	3.02	2.70
MnO	0.07	0.07	0.04	0.04	0.07	0.04	0.04	0.04	0.05	0.13	0.10	0.05	0.14
CaO	0.08	0.88	0.21	0.20	1.59	0.03	0.15	bd	0.15	0.19	0.50	0.28	0.88
Na_2O	1.28	2.78	1.41	1.46	1.32	1.20	1.48	0.15	0.32	0.27	2.04	1.58	1.98
K_2O	3.53	2.79	3.47	3.67	2.16	3.80	4.10	4.67	4.67	4.24	3.48	3.97	3.32
P_2O_5	0.06	0.09	0.14	0.12	0.98	0.11	0.14	0.16	0.17	0.21	0.20	0.16	0.27
LOI	9.03	6.08	5.14	5.64	6.45	8.95	4.80	6.18	4.37	4.02	4.27	4.17	3.30
Total	99.72	99.69	99.70	99.67	99.73	99.77	99.70	99.77	99.41	99.15	99.73	99.72	99.77
Be	1.74	2.59	2.81	2.97	2.04	2.67	3.21	3.09	3.85	2.88	2.36	3.52	2.97
Rb	120	104	130	150	93.6	157	184	196	183	165	137	151	146
Sr	38.0	180	84.6	87.0	66.0	90.6	82.2	88.4	89.7	127	68.0	52.5	151
Cs	5.34	3.96	7.57	10.4	8.91	7.60	7.31	8.78	11.9	10.3	4.68	6.35	7.28
Ba	686	576	707	893	418	1486	749	865	693	805	618	697	1405

La	46.2	29.0	45.3	45.8	35.2	40.3	44.6	38.1	40.8	43.8	36.8	32.6	41.4
Ce	99.8	58.4	83.4	83.3	72.3	75.0	90.2	55.0	82.6	85.3	73.2	43.9	74.4
Pr	12.8	7.05	10.9	11.0	8.91	9.10	11.0	10.1	10.2	9.80	8.90	7.48	8.97
Nd	50.8	26.5	42.6	44.2	34.1	35.4	43.0	39.0	39.1	39.1	34.9	29.2	36.5
Sm	10.9	5.57	8.50	9.15	8.04	6.93	8.32	8.51	7.95	7.67	6.76	5.75	7.22
Eu	2.33	1.30	1.72	1.80	1.60	1.40	1.85	1.78	1.56	1.51	1.45	1.14	1.59
Gd	8.41	4.56	7.20	7.10	8.76	5.03	6.64	7.28	6.79	6.20	5.62	4.73	5.96
Tb	1.29	0.73	1.17	1.19	1.27	0.77	0.97	1.12	1.06	0.97	0.85	0.73	0.91
Dy	6.73	4.16	6.90	7.56	8.10	4.70	5.64	6.73	6.32	5.64	4.85	4.55	5.83
Ho	1.32	0.89	1.44	1.64	1.60	0.87	1.11	1.23	1.25	1.10	1.09	0.99	1.04
Er	3.46	2.30	3.91	4.35	4.42	2.30	3.26	3.40	3.54	3.15	2.89	2.46	3.14
Tm	0.54	0.37	0.57	0.63	0.70	0.35	0.48	0.54	0.53	0.48	0.43	0.39	0.49
Yb	3.40	2.51	4.02	4.33	4.55	2.79	3.20	3.76	3.70	3.20	3.02	2.81	3.25
Lu	0.54	0.44	0.59	0.67	0.68	0.4	0.49	0.62	0.58	0.49	0.40	0.40	0.42
Th	9.86	9.59	12.84	13.2	8.83	14.4	12.3	12.3	14.0	13.3	10.9	12.7	12.0
U	5.46	3.79	4.60	4.76	3.34	6.21	2.80	4.10	4.17	3.36	3.04	3.20	4.00
Sc	18.6	17.6	18.2	21.1	na	na	20.5	20.4	21.7	21.4	18.7	17.2	14.9
V	162	126	168	181	110	139	128	159	131	117	112	119	239

Continued

Table A13 Analyses of Lower Cambrian shales. Units VI–XII. SHD. CIZ.—cont'd

Sample	CR-1	FG-5	SSA-3	SSA-4	SSA-19	SSA-25	SSA-24	SSA-28	MC-25	MC-27	CR-5	FG-1	SSA-26
Unit	VI	VI	VI	VI	VI	VI	VII	VII	VII	VII	VIII	VIII	VIII
Cr	100	96	124	127	80	93	111	133	106	102	100	96	150
Co	43	29	19	25	28	113	13	19	8	9	23	11	19
Ni	63	63	62	65	61	100	45	37	30	42	50	41	50
Cu	48	51	43	60	33	66	29	27	18	18	40	49	68
Zn	59	74	135	135	173	39	123	105	73	252	121	98	125
Ga	24.4	2.40	27.2	29.4	19.9	26.0	32.4	30.5	28.1	28.6	26.2	26.9	24.1
Ge	1.44	1.57	1.52	1.77	1.24	1.17	1.31	1.58	1.75	1.77	1.72	1.46	1.53
As	47.4	10.6	0.66	1.65	145	83.5	4.15	14.4	20.2	2.28	14.4	21.1	0.59
Y	36.8	25.0	40.8	45.3	51.2	23.6	36.3	33.8	36.0	32.1	28.9	24.8	35.6
Zr	154	155	184	211	194	150	174	229	193	176	149	148	159
Nb	10.7	9.77	13.2	14.1	11.0	13.1	13.5	14.6	13.3	13.9	11.4	12.7	11.6
Mo	12.6	7.16	3.38	4.00	1.44	16.6	0.09	0.47	0.55	bd	0.25	0.19	0.52
Sn	3.40	2.87	3.88	4.04	2.53	5.30	4.87	5.08	4.77	4.2	3.65	4.31	3.53
Sb	3.95	3.55	3.57	7.31	6.07	5.25	1.01	9.56	1.93	0.69	1.48	0.55	1.86
Hf	3.87	4.02	4.93	5.25	4.51	4.05	4.47	6.20	5.44	4.85	4.23	4.16	4.05
Ta	0.95	0.95	1.21	1.27	0.99	1.22	1.18	1.35	1.29	1.32	1.06	1.18	1.03
Pb	27.6	23.8	39.1	35.8	28.3	38.0	11.4	9.85	35.1	21.0	15.2	20.9	35.2
Bi	0.53	0.50	0.59	0.66	0.53	0.99	0.23	0.06	0.88	0.21	0.31	0.69	0.49

Sample	SSA-1	SSA-20	SSA-29	SSA-30	SSA-17	SSA-18	SSA-15	GUA-31	GUA-32	SSA-16	GUB-20	GUB-21	GUB-22	GUB-23	VCA-2
Unit	IX	IX	IX	X	X	X	XI	XI	XI	XII	XII	XII	XII	XII	XII
SiO_2	58.95	56.92	58.11	59.16	58.86	56.09	59.29	58.61	61.13	60.15	60.38	61.35	57.37	59.05	61.2
TiO_2	0.72	0.88	0.92	0.81	0.67	0.61	0.88	0.73	0.83	0.86	0.77	0.79	0.87	0.83	0.77
Al_2O_3	19.76	20.66	21.13	20.58	19.3	18.19	20.32	19.92	18.92	20.34	18.45	18.42	20.97	18.67	18.17
Fe_2O_3	6.95	7.43	6.63	7.29	5.66	7.94	7.43	7.52	6.56	6.94	6.85	6.58	7.15	7.61	7.29
MgO	2.84	2.97	2.42	2.51	2.85	2.20	2.27	3.03	2.82	2.28	3.02	2.71	2.85	3.22	2.77
MnO	0.05	0.07	0.04	0.06	0.04	0.03	0.08	0.08	0.05	0.08	0.07	0.07	0.11	0.09	0.08
CaO	0.19	0.42	0.03	0.02	0.36	0.67	0.03	0.58	0.32	0.15	0.39	0.74	0.87	0.49	0.76
Na_2O	1.77	1.79	0.54	0.11	1.71	1.46	0.18	1.24	0.09	0.27	1.32	1.77	1.95	2.03	1.97
K_2O	3.66	4.19	3.92	4.64	3.67	3.56	4.42	3.99	4.96	4.11	3.89	3.89	4.18	3.44	3.31
P_2O_5	0.19	0.23	0.22	0.21	0.09	0.15	0.13	0.11	0.2	0.17	0.18	0.22	0.17	0.17	0.13
LOI	4.68	4.03	5.86	4.42	6.67	8.79	4.66	4.11	4.03	4.35	4.22	3.45	3.78	3.97	3.45
Total	99.76	99.59	99.82	99.81	99.88	99.69	99.69	99.92	99.91	99.7	99.53	100.0	100.3	99.56	99.9
Be	2.67	3.59	3.66	3.21	3.19	2.86	3.68	4.01	4.59	3.47	3.26	3.29	4.14	3.55	3.71
Rb	160	178	171	197	160	162	191	170	202	171	166	163	176	145	139
Sr	85.4	89	53.4	49.2	94.3	105	63.7	94.3	41.7	80.0	71.0	110	169	170	118
Cs	7.62	6.86	6.42	7.04	8.71	8.48	8.44	7.57	8.21	8.73	5.41	7.41	5.79	6.70	9.33
Ba	682	851	672	758	861	394	736	868	915	637	738	770	740	744	598
La	40.9	35.4	49.3	41.5	30.0	22.4	51.8	37.0	29.5	45.0	37.0	45.4	56.0	47.0	40.5

Continued

Table A13 Analyses of Lower Cambrian shales. Units VI–XII. SHD. CIZ.—cont'd

Sample	SSA-1	SSA-20	SSA-29	SSA-30	SSA-17	SSA-18	SSA-15	GUA-31	GUA-32	SSA-16	GUB-20	GUB-21	GUB-22	GUB-23	VCA-2
Unit	IX	IX	IX	X	X	X	XI	XI	XI	XII	XII	XII	XII	XII	XII
Ce	78.8	68.7	95.1	80.4	59.3	44.3	104	73.9	52.8	90.8	78.4	80.9	101	89	69.5
Pr	10.1	8.67	12.4	9.96	7.05	5.46	12.5	8.89	7.62	10.6	8.74	10.5	12.5	10.7	9.82
Nd	38.8	33.1	48.5	38.4	27.9	21.2	48.7	31.9	27.5	40.5	32.5	40.0	46.6	40.8	35.8
Sm	7.64	6.83	11.2	7.80	5.49	4.33	10.0	6.39	5.47	7.60	6.46	7.96	8.81	7.80	7.00
Eu	1.66	1.51	2.26	1.68	1.32	1.03	2.01	1.23	1.05	1.54	1.23	1.59	1.80	1.58	1.39
Gd	6.77	5.68	10.2	6.69	4.68	3.80	8.05	4.56	4.14	6.76	5.30	6.47	7.20	6.27	5.24
Tb	1.06	0.92	1.50	1.02	0.79	0.59	1.28	0.72	0.71	1.05	0.82	0.98	1.08	0.95	0.79
Dy	6.14	5.70	8.40	6.33	4.78	3.73	6.93	4.12	4.10	5.98	4.82	5.60	6.24	5.37	4.77
Ho	1.35	1.18	1.70	1.26	0.91	0.75	1.51	0.88	0.82	1.33	0.95	1.07	1.17	1.02	0.94
Er	3.11	3.01	4.72	3.24	2.70	2.35	3.86	2.36	2.35	3.60	2.69	3.06	3.27	2.85	2.71
Tm	0.53	0.50	0.66	0.54	0.46	0.37	0.60	0.39	0.34	0.53	0.40	0.46	0.46	0.42	0.42
Yb	3.47	3.52	4.72	3.49	3.16	2.56	4.09	2.42	2.33	3.53	2.78	3.15	3.16	2.86	2.82
Lu	0.52	0.54	0.81	0.55	0.49	0.42	0.60	0.42	0.37	0.54	0.43	0.49	0.48	0.45	0.43
Th	11.5	15.7	15.0	13.4	10.7	11.4	14.8	14.8	17.1	14.1	12.7	13.8	14.5	13.2	15.8
U	2.84	4.08	3.78	3.67	3.69	3.77	4.26	3.46	3.73	3.65	2.34	3.68	2.70	2.22	3.39
Sc	19.6	19.1	19.6	20.9	19.7	na	20.5	18.7	16.5	19.9	16.7	16.2	19.8	18.6	14.9
V	118	130	125	122	137	124	132	126	108	121	109	117	121	109	112
Cr	105	109	101	98	105	90	114	97	88	103	101	102	117	107	97

Co	12	14	14	8	24	32	12	28	16	10	22	18	19	22	16
Ni	31	48	38	43	73	82	26	53	43	39	49	46	54	54	44
Cu	18	28	29	57	44	61	24	47	12	16	38	35	32	31	33
Zn	110	114	108	100	117	64	112	125	111	93	123	110	113	127	111
Ga	29.2	29.9	29.2	27.9	28.0	26.7	31.5	27.8	25.9	29.0	27.8	27.6	31.5	28.0	25.6
Ge	1.75	1.18	1.45	1.39	1.34	1.11	1.63	1.26	1.32	1.96	1.79	1.86	2.06	2.06	1.31
As	7.32	9.88	2.75	30.7	3.18	185	22.4	25.3	9.08	24.1	8.16	12.1	13.7	4.19	19.4
Y	34.2	35.2	48.6	35.7	26.5	23.1	39.0	24.9	23.9	80.0	28.7	32.0	36.0	30.1	26.9
Zr	155	185	206	170	152	125	197	137	199	168	159	197	161	162	192
Nb	12.6	16.6	15.7	13.6	10.9	10.2	15.1	12.3	14.7	14.7	13.7	14.4	15.8	14.3	13.3
Mo	0.49	0.21	0.08	0.31	2.75	7.91	1.91	0.62	0.06	0.18	0.44	0.39	bd	0.37	0.10
Sn	4.38	5.83	5.27	4.71	4.04	4.20	4.81	4.10	4.18	4.66	4.78	5.19	5.53	4.28	3.54
Sb	0.78	1.45	1.26	3.02	1.77	3.87	10.9	2.20	1.13	23.5	1.05	2.92	2.46	0.68	0.64
Hf	4.10	5.03	5.97	4.83	4.01	3.26	5.37	3.49	4.95	4.69	4.26	5.23	4.31	4.32	4.77
Ta	1.13	1.42	1.44	1.22	1.04	0.98	1.41	1.09	1.26	1.35	1.23	1.31	1.37	1.23	1.20
Pb	12.3	14.0	8.43	53.4	25.0	37	28.8	42.6	12.3	12.5	18.0	16.2	16.1	15.3	19.4
Bi	0.27	0.27	0.32	0.89	0.60	0.98	0.42	1.01	0.33	0.07	0.23	0.20	0.13	0.27	0.25

na: not analyzed; bd: below detection limit. New data in blue.

Data after Ugidos, J.M., Valladares, M.I., Barba, P., Ellam, R.M., 2003. The Upper Neoproterozoic-Lower Cambrian of the Central Iberian Zone, Spain: chemical and isotopic (Sm-Nd) evidence that the sedimentary succession records an inverted stratigraphy of its source. Geochim. Cosmochim. Acta 67, 2615–2629, Erratum: 2004, Geochim. Cosmochim. Acta 68, 1953.

Table A14 Analyses of Ediacaran shales. NAS and MND. WALZ.

Sample	TIN-4	TIN-5	VIL-2	PS-1	LR-2	RIE-1	CN-2	GED-4	GED-5	GED-6	GED-8	GED-9	VIL-3	RIE-2	SAR-3	SAR-5	SAR-8	OEN-6
	T	T	T	T	T	T	A	A	A	A	A	A	A	A	M	M	M	M
SiO_2	56.42	53.83	55.79	55.54	60.57	58.20	59.23	59.44	60.98	56.86	57.98	59.33	60.93	56.55	56.64	61.51	60.73	59.52
TiO_2	0.96	0.99	0.95	0.86	0.88	0.82	0.88	0.94	0.78	0.93	0.89	0.84	0.90	0.93	1.06	1.09	1.09	0.97
Al_2O_3	20.23	21.22	19.54	17.61	17.91	18.90	18.11	18.63	18.15	20.35	19.20	18.44	17.93	19.51	20.46	19.67	19.57	19.46
Fe_2O_3	7.81	6.74	8.54	8.19	7.41	7.72	7.89	7.95	6.73	7.93	7.42	7.42	7.05	8.60	7.56	8.00	8.44	7.64
MgO	3.18	4.89	3.59	3.44	2.97	3.36	3.66	3.21	2.96	3.52	3.44	3.31	3.05	3.42	3.26	1.79	2.39	2.86
MnO	0.04	0.03	0.07	0.07	0.08	0.07	0.08	0.05	0.04	0.05	bd	0.06	0.07	0.06	0.03	0.08	0.04	0.04
CaO	0.29	0.41	0.81	0.75	0.84	0.88	1.11	0.35	0.56	0.25	0.42	1.00	0.71	0.62	0.28	0.36	0.13	0.19
Na_2O	2.32	0.71	2.59	1.82	2.63	2.19	2.56	2.26	2.41	2.79	2.81	2.46	2.44	2.72	0.90	0.67	0.72	1.63
K_2O	3.63	5.07	3.29	3.05	3.40	3.70	3.25	3.59	3.47	3.54	3.27	3.46	3.49	3.47	4.89	3.02	3.08	4.00
P_2O_5	0.28	0.28	0.26	0.26	0.22	0.22	0.31	0.25	0.29	0.25	0.27	0.26	0.26	0.23	0.22	0.30	0.16	0.21
LOI	4.86	5.69	4.45	9.84	3.54	3.82	3.25	3.76	3.50	4.02	4.15	3.29	3.62	3.98	5.58	3.89	4.46	4.19
Total	100.0	99.86	99.88	100.4	100.5	99.88	100.3	100.4	99.87	100.5	99.85	99.87	100.5	100.1	100.9	100.4	100.8	100.7
Be	2.94	3.31	3.06	2.61	2.58	3.43	2.55	2.71	2.83	3.05	2.89	2.55	2.79	2.94	3.94	2.66	2.75	3.05
Rb	123	156	113	103	119	132	93.3	119	118	122	103	110	116	117	196	119	128	156
Sr	94.8	43.7	96.0	123	128	92.0	113	89.3	90.6	109	116	107	108	96.0	78.9	114	138	53.3
Cs	6.46	6.53	4.11	14.8	5.63	5.88	1.90	4.55	4.83	5.26	3.32	2.59	3.77	4.21	5.82	8.61	9.01	10.1
Ba	851	501	671	782	561	601	629	710	585	777	586	583	526	589	953	498	494	636
La	47.2	43.1	40.8	35.6	41.3	30.1	37.2	41.8	35.2	48.3	44.0	42.6	40.2	50.7	46.1	43.2	36.3	44.5

Table A14 Analyses of Ediacaran shales. NAS and MND. WALZ.—cont'd

Sample	TIN-4	TIN-5	VIL-2	PS-1	LR-2	RIE-1	CN-2	GED-4	GED-5	GED-6	GED-8	GED-9	VIL-3	RIE-2	SAR-3	SAR-5	SAR-8	OEN-6
	T	T	T	T	T	T	A	A	A	A	A	A	A	A	M	M	M	M
Ce	93.3	87.1	85.2	75.6	84.0	59.7	74.0	84.3	71.1	95.7	86.6	84.2	82.3	105	92.7	91.4	74.6	92.0
Pr	11.9	11.0	10.7	8.95	10.3	7.86	9.75	10.5	8.82	11.9	10.9	10.5	10.2	12.4	11.6	10.6	8.84	11.3
Nd	47.3	42.4	42.5	35.9	40.4	32.3	39.3	40.4	34.3	46.0	42.1	40.8	39.2	47.4	45.1	40.9	35.4	43.9
Sm	9.80	8.59	8.64	6.83	7.77	6.82	8.50	8.13	7.05	9.37	8.68	8.34	8.05	9.34	9.20	8.22	7.44	9.13
Eu	1.97	1.69	1.96	1.53	1.65	1.47	1.67	1.68	1.43	2.04	1.70	1.99	1.59	1.86	1.80	1.70	1.59	1.81
Gd	9.05	7.57	7.35	6.32	6.38	6.09	7.44	6.72	6.08	7.95	7.37	7.22	6.85	7.92	8.33	7.15	6.43	8.04
Tb	1.44	1.18	1.11	0.96	1.04	0.96	1.15	1.03	0.95	1.21	1.16	1.10	1.07	1.15	1.38	1.12	1.01	1.25
Dy	8.68	6.90	6.66	5.44	5.73	5.93	6.79	6.06	5.60	6.90	6.85	6.60	6.08	6.53	8.54	6.59	6.11	7.45
Ho	1.79	1.40	1.39	1.10	1.21	1.09	1.32	1.17	1.10	1.31	1.32	1.23	1.15	1.32	1.71	1.27	1.25	1.43
Er	5.11	3.99	3.77	2.92	3.34	2.99	3.76	3.35	3.16	3.66	3.69	3.44	3.23	3.69	4.92	3.65	3.55	4.09
Tm	0.74	0.58	0.58	0.43	0.50	0.49	0.55	0.50	0.46	0.53	0.54	0.51	0.48	0.58	0.73	0.56	0.55	0.61
Yb	4.91	4.11	3.53	3.05	3.40	3.22	3.86	3.44	3.19	3.71	3.67	3.40	3.21	3.88	4.78	3.67	3.84	4.04
Lu	0.75	0.62	0.57	0.49	0.52	0.51	0.57	0.53	0.49	0.57	0.57	0.52	0.50	0.61	0.72	0.57	0.59	0.63
Th	13.1	13.5	13.3	9.86	13.8	12.1	11.2	12.0	11.4	13.4	12.6	11.8	12.6	17.0	15.6	14.1	13.0	14.4
U	3.48	3.18	3.84	2.89	3.88	3.34	3.71	3.15	3.42	3.72	3.31	3.00	3.46	4.57	3.73	3.96	4.06	3.94
Sc	22.6	25.0	23.2	21.0	18.8	19.9	20.9	21.3	19.3	22.8	21.4	19.7	18.5	22.5	23.3	20.1	20.4	21.9
V	134	143	162	145	118	137	125	123	120	138	125	115	113	142	146	122	126	128
Cr	102	105	117	102	103	106	87	98	96	107	101	96	100	111	117	115	126	105

Continued

Table A14 Analyses of Ediacaran shales. NAS and MND. WALZ.—cont'd

Sample	TIN-4	TIN-5	VIL-2	PS-1	LR-2	RIE-1	CN-2	GED-4	GED-5	GED-6	GED-8	GED-9	VIL-3	RIE-2	SAR-3	SAR-5	SAR-8	OEN-6
	T	T	T	T	T	T	A	A	A	A	A	A	A	A	M	M	M	M
Co	11	20	29	33	12	14	20	21	19	17	13	13	21	28	11	9	28	10
Ni	39	52	69	58	47	51	46	42	49	40	44	43	47	65	39	39	74	35
Cu	28	12	63	47	43	43	39	36	36	25	26	37	26	55	31	12	45	23
Zn	129	65	78	140	115	129	119	126	121	110	41	114	93	73	131	114	183	128
Ga	28.0	29.0	30.0	25.4	27.5	27.9	24.5	24.6	24.4	27.7	26.1	24.5	25.7	28.6	31.2	26.5	26.8	27.8
Ge	1.71	1.77	1.81	1.69	1.73	1.89	1.64	1.66	1.67	1.83	1.60	1.54	1.51	2.11	2.07	2.14	2.00	1.88
As	15.1	7.38	51.5	52.5	18.8	12.7	10.2	17.7	15.7	17.4	15.7	11.5	12.4	33.6	15.1	16.1	16.9	21.7
Y	57.9	41.5	39.0	31.0	34.0	35.0	37.9	33.4	31.7	37.7	37.6	36.6	33.7	37.0	53.7	37.1	36.7	43.2
Zr	215	205	221	185	206	184	201	199	196	213	202	194	193	197	227	265	289	208
Nb	13.3	12.8	13.7	11.6	13.2	11.7	12.6	12.4	12.3	13.1	12.7	12.0	12.3	13.6	14.7	14.9	15.7	13.4
Mo	bd	bd	0.43	0.61	0.20	bd	0.41	bd	0.96	bd	bd	bd	0.68	0.88	bd	bd	0.60	0.58
Sn	3.52	3.80	3.34	2.90	3.55	3.25	3.66	3.19	3.13	3.63	3.31	3.06	3.33	3.48	4.11	3.54	3.53	3.51
Sb	3.62	1.80	9.44	3.06	2.60	2.38	1.08	1.39	bd	1.37	bd	bd	1.82	17.5	17.0	1.03	0.82	1.06
Hf	5.74	5.61	6.07	4.69	5.50	4.58	5.63	5.39	5.36	5.92	5.75	5.40	5.59	5.58	6.06	7.08	7.24	5.74
Ta	1.20	1.19	1.20	1.02	1.20	0.98	1.17	1.24	1.16	1.25	1.22	1.15	1.14	1.27	1.30	1.35	1.41	1.22
Pb	15.2	9.66	18.6	11.5	16.4	19.2	25.3	24.3	8.56	7.69	2.06	8.55	4.32	8.03	11.4	20.6	22.2	12.0
Bi	0.25	0.27	0.46	0.25	0.27	0.17	0.41	0.38	0.22	0.31	0.22	0.21	0.27	0.85	0.13	0.29	0.31	0.23

na: not analyzed; bd: below detection limit. T: Tineo gp. A: Allande gp. M: MND. New data in blue.

Data after Ugidos, J.M., Sánchez-Santos, J.M., Barba, P., Valladares, M.I., 2010. Upper Neoproterozoic series in the Central Iberian, Cantabrian and West Asturian Leonese zones (Spain): geochemical data and statistical results as evidence for a shared homogenised source area. Precamb. Res. 178, 51–58.

Table A15 Analyses of Upper Ediacaran sandstones. NAS. WALZ.

Sample	GED-7	GED-11	GED-12	CN-5	CN-6	CN-11
SiO_2	63.62	65.70	65.16	63.71	69.69	65.67
TiO_2	1.22	0.98	0.87	0.92	0.90	0.92
Al_2O_3	14.55	15.00	15.38	16.48	13.67	14.76
Fe_2O_3	6.68	5.94	6.07	6.29	5.17	5.87
MgO	2.74	2.29	3.00	2.65	1.96	2.32
MnO	0.07	0.07	0.08	0.09	0.08	0.07
CaO	2.03	1.49	0.94	1.83	1.90	1.67
Na_2O	3.57	3.81	3.21	2.59	3.99	3.73
K_2O	1.57	1.97	2.23	2.96	1.78	2.23
P_2O_5	0.28	0.23	0.21	0.28	0.21	0.20
LOI	3.55	2.66	3.48	2.72	1.59	2.24
Total	99.88	100.1	100.6	100.5	100.9	99.66
Be	1.84	2.11	2.22	2.63	1.98	2.01
Rb	53.0	62.0	70.0	113	68.1	75.8
Sr	215	193	150	220	225	259
Cs	1.97	2.26	2.23	8.34	4.39	4.33
Ba	362	485	493	703	464	544
La	50.0	38.4	33.7	36.6	37.5	37.0
Ce	104	79.0	69.6	74.9	76.7	76.3
Pr	12.2	9.35	8.32	9.22	8.97	9.01
Nd	46.3	36.1	31.7	35.9	34.2	34.4
Sm	8.95	7.31	6.37	7.34	6.70	6.79
Eu	1.97	1.55	1.49	1.57	1.43	1.51
Gd	7.30	6.24	5.42	6.28	5.54	5.61
Tb	1.11	0.95	0.85	0.98	0.84	0.87
Dy	6.43	5.60	5.11	5.79	4.97	5.07
Ho	1.23	1.09	0.99	1.12	0.97	0.99
Er	3.57	3.16	2.90	3.22	2.82	2.84

Continued

Table A15 Analyses of Upper Ediacaran sandstones. NAS. WALZ.—cont'd

Sample	GED-7	GED-11	GED-12	CN-5	CN-6	CN-11
Tm	0.53	0.47	0.44	0.48	0.42	0.42
Yb	3.72	3.27	2.93	3.24	2.94	2.90
Lu	0.59	0.49	0.45	0.50	0.46	0.45
Th	17.9	11.8	9.61	10.8	11.2	10.9
U	4.83	3.15	2.84	3.22	2.79	2.82
Sc	na	na	na	na	na	na
V	122	112	104	117	95	101
Cr	164	114	93	92	95	113
Co	25	23	24	32	30	26
Ni	37	38	40	44	30	39
Cu	34	35	19	31	15	26
Zn	77	96	90	144	83	129
Ga	19.5	20.4	21.2	23.2	17.5	19.3
Ge	1.44	1.57	1.32	1.69	1.46	1.56
As	16.2	10.3	10.1	17.7	10.5	17.1
Y	35.5	32.0	29.0	33.0	27.0	28.0
Zr	591	334	213	204	307	276
Nb	14.8	12.3	11.5	12.2	11.0	11.6
Mo	bd	bd	bd	0.56	bd	bd
Sn	2.56	2.60	2.52	2.63	2.25	2.36
Sb	bd	0.56	1.17	0.91	0.94	1.43
Hf	14.9	8.44	5.79	5.59	7.60	7.03
Ta	1.43	1.13	1.06	1.19	1.16	1.09
Pb	3.24	24.5	3.69	12.8	11.6	11.9
Bi	0.17	0.37	bd	bd	bd	bd

na: not analyzed; bd: below detection limit. New data.

Table A16 Analyses of the underlying and overlying Lower Cambrian shales. NAS and MND. WALZ.

Sample	TIN-2	TIN-3	VIL-5	GED-3	TIN-1	BOA-1	OEN-2	SAR-4	SAR-6.	SAR-7	SAR-9	OEN-1	OEN-3
	NASD	NASD	NASD	NASD	NASD	NASD	MND	MND	MND	MND	MND	MND	MND
	UN	UN	UN	OV	OV	OV	OV	UN	UN	UN	UN	UN	UN
SiO_2	60.95	60.55	59.38	55.59	56.56	52.52	62.10	57.19	58.15	57.86	56.55	64.13	58.36
TiO_2	0.76	0.72	0.80	0.88	0.83	1.00	1.00	1.10	0.89	1.14	1.13	0.88	1.26
Al_2O_3	21.77	18.38	19.05	21.79	20.11	22.82	17.32	23.52	20.47	22.49	23.24	16.65	22.04
Fe_2O_3	4.67	7.84	7.24	6.72	7.25	6.97	6.97	7.63	7.16	8.02	8.05	6.30	7.25
MgO	1.23	2.52	3.20	4.31	3.52	3.61	2.71	0.76	3.14	1.89	1.81	2.74	1.90
MnO	bd	bd	0.05	bd	bd	0.03	0.05	0.03	0.02	0.13	0.07	0.09	0.11
CaO	0.00	0.17	0.33	0.21	0.00	0.17	0.31	0.09	0.00	0.05	0.05	0.58	0.31
Na_2O	0.28	0.23	2.40	0.08	0.70	0.16	2.40	0.24	0.36	0.39	1.20	2.58	0.36
K_2O	6.14	5.51	3.78	5.59	5.44	6.55	4.46	5.83	4.93	3.93	3.84	3.18	4.59
P_2O_5	0.14	0.21	0.18	0.21	0.18	0.18	0.22	0.12	0.15	0.07	0.07	0.23	0.27
LOI	3.80	3.75	3.51	4.41	5.24	4.91	3.14	4.00	5.31	4.07	4.43	2.98	4.03
Total	99.74	99.88	99.93	99.79	99.83	98.92	100.7	100.5	100.6	99.99	100.4	100.3	100.5
Be	3.06	3.12	3.49	3.84	4.08	5.08	3.50	3.23	3.45	3.19	3.40	2.57	2.96
Rb	179	201	151	150	196	168	156	222	159	173	158	128	194
Sr	24.5	30.6	125	50.5	30.5	75.6	82.3	49.7	34.7	128	190	93.5	71.9
Cs	9.51	15.3	7.10	3.50	6.90	2.77	6.21	9.86	8.33	10.9	10.6	5.50	16.1
Ba	278	886	545	217	563	1101	567	498	696	646	578	422	527

Continued

Table A16 Analyses of the underlying and overlying Lower Cambrian shales. NAS and MND. WALZ.—cont'd

Sample	TIN-2	TIN-3	VIL-5	GED-3	TIN-1	BOA-1	OEN-2	SAR-4	SAR-6.	SAR-7	SAR-9	OEN-1	OEN-3
	NASD	NASD	NASD	NASD	NASD	NASD	MND	MND	MND	MND	MND	MND	MND
	UN	UN	UN	OV	OV	OV	OV	UN	UN	UN	UN	UN	UN
La	43.0	34.5	41.3	61.9	43.3	65.8	48.5	46.1	25.1	54.9	54.6	42.1	47.9
Ce	87.5	72.0	78.2	120	85.5	127	93.1	95.0	44.0	111	106	81.9	102
Pr	10.8	9.07	9.82	14.0	10.1	15.1	10.6	11.0	6.67	13.0	12.6	10.2	12.0
Nd	41.5	34.9	36.8	51.2	38.0	57.1	39.5	41.0	27.5	49.8	49.6	39.0	46.9
Sm	8.66	7.21	7.15	9.15	7.53	10.6	7.47	8.01	6.25	9.85	9.17	7.86	9.74
Eu	1.57	1.52	1.45	1.81	1.50	2.19	1.46	1.63	1.46	2.02	1.92	1.56	2.10
Gd	7.23	6.04	5.99	7.23	6.24	7.52	6.25	6.76	6.51	7.74	6.70	6.50	8.39
Tb	1.13	0.93	0.95	1.14	0.99	1.11	0.99	1.10	1.13	1.21	1.09	1.04	1.23
Dy	6.61	5.36	5.50	6.68	5.85	6.33	5.93	6.53	6.96	7.16	6.80	6.05	7.28
Ho	1.30	1.07	1.07	1.28	1.14	1.24	1.18	1.27	1.39	1.42	1.38	1.15	1.40
Er	3.71	3.11	3.13	3.70	3.24	3.63	3.52	3.68	4.02	3.99	3.99	3.33	4.00
Tm	0.56	0.47	0.46	0.55	0.48	0.56	0.53	0.56	0.60	0.60	0.61	0.51	0.61
Yb	3.94	3.33	3.13	3.83	3.35	3.77	3.59	3.86	4.11	4.12	4.17	3.42	4.12
Lu	0.60	0.51	0.49	0.56	0.51	0.59	0.56	0.59	0.62	0.64	0.64	0.52	0.64
Th	14.4	10.9	12.8	16.9	15.2	16.6	20.4	16.2	12.3	15.9	14.5	13.1	15.5
U	4.14	3.68	2.95	4.08	3.34	7.14	3.85	4.28	3.73	4.17	4.44	3.22	4.33
Sc	19.7	18.7	19.7	20.5	20.2	24.3	16.6	23.5	21.6	23.3	24.1	16.2	23.5
V	101	173	116	111	112	143	89	151	132	146	144	90	144

Cr	85	106	100	119	96	122	96	108	111	128	127	94	126
Co	8	21	30	24	22	15	24	8	10	16	24	17	6
Ni	18	52	48	55	46	37	45	32	43	59	52	38	44
Cu	4	bd	26	5	124	46	4	5	10	42	29	24	10
Zn	31	65	95	31	85	50	87	58	127	114	130	117	98
Ga	28.1	25.0	27.5	31.1	27.2	31.5	24.6	33.2	27.0	31.1	31.0	23.0	30.2
Ge	1.67	2.08	1.75	1.73	1.97	2.17	1.61	2.16	2.14	2.05	1.83	1.77	1.93
As	7.15	7.31	1.21	11.7	2.21	36.2	bd	bd	13.8	13.6	5.62	3.24	13.3
Y	36.7	32.1	32.1	37.5	32.9	36.9	35.7	37.3	41.5	40.6	39.4	34.6	40.9
Zr	181	185	155	179	161	176	448	226	200	212	221	222	325
Nb	12.7	11.1	12.4	17.7	14.4	14.1	18.2	16.4	12.1	17.6	16.2	14.3	17.2
Mo	bd	bd	bd	bd	bd	bd	bd	0.92	bd	bd	1.03	bd	bd
Sn	6.68	3.11	4.09	4.01	4.68	3.96	4.33	4.74	4.13	4.56	4.44	3.66	4.08
Sb	bd	1.18	2.48	bd	bd	2.40	0.37	3.74	0.39	2.82	2.50	0.53	0.76
Hf	5.29	5.02	4.45	5.04	4.54	4.81	10.9	6.15	5.51	6.25	6.26	6.01	8.48
Ta	1.23	1.07	1.22	1.53	1.35	1.31	1.64	1.48	1.12	1.59	1.42	1.28	1.51
Pb	1.88	2.04	4.22	bd	2.88	10.4	5.41	16.0	5.57	11.3	33.3	11.7	7.26
Bi	0.24	0.10	0.30	0.27	0.35	bd	0.10	bd	0.14	0.31	0.31	0.31	0.21

na: not analyzed; bd: below detection limit. New data. UN: underlying shales. OV: overlying shales.

References

Ábalos, B., 2001. Nuevos datos microestructurales sobre la existencia de deformaciones precambricas en la Sierra de la Demanda (Cordillera lberica). Geogaceta 30, 3–6.

Abd El-Rahman, Y., Polat, A., Fryer, B.J., Dilek, Y., El-Sharkawy, M., Sakran, S., 2010. The provenance and tectonic setting of the Neoproterozoic Um Hassa Graywacke Member, Wadi Hammamat area, Egypt: evidence from petrography and geochemistry. J. Afr. Earth. Sci. 58, 185–196.

Abd El-Rahman, Y., Abu Ambar, M., Li, X.H., Li, J., Ling, X.-X., Wu, L.-G., Masoud, A.E., 2019. The evolution of the Arabian-Nubian Shield and survival of its zircon U-Pb-H-O isotopic feature: a tale from the Um Had Conglomerate, central Eastern Desert, Egypt. Precambrian Res. 320, 46–62.

Abdelsalam, M.G., Stern, R.J., 1996. Mapping precambrian structures in the Sahara Desert with SIR-C/X-SAR radar: the Neoproterozoic Keraf suture, NE Sudan. J. Geophys. Res. 101, 23063–23076.

Abdelsalam, M.G., Liégeois, J.-P., Stern, R.J., 2002. The Saharan Metacraton. J. Afr. Earth. Sci. 34, 119–136.

Abu Sharib, A.S.A.A., Maurice, A.E., Abd El-Rahman, Y.M., Sanislav, I.V., Schulz, B., Bakhit, B.R., 2019. Neoproterozoic arc sedimentation, metamorphism and collision: evidene for the northern tip of the Arabian-Nubian Shield and implication for the terminal collision between East and West Gondwana. Gondw. Res. 66, 13–42.

Álvarez Nava, H., García Casquero, J.L., Gil, A., Hernández Hurroz, J., Lorenzo, S., López Díaz, F., Mira, M., Monteserín, V., Nozal, F., Pardo, M.V., Piart, J., Robles, R., Santamaría, J., Solé, J., 1988. Unidades litoestratigráficas de los materiales precámbrico-cámbricos de la mitad suroriental de la Zona Centro-Ibérica. In: Vera, J.A. (Ed.), II Congreso Geológico de España. vol. 1. Universidad de Granada, Granada, pp. 19–22.

Álvaro, J.J., Blanc-Valleron, M.M., 2002. Stratigraphic and structural framework of the Neoproterozoic Paracuellos group, Iberian chains, NE Spain. Bull. Soc. Geol. Fr. 173, 219–227.

Álvaro, J.J., Bauluz, B., Gil Imaz, A., Simón, J.L., 2008. Multidisciplinary constraints on the Cadomian compression and early Cambrian extension in the Iberian chains, NE Spain. Tectonophysics 461, 215–227.

Amidon, W.H., Burbank, D.W., Gehrels, G.E., 2005. Construction of detrital mineral populations: insights from mixing of U-Pb zircon ages in Himalayan rivers. Basin Res. 17, 463–485.

Amthor, J.E., Grotzinger, J.P., Schröder, S., Bowring, S.A., Ramezani, J., Martin, M.W., Matter, A., 2003. Extinction of *Cloudina* and *Namacalathus* at the Precambrian-Cambrian boundary in Oman. Geology 31, 431–434.

Andersen, T., 2005. Detrital zircons as tracers of sedimentary provenance: limiting conditions from statistics and numerical simulation. Chem. Geol. 216, 249–270.

Andersen, T., 2013. Age, Hf isotope and trace element signatures of detrital zircons in the Mesoproterozoic Eriksfjord sandstone, southern Greenland: are detrital zircons reliable guides to sedimentary provenance and timing of deposition? Geol. Mag. 150, 426–440.

Andersen, T., Kristoffersen, M., Elburg, M., 2016. How far can we trust provenance and crustal evolution information from detrital zircons? A south African case study. Gondw. Res. 34, 129–148.

Armstron-Altrin, J.S., Verma, S.P., 2005. Critical evaluation of six tectonic setting discrimination diagrams using geochemical data of Neogene sediments from known tectonic settings. Sediment. Geol. 177, 115–129.

Armstrong-Altrin, J.S., 2015. Evaluation of two multidimensional discrimination diagrams from beach and deep-sea sediments from the Gulf of Mexico and their application to Precambrian clastic sedimentary rocks. Int. Geol. Rev. 57, 1446–1461.

Avigad, D., Gvirtzman, Z., 2009. Late Neoproterozoic rise and fall of the northern Arabian-Nubian shield: the role of lithospheric mantle delamination and subsequent thermal subsidence. Tectonophysics 477, 217–228.

Avigad, D., Weissbrod, T., Gerdes, A., Zlatkin, O., Ireland, T.R., Morag, N., 2015. The detrital zircon U-Pb-Hf fingerprint of the northern Arabian-Nubian shield as reflected by a late Ediacaran arkosic wedge (Zenifim formation; subsurface Israel). Precambrian Res. 266, 1–11.

Banfield, J.F., Eggleton, R.A., 1989. Apatite replacement and rare earth mobilization, fractionation, and fixation during weathering. Clays Clay Miner. 37, 113–127.

Barba, P., Fernández, L.P., 1991. Mapa Geológico de España E. 1:50.000. Inst. Geol. Min. Esp. Memoria explicativa de la Hoja 102 (Barrios de Luna), 130 pp.

Barbero, L., Villaseca, C., Rogers, G., Brown, P.E., 1995. Geochemical and isotopic disequilibrium in crustal melting: an insight from the anatectic granitoids from Toledo, Spain. J. Geophys. Res. 100, 15,745–15,765.

Bau, M., 1999. Scavening of dissolved yttrium and rare earths by precipitating iron oxyhydroxide: experimental evidence for Ce oxidation, Y-Ho fractionation, and lanthanide tetrad effect. Geochim. Cosmochim. Acta 63, 67–77.

Bauluz, B., Mayayo, M.J., Fernández-Nieto, C., González López, J.M., 2000. Geochemistry of Precambrian and Paleozoic siliciclastic rocks from the Iberian range (NE Spain): implications for source-area weathering, sorting, provenance and tectonic setting. Chem. Geol. 168, 135–150.

Beetsma, J.J., 1995. The Late Proterozoic/Paleozoic and Hercynian crustal evolution of the Iberian Massif, N Portugal. Published Ph.D. thesis, Vrije University, Amsterdam. 223 pp.

Bentor, Y.K., 1985. The crustal evolution of the Arabo-Nubian massif with special reference to the Sinai peninsula. Precambrian Res. 28, 1–74.

Bhatia, M.R., 1983. Plate tectonics and geochemical composition of sandstones. J. Geol. 91, 611–627.

Bhatia, M.R., Crook, K.A.W., 1986. Trace element characteristics of graywackes and tectonic discrimination of sedimentary basins. Contrib. Mineral. Petrol. 92, 181–193.

Birgul, O., 1981. Scandium-iron correlation in clay minerals. Earth Planet. Sci. Lett. 55, 450–452.

Bock, B., McLennan, S.M., Hanson, G.N., 1994. Rare earth elements reditribution and its effects on the neodymium isotope system in the Austin Glen member of the Normenskill formation, New York, USA. Geochim. Cosmochim. Acta 58, 5245–5253.

Bock, B., McLennan, S.M., Hanson, G.N., 1998. Geochemistry and provenance of the middle Ordovician Austin Glen member (Normanskill formation) and the Taconic Orogeny in New England. Sedimentology 45, 635–655.

Bouch, J.E., Hole, M.J., Trewin, N.H., Morton, A.C., 1995. Low temperature aqueous mobility of the rare-earth elements during sandstone diagenesis. J. Geol. Soc. Lond. 152, 895–898.

Bouyx, E., 1970. Contribution a l'etude des Formations Anteordoviciennes de la Meseta Méridionale (Ciudad Real et Badajoz). Mem. Inst. Geol. Min. Esp. 73, 1–293.

Braun, J.-J., Pagel, M., Herbillon, A., Rosin, C., 1993. Mobilization and redistribution of REEs and thorium in a syenitic lateritic profile: a mass balance study. Geochim. Cosmochim. Acta 57, 4419–4434.

Brown, D.J., Helmke, P.A., Clayton, M.K., 2003. Robust geochemical indices for redox and weathering on a granitic laterite landscape in Central Uganda. Geochim. Cosmochim. Acta 67, 2711–2723.

Calvet, F., Salas, R., 1988. Tipos de plataformas carbonatadas del Precámbrico terminal de la Zona Centro Ibérica. In: Vera, J.A. (Ed.), II Congreso Geológico de España, vol. 1. Universidad de Granada, pp. 59–62.

Carignan, J., Hild, P., Mévelle, G., Morel, J., Yeghicheyan, D., 2001. Routine analyses of trace elements in geological samples using flow injection and low pressure online liquid chromatography coupled to ICP-MS: a study of reference materials BR, DR-N, UB-N, AN-G and GH. Geostand. Newslett. 25, 187–198.

Carpentier, M., Chauvel, C., Mauri, R.C., Matielli, N., 2009. The "zircon effect" as recored by the chemical and Hf isotopic compositions of Lesser Antilles forearc sediments. Earth Planet. Sci. Lett. 287, 86–99.

Carrington da Costa, J., 1950. Noticia sobre uma carta geológica do Buçaco, de Nery Delgado. Serviços Geológicos de Portugal, Lisboa, pp. 1–27.

Castiñeiras, P., Villaseca, C., Barbero, L., Martín Romera, C., 2008. SHRIMP U–Pb zircon dating of anatexis in high-grade migmatite complexes of Central Spain: implications in the Hercynian evolution of Central Iberia. Int. J. Earth Sci. 97, 35–50.

Cawood, P.A., Hawkesworth, C.J., Dhuime, B., 2014. The continental record and the generation of continental crust. Geol. Soc. Am. Bull. 125 (1–2), 14–32.

Cohen, K.M., Finne, S.C., Gibbard, P.L., Fan, J.X., 2013. The ICS international Chronostratigraphic chart. Episodes 36, 199–204.

Colchen, M., 1974. Geologie du la Sierra de la Demanda (Burgos-Logroño) Espagne. Mem. Inst. Geol. Min. Esp. 85, 1–436.

Comte, P., 1959. Recherches sur les terrains anciens de la Cordillere Cantabrique. Mem. Inst. Geol. Min. Esp. 60, 1–440.

Condie, K.C., 1991. Another look at rare earth elements in shales. Geochim. Cosmochim. Acta 55, 2527–2531.

Condie, K.C., 1993. Chemical composition and evolution of the upper continental crust: contrasting results from surface samples and shales. Chem. Geol. 104, 1–37.

Condie, K.C., Wronkiewciz, D.J., 1990. The Cr/Th in Precambrian pelites from the Kaapvaal craton as an index of craton evolution. Earth Planet. Sci. Lett. 97, 256–267.

Condie, K.C., Dengate, J., Cullers, R.L., 1995. Behaviour of rare earth elements in a paleoweathering profile on granodiorite in the front range, Colorado, USA. Geochim. Cosmochim. Acta 59, 279–294.

Corrales, I., Valladares, M.I., 1980. Facies carbonatadas del Cámbrico de Salamanca. In: Stvdia Geológica Salmanticensia XVI. Universidad de Salamanca, pp. 95–102.

Corrales, I., Manjón, M., Valladares, M.I., 1974. La serie carbonatada de Navarredonda de la Rinconada (Salamanca, España). In: Stvdia Geológica Salmanticensia VII. Universidad de Salamanca, pp. 85–91.

Cortijo, I., Martí Mus, M., Jensen, S., Palacios, T., 2010. A new species of *Cloudina* from the terminal Ediacaran of Spain. Precambrian Res. 176, 1–10.

Cortijo, I., Martí Mus, M., Jensen, S., Palacios, T., 2015. Late Ediacaran skeletal body fossil assemblage from the Navalpino anticline, Central Spain. Precambrian Res. 267, 186–195.

Cotten, J., Le Dez, A., Bau, M., Caroff, M., Maury, R.C., Dulsky, P., Fourcade, S., Bohn, M., Brousse, R., 1995. Origin of anomalous rare-earth element and yttrium enrichments in subaerially exposed basalts: evidence from French Polynesia. Chem. Geol. 119, 85–91.

Cox, R., Lowe, D.R., 1995. A conceptual review of regional-scale controls on the composition of clastic sediment and the co-evolution of continental blocks and their sedimentary cover. J. Sediment. Res. A65, 1–12.

Cox, R., Lowe, D.R., Cullers, R.L., 1995. The influence of sediment recycling and basement composition on evolution of mudrock chemistry in the southwestern United States. Geochim. Cosmochim. Acta 59, 2919–2940.

Crespo, V., Rey, J., 1971. Contribución al estudio del Valle de Alcudia. Bol. Geol. Min. Esp. 82 (6), 512–515.

Crichton, J.G., Condie, K.C., 1993. Trace elements as source indicators in cratonic sediments: a case study from the early Proterozoic Libby Creek Group, Southeastern Wyoming. J. Geol. 101, 319–332.

Cullers, R.L., 1988. Mineralogical and chemical changes of soil and stream sediment formed by intense weathering of the Danburg granite, Georgia, U.S.A. Lithos 21, 301–314.

Cullers, R.L., Podkovyrov, V.N., 2000. Geochemistry of the Mesoproterozoic Lakhanda shales in southeastern Yakutia, Russia: implications for mineralogical and provenance control, and recycling. Precambrian Res. 104, 77–93.

Cullers, R.L., Chaudhuri, S., Kilbane, N., Koch, R., 1979. Rare earths in size fractions and sedimentary rocks of Pennsylvanian-Permian age from the mid-continent of the USA. Geochim. Cosmochim. Acta 43, 1285–1302.

Díaz García, F., 2006. Geometry and regional significance of Neoproterozoic (Cadomian) structures of the Narcea antiform, NW Spain. J. Geol. Soc. Lond. 163, 499–508.

Diez Balda, M.A., 1980. La sucesión estratigráfica del Complejo esquisto-grauváquico al Sur de Salamanca. Estud. Geol. 36, 131–138.

Diez Balda, M.A., 1986. El Complejo Esquisto Grauváquico, las series paleozoicas y la estructura hercínica. In: Acta Salmanticensia Ciencias. vol. 52. Universidad de Salamanca. 162 pp.

Diez Balda, M.A., Vegas, R., González Lodeiro, F., 1990. Authoctonous sequences in the central Iberian zone: structure. In: Dallmeyer, R.D., Martínez García, E. (Eds.), Pre-Mesozoic Geology of Iberia. Springer, Berlin, pp. 172–188.

Dozy, J.J., 1983. The geology of the region to the southeast of Lugo (NW Spain). Leidse Geol. Meded. 52, 513–524.

Elderfield, H., Upstill-Goddard, R., Sholkovitz, E.R., 1990. The rare earth elements in rivers, stuaries and coastal sea waters: processes affecting crustal input of elements to the ocean and their significance to the composition of sea water. Geochim. Cosmochim. Acta 54, 971–991.

Eliwa, H., Breitkreuz, C., Khalaf, I., El Gameel, K., 2010. Depositional styles of early Ediacaran terrestrial volcanosedimentary succession in Gebel El Urf area, N Eastern Desert, Egypt. J. Afr. Earth Sci. 57, 328–344.

Eriksson, K.A., Taylor, S.R., Korsch, R.J., 1992. Geochemistry of 1.8-1.67 Ga mudstones and siltstones from the Mount Isa Inlier, Queensland, Australia: provenance and tectonic implications. Geochim. Cosmochim. Acta 56, 899–909.

Fedo, C.M., Nesbitt, H.W., Young, G.M., 1995. Unraveling the effects of potasium metasomatism in sedimentary rocks and paleosols, with implications for paleoweathering conditions and provenance. Geology 23, 921–924.

Feng, R., Kerrich, R., Maas, R., 1993. Geochemical, oxygen, and neodymium isotope compositions of metasediments from the Abitibi greenstone belt and Pontiac subprovince, Canada: Evidence for ancient crust and Archean terrane yuxtaposition. Geochim. Cosmochim. Acta 57, 641–658.

Fernández-Suárez, J., Gutiérrez-Alonso, G., Jenner, G.A., Jackson, S.E., 1998. Geochronolgy and geochemistry of the Pola de Allande granitoids (northern Spain): their bearing on the Cadomian-Avalonian evolution of Northwest Iberia. Can. J. Earth Sci. 35, 1439–1453.

Fernández-Suárez, J., Gutiérrez-Alonso, G., Jenner, G.A., Tubrett, M.N., 2000. New ideas on the Proterozoic–early Palaeozoic evolution of NW Iberia: insights from U–Pb detrital zircon ages. Precambrian Res. 102, 185–206.

Fernández-Suárez, J., Gutiérrez-Alonso, G., Pastor-Galán, D., Hofmann, M., Murphy, J.B., Linnemann, U., 2014. The Ediacaran-early Cambrian detrital zircon record of NW Iberia: possible sources and paleogeographic constraints. Int. J. Earth Sci. 103, 1335–1357.

Floyd, P.A., Shail, R., Leveridge, B.E., Franke, W., 1991. Geochemistry and provenance of Rhenohercynian synorogenic sandstones: implications for tectonic environment discrimination. In: Morton, A.C., Todd, S.P., Haughton, P.D.W. (Eds.), Developments in sedimentary provenance studies. Geological Society, London, Special Publications, vol. 57, pp. 173–188.

Fuenlabrada, J.M., Pieren, A.P., Díez Fernández, R., Sánchez Martínez, S., Arenas, R., 2016. Geochemistry of the Ediacaran-early Cambrian transition in Central Iberia: tectonic setting and isotopic sources. Tectonophysics 681, 15–30.

Garrels, R.M., 1988. Sediment cycling during earth history. In: Lerman, M., Meybeck, M. (Eds.), Physical and Chemical Weathering in Geochemical Cycles. Kluwer Academic Publishers, New York, pp. 341–355.

Garzanti, E., 1986. Source rocks versus sedimentary control on the mineralogy of deltaic volcanic arenites (upper Triassic, northern Italy). J. Sediment. Petrol. 56, 267–275.

Geyer, G., Landing, E., 2004. A unified lower-middle Cambrian chronostratigraphy for West Gondwana. Acta Geol. Pol. 54, 179–218.

Grenne, T., Pedersen, R.B., Bjerkgard, T., Braathen, A., Selassie, M.G., Worku, T., 2003. Neoproterozoic evolution of Western Ethiopia: igneous geochemistry, isotope systematics and U-Pb ages. Geol. Mag. 140, 373–395.

Grotzinger, J.P., Bowring, S.A., Saylor, B.Z., Kaufman, A.J., 1995. Biostratigraphic and geochronologic constraints on early animal evolution. Science 270, 598–604.

Grotzinger, J.P., Adams, E.W., Schröder, S., 2003. Microbial-metazoan reefs of the terminal Proterozoic Nama group (c. 550-543 Ma), Namibia. Geol. Mag. 142, 499–517.

Gutiérrez-Alonso, G., 1996. Strain partitioning in the footwall of the Somiedo Nappe: structural evolution of the Narcea tectonic window, NW Spain. J. Struct. Geol. 18, 1217–1229.

Gutiérrez-Alonso, G., Fernández-Suárez, J., Jeffries, T.E., Jenner, G.A., Tubrett, M.N., Cox, R., Jackson, S.E., 2003. Terrane accretion and dispersal in the northern Gondwana margin. An early Paleozoic analogue of a long-lived active margin. Tectonophysics 365, 221–232.

Gutiérrez-Alonso, G., Fernández-Suárez, J., Jeffries, T.F., 2004. Age and setting of the upper Neoproterozoic Narcea Antiform volcanic rocks (NW Iberia). Geogaceta 35, 79–83.

Hargrove, U.S., Stern, R.J., Kimura, J.-I., Manton, W.I., Johnson, P.R., 2006. How juvenile is the Arabian-Nubian shield? Evidence from Nd isotopes and pre-Neoproterozoic inherited zircon in the Bi'r Umq suture zone, Earth Planet. Sci. Lett. 252, 308–326.

Harms, U., Schandelmeier, H., Darbyshire, D.P.F., 1990. Pan-African reworked early/middle Proterozoic crust in the NE Africa west of the Nile: Sr and Nd isotope evidence. J. Geol. Soc. Lond. 147, 859–872.

Haughton, P.D.W., 1988. A cryptic Caledonian flysch terrane in Scotland. J. Geol. Soc. Lond. 145, 685–703.

Henderson, B.J., Collins, W.J., Murphy, J.B., Gutiérrez-Alonso, G., Hand, M., 2016. Gondwanan basement terranes of the Variscan-Appalachian orogen: Baltican, Saharan and west African hafnium isotopic fingerprints in Avalonian, Iberian and the Armorican Terranes. Tectonophysics 681, 278–304.

Holail, H.M., Moghazi, A.M., 1998. Provenance, tectonic setting and geochemistry of greywackes and siltstones of the late Precambrian Hammamat Group, Egypt. Sediment. Geol. 116, 227–250.

Hu, Z., Gao, S., 2008. Upper crustal abundances of trace elements: a revision and update. Chem. Geol. 253, 205–221.

Jahn, B.M., Condie, K.C., 1995. Evolution of the Kaapvaal Craton as viewed from geochemical and Sm-Nd isotopic analyses of intracratonic pelites. Geochim. Cosmochim. Acta 59, 2239–2258.

Jensen, S., Palacios, T., 2016. The Ediacaran-Cambrian trace fossil record in the Central Iberian Zone, Iberian Península. Comunicações Geológicas 103 (Especial I), 83–92.

Jensen, S., Palacios, T., Martí Mus, M., 2007. A brief review of the fossil record of the Ediacaran-Cambrian transition in the area of Montes de Toledo-Guadalupe, Spain. In: Vickers-Rich, P., Komarovwer, P. (Eds.), The rise and fall of the Ediacaran biota. Geological Society, London, Special Publications, vol. 286, pp. 223–235.

Jensen, S., Palacios, T., Martí Mus, M., 2010. Revised biochronology of the lower Cambrian of the central Iberian zone, southern Iberian massif, Spain. Geol. Mag. 147, 690–703.

Johnson, P.R., 2003. Post-amalgamation basins of the NE Arabian shield and implications for Neoproterozoic tectonism in the northern east African Orogen. Precambrian Res. 123, 321–337.

Johnson, P.R., Andresen, A., Collins, A.S., Fowler, A.R., Fritz, H., Ghebreab, W., Kusky, T., Stern, R.J., 2011. Late Cryogenian-Ediacaran history of the Arabian-Nubian shield: a review of depositional, plutonic, structural, and tectonic events in the closing stages of the northern east African Orogen. J. Afr. Earth Sci. 61, 167–232.

Johnsson, M.J., 1993. The system controlling the composition of clastic sediments. In: Johnsson, M.J., Basu, A. (Eds.), Processes Controlling the Composition of Clastic Sediments. The Geological Society of America. Special Paper, vol. 284, pp. 21–40.

Khalaf, E.A., 2010. Stratigraphy, facies architecture, and palaeoenvironment of Neoproterozoic volcanics and volcaniclastic deposits in Fatira area, central Eastern Desert, Egypt. J. Afr. Earth Sci. 58, 405–426.

Knoll, A.H., Walter, M.R., Narbonne, G.M., Christie-Blick, N., 2004. A new period for the geologic time scale. Science 305, 621–622.

Kolodner, K., Avigad, D., McWilliams, M., Wooden, J.L., Weissbrod, T., Feinstein, S., 2006. Provenance or north Gowndwana Cambrian-Ordovician sandstone: U-Pb SHRIMP dating of detrital zircons from Israel and Jordan. Geol. Mag. 143, 367–391.

Küster, D., Liégeois, J.P., Matukov, D., Sergeev, S., Lucassen, F., 2008. Zircon geochronology and Sr, Nd, Pb isotope geochemistry of granitoids from Bayuda Desert and Sabaloka (Sudan): evidence for a Bayudan event (920-900 Ma) preceding the pan-African orogenic cycle (860-590 Ma) at the eastern boundary of the Saharan Metacraton. Precambrian Res. 164, 16–39.

Leake, B.E., 1964. The chemical distinction between ortho- and para-amphibolites. J. Petrol. 5, 238–254.

Lev, S.M., McLennan, S.M., Hanson, G.N., 1999. Mineralogic control on REE mobility during black-shale diagenesis. J. Sediment. Res. 69, 1071–1082.

Liñan, E., Gámed-Vintaned, J.A., 1993. Lower Cambrian palaeogeography of the Iberian Península and its relations with some neighbouring European areas. Bull. Soc. Geol. Fr. 164, 831–842.

Liñán, E., Tejero, R., 1988. Las formaciones precámbricas del antiforme de Paracuellos (Cadenas Ibéricas). Bol. Real Soc. Esp. Hist. Nat. 84, 39–49.

Liñán, E., Gozalo, R., Palacios, T., Gámed-Vintaned, J.A., Ugidos, J.M., Mayoral, E., 2002. Cambrian. In: Gibbons, W., Moreno, T. (Eds.), The Geology of Spain. Geological Society, London, pp. 17–29.

Liñán, E., Gámez-Vintaned, J.A., San José, M.A., Tejero, R., 2004. El sustrato prepaleozoico. In: Vera, J.A. (Ed.), Geología de España. Soc. Geol. España, Inst. Geol. Min. Esp, pp. 470–471.

López-Díaz, F., 1994. Estratigrafía de los materiales anteordovícicos del anticlinal de Navalpino (Zona Centroibérica). Rev. Soc. Geol. Esp. 7, 31–45.

López-Guijarro, R., Armendáriz, M., Quesada, C., Fernández-Suárez, J., Murphy, J.B., Pin, C., Bellido, F., 2008. Ediacaran-Palaeozoic tectonic evolution of the Ossa Morena and central Iberian zones (SW Iberia) as revealed by Sm-Nd isotope systematics. Gondw. Res. 461, 202–214.

Lotze, F., 1961. Das Kambrium Spaniens. Teil 1: Stratigraphie. Abhandlungen der Gesellschaft der Wissenschaften zu Göttingen (Mathematisch-Physikalische Klasse). 6, 283–501 (translated into Spanish by Gómez de Llarena, J., 1970. El Cámbrico de España. Mem. Inst. Geol. Min. España 70, 1–256).

Mack, G.H., 1984. Exceptions to the relationship between plate tectonic and sandstone composition. J. Sediment. Petrol. 54, 212–220.

Malusà, M.G., Carter, A., Limonchelli, M., Villa, I.M., 2013. Bias in detrital zircon geochronology and thermochronometry. Chem. Geol. 359, 90–107.

Malusà, M.G., Resentini, A., Garzanti, E., 2016. Hydraulic sorting and mineral fertility bias in detrital geochronology. Gondw. Res. 31, 1–19.

Martín Parra, L.M., 1989. Mapa Geológico de España E. 1:50.000. Memoria explicativa de la Hoja 128 (Riello). Inst. Tecnol. Geomin. España. 119 pp.

Martínez-Catalán, J.R., 1985. Estratigrafía y estructura del domo de Lugo (sector oeste de la zona Asturoccidental-leonesa). Corpus Geol. Gallaeciae (2ª serie). 291 pp.

Martínez-Catalán, J.R., 2011. Are the oroclines of the Variscan belt related to late Variscan strike-slip tectonics? Terra Nova 23, 241–247.

Matte, P., 1968. La structure de la virgation hercynienne de Galice (Spagne). Rev. Geol. Alp. 44, 1–128.

McCann, T., 1991. Petrological and geochemical determination of provenance in the southern Welsh Basin. In: Morton, A.C., Todd, S.P., Haughton, P.D.W. (Eds.), Developments in Sedimentary Provenance Studies. Geological Society, London, Special Publications, vol. 57, pp. 215–230.

McCulloch, M.T., Wasserburg, G.J., 1978. Sm-Nd and Rb-Sr chronology of continental crust formation. Science 200, 1003–1011.

McDaniel, D.K., Hemming, S.R., McLennan, S.M., Hanson, G.N., 1994. Resetting of neodymium isotopes and redistribution of REEs during sedimentary processes: the early Proterozoic Chelmsford formation, Sudbury Basin, Ontario, Canada. Geochim. Cosmochim. Acta 58, 931–941.

McLennan, S.M., 1989. Rare earth elements in sedimentary rocks: influence of provenance and sedimentary processes. In: Lipin, B.R., McKay, G.A. (Eds.), Reviews in Mineralogy. vol. 21, pp. 169–200.

McLennan, S.M., Hemming, S., 1992. Samarium/neodymium elemental and isotopic systematics in sedimentary rocks. Geochim. Cosmochim. Acta 56, 887–898.

McLennan, S.M., Taylor, S.R., 1991. Sedimentary rocks and crustal evolution: tectonic setting and secular trends. J. Geol. 99, 1–21.

McLennan, S.M., Taylor, S.R., McCulloch, M.T., Maynard, J.B., 1990. Geochemical and Nd-Sr isotopic composition of deep-sea turbidites: crustal evolution and plate tectonic associations. Geochim. Cosmochim. Acta 54, 2015–2050.

McLennan, S.M., Hemming, S., McDaniel, D.K., Hanson, G.N., 1993. Geochemical approaches to sedimentation, provenance, and tectonics. In: Johnsson, M.J., Basu, A. (Eds.), Processes Controlling the Composition of Clastic Sediments. vol. 284. The Geological Society of America. Special Paper, pp. 21–40.

McLennan, S.M., Taylor, S.R., Hemming, S.R., 2006. Composition, differentiation, and evolution of continental crust: constraints from sedimentary rocks and heat flow. In: Brown, M., Rushmer, T. (Eds.), Evolution and Differentiation of the Continental Crust. Cambridge University Press, Cambridge, pp. 92–134.

Miller, R.G., O'Nions, R.K., Hamilton, P.J., Welin, E., 1986. Crustal residence ages of clastic sediments, orogeny and continental evolution. Chem. Geol. 57, 87–99.

Milodowski, A.E., Zalasiewicz, J.A., 1991. Redistribution of rare earth elements during diagenesis of turbidite/hemipelagite mudrock sequences of Llandovery age from central Wales. In: Morton, A.C., Todd, S.P., Haughton, P.D.W. (Eds.), Developments in Sedimentary Provenance Studies, vol. 57. Geological Society, London, Special Publication, pp. 101–124.

Moecher, D.P., Samson, S.D., 2006. Differential zircon fertility of source terranes and natural bias in the detrital zircon record: implications for sedimentary analysis. Earth Planet. Sci. Lett. 247, 252–266.
Moghazi, A.K.M., 1999. Magma source and evolution of late Neoproterozoic granitoids in the Gabal El-Urf area, Eastern Desert, Egypt: geochemical and Sr-Nd isotope constraints. Geol. Mag. 136, 285–300.
Moghazi, A.M., 2003. Geochemistry and Petrogenesis of a high-K calc-alkaline Dokhan volcanic suite, South Safaga area, Egypt: the role of late Neoproterozoic crustal extension. Precambrian Res. 125, 161–178.
Moghazi, A.K.M., Ali, K.A., Wilde, S.A., Zhou, Q., Andersen, T., Andresen, A., Abu El-Enen, M.M., Stern, R.J., 2012. Geochemistry, geochronology, and Sr–Nd isotopes of the late Neoproterozoic Wadi Kid volcano-sedimentary rocks, Southern Sinai, Egypt: implications for tectonic setting and crustal evolution. Lithos 154, 147–165.
Morag, N., Avigad, D., Gerdes, A., Belousova, E., Harlavan, Y., 2011. Crustal evolution and recycling in the northern Arabian-Nubian shield: new perspectives from zircon Lu-Hf and U-Pb systematics. Precambrian Res. 186, 101–116.
Morag, N., Avigad, D., Gerdes, A., Harlavan, Y., 2012. 1000-500 Ma crustal evolution in the northern Arabian-Nubian shield revealed by U-Pb-Hf of detrital zircons from the late Neoproterozoic sediments (Elat area, Israel). Precambrian Res. 208-211, 197–212.
Morton, A.C., Hallsworth, C.R., 1999. Processes controlling the composition of heavy mineral assemblages in sandstones. Sediment. Geol. 124, 3–29.
Moussa, E.M.M., Stern, R.J., Manton, W.I., Ali, K.A., 2008. SHRIMP zircon dating and Sm/Nd isotopic investigations of Neoproterozoic granitoids, Eastern Desert, Egypt. Precamb. Res. 160, 341–356.
Nägler, T.F., Schäfer, H.J., Gebauer, D., 1995. Evolution of the Western European continental crust: implications from Nd and Pb isotopes in Iberian sediments. Chem. Geol. 121, 345–357.
Naidoo, T., Zimmermann, U., Vervoort, J., Tait, J., 2018. Evidence of early Archean crust in Northwest Gondwana, from U-Pb and Hf isotope analysis of detrital zircon, in Ediacaran supracrustal rocks of northern Spain. Int. J. Earth Sci. 107, 409–429.
Narbonne, G.M., Xiao, S., Shields, G., 2012. The Ediacaran period. In: Gradstein, F., Ogg, J., Schmidt, M.D., Ogg, G. (Eds.), Geologic Timescale. Elsevier, pp. 427–449.
Navidad, M., Castiñeiras, P., 2011. Early Ordovician magmatism in the northern Central Iberian Zone (Iberian Massif): new U–Pb (SHRIMP) ages and isotopic Sr–Nd data. In: Gutiérrez-Marco, J.C., Rábano, I., García-Bellido, D. (Eds.), Ordovician of the World. vol. 14. Cuadernos del Museo Geolminero de España, pp. 391–398.
Nesbitt, H.W., Markovicks, G., 1997. Weathering of granodioritic crust, long-term storage of elements in weathering profiles, and petrogenesis of siliciclastic sediments. Geochim. Cosmochim. Acta 61, 1653–1670.
Nesbitt, H.W., Wilson, R.E., 1992. Recent chemical weathering of basalts. Am. J. Sci. 292, 740–777.
Nesbitt, H.W., Young, G.M., 1982. Early Proterozoic climates and plate motions inferred from major elements chemistry of lutites. Nature 299, 715–717.
Nesbitt, H.W., Markovics, Ç.G., Price, R.C., 1980. Chemical processes affecting alkalis and alkaline earths during continental weathering. Geochim. Cosmochim. Acta 44, 1659–1666.
Öhlander, B., Ingri, J., Land, M., Schöberg, H., 2000. Change of Sm-Nd isotope composition during weathering of till. Geochim. Cosmochim. Acta 64, 813–820.
Ohr, M., Halliday, A.N., Peator, D.R., 1991. Sr and Nd isotopic evidence for punctuated clay diagenesis, Texas Gulf Coast. Earth Planet. Sci. Lett. 105, 110–126.
Orejana, D., Merino Martínez, E., Villaseca, C., Andersen, T., 2015. Ediacaran-Cambrian paleogeography and geodynamic setting of the central Iberian zone: constraints from coupled U-Pb-Hf isotopes of detrital zircons. Precambrian Res. 261, 234–251.

Ortega, E., González Lodeiro, F., 1986. La discordancia intraalcudiense en el dominio meridional de la zona centroibérica. Breviora Geológica Astúrica 3, 27–32.

Ovtracht, A., Tamain, G., 1970. Tectonique en Sierra Morena (España). C.R. Acad. Sci. Paris (Ser. D) 270, 2634–2636.

Padel, M., Álvaro, J.J., Casas, J.M., Clausen, S., Poujol, M., Sánchez-García, T., 2018. Cadomian volcanosedimentary complexes across the Ediacaran-Cambrian transition of the Eastern Pyrenees, southwestern Europe. Int. J. Earth Sci. 107, 1579–1601.

Palacios, T., Vidal, G., 1992. Lower Cambrian achritarchs from northern Spain: the Precambrian-Cambrian boundary and biostratigraphic implications. Geol. Mag. 129, 421–436.

Parnell, J., Mark, D.F., Frei, R., Fallick, A.E., Ellam, R.M., 2014. $^{40}Ar/^{39}Ar$ dating of exceptional concentration of metals by weathering of Precambrian rocks at the Precambrian-Cambrian boundary. Precambrian Res. 246, 54–63.

Pastor-Galán, D., Gutiérrez-Alonso, G., Murphy, J.B., Fernández-Suárez, J., Hoffmann, M., Linnemann, U., 2013. Provenance analysis of the Paleozoic sequences of the northern Gondwana margin in NW Iberia: passive margin to Variscan collision and orocline development. Gondw. Res. 23, 1089–1103.

Patchett, P.J., White, W.M., Feldmann, H., Kienlinczuk, S., Hofmann, A.W., 1984. Hafnium/rare earth element fractionation in the sedimentary system and crustal recycling into the Earth's mantle. Earth Planet. Sci. Lett. 69, 365–378.

Pe-Piper, G., Triantafyllidis, S., Piper, D.J.W., 2008. Geochemical identification of clastic sediment provenance from known sources of similar geology: the cretaceous scotian basin. Can. J. Sediment. Res. 78, 595–607.

Pereira, M.F., Linnemann, U., Hofmann, M., Chichorro, M., Solá, A.R., Medina, J., Silva, J.B., 2012. The provenance of late Ediacaran and early Ordovician siliciclastic rocks in the southwest central Iberian zone: constraints from detrital zircon data on northern Gondwana margin evolution during the late Neoproterozoic. Precambrian Res. 192–195, 166–189.

Pérez Estaún, A., 1978. Estratigrafía y estructura de la rama Sur de la Zona Asturoccidental-leonesa. Mem. Inst. Geol. Min. Esp. 92, 1–150.

Pérez Estaún, A., Bastida, F., Martínez-Catalán, J.R., Gutiérrerz Marco, J.C., Marcos, A., Pulgar, J.A., 1990. West Asturian Leonese zone. Stratigraphy. In: Dallmeyer, R.D., Martínez García, E. (Eds.), Pre-Mesozoic Geology of Iberia. Springer, Berlin, pp. 92–102.

Pieren, A.P., García-Hidalgo, J.F., 1999. The upper Alcudian from the Alcudian anticline revisited (Ciudad Real, Central Spain). XV Reunión de Geología del Oeste Peninsular. J. Conf. Abstr. 4 (3), 1018–1019.

Pin, C., Santos Zalduegui, J.F., 1997. Sequential separation of LREE, Th and U by miniaturized extraction chromatography: application to isotopic analyses of silicate rocks. Anal. Chim. Acta 339, 79–89.

Pourmand, A., Daupas, N., Ireland, J.T., 2012. A novel extraction chromatography and MC-ICP-MS technique for rapid analysis of REE, Sc and Y: revising CI-chondrite and post-Archean Australian shale. Chem. Geol. 291, 38–54.

Ramírez Merino, J.I., Olivé Davó, A., Álvaro López, M., Hernández Samaniego, A., 1990. Mapa Geológico de España E. 1:50.000. Memoria explicativa de la Hoja 241 (Anguiano). Inst. Tecnol. Geomin. España, 1–63.

Rasmussen, B., Buick, R., Taylor, W.R., 1998. Removal of oceanic REE by authigenic precipitation of phosphatic minerals. Earth Planet. Sci. Lett. 164, 135–149.

Rodriguez Alonso, M.D., 1985. El complejo esquisto-grauváquico y el Paleozoico en el centro-oeste español. Acta Salmanticensia Ciencias 51, 1–174.

Rodríguez Alonso, M.D., Peinado, M., López-Plaza, M., Franco, P., Carnicero, A., Gonzalo, J.C., 2004. Neoproterozoic-Cambrian synsedimentary magmatism in the central Iberian zone (Spain): geology, petrology and geodynamic significance. Int. J. Earth Sci. 93, 897–920.

Rodríguez-Fernández, L.R., Oliveira, J.T. (Eds.), 2015. Mapa Geológico de España y Portugal E 1:1.000.000. Instituto Geológico Minero de España (IGME) y Laboratório Nacional de Energía e Geologia de Portugal (LNGP).

Rojo-Pérez, E., Arenas, R., Fuenlabrada, J.M., Sánchez Martínez, S., Martín Parra, L.M., Matas, J., Pieren, A.P., Diez Fernández, R., 2019. Contrasting isotopic sources (Sm-Nd) of late Ediacaran series in the Iberian Massif: implications for the Central-Iberian Ossa Morena boundary. Precambrian Res. 324, 194–207.

Rollinson, H., 1993. Using geochemical data: evaluation, presentation, interpretation. Longman, New York, pp. 1–352.

Roser, B.P., Korsch, R.J., 1988. Provenance signatures of sandstone-mudstone suites determined using discriminant function analysis of major element data. Chem. Geol. 67, 119–139.

Rubio-Ordóñez, A., Barba, P., Cuesta, A., Gallastegui, G., Suárez, O., Ugidos, J.M., Valladares, M.I., 2004. Los cantos vocánicos del conglomerado basal de la Fm. Herrería: Evidencias de un vulcanismo Neoproterozoico en la base del Cámbrico. Geogaceta 36, 11–14.

Rubio-Ordóñez, A., Gutiérrez Alonso, G., Valverde-Vaquero, P., Cuesta, A., Gallastegui, G., Gerdes, A., Cárdenes, V., 2015. Arc-related Ediacaran magmatism along the northern margin of Gondwana. Geochronology and isotopic geochemistry from northern Iberia. Gondwana Res. 27, 216–227.

Ryan, K.M., Williams, D.M., 2007. Testing the reliability of discrimination diagrams for determining the tectonic depositional environment of ancient sedimentary basins. Chem. Geol. 242, 103–125.

Samuel, M.D., Be'eri-Shlevin, Y., Azer, M.K., Whitehouse, M.J., Moussa, H.E., 2011. Provenance of conglomerate clasts from the volcano-sedimentary sequence at Wadi Rutig in southern Sinai, Egypt, as revealed by SIMS U-Pb dating of zircon. Gondw. Res. 20, 450–464.

San José, M.A., Pieren, A.P., García Hidalgo, F.J., Vilas, L., Herranz, P., Pelaez, J.R., Perejón, A., 1990. Central Iberian Zone. Ante-Ordovician stratigraphy. In: Dallmeye, R.D., Martínez-García, E. (Eds.), Pre-Mesozoic geology of Iberia. Springer-Verlag, Berlin, pp. 147–159.

Sandler, A., Teutsch, N., Avigad, D., 2012. Sub-Cambrian pedogenesis recorded in weathering profiles of the Arabian-Nubian shield. Sedimentology 59, 1305–1320.

Santamaría, J., 1995. Los yacimientos de fosfato sedimentario en el límite Precámbrico-Cámbrico del Anticlinal de Valdelacasa (Zona Centro Ibérica). PhD thesis, Universidad Autónoma de Barcelona. 233 pp.

Shalaby, A., Stüwe, K., Fritz, H., Makroum, F., 2006. The El Mayah molasse basin in the Eastern Desert of Egypt. J. Afr. Earth Sci. 45, 1–15.

Shaw, J., Johnston, S.T., Gutiérrez-Alonso, G., 2012. Oroclines of the Variscan orogen of Iberia: Paleocurrent analysis and paleogeographic implications. Earth Planet. Sci. Lett. 329-330, 60–70.

Simón, J., 2017. Trace fossils and dubiofossils from the Ediacaran and Cambrian of the Alcudia anticline, Spain. Estudios Geológicos 73 (2), 1–11.

Stephan, T., Kroner, U., Romer, R.L., 2019. The pre-orogenic detrital zircon record of the Peri-Gondwanan crust. Geol. Mag. 156, 281–307.

Stern, R.J., 2002. Crustal evolution of the East African Orogen: a neodymium isotopic perspective. J. Afr. Earth Sci. 34, 109–117.

Stern, R.J., Ali, K.A., Liégeois, J.P., Johnson, P., Wiescek, F., Kattan, F., 2010. Distribution and significance of pre-Neoproterozoic zircons in juvenile Neoproterozoic igneous rocks of the Arabian-Nubian shield. Am. J. Sci. 310, 791–811.

Talavera, C., Montero, P., Martínez Poyatos, D., Williams, I.S., 2012. Ediacaran to lower Ordovician age for rocks ascribed to the schist-Graywacke complex (Iberian massif, Spain): evidence from detrital zircon SHRIMP U-Pb geochronology. Gondw. Res. 22, 928–942.

Talavera, C., Montero, P., Bea, F., González Lodeiro, F., Whitehouse, M., 2013. U-Pb zircon geochronology of the Cambro-Ordovician metagranites and metavolcanic roks of central and NW Iberia. Int. J. Earth Sci. 102, 1–23.

Talavera, C., Martínez Poyatos, D., González Lodeiro, F., 2015. SHRIMP U-Pb geochronological constraints on the timing of the intra-Alcudian (Cadomian) angular unconformity in the Central Iberian Zone (Iberian Massif, Spain). Int. J. Earth Sci. 104, 1739–1757.

Tassinari, C.C.G., Medina, J., Pinto, M.S., 1996. Rb-Sr and Sm-Nd geochronology and isotope geochemistry of Central Iberian metasedimentary rocks (Portugal). Geologie en Mijnbouw 75, 69–79.

Teixeira, R.J.S., Neiva, A.M.R., Silva, P.B., Gomes, M.E.P., Andersen, T., Ramos, J.M.F., 2011. Combined U-Pb geochronology and Lu-Hf isotope systematics by LAM-ICPMS of zircons from granites and metasedimentary rocks of Carrazeda de Ansiaes and Sabugal areas, Portugal, to constrain granite sources. Lithos 125, 121–334.

Teklay, M., Kröner, A., Mezger, K., Oberhänsli, R., 1998. Geochemistry, Pb–Pb single zircon ages and Nd–Sr isotope composition of Precambrian rocks from southern and eastern Ethiopia: implications for crustal evolution in east Africa. J. Afr. Earth Sci. 26, 207–227.

Thirlwall, M.F., 1991. Long-term reproductivility of multicollector Sr and Nd isotope ratio analysis. Chem. Geol. 94, 85–104.

Torsvik, T.H., Smethurst, M.A., Meert, J.G., Van der Voo, R., McKerrow, W.S., Brasier, M.D., Sturte, B.A., Walderhaug, H.J., 1996. Continental break-up and collision in the Neoproterozoic and Paleozoic—a tale of Baltic and Laurentia. Earth Sci. Rev. 40, 229–258.

Ugidos, J.M., Valladares, M.I., Recio, C., Rogers, G., Fallick, A.E., Stephens, W.E., 1997a. Provenance of upper Precambrian-lower Cambrian shales in the central Iberian zone, Spain: evidence from a chemical and isotopic study. Chem. Geol. 136, 55–70.

Ugidos, J.M., Armenteros, I., Barba, P., Valladares, M.I., Colmenero, J.R., 1997b. Geochemistry and petrology of recycled orogen-derived sediments: a case study from upper Precambrian siliciclastic rocks of the central Iberian Zone, Iberian Massif, Spain. Precambrian Res. 84, 163–180.

Ugidos, J.M., Valladares, M.I., Barba, P., Ellam, R.M., 2003. The Upper Neoproterozoic-Lower Cambrian of the Central Iberian Zone, Spain: Chemical and isotopic (Sm-Nd) evidence that the sedimentary succession records an inverted stratigraphy of its source. Geochim. Cosmochim. Acta 67, 2615–2629. Erratum: 2004, Geochim. Cosmochim. Acta 68, p. 1953.

Ugidos, J.M., Stephens, W.E., Carnicero, A., Ellam, R.M., 2009. A reactive assimilation model for regional-scale cordierite-bearing granitoids: geochemical evidence from the late Variscan granites of the central Iberian zone, Spain. Earth Environ. Sci. Trans. R. Soc. Edinb. 99, 1–26.

Ugidos, J.M., Sánchez-Santos, J.M., Barba, P., Valladares, M.I., 2010. Upper Neoproterozoic series in the Central Iberian, Cantabrian and West Asturian Leonese zones (Spain): geochemical data and statistical results as evidence for a shared homogenised source area. Precambrian Res. 178, 51–58.

Ugidos, J.M., Barba, P., Valladares, M.I., Suárez, M., Ellam, R.M., 2016. The Ediacaran-Cambrian transition in the Cantabrian zone (North Spain): sub-Cambrian weathering, K-metasomatism and provenance of detrital series. J. Geol. Soc. Lond. 173, 603–615.

Valladares, M.I., 1995. Siliciclastic-carbonate slope apron in an immature tensional margin (upper Precambrian-lower Cambrian), central Iberian zone, Salamanca, Spain. Sediment. Geol. 94, 165–186.
Valladares, I., Ugidos, J.M., Recio, C., 1993. Criterios geoquímicos de correlación y posible área fuente de las pelitas del Precambrico Superior-Cámbrico Inferior en la Zona Centro Ibérica (Macizo Ibérico. España). Rev. Soc. Geol. Esp. 6 (1/2), 37–45.
Valladares, M.I., Barba, P., Colmenero, J.R., Armenteros, I., Ugidos, J.M., 1998. La sucesión sedimentaria del Precámbrico superior–Cámbrico inferior en el sector central de la Zona Centro Ibérica: Litoestratigrafía, geoquímica y facies sedimentarias. Rev. Soc. Geol. Esp. 11, 271–283.
Valladares, M.I., Barba, P., Ugidos, J.M., Colmenero, J.R., Armenteros, I., 2000. Upper Neoproterozoic-lower Cambrian sedimentary successions in the Central Iberian Zone (Spain): sequence stratigraphy, petrology and chemostratigraphy. Implications for other European zones. Int. J. Earth Sci. 89, 2–20.
Valladares, M.I., Ugidos, J.M., Barba, P., Colmenero, J.R., 2002a. Contrasting geochemical features of the central Iberian zone shales (Iberian massif, Spain): implications for the evolution of Neoproterozoic-lower Cambrian sediments and their sources in other Peri-Gondwanan areas. Tectonophysics 352, 121–132.
Valladares, M.I., Barba, P., Ugidos, J.M., 2002b. Precambrian. In: Gibbons, W., Moreno, T. (Eds.), The Geology of Spain. Geological Society, London, pp. 7–16.
Valladares, M.I., Ugidos, J.M., Barba, P., Fallick, A.E., Ellam, R.M., 2006. Oxygen, carbon and strontium isotope records of Ediacaran carbonates in Central Iberia (Spain). Precambrian Res. 147, 354–365.
van der Bosch, W.J., 1969. Geology of the Luna-Sil region, Cantabrian Mountains (Spain). Leidse Geol. Meded. 44, 137–225.
Van Wagoner, J.C., Posamentier, H.W., Mitchum, R.M., 1988. An overview of the fundamentals of sequence stratigraphy and key definitions. In: Wilgus, C.K., Hasstings, B.S., Kendall, C.G.S.C., Posamentier, H.W., Ross, C.A., Van Wagoner, J.C. (Eds.), Sea-Level Changes: An Integrated Approach. 42. SEMP, Spec. Pub, pp. 39–45.
Veizer, J., Jansen, S.L., 1985. Basement and recycling, 2. Time dimension to global tectonics. J. Geol. 93, 625–643.
Velvel, M.A., 1985. Geochemical mass balances and weathering rates in forested watersheds of the southern blue ridge. Am. J. Sci. 285, 904–930.
Vidal, G., Palacios, T., Gámez-Vintaned, J.A., Díez Balda, M.A., Grant, S.W.F., 1994. Neoproterozoic-early Cambrian geology and paleontology of Iberia. Geol. Mag. 131, 729–765.
Villaseca, C., Merino, E., Oyarzun, R., Orejana, D., Pérez-Soba, C., Chicharro, E., 2014. Contrasting chemical and isotopic signatures from Neoproterozoic metasedimentary rocks in the central Iberian zone (Spain) of pre-Variscan Europe: implications for terrane analysis and early Ordovician magmatic belts. Precambrian Res. 245, 131–145.
Waggoner, B., 1999. Biogeographic analyses of the Ediacara biota: a conflict with paleotectonic reconstructions. Paleobiology 25 (4), 440–458.
Waggoner, B., 2003. The Ediacaran biotas in space and time. Integr. Comp. Biol. 43, 104–113.
Wasserburg, G.J., Jacobsen, S.B., DePaolo, D.J., McCulloch, M.T., Wen, T., 1981. Precise determination of Sm/Nd ratios, Sm and Nd isotopic abundances in standard solutions. Geochim. Cosmochim. Acta 45, 2311–2323.
Wilde, S.A., Youssef, K., 2002. A re-evaluation of the origin and setting of the late Precambrian Hammamat group based on SHRIMP U-Pb dating of detrital zircons from Gebel umm Tawat, north Eastern Desert, Egypt. J. Geol. Soc. Lond. 159, 595–604.

Winchester, J.A., Max, M.D., 1989. Tectonic setting discrimination in clastic sequences: an example from the late Proterozoic Erris group, NW Ireland. Precambrian Res. 45, 191–201.

Woldemichael, B.W., Kimura, J.I., Dunkley, D.J., Tani, K., Ohira, H., 2010. SHRIMP U–Pb zircon geochronology and Sr–Nd isotopic systematics of the Neoproterozoic Ghimbi-Nedjo mafic to intermediate intrusions of Western Ethiopia: a record of passive margin magmatism at 855 Ma? Int. J. Earth Sci. 99, 1773–1790.

Wood, R.A., 2011. Paleoecology for the earliest skeletal metazoan communities: implications for early biomineralization. Earth Sci. Rev. 106, 184–190.

Yang, S., Zhang, F., Wang, Z., 2012. Grain size distribution and age population of detrital zircons from the Changjiang (Yangtze) river system. China. Chem. Geol. 296–297, 26–38.

Yeshanew, F.G., Pease, V., Abdelsalam, M.G., Whitehouse, M.J., 2017. Zircon U-Pb ages, $\delta^{18}O$ and whole-rock Nd isotopic compositions of the Dire Dawa Precambrian basement, eastern Ethiopia: implications for the assembly of Gondwana. J. Geol. Soc. Lond. 174, 142–156.

Zamarreño, I., Vegas, R., Moreno, F., 1976. El nivel carbonatado de los Navalucillos y su posición en la sucesión Cámbrica de los Montes de Toledo occidentales (Centro de España). Breviora Geológica Astúrica 20, 56–64.

Zeck, H.P., Wingate, M.T.D., Pooley, G.D., Ugidos, J.M., 2004. A sequence of pan-African and Hercynian events recorded in zircons from an orthogneiss from the Hercynian Belt of Western Central Iberia—an ion microprobe U-Pb study. J. Petrol. 45, 1613–1629.

Zhao, J.X., McCulloch, M.T., Bennett, V.C., 1992. Sm-Nd and U-Pb zircon isotopic constraints on the provenance of sediments from the Amadeus Basin, Central Australia: evidence for REE fractionation. Geochim. Cosmochim. Acta 56, 921–940.

Zhuravlev, A.Y., Liñán, E., Gámez-Vintaned, J.A., Debrenne, F., Fedorov, A.B., 2012. New finds of skeletal fossils in the terminal Neoproterozoic of the Siberian platform and Spain. Acta Paleontol. Pol. 57, 205–224.

Zimmermann, U., Andersen, T., Madland, M.V., Larsen, I.S., 2015. The role of U-Pb ages of detrital zircons in sedimentology—an alarming case study for the impact of sampling for provenance interpretation. Sediment. Geol. 320, 38–50.

CHAPTER FIVE

Calibrating $\delta^{13}C$ and $\delta^{18}O$ chemostratigraphic correlations across Cambrian strata of SW Europe and Morocco, West Gondwana

J. Javier Álvaro*
Instituto de Geociencias (CSIC-UCM), Madrid, Spain
*Corresponding author: e-mail address: jj.alvaro@csic.es

Contents

Abstract

The Cambrian remains one of the least constrained Phanerozoic periods from the International Geological Timescale. Several global boundaries of series and stages are

Stratigraphy & Timescales, Volume 5
ISSN 2468-5178
https://doi.org/10.1016/bs.sats.2020.08.003

still awaiting formal definition and ratification. Particularly in West Gondwana, a major obstacle to the development of a common regional Cambrian timescale and its global correlation is the endemic character of its shelly fauna (mainly trilobites, archaeocyaths, echinoderms and brachiopods) due to the rare exposure of fossiliferous slope-to-basinal strata bearing cosmopolitan taxa. Over the past two decades, there has been a growing interest in the application of isotopic chemostratigraphic methods to regional studies, in an effort to solve and calibrate the timescale resolution in the Cambrian Atlas–Ossa-Morena rift axes and their lateral flanks preserved in Morocco, the Iberian Peninsula and South France. A critical re-appraisal of the resulting chemostratigraphic methods and results applied to the recognition and regional correlation of the Ediacaran–Cambrian (BACE event), Cambrian Series 2–Miaolingian (ROECE) and Miaolingian–Furongian (SPICE) boundaries is presented here. The reliability of isotopic data is mainly controlled by (i) the preferential analysis of bulk rock in carbonates and shales with total organic contents (TOC) lesser than 1 wt%; (ii) the unequal record of diagenetic (e.g., early vs. late diagenetic dolomitization) and metamorphic processes related to Variscan tectono-thermal events; and (iii) dating uncertainty of chemostratigraphic fluctuations by chronostratigraphic and radiometric constraints.

1. Introduction

Fluctuations in the oceanic stable isotope composition of dissolved inorganic carbon (DIC) are currently used as indicators of environmental changes in oceanic waters and substrates (e.g., Weissert et al., 2008). However, due to the diagenetic and metamorphic modifications recorded in marine carbonate and shale strata, the original composition of DIC in seawater is difficult to interpolate in Cambrian case studies. Carbon and oxygen isotope values are affected by thermal burial effects and interaction with circulating diagenetic, hydrothermal and metamorphic fluids. As a result, the absolute isotopic values are commonly modified through time. However, if they are affected in a similar way throughout the whole study sections, the $\delta^{13}C$ fluctuations (not the absolute values) are not predominantly induced by diagenetic alteration and the resulting fluctuations are useful for chemostratigraphic correlation (e.g., Immenhauser et al., 2008; Veizer et al., 1997, 1999).

This has been the procedure followed for worldwide Cambrian chemostratigraphic correlations based on carbon isotopes: the chemostratigraphic fluctuations are somewhat depleted in orogenic belts, such as those preserved in the Variscan Orogen of south-western Europe and the Moroccan High Atlas, by comparison with those somewhat protected from their tectonothermal effects, such as the Anti-Atlas and the worldwide referenced

South China continent (Álvaro et al., 2008c; Maloof et al., 2005; Wotte et al., 2007; Zhu et al., 2004). In areas strongly affected by the Variscan orogeny, two subsequent chemostratigraphic controls are currently neglected: (i) carbon isotope variations are predominantly studied in either bulk carbonate material or shales with total organic contents (TOC) lesser than 1 wt%; and (ii) chemostratigraphic methods are applied to stratigraphic logs with poor chronostratigraphic (based on their endemic fossil record) and radiometric constrains. As a result, carbon isotope stratigraphy is, in some cases, based on the correlation of similar fluctuations ("wiggle matching" or "blind correlation" effects) within a dating uncertainty (Weissert et al., 2008). This is particularly problematic in active rifting systems, where the shallow-water carbon isotope pool of DIC is affected to varying degrees by local carbon cycle processes, *syn*depositional diagenesis and fractionation effects associated with varying contributions of aragonite, low- and high-magnesium calcite, and dolomite (Immenhauser et al., 2008). Moreover, the shape of shallow-water carbon isotope curves is strongly modulated by sharp modifications in accommodation space, many of them tectonically enhanced by the breakdown of rift branches into horst-and-graben palaeotopographies. During the pre-rift, *syn*-rift and break-up events, subaerial exposure and re-flooding of the basement palaeorelief generate episodes of non-deposition, erosion, reworking and bypass of sediments (reports from the Atlas–Ossa-Morena Rift occur in Álvaro and Subías, 2011; Álvaro, 2013; Álvaro and Vizcaïno, 2018). As a consequence, the carbon isotope signal may be commonly lost in unconformity surfaces or condensed in rift-related strata. Hence "wiggle matching" between carbon isotope curves, with locally different absolute values but similar gradients and fluctuations, allows multiple stratigraphic interpretations (Strasser, 2015).

This paper synthesizes the carbon and oxygen isotope data from Cambrian carbonate and shale strata of Morocco, Spain and southern France published in recent years. A chemostratigraphic correlation of the interval across rift axes and fringing stable platform-block transects is provided, on the basis of chronostratigraphic charts and radiometric dating, where available. A critical re-appraisal of the reliability of isotopic values ($\delta^{13}C_{carb}$ vs. $\delta^{18}O_{carb}$ and $\delta^{13}C_{org}$ vs. TOC) is discussed.

2. Geological and geodynamic background

The Cambrian palaeogeographic margin of West Gondwana selected here includes the so-called Atlas–Ossa-Morena Rift and its platform-block flanks fringing the Gondwanan coastline, which is exposed at present in Morocco, the Iberian Peninsula and South France (Álvaro et al., 2014c, 2018b; Sánchez-García et al., 2019). In Morocco, the Anti-Atlas Ranges lie south of a major inherited swarm of faults, known as the South Atlas Fault, which was active during late Ediacaran and Cambrian times, separating a shallow-water (or epeiric) part of the Atlas rift transect from its basinal counterpart or northern offshore High Atlas-Coastal Meseta rift transect (Álvaro et al., 2014c). The South Atlas Fault also marks an inherited structural contact between a northern domain, with sparsely fossiliferous and Variscan–controlled highly deformed and metamosphosed rocks, and a southern domain with unmetamorphosed, fossiliferous rocks that have recorded only a weak Variscan–Alleghanian (Late Devonian–Carboniferous) deformation (Michard et al., 2008).

In south-western Europe, the Variscan Ibero-Armorican Arc separates two branches: (i) the western branch comprises the Iberian Massif and its lateral prolongation into the Iberian Chains; and (ii) the eastern branch includes the Pyrenees and the southern edge of the French Massif Central, or Montagne Noire (Fig. 1A). In the Iberian Massif, the main Variscan tectonostratigraphic domains, commonly used to distinguish Cambrian palaeogeographic transects of the peri-Gondwana margin are, from NE to SW, the Cantabrian, West Asturian-Leonese, Central Iberian and Ossa-Morena Zones. The contact between the Cantabrian and the West Asturian-Leonese Zones is extended to the SE into the Datos Fault (Iberian Chains; Álvaro et al., 2018a). Both the South Portuguese (lacking exposed Cambro–Ordovician rocks) and the Galicia-Trás-os-Montes Zones amalgamated against this margin of Gondwana as a result of the Variscan orogeny (Fig. 1A).

As stated above, an active network of Cambrian rift axes developed in West Gondwana, linking the Ossa-Morena Zone to the south-western High Atlas-Coastal Meseta and Anti-Atlas rift transects, forming the so-called Atlas–Ossa-Morena Rift. In the Iberian Massif, a proximal-distal gradient of the proximal stable margin can be recognized following a NE-SW gradient formed by the Cantabrian, West Asturian-Leonese and Central Iberian Zones, which were bounded to the NE by the Cantabro-Ebroan Source area (Fig. 1A), at present covered by the Cenozoic Ebro Basin, and to the SW by the (marginal) Ossa-Morena rift axes. In Morocco, source areas were located eastward of the preserved eastern edges of the Anti-Atlas, High Atlas and Coastal Meseta (Álvaro et al., 2014c; Álvaro and Debrenne, 2010; Gasquet et al., 2005; Michard et al., 2008; Pouclet et al., 2018). A latest Ediacaran–to–Fortunian pre-rift unconformity lies

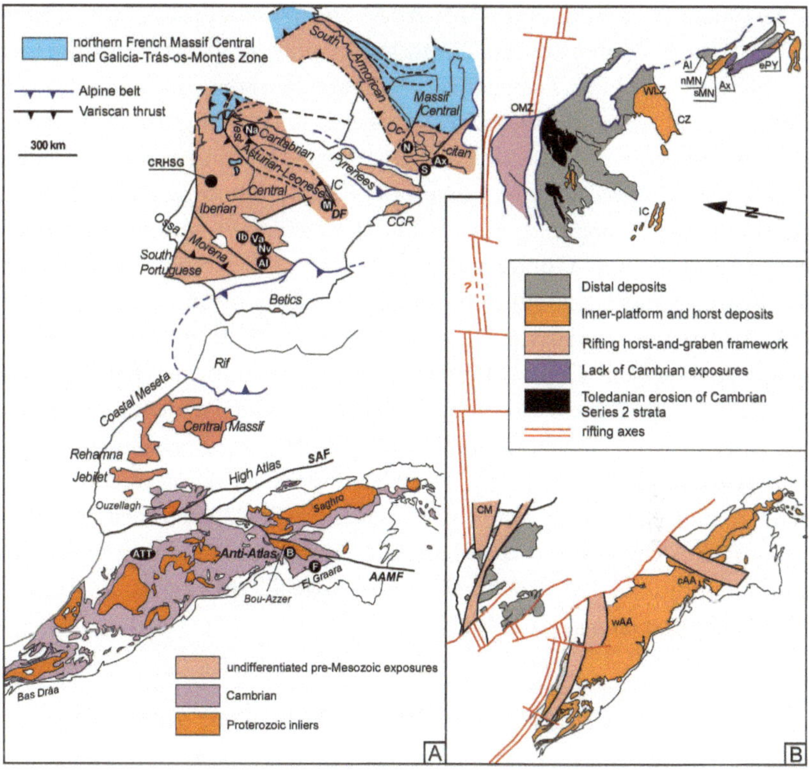

Fig. 1 (A) Geological setting of the study area in Morocco, the Iberian Peninsula and South France; abbreviations: *Al* Alcudia valley, *ATT* Amouslek-Tazzemourt-Tiout villages and Oued Sdas, *Ax* Axial Montagne Noire, *B* Boho volcano, *CCR* Coastal Catalan Range, *CRHSG* Ciudad Rodrigo-Hurdes-Sierra de Gata domain, *DF* Datos Fault, *E* Esla nappe, *F* Foum Zguid, Ib Ibor anticline, *IC* Iberian Chains, *M* Murero, *N* northern Montagne Noire (Albigeois), *Na* Narcea anticline, *Nv* Navalpino anticline, *S* southern Montagne Noire (Minervois and Pardailhan nappes) and *Va* Valdelacasa anticline. (B) Tentative palaeogeographic reconstruction of the Cambrian Atlas–Ossa-Morena Rift and flanking areas discussed in the text; abbreviations: *Al* Albigeois, *Ax* Axial Montagne Noire, *CZ* Cantabrian Zone, *ePY* eastern Pyrenees, *IC* Iberian Chains, *nMN* northern Montagne Noire, *OMZ* Ossa-Morena Zone, *sMN* southern Montagne Noire and *WALZ* West Asturian-Leonese Zone. *Based on Álvaro, J.J., Zamora, S., Clausen, S., Vizcaïno, D., Smith, A.B., 2013. The role of abiotic factors in the Cambrian Substrate Revolution: a review from the benthic community replacements of West Gondwana. Earth-Sci. Rev. 118, 69–82. Álvaro, J.J., Bauluz, B., Gil-Imaz, A., Ubide, T., 2018a. 40Ar/39Ar chronological constraints on syn- and post–Variscan biotite porphyroblasts from the Iberian chains, NE Spain. J. Iber. Geol. 44, 655–670. Álvaro, J.J., Casas, J.M., Clausen, S., Quesada, C., 2018b. Early Palaeozoic geodynamics in NW Gondwana. J. Iber. Geol. 44, 551–565. Pouclet, A., Álvaro, J.J., Bardintzeff, J.M., Gil Imaz, A., Monceret, E., Vizcaïno, D., 2017. Cambrian–early Ordovician volcanism across the south Armorican and Occitan domains of the Variscan Belt in France: continental break-up and rifting of the northern Gondwana margin. Geosci.*

(Continued)

at the Ouarzazate/Adoudou (Anti-Atlas) and Malcocinado/Torreárboles-Bodonal (Ossa-Morena Zone) lithostratigraphic contacts (Álvaro et al., 2014b). The overlying Terreneuvian–Miaolingian strata are punctuated by several *syn*-rift unconformities, recognizable as both angular discordances and paraconformities, and associated with tilting/uplifting events, and volcanic and hydrothermal activity. For about 60 m.y., the rifting processes propagated "zip-like" along the main axes and then widened cratonward affecting inner parts of the peri-Gondwana margin, such as the Central Iberian Zone (Sánchez-García et al., 2019). A Furongian break-up (or rift/drift) unconformity is well represented in the Ossa-Morena (Main Rifting Volcanic Group/Barriga contact) and Central Iberian (Cortijos/Armorican Quartzite contact) Zones and in the Anti-Atlas (Tabanite/Lower Fezouata contact) of Morocco, incompletely recorded in the Cantabrian Zone and absent in the remaining sectors of the Iberian Massif. The break-up unconformity, related to thermal doming, uplift and denudation of elevated areas, is regionally known as the Toledanian Phase and marks the beginning of passive-margin conditions (Álvaro et al., 2007, 2018b; Álvaro and Vizcaïno, 2018; Sánchez-García et al., 2019).

3. Methodological problems—Sampling strategy

Two methodological problems, related to the selection of both bulk rock for isotopic analysis of carbonate samples bearing complex diagenetic and metamorphic processes, and shale samples with very low TOC content, are discussed below.

3.1 Isotopic analysis of bulk rock in carbonates

In the Variscan tectonostratigraphic units described above, the Cambrian carbonate strata selected for carbon and oxygen isotopic analysis are in need of careful diagenetic, metamorphic and hydrothermal evaluation.

Fig. 1—Cont'd *Front. 8, 25–64. Pouclet, A., El Hadi, H., Álvaro, J.J., Bardintzeff, J.M., Benharref, M., Fekkak, A., 2018. Review of the Cambrian volcanic activity in Morocco: geochemical fingerprints and geotectonic implications for the rifting of West Gondwana. Int. J. Earth Sci. 107, 2101–2123. Sánchez-García, T., Chichorro, M., Solá, R., Álvaro, J.J., Díez Montes, A., Bellido F., Ribeiro, M.L., Quesada, C., Lopes, J.C., Dias da Silva, I., González-Clavijo, E., Gómez Barreiro, J., López-Carmona, A., 2019. The Ediacaran–early Ordovician rift stage in the Gondwanan units of the Iberian Massif. In: Quesada, C., Oliveira, J.T. (Eds.), The Geology of Iberia: A Geodynamic Approach. Regional Geology Review Series, vol. 2. Springer, Heidelberg, pp. 27–74.*

Diagenesis and metamorphism are inexorable severe problems because of the interchange between interstitial waters and carbonate rocks. This is particularly evident for negative carbon-isotope shifts, which may either reflect a real change in ocean chemistry or represent diagenetic artifacts caused by the introduction of isotopically light cements (Marshall, 1992).

Introduction of meteoric water causes major isotopic readjustment of carbonate phases, as the freshwater influence is commonly responsible for depleting $\delta^{13}C_{carb}$ values. The influence of meteoric waters can be related to: (i) contemporaneous effects of subaerially exposed substrates during relative sea-level falls, which may be synchronously distributed worldwide if related to eustatic fluctuations (Allen and Matthews, 1982; Immenhauser et al., 2008; Maliva et al., 1995); (ii) short-term uplift and tilting related to *syn*-rift and break-up events (the latter named Toledanian Phase in the Anti-Atlas and the Ossa-Morena and Central Iberian Zones); and (iii) late-diagenetic processes associated with general exhumation of the Cambrian basement and tectono-thermal events associated with the Variscan orogeny, mainly reflected as meteoric dedolomitization processes (Álvaro et al., 2020; Álvaro and González-Acebrón, 2019). Other contemporaneous processes related to syn-rift and break-up events in the Atlas–Ossa-Morena Rift are related to hydrothermal alteration in the proximity of polymetallic ore stockwork deposits, whereas further diagenetic processes, such as recrystallization and dolomitization, progressively increase at higher temperatures in burial-diagenetic environments (Álvaro, 2013; Álvaro and Subías, 2011). The adverse effect of sediment burial, controlling pressure and temperature as well as percolating fluids, is currently considered to reset a large portion of the carbonate isotopic archive.

In marine carbonates, it is widely accepted that carbon isotopes are much less prone to diagenetic alteration than oxygen isotopes, because pore-water in the sediments generally contains little organic matter, the carbon isotope system is rock-dominated, and carbon isotopes show no significant temperature-controlled fractionation during burial (Marshall, 1992). In carbonates, $\delta^{18}O$ values currently relate to: (i) the evaporation/precipitation balance in seawater and palaeotemperature, both potentially dependent on latitude and climate patterns; and (ii) the influence of diagenetic, metamorphic and hydrothermal fluids. The fluids are responsible for solution and precipitation of calcite (including late-diagenetic dedolomite) and dolomite, each cement characterized by its own $\delta^{13}C/\delta^{18}O$ ratio. In general, positive correlation (quantified with linear correlation coefficients) between carbon and oxygen values mainly indicates the infiltration of fluids containing isotopically distinct carbon and oxygen datasets. The absence of such a positive correlation may suggest that the carbon isotopic values remained preferentially unchanged during early or burial diagenesis. Other chemical analyses, such as Sr isotopes and Mn/Sr ratios, are also useful for the evaluation of diagenetic changes in carbonate rocks. For instance, limestones deposited under marine environments with Mn/Sr ratios <2 indicate that they were not subjected to significant diagenesis (Jacobsen and Kaufman, 1999; Kaufman and Knoll, 1995).

Microbial carbonates, displaying stromatoid and thromboid textures currently known as stromatolites and thrombolites, are also problematic for chemostratigraphic comparisons. These are overabundant in Terreneuvian (Adoudou and Lie-de-vin formations) and Cambrian Series 2 (Igoudine, Amouslek and Issafen formations) in the Anti-Atlas, due to the nucleation and growth of microbial and microbial-archaeocyathan reefs (Álvaro et al., 2006a, 2008c, 2014c; Álvaro and Debrenne, 2010; Monninger, 1979). $\delta^{13}C$ and $\delta^{18}O$ values of microbial carbonates can be biased by CO_2 release from organic impurities, which becomes critical for carbonate isotope analyses from bulk sediments containing comparably high amounts of microbial (organic) content. Although several methods have been developed to remove organic matter prior to analyses, pre-treatment methods cause isotopic bias in themselves (Oehlerich et al., 2013).

In summary, given the bulk rock multicomponent nature of many of the Cambrian carbonate samples analyzed for chemostratigraphic purposes detailed below, secondary signals entrained by diagenetic, metamorphic and hydrothermal processes should be distinguished before selecting micritic rock vs. bulk rock for isotopic analysis.

3.2 Isotopic analysis of organic matter in TOC-poor shales

In addition to carbonates, carbon isotopic changes in the composition of the organic matter preserved in shales is also useful for chemostratigraphic correlation. Controls on the $\delta^{13}C$ composition of preserved organic matter are considerably more complex than those on carbonate and can reflect, in addition to DIC composition of seawater, changes in the marine/terrestrial organic carbon ratio, modification in primary producers with different selectivity against ^{13}C, or post-burial bacterial reworking of the organic matter (Zhang et al., 2018). For sediments containing relatively high (5–10%) TOC levels, pyrolysis and thermal analysis are applied. However, many of these techniques are not sensitive enough to be applied directly to the bulk material of TOC-poor (<1%) shales, necessitating the use of extractions to separate organic-matter compounds (Fox et al., 2017). As standard minimum requirement, shales targeted for chemostratigraphic analysis should contain at least >1% TOC. Many ancient shales that once contained >1% TOC may now have <1% TOC because of organic content loss during thermal maturation or metamorphism (Raiswell and Berner, 1987).

Paired carbonate and organic ($\delta^{13}C_{carb}$ vs. $\delta^{13}C_{org}$) records can exhibit coupled negative excursions resulting from multiple episodes of both meteoric alteration of the carbonate $\delta^{13}C$ record, and coeval contributions of isotopically depleted organic matter to the sedimentary record. Classically, covariant $\delta^{13}C_{carb}$ and $\delta^{13}C_{org}$ records are interpreted as evidence that both the carbonate and organic matter were originally produced in the surface waters of the ocean, and that they have retained their original $\delta^{13}C$ composition. In contrast, decoupled $\delta^{13}C_{carb}$ and $\delta^{13}C_{org}$ records have been interpreted as evidence for diagenetic alteration, or for local *syn*-sedimentary processes making the $\delta^{13}C_{org}$ record noisy (Maloof et al., 2010; Oehlert and Swart, 2014). Recently, Lin et al. (2019) have highlighted the decoupling of organic and inorganic carbon isotopes in Terreneuvian and Cambrian Series 2 sections. Although the origin of decoupled $\delta^{13}C_{carb}$ and $\delta^{13}C_{org}$ remains unresolved, possible causes may be linked to significant chemoautotrophic-methanotrophic biomass contribution to bulk organic matter in anoxic-euxinic environments (Hayes et al., 1999), contribution of dissolved organic carbon in the ocean (Rothman et al., 2003), recycled organic matter (Jiang et al., 2010), influence of detrital organic carbon (Dilliard et al., 2007), and/or diagenetic alteration of organic matter and/or amplification of detrital organic carbon isotope signature in low-TOC carbonates (Jiang et al., 2012).

4. Chronostratigraphic and radiometric constraints on Cambrian Series boundary intervals

The $\delta^{13}C_{cab}$ and $\delta^{13}C_{org}$ chemostratigraphic analyses documented below have been made in Cambrian strata of the Anti-Atlas and High Atlas in Morocco (Álvaro et al., 2008c; Maloof et al., 2005), the Cantabrian (Wotte et al., 2007) and Central Iberian Zones (Valladares et al., 2000, 2006) and the Iberian Chains (Gozalo et al., 2013) in Spain, and the southern Montagne Noire in France (Álvaro et al., 2008c; Wotte et al., 2007). An updated revision of the Cambrian Series boundaries, based on chronostratigraphic and radiometric data, will allow improving a better control on the chemostratigraphic logs studied so far in Morocco and SW Europe.

4.1 Ediacaran–Terreneuvian transition

A pronounced negative carbon isotope shift (the so-called Basal Cambrian Isotope Excursion or BACE event; Zhu et al., 2004), nearly coincident with the Ediacaran–Cambrian boundary at ca. 543–542 Ma, has been recognized worldwide (e.g., Bowring et al., 1993; Grotzinger et al., 1995). The presence of carbonate successions potentially crossing the Ediacaran–Cambrian boundary interval in the Anti-Atlas (Adoudou Formation) and the Central Iberian Zone (Ibor Group) has encouraged the search of this chemostratigraphic shift. The relative synchronicity of the first appearance of *Treptichnus pedum* and the negative $\delta^{13}C$ excursion can be tested in the Central Iberian Zone.

In the Anti-Atlas and High Atlas Ranges of Morocco, the location of the Ediacaran–Cambrian boundary has been an everlasting matter of discussion due to the interplay of carbonate productivity and volcanic record in the Atlas rift branches. The Adoudou (Tabia and Tifnout members) and Lie-de-vin formations, sandwiched between the Ouarzazate Supergroup dated from about 615–548 Ma (Gasquet et al., 2005, 2008; Thomas et al., 2002; Walsh et al., 2012) and the Igoudine Formation bearing the oldest (Cambrian Epoch 2) trilobites and archaeocyaths of the Anti-Atlas (Álvaro et al., 2006a; Debrenne and Debrenne, 1995; Geyer, 2019), are devoid of chronostratigraphically significant fossils. The Ediacaran–Cambrian boundary has been tentatively correlated with carbon isotope signatures at the base of the Tamjout Bed (Tifnout Member) in the western Anti-Atlas (Maloof et al., 2005) (Fig. 2). However, based on zircon data, this chemostratigraphic shift is not lithostratigraphically isochronous throughout the entire Anti-Atlas. The diachronism is based on the record of: (i) reworked zircon grains from a pyroclastic breccia encased in the Tabia Member, northwest of the Bou-Azzer inlier, which gave a weighted mean age of 541 ± 6 Ma (Blein et al., 2014; Chèvremont et al., 2013); hence the base of the Adoudou Formation post-dated the magmatic sources of these zircon populations; (ii) the youngest detrital zircon grains yielded by a sandstone bed from the upper volcanic Unit of the Ouarzazate Supergroup in the Bas Draâ inlier and dated at ca. 541 Ma (Karaoui et al., 2015); (iii) the youngest rhyolitic dyke encased in the Ouarzazate Supergroup from the northern Saghro inlier, dated

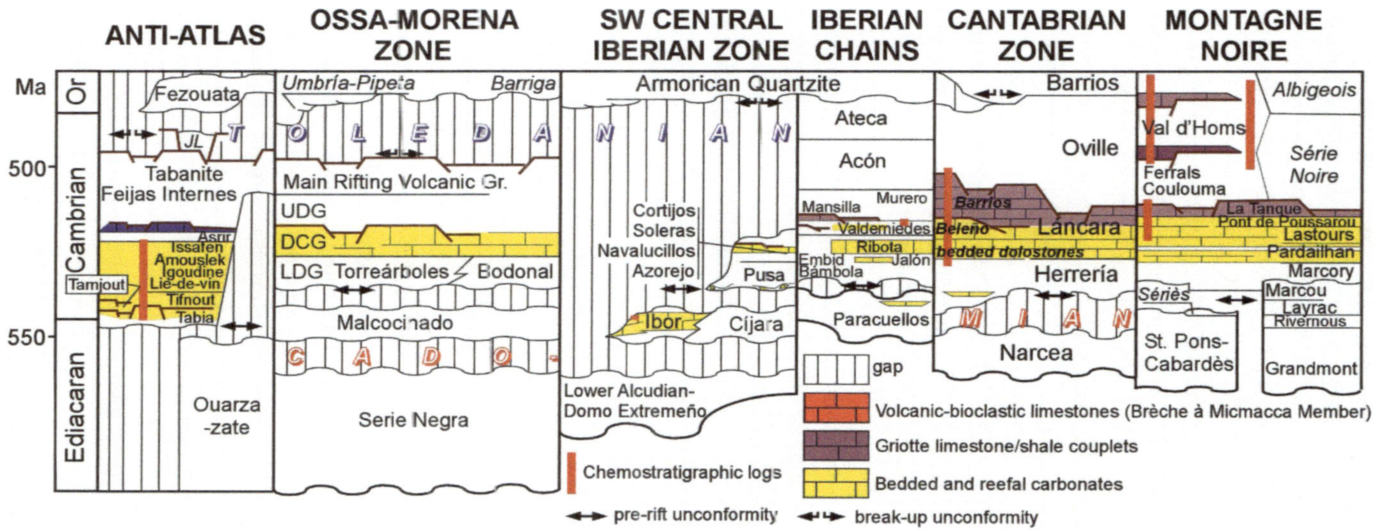

Fig. 2 Stratigraphic setting of the chemostratigraphic logs described in the text; abbreviations: *DCG* Detrital-Carbonate Group, *Gr.* Group, *JL* Jbel Lmgaysmat Formation (Tabanite Group), *LDG* Lower Detrital Group, *Or* Ordovician and *UDG* Upper Detrital Group. *Modified from Álvaro, J.J., Ferretti, A., González-Gómez, C., Serpagli, E., Tortello, M.F., Vecoli, M., Vizcaïno, D., 2007. A review of the late Cambrian (Furongian) palaeogeography in the western Mediterranean region, NW Gondwana. Earth-Sci. Rev. 85, 47–81. Álvaro, J.J., Bauluz, B., Clausen, S., Devaere, L., Gil Imaz, A., Monceret, E., Vizcaïno, D., 2014a. Stratigraphic review of the Cambrian–lower Ordovician volcanosedimentary complexes from the northern Montagne noire, France. Stratigraphy 11, 83–96. Álvaro, J.J., Bellido, F., Gasquet, D., Pereira, F., Quesada, C., Sánchez-García, T., 2014b. Diachronism of late Neoproterozoic-Cambrian arc-rift transition of North Gondwana: a comparison of Morocco and the Iberian Ossa-Morena zone. J. Afr. Earth Sci. 98, 113–132. Álvaro, J.J., Shields, G.A., Ahlberg, P., Jensen, S., Palacios, T., 2016. Ediacaran–Cambrian phosphorites from the western margins of Gondwana and Baltica. Sedimentology 63, 350–377. Álvaro, J.J., Casas, J.M., Clausen, S., Quesada, C., 2018b. Early Palaeozoic geodynamics in NW Gondwana. J. Iber. Geol. 44, 551–565. Álvaro, J.J., Bauluz, B., Gil-Imaz, A., Ubide, T., 2018a. 40Ar/39Ar chronological constraints on syn- and post–Variscan biotite porphyroblasts from the Iberian chains, NE Spain. J. Iber. Geol. 44, 655–670. Álvaro, J.J., Cortijo, I., Jensen, S., Lorenzo, S., Palacios, T., Pieren, A.P., 2019. Updated stratigraphic framework and biota of the Ediacaran and Terreneuvian in the Alcudia-Toledo Mountains of the central Iberian zone, Spain. Estud. Geol. 75(2), e093.*

at 543±9 Ma (Gasquet et al., 2005); and (iv) the magmatic zircon populations from the Boho volcano lavas onlapped by the upper part of the Tamjout Bed and the overlying Lie-de-vin Formation giving similar 529±3 and 531±5 Ma ages (Álvaro et al., 2006b, 2014c; Ducrot and Lancelot, 1977; Gasquet et al., 2005).

Based on these radiometric data, the Adoudou Formation should be placed in the Fortunian Stage, and the whole Taroudant Group (Adoudou and Lie-de-vin formations) in the Terreneuvian Series (Pouclet et al., 2018). This dating is supported by further U—Pb ages, such as the Azegour dacite from the Ouzellagh inlier (ca. 532; Pouclet et al., 2008), the Aghbar trachytic sills of the Bou Azzer-El Graara inlier (ca. 531 Ma; Gasquet et al., 2005), a felsic tuff from the Tifnout Member at Oued Sdas (ca. 525.3 Ma; Maloof et al., 2010), and several felsic tuffs from the Lie-de-vin Formation in the vicinity of Tiout and Oued Sdas, ranging from 525 to 520 Ma (Compston et al., 1992; Landing et al., 1998; Maloof et al., 2005).

In the Iberian Peninsula, the Precambrian–Cambrian boundary has been traditionally placed at distinct angular discordances and paraconformities related to the onset of Cadomian deformation. In the Narcea Antiform, which separates the Cantabrian and West Asturian-Leonese Zones, the Ediacaran Narcea Group is unconformably overlain by the lowermost Cambrian Herrería and Cándana formations (Fig. 2). Both the Narcea Group (up to 2000 m of monotonous greywackes and siltstones) and the Cándana Formation (up to 2500 m of sandstones and slates) are rather barren of fossils and volcanosedimentary interbeds, precluding any appropriate identification of the Ediacaran–Cambrian boundary. The Herrería Formation, 500–1400 m thick, consists of conglomerates, sandstones and shales with subsidiary phosphatic carbonate and paleosol interbeds (Álvaro, et al., 2003b; Aramburu et al., 2004). The lower member, up to 300 m thick, comprises a monotonous variegated shale interval interrupted by conglomeratic channels, sandstone bars and phosphatic carbonate interbeds. Palacios and Vidal (1992) reported the occurrence of *Treptichnus pedum* about 4 m above the basal angular discordance. The Narcea/Herrería contact represents an inherited palaeorelief, associated with the Cadomian orogeny, partly filled with gravel tuffaceous litharenites that reflect the reworking of late Ediacaran felsic volcanics (Álvaro et al., 2016). Some reworked clasts lining the Narcea/Herrería contact have yielded U—Pb zircon ages of ca. 575–560 Ma (Rubio-Ordóñez et al., 2015). Acritarchs from the lower Herrería member allows recognition of the *Granomarginata–Leiosphaeridia* Zone, those from the middle member the *Asteridium–Comasphaeridium*

Zone, and those from the upper member (which has also yielded trilobites; Liñán et al., 2015) the *Skiagia–Fumbriaglomerella* Zone (Palacios and Vidal, 1992; updated in Palacios et al., 2020). The three acritarch-based zones broadly represent the Fortunian, Cambrian Age 2 (Terreneuvian) and Cambrian Age 3 (Cambrian Epoch 2), respectively (Palacios et al., 2020).

The Datos Fault of the Iberian Chains represents the southeastern prolongation of the Cantabrian/West Asturian-Leonese tectonostratigraphic contact (Álvaro et al., 2018a). The contact of the Paracuellos Group and the overlying Bámbola Formation marks a tectono-thermal discontinuity associated with a Cadomian compressive deformation and a low-grade, intermediate-pressure metamorphism, absent in the Bámbola and overlying strata. The contact represents stratigraphically an erosive unconformity related to either the onset of alluvial plain sequences or synsedimentary faults (Álvaro et al., 2008a; Álvaro and Blanc-Valleron, 2002). The earliest ichnofossils with Cambrian significance occur in the upper part of the Bámbola Formation (Liñán et al., 1993), but recent re-assignations of the shelly microfossils yielded by the Codos phosphorite Bed of the Paracuellos Group suggest an earliest Cambrian age for them (Álvaro et al., 2016; Zhuravlev et al., 2012) (Fig. 2). As a result, the Ediacaran–Cambrian boundary lies within the Paracuellos Group, below the Cadomian unconformity that marks the Paracuellos/ Bámbola contact. The high accumulation rates related to the deposition of the alluvial-to-deltaic Bámbola Formation (reflecting the Cadomian aftermath) explain the delayed first occurrence of *Treptichnus pedum*.

The Central Iberian Zone offers a distinct stratigraphic framework for the Ediacaran–Cambrian transition, based on the onset of a retro-arc basin recording denudation of a neighboring Cadomian arc (Álvaro et al., 2019, 2020). Another stratigraphic peculiarity of the Central Iberian Zone is the presence of a Furongian gap; in some cases, the involved unconformity, capped by the Lower Ordovician Armorican Quartzite, is scouring and eroding the entire Cambrian and part of the uppermost Ediacaran. Two sectors of the Central Iberian Zone have received special attention to identify the stratigraphic setting of the Ediacaran–Cambrian boundary, the Toledo Mountains (including the Valdelacasa, Navalpino and Ibor anticlines and the Alcudia valley; for a recent revision, see Álvaro et al., 2019) and the Ciudad Rodrigo-Hurdes-Sierra de Gata domain (CRHSG in Fig. 1; Rodríguez Alonso et al., 2004).

A recent revision of the Ediacaran–Cambrian boundary interval in the Toledo Mountains (Álvaro et al., 2019) has recognized: (i) a latest Ediacaran biota in the Villarta Formation (Ibor Group), composed of the frame-building microfossil *Cloudina* (*C. carinata*, *C. hartmanae* and probably *C. xuanjiangpingensis*) associated with reefal dwellers, such as *Sinotubulites* (*S. baimatuoensis*) and *Protolagena* sp. (a problematic flask-shaped microfossil; Cortijo et al., 2010, 2015; Álvaro et al., 2020), interbedded and overlain by shales bearing vendotaenids and sabelliditids (Vidal et al., 1994a,b); and (ii) the occurrence of *Treptichnus pedum* on the basal shales of the overlying Arrocampo Formation cropping out in the Navalpino anticline (Álvaro et al., 2019; Jensen et al., 2007, 2010; Jensen and Palacios, 2016) (Fig. 2). Its first occurrence is strongly facies-controlled due to the predominance of carbonates in the underlying Villarta Formation, where earliest diagenetic lithification precludes its potential preservation. The lower member of the Villarta Formation contains *Cloudina*-thromboid reef complexes interbedded with simple ichnofossils that penetrate some 7 mm into the sediment pointing to a latest Ediacaran age (Álvaro et al., 2020). Due to the lack of radiometric ages across the Villarta and Arrocampo formations, the base of the Cambrian should be placed somewhere throughout the middle and upper members of the Villarta Formation, before the delayed first occurrence of the index ichnofossil *T. pedum*. Therefore, the Ediacaran–Cambrian boundary interval lies above the Lower Alcudian-Domo Extremeño/Ibor contact discordance, which marks the major Cadomian tectono-thermal episode, but below the Ibor/Pusa (pre-rift) unconformity (Álvaro et al., 2019) (Fig. 2).

In the CRHSG domain, four unnamed lithostratigraphic units (labeled I to IV) have been distinguished below the occurrence of Cambrian fossils (Valladares et al., 2000). They represent slope-related and base-of-slope environments capped by a major erosive unconformity. The latter is covered by megabreccia lenses, similar to those described in the Toledo Mountains (e.g., the Fuentes, Membrillar and Navalpino megabreccia beds) highlighting the pre-rift Ibor/Pusa contact. Allochthonous *Cloudina* shells have been described within some limestone boulders of the Unit V megabreccia (Vidal et al., 1994a,b).

In southern France, the Sériès Tuffs of the Montagne Noire Axial Zone and the Rivernous and Layrac formations of the northern Montagne Noire have yielded U—Pb zircon ages of ca. 542–539 Ma (Álvaro et al., 2014a; Lescuyer and Cocherie, 1992; Padel et al., 2017; Pouclet et al., 2017) reflecting the onset of Cadomian magmatic events across the Ediacaran–Cambrian boundary interval (Fig. 2). Carbonate production took place in the volcanic event aftermath, as documented by the presence of abundant microfossil remains in the overlying *Heraultia* Member representing the *Watsonella crosbyi–Oelandiella korobkovi* interval Zone, correlated with the base of Cambrian Age 2 (broad Tommotian) throughout Siberia, China, Mongolia and Avalonia (Devaere et al., 2013).

4.2 Cambrian Epoch 2–Miaolingian transition

Worldwide, a negative $\delta^{13}C_{carb}$ isotope shift, known as the Redlichiid-Olenellid Extinction Carbon isotope Excursion or ROECE (Zhu et al., 2004), is considered to reflect global transgression and related redox changes (Hough et al., 2006; Zhuravlev and Wood, 1996), succeeding the global Hawke Bay/Toyonian regression that marks the end of Cambrian Epoch 2. ROECE is associated with the extinction of trilobite redlichiids and oleneliids and the disappearance of microbial-archaeocyathan reefs (Montañez et al., 2000). In the study area, three tectonostratigraphic units have been selected for chemostratigraphic analysis across the Cambrian Epoch 2–Miaolingian (former "lower–middle Cambrian") transition: the Cantabrian Zone, the southern Montagne Noire and the Iberian Chains.

4.2.1 Cantabrian Zone

A carbonate and mixed (carbonate-siliciclastic) deposition characterizes the Cambrian Epoch 2–Miaolingian transition in the Cantabrian Mountains. Uppermost Cambrian Series 2 facies associations are characterized by peritidal, microbial and archaeocyathan-microbial reefs, ooidal-dominated shoals and scattered evaporitic pseudomorphs (Álvaro et al., 2000a; Aramburu et al., 2004; Zamarreño, 1972) (Fig. 2). In the Esla nappe, a peritidal-dominant ramp (bedded dolostones of the lower Láncara Member) was severely affected by tectonic instability across the series boundary, which led to (i) the occurrence of low-angle, ossicle-rich shoals (gray limestone facies of the lower Láncara Member) protecting back-shoal microbial-archaeocyathan patch-reefs, subsequently affected by another episode of tectonic breackdown leading to (iii) widespread development of low-angle shoal barriers rife of disarticulated skeletons (Beleño facies), underlain by a graveled lag that includes clasts derived from the underlying facies; another generalized breakdown of the horst-and-graben system led to the final deposition of (iv) variegated centimeter-thick, bioclastic packstone-wackestone limestones and shales, representing deposition on the top of palaeohorsts ("griotte" or Barrios facies from the upper Láncara Member), laterally flanked by offshore-dominant clayey substrates on the bottom of (half-) grabens (Álvaro et al., 2000a; Aramburu et al., 2004; Zamarreño, 1972). Chronostratigraphically, the regional "lower–middle Cambrian boundary" unconformity is bracketed between: (i) the fossil assemblage yielded by the gray limestone facies, including trilobites (endemic species of the genera *Berabichia* and *Kingaspis*; Álvaro, 2007), skeletonized microfossils (phosphatic tubes; Clausen and Álvaro, 2006) and archaeocyaths (genera *Archaeocyathus* and *Pycnoidocyathus*; Perejón et al., 2012); and (ii) that from the overlying Beleño facies, which comprise trilobites (e.g., *Acadoparadoxides*) and microfossils such as hyoliths, hexactinellid and heteractinid sponge spicules, chancelloriid sclerites, cambroclaves, probable eoconchariids and echinoderm ossicles (Álvaro, 2007; Clausen and Álvaro, 2006).

The southern Montagne Noire represents the lateral prolongation of the Cantabrian peri-Gondwana margin. Lithologies and facies are similar, though slightly diachronous. Three lithostratigraphic units are involved in the series boundary, from bottom to top, (i) the upper Lastours Member, 40–150 m thick and comprising alternations of bedded carbonates with thromboid reefs recording episodes of subaerial exposure and variegated shales; (ii) the Pont de Poussarou Formation, 20–80 m thick and composed of massive to bedded limestones and marbles; and (iii) the La Tanque Formation, 10–50 m thick and consisting of centimeter-thick alternations of reddish-to-purple, nodular and bedded limestone/shale couplets (griotte facies) (Fig. 2). Due to the lack of radiometric ages, the series boundary should be chronostratigraphically located between the upper member of the Lastours Formation, which has yielded the trilobite *Ferralsia blayaci* (congeneric with another species from the Görlitz Synclinorium in Germany and the Ossa-Morena Zone in the Iberian Peninsula; Álvaro et al., 1998a; Geyer and Elicki, 1995) and the top of the La Tanque Formation rich in trilobites such as *Badulesia* (Álvaro et al., 1998b) (Fig. 2).

In the Iberian Chains, the stratigraphic triad that reflects the same palaeogeographic evolution than that recorded in the Cantabrian Zone and the southern Montagne Noire is composed of the Daroca, Valdemiedes and Mansilla formations (Fig. 2). The Daroca Formation comprises a highstand-to-transgressive composite systems tract with coastal-plain deposits recording subaerial exposure (Hawke Bay/Toyonian Regression; Álvaro and Vennin, 1998). The Valdemiedes Formation reflects the establishment of peritidal environments locally truncated by the input of storm-induced skeletal remains. The base of the regional Leonian Stage (marking the base of the regional "middle Cambrian"; Sdzuy et al., 1999) lies within the Valdemiedes aggradational systems tract, both in peritidal environments and offshore-dominant bottom grabens (such as that represented by the Valdemiedes rivulet stratotype). Carbonate/shale cyclic couplet sedimentation took place across the Valdemiedes and Mansilla (griotte facies) formations, reflecting the interplay of two variables: an episodic tectonic activity and a cyclic Milankovich-like orbital forcing. A quantitative analysis of the tectonically induced subsidence recorded in the Iberian Platform revealed two main tilting disturbances, whereas other minor pulses were recorded only locally on some parts of the area (Álvaro and Vennin, 1996). The successive tectonically induced disturbances caused complex and varied subsidence patterns across platform regions. In addition, a multifold cyclicity has been reported from the griotte exposures. Spectral analyses of the high-frequency cycle-stacking patterns have revealed the interplay of a two-fold hierarchy, which results in a cyclic arrangement of carbonate/shale couplets close to the obliquity/precessional ratio predicted for early Paleozoic time (Álvaro et al., 2000b). The Valdemiedes rivulet of the western Iberian Chains represents the stratotype of the regional "lower–middle Cambrian" boundary, marked by the first occurrence of *Acadoparadoxides mureroensis* in a monotonous succession of marly shales (Valdemiedes Formation; Álvaro et al., 1993).

4.3 Miaolingian–Furongian transition

Another worldwide positive $\delta^{13}C_{carb}$ isotope shift, known as the Steptoean Positive Carbon Isotope Excursion or SPICE (Zhu et al., 2004), has been considered in the study area as a complementary approach to recognize the base of the Furongian. The peak of the SPICE event coincides with a time of maximum regression, and the subsequent increased weathering and erosion rates during relative sea-level fall, which would increase the burial fraction of organic carbon, triggering a major increase in atmospheric O_2 (Saltzman et al., 1998, 2000, 2011).

The Furongian–basal Tremadocian Toledanian Phase, responsible for uplift and erosion along the rift axes of the Atlas–Ossa-Morena Rift, has controlled the presence/absence of Furongian sediments. The phase has been interpreted as different ways, such as the onset of a break-up (or rift/drift) unconformity (Álvaro et al., 2018b; Sánchez-García et al., 2019). In the Ossa-Morena and Central Iberian Zones, the phase is marked by an angular discordance separating a palaeorelief formed by variably tilted

Ediacaran–Cambrian Series 2 strata from an overlying passive-margin succession. The involved gap includes, at least, most of the Furongian and basal Ordovician. Peri-Gondwanan margins sandwiched between the rift axes and Gondwana (e.g., the Cantabrian Zone, its lateral prolongation into the Iberian Chains, and the Montagne Noire) have preserved Furongian strata (Fig. 2).

In the Anti-Atlas, the Furongian gap lies at the Azlag/Lower Fezouata contact, except in the vicinity of Foum Zguid (central Anti-Atlas), where Álvaro and Vizcaïno (2018) have described several unconformities (joined laterally into one single angular discordance or paraconformity capping the Miaolingian Azlag Formation) associated with silcretization processes related to uplift, denudation and weathering of an exposed landscape, hydrothermal mineralization and final blanketing with alluvial conglomerates (lower member of the Lower Fezouata Formation).

Far from these uplifting areas, the Furongian is dominated by shale deposition (e.g., upper part of the Valtorres Formation in the Iberian Chains; Shergold and Sdzuy, 1991), locally interrupted by griotte facies in the southern Montagne Noire (Val d'Homs Formation; Álvaro and Vizcaïno, 1998; Álvaro et al., 2003a, 2007).

5. Chemostratigraphic resolution of the BACE, ROECE and SPICE events

As stated above, three Cambrian chemostratigraphic shifts have been targeted in West Gondwana: the BACE shift close to the Ediacaran–Cambrian boundary, the ROECE shift close to the Cambrian Series 2–Miaolingian boundary, and the SPICE shift close to the Miaolingian–Furongian boundary (Zhu et al., 2004).

5.1 Looking for the BACE shift in the Central Iberian Zone

A chemostratigraphic analysis of the C, O and Sr isotopic and chemical (Ca, Mg, Sr, Mn and Fe) record has been made in the Ciudad Rodrigo-Hurdes-Sierra de Gata domain (CRHSG). The selected level is a carbonate-dominant unit, about 25 m thick and including bedded and slumped-brecciated carbonates. The carbonate level is embedded in a lenticular slope-related deposit, up to 3 km long and 260 m thick, dominated by black shales and named Unit IV by Valladares et al. (2000, 2006). Unit IV, up to 500 m thick, represents a lateral equivalent of the Ibor Group (Castañar, Villarta and Arrocampo formations). The laterally equivalent

Unit IV in the CRHSG, and the Ibor Group and Cíjara Formation of the Toledo Mountains are unconformably capped by a discontinuous level of megabreccias bearing carbonate blocks containing (allochthonous) *Cloudina* remains, which mark the beginning of rifting conditions in this inherited Cadomian fore-arc basin (Álvaro et al., 2019).

5.1.1 Diagenetic and metamorphic overprint

As described by Valladares et al. (2000), the carbonate level selected for chemostratigraphic analysis belongs to a lenticular slope-related deposit encased in a monotonous succession of black shales infilling a retro-arc basin; 25 m of the about 500 m-thick Unit IV were checked chemostratigraphically. The carbonate level consists of alternating bedded limestones (where bulk rock microsamples were selected from "homogeneous carbonate samples") and calcareous breccia lenses. The Ibor Group and its lateral lithostratigraphic equivalents are sandwiched between a lower gap, representative of the so-called "main Cadomian folding event," and an upper gap marking the "pre-rift unconformity." Both the pre-rift and break-up (related to the Toledanian Phase) unconformities point to uplift and tilting processes associated with forced regressive trends responsible for the influence of early-diagenetic dissolution processes and occlusion of secondary porosity with meteoric cements. No diagenetic control was made to control their effects.

Recently, Álvaro et al. (2020) analyzed the different carbon and oxygen isotopic values of latest Ediacaran *Cloudina*-microbial reefal and peri-reefal sediments from the Toledo Mountains. Paragenetic relationships reflect the onset of several phases of neomorphism and precipitation of calcite and dolomite cements, in some cases induced by polyphase fissuring and hydrothermal influx controlled by the latest episodes of Cadomian activity and pre-rift events. The most significant phase of fissuring and solution was related to the Toledanian Phase, which is highlighted by the precipitation of ferroan dolomite and ankerite, with depleted carbon isotopic values and influenced by meteoric waters. The diagenetic complexity recognized in these carbonates reflects the potential problems derived by geochemical analysis of bulk rock.

5.1.2 Chemostratigraphic dataset

Valladares et al. (2006) selected 35 samples along 25 m of carbonates from the entire Unit IV (about 500 m) along the Pastores section for chemostratigraphic analysis (Fig. 3A). The $\delta^{13}C$ (V-PDB) values range from −8.7

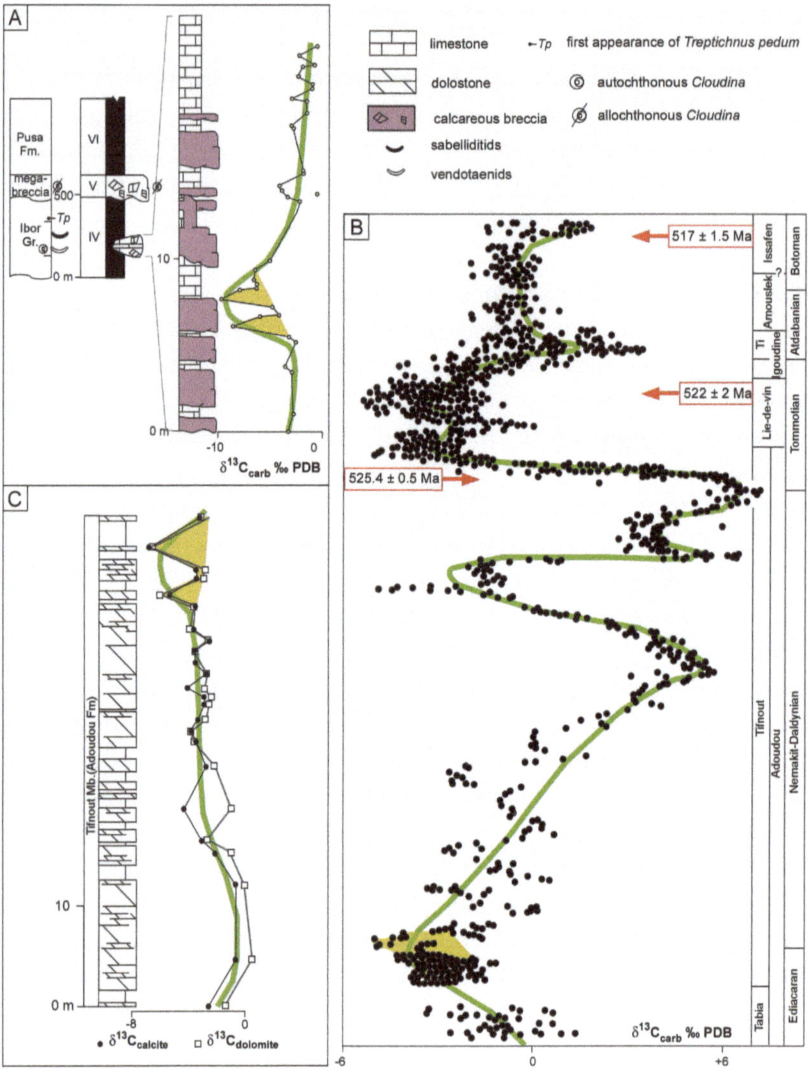

Fig. 3 Updated revision of the Cambrian chemostratigraphic logs crossing the Ediacaran–Cambrian transition in Morocco and Spain. (A). $\delta^{13}C_{carb}$ values from the Pastores section, Central Iberian Zone. (B) $\delta^{13}C_{carb}$ and $\delta^{13}C_{dolomite}$ values described at Tizi n'Tinififft jbel onlapping the southern Saghro inlier, central Anti-Atlas. (C) $\delta^{13}C_{carb}$ values described in the Anti-Atlas and High Atlas of Morocco; conversion equation between $\delta^{18}O_{SMOW}$ and $\delta^{18}O_{PDB}$, for Valladares et al.'s (2006) dataset, is $\delta^{18}O_{SMOW} = 1.03086\ \delta^{18}O_{PDB} + 30.86$. *Panel (A) modified from Valladares, M.I., Ugidos, J.M., Barba, P., Fallick, A.E., Ellam, R.M., 2006. Oxygen, carbon and strontium isotope records of Ediacaran carbonates in Central Iberia (Spain). Precambrian Res. 147, 354–365. Panel (B) modified from Álvaro, J.J., Macouin, M., Ezzouhairi, H., Ribeiro, M.L., Ader, M., Charif, A., Ait-Ayad, N., 2008c. Late Neoproterozoic carbonate productivity in a rifting context: the Adoudou formation and its associated bimodal volcanism onlapping the western Saghro inlier, Morocco. In: Ennih, N., Liégeois, J.P. (Eds.), The boundaries of the west African Craton. Geological Society London Special Publications, vol. 297. pp. 285–302. Panel (C) Modified from Maloof, A.C., Schrag, D.P., Crowley, J.L., Bowring, S.A., 2005. An expanded record of early Cambrian carbon cycling for the anti-atlas margin, Morocco. Can. J. Earth Sci. 42, 2195–2216.*

to −0.5‰, and the $\delta^{18}O$ values from +15.4 to +18.9‰ relative to V-SMOW (or from −15 to −8.8‰ relative to V-PDB). A negative peak is documented through about 4m, characterized by a shift of about 3–4‰, reaching a minimum of −7.7‰ that coincides with a $\delta^{18}O$ minimum of −14.2‰.

The low $\delta^{18}O$ values recorded in the carbonate samples suggest that alteration affected the complete succession, indicating substantial post-depositional oxygen isotopic mobility. The Mg/Ca ratios exclude dolomitization processes, but the relatively low abundance of Mn (61–340ppm) and low Mn/Sr ratios (0.050–0.591) suggest that the dataset was irregularly altered. In some cases, such as the stratigraphic interval between 6 and 9.7m, positive excursions of Mn/Sr and $^{87}Sr/^{86}Sr$ ratios coincide with negative excursions of $\delta^{13}C$ and $\delta^{18}O$, suggesting alteration of the sampled carbonates. Another result is the absence of covariation between Mg/Ca and the Mn/Sr, $\delta^{18}O$, $\delta^{13}C$, and $^{87}Sr/^{86}Sr$ parameters, consistently suggesting that neither neomorphism nor dolomitization processes did seriously affect these carbonates. Finally, the analyzed 25m-thick carbonate package contrasts with the ca. 500m measured by the entire Univ IV, devoid of chemostratigraphic analysis. An additional analysis of the whole Unit IV would strengthen the identification of the BACE event.

5.2 Terreneuvian–Cambrian Epoch 2 chemostratigraphic standard from the Anti-Atlas

Although several geographical areas have been selected in the Anti-Atlas and High Atlas for chemostratigraphic analysis of Terreneuvian–Cambrian Series 2 carbonates (Maloof et al., 2005: Fig. 1), two of them have yielded the most complete dataset: (i) the Adoudou-to-Issafen strata fringing the Cenozoic Oued Souss, and linking Amouslek, Tazemmourt and Tiout villages and Oued Sdas in the western Anti-Atlas (Maloof et al., 2005) (Fig. 3B); and (ii) the Adoudou strata at the Tizi n'Tinififft jbel onlapping the southern Saghro inlier in the central Anti-Atlas (Álvaro et al., 2008c) (Fig. 3C). Cambrian exposures mainly occur as monoclinal strata onlapping an inherited palaeorelief formed by the volcanosedimentary Ouarzazate Supergroup. The palaeorelief formed as a result of two main events: (i) the so-called Panafrican-3 (Álvaro et al., 2014c; Walsh et al., 2012), dated at about 564–560Ma and stratigraphically marked by the intra-Ouarzazate Mançour/Imlas groups contact (Walsh et al., 2012); and (ii) the Ouarzazate/Adoudou contact, representing a pre-rift unconformity preserved as angular discordant to paraconformable discontinuities associated with the onset of normal faults bounding (half-)grabens (Álvaro et al., 2008c, 2014c). As explained above, the pre-rift (Ouarzazate/Adoudou) unconformity is diachronous throughout the Anti-Atlas, as a result of which the beginning of carbonate productivity can range from latest Ediacaran to Fortunian.

5.2.1 Diagenetic and metamorphic overprint

The palaeothermal conditions recorded in the Cambrian of the Anti-Atlas can be inferred based on the study of the illite crystallinity (IC) index of shales. Their metamorphism is of low to very low degree, increasing from NE to SW: whereas in the eastern Anti-Atlas diagenetic and anchizonal IC-values are predominant, in the western and central Anti-Atlas Cambrian shales display epizonal conditions (Ruiz et al., 2008). The geothermal gradient has been interpreted as the result of both burial (the epizone-anchizone and anchizone-diagenesis boundaries are separated by a 2.5–3 km thick sedimentary pile generating a 33–27 °C palaeogeothermal gradient) and tectonic thickening of a sedimentary pile enhanced by the Variscan orogeny; the Variscan shortening in the western Anti-Atlas varies between 15 and 25% (Robert-Charrue and Burkhard, 2008). Shortening was accommodated by polyharmonic buckle folding of the cover in a thick-skinned fashion without development of any significant thrust/duplex systems (Burkhard et al., 2006). The anchizone/epizone boundary is reached at depths of about 10–12 km and temperatures close to 300 °C (Frey and Robinson, 1999).

The main diagenetic and hydrothermal processes recorded in the Terreneuvian–Cambrian Series 2 carbonates are related to: (i) intra-Cambrian *syn*-rift and pre-rift episodes of basement uplift and tilting, related to crystal neomorphism, precipitation of successive calcite and dolomite cements progressively depleted in $\delta^{13}C$ and $\delta^{18}O$ values (reflecting the influence of both meteoric waters and elevated temperatures of precipitation); (ii) the hydrothermal influence recorded in the Taguedit (Tabia Member) and Tamjout Beds (Tifnout Member) of the Adoudou Formation; and (iii) late-diagenetic burial and Variscan influences (Álvaro, 2013; Álvaro and González-Acebrón, 2019).

The sector of the western Anti-Atlas linking Amouslek, Tazemmourt and Tiout villages was relatively preserved by sharp syn-rift episodes during Terreneuvian times. As a result, the dolostones of the Adoudou and

Lie-de-vin formations and the lower member of the Igoudine Formation are relatively homogeneous and range petrographically from sparry dolostones to argillaceous dolosiltstones (Monninger, 1979). An earliest diagenetic, microsparry dolomite, locally neomorphosed to sparry dolomite, comprises the bulk of the strata. The presence of ferroan dolomite is related to faults; subsidiary clusters of coarse sparry dolomite are associated with (post-stylolitization) burial, and dedolomitization processes associated with vadose conditions are probably related to Variscan uplift (Monninger, 1979).

However, a significant influence of hydrothermal events occurs in the Taguedit (Tabia Member) and Tamjout Beds (Tifnout Member) of the Adoudou Formation. A syn-rift unconformity is recognized bounding the Taguedit Bed, reflecting a short-term episode of extensive fault-controlled hydrothermal activity (mainly veining and dissemination of Cu—Ag sulfides subsequently reworked and preserved as capping lags), carbonate production and phosphogenesis followed by sealing of the palaeorelief by the microbial carbonate-dominated platforms of the Tifnout Member (Álvaro, 2013; Álvaro and Subías, 2011). The Tamjout dolostone Bed, which marks the base of the Tifnout Member in the western Anti-Atlas, hosts synsedimentary polymetallic (Pb–Cu–Fe) stockworks that affected the porosity and subsequent occlusion by successive generations of hydrothermal dolomite (Álvaro and González-Acebrón, 2019; Oummouch et al., 2017).

The overlying Cambrian Series 2 carbonates of the Tiout Member (Igoudine Formation) and Amouslek and Issafen formations are dominated by limestone strata. Their facies associations represent different horst-and-graben settings related to rift axes punctuated by microbial (thromboid-dominated)-archaeocyathan reef complexes (Álvaro et al., 2006a; Álvaro and Clausen, 2005). Sharp modifications in facies and thickness accompanied by changes in frame-building styles (reef vs. mud-mound complexes; Álvaro and Debrenne, 2010) reflect rifting changes in accommodation space lacking distinct hydrothermally induced overprints. As a result, dolomitization processes are scarce and calcite cements are mainly related to occlusion of primary porosity.

5.2.2 Chemostratigraphic dataset

In the western Anti-Atlas, Maloof et al. (2005) analyzed the $\delta^{13}C$ values of the Tifnout Member (Adoudou Formation) and the Lie-de-vin, Igoudine,

Amouslek and Issafen formations, completing stratigraphically chemostratigraphic data by Tucker (1986) and Magaritz et al. (1991) from the Tiout reference section. Samples were collected at 0.2–1.5 m intervals through 2.5 km of nearly continuous carbonates from 15 measured stratigraphic logs. Clean dolostones and limestones without siliciclastic components or secondary veining or cleavage were targeted. In the Tabia Member, $\delta^{13}C$ varies between −4 and 0‰; the lower Tifnout Member (Tamjout Bed) shows near uniformity across the Anti-Atlas, descending rapidly from −2 to a low of −6‰ (correlated with BACE or Ediacaran–Nemakyt-Daldynian boundary), then increasing to a high of +7‰ in the upper Tifnout Member (correlated with the Nemakit-Daldynian–Tommotian boundary in Maloof et al., 2005; but replaced below the same boundary in Maloof et al., 2010). In the uppermost Tifnout Member, $\delta^{13}C$ falls again to −3‰ and stays between −3.5 and + 0.5‰ for most of the Lie-de-Vin Formation. $\delta^{13}C$ rises to a peak of +3.5‰ in the uppermost Lie-de-Vin and lower Igoudine Member and then stabilizes at −0.5‰ for the remainder upper Igoudine Member and Amouslek and Issafen formations. Oxygen isotopes show a mean value of −6 to −7‰, with a negative shift up to −14‰ below the base of the Tata Group, and remain highly depleted until the Issafen Formation. The evolving oxygen-composition of seawater includes values of −4 to −10‰ in the Neoproterozoic (Jacobsen and Kaufman, 1999) dropping to a nadir in the late Cambrian Epoch 2 to Furongian of −10 to −14‰ (Veizer et al., 1999).

In the central Anti-Atlas, Álvaro et al. (2008c) analyzed 23 carbonate samples spanning a mixed (carbonate-siliciclastic) stratigraphic interval of 50 m from the Adoudou Formation. Based on partial earliest diagenetic dolomitization of these carbonates, carbon isotope values were analyzed in drilled microsparry calcite and dolomite microsamples. $\delta^{13}C_{calcite\ PDB}$ and $\delta^{13}C_{dolomite\ PDB}$ values ranged from −6.6 to −0.6‰ and from −6.4 to +0.4‰, respectively; and $\delta^{18}O_{calcite\ PDB}$ and $\delta^{18}O_{dolomite\ PDB}$ ones from −5.5 to −0.6‰ and from −8.7 to −3.7‰, respectively. Despite the lack of covariance between $\delta^{13}C$ and $\delta^{18}O$ values, reflecting the lack of diagenetic processes affecting preferentially the $\delta^{13}C$ values in calcite or dolomite, a negative shift of about 3.2–3.6‰ is located in $\delta^{13}C_{calcite}$ and $\delta^{13}C_{dolomite}$ values about 3 m below the Adoudou/Tikirt contact, after which the values sharply decrease to background averages. The negative $\delta^{13}C$ shift is coeval with a positive shift of about 3.8–3.9‰ in $\delta^{18}O_{calcite}$ and $\delta^{18}O_{dolomite}$ values probably reflecting a diagenetic signal. As a result, Álvaro et al. (2008c) tentatively proposed this negative shift in $\delta^{13}C_{calcite}$ and $\delta^{13}C_{dolomite}$ values as representative of the $\delta^{13}C$ excursion to −6‰ located at the base of the Tifnout Member by Maloof et al. (2005).

Both the negative Ediacaran–Nemakit-Daldynian (BACE) and the positive Nemakit-Daldynian–Tommotian boundaries have been calibrated in the Adoudou Formation of the Anti-Atlas (Maloof et al., 2005), though the latter, based on further radiometric ages, was dated at about 525.343 ± 0.088 Ma and preceding the Nemakit-Daldynian–Tommotian boundary dated at 524.837 ± 0.092 Ma in Maloof et al. (2010).

5.3 Looking for the ROECE shift across the regional "lower–middle Cambrian" transition

Four sections have been selected in south-western Europe to check the potential record of the ROECE event. The base of the traditional "middle Cambrian" was regionally placed at the first occurrence of the trilobite *Acadoparadoxides mureroensis* proposing the Valdemiedes rivulet section of the Iberian Chains as stratotype for the regional Bilbilian–Leonian boundary (Álvaro et al., 1993; Sdzuy et al., 1999). This log represents a graben with infill dominated by homogeneous shales deposited on offshore substrates. Three similar series boundary sections were tested in carbonate-dominant successions, in the Esla and Somiedo nappes of the Cantabrian Zone (Álvaro et al., 2000a) and in the Pardailhan nappe of the southern Montagne Noire (Álvaro et al., 1998b, 1999) (Fig. 4). From a biostratigraphic point of view, the setting of the series boundary is not well constrained in the three sections, and the boundary was placed at an erosive unconformity.

5.3.1 *Diagenetic and metamorphic overprint*

The Cantabrian Zone and its southeastern prolongation into the Iberian Chains display a thin-skinned type of structural geometry, in which the basement was not involved in the deformation, but the framework was complicated by the existence of several detachment levels. Metamorphism reached the transition from anchizone to low grade in the innermost thrust units (Bastida et al., 1999). In contrast, the West Asturian-Leonese Zone and its lateral prolongation in the Iberian Chains show thick-skinned tectonics with partial involvement of the basement. The distribution of the dynamothermal metamorphic zones points to a westward increase from epizonal to catazonal metamorphic conditions. Low metamorphic conditions occur in

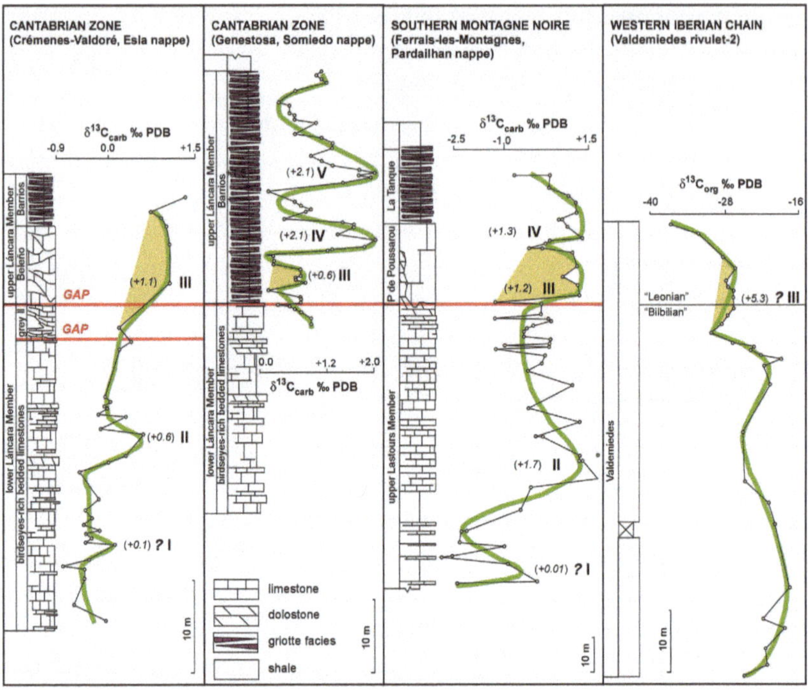

Fig. 4 Updated revision of the Cambrian chemostratigraphic logs crossing the regional "lower–middle Cambrian" boundary interval in the Cantabrian Zone (Esla and Somiedo nappes), southern Montagne Noire and western Iberian Chain transition in Morocco and Spain. *Modified from Wotte, T., Álvaro, J.J., Shields, G.A., Brasier, M., Veizer, J., 2007. C-, O- and Sr–isotope stratigraphy across the lower–middle Cambrian transition of the Cantabrian Mountains (Spain) and the Montagne noire (France), West Gondwana. Palaeogeography, Palaeoclimatology, Palaeoecology 256, 47–70 and Gozalo, R., Dies Álvarez, M.E., Gámez Vintaned, J.A., Zhuravlev, A.Yu., Bauluz, B., Subías, I., Chirivella Martorell, J.B., Mayoral, E., Gursky, H.J., Andrés, J.A., Liñán, E., 2013. Proposal of a reference section and point for the Cambrian series 2–3 boundary in the Mediterranean subprovince in Murero (NE Spain) and its intercontinental correlation. Geochem. J. 48, 142–155.*

the eastern belts of the Iberian Chains fringing the Datos Fault and the eastern belt of the West Asturian-Leonese Zone flanking the Narcea Antiform, where Paleozoic exposures comprise chlorite-to-biotite zones (Álvaro et al., 2018a; Bauluz et al., 1998; García-López et al., 2007). The southern Montagne Noire, which represents the northeastern prolongation of the Cantabrian Zone, comprises a framework of south-facing thrust slices and nappes. IC analysis combined with estimates of the conodont alteration index in Carboniferous samples allowed determining anchizonal to epizonal conditions (Engel et al., 1981; Wiederer et al., 2002).

Therefore, the sections reported below (e.g., Esla, Somiedo and Pardailhan nappes and Valdemiedes rivulet) display both anchizonal-to-epizonal conditions and chlorite isograds, pointing to burial depths of about 10–12 km associated with Variscan low-grade metamorphic conditions taking place at approximately 200–320 °C and relatively low pressure conditions (Frey and Robinson, 1999).

As stated above, in the Esla nappe of the Cantabrian Zone, the carbonates of the Láncara Formation are punctuated by three tectonically induced discontinuities, marking: (i) the bedded dolostone/Beleño facies contact within the lower Láncara Member; (ii) the lower/upper Láncara member contact, and (iii) the Beleño/Barrios facies contact of the upper Láncara Member. The discontinuities separate, from bottom to top, a highstand systems, an aggradational and a transgressive systems tracts (Álvaro and Clausen, 2005). The main diagenetic processes recorded in these carbonates include an earliest dolomitization event in the bedded dolostones of the lower member, as well as a conspicuous neomorphism and occlusion of primary porosities (including fenestral pores and birdseyes) by mosaics of sparry calcite (Álvaro et al., 2000a; Zamarreño, 1972) (Fig. 4).

In the upper member of the Lastours Formation and the Pont de Poussarou and La Tanque formations, a similar distribution of unconformities marking tilting episodes is recognized separating the three reported lithostratigraphic units (Álvaro et al., 1999). Again, a mixture of microbial carbonates and ooidal shoals dominate the highstand systems tract, a strongly neomorphosed low-angle skeletal shoal complex the aggradational systems tract, and the La Tanque griotte facies, deposited on palaeohorsts, the transgressive systems tract. Neomorphism and occlusion of primary pores by calcite dominate the main early diagenetic processes (Álvaro and Clausen, 2005).

5.3.2 Chemostratigraphic dataset

Wotte et al. (2007) analyzed the carbonate-dominated sections of the Cantabrian Mountains (northern Spain) and Montagne Noire (southern France). They recognized a distinct positive shift in the uppermost limestones of the lower Láncara Member (+0.6/+1.7‰) and two positive shifts (peaking at +1.1/+1.2‰ and +1.3‰, respectively) in the Barrios facies of the upper Láncara Member, which were recognized in Ferrals-les-Montagnes (Montagne Noire). As stated by the authors, the marked negative $\delta^{13}C_{carb}$ excursion reported from the Cambrian Epoch 2–Miaolingian boundary in China and Laurentia is absent, though this may be related to the erosive unconformity bearing an undated gap marking the lower/upper Láncara members contact (Álvaro et al., 2000a). A regional correlation of the series boundary was only suggested between the Esla and Somiedo nappes (Cantabrian Mountains) and the Pardailhan nappe (Montagne Noire), where the analysis was made (Fig. 4).

Subsequently, Gozalo et al. (2013) documented a chemostratigraphic test in a contemporaneous single shale-dominated section from the Iberian Chains (Fig. 4). In the Valdemiedes rivulet section, they recognized a broad negative trend in $\delta^{13}C_{org}$ from −16 to −38.5‰, with a small negative peak of c. 7‰ underlying the regional base of the "middle Cambrian" (or Leonian stage; Sdzuy et al., 1999). Despite the lack of further arguments, they correlated this negative shift with the global negative shift known as the ROECE event (Zhu et al., 2004), close to the FAD of *Ovatoryctocara granulata* in Laurentia and Siberia. Due to the lack of any positive $\delta^{13}C_{org}$ trend in the Valdemiedes rivulet, Gozalo et al. (2013) arbitrarily selected one of the Leonian positive $\delta^{13}C_{carb}$ shifts described by Wotte et al. (2007) in the Cantabrian Mountains to recognize another positive shift close to the global first occurrence of *Oryctocephalus indicus*, which marks the base of the Miaolingian. As a result, although this blind dating is a risk hypothesis in need of confirmation, due to the lack of cosmopolitan fossils available to calibrate both blind chemostratigraphic correlations, it has been subsequently used as a global correlation method to support the first occurrence of *Acadoparadoxides mureroensis* as a GSSP candidate (Gozalo et al., 2007, 2013).

In the absence of either well-constrained radiometric data or cosmopolitan, biostratigraphically significant taxa, carbon isotope stratigraphy cannot be used alone for worldwide correlation. This is the case of the chemostratigraphic dating made in Iberia by Gozalo et al. (2007, 2013) across the Cambrian Series 2–Miaolingian boundary interval. The chemostratigraphic analysis made by Wotte et al. (2007) was only valid for regional correlation throughout the Cantabrian Zone and the southern Montagne Noire. The lack of the ROECE shift in the Iberian margin of West Gondwana may be related to the onset of a distinct unconformity marking the regional "lower–middle Cambrian boundary." The reliability of the involved $\delta^{13}C_{org}$ values from the Iberian Chains is discussed below.

5.4 Looking for the SPICE shift throughout the regional "middle–upper Cambrian" transition

The presence of carbonate interbeds in the Miaolingian–to–Tremadocian Val d'Homs Formation of the southern Montagne Noire allowed extraction of chronostratigraphically significant trilobites, linguliformean brachiopods and conodonts (Álvaro et al., 2003a, 2007; Feist and Courtessole, 1984; González-Gómez, 2005; Serpagli et al., 2007). The lack of the Toledanian Phase and the record of a rich shelly fauna encouraged a chemostratigraphic analysis of the carbonate interbeds (Álvaro et al., 2008b).

5.4.1 Diagenetic overprint

Two sections were analyzed in the southern Montagne Noire: Sallèles-Cabardès (Minervois nappe) for both $\delta^{13}C_{carb}$ and $\delta^{13}C_{org}$ interbedded limestone and shale samples, and Campelou (Pardailhan nappe) for $\delta^{13}C_{org}$ shale samples. The mineralogical composition of shales, dominated by illite and chlorite with subsidiary feldspar and calcite, was analyzed. No significant trends or variations of the mineral associations were observed at Sallèles-Cabardès, whereas a gradual increase in the illite + chlorite/quartz + feldspar ratio was identified at Campelou and interpreted as a result of both weathering (feldspar transforming into illite) and decreased siliciclastic input related to transgressive conditions. The IC values reflect late diagenesis-to-anchizone conditions, which fits well with the anchizonal degree supported by CAI values of contemporaneous conodonts (Serpagli et al., 2007).

The transgressive Miaolingian–middle Tremadocian mixed Val d'Homs Formation and the laterally equivalent siliciclastic La Gardie Formation are characterized by: (i) a progressive increase of $\delta^{13}C_{carb}$ (from −3 to −1‰) in limestone interbeds coeval with a decrease in $\delta^{13}C_{org}$ (from −22 to −27.5‰) in shales; and (ii) a large-scale increase in organic matter content reflecting a three-step trend from a stable impoverished content, less than 0.15 wt%, a subsequent fluctuation reaching peaks of 0.15–0.25 wt% that rapidly decrease close to the base of the prograding La Dentelle shoal complexes, and a late fluctuation reaching a peak of 0.20 wt% (Álvaro et al., 2008b) (Fig. 5).

Fluctuations of $\delta^{18}O$ values in the limestone interbeds range from −9 to −16‰, whereas similar analyses compiled from Cambrian tropical low-Mg calcite-walled brachiopods (which probably yield the closer proxy to real seawater chemistry) vary from −7 to −12‰ (Wadleigh and Veizer, 1992). However, as stated by Bruckschen et al. (1999), these differences must not be exclusively related to diagenetic processes, because of latitudinal shifts: $\delta^{18}O$ values in modern tropical brachiopods at shelf depths are ~4‰, and, for specimens sampled in temperate climates, they span ~8‰. Oxygen isotopic differences can also be triggered by metabolic effects or seasonality (Grossman et al., 1991) and may reflect superimposed vital fractionation

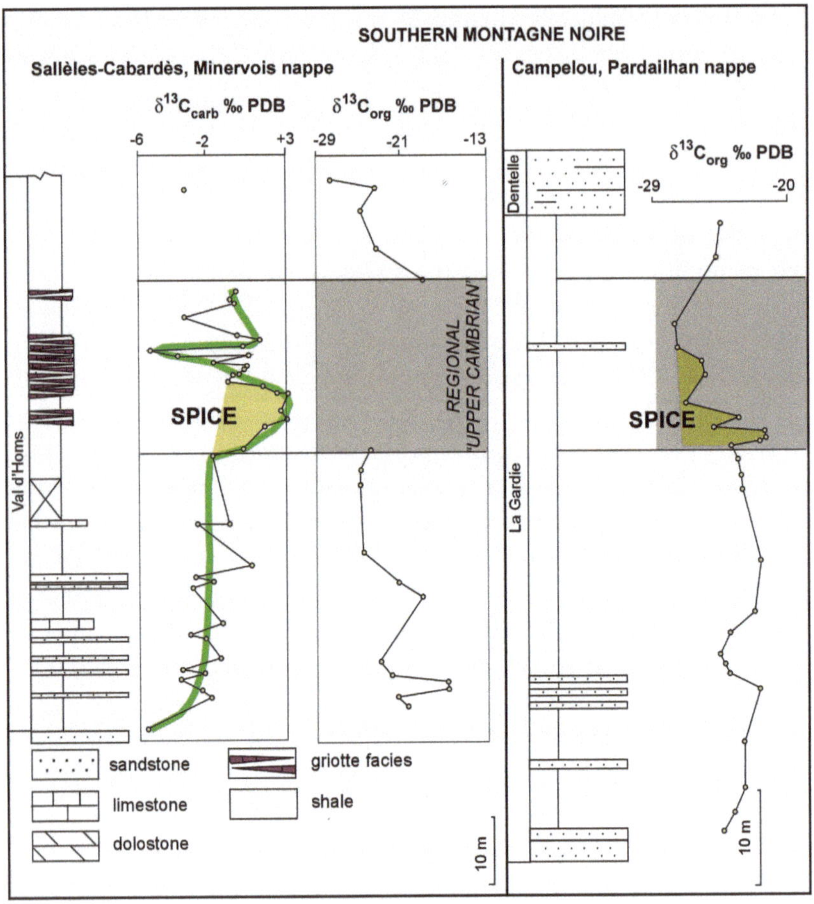

Fig. 5 Updated revision of the Cambrian chemostratigraphic logs crossing the regional "middle–upper Cambrian" boundary interval in the southern Montagne Noire (Minervois and Pardailhan nappes). *Modified from Álvaro, J.J., Bauluz, B., Pierre, C., Subías, I., Vizcaïno, D., 2008b. Carbon chemostratigraphy of the Cambrian–Ordovician transition in a midlatitude mixed platform, Montagne Noire, France. Geol. Soc. Am. Bull. 120, 962–975.*

effects (Qing and Veizer, 1994) or recrystallization of the shell calcite after deposition (Veizer et al., 1986). On average, the difference between bulk rock and skeletal calcite (e.g., calcite-walled brachiopods) $\delta^{18}O$ values amounts to ~2‰ (Schrag et al., 1995; Veizer et al., 1999). The $\delta^{18}O$ data from Sallèles-Cabardès show more oscillations than the $\delta^{13}C$ data; this fact and their depleted character can be reasonably ascribed to diagenetic alteration. This is consistent with the burial history of the Minervois and Pardailhan nappes in the southern Montagne Noire, since the anchizonal degree could be achieved by increasing lithostatic pressure related to temperatures close to 250 °C (Frey and Robinson, 1999). Micritic recrystallization would have also caused $\delta^{18}O$ values to decrease. Therefore, the depleted and variable values of $\delta^{18}O$ in Sallèles-Cabardès reflect a diagenetic resetting and post-depositional alteration of the Val d'Homs limestone samples, probably primarily controlled by thermal burial temperatures.

(i) Carbon isotopes from carbonates (Fig. 5). As explained above, the limestone interbeds of Sallèles-Cabardès exhibit a progressive increase of $\delta^{13}C_{carb}$ values from −3 to −1‰, interrupted by a positive shift peak at +3.02‰. These values are slightly depleted in comparison with the background $\delta^{13}C_{carb}$ values analyzed in Cambrian tropical brachiopods (ranging from −2.2 to +1.0‰; Wadleigh and Veizer, 1992). Possible controls on ^{13}C depletion can be related to three factors: the proportion of oxidized kerogen incorporated in the host limestone (Schumacher, 1996), the influence of riverine carbon in nearshore environments (Koch et al., 1992), and diagenesis. The carbon isotope value of marine organic matter is ~25‰ more negative than inorganic bicarbonate (Marshall, 1992); thus, degradation of organic matter has the potential to significantly alter the carbon isotope composition of marine carbonate sediment (Glumac and Walker, 1998). The condensed "griotte" limestones of Sallèles-Cabardès are reddish-colored, a feature influenced by the oxidation state of iron and manganese. They have been sedimentologically interpreted as deposited on the tops of tectonically induced palaeohighs, well oxygenated and submitted to the action of waves (Álvaro et al., 2003a). Because organic-matter preservation is directly related to rapid burial (preserving it from oxidation and scavengers), the scarcity (or virtual absence) of kerogen in these limestone interbeds may be related to irreversible, synsedimentary to early diagenetic oxidation of organic matter. In addition, $\delta^{13}C$ values of riverine carbon are shifted to more negative values than oceanic values, which may have somewhat impacted the carbon isotopic values of the shoreface-to-offshore griotte limestones. The wide variability of $\delta^{18}O$ values and lack of covariance between $\delta^{13}C$ and $\delta^{18}O$ values (coefficient of correlation = 0.23) suggest that the studied limestones were not significantly submitted either to meteoric waters or to marine-meteoric waters of the mixing zone (Allen and Matthews, 1982). As a result, despite the depleted character of the ^{18}O and ^{13}C curves, the lack of

$\delta^{13}C_{carb}/\delta^{18}O_{carb}$ covariance in Sallèles-Cabardès suggests similar diagenetic behaviors for the whole section. The absolute isotopic values were probably depleted by burial temperature, but they were affected in a similar way throughout the whole section. The $\delta^{13}C_{carb}$ fluctuations (not the absolute values) were not predominantly induced by diagenetic alteration. Álvaro et al. (2008b) considered that, despite the depleted isotope carbon values of the Sallèles-Cabardès $\delta^{13}C$ curve, its baseline and excursion are well constrained, and potentially preserved the original chemostratigraphic trends.

(ii) Carbon isotopes from shales (Fig. 5). Two apparent paradoxes displayed in the Campelou section need discussion: the gradual increase of $\delta^{13}C_{carb}$ values in the palaeohorsts is coeval with a gradual decrease in $\delta^{13}C_{org}$ values in the fringing lows; and a TOC/$\delta^{13}C_{org}$ covariance (coefficient of correlation = 0.9). The difference $\delta^{13}C_{carb} - \delta^{13}C_{org} = \Delta\delta^{13}C$ is generally considered to separate secondary fractionation and source effects from secular variations in the $\delta^{13}C_{DIC}$ signal. Constant $\Delta\delta^{13}C$ values (e.g., approaching 28.5 ± 2.0‰ for the Neoproterozoic in Knoll et al., 1986) were attributed to global changes in $\delta^{13}C_{DIC}$, whereas changes in $\Delta\delta^{13}C$ are currently interpreted as changes in the origin of the organic matter, fractionation processes, or diagenesis (Hayes et al., 1989; Magaritz et al., 1992). Although the analysis of $\delta^{13}C_{carb}$ and $\delta^{13}C_{org}$ values could not be made in the same samples of the Sallèles-Cabardès section, their respective background trends indicate an upsection modification of $\Delta\delta^{13}C$ values. The direct TOC/$\delta^{13}C_{org}$ relationship explains the modification in $\Delta\delta^{13}C$ because the upward increase in TOC concentration controlled the coeval decrease in $\delta^{13}C_{org}$ values. This relationship has also been observed in some Neoroterozoic–Lower Paleozoic strata and can be related to changes in the degree of biodegradation of organic matter, sulfate reduction, and methanogenesis before or soon after burial (Hayes et al., 1983; Kaufman and Knoll, 1995).

There is another explanation that can clarify the fact that C_{carb} and C_{org} production and burial may have been dominated by decoupled processes in the development of water-column stratification and bottom anoxia affecting shelf biota (Bartley and Kah, 2004): although biological productivity and carbonate production are primarily attributed to benthic shelly metazoans, resulting in strong coupling of C_{org} and C_{carb} production and burial, during an environmental crisis (e.g., related to water-column stratification), open-water production dominated by non-skeletal plankton would provide a potential source of C_{org} production and burial uncoupled from C_{carb}

production and burial. The difference in $\delta^{13}C_{DIC}$ between surface and bottom waters, related to the "photic pump" effect, may have strongly affected ^{13}C abundance in surface and deep waters without necessarily altering average seawater isotopic composition (Calver, 2000). However, despite the uncoupled C_{org}/C_{carb} signatures exhibited in comparisons of the shallower Sallèles-Cabardès and the deeper Campelou sections, the chemostratigraphic signal related to a synchronic positive shift of $\delta^{13}C_{carb}$ and $\delta^{13}C_{org}$ values located close to the base of the regional upper Cambrian is still preserved in both sections.

Other overprinting modifications can be considered. The thermally induced loss of hydrocarbons, which can represent an important factor affecting the stable isotopic composition of organic carbon (Hayes et al., 1983; Strauss et al., 1992), may have been significant in the Campelou area because a ^{13}C enrichment is generally observed with increasing thermal alteration. Thus, the higher $\delta^{13}C_{org}$ values might be explained by a thermally driven loss of hydrocarbons in TOC-poor samples, as pointed out by Joachimski (1997). In summary, the thermal burial temperatures achieved during the very low metamorphic degree of the Minervois and Pardailhan nappes probably affected the Cambrian strata in a homogeneous way, leading to a depletion of $\delta^{13}C_{carb}$ and $\delta^{18}O_{carb}$ values, and an increase of $\delta^{13}C_{org}$ values in shales.

5.4.2 Chemostratigraphic dataset

The Miaolingian chemostratigraphic data from the shale-dominated basinal part of the Montagne Noire platform (Campelou section) reflect the onset of a negative $\delta^{13}C_{org}$ shift, ~25 m thick, from a background of −22 to a negative peak of −24.5‰. Although it was the first time that a negative mid Miaolingian excursion is tentatively identified on the basis of $\delta^{13}C_{org}$ values, the shift fits well with the aforementioned patterns of the DICE $\delta^{13}C_{carb}$ excursion. However, this chemostratigraphic correlation should be tested in other subtropical shale-dominated platforms bearing and lacking the index fossil *P. atavus* that marks the base of the Miaolingian. The DICE identification in the Montagne Noire remains tentative, as suggested by Álvaro et al. (2008b).

In addition, the regional "upper Cambrian" chemostratigraphic data from nearshore and deeper parts of the Montagne Noire display a common positive isotope excursion: (i) the limestone interbeds of Sallèles-Cabardès exhibit a progressive increase of $\delta^{13}C_{carb}$ values from −3 to −1‰, interrupted by a positive shift peak at +3.02‰; and (ii) the biostratigraphically coeval shift in the 10-m-thick Campelou section is characterized by a sharp increase in $\delta^{13}C_{org}$ of 2.3‰ followed by a sharp decrease of 4.4‰, reaching a peak close to −21‰. In both sectors of the platform, the regional "upper Cambrian" excursion is unaffected by the influence of the La Dentelle regression, and mimics the SPICE event. Datasets from micritic analysis of the SPICE excursion in carbonate successions (Saltzman et al., 1998, 2000) yield parallel trends, but they broadly rise from −0.5 to +1.5‰ and peak at +4.5 to +5‰. Therefore, the $\delta^{13}C$ values from Sallèles-Cabardès are ~3‰ more depleted in comparison to published values from micrites of carbonate-dominant platforms. Fluctuations of $\delta^{18}O$ values in the limestones of Sallèles-Cabardès range from −9 to −16‰. The $\delta^{18}O$ data sets from the SPICE excursion analyzed in the micritic fraction of Furongian carbonate-dominated successions range from −7.8 to −13.1‰ (Nevada), −6.2 to −10.2‰ (Malyi Karatau region, Kazakhstan), or −5.4 to −10.8‰ (Hunan, China) (Saltzman et al., 1998, 2000). As a result, the measured $\delta^{18}O$ values in Sallèles-Cabardès are also shifted ~3–4‰ more negatively in comparison to the reported values from micrites of carbonate platforms. Despite the "depleted character" of the $\delta^{13}C$ SPICE excursion in the Montagne Noire, other "depleted SPICE excursions" have been described in dolomitized and highly altered rocks hosting ore deposits. The survival of the positive carbon isotope excursions in Montagne Noire suggests that the SPICE excursion represents a perturbation in the global (not regional) cycling of carbon. The proposed SPICE shift at Sallèles-Cabardès differs from other coeval shifts in carbonate-dominated platforms in its asymmetric shape: after the positive peak, the carbon isotope values decrease sharply until values about −5.3‰, and then they finally increase gradually to reach background values. The stratigraphically condensed character of the griotte limestone in Sallèles-Cabardès (the whole Furongian would be represented by ~25 m in thickness) suggests that other later Furongian chemostratigraphic shifts can be included in these strata. A similar negative $\delta^{13}C_{org}$ excursion, predating the Cambrian–Ordovician boundary, is also identifiable at Campelou, ~15 m above the end of the SPICE event: the negative shift is ~7 m thick and reaches a negative peak at −27.5‰. Although no further chemostratigraphic excursions were proposed by Zhu et al. (2006), Miller et al. (2006) proposed another negative $\delta^{13}C_{carb}$ excursion (coined the HERB event) close to the base of the highest Furongian (still unnamed) stage, the latter of which is proposed at the first occurrence of the conodont *Cordylodus andresi*. Evidence of the HERB event has been found in Australia, Newfoundland, and western

North America (Miller et al., 2006, and references therein), where it is characterized by a negative peak at −3 to −4‰, but the shift is in need of worldwide confirmation.

The SPICE excursion can be used for chemostratigraphic correlation of the Furongian base in the mid-latitude Montagne Noire platform, where chronostratigraphically significant trilobites, useful for interregional correlation, are absent. Because the first appearance of the index trilobite *Palaeadotes latefalcata*, which marks the base of the regional "upper Cambrian," occurs stratigraphically below the SPICE excursion, the base of the regional "upper Cambrian" is still slightly pre–Furongian in age.

6. Reliability of isotopic data: A critical re-appraisal

As stated above, bulk carbonate stable isotope signatures recorded by shallow-water limestone strata, provide evidence for a differential multistage diagenetic alteration. In our case study, these signatures would be also affected by the tectonically induced breakdown of the basement controlled by rifting episodes. This is revealed by bivariate plots of $\delta^{18}O_{carb}$ vs. $\delta^{13}C_{carb}$ values showing clustering of isotopic fields, and covariance analysis of the datasets (r and n values are encased in the following Figs. 6–8) determining the relationship between isotope records and quantified with Pearson's regression analysis.

A $\delta^{13}C_{carb}$ vs. $\delta^{18}O_{carb}$ bivariate plot for carbonates deposited across the Ediacaran–Cambrian transition in the western and central Anti-Atlas and the Central Iberian Zone allows distinction of two separate isotopic fields (Fig. 6). A distinct cluster includes the synsedimentary and early-diagenetic isotopic values from the Adoudou-to-Issafen formations in the Anti-Atlas, with $\delta^{13}C_{carb}$ values broadly ranging from +5 to −6‰, and $\delta^{18}O_{carb}$ values from −2 to −10‰. In the Adoudou Formation, $\delta^{13}C$ values vary between −4 and 0‰ in the Tabia Member, and between −6 and +7‰ in the Tifnout Member, except for some anomalous data from the Tagrara inlier (Álvaro et al., 2008c; Maloof et al., 2005). Hydrothermally affected carbonates of the Taguedit Bed (Tabia Member) display $\delta^{13}C$ values from −0.2 to −3.7‰, whereas their hydrothermally induced karstic infill (speleothems) ranges from −8.2 to −8.6‰ (Álvaro, 2013). The Adoudou Formation displays mean values of $\delta^{18}O$ close to −6/−7‰, although some isolated data from the Tabia Member at the Tagrara inlier is as low as −14‰ reflecting influence by hydrothermal alteration (Maloof et al., 2005).

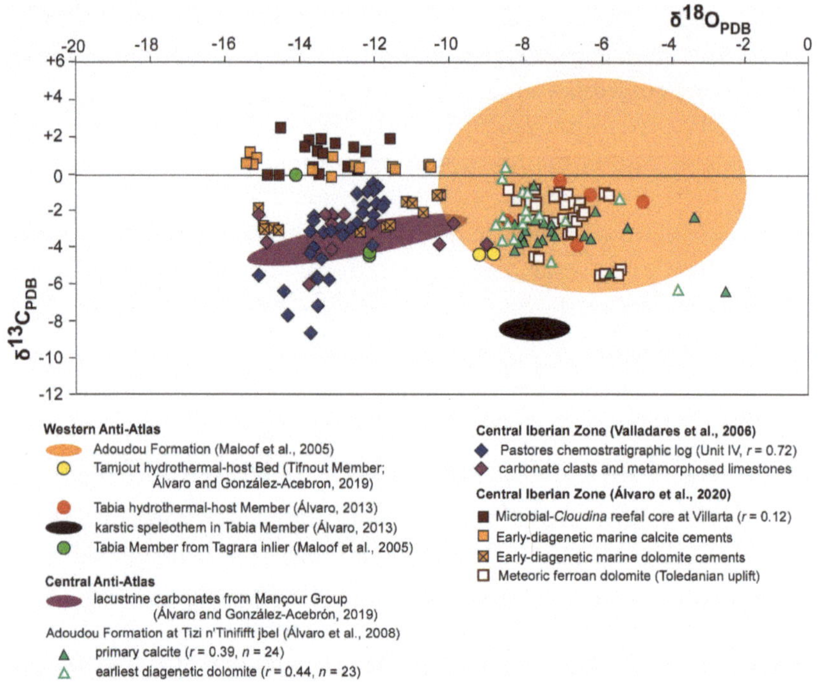

Fig. 6 Bivariate $\delta^{13}C_{carb}$ vs. $\delta^{18}O_{carb}$ plot of the Ediacaran–Cambrian transition in the Central Iberian Zone and the Terreneuvian–Cambrian Series 2 of the Anti-Atlas and High Atlas, Morocco. *Based on Maloof, A.C., Schrag, D.P., Crowley, J.L., Bowring, S.A., 2005. An expanded record of early Cambrian carbon cycling for the anti-atlas margin, Morocco. Can. J. Earth Sci. 42, 2195–2216, Valladares, M.I., Ugidos, J.M., Barba, P., Fallick, A.E., Ellam, R.M., 2006. Oxygen, carbon and strontium isotope records of Ediacaran carbonates in Central Iberia (Spain). Precambrian Res. 147, 354–365, Álvaro, J.J., Bauluz, B., Gil Imaz, A., Simón, J.L., 2008a. Multidisciplinary constraints about Cadomian compression and early Cambrian extension in the Iberian chains, NE Spain. Tectonophysics 461, 215–227, Álvaro, J.J., Bauluz, B., Pierre, C., Subías, I., Vizcaïno, D., 2008b. Carbon chemostratigraphy of the Cambrian–Ordovician transition in a midlatitude mixed platform, Montagne Noire, France. Geol. Soc. Am. Bull. 120, 962–975, Álvaro, J.J., Macouin, M., Ezzouhairi, H., Ribeiro, M.L., Ader, M., Charif, A., Ait-Ayad, N., 2008c. Late Neoproterozoic carbonate productivity in a rifting context: the Adoudou formation and its associated bimodal volcanism onlapping the western Saghro inlier, Morocco. In: Ennih, N., Liégeois, J.P. (Eds.), The boundaries of the west African Craton. Geological Society London Special Publications, vol. 297. pp. 285–302, Álvaro, J.J., Cortijo, I., Jensen, S., Martí Mus, M., Palacios, T., 2020. Cloudina-microbial reef resilience to substrate instability in a Cadomian retro-arc basin of the Iberian Peninsula. Precambrian Res. 336, 105479, Álvaro, J.J., 2013. Late Ediacaran syn-rift/post-rift transition and related fault-driven hydrothermal systems in the anti-Atlas Mountains, Morocco. Basin Res. 25, 348–360, and Álvaro, J.J., González-Acebrón, L., 2019. Sublacustrine hydrothermal seeps and silicification of microbial bioherms in the Ediacaran Oued Dar'a caldera, anti-atlas, Morocco. Sedimentology 66, 2048–2071.*

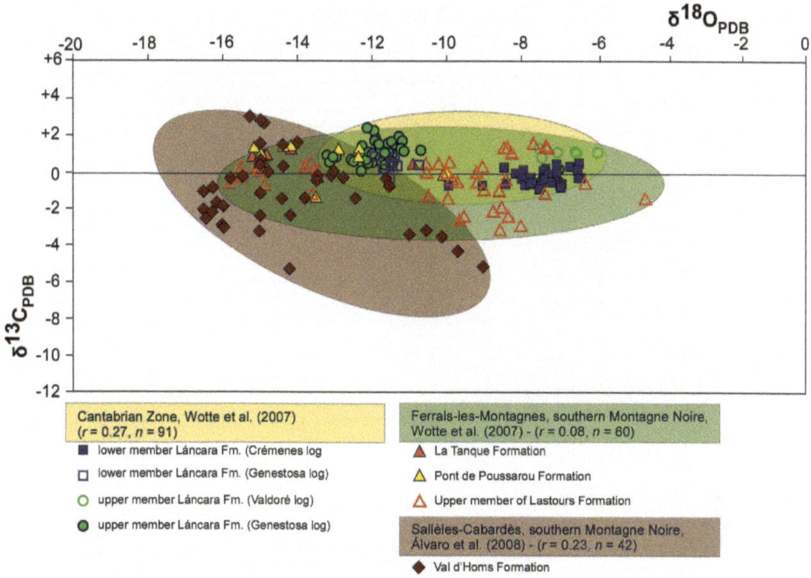

Fig. 7 Bivariate $\delta^{13}C_{carb}$ vs. $\delta^{18}O_{carb}$ plot of the regional "lower–middle Cambrian" transition from the southern Montagne Noire and the regional "middle–upper Cambrian" transition from the southern Montagne Noire (Minervois and Pardailhan nappes). *Based on Álvaro, J.J., Bauluz, B., Pierre, C., Subías, I., Vizcaïno, D., 2008b. Carbon chemostratigraphy of the Cambrian–Ordovician transition in a midlatitude mixed platform, Montagne Noire, France. Geol. Soc. Am. Bull. 120, 962–975. Wotte, T., Álvaro, J.J., Shields, G.A., Brasier, M., Veizer, J., 2007. C-, O- and Sr–isotope stratigraphy across the Lower–Middle Cambrian transition of the Cantabrian Mountains (Spain) and the Montagne Noire (France), West Gondwana. Palaeogeogr. Palaeoclimatol. Palaeoecol. 256, 47–70.*

Therefore, excluding isotopic data from karstic speleothems from the Tabia Member, all the Neoproterozoic–Terreneuvian data plot in a continuous domain of the $\delta^{13}C/\delta^{18}O$ field, with partial overlapping areas from higher values of post–Panafrican (Adoudou Group) to lower values of pre–Panafrican and syn–Panafrican (Taghdout Group and Mançour lacustrine carbonates). This reflects a possible overprint of the isotopic signal by burial trend and the influence of hydrothermal processes related to the onset of the Panafrican Orogeny (Álvaro and González-Acebrón, 2019).

In the Central Iberian Zone, another distinct cluster includes the synsedimentary and early-diagenetic dataset from the Ibor Group (including Unit IV of Valladares et al., 2006), with $\delta^{13}C_{carb}$ values broadly ranging from +3 to −9‰, and $\delta^{18}O_{carb}$ values from −9 to −16‰. The meteoric ferroan dolomite reflecting the influence of the Toledanian break-up event shows

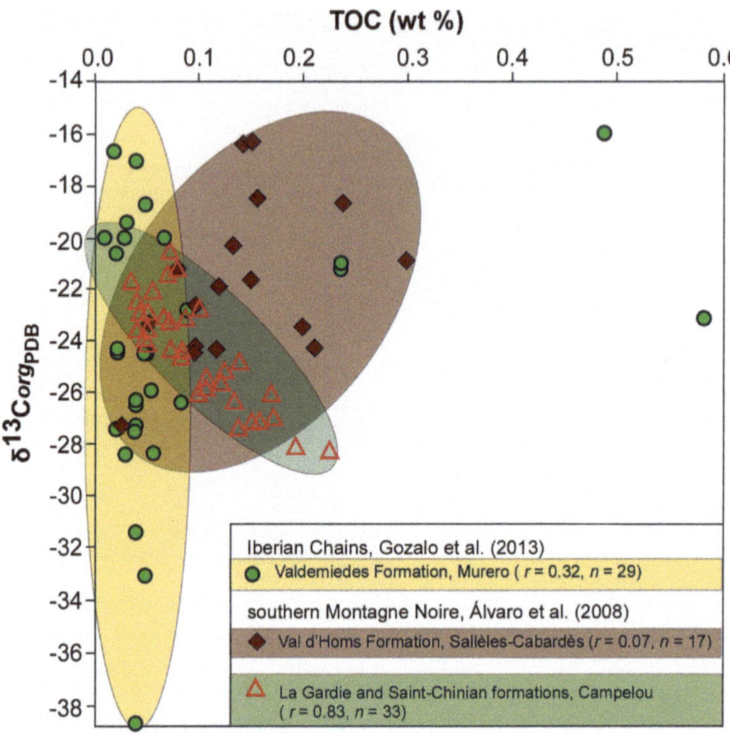

Fig. 8 Bivariate $\delta^{13}C_{org}$ vs. TOC plot of the regional "lower–middle Cambrian" transition in the Iberian Chains and the regional "middle–upper Cambrian" transition from the southern Montagne Noire (Minervois and Pardailhan nappes). *Based on Álvaro, J.J., Bauluz, B., Pierre, C., Subías, I., Vizcaïno, D., 2008b. Carbon chemostratigraphy of the Cambrian–Ordovician transition in a midlatitude mixed platform, Montagne Noire, France. Geol. Soc. Am. Bull. 120, 962–975. Gozalo, R., Dies Álvarez, M.E., Gámez Vintaned, J.A., Zhuravlev, A.Yu., Bauluz, B., Subías, I., Chirivella Martorell, J.B., Mayoral, E., Gursky, H.J., Andrés, J.A., Liñán, E., 2013. Proposal of a reference section and point for the Cambrian series 2–3 boundary in the Mediterranean subprovince in Murero (NE Spain) and its intercontinental correlation. Geochem. J. 48, 142–155.*

less depleted $\delta^{18}O_{carb}$ values, falling within the Anti-Atlas cluster. The sharp difference in $\delta^{18}O_{carb}$ values that allows distinction of the Anti-Atlas and Central Iberian clusters reflects the poor vs. strong influence of Variscan tectono-thermal effects, which preferentially depleted the $\delta^{18}O_{carb}$ values in the Iberian Massif. Another depletion in $\delta^{18}O_{carb}$ values is highlighted in the lacustrine carbonates embedded in the Ediacaran Ouarzazate Supergroup of the Anti-Atlas reflecting an important influence of meteoric waters and a relatively high $\delta^{13}C_{carb}/\delta^{18}O_{carb}$ covariance (Álvaro and González-Acebrón, 2019).

The bivariate plot of the carbonates crossing the regional "lower–middle Cambrian" boundary (Fig. 7) (Wotte et al., 2007) shows a single cluster with wider isotopic distributions recorded in the southern Montagne Noire (with $\delta^{13}C_{carb}$ values broadly ranging from +2 to −3‰, and $\delta^{18}O_{carb}$ values from −4 to −16‰) than in the Cantabrian Zone (ranging from +2 to −2‰, and from −13 to −6‰, respectively). The bivariate plot of the carbonates crossing the regional "middle–upper Cambrian" boundary in the southern Montagne Noire (Fig. 7) shows a larger cluster, with $\delta^{13}C_{carb}$ values broadly ranging from +4 to −6‰, and $\delta^{18}O_{carb}$ values from −9 to −17‰. Covariance indexes are low in all the dataset.

Finally, the $\delta^{18}O_{org}$ vs. TOC bivariate plot (Fig. 8) for shales deposited across the "lower–middle Cambrian" (Gozalo et al., 2013) and "middle–upper Cambrian" (Álvaro et al., 2008b) boundary intervals reflect (i) a dramatic impoverishment of TOC values for the Valdemiedes rivulet stratotype with 24 of the 29 samples yielding TOC values <1%; (ii) an appropriate relationship of data in the Sallèles-Cabardès mixed Val d'Homs Formation, with $\delta^{13}C_{org}$ ranging from −27 to −16‰ and TOC contents from 0.02 to 0.3%; and (iii) a linear distribution of values in the Campelou section with a high covariance index (r=0.83) reflecting the $\delta^{13}C_{org}$/TOC relationship pattern.

7. Conclusions

This work summarizes a critical re-appraisal of carbon and oxygen isotopic datasets from Cambrian carbonate and shale successions along the axes and flanks of the Atlas–Ossa-Morena Rift, West Gondwana. The primary methodological problems found in previous works are shared by similar Cambrian exposures affected by diagenetic, metamorphic, hydrothermal and short-term uplifts associated with pre-rift, *syn*-rift and break-up events punctuating the evolution of the rift. As a result, early-diagenetic dissolution processes and the subsequent occlusion of the involved secondary porosity by meteoric cements point to complex multi-stage parageneses. Analysis of bulk rock and homogeneous (micro-)sparry textures in carbonate samples bearing so complex diagenetic processes are not adequate. Isotopic analyses made in shales with TOC content <1%, strongly affected by pyrolysis and thermal analysis, are not sensitive enough to be applied.

The chemostratigraphic BACE shift broadly marking the Ediacaran–Cambrian boundary is well constrained in the Anti-Atlas and High Atlas due to the number of sections analyzed. However, the chemostratigraphic anomaly is also related to the pre-rift unconformity marking the base of the Adoudou Formation and the syn-rift episodes locally punctuating the contacts of the Taguedit Bed, the Tabia Member and the Tamjout Bed (Adoudou Formation). The hydrothermally induced precipitation of secondary calcite and dolomite is highlighted there by covariance of $\delta^{13}C/\delta^{18}O$ datasets, strong modifications in $\delta^{18}O$ values (controlled by diagenetic palaeotemperatures) and its association with other hydrothermal byproducts. The BACE is tentatively recognized in the Central Iberian Zone of Spain, though further correlations are needed regionally to constrain this shift throughout the complete succession bracketed between the latest Ediacaran fossils and the first occurrence of *Treptichnus pedum*.

Some chemostratigraphic shifts cannot be correlated with global anomalies. The record of an erosive unconformity marking the regional "lower–middle Cambrian" boundary in the Cantabrian Zone and the southern Montagne Noire probably precludes recognition of the ROECE. However, until four minor fluctuations in $\delta^{13}C$ values crossing the series transition are useful for regional correlation and represent similar environmental conditions throughout horst-and-graben palaeotopographies fringing the main Atlas–Ossa-Morena rift axes. Finally, despite the absence of Furongian strata in the vicinity of shoulders fringing the main rift axes, the episodic preservation of carbonate production in the southern Montagne Noire has allowed identification of the SPICE event on the top of lateral horsts recording alternations of carbonate and shale (griotte facies). Other chemostratigraphic shifts are still in need of chronostratigraphic and radiometric constraints before global correlation.

Since the petrographic and geochemical features of the carbonate phases described above clearly indicate a rather complex multi-stage diagenetic influence, future research should establish a best-fit chemostratigraphic framework by separating micritic (or microsparitic) sedimentary vs. early-diagenetic calcite $\delta^{13}C_{carb}/\delta^{18}O_{carb}$ datasets. Analysis of $\delta^{13}C_{org}$ in shales should encounter samples with TOC content $>1\%$, a problem related to kerogen preservation commonly affected by diagenetic and metamorphic resetting.

Acknowledgments

This paper is a contribution to project CGL2017-87631-P from Spanish Ministry of Science and Innovation.

References

Allen, J.R., Matthews, R.K., 1982. Isotopic signature associated with early meteoric diagenesis. Sedimentology 29, 791–817.

Álvaro, J.J., 2007. New ellipsocephalid trilobites from the lower Cambrian member of the Láncara formation, Cantabrian Mountains, northern Spain. Mem. Assoc. Australas. Paleontol. 34, 29–41.

Álvaro, J.J., 2013. Late Ediacaran syn-rift/post-rift transition and related fault-driven hydrothermal systems in the Anti-Atlas Mountains, Morocco. Basin Res. 25, 348–360.

Álvaro, J.J., Blanc-Valleron, M.M., 2002. Stratigraphic and structural framework of the Neoproterozoic Paracuellos Group, Iberian Chains, NE Spain. Bull. Soc. Géol. Fr. 173 (3), 27–35.

Álvaro, J.J., Clausen, S., 2005. Major geodynamic and sedimentary constraints on the chronostratigraphic correlation of the Lower–Middle Cambrian transition in the western Mediterranean region. Geol. J. 9, 145–160.

Álvaro, J.J., Debrenne, F., 2010. The Great Atlasian Reef Complex: an early Cambrian subtropical fringing belt that bordered West Gondwana. Palaeogeogr. Palaeoclimatol. Palaeoecol. 294, 120–132.

Álvaro, J.J., González-Acebrón, L., 2019. Sublacustrine hydrothermal seeps and silicification of microbial bioherms in the Ediacaran Oued Dar'a caldera, Anti-Atlas, Morocco. Sedimentology 66, 2048–2071.

Álvaro, J.J., Subías, I., 2011. Interplay of phosphogenesis and hydrothermalism in the latest Ediacaran rift of the High Atlas, Morocco. J. Afr. Earth Sci. 59, 51–60.

Álvaro, J.J., Vennin, E., 1996. Tectonic control on Cambrian sedimentation in South-Western Europe. Eclogae Geol. Helv. 89, 935–948.

Álvaro, J.J., Vennin, E., 1998. Stratigraphic signature of a terminal early Cambrian regressive event in the Iberian Peninsula. Can. J. Earth Sci. 35, 402–411.

Álvaro, J.J., Vizcaïno, D., 1998. Révision biostratigraphique du Cambrien moyen du versant méridional de la Montagne Noire (Languedoc, France). Bull. Soc. Géol. Fr. 169, 233–242.

Álvaro, J.J., Vizcaïno, D., 2018. The Furongian break-up (rift/drift) unconformity in the Anti-Atlas, Morocco. J. Iber. Geol. 44, 567–587.

Álvaro, J.J., Gozalo, R., Liñán, E., Sdzuy, K., 1993. The palaeogeography of northern Iberia at the Lower–Middle Cambrian transition. Bull. Soc. Géol. Fr. 164, 843–850.

Álvaro, J.J., Liñán, E., Vizcaïno, D., 1998a. Biostratigraphical significance of the genus *Ferralsia* (lower Cambrian, Trilobita). Geobios 31, 499–504.

Álvaro, J.J., Chauvel, J.J., Courjault-Radé, P., Dabard, M.P., Debrenne, F., Feist, R., Pillola, G.L., Vennin, E., Vizcaïno, D., 1998b. Nouveau découpage stratigraphique du Cambrien des nappes de Pardailhan et du Minervois (versant sud de la Montagne Noire, France). Géol. Fr. 1998 (2), 3–12.

Álvaro, J.J., Vennin, E., Vizcaïno, D., 1999. Depositional controls on Early Cambrian microbial carbonates from the Montagne Noire, southern France. Trans. R. Soc. Edinb. Earth Sci. 89, 135–143.

Álvaro, J.J., Vennin, E., Moreno-Eiris, E., Perejón, A., Bechstädt, T., 2000a. Sedimentary patterns across the Lower–Middle Cambrian transition in the Esla nappe (Cantabrian Mountains, northern Spain). Sediment. Geol. 137, 43–61.

Álvaro, J.J., Vennin, E., Muñoz, A., Sánchez-Valverde, B., Ojeda, J.L., 2000b. Interplay of orbital forcing and tectonic pulses in the Cambrian Iberian platform, NE Spain. Int. J. Earth Sci. 89, 366–376.

Álvaro, J.J., González-Gómez, C., Vizcaïno, D., 2003a. Paleogeographic patterns of the Cambrian–Ordovician transition in the southern Montagne Noire (France): preliminary results. Bull. Soc. Géol. Fr. 174 (3), 23–31.

Álvaro, J.J., Van Vliet-Lanoë, B., Vennin, E., Blanc-Valleron, M.M., 2003b. Lower Cambrian paleosols form the Cantabrian Mountains (northern Spain): a comparison with Neogene–Quaternary estuarine analogues. Sediment. Geol. 163, 67–84.

Álvaro, J.J., Clausen, S., El Albani, A., Chellai, E.H., 2006a. Facies distribution of lower–Cambrian cryptic microbial and epibenthic archaeocyathan-microbial communities in the western Anti-Atlas, Morocco. Sedimentology 53, 35–53.

Álvaro, J.J., Ezzouhairi, H., Vennin, E., Ribeiro, M.L., Clausen, S., Charif, A., Aït-Ayad, N., Moreira, M.E., 2006b. The Early–Cambrian Boho volcano of El Graara massif, Morocco: petrology, geodynamic setting and coeval sedimentation. J. Afr. Earth Sci. 44, 396–410.

Álvaro, J.J., Ferretti, A., González-Gómez, C., Serpagli, E., Tortello, M.F., Vecoli, M., Vizcaïno, D., 2007. A review of the Late Cambrian (Furongian) palaeogeography in the western Mediterranean region, NW Gondwana. Earth-Sci. Rev. 85, 47–81.

Álvaro, J.J., Bauluz, B., Gil Imaz, A., Simón, J.L., 2008a. Multidisciplinary constraints about Cadomian compression and early Cambrian extension in the Iberian Chains, NE Spain. Tectonophysics 461, 215–227.

Álvaro, J.J., Bauluz, B., Pierre, C., Subías, I., Vizcaïno, D., 2008b. Carbon chemostratigraphy of the Cambrian–Ordovician transition in a midlatitude mixed platform, Montagne Noire, France. Geol. Soc. Am. Bull. 120, 962–975.

Álvaro, J.J., Macouin, M., Ezzouhairi, H., Ribeiro, M.L., Ader, M., Charif, A., Ait-Ayad, N., 2008c. Late Neoproterozoic carbonate productivity in a rifting context: the Adoudou formation and its associated bimodal volcanism onlapping the western Saghro inlier, Morocco. In: Ennih, N., Liégeois, J.P. (Eds.), The Boundaries of the West African Craton. Geol. Soc., London, Spec. Publ., 297, pp. 285–302.

Álvaro, J.J., Bauluz, B., Clausen, S., Devaere, L., Gil Imaz, A., Monceret, E., Vizcaïno, D., 2014a. Stratigraphic review of the Cambrian–Lower Ordovician volcanosedimentary complexes from the northern Montagne Noire, France. Stratigraphy 11, 83–96.

Álvaro, J.J., Bellido, F., Gasquet, D., Pereira, F., Quesada, C., Sánchez-García, T., 2014b. Diachronism of late Neoproterozoic-Cambrian arc-rift transition of North Gondwana: a comparison of Morocco and the Iberian Ossa-Morena Zone. J. Afr. Earth Sci. 98, 113–132.

Álvaro, J.J., Benziane, F., Thomas, A.R., Walsh, G.J., Yazidi, A., 2014c. Neoproterozoic–Cambrian stratigraphic framework of the Anti-Atlas and Ouzellagh promontory (High Atlas), Morocco. J. Afr. Earth Sci. 98, 19–33.

Álvaro, J.J., Shields, G.A., Ahlberg, P., Jensen, S., Palacios, T., 2016. Ediacaran–Cambrian phosphorites from the western margins of Gondwana and Baltica. Sedimentology 63, 350–377.

Álvaro, J.J., Bauluz, B., Gil-Imaz, A., Ubide, T., 2018a. $^{40}Ar/^{39}Ar$ chronological constraints on syn- and post–Variscan biotite porphyroblasts from the Iberian Chains, NE Spain. J. Iber. Geol. 44, 655–670.

Álvaro, J.J., Casas, J.M., Clausen, S., Quesada, C., 2018b. Early Palaeozoic geodynamics in NW Gondwana. J. Iber. Geol. 44, 551–565.

Álvaro, J.J., Cortijo, I., Jensen, S., Lorenzo, S., Palacios, T., Pieren, A.P., 2019. Updated stratigraphic framework and biota of the Ediacaran and Terreneuvian in the Alcudia-Toledo Mountains of the Central Iberian Zone, Spain. Estud. Geol. 75 (2), e093.

Álvaro, J.J., Cortijo, I., Jensen, S., Martí Mus, M., Palacios, T., 2020. *Cloudina*-microbial reef resilience to substrate instability in a Cadomian retro-arc basin of the Iberian Peninsula. Precambrian Res. 336, 105479.

Aramburu, C., Méndez-Bedia, I., Arbizu, M., García-López, S., 2004. Zona Cantábrica. Estratigrafía. La secuencia preorogénica. In: Vera, J. (Ed.), Geología de España. Sociedad Geológica de España & Instituto Geológico y Minero de España, Madrid, pp. 27–34.

Bartley, J.K., Kah, L.C., 2004. Marine carbon reservoir C_{org}–C_{carb} coupling, and the evolution of the Proterozoic carbon cycle. Geology 32, 129–132.

Bastida, F., Brime, C., García-López, S., Sarmiento, G.N., 1999. Tectonothermal evolution in a region with thin-skinned tectonics: the western nappes of the Cantabrian Zone (Variscan belt of NW Spain). J. Iber. Geol. 88, 34–48.

Bauluz, B., Fernández Nieto, C., González López, J.M., 1998. Diagenesis-very low grade metamorphism of clastic Cambrian and Ordovician sedimentary rocks in the Iberian Range (Spain). Clay Miner. 33, 373–393.

Blein, O., Baudin, T., Chèvremont, P., Soulaimani, A., Admou, H., Gasquet, D., Cocherie, A., Egal, E., Youbi, N., Razin, P., Bouabdelli, N., Gombert, P., 2014. Geochronological constraints on the polycyclic magmatism in the Bou Azzer-El Graara inlier (Central Anti-Atlas, Morocco). J. Afr. Earth Sci. 99, 287–306.

Bowring, S.A., Grotzinger, J.P., Isachsen, C.E., Knoll, A.H., Pelechaty, S.M., Kolosov, P., 1993. Calibrating rates of early Cambrian evolution. Science 261, 1293–1298.

Bruckschen, P., Oesmann, S., Veizer, J., 1999. Isotope stratigraphy of the European carboniferous: proxy signals for ocean chemistry, climate and tectonics. Chem. Geol. 161, 127–163.

Burkhard, M., Caritg, S., Helg, U., Robert-Charrue, C., Soulaimanai, A., 2006. Tectonics of the Anti-Atlas of Morocco. C. R. Geosci. 338, 11–24.

Calver, C.R., 2000. Isotope stratigraphy of the Ediacaran (Neoproterozoic III) of the Adelaide rift complex, Australia, and the overprint of water column stratification. Precambrian Res. 100, 121–150.

Chèvremont, P., Blein, O., Razin, P., Baudin, T., Barbanson, L., Gasquet, D., Soulaimani, A., Admou, H., Youbi, N., Bouabdelli, M., Anzar-Conseil, 2013. Carte géologique du Maroc au 1/50.000, feuille Bou Azzer. In: Notes et Mémoires du Service géologique du Maroc 535, Notice explicative 535 bis.

Clausen, S., Álvaro, J.J., 2006. Skeletonized microfossils from the Lower–Middle Cambrian transition of the Cantabrian Mountains, northern Spain. Acta Palaeontol. Pol. 51, 223–238.

Compston, W., Williams, I.S., Kirschvink, J.L., Zichao, Z., Guogan, M., 1992. Zircon U–Pb ages for the Early Cambrian time-scale. J. Geol. Soc. London 149, 171–184.

Cortijo, I., Martí Mus, M., Jensen, S., Palacios, T., 2010. A new species of *Cloudina* from the terminal Ediacaran of Spain. Precambrian Res. 176, 1–10.

Cortijo, I., Martí Mus, M., Jensen, S., Palacios, T., 2015. Late Ediacaran skeletal body fossil assemblage from the Navalpino anticline, Central Spain. Precambrian Res. 267, 186–195.

Debrenne, F., Debrenne, M., 1995. Archaeocyaths of the Lower Cambrian of Morocco. Beringeria 2, 121–145. Special Issue.

Devaere, L., Clausen, S., Steiner, M., Álvaro, J.J., Vachard, D., 2013. Chronostratigraphic and palaeogeographic significance of an early Cambrian microfauna from the *Heraultia* limestone, northern Montagne Noire, France. Palaeontol. Electron. 16 (17A), 1–91.

Dilliard, K.A., Pope, M.C., Coniglio, M., Hasiotis, S.T., Lieberman, B.S., 2007. Stable isotope geochemistry of the lower Cambrian Sekwi formation, Northwest Territories, Canada: implications for ocean chemistry and secular curve generation. Palaeogeogr. Palaeoclimatol. Palaeoecol. 256, 174–194.

Ducrot, J., Lancelot, J.R., 1977. Problème de la limite Précambrien–Cambrien: étude radiochronologique par la méthode U–Pb sur zircon du Jbel Boho (Anti-Atlas marocain). Can. J. Earth Sci. 14, 2771–2777.

Engel, W., Feist, R., Franke, W., 1981. Le Carbonifère anté–Stéphanien de la Montagne Noire: rapports entre mise en place des nappes et sédimentation. Bull. BRGM 4, 341–389. série 2.

Feist, R., Courtessole, R., 1984. Découverte de Cambrien supérieur à trilobites de type est-asiatique dans la Montagne Noire (France méridionale). C. R. Acad. Sci. 298, 177–182. Paris (2e Série).
Fox, P.M., Nico, P.S., Tfaily, M.M., Heckman, K., Davis, J.A., 2017. Characterization of natural organic matter in low-carbon sediments: extraction and analytical approaches. Org. Geochem. 114, 12–22.
Frey, M., Robinson, D., 1999. Low-Grade Metamorphism. Blackwell Science, Oxford.
García-López, S., Brime, C., Valín, M.L., Sanz-López, J., Bastida, F., Aller, J., Blanco-Ferrera, S., 2007. Tectonothermal evolution of a foreland fold and thrust belt: the Cantabrian Zone (Iberian Variscan belt, NW Spain). Terra Nova 19, 469–475.
Gasquet, D., Levresse, G., Cheilletz, A., Azizi-Samir, M.R., Mouttaqi, A., 2005. Contribution to a geodynamic reconstruction of the Anti-Atlas (Morocco) during Pan-African times with the emphasis on inversion tectonics and metallogenic activity at the Precambrian–Cambrian transition. Precambrian Res. 140, 157–182.
Gasquet, D., Ennih, N., Liegeois, J.P., Soulaimani, A., Michard, A., 2008. The Pan-African belt. In: Michard, A., Chalouan, A., Saddiqi, O., de Lamotte, D. (Eds.), Continental Evolution: The Geology of Morocco Structure, stratigraphy, and tectonics of the African-Atlantic-Mediterranean triple junction. Lect. Notes Earth Sci., 116, pp. 33–64.
Geyer, G., 2019. The earliest known West Gondwanan trilobites from the Anti-Atlas of Morocco, with a revision of the Family Bigotinidae Hupé, 1953. Fossils Strata 64, 55–153.
Geyer, G., Elicki, O., 1995. The Lower Cambrian trilobites from the Görlitz Synclinorium (Germany)—review and new results. Paläontol. Z. 69, 87–119.
Glumac, B., Walker, K.R., 1998. A Late Cambrian positive carbon-isotope excursion in the southern Appalachians: relation to biostratigraphy, sequence stratigraphy, environments of deposition, and diagenesis. J. Sediment. Res. 68, 1212–1222.
González-Gómez, C., 2005. Linguliformean brachiopods of the Middle–Upper Cambrian transition from the Val d'Homs formation, southern Montagne Noire, France. J. Paleo. 79, 29–47.
Gozalo, R., Liñán, E., Dies Álvarez, M.E., Gámez Vintaned, J.A., Mayoral, E., 2007. The Lower–Middle Cambrian boundary in the Mediterranean subprovince. In: Linnemann, U., Nance, R.D., Kraft, P., Zulauf, G. (Eds.), The Evolution of the Rheic Ocean: From Avalonian-Cadomian active margin to Alleghenian-Variscan collision. Geol. Soc.Am., Spec. Pap., 423, pp. 359–373.
Gozalo, R., Dies Álvarez, M.E., Gámez Vintaned, J.A., Zhuravlev, A.Y., Bauluz, B., Subías, I., Chirivella Martorell, J.B., Mayoral, E., Gursky, H.J., Andrés, J.A., Liñán, E., 2013. Proposal of a reference section and point for the Cambrian Series 2–3 boundary in the Mediterranean subprovince in Murero (NE Spain) and its intercontinental correlation. Geochem. J. 48, 142–155.
Grossman, E.L., Zhang, C., Yancey, T.E., 1991. Stable isotope stratigraphy of brachiopods from Pennsylvanian shales in Texas. Geol. Soc. Am. Bull. 103, 953–965.
Grotzinger, J.P., Bowring, S.A., Saylor, B.Z., Kaufman, A.J., 1995. Biostratigraphic and geochronologic constraints on early animal evolution. Science 270, 598–604.
Hayes, J.M., Kaplan, I.R., Wedeking, K.W., 1983. Precambrian organic geochemistry, preservation of the record. In: Schopf, J.W. (Ed.), The Earth's Earliest Biosphere: Its Origin and Evolution. Princeton University Press, Princeton, pp. 93–134.
Hayes, J.M., Popp, B.N., Takigiku, R., Johnson, M.W., 1989. An isotopic study of biogeochemical relationships between carbonates and organic carbon in the Greenhorn Formation. Geochim. Cosmochim. Acta 53, 2961–2972.
Hayes, J.M., Strauss, H., Kaufman, A.J., 1999. The abundance of ^{13}C in marine organic matter and isotopic fractionation in the global biogeochemical cycle of carbon during the past 800 ma. Chem. Geol. 161, 103–125.

Hough, M.L., Shields, G.A., Evins, L.Z., Strauss, H., Henderson, R.A., Mackenzie, S., 2006. A major sulphur isotope event at 510 Ma: a possible anoxia-extinction-volcanism connection during the Early–Middle Cambrian transition? Terra Nova 18, 257–263.

Immenhauser, A., Holmden, C., Patterson, W.P., 2008. Interpreting the carbon-isotope record of ancient shallow epeiric seas: lessons from the recent. In: Pratt, B.R., Holmden, C. (Eds.), Dynamics of Epeiric Seas. Geol. Assoc. Canada, Spec. Publ., 48, pp. 135–174.

Jacobsen, S., Kaufman, A., 1999. The Sr, C, and O isotopic evolution of Neoproterozoic seawater. Chem. Geol. 161, 37–57.

Jensen, S., Palacios, T., 2016. The Ediacaran–Cambrian trace fossil record in the Central Iberian Zone, Iberian Peninsula. Comun. Geol. 103, 83–92.

Jensen, S., Palacios, T., Martí Mus, M., 2007. A brief review of the fossil record of the Ediacaran–Cambrian transition in the area of Montes de Toledo-Guadalupe (Spain). In: Vickers-Rich, P., Komarover, P. (Eds.), The Rise and Fall of the Ediacaran Biota. Geol. Soc., London, Spec. Publ., 286, pp. 223–235.

Jensen, S., Palacios, T., Martí Mus, M., 2010. Revised biochronology of the Lower Cambrian of the Central Iberian Zone, southern Iberian massif. Geol. Mag. 147, 690–703.

Jiang, G., Wang, X., Shi, X., Zhang, S., Xiao, S., Dong, J., 2010. Organic carbon isotope constraints on the dissolved organic carbon (DOC) reservoir at the Cryogenian–Ediacaran transition. Earth Planet. Sci. Lett. 299, 159–168.

Jiang, G.Q., Wang, X.Q., Shi, X.Y., Xiao, S., Zhang, S.H., Dong, J., 2012. The origin of decoupled carbonate and organic carbon isotope signatures in the early Cambrian (ca. 542–520 ma) Yangtze platform. Earth Planet. Sci. Lett. 317–318, 96–110.

Joachimski, M.M., 1997. Comparison of organic and inorganic isotope patterns across the Frasnian–Famennian boundary. Palaeogeogr. Palaeoclimatol. Palaeoecol. 132, 133–145.

Karaoui, B., Breitkreuz, C., Mahmoudi, A., Youbi, N., Hofmann, M., Gärtner, A., Linnemann, U., 2015. U–Pb zircon ages from volcanic and sedimentary rocks of the Ediacaran Bas Draâ inlier (Anti-Atlas Morocco): chronostratigraphic and provenance implications. Precambrian Res. 263, 43–58.

Kaufman, A.J., Knoll, A.H., 1995. Neoproterozoic variations in the carbon isotopic composition of seawater: stratigraphic and biogeochemical implications. Precambrian Res. 73, 27–49.

Knoll, A.H., Hayes, J.M., Kaufman, A.J., Swett, K., Lambert, I.B., 1986. Secular variation in carbon isotope ratios from Upper Proterozoic successions of Svalbard and East Greenland. Nature 321, 832–838.

Koch, P.L., Zachos, J.C., Gingerich, P.D., 1992. Correlation between isotope records in marine and continental carbon reservoirs near the Palaeocene/Eocene boundary. Nature 358, 319–322.

Landing, E., Bowring, S.A., Davidek, K.L., Westrop, S.R., Geyer, G., Heldmaier, W., 1998. Duration of the Early Cambrian: U–Pb ages from Avalon and Gondwana. Can. J. Earth Sci. 35, 329–338.

Lescuyer, J.L., Cocherie, A., 1992. Datation sur monozircons des métadacites de Sériès: arguments pour un âge protérozoïque terminal des schistes X de la Montagne Noire (Massif central français). C. R. Acad. Sci. Paris 314, 1071–1077.

Lin, J.P., Sundberg, F.A., Jiang, G.Q., Montañez, I.P., Wotte, T., 2019. Chemostratigraphic correlations across the first major trilobite extinction and faunal turnovers between Laurentia and South China. Sci. Rep. 9, 17392.

Liñán, E., Perejón, A., Sdzuy, K., 1993. The Lower–Middle Cambrian stages and stratotypes from the Iberian Peninsula: a revision. Geol. Mag. 130, 817–833.

Liñán, E., Gámez Vintaned, J.A., Gozalo, R., 2015. The middle lower Cambrian (Ovetian) *Lunagraulus* n. gen. from Spain and the oldest trilobite records. Geol. Mag. 152, 1123–1136.

Magaritz, M., Kirschvink, J., Latham, A., Zhuralev, A., Rozanov, A., 1991. Precambrian/ Cambrian boundary problem: carbon isotope correlations for Vendian and Tommotian time between Siberia and Morocco. Geology 19, 847–850.

Magaritz, M., Krishnamurthy, R.V., Holser, T., 1992. Parallel trends in organic and inorganic carbon isotopes across the Permian/Triassic boundary. Am. J. Sci. 292, 727–739.

Maliva, R.G., Dickson, J.A.D., Smalley, P.C., Oxtoby, N.H., 1995. Diagenesis of the Machar field (British North-Sea) chalk: evidence for decoupling of diagenesis in fractures and the host rock. J. Sediment. Res. 65, 105–111.

Maloof, A.C., Schrag, D.P., Crowley, J.L., Bowring, S.A., 2005. An expanded record of Early Cambrian carbon cycling for the Anti-Atlas margin, Morocco. Can. J. Earth Sci. 42, 2195–2216.

Maloof, A.C., Ramezani, J., Bowring, S.A., Fike, D.S., Porter, S.M., Mazouad, M., 2010. Constraints on early Cambrian carbon cycling from the duration of the Nemakit-Daldynian–Tommotian boundary $\delta^{13}C$ shift, Morocco. Geology 38 (7), 623–626.

Marshall, J.D., 1992. Climatic and oceanographic isotopic signals from the carbonate rock record and their preservation. Geol. Mag. 129, 143–160.

Michard, A., Hoepfner, C., Soulaimani, A., Baidder, L., 2008. The Variscan belt. In: Michard, A., Saddiqi, O., Chalouan, A., Frizon de Lamotte, D. (Eds.), Continental Evolution: The Geology of Morocco. Lecture Notes in Earth Sciences, 116. Springer, Berlin, pp. 65–132.

Miller, J.F., Ethington, R.L., Evans, K.R., Holmer, L.E., Coch, J.D., Popov, L.E., Repetski, J.E., Ripperdan, R.L., Taylor, J.F., 2006. Proposed stratotype for the base of the highest Cambrian stage at the first appearance datum of *Cordylodus andresi*, Lawson cow section, Utah, USA. Palaeoworld 15, 384–405.

Monninger, W., 1979. The section of Tiout (Precambrian/Cambrian boundary beds, Anti-Atlas, Morocco): an environmental model. Arb. Paläont. Inst. Würzburg 1, 1–289.

Montañez, I.P., Osleger, D.A., Banner, J.L., Mack, L.E., Musgrove, M., 2000. Evolution of the Sr and C isotope composition of Cambrian oceans. GSA Today 10, 1–7.

Oehlerich, M., Baumer, M., Lücke, A., Mayr, C., 2013. Effects of organic matter on carbonate stable isotope ratios ($\delta^{13}C$, $\delta^{18}O$ values)—implications for analyses of bulk sediments. Rapid Commun. Mass Spectrom. 27 (6), 707–712.

Oehlert, A., Swart, P.K., 2014. Interpreting carbonate and organic carbon isotope covariance in the sedimentary record. Nat. Commun. 5, 4672. https://doi.org/10.1038/ncomms5672.

Oummouch, A., Esaifi, A., Zayane, R., Maddi, O., Zouhait, M., Maacha, L., 2017. Geology and metallogenesis of the sediment-hosted Cu–Ag deposit of Tizert (Igherm inlier, Anti-Atlas Copperbelt, Morocco). *Geofluids*, ID 7508484.

Padel, M., Álvaro, J.J., Clausen, S., Guillot, F., Poujol, M., Chichorro, M., Monceret, E., Pereira, M.F., Vizcaïno, D., 2017. U–Pb laser ablation ICP–MS zircon dating across the Ediacaran–Cambrian transition of the Montagne Noire, southern France. C. R. Geosci. 349, 380–390.

Palacios, T., Vidal, G., 1992. Lower Cambrian acritarchs from northern Spain: the Precambrian–Cambrian boundary and biostratigraphic implications. Geol. Mag. 129, 421–436.

Palacios, T., Högström, A.E.S., Ebbestad, J.O.R., Agić, H., Høyberget, M., Jensen, S., Meinhold, G., Taylor, W.L., 2020. Acritarchs from the Duolbagáisá formation (Cambrian Series 2, Miaolingian) on the Digermulen Peninsula, Finnmark, Arctic Norway: towards a high resolution Cambrian chronostratigraphy. Geol. Mag. https://doi.org/10.1017/S001675682000031X.

Perejón, A., Moreno-Eiris, E., Bechstädt, T., Menéndez, S., Rodríguez-Martínez, M., 2012. New Bilbilian (early Cambrian) archaeocyath-rich thrombolitic microbialite from the Láncara formation (Cantabrian Mts., northern Spain). J. Iber. Geol. 38 (2), 313–330.

Pouclet, A., Ouazzani, H., Fekkak, A., 2008. The Cambrian volcanosedimentary formations of the westernmost High Atlas (Morocco): their place in the geodynamic evolution of the West African Palaeo-Gondwana northern margin. In: Ennih, N., Liégeois, J.P. (Eds.), The Boundaries of the West African Craton. 297. Geol. Soc., London, Spec. Publ, pp. 303–327.

Pouclet, A., Álvaro, J.J., Bardintzeff, J.M., Gil Imaz, A., Monceret, E., Vizcaïno, D., 2017. Cambrian–early Ordovician volcanism across the South Armorican and Occitan domains of the Variscan Belt in France: continental break-up and rifting of the northern Gondwana margin. Geosci. Front. 8, 25–64.

Pouclet, A., El Hadi, H., Álvaro, J.J., Bardintzeff, J.M., Benharref, M., Fekkak, A., 2018. Review of the Cambrian volcanic activity in Morocco: geochemical fingerprints and geotectonic implications for the rifting of West Gondwana. Int. J. Earth Sci. 107, 2101–2123.

Qing, H., Veizer, J., 1994. Oxygen and carbon isotope composition of Ordovician brachiopods: implications for coeval seawater. Geochim. Cosmochim. Acta 58, 4429–4442.

Raiswell, R., Berner, R.A., 1987. Organic carbon losses during burial and thermal maturation of normal marine shales. Geology 15, 853–856.

Robert-Charrue, C., Burkhard, M., 2008. Inversion tectonics, interference pattern and extensional fault-related folding in the Eastern Anti-Atlas, Morocco. Swiss J. Geosci. 101, 397–408.

Rodríguez Alonso, M.D., Peinado, M., López-Plaza, M., Franco, P., Carnicero, A., Gonzalo, J.C., 2004. Neoproterozoic–Cambrian synsedimentary magmatism in the Central Iberian Zone (Spain): geology, petrology and geodynamic significance. Int. J. Earth Sci. 93, 897–920.

Rothman, D.H., Hayes, J.M., Summons, R.E., 2003. Dynamics of the neoproterozoic carbon cycle. PNAS 100 (14), 8124–8129.

Rubio-Ordóñez, A., Gutiérrez-Alonso, G., Valverde-Vaquero, P., Cuesta-Fernández, A., Gallastegui, G., Gerdes, A., Cárdenes, V., 2015. Arc-related Ediacaran magmatism along the northern margin of Gondwana: geochronology and isotopic geochemistry from northern Iberia. Gondw. Res. 27, 216–227.

Ruiz, G.M.H., Helg, U., Negro, F., Adatte, T., Burkhard, M., 2008. Illite crystallinity patterns in the Anti-Atlas of Morocco. Swiss J. Geosci. 101 (2), 387–395.

Saltzman, M.R., Runnegar, B., Lohmann, K.C., 1998. Carbon isotope stratigraphy of Upper Cambrian (Steptoean stage) sequences of the eastern Great Basin: record of a global oceanographic event. Geol. Soc. Am. Bull. 110, 285–297.

Saltzman, M.R., Ripperdan, R.L., Brasier, M.D., Lohmann, K.C., Robison, R.A., Chang, W.T., Peng, S., Ergaliev, E.K., Runnegar, B., 2000. A global carbon isotope excursion (SPICE) during the Late Cambrian: relation to trilobite extinctions, organic-matter burial and sea level. Palaeogeogr. Palaeoclimatol. Palaeoecol. 162, 211–223.

Saltzman, M.R., Young, S.A., Kumo, L.R., Gill, B.C., Lyons, T.W., Runnegar, B., 2011. Pulse of atmospheric oxygen during the late Cambrian. PNAS 108 (10), 3876–3881.

Sánchez-García, T., Chichorro, M., Solá, R., Álvaro, J.J., Díez Montes, A., Bellido, F., Ribeiro, M.L., Quesada, C., Lopes, J.C., Dias da Silva, I., González-Clavijo, E., Gómez Barreiro, J., López-Carmona, A., 2019. The Ediacaran–Early Ordovician rift stage in the Gondwanan units of the Iberian Massif. In: Quesada, C., Oliveira, J.T. (Eds.), The Geology of Iberia: A Geodynamic Approach. Regional Geology Review Series, 2. Springer, Heidelberg, pp. 27–74.

Schrag, D.P., DePaolo, D.J., Richter, F.M., 1995. Reconstructing past sea surface temperatures: correcting for diagenesis of bulk marine carbonate. Geochim. Cosmochim. Acta 59, 2265–2278.

Schumacher, D., 1996. Hydrocarbon-induced alteration of soils and sediments. In: Schumacher, D., Abrams, M.A. (Eds.), Hydrocarbon Migration and its Near-Surface Expression. Am. Ass. Petrol. Geol., 66, pp. 71–89.
Sdzuy, K., Liñán, E., Gozalo, R., 1999. The Leonian stage (early Middle Cambrian): a unit for Cambrian correlation in the Mediterranean subprovince. Geol. Mag. 136, 39–48.
Serpagli, E., Ferretti, A., Vizcaïno, D., Álvaro, J.J., 2007. A new Early Ordovician conodont genus from the southern Montagne Noire, France. Palaeontology 50 (6), 1447–1457.
Shergold, J.H., Sdzuy, K., 1991. Late Cambrian trilobites from the Iberian Mountains, Zaragoza Province, Spain. Beringeria 4, 193–235.
Strasser, A., 2015. Hiatuses and condensation: an estimation of time lost on a shallow carbonate platform. Dep. Rec. 1, 91–117.
Strauss, H., DesMarais, D.J., Hayes, J.M., Summons, R.E., 1992. The carbon-isotope record. In: Schopf, J.W., Klein, C. (Eds.), The Proterozoic Biosphere: A Multidisciplinary Study. Cambridge University Press, Cambridge, pp. 117–128.
Thomas, R.J., Chevallier, L.P., Gresse, P.G., Harmer, R.E., Eglington, B.M., Armstrong, R.A., de Beer, C.H., Martini, J.E.J., de Kock, G.S., Macey, P.H., Ingram, B.A., 2002. Precambrian evolution of the Sirwa Window, Anti-Atlas Orogen, Morocco. Precambrian Res. 118, 1–57.
Tucker, M., 1986. Carbon isotope excursions in Precambrian/Cambrian boundary beds, Morocco. Nature 319, 48–49.
Valladares, M.I., Barba, P., Ugidos, J.M., Colmenero, J.R., Armenteros, I., 2000. Upper Neoproterozoic–Lower Cambrian sedimentary successions in the Central Iberian Zone (Spain): sequence stratigraphy, petrology and chemostratigraphy. Implications for other European zones. Int. J. Earth Sci. 89, 2–20.
Valladares, M.I., Ugidos, J.M., Barba, P., Fallick, A.E., Ellam, R.M., 2006. Oxygen, carbon and strontium isotope records of Ediacaran carbonates in Central Iberia (Spain). Precambrian Res. 147, 354–365.
Veizer, J., Fritz, P., Jones, B., 1986. Geochemistry of brachiopods: oxygen and carbon isotopic records of Paleozoic oceans. Geochim. Cosmochim. Acta 50, 1679–1696.
Veizer, J., Bruckschen, P., Pawellek, F., Diener, A., Podlaha, O.G., Carden, G.A.F., Jasper, T., Korte, C., Strauss, H., Azmy, K., Ala, D., 1997. Oxygen isotope evolution of Phanerozoic seawater. Palaeogeogr. Palaeoclimatol. Palaeoecol. 132, 159–172.
Veizer, J., Ala, D., Azmy, K., Bruckschen, P., Buhl, D., Bruhn, F., Carden, G., Diener, A., Ebneth, S., Godderis, Y., Jasper, T., Korte, C., Pawellek, F., Podlaha, O., Strauss, H., 1999. $^{87}Sr/^{86}Sr$, $\delta^{13}C$ and $\delta^{18}O$ evolution of Phanerozoic seawater. Chem. Geol. 161, 59–88.
Vidal, G., Jensen, S., Palacios, T., 1994a. Neoproterozoic (Vendian) ichnofossils from lower Alcudian strata in Central Spain. Geol. Mag. 131, 169–179.
Vidal, G., Palacios, T., Gámez-Vintaned, J.A., Díaz Balda, M.A., Grant, S.W.F., 1994b. Neoproterozoic–early Cambrian geology and palaeontology of Iberia. Geol. Mag. 131, 729–765.
Wadleigh, M.M., Veizer, J., 1992. $^{18}O/^{16}O$ and $^{13}C/^{12}C$ in lower Paleozoic brachiopods: implications for the isotopic composition of sea water. Geochim. Cosmochim. Acta 56, 431–443.
Walsh, G.J., Benziane, F., Aleinikoff, J.N., Harrison, R.W., Yazidi, A., Burton, W.C., Quick, J.E., Saadane, A., 2012. Neoproterozoic tectonic evolution of the Jebel Saghro and Bou Azzer-El Graara inliers, eastern and central Anti-Atlas, Morocco. Precambrian Res. 216–219, 23–62.
Weissert, H., Joachimski, M.M., Sarnthein, M., 2008. Chemostratigraphy. Newsl. Stratigr. 42, 145–179.

Wiederer, U., Königshof, P., Feist, R., Franke, W., Doublier, M., 2002. Low-grade metamorphism in the Montagne Noire (S-France): Conodont Alteration Index (CAI) in Palaeozoic carbonates and implications for the exhumation of a hot metamorphic core complex. Schweiz. Mineral. Petrogr. Mitt. 82, 393–407.

Wotte, T., Álvaro, J.J., Shields, G.A., Brasier, M., Veizer, J., 2007. C-, O- and Sr–isotope stratigraphy across the Lower–Middle Cambrian transition of the Cantabrian Mountains (Spain) and the Montagne Noire (France), West Gondwana. Palaeogeogr. Palaeoclimatol. Palaeoecol. 256, 47–70.

Zamarreño, I., 1972. Las litofacies carbonatadas del Cámbrico de la Zona Cantábrica (NW España) y su distribución paleogeográfica. Trab. Geol., Univ. Oviedo 5, 1–118.

Zhang, Y.Y., He, Z.L., Jiang, S., Lu, S.F., Xiao, D.S., Chen, G.H., Zhao, J.H., 2018. Controls on the organic carbon content of the lower Cambrian black shale in the southeastern margin of Upper Yangtze. Pet. Sci. 15, 709–721.

Zhu, M.Y., Zhang, J.M., Li, G.X., Yang, X., 2004. Evolution of C isotopes in the Cambrian of China: implications for Cambrian subdivision and trilobite mass extinctions. Geobios 37, 287–310.

Zhu, M.Y., Babcock, L.E., Peng, S.C., 2006. Advances in Cambrian stratigraphy and paleontology: integrating correlation techniques, paleobiology, taphonomy and paleoenvironmental reconstruction. Palaeoworld 15, 217–222.

Zhuravlev, A.Y., Wood, R.A., 1996. Anoxia as the cause of the mid–Early Cambrian (Botomian) extinction event. Geology 24, 311–314.

Zhuravlev, A., Liñán, E., Gámez Vintaned, J.A., Debrenne, F., Fedorov, A.B., 2012. New finds of skeletal fossils in the terminal Neoproterozoic of the Siberian Platform and Spain. Acta Palaeontol. Pol. 57, 205–224.

CHAPTER SIX

Ordovician $\delta^{13}C$ chemostratigraphy: A global review of major excursions and their ties to graptolite and conodont biostratigraphy

Stig M. Bergström[a,*], Mats E. Eriksson[b], and Per Ahlberg[b]
[a]School of Earth Sciences, The Ohio State University, Columbus, OH, United States
[b]Department of Geology, Lund University, Lund, Sweden
*Corresponding author: e-mail address: bergstrom.1@osu.edu

Contents

Stratigraphy & Timescales, Volume 5
ISSN 2468-5178
https://doi.org/10.1016/bs.sats.2020.09.001

Abstract

Recent studies on $\delta^{13}C$ chemostratigraphy have produced a wealth of new information of major importance for clarification of local and global stratigraphic relationships. Currently, approximately 15 named positive and negative isotopic excursions have been identified in the Ordovician System. Some of these are significant and geographically widespread, whereas others are poorly developed and regional, thus hindering unambiguous usage. Because isotopic age data useful for precise excursion dating are rarely available in the Ordovician, biostratigraphic information from geographically widespread fossil groups such a graptolites and conodonts are of critical importance for correct identification of isotopic excursions. This study is a global review of the relations between Ordovician $\delta^{13}C$ excursions and graptolite and conodont biostratigraphic units with special emphasis on Baltoscandia, North America, and Argentina, three large regions that have been centers of Ordovician chemostratigraphic research. In this report we review for each excursion its proper nomenclatural designation, relations to biostratigraphic units, its known geographic range, and key study localities.

1. Introduction

Ever since the initial development of the global Ordovician chronostratigraphy, a classical problem has been the provincial differentiation of the shelly, graptolite, and conodont faunas that has made particularly long-distance correlations difficult, and in many cases uncertain. This provincial biogeographic differentiation applies at a very significant extent also to acid-resistant microfossils, such as chitinozoans and acritarchs (e.g., Molyneux et al., 2013; Vandenbroucke et al., 2013). The relatively recent development of $\delta^{13}C$ chemostratigraphy has led to a breakthrough in global correlation in the Ordovician, resulting in clarification of many previously uncertain, unclear, or misunderstood stratigraphic relationships at both the local and regional level. However, apart from the prominent Hirnantian excursion (HICE), the Ordovician carbon isotope curve perturbations do not carry their own identification marks in terms of amplitude and curve architecture, and for correct identification they must be integrated into some biostratigraphic or geochronologic framework, preferably based on graptolites and/or conodonts. Whereas recent investigations have made the major features of the Ordovician $\delta^{13}C$ curve reasonably well known (e.g., Bergström et al., 2020a), the precise relations between significant perturbations in this generalized curve and biostratigraphic units are in many cases still incomplete and a subject of current exploration. Furthermore, there is a need to better define significant excursions and when appropriate, establish suitable reference localities. The purpose of the present study is to discuss these matters in an effort to tighten up the relations between the Ordovician $\delta^{13}C$ chemostratigraphy and biostratigraphy (Fig. 1). As far as we are aware, this study is the first to deal with all the named Ordovician isotopic excursions in a global context. Our discussions below are based on two general assumptions. First, we interpret the excursions as in general covering broadly the same geochronological interval, even though it has become evident that the stratigraphical pattern of the isotope curves can vary laterally (Edwards and Saltzman, 2016; Henderson et al., 2018) and $\delta^{13}C$ events may be facies dependent and diachronous (Barili et al., 2018; Schiffbauer et al., 2017). Second, these excursions have a regional, often global, occurrence as far as strata is available for study. In our review, we focus on key sections in which it is possible to establish direct ties between $\delta^{13}C$ positive and negative isotopic perturbations and conodont and/or graptolite biostratigraphic units. It is our hope that this study will be useful to both biostratigraphers and chemostratigraphers to correctly identify the isotopic perturbations they encounter in their study sections.

2. The Cambrian/Ordovician boundary interval

A minor positive $\delta^{13}C_{carb}$ excursion with a local magnitude of ca 1‰ above baseline values has been recorded from the Cambrian/Ordovician boundary interval in several parts of the world. In the succession in the Precordillera of Argentina, it was first recognized by Buggisch et al. (2003) and later designated the Cambrian-Ordovician Transition Isotopic Carbon Excursion (COTICE) by Lehnert et al. (2014). The COTICE is located above the HEllnmaria-Red Tops Boundary Event (HERB) or Top of Cambrian Excursion (TOCE) that is a high-amplitude negative carbon excursion well below the top of the Cambrian Stage 10 (Azmy, 2019; Miller et al., 2014, 2015; Ripperdan, 2002; Zhu et al., 2006). Although COTICE has been illustrated as extending into the lowermost Tremadocian Stage in the La Silla Formation and ranging through the *Cordylodus angulatus* Conodont Zone in the Precordillera (e.g., Lehnert et al., 2014), it should be noted that the position of the base of the Ordovician remains somewhat uncertain in the Argentine sections investigated. In the biostratigraphically well-studied Great Basin successions in western USA, the top of the COTICE coincides with the base of the Ordovician (Fig. 2) and there is currently no firm evidence that this excursion extends into the lowermost Ordovician in the well-studied sections in southern Sweden (cf., e.g., Ahlberg et al., 2019; Terfelt et al., 2014). In North China and western

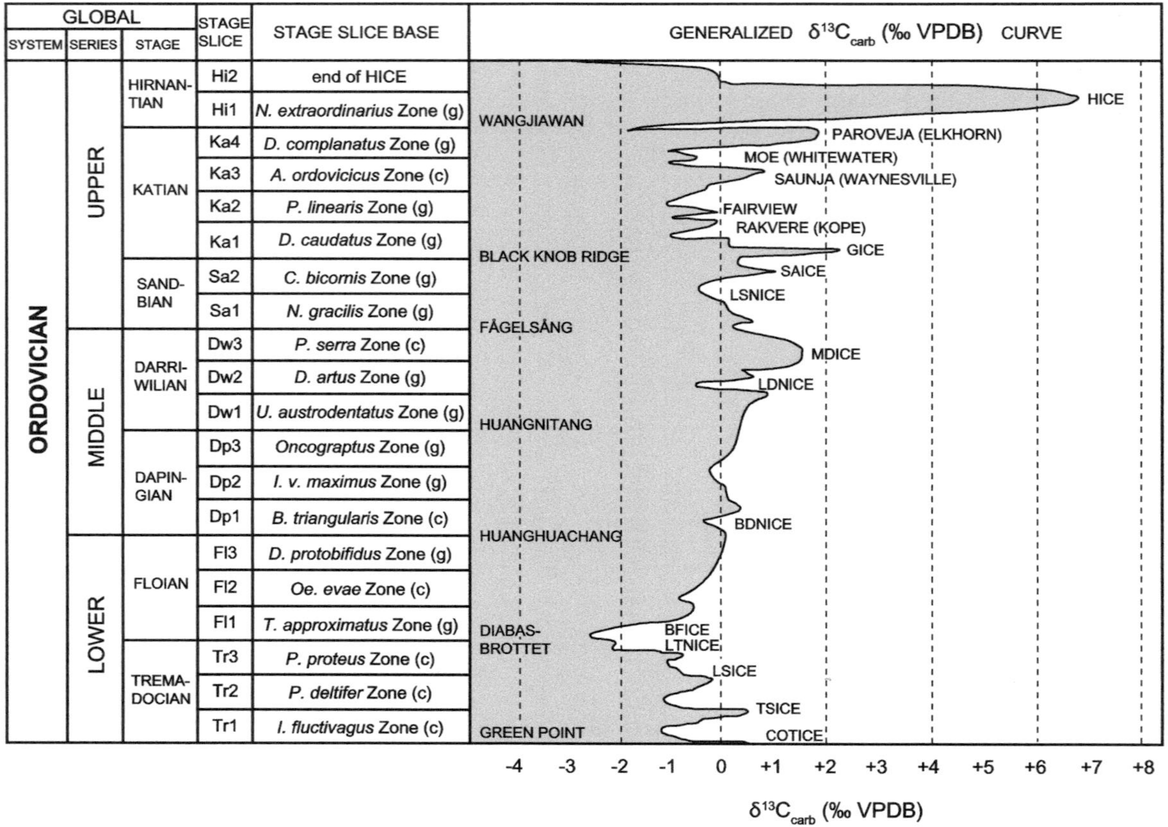

Fig. 1 General review of the named Ordovician $\delta^{13}C$ excursions discussed in the text and their relations to major stratigraphic units. Also listed are the names of the global stage GSSPs, the designations of stage slices, and the fossil zone that defines the base of each stage slice. c, conodont; g, graptolite. *Figure modified from Bergström, S.M., Chen, X., Gutiérrez-Marco, J.C., Dronov, A., 2009. The new chronostratigraphic classification of the Ordovician System and its relations to major regional series and stages and to $\delta^{13}C$ chemostratigraphy. Lethaia 42, 97–107, fig. 2.*

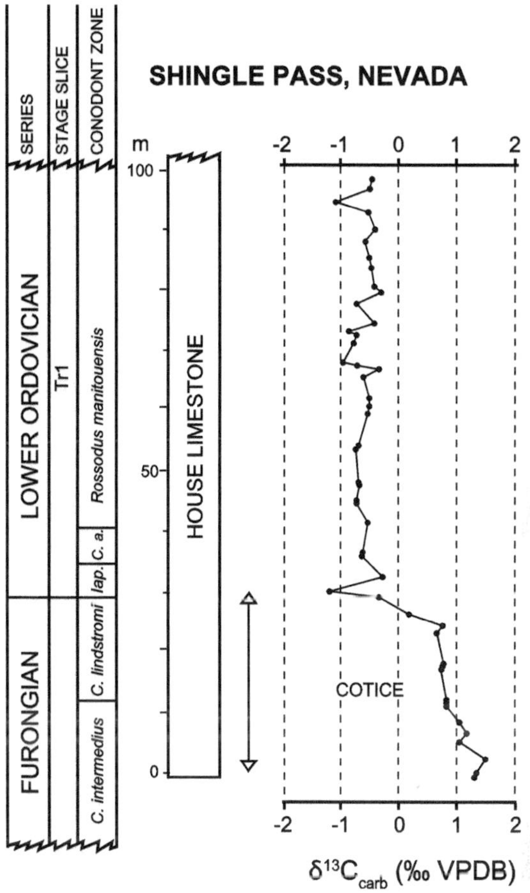

Fig. 2 The $\delta^{13}C_{carb}$ chemostratigraphy of the Cambrian/Ordovician boundary interval in the House Limestone at Shingle Pass, Nevada, USA. The curve is based on isotopic values published by Edwards and Saltzman (2014, table 1). Note the absence of a notable isotopic excursion in the basalmost Ordovician in this important section. *Iap.*, *Iapetognathus fluctivagus*, formerly considered the index of the lowermost conodont zone in the Ordovician; *C. a.*, *Cordylodus angulatus*, the next younger conodont zone. As in the following figures, arrowed vertical line represents the approximate range of a particular excursion. COTICE refers to the Cambrian-Ordovician Transition Isotopic Excursion of Lehnert et al. (2014) whose uppermost falling limb may extend into the very basalmost Ordovician in some stratigraphically complete successions.

Newfoundland, a short-ranging positive $\delta^{13}C_{carb}$ excursion, the HSS Spike of Wang et al. (2019), is located slightly below the base of the Ordovician. It has been recorded from the upper *Cordylodus caboti* and lower *C. intermedius* zones and may partly correspond to the COTICE. Another positive excursion, no. 3 of Wang et al. (2019), occurs in the basalmost Ordovician and has a similar amplitude as the COTICE, and could thus also correlate to the latter. The base and top of the COTICE are not tied to graptolite biostratigraphy, this carbon excursion being located below the base of the Ordovician graptolite zonation. Collectively, the isotopic data from vastly different regions provide inconconclusive evidence as to the precise definition and placement of the COTICE. Further studies are needed to show whether or not this excursion can be used outside of its key area of Argentina.

3. The Lower Ordovician

Whereas the Middle and Upper Ordovician $\delta^{13}C$ chemostratigraphy has been dealt with in numerous papers, relatively few such studies have been focused on the Lower Ordovician. Important pioneer studies were carried out on the rather thick and stratigraphically virtually complete, carbonate-dominated succession in the Precordillera of Argentina (e.g., Buggisch et al., 2003; Lehnert et al., 2014; Thompson and Kah, 2012). Similar chemostratigraphic work has also been performed in the thick carbonate succession in the Great Basin of western North America (e.g., Edwards and Saltzman, 2014) and western Newfoundland (Azmy et al., 2014; Thompson and Kah, 2012). Some chemostratigraphic work has been carried out on the stratigraphically very condensed Lower Ordovician successions in Baltoscandia (e.g., Bergström et al., 2020c; Lehnert et al., 2014; Lindskog et al., 2019; Rasmussen et al., 2016; Wu et al., 2015) and the shallow marine carbonate successions of the Yangtze Platform (Wu et al., 2020). Only very limited information is currently available from other parts of the world.

Several positive and negative excursions have been recognized in the Lower Ordovician. In a summary paper, Lehnert et al. (2014) defined and named six excursions based mainly on the successions in the Precordillera of Argentina (Fig. 3). Each of these excursions will be discussed below. Unfortunately, very little work has been done in this region on the precise relations between these excursions and the graptolite biostratigraphy, virtually all of these excursions being recognized only in carbonate facies with no direct graptolite control. However, this more than 500 m thick succession without significant stratigraphic gaps is important because its chemostratigraphy has been tied to a very useful conodont biostratigraphy.

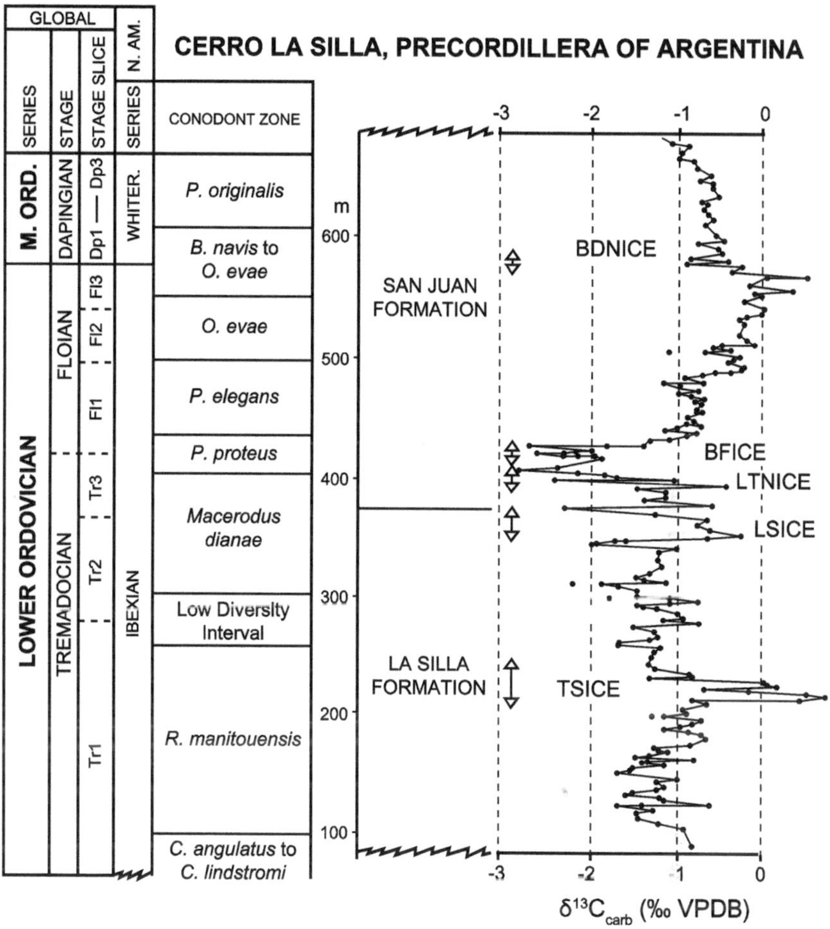

Fig. 3 Tremadocian and Floian $\delta^{13}C_{carb}$ excursions through the Precordilleran Cerro La Silla succession. Note the TSICE in the Tremadocian *Rossodus manitouenis* Conodont Zone, the LSICE in the *Macerodus dianae* Conodont Zone, and the negative LTNICE in the *Paroistodus proteus* Conodont Zone around the Tremadocian-Floian boundary. The basal Middle Ordovician (Dapingian) BDNICE is present in the *Baltoniodus navis* Conodont Zone in the Volkhov Formation of the St. Petersburg region of Russia (Zaitsev and Pokrovsky, 2014). In the Precordillera as well as in the Great Basin in western USA, the BDNICE occurs in an interval with the first appearance of the biostratigraphically important conodont *Tripodus laevis. Modified after Lehnert, O., Meinhold, G., Wu, R.C., Calner, M., Joachimski, M.M., 2014. δ13C chemostratigraphy in the upper Tremadocian through lower Katian (Ordovician) carbonate succession of the Siljan district, Central Sweden. Estonian J. Earth Sci. 63, 277–286.*

3.1 The Top Skullrockian positive Isotopic Carbon Excursion (TSICE)

This positive isotopic excursion (Fig. 3), which is the largest in the Tremadocian and has about the same magnitude as COTICE has in the Precordillera of Argentina, is located in Stage Slice Tr1 of Bergström et al. (2009). In terms of biostratigraphy, it is present in the upper portion of the *Rossodus manitouensis* Conodont Zone in the Precordillera (Lehnert et al., 2014), in the broadly coeval to slightly younger interval in the Great Basin (Fig. 4), and in the St. George Group of western Newfoundland. It should be noted that in both the Precordillera and the Great Basin, the *R. manitouensis* Zone is overlain by a biostratigraphically poorly controlled interval (referred to as the Low Diversity Interval) that currently makes it difficult to properly define the top of the *R. manitouensis* Zone. In the Ibex area of the Great Basin, the TSICE appears to range into The Low Diversity Interval (Fig. 4) but if this is only an artifact of poor biostratigraphy remains uncertain. The positive C1 excursion that was recently described by Wu et al. (2020) from the lower *R. manitouensis* Zone of the Yangtze Platorm, indicates that the TSICE may also be present in South China, as these excursions are most likely the same. As far as we are aware, the TSICE has not yet been identified in graptolite-bearing strata anywhere in the world but based on available correlations, it should range in an interval from the *Adelograptus victoriae* to the *Araneograptus murrayi* graptolite zones.

3.2 The Late Stairsian positive Isotopic Carbon Excursion (LSICE)

This is a short and minor excursion (~1‰ above baseline values) in the middle Tremadocian Stage (Stage Slice Tr2 of Bergström et al., 2009). In the Precordillera of Argentina, it is recorded from the upper *Macerodus dianae* Conodont Zone (previously referred to as the *Paroistodus deltifer* Zone, see Albanesi and Ortega, 2002; Fig. 3). Although indistinct, this excursion appears to occur also in the Great Basin in western North America, and if so it is present in the upper *M. dianae* and very lowermost *Diaphorodus deltatus*/*Oneotodus costatus* Conodont zones in the Parker Spring Formation (Fig. 4). Edwards and Saltzman (2014, fig. 9) did not identify and label this excursion in their correlation chart. Clearly there is a need for further investigation before this putative minor excursion can be positively identified and used.

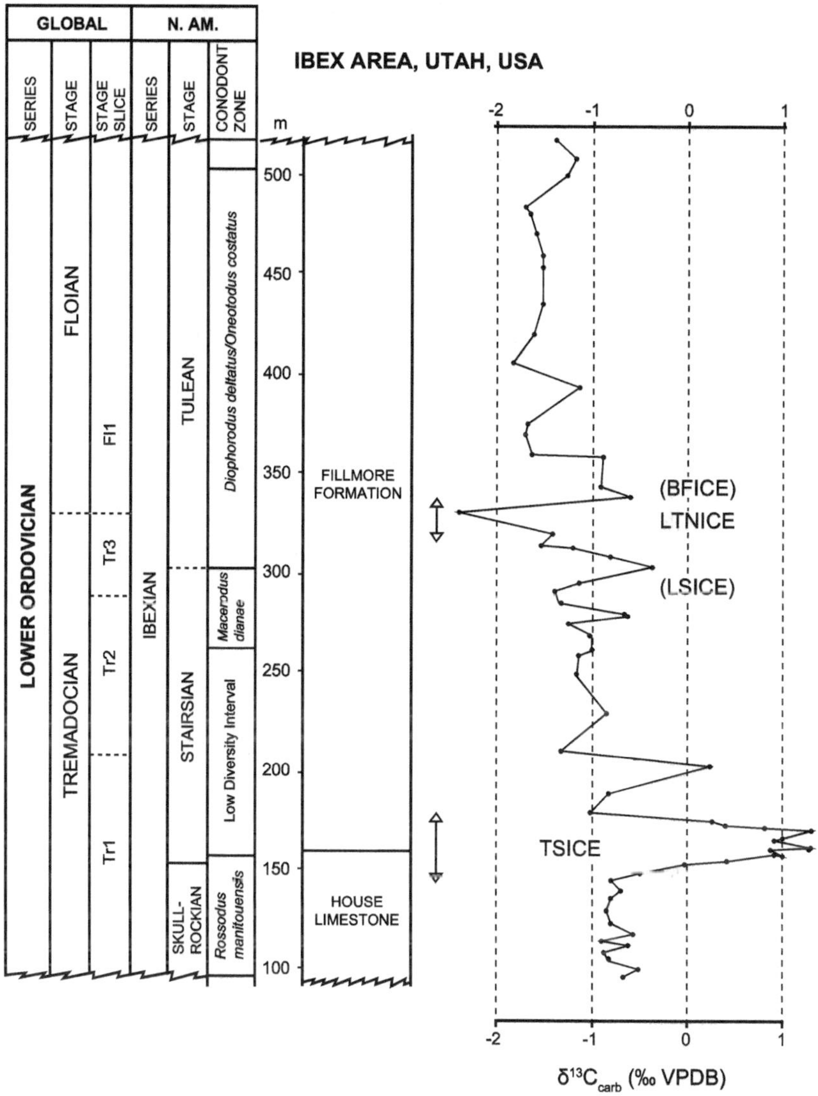

Fig. 4 $\delta^{13}C_{carb}$ chemostratigraphy through the North American Ibexian Series in the Ibex area, Utah. The curve is based on isotopic values published by Edwards and Saltzman (2014, table 2). Note the conspicuous TSICE and the poor development of the LSICE, LTNICE, and BFICE. Also note the baseline isotopic values through much of the lower Floian interval that make the Ibex isotopic curve different in appearance from that of the coeval interval in the Precordillera.

3.3 The Late Tremadocian Negative Isotopic Carbon Excursion (LTNICE)

This is a short negative carbon isotopic excursion in the uppermost *M. dianae* Conodont Zone and lowermost *Paroistodus proteus* Conodont Zone in the lowermost San Juan Limestone of the Precordillera of Argentina (Lehnert et al., 2014) where it reaches 1.5‰ below baseline values (Fig. 3). It may also be present in the middle part of the Fillmore Formation of the Great Basin (Fig. 4). Another possible record is from the western Newfoundland succession (Azmy and Lavoie, 2009) based on the correlation between the Precordillera and that region proposed by Edwards and Saltzman (2014, fig. 9). Lehnert et al. (2014) noted the occurrence of this negative excursion with question in two drill cores from southern Sweden but very few details about this potential occurrence are currently available. It may also occur in the Xiangshuidong section on the Yangtze Platform of South China, where it seems to be represented by a negative interval between the positive C2 and C3 excursions of Wu et al. (2020). In a recent study on the global stratotype of the Floian Stage at Diabasbrottet in southwestern Sweden, Bergström et al. (2020c) were unable to recognize this excursion in the Tremadocian-Floian boundary interval. Like some of the above named excursions (COTICE and LSICE), the LTNICE seems to be of highly questionable applicability, particularly in light of the fact that most of the authors responsible for naming this excursion (Lehnert et al., 2014) seem to themselves have abandoned its usage (Wu et al., 2020).

3.4 The Basal Floian positive Isotopic Carbon Excursion (BFICE)

In the Precordilleran succession, this is a very minor isotopic excursion (<1‰) in the upper *P. proteus* Conodont Zone (Fig. 3). This excursion is weakly developed also in the Great Basin successions but difficult to recognize with confidence (Fig. 4). It was not recorded in the recent study on the chemostratigraphy of the GSSP of the Floian Stage at Diabasbrottet in Sweden (Bergström et al., 2020c), but Lehnert et al. (2014, fig. 4) recorded it (although seemingly of inversed excursion direction) in two drill cores from southern Sweden. As for the LTNICE, the usage of the BFICE appears to have been abandoned even by the designation authors (see Wu et al., 2020). Thus, the BFICE is still poorly known and further studies are needed to establish if it is a globally identifiable perturbation in the Ordovician carbon isotope curve. Perhaps it would be preferable to consider the entire negative interval around the base of the Floian Stage, which is the most prominent negative interval in the Lower Ordovician, as one excursion which could be referred to as the Proteus Negative Isotopic Carbon Excursion (PNICE). This new excursion designation would include both the LTNICE and the BFICE of Lehnert et al. (2014).

4. Middle Ordovician

Recent research in Baltoscandia, North America and Argentina has resulted in several useful ties between $\delta^{13}C$ chemostratigraphy and biostratigraphy in this portion of the Ordovician System. Direct ties between standard trilobite zones and chemostratigraphy have also been established in Darriwilian successions in Sweden (Lindskog et al., 2019). Of particular importance are the ties between graptolite biostratigraphy and chemostratigraphy discovered by work in graptolite shale successions in southern Sweden (Bergström et al., 2018, 2020b) in which the graptolite biostratigraphy has been recently revised (Maletz and Ahlberg, 2020; Maletz et al., 2020). It is outside the scope of this review to deal with all the chemostratigraphically studied Middle Ordovician successions around the world and we herein focus on those of particular significance in clarifying the relations between isotope chemostratigraphy and biostratigraphy.

4.1 The Basal Dapingian Negative Isotopic Carbon Excursion (BDNICE)

In the Precordilleran succession, this excursion is marked by a rather minor negative shift (<1‰) in the $\delta^{13}C$ curve in the *Baltoniodus navis* Conodont Zone (Fig. 3). This shift starts just above the first appearance of the North American index conodont *Tripodus laevis*. As noted by Lehnert et al. (2014), this stratigraphic level approximately coincides with the base of the Dapingian Global Stage. This negative trend continues through most of the Dapingian in the Precordillera, Great Basin (Fig. 5), Newfoundland, and Baltoscandia, to end in the lowermost Darriwilian. In the classical Kinnekulle succession in southern Sweden (e.g., Lindskog et al., 2019, fig. 3), the isotopic values through the lower-middle Dapingian straddles baseline values with minor fluctuations and show no distinctive trend until the middle of the Darriwilian *Baltoniodus norrlandicus* Conodont Zone, where there is a minor negative trend that continues to near the base of the *Lenodus variabilis* Conodont Zone that has been referred to as the Lower Darriwilian Negative Isotopic Carbon Excursion (LDNICE; Lehnert et al. 2014). Lehnert et al. (2014) recorded a roughly similar negative trend in the Mora 001 drill core from the Siljan area of south-central Sweden but unfortunately, no conodont biostratigraphy of that core succession has been published.

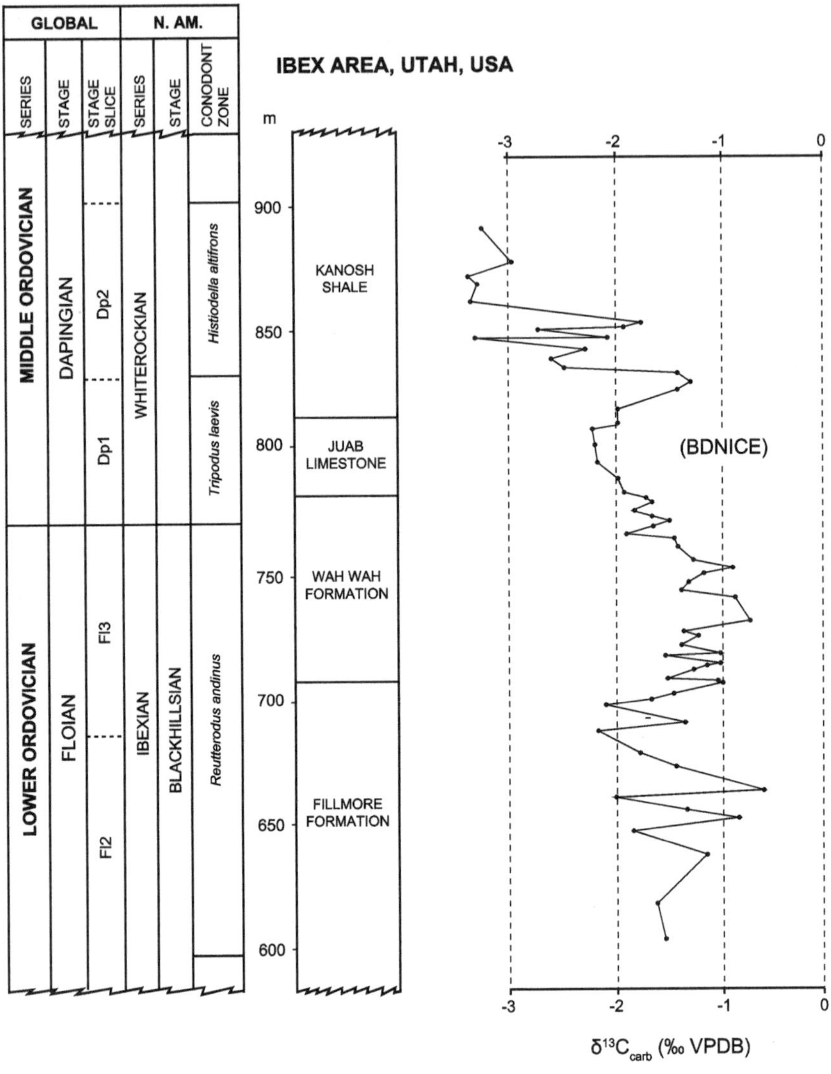

Fig. 5 $\delta^{13}C_{carb}$ isotopic curve through the Lower-Middle Ordovician (Floian-Dapingian) boundary interval in the Ibex area, Utah. Note the poor development of the BDNICE and the "noisy" isotopic curve in the upper Floian *Reutterodus andinus* Conodont Zone. Isotopic values from Edwards and Saltzman (2014, table 2).

A similar negative excursion is also present in the Dapingian-lower Darriwilian interval in Estonia (cf. Bauert et al., 2014) and the St. Petersburg region of Russia (e.g., Rasmussen et al., 2016). It may also correlate with the C5 negative shift in the Yangtze Platform, South China (Wu et al., 2020). The BDNICE and LDNICE also belong to those putative isotopic excursions that are poorly documented and need to be further assessed.

4.2 The Middle Darriwilian positive Isotopic Carbon Excursion (MDICE)

This is one of the most widely identified $\delta^{13}C$ excursions in the Ordovician, being recorded from numerous localities in Baltocandia, North America, Argentina, and China (cf. Schmitz et al., 2010; Wu et al., 2017 and references therein). The vast majority of these occurrences are in carbonate facies but there are also a few records from graptolite shale successions, especially from two drill cores in southern Sweden (Bergström et al., 2018, 2020b). In the latter, the isotopic values are more subdued than in carbonate facies but the excursion is still recognizable (Fig. 6).

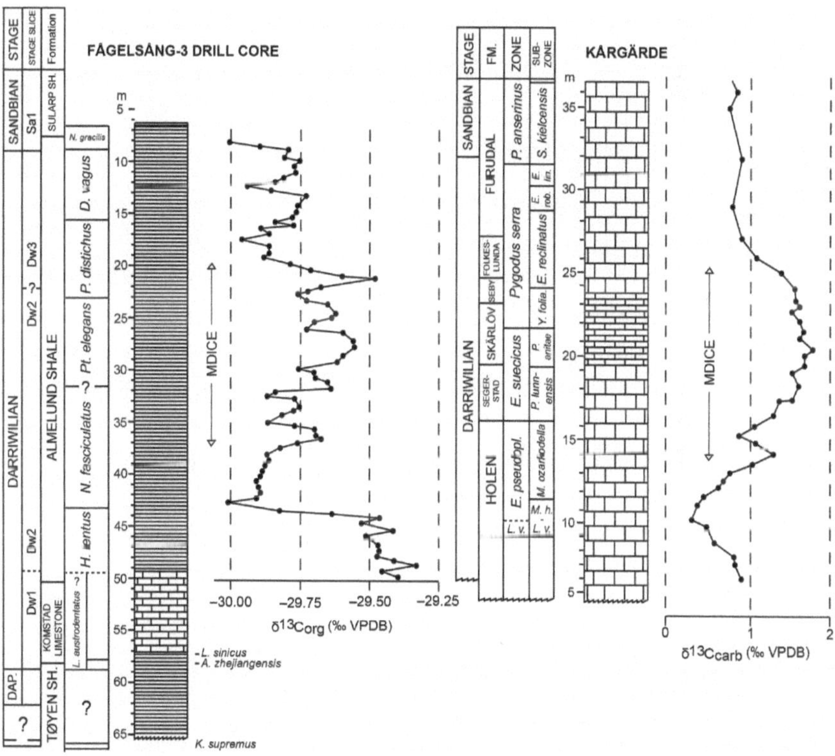

Fig. 6 The MDICE in shale facies in the Fågelsång-3 drill core (3-point running average) from southernmost Sweden and in the carbonate facies of the Kårgärde section from south-central Sweden. Note the range of the MDICE in terms of graptolite zones at Fågelsång and in conodont zones and subzones at Kårgärde. *Figure modified from Bergström, S.M., Ahlberg, P., Maletz, J., Lundberg, F., Joachimski, M.M. 2018. Darriwilian (Middle Ordovician) chemostratigraphy linked to graptolite, conodont and trilobite biostratigraphy in the Fågelsång-3 drill core, Scania, Sweden. GFF 140, 229–240, fig. 7.*

Whereas MDICE peak values in carbonate facies commonly reach 1–2‰ above baseline values, corresponding ones in shale facies often are only <0.5‰ above the baseline values. This obviously complicates the recognition of the precise beginning of the MDICE but in the Fågelsång-3 drill core from Scania, southern Sweden (Bergström et al., 2018), the MDICE starts in the middle Darriwilian *Nicholsonograptus fasciculatus* Graptolite Zone and ends in the upper Darriwilian *Pseudamplexograptus distichus* Graptolite Zone, hence an interval covering parts of Dw2 and Dw3 Stage Slices of Bergström et al. (2009). The range of the MDICE in the conodont zone succession in calcareous facies is more clear as illustrated by the Kårgärde section in south-central Sweden in which the MDICE ranges from the topmost part of the *Eoplacognathus pseudoplanus* Zone to the middle part of the *Eoplacognathus reclinatus* Subzone of the *Pygodus serra* Zone (cf. Bergström et al., 2018, fig. 7; see also Wu et al., 2017, fig. 8).

5. Upper Ordovician

Based on a considerable number of investigations in especially Baltoscandia and North America, the Upper Ordovician $\delta^{13}C$ chemostratigraphy can now be considered comparatively well established. Most major isotopic excursions have been recorded in Baltoscandia and North America (e.g., Bergström et al., 2014, 2016) and a very significant amount of work has been carried out in key successions around the world. As will be discussed below, most significant isotopic excursions are now rather closely tied to graptolite and conodont biostratigraphy. Among the nine $\delta^{13}C$ excursions now generally recognized in the Sandbian-Hirnantian interval, three are of particular significance, namely the GICE, MOE, and HICE. However, in the text below, we discuss the biostratigraphic correlation of all the eight Upper Ordovician excursions recognized herein.

5.1 The Lower Sandbian Negative Isotopic Carbon Excursion (LSNICE)

In an interval not far above the MDICE there is a distinctive negative excursion. Originally known as the Upper Kukruse Low (Kaljo et al., 2004), it is now referred to as the Lower Sandbian Negative Isotopic Carbon Excursion (LSNICE) (Bauert et al., 2014). This perturbation was originally described from the upper portion of the regional Kukruse Stage in Estonia where it occurs in the *Baltoniodus variabilis* Subzone of the *Amorphognathus tvaerensis* Conodont Zone (Bauert et al., 2014). At the GSSP of the Sandbian Stage at Fågelsång in southern Sweden, only the lowermost part of the excursion interval is available at the E14a and E14b natural outcrops and in the Fågelsång-3 drill core (Bergström et al., 2018, 2020c). In these successions, the excursion occurs in the lower Sularp Shale in an interval referable to the *Nemagraptus gracilis* Graptolite Zone (Fig. 7). Recently, it has also been recorded in the uppermost part of the Röstånga-2 drill core from Scania (Bergström et al., 2020b). Little is currently known about the occurrence of this negative excursion outside Baltoscandia and it has not yet been clearly identified in Argentina or China. However, it is possible that the prominent but poorly dated negative excursion (Fig. 8) above the MDICE in the Clear Spring section in Maryland (Leslie et al., 2011) may represent the LSNICE (cf. Young et al., 2016). It could be argued that the LSNICE just represents the falling limb of the MDICE, but the LSNICE covers an interval slightly above the end of the MDICE.

5.2 The Sandbian Positive Isotopic Carbon Excursion (SAICE)

Some 15 years ago, a positive $\delta^{13}C$ excursion was recognized in the topmost part of the Sandbian succession of the Copenhagen Formation in Nevada (Saltzman and Young, 2005). This perturbation was later recorded also in Maryland by Leslie et al. (2011) who named it the Sandbian Isotopic Carbon Excursion (SAICE). As far as we are aware, this excursion has not yet been identified in graptolite-bearing strata anywhere in the world. In Nevada, it occurs in the uppermost portion of the *Amorphognathus tvaerensis* Conodont Zone, but it is still not clear if it starts in the upper *Baltoniodus gerdae* Subzone or lower *Baltoniodus alobatus* Subzone. In the Solberga-1 drill core of the Siljan area, south-central Sweden, the base of the SAICE falls near the base of the *B. gerdae* Subzone (Bergström et al., 2016, fig. 4) but the conodont biostratigraphy is not yet firmly established in that drill core. In a recent study of the extensive Evans Ferry succession in the eastern Appalachians of Tennessee, Kozic et al. (2019, fig. 6) recorded the start of the SAICE from an interval near the base of the *B. gerdae* Subzone but this level is considerably above the last occurrence of *Eoplacognathus elongatus,* a common associate of *B. gerdae,* as recorded by Carnes (1975) in a comprehensive study of the conodont biostratigraphy of this particular section. The subzonal index *B. gerdae* has not been recorded from the Evans Ferry section in Tennessee but is present in thrust belts farther to the east in Tennessee (Bergström and Carnes, 1976). It may be noted that the apparent absence of the SAICE in some successions in Baltoscandia (especially in

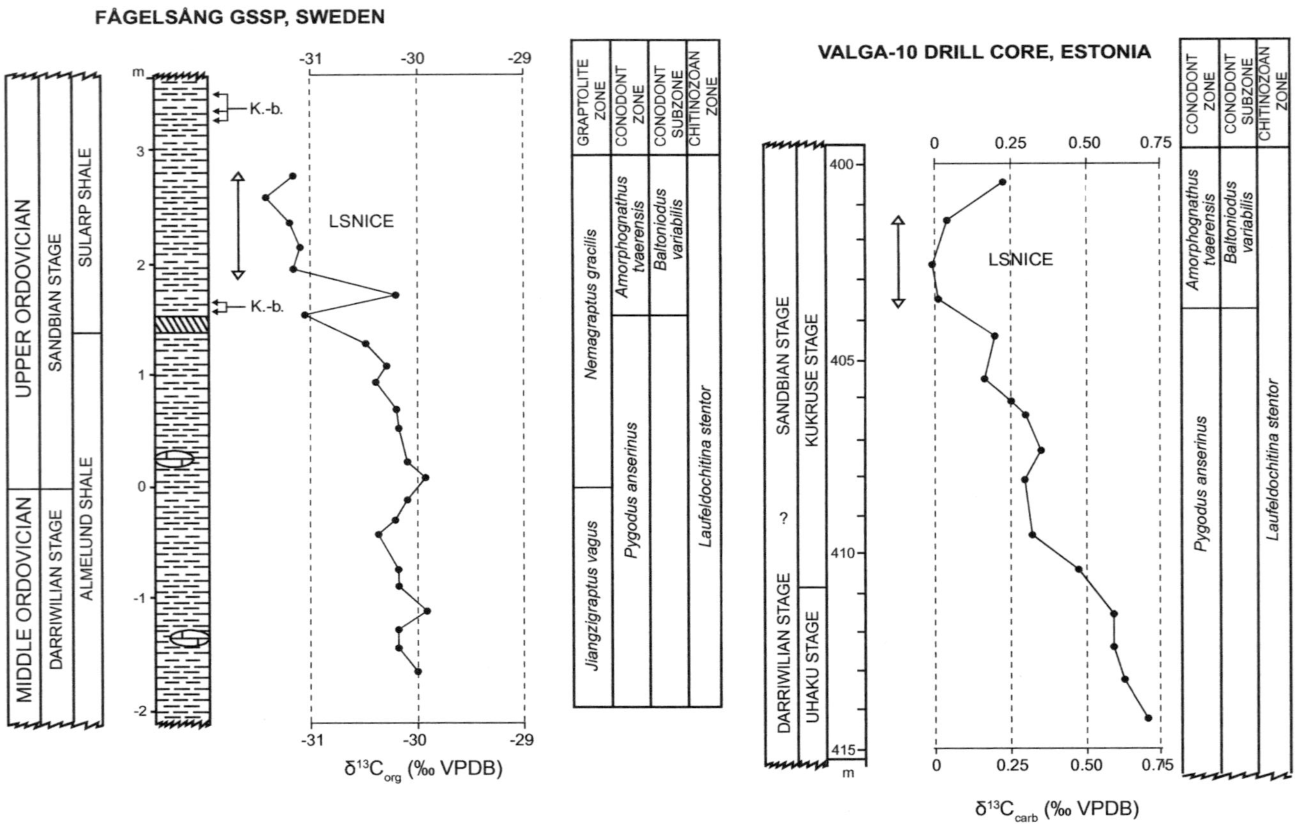

Fig. 7 Comparison of the short-ranging lower Sandbian LSNICE in the shale facies of the Fågelsång area, Sweden, and in the carbonate facies of the Valga-10 drill core in Estonia. Both occurrences are in the *Amorphognathus tvaerensis* Conodont Zone. *Figure slightly modified from Bergström, S.M., Schmitz, B., Terfelt, F., Eriksson, M.E., Ahlberg, P., 2020a. The $\delta^{13}C_{org}$ chemostratigraphy of Ordovician global stage stratotypes: geochemical data from the Floian and Sandbian GSSPs in Sweden. GFF 142, 23–32, fig. 6.*

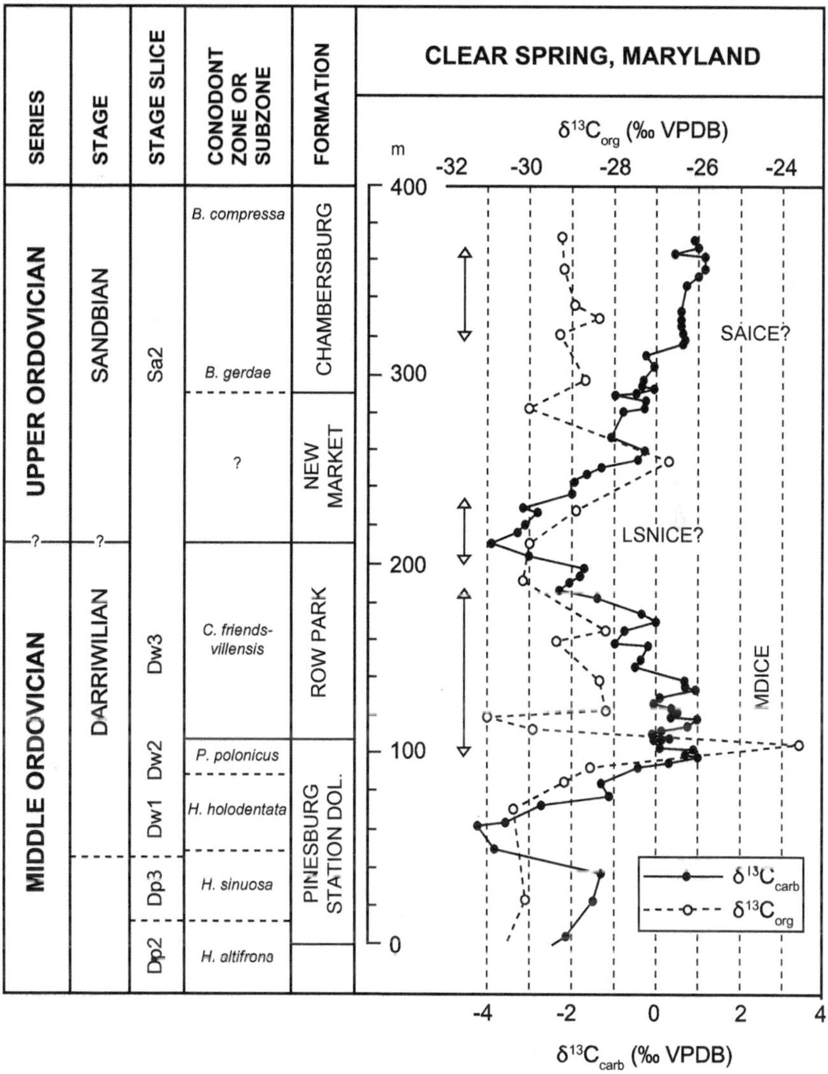

Fig. 8 $\delta^{13}C_{org}$ and $\delta^{13}C_{carb}$ curves from the Middle-Upper Ordovician succession at Clear Spring, Maryland, eastern USA. *Figure modified after Leslie, S.A., Saltzman, M.R., Bergström, S.M., Repetski, J.E., Howard, A., Seaward, A.M., 2011. Conodont biostratigraphy and stable isotope stratigraphy across the Ordovician Knox/Beekmantown unconformity in the central Appalachians. In: J.C. Guitiérrez-Marco, Rábano, I., García-Bellido, D. (Eds.), Ordovician of the World. 11th International Symposium on the Ordovician System. Cuadernos del Museo Geominero 14, Instituto Geológico y Minero de España, Madrid, 301–308 and Young, S.A., Gill, B.C., Edwards, C.T., Saltzman, M.R., Leslie, S.A., 2016. Middle-late Ordovician (Darriwilian–Sandbian) decoupling of global sulfur and carbon cycles: isotopic evidence from eastern and southern Laurentia. Palaeogeogr. Palaeoclimatol. Palaeoecol. 458, 118–132.*

Estonia) may be due to the presence of a locally developed stratigraphic gap in an interval of variable size between the Kukruse and Haljala regional stages (cf. Paiste et al., 2020). Whether or not this gap is present also in some Swedish carbonate successions is currently not clear.

The relations of the SAICE to the excursions recognized in the uppermost Sandbian Black River Group of New York State and coeval strata in the Midcontinent remain uncertain (Bergström et al., 2010a) and in need of further study. Although currently not recorded from a section with good graptolite control anywhere in the world, it is likely to be in strata coeval with part of the upper *Climacograptus bicornis* Zone.

5.3 The Guttenberg positive Isotopic Carbon Excursion (GICE)

Along with the HICE, the GICE is the most widely recorded $\delta^{13}C$ excursion in the Ordovician, being known from many localities not only in North America and Baltoscandia but also China, western Siberia, and Malaysia. For a general review of this excursion, see Bergström et al. (2010a). At most localities, the maximum $\delta^{13}C$ values are 1–2‰ above baseline values but there is a great deal of variation. Up to three peaks have been recognized in the GICE curves by some authors and used for local correlation but regional reviews indicate that such peaks are not consistent and hence of limited use for long-distance correlations.

The GICE was first described from Katian sections in Iowa in the Mississippi Valley (Hatch et al., 1987), where it occurs a little above the very widespread Deicke-Millbrig K-bentonite complex (e.g., Bergström et al., 2010a). In that region, as well as elsewhere in the North American Midcontinent, there is no direct graptolite control in the GICE successions that are dominated by carbonates. In New York State, the base of the GICE corresponds to a level in the lower-middle Napanee Formation that is regarded to be near the boundary between the *C. bicornis* and *Diplacanthograptus caudatus* zones (Barta et al., 2007), but direct correlation between the graptolite succession and the carbon chemostratigraphy has not been established in that region. The GICE was identified in the GSSP section of the global Katian Stage at Black Knob Ridge in southeastern Oklahoma (Goldman et al., 2007), where the base of the excursion falls at, or near, the base of the *Diplacanthograptus caudatus* Zone, which by definition is the base of the Katian Stage. At this section, the top of the GICE is within the *D. caudatus* Zone. The GICE curve at this locality shows a remarkable similarity to that of the GICE in the Röstånga-1 drill core from Sweden (cf. Bergström et al., 2016, fig. 8), which unfortunately has less precise graptolite biostratigraphy of the GICE interval although the end of the GICE appears to fall within the *Dicranograptus clingani* Zone.

In the North American Midcontinent conodont biostratigraphy, the base of the GICE is typically located in the upper part of the *Phragmodus undatus* Conodont Zone a short distance above the widespread Deicke-Millbrig K-bentonite beds and the excursion ranges to near the top of the overlying *Plectodina tenuis* Conodont Zone (Fig. 9). In some successions it appears to continue into the lower portion of the overlying *Belodina confluens* Zone (e.g., Bergström et al., 2010a). Elsewhere in the world, using the Atlantic conodont zone biostratigraphy, as is shown particularly well in sections of the Pagoda Limestone on the Yangtze Platform of China, the GICE ranges from the topmost portion of the *Amorphognathus tvaerensis* Conodont Zone into the lower part of the *A. superbus* Conodont Zone (e.g., Fan et al., 2015).

5.4 The RAKVERE (KOPE) positive carbon isotopic excursion

This $\delta^{13}C$ perturbation in the lower part of the Katian Stage was described as the RAKVERE excursion in Estonia (Ainsaar et al., 2010) a few weeks before a coeval excursion was recognized in North America, where it was named the KOPE excursion (Bergström et al., 2010b). Hence, the designation RAKVERE has priority over KOPE and the former designation is used herein. This is not a very prominent excursion (Fig. 9), the peak isotopic values commonly being no larger than 1‰ above the baseline values. No reference section for this excursion has been proposed previously but in view of the fact that in the Cincinnati region of Ohio and adjacent states, there is better graptolite and conodont biostratigraphic control of the excursion interval than anywhere in Baltoscandia, it would be preferable to select a reference section in the Cincinnati area. In that region (Fig. 9), the base of this excursion is in what was classified as the middle portion of the *Climacograptus spiniferus* Zone and the top in the lower-middle *G. pygmaeus* Zone (cf. Bergström et al., 2010b, fig. 1). In terms of a later, less detailed, graptolite zone classification, both the base and the top of this excursion are in the *D. caudatus* Graptolite Zone. In terms of the North American Midcontinent conodont zone classification (Sweet, 1984), the RAKVERE excursion is present in the middle and upper parts of the *Belodina confluens* Zone. In the Atlantic conodont zone classification, it is present in the middle portion of the *A. superbus* Zone.

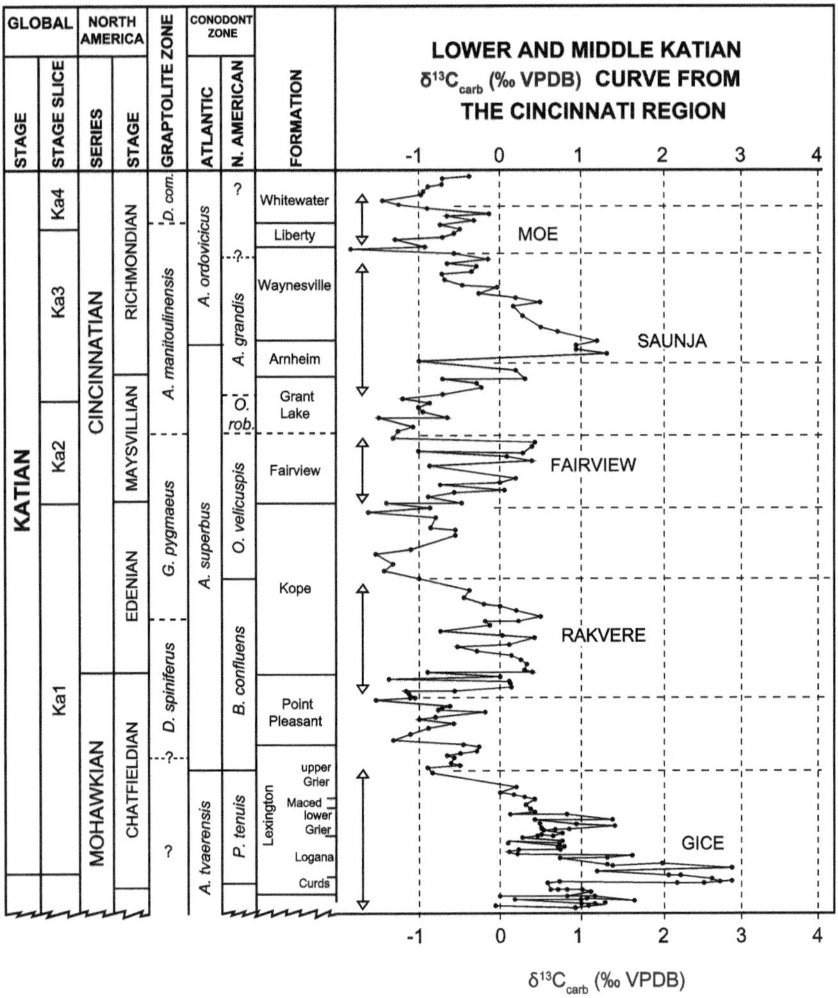

Fig. 9 Summary diagram showing the relations between five $\delta^{13}C$ excursions recognized in the Cincinnati region of Ohio, Kentucky, and Indiana, and local formations, graptolite and conodont zones, and North American regional series and stages. Note that the RAKVERE, SAUNJA, and MOE isotopic excursions were previously known as the KOPE, WAYNESVILLE, and WHITEWATER excursions, respectively. *Figure modified after Bergström, S.M., Schmitz, B., Saltzman, M.R., Huff, W.D., 2010a. The Upper Ordovician Guttenberg $\delta^{13}C$ excursion (GICE) in North America and Baltoscandia: Occurrence, chronostratigraphic significance, and paleoenvironmental relationships. Geol. Soc. Am. Spec. Pap. 466, 37–67 and Bergström, S.M., Young, S., Schmitz, B., 2010b. Katian (Upper Ordovician) $\delta^{13}C$ chemostratigraphy and sequence stratigraphy in the United States and Baltoscandia: a regional comparison. Palaeogeogr. Palaeoclimatol. Palaeoecol. 296, 217–234, fig. 3.*

5.5 The FAIRVIEW positive isotopic carbon excursion

Bergström et al. (2010b) first recorded this positive excursion in the Cincinnati region in the eastern Midcontinent of the USA where it ranges through the Fairview Formation (Fig. 9). It exhibits peak $\delta^{13}C_{carb}$ values of ~1.5‰ above the baseline values. Biostratigraphically, it is present in the upper part of the *Oulodus velicuspis* Conodont Zone and in the upper third of the *G. pygmaeus* Graptolite Zone. Because it has not been recorded with certainty in the middle Katian successions in the Upper Mississippi Valley, Oklahoma (cf. Bergström et al., 2010b), and Baltoscandia, its status as a significant regional excursion is in need of confirmation.

5.6 The SAUNJA (WAYNESVILLE) positive carbon isotopic excursion

As was the case with several Middle and Upper Ordovician excursions, this positive perturbation in the isotope curve was first recognized in the East Baltic region, where it was formally named the SAUNJA excursion by Ainsaar et al. (2010) a few weeks before Bergström et al. (2010b) described an apparently coeval positive perturbation from the Cincinnati region of the eastern North American Midcontinent as the WAYNESVILLE excursion. In the latter region, the base of this excursion is poorly controlled by graptolites but this level is commonly referred to the upper portion of the *Amplexograptus manitoulinensis* Zone (cf. Bergström et al., 2010b, fig. 1). Notably, at least in western Ohio (Fig. 9), the base of this excursion appears to coincide approximately with the base of the Atlantic *Amorphognathus ordovicicus* Conodont Zone. In Baltoscandia this excursion occurs in the *Pleurograptus linearis* Graptolite Zone in the Röstånga-1 drill core (Bergström et al., 2016, fig. 3) and coeval strata in carbonate facies of the Baltoscandic Nabala Regional Stage (Ainsaar et al., 2010).

5.7 The MOE (WHITEWATER) positive isotopic carbon excursion

As was the case with several Ordovician excursions, the Baltoscandic name MOE Excursion (Ainsaar et al., 2010) has priority over the coeval North American WHITEWATER Excursion (Bergström et al., 2010b) by publication date. This is one of the geographically most widely recorded excursions in the Ordovician with a magnitude of up to 2‰ or more above baseline values. It is of special interest in North America, where it is a widespread marker of a level a little above the base of the regional Richmondian

Stage (Bergström et al., 2020a). Typically, this excursion occurs in the *Dicellograptus complanatus* (*D. ornatus*) Graptolite Zone. In terms of conodont biostratigraphy, it is present in the lower *Aphelognathus divergens* Midcontinent Conodont Zone and the lower half of the Atlantic *A. ordovicicus* Conodont Zone (Bergström et al., 2020a). In Baltoscandia, it is present in the upper Vormsi and lower Pirgu regional stages. As far as we are aware, this prominent excursion has not yet been recorded in graptolite facies in China and other parts of Asia or in South America.

5.8 The PAROVEJA positive isotopic carbon excursion

As shown in a recent review by Bergström et al. (2020a), this positive carbon isotope excursion occurs in the Stage Slice Ka4 in the uppermost Katian Stage. It is known as the PAROVEJA excursion in Baltoscandia (Ainsaar et al., 2010) and was previously known in North America as the ELKHORN excursion (Bergström et al., 2010b). As shown by Bergström et al. (2020a), these names refer to the same isotopic perturbation. Because the designation PAROVEJA has a few weeks priority in terms of publication date it should replace the term ELKHORN in North America and elsewhere. No reference section for this unit has yet been selected but such a succession could preferably be one in the Great Basin of USA, where both graptolite and conodont control are available at, for instance, the Martin Ridge succession, Monitor Range, Nevada (Fig. 10). Compared with the next younger excursion, the HICE, the magnitude of the PAROVEJA is rather modest and generally amounts to 1–2‰ above baseline values. In stratigraphically seemingly complete successions both the base and the top of this excursion curve are smoothly transitional from the baseline values which may complicate establishment of its precise vertical range. The only locality known to us with some graptolite control of the base of the PAROVEJA is at Martin Ridge where this level appears to at least broadly coincide with the base of the *Paraorthograptus pacificus* Zone. In stratigraphically relatively complete successions, the top of the PAROVEJA is within this graptolite zone in the uppermost Katian Stage Slice Ka4 and separated from the HICE by a relatively narrow stratigraphical interval. In terms of Atlantic conodont biostratigraphy (Bergström, 1971), the entire PAROVEJA is within the upper part of the *Amorphognathus ordovicicus* Zone but it should be noted that there is no precise conodont or graptolite biostratigraphic control of the range of PAROVEJA in Baltoscandia. In the Upper Ordovician North American conodont biostratigraphy introduced

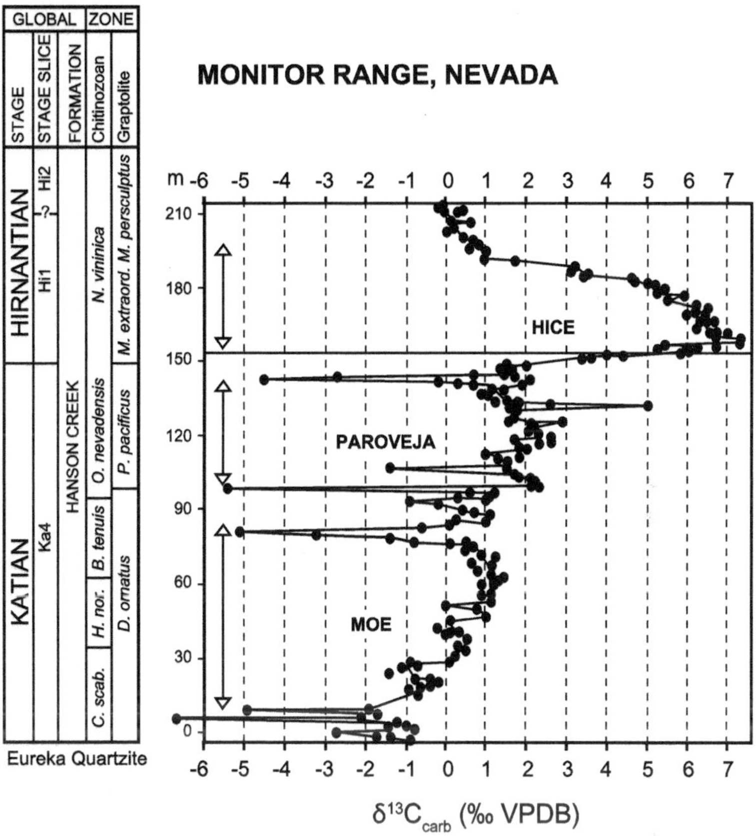

Fig. 10 $\delta^{13}C_{carb}$ curve through the more than 210 m thick upper Katian and Hirnantian succession at Monitor Range, Nevada, USA, showing the vertical ranges of the MOE, PAROVEJA, and HICE isotopic excursions and their relations to chitinozoan and graptolite zones. *Figure modified after Bergström, S.M., Kleffner, M., Eriksson, M.E., 2020b (published online 2019). Upper Katian (Upper Ordovician) trans-Atlantic $\delta^{13}C$ chemostratigraphy: the geochronological equivalence of the ELKHORN and PAROVEJA excursions and its implications. Lethaia 53, 199–216, fig. 10.*

by Sweet (1979), most of the PAROVEJA interval in the important succession at Barn Hills, Utah (cf. Bergström et al., 2020a) falls in the *Aphelognathus shatzeri* Zone. Apart from its occurrence in western Ohio (Bergström et al., 2020a) little is known about the PAROVEJA in central United States where it should be present in the Maquoketa Group and equivalent shaly strata that have not yet been the subject of detailed chemostratigraphic study. As far as we are aware, this excursion has not yet been identified in successions outside Baltoscandia and North America.

5.9 The Hirnantian positive Isotopic Carbon Excursion (HICE)

The youngest $\delta^{13}C$ excursion in the Ordovician is the Hirnantian carbon isotopic excursion. This excursion, one of the most prominent in the Lower Paleozoic, has been documented from the lower to middle Hirnantian Stage at numerous localities in North America, Europe, and Asia. For aspects of the development of this perturbation at many global localities, see Bergström et al. (2014) and Bergström and Goldman (2019).

The GSSP of the global Hirnantian Stage at Wangjiwan in the Yichang region of the Hubei Province in China is not an ideal succession for the study of the relations between chemostratigraphy and biostratigraphy. Whereas two graptolite zones have been recognized in the Hirnantian (e.g., Melchin and Holmden, 2006), the middle part of the Hirnantian is developed in carbonate facies at many other localities and lacks graptolites. Furthermore, the conodont biostratigraphy of the Hirnantian GSSP is essentially unknown in the absence of diagnostic conodonts. Another problem is that the most detailed $\delta^{13}C_{carb}$ and $\delta^{13}C_{org}$ curves from this very condensed section (Gorjan et al., 2012) are not typical in some respects, especially in the end interval of the HICE that is not well defined and biostratigraphically controlled. However, currently available data suggest that the base of the HICE is at about the level of the base of the *Metabolograptus extraordinarius* Graptolite Zone (cf. Bergström and Goldman, 2019, fig. 8: 5). The graptolite correlation of the top of the HICE in this section remains unclear but may be somewhere in the lower *Metabolograptus persculptus* Zone.

The biostratigraphically better controlled, and muck thicker, succession on Martin Ridge, Monitor Range, Nevada has an excellent development of the HICE with $\delta^{13}C_{org}$ and $\delta^{13}C_{carb}$ values up to 3‰ and 6‰, respectively, above baseline values (Fig. 10) but the top of the excursion interval is marked by a significant unconformity, and it is possible that also the base of the Hirnantian coincides with a small unconformity (Bergström et al., 2014). The HICE is well developed with maximum $\delta^{13}C_{org}$ values of between −25‰ and −26‰ (Bergström et al., 2014, fig. 13). The Martin Ridge succession is similar to the Hirnantian GSSP sequence in that the base of the Hirnantian and the HICE approximately coincide with the base of the *M. extraordinarius* Zone (e.g., Bergström et al., 2014). This graptolite correlation of the base of the HICE agrees with that in the important Vinini Creek succession in Nevada (e.g., Bergström et al., 2014, fig. 13; Štorch et al., 2011). The graptolite correlation of the top of the HICE is less precise in these sections. At Martin Ridge, it falls well above the base of the

M. persculptus Zone but the top of this zone is cut off by an unconformity. At Vinini Creek, the top part of the HICE appears to be missing, the HICE interval being directly overlain by Silurian strata.

The conodont biostratigraphy of the Martin Ridge and Vinini Creek sections have been investigated by Sweet (2000) who reported that the base of the HICE interval is not marked by a significant change in the conodont fauna, which represents the *Amorphognathus ordovicicus* Zone. At Martin Ridge most of the HICE interval has not yielded diagnostic conodonts (cf. Bergström et al., 2014, fig. 13) but at a stratigraphic level a little below the top of the HICE, a new conodont fauna appears that is characteristic of the *Ozarkodina hassi* Zone, which elsewhere ranges from the topmost Hirnantian into the basal Silurian. Unfortunately, the absence of conodonts in much of the HICE interval at Martin Ridge makes it impossible to establish the precise relations between the HICE and the conodont biostratigraphy in much of the excursion interval.

Several sections with excellent chemostratigraphy in the HICE interval have been investigated in Baltoscandia (e.g., Bergström et al., 2014) but the common absence of graptolites and conodonts in the Hirnantian interval prevents reliable correlation between biostratigraphy and chemostratigraphy. An exception is the succession at Röstånga in southern Sweden, which has good late Hirnantian graptolite biostratigraphy but no conodont biostratigraphy. As shown by Bergström et al. (2014), the base of the HICE is not controlled biostratigraphically but the top of the HICE interval corresponds approximately to the base of the latest Hirnantian *Akidograptus avitus* Faunal Interval of Koren´ et al. (2003; Fig. 11 herein). At this locality, this topmost Ordovician interval has a thickness of a little more than 3 m and is conformably overlain by Silurian strata containing graptolites of the basal Silurian *Akidograptus ascensus* Zone (Maletz et al., 2014).

The $\delta^{13}C_{org}$ chemostratigraphy of the Ordovician/Silurian boundary GSSP at Dob's Linn in Southern Scotland described by Underwood et al. (1997) appears to be post-HICE and hence of limited interest in the present study. More important information about the relations between the HICE and graptolite and conodont biostratigraphy is provided by the succession along Mirno Creek in eastern Siberia (e.g., Bergström and Goldman, 2019; Kaljo and Martma, 2011; Koren´ and Sobolevskaya, 2008; Koren' et al., 1988; Zhang and Barnes, 2007). In this important but remotely located succession, most of the HICE has not conspicuously high isotopic values but the base of this major excursion falls just below the base of the *M. extraordinarius* Zone and its top in the middle part of the *M. persculptus*

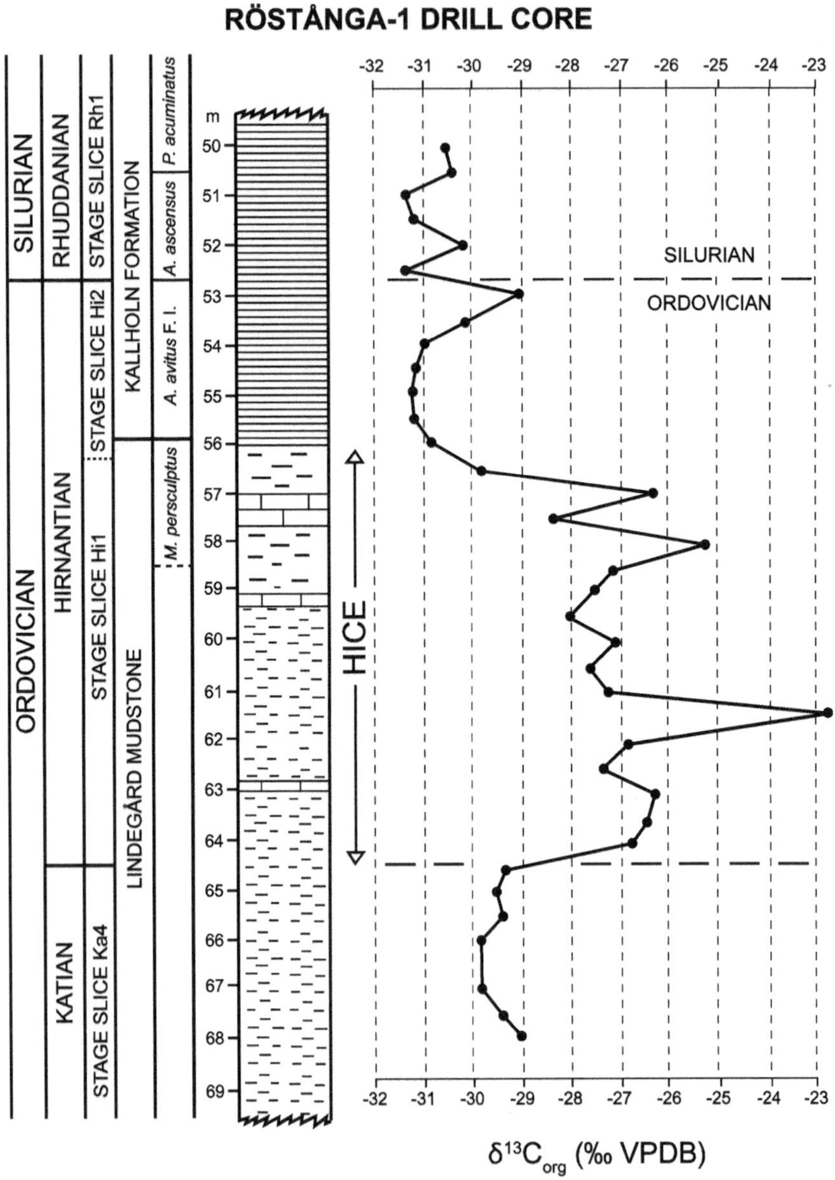

Fig. 11 Relations between graptolite zones and the $\delta^{13}C_{org}$ curve through the Hirnantian and lowermost Rhuddanian in the Röstånga-1 drill core from southernmost Sweden. Note that the HICE ends at the top of an interval classified as the *M. persculptus* Graptolite Zone. *Figure slightly modified after Bergström, S.M., Eriksson, M.E., Young, S.A., Ahlberg, P., Schmitz, B., 2014: Hirnantian (latest Ordovician) $\delta^{13}C$ chemostratigraphy in southern Sweden and globally: a refined integration with the graptolite and conodont zone successions. GFF 136, 355–386, fig. 2.*

Zone. About a dozen Mirno Creek conodont species were reported by Zhang and Barnes (2007), but because the currently known Hirnantian conodont fauna from that locality includes no zone index species or other taxa with restricted ranges, the relations between conodont biostratigraphy and $\delta^{13}C$ chemostratigraphy remain unclear.

In summary, whereas the biostratigraphy of several sections indicates that the base of the HICE coincides approximately with the base of the *M. extraordinarius* Zone, the top of this major excursion is more difficult to precisely tie to the graptolite biostratigraphy although it appears that it corresponds to a level in the lower to middle part of the *M. persculptus* Zone. In the Röstånga succession, this horizon corresponds to a level near the base of the highest Hirnantian *A. avitus* Faunal Interval.

6. Concluding remarks

Although having been studied for only a little more than a quarter of a century, Ordovician $\delta^{13}C$ chemostratigraphy has now reached a relatively mature stage. In this report we deal with the biostratigraphic relations of four Lower Ordovician, two (three if counting LDNICE) Middle Ordovician, and nine Upper Ordovician named isotope excursions. Based on information from key localities or important regions, we have tried to establish the approximate beginning and end of each of these excursions in terms of conodont and/or graptolite biostratigraphy. Although there remains some uncertainty regarding the precise biostratigraphic position of a couple of $\delta^{13}C$ excursions, we believe that at least in our investigated key successions, the correlation between Ordovician biostratigraphy and isotope chemostratigraphy is now reasonably well established.

A summary of our conclusions is presented in Fig. 12, which is based on the study successions in North America, Baltoscandia, and the Precordillera of Argentina. In this figure intervals with poor, if any, ties between biostratigraphy and isotopic chemostratigraphy are left blank, hence the illustrated local conodont and graptolite zone successions are not complete. The gaps shown represent intervals within which additional work is needed. For instance, the precise relations between $\delta^{13}C$ chemostratigraphy and graptolite biostratigraphy in the Lower Ordovician are incompletely known even in our study regions.

It is also important to note that little, if any, Ordovician chemostratigraphic work has been carried out in large parts of the world, such as the Artic regions, much of Asia outside China, Australia, Africa, the

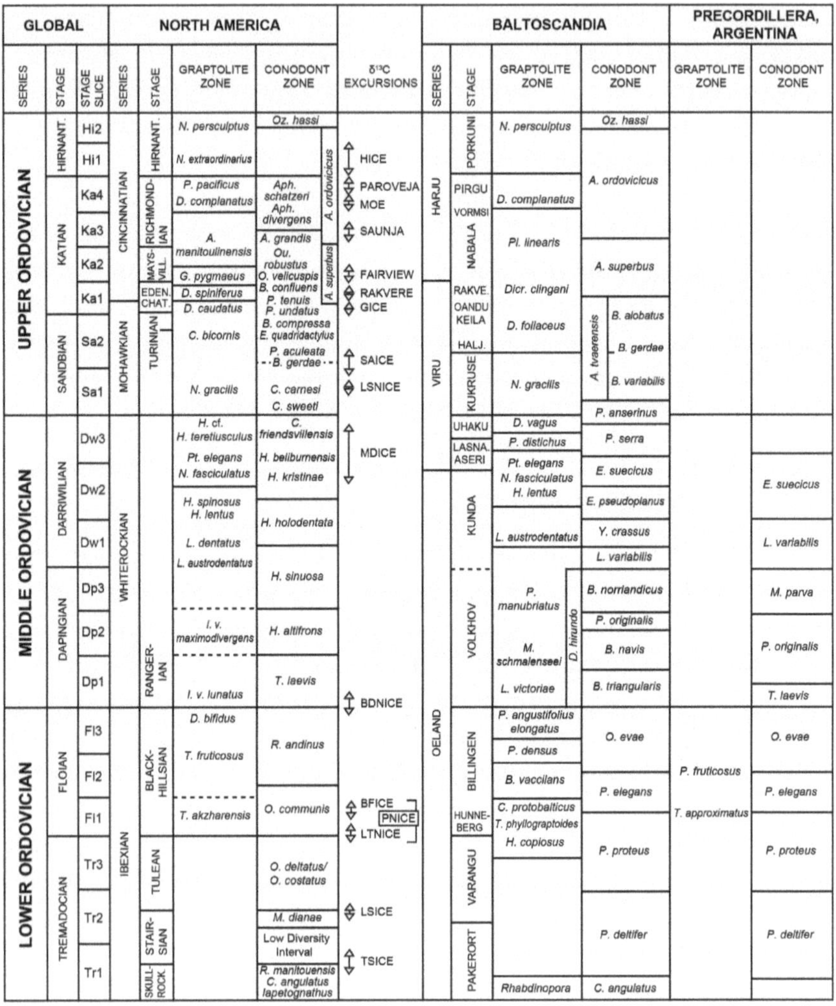

Fig. 12 The relations between the Ordovician $\delta^{13}C$ isotopic excursions and graptolite and conodont zones, stages, and stage slices in North America, Baltoscandia, and the Precordillera of Argentina. Stratigraphical intervals without direct ties between $\delta^{13}C$ chemostratigraphy and faunal zones are left blank. Arrowed vertical lines in the excursion column indicate approximate excursion ranges. Note that isotopic excursions are more common in the Upper Ordovician than in the Lower and Middle Ordovician. Also note that whereas some of the named excursions are prominent, well-known, and have been recognized in many regions of the world, others are rather insignificant in amplitude and have only local records from one or a few localities, and therefore remain subject of further studies in order to assess their significance and potential applicability (see main text for discussion).

Mediterranean region, and even in a geologically classical area such as the British Isles. Hence, Fig. 12 should be regarded as a kind of progress report representing the status of current knowledge and not as a definite final summary.

Acknowledgments

M.E.E. acknowledges funding from the Swedish Research Council (VR), and P.A. and S.M.B. thank the Crafoord Foundation and the Gyllenstierna Krapperup's Foundation for funding their research. We thank Leo Ainsaar for valuable comments.

References

Ahlberg, P., Lundberg, F., Erlström, M., Calner, M., Lindskog, A., Dahlqvist, P., Joachimski, M.M., 2019. Integrated Cambrian biostratigraphy and carbon isotope chemostratigraphy of the Grönhögen-2015 drill core, Öland, Sweden. Geol. Mag. 156, 935–949.

Ainsaar, L., Kaljo, D., Martna, T., Meidla, T., Männik, P., Nõlvak, J., Tinn, O., 2010. Middle and Upper Ordovician carbon isotope chemostratigraphy in Baltoscandia: a correlation tool and clues to environmental history. Palaeogeogr. Palaeoclimatol. Palaeoecol. 294, 189–201.

Albanesi, G.L., Ortega, G., 2002. Advances on conodont-graptolite biostratigraphy of the Ordovician system of Argentina. In: Aceñolaza, F.G. (Ed.), Aspects of the Ordovician System in Argentina. vol. 16. INSUGEO Serie Correlación Geológico, pp. 145–166.

Azmy, K., 2019. Carbon-isotope stratigraphy of the uppermost Cambrian in eastern Laurentia: implications for global correlation. Geol. Mag. 156, 759–771.

Azmy, K., Lavoie, D., 2009. High-resolution isotope stratigraphy of the Lower Ordovician St. George group of western Newfoundland, Canada: implications for global correlation. Can. J. Earth Sci. 46, 403–423.

Azmy, K., Stouge, S., Brand, U., Bagnoli, G., Ripperdan, R., 2014. High-resolution chemostratigraphy of the Cambrian–Ordovician GSSP: enhanced global correlation. Palaeogeogr. Palaeoclimatol. Palaeoecol. 409, 135–144.

Barili, R., Neilson, J.E., Brasier, A.T., Goldberg, K., Bardola, T.P., de Ros, L.F., Leng, M., 2018. Carbon isotopes, stratigraphy, and environmental change: the Middle–Upper Cambrian positive excursion (SPICE) in Port au Port Group, western Newfoundland, Canada. Can. J. Earth Sci. 55, 1209–1222.

Barta, N.C., Bergström, S.M., Saltzman, M.R., Schmitz, B., 2007. First record of the Ordovician Guttenberg $\delta^{13}C$ excursion (GICE) in New York state and Ontario: local and regional chronostratigraphic implications. Northeast. Geol. Environ. Sci. 29, 276–298.

Bauert, H., Ainsaar, L., Põldsaar, K., Sepp, S., 2014. $\delta^{13}C$ chemostratigraphy of the Middle and Upper Ordovician succession in the Tartu-453 drillcore, southern Estonia, and the significance of the HICE. Estonian J. Earth Sci. 63, 195–200.

Bergström, S.M., 1971. Conodont biostratigraphy of the Middle and Upper Ordovician of Europe and eastern North America. Geol. Soc. Am. Mem. 127, 83–159.

Bergström, S.M., Carnes, J.B., 1976. Conodont biostratigraphy and paleoecology of the Holston Formation (Middle Ordovician) and associated strata in eastern Tennessee. Geol. Assoc. Can. Spec. Pap. 15, 27–57.

Bergström, S.M., Goldman, D., 2019. $\delta^{13}C$ chemostratigraphy of the Ordovician–Silurian boundary interval. In: Sial, A.N., Gaucher, C., Ramkumar, M., Ferreira, V.P. (Eds.), Chemostratigraphy Across Major Chronological Boundaries. vol. 240. American Geophysical Union and John Wiley & Sons Geophysical Monographs, pp. 145–158.

Bergström, S.M., Chen, X., Gutiérrez-Marco, J.C., Dronov, A., 2009. The new chronostratigraphic classification of the Ordovician System and its relations to major regional series and stages and to $\delta^{13}C$ chemostratigraphy. Lethaia 42, 97–107.

Bergström, S.M., Schmitz, B., Saltzman, M.R., Huff, W.D., 2010a. The upper Ordovician Guttenberg $\delta^{13}C$ excursion (GICE) in North America and Baltoscandia: occurrence, chronostratigraphic significance, and paleoenvironmental relationships. Geol. Soc. Am. Spec. Pap. 466, 37–67.

Bergström, S.M., Young, S., Schmitz, B., 2010b. Katian (Upper Ordovician) $\delta^{13}C$ chemostratigraphy and sequence stratigraphy in the United States and Baltoscandia: a regional comparison. Palaeogeogr. Palaeoclimatol. Palaeoecol. 296, 217–234.

Bergström, S.M., Eriksson, M.E., Young, S.A., Ahlberg, P., Schmitz, B., 2014. Hirnantian (latest Ordovician) $\delta^{13}C$ chemostratigraphy in southern Sweden and globally: a refined integration with the graptolite and conodont zone successions. GFF 136, 355–386.

Bergström, S.M., Eriksson, M.E., Schmitz, B., Young, S.A., Ahlberg, P., 2016. Upper Ordovician $\delta^{13}C_{org}$ chemostratigraphy, K-bentonite stratigraphy, and biostratigraphy in southern Scandinavia: a reappraisal. Palaeogeogr. Palaeoclimatol. Palaeoecol. 454, 175–188.

Bergström, S.M., Ahlberg, P., Maletz, J., Lundberg, F., Joachimski, M.M., 2018. Darriwilian (middle Ordovician) chemostratigraphy linked to graptolite, conodont and trilobite biostratigraphy in the Fågelsång-3 drill core, Scania, Sweden. GFF 140, 229–240.

Bergström, S.M., Kleffner, M., Eriksson, M.E., 2020a. Upper Katian (Upper Ordovician) trans-Atlantic $\delta^{13}C$ chemostratigraphy: the geochronological equivalence of the ELKHORN and PAROVEJA excursions and its implications. Lethaia 53, 199–216 (published online 2019).

Bergström, S.M., Ahlberg, P., Maletz, J., Lundberg, F., Joachimski, M.M., 2020b. Integration of Darriwilian (Middle Ordovician) $\delta^{13}C_{org}$ chemostratigraphy with graptolite biostratigraphy in the classical Röstånga area in northwestern Scania (southern Sweden). Estonian J. Earth Sci. 69, 121–133.

Bergström, S.M., Schmitz, B., Terfelt, F., Eriksson, M.E., Ahlberg, P., 2020c. The $\delta^{13}C_{org}$ chemostratigraphy of Ordovician global stage stratotypes: geochemical data from the Floian and Sandbian GSSPs in Sweden. GFF 142, 23–32.

Buggisch, W., Keller, M., Lehnert, O., 2003. Carbon isotope record of late Cambrian to early Ordovician carbonates of the Argentine Precordillera. Palaeogeogr. Palaeoclimatol. Palaeoecol. 195, 357–373.

Carnes, J.B., 1975. Conodont Biostratigraphy in the Lower Middle Ordovician of the Western Appalachian Thrust-Belts in Northeastern Tennessee. Unpublished Ph.D. dissertation, The Ohio State University. 291 pp.

Edwards, C.T., Saltzman, M.R., 2014. Carbon isotope ($\delta^{13}C_{carb}$) stratigraphy of the Lower–Middle Ordovician (Tremadocian–Darriwilian) in the Great Basin, western United States: implications for global correlation. Palaeogeogr. Palaeoclimatol. Palaeoecol. 399, 1–20.

Edwards, C.T., Saltzman, M.R., 2016. Paired carbon isotopic analysis of Ordovician bulk carbonate ($\delta^{13}C_{carb}$) and organic matter ($\delta^{13}C_{org}$) spanning the great Ordovician biodiversification event. Palaeogeogr. Palaeoclimatol. Palaeoecol. 458, 102–117.

Fan, R., Bergström, S.M., Lu, Y.Z., Zhang, X., Zhang, S., Li, X., Deng, S.H., 2015. Upper Ordovician carbon isotope chemostratigraphy on the Yangtze platform, southwestern China: implications for the correlation of the Guttenberg $\delta^{13}C$ excursion (GICE) and paleoceanic change. Palaeogeogr. Palaeoclimatol. Palaeoecol. 433, 81–90.

Goldman, D., Leslie, S.A., Nõlvak, J., Young, S., Bergström, S.M., Huff, W.D., 2007. The Global Stratotype Section and Point (GSSP) for the base of the Katian stage of the upper Ordovician series at Black Knob Ridge, southeastern Oklahoma, USA. Episodes 30, 258–270.

Gorjan, P., Kaiho, K., Fike, D.A., Chen, X., 2012. Carbon- and sulfur-isotope geochemistry of the Hirnantian (Late Ordovician) Wangjiawan (Riverside) section, South China: global correlation and environmental event interpretation. Palaeogeogr. Palaeoclimatol. Palaeoecol. 337–338, 14–22.

Hatch, J.R., Jacobson, S.R., Witzke, B.J., Risatti, J.B., Anders, D.E., Watney, W.L., Newell, K.D., Vuletich, A.K., 1987. Possible late Middle Ordovician organic carbon isotope excursion: Evidence from Ordovician oils and hydrocarbon source rocks, Mid-continent and east-Central United States. Am. Assoc. Pet. Geol. Bull. 71, 1342–1354.

Henderson, M.A., Serra, F., Feltes, N.A., Albanesi, G.L., Kah, L.C., 2018. Paired isotope records of carbonate and organic matter from the Middle Ordovician of Argentina: intra-basinal variation and effects of the marine chemocline. Palaeogeogr. Palaeoclimatol. Palaeoecol. 490, 107–130.

Kaljo, D., Martma, T., 2011. Carbon isotope trend in the Mirny Creek area, NE Russia, its specific features and possible implications of the uppermost Ordovician stratigraphy. In: Guitiérrez-Marco, J.C., Rábano, I., García-Bellido, D. (Eds.), *Ordovician* of the World. 11th International Symposium on the Ordovician System. Cuadernos del Museo Geominero 14, Instituto Geológico y Minero de España, Madrid, pp. 267–273.

Kaljo, D., Hints, L., Martma, T., Nõlvak, J., Oraspõld, A., 2004. Late Ordovician carbon isotope trend in Estonia, its significance in stratigraphy and environmental analysis. Palaeogeogr. Palaeoclimatol. Palaeoecol. 210, 165–185.

Koren′, T.N., Sobolevskaya, R.F., 2008. The regional stratotype section and point for the base of the Hirnantian Stage (the uppermost Ordovician) at Mirny Creek, Omulev Mountains, Northeast Russia. Estonian J. Earth Sci. 57, 1–10.

Koren′, T., Oradovskaya, M.M., Sobolevskaya, R.F., 1988. The Ordovician-Silurian boundary beds of the north-east USSR. Bull. Br. Mus. Nat. Hist. 43, 133–138.

Koren′, T.N., Ahlberg, P., Nielsen, A.T., 2003. The post-*persculptus* and pre-*ascensus* graptolite fauna in Scania, south-western Sweden: Ordovician or Silurian? In: Ortega, G., Aceñolaza, G.F. (Eds.), Proceeding of the 7th International Graptolite Conference & Field Meeting of the Subcommission on Silurian Stratigraphy. Serie Correlación Geológica, vol. 18. Instituto Superior de Correlación Geológica (INSUGEO), pp. 133–138.

Kozic, N.P., Young, S.A., Bowman, C.N., Saltzman, M.R., Them II, T.R., 2019. Middle–Upper Ordovician (Darriwilian–Sandbian) paired carbon and sulphur isotope stratigraphy from the Appalachian Basin, USA: implications for dynamic redox conditions spanning the peak of the great Ordovician biodiversification event. Palaeogeogr. Palaeoclimatol. Palaeoecol. 520, 188–202.

Lehnert, O., Meinhold, G., Wu, R.C., Calner, M., Joachimski, M.M., 2014. $\delta^{13}C$ chemostratigraphy in the upper Tremadocian through lower Katian (Ordovician) carbonate succession of the Siljan district, Central Sweden. Estonian J. Earth Sci. 63, 277–286.

Leslie, S.A., Saltzman, M.R., Bergström, S.M., Repetski, J.E., Howard, A., Seaward, A.M., 2011. Conodont biostratigraphy and stable isotope stratigraphy across the Ordovician Knox/Beekmantown unconformity in the central Appalachians. In: Guitiérrez-Marco, J.C., Rábano, I., García-Bellido, D. (Eds.), *Ordovician* of the World. 11th International Symposium on the Ordovician System. Cuadernos del Museo Geominero 14, Instituto Geológico y Minero de España, Madrid, pp. 301–308.

Lindskog, A., Eriksson, M.E., Bergström, S.M., Young, S.A., 2019. Lower–Middle Ordovician carbon and oxygen isotope chemostratigraphy at Hällekis, Sweden: implications for regional to global correlation and palaeoenvironmental development. Lethaia 52, 204–219.

Maletz, J., Ahlberg, P., 2020. Graptolite biostratigraphy of the Ordovician Almelund and Sularp Shale formations of the Fågelsång-3 drill core, Scania, Sweden. GFF 142, 33–51.

Maletz, J., Ahlberg, P., Suyarkova, A., Loydell, D.K., 2014. Silurian graptolite biostratigraphy of the Röstånga-1 drill core, Scania—a standard for southern Scandinavia. GFF 136, 175–178.

Maletz, J., Ahlberg, P., Lundberg, F., 2020. Ordovician graptolite biostratigraphy of the Röstånga-2 drill core (Scania, southern Sweden). GFF 142, 206–222. https://doi.org/10.1080/11035897.2020.1739743.

Melchin, M.J., Holmden, C., 2006. Carbon isotope chemostratigraphy in Arctic Canada: sea-level forcing of carbonate platform weathering and implications for Hirnantian global correlation. Palaeogeogr. Palaeoclimatol. Palaeoecol. 234, 186–200.

Miller, J.F., Evans, K.R., Freeman, R.L., Ripperdan, R.L., Taylor, J.F., 2014. The proposed GSSP for the base of Cambrian stage 10 at the first appearance datum of the conodont *Eoconodontus notchpeakensis* (Miller, 1969) in the House Range, Utah, USA. GFF 136, 189–192.

Miller, J.F., Ripperdan, R.L., Loch, J.D., Freeman, R.L., Evans, K.R., Taylor, J.F., Tolbart, Z.C., 2015. Proposed GSSP for the base of Cambrian stage 10 at the lowest occurrence of *Eoconodontus notchpeakensis* in the House Range, Utah, USA. Ann. Paléontol. 101, 199–211.

Molyneux, S.G., Delabroye, A., Wicander, R., Servais, T., 2013. Biogeography of early to mid Palaeozoic (Cambrian–Devonian) marine phytoplankton. In: Harper, D.A.T., Servais, T. (Eds.), Early Palaeozoic Biogeography and Palaeogeography. Memoirs, vol. 38. Geological Society, London, pp. 365–397.

Paiste, T., Männik, P., Nõlvak, J., Meidla, T., 2020. The lower boundary of the Haljala regional stage (Sandbian, Upper Ordovician) in Estonia. Estonian J. Earth Sci. 69, 76–90.

Rasmussen, C.M.Ø., Ullman, C.V., Jakobsen, K.G., Lindskog, A., Hansen, J., Hansen, T., Eriksson, M.E., Dronov, A., Frei, R., Korte, C., Nielsen, A.T., Harper, D.A.T., 2016. Onset of main Phanerozoic marine radiation sparked by emerging mid Ordovician icehouse. Sci. Rep. 6, 18884. https://doi.org/10.1038/srep18884.

Ripperdan, R.L., 2002. The HERB event: end of the Cambrian carbon cycle paradigm? Geol. Soc. Am. 34 (6), 413. Abstracts with Programs.

Saltzman, M.R., Young, S.A., 2005. Long-lived glaciation in the late Ordovician? Isotopic and sequence-stratigraphic evidence from western Laurentia. Geology 33, 109–112.

Schiffbauer, J.D., Huntley, J.W., Fike, D.A., Jeffrey, M.J., Gregg, J.M., Shelton, K.L., 2017. Decoupling biogeochemical records, extinction, and environmental change during the Cambrian SPICE event. Sci. Adv. 3 (3), e1602158.

Schmitz, B., Bergström, S.M., Xiaofeng, W., 2010. The middle Darriwilian (Ordovician) $\delta^{13}C$ excursion (MDICE) discovered in the Yangtze Platform succession in China: implications of its first recorded occurrence outside Baltoscandia. J. Geol. Soc. Lond. 167, 249–259.

Štorch, P., Mitchell, C.E., Finney, S.C., Melchin, M.J., 2011. Uppermost Ordovician (upper Katian–Hirnantian) graptolites of north-Central Nevada, U.S.A. Bull. Geosci. 86, 301–386.

Sweet, W.C., 1979. Late Ordovician conodonts and biostratigraphy of the western Midcontinent Province. Brigham Young Univ. Geol. Stud. 26, 45–86.

Sweet, W.C., 1984. Graphic correlation of upper middle and upper Ordovician rocks, north American Midcontinent Province, USA. In: Bruton, D.L. (Ed.), Aspects of the Ordovician System. Palaeontological Contributions from the University of Oslo, vol. 295. Universitetsforlaget, Oslo, pp. 23–35.

Sweet, W.C., 2000. Conodonts and biostratigraphy of upper Ordovician strata along a shelf to basin transect in Central Nevada. J. Paleo. 74, 1148–1160.

Terfelt, F., Eriksson, M.E., Schmitz, B., 2014. The Cambrian–Ordovician transition in dysoxic facies in Baltica—diverse faunas and carbon isotope anomalies. Palaeogeogr. Palaeoclimatol. Palaeoecol. 394, 59–73.

Thompson, C.K., Kah, L.C., 2012. Sulfur isotope evidence for widespread euxinia and a fluctuating oxycline in Early to Middle Ordovician greenhouse oceans. Palaeogeogr. Palaeoclimatol. Palaeoecol. 313–314, 189–214.

Underwood, C.J., Crowley, S., Marshall, J.D., Brenchley, P.J., 1997. High-resolution carbon isotope stratigraphy of the basal Silurian stratotype (Dob's Linn, Scotland) and its global correlation. Journal of the Geological Society, London 154, 709–718.

Vandenbroucke, T.R.A., Armstrong, H.A., Williams, M., Paris, F., Sabbe, K., Zalasiewicz, J.A., 2013. Late Ordovician zooplankton maps and the climate of the early Palaeozoic icehouse. In: Harper, D.A.T., Servais, T. (Eds.), Early Palaeozoic Biogeography and Palaeogeography. Memoirs, vol. 38. Geological Society, London, pp. 399–405.

Wang, X.-F., Stouge, S., Maletz, J., Bagnoli, G., Qi, Y.-P., Raevskaya, E.G., Wang, C.-S., Yan, C.-B., 2019. Correlating the global Cambrian–Ordovician boundary: precise comparison of the Xiaoyangqiao section, Dayangcha, North China with the Green Point GSSP section, Newfoundland, Canada. Palaeoworld 28, 243–275.

Wu, R.C., Calner, M., Lehnert, O., Peterffy, O., Joachimski, M.M., 2015. Lower–Middle Ordovician $\delta^{13}C$ chemostratigraphy of western Baltica (Jämtland, Sweden). Palaeoworld 24, 110–122.

Wu, R.C., Calner, M., Lehnert, O., 2017. Integrated conodont biostratigraphy and carbon isotope chemostratigraphy in the lower–middle Ordovician of southern Sweden reveals a complete record of the MDICE. Geol. Mag. 154, 334–353.

Wu, R.C., Liu, J.B., Calner, M., Gong, F.Y., Lehnert, O., Luan, X.C., Li, L., Zhan, R.B., 2020. High-resolution carbon isotope stratigraphy of the Lower and Middle Ordovician succession of the Yangtze Platform, China: implications for global correlation. J. Geol. Soc. 177, 537–549. https://doi.org/10.1144/jgs2018-208.

Young, S.A., Gill, B.C., Edwards, C.T., Saltzman, M.R., Leslie, S.A., 2016. Middle–Late Ordovician (Darriwilian–Sandbian) decoupling of global sulfur and carbon cycles: isotopic evidence from eastern and southern Laurentia. Palaeogeogr. Palaeoclimatol. Palaeoecol. 458, 118–132.

Zaitsev, A.V., Pokrovsky, B.G., 2014. Carbon and oxygen isotope compositions of lower–middle Ordovician carbonate rocks in the northwestern Russian Platform. Lithol. Miner. Resour. 49, 272–279.

Zhang, S., Barnes, C.R., 2007. Late Ordovician to early Silurian conodont faunas from the Kolyma terrane, Omulev Mountains, Northeast Russia and their paleobiogeographic affinity. J. Paleo. 81, 490–512.

Zhu, M.Y., Babcock, L.E., Peng, S.C., 2006. Advances in Cambrian stratigraphy and paleontology: integrating correlation techniques, paleobiology, taphonomy and paleoenvironmental reconstruction. Palaeoworld 15, 217–222.

CHAPTER SEVEN

Ordovician carbon isotope stratigraphy of China: A synthesis

Chao Li[a,b] and Yuandong Zhang[a,b,*]
[a]State Key Laboratory of Palaeobiology and Stratigraphy, Nanjing Institute of Geology and Palaeontology and Center for Excellence in Life and Paleoenvironment, Chinese Academy of Sciences, Nanjing, China
[b]University of Chinese Academy of Sciences, Beijing, China
[*]Corresponding author: e-mail address: ydzhang@nigpas.ac.cn

Contents

Abstract

In this paper we review the published carbon isotope records ($\delta^{13}C_{carb}$) of the Ordovician System from Chinese terranes, principally South China, North China, Tarim and Xizang (Tibet). All the $\delta^{13}C_{carb}$ values from a number of sections in each terrane and published by different authors are calibrated to integrated graptolite and conodont biostratigraphic zones, and are correlated with each other. The $\delta^{13}C_{carb}$ curves for each block or terrane are interpreted by referring to the specific geological backgrounds and are discussed with their implications. The composite curve of $\delta^{13}C_{carb}$ from Tarim agrees well with the global composite curve, but with a significantly prolonged GICE and a lack of data for the late Katian and Hirnantian. In South China, where the published $\delta^{13}C_{carb}$ values of different facies are reviewed separately, the results show that the composite curves of three facies are generally well correlated, especially for the Tremadocian and early Floian. The MDICE excursions are not convincingly confirmed in the Middle Yangtze Platform and the Jiangnan Slope, suggesting the likely presence of hiatuses within a succession of uniform facies, or that the excursions were missed due to inadequate sampling, a common problem for condensed sections. The GICE excursions have been recognized in all the facies, but show a significant

Stratigraphy & Timescales, Volume 5
ISSN 2468-5178
https://doi.org/10.1016/bs.sats.2020.08.010

prolongation relative to that in the United States. The HICE can be steadily identified in the Upper and Middle Yangtze Platform, but is distinguished by overall low values of $\delta^{13}C_{carb}$, ca. 5‰ lower than Baltoscandia and Anticosti Island, suggesting the likelihood of missing peaks due to tectonic disturbances.

1. Introduction

In the past 20 years, the Ordovician carbon isotope chemostratigraphy of China has achieved substantial progress, as evidenced by a number of publications addressing the subdivisions and correlations of varying stratigraphic intervals, and/or different blocks or terranes. Some of these papers calibrated the $\delta^{13}C$ values and curves of a specific area or section to different or sometime contradictory biostratigraphic zonations or chronostratigraphic schemes, which introduced confusion and lowered the stratigraphic resolution. This calls for a synthesis and a careful review of these data with a constraint of widely agreed, integrated conodont and graptolite biostratigraphic zonations. Reliable $\delta^{13}C$ curves from a specific terrane or region may be used as a helpful tool for refining the intra- and inter-regional stratigraphic correlation, especially in the case of drill cores or where biostratigraphically diagnostic fossils are rare or absent. Therefore, we take this chance to conduct a review on these published carbon isotope results of the Ordovician from the several presumably discrete terranes or blocks comprising China.

Geologically, China embraces eight major blocks or terranes, including South China, North China, Jungar, Tarim-Yiri, Qaidam, Tibet, Indo-China and Sibumasu (Chen et al., 1995; Zheng et al., 2013; Fig. 1). In these blocks, Ordovician successions were developed and exposed to a different extent, e.g., in South China they are extensively distributed, and are represented by platform, slope and basin facies, while in Jungar the Ordovician is poorly represented by scarce and intermittent outcrops that have been severely tectonically altered. There are three substantial orogenic belts between or attached to some of these blocks, where the Ordovician records are poorly preserved or absent, namely the Altai-Hinggan Orogenic Belt (or Central Asian Orogenic Belt) in the north extending into Mongolia and eastern Siberia, and the Songpan-Garze Orogenic Belt between Tibetan terranes and the Qaidam block, and the Qinling-Dabie Orogenic Belt between South China and North China (Fig. 1).

The location of these major blocks or terranes during the Ordovician has been a big question addressed by a number of publications. Nevertheless, it has become a consensus that they scattered in the northeastern

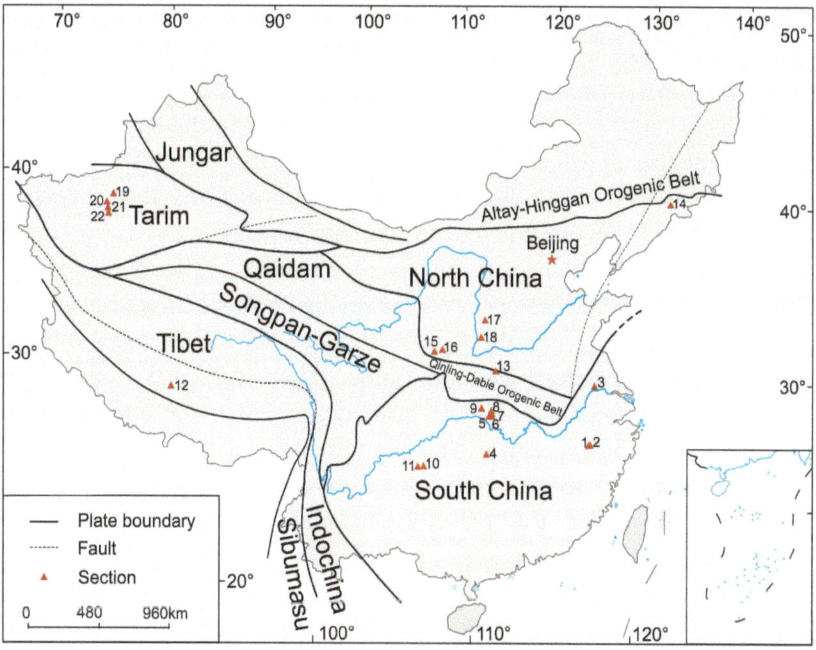

Fig. 1 Index map for the locations of the sections discussed in the present paper. Sections: 1. Huangnitang, Changshan County, Zhejiang Province; 2. Putangkou, Changshan County, Zhejiang Province; 3. Ganggangshan, Jurong County, Jiangsu Province; 4. Maocaopu, Taoyuan County, Hunan Province; 5. Huanghuachang, Yichang City, Hubei Province; 6. Puxihe, Yichang City, Hubei Province; 7. Wangjiawan, Yichang City, Hubei Province; 8. Zhenjin, Yuan'an County, Hubei Province; 9. Gudongkou, Xingshan County, Hubei Province; 10. Daijiagou, Tongzi County, Guizhou Province; 11. Honghuayuan, Tongzi County, Guizhou Province; 12. Xainza, Nagqu, Tibet; 13. Sigang, Neixiang County, Henan Province; 14. Dayangcha, Baishan, Jilin Province; 15. Qishan, Shaanxi Province; 16. Tiewadian, Jinyang County, Shaanxi Province; 17. Linfen, Shanxi Province; 18. Hejin, Shanxi Province; 19. Dawangou, Kalpin, Xinjiang Autonomous Region; 20. Kalpin Cement Plant, Kalpin, Xinjiang Autonomous Region; 21. Dabantag No.2, Bachu, Xinjiang Autonomous Region; 22. Dabantag No.3, Bachu, Xinjiang Autonomous Region.

Peri-Gondwanan region, possibly separated by narrow oceans. North China, a coherent craton as widely agreed and evidenced by the sedimentary and tectonic records in the Paleozoic, is significantly different from the others with its Ordovician successions dominated by warm-water carbonates and evaporites, interrupted by an extensive hiatus and terminated no later than the Early Katian, as a consequence of the profound Huaiyuan Epeirogeny that lasted until early Carboniferous times (Zhen et al., 2016).

North China has been palaeogeographically reconstructed as straddling the equator, or located in the low latitudes of the southern hemisphere and close to the Siberia Block (Wang et al., 2016). The amalgamation of these Chinese blocks into a single unity likely commenced in the late Ordovician (known as the Kwangsian Orogeny; Chen et al., 2012, 2014) and finally ended probably in the late Triassic (known as the Indosinian Orogeny), but whose complex history is beyond the scope of present paper.

Studies on the Ordovician carbon isotope records of China date back to 1990s, when Wang and Yang (1994) for the first time obtained the continuous curve of $\delta^{13}C_{carb}$ for the Cambrian and Ordovician systems in the Kalpin area, Tarim, northwestern China. Their carbon isotope work, though based on loose samples and primarily aimed at the interpretations of transgressions and regressions and the potential relationship with biotic extinctions, revealed some significant positive excursions that were later confirmed and named the MDICE and GICE elsewhere. In South China, among the earliest studies of Ordovician carbon isotopes records was the $\delta^{13}C_{carb}$ curve of Cambrian-Ordovician transition interval by Feng et al. (2000), who attempted to use it to address marine primary productivity, rather than for stratigraphic correlation.

Substantial chemostratigraphic work on the Ordovician of China was initiated during 2000–2010, when the carbon isotope record was widely appreciated as an effective and reliable tool to improve the resolution of stratigraphic correlation for most systems of the Paleozoic. These early works, mainly focusing on calibration for specific short time intervals or geological events, were then followed by a large number of studies through the Ordovician during 2010–2020, aimed at generating continuous curves through the entire Ordovician for the distinct blocks or terranes comprising China, e.g., South China, Tarim, North China, etc. Wang et al. (2005) published the $\delta^{13}C$ records across the base of the Dapingian Stage at Huanghuachang section, Yichang, South China, where no significant excursions but just some fluctuations were recorded during the interval. Young et al. (2008) and Bergström et al. (2009a,b) pioneered the chemostratigraphic work on the Sandbian-Katian interval of South China, and successfully identified the Guttenberg Carbon Isotope Excursion (GICE) in a number of sections in the Yangtze Platform and Jiangnan Slope regions. Zhang et al. (2010) published carbon isotope records through the Ordovician to early Silurian from the Honghuayuan section, Tongzi County, Guizhou Province, focusing mainly on inorganic carbons with an aim to address potential links to the Ordovician climatic cooling and the onset of the Great Ordovician Biodiversification Event (GOBE), and were able to identify a magnificent positive excursion of $\delta^{13}C_{org}$ in the early Floian to the end of Darriwilian, which was not supported by inorganic carbon records. Fan et al. (2009) focused on the carbon isotope records of the Hirnantian Stage in Yichang area, and revealed a positive shift of $\delta^{13}C_{org}$ by 2‰, for which they proposed a two-peak pattern covering the Hirnantian Isotope Carbon Excursion (HICE).

Munnecke et al. (2011) closely sampled 11 sections of the Ordovician in South China, which included both shallow-water and deep-water settings, and were able to present a complete sequence of $\delta^{13}C_{carb}$ through the entire Ordovician. It is noteworthy that they were able to identify, based on the chemostratigraphic results, a few significant stratigraphic gaps across the blocks, e.g., the Sandbian and Hirnantian gaps in the Yangtze Region, which was confirmed by subsequent conodont biostratigraphic results from some other sections (Fan et al., 2013; Ma et al., 2015). Recently, Wu et al. (2020) documented the $\delta^{13}C_{carb}$ records from the basal Ordovician to the earliest Darriwilian in Gudongkou section, Yichang, and identified five shifts of $\delta^{13}C_{carb}$ that show potential for international correlation, and revealed that $\delta^{13}C_{carb}$ became heavier with increasing depositional depth.

In North China, Ordovician carbon isotope works were not conducted until recent years, and have been concentrated on the western and southwestern slope facies areas of the block (Guo et al., 2014), largely due to the wide extent of a great hiatus on the platform region where the Ordovician successions are substantially incomplete, and the pervasive occurrences of evaporites resulting from the hot and arid climate in the tropical or arid belts (Boucot et al., 2013).

In Tarim, there have been quite a few carbon isotope chemostratigraphic studies for the Ordovician in recent years, largely driven by the demands from petroleum companies on age determination and precise correlation for the drill cores, which generate material that is insufficient for macro fossils in most cases, and commonly inadequate for micro fossils either. Carbon isotope chemostratigraphy has been widely used as a helpful tool complementary to the lithostratigraphic and biostratigraphic works. Among the most substantial chemostratigraphic studies of the Ordovician in Tarim is that by Zhang and Munnecke (2016), who presented the longest and most continuous curves through the Ordovician, based on nine sections and drill cores. In the Jungar, which likely comprised the eastern part of Kazakhstania, or a separate region with abundant volcanic deposits, the Ordovician outcrops are limited geographically and chronologically, and are not suitable for chemostratigraphic studies.

For the Qaidam Block, no carbon isotope chemostratigraphic studies have been reported yet, though there are some Ordovician successions in the region, which may be suitable for such a work. In the early Paleozoic, the block experienced severe tectonic collisions and compressions from the north side, where several narrow accretionary belts, with embedded ophiolites and oceanic arc complex, were attached to the block (Song et al., 2017), and also from the south side, where the Songpan-Garze Block of cryptic provenance is located.

In Tibet, which comprises several parallel and likely discrete blocks (Qiangtang terrane, Lhasa terrane, Himalaya Orogen), only a few chemostratigraphic studies have been attempted for some specific intervals of the Ordovician (Yu et al., 2019; Yuan et al., 2018). The Ordovician strata have presumably been seriously altered by the collision of the Indian Continent with Eurasia, and whether the chemostratigraphic results can be used for precise stratigraphic correlations is yet to be assessed.

2. South China

South China Block is tectonically bounded to the north with North China through the Qinling-Dabie Orogenic Belt and Tan-Lu Fault, to the west by the Longmenshan Fault with the Songpan-Garze Belt, and to the southwest by the Ailaoshan-Songda Suture with the Indochina Block (central Yunnan Province and northern Viet Nam) (Zhang et al., 2013; Figs. 1 and 2). The southern and eastern boundaries of South China are not yet well-defined, but are commonly placed at the Zhenghe-Dapu Fault Zone extending southwestwards from Zhenghe in the northern Fujian Province to Dapu in the eastern Guangdong Province (Shu, 2012). South China has been geographically reconstructed as located in low latitudes of the Southern Hemisphere, being close to the paleo-equator in the Ordovician (Torsvik and Cocks, 2013).

In the northwestern and southwestern parts of South China there existed several subaerial lands, separate or connected, and there was a general decline in elevation from northwest to southeast across the entire block. Ordovician strata are extensively distributed across South China, which are classified into three basic facies, from northwest to southeast: the Yangtze Platform, Jiangnan Slope and Zhujiang Basin (Zhang, 2016; Fig. 2). This pattern of sedimentary facies differentiation remained consistent until the late

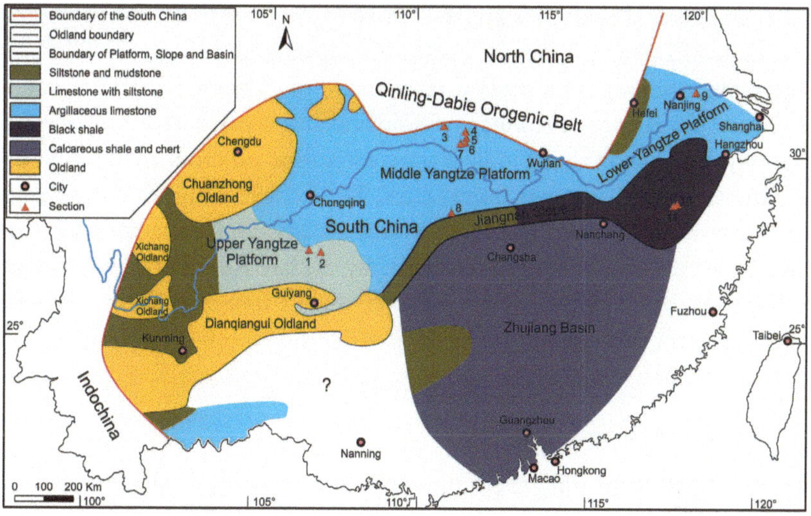

Fig. 2 Lithofacies map of South China during the middle and late Darriwilian. Sections: 1. Daijiagou; 2. Honghuayuan; 3. Gudongkou; 4. Zhenjin; 5. Wangjiawan; 6. Puxihe; 7. Huanghuachang; 8. Maocaopu; 9. Ganggangshan; 10. Putangkou; 11. Huangnitang. *Modified from Zhang, L.N., 2016. Quantitative Palaeogeographic Reconstruction of South China in the Ordovician Based on Database and GIS. (PhD Thesis. In Chinese with English abstract).*

Ordovician, when the Kwangsian Orogeny initiated, which eventually led to extensive uplift of the southern part of South China. At that point the western and southern subaerial lands partly surrounded a restricted epicontinental sea (Yangtze Sea) opening to the north.

The Ordovician successions on the Yangtze Platform are dominated by carbonates with some silisiclastic rocks, while the Jiangnan Slope is dominated by siliciclastic rocks intercalated with some carbonates (Fig. 2). The basin (Zhujiang Basin) was dominated by siliciclastic rocks. The Lower and Middle Ordovician $\delta^{13}C$ stratigraphic studies in South China have been carried out mostly for the Yangtze Platform and Jiangnan Slope (e.g., Fan et al., 2015; Munnecke et al., 2011; Wu et al., 2020). In the late Ordovician, as a consequence of the initiation of the Kwangsian Orogeny from the south, the Zhujiang Basin and Jiangnan Slope were partly uplifted to become subaerial (Chen et al., 2014). The Yangtze platform was changed into a semi-restricted Yangtze sea, with black shales dominating the middle Katian to Hirnantian (Zhang and Munnecke, 2016).

In this paper, we review separately the $\delta^{13}C$ records of the three different facies in South China (Upper Yangtze Platform, Middle Yangtze Platform, and Jiangnan Slope). The studied sections on the Upper Yangtze Platform are located close to the Dianqiangui Oldland and represented an inner shelf facies, while those on the Middle Yangtze Platform an outer shelf facies, and the Jiangnan Slope of upper slope facies. The Lower Yangtze Platform was largely connected with the Middle Yangtze Platform during the Early and Middle Ordovician, but became isolated by the late Katian, and for this reason the selected late Katian-Hirnantian $\delta^{13}C_{org}$ curve from this region was combined with those from the nearby sections of the Jiangnan Slope.

2.1 Upper Yangtze Platform

The carbon isotope chemostratigraphic studies in the Upper Yangtze Platform have been concentrated in Guizhou, Sichuan and Chongqing provinces. All these localities of studied sections were close to the Qianzhong Oldland or Dianqiangui Oldland during the Ordovician, and consequently the successions are typified by abundant mudstones resulting from terrestrial detrital input and intercalated carbonates (Zhan and Jin, 2007). The Lower and Middle Ordovician in the Upper Yangtze Platform, such as Tungtzu, Hunghuayuan and Shihtzupu formations, contain carbonates as well as many shale layers, a sequence suitable for both $\delta^{13}C_{carb}$ and $\delta^{13}C_{org}$ analyses.

Among the studied sections, the Honghuayuan section in northern Guizhou Province is one of the classic sections for the Ordovician of the Upper Yangtze Platform. The section exposes all the lithostratigraphic units of the Ordovician, i.e., the Tungtzu, Hunghuayuan, Meitan, Shihtzupu, Pagoda, Chientsaokou and Wufeng formations (Fig. 3). Zhang et al. (2009, 2010), Munnecke et al. (2011) conducted analyses of the $\delta^{13}C_{carb}$ and $\delta^{13}C_{org}$ records in this section, respectively. Some other nearby sections, such as Yanmen Power Station, Daijiagou and Dajiaosi sections in the northern Guizhou Province, were also sampled for $\delta^{13}C_{carb}$ studies (Munnecke et al., 2011). In the Sichuan and Chongqing Provinces, sections of the Lower and Middle Ordovician suitable for $\delta^{13}C$ analysis are relatively rare, due to the tremendous input of terrestrial clastics probably from the western subaerial lands. Nevertheless, the successions of the Upper Ordovician, including Pagoda, Linhsiang/Chientsaokou, and Wufeng (including Kuanyinchiao Bed) formations, are well-developed in the area and are suitable for $\delta^{13}C_{carb}$ and $\delta^{13}C_{org}$ studies. Fan et al. (2015) carried

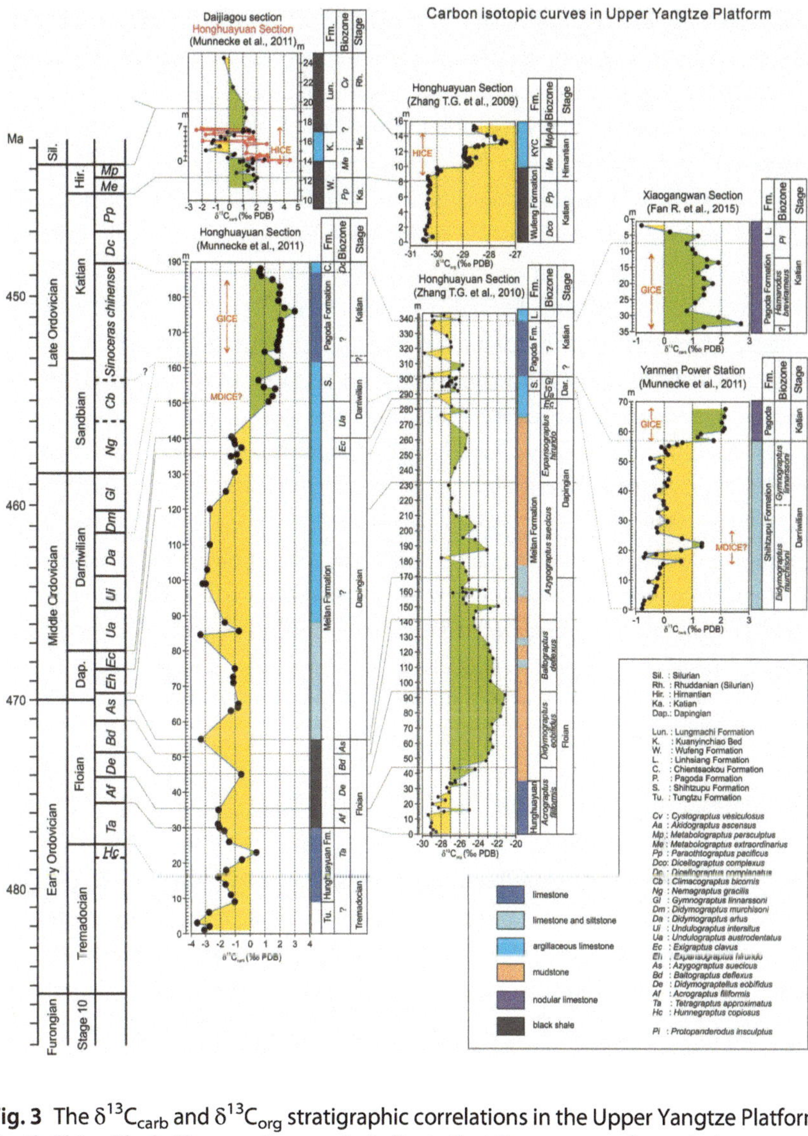

Fig. 3 The $\delta^{13}C_{carb}$ and $\delta^{13}C_{org}$ stratigraphic correlations in the Upper Yangtze Platform, South China Block. The question marks indicate that the concerned correlations are not sure. The dashed lines for the boundary of biozones means they are not well-supported, but placed tentatively.

out studies of the stable carbon isotope in the carbonates of the Pagoda and Linhsiang formations in many sections in the area, including Xiaogangwan (Chongqing), Liangcun (Guizhou), Zhuazhuayan (Sichuan), Shatan (Sichuan) sections.

By assessing the quality of the sections, the sampling density, and potential diagenetic and tectonic alteration, herein we select the best $\delta^{13}C_{carb}$ analyses from among the previous carbon isotope chemostratigraphic studies in the Upper Yangtze Platform, including the Honghuayuan, Daijiagou, Yanmen Power Station and Xiaogangwan sections (Fig. 3). Because of the relatively low sampling density across the Sandbian-Katian boundary in the Honghuayuan section, we adopt the $\delta^{13}C_{carb}$ values of the contemporary interval in the Yanmen Power Station section. The $\delta^{13}C_{carb}$ values of the Pagoda and Linhsiang formations from many sections in Sichuan and Chongqing provinces are consistent and rather similar (Fan et al., 2015), and thus we selected the data from only one section, the Xiaogangwan section, for a discussion and correlation (Fig. 3).

The $\delta^{13}C$ values of Lower and Middle Ordovician were best recorded in the Honghuayuan section (Munnecke et al., 2011; Zhang et al., 2010). In this section, the $\delta^{13}C_{carb}$ values begin from ca. −3‰ in the upper Tungtzu Formation and remain negative until the lower part of the Middle Ordovician. The $\delta^{13}C_{carb}$ results showed a small positive excursion in the *Tetragraptus approximatus* Biozone (lower Floian) with a peak value of +0.45‰ (Fig. 3), which is followed by a decrease to ca. −2‰ in the *Acrograptus filiformis* Biozone. The $\delta^{13}C_{org}$ values begin from ca. −29‰ in the Hunghuayuan Formation, and then display a positive excursion in the lower part of the Meitan Formation peaking at −21‰ in the uppermost of the *Didymograptellus eobifidus* Biozone, and then a drop to the −28‰ at the base of the Dapingian (Fig. 3; Zhang et al., 2010). However, this $\delta^{13}C_{org}$ excursion is not coupled with the values of $\delta^{13}C_{carb}$, which show a one-sample positive peak (−0.58‰) near the top of the *Didymograptellus eobifidus* Biozone followed by a decline to the one-sample negative value (−3.31‰) in the basal Dapingian (Fig. 3).

Both the values of $\delta^{13}C_{carb}$ and $\delta^{13}C_{org}$ show a positive excursion in the middle Meitan Formation (lower Dapingian) with the peak values of −0.8‰ and −23‰ for $\delta^{13}C_{carb}$ and $\delta^{13}C_{org}$, respectively, in the Honghuayuan section (Fig. 3). The $\delta^{13}C_{carb}$ values in the Honghuayuan section begin to increase in the upper Meitan Formation, and become positive in the Shihtzupu Formation (Fig. 3). It is noteworthy that the $\delta^{13}C_{carb}$ values reach a small peak in the middle of Shihtzupu Formation (*Didymograptus murchisoni* Biozone) in both the Honghuayuan section and the Yanmen Power Station section (peak values at +1.72‰ and +1.34‰, respectively), suggesting a possible MDICE.

The recognition of the base of the Katian in the Upper Yangtze Platform is slightly problematic due to the dominance of carbonates of the Pagoda Formation. This problem causes the difficulty in identifying the GICE in the region. Nevertheless, an excursion to maximum values of ca. 3‰ has been identified in the Honghuayuan and Xiaogangwan sections (Fan et al., 2015; Munnecke et al., 2011), which is constrained by the conodont *Hamarodus brevirameus* Biozone in the latter section, and may be correlated to the GICE in the Middle Yangtze Platform (Bergström et al., 2009a,b). Surprisingly, the $\delta^{13}C_{carb}$ values fluctuate gently around +2‰ through most of the Pagoda Formation in the Honghuayuan section, spanning most of the early Katian.

For the late Katian and Hirnantian, the available $\delta^{13}C_{org}$ values showed a two-step positive excursion beginning at the base of Hirnantian (about −30.3‰) with a peak value of ca. −27.4‰ in the Kuanyinchiao Bed (Zhang et al., 2009), well correlated with the HICE recognized elsewhere. Comparably, the $\delta^{13}C_{carb}$ results also showed a positive excursion in the early Hirnantian with a peak value of +2.6‰, +4.4‰, in the Daijiagou and Honghuayuan sections respectively (Munnecke et al., 2011). It is interesting to note that the likely HICE is followed by a significant negative excursion within the Kuanyinchiao Bed in both the Honghuayuan and Daijiagou sections (Fig. 3), a phenomenon also shared with that in the Middle Yangtze Platform (Chen et al., 2017a).

2.2 Middle Yangtze Platform

On the Middle Yangtze Platform, especially in the classic Yichang area for the studies of Ordovician System, the Ordovician successions are chronostratigraphically well constrained by two GSSPs, for the bases of Dapingian and Hirnantian stages (Chen et al., 2006; Wang et al., 2009). The well-established biostratigraphic sequences have facilitated a number of carbon isotope studies on the Ordovician, using both $\delta^{13}C_{carb}$ and $\delta^{13}C_{org}$ (e.g., Bergström et al., 2009a,b; Fan et al., 2009; Ma et al., 2015; Wang et al., 1993, 2005; Wu et al., 2020). The Ordovician successions in this region are mostly dominated by carbonates, particularly in the Early to the early Late Ordovician. The scarcity of siliciclastic rocks makes it difficult to obtain reliable $\delta^{13}C_{org}$ values in the region, except Young et al. (2008) who used $\delta^{13}C_{org}$ from the Puxihe section to successfully identify the GICE. However, from the late Katian onwards the black shale of the

Wufeng Formation dominated the Middle Yangtze area, of which the $\delta^{13}C_{org}$ values have been used for a precise correlation of upper Katian and Lower Hirnantian rocks (Fan et al., 2009). Subsequently, late Katian $\delta^{13}C_{carb}$ records were successfully obtained from the Wufeng Formation in the Riverside section at Wangjiawan, showing a prominent excursion corresponding to the HICE (Gorjan et al., 2012).

Many classic sections of the Ordovician in the Yichang area, most of them well exposed, have been sampled for $\delta^{13}C_{carb}$ and $\delta^{13}C_{org}$ analyses, including Wangjiawan (Chen et al., 2017a; Fan et al., 2009; Gorjan et al., 2012), Huanghuachang (Feng et al., 2000; Wang et al., 2005), Jieling

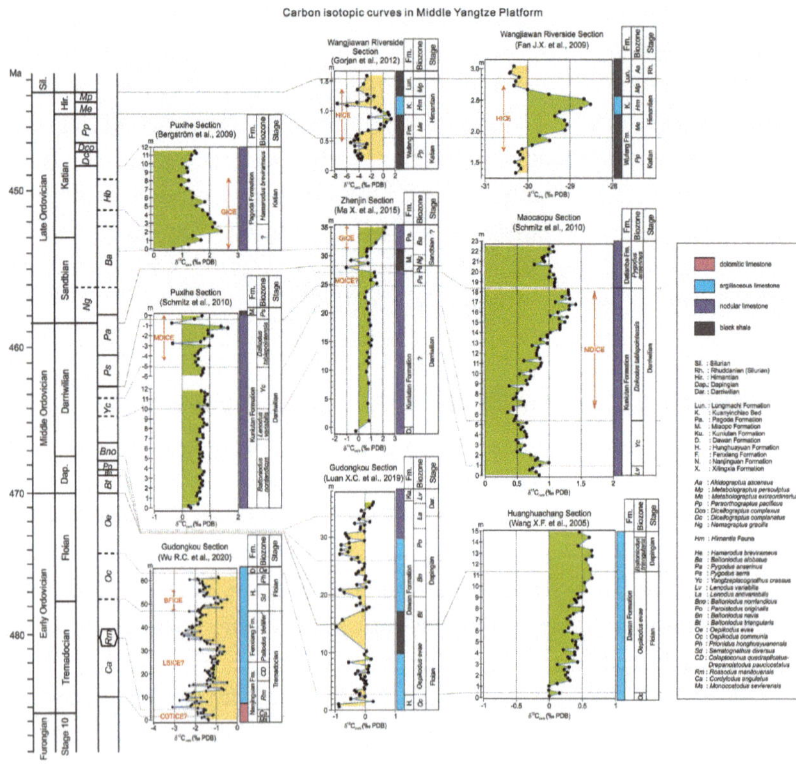

Fig. 4 The correlations of $\delta^{13}C_{carb}$ and $\delta^{13}C_{org}$ among sections on the Middle Yangtze Platform, South China. The question marks indicate that the concerned correlations are not sure. The dashed lines for the boundary of biozones imply they are not well-supported, but placed tentatively.

(Ma et al., 2015), and Xiangshuidong (Luan et al., 2019; Wu et al., 2020). Herein, we choose and illustrate some of them for discussions (Fig. 4).

In the Middle Yangtze Platform, Lower Ordovician $\delta^{13}C_{carb}$ values have been obtained from the Huanghuachang, Gudongkou and Xiangshuidong sections, most of the excursions being negative (Feng et al., 2000; Wu et al., 2020). A positive excursion, peaking at ca. −0.5‰, was recognized at a horizon corresponding to the *Cordylodus angulatus* Conodont Biozone in the Gudongkou section, and was correlated with the COTICE named in the Precordillera by Lehnert et al. (2014). Another two positive excursions were recognized in the higher horizons: the first at the boundary between the Nanjinguan and Fenxiang formations (with peak value of about −0.6‰, upper Tremadocian), and second within the lower Hunghuayuan Formation (with peak value of about −0.5‰, basal Floian). These two excursions could also be correlated with the LSICE and BFICE named by Lehnert et al. (2014).

At the boundary between the Hunghuayuan and Dawan formations (middle Floian), a sharp $\delta^{13}C_{carb}$ rise, from −1.8‰ to 0‰, occurred in the Gudongkou section (Fig. 4), which coincided with a rise at the boundary between the Hunghuayuan and Zitai formations in marginal platform setting (Luan et al., 2019; Wu et al., 2020). In the Gudongkou, Xiangshuidong and Daling sections, a negative excursion was recognized in the lower Dapingian (*Baltoniodus triangularis* Biozone) and correlated with the BDNICE, a negative excursion named by Lehnert et al. (2014). Wu et al. (2020) noted that this negative excursion has been also recorded in the Honghuayuan section and partly in the Huanghuachang section.

From the Dapingian to Darriwilian (Dawan, Zitai and Kuniutan formations), the $\delta^{13}C_{carb}$ values in the Middle Yangtze Platform became commonly positive and fluctuated within the range 0‰ to +1‰, except for the Dawan Formation in the Gudongkou section (Fig. 4; Luan et al., 2019; Wang et al., 2005; Wu et al., 2020). In the upper Kuniutan Formation, of age corresponding to *Yangtzeplacognathus crassus* to *Pygodus serra* biozones, the $\delta^{13}C_{carb}$ values showed a slightly positive excursion with a peak value at +1.5‰ (Fig. 4; Ma et al., 2015), which may be correlated to the MDICE identified in the Puxihe and Maocaopu sections, South China (Schmitz et al., 2010) and elsewhere.

The Miaopo Formation of Sandbian age is dominated by black shale interbedded with a few limestone layers (Ma et al., 2015). The $\delta^{13}C_{carb}$ values from the interbedded limestones exhibited a slightly negative excursion, with the lowermost value of −1‰ in the Zhenjin section, across the base of the *Nemagraptus gracilis* Biozone (Fig. 4). This negative excursion could be correlated to the Upper Kukruse Low named in the lower Sandbian of the Baltoscandia by Ainsaar et al. (2010). In the lower Pagoda Formation, a positive excursion (with a peak value of +2.4‰) has been widely recognized in many sections in the Middle Yangtze Platform and has been correlated as the GICE (Fig. 4; Bergström et al., 2009b; Ma et al., 2015). Young et al. (2008) carried out $\delta^{13}C_{org}$ analyses of the Pagoda Formation in the Puxihe section and identified the GICE in the lower Pagoda Formation, which increases from ca. −29‰ to peak value of −26.75‰ (Fig. 3, Young et al., 2008). In the Wangjiawan section, a HICE excursion of both $\delta^{13}C_{carb}$ and $\delta^{13}C_{org}$ can be recognized, beginning in the *Paraorthograptus pacificus* Biozone (from ca. −4‰ for $\delta^{13}C_{carb}$ and −30‰ for $\delta^{13}C_{org}$) and reaching peak values (ca. +1.3‰ for $\delta^{13}C_{carb}$ and −28.6‰ for $\delta^{13}C_{org}$) around the base of the Kuanyinchiao Bed (*Metabolograptus extraordinarius* Biozone). In addition to the Wangjiawan section, a HICE of $\delta^{13}C_{org}$ was also recognized in the Huanghuachang section (Wang et al., 1993).

2.3 Lower Yangtze Platform and Jiangnan Slope

The Lower Yangtze Platform in the Early and Middle Ordovician shared the same outer shelf facies with the Middle Yangtze Platform, and hence the $\delta^{13}C_{carb}$ values from Daling, Anhui Province (Fig. 2) showed a similar pattern to the Middle Yangtze Platform (Gong et al., 2018; Luan et al., 2019; Wu et al., 2020). The Lower and Middle Ordovician of the Jiangnan Slope is dominated by siliciclastic rocks intercalated with argillaceous limestones (Zhang et al., 2007). Previous $\delta^{13}C$ studies were concentrated in the JCY area (Jiangshan and Changshan in western Zhejiang Province, and Yushan in northeastern Jiangxi Province) (Munnecke et al., 2011). In the Late Ordovician, affected by the Kwangsian Orogeny, the Lower Yangtze Platform became disconnected from the Middle and Upper Yangtze Platform, and instead changed to grade into northwestern Zhejiang and northeastern Jiangxi, which belonged to the Jiangnan Slope in the Early and Middle Ordovician (Chen et al., 2018). The Upper Ordovician of this region is dominated by carbonates in the early to middle Katian and mainly siliciclastic rocks in the late Katian to Hirnantian (Zhang et al., 2007).

Here, we choose the Huangnitang and Putangkou sections in the JCY area to represent the Tremadocian to lower Katian of the Jiangnan Slope facies (Munnecke et al., 2011), and the Ganggangshan section near Nanjing to represent the upper Katian to Hirnantian (Botting et al., 2018; Fig. 5).

In the Huangnitang section, the $\delta^{13}C_{carb}$ values of the interval from stage 10 of Cambrian to the *Anisograptus matanensis* Biozone of the Tremadocian fluctuate within +1‰ to +2‰, and dropped to ca. −3‰ in the middle Tremadocian (Fig. 5). These positive values might represent an excursion

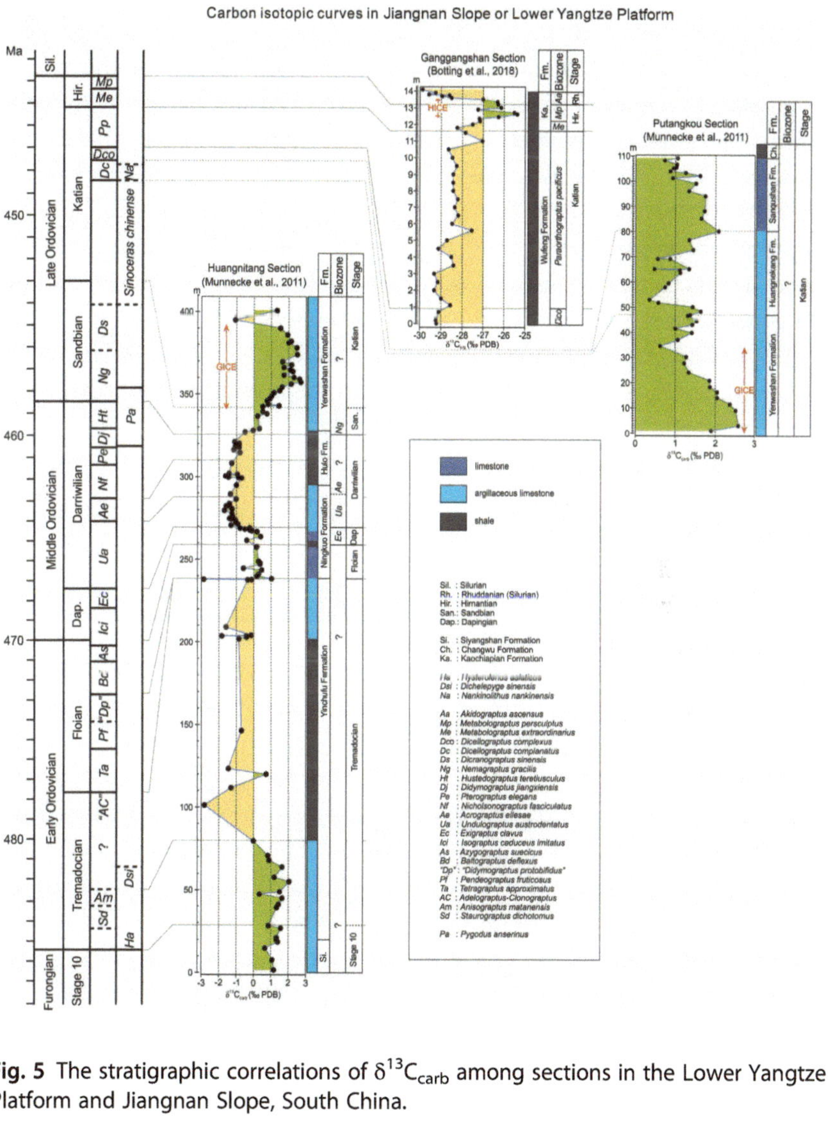

Fig. 5 The stratigraphic correlations of $\delta^{13}C_{carb}$ among sections in the Lower Yangtze Platform and Jiangnan Slope, South China.

correlated to the COTICE. The $\delta^{13}C_{carb}$ values then fluctuated sharply in the upper Tremadocian and finally increased to about 0‰ in the lower Ningkuo Formation (Fig. 5). A sharply negative shift from 0.39‰ to a minimum of −1.7‰ has been recorded in the *Undulograptus austrodentatus* Biozone (lower Darriwilian) in the middle Ningkuo Formation (Fig. 5), a negative excursion not recorded in other sections in South China (Figs. 3 and 4). Negative values of $\delta^{13}C_{carb}$ remain fluctuating between −1‰ and −2‰ in the upper Ningkuo Formation and Hulo Formation. From the beginning of the Sandbian, the $\delta^{13}C_{carb}$ values began to increase progressively and eventually reached their peak value of +2.7‰ in the middle Yenwashan Formation (Fig. 5). This positive excursion in the Yenwashan Formation was recognized as the GICE (Bergström et al., 2009b; Munnecke et al., 2011). After the GICE, the $\delta^{13}C_{carb}$ values displayed some small fluctuations between 0‰ and +2‰ in the Putangkou section (Fig. 5). The strata overlying the Sanqushan Formation in the JCY area are dominated by siliciclastic rocks in a thickness of thousands of meters, and are not suitable for $\delta^{13}C_{carb}$ or $\delta^{13}C_{org}$ analyses.

In the Ganggangshan section near Nanjing, $\delta^{13}C_{org}$ values of late Katian to Hirnantian have been made available for chemostratigraphic correlation (Fig. 5), which showed an increase from ca. −29‰ in the latest Katian (upper *Paraorthograptus pacificus* Biozone) to a peak value of −25.38‰ recorded from the lower *Metabolograptus persculptus* Biozone, indicating an excursion representing the HICE (Botting et al., 2018; Fig. 5).

3. North China

The North China Block is bounded to the south by the Qinling-Dabie Orogenic Belt and the Tan-Lu Fault with South China, and to the north by the Altai-Hingan Orogenic Belt with Siberia (Fig. 1). It was palaeogeographically reconstructed to be located near the paleo-equator during the Ordovician (Torsvik and Cocks, 2013). The Lower Ordovician and Middle Ordovician are dominated by carbonates, commonly limestones and dolomitic limestones. During the late Middle and Late Ordovician, the main part of North China was uplifted to be subaerial as a consequence of the dramatic Huaiyuan Epeirogeny, and only in the southwestern marginal belt, i.e., the Ordos Basin, some muddy limestone with shale were deposited prior to middle Katian (Chen et al., 2013; Guo et al., 2014; Zhen et al., 2016).

In the eastern part of North China, the carbon isotope data of the Dayangcha section, Jilin Province are of significance for the boundary interval of the Cambrian and Ordovician (Ripperdan et al., 1993). The $\delta^{13}C_{carb}$ values displayed a possible positive excursion from 0‰ to ca. +1.4‰ in the unnamed stage 10 of Cambrian, and another positive excursion from 0‰ to ca. 1‰ in the earliest Tremadocian, corresponding roughly to the *Cordylodus lindstromi* Biozone, which might be correlated to the COTICE (Fig. 6).

In the Hejin and Linfen sections of the Ordos Basin, the $\delta^{13}C_{carb}$ values display a great decrease from +0.5‰ in the topmost Tremadocian to ca. −5‰ in the middle Floian, the lowermost values coinciding with the boundary between the Liangjiashan and Majiagou formations (Fig. 6). In the early Darriwilian, corresponding to the top part of the Majiagou Formation, there appears to be another negative excursion of $\delta^{13}C_{carb}$ from 0‰ to ca. −4‰, which may be indicative of LDNICE. In the Tiewadian section, Shaanxi Province, a positive excursion of ca. 3‰ was recognized in the middle Kelimoli Formation, which is of middle-late Darriwilian age (*Pterograptus elegans* Biozone; Chen et al., 2017b) and may be correlated with the MDICE (Fig. 6). The $\delta^{13}C_{carb}$ records of Sandbian to Katian are obtained from the Qishan section, Shaanxi Province, which show a scattering between −1‰ and +2‰, and from which it is difficult to extract any useful signatures.

Some carbon isotope records of Katian age were also reported from the Sigang section, Neixiang County, Henan Province, which belongs to the southern margin of North China Block and was likely seriously affected by the multi-episodic collisions with South China during the late Paleozoic to Triassic (Fig. 1; Jing et al., 2019). The $\delta^{13}C_{carb}$ values formed a negative excursion in the *Belodina confluens-Oulodus ulrichi* Biozone, and then fluctuated mainly between −3‰ and +1‰ in the *Oulodus robustus* to *Apelognathus divergens* biozones (Fig. 6). Jing et al. (2019) were able to recognize tentatively four to five positive excursions in the section: Kope, Fairview, Waynesville, Whitewater and Elkhorn.

4. Tarim

Tarim is bounded to the north by the Tianshan Mountains with the Jungar Basin and to the south by the West Kunlun-Altun Mountain range of the Qinghai–Tibet Plateau (Fig. 1). The Ordovician strata of Tarim are extensively distributed and yield rich petroleum resources, but most of

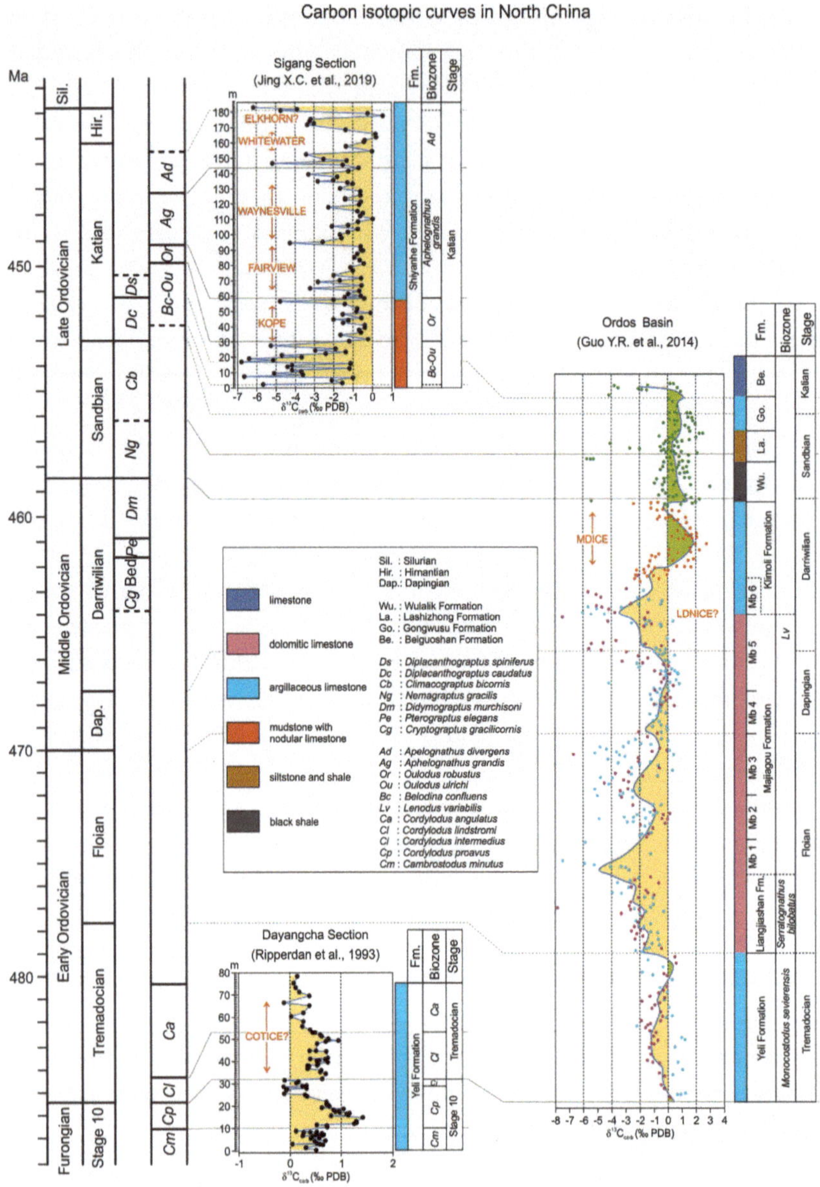

Fig. 6 The stratigraphic correlations of $\delta^{13}C_{carb}$ among sections in North China Block. Note that the $\delta^{13}C_{carb}$ curve from the Ordos Basin is a composite curve which was correlated to chronostratigraphic stages by the authors, and the samples in green, orange, purple and blue colors are from Qishan, Tiewadian, Hejin, and Linfen sections, respectively (Guo et al., 2014), and the Member 6 of the Majiagou Formation is contemporaneous with lower part of the Kelimoli Formation.

the deposits are covered by the vast and thick Taklimakan Desert, leaving some outcrops in the northwestern peripheral region and the Kuruktag area in the eastern part (Zhang and Munnecke, 2016).

The Ordovician strata in Tarim are of several different facies. The central Tarim was mainly a platform, while the western and eastern parts of Tarim were slope and basin, respectively. Since only a few sections are exposed, the distributions of these facies have not yet been clearly understood (Chen et al., 2012). In the western Tarim, the Ordovician successions are mainly composed of carbonates, sandstones, and some cherts and black shale, representing a slope facies or an intra-platform depression (Zhang and Munnecke, 2016). In the Bachu area, and the central Tarim under the Taklimakan Desert, the Ordovician sequence is dominated by carbonates of shallow-water facies, with several types of reefs (Zhang et al., 2015). In the eastern Tarim, siliciclastic rocks dominate the lithofacies during the entire Ordovician. In the late Katian, the Tarim Block was mostly uplifted to be subaerial, and the late Katian to Hirnantian records are largely absent (Chen et al., 2013).

Several outcrop sections and industrial wells in Tarim have been sampled for $\delta^{13}C$ stratigraphic studies (e.g., Jing et al., 2008; Liu et al., 2016; Zhang and Munnecke, 2016; Zhao et al., 2017). In order to understand and assess the potential effects of different facies on the carbon isotope records, we select some sections from the Kalpin area in the northwestern part, where the Ordovician is of likely slope or deep-water facies, and some sections from the Bachu area, where represents a shallow-water, platform facies (Fig. 1).

The carbon isotope signatures of the Cambrian-Ordovician boundary were recorded in the Kalpin Cement Plant section (Fig. 7; Jing et al., 2008). The $\delta^{13}C_{carb}$ values of the Lower Ordovician began with ca. -1‰ and increased to the peak value of +0.4‰ in the *Teridontus huanghuachangensis* Biozone. This positive excursion may coincide roughly with the Cambrian-Ordovician boundary, which is followed by a decrease to values lower than −1‰ in the middle Tremadocian (Jing et al., 2008). In the Dabantag No.2 section, the $\delta^{13}C_{carb}$ records varied between −2‰ and −1‰ in the Tremadocian and reached a small peak value of −0.47‰ in the uppermost Tremadocian (Zhang and Munnecke, 2016; Fig. 7). For most of the Floian and Dapingian, the $\delta^{13}C_{carb}$ values are negative and fluctuate between −1‰ and −3‰, until a sharp, steady increase back to around 0‰ in the upper Dapingian and lower Darriwilian. In the Dabantag No. 3 section, there seems to be a gentle positive excursion in

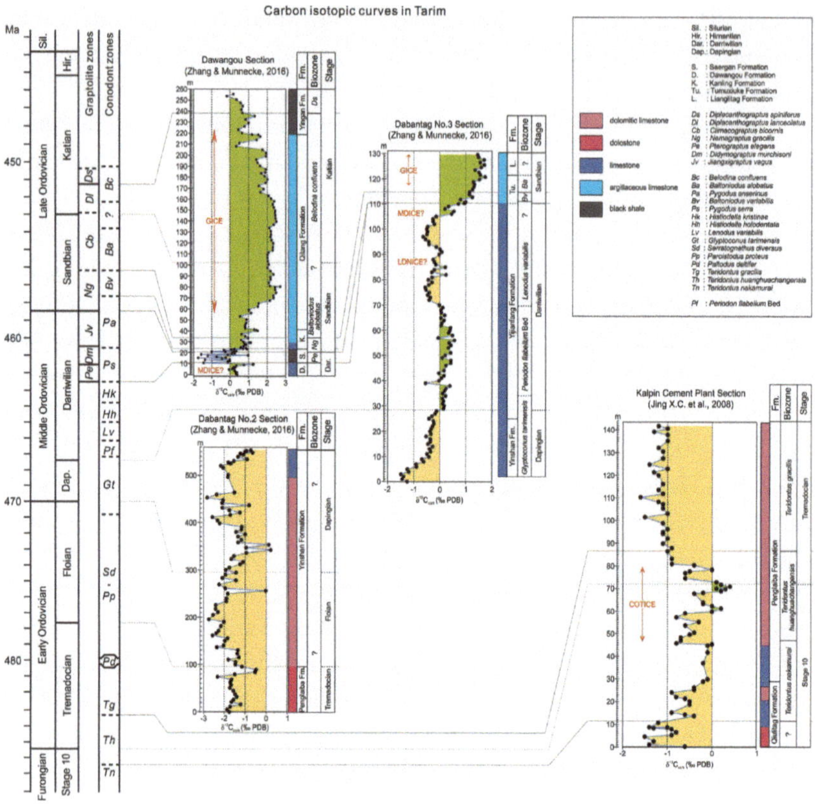

Fig. 7 The stratigraphic correlations of $\delta^{13}C_{carb}$ among the Ordovician sections in the Tarim Block.

the earliest Darriwilian. In the top part of the Dawangou Formation in the Dawangou section, and the upper Yijianfang Formation in the Dabantag No. 3 section, a small positive excursion was recognized as the MDICE in the middle Darriwilian (Zhang and Munnecke, 2016). Based on conodont biostratigraphy, it seems that the upper part of the MDICE in the Kalpin and Bachu areas are absent due to occurrence of a likely hiatus between the Dawangou Formation and the overlying black shale of the Saergan Formation.

In the Dawangou section, the upper Darriwilian to the basal Sandbian were represented by black shales with some thin layers of cherty limestone of the Saergan Formation. The $\delta^{13}C_{carb}$ values in the Saergan Formation varied between −2‰ and 1‰, somewhat surprisingly (Fig. 7). In the upper

Sandbian to the lower Katian in the Dawangou and Dabantag No.3 sections, a conspicuous and well-supported positive excursion of $\delta^{13}C_{carb}$ to a maximum value of +2.5‰ occurs, which was correlated to the GICE as identified elsewhere (Zhang and Munnecke, 2016). After the middle Katian, most part of the Tarim was uplifted and consequently only few strata are available for $\delta^{13}C_{carb}$ analysis.

5. Xizang (Tibet)

The Xizang Region, as used herein, is composed of three major blocks, i.e., the Qiangtang, Lhasa terranes and the Himalaya Orogen (Fig. 1). In the southern-central Xizang, the Ordovician strata prior to the early Katian are composed of limestones with some dolomitic limestones and siltstones, while the upper Katian and Hirnantian are dominated by shales (Zhen et al., 2020). The lithostratigraphic and biostratigraphic studies of Xizang started in the 1960s, but no carbon isotope stratigraphic studies have been carried out until recently, when Yuan et al. (2018) reported briefly a $\delta^{13}C_{carb}$ curve for the upper Ordovician in the Xainza area. The $\delta^{13}C_{carb}$ values show a single-sample positive excursion in the lower Gangmusang Formation, a case to be confirmed by more sections for geological implications. The Katian is typified by a steady decrease of $\delta^{13}C_{carb}$ from ca. +1.5‰ to +0.7‰ in the Gangmusang Formation, which is followed by a positive excursion to a maximum of ca. +2.5‰ in in the lower Dewukaxia Formation, which was recognized as the HICE (Yuan et al., 2018; Fig. 8). Surprisingly, the HICE was recorded within a shale-dominated succession (Fig. 8).

Recently, Yu et al. (2019, 2020 under review) reported carbon isotope results in the Alai Formation and lower Jiaqu Formation at Jiacun section, Nyalam, Southern-central Xizang (Figs. 1 and 9). With the constraint of well-established biostratigraphy, a positive excursion in the *Histiodella kristinae* to *Pygodus serra* biozones was recognized as the MDICE (Yu et al., 2020 under review). The MDICE here began in the middle Alai Formation and reached its peak value of ca. +2.3‰ in the lower Jiaqu Formation (Fig. 8).

6. Inter-regional correlation and discussion

We try herein to combine the usable carbon isotope values of $\delta^{13}C_{carb}$ for each block or region of specific facies into a composite curve using a

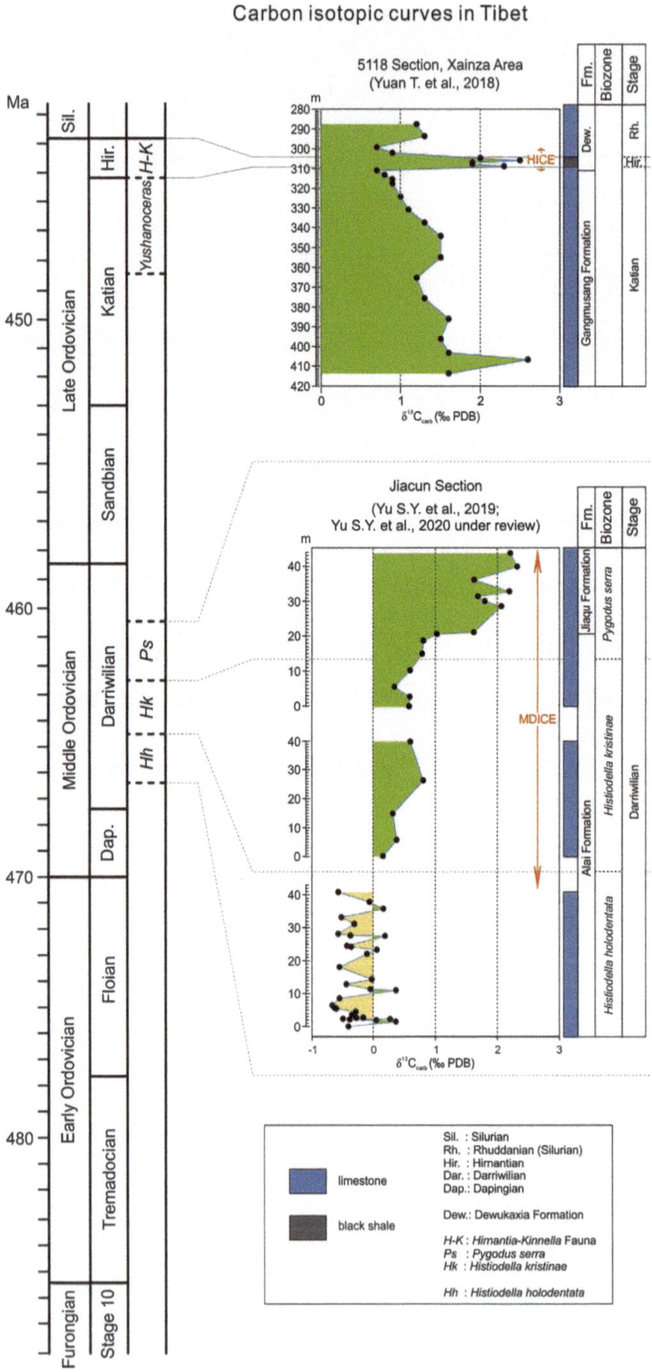

Fig. 8 The $\delta^{13}C_{carb}$ records the Ordovician in Xizang (Tibet).

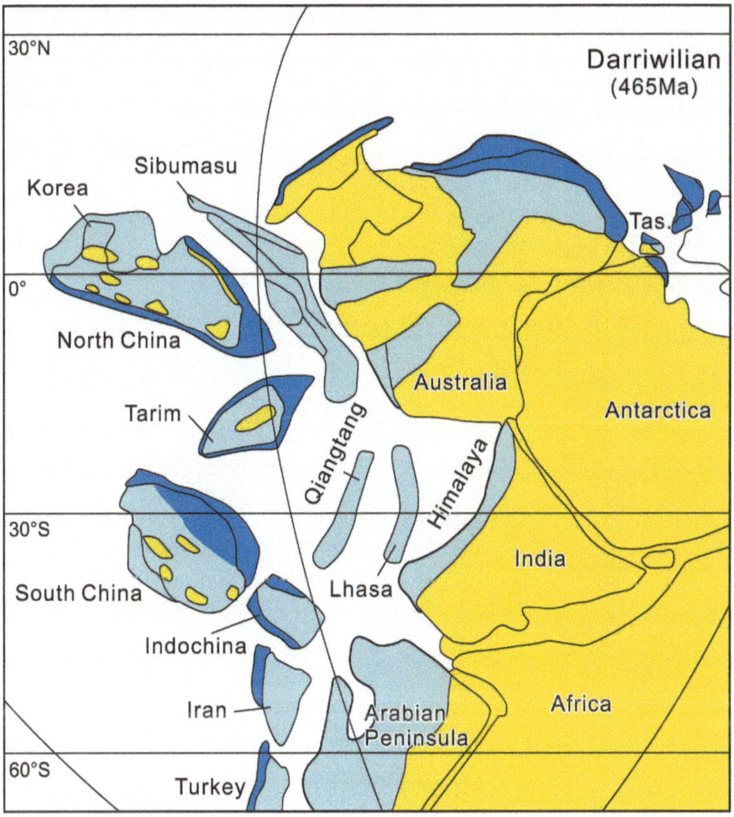

Fig. 9 Palaeogeographic reconstruction of Chinese terranes in Darriwilian Stage of Ordovician. *After Yu, S.Y., Fang, X., Munnecke, A., Li, W., Zhen, Y., Li, Y., Wang, Z., Zhang, Y.D., 2019. First documentation of Middle Ordovician warm-water carbonates in the Mount Jolmo Lungma (Mount Everest) area, southern Xizang (Tibet), China, and its paleogeographic implications. Palaeogeogr. Palaeoclimatol. Palaeoecol. 530, 136–151.*

moving average of three successive values, against the global official chronostratigraphic divisions of the Ordovician System. These composite curves, from Tarim, the Upper Yangtze Region, Middle Yangtze Region, and the Jiangnan Slope-Lower Yangtze Region, then are correlated with each other and with the global composite curve, so as to show whether they are of similar trend and display correlatable excursions (Fig. 10). For significant discrepancies, we check the validities of the values involved and consider likely controlling factors, or otherwise analyse and discuss whether it is possible to reveal any potential geological signatures.

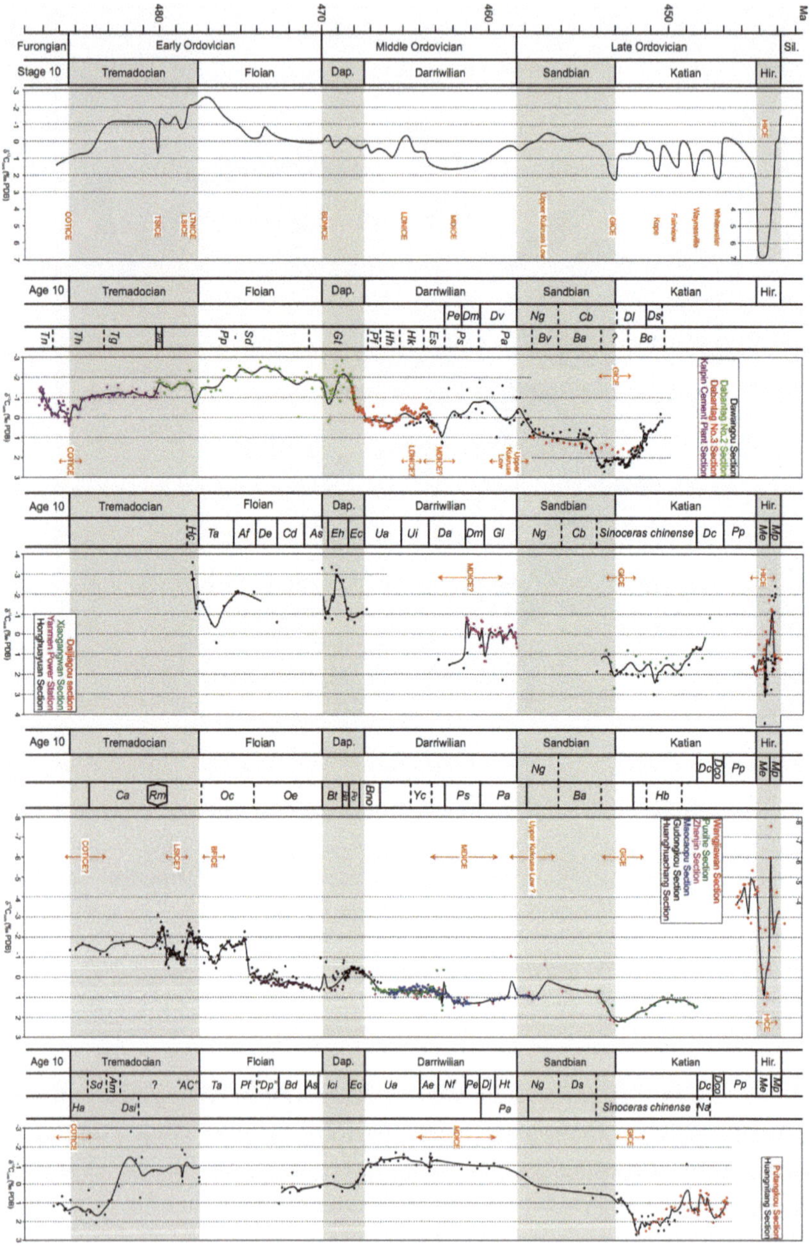

Fig. 10 See figure legend on opposite page.

The composite curve of $\delta^{13}C_{carb}$ for the Tarim Block is composed of the carbon isotope records from the Dawanggou, Dabantag No.2, Dabantag No.3 and the Kalpin Cement Plant sections (Fig. 10). The curve covers fairly continuously the large time interval of the Ordovician, from latest Cambrian to early Katian. The absence of late Katian and Hirnantian values are due to the presences of two stratigraphic gaps within the interval, one between the Yingan Formation (early Katian) and the overlying Terekawat Formation (late Katian), and the other between the Terekawat Formation and the overlying Kalpintag Formation (early Silurian) (Zhang et al., 2020). The Terekawat Formation is dominated by sandstones, siltstones and shales, and does not contain sufficient carbonates for carbon isotope analysis.

The available composite curve matches largely with the global composite curve. From the basal Tremadocian to Dapingian, the Tarim curve shows a steadily decreasing trend of $\delta^{13}C_{carb}$, with the most negative values approaching −3‰ (Fig. 10). In the Darriwilian, a short middle Darriwlian positive excursion, followed by a subsequent decrease which are supported by considerably scattered values, is shown to be likely correlated to the MDICE recognized in Baltoscandia (Ainsaar et al., 2010). If this minor positive shift represents the lower limb of MDICE as recognized elsewhere, which is to be confirmed by future further works, it seems quite likely that the peak and the upper limb are truncated by a potential disconformity yet to be confirmed by conodont biostratigraphy.

The Sandbian interval begins with an increase of $\delta^{13}C_{carb}$ from the preceding, largely negatives in Darriwilian, and is represented by various values gently fluctuating around 1‰. From the latest Sandbian to early Katian, an interval corresponding to the Qilang and Yingan formations (dominated by

Fig. 10 A correlation of the composite curves for each block or region of specific facies in China, and a comparison with the global composite curve. Data sources: 1. Global composite curve: the Darriwilian and late Ordovician segment is after Ogg et al. (2016), while the Tremadocian to early Darriwilian segment is after Lehnert et al. (2014); 2. Tarim: Dawangou, Dabantag No.2 and Dabantag No.3 sections (Zhang and Munnecke, 2016), Kalpin Cement Plant section (Jing et al., 2008); 3. Upper Yangtze Platform: Daijiagou, Yanmen Power Station and Honghuayuan sections (Munnecke et al., 2011), Xiaogangwan (Fan et al., 2015); 4. Middle Yangtze Platform: Wangjiawan section (Gorjan et al., 2012), Puxihe section (Bergström et al., 2009; Schmitz et al., 2010), Zhenjin section (Ma et al., 2015), Maocaopu section (Schmitz et al., 2010), Gudongkou section (Luan et al., 2018; Wu et al., 2020), Huanghuachang section (Wang et al., 2005); 5. Jiangnan Slope: Putangkou and Huangnitang (Munnecke et al., 2011).

thin- to medium-bedded limestone alternating with shale), there is an obvious positive excursion, which was correlated with the globally well-identified GICE (Zhang and Munnecke, 2016). However, the GICE in Tarim appears more protracted than in the United States (Bergström et al., 2010; Young et al., 2008).

In the Upper Yangtze Platform, where the Ordovician successions are dominated by siliciclastic deposits with some carbonates, the composite curve is composed of five discrete segments (Fig. 10). This discontinuity is due to the absence of carbonates suitable for carbon isotope analysis in some specific intervals, and the presence of stratigraphic gaps including that between the Shihtzupu and Pagoda formations and spanning most of the Sandbian. Overall, the available values of Floian and Dapingian $\delta^{13}C_{carb}$ are mostly negative, and are comparable with those records from elsewhere. The Darriwilian $\delta^{13}C_{carb}$ of the Shihtzupu Formation does not show a clear positive excursion, possibly representing the aftermath of the MDICE. During the early Katian, $\delta^{13}C_{carb}$ experienced a lasting prominent positive excursion, which may be correlated to the GICE, or perhaps also those of Kope and Fairview. The latter case seems quite likely but remains to be confirmed, if those stratigraphic records with negative $\delta^{13}C_{carb}$ have been missing due to the shallow-water setting of the region and the fluctuation of sea levels.

The $\delta^{13}C_{carb}$ data of the Hirnantian Stage from the Daijiagou and Honghuayuan sections, Tongzi County, Guizhou Province (Munnecke et al., 2011), represent probably only the upper limb of the HICE, as evidenced by only a few values over +3‰ being recorded, and followed by a subsequent decline, and the presence of conglomerates indicative of likely erosion at the base of the Kuanyinchiao Bed. A minor disconformity was suggested for the early Hirnantian (Munnecke et al., 2011), which may account for the lower values $\delta^{13}C_{carb}$ than those elsewhere (Brenchley et al., 2003; Jones et al., 2011; Melchin et al., 2013).

The $\delta^{13}C_{carb}$ values of the Ordovician on the Middle Yangtze Platform are fairly consistent with little variation (Fig. 10). The Tremadocian and early Floian $\delta^{13}C_{carb}$ values are all negative, within 0 to −3‰, with one or two possible positive spikes in late Tremadocian and another one in early Floian. The first spike may be correlated with the TSICE identified in Precordillera of Argentina (Buggisch et al., 2003; Lehnert et al., 2014). It is surprising that no obvious or complete MDICE has been recognized in this region, nor even across the entire South China plate, despite Schmitz et al. (2010) identifying a possible excursion of middle Darriwilian age at

Maocaopu section, Taoyuan, Hunan Province, which is distinguished by a positive shift of less than 1‰ truncated by a disconformity identified by conodont biostratigraphic work. They were also able to identify a spike of MDICE at the Puxihe section, Yichang. Nevertheless, more multidisciplinary approaches will be required to decipher the reason for the absence and incompleteness of the MDICE in South China. An obvious late Sandbian to early Katian positive excursion has been widely identified in the region, and correlated to the GICE (Bergström et al., 2009b; Young et al., 2008). This excursion, however, exhibits a prolongation until the *Hamarodus brevirameus* Biozone (Fig. 10).

The $\delta^{13}C$ changes at Wangjiawan section near Yichang have attracted a lot of attention and there have been studies of organic and inorganic carbon isotopes across the bases of the Hirnantian and the Silurian, due to its precise biostratigraphic and chronostratigraphic constraints as the GSSP for the base of the Hirnantian Stage. Fan et al. (2009) identified a positive excursion of $\delta^{13}C_{org}$ with two peaks, the first, minor peak in the *Metabolograptus extraordinarius* Biozone (topmost Wufeng Shale), and the second, higher peak within the Kuanyinchiao Bed bearing the *Hirnantia* Fauna. Based on samples collected also from the Riverside section at Wangjiawan, ca. 200 m from the GSSP, Gorjan et al. (2012) were able to recognize a significant positive excursion of $\delta^{13}C_{carb}$ by a ca. 6‰ shift across the base boundary, though their overall values are 5‰ lower than elsewhere. Coincidently their curves also displayed two probable peaks in a single excursion. It is noteworthy that a sharp, significant negative excursion with a $\delta^{13}C_{org}$ shift of 9–13‰, in the aftermath of the HICE, was identified by Gorjan et al. (2012) and Tu et al. (2012), which corresponds to the Kuanyinchiao Bed. The negative shift has been interpreted as possible evidence of a mass release of methane and/or other organic matter (Tu et al., 2012).

The $\delta^{13}C_{carb}$ curve of the Jiangnan Slope presented in the present paper, composed of the Huangnitang and Putangkou sections, is basically continuous except for the absences of data in two intervals (Fig. 10). The first interval of $\delta^{13}C_{carb}$ absence is in the Floian, which is due to the stratigraphic gap at Huangnitang (Munnecke et al., 2011; Zhang et al., 2007), and the second absence of data is in late Katian and Hirnantian due to the increasing dominance of siliciclastics as a consequence of the Kwangsian Orogeny (Chen et al., 2012). However, the other intervals are fairly well-supported by carbon isotope data. When compared with the curves from the Middle Yangtze Region and Tarim, there is a discrepancy of $\delta^{13}C_{carb}$ values for the Darriwilian—consistently negative in the Huangnitang section while their coeval values in the other two regions are largely positive. Additionally, no MDICE has been identified for the region, a case reminiscent of the composite curve of Monitor-Antelope Ranges in Nevada (Kaljo et al., 2007). Whether or not this discrepancy results from facies differentiation is a question to be further addressed in the future.

An early Katian positive excursion of $\delta^{13}C_{carb}$ is well supported by data from both Huangnitang and Putangkou sections, and may be correlated with the widely-recognized GICE (Fig. 10). However, as calibrated by conodont biostratigraphy, the excursion appears somewhat delayed and spans longer than elsewhere. This excursion is followed by two minor positive excursions of middle to late Katian age, which may be correlated to Waynesville and Whitewater excursions. A more precise correlation is, however, not possible at the moment due to the rough biostratigraphic constraint, and will require more data from additional sections.

The Ordovician $\delta^{13}C_{carb}$ values from North China exhibit considerable scatter, possibly due to the deposits being formed in mostly shallow-water, lagoonal marine environments, and an arid climate. Though a few excursions such as MDICE have been identified tentatively (Guo et al., 2014), they are not comparable with curves of Chinese other regions discussed here, or the global composite curve.

To summarize, the overall patterns of Ordovician $\delta^{13}C_{carb}$ values from Chinese blocks are comparable with each other and with the global composites by Bergström et al. (2009a,b, 2012) and Lehnert et al. (2014). However, some discrepancies or even decoupled segments are revealed between them, including mainly: (1) the Floian $\delta^{13}C_{carb}$ values in the Tarim being lower overall than elsewhere; (2) the Darriwilian $\delta^{13}C_{carb}$ values on the Yangtze Platform largely fluctuate gently without a clear MDICE, and are even largely negative in the Jiangnan Slope facies, suggesting the possibility that some critical intervals may have been missed in sampling or there may be an extensive, short hiatus during the middle Darriwilian across the platform; (3) the GICE has been widely recognized in South China and Tarim, but appears to be more prolonged than in the global composite curve.

7. Conclusions

The carbon isotope records, particularly $\delta^{13}C_{carb}$, of the Chinese Ordovician terranes, including South China, North China, Tarim and Xizang (Tibet), are reviewed. All published $\delta^{13}C_{carb}$ values from a number of sections in each terrane are calibrated by graptolite and conodont stratigraphic zones, and correlated with each other. The results of carbon isotope chemostratigraphy for each block or terrane are interpreted with reference to the geological contexts and are discussed with their implications. A composite $\delta^{13}C_{carb}$ curve is achieved by tentatively integrating some of the discussed curve segments in each block or terrane for inter-regional correlation and discussion, aiming to assess whether the $\delta^{13}C_{carb}$ curve can be used for the correlation of Ordovician stratigraphic successions across different facies, or among blocks with different geological histories. The composite curve of each terrane or region, once proved to be consistent and truly reflecting the geological signatures, may be used for refining the resolution of regional stratigraphic correlation significantly, especially in the cases when biostratigraphically diagnostic fossils are rare or absent (e.g., in the drill cores of Tarim and Sichuan basins), and, in turn, improve our understanding of the geological history of the region or terrane.

Overall, the composite curve of $\delta^{13}C_{carb}$ from Tarim agrees well with the global composite curve. The Tremadocian and Floian is represented by a steady decrease of $\delta^{13}C_{carb}$ with the most negative values approaching −3‰. A small positive spike is identified in the early Dapingian, which seems to be reflecting true signatures of the deposits. A minor and short positive excursion of middle Darriwilian age in the uppermost Dawangou Formation at the Dawangou section may be correlated to the MDICE. The late Sandbian to early Katian positive excursion is correlated to the GICE, but with an obvious prolongation into *Diplacanthograptus spiniferus* Biozone.

In South China, we integrate composite curves separately for different facies, i.e., Upper Yangtze Platform, Middle Yangtze Platform and Jiangnan Slope, to reveal whether the $\delta^{13}C_{carb}$ values are facies-dependent. The results show that the composite curves of the three facies are generally correlative, particularly for the Tremadocian and early Floian. The GICE excursions are readily recognized in all the facies, but showing a significant prolongation compared with that in the United States. The MDICE excursions are not convincingly confirmed in the Middle Yangtze Platform and the Jiangnan Slope, suggesting the likely presence of hiatuses within a uniform facies succession, or that the precise intervals are missed due to inadequate sampling (common situation in condensed sections).

A sharp positive excursion representing the HICE can be clearly identified in the Upper and Middle Yangtze Platform, but typified by overall low values of the $\delta^{13}C_{carb}$, ca. 5‰ lower than in Baltoscandia and Anticosti

Island. It seems likely that the peaks of the HICE are missing or truncated in the Upper Yangtze Platform, due to the integrated consequences of the active Kwangsian Orogeny and the potential sea-level drop in response to the global Hirnantian glaciation.

Acknowledgments

We are indebted to Mr. Cheng Junfeng for permission to refer to some data of his unpublished thesis on the Ordovician chemostratigraphy of Middle Yangtze Region, to Dr. Zhang Linna for permission to refer to her unpublished palaeogeographic maps, to Mr. Yu Shenyang for sharing his unpublished data about Xizang (Tibet), and to Dr. Fang Xiang for helpful suggestions in the preparation of this manuscript. We are grateful to Dr. Joseph Botting for polishing the language. This study was financially supported by Chinese Academy of Sciences (Grant number XDB26000000), The Second Tibetan Plateau Scientific Expedition and Research Program (No. 2019QZKK0706), Chinese Ministry of Science and Technology (Grant number 2017ZX05036-001-004), and National Natural Science Foundation of China (Grant number 41772005). This paper is a contribution to the IGCP653 project (the Onset of the Great Ordovician Biodiversification Event), and the IGCP668 project (Equatorial Gondwanan History and Early Paleozoic Evolutionary Dynamics).

References

Ainsaar, L., Kaljo, D., Martma, T., Meidla, T., Männik, P., Nõlvak, J., Tinn, O., 2010. Middle and Upper Ordovician carbon isotope chemostratigraphy in Baltoscandia: a correlation standard and clues to environmental history. Palaeogeogr. Palaeoclimatol. Palaeoecol. 294, 189–201.

Bergström, S.M., Chen, X., Gutiérrez-Marco, J.C., Dronov, A., 2009a. The new chronostratigraphic classification of the Ordovician System and its relations to major regional series and stages and to $\delta^{13}C$ chemostratigraphy. Lethaia 42, 1–11.

Bergström, S.M., Chen, X., Schmitz, B., Young, S., Rong, J.Y., Saltzman, M.R., 2009b. First documentation of the Ordovician Guttenberg $\delta^{13}C$ excursion (GICE) in Asia: chemostratigraphy of the Pagoda and Yanwashan formations in southeastern China. Geol. Mag. 146, 1–11.

Bergström, S.M., Young, S., Schmitz, B., 2010. Katian (Upper Ordovician) $\delta^{13}C$ chemostratigraphy and sequence stratigraphy in the United States and Baltoscandia: a regional comparison. Palaeogeogr. Palaeoclimatol. Palaeoecol. 296 (3/4), 217–234.

Bergström, S.M., Lehnert, O., Calner, M., Joachimski, M.M., 2012. A new upper Middle Ordovician–Lower Silurian drillcore standard succession from Borenshult in Östergötland, southern Sweden: 2. Significance of $\delta^{13}C$ chemostratigraphy. GFF 134 (1), 39–63.

Botting, J.P., Muir, L.A., Wang, W., Qie, W., Tan, J., Zhang, L., Zhang, Y.D., 2018. Sponge-dominated offshore benthic ecosystems across South China in the aftermath of the end-Ordovician mass extinction. Gondw. Res. 61, 150–171.

Boucot, A.J., Chen, X., Scotese, C.R., Morley, R.J., 2013. Phanerozoic paleoclimate: an atlas of lithologic indicators of climate. In: SEPM Concepts in Sedimentology and Paleontology No. 11. SEPM (Society for Sedimentary Geology), pp. 1–478.

Brenchley, P.J., Carden, G.A., Hints, L., Kaljo, D., Marshall, J.D., Martma, T., Meidla, T., Nõlvak, J., 2003. High-resolution stable isotope stratigraphy of Upper Ordovician sequences: constraints on the timing of bioevents and environmental changes associated with mass extinction and glaciation. Geol. Soc. Amer. Bull. 115, 89–104.

Buggisch, W., Keller, M., Lehnert, O., 2003. Carbon isotope record of the Late Cambrian and Early Ordovician carbonates of the Argentine Precordillera. Palaeogeogr. Palaeoclimatol. Palaeoecol. 195, 357–373.

Chen, X., Rong, J.Y., Wang, X.F., Wang, Z.H., Zhang, Y.D., Zhan, R.B., 1995. Correlation of the Ordovician Rocks of China. International Union of Geological Sciences Publication, no. 31, pp. 1–104.

Chen, X., Rong, J.Y., Fan, J.X., Zhan, R.B., Mitchell, C.E., Harper, D.A.T., Melchin, M.J., Peng, P., Finney, S.C., Wang, X., 2006. The global boundary stratotype section and point (GSSP) for the base of the Hirnantian Stage (the uppermost of the Ordovician System). Episodes 29, 183–196.

Chen, X., Zhang, Y.D., Fan, J.X., Tang, L., Sun, H.Q., 2012. Onset of the Kwangsian Orogeny as evidenced by biofacies and lithofacies. Sci. China Earth Sci. 55 (10), 1592–1600.

Chen, X., Bergström, S.M., Zhang, Y.D., Wang, Z.H., 2013. A Katian (Late Ordovician) regional tectonic event in China. Sci. China Earth Sci. 58 (34), 4292–4299.

Chen, X., Fan, J.X., Chen, Q., Tang, L., Hou, X.D., 2014. Toward a stepwise Kwangsian Orogeny. Sci. China Earth Sci. 57, 379–387.

Chen, C., Wang, J.S., Algeo, T.J., Wang, Z., Tu, S., Wang, G.Z., Yang, J.X., 2017a. Negative $\delta^{13}C_{carb}$ shifts in Upper Ordovician (Hirnantian) Guanyinqiao Bed of South China linked to diagenetic carbon fluxes. Palaeogeogr. Palaeoclimatol. Palaeoecol. 487, 430–446.

Chen, X., Zhang, Y.D., Goldman, D., Bergström, S.M., Fan, J.X., Wang, Z., Finney, S.C., Chen, Q., Ma, X., 2017b. Darriwilian to Katian (Ordovician) Graptolites from Northwest China. Zhejiang University Press, Hangzhou.

Chen, Q., Fan, J.X., Zhang, L.N., Chen, X., 2018. Paleogeographic evolution of the Lower Yangtze region and the break of the "platform-slope-basin" pattern during the Late Ordovician. Sci. China Earth Sci. 61, 625–636.

Fan, J.X., Peng, P.A., Melchin, M.J., 2009. Carbon isotopes and event stratigraphy near the Ordovician-Silurian boundary, Yichang, South China. Palaeogeogr. Palaeoclimatol. Palaeoecol. 276, 160–169.

Fan, R., Lu, Y.Z., Zhang, X.L., Zhang, S.B., Duan, W.Z., Li, X., Deng, S.H., 2013. New understanding of the contact relationship between Shihtzupu Formation and Pagoda Formation in Sichuan Basin. Acta Geol. Sinica 87, 321–329.

Fan, R., Bergström, S.M., Lu, Y.Z., Zhang, X.L., Zhang, S.B., Li, X., Deng, S.H., 2015. Upper Ordovician carbon isotope chemostratigraphy on the Yangtze Platform, Southwestern China: implications for the correlation of the Guttenberg $\delta^{13}C$ excursion (GICE) and paleoceanic change. Palaeogeogr. Palaeoclimatol. Palaeoecol. 433, 81–90.

Feng, H.Z., Liu, J.R., Shi, G.J., 2000. Records of carbon and oxygen isotopes from the Cambrian and Lower Ordovician carbonates in Yichang Area, Hubei province. Geol. J. China Univ. 6, 106–115.

Gong, F.Y., Wu, R.C., Zhan, R.B., Luan, X.C., 2018. Upper Ordovician carbon isotope chemostratigraphy in Shitai, Anhui province. J. Stratigr. 41, 401–410 (In Chinese with English abstract).

Gorjan, P., Kaiho, K., Fike, D.A., Xu, C., 2012. Carbon- and sulfur-isotope geochemistry of the Hirnantian (Late Ordovician) Wangjiawan (Riverside) section, South China: global correlation and environmental event interpretation. Palaeogeogr. Palaeoclimatol. Palaeoecol. 337–338, 14–22.

Guo, Y.R., Zhao, Z.Y., Xu, W.L., Shi, X.Y., Gao, J.R., Bao, H.P., Liu, J.B., Zhang, Y.L., Zhang, Y.Q., 2014. Sequence stratigraphy of the Ordovician system in the Ordos Basin. Acta Sedimentol. Sin. 32, 44–60 (In Chinese with English abstract).

Jing, X.C., Deng, S.H., Zhao, Z.J., Lu, Y.Z., Zhang, S.B., 2008. Carbon isotope composition and correlation across the Cambrian-Ordovician boundary in Kalpin Region of the Tarim Basin, China. Sci. China, Ser. D Earth Sci. 51, 1317–1329.

Jing, X.C., Stouge, S., Tian, Y.F., Wang, X.L., Zhou, H.R., 2019. Katian (Upper Ordovician) carbon isotope chemostratigraphy in the Neixiang area, central China: implications for intercontinental correlation. Geol. Mag. 156, 2053–2066.

Jones, D.S., Fike, D.A., Finnegan, S., Fischer, W.W., Schrag, D.P., McCay, D., 2011. Terminal Ordovician carbon isotope stratigraphy and glacioeustatic sea-level change across Anticosti Island (Quebec, Canada). Geol. Soc. Am. Bull. 123, 1645–1664.

Kaljo, D., Martma, T., Saardre, T., 2007. Post-Hunnebergian Ordovician carbon isotope trend in Baltoscandia, its environmental implications and some similarities with that of Nevada. Palaeogeogr. Palaeoclimatol. Palaeoecol. 245, 138–155.

Lehnert, O., Meinhold, G., Wu, R.C., Calner, M., Joachimski, M.M., 2014. $\delta^{13}C$ chemostratigraphy in the upper Tremadocian through lower Katian (Ordovician) carbonate succession of the Siljan District, Central Sweden. Est. J. Earth Sci. 63, 277–286.

Liu, C.G., Qi, L.X., Liu, Y.L., Luo, M.X., Shao, X.M., Luo, P., Zhang, Z.L., 2016. Positive carbon isotope excursions: global correlation and genesis in the Middle–Upper Ordovician in the northern Tarim Basin, Northwest China. Pet. Sci. 13, 192–203.

Luan, X.C., Wu, R.C., Zhan, R.B., Liu, J.B., 2019. The Zitai Formation in South China: unique deeper-water marine red beds in terms of lithology, distribution and $\delta^{13}C_{carb}$ chemostratigraphy. Palaeoworld 28, 198–210.

Ma, X., Wang, Z.H., Zhang, Y.D., Song, Y.Y., Fang, X., 2015. Carbon isotope records of the Middle–Upper Ordovician transition in Yichang area, South China. Palaeoworld 24, 136–148.

Melchin, M.J., Mitchell, C.E., Holmden, C., Štorch, P., 2013. Environmental changes in the Late Ordovician– early Silurian: review and new insights from black shales and nitrogen isotopes. Geol. Soc. Amer. Bull. 125, 1635–1670.

Munnecke, A., Zhang, Y.D., Liu, X., Cheng, J.F., 2011. Stable carbon isotope stratigraphy in the Ordovician of South China. Palaeogeogr. Palaeoclimatol. Palaeoecol. 307, 17–43.

Ogg, J.G., Ogg, G.M., Gradstein, F.M., 2016. A Concise Geologic Time Scale. Elsevier, pp. 1–240.

Ripperdan, R.L., Magaritz, M., Kirschvink, J.L., 1993. Carbon isotope and magnetic polarity evidence for non-depositional events within the Cambrian-Ordovician Boundary section near Dayangcha, Jilin Province, China. Geol. Mag. 130, 443–452.

Schmitz, B., Bergström, S.M., Wang, X.F., 2010. The middle Darriwilian (Ordovician) $\delta^{13}C$ excursion (MDICE) discovered in the Yangtze Platform succession in China: implications of its first recorded occurrences outside Baltoscandia. J. Geol. Soc. London 167, 249–259.

Shu, L.S., 2012. An analysis of principal features of tectonic evolution in South China Block. Geol. Bull. China 31 (7), 1035–1053.

Song, S.G., Yang, L.M., Zhang, Y.Q., Niu, Y.L., Wang, C., Su, L., Gao, Y.L., 2017. Qi-Qin Accretionary Belt in Central China Orogen: accretion by trench jam of oceanic plateau and formation of intra-oceanic arc in the Early Paleozoic Qin-Qi-Kun Ocean. Sci. Bull. 62, 1035–1038.

Torsvik, T.H., Cocks, L.R.M., 2013. New global palaeogeographical reconstructions for the Early Palaeozoic and their generation. Geol. Soc. Mem. 38, 5–24.

Tu, S., Wang, Z., Wang, J.S., 2012. Interpretation for the high resolution stable carbon and oxygen isotope records across Ordovician–Silurian boundary from Wangjiawan, South China. Earth Sci. J. China Univ. Geosci. 37 (Suppl. 2), 165–174.

Wang, Z.Z., Yang, J.D., 1994. Features of the carbon isotope changes in the Early Palaeozoic rocks of the Kalpin area, Xinjiang and their significance. J. Stratigr. 18, 45–52.

Wang, K., Orth, C.J., Attrep, M., Chatterton, B.D.E., Wang, X., Li, J.j., 1993. The great latest Ordovician extinction on the South China Plate: chemostratigraphic studies of the Ordovician-Silurian boundary interval on the Yangtze platform. Palaeogeogr. Palaeoclimatol. Palaeoecol. 104, 61–79.

Wang, X.F., Stouge, S., Erdtmann, B.D., Chen, X.H., Li, Z.H., Wang, C.S., Zeng, Q.L., Zhou, Z.Q., Chen, H.M., 2005. A proposed GSSP for the base of the Middle Ordovician Series: the Huanghuachang section, Yichang, China. Episodes 28, 105–117.

Wang, X.F., Stouge, S., Chen, X.H., Li, Z.H., Wang, C.S., Finney, S.C., Zeng, Q.L., Zhou, Z.Q., Chen, H.M., Erdtmann, B.-D., 2009. The global stratotype section and point for the base of the middle ordovician series and the third stage (Dapingian). Episodes 32, 96–113.

Wang, H.H., Li, J.H., Zhang, H.T., Xu, L., Li, W.B., 2016. The absolute paleoposition of the North China Block during the Middle Ordovician. Sci. China Earth Sci. 59, 573–582.

Wu, R.C., Liu, J., Calner, M., Gong, F.Y., Lehnert, O., Luan, X.C., Li, L., Zhan, R.B., 2020. High-resolution carbon isotope stratigraphy of the Lower and Middle Ordovician succession of the Yangtze Platform, China: implications for global correlation. J. Geol. Soc. London 177, 537–549.

Young, S.A., Saltzman, M.R., Bergström, S.M., Leslie, S.A., Chen, X., 2008. Paired $\delta^{13}C_{carb}$ and $\delta^{13}C_{org}$ records of Upper Ordovician (Sandbian-Katian) carbonates in North America and China: implications for paleoceanographic change. Palaeogeogr. Palaeoclimatol. Palaeoecol. 270, 166–178.

Yu, S.Y., Fang, X., Munnecke, A., Li, W.J., Zhen, Y.Y., Li, Y., Wang, Z.H., Zhang, Y.D., 2019. First documentation of Middle Ordovician warm-water carbonates in the Mount Jolmo Lungma (Mount Everest) area, southern Xizang (Tibet), China, and its paleogeographic implications. Palaeogeogr. Palaeoclimatol. Palaeoecol. 530, 136–151.

Yu, S.Y., Fang, X., Zhang, Y.D., Zhen, Y.Y., Li, W.J., Li, Y., Munnecke, A., 2020. First record of the Middle Darriwilian $\delta^{13}C$ excursion (MDICE) in southern Xizang (Tibet), China, and its implications. Carbonates Evaporites. under review.

Yuan, T., Yi, H.S., Zhang, S., Cai, Z.H., Li, G.J., 2018. Carbon Isotope Excursions and Paleo-Oceanography of the Ordovician–Silurian Boundary Carbonate Rocks from the Xainza Area, Tibet. Acta Geol. Sin. 92, 2052–2054.

Zhan, R.B., Jin, J.S., 2007. Ordovician-Early Silurian (Llandovery) Stratigraphy and Palaeontology of the Upper Yangtze Platform, South China. Science Press, Beijing.

Zhang, L.N., 2016. Quantitative Palaeogeographic Reconstruction of South China in the Ordovician Based on Database and GIS. (PhD Thesis. In Chinese with English abstract).

Zhang, Y.D., Munnecke, A., 2016. Ordovician stable carbon isotope stratigraphy in the Tarim Basin, NW China. Palaeogeogr. Palaeoclimatol. Palaeoecol. 458, 154–175.

Zhang, Y.D., Chen, X., Yu, G.H., Goldman, D., Liu, X., 2007. Ordovician and Silurian Rocks of Northwestern Zhejiang and Northeast Jiangxi Provinces, SE China. University of Science and Technology of China Press, Hefei.

Zhang, T.G., Shen, Y.A., Zhan, R.B., Shen, S.Z., Chen, X., 2009. Large perturbations of the carbon and sulphur cycle associated with the Late Ordovician mass extinction in South China. Geology 37, 299–302.

Zhang, T.G., Shen, Y., Algeo, T.J., 2010. High-resolution carbon isotopic records from the Ordovician of South China: links to climatic cooling and the Great Ordovician Biodiversification Event (GOBE). Palaeogeogr. Palaeoclimatol. Palaeoecol. 289, 102–112.

Zhang, G.W., Guo, A.L., Wang, Y.J., Li, S.Z., Dong, Y.P., Liu, S.F., He, D.F., Cheng, S.Y., Lu, R.K., Yao, A.P., 2013. Tectonics of South China continent and its implications. Sci. China Earth Sci. 56, 1804–1828.

Zhang, Y.Y., Wang, J., Munnecke, A., Li, Y., 2015. Ramp morphology controlling the facies differentiation of a Late Ordovician reef complex at Bachu, Tarim Block, NW China. Lethaia 48 (4), 509–521.
Zhang, Y.D., Zhan, R.B., Wang, Z.H., Yuan, W.W., Fang, X., Liang, Y., Yan, K., Wang, Y.J., Liang, K., Zhang, J.P., Chen, T.E., Chen, Q., Jeon, J.W., Ma, X., Li, W.J., Wu, X.J., 2020. Ordovician stratigraphy and index fossils of China. Zhejiang University Press, Hangzhou. in press.
Zhao, Z.J., Wang, J.G., Wang, P.X., 2017. Carbon and oxygen isotope variation and its implication for marine sequence: a case study of Ordovician in Tarim Basin. Pet. Res. 2, 281–290.
Zhen, Y.Y., Zhang, Y.D., Wang, Z.H., Percival, I.G., 2016. Huaiyuan Epeirogeny—Shaping Ordovician stratigraphy and sedimentation on the North China Platform. Palaeogeogr. Palaeoclimatol. Palaeoecol. 448, 363–370.
Zhen, Y.Y., Zhang, Y.D., Harper, D.A.T., Zhan, R.B., Fang, X., Wang, Z.H., Yu, S.Y., Li, W.J., 2020. Ordovician successions in southern-central Xizang (Tibet), China—refining the stratigraphy of the Himalayan and Lhasa terranes. Gondw. Res. 83, 372–389.
Zheng, Y.F., Xiao, W.J., Zhao, G.C., 2013. Introduction to tectonics of China. Gondw. Res. 23 (4), 1189–1206.

CHAPTER EIGHT

Comparative carbon isotope chemostratigraphy of major Late Devonian biotic crises

Agnieszka Pisarzowska and Grzegorz Racki*
Institute of Earth Sciences, University of Silesia in Katowice, Sosnowiec, Poland
*Corresponding author: e-mail address: racki@us.edu.pl

Contents

Stratigraphy & Timescales, Volume 5
ISSN 2468-5178
https://doi.org/10.1016/bs.sats.2020.08.001

Abstract

Changes in the climate-sensitive carbon cycle during the major and second-order Late Devonian global events were more complex than usually expected, i.e., they were not characterized only by positive $\delta^{13}C_{carb}$ and $\delta^{13}C_{org}$ excursions up to 5.5‰. In fact, the carbon (and oxygen) isotope signatures in lime mudstones are less susceptible to diagenetic bias and therefore several negative trends and spikes in different facies of distant continents are re-interpreted as primary biogeochemical signals. In addition to significant intra- and inter-regional variability resulting from diverse local carbon cycling factors, the initial phase of both Kellwasser Crisis events [Frasnian-Famennian (F-F) mass extinction] and end-Devonian Hangenberg Crisis confirms the predicted causal relationship between large-scale volcanic degassing (mainly in large igneous provinces?) and the response of global carbon cycle in ^{13}C depleted signals. Potential volcanic signals were confirmed in more than 40% of the 73 F-F sites analyzed for C-isotope stratigraphy, as summarized in the global database. Lowered $\delta^{13}C$ values characterize also onset of other less known Late Devonian events, exemplified particularly by well-documented Frasnian Middlesex/*punctata* Event. Thus, consistently recognized worldwide, paired negative $\delta^{13}C$ and $\delta^{18}O$ excursions, even only in carbonate signatures, can be cautiously regarded as a primary volcano-climatic (greenhouse) signal and additional proxy of the volcanic trigger. In addition, the repeated negative $\delta^{13}C$ spike near the F-F boundary correlates with Hg anomalies considered to be the signature of the arc magmatism acme (and alkaline/carbonatite/kimberlite paroxysm?). However, both the carbon cycle response and time relationships between disturbances in C and Hg cycles during mass extinctions were certainly complex and required high-resolution studies in the most stratigraphically extended series and advanced quantitative modeling.

1. Introduction

On geological timescales, the long-term global carbon cycle is principally stable and controlled by factors that moderate atmospheric CO_2 level. Large-scale turnovers in the biosphere at significant stratigraphic boundaries, related with catastrophic environmental changes, were immanently correlated with major perturbations in the steady-state cycle (Archer, 2010; Sarmiento and Gruber, 2006; Sundquist and Visser, 2011, recorded in rapid variations in the $^{13}C/^{12}C$ ratio of total dissolved inorganic carbon (DIC) in the oceanic reservoir (Saltzman and Thomas, 2012; Schobben et al., 2019a; Walliser, 1996).

The Late Devonian epoch was particularly eventful, which is why the sudden biogeochemical changes were especially common then in the carbon sources and sinks in the quickly evolving Earth system. In fact, the $\delta^{13}C$ shifts and excursions, conceptualized as the C-isotope chemostratigraphy (Weissert et al., 2008), offer substantial indications as to the nature of several Devonian global events, especially that the biotic patterns were not always clear (Becker et al., 2016a; Buggisch and Joachimski, 2006; Racki, 2005; Saltzman, 2005). The principal $\delta^{13}C$ curves from carbonate (brachiopod calcite and bulk rock) and organic matter analyses are well known for the most highlighted Late Devonian crisis interval corresponding to the Frasnian-Famennian boundary (F-FB) transition (Joachimski and Buggisch, 1993; Van Geldern et al., 2006), but their intra- and inter-regional variation is an emerging challenge (Hillbun et al., 2015, 2016; Ma et al., 2016; Racki et al., 2019; Whalen et al., 2017). Other Late Devonian global changes are less known (e.g., Middlesex/*punctata* Event) and/or even scarcely documented in the biogeochemical archives (e.g., Azmy et al., 2012; Becker et al., 2016a; Buggisch and Joachimski, 2006; Myrow et al., 2011, 2013; Racki, 2005; Saltzman, 2005; Van Geldern et al., 2006; Zhang et al., 2019).

This contribution focuses on the C-isotope systematics during the major multiphase biotic crises near the F-F boundary [Kellwasser (KW) Crisis], and the end-Devonian global change [Hangenberg (HB) Crisis], in particular in the context of recently proposed volcanic scenarios (Pisarzowska et al., 2020a; Racki, 2020b; Racki et al., 2018). Three other second-order Late Devonian global events are accounted as well. The origin of negative $\delta^{13}C$ shifts is reviewed, especially because of their usually approved diagenetic origin. As concluded by Schobben et al. (2019a; see also Table 1), in contrast to the previous paradigm, only volcanic degassing may have provided enough amount of the isotopically light carbon which effectively influenced the carbon cycle toward negative carbon isotope spikes in times of the ecosystem collapse. The question was noted for Late Devonian $\delta^{13}C$ events by Zhang et al. (2019, pp. 5, 10–11), but the LIP—controlled $\delta^{13}C$ excursions were thought as only regional environmental change, and "can not be regarded as global signal of the carbon cycle" (conversely to "heavy carbon" events). In fact, the distinction between global and local signals is a key issue in carbon stratigraphy poorly recognized in the Devonian signatures (but see e.g., Late Permian case in Bagherpour et al., 2020). The potential volcanic signals were searched in particular in all F-F sites (Supplementary Material 1 in the online version at https://doi.org/10.1016/bs.sats.2020.08.001), analyzed for C-isotope stratigraphy, and summarized in a global database prepared for this article.

In addition, a meaning of commonly omitted oxygen isotope ratios in bulk carbonate samples is critically accounted for the F-F interval. With the results discussed here it is hoped that the immanent uncertainty regarding the key biogeochemical perturbations may be better resolved not only for Late Devonian biotic crises, because the similar volcanism-related feedbacks are recently debated for many other Phanerozoic mass extinctions and global changes (see current reviews in Clapham and Renne, 2019; Elewa and Abdelhady, 2020; Ernst and Youbi, 2017; Font and Bond, 2020; Hu et al., 2020; Racki, 2020a; Rampino and Self, 2015).

2. Late Devonian global events and major biotic crises

Only two first-order global events, with catastrophic biotic consequences, are summarized below. The minor global Middlesex/*punctata*, *Annulata* and Dasberg events are characterized by Walliser (1996), House (2002), Racka et al. (2010), Hartenfels and Becker (2009), Becker et al. (2016a), Hartenfels et al. (2016), and Pisarzowska et al. (2020b).

2.1 Kellwasser Crisis

The biosphere recovery after the latest Ordovician mass extinction is perfectly exemplified by the Phanerozoic acme in the Middle to Late Devonian reef development, formed by coral–sponge (stromatoporoid) communities on vast carbonate shelves (Copper, 2011). The rapid collapse of the diverse metazoan-controlled ecosystems and effective carbonate factory is the leading issue in explaining this large-scale ecological disturbance in the Paleozoic biosphere. Thus, the major intra-Devonian extinction across the broadly defined F-F transition is the reason for great scientific interest. The term "Kellwasser Crisis" of Schindler (1990, 1993), based on classic sections of the Rhine Slate Mountains and Harz Mountains, is universally applied to label both the Late Frasnian Kellwasser black shale horizons (lower, LKW; upper, UKW) and the corresponding worldwide anoxic episodes (LKE and UKE, respectively; Fig. 1A, note the etymological connotation in House, 2002, p. 9). On other hand, an array of well-documented Late Devonian transgressive-anoxic events, defined in the conceptual framework of Walliser's (1996) global event stratigraphy (modified by Becker et al., 2016a), includes particularly significant Hangenberg (HB) Crisis in the Devonian-Carboniferous (D-C) boundary beds (see summary in Kaiser et al., 2015; Lakin et al., 2016).

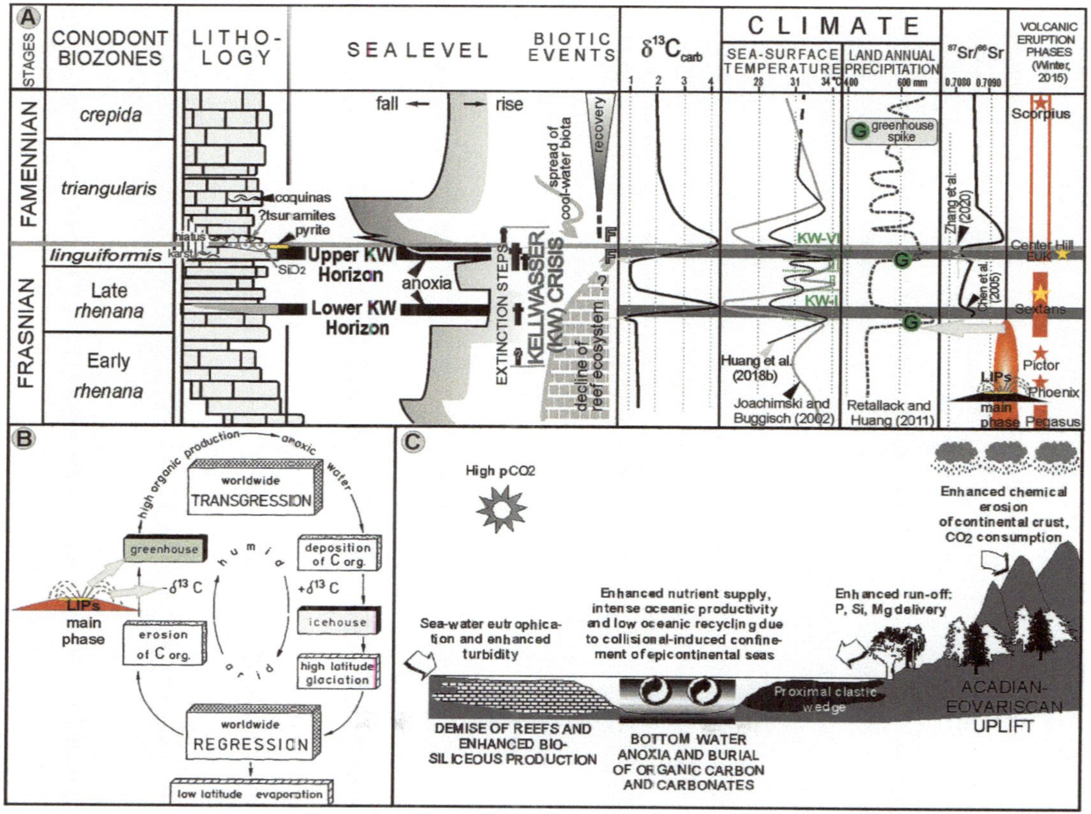

Fig. 1 See figure legend on next page.

In light of updated analysis of global paleobiodiversity changes of Stanley (2016), both the KW and HB crises are shown as marked by relatively low species losses, despite so severe and long lasting ecosystem consequences. In particular, only ~40% drop is attributed to the KW Crisis, mainly due to a depressed speciation rate (Bambach, 2006). Therefore, the F-F mass extinction sensu *lato* should be viewed as a major biotic or ecological crisis, but not a true "biodiversity crisis" (see discussion in Racki, 2020b). This noticeably subdued aspect is confirmed by higher-resolution summary of invertebrate biodiversity changes of Fan et al. (2020, p. 274), that reveals that: "the Frasnian–Famennian event to be embedded within a protracted diversity decline that began in the Eifelian (392.72 Ma) and continued to ~368.24 Ma with no subsequent major recovery. No discrete Frasnian–Famennian event was evident (...)."

The hypothetically isochronous marker intervals of the KW black shales (KW BS), which are time-specific facies as defined in the standard German localities (Gereke and Schindler, 2012; Schindler, 1993; Walliser, 1996), are correlated only approximately with the concomitant biogeochemical perturbations in C cycling recorded in carbon isotope excursions (CIEs; Racki et al., 2019), as reviewed below. Based on high-resolution U-Pb dating of the volcanogenic ash horizon, the astronomically constrained relative timescale indicates that the crucial KW interval (=from the base of LKW BS to the top of UKW BS) lasted 0.57 Ma, and the revised date of 371.9 Ma for the F-F boundary (Da Silva et al., 2020; Percival et al., 2018). However, note alternative astronomical time calibration with date of 373.3 ± 0.49 Ma to the F-FB, assessed by Ma et al. (2020).

Fig. 1 (A) Composite sedimentary and geochemical records of the stepwise Kellwasser Crisis, to show generalized eustatic, biotic and volcanic events and climatic variations (modified fig. 2 from Racki, 2020b; compiled mainly from fig. 1 in Schindler, 1993, and fig. 2 in Joachimski and Buggisch, 2002; EUK—earlier UKW eruptions). The land annual precipitation and greenhouse/CO_2 spikes compiled from Retallack and Huang (2011, fig. 2); for updated Famennian conodont zonation see Spalletta et al. (2017). The schematized $\delta^{13}C$ curve corresponds to the "German standard" herein, but updated for volcanically induced negative spikes. Six climate cycles KW I-VI on the basis of the temperature curve of Huang et al. (2018b; Fig. 1A) correspond to the cyclicity proposed by Buggisch (1991). (B and C) Two key Earth-born multicause Kellwasser Crisis scenarios considered a conceptual extension of the volcanic scenario of Racki (2020b): (B) autocyklic model of Buggisch (1991, fig. 11; used by permission from Springer Verlag), with added volcanic trigger of climate warming, and (C) endogenous/tectonic scenario of Averbuch et al. (2005, fig. 8; used by permission from John Wiley and Sons).

Recent studies have emphasized that a complex combination of profound sea-level changes and climatic variations, including the concomitant imbalance of nutrient sources and the related anoxia levels, was the key factor in the multicausal scenario for the F-F mass extinction (see review in Carmichael et al., 2019; Ma et al., 2016; Fig. 1B-C). A combination of the broadly synchronous processes and circumstances can be shown as a key to the KW Crisis timespan, i.e., twice intermittent Devonian sea-level highstand and similarly punctuated extreme hyperthermal mode (up to 34 °C), paired with progressive tectonic uplift in the tropical zone. Greenhouse gases overload have been suggested to cause the extreme warming, which has deoxygenated the oceans, resulting in oscillating oxygen deficiency in water columns (Keeling et al., 2010; see Racki, 2020b for further discussion). Noteworthy, as modeled by Goddéris and Joachimski (2004), the ~3‰ positive $\delta^{13}C$ excursions during the KW events (KEs) would have paired with bioproductivity peak that mediated a short-term drop in atmospheric CO_2 content (from ~3000 to 1560 ppm), and this organic matter sink was sufficient to a transient 5–7 °C cooling evidenced by $\delta^{18}O$ systematics (Huang et al., 2018a; Fig. 1).

Furthermore, Da Silva et al. (2020, p. 10) recognized that the "the obliquity, under a low-eccentricity orbit make the Earth system more prone to developing marine anoxia." The decisive control of orbital forcing is therefore highlighted (De Vleeschouwer et al., 2017). This hypothesis is noticeable in the conceptual framework of the press-pulse model (Arens and West, 2008), such that the orbitally tuned F-F climatic rhythms and resulted eustatic turning point in the highstand peak were a decisive trigger in the previously stressed biosphere. A large-scale environmental instability and event cascade near the F-F boundary is clear in light of common hiatus or stratigraphic condensation, and diversity of concomitant "litho-events," such as grained deposits (from minor coquinas to coarse-grained breccias), and chert and pyrite horizons (see Becker et al., 2018; Du et al., 2008; Gereke and Schindler, 2012; Racki, 2020b; Racki et al., 2002; Schindler, 1993; Figs. 1A and 2).

Volcanic paroxysms coupled with the Eovariscan tectonism have recently been justified by the synchronous, worldwide, euryfacies distribution of anomalous mercury (Hg) enrichments (Racki, 2020b; Racki et al., 2018). The diverse intrusive and effusive outbursts in large igneous provinces (LIPs) in eastern Siberia and on the East European Platform (Ernst et al., 2020) were likely combined with augmented arc volcanism and hydrothermal

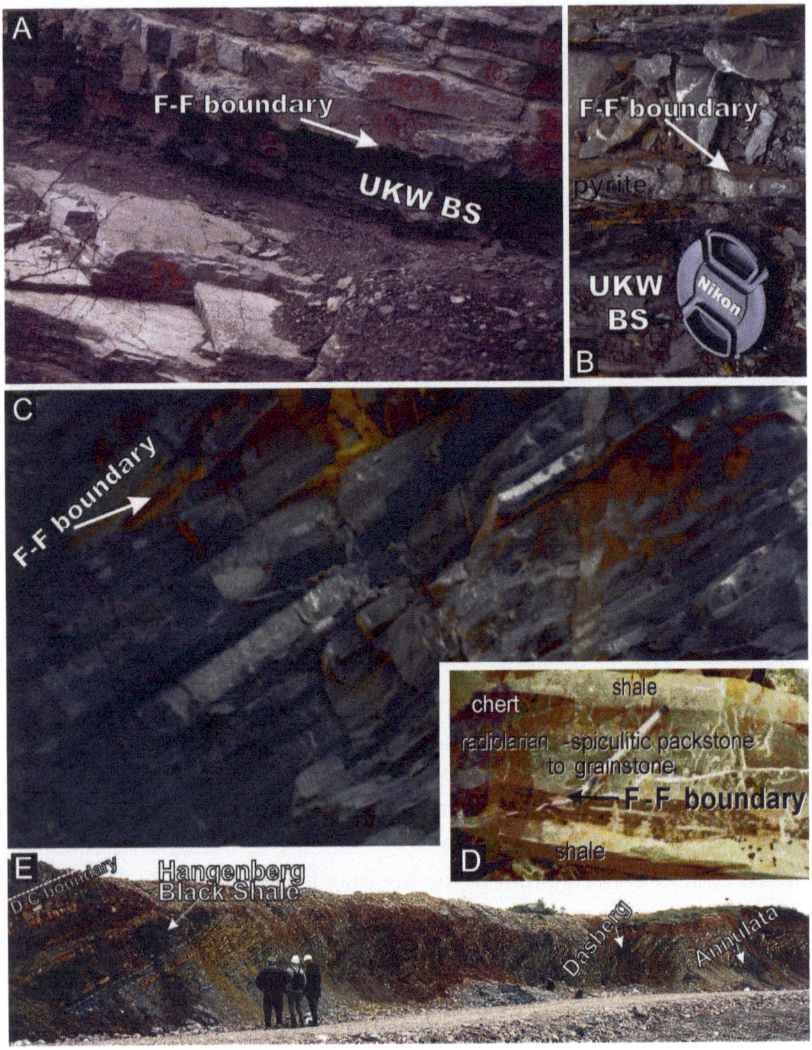

Fig. 2 Sedimentary markers of the Kellwasser (KW) levels (A–D; Fig. 1A), and upper Famennian black shales (E) in Germany and Poland. (A) General view of Steinbruch Schmidt quarry, Rhine Slate Mountains, one of the first succession studied for stable isotopes (McGhee et al., 1986; Fig. 7); the UKW black shales and limestones are ~0.5 m thick (courtesy of E. Schindler). (B) Pyrite horizon between of Upper KW black shale and conglomeratic limestone; Silberberg site, northern Bavaria (Fig. 5C; photo by K. Owocki). (C and D) The F-F passage beds include dark-colored rhythmic marly unit and thick-bedded cherty unit limestones (C) in the section Kx at active Kowala Quarry near Kielce in 1999 (Racki et al., 2002), and the chert-bearing F-F boundary layer (in available only bleached appearance, D; after Racki, 2005, fig. 9B and C). (E) Upper Famennian black shale horizons in basin-type marly succession, erosively overlain by Zechstein conglomerates, at Kowala Quarry in 2011 (courtesy of M. Rakociński).

activities. The updated volcanic press-pulse model for the KW Crisis, based on concepts of Buggisch (1991), Racki (1998), Veimarn and Korneeva (2007), and Winter (2015), is proposed by Racki (2020b) with reference to the recent timing of LIPs and arc magmatism and the revised date of 371.9 Ma for the F-F boundary.

The Late Frasnian global changes were initiated essentially by the pre-KW effusive activity in the LIPs, which caused extreme stress in the global ecosystem. However, at least two decisive pulses of kimberlite/carbonatite eruptions and/or sill-type intrusions, in the host-rock massifs capable of mass devolatilization, finally resulted in the end-Frasnian ecological catastrophe. Many lines of evidence support the proposed coupling of the rapid climate oscillations with diverse feedbacks driven by volcanism and tectonics (Racki, 2020a, b, and references therein; also e.g., Horton, 2015; Jenkyns et al., 2017; Jerram et al., 2016; Patterson and Francis, 2013; Zeng et al., 2018). In addition to usually quoted sulfur outgassing, the icehouse loop includes volcanogenic fertilization in seas and the resulting bioproductivity pulses (biological pump), as well as the weathering of newly erupted and uplifted or exhumed trap series (weathering/chemical pump), and ice/ash albedo effects. However, the eventually anti-greenhouse consequence of rapidly expanding intertidal–estuarine and riparian woodlands during brief CO_2-greenhouse episodes was another noteworthy and specific terrestrial feedback with principal biogeochemical implications (see "the R&S hypothesis" below), that was multiply effective during the Late Devonian global changes (Retallack and Huang, 2011). Glacioustatic sea-level falls of 35–45 m (McClung et al., 2016) are therefore an obvious consequence of CO_2 sink due to rapidly increased marine and terrestrial bioproduction, combined with accelerated chemical weathering in tropical zones due to the developing Eovariscan orogeny (compare McKenzie and Jiang, 2019).

2.2 Hangenberg Crisis

The Hangenberg Event (HE) was defined in the Rhenish Massif (Germany) and occurred during the time interval from the middle of the *Siphonodella praesulcata* conodont Zone to the middle of the *Siphonodella sulcata* Zone (Fig. 3; Becker, 1993, 1996; Kaiser et al., 2006, 2011; Walliser, 1984, see also alternative biozonation after Corradini et al., 2017). The HE is classically recognized within a lithology sequence of recording eustatic changes (Fig. 3), and begins with the deposition of a transgressive organic-rich shale, with the type section being the Hangenberg Black Shale (HBS) in Germany

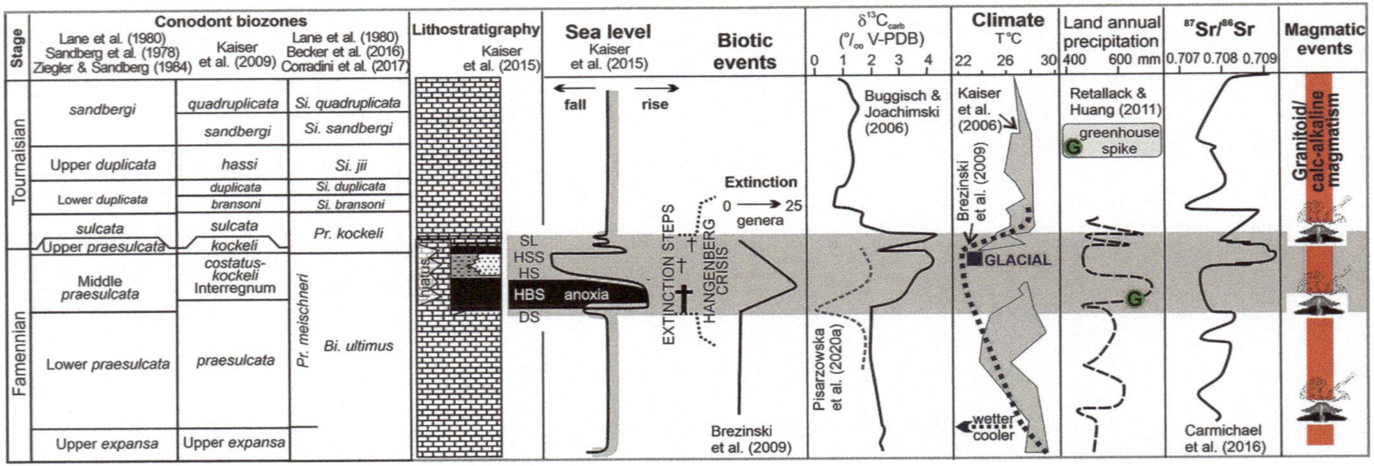

Fig. 3 Composite sedimentary and geochemical records of the Hangenberg Crisis, to show biostratigraphy, lithostratigraphy event stratigraphy and chemostratigraphy of the Devonian-Carboniferous boundary beds (Kaiser et al., 2009; Lane et al., 1980; Sandberg et al., 1978; Ziegler and Sandberg, 1984). The HB Crisis shown with lithostratigraphic units from the type area in the Rhenish Massif (after Becker et al., 2016b; Kaiser et al., 2015). The major climatic trend line after Brezinski et al. (2009) shows only warmer and colder intervals. Magmatic events after Tessalina et al. (2017), volcanic events after Marynowski and Filipiak (2007). Abbreviations: *Bi.*—*Bispathodus*, Pr.—*Protognathodus*, Si.—*Siphonodella*, DS—Drewer Sandstone, HBS—Hangenberg Black Shale, HS—Hangenberg Shale, HSS—Hangenberg Sandstone, SL—Stockum Limestone.

(see e.g. Kaiser et al., 2015). The deposition of the HBS durated ~50–190 ka (Myrow et al., 2014) and corresponds to ocean-wide marine anoxia (to euxinia) most visible in the deeper facies (Kaiser et al., 2015; Pisarzowska et al., 2020a) and short-term greenhouse to hothouse conditions. The HBS is in turn overlain by regressive units of gray shales (Hangenberg Shale, HS), sandstones (Hangenberg Sandstone, HSS), and limestones (Stockum Limestone, SL; Caplan and Bustin, 1999; Kaiser et al., 2015). Locally, the HBS is preceded by deposition of the Drewer Sandstone (DS). However, some locations that span the D-C boundary (DCB) do not contain black shale unit, and many studies show the presence an unconformity (Cole et al., 2015), which is used to define the D-C boundary. In the absence of appropriate fossil assemblages and visible black shale horizon or unconformities, isotope chemostratigraphy may be required to constrain both the HE and the D-C boundary.

The Hangenberg biotic crisis is characterized by the increasing $\delta^{13}C$ and $\delta^{34}S$ values of marine sulfate (Emsbo, 2017), $\delta^{98}Mo$ ratio (Dahl et al., 2010; Pisarzowska et al., 2020a; Yang, 2019), and the presence of a significant $^{87}Sr/^{86}Sr$ excursion (Fig. 3), attributed to the increased nutrient fluxes accompanying accelerated continental weathering and storage of organic carbon and pyrite sulfur. Numerous mechanisms (and triggers) were proposed as a reason of the ecological crisis including large eustatic fluctuations, upwelling of anoxic bottom waters, oceanic shallow water eutrophication, increased weathering rates, evolutionary impact of land plants, forest wildfires and changes to biogeochemical cycling (see review in Kaiser et al., 2015; McGhee, 2018). The terminal Devonian was a time sustained climatic instability combined with rapid changes in CO_2 (Carmichael et al., 2016). It is assumed that enhanced bioproductivity during warming seawater temperatures contributed to high organic carbon burial rates (HBS and equivalents). The global cooling at the end of the Devonian is correlated with regression, deposition of the HSS or unconformity (Fig. 3), and has been proposed as a consequence of the CO_2 drawdown suggested by increasing carbonate $\delta^{13}C$ values in many sections (Azmy et al., 1998; Joachimski and Buggisch, 2002; Kaiser et al., 2006).

The impact of magmatic activity (especially large igneous provinces and/or arc volcanism) on global climate recently is increasingly postulated in context of mass extinctions. Kaiser et al. (2015, p. 412) suggested "an ultimate volcanogenic trigger of the warming," with decisive interference of Milankovitch cycles in the Hangenberg Crisis interval. Similarity with the volcanic F-F scenario is hence understandable, supported by the multiple extreme Hg enrichments placed preferentially in the HB black shale

(Kalvoda et al., 2019; Paschall et al., 2019; Pisarzowska et al., 2020a; Racki, 2020b; Rakociński et al., 2020). The current LIP dates appear unexpectedly to rather preclude intensive flood basalt eruptions in the D-C transition (Ernst et al., 2020; Percival et al., 2019), although precise final timing of the East European LIPs is conjectural (Racki, 2020b). Thus, Kaiser et al. (2015) suggest a lost oceanic volcanic province in the giant Panthalassan oceanic domain. However, a major and short-lived mafic magmatism in the Armorican Variscan internal domain (France) during the Devonian-Carboniferous transition (ca. 360 Ma) is interpreted by Pochon et al. (2016, p. 2440) as "a within-plate type like alkaline continental flood basalts." In addition, this interval corresponds to a geodynamic shift from the Gondwana-Armorica plate subduction to collisional regimes that was associated with a large-scale metallogenic event on the entire Variscan belt scale (Lescuyer et al., 1997; Pochon et al., 2019). Additionally, the late stage of the "Magnitogorsk arc—Laurussia continent" collision and a beginning of large-scale subduction-related granitoid-magmatism coincident with a giant Volcanic-Hosted Massive Sulphide deposits in Urals during the latest Famennian (Tessalina et al., 2017). In fact, hydrothermal and mineralization signatures, as well as volcanoclastic deposits were reported worldwide from the D-C transition (reviewed by Emsbo et al., 2018; Kaiser et al., 2015; Pisarzowska et al., 2020a).

3. Materials and methods

Carbon isotope records based on organic matter ($\delta^{13}C_{org}$) and carbonate rocks ($\delta^{13}C_{carb}$) were used for correlating intra- and inter-regional variations. Changes in the global carbon cycle have been revealed by several published carbon isotope studies in the Late Devonian and around the D-C boundary in Europe, North America and Asia (Buggisch and Joachimski, 2006; Cole et al., 2015; Cramer et al., 2008; Day et al., 2013; Kaiser et al., 2006, 2008, 2015; Königshof et al., 2012; Kumpan et al., 2015; Qie et al., 2015; Paschall et al., 2019).

For new F-F and D-C sections, $\delta^{13}C_{carb}$ and $\delta^{13}C_{org}$ data were measured in the same rock samples. The rock pieces were powdered and those that were intended for $\delta^{13}C_{org}$ analyses were acidified with an excess of 10% HCl and kept at 60 °C for at least 8 h. Then samples were rinsed with ultrapure (>18 MΩ) deionized water to remove acid and were oven-dried at 60 ° C. Analyses of sedimentary organic carbon ($\delta^{13}C_{org}$) isotope signatures were carried out using a Thermo Flash EA 1112HT elemental analyzer connected to a Thermo Delta V Advantage isotope ratio mass spectrometer in a

Continuous Flow system at IGS PAS (Warsaw, Poland) and (for samples above the HBS from Kowala) a Carlo-Erba elemental analyzer connected to a ThermoFinnigan Delta plus mass spectrometer at GeoZentrum Nordbayern, University of Erlangen-Nürnberg (Erlangen, Germany). Isotopic values for carbon were reported in per mil (‰) relative to the Vienna PeeDee Belemnite (VPDB) standard and calibrated using certified international standards. The precision of $\delta^{13}C$ analyses was ±0.33‰. Samples for $\delta^{13}C$ analyses in carbonate were reacted with 100% H_3PO_4 at 70 °C. Isotope measurements were conducted with a KIEL IV Device that was connected on-line to a FinniganMAT Delta plus isotope ratio mass spectrometer (IGS PAS, Warsaw, Poland). The results were reported relative to the PDB standard by using the NBS-19 reference sample. The precision was better than ±0.03‰ ($\delta^{13}C$).

4. Carbon isotope records and diagenesis

The $\delta^{13}C$ values of the ocean dissolved inorganic carbon (DIC) have been various over the geologic time (Saltzman and Thomas, 2012; Sundquist and Visser, 2011). The $\delta^{13}C$ composition of DIC in the oceans is linked to the $\delta^{13}C$ of CO_2 in the atmosphere through exchange between the atmosphere and surface ocean and through circulation between surface and deep waters in the oceans (Maslin and Swann, 2005). Therefore, a change in carbon isotope composition of the oceanic DIC reservoir is reflected in the isotopic composition of marine carbonates and also other components of the carbon cycle, such as organic matter in marine and terrestrial sediments (Hayes et al., 1999), plant material (e.g., Robinson and Hesselbo, 2004), and carbonate nodules in soils (Ekart et al., 1999).

The lithospheric organic carbon reservoir is dominated by dispersed organic matter. The isotope composition of sedimentary organic carbon may serve as signatures of both the origin of the organic materials and the specific factors influencing that primary signature, such as temperature, nutrient levels, and productivity. Variations in the $\delta^{13}C_{org}$ commonly are interpreted as changes in relative contributions of terrigenous and marine inputs to basin, productivity, burial rate as well as diagenetic alterations.

4.1 Inorganic versus organic carbon negative excursions

Bulk rock carbonate samples are often composed of multiple components, which have specific reactivity and can be stabilized to less reactive form after their initial deposition. Interaction with meteoric water, marine-derived or hydrothermal fluids could lead to diagenetic alteration of carbonate rock constituents (and isotope signature; see Schobben et al., 2019a, b). The most common reason for alter $\delta^{13}C$ record is meteoric diagenesis (Marshall, 1992). In general, meteoric water diagenesis results in lowered $\delta^{13}C$ values of carbonates due to incorporation of ^{12}C soil-derived from oxidized organic matter (Allan and Matthews, 1982; Joachimski, 1994; Lohmann, 1988). Diagenetic factors such as thermal maturation and migration of hydrocarbons can affect the $\delta^{13}C_{org}$ (Dolenec et al., 2004; Hayes et al., 1989). Thermal alteration (and a thermally driven loss of hydrocarbons) tends to elevate the $\delta^{13}C$ in residual organic matter. Additionally, selective preservation of some resistant constituents of organic matter relative to more degradable compounds may be reflected by positive $\delta^{13}C_{org}$ value (and the negative correlation between the $\delta^{13}C_{org}$ and TOC, compare in Marynowski et al., 2017; Meyers, 1994). The contribution of remineralized organic carbon during diagenesis recrystallization may be reason of lowering the C-isotopic composition of carbonate (noticed in organic-rich shale horizons; Rosales et al., 2001; Saltzman, 2002). A separate problem is a reworking of carbon isotope events by incorporation of limestones formed during the excursion into younger sediments (compare Buggisch and Joachimski, 2006).

Therefore, negative C-isotope excursions are interpreted most commonly as a signature of diagenetic alteration. However, there are several primary processes (Table 1), and inherent changes in the global carbon cycle with geologic time and biotic evolution recorded as a negative carbon isotope anomaly. The carbon isotope signatures of several Jurassic and Cretaceous oceanic anoxic events (OAE) commonly display both positive and distinct negative excursions (Sial et al., 2019 and reference therein). For example, a fast input of huge amount of ^{12}C-enriched carbon into the atmosphere-ocean from lithospheric sources could be a reason of the Paleocene-Eocene Thermal Maximum Events (Sluijs et al., 2007), and Permian-Triassic and Triassic-Jurassic mass extinctions event (e.g., Korte et al., 2019; Retallack and Jahren, 2008).

5. Volcanism and global carbon cycle

The generally stable chemostratigraphic $\delta^{13}C$ record was interrupted during biocrisis intervals by high amplitude (up to 8‰) positive or negative CIEs. These prime perturbations of the long-term carbon cycling are attributed to large-scale fluxes of gaseous carbon to the surface reservoirs and/or turnovers in the climate-sensitive carbon cycling. In the leading case of enormous negative CIE during end-Permian mass extinction (EPME), a

Table 1 Primary and diagenetic causes of negative $\delta^{13}C$ excursions.

Isotope signature	Hypothesis	Author(s)
$\delta^{13}C$	Massive release of methane from gas hydrates	Dickens et al. (1995), Hesselbo et al. (2000), and Kemp et al. (2005)
$\delta^{13}C$	Thermal release of methane from organic carbon-rich strata by intruded magmas	Svensen et al. (2004) and McElwain et al. (2005)
$\delta^{13}C$	Combustion of terrestrial carbon deposits accumulated in swamps	Kurtz et al. (2003)
$\delta^{13}C$	Asteroid or comet impacts	Wang et al. (1996) and Glikson (2005)
$\delta^{13}C$	Drastic reduction of organic carbon burial in the aftermath of extinction	Grard et al. (2005)
$\delta^{13}C$	Increase in organic-carbon weathering	Bickert et al. (1997), Holser et al. (1989), Munecke et al. (2003), Buggisch and Joachimski (2006), Wang (2007), and Xie et al. (2007)
$\delta^{13}C$	Collapse of the biological pump	Zachos et al. (1989) and Rampino and Caldeira (2005)
$\delta^{13}C$	Overturn of a stratified ocean with isotopically light deep waters and recycling of respired CO_2 in stratified seas	Knoll et al. (1996), Schouten et al. (2000), and Van Breugel et al. (2005a)
$\delta^{13}C$	Episodes of intense volcanic activity/input of mantle-derived carbon	Renne et al. (1995), Grard et al. (2005), Payne and Kump (2007), Herrmann et al. (2010), and Schobben et al. (2019a, b)
$\delta^{13}C$	Extensive wildfires	Takahashi et al. (2010)
$\delta^{13}C$	Episodes of hydrothermal activities in the basin	Chen et al. (2009) and Wang et al. (2012)
$\delta^{13}C$	Extensive freshwater plumes caused by melting of ice caps during catastrophic deglaciation	Shields (2005)
$\delta^{13}C$	Pulses in the remineralization or bioremineralization of a large oceanic organic carbon pool	Rothman et al. (2003) and Van Breugel et al. (2005b)

Continued

Table 1 Primary and diagenetic causes of negative $\delta^{13}C$ excursions.—cont'd

Isotope signature	Hypothesis	Author(s)
$\delta^{13}C_{org}$	Water column bacterial sulfate reduction cause seawater alkalinity	Railsback et al. (1990) and Kaufman (2019)
$\delta^{13}C_{org}$	Changes in the dominant type of organic matter	Kuypers et al. (2001)
$\delta^{13}C_{org}$	Specific microbial activity near the chemocline boundary	Uveges et al. (2018)
$\delta^{13}C$	Diagenetic alteration: – meteoric water diagenesis – remineralisation of organic carbon	Allan and Matthews (1982), Lohmann (1988), Coniglio (1989), Joachimski et al. (2001, 2002), and Le Guerroué et al. (2006)
$\delta^{13}C_{carb}$	Deposition or early diagenesis in brackish or nonmarine fluids	Myrow et al. (2015)
$\delta^{13}C_{carb}$	Carbonate precipitation/early cementation of deepwater carbonates in marine bottom waters depleted in ^{13}C	Bickert et al. (1997) and Munecke et al. (2003)

plethora of scenarios have been postulated (see summary in Schobben et al., 2019a, b; compare Table 1), in particular (a) a rapid oceanic overturn that caused an incursion of ^{12}C-enriched deep anoxic water masses into photic zone, (b) massive release of ^{13}C-depleted carbon as a result of exposure and weathering of organic-rich shelf sediments due to a sea-level fall, and (c) a suppression of pelagic marine productivity that resulted in an influx of isotopically light carbon to the oceanic DIC pool (cf. the "Strangelove ocean" model of the impact induced end-Cretaceous mass extinction; Zachos et al., 1989). As shown by Schobben et al. (2019a, b), all these causal mechanisms are inadequate in light of current geological evidence or quantitative simulations.

The negative CIEs are markedly associated with independent evidence of strong greenhouse effect, spikes in atmospheric CO_2 content and ocean acidification (Black and Gibson, 2019; Knoll, 2013). Therefore, in the conceptual frame of modern volcanic greenhouse scenario (Ernst and Youbi, 2017; Font and Bond, 2020; Jones et al., 2016; Racki, 2020b), an explanation for these carbon cycle disruption focuses on rapid release of isotopically light carbon, with usually implied $\delta^{13}C$ values between 5‰ and −8‰ in LIP magmas (Black and Gibson, 2019). However, this value range of magmatic carbon isotopes is in fact conjectural and could be lower from the standard due to heterogeneity of mantle carbon, its magmatic fractionation or recycling back of surface-derived carbon into the mantle (Black and Gibson, 2019; Sobolev et al., 2011; Werner et al., 2019). Thus, the initial negative EPME CIE is now elucidated by light carbon input from both direct (volcanogenic) and indirect (thermogenic) devolatilization in the giant Siberian volcanic system, with the latter outgassing type likely having the greatest control on the carbon cycle disruption (Burgess et al., 2017; Jerram et al., 2016; Jones et al., 2016). Importantly, the extremely voluminous emission from LIPs was promoted by recurring episodes of eruptions/intrusions with very different durations on timescales <100 kA over a prolonged timespan (<2 mA) of the of LIP cycle (Black and Gibson, 2019). These degassing pulses were difficult to buffering by the short- and long-term biogeochemical processes (see review in Sarmiento and Gruber, 2006), therefore having the potentially catastrophic ecosystem consequences (McKenzie and Jiang, 2019; Racki, 2020a).

On the other hand, as claimed by Wong et al. (2019, p. 1) in context of modeled carbon degassing budget over the past 200 mA, the volcanic devolatilization process is "poorly correlated with atmospheric CO_2, implying that surface carbon cycling processes play a significant role in modulating carbon concentrations and/or there is a long-term crustal or lithospheric storage of carbon which modulates the outgassing flux." Thus, there may be a balance between the ability for carbon sink during the LIP weathering and the effectivity of igneous degassing," and "climate response to S & C emissions likely to be unique for each individual LIP" (Jones et al., 2016). The timescales of anti-greenhouse effect may depend, among others, on the extent of extrusive versus intrusive activity and/or subaerial versus submarine volcanism in a given LIP, participation of volatile-rich alkaline and carbonatite magmas and the potential for water–rock reactions in hydrothermal systems (Black and Gibson, 2019). For the carbon budget in arc volcanism, amount of carbon in subducted rock masses is a fundamental constraint (Werner et al., 2019).

As stressed finally by Schobben et al. (2019a, p. 337), "carbon isotopes are not a panacea for understanding all aspects of mass extinctions. Most, perhaps all, extinction crises coincide with large-scale volcanism and disturbance to the long-term carbon cycle. But the associated carbon gas emissions might have left little imprint on $\delta^{13}C$ systematics. Additionally, failure to account for the limitations inherent to certain isotope recording mediums might lead to erroneous interpretations of carbon isotope excursion. Our evaluation further emphasizes the need for understanding global carbon reservoirs, fluxes, interconnections, and their respective carbon isotope compositions… Looking forward, the release of greenhouse gasses by intrusive volcanism and through positive feedbacks (as for methane hydrates) provides a tantalizing clue in the search for the smoking gun of ancient extinction events." Additional fundamental questions are given by Black and Gibson (2019, p. 323), in particular: "How do alkaline and carbonatite magmas relate to and physically interact with tholeiitic magmas, and what do alkaline magmas tell us about the influence of CO_2 and other volatiles on the petrologic evolution of LIP magmas? … On the 10^5–10^6 year timescales of silicate weathering, does LIP emplacement lead to prolonged warming, transient warming followed by cooling as silicate weathering proceeds, negligible climate change, or predominant cooling?"

5.1 Tested hypothesis

Negative $\delta^{13}C$ shifts are overall attributed to post-depositional alterations of primary signal in case of Late Devonian isotope archives (Carmichael et al., 2019; Joachimski, 1997; Joachimski et al., 2002; Van Geldern et al., 2006; see below), conversely to the similar primary biogeochemical markers of other mass extinctions. As summarized above, only volcanic degassing and related positive feedbacks are appropriate processes to explain this major biogeochemical anomaly (independently of other contributing phenomena as a productivity collapse). Combining the conclusion of Schobben et al. (2019a) with model proposed by Racki (2020b), mostly implied from concept of Retallack and Huang (2011) based on pedogenic record, the tested R&S (Retallack/Racki—Schobben) hypothesis is formulated with reference to two associated pathways (Fig. 4):

(a) Igneous outbursts (both in LIPs and volcanic arcs) → massive CO_2 degassing → greenhouse spike → terrestrial vegetation acme and CO_2 storage → cooling → marine productivity acme → expanding anoxia → KW Event, as defined climatically by Joachimski and Buggisch (2002) from oxygen isotopes in biogenic apatites.

(b) Igneous outbursts (both in LIPs and volcanic arcs) → massive CO_2 degassing → —large-scale fluxes of ^{12}C to surface carbon reservoirs → initial negative $\delta^{13}C$ excursion → increasing marine productivity and the trend reversal toward positive $\delta^{13}C$ → dominant positive $\delta^{13}C$ anomaly, as demonstrated worldwide for the KW BS and age-equivalent strata (Carmichael et al., 2019; Ma et al., 2016).

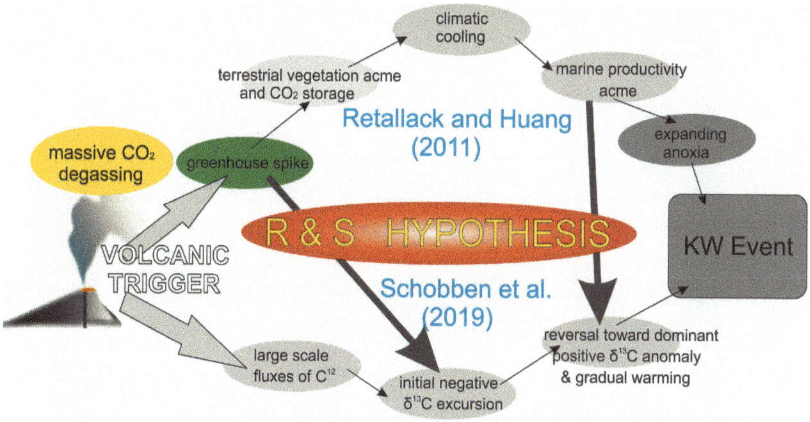

Fig. 4 Schematic representation of the R&S hypothesis as a starting point for the search for negative $\delta^{13}C$ shifts in the initial phase of Kellwasser anoxic events, assumed as the signature of volcanically induced hyperthermal peaks (Fig. 1; for general volcanic catalysm model see Jones et al., 2016; Ernst and Youbi, 2017; Racki, 2020b; Zhang et al., 2020a).

The thesis under test can be recapitulated as follows: levels enriched in ^{12}C correspond to the initial part of the Late Devonian global events (as a record of volcanic warming, recorded in greenhouse pedogenic episodes and oxygen isotope trends; Fig. 1). Consequently, the lightest $\delta^{13}C$ (and $\delta^{18}O$; see below) values or multiple negative spikes are searched in the broad basal intervals of global events, especially in the announced above world F-F database (Supplementary Material 1 in the online version at https://doi.org/10.1016/bs.sats.2020.08.001). Notably, magmatic outgassing pulses (modulated by sill intrusions?—Racki, 2020b) and clathrate dissociation and methane release (Chen et al., 2002; Gharaie et al., 2007) are supposed as an ultimate driving mechanism in the F-F volcanic scenarios, as guided by negative $\delta^{13}C$ events. Such hypothetical feedback was already outlined by Chen et al. (2005) and Zhang et al. (2020a). Because all Late Devonian global events are considered by Retallack and Huang (2011) as the greenhouse crises, the C-isotope curves from KW and HB intervals are below re-examined in the context of the R&S hypothesis (see Figs. 5–8, 11, and 16).

Of course, quoting a variety of limitations in real successions (diagenetic impact, cryptic hiatuses, approximate dating, insufficient sampling, possible diachronism of the BS bases, etc.), only a wide stratigraphic correlation is expected. Thus, only the thickest, more complete and refined stratigraphic

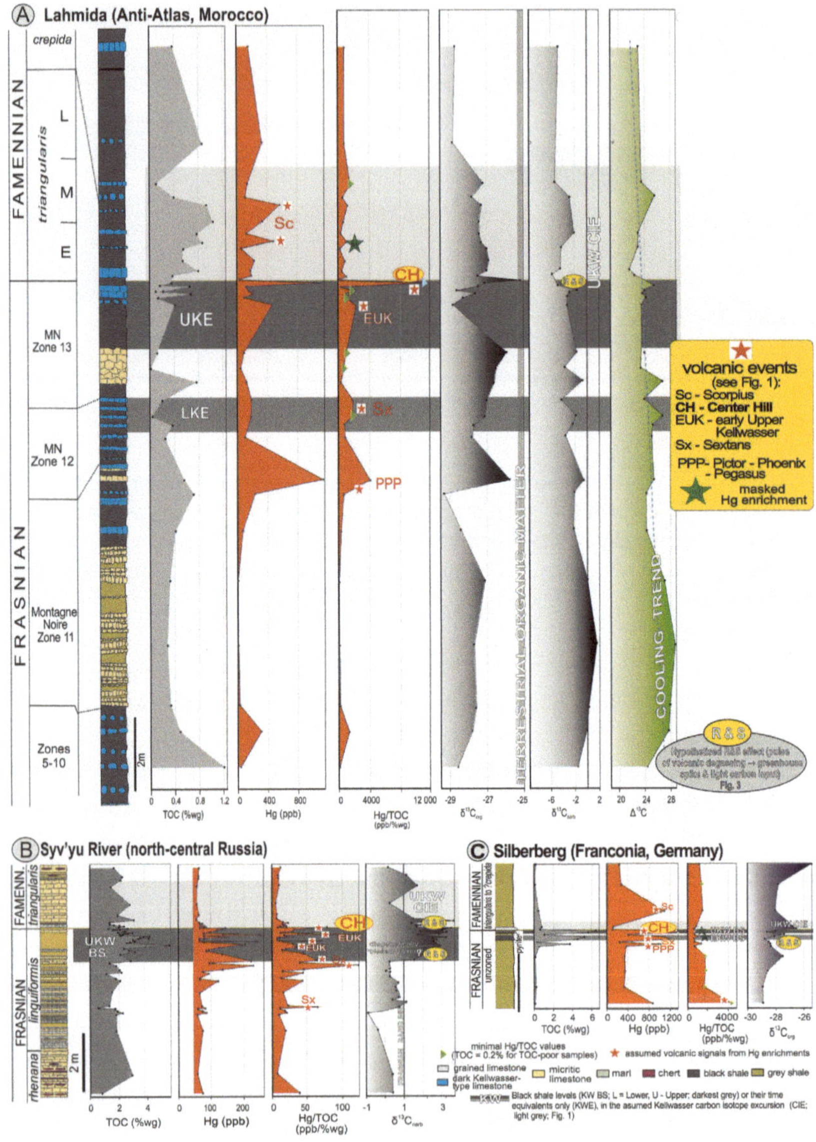

Fig. 5 Three reference F-F sections located in Morocco (after Dopieralska, 2003; A), north-central Russia (after Yudina et al., 2002; B), and Germany (see SM 1 in Racki, 2020b; C), with the different-resolution $\delta^{13}C$ and Hg signatures (partly after Racki, 2020b, fig. 16). Note the same thickness scale for all sections, and therefore the most complete KW record in the Lahmida site, but this advantage is balanced by the dense sampling of the Urals section. Masked Hg enrichment refers to volcanic Hg supply overwhelmed in Hg/TOC ratios by surplus organic matter under conditions of extreme bioproduction bloom (Racki, 2020b).

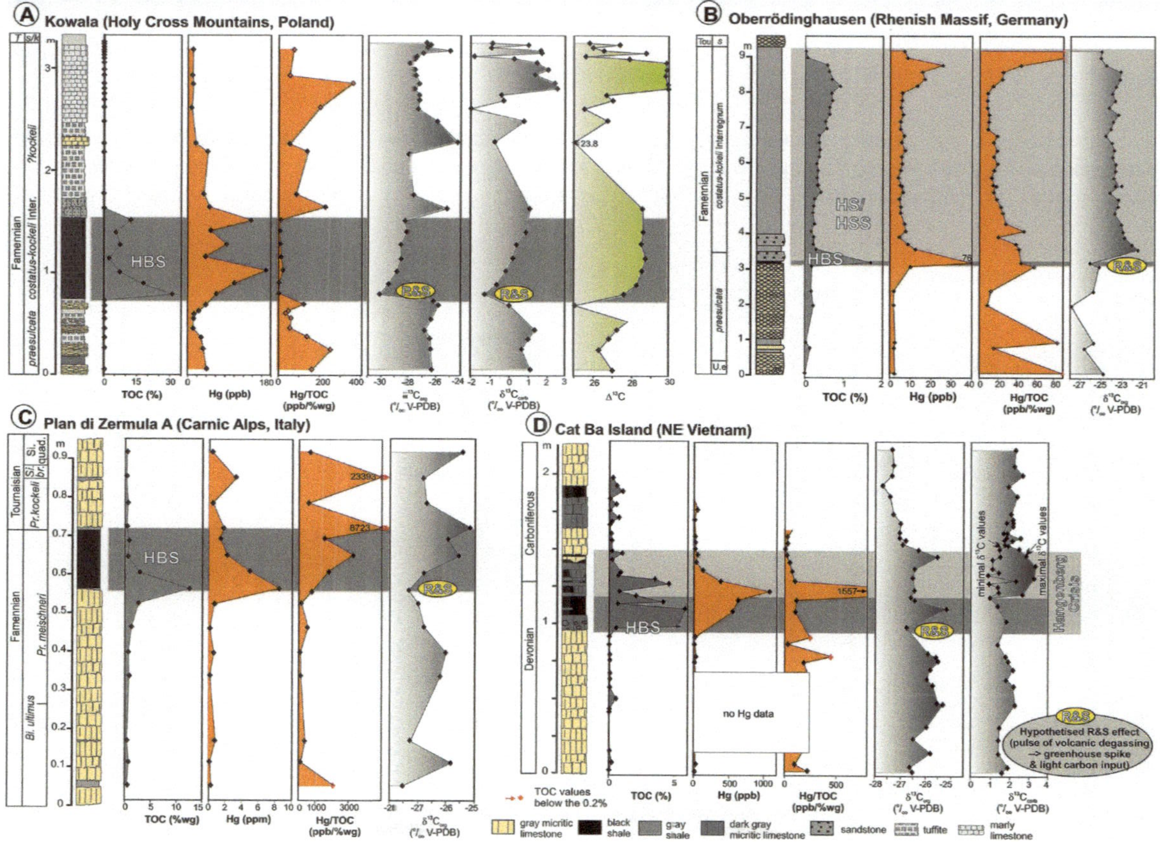

Fig. 6 Reference Devonian-Carboniferous boundary sections located in Poland (partially after Pisarzowska et al., 2020a; A), Rhenish Massif (B), Carnic Alps (Pisarzowska et al., 2020a; C) and northern-east Vietnam (Paschall et al., 2019; D) with the high-resolution $\delta^{13}C$, TOC, and Hg records with marked evidence for the hypothetical R&S correlation. The equivalent to the Hangenberg Black Shale (s. l.) in Cat Ba Island after Komatsu et al. (2014) and Shizuya et al. (2020). For explanations see Fig. 3.

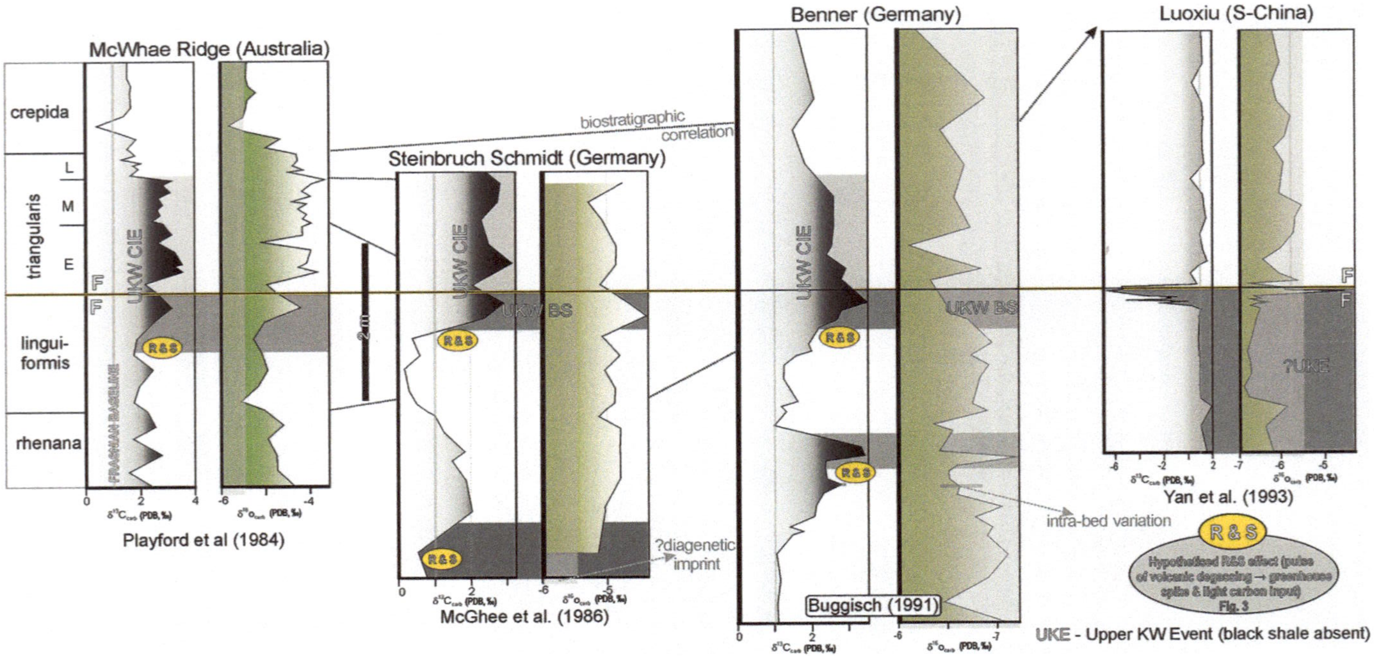

Fig. 7 Inter-regional correlation of the paired $\delta^{13}C$ and $\delta^{18}O$ profiles across the KW Crisis interval, based on first reports from various continents (see Fig. 10). Potential diagenetic bias is marked for ^{18}O-depleted values (light gray fields), but note an alleged cooling pulse recorded by abrupt positive $\delta^{18}O$ shifts in the most KW levels studied (compare Fig. 16) (Playford et al., 1984).

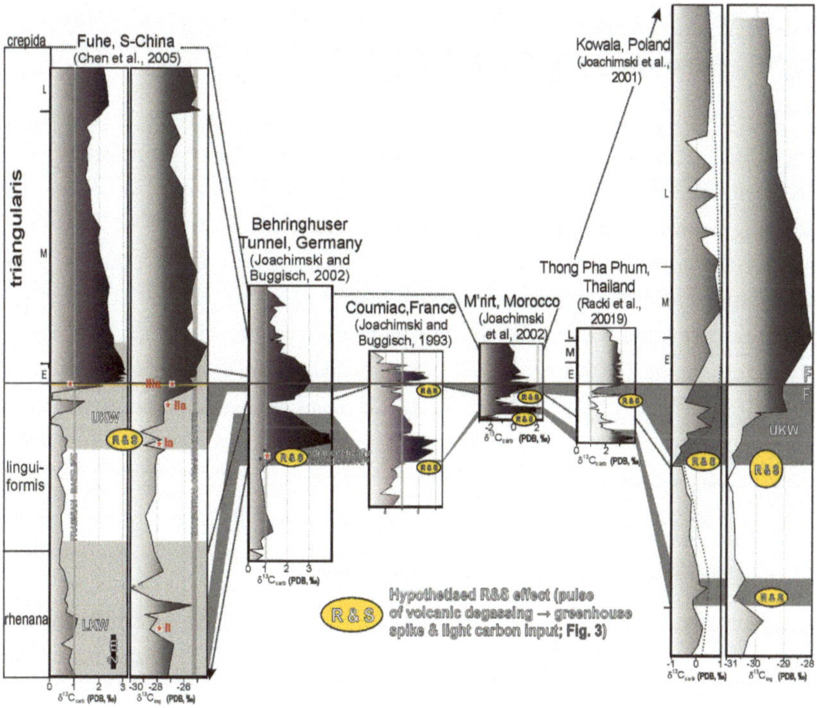

Fig. 8 Inter-regional correlation of the $\delta^{13}C$ record across the KW Crisis interval, based on reference lime mudstone dominated sections from various continents (see Fig. 10), with marked evidence for the hypothetical R&S correlation (asterisked are volcanic signals assumed by Chen et al., 2005; Zhang et al., 2020a, b). Note very different stratigraphic resolution and sample density for particular successions, the expected positive shift after the "positive smoothing" of highly fluctuating $\delta^{13}C$ values at Kowala (see Joachimski et al., 2001, fig. 8; compare Fig. 16), as well as variable $\delta^{13}C$ backgrounds. Late Devonian averaged $\delta^{13}C$ value of terrestrial organic matter after Strauss and Peters-Kottig (2003).

records can provide conclusive data (see examples in Racki, 2020b; Schobben et al., 2019b). In the context, Wang et al. (1996, table 1) and Whalen et al. (2017, table 1) show the advantages of the Chinese, and especially North American and Australian successions, over the condensed German standard archives because sedimentation rates are at least one order higher (see more details in Table S1, Supplementary Material 1 in the online version at https://doi.org/10.1016/bs.sats.2020.08.001).

6. New analytical sections

This contribution presents mostly results of high-budget project "Devonian deep-water marine realm as a key to elucidate global ecosystem perturbations" from the National Science Centre—Poland (to the second Author), realized in 2013–2018. Among many F-F and D-C localities recently studied or re-studied (see Racki, 2020b), new $\delta^{13}C$ data are presented from selected successions (Figs. 5, 6, and 11). Jointly with comprehensive literature data (Table S1 in Supplementary Material 1 in the online version at https://doi.org/10.1016/bs.sats.2020.08.001; based partly on table S1 of Carmichael et al., 2019), this dataset allows for a refined insight into the temporal and spatial pattern of the major biogeochemical perturbations, mostly at the intra-zonal scale.

6.1 Frasnian-Famennian (F-F) localities

New C-isotope data (Appendix A) are presented from two additional deep-water reference localities: (1) the Lahmida section (Rheris Basin) near Erfoud in the Anti-Atlas, Morocco and (2) the Silberberg section near Wallenfels in the Frankenwald (Bavaria), Germany. Both sections are characterized by high-scale Hg spikes in the KW interval, even if different constraints determine these chemostratigraphic patterns, as discussed in Racki (2020b).

The Lahmida succession accumulated on the deep-water Rheiris shelf basin which stretched to north from Tafilalt Platform. The investigated section stratigraphically extends from the Lower Frasnian to the Middle Famennian *rhomboidea* Zone (see Dopieralska, 2003). The argillaceous-lime succession, including dark gray shales, represents partly a regional variety of Kellwasser-type facies (for more details about geology, see Dopieralska, 2003). Analyzed 44 samples were collected by Michał Rakociński and Leszek Marynowski in 2015, guided by Zdzisław Bełka.

The Silberberg section comprises deep-water greenish-gray slate-type strata containing two black-shale horizons (Gereke, 2007, p. 75). This section is referred to the Saxothuringian Ocean but has not yet been described (R.T. Becker, mail communication, 2017). In an unpublished lecture authored by Tragelehn (2012), the 20-cm thick UKW level was reported, with rare conodonts and tentaculitoids, as well as the Famennian conodont zones above. The co-occurrence of both KW BSs is approved due to distinct geochemical differences between the organic-rich shales, succeeded by unique cm-thick pyrite intercalation and conglomeratic limestone layer in the basal Famennian (Fig. 2B).

While the particularly condensed oceanic Bavarian succession well corresponds to the standard chemostratigraphic pattern (Fig. 1A), marked by two distinctive positive $\delta^{13}C_{org}$ anomalies of ~2‰ in the both KW BS intervals, the stratigraphically extended Moroccan record is far more atypical, especially in the carbonate signature, paired with a prominent negative $\delta^{13}C_{org}$ excursion in the upper UKW interval.

Carbon and oxygen isotopes were analyzed bed-by-bed, at decimeter resolution (Joachimski, unpublished; Fig. 16), through a 16 m rhythmic marly succession that straddles the F-F boundary at Kowala Quarry near Kielce, Holy Cross Mountains, southern Poland (Racki, 2005; Racki et al., 2002). Lower-resolution but comprehensive carbon isotope systematics from this well-known locality were earlier presented by Joachimski et al. (2001) from another ephemeral Kowala section (Fig. 8) in the active quarry.

6.2 Annulata and Hangenberg localities

New C-isotope data are presented for the *Annulata* Event and the D-C boundary sections (Figs. 6, 11, and 15) in the northern part of the Rhenish Massif (Oberrödinghausen, Drewer II—only the D-C boundary interval) and a section in the Holy Cross Mountains (Kowala).

Oberrödinghausen railway cut section is situated on the northern flank of the Remscheid-Altena anticline; it constitutes a part of the late Famennian-early Carboniferous turbidite-influenced basin (Bábek et al., 2016). The Drewer section located in a non-active quarry situated on the Belecke anticline was deposited on a deep intrabasinal swell with starved condensed basin facies (Korn, 2010; Korn and Weyer, 2003; Paproth, 1986). Both sections comprise the nodular pelagic limestones of the Wocklum Limestone member (upper Famennian, *expansa* to *praesulcata* conodont Zone). Distinct HBS occurs in the Drewer locality, while in Oberrödinghausen, the horizon is less visible (probably by weathering). The overlying green and silty shale (HS in Oberrödinghausen) and sandstone (HSS in Drewer) represent the regressive phase of the HE. Due to a distal setting, the equivalent of the Hangenberg Sandstone in Oberrödinghausen is represented by greenish-gray shales with minor content of silt and fine sand (Kaiser, 2005).

Within the higher part of the Upper *trachytera* Zone, the Oberrödinghausen section contains the Lower and Upper *Annulata* Black Shale beds. The lower shale is 25 cm thick. A few cm-thick upper shales lies

43 cm above the lower horizon. In the Kowala D-C section thin-bedded marly limestones intercalated with shales and volcanic ash layers (~1-m thick) lie directly below black bituminous shales (~1.0-m thick equivalent of the HBS; Fig. 2E) enriched in volcanic material. Light gray sandy-limestone, green volcanic ash layers and marly shales occur above the black shale horizon.

The $\delta^{13}C_{org}$ signatures in all sections show similar pattern (Figs. 6 and 11). The negative anomaly, with amplitudes in the range 0.5‰ (Oberrödinghausen), 1‰ (Drewer) to 3‰ (Kowala), occurred in the HBS horizon and was followed by a positive shift in the lower part of the Upper *praesulcata* Zone. The Hg concentrations in the Rhenish Massif and the Holy Cross Mountains are low, and only a minor Hg-enrichment corresponds to the organic-rich HBS (Fig. 6, fig. 4 in Pisarzowska et al., 2020a).

7. Carbon isotope chemostratigraphy of the Kellwasser Crisis

McGhee (1996, p. 224) summarized knowledge of carbon isotopes as follows: "... mystery of the Frasnian-Famennian carbon isotope anomalies remain unresolved." In fact, literature data then showed diverging $\delta^{13}C$ excursions across the major crisis level (Fig. 7), ranging from positive (Europe, Australia, USA) to negative, Strangelove Ocean type (Canada, China). Importantly, the biogeochemical anomaly is not limited to anoxic sediments, as stressed by Joachimski and Buggisch (1993), but McGhee (1996) was uncertain about the global scale of postulated heavy carbon events, i.e., as a reflection of the signals from marine DIC reservoir. As alternative, he considered local upwelling of deeper oceanic water masses (McGhee et al., 1986).

7.1 Reference sections

After benchmark paper of Joachimski et al. (2002), a positive excursion in $\delta^{13}C$ in both carbonates and organic carbon has been highlighted as one of the diagnostic chemostratigraphic features for the both KW CIEs, with the major spike mostly near the top or just above of the KW BS levels (Figs. 5, 7, and 8). The C-isotope ratios show a sharp rise of up to 4.5‰ (LKW) and 5.5‰ (UKW) over a general Frasnian background value of +1‰ (Joachimski et al., 2002, table 1), as manifested in the reference German localities at Benner and Steinbruch Schmidt, as well as in more extended Behringhuser Tunnel succession (Figs. 7 and 8). The negative $\delta^{13}C$ shifts up to −7.6‰ reported from China and Poland (Kowala) were interpreted as a diagenetic artifact because they were measured on carbonate from organic carbon-rich rocks, i.e., "the potential contribution of isotopically light carbon during the diagenetic stabilization of the carbonates" (Joachimski et al., 2002, p. 100). In fact, the suspected light carbon events from China were not confirmed by $\delta^{13}C$ data neither from brachiopod calcites nor from organic matter (Fig. 7).

In recent review of C-isotope negative excursions, Zhang et al. (2019) stated that "These abnormal records are probably local phenomenon and we admitted that they could potentially reflect possible thermogenic effects of LIPs or LIP-derived 'Strangelove Ocean' on a regional scale. But this speculation obviously needs further investigations. The influence of late diagenesis must be excluded at the stratigraphic horizons where negative CIEs develop, especially for C-org rich rock samples." However, already, Chen et al. (2005) interpreted ^{13}C-depleted trends in the both KW CIEs in extended Chinese successions (Fig. 8) as primary ones and induced by volcanic degassing. Notably, the light carbon events in Chinese successions are documented both in carbonate and organic carbon archives (Chen et al., 2005; Xu et al., 2012), and therefore simple diagenetic artifacts are precluded. The growing diversity of C-isotope curves during the KW Crisis in different facies of distant continents becomes evident in the light of incoming data with more or less "odd" $\delta^{13}C$ time series (Fig. 9), summarized

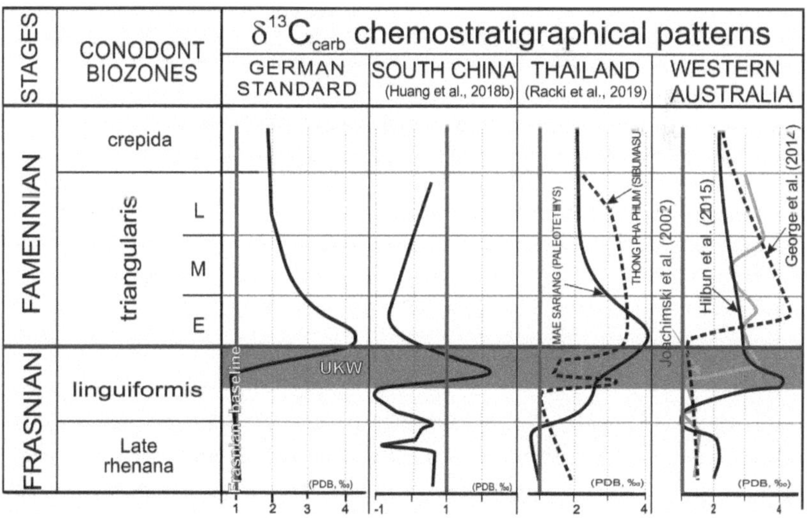

Fig. 9 Schematized carbon isotope time series of F-F boundary beds from S China, Thailand and Western Australia (modified fig. 9 from Racki et al., 2019) against the German standard (Fig. 1A). Only selected atypical Australasian $\delta^{13}C_{carb}$ profiles are shown (see full data in Table S1; Supplementary Material 1).

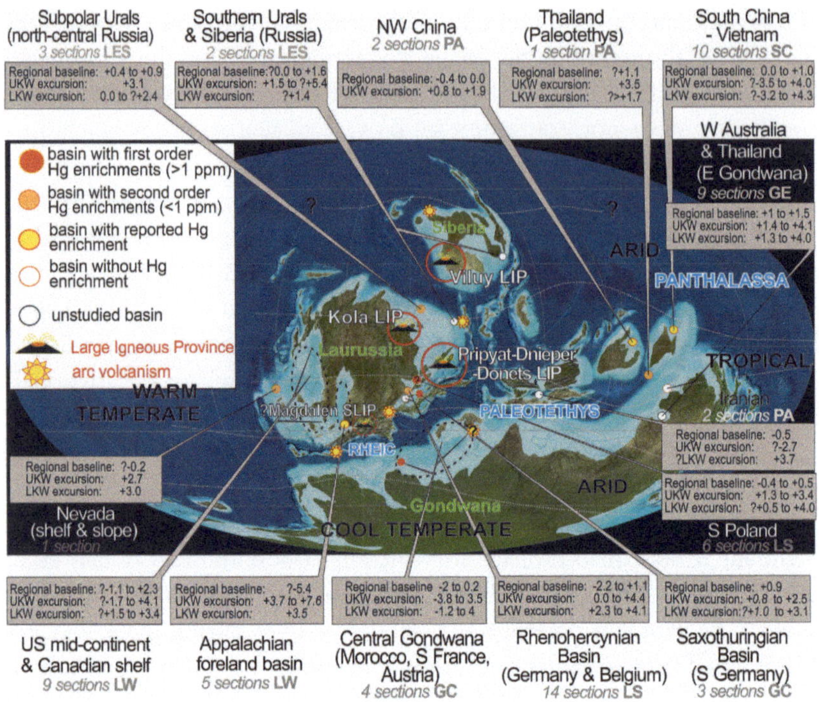

Fig. 10 Global paleogeographic and climate zones reconstruction for the Late Devonian (after Blakey, 2016a; Scotese, 2000; http://www.scotese.com/ldevclim.htm), and location of selected 14 sedimentary basins (LES, PA, etc.—domain codes in Table 2), characterized in terms of the C-isotope patterns (more details in SM 1) against the overall Hg chemostratigraphy and LIPs and volcanic arcs (see data in Racki, 2020b). The synoptic $\delta^{13}C$ values in particular basins/domains refer mostly to carbonate-based data (italicized are values from organic matter), and baselines refer to averaged values of the pre-KW (rarely inter-KWEs) curves (table 1 in Joachimski et al., 2002). Note accumulation of the data in the southern warm temperate and tropical climatic belts, and still existing paucity of studies in higher-latitude domains (see Carmichael et al., 2019; Racki, 2005).

by Hillbun et al. (2016), Ma et al. (2016), Whalen et al. (2017), Chang et al. (2019), and Racki et al. (2019). The patterns are not easy to explain based on diagenetic bias and/or ambiguous timing alone, as assumed by Zhang et al. (2019).

For example, among two $\delta^{13}C_{carb}$ profiles reported from west Thailand by Racki et al. (2019), the more condensed isotope signature in Mae Sariang generally fulfills the European standard, while the peculiar Thong Pha Phum curve is similar to "atypical" Australian signatures (Fig. 9), what agrees with the predicted geotectonic affinity of the Sibumasu terrane. The first-order trends and peak $\delta^{13}C$ values are certainly irregularly arranged in the Australian F-F sections, distinguished by absence of anoxia (George et al., 2014; Hillbun et al., 2015, 2016). In addition, similar differences are exhibited by some Chinese basin successions, well exemplified by the carbonate-biosiliceous succession at Nandong (Huang et al., 2018b; see below). The preliminary summary of global database conclusively shows that many stratigraphically extended sections from China, Australia and North America exhibit diversity of $\delta^{13}C$ secular patterns in both KW intervals, and especially occurrences of "light carbon" episodes in different facies and paleogeographic settings (see below).

7.2 Intra- and inter-regional variation

Unsurprisingly, in the recent most review, Carmichael et al. (2019, p. 7) concluded that "Outside of Laurussia and parts of Gondwana, the $\delta^{13}C$ patterns across the UKE and LKE appear to be variable, and at times contradictory." The authors quote the three main regions from several Devonian continents, developed in various epeiric to pelagic facies, with atypical $\delta^{13}C_{carb}$ and $\delta^{13}C_{org}$ signals at least in some localities (Carmichael et al., 2019, pp. 5–6; Figs. 9 and 10), updated in the world database (Supplementary Material 1 in the online version at https://doi.org/10.1016/bs.sats.2020.08.001; see below):

1) South China (shelf basins): highly variable CIEs, also including one or two negative excursions in $\delta^{13}C$ in many sites (e.g., Chen et al., 2005; Gharaie et al., 2007; Huang et al., 2018b), and "where there is no pronounced excursion at all" (Whalen et al., 2015).
2) Appalachian foreland basin (Lash, 2017; Uveges et al., 2018), and North American cratonic basins (Uveges et al., 2018): in addition to standard $\delta^{13}C$ profiles, others "show negative $\delta^{13}C$ excursions or show excursions that are somewhat offset from the UKE and LKE intervals (...) or do not show significant excursions at all." The most peculiar is coupling of positive $\delta^{13}C_{carb}$ CIEs with negative $\delta^{13}C_{org}$ anomalies in the carbonate-shale/mudstone succession of Abitibi River, Ontario, Canada (Levman and Von Bitter, 2002). In Western North America (Goodfellow et al., 1989; Wang et al., 1996; Whalen et al., 2015), the profiles are "even more variable," with positive or negative excursions or no excursions.
3) Lennard Shelf and Canning Basin of Australia: as stressed above, this reef-dominated carbonate platform provides "somewhat contradictory

information" (Fig. 9), especially for the LKW record (no excursion in McWhae Ridge section; Joachimski et al., 2002). Furthermore, the positive shifts frequently pre-date the expected biostratigraphic location (George et al., 2014; Hillbun et al., 2015; Stephens and Sumner, 2003). Similar conclusions can be provided for far less recognized C-isotope signatures in an island arc of South Urals (Mizens et al., 2015) and Central Asian Orogenic Belt (shallow water facies of NW China; Carmichael et al., 2014; Wang et al., 2016). Also one of the Moroccan sections reveals various CIEs in the unexpected biostratigraphic location (Lahmida—Fig. 5A).

Other specificity is manifested in the regional Frasnian $\delta^{13}C_{carb}$ baselines (see Fig. 10 and Table S1 in Supplementary Material 1 in the online version at https://doi.org/10.1016/bs.sats.2020.08.001), which can shift toward negative values in relation to the German standard (+1‰), even if it actually fluctuates between 0‰ and +1.5‰ (Fig. 7; table 1 in Joachimski et al., 2002). This separateness is characteristic especially the both Moroccan basin successions at Lahmida and M'rirt, with −2.7‰ to −2.2‰ background values, respectively (Figs. 5A and 8). Also the deep-water Harz and Ardennes successions exhibit the baseline between −2.2‰ (Sinsin) and −0.8‰ (Kellwassertal—Azmy et al., 2012; Joachimski et al., 2002; Kaiho et al., 2013). The most anomalously ^{13}C-depleted value is shown for east Canadian foreland section (−5.4‰; Levman and Von Bitter, 2002), distinguished also by the largest UKW CIE of +7.6‰.

In fact, various isotopic signatures from epicontinental and semi-restricted shelf basins may not reveal open-ocean isotope ratios (e.g., Brand et al., 2009; Dopieralska et al., 2006; Joachimski et al., 2009; Pisarzowska and Racki, 2012), and noticeable intra- and inter-regional differentiation of $\delta^{13}C$ profiles had been reported from Ordovician epeiric seas (Panchuk et al., 2006). In the context, Buggisch and Joachimski (2006, p. 78) noted: "A systematic difference in $\delta^{13}C$ is observed for time-equivalent samples from shallow water, slope and basinal sections with deep-water carbonates being depleted in ^{13}C relative to shallow water carbonates…" This difference can be explained either by (1) a higher contribution of aragonite to shallow water sediments or (2) carbonate precipitation/early cementation of deepwater carbonates in marine bottom waters depleted in ^{13}C. On the other hand, Brand et al. (2009) quoted an unusually highly positive $\delta^{13}C$ value (above 4‰) for Middle Devonian open-ocean brachiopod calcite from south China. Thus, the more refined study of the surprising baseline discrepancies, especially within one pelagic setting, is one of key challenges in future studies.

Zhang et al. (2019, p. 6) assumed that "Previous discrepancy in timing of the positive CIE is likely caused by imprecise conodont dating during the

Upper Kellwasser event or they just only reflect local variations (…).” It seems more likely to include broad regional specificities in the discussion due to these surprisingly frequent discrepancies.

7.3 Epeiric versus oceanic signatures

To explain such a great paleogeographic variability of imbalance in the regional marine carbon pool due to extreme stresses during the KW Crisis, Carmichael et al. (2019) and Racki et al. (2019) invoke, in addition to possible post-depositional alteration, a variety of primary local controls. In particular, negative $\delta^{13}C_{org}$ excursions within black shale units commonly occur in the Appalachian basin, and this decoupling from the global $\delta^{13}C_{carb}$ pattern is causally linked by Uveges et al. (2018) with evolving chemocline and the water-column stratification. This anomalous pattern was modulated by “chemocline-derived production where microbial communities with varying pathways of carbon fixation, assimilated ^{13}C-depleted DIC (…) or organic compounds such as methane (…)" (Uveges et al., 2018, p. 10). Thus, “elevated utilization of isotopically depleted nutrients, incomplete utilization of highly concentrated nutrients either at/below a shallow chemocline/in surface waters during episodic overturn events” (Uveges et al., 2018, p. 11) may exert primary control in shelf settings.

The biogeochemical shifts indicated a great lateral inconsistency in the rate of primary production and biological pumping (seawater aging), paired with the transfer of weathered carbonate and sedimentary organic matter, especially in heterogeneous epeiric basins (Huang et al., 2018b; Immenhauser et al., 2002, 2008; Panchuk et al., 2006). In addition, variable values of $\delta^{13}C_{carb}$ were correlated with bathymetric changes (e.g., Immenhauser et al., 2002; Katz et al., 2007; Yans et al., 2007), and the isotopic trends are used to improved analysis of depositional systems through diversified facies of Australian carbonate shelf slopes (Hillbun et al., 2016). In case of $\delta^{13}C_{org}$ record in the Appalachian foreland succession, Sageman et al. (2003) quote a key role of such factors as variations in organic matter source, primary production rates, and basin circulation (i.e., upward advection of water mass rich in respired CO_2).

Therefore, oceanic C-isotope trends may be different from these commonly recorded from concomitant carbonate factories of vast shelves. Particularly in Silurian pelagic regimes, CIEs were astonishingly minor, approving the dominant burial loci of organic carbon in epeiric environments (see discussion in Loydell, 2007). This pattern is not confirmed in the Late

Devonian settings (see Fig. 10), and Buggisch and Joachimski (2006) stressed that comparable amplitudes of the $\delta^{13}C$ excursions were recognized in shallow- and deep-water facies of different palaeogeographic domains. In the Thai section, in fact, the $\delta^{13}C$ time series of condensed pelagic limestones at Mae Sariang seamount agrees with the German chemostratigraphic reference (Racki et al., 2019; Fig. 9). On the other hand, in the deep-water carbonate-biosiliceous succession at Nandong, South China, the positive UKW CIE is narrowed to the *linguiformis* Zone, while negative $\delta^{13}C$ signals distinguish the LKW interval (Huang et al., 2018b; Fig. 9). Furthermore, truly oceanic, radiolarite F-F successions still are the challenge in the context, because the only sampled biosiliceous succession at Bancheng, Guangxi, south China (Zhang et al., 2008) appears to include totally weathered, 'bleached' cherty lithologies.

7.4 Chemostratigraphic pattern and possible volcanic impact

In light of the world database, among 73 localities studied in terms of C-isotope stratigraphy (Table S1 in Supplementary Material 1 in the online version at https://doi.org/10.1016/bs.sats.2020.08.001), the predicted $\delta^{13}C$ minima and negative shift around the basal parts of the both KW levels, as well as near the F-F boundary (correlated with the CH eruptive event; Fig. 1) were less or more distinctly recognizable in more of 38% curves reported in literature (Table 2). Of course, diagenetic bias may partly obscure this statistics, but recurrence of the negative $\delta^{13}C$ spikes in the awaited interval near the F-F boundary, in particular in Chinese sections, suggests rather diagenetic strengthening of the volcanism-driven biogeochemical changes. This diagenetic enhancement of the volcanic signal is awaited especially in case of negative $\delta^{13}C$ spikes in limestones not very rich in OM, but correlated with "heavy" $\delta^{18}O$ values, e.g., $\delta^{13}C = -5.91‰$, $\delta^{18}O = -6.03‰$, TOC = 0.83% at Nandong (sample K68–1 in table S3; Huang et al., 2018b). The highest frequency is found in Lower KW event (47.1%), but all three signals occur worldwide at least in six from seven distinguished domains. Note in the context that inter-KW baseline commonly shows more or less distinctive negative trend, with a minimum just below the onset of UKW CIE (see Fig. 1, and also fig. 5 in Da Silva et al., 2020).

The prominent examples of applicability of the R&S hypothesis are provided particularly by extended chemostratigraphic time series:

1) The south Chinese basin successions, in particular Fuhe, Baisha, Nandong and Dongcun, reveal the decreasing $\delta^{13}C_{carb}$ and $\delta^{13}C_{org}$ values up to 2‰ not only in the basal parts and prior to the KW intervals,

Table 2 The frequency of occurrence of potentially volcanism-induced negative minima in the KW CIEs (Fig. S1 in Supplementary Material 1 in the online version at https://doi.org/10.1016/bs.sats.2020.08.001) in various paleogeographic domains, as predicted for LKW and UKW anomalies in the light of the R&S hypothesis (Fig. 4).

			UKW	
Paleogeographic domains (see codes in Fig. 10)	**Number of localities LKW/UKW**	**Initial negative $\delta^{13}C_{carb}$ change LKW**	**Initial negative $\delta^{13}C_{carb}$ change**	**Negative $\delta^{13}C_{carb}$ change near the F-FB**
LS: Laurussia—South	15/17	6	8	3
LES: Laurussia East—Siberia	3/4	1	1	4
LW: Laurussia West	9/14	4	7	8
GC: Gondwana—North-Central including Saxothuringian Basin	6/7	4	4	1
GE: Gondwana—North Eeast	8/9	3	3	4
SC: South China	8/10	5	6	5
PA: Paleotethys—Central Asia (also microcontinents and arcs)	2/5	1	1	0
Summary	51/66	24 (47.1%)	30 (45.4%)	25 (37.9%)

Based on review of 73 localities recorded in Table S1 (in Supplementary Material 1 in the online version at https://doi.org/10.1016/bs.sats.2020.08.001; diagenetically imprinted signatures are omitted).

but also −2.7‰ $\delta^{13}C$ spikes at the F-F boundary (Chen et al., 2005; Huang et al., 2018b; Xu et al., 2012). On the other hand, the depleted oscillating LKW signature (with minimum value of −3.2‰) in the most deep-water Nandong sections certainly may require a more complex explanation than simple volcanic driver.

2) At Kowala quarry section, the lower-resolution $\delta^{13}C$ curve exhibits distinctive minima in inorganic and organic archives near the bottom of supposed KW levels (Joachimski et al., 2001; Fig. 8). The more refined 1993 succession records additional negative CIEs, in particular near the F-F boundary (Fig. 16).

3) In the section at the Bruns Quarry, Iowa, a distinct decreasing trend peaked at the bottom LKW minimum (below −2‰), quickly reversed within the LKW level to a broad positive excursion with a prominent amplitude of 3.4‰ (Day and Witzke, 2017, fig. 14).

The highlighted attributes of $\delta^{13}C$ curves, related to the volcanic-biogeochemical R&S hypothesis, are detectable in more shortened stratigraphic records, well exemplified by Moroccan sections, but also by the "classic" Steinbruch Schmidt and Coumiac localities (Figs. 5 and 7). In case of several North American shelf sites, the reported $\delta^{13}C$ changeover at the F-F boundary was certainly sharpened by depositional hiatuses (e.g., Goodfellow et al., 1989).

8. Devonian-Carboniferous (D-C) carbon isotope chemostratigraphy

In opinion of some authors (Brand et al., 2004; Buggisch and Joachimski, 2006; Cramer et al., 2008; Kaiser, 2005; Kaiser et al., 2008; Myrow et al., 2011), the Hangenberg Event was associated with the positive $\delta^{13}C$ excursion. The anomaly in carbonates was reported from Moravian Karst (Lesni lom Quarry, Kumpan et al., 2014a), Carnic Alps (Grüne Schneid; Kaiser et al., 2006), Graz (Trolp; Kaiser et al., 2008), Namur-Dinant Basin (Gendron-Celles and Anseremme, Buggisch and Joachimski, 2006; Kumpan et al., 2014b), Iowa and Utah (western Illinois basin and Chaffee Group of Colorado; Day et al., 2013; Myrow et al., 2013). The $\delta^{13}C_{carb}$ values at the D-C boundary in the stratotype section at La Serre (Montagne Noire, France) exhibit a wide range of values (from −4‰ to +5‰; Brand et al., 2004; Buggisch and Joachimski, 2006) due to presence of intra- and extraclasts which indicate a disturbance of the original signal by redeposition (Buggisch and Joachimski, 2006). The heavy carbon events are usually interpreted to reflect enhanced burial of organic carbon and drawdown of atmosphere CO_2 (Brand et al., 2004; Buggisch and Joachimski, 2006; Cramer et al., 2008; Kaiser et al., 2006, 2008).

The $\delta^{13}C_{carb}$ and $\delta^{13}C_{org}$ records in the unaltered diagenetically Kowala basin succession show a sharp negative spike of 2.5‰ and 3‰, respectively, below a background value of +1‰ and −27‰, respectively, in the lowermost, the strongest TOC-enriched part of the HBS horizon. Positive $\delta^{13}C_{carb}$ anomaly up to 2.5‰ occurs 1 m above the HBS, just below the DCB (see also Matyja et al., 2020). Positive $\delta^{13}C_{org}$ excursion (~2.5‰) slightly precedes the $\delta^{13}C_{carb}$ anomaly (Figs. 6 and 11).

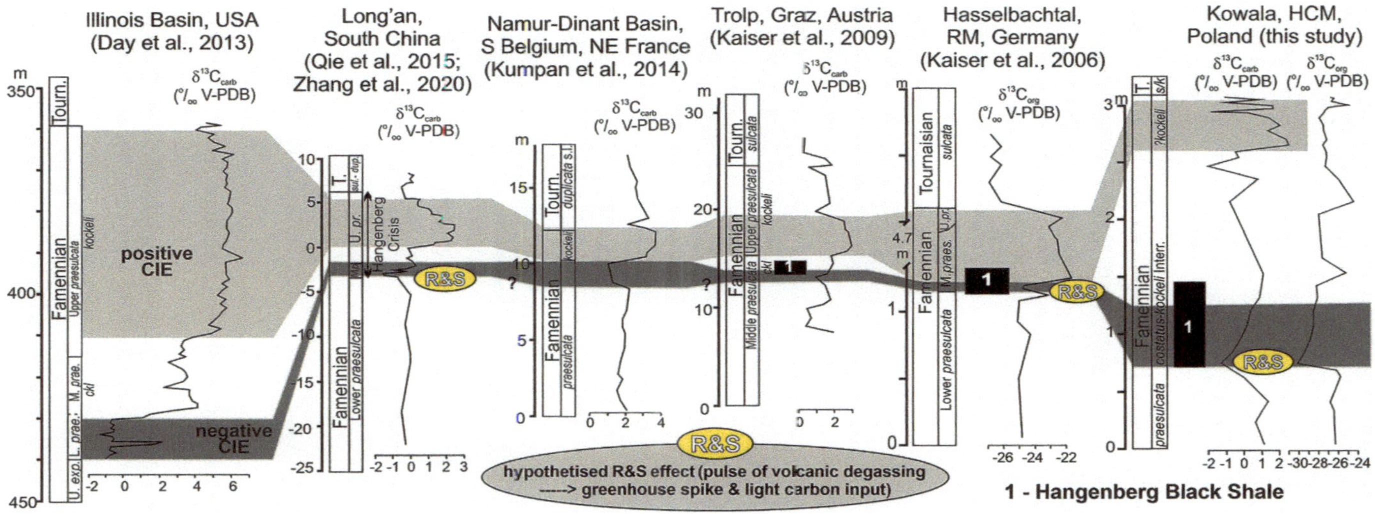

Fig. 11 Chemostratigraphic correlation based $\delta^{13}C_{carb}$ and $\delta^{13}C_{org}$ systematics of the end-Devonian sections in the North America, South China and Europe. Dark-gray marks $\delta^{13}C$ negative anomaly, light gray—$\delta^{13}C$ positive anomaly. The assumed R&S correlation is marked only for three most refined chemostratigraphic records at Hasselbachtal, Kowala and Long'an. Abbreviation: ckl—*costatus-kockeli* Interregnum, M.—Middle, *p., pr., praesul.—praesulcata*, T., Tou., Tourn.—Tournaisian, U.—Upper, *sul.—dup.—sulcata—duplicata, s/k—sulcata/kockeli*.

8.1 Intra- and inter-regional variation

Vertical changes in $\delta^{13}C_{carb}$ and/or $\delta^{13}C_{org}$ can be used for both regional and inter-regional correlations of the HE (compare Kumpan et al., 2014b). Intra- and inter-regional diversity (Fig. 12), stratigraphic gaps or unprecise biostratigraphy and less denser sampling make that precise supraregional correlation of curves difficult or even impossible (e.g., Kaiser et al., 2015; Matyja et al., 2020; Myrow et al., 2014; Qie et al., 2015). The authors quote the eight main regions from several Devonian continents, developed mostly in carbonate epeiric to pelagic facies, with atypical C-isotope signals at least in some localities:

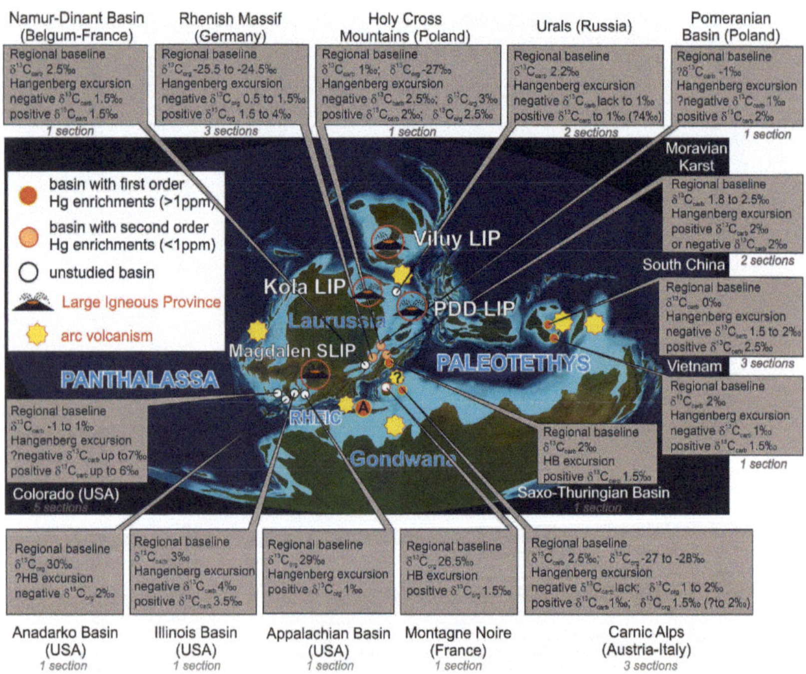

Fig. 12 Paleogeographic reconstruction for the Late Devonian (after Blakey, 2016b), and approximate location of the investigated D-C basins characterized in terms of C-isotope stratigraphy and reliability of the signatures, against the mercury data (see discussion in Pisarzowska et al., 2020a). (A) Giant Almadén mercury deposit (Spain). Reference data for the CIEs provided in Table S2 in Supplementary Material 2. Location of the D/C arc volcanism activity after Acquafredda et al. (1994), Brown et al. (2006), Calpron et al. (2007), Rosa et al. (2008), Wang et al. (2015), Carmichael et al. (2016), and Cimen (2018).

1) South China and Vietnam (near-shore and shelf marginal basins): a negative $\delta^{13}C_{carb}$ shift near the base of the Middle *praesulcata* Zone, and high $\delta^{13}C_{carb}$ values in the Upper *praesulcata* (Paschall et al., 2019; Qie et al., 2015; Zhang et al., 2020b).
2) Urals (carbonate platform and near-shore basin): in Pechora Plate, Subpolar Urals a negative shifts in $\delta^{13}C_{org}$ in conodont organic matter, weak positive $\delta^{13}C_{carb}$ in the *praesulcata* Zone (Zhuravlev and Sobolev, 2019). Eastern Urals (isolated carbonate platform): weak negative $\delta^{13}C_{carb}$ shift in the *praesulcata* Zone and a positive shift in the lower part of the *sulcata* Zone (Mizens et al., 2015).
3) Appalachian basin: no significant $\delta^{13}C_{org}$ excursion or a positive CIE (Martinez et al., 2019).
4) Illinois basin (carbonate platform): Day et al. (2013) adopted a value of −1‰ (the lowest $\delta^{13}C_{carb}$ values recorded in the *praesulcata* Zone) as the Late Famennian baseline for the Illinois section. An abrupt positive excursion to 4‰ marks the onset of the Hangenberg CIE and the $\delta^{13}C_{carb}$ values increase up to 6‰.
5) Rhenohercynian basin (neritic and pelagic facies): a negative $\delta^{13}C_{carb}$ and $\delta^{13}C_{org}$ anomaly at the base of the HBS (Pisarzowska et al., 2020a; see also Kaiser et al., 2006; Kalvoda et al., 2019) or just below DCB (Kumpan et al., 2014a, b), increase in $\delta^{13}C_{carb}$ "leading to the top of the *costatus—kockeli* Interregnum and continuing with the highest values [...] into the *kockeli* Zone" (Matyja et al., 2020; see also Kaiser et al., 2006, 2015; Trela and Malec, 2007). In shallow carbonate ramp (Pomeranian Basin), a beginning of the (only) positive CIE occurs in the uppermost Famennian (Matyja et al., 2015), although the ^{12}C-enriched values correspond to the highest TOC values below the positive CIE.
6) Saxothuringian basin (pelagic facies): a positive $\delta^{13}C_{carb}$ excursion "is observed at the DCB, although a gap must be considered, which has influence on the isotopic record" (Matyja et al., 2020).
7) Proto-Alps (tips of Gondwana): two $\delta^{13}C_{carb}$ shifts are recognized in Eastern Alps (pelagic depositional environments of a distal shelf region), "the older is situated at the level of the Lower to Middle *praesulcata* Zone, the younger one is found at the level of the Upper *praesulcata* Zone" (Bojar et al., 2013). Carnic Alps (shallower and offshore pelagic basinal facies) are characterized by positive $\delta^{13}C_{carb}$ excursion in the Middle and Upper *praesulcata* Zones, preceded by a minor decrease in $\delta^{13}C_{carb}$ (Kaiser, 2005). "A negative shift in $\delta^{13}C_{org}$ values across the HBS [...], and a slightly increasing trend in values within the HBS" (Pisarzowska et al., 2020a).

8) Iberian margin (Gondwana) (offshore to near-shore environments): "drastic and sudden excursion of the curve to heavier $\delta^{13}C_{carb}$ [brachiopod calcite—based] values during the uppermost Middle S. *praesulcata* Subzone followed subsequently by a return to more 'normal' values" (Brand et al., 2004). No data for the basal part of the HE.

The intra-regional correlation of the $\delta^{13}C_{org}$ curves of the HE, e.g., in the Rhenish Massif has revealed difference in magnitude of negative and positive CIEs (Fig. 13). However, despite lithological and biostratigraphic

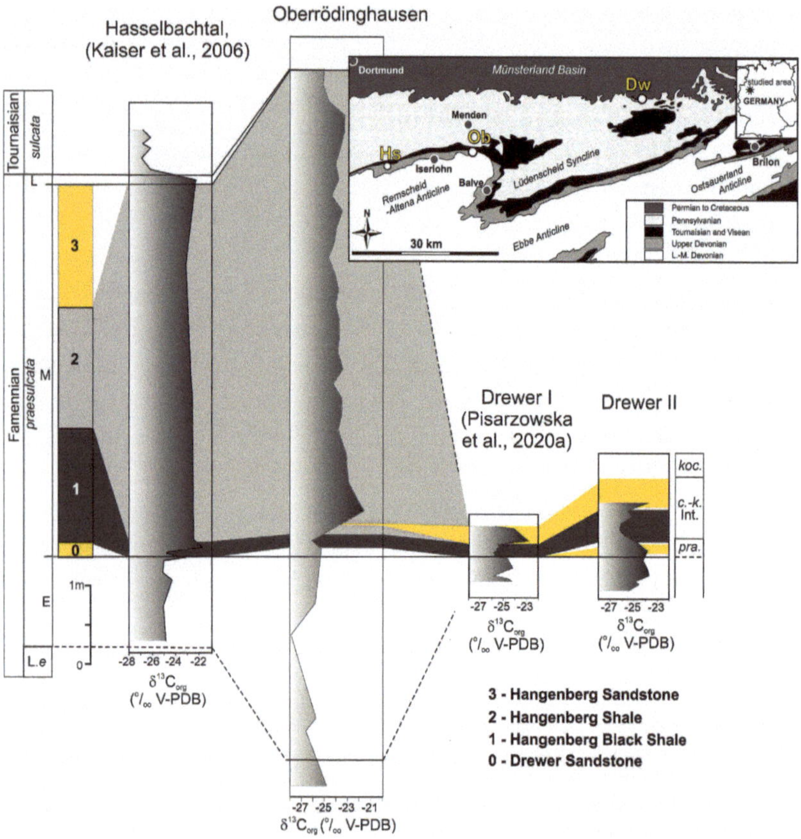

Fig. 13 Intra-regional variations of biozones, lithostratigraphic units and the $\delta^{13}C_{org}$ signatures in the Rhenish Massif sections. Conodont zonation for the Oberrödinghausen after Kaiser (2005), biostratigraphy of the Drewer sections after Korn et al. (1994) and Becker et al. (2016c). Note the same thickness scale for all sections. Geological map of the northern Rhenish Massif with the position of the studied sections (Hs—Hasselbachtal, Ob—Oberrödinghausen, Dw—Drewer) after Kumpan et al. (2015).

constraints, the carbon isotope chemostratigraphic correlation (particularly based on organic matter) of the HE is possible.

8.2 Epeiric versus oceanic signatures

Carbon isotope ratios can be different among various environments at the same time as they are affected among others by the bathymetric gradient (Brand et al., 2009; Kumpan et al., 2014b; see the F-F examples above). The worldwide change from deep water to shallow water related to the Hangenberg Event is well reflected in the lithology. The HBS and its equivalents occur from near-shore to basinal facies, in opposite to shallow carbonate platform settings.

The $\delta^{13}C_{org}$ values from the Rhenish Massif are generally higher than those from the other regions. These differences are not related to bathymetry gradient, but may be related to the dominant type of organic matter. In the case of the Rhenish Massif, input of terrestrial origin of organic matter during sea-level shallowing (Drewer and HB Sandstones) cannot be excluded. Green shale in Hasselbachtal which under- and overlying the HBS contain higher contribution of terrestrial organic matter than ^{12}C-enriched, mostly amorphous HBS organic matter (see Kaiser et al., 2006). A significant increase in terrestrial biomass input associated with coarser, clastic rocks and a shift to more ^{13}C-enriched $\delta^{13}C_{org}$ values in the Appalachian Basin (Martinez et al., 2019). The synchronous positive $\delta^{13}C_{org}$ and $\delta^{13}C_{carb}$ excursion (~2‰) in near the base of the HB Black Shale in European part of Laurussia and the most complete Appalachian section suggest that the positive excursion "could provide a useful tool to identify the onset of HB in other geographical marine localities on productive continental margins near the Devonian-Carboniferous boundary" (Martinez et al., 2019).

A distinct lateral $\delta^{13}C_{carb}$ gradient is observed in the South China (Qie et al., 2015). The positive $\delta^{13}C_{carb}$ shift is larger in shallow- than deep-water environments. Kumpan et al. (2014b) recognized a similar relationship in in the Namur-Dinant Basin. However, distinct negative vertical $\delta^{13}C_{DIC}$ gradient occurs previous to and during the Hangenberg Crisis, probably reflecting the contribution of ^{12}C-enriched DIC from organic matter remineralization on the restricted carbonate platform (Qie et al., 2015).

8.3 Chemostratigraphic pattern and possible volcanic impact on carbon cycling

Several studies (e.g., Bojar et al., 2013; Carmichael et al., 2016; Matyja et al., 2020) suggested that the positive $\delta^{13}C$ excursion cannot be used to denote the HE, because in some locations that cross the D-C boundary it is often ambiguous or invisible. For example, in Chaffee Group of Colorado (Myrow et al., 2013) and South China (Liu et al., 2016 and review in Kaiser et al., 2015) the DCB was marked by negative $\delta^{13}C_{carb}$ anomaly, while the positive excursion is not reported from Milles locality, Pyrenees (Buggisch and Joachimski, 2006). Such atypical records were commonly explained by the presence of an unconformity within the section or obscured by diagenetic processes during, e.g. local lowstands or weathering (Buggisch and Joachimski, 2006; Myrow et al., 2011). Cramer et al. (2008) observed that in the Louisiana Limestone (the US mid-continent), same as in the European sections, only the very beginning of the rising limb of the Hangenberg CIE is recorded within the HBS, while the majority of the excursion is recorded in the overlying rocks (corresponding to the HS and SL; see Fig. 3) deposited during the highstand phase of the very late Famennian post-glacial sea-level rise.

Carbon isotope chemostratigraphy is recently summarized by Pisarzowska et al. (2020a). Conversely to previous C-isotope studies of the D-C transition, a positive excursion connected with the Hangenberg Event is preceded by a negative, short-term excursion (Fig. 11) in all sections within the deposits associated with the HBS. These negative CIE is mainly associated with organic C-rich deposits (TOC > 3%). However, the $\delta^{13}C$ profiles exhibit distinctive minima in inorganic and organic signatures also in sections where the HE is not marked in the form of HBS deposits (Fig. 11). Furthermore, the paired negative signals in both $\delta^{13}C_{carb}$ and $\delta^{13}C_{org}$ are confirmed in some localities in Laurussia and China (Figs. 11 and 12; see table 1 in Pisarzowska et al., 2020a). The driver of the negative CIEs is seen in specific C-isotope fractionation under high $[CO_2]_{aq}$ conditions above the chemocline, supplemented by augmented chemoautotrophic carbon fixation near the chemocline boundary (Uveges et al., 2018; see above). However, the volcanic-biogeochemical R&S hypothesis (Fig. 3) is at least an additional explanation for the $\delta^{13}C$ minima in carbonate and organic matter records near the base of HBS (Fig. 11). This interpretation is favored because of the abundant volcanic signatures associated with the anoxic level (see Kaiser et al., 2015; Myrow et al., 2014; Pisarzowska et al., 2020a), and a greenhouse spike recorded in paleosols (Retallack and Huang, 2011).

9. Other Late Devonian global events

9.1 Middlesex/punctata Event

The isotope *punctata* Event (term of Yans et al., 2007) consists of four-step changes in the global carbon cycle, therefore this multiphase biogeochemical perturbation was collectively called "Early-Middle Frasnian (E–MF) perturbation" by Racki et al. (2008; see also Pisarzowska and Racki, 2012). The extended $\delta^{13}C$ excursion interval coincided with the global Timan, Middlesex and the Lower Rhinestreet events (sensu Becker et al., 2016a; House, 2002; House et al., 2000; House and Kirchgasser, 1993; Racki et al., 2008). The major positive isotope excursion of as much as 6–8‰ (mainly ~3‰), worldwide recorded in carbonate and organic carbon, begins near the end of the Early Frasnian *transitans* Zone (=beginning of the *Ancyrodella nodosa* Zone, see discussion in Pisarzowska et al., 2020b), at the level of the sedimentologically defined the Middlesex Event (Lash, 2019). The most characteristic, extended plateau in $\delta^{13}C$ values, occurs largely in the Middle Frasnian *punctata* Zone (Fig. 14). The positive carbon isotope excursion is preceded by the noteworthy negative $\delta^{13}C$ excursion (mainly as much as 2–3‰), well documented in many successions worldwide (Event II—Pisarzowska and Racki, 2012; Racki et al., 2008; Fig. 14).

Proximal-distal $\delta^{13}C$ gradients documented in the South Polish carbonate shelf and the Western Canada Sedimentary Basin (Śliwiński et al., 2011; fig. 11), disclosed a larger magnitude of the $\delta^{13}C$ positive excursion in shallow water, platform settings than deeper-water, basinal successions (see fig. 7 in Pisarzowska and Racki, 2012; see also the carbon isotope patterns in biostromal and lagoonal facies the Belgian Frasnian sedimentary basin in da Silva and Boulvain, 2008, fig. 6). The observed differences in the carbon isotope patterns between Laurussian foreslope and basin settings might be result of local productivity, the regional and global pattern of ocean circulation, and local to regional effects (Saltzman and Thomas, 2012). In addition the gradients in biogeochemical processes working within a depositional environment and ecosystem (Immenhauser et al., 2008) could be reflected in the isotope records.

Increasing attention in generating the negative excursions of the EMF carbon isotope perturbation is directed to extraterrestrial impact (Lash, 2019), and also to eruptive episodes connected with Devonian large igneous provinces ca. 380 Ma (Racki, 2020b). The second causal link is also suggested by moderate Hg enrichment and the initial negative $\delta^{13}C$ excursion (Pisarzowska et al., 2020b), understandable in light of the R&S hypothesis (4).

9.2 Annulata Event

The *Annulata* Event, recorded in the Late Famennian Upper *trachytera* Zone, was associated with global hypoxia to anoxia (Hartenfels and Becker, 2016; Racka et al., 2010; Walliser, 1996; Wilde and Berry, 1984) and an

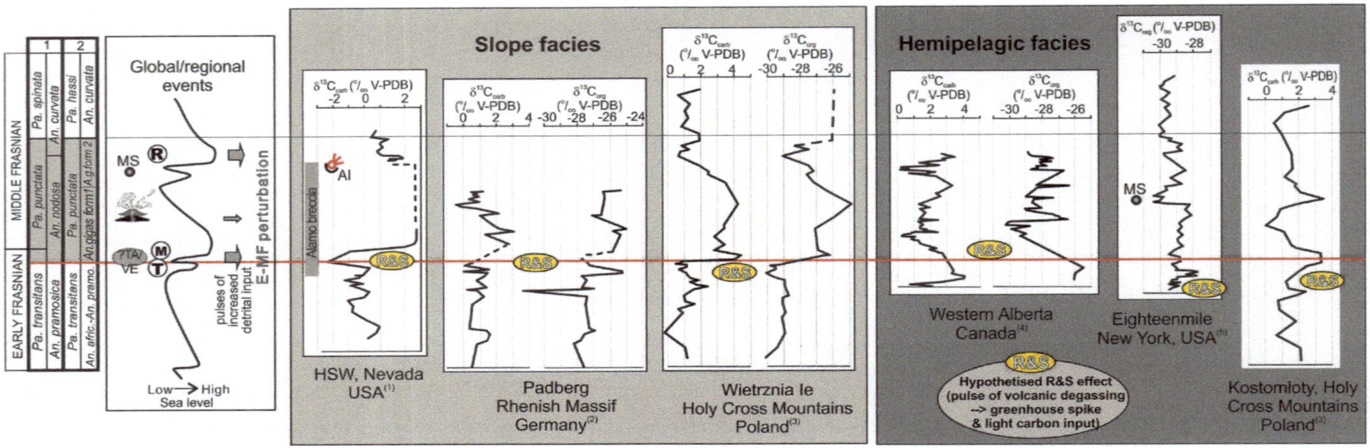

Fig. 14 Effect of the bathymetric gradient on the C-isotope records in the Laurussian shelf basins during the Middlesex/*punctata* Event plotted against global biotic and eustatic events (Pisarzowska et al., 2020b, modified) and assumed volcanic control of the isotope signals in light of the R&S hypothesis. (1) Morrow et al. (2009), (2) Pisarzowska et al. (2020b), (3) Pisarzowska and Racki (2012), (4) Śliwiński et al., 2011, (5) Lash (2019). Conodont zonation based on the *Ancyrodella* index species (1) revised in Pisarzowska et al. (2020b) and (2) proposed by Racki and Bultynck (1993). Conodont zones: *An.*—*Ancyrodella*, *An. afric.*—*An. pramo.*—*Ancyrodella africana*—*Ancyrodella pramosica*, *A. g.*—*Ancyrodella gigas*, *Pa.*—*Palmatolepis*. AI—position of Alamo impact after Morrow et al. (2009), MS—position of the microtektite-like spherules described by Lash (2019). TA—tectonic activation, VE—volcanic eruption, T—Timan Event, M—Middlesex Event.

interglacial sea-level rise (Joachimski et al., 2009; Sandberg et al., 2002). This event is not marked by rapid biotic changes, only negligible changes in conodont assemblages and a radiation event of the ammonoid *Platyclymenia annulata* (Hartenfels and Becker, 2009; House, 1985; Walliser, 1996). In the European part of Laurussia this event is recorded in the form of two black shale beds (Lower and Upper *Annulata* Black Shales), with abundant ammonites, ostracods and bivalves (Korn, 2004; Racka et al., 2010).

Previous studies have revealed a very muted carbon isotopic signal (Buggisch and Joachimski, 2006; Cole et al., 2015; Myrow et al., 2011). Chemostratigraphic correlation has been difficult due to a limited number of $\delta^{13}C$ data points. New C-isotope data are presented for the Holy Cross Mountains (Kowala) and the Rhenish Massif, and compared with literature data (Fig. 15). Except the Effenberg locality, where data are sparse and record sharp negative shifts directly below and above the *Annulata* BS (Myrow et al., 2011), decrease of $\delta^{13}C_{carb}$ and $\delta^{13}C_{org}$ values coincides with the Lower and Annulata Black Shale horizons, as predicted for the volcanically triggered greenhouse/interglacial episode by the R&S hypothesis (Racki, 2020b; Retallack and Huang, 2011; Fig. 3). Above the Upper *Annulata* Shale, the $\delta^{13}C$ values show positive CIE in all regions (see also the $\delta^{13}C_{carb}$ data from Oese, Rhenish Massif, fig. 8 in Myrow et al., 2011).

9.3 Dasberg Event

The Dasberg Event occurs in black shale level of the upper part of the Lower *expansa* Zone (Becker, 1993; Hartenfels and Becker, 2009). This event is associated with eustatic sea-level rise IIf sensu Johnson et al. (1985), linked causally to termination of glacial episodes (Sandberg et al., 2002), an extreme change in oceanic circulation patterns (Stock and Sandberg, 2019), radiation of the Gonioclymeniina (House, 2002), but not an extinction event (Hartenfels and Becker, 2009).

Geochemical record of the Dasberg Event is poorly documented globally. Carbon isotope data from carbonate facies of Europe and Morocco show ambiguous positive anomaly with slight change across the event boundaries (Buggisch and Joachimski, 2006; Kaiser et al., 2008; Myrow et al., 2011). The carbon isotope shift in Lower *expansa* Zone in Utah and Colorado sections with ~3‰ magnitude was tentatively correlated to the Dasberg Event (Myrow et al., 2011). However, re-correlation of the carbon isotope curves in the same sections shows that the positive $\delta^{13}C_{carb}$

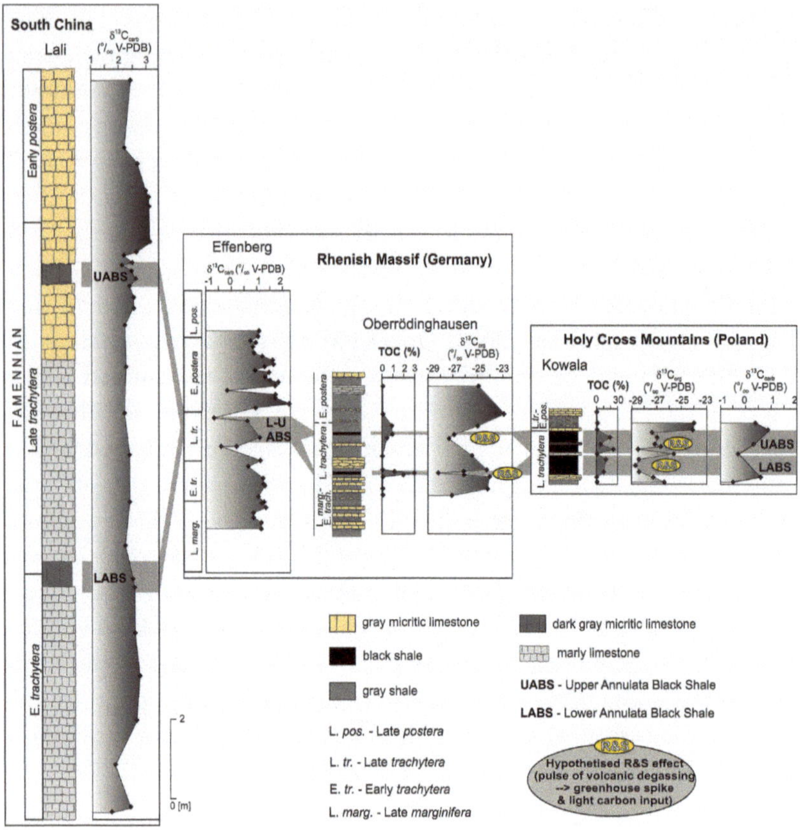

Fig. 15 Correlation of $\delta^{13}C_{carb}$ and $\delta^{13}C_{org}$ curves of selected section of the *Annulata* Event from Europe (Effenberg after Myrow et al., 2011) and South China (Zhang et al., 2019) and their assumed volcanic control in light of the R&S hypothesis. The Effenberg section contains conspicuous black shale marker beds, including the Lower and Upper *Annulata* Black Shale (ABS).

anomaly in both Utah and Colorado is preceded by a distinct, broad negative excursion (Stock and Sandberg, 2019, fig. 10). In the North American sections, therefore, the Dasberg Event is a time interval marked by several changes in the carbon isotope curve (multiphase event; Stock and Sandberg, 2019). The high-resolution $\delta^{13}C_{org}$ record of the Dasberg Event in deeper-water facies of the Laurussia (Holy Cross Mountains, Poland and Rhenish Massif, Germany, and Montana, USA; A. Pisarzowska, unpublished data) documents a negative up to 3‰ excursion, completed in a return to the higher $\delta^{13}C_{org}$ values just above the Dasberg BS interval.

10. Oxygen isotopes in biogenic apatite and whole-rock carbonate records

Diagenetically resistant brachiopod calcites or even only conodont apatites are recently utilized in recent Paleozoic paleotemperature studies (see review in Grossman, 2012), while the routinely measured $\delta^{18}O$ values in whole-rock samples are totally ignored since 1990s as an artifact of diagenesis. In the thoroughly studied Kowala succession (Racki, 2005; Figs. 2C and D, 6, and 16), due to the severe post-depositional resetting recorded in the $\delta^{13}C$ scatter (Joachimski et al., 2001, 2002), O-isotope signature is a priori uncertain. However, low burial temperatures characterize this site, and therefore the $\delta^{18}O_{phosp}$ curve, published in Joachimski et al. (2009, fig. 5), is highly trustworthy. It proves the ~1.7‰ increase, culminating exactly at the F-F boundary. This worldwide isotope event (Fig. 1A) is thought as record of a rapid cooling pulse (up to 9 °C in the localized Kowala basin; Racki et al., 2002) in the critical time of the KW Crisis (Joachimski and Buggisch, 2002; see also $\Delta^{13}C$ data in Chen et al., 2005; Xu et al., 2012, Fig. 6A). However, at least three "icehouse" spikes broke the extreme greenhouse mode at the start of the UKW event, which was recognized in the high-resolution signature $\delta^{18}O_{phosp}$ in the extended Fuhe succession in southern China by Huang et al. (2018b) and Zhang et al. (2020a). Moreover, Girard et al. (2020) implied palaeogeographic differences in Famennian climate evolution based on $\delta^{18}O_{phosp}$ data from Montagne Noire and Saxony-Thuringia, possibly due to active tectonic control of oceanographic conditions (see other factors in epeiric seas in Immenhauser et al., 2003).

If only bulk samples representing the heaviest "least-altered" $\delta^{18}O_{carb}$ values (i.e., above −5.5‰ after Van Geldern et al., 2006) are extracted by the simple graphic procedure of 'positive smoothing' (Fig. 17; see e.g., Metzger and Fike, 2013), selectively approximated curves exhibit matching signals independently of the sampled source mineralogy (Fig. 16). In fact, when positively smoothed, the $\delta^{18}O_{carb}$ dataset from Kowala includes almost exclusively values higher than −5.5‰, and explicitly matches the positive $\delta^{18}O_{phosp}$ shift of ~1.5‰ increase toward the F-F boundary, subsequent to generally stable values at around −4.8‰ in the 10-m thick upper Frasnian succession. The only difference of note is the cooling signature with its peak in a lower position, 0.5 m below the F-F boundary. Furthermore, a subordinate negative excursion, reversing this increasing terminal Frasnian

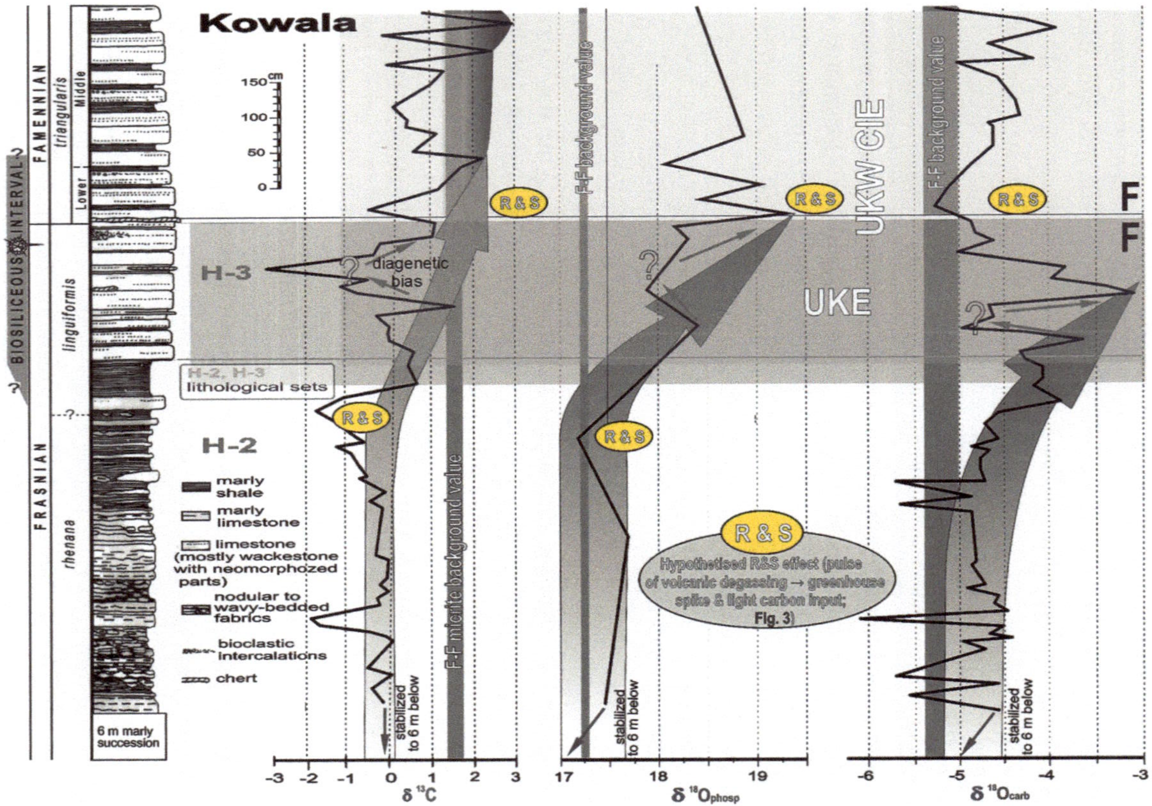

Fig. 16 See figure legend on next page.

$\delta^{18}O$ trend, is clearly seen in carbonate and phosphate archives, approximately close to beginning of the UKW event, as well as crudely correlated with the $\delta^{13}C_{carb}$ systematics (according with the R&S testing; Fig. 16). Thus, as shown by this cautious comparative analysis of corresponding isotope-rock time series, $\delta^{18}O_{carb}$ archives may still preserve an echo of seawater chemistry.

The conspicuous diagenetic instability of the precursor carbonate muds may be suggested because of mass abundance of susceptible opaline skeletons (radiolarians, silicisponges) and chert formation near the F-F boundary only (Racki et al., 2002: Fig. 2D). Nevertheless, the isotopic ratios still indicate a primary marine signal exceeding −5.5‰, concurrent with the KW climatic fluctuations (Fig. 1A). Furthermore, response of the biogeochemical cycle to orbital forcing is another test of a preservation of the primary oceanic signal, well known for highly sensitive carbon cycling and climate oscillations in volcanically stressed Late Cretaceous oceanic ecosystem (Barnet et al., 2019; Batenburg et al., 2016). As concluded by De Vleeschouwer et al. (2013, p. 31 and fig. 3K therein) for the Kowala succession, "the elevated spectral power observed for both $\delta^{13}C$ and $\delta^{18}O$ around frequency 0.027cycles/cm occurs exactly at the frequency at which one expects the influence of 18-kyr precession. The recognition of this 18-kyr period in environmental proxies such as $\delta^{13}C$ and $\delta^{18}O$ demonstrates the importance of precessional forcing on the Late-Devonian tropical climate." Thus, the preserved influence of monsoonal activity in the equivocal O-isotope systematics is a decisive argument against its strongly distorted nature due to the dominating influence of ^{16}O enriched fluids, attributed to post-depositional processes (Joachimski et al., 2004; Van Geldern et al., 2006).

Fig. 16 Oxygen isotope ratios of conodonts and bulk carbonates from the F-F boundary section exposed in 1993 at Kowala, against the composite lithological succession and C-isotope profile (see Joachimski et al., 2001, 2002; Racki et al., 2002, Fig. 8; for data see www.researchgate.net/publication/322386959_Data_Repository_Oxygen_isotope_ratios_in_Devonian_whole-rock_carbonate_record_no_much_more_than_an_artifact_of_diagenesis), and their assumed volcanic control in light of the R&S hypothesis. Upper Frasnian isotope ratios plot as a flat distribution throughout 10m succession in each case, while positively smoothed records, plotted as arrows, show comparable shift toward anomalously heavy values across the F-F boundary interval, independent of the sampled source mineralogy. Note the different regional and global $\delta^{18}O_{carb}$ and $\delta^{13}C_{carb}$ background levels (after Buggisch and Joachimski, 2006, fig. 6; Van Geldern et al., 2006, fig. 6; Joachimski et al., 2009, fig. 4), especially significant for heavy $\delta^{18}O$ signals.

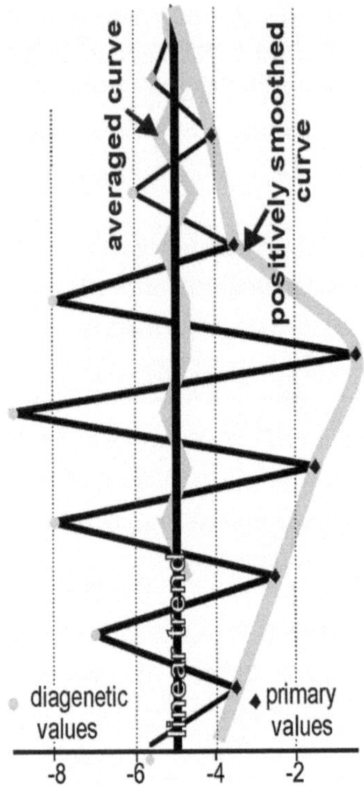

Fig. 17 Schematic presentation of differentiation between averaged trend (by three moving mean) and positively smoothed approximation of idealized "rhytmically" varied $\delta^{18}O_{carb}$ profile, generated by the Microsoft Excel. The original positive excursion is obscured regularly by post-depositional alteration phenomena that tend to shift $\delta^{18}O_{carb}$ values downwards. The averaged curve is shaped by these negative outliers, and therefore gives a false picture of the stabilized secular pattern around the linear trend. The alternative graphic approach (positive smoothing), highlighting only positively skewed data points reflecting original values, is a simply factual signal re-exposure (see the Kowala C-isotope curves in Fig. 8; Joachimski et al., 2001, fig. 8).

From the review of 33 sites studied in terms of the $\delta^{18}O_{carb}$ systematics (see selected 14 examples in Table 3; more details in Supplementary Material 1 in the online version at https://doi.org/10.1016/bs.sats.2020.08.001), it is clear that similar correspondences of climate signals occur in many other successions on a global scale and in different facies. The most representative examples are from South Chinese sections such the basinal Baisha succession (Chen et al., 2005), but also from shallow water succession at Trout River, Canada (Geldsetzer et al., 1993). Obviously, the significant intra- and inter-

Table 3 Summary list of selected localities studied for oxygen isotopes in the KW interval, with reference to their facies setting, stratigraphic reliability, $\delta^{18}O_{carb}$ data characteristics and climatic interpretation.

Locality	Facies setting	$\delta^{18}O_{carb}$ values rangea/their median value/*SR*	Interpreted climate setting of the UKW CIE	Sources
Kowala Quarry (active), Holy Cross Mountains, Poland	Deep shelf (partly restricted basin)	−6.1 to −3.1 **−4.6** *SR*=*0.2*	Two warming/cooling cycles; the F-FB in warming interval	This paper (Fig. 10)
Dębnik Z17 trench, Cracow region, Poland	Shallow shelf (ramp-style)	−7.7a to −5.0 **−5.5** *SR*=*0.15*	Overall cooling trend in the F-FB transition	This paper
Schmidt Quarry, Rheinisches Schiefergebirge, Germany	Submarine rise within shelf	−5.3 to −4.4 **−4.9** *SR*=*0.2*	Cooling to warming change near the F-FB	McGhee et al. (1986)
Xiangtian, Yangshuo Basin, Guangxi, China	Deep shelf basin	−7.2 to −5.3 **−6.5** *SR*=*0.15*	?Two cooling/warming cycles	Wang et al. (1991)
Baisha, Yangshuo Basin, Guangxi, China	Deep shelf basin	−7.0 to −4.3/ **−6.3** *SR*=*0.5*	Five warming/cooling cycles; the extreme warming interrupted by icehouse spike near the F-FB	Chen et al. (2005)
Fuhe (=Yangdi), Yangshuo Basin, Guangxi, China	Deep shelf basin	−6.7 to −5.2/ **−5.9** *SR*=*0.2*	Fluctuating cooling trend with icehouse spike near the F-FB	Chen et al. (2005)
Luoxiu, Guangxi, China	Deep shelf basin	−6.7 to −5.74.5/ **−6.2** *SR*=*0.15*	Cooling to warming shift near the F-FB	Yan et al. (1993)
Yangdi, Guilin Platform, Guangxi, China	Shallow shelf—carbonate platform	−6.7 to −2.0/ **−5.7** *SR*=*0.45*	Two distinctive warming/cooling cycles before the F-FB; one weak cycle after the F-FB	Gong et al. (2005)

Continued

Table 3 Summary list of selected localities studied for oxygen isotopes in the KW interval, with reference to their facies setting, stratigraphic reliability, $\delta^{18}O_{carb}$ data characteristics and climatic interpretation.—cont'd

Locality	Facies setting	$\delta^{18}O_{carb}$ values rangea/their median value/*SR*	Interpreted climate setting of the UKW CIE	Sources
Liujing, Guangxi, China	Shallow shelf slope	−5.9 to −4.1/ **−4.6** *SR* = *0.2*	Broad cooling trend interrupted by a warming episode prior to the F-FB	Gharaie et al. (2007)
Nandong, Guangxi, China	Deep shelf basin	−6.7 to −2.9/ **−5.8** *SR* = *0.6*	Fluctuating shift to greenhouse near the F-FB	Huang et al. (2018b)
Pershino, southern Urals, Russia	Oceanic island arc—isolated carbonate platform	−7.6 to—4.2/ **−6.5** *SR* = *1.1*	Weak cooling pulse near the F-FB	Mizens et al. (2015)
Syv'yu River, Subpolar Urals, Russia	Deep shelf slope	−7.6 to −5.1[a]/ **−6.1** *SR* = *0.1*	Fluctuating trend from cooling to warming, with ?two greenhouse spikes in the onset of UKW and near the F-FB	Yudina et al. (2002)
Kosoy Utyos, Kuznetsk Basin, Siberia, Russia	Deep shelf slope	−9.4 to −7.1/ **−7.8** *SR* = *0.6*	Cooling trend	Izokh et al. (2009)
Trout River, Northwest Territories, Canada	Shallow shelf back-reef	−9.3 to −6.7/ **−7.9** *SR* = *0.6*	Three cooling-warming cycles, with icehouse spike near the F-FB	Geldsetzer et al. (1993)

[a]Values above 8‰ are omitted.
SR—stratigraphic resolution refers to average sampling interval.

regional variation of the $\delta^{18}O_{carb}$ signatures reflects both the original seawater differentiation, as well as at least partially sampling pattern and "noisy" diagenetic imprint (see the Polar Urals locality; Fig. 6B and Table A3 in Appendix A; Yudina et al., 2002). In fact, the "icehouse" setting of the

F-F biocrisis was anticipated already in the first O-isotope studies of F-F sections, summarized by McGhee (1996; Fig. 7) and Streel et al. (2000), well exemplified by shift from ~−5‰ baseline up to −4.3‰ in the Rhenish Steinbruch Schmidt section (McGhee et al., 1986). This case is well exemplified by the stratigraphically extended South Chinese sections (Chen et al., 2005; Gharaie et al., 2007), and it is correspondingly re-interpreted herein for the Polar Urals locality (Fig. 6B; Yudina et al., 2002). This trend may also be recognized in the high-latitude basin of southern West Siberia (from ~−8.5‰ baseline up to −7.1‰ in the Kosoy Utyo section; Izokh et al., 2009), where the $\delta^{18}O_{carb}$ values are surprisingly 3‰ lower than those of the world's coeval profiles. Even some sparry-rich foreslope lithologies may still preserve the key $\delta^{18}O$ signal, as shown by introductorily studied Holy Cross sections at Psie Górki (from at least −4.9‰ to −4.5‰) and Kostomłoty (from at least −4.9‰ to −4.2‰). Surprisingly, the great warming at the beginning of the LKW and the transition to final cooling is only recognizable in single sections (Fuhe, ?Pershino—see Supplementary Material 1 in the online version at https://doi.org/10.1016/bs.sats.2020.08.001), but in others (Baisha, Nandong) it is possible to decipher the pre-KW change from cooling to greenhouse mode.

Diagenetic processes may take place in such a way that the ratio of primary to secondary calcite is stable overall throughout the succession (Álvaro et al., 2008; Li et al., 2006; Metzger and Fike, 2013). When the local $\delta^{18}O_{carb}$ trend is uniformly negatively skewed to global chemostratigraphic patterns, particularly exposed after the curve positive smoothing, the signature is reasonably assumed as tracer of seawater isotopic composition, with a post-depositional bias influencing the succession in a generally expected way. Interestingly, when the oxygen isotope curve co-varies with the carbon curve, a diagenetic resetting in meteoric setting was usually presumed (Marshall, 1992; Oehlert and Swart, 2019). However, the $\delta^{18}O_{phosp}$ and $\delta^{13}C_{carb}$ curves are paired in distinct excursions during the KW episodes, and thought as the record of high rates of organic matter burial and resulting drawdown of atmospheric CO_2 levels, finalized in the cooling episodes (Figs. 1A, 4, and 16). This case is modeled for the Devonian isotope perturbations by Simon et al. (2007), and mostly proven by increasing $\delta^{18}O_{phosp}$ values by Elrick et al. (2009) and Joachimski et al. (2009). The similar data on transient cooling are presented according to comparative analysis of $\delta^{18}O_{carb}$ and $\delta^{18}O_{phosp}$ signals in the Early-Middle Frasnian interval (*punctata* Event; Pisarzowska and Racki, 2012; see above). Furthermore, the same feedback is evidenced in many other organic-rich deposits and implied cooling episodes (Azmy et al., 1998; Małkowski et al., 2009; Mii et al., 1999; Saltzman, 2002; among others).

In brief, care must be taken to prevent quick rejection of $\delta^{18}O_{carb}$ datasets. It is particularly unlikely that the major chemostratigraphic variation is only artifact of post-depositional resetting when: (1) the incorporated $\delta^{18}O_{carb}$ values overlap the ranges of coeval brachiopod calcites and early cements, and/or (2) they create a positive shift from the reliable isotope baseline, (3) the approximated shifts are reproduced synchronously in a diversity of depositional and diagenetic settings, (4) they preferentially occur in early lithified deposits that originated below the storm wave actions, and (5) the precession cyclicity was deciphered in the high-resolution isotope record. Thus, the popular idea in chemostratigraphy that bulk carbonate-derived $\delta^{18}O$ records represent merely a diagenetically-controlled information noise needs to be critically re-examined in each case, because a first-order climatic trend may be preserved more frequently than usually assumed.

11. Discussion

The expected time correlation between minima in the C-isotope curves and the basal portions of KW and HB BS levels (and other Late Devonian global events) is proved in surprisingly many sections around the world, even if other factors of the ^{13}C depletion (or at least its enhancement), including a diagenetic bias, cannot be excluded in more organic-rich horizons. In addition, recurrent $\delta^{13}C$ negative spike near the F-F boundary clearly mimics the Hg chemostratigraphic signal, referred primarily to the amplified arc magmatism (CH event in Racki et al., 2018; see discussion in Racki, 2020b).

Regarding time correlation between $\delta^{13}C$ minima and Hg enrichments, this link is not very robustly shown in the current chemostratigraphical record of the KW Crisis (see Fig. 5). Only a very broad correspondence may be detected in the Uralian (UKW) or Moroccan (LKW) records. In case of the Hangenberg interval, Hg enrichments in basal parts of the black shale are more commonly present (Fig. 6; Pisarzowska et al., 2020a), exemplified by the Bavarian basin succession at Losau (see fig. 18A in Racki, 2020b). In addition, moderate Hg enrichments, correlated with negative $\delta^{13}C$ excursion (event II of Pisarzowska and Racki, 2012) have also been recognized in lower interval of the Early-Middle Frasnian Middlesex/*punctata* Event in the

Padberg section of the Rhenish region (Pisarzowska et al., 2020b). To sum up, the predicted causal link between large-scale volcanic degassing (from LIPs?) and the response of carbon cycling is confirmed for the initial steps not only of the both major Late Devonian biotic crises.

On the other hand, not very strong δ^{13}C-Hg correspondence exists also in case of other, better studied mass extinction intervals, even if the end-Cretaceous isotope events were primarily shaped by the extraterrestrial impact (Sial et al., 2020). In many sections straddling the most apocalyptic Permian-Triassic boundary, the anti-correlative δ^{13}C-Hg variation is indeed manifested, but evidently obscured in the stratotype Meishan section, especially in the $\delta^{13}C_{org}$ profile (fig. 9 in Sial et al., 2019; see also e.g., Shen et al., 2019).

The temporal links between eruptive episodes in the Central Atlantic LIP and the geochemical records are best established for the Triassic-Jurassic (T-J) transition, and only partial correspondence is established. The large negative Marshi CIE had no link with Hg signal (Lindström et al., 2019; Fig. 18). The major shift δ^{13}C values downwards, as the opening marker of this biotic crisis, cannot be directly link with the Hg signal, maybe partly because of dominating intrusive degassing and some delay (lag time) in

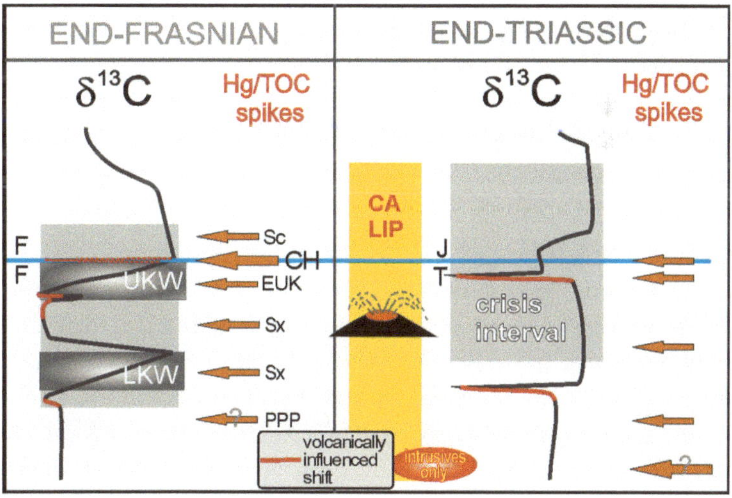

Fig. 18 Comparison of C-isotope and Hg records between the F-F and end-Triassic biotic crises. The T-J geochemical patterns are located relative to a well-dated range of volcanic activity in the Central Atlantic province (CA LIP), based on figs. 1 and 5 (Danish Stenlille-1 locality) in Lindström et al. (2019).

the biogeochemical response. The similar time relations may be assumed at least for the UKW initial phase (Racki, 2020b). So, temporal relationships between perturbations in C and Hg cycles seem to be complex and mostly unpredictable, at least in the light of current data.

As reviewed by Hu et al. (2020), hyperthermal global events in the Mesozoic and Cenozoic, characterized by overall positive CIEs, were triggered by the LIP volcanism in deep-sea domains, i.e., the supplied heat and nutrients loads were buffered by water masses, thus having a more significant impact on deep-sea biota than on shallow marine and terrestrial fauna and flora. The pulsed emplacement of immense oceanic plateau resulted in significant climate shifts (from super-greenhouse to prolonged cooling interludes) and various oceanographic disturbances, exemplified primarily by the Aptian biocalcification crisis (Erba et al., 2015; Ernst and Youbi, 2017; Jenkyns et al., 2017).

Thus, seafloor magmatic activity can also be considered in the case of Late Devonian biotic crises (Kaiser et al., 2015). Nevertheless, a combined continental-oceanic volcanism in the LIPs and arc belts is more probable owing to a major collapse of epeiric carbonate factory. A similar geodynamic setting, associated with evolving Siberian Superplume activity (Black and Gibson, 2019; Burgess et al., 2017; Jerram et al., 2016), but notably paired with progressive arc volcanism in the expanding subduction zones, lastly shown for the EPME (Chen and Xu, 2019).

12. Conclusions and implications

Changes in the climate-sensitive carbon cycle during major and second-order Late Devonian global events were more complex than usually expected, i.e., they were not characterized only by positive $\delta^{13}C_{carb}$ and $\delta^{13}C_{org}$ excursions. In fact, the carbon (and oxygen) isotope signatures in lime mudstones are not very susceptible to diagenetic bias and therefore several negative trends and spikes in different facies of distant continents are re-interpreted as primary biogeochemical signals. In addition to significant intra- and inter-regional variability resulting from diverse local carbon cycling factors, the initial phase of both Kellwasser and Hangenberg Crises confirms the predicted causal relationship between large-scale volcanic degassing (mainly in large igneous provinces?) and the response of global carbon cycle in ^{13}C depleted signals. Lowered $\delta^{13}C$ values characterize also other Late Devonian events, exemplified particularly by well-documented Frasnian Middlesex/*punctata* Event. Thus, consistently approved, paired negative $\delta^{13}C$ and $\delta^{18}O$ excursions, even only in carbonate signatures, can be cautiously regarded as a primary volcano-climatic (greenhouse) signal and additional proxy of the volcanic trigger. Current data suggest such oscillatory interactions and multiple climate changes caused by volcanic pulsating paroxysms, at least during the Upper Kellwasser and Hangenberg times. In addition, the repeated negative $\delta^{13}C$ spike near the F-F boundary correlates with Hg anomalies considered to be the signature of the arc magmatism acme (and alkaline/carbonatite/kimberlite paroxysm?; see Racki, 2020b). However, both the carbon cycle response and time relationships between disturbances in C and Hg cycles during mass extinctions were certainly complex, as predicted for the biogeochemical interactions by Jones et al. (2016).

Following key challenges are recognized for future high-resolution studies, especially in the most stratigraphically extended successions:

1. Truly oceanic biosiliceous successions still await basic recognition of geochemical patterns.
2. The standard paired $\delta^{13}C_{carb}$ and $\delta^{18}O_{carb}$ analysis of deeper-water lime mudstones appears to have a higher than usually approved potential to preserve the primary isotope ratios, as a record of volcanically-driven climatic changes.
3. The surprising large-scale shifts in regional background $\delta^{13}C_{carb}$ values, even within one pelagic setting, need refined explanation.
4. In context of amplified primary production assumed for black shale deposition, low nutritional quality, and possible toxicity is considered for dominating picoplankton and bacterioplankton (cyanobacteria and green sulfur bacteria; Joachimski et al., 2001; Song et al., 2020; van de Schootbrugge and Gollner, 2013), what needs more proper study in the key context of evolving carbon cycling and trophic webs during Late Devonian global events.
5. The predicted causal link between large-scale volcanic degassing and the response of carbon cycling recorded in depleted C-isotope ratios in Late Devonian signatures presents an attractive perspective for confirmation of the biogeochemical feedback in the initial steps of older Devonian greenhouse (to icehouse) crises, listed by Retallack and Huang (2011).

6. Age relations between disturbances in C and Hg cycles during major extinction events are indeed unclear, especially due to their control through variable degassing in intrusive and extrusive LIP phases. The key issue requires high-resolution studies and advanced quantitative modeling.

Acknowledgments

The study based on results of the MAESTRO grant 2013/08/A/ST10/00717 from the National Science Centre—Poland (to G. Racki), as well as on grants 2011/03/B/ST10/04602 and 2014/15/B/ST10/03705 (to A. Pisarzowska). Many peoples participated in field work program during the MAESTRO grant realization (see SM 1 and SM 3 in Racki, 2020b), and especially participation of Michał Rakociński (also kindly provided the draft of part of Fig. 5), Leszek Marynowski, Zdzisław Bełka, Thomas Becker, Sven Hartenfels and Harald Tragelehn was significant for the present work. Michael Joachimski kindly provided in 1993 the isotope measures for the F-F section at Kowala, unpublished till now (Fig. 16). Photos for Fig. 2 are made by Eberhard Schindler (A), Krzysztof Owocki (B), and Michał Rakociński (E). The text has been checked by Cambridge Proofreading & Editing LLC.

Appendix A

Selected geochemical data for the Frasnian-Famennian sections presented in Fig. 5 (see also Supplementary Material 1 in the online version at https://doi.org/10.1016/bs.sats.2020.08.001; for methods see Racki, 2020b).

See Tables A1–A3.

Table A1 Hg, TOC, Al_2O_3, Mo contents, Hg/TOC, Hg/Al, Hg/Mo ratios and $\delta^{13}C$ data from F-F succession at Lahmida, and assumed volcanic signals (CH, Sx, PPP) based on the Hg data (red ellipses, see Fig. 1A).

Stage	Sample	Hg	TOC	Al_2O_3	Mo	Hg/TOC	Hg/Al	$\delta^{13}C_{org}$	$\delta^{13}C_{carb}$
	Method	ICP-MS	Eltra CS-500	ICP-ES	ICP-MS				
		(ppb)	(%)	(%)	(ppm)	(ppb/%)	(ppb/%)	‰	‰
	Max detection limit (MDL)	10	0.01	0.01	0.1				
	LA 44	80	0.61	4.83	1.0	132.2	31.3	-27,2	no data
	LA 43/44	40	0.68	5.23	1.3	58.4	14.5	-28,0	-2,5
	LA 42	270	0.83	5.85	4.4	327.1	87.2	-28,8	-3,2
	LA 41/42	270	1.14	21.90	5.8	236.0	23.3	-28,3	no data
	LA 40T	50	0.59	3.37	0.3	85.4	28.0	-28,9	-3,9
	LA 38	100	0.42	2.93	0.9	240.2	64.5	-28,6	-1,5
	LA 35/36B	250	0.83	20.60	0.7	300.4	22.9	-28,2	no data
	LA 34	250	0.75	5.50	0.3	332.8	85.9	-28,5	-4,0
	LA 32/33	150	0.46	20.72	1.4	323.7	13.7	-27,7	no data
	LA 32M	140	0.38	2.61	1.5	367.8	101.4	-28,7	-4,6
	LA 30	400	0.86	3.58	1.6	467.0	211.1	-28,8	-9,5
	LH 29	170	0.11	1.73	4.6	1565.0	185.7	-27,2	-5,3
	LH 28	110	0.40	5.41	1.1	276.8	38.4	-27,8	-2,9
	LH 27A2	420	0.93	21.04	1.9	449.2	37.7	-27,3	no data
	LA 27/28	260	1.05	20.70	2.7	246.6	23.7	-27,4	no data
	LH 27	120	0.77	4.97	4.9	156.4	45.6	-27,4	-2,2
	LA 26/27	360	0.86	21.02	2.3	418.7	32.4	-27,3	no data
	LH 26	60	0.55	2.43	3.3	108.8	46.7	-27,0	-4,3
	LA 25/26	190	0.54	18.85	2.5	350.6	19.1	-26,9	no data
	LH 25T	180	0.79	3.10	0.9	228.1	109.7	-27,0	-4,4
FAMEN.	LH 25B	210	0.37	2.57	2.3	572.0	154.4	-27,5	-5,8
FRASN.	LA 24/25S	110	0.35	3.17	1.6	313.8	65.6	-25,2	no data
	LA 24/2 CH	1530	0.16	2.33	0.9	9759.0	1240.8	-27,1	no data
	LH 24T	110	0.65	2.79	2.0	169.7	74.5	-27,5	-4,9
UKW	LH 24B ?	190	0.10	0.96	0.9	1975.3	374.0	-28,1	-1,5
	LH 23T	250	0.67	1.99	4.4	371.6	237.4	-28,6	-3,0
	LH 23B	90	0.13	4.24	9.9	713.7	40.1	-27,6	-3,3
	LH 22 ?	490	0.41	2.08	15.0	1201.8	445.2	-28,8	-3,3
	LA 21A	290	<0.01	3.97	1.6	29000.0	138.0	no data	no data
	LH 21	120	0.11	3.21	3.1	1100.4	70.6	-25,9	-1,7
	LH 20	90	0.01	2.41	1.2	7558.3	70.6	-26,5	-3,7
	LH 19	230	0.76	2.49	1.3	302.0	174.6	-26,6	-0,7
LKW	LH 18 ?Sx	250	0.20	0.93	5.0	1225.0	508.0	-27,0	-4,2
	LA 17/18T	260	0.03	1.04	1.6	9512.9	472.4	-27,1	-1,8
	LA 17/18B	170	0.37	1.61	1.5	455.8	199.5	-27,7	-2,3
	LA 17M	70	0.24	2.75	2.0	296.4	48.1	-27,8	-3,6
	LA 14 PPP	1050	0.56	3.49	4.6	1867.4	568.5	-25,8	-0,5
	LA 13T	200	0.71	1.05	1.7	282.5	359.9	-29,3	-2,2
	LA 12	50	0.41	3.78	0.5	121.2	25.0	-28,4	-2,3
	LA 10	<10	0.32	2.09	0.5	31.3	9.0	-27,1	0,1
	LA 8	<10	0.28	1.71	<0.1	36.2	11.1	-27,7	1,6
	LA 6	30	0.33	2.29	0.2	90.4	24.8	-27,6	0,2
	LA 5	320	0.48	1.55	3.1	670.2	390.1	-27,8	-0,1
	LA 4A	30	1.21	1.84	1.6	24.9	30.8	-28,6	-1,5

KW intervals are grayish.

Table A2 Hg, TOC, Al_2O_3, Mo contents, Hg/TOC, Hg/Al, Hg/Mo ratios and $\delta^{13}C_{org}$ data from F-F succession at Silberberg, and assumed volcanic signals based on the Hg data (red ellipses, see Fig. 1A).

Stage	Sample	Hg	TOC	Al_2O_3	$\delta^{13}C_{org}$	Mo	Hg/TOC	Hg/Al
	Method	ICP-MS	Eltra CS-	ICP-ES		ICP-		
		(ppb)	%	%	‰	(ppm	(ppb/%)	(ppb/%)
	MDL	10	0.01	0.01		0.1		
	Sb 17	240	0.15	18.34	-25.43	<0.1	1631.9	24.7
	Sb 16	280	0.11	18.64	-27.94	<0.1	2441.0	28.4
	Sb 15 Sc	1140	0.28	18.87	-28.14	<0.1	4110.4	114.2
	Sb 14	120	0.59	2.49	-27.00	0.4	201.7	91.1
FAMENNIAN	Sb 13	110	0.47	2.61	-26.75	0.5	232.7	79.6
FRASNIAN	Sb 12 ?CH	1340	6.46	10.14	-28.13	28.1	207.3	249.7
UKW	Sb 11	1380	5.54	10.95	-28.70	22.5	249.2	238.2
	Sb 10	410	0.66	18.28	-27.55	1.7	617.2	42.4
LKW	Sb 9	1100	5.28	8.68	-28.18	14.7	208.2	239.5
	Sb 8 ?Sx	1200	4.51	9.06	-28.85	20.6	265.9	250.3
	Sb 7	250	0.25	16.21	-28.59	0.5	981.3	29.1
	Sb 6	1060	3.72	8.92	-28.45	14.5	285.1	224.6
	Sb 5	380	0.44	15.19	-28.42	0.2	867.3	47.3
	Sb 4	350	0.14	18.21	-27.63	<0.1	2415.1	36.3
	Sb 3	360	0.17	18.28	-28.64	0.1	2143.5	37.2
	Sb 2	340	0.17	17.62	-28.99	0.2	1977.4	36.5
	Sb 1 ?PPP	880	0.19	17.87	-28.97	<0.1	4633.7	93.1

Table A3 Carbon and oxygen isotope and TOC data for Siv'yu section, Russia (for other geochemical data, see Racki et al., 2018; Yudina et al., 2002).

Sample	$\delta^{13}C_{carb}$ (‰)	$\delta^{18}C_{carb}$ (‰)	TOC (%)
Sv96-S12	0.12	−5.68	1.64
Sv96-S10	1.12	−5.76	2.24
Sv96-S7	1.69	−5.9	1.46
Sv96-S4	0.25	−6.76	2.79
Sv96132	1.48	−6.1	3.09
CB 99355	3.6	−5.78	1.39
CB 99353	1.3	−10.13	1.64
CB 99352	1.35	−9.08	2.04
CB 99351	1.73	−7.57	2.4
CB 99350	1.84	−7.27	No data
CB 99349	1.6	−10.38	2.0
CB 99348	1.61	−10.13	2.66
Sv96-138/1	1.67	−8.99	2.13

Table A3 Carbon and oxygen isotope and TOC data for Siv'yu section, Russia (for other geochemical data, see Racki et al., 2018; Yudina et al., 2002).—cont'd

Sample	$\delta^{13}C_{carb}$ (‰)	$\delta^{18}C_{carb}$ (‰)	TOC (%)
CB 99346	3.48	−6.29	2.11
CB 99344	3.10	−6.5	2.47
CB 99343	3.06	−6.83	No data
CB 99342	3.0	−6.47	1.97
CB 99340	1.85	−6.83	3.05
CB 99338	3.24	−5.89	2.65
CB 99335	2.89	−6.23	2.17
CB 99334	1.72	−6.21	4.18
CB 99330	3.24	−6.24	3.35
CB 99329	2.5	−5.92	3.0
CB 99328	2.72	−5.94	3.13
CB 99326	2.5	−5.92	2.58
CB 99324	2.96	−5.76	No data
CB 99318	1.51	−5.68	2.84
Cb 99317	0.19	−6.07	1.36
CB 99315	0.24	−5.95	No data
CB 99313	−0.01	−6.21	2.24
CB 99312	0.24	−6.08	1.16
CB 99310	0.23	−6.03	1.81
CB 99308	0.67	−6.08	No data
CB 99306	0.54	−6.39	No data
CB 99305	0.66	−6.48	1.55
CB 99304	−1.88	−10.93	0.83
CB 99303	−1.21	−11.77	No data
CB 99302	0.15	−6.02	No data
CB 99229	0.78	−6.2	1.75
CB 99227	0.38	−5.93	2.28

Continued

Table A3 Carbon and oxygen isotope and TOC data for Siv'yu section, Russia (for other geochemical data, see Racki et al., 2018; Yudina et al., 2002).—cont'd

Sample	$\delta^{13}C_{carb}$ (‰)	$\delta^{18}C_{carb}$ (‰)	TOC (%)
CB 99226	1.1	−5.42	1.83
Syv96-91	0.71	−5.77	2.92
CB 99222	−0.1	−5.61	0.83
CB 99221	−0.98	−5.07	No data
CB 99220	0.35	−6.47	No data
Sv96-S12	0.37	−7.53	No data
Sv96-S10	2.81	−5.83	No data
Sv96-S7	0.67	−7.08	No data

Appendix B

Selected geochemical data for the upper Famennian sections presented in Figs. 6 and 15.

Table B1 Geochemical data from Oberrödinghausen section (OB, OBH) for the Hangenberg Crisis and the *Annulata* Event intervals.

Stage	Sample	Hg	TOC	Al_2O_3	Mo	Hg/TOC	Hg/Al	$\delta^{13}C_{org}$
	Method	AAS	Thermo Flash	XRF	LAICP-MS			
		(ppb)	(%)	(%)	(ppm)	(ppb/%)	(ppb/%)	‰
	Max detection limit (MDL)	2	0.01	0.01	0.1			
	OBH 920	7	0.02			348		-25.0
HS	OBH 900	9	0.06	19.50	1.2	142	0.44	-25.1
HS	OBH 880	26	0.58			44		-24.6
HS	OBH 860	16	0.62			25		-23.2
HS	OBH 840	17	0.69	14.40	2.6	24	1.17	-23.3
HS	OBH 820	13	0.90			15		-23.7
HS	OBH 800	6	0.68			9		-24.0
HS	OBH 780	6	0.70	11.90	1.2	8	0.49	-23.7
HS	OBH 760	5	0.63			8		-24.3
HS	OBH 740	6	0.63			10		-23.4
HS	OBH 720	5	0.52	11.10	0.6	10	0.45	-23.9
HS	OBH 700	5	0.56			9		-24.1
HS	OBH 680	7	0.36			19		-24.6
HS	OBH 660	6	0.37	16.20	0.4	15	0.35	-24.0
HS	OBH 640	7	0.39			19		-23.8
HS	OBH 620	6	0.37			15		-23.9
HS	OBH 600	6	0.30	17.40	0.6	19	0.34	-23.6
HS	OBH 580	5	0.33			16		-23.7
HS	OBH 560	4	0.31			13		-23.8
HS	OBH 540 + 2cm	5	0.34	16.60	0.8	15	0.30	-23.1
HS	OBH 540 - 2cm	6	0.32			18		-23.7
HS	OBH 520	6	0.41			15		-24.2
HS	OBH 500	5	0.25	15.90	1.0	19	0.29	-23.4
HS	OBH 480	5	0.23			23		-23.4
HS	OBH 460	5	0.21	16.20	0.2	25	0.33	-23.7
HS	OBH 430	7	0.25			29		-23.1
HS	OBH 410	8	0.16	15.60	1.2	48	0.49	-22.9
HS	OBH 390	5	0.23			20		-22.8
HS	OBH 370	9	0.22			39		-22.1
HS	OBH 350	12	0.29	17.10	2.8	43	0.72	-21.5
HBS	OBH 320	76	1.70	20.90	8.4	45	3.65	-26.1
	OBH 310	10	0.17			59		-25.2
	OBH 240	2	0.16			12		-25.8
	OBH 200	2	0.22	5.07	0.4	9	0.38	-27.9
	OBH 090	3	0.03	16.70	0.8	84	0.15	-25.7
	OBH 080	2	0.13			16		-26.6
	OBH 0	2	0.01			232		-24.8
	OB13		0.21					-24.9
	OB12		0.17					-22.8
	OB11		0.79					-25.0
UAN	OB10		0.87					-26.9
	OB9		0.47					-27.3
	OB8		0.09					-25.4
	OB7		0.12					-24.3
LAN	OB6		1.20					-26.3
LAN	OB5		2.91					-28.3
LAN	OB4		1.95					-26.2
	OB3		0.18					-24.0

Light gray denotes the Hangenberg Crisis; dark gray indicates the *Annulata* Event.
LAN—Lower Annulata Event, UAN—Upper Annulata Event, HBS—Hangenberg Black Shale, HS—Hangenberg Shale.

Table B2 Geochemical data in the samples of Kowala (KQ) section for the D-C interval and the *Annulata* Event.

Stage	Sample	Hg	TOC	Al_2O_3	Mo	Hg/TOC	Hg/Al	$\delta^{13}C_{org}$	$\delta^{13}C_{carb}$
	Method	AAS	Thermo Flash	XRF	LAICP-MS/ ICP-MS				
		(ppb)	(%)	(%)	(ppm)	(ppb/%)	(ppb/%)	‰	‰
	Max detection limit (MDL)	10	0.01	0.01	0.1/2				
	KQ 132		0.1					-26.8	-0.9
	KQ 131		0.1					-26.4	0.8
	KQ 130		0.3					-26.9	-0.8
	KQ 129	12	0.2			75		-24.9	1.6
	KQ 128 top		0.1					-27.0	1.7
	KQ 127b top		0.2					-27.4	-1.7
	KQ 127a base		0.2					-27.5	0.1
	KQ 126		0.2					-27.9	1.4
	KQ 125 base		0.2					-27.4	2.0
	KQ 124	8	0.2			51		-27.5	1.2
	KQ 123b top	11	0.03			367		-27.0	2.4
	KQ 122		0.1					-27.7	2.5
	KQ 121		0.1					-27.3	-0.6
	KQ 120		0.1					-27.3	-0.3
	KQ 119	7	0.03			208		-27.5	-2.0
	KQ 117		0.1					-26.0	0.8
	KQ 114	16	0.3			55		-24.6	-0.8
	KQ113B	43	0.3	19.36	16.0	134	2.2	-27.9	
	KQ 112	34	0.4	17.83	0.5	92	1.9	-27.5	
	KQ 111	46	0.2	7.22	4.4	242	6.4	-25.2	1.1
HBS	KQ110G	138	12.5	9.78	11.7	11	14.1	-28.2	b.d.
HBS	KQ110F	50	5.7	9.12	7.9	9	5.5	-28.1	0.9
HBS	KQ110E	83	7.6	15.50	28.0	11	5.4	-28.5	0.4
HBS	KQ110D	38	2.6	19.00	14.2	14	2.0	-28.5	0.2
HBS	KQ110C	169	7.2	16.00	24.1	23	10.6	-28.9	-0.2
HBS	KQ110B	99	18.2	6.94	34.8	5	14.3	-29.4	-0.7
HBS	KQ110A	63	30.9	6.12	161.2	2	10.3	-30.1	-1.3
	KQ 109G	38	0.5	8.97	19.0	79	4.2	-26.2	
	KQ 109F	24	0.3	3.06	72.0	80	7.8	-25.3	0.0
	KQ 109D	13	0.4	8.41	3.3		1.5	-29.6	
	KQ 109C	11	0.2	13.31	1.9	31	0.8	-26.4	1.3
	KQ 109B	28	0.2	3.22	3.0	56	8.7	-26.2	0.9
	KQ 109A	38	0.2	16.33	0.6	189	2.3	-26.4	
	KQ 108B	34	0.1	2.58	0.8	253	13.2	-25.7	0.7
	KQ 106	40	0.2	4.75	0.5	240	8.4	-25.8	1.1
	KQ 11/K8		0.35	1.76	**3**				
	KQ 10/7		1.3	10.23	**4**			-23.6	
	KQ 9/6		0.98	4.08	**4**			-24.1	
UAN	KQ 8/5ST							-27.4	
UAN	KQ 8/5		19.05	4.18	**24**			-27.0	
UAN	KQ 7/5		5.81	1.21	**6**			-27.2	
UAN	KQ 6/5SR							-26.8	
UAN	KQ 6/5		23.26	5.91	**61**			-28.5	
	KQ 5/4		0.71	6.43	**<2**			-25.7	
LAN	KQ 4/3		12.56	6.69	**9**			-28.4	
LAN	KQ 3/3SR							-28.7	
LAN	KQ 3/3		10.17	8.66	**3**			-28.3	
	KQ 2/2		0.06	3.04	**<2**			-27.3	

Light gray denotes the Hangenberg Crisis; dark gray indicates the *Annulata* Event.
b.d.—below detection; for explanations see Table 1.

References

Acquafredda, P., Lorenzoni, S., Lorenzoniet, E.Z., 1994. Palaeozoic sequences and evolution of the Calabrian-Peloritan Arc (Southern Italy). Terra Nova 6, 582–594.

Allan, J.R., Matthews, R.K., 1982. Isotope signatures associated with early meteoric diagenesis. Sedimentology 29, 797–817.

Álvaro, J.J., Bauluz, B., Subías, I., Pierre, C., Vizcaïno, D., 2008. Carbon chemostratigraphy of the Cambrian-Ordovician transition in a midlatitude mixed platform, Montagne Noire, France. GSA Bull. 120, 962–975.

Archer, D., 2010. The Global Carbon Cycle. Princeton University Press, Princeton, NJ, p. 202.

Arens, N., West, I.D., 2008. Press-pulse: a general theory of mass extinction? Paleobiology 34, 456–471.

Averbuch, O., Tribovillard, N., Devleeschouwer, X., Riquier, L., Mistiaen, B., van Vliet-Lanoe, B., 2005. Mountain building-enhanced continental weathering and organic carbon burial as major causes for climatic cooling at the Frasnian–Famennian boundary (c. 376 Ma)? Terra Nova 17, 25–34.

Azmy, K., Veizer, J., Bassett, M.G., Copper, P., 1998. Oxygen and carbon isotopic composition of Silurian brachiopods: implications for coeval seawater and glaciations. GSA Bull. 110, 1499–1512.

Azmy, K., Poty, E., Mottequin, B., 2012. Biochemostratigraphy of the upper Frasnian in the Namur–Dinant Basin, Belgium: implications for a global Frasnian–Famennian pre-event. Palaeogeogr. Palaeoclimatol. Palaeoecol. 313–314, 93–106.

Bábek, O., Kumpan, T., Kalvoda, J., Matys Grygar, T., 2016. Devonian/Carboniferous boundary glacioeustatic fluctuations in a platform-to-basin direction: a geochemical approach of sequence stratigraphy in pelagic settings. Sediment. Geol. 337, 81–99.

Bagherpour, B., Bucher, H., Vennemann, T., Schneebeli-Hermann, E., Yuan, D.X., Leu, M., Zhang, C., Shen, S.Z., 2020. Are Late Permian carbon isotope excursions of local or of global significance? GSA Bull. 132, 521–544.

Bambach, R.K., 2006. Phanerozoic biodiversity mass extinctions. Annu. Rev. Earth Planet. Sci. 34, 127–155.

Barnet, J.S.K., Littler, K., Westerhold, T., Kroon, D., Leng, M.J., Bailey, I., Röhl, U., Zachos, J.C., 2019. A high-fidelity benthic stable isotope record of late Cretaceous–early Eocene climate change and carbon-cycling. Paleoceanogr., Paleoclimatol. 34, 672–691.

Batenburg, S.J., De Vleeschouwer, D., Sprovieri, M., Hilgen, F.J., Gale, A.S., Singer, B.S., Koeberl, C., Coccioni, R., Claeys, P., Montanari, A., 2016. Orbital control on the timing of oceanic anoxia in the Late Cretaceous. Clim. Past 12, 1995–2009.

Becker, R.T., 1993. Anoxia, eustatic changes, and Upper Devonian to lowermost Carboniferous global ammonoid diversity. In: House, M.R. (Ed.), The Ammonoidea: Environment, Ecology and Evolutionary Change. Syst. Assoc. Spec., Vol. 47, pp. 115–163.

Becker, R.T., 1996. New faunal records and holostratigraphic correlation of the Hasselbachtal D/C-boundary auxiliary stratotype (Germany). Ann. Soc. Géol. Belgique 117, 19–45.

Becker, R.T., Königshof, P., Brett, C.E., 2016a. Devonian climate, sea level and evolutionary events: an introduction. In: Becker, R.T., Königshof, P., Brett, C.E. (Eds.), Devonian Climate, Sea Level and Evolutionary Events, pp. 1–10. Geol. Soc., Lond. Spec. Publ. 423.

Becker, R.T., Kaiser, S., Aretz, M., 2016b. Review of chrono-, litho- and biostratigraphy across the global Hangenberg Crisis and Devonian–Carboniferous Boundary. In: Becker, R.T., Königshof, P., Brett, C.E. (Eds.), Devonian Climate, Sea Level and Evolutionary Events, pp. 355–386. Geol. Soc., Lond. Spec. Publ. 423.

Becker, R.T., Hartenfels, S., Weyer, D., Kumpan, T., 2016c. The famennian to lower Viean at Drewer (northern Rhenish Massif). Münst. Forsch. Geol. Paläont. 108, 158–178.
Becker, R.T., Aboussalam, Z., Hartenfels, S., 2018. The Frasnian-Famennian boundary mass extinction–widespread seismic events, the timing of climatic pulses, "pelagic death zones", and opportunistic survivals. In: The Fossil Week, Abstract Book. 5th International Paleontological Congress—Paris, 9th–13th July 2018, p. 10.
Bickert, T., Pätzold, J., Samtleben, C., Munneke, A., 1997. Paleoenvironmental changes in the Silurian indicated by stable isotopes in brachiopod shells from Gotland, Sweden. Geochim. Cosmochim. Acta 61, 2717–2730.
Black, B.A., Gibson, S.A., 2019. Deep carbon and the life cycle of large igneous provinces. Elements 15, 319–324.
Blakey, R., 2016a. Devonian—370 Ma, Global Paleogeography and Tectonics in Deep Time Series. Deep Time Maps™ Paleogeography.
Blakey, R., 2016b. Devonian—360 Ma, Global Paleogeography and Tectonics in Deep Time Series. Deep Time Maps™ Paleogeography.
Bojar, A.V., Neubauer, F., Koeberl, C., 2013. Geochemical record of Late Devonian to Early Carboniferous events, Palaeozoic of Graz, Eastern Alps, Austria. In: Gąsiewicz, A., Słowakiewicz, M. (Eds.), Palaeozoic Climate Cycles: Their Evolutionary and Sedimentological Impact, pp. 87–108. Geol. Soc., Lond. Spec. Publ. 376.
Brand, U., Legrand-Blain, M., Streel, M., 2004. Biochemostratigraphy of the Devonian–Carboniferous boundary global stratotype section and point, Griotte Formation, La Serre, Montagne Noire, France. Palaeogeogr. Palaeoclimatol. Palaeoecol. 205, 337–357.
Brand, U., Tazawa, J., Sano, H., Azmy, K., Lee, X., 2009. Is mid-late Paleozoic oceanwater chemistry coupled with epeiric seawater isotope records? Geology 237, 823–826.
Brezinski, D.K., Cecil, C.B., Skema, V.W., Kertis, C.A., 2009. Evidence for long-term climate change in Upper Devonian strata of the central Appalachians. Palaeogeogr. Palaeoclimatol. Palaeoecol. 284, 315–325.
Brown, D., Spadea, P., Puchkov, V., Alvarez-Marron, J., Herrington, R., Willner, A.P., Hetzel, R., Gorozhanina, Y., Juhlin, C., 2006. Arc-continent collision in the Southern Urals. Earth Sci. Rev. 79, 261–287.
Buggisch, W., 1991. The global Frasnian-Famennian "Kellwasser Event". Geol. Rundsch. 80, 49–72.
Buggisch, W., Joachimski, M.M., 2006. Carbon isotope stratigraphy of the Devonian of Central and Southern Europe. Palaeogeogr. Palaeoclimatol. Palaeoecol. 240, 68–88.
Burgess, S.D., Muirhead, J.D., Bowring, S.A., 2017. Initial pulse of Siberian Traps sills as the trigger of the end-Permian mass extinction. Nat. Commun. 8 (164), 1–6.
Calpron, M., Nelson, J.L., Murphy, D.C., 2007. Northern Cordilleran terranes and their interactions through time. GSA Today 17, 4–10.
Caplan, M.L., Bustin, R.M., 1999. Devonian–Carboniferous Hangenberg mass extinction event, widespread organic-rich mudrock and anoxia: causes and consequences. Palaeogeogr. Palaeoclimatol. Palaeoecol. 148, 187–207.
Carmichael, S.K., Waters, J.A., Suttner, T.J., Kido, E., DeReuil, A.A., 2014. A new model for the Kellwasser anoxia events (Late Devonian): shallow water anoxia in an open oceanic setting in the Central Asian Orogenic Belt. Palaeogeogr. Palaeoclimatol. Palaeoecol. 399, 394–403.
Carmichael, S.K., Waters, J.A., Batchelor, C.J., Coleman, D.M., Suttner, T.J., Kido, E., Moore, L.M., Chadimová, L., 2016. Climate instability and tipping points in the Late Devonian: detection of the Hangenberg event in an open oceanic island arc in the Central Asian Orogenic Belt. Gondwana Res. 32, 213–231.
Carmichael, S.K., Waters, J.A., Königsof, P., Suttner, T.J., Kido, E., 2019. Paleogeography and paleoenvironments of the Late Devonian Kellwasser event: a review of its sedimentological and geochemical expression. Glob. Planet. Change 183 (102984), 1–28.

Chang, J., Bai, Z., Sun, Y., Peng, Y., Qin, S., Shen, B., 2019. High resolution bio- and chemostratigraphic framework at the Frasnian-Famennian boundary: implications for regional stratigraphic correlation between different sedimentary facies in South China. Palaeogeogr. Palaeoclimatol. Palaeoecol. 531A (108299), 1–14.

Chen, J., Xu, Y.G., 2019. Establishing the link between Permian volcanism and biodiversity changes: insights from geochemical proxies. Gondwana Res. 75, 68–96.

Chen, D.Z., Tucker, M.E., Shen, Y.A., Yans, J., Preat, A., 2002. Carbon isotope excursions and sea-level change: implications for the Frasnian–Famennian biotic crisis. J. Geol. Soc. Lond. 159, 623–626.

Chen, D.Z., Qing, H.R., Li, R.W., 2005. The Late Devonian Frasnian–Famennian (F/F) biotic crisis: insights from $\delta^{13}C_{carb}$, $\delta^{13}C_{org}$ and $^{87}Sr/^{86}Sr$ isotopic systematics. Earth Planet. Sci. Lett. 235, 151–166.

Chen, D.Z., Wang, J.G., Qing, H.R., Yan, D.T., Li, R.W., 2009. Hydrothermal venting activities in the Early Cambrian, South China: petrological, geochronological and stable isotopic constrains. Chem. Geol. 258, 168–181.

Cimen, O., 2018. The new findings on the Late Devonian volcanism in the Eastern Taurides (Develi, Kayseri): preliminary data. Geol. Bull. Turkey 61, 75–90.

Clapham, M.E., Renne, P.R., 2019. Flood basalts and mass extinctions. Annu. Rev. Earth Planet. Sci. 47, 275–303.

Cole, D., Myrow, P.M., Fike, D.A., Hakim, A., Gehrels, G.E., 2015. Uppermost Devonian (Famennian) to Lower Mississippian events of the western U.S.: stratigraphy, sedimentology, chemostratigraphy, and detrital zircon geochronology. Palaeogeogr. Palaeoclimatol. Palaeoecol. 427, 1–19.

Coniglio, M., 1989. Neomorphism and cementation in ancient deepwater limestones, Cow Head Group (Cambro-Ordovician), western Newfoundland. Canada. Sediment. Geol. 65, 15–33.

Copper, P., 2011. 100 million years of reef prosperity and collapse: Ordovician to Devonian interval. In: Stanley, G.D. (Ed.), Corals and Reef Crises, Collapse and Change, pp. 15–32. Paleont. Soc. Papers 17.

Corradini, C., Spalletta, C., Mossoni, A., Matyja, H., Over, D.J., 2017. Conodonts across the Devonian/Carboniferous boundary: a review and implication for the redefinition of the boundary and a proposal for an updated conodont zonation. Geol. Mag. 154, 888–982.

Cramer, B., Saltzman, M., Day, J.E., Witzke, B.J., 2008. Record of the Late Devonian Hangenberg global positive carbon-isotope excursion in an epeiric sea setting: carbonate production, organic-carbon burial and paleoceanography during the Late Famennian. In: Pratt, B.R., Holmden, C. (Eds.), Dynamics of Epeiric Seas, pp. 103–118. GAC Spec. Pap. 48.

da Silva, A.C., Boulvain, F., 2008. Carbon isotope lateral variability in a Middle Frasnian carbonate platform (Belgium): significance of facies, diagenesis and sea-level history. Palaeogeogr. Palaeoclimatol. Palaeoecol. 269, 189–204.

Da Silva, A.C., Sinnesael, M., Claeys, P., Davies, J.H.F.L., Winter, N.J.d., Percival, L.M.E., Schaltegger, U., De Vleeschouwer, D., 2020. Anchoring the Late Devonian mass extinction in absolute time by integrating climatic controls and radio-isotopic dating. Sci. Rep. 10 (12940), 1–12.

Dahl, T.W., Hammarlund, E.U., Anbar, A.D., Bond, D.P.G., Gill, B.C., Gordon, G.W., Knoll, A.H., Nielsen, A.T., Schovsbo, N.H., Canfield, D.E., 2010. Devonian rise in atmospheric oxygen correlated to the radiations of terrestrial plants and large predatory fish. Proc. Natl. Acad. Sci. U.S.A. 107, 17911–17915.

Day, J., Witzke, B.J., 2017. Chapter Six—Upper Devonian biostratigraphy, event stratigraphy, and Late Frasnian Kellwasser extinction bioevents in the Iowa Basin: Western Euramerica. In: Montenari, M. (Ed.), Stratigraphy & Timescales. Vol. 2, pp. 243–332.

Day, J., Witzke, B.J., Rowe, H., Elwood, B., 2013. Magnetic susceptibility and carbon isotope stratigraphies through the Devonian-Carboniferous boundary interval in the

western Illinois basin—central North America. In: Whalen, M., et al. (Eds.), Geophysical and Geochemical Techniques: A Window on the Palaeozoic World, IGCP 580–596, 27 August–1 September 2013, Programme with Abstracts.
De Vleeschouwer, D., Rakociński, M., Racki, G., Bond, D.P.G., Sobień, K., Claeys, P., 2013. The astronomical rhythm of Late-Devonian climate change 1023 (Kowala section, Holy Cross Mountains, Poland). Earth Planet. Sci. Lett. 365, 25–37.
De Vleeschouwer, D., da Silva, A.C., Sinnesael, M., Chen, D.Z., Day, J.E., Whalen, M.T., Guo, Z.H., Claeys, P., 2017. Timing and pacing of the Late Devonian mass extinction event regulated by eccentricity and obliquity. Nat. Commun. 8 (2268), 1–11.
Dickens, G.R., O'Neil, J.R., Rea, D.K., Owen, R.M., 1995. Dissociation of oceanic methane hydrate as a cause of the carbon isotope excursion at the end of the Paleocene. Paleoceanography 10, 965–971.
Dolenec, T., Ogorelec, B., Dolenec, M., Lojen, S., 2004. Carbon isotope variability and sedimentology of the Upper Permian carbonate rocks and changes across the Permian-Triassic boundary in the Masore section (Western Slovenia). Facies 50, 287–299.
Dopieralska, J., 2003. Neodymium isotopic composition of conodonts as a palaeoceanographic proxy in the Variscan oceanic system. Justus-Liebig-University, Giessen, p. 111. PhD thesis.
Dopieralska, J., Bełka, Z., Haack, U., 2006. Geochemical decoupling of water masses in the Variscan oceanic system during Late Devonian times. Palaeogeogr. Palaeoclimatol. Palaeoecol. 240, 108–119.
Du, Y.S., Gong, Y.M., Zeng, X.W., Huang, H.W., Yang, J.H., Zhang, Z., Huanh, Z.Q., 2008. Devonian Frasnian-Famennian transitional event deposits of Guangxi, South China and their possible tsunami origin. Sci. China D–Earth Sci. 5, 1570–1580.
Ekart, D.D., Cerling, T.E., Montanez, I.P., Tanopr, N.J., 1999. A 400 million year carbon isotope record of pedogenic carbonate: Implications for paleoatmospheric carbon dioxide. Am. J. Sci. 299, 805–827.
Elewa, A.M.T., Abdelhady, A.A., 2020. Past, present, and future mass extinctions. J. Afr. Earth Sci. 162, 103678.
Elrick, M., Berkyová, S., Klapper, G., Sharp, Z., Joachimski, M., Frýda, J., 2009. Stratigraphic and oxygen isotope evidence for My-scale glaciation driving eustasy in the Early–Middle Devonian greenhouse world. Palaeogeogr. Palaeoclimatol. Palaeoecol. 276, 170–181.
Emsbo, P, 2017. Sedex brine expulsions to Paleozoic basins may have changed global marine $^{87}Sr/^{86}Sr$ values, triggered anoxia, and initiated mass extinctions. Ore Geol. Rev. 86, 474–486.
Emsbo, P., Premo, W.R., McLaughlin, P.I., Neymark, L.A., Vandenbroucke, T.R.A., Day, J.E., du Bray, E.A., Manning, A.H., Bancroft, A.M., 2018. Impact of sedimentary-exhalative hydrothermal systems on marine chemistry and mass extinctions: applications for ore genesis research and mineral exploration. In: Arribas, A.M., Mauk, J.L. (Eds.), Metals, Minerals, and Society, pp. 75–87. Soc. Econ. Geol., Spec. Publ. 21.
Erba, E., Duncan, R.A., Bottini, C., Tiraboschi, D., Weissert, H., Jenkyns, H.C., Malinverno, A., 2015. Environmental consequences of Ontong Java Plateau and Kerguelen Plateau volcanism. In: Neal, C.R., Sager, W.W., Sano, T., Erba, E. (Eds.), The Origin, Evolution, and Environmental Impact of Oceanic Large Igneous Provinces, pp. 271–303. GSA Spec. Pap. 511.
Ernst, R.E., Youbi, N., 2017. How Large Igneous Provinces affect global climate, sometimes cause mass extinctions, and represent natural markers in the geological record. Palaeogeogr. Palaeoclimatol. Palaeoecol. 478, 30–52.
Ernst, R.E., Rodygin, S.A., Grinev, O.M., 2020. Age correlation of Large Igneous Provinces with Devonian biotic crises. Glob. Planet. Change 185 (103097), 1–13.

Fan, J., Shen, S., Erwin, D.E., 17 others, 2020. A high-resolution summary of Cambrian to Early Triassic marine invertebrate biodiversity. Science 367, 272–277.

Font, E., Bond, D.P.G., 2020. Volcanism and Mass Extinction. Reference Module in Earth Systems and Environmental Sciences, 1–11. https://doi.org/10.1016/b978-0-12-409548-9.12108-6.

Geldsetzer, H.H.J., Goodfellow, W.D., McLaren, D.J., 1993. The Frasnian-Famennian extinction event in a stable cratonic shelf setting: Trout River, Northwest Territories, Canada. Palaeogeogr. Palaeoclimatol. Palaeoecol. 104, 81–95.

George, A.D., Chow, N., Trinajstic, K., 2014. Oxic facies and the Late Devonian mass extinction, Canning Basin, Australia. Geology 42, 327–330.

Gereke, M., 2007. Die oberdevonische Kellwasser-Krise in der Beckenfazies von Rhenohercynikum und Saxothuringikum (spätes Frasnium/frühestes Famennium, Deutschland). Köln. For. Geol. Paläont. 17. 228 p.

Gereke, M., Schindler, E., 2012. Time-specific facies and biological crises—the Kellwasser event interval near the Frasnian/Famennian boundary (Late Devonian). Palaeogeogr. Palaeoclimatol. Palaeoecol. 367–368, 19–29.

Gharaie, M.H.M., Matsumoto, R., Racki, G., Kakuwa, Y., 2007. Chemostratigraphy of Frasnian-Famennian transition: possibility of methane hydrate dissociation leading to mass extinction. In: Monechi, S., Coccioni, R., Rampino, M. (Eds.), Large Ecosystem Perturbations: Causes and Consequences, pp. 109–125. GSA Spec. Pap. 424.

Girard, C., Cornée, J.J., Joachimski, M.M., Charruault, A.L., Dufour, A.B., Renaud, S., 2020. Paleogeographic differences in temperature, water depth and conodont biofacies during the Late Devonian. Palaeogeogr. Palaeoclimatol. Palaeoecol. 549 (108852), 1–12.

Glikson, A., 2005. Asteroid/comet impact clusters, flood basalts and mass extinctions: Significance of isotopic age overlaps, Earth Planet. Sci. Lett. 236, 933–937.

Goddéris, Y., Joachimski, M.M., 2004. Global change in the Late Devonian: modelling the Frasnian-Famennian short-term carbon isotope excursions. Palaeogeogr. Palaeoclimatol. Palaeoecol. 202, 309–329.

Gong, Y., Xu, R., Tang, Z., Si, Y., Li, B., 2005. Relationships between bacterial-algal proliferating and mass extinction in the Late Devonian Frasnian-Famennian transition: Enlightening from carbon isotopes and molecular fossils. Sci. China Ser. D, Earth Sci. 48, 1656–1665.

Goodfellow, W.D., Geldsetzer, H.H.J., McLaren, D.J., Orchard, M.J., Klapper, G., 1989. The Frasnian–Famennian extinction: current results and possible causes. In: McMillan, N.J., Embry, A.F., Glass, D.J. (Eds.), Devonian of the World, Vol. 3, pp. 9–21. Can. Soc. Petrol. Geol. Mem. 14.

Grard, A., François, L.M., Dessert, C., Dupré, B., Goddéris, Y., 2005. Basaltic volcanism and mass extinction at the Permo-Triassic boundary: environmental impact and modeling of the global carbon cycle. Earth Planet Sci. Lett. 234, 207–221.

Grossman, E.L., 2012. Applying oxygen isotope paleothermometry in deep time. In: Ivany, L.C., Huber, B.T. (Eds.), Reconstructing Earth's Deep-Time Climate—The State of the Art in 2012, pp. 39–67. Paleontological Society Short Course, November 3, 2012. Paleontological Society Papers 18.

Hartenfels, S., Becker, R.T., 2009. Timing of the global Dasberg Crisis—implications for Famennian eustasy and chronostratigraphy. In: Over, D.J. (Ed.), Studies in Devonian Stratigraphy: Proceedings of the 2007 International Meeting of the Subcommission on Devonian Stratigraphy and IGCP 499, pp. 71–97. Palaeontogr. Am. 63.

Hartenfels, S., Becker, R.T., 2016. Famennian sedimentation, faunas, and event stratigraphy at Effenberg Quarry (Remscheid-Altena Anticline, Rhenish Massif). Münster. Forsch. Geol. Paläont. 108, 141–157.

Hartenfels, S., Becker, R.T., Aboussalam, Z.S., 2016. Givetian to Famennian stratigraphy, Kellwasser, Annulata and other events at Beringhauser Tunnel (Messinghausen Anticline, eastern Rhenish Massif). Münst. Forsch. Geol. Paläont. 108, 196–219.

Hayes, J.M., Popp, B.N., Takigiku, R., Johnson, M.W., 1989. An isotopic study of biogeochemical relationships between carbonates and organic carbon in the Greenhorn Formation. Geochim. Cosmochim. Acta 53, 2961–2972.

Hayes, J.M., Strauss, H., Kaufmann, A.J., 1999. The abundance of 13C in marine organic matter and isotopic fractionation in the global biogeochemical cycle of carbon during the past 800 Ma. Chem. Geol. 161, 103–125.

Herrmann, A.D., Macleod, K.G., Leslie, S.A., 2010. Did a volcanic mega-eruption cause global cooling during the Late Ordovician? Palaios 25, 831–836.

Hesselbo, S.P., Gröcke, D.R., Jenkyns, H.C., Bjerrum, C.J., Farrimond, P., Morgans Bell, H.S., Green, O.R., 2000. Massive dissociation of gas hydrate during a Jurassic oceanic anoxic event. Nature 406, 392–395.

Hillbun, K., Playton, T.E., Tohver, E., Ratcliffe, K., Trinajstic, K., Roelofs, B., Caulfield-Kerney, S., Wray, D., Haines, P., Hocking, R., Katz, D., Montgomery, P., Ward, P., 2015. Upper Kellwasser carbon isotope excursion pre-dates the F–F boundary in the Upper Devonian Lennard Shelf carbonate system, Canning Basin, Western Australia. Palaeogeogr. Palaeoclimatol. Palaeoecol. 438, 180–190.

Hillbun, K., Playton, T.E., Katz, D., Toher, E., Trinajstic, K., Haines, P., Hocking, R., Roelofs, B., Montgomery, P., 2016. Correlation and sequence stratigraphic interpretation of Upper Devonian carbonate slope facies using carbon isotope chemostratigraphy, Lennard shelf, Canning Basin, western Australia. In: Playton, T.E., Kerans, C., Weissenberger, J.A.W. (Eds.), New Advances in Devonian Carbonates: Outcrop Analogs, Reservoirs, and Chronostratigraphy, pp. 302–318. SEPM Spec. Publ. 107.

Holser, W.T., Schönlaub, H.P., Attrep, M.J., Boeckelmann, K., Klein, P., Magaritz, M., Mauritsch, H., Pak, E., Schramm, J.M., Stattegger, K., Schmoller, R., 1989. A unique geochemical record at the Permian/Triassic boundary. Nature 337, 39–44.

Horton, F., 2015. Did phosphorus derived from the weathering of large igneous provinces fertilize the Neoproterozoic ocean? Geochem. Geophys. Geosyst. 16, 1723–1738.

House, M.R., 1985. Correlation of mid-Palaeozoic ammonoids evolutionary events with global sedimentary pertubations. Nature 313, 17–22.

House, M.R., 2002. Strength, timing, setting and cause of mid-Palaeozoic extinctions. Palaeogeogr. Palaeoclimatol. Palaeoecol. 181, 5–25.

House, M.R., Kirchgasser, W.T., 1993. Devonian goniatite biostratigraphy and timing of facies movements in the Frasnian of eastern North America. In: Hailwood, E.A., Kidd, R.B. (Eds.), High Resolution Stratigraphy, pp. 267–292. Geol. Soc. Spec. Publ. 70.

House, M.R., Menner, V.V., Becker, R.T., Klapper, G., Ovnatanova, N.S., Kuźmin, V., 2000. Reef episodes, anoxia and sea-level changes in the Frasnian of the southern Timan (NE Russian platform). In: Insalaco, E., Skelton, P.W., Palmer, T.J. (Eds.), Carbonate Platform Systems: Components and Interactions, pp. 147–176. Geol. Soc., Lond. Spec. Publ. 178.

Hu, X., Li, J., Han, Z., Li, Y., 2020. Two types of hyperthermal events in the Mesozoic-Cenozoic: environmental impacts, biotic effects, and driving mechanisms. Sci. China Earth Sci. 63, 1041–1058. https://doi.org/10.1007/s11430-019-9604-4.

Huang, C., Michael, M.J., Gong, Y.M., 2018a. Did climate changes trigger the Late Devonian Kellwasser Crisis? Evidence from a high-resolution conodont record from South China. Earth Planet. Sci. Lett. 495, 174–184.

Huang, C., Song, J., Shen, J., Gong, Y.M., 2018b. The influence of the Late Devonian Kellwasser events on deep-water ecosystems: evidence from palaeontological and geochemical records from South China. Palaeogeogr. Palaeoclimatol. Palaeoecol. 504, 60–74.

Immenhauser, A., Kenter, J.A.M., Ganssen, G., Bahamonde, J.R., Van Vliet, A., Saher, M.H., 2002. Origin and significance of isotope shifts in Pennsylvanian carbonates (Asturias, NW Spain). J. Sediment. Res. 72, 82–94.

Immenhauser, A., Della Porta, G., Kenter, J.A.M., Bahamonde, J.R., 2003. An alternative model for positive shifts in shallow-marine carbonate $\delta^{13}C$ and $\delta^{18}O$. Sedimentology 50, 953–959.

Immenhauser, A., Holmden, C., Patterson, W.P., 2008. Interpreting the carbon-isotope record of ancient shallow epeiric seas: lessons from the recent. In: Holmden, C., Pratt, B.R. (Eds.), Dynamics of Epeiric Seas: Sedimentological, Paleontological and Geochemical Perspectives, pp. 137–174. GAC Spec. Publ. 48.

Izokh, O.P., Izokh, N.G., Ponomarchuk, V.A., Semenova, D.V., 2009. Carbon and oxygen isotopes in the Frasnian-Famennian section of the Kuznetsk basin (southern West Siberia). Russian Geol. Geophys. 50, 610–617.

Jenkyns, H.C., Dickson, A.J., Ruhl, M., van den Boorn, S.H.J.M., 2017. Basalt–seawater interaction, the Plenus Cold Event, enhanced weathering and geochemical change: deconstructing Oceanic Anoxic Event 2 (Cenomanian–Turonian, Late Cretaceous). Sedimentology 64, 16–43.

Jerram, D.A., Svensen, H.H., Planke, S., Polozov, A.G., Torsvik, T.H., 2016. The onset of flood volcanism in the north-western part of the Siberian traps: explosive volcanism versus effusive lava flows. Palaeogeogr. Palaeoclimatol. Palaeoecol. 441, 38–50.

Joachimski, M.M., 1994. Subaerial exposure and deposition of shallowing upward sequences: evidence from stable isotopes of Purbeckian peritidal carbonates (basal Cretaceous), Swiss and French Jura Mountains. Sedimentology 41, 805–824.

Joachimski, M.M., 1997. Comparison of organic and inorganic carbon isotope patterns across the Frasnian-Famennian boundary. Palaeogeogr. Palaeoclimatol. Palaeoecol. 132, 133–145.

Joachimski, M.M., Buggisch, W., 1993. Anoxic events in the late Frasnian—causes of the Frasnian-Famennian faunal crisis? Geology 21, 675–678.

Joachimski, M.M., Buggisch, W., 2002. Conodont apatite $\delta^{18}O$ signatures indicate climatic cooling as a trigger of the Late Devonian mass extinction. Geology 30, 711–714.

Joachimski, M.M., Ostertag-Henning, C., Pancost, R.D., Strauss, H., Freeman, K.H., Littke, R., Damsté, J.S., Racki, G., 2001. Water column anoxia, enhanced productivity and concomitant changes in $\delta^{13}C$ and $\delta^{34}S$ across the Frasnian-Famennian boundary (Kowala—Holy Cross Mountains/Poland). Chem. Geol. 175, 109–131.

Joachimski, M.M., Pancost, R.D., Freeman, K.H., Ostertag-Henning, C., Buggisch, W., 2002. Carbon isotope geochemistry of the Frasnian-Famennian transition. Palaeogeogr. Palaeoclimatol. Palaeoecol. 181, 91–109.

Joachimski, M.M., Van Geldern, R., Breisig, S., Buggisch, W., Day, J., 2004. Oxygen isotope evolution of biogenic calcite and apatite during the Middle and Late Devonian. Int. J. Earth Sci. 93, 542–553.

Joachimski, M.M., Breisig, S., Buggisch, W., Talent, J.A., Mawson, R., Gereke, M., Morrow, J.R., Day, J., Weddige, K., 2009. Devonian climate and reef evolution: insights from oxygen isotopes in apatite. Earth Planet. Sci. Lett. 284, 599–609.

Johnson, J.G., Klapper, G., Sandberg, C.A., 1985. Devonian eustatic fluctuations in Euramerica. GSA Bull. 96, 567–587.

Jones, M.T., Jerram, D.A., Svensen, H.H., Grove, C., 2016. The effects of large igneous provinces on the global carbon and sulphur cycles. Palaeogeogr. Palaeoclimatol. Palaeoecol. 441, 4–21.

Kaiho, K., Yatsu, S., Oba, M., Gorjan, P., Casier, J.-G., Ikeda, M., 2013. A forest fire and soil erosion event during the Late Devonian mass extinction. Palaeogeogr. Palaeoclimatol. Palaeoecol. 392, 272–280.

Kaiser, S., 2005. Mass Extinctions, Climatic and Oceanographic Changes at the Devonian/Carboniferous Boundary. Unpublished PhD thesis, Ruhr-Universität, Bochum, Germany. 160 p.

Kaiser, S.I., Steuber, T., Becker, R.T., Joachimski, M.M., 2006. Geochemical evidence for major environmental change at the Devonian–Carboniferous boundary in the Carnic Alps and the Rhenish Massif. Palaeogeogr. Palaeoclimatol. Palaeoecol. 240, 146–160.

Kaiser, S.I., Steuber, T., Becker, R.T., 2008. Environmental change during the Late Famennian and Early Tournaisian (Late Devonian–Early Carboniferous): implications from stable isotopes and conodont biofacies in southern Europe. Geol. J. 43, 241–260.

Kaiser, S.I., Becker, R.T., Spalletta, C., Steuber, T., 2009. High-resolution conodont stratigraphy, biofacies, and extinctions around the Hangenberg Event in pelagic successions from Austria, Italy, and France. Palaeontogr. Am. 63, 99–143.

Kaiser, S.I., Becker, R.T., Steuber, T., Aboussalam, S.Z., 2011. Climate-controlled mass extinctions, facies, and sea-level changes around the Devonian–Carboniferous boundary in the eastern Anti-Atlas (SE Morocco). Palaeogeogr. Palaeoclimatol. Palaeoecol. 310, 340–364.

Kaiser, S.I., Aretz, M., Becker, R.T., 2015. The global Hangenberg Crisis (Devonian–Carboniferous transition): review of a first-order mass extinction. In: Becker, R.T., Königshof, P., Brett, C.E. (Eds.), Devonian Climate, Sea Level and Evolutionary Events, pp. 387–437. Geol. Soc., Lond. Spec. Publ. 423.

Kalvoda, J., Kumpan, T., Qie, W., Frýda, J., Bábek, O., 2019. Mercury spikes at the Devonian-Carboniferous boundary in the eastern part of the Rhenohercynian zone (central Europe) and in the South China block. Palaeogeogr. Palaeoclimatol. Palaeoecol 531 (A; 109221), 1–12.

Katz, D., Buoniconti, M., Montãnez, I., Swart, P., Eberli, G., Smith, L., 2007. Timing and local perturbations to the carbon pool in the lower Mississippian Madison Limestone, Montana and Wyoming. Palaeogeogr. Palaeoclimatol. Palaeoecol. 256, 231–253.

Kaufman, A.J., 2019. The Ediacaran-Cambrian transition: a resource-based hypothesis for the rise and fall of the Ediacara Biota. In: Sial, A.N., Gaucher, C., Ramkumar, M., Ferreira, V.P. (Eds.), Chemostratigraphy Across Major Chronological Boundaries, pp. 115–142. Geophysical Monograph 240.

Keeling, R.F., Körtzinger, A., Gruber, N., 2010. Ocean deoxygenation in a warming world. Annu. Rev. Mar. Sci. 2, 463–493.

Kemp, D.B., Coe, A.L., Cohen, A.S., Schwark, L., 2005. Astronomical pacing of methane release in the Early Jurassic period. Nature 437, 396–399.

Knoll, A.H., 2013. Systems paleobiology. GSA Bull. 125, 3–13.

Knoll, A.H., Bambach, R.K., Canfield, D.E., Grotzinger, J.P., 1996. Comparative earth history and Late Permian mass extinction. Science 273, 452–457.

Komatsu, T., Kato, S., Hirata, K., Takashima, R., Ogata, Y., Oba, M., Naruse, H., Ta, P.H., Nguyen, P.D., Dang, H.T., 2014. Devonian–Carboniferous transition containing a Hangenberg Black Shale equivalent in the Pho Han Formation on Cat Ba Island, northeastern Vietnam. Palaeogeogr. Palaeoclimatol. Palaeoecol. 404, 30–43.

Königshof, P., Savage, N.M., Lutat, P., Sardsud, A., Dopieralska, J., Belka, Z., Racki, G., 2012. Late Devonian sedimentary record of the Paleotethys Ocean—the Mae Sariang section, northwestern Thailand. J. Asian Earth Sci. 52, 146–157.

Korn, D., 2004. The mid-Famennian ammonoid succession in the Rhenish Mountains: the "*annulata* Event" reconsidered. Geol. Q. 48, 245–252.

Korn, D., 2010. Lithostratigraphy and biostratigraphy of the Kulm succession in the Rhenish Mountains. Z. Dtsch. Ges. Geowiss. 161, 431–453.

Korn, D., Weyer, D., 2003. High resolution stratigraphy of the Devonian-Carboniferous transitional beds in the Rhenish Mountains. Mitt. Museum Naturk. Berlin, Geowissensch. Reihe 6, 79–124.

Korn, D., Clausen, C.D., Belka, Z., Leuteritz, K., Luppold, F.W., Fiest, R., Weyer, D., 1994. Die Devon/Karbon-Grenze bei Drever (Rheinisches Schiefergebirge). Geol. Paläontol. Westf. 29, 97–147.

Korte, C., Ruhl, M., Palfy, J., Ullmann, C.V., Hesselbo, S.P., 2019. Chemostratigraphy across the triassic–jurassic boundary. In: Sial, C.N., Gaucher, C., Ramkumar, M., Ferreira, V.P. (Eds.), Chemostratigraphy Across Major Chronological Boundaries, pp. 185–210. Geophysical Monograph 240.

Kumpan, T., Bábek, O., Kalvoda, J., Frýda, J., Matys Grygar, T., 2014a. A high-resolution, multiproxy stratigraphic analysis of the Devonian–Carboniferous boundary sections in the Moravian Karst (Czech Republic) and a correlation with the Carnic Alps (Austria). Geol. Mag. 151, 201–215.

Kumpan, T., Bábek, O., Kalvoda, J., Matys Grygar, T., Frýda, J., 2014b. Sea-level and environmental changes around the Devonian–Carboniferous boundary in the Namur–Dinant Basin (S Belgium, NE France): a multi-proxy stratigraphic analysis of carbonate ramp archives and its use in regional and interregional correlations. Sediment. Geol. 311, 43–59.

Kumpan, T., Bábek, O., Kalvoda, J., Matys Grygar, T., Frýda, J., Becker, R.T., Hartenfels, S., 2015. Petrophysical and geochemical signature of the Hangenberg Events: an integrated stratigraphy of the Devonian–Carboniferous boundary interval in the Northern Rhenish Massif (Avalonia, Germany). Bull. Geosci. 90, 667–694.

Kurtz, A.C., Kump, L.R., Arthur, M.A., Zachos, J.C., Paytan, A., 2003. Early Cenozoic decoupling of the global carbon and sulfur cycles. Paleoceanography 18, 1–14.

Kuypers, M.M.M., Blokker, P., Erbacher, J., Kinkel, H., Pancost, R.D., Schouten, S., Sinninghe Damste, J.S., 2001. Massive expansion of marine archaea during a mid-Cretaceous oceanic anoxic event. Science 293, 92–94.

Lakin, J., Marshall, J., Troth, I., Harding, I., 2016. Greenhouse to icehouse: a biostratigraphic review of latest Devonian–Mississippian glaciations and their global effects. In: Becker, R.T., Königshof, P., Brett, C.E. (Eds.), Devonian Climate, Sea Level and Evolutionary Events, pp. 439–464. Geol. Soc., Lond. Spec. Publ. 423.

Lane, H.R., Sandberg, C.A., Ziegler, W., 1980. Taxonomy and phylogeny of some Lower Carboniferous conodonts and preliminary standard post-Siphonodellazonation. Geol. Palaeont. 14, 117–164.

Lash, G.G., 2017. A multiproxy analysis of the Frasnian Famennian transition in western New York State, U.S.A. Palaeogeogr. Palaeoclimatol. Palaeoecol. 473, 108–122.

Lash, G.G., 2019. A global biogeochemical perturbation during the Middle Frasnian punctata Event: evidence from muted carbon isotope signature in the Appalachian Basin, New York State (USA). Glob. Planet. Change 177, 239–254.

Le Guerroué, E., Allen, P.A., Cozzi, A., 2006. Chemostratigraphic and sedimentological framework of the largest negative carbon isotopic excursion in Earth history: the Neoproterozoic Shuram Formation (Nafun Group, Oman). Precambr. Res. 146, 68–92.

Lescuyer, J.L., Leistel, J.M., Marcoux, E., Milési, J.P., Thiéblemont, D., 1997. Late devonian-early carboniferous peak sulphide mineralization in the Western Hercynides. Miner. Deposita 33, 208–220.

Levman, B.G., Von Bitter, P.H., 2002. The Frasnian-Famennian (mid-Late Devonian) boundary in the type section of the Long Rapids Formation, James Bay Lowlands, northern Ontario, Canada. Can. J. Earth Sci. 39, 1795–1818.

Li, X.H., Jenkyns, H.C., Wang, C.S., Hu, X.M., Chen, X., Wei, Y.H., Huang, Y.J., Cui, J., 2006. Upper Cretaceous carbon- and oxygen-isotope stratigraphy of hemipelagic carbonate facies from southern Tibet, China. J. Geol. Soc. 163, 375–382.

Lindström, S., Sanei, H., van de Schootbrugge, B., Pedersen, G.K., Lesher, C.E., Tegner, C., Heunisch, C., Dybkjær, K., Outridge, P.M., 2019. Volcanic mercury and mutagenesis in land plants during the end-Triassic mass extinction. Sci. Adv. 5 (eaaw4018), 1–13.

Liu, J., Qie, W., Algeo, T.J., Yao, L., Huang, J., Luo, G., 2016. Changes in marine nitrogen fixation and denitrification rates during the end-Devonian mass extinction. Palaeogeogr. Palaeoclimatol. Palaeoecol. 448, 195–206.

Lohmann, K.C., 1988. Geochemical patterns of meteoric diagenetic systems and their application to studies to palaeokarst. In: James, N.P., Choquette, P.W. (Eds.), Palaeokarst. Springer-Verlag, Berlin, pp. 55–80.

Loydell, D.K., 2007. Early Silurian positive $\delta^{13}C$ excursions and their relationship to glaciations, sea-level changes, and extinction events. Geol. J. 42, 531–546.

Ma, X.P., Gong, Y.M., Chen, D.Z., Racki, G., Chen, X.Q., Liao, W.H., 2016. The Late Devonian Frasnian–Famennian event in South China—patterns and causes of extinctions, sea level changes, and isotope variations. Palaeogeogr. Palaeoclimatol. Palaeoecol. 448, 224–244.

Ma, K., Hinnov, L.A., Zhang, X., Gong, Y., 2020. Astronomical time calibration of the Upper Devonian Lali section, South China. Glob. Planet. Change 193, 103267.

Małkowski, K., Racki, G., Drygant, D., Szaniawski, H., 2009. Carbon isotope stratigraphy across the Silurian–Devonian transition in Podolia, Ukraine: evidence for a global biogeochemical perturbation. Geol. Mag. 146, 674–689.

Marshall, J.D., 1992. Climatic and oceanographic isotopic signals from the carbonate rock record and their preservation. Geol. Mag. 129, 143–160.

Martinez, A., Boyer, D.L., Droser, M.L., Barrie, C., Love, G.D., 2019. A stable and productive marine microbial community was sustained through the end-Devonian Hangenberg Crisis within the Cleveland Shale of the Appalachian Basin, United States. Geobiology 17, 27–42.

Marynowski, L., Filipiak, P., 2007. Water column euxinia and wildfire evidence during deposition of the Upper Famennian Hangenberg event horizon from the Holy Cross Mountains (central Poland). Geol. Mag. 144, 569–595.

Marynowski, L., Pisarzowska, A., Rakociński, M., Derkowski, A., Środoń, J., Szaniawski, R., Cohen, A., 2017. Influence of palaeoweathering on trace metal concentrations and environmental proxies in black shales. Palaeogeogr. Palaeoclimatol. Palaeoecol. 472, 177–191.

Maslin, M.A., Swann, G.E.A., 2005. Isotopes in marine sediments. In: Leng, M.J. (Ed.), Isotopes in Paleoenvironmental Research. Springer, Dordrecht, Netherlands, pp. 227–290.

Matyja, H., Sobień, K., Marynowski, L., Stepień-Sałek, M., Małkowski, K., 2015. The expression of the Hangenberg Event (latest Devonian) in arelatively shallow-marine succession (Pomeranian Basin, Poland): the results of a multi-proxy investigation. Geol. Mag. 152, 400–428.

Matyja, H., Woroncowa-Marcinowska, T., Filipiak, P., Brański, P., Sobień, K., 2020. The Devonian/Carboniferous boundary interval in Poland: multidisciplinary studies in pelagic (Holy Cross Mountainsand Sudetes) and ramp (Western Pomerania) successions. Palaeobiodiv. Palaeoenv. https://doi.org/10.1007/s12549-020-00442-3. in press.

McClung, W.S., Cuffey, C.A., Eriksson, K.A., Terry Jr., D.O., 2016. An incised valley fill and lowstand wedges in the Upper Devonian Foreknobs Formation, central Appalachian Basin: implications for Famennian glacioeustasy. Palaeogeogr. Palaeoclimatol. Palaeoecol. 446, 125–143.

McElwain, J.C., Wade-Murphy, J., Hesselbo, S.P., 2005. Changes in carbon dioxide during an oceanic anoxic event linked to intrusion into Gondwana coals. Nature 435, 479–482.

McGhee, G.R., 1996. The Late Devonian Mass Extinction. The Frasnian-Famennian Crisis. Columbia University Press, New York. 303 p.

McGhee, G.R., 2018. Carboniferous Giants and Mass Extinction: The Late Paleozoic Ice Age World. Columbia University Press. 304 p.

McGhee, G.R., Orth, C.J., Quintana, L.R., Gilmore, J.S., Olsen, E.J., 1986. Geochemical analyses of the Late Devonian "Kellwasser Event" stratigraphic horizon at Steinbruch Schmidt (F.R.G.). In: Walliser, O.H. (Ed.), Global Bio-Events. Springer, Berlin, pp. 219–224.

McKenzie, N.R., Jiang, H.H., 2019. Earth's outgassing and climatic transitions: the slow burn towards environmental "catastrophes"? Elements 15, 325–330.

Metzger, J.G., Fike, D.A., 2013. Techniques for assessing spatial heterogeneity of carbonate $\delta^{13}C$ values: implications for craton-wide isotope gradients. Sedimentology 60, 1405–1431.

Meyers, P.A., 1994. Preservation of elemental and isotopic source identification of sedimentary organic matter. Chem. Geol. 114, 289–302.

Mii, H., Grossman, E.L., Yancey, T.E., 1999. Carboniferous isotope stratigraphies of North America; implications for Carboniferous paleoceanography and Mississippian glaciations. GSA Bull. 111, 960–973.

Mizens, G.A., Kuleshov, V.N., Stepanova, T.I., Kucheva, N.A., 2015. Evidence for global Famennian and Tournaisian geologic events in the section of an isolated carbonate platform in the eastern Urals. Russ. Geol. Geophys. 56, 1531–1543.

Morrow, J.R., Sandberg, C.A., Malkowski, K., Joachimski, M.M., 2009. Carbon isotope chemostratigraphy and precise dating of middle Frasnian (lower Upper Devonian) Alamo Breccia, Nevada, USA. Palaeogeogr. Palaeoclimatol. Palaeoecol. 282, 105–118.

Munecke, A., Samtleben, C., Bickert, T., 2003. The Ireviken Event in the lower Silurian of Gotland, Sweden—relation to similar Palaeozoic and Proterozoic events. Palaeogeogr. Palaeoclimatol. Palaeoecol. 195, 99–124.

Myrow, P.M., Strauss, J.V., Creveling, J.R., Sicard, K.R., Ripperdan, R., Sandberg, C.A., Hartenfels, S., 2011. A carbon isotopic and sedimentological record of the latest Devonian (Famennian). Palaeogeogr. Palaeoclimatol. Palaeoecol. 306, 147–159.

Myrow, P.M., Hanson, A., Phelps, A.S., Creveling, J.R., Strauss, J.V., Fike, D.A., Ripperdan, R.L., 2013. Latest Devonian (Famennian) global events in western Laurentia: variations in the carbon isotopic record linked to diagenetic alteration below regionally extensive unconformities. Palaeogeogr. Palaeoclimatol. Palaeoecol. 386, 194–209.

Myrow, P.M., Ramezani, J., Hanson, A.E., Bowring, S.A., Racki, G., Rakociński, M., 2014. High-precision U–Pb age and duration of the latest Devonian (Famennian) Hangenberg event, and its implications. Terra Nova 26, 222–229.

Myrow, P.M., Cole, D., Johnston, D.T., Fike, D.A., Hakim, A., 2015. Passive transgression: remarkable preservation and spatial distribution of uppermost Devonian (Famennian) marginal and nearshore marine facies and fauna of Western Laurentia. Palaios 30, 490–502.

Oehlert, A.M., Swart, P.K., 2019. Rolling window regression of $\delta^{13}C$ and $\delta^{18}O$ values in carbonate sediments: implications for source and diagenesis. Deposit. Rec. 5, 613–630.

Panchuk, K.M., Holmden, C.E., Leslie, S.S., 2006. Local controls on carbon cycling in the Ordovician Midcontinent region of North America with implications for carbon isotope secular curves. J. Sediment. Res. 76, 200–211.

Paproth, E., 1986. An introduction to a field trip to the Late Devonian outcrops in the northern Rhenishes Schiefergebirge (Federal Republic of Germany). Ann. Soc. Géol. Belgique 109, 275–284.

Paschall, O., Carmichael, S., Königshof, P., Waters, J., Ta, H.P., Komatsu, T., Dombrowski, A., 2019. The Devonian-Carboniferous boundary in Vietnam: sustained ocean anoxia with a volcanic trigger for the Hangenberg Crisis? Glob. Planet. Change 175, 64–81.

Patterson, M.V., Francis, D., 2013. Kimberlite eruptions as triggers for early Cenozoic hyperthermals. Geochem. Geophys. Geosyst. 14, 448–456.

Payne, J.L., Kump, L.R., 2007. Evidence for recurrent Early Triassic massive volcanism from quantitative interpretation of carbon isotope fluctuations. Earth Planet. Sci. Lett. 256, 264–277.

Percival, L.M.E., Davies, J.H.F.L., Schaltegger, U., De Vleeschouwer, D., Da Silva, A.C., Föllmi, K.B., 2018. Precisely dating the Frasnian–Famennian boundary: implications for the cause of the Late Devonian mass extinction. Sci. Rep. 8 (9578), 1–10.

Percival, L.M.E., Selby, D., Bond, D.P.G., Rakociński, M., Racki, G., Adatte, T., Spangenberg, J.E., Föllmi, K.B., 2019. Osmium-isotope evidence for pulses of extreme continental weathering associated with multiple Late Devonian climate perturbations. Palaeogeogr. Palaeoclimatol. Palaeoecol. 524, 240–249.

Pisarzowska, A., Racki, G., 2012. Isotopic geochemistry across the Early-Middle Frasnian transition (Late Devonian) on the South Polish carbonate shelf: a reference for the global *punctata* Event. Chem. Geol. 334, 199–220.

Pisarzowska, A., Rakociński, M., Marynowski, L., Szczerba, M., Thoby, M., Paszkowski, M., Perri, M.C., Spalletta, C., Schönlaub, H.P., Kowalik, N., Gereke, M., 2020a. Large environmental disturbances caused by magmatic activity during the Late Devonian Hangenberg Crisis. Glob. Planet. Change 188 (103155), 1–24.

Pisarzowska, A., Becker, R.T., Aboussalam, Z.S., Szczerba, M., Sobień, K., Kremer, B., Owocki, K., Racki, G., 2020b. Middlesex/*punctata* Event in the Rhenish Basin (Padberg section, Sauerland, Germany)–geochemical clues to the early-middle Frasnian global biogeochemical perturbation. Glob. Planet. Change 191 (103211), 1–14.

Playford, P.E., McLaren, D.J., Orth, C.J., Gilmore, J.S., Goodfellow, W.D., 1984. Iridium anomaly in the Upper Devonian of the Canning Basin, Western Australia. Science 226, 437–439.

Pochon, A., Poujol, M., Gloaguen, E., Branquet, Y., Cagnard, F., Gumiaux, C., Gapais, D., 2016. U-Pb LA-ICP-MS dating of apatite in mafic rocks: Evidence for a major magmatic event at the Devonian-Carboniferous boundary in the Armorican Massif (France). Am. Miner. 101, 2430–2442.

Pochon, A., Branquet, Y., Gloaguen, E., et al., 2019. A Sb$\pm$Au mineralizing peak at 360 Ma in the Variscan belt. BSGF–Earth Sci. Bull. 190 (4), 1–12.

Qie, W., Liu, J., Chen, J., Wang, X., Mii, H.-S., Zhang, X., Huang, X., Yao, L., Algeo, T.J., Luo, G., 2015. Local overprints on the global carbonate $\delta^{13}C$ signal in Devonian-Carboniferous boundary successions of South China. Palaeogeogr. Palaeoclimatol. Palaeoecol. 418, 290–303.

Racka, M., Marynowski, L., Filipiak, P., Sobstel, M., Pisarowska, A., Bond, P.G., 2010. Anoxic annulata events in the late Famennian of the Holy Cross Mountains (Southern Poland): geochemical and palaeontological record. Palaeogeogr. Palaeoclimatol. Palaeoecol. 297, 549–575.

Racki, G., 1998. Frasnian–Famennian biotic crisis: undervalued tectonic control? Palaeogeogr. Palaeoclimatol. Palaeoecol. 141, 177–198.

Racki, G., 2005. Toward understanding Late Devonian global events: few answers, many questions. In: Over, D.J., Morrow, J.R., Wignall, P.B. (Eds.), Understanding Late Devonian and Permian-Triassic Biotic and Climatic Events: Towards an Integrated Approach, pp. 5–36. Developments in Paleontology and Stratigraphy 20.

Racki, G., 2020a. A volcanic scenario for the F-F major biotic crisis and other Late Devonian global changes: more answers than questions? Glob. Planet. Change 189 (103174), 1–29.

Racki, G., 2020b. Volcanism as a prime cause of mass extinction: retrospectives and perspectives. In: Adatte, T., Bond, D.P.G., Keller, G. (Eds.), Mass Extinctions, Volcanism, and Impacts: New Developments, pp. 1–34. GSA Spec. Pap. 544.

Racki, G., Bultynck, P., 1993. Conodont biostratigraphy of the middle to upper devonian boundary Beds in the Kielce area of the Holy Cross Mts. Acta Geol. Pol. 44, 1–25.

Racki, G., Racka, M., Matyja, H., Devleeschouwer, X., 2002. The Frasnian/Famennian boundary interval in the South Polish-Moravian shelf basins: integrated event-stratigraphical approach. Palaeogeogr. Palaeoclimatol. Palaeoecol. 181, 251–297.

Racki, G., Joachimski, M.M., Morrow, J.R., 2008. A major perturbation of the global carbon budget in the Early–Middle Frasnian transition (Late Devonian). Palaeogeogr. Palaeoclimatol. Palaeoecol. 269, 127–129.

Racki, G., Rakocinski, M., Marynowski, L., Wignall, P.B., 2018. Mercury enrichments and the Frasnian-Famennian biotic crisis: a volcanic trigger proved? Geology 46, 543–546.

Racki, G., Königshof, P., Belka, Z., Dopieralska, J., Pisarzowska, A., 2019. Diverse depositional and geochemical signatures of the Frasnian-Famennian global event in western Thailand reveal Palaeotethyan vs. Western Australian geotectonic affinities. J. Asian Earth Sci. X 2 (100010), 1–23.

Railsback, L.B., Ackerly, S.C., Anderson, T.F., Cisne, J.L., 1990. Palaeontological and isotope evidence for warm saline deep waters in Ordovician oceans. Nature 343, 156–159.

Rakociński, M., Marynowski, L., Pisarzowska, A., Bełdowski, J., Siedlewicz, G., Zatoń, M., Perri, M.C., Spalletta, C., Schönlaub, H.-P., 2020. Volcanic related methylmercury poisoning as the possible driver of the end-Devonian Mass Extinction. Scientific Reports 10 (7344), 1–8.

Rampino, M.R., Caldeira, K., 2005. Major perturbation of ocean chemistry and a 'Strangelove Ocean' after the end-Permian mass extinction. Terra Nova 17, 554–559.

Rampino, M.R., Self, S., 2015. Large Igneous Provinces and biotic extinctions. In: Sigurdsson, H. (Ed.), The Encyclopedia of Volcanoes, second ed. Elsevier, Amsterdam, pp. 605–644.

Renne, P.R., Black, M.T., Zhang, Z.C., Richards, M.A., Basu, A.R., 1995. Synchrony and causal relations between Permian–Triassic boundary crises and Siberian flood volcanism. Science 169, 1413–1416.

Retallack, G.J., Huang, C.M., 2011. Ecology and evolution of Devonian trees in New York, USA. Palaeogeogr. Palaeoclimatol. Palaeoecol. 299, 110–128.

Retallack, G.J., Jahren, A.H., 2008. Methane release from igneous intrusion of coal during Late Permian extinction events. J. Geol. 116, 1–20.

Robinson, S.A., Hesselbo, S.P., 2004. Fossil-wood carbon-isotope stratigraphy of the non-marine Wealden Group (Lower Cretaceous, southern England). J. Geol. Soc. 161, 133–145.

Rosa, D.R.N., Finch, A.A., Andersen, T., Inverno, C.M.C., 2008. U-Pb geochronology of felsic volcanic rocks hosted in the Gafo Formation, South Portuguese Zone: the relationship with Iberian Pyrite Belt magmatism. Miner. Mag. 72, 1129–1144.

Rosales, I., Quesada, S., Robles, S., 2001. Primary and diagenetic isotopic signals in fossils and hemipelagic carbonates: the Lower Jurassic of northern Spain. Sedimentology 48, 1149–1169.

Rothman, D.H., Hayes, J.M., Summons, R.E., 2003. Dynamics of the neoproterozoic carbon cycle. Proc. Natl. Acad. Sci. U.S.A. 10, 8124–8129.

Sageman, B.B., Murphy, A.E., Werne, J.P., Ver Straeten, C.A., Hollander, D.J., Lyons, T.W., 2003. A tale of shales: the relative roles of production, decomposition, and dilution in the accumulation of organic-rich strata, Middle–Upper Devonian, Appalachian Basin. Chem. Geol. 195, 229–273.

Saltzman, M.R., 2002. Carbon and oxygen isotope stratigraphy of the Lower Mississippian (Kinderhookian–lower Osagean), western United States: implications for seawater chemistry and glaciation. GSA Bull. 114, 96–108.

Saltzman, M.R., 2005. Phosphorus, nitrogen, and the redox evolution of the Paleozoic oceans. Geology 33, 573–576.

Saltzman, M.R., Thomas, E., 2012. Chapter 11. Carbon isotope stratigraphy. In: Gradstein, F.M., Ogg, J.G., Schmitz, M., Ogg, G. (Eds.), The Geologic Time Scale 2012. Elsevier, Amsterdam, pp. 207–232.

Sandberg, C.A., Ziegler, W., Leuteritz, K., Brill, S.M., 1978. Phylogeny, speciation, and zonation of Siphonodella (Conodonta, Upper Devonian and Lower Carboniferous). Newslett. Stratigr. 7, 102–112.

Sandberg, C.A., Morrow, J.R., Ziegler, W., 2002. Late Devonian sea-level changes, catastrophic events, and mass extinctions. In: Koeberl, C., MacLeod, K.G. (Eds.), Catastrophic Events and Mass Extinctions: Impacts and Beyond, pp. 473–487. GSA Spec. Pap. 356.

Sarmiento, J.L., Gruber, N., 2006. Ocean Biogeochemical Dynamics. Princeton University Press, Princeton, NJ. 503 p.

Schindler, E., 1990. Die Kellwasser Krise (hohe Frasne-Stufe Ober-Devon). Göttinger Arb. Geol. Pälontol. 46, 1–115.

Schindler, E., 1993. Event-stratigraphic markers within the Kellwasser Crisis near the Frasnian/Famennian boundary (Upper Devonian) in Germany. Palaeogeogr. Palaeoclimatol. Palaeoecol. 104, 115–125.

Schobben, M., Schootbrugge, B.v.d., Wignall, P.B., 2019a. Interpreting the carbon isotope record of mass extinctions. Elements 15, 331–337.

Schobben, M., Heuer, F., Tietje, M., Ghaderi, A., Korn, D., Korte, C., Wignall, P.B., 2019b. Chemostratigraphy across the Permian-Triassic boundary: the effect of sampling strategies on carbonate carbon isotope stratigraphic markers. In: Sial, A.N., Gaucher, C., Ramkumar, M., Ferreira, V.P. (Eds.), Chemostratigraphy Across Major Chronological Boundaries. Geophysical Monograph, Vol. 240, pp. 159–181.

Schouten, S., van Kaam-Peters, H.M.E., Rijpstra, W.I.C., Schoell, M., Sinninghe Damsté, J.S., 2000. Effects of an oceanic anoxic event on the stable carbon isotopic composition of early Toarcian carbon. Am. J. Sci. 300, 1–22.

Scotese, C.R., 2000. PALAEOMAP Project. Climate history. http://www.scotese.com/ldevclim.htm.

Shen, J., Algeo, T.J., Planavsky, N.J., Yu, J.X., Feng, Q.L., Song, H.J., Song, H.Y., Rowe, H., Zhou, L., Chen, J.B., 2019. Mercury enrichments provide evidence of early Triassic volcanism following the end-Permian mass extinction. Earth Sci. Rev. 195, 191–212.

Shields, G.A., 2005. Neoproterozoic cap carbonates: a critical appraisal of existing models and the plumeworld hypothesis. Terra Nova 17, 299–310.

Shizuya, A., Oba, M., Ando, T., Ogata, Y., Takashima, R., Nishi, H., Komatsu, T., Nguyen, P.D., 2020. Variations in trace elements, isotopes, and organic geochemistry during the Hangenberg Crisis, Devonian–Carboniferous transition, northeastern Vietnam. Island Arc. 29, e12337.

Sial, A.N., Gaucher, C., Ramkumar, M., Ferreira, V.P., 2019. Chemostratigraphy as a formal stratigraphic method. In: Sial, A.N., Gaucher, C., Ramkumar, M., Ferreira, V.P. (Eds.), Chemostratigraphy Across Major Chronological Boundaries. Geophysical Monograph, Vol. 240, pp. 3–26.

Sial, A.N., Chen, J., Lacerda, L.D., Korte, C., Spangenberg, J.E., Silva-Tamayo, J.C., Gaucher, C., Ferreira, V.P., Barbosa, J.A., Pereira, N.S., Benigno, A.P., 2020. Globally enhanced Hg deposition and Hg isotopes in sections straddling the Permian–Triassic boundary: link to volcanism. Palaeogeogr. Palaeoclimatol. Palaeoecol. 540 (109537), 1–20.

Simon, L., Goddéris, Y., Werner, B., Strauss, H., Joachimski, M.M., 2007. Modeling the carbon and sulfur isotope compositions of marine sediments: climate evolution during the Devonian. Chem. Geol. 246, 19–38.

Śliwiński, M.G., Whalen, M.T., Newberry, R.J., Payne, J.H., Day, J., 2011. Stable isotope ($\delta^{13}C_{carb \& org}$, $\delta^{15}N_{org}$) and trace element anomalies during the Late Devonian 'punctata Event' in the Western Canada Sedimentary Basin. Palaeogeogr. Palaeoclimatol. Palaeoecol. 307, 245–271.

Sluijs, A., Bowen, G.J., Brinkhuis, H., Lourens, L.J., Thomas, E., 2007. The Palaeocene-Eocene Thermal maximum super greenhouse: biotic and geochemical signatures, age models and mechanisms of global change. In: Williams, M., Haywood, A.M., Gregory, F.J., Schmidt, D.N. (Eds.), Deep Time Perspectives on Climate Change: Marrying the Signal From Computer Models and Biological Proxies. Micropalaeontological Society, Special Publications, Geological Society, London, pp. 323–351.

Sobolev, S.V., Sobolev, A.V., Kuzmin, D.V., Krivolutskaya, N.A., Petrunin, A.G., Arndt, N.T., Radko, V.A., Vasiliev, Y.R., 2011. Linking mantle plumes, large igneous provinces and environmental catastrophes. Nature 477, 312–316.

Song, Y., Gilleaudeau, G.J., Algeo, T.J., Over, D.J., Lyons, T.W., Anbar, A.D., Xie, S., 2020. Biomarker evidence of algal-microbial community changes linked to redox and salinity variation, Upper Devonian Chattanooga Shale (Tennessee, USA). GSA Bull. in press https://doi.org/10.1130/B35543.1.

Spalletta, C., Perri, M.C., Over, D.J., Corradini, C., 2017. Famennian (Upper Devonian) conodont zonation: revised global standard. Bull. Geosci. 92, 31–57.

Stanley, S.M., 2016. Estimates of the magnitudes of major marine mass extinctions in Earth history. Proc. Natl. Acad. Sci. USA 113, 6325–6334.

Stephens, N.P., Sumner, D.Y., 2003. Late Devonian carbon isotope stratigraphy and sea level fluctuations, Canning Basin, Western Australia. Palaeogeogr. Palaeoclimatol. Palaeoecol. 191 (2), 203–219.

Stock, C.W., Sandberg, C.A., 2019. Latest Devonian (Famennian, expansa Zone) conodonts and sponge-microbe symbionts in Pinyon Peak Limestone, Star Range, southwestern Utah, lead to reevaluation of global Dasberg Event. Palaeogeogr. Palaeoclimatol. Palaeoecol. 534, 109271.

Strauss, H., Peters-Kottig, W., 2003. The Paleozoic to Mesozoic carbon cycle revisited: the carbon isotopic composition of terrestrial organic matter. Geochem. Geophys. Geosyst. 4 (10–1083), 1–15.

Streel, M., Caputo, M.V., Loboziak, S., Melo, J.H.G., 2000. Late Frasnian–Famennian climates based on palynomorph analyses and the question of the Late Devonian glaciations. Earth Sci. Rev. 52, 121–173.

Sundquist, E.T., Visser, K., 2011. The geologic history of the carbon cycle. In: Holland, H.D., Turekian, K.K. (Eds.), Geochemistry of Earth Surface Systems From the Treatise on Geochemistry. Elsevier & Academic Press, Amsterdam, pp. 341–387.

Svensen, H., Planke, S., Malthe-Sørenssen, A., Jamtveit, B., Myklebust, R., Eidem, T.R., Rey, S.S., 2004. Release of methane from a volcanic basin as a mechanism for initial Eocene global warming. Nature 429, 542–545.

Takahashi, S., Kaiho, K., Oba, M., Kakegawa, T., 2010. A smooth negative shift of organic carbon isotope ratios at an end-Permian mass extinction horizon in central pelagic. Panthalassa Palaeogeogr. Palaeoclimatol. Palaeoecol. 292, 532–539.

Tessalina, S.G., Jourdan, F., Belogub, E.V., 2017. Significance of Late Devonian–Lower Carboniferous ages of hydrothermal sulphides and sericites from the Urals Volcanic-Hosted Massive Sulphide deposits. Ore Geol. Rev. 85, 131–139.

Tragelehn, H., 2012. The Devonian to Lower Carboniferous of Franconia—stratigraphy, facies history and tectonics. In: Unpublished lecture. Univ, Münster.

Trela, W., Malec, J., 2007. Zapis $\delta^{13}C$ w osadach pogranicza dewonu i karbonu w południowej części Gór Świętokrzyskich. Przegl. Geol. 55, 411–415.

Uveges, B.T., Junium, C.K., Boyer, D.L., Cohen, P.A., Day, J.E., 2018. Biogeochemical controls on black shale deposition during the Frasnian-Famennian biotic crisis in the Illinois and Appalachian Basins, USA, inferred from stable isotopes of nitrogen and carbon. Palaeogeogr. Palaeoclimatol. Palaeoecol. 531A (108787), 1–14.

Van Breugel, Y., Schouten, S., Paetzel, M., Nordeide, R., Sinninghe Damste, J.S., 2005a. The impact of recycling of organic carbon on the stable carbon isotopic composition of dissolved inorganic carbon in a stratified marine system (Kyllaren fjord, Norway). Org. Geochem. 36, 1163–1173.

Van Breugel, Y., Schouten, S., Paetzel, M., Ossebaar, J., Sinninghe Damste, J.S., 2005b. Reconstruction of $\delta^{13}C$ of chemocline CO_2 (aq) in past oceans and lakes using the $\delta^{13}C$ of fossil isorenieratene. Earth Planet. Sci. Lett. 235, 421–434.

van de Schootbrugge, B., Gollner, S., 2013. Altered primary productivity during mass-extinction events. In: Bush, A.M., Pruss, S.B., Payne, J.L. (Eds.), Ecosystem Paleobiology and Geobiology, pp. 87–114. Paleont. Soc. Pap. 19.

Van Geldern, R., Joachimski, M.M., Jansen, U., Alvarez, F., Yolkin, E.A., Ma, X.-P., 2006. Carbon, oxygen and strontium isotope records of Devonian brachiopod shell calcite. Palaeogeogr. Palaeoclimatol. Palaeoecol. 240, 47–67.

Veimarn, A.B., Korneeva, G.A., 2007. Global geological events at Frasnian/Famennian boundary. Biul. Mosk. Obsch. Ispyt. Prirody, Otd. Geol. 82 (1), 48–68. In Russian.

Walliser, O.H., 1984. Pleading for a natural D/C boundary. Courier Forschungsinstitut Senckenberg 67, 241–246.

Walliser, O.H. (Ed.), 1996. Global Events and Event Stratigraphy in the Phanerozoic. Springer, Berlin. 333 p.

Wang, C., 2007. Anomalous hopane distributions at the Permian–Triassic boundary, Meishan, China—evidence for the end-Permian marine ecosystem collapse. Org. Geochem. 38, 52–66.

Wang, K., Orth, C.J., Attrep, M., Chatterton, B.D.E., Hou, H., Geldsetzer, H.H.J., 1991. Geochemical evidence for a catastrophic biotic event at the Frasnian/Famennian boundary in south China. Geology 19, 776–779.

Wang, K., Geldsetzer, H.H.J., Goodfellow, W.D., Krouse, H.R., 1996. Carbon and sulfur isotope anomalies across the Frasnian–Famennian extinction boundary, Alberta, Canada. Geology 24, 187–191.

Wang, J., Chen, D., Yan, D., Wei, H., Xiang, L., 2012. Evolution from an anoxic to oxic deep ocean during the Ediacaran–Cambrian transition and implications for bioradiation. Chem. Geol. 306–307, 129–138.

Wang, Z.-W., Pei, F.-P., Xu, W.-L., Cao, H.-H., Wang, Z.-J., 2015. Geochronology and geochemistry of Late Devonian and early Carboniferous igneous rocks of central Jilin Province, NE China: implications for the tectonic evolution of the eastern Central Asian Orogenic Belt. J. Asian Earth Sci. 97, 260–278.

Wang, Z., Becker, R.T., Aboussalam, Z.S., Hartenfels, S., Joachimski, M.M., Gong, Y.M., 2016. Conodont and carbon isotope stratigraphy near the Frasnian/Famennian (Devonian) boundary at Wulankeshun, Junggar Basin, NW China. Palaeogeogr. Palaeoclimatol. Palaeoecol. 448, 279–297.

Weissert, H., Joachimski, M., Sarnthein, M., 2008. Chemostratigraphy. Newslett. Strat. 42, 145–179.

Werner, C., Fischer, T.P., Aiuppa, A., et al., 2019. Carbon dioxide emissions from subaerial volcanic regions: two decades in review. In: Orcutt, B., Dasgupta, R., Daniel, I. (Eds.), Deep Carbon: Past to Present. Cambridge University Press, Cambridge, pp. 188–236.

Winter, J., 2015. Vulkanismus und Kellwasser-Krise—Zirkon-Tephrostratigrafie, Identifizierung und Herkunft distaler Fallout-Aschenlagen (Oberdevon, Synklinorium von Dinant, Rheinisches Schiefergebirge, Harz). Z. Dt. Ges. Geowiss. 166, 227–251.

Whalen, M., Śliwiński, M.G., Payne, J.H., Day, J., Over, J., Chen, D., Da Silva, A.C., Da Silva, A.C., et al., 2015. Chemostratigraphy and magnetic susceptibility of the Late Devonian Frasnian–Famennian transition in western Canada and southern China: implications for carbon and nutrient cycling and mass extinction. In: Magnetic Susceptibility Application: A Window onto Ancient Environments and Climatic Variations, pp. 37–72. Geol. Soc., London Spec. Publ. 414.

Whalen, M., De Vleeschouwer, D., Payne, J.H., Day, J., Over, J., Claeys, P., 2017. Pattern and timing of the Late Devonian biotic crisis in Western Canada: insights from carbon isotopes and astronomical calibration of magnetic susceptibility data. In: Playton, T.E., Kerans, C., Weissenberger, J.A.W. (Eds.), New Advances in Devonian Carbonates: Outcrop Analogs, Reservoirs, and Chronostratigraphy, pp. 185–201. SEPM Spec. Publ. 107.

Wilde, P., Berry, W., 1984. Destabilization of the oceanic density structure and its significance to marine "extinction" events. Palaeogeogr. Palaeoclimatol. Palaeoecol. 48, 143–162.

Wong, K., Mason, E., Brune, S., East, M., Edmonds, M., Zahirovic, S., 2019. Deep carbon cycling over the past 200 million years: a review of fluxes in different tectonic settings. Front. Earth Sci. 7 (263), 1–22.

Xie, S.C., Pancost, R.D., Huang, X.Y., Jiao, D., Lu, L.Q., Huang, J.H., et al., 2007. Molecular and isotopic evidence for episodic environment change across the Permo/Triassic boundary at Meishan in South China. Glob. Planet. Change 55, 56–65.

Xu, B., Gu, Z., Wang, C., Hao, Q., Han, J., Liu, Q., Wang, L., Lu, Y., 2012. Carbon isotopic evidence for the associations of decreasing atmospheric CO_2 level with the Frasnian-Famennian mass extinction. J. Geoph. Res. Biogeosci. 117 (G01032), 1–12.

Yan, Z., Hou, H.F., Ye, L.F., 1993. Carbon and oxygen isotope event markers near the Frasnian-Famennian boundary, Luoxiu section, South China. Palaeogeogr. Palaeoclimatol. Palaeoecol. 104, 97–104.

Yang, S., 2019. Reconstruction of Local and Global Marine Paleoredox Conditions during Deposition of the Devonian-Mississippian Exshaw Formation (Black Shale Member) and the Evaluation of Hydrocarbon Maturation Effects. Unpublished Master thesis, University of Waterloo, Ontario, Canada.

Yans, J., Corfield, R.M., Racki, G., Preat, A., 2007. Evidence for perturbation of the carbon cycle in the Middle Frasnian punctata Zone (Late Devonian). Geol. Mag. 144, 263–270.

Yudina, A.B., Racki, G., Savage, N.S., Racka, M., Małkowski, K., 2002. The Frasnian–Famennian events in deep-shelf succession, Subpolar Urals: biotic, depositional and geochemical records. Acta Palaeont. Pol. 47, 355–372.

Zachos, J.C., Arthur, M.A., Dean, W.E., 1989. Geochemical evidence for suppression of pelagic marine productivity at the Cretaceous/Tertiary boundary. Nature 337, 61–64.

Zeng, Z., Pike, M., Tice, M.M., Kelly, C., Marcantonio, F., Xu, G., Maulana, I., 2018. Iron fertilization of primary productivity by volcanic ash in the Late Cretaceous (Cenomanian) Western Interior Seaway. Geology 46, 859–862.

Zhang, N., Xia, W.C., Dong, Y.X., Shang, H.J., 2008. Conodonts and radiolarians from pelagic cherts of the Frasnian–Famennian boundary interval at Bancheng, Guangxi, China: global recognition of the upper Kellwasser event. Mar. Micropaleont. 67, 180–190.

Zhang, X.S., Joachimski, M.M., Over, D.J., Ma, K.Y., Huang, C., Gong, Y.M., 2019. Late Devonian carbon isotope chemostratigraphy: a new record from the off-shore facies from South China. Glob. Planet. Change 182 (103024), 1–18.

Zhang, L., Chen, D., Huang, T., Yu, H., Zhou, X., Wang, J., 2020a. An abrupt oceanic change and frequent climate fluctuations across the Frasnian–Famennian transition of Late Devonian: constraints from conodont Sr isotope. Geol. J. 55, 4479–4492.

Zhang, F., Dahl, T.W., Lenton, T.M., Luo, G., Shen, S.Z., Algeo, T.J., Planavsky, N., Liu, J., Cui, Y., Qie, W., Romaniello, S.J., Anbar, A.D., 2020b. Extensive marine anoxia associated with the Late Devonian Hangenberg Crisis. Earth Planet. Sci. Lett. 533, 115976.

Zhuravlev, A.V., Sobolev, D.B., 2019. Devonian-Carboniferous boundary in the East of the Pechora Plate (Kamenka River and Vangyr River sections). Vestnik IG Komi SC UB RAS 10, 16–22.

Ziegler, W., Sandberg, C.A., 1984. Palmatolepis-based revision of upper part of standard Late Devonian conodont zonation. In: Clark, D.L. (Ed.), Conodont Biofacies and Provincialism, pp. 179–194. GSA Spec. Pap. 196.

Further reading

Berner, R.A., 2004. The Phanerozoic Carbon Cycle: CO_2 and O_2. Oxford Univ, Press, Oxford, p. 150.

Brune, S., Williams, S.E., Müller, R.D., 2017. Potential links between continental rifting, CO_2 degassing and climate change through time. Nat. Geosci. 10, 941–946.

Oppenheimer, C., 2011. Volcanic degassing. In: Holland, H.D., Turekian, K.K. (Eds.), Geochemistry of Earth Surface Systems From the Treatise on Geochemistry. Elsevier & Academic Press, Amsterdam, pp. 1–44.

Orcutt, B., Dasgupta, R., Daniel, I. (Eds.), 2020. Deep Carbon: Past to Present. Cambridge University Press, Cambridge, p. 669.

Racki, G., 2012. The Alvarez impact theory of mass extinction; limits to its applicability and the "great expectations syndrome". Acta Palaeont. Pol. 57, 681–702.

Ruhl, M, Kürschner, W.M., 2011. Multiple phases of carbon cycle disturbance from large igneous province formation at the Triassic-Jurassic transition. Geology 39, 431–434.

Wieczorek, R., Fantle, M.S., Kump, L.R., Ravizza, G., 2013. Geochemical evidence for volcanic activity prior to and enhanced terrestrial weathering during the Paleocene Eocene Thermal Maximum. Geochim. Cosmochim. Acta 119, 391–410.

CHAPTER NINE

Carbon isotope stratigraphy of the Tournaisian (Lower Mississippian) successions of NE Europe

Andrey V. Zhuravlev*, Artem N. Plotitsyn, Denis A. Gruzdev, and Irina V. Smoleva
N.P. Yushkin Institute of Geology Komi SC UrB RAS, Syktyvkar, Russia
*Corresponding author: e-mail address: micropalaeontology@gmail.com

Contents

Abstract

Biostratigraphically constrained $\delta^{13}C_{carb}$ variations in the Tournaisian successions of NE Europe (Komi Republic, Russia) are considered. Eight reference sections, characterizing platforms, slopes, and shelf depression facies, comprise the basis of this study. Of 412 studied samples of bulk carbonates, 268 samples passed all of the screening tests and were used to analyze the carbonate carbon isotope record. The established regional $\delta^{13}C_{carb}$ data allowed the recognition of the distinct, terminal Famennian and mid-Tournaisian carbon isotope events and two minor positive shifts in the sulcata Zone and Upper duplicata—lowermost quadruplicata interval. The age of the isotope shifts is underpinned by conodont biostratigraphy. These isotope events and shifts are traced through the facies gradient from the shallow-water carbonate platform to deep-water

Stratigraphy & Timescales, Volume 5
ISSN 2468-5178
https://doi.org/10.1016/bs.sats.2020.08.007

shelf depression. Because the regional $\delta^{13}C_{carb}$ variations match those known in different regions (Europe, South China, and North America), the recognized C-isotope shifts can be considered to be reliable isotope-stratigraphic markers for both regional and world-wide correlation.

1. Introduction

Variations in the isotopic carbon composition of sedimentary carbonates are usefsul proxies for paleoecological perturbations (changes in carbon cycle) and can be used as reliable stratigraphic markers (Saltzman and Thomas, 2012). The Tournaisian was a time of the post-Hangenberg biotic recovery and transfer from the green-house to ice-house climate of the Late Paleozoic Ice Age (Lakin et al., 2016; Mii et al., 1999). This time period corresponds to the C1 isotopic stage described by Mii et al. (1999). The climate changes, which started in the Late Famennian as the initial phases of the glaciation in Gondwana (Caputo et al., 2008; Lakin et al., 2016; Qie et al., 2019), affected global circulation, ocean chemistry, biota, and sedimentation. Changes in the climate led to significant perturbations in the carbon cycle (Qie et al., 2011; Saltzman, 2002; Saltzman et al., 2004; Yao et al., 2015). These dramatic fluctuations are reflected in the carbon isotope record and are known as the Hangenberg Event (HICE), Mid-Aikuanian Event (TICE) and two episodes in the Early and Middle Tournaisian, named P1 and P2 (Qie et al., 2011).

Carbonate carbon isotope variations have been well-studied around the Devonian/Carboniferous boundary and preceding Hangenberg Event (Buggisch and Joachimski, 2006; Kaiser et al., 2006, 2008; Qie et al., 2011, 2019). A positive shift in $\delta^{13}C_{carb}$ (HICE) is characteristic of the terminal Famennian (Cramer et al., 2008). This twofold shift corresponds to the Middle-Upper praesulcata Zones (Buggisch and Joachimski, 2006; Kaiser et al., 2006, 2008; Kumpan et al., 2014; Lakin et al., 2016; Qie et al., 2019). Continuously decreasing $\delta^{13}C_{carb}$, interrupted by short-term positive anomalies, can be observed in the lower Tournaisian sulcata through the sandbergi conodont Zones (Bruckschen and Veizer, 1997; Kaiser et al., 2008; Qie et al., 2011). The positive trend in the Middle Tournaisian was detected worldwide and is terminated by a huge, positive carbon isotope excursion near the Middle-Upper Tournaisian boundary (Upper crenulata—isosticha and Lowe typicus conodont Zones) (Cole et al., 2015; Liu et al., 2018; Saltzman, 2002). This excursion (TICE) was one of the largest in the Phanerozoic (Bruckschen and Veizer, 1997; Cheng et al., 2020; Saltzman, 2002; Saltzman et al., 2004; Yao et al., 2015).

The Tournaisian carbon isotope trends were distinguished in different regions (Buggisch and Joachimski, 2006; Kaiser et al., 2006, 2008; Qie et al., 2011, 2019; Saltzman, 2002; Saltzman and Thomas, 2012; Saltzman et al., 2004). Generally, manifestations of the terminal Famennian—Tournaisian isotope excursions are well studied in North America, western and central Europe and China, however, much less is known in eastern Europe (e.g., Erofeevskiy and Zhuravlev, 2019; Saltzman et al., 2004; Sobolev, 2011).

Some data on the carbon isotope stratigraphy of the north-eastern Europe region (Pechora Plate, North of Urals and Pai-Khoi) were considered earlier. Information regarding carbonate carbon isotope composition has been mentioned with respect to the Devonian-Carboniferous boundary in the Kozhim River section (Sobolev, 2011; Sobolev et al., 2002), the Syvyu River section (Plotitsyn et al., 2014), the Izyayol River section (Plotitsyn and Gruzdev, 2016), the Kamenka River section (Zhuravlev and Sobolev, 2019), and the Vangyr River section (Sobolev, 2011; Zhuravlev and Sobolev, 2019). The TICE isotope event was detected in the Kozhim River section and Izyayu River section (Saltzman et al., 2004) and was later traced in the Kochmes-1 borehole and Kozhva River section (Zhuravlev and Vevel, 2018; Zhuravlev et al., 2019). In the Malaya Usa River section, changes in the carbon isotope composition were described with respect to the Mid-Tournaisian (Lower Alum Shale) Event (Plotitsyn et al., 2018).

The first isotope stratigraphic framework of the Tournaisian of the region was elaborated by Sobolev (2011) on the basis of the Kozhim River and Vangyr River sections. Sobolev mentioned two positive excursions in $\delta^{13}C_{carb}$ in the lower part of the Tournaisian but a generally negative trend up to the base of the quadruplicata conodont Zone. Recent revision of the conodont biostratigraphy of the Kozhym River section (Plotitsyn and Vevel, 2019) demonstrated that both of the Tournaisian positive isotope shifts mentioned by Sobolev (2011) correspond to the sulcata conodont Zone and probably reflect the P1 isotope excursion described by Qie et al. (2011) and/or diagenetic overprints.

The present study is focused on the compilation of the biostratigraphically constrained regional carbonate carbon isotope record for NE Europe (Pechora Plate and North of Urals), that occupied the north-eastern part of the Laurussia paleocontinent in Mississippian times.

2. Geological setting

The studied sections are located in North-Eastern Europe (Komi Republic, Russia) and paleogeographically correspond to the shelf basin of North-Eastern Laurussia (Fig. 1). The sections comprise the stratigraphic interval from the terminal Famennian through to the lower part of the Upper Tournaisian (Fig. 2). The biostratigraphy of the sections is dominated by conodonts, their key species are illustrated in Fig. 3.

2.1 Lithology and biostratigraphy of the shallow-water carbonate platform successions

2.1.1 Kamenka River and Kozhva River sections

The shallow-water, clayey-carbonate facies of the Tournaisian outcrop in the southern part of the Pechora-Kozhva Uplift (Fig. 1). A number of sections are located on the banks of the Kamenka River and Kozhva River, situated c. 25 km south-west of the town of Pechora. The Famennian-Mississippian sections in this area have been known since the middle of the twentieth century and were studied with respect to their lithology, facies, isotope stratigraphy, and foraminifer and conodont biostratigraphy (Tchernov, 1960; Vevel et al., 2012; Zhuravlev, 2017; Zhuravlev and Sobolev, 2019; Zhuravlev et al., 2019).

The terminal Famennian—Middle Tournaisian sections are situated on the banks of the Kamenka River (loc. 121, N 65° 04′ 27.4″ E 56° 42′ 50.9″; loc. 122, N 65° 04′ 21.6″ E 56° 43′ 13.7″; locs. 111A and 111, N 65° 01′ 40.3″ E 56° 42′ 09.3″) (Fig. 4) and the Middle-Upper Tournaisian sections are located on the banks of the Kozhva River (loc. KV4, N 65° 06′ 26.9″ E 56° 41′ 03.1″ and loc. KV3, N 65° 05′ 49.35″ E 56°41′ 47.98″) (Fig. 5).

The terminal Famennian and Lower Tournaisian interval, approximately corresponding to the praeluclata—sandbergi Zones, is represented by light-gray, wavy laminated, bioclastic limestones (mainly pack and wackestones) intercalated with gray, limey clays. The Middle Tournaisian (crenulata—isosticha Zones) is similar in lithology and demonstrates a gradual upward decrease in clay content. The upper part of the Tournaisian probably corresponds to the typicus—anchoralis Zones and comprises bioclastic limestones, grading into clayey limestones and terrigenous unit that are poorly exposed and only known in boreholes.

The biostratigraphy of the sections under consideration is based on the regional, shallow-water conodont zonation of the uppermost Famennian

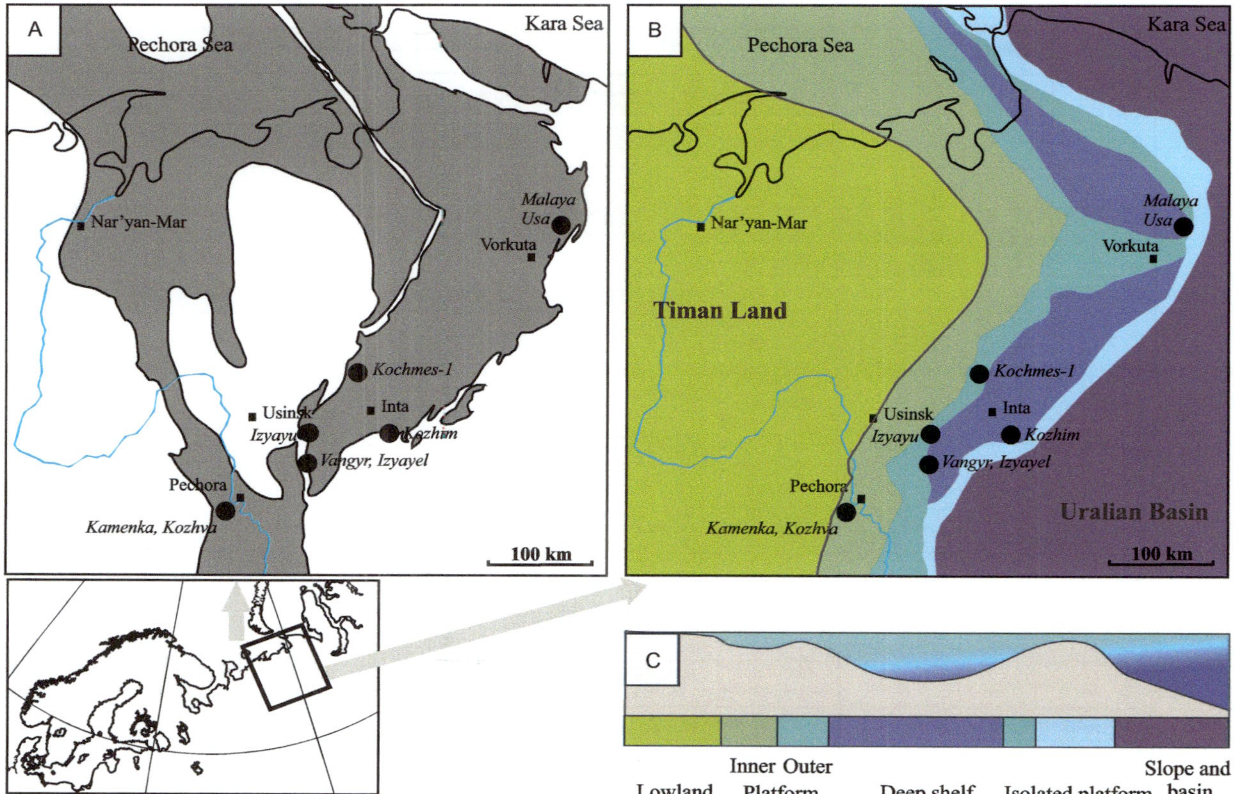

Fig. 1 Location of studied sections. (A) Scheme of recent distribution of the Tournaisian in the Pechora Plate and North of Urals. (B) Paleogeographic map of the region. (C) Facies profile for the paleogeographic map.

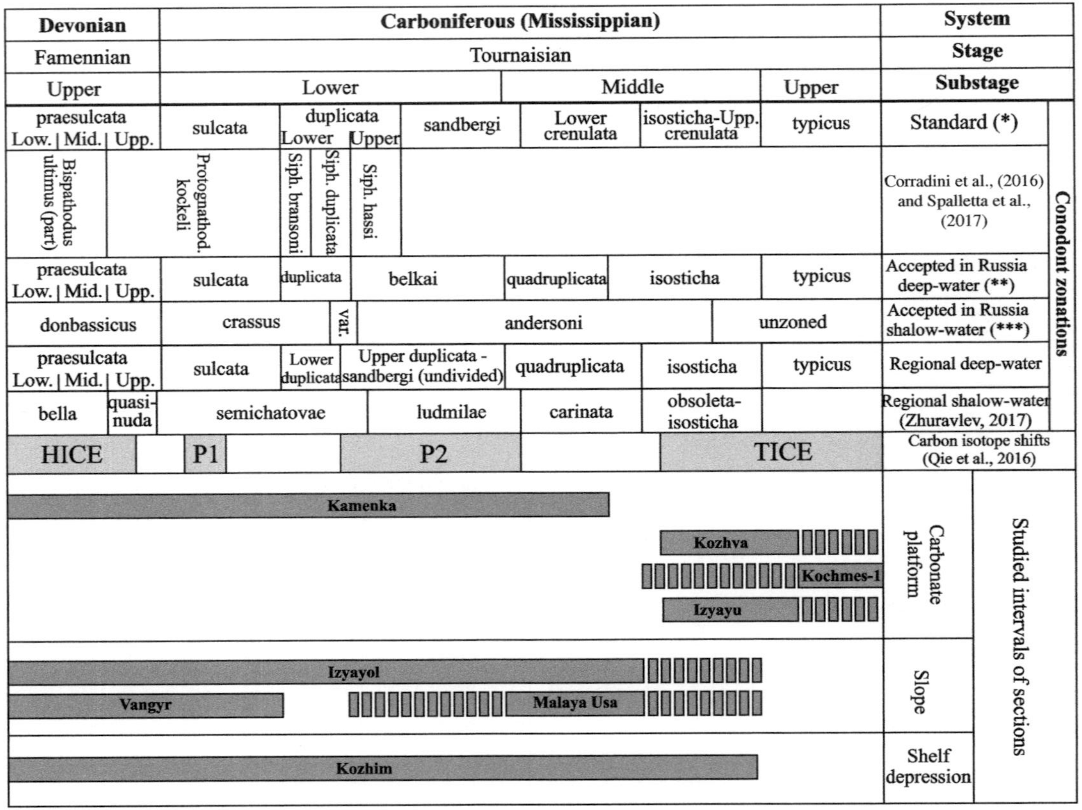

Fig. 2 Correlation of biostratigraphic units, global carbon isotope shifts, and ranges of studied sections. * Sandberg et al. (1978); ** Sostoyanie izuchennosti… (2008); *** Barskov et al. (1984).

Fig. 3 The stratigraphically important conodont species from the uppermost Famennian—Tournaisian successions of NE Europe. 1—*Siphonodella praesulcata* Sandberg, 1972, specimen 512/1-25, sample S-40, Syvyu River section, Subpolar Urals; 2—*Siphonodella sulcata* (Huddle, 1934), specimen 512/2-46, sample 58g/14, Kozhim River section, Subpolar Urals; 3—*Siphonodella duplicata* (Branson et Mehl, 1934),

(Continued)

through to the Middle Tournaisian, which is correlated with the standard conodont zonation (Zhuravlev, 2017). The latest Famennian part of the succession is composed of the Siphonodella bella and Siphonodella quasinuda regional conodont zones (Zhuravlev, 2017). The base of the Siphonodella bella Zone is defined by the first appearance datum (FAD) of the index species. Conodont associations in the zone include typical Famennian forms, such as *Pseudopolygnathus graulichi* Bouckaert et Groessens and *Polygnathus taxophorus* Cooper; numerous, long-living polygnathids and spathognathids also occur, including *Neopolygnathus communis* (Branson et Mehl) and *Polygnathus parapetus* Druce. The lower boundary of the Siphonodella quasinuda Zone corresponds to the FAD of the index species. The conodont associations in the zone include the Famennian forms, such as *Pseudopolygnathus graulichi*, *Mehlina strigosa* (Branson et Mehl), and *Polygnathus taxophorus* Cooper. These species become extinct at the top of the zone. Transient Famennian-Tournaisian taxa, such as *Pandorinellina nota* Kononova et Migdisova, *Siphonodella bella* Kononova et Migdisova, and *Siphonodella praesulcata* Sandberg, occur as well.

The Lower Tournaisian part of the succession comprises Siphonodella semichatovae and Siphonodella ludmilae regional conodont zones (Zhuravlev, 2017). The base of the Siphonodella semichatovae Zone coincides

Fig. 3—Cont'd specimen 512/7-175, sample GER3-100, Izyayol River section, Cis-Urals; 4—*Siphonodella wilberti* Bardasheva, Bardashev, Weddige, Ziegler, 2004, specimen 512/5-72, sample GER4-1, Izyayol River section, Cis-Urals; 5—*Siphonodella belkai* Dzik, 1997, specimen 512/6-14, sample 81a/14, Kozhim River section, Subpolar Urals; 6—*Siphonodella quadruplicata* (Branson et Mehl, 1934), specimen 512/4-53, sample 80o/14, Kozhim River section, Subpolar Urals; 7—*Siphonodella nandongensis* Li, 2014, specimen 512/1-39, sample mu3-13, Malaya Usa River section, Polar Urals; 8—*Siphonodella crenulata* (Cooper, 1939), specimen 512/4-31, sample 85d/14, Kozhim River section, Subpolar Urals; 9—*Siphonodella isosticha* (Cooper, 1939), specimen 512/6-35, sample 95b/14, Kozhim River section, Subpolar Urals; 10—*Protognathodus kockeli* (Bischoff, 1957), specimen 512/7-123, sample GER3-92f, Izyayol River section, Cis-Urals; 11—*Siphonodella bella* Kononova et Migdisova, 1984, sample 121-13/16, Kamenka River section, Pechora Plate; 12—*Siphonodella quasinuda* Gagiev, Kononova et Pazuhin, 1987, sample 121-15/16, Kamenka River section, Pechora Plate; 13—*Siphonodella semichatovae* Kononova et Lipnjagov, 1976, specimen 512/8-20, sample 121-12/11, Kamenka River section, Pechora Plate; 14—*Siphonodella ludmilae* Zhuravlev et Plotitsyn, 2017, specimen 703/17, sample 121-7/18, Kamenka River section, Pechora Plate; 15—*Siphonodella carinata* Zhuravlev, 2018, specimen 517/2, sample 122-6/16, Kamenka River section, Pechora Plate; 16—*Pseudopolygnathus graulichi* Bouckaert et Groessens, 1977, specimen 512/8-16, sample 121-4/16, Kamenka River section, Pechora Plate.

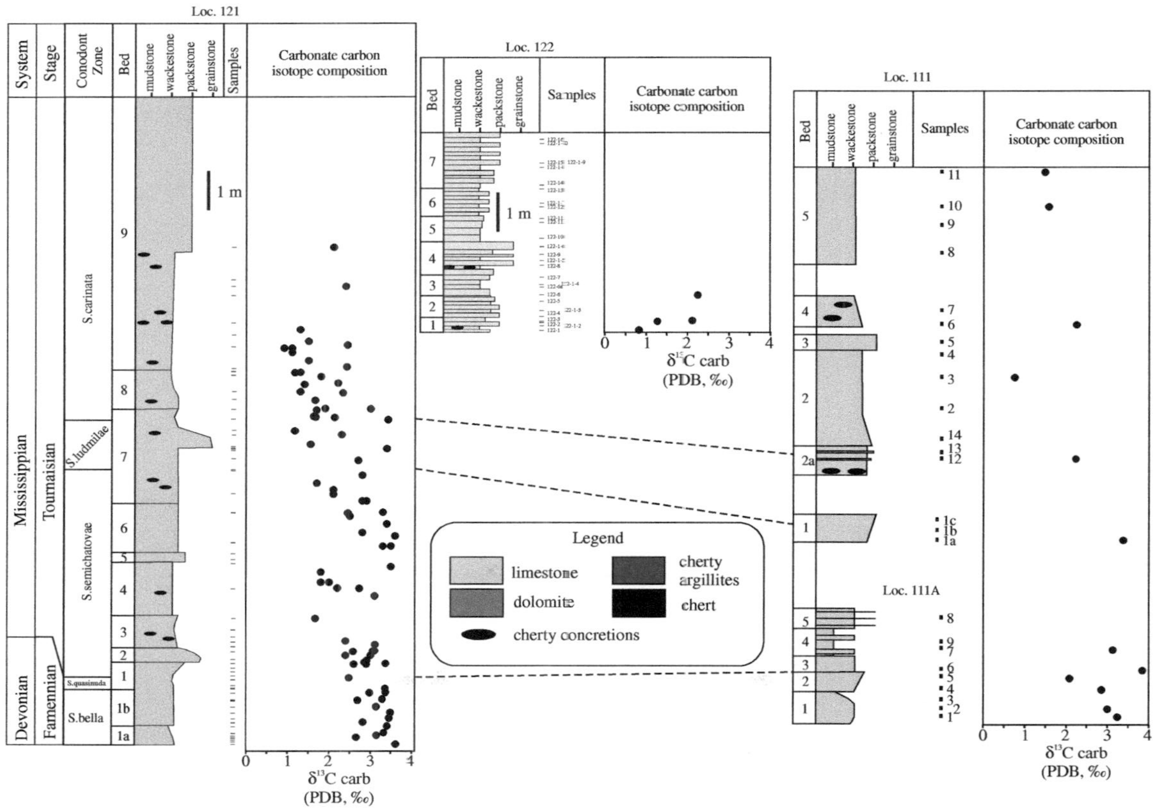

Fig. 4 Logs and screened carbon isotope data of the Kamenka River sections.

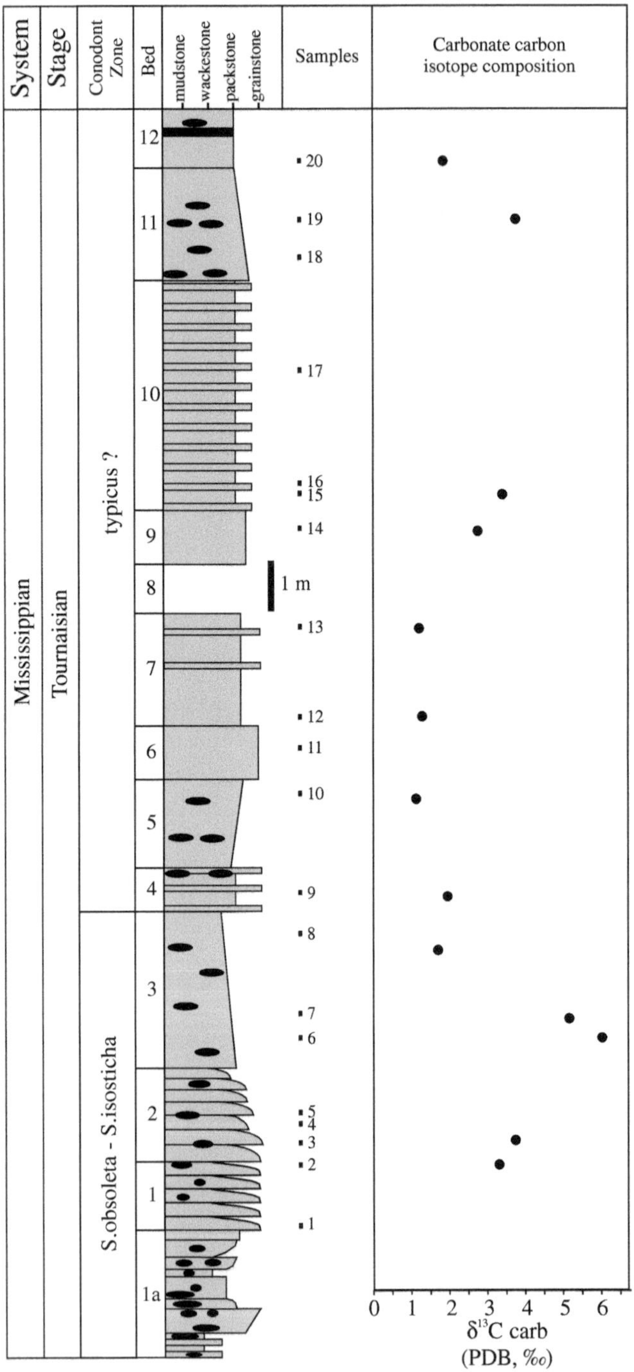

Fig. 5 Log and screened carbon isotope data of the Kozhva River section. See Fig. 4 for legend.

with the Devonian-Carboniferous boundary. It is marked by FADs of the index species, *Siphonodella sulcata* (Huddle), and *Patrognathus crassus* Kononova et Migdisova. Conodont associations of the zone include *Siphonodella bella*, *S. quasinuda* Gagiev, Kononova et Pazuhin, *S. sulcata*, the Tournaisian representatives of *Patrognathus*, as well as *Neopolygnathus communis* (Branson et Mehl), *Polygnathus parapetus* Druce, *Hindeodus penescitulus* Rexroad et Collinson, *H. crassidentatus* (Branson et Mehl), *Ligonodina discreta* (Austin et Husri), etc. This zone is correlated with the Siphonodella sulcata and Lower Siphonodella duplicata standard zones (Zhuravlev, 2017). The lower boundary of Siphonodella ludmilae Zone is defined by the FAD of the index species. Conodont associations of the zone are similar to those of the Siphonodella semichatovae Zone, excluding the appearance of *Patrognathus andersoni* Klapper. The zone is correlated with the Upper Siphonodella duplicata and Siphonodella sandbergi standard conodont zones (Zhuravlev, 2017).

The Middle Tournaisian is represented by the Siphonodella carinata regional conodont Zone and unzoned intervals with rare *Siphonodella obsoleta* Hass and *S. isosticha* (Cooper). The base of the Siphonodella carinata Zone corresponds to the FAD of the index species and the appearance of *Hindeodus cristulus* (Youngquist et Miller) (Zhuravlev, 2017). Conodont associations of the zone are close to those of the Siphonodella ludmilae Zone. The Siphonodella carinata regional conodont Zone correlates with the Lower crenulata standard conodont Zone (Zhuravlev, 2017). Most of the shallow-water siphonodellids became extinct at the upper limit of the Siphonodella carinata Zone. The overlaying part of the Middle Tournaisian, cropping out in the Kozhva River section, contains poor conodont associations, composed of rare *Neopolygnathus communis* (Branson et Mehl), *Siphonodella obsoleta* Hass, *S.* aff. *isosticha* (Cooper), *Polygnathus inornatus* Branson, *P. lacinatus* Huddle, *Bispathodus stabilis* (Branson et Mehl), and *Hindeodus scitulus* (Hinde). The Middle-Upper Tournaisian transition is marked by the disappearance of the last siphonodellids. The unzoned interval is roughly correlated with the Upper crenualata—isosticha Zone and lower parts of the typicus Zone (Zhuravlev et al., 2019).

Diagenetic changes in the Kamenka River succession are quite weak. They are represented by recrystallization of the micritic components of the rock, mainly developed around the clay layers. Locally scattered silicification occurs as well.

2.1.2 Izyayu River section and the Kochmes-1 borehole

Additional data on the Middle-Upper Tournaisian shallow-water successions were collected in the Izyayu River section (Saltzman et al., 2004) and borehole Kochmes-1 (Zhuravlev and Vevel, 2018). These successions are represented by light-gray, wavy laminated pack and grainstones, which lack conodonts but contain numerous foraminifera and calcareous algae (Zhuravlev and Vevel, 2018). The presence of foraminifera *Chernyshinella glomiformis* (Lipina), *Eoforschia gigantea minoris* (Lipina), large *Tournayella*, and green algae *Kamaena delicata* Antropov, *K. minuta* R. Ivanova, *Exvotarisella index* (Ehrenberg, emend. Moeller), *Crassikamaena foraminosa* Brenckle, *Aphanocapites granulosus* Maslov allows rough correlation with the Upper crenualata—isosticha Zone and typicus Zone (Zhuravlev and Vevel, 2018). The deposits demonstrate low-grade diagenetic alterations, including local recrystallization and scattered silicification.

2.2 Lithology and biostratigraphy of the deep-water shelf successions

2.2.1 Kozhim River section (central part of the shelf depression)

The Kozhim River key section is located on the right bank of the Kozhim River (Subpolar Urals) approximately 40 km SSE of the town of Inta (N65° 41′10″ E59° 46′3″) (Fig. 6). The Upper Famennian part (= expansa and the lower part of the praesulcata conodont zones) of the Kozhim key section is lithologically characterized by gray, nodular, slightly wavy, massive and horizontally-layered micritic limestones, and dark gray or black, siliceous argillites (up to 6% of TOC). A few beds of litho and bioclastic limestones with gradation fabrics (calciturbidites) are noted. The Devonian/Carboniferous boundary is located between Famennian siliceous and calcareous-siliceous, black argillites (interpreted as "Hangenberg black shales", see Sedaeva et al., 2010) and massive, wavy-laminated Tournaisian carbonate sandstones, comprising four gradation cyclites. The black shale part of the section is conditionally compared with the lower part of the Middle praesulcata Zone. The role of injection of turbidite flows of carbonate material increases in the Lower and Middle Tournaisian (the sulcata-isosticha zones). As a result, a predominance of calciturbidites among the carbonate rocks is noted. From the base of the isosticha Zone, the role of clayey and siliceous material in sedimentation increases. The Upper Tournaisian (upper part of the isosticha Zone, typicus and anchoralis conodont zones) is represented by thin-bedded, dark gray or black shales (calcareous argillites) containing layers of clayey and cherty limestones

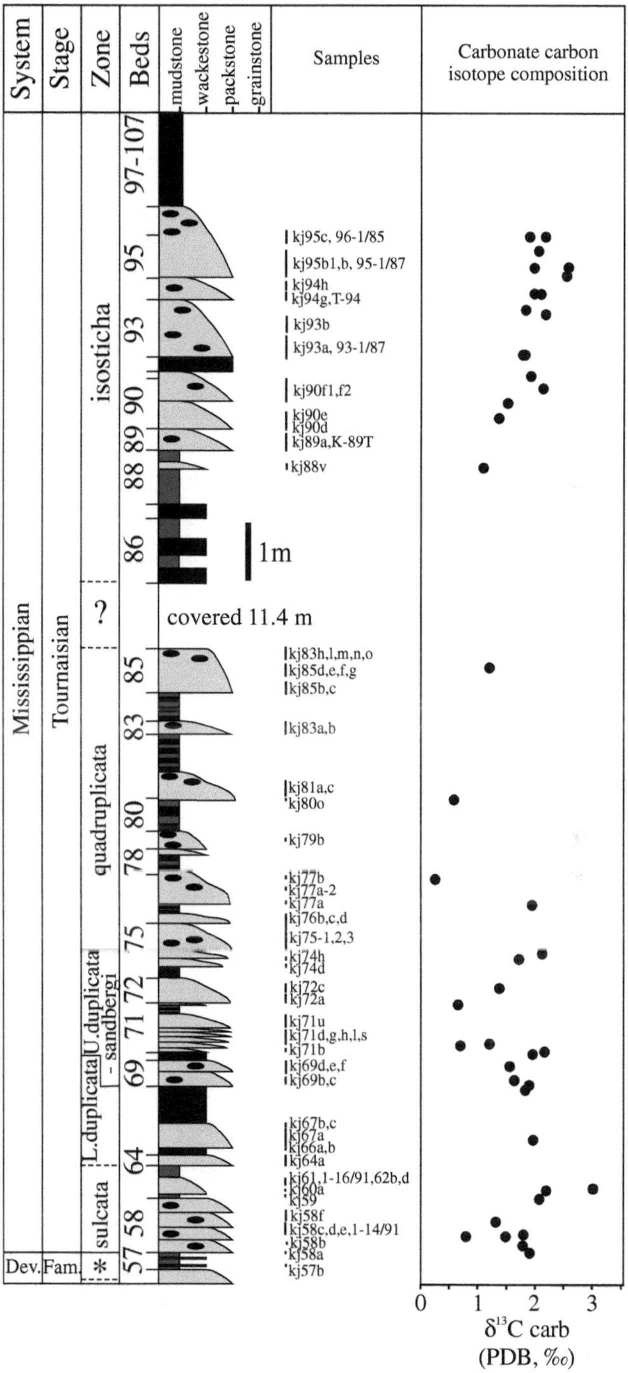

Fig. 6 Log and screened carbon isotope data of the Kozhim River section. See Fig. 4 for legend. * Praesulcata Zone.

and cherts (for more detailed lithological information, see Sobolev et al., 2002 and Plotitsyn and Vevel, 2019). Diagenetic modifications of the rocks are represented by strong silicification (including cherty concretions), carbonate recrystallization, and pyritization. Lithologically similar, deep-water successions were previously described in the Syvyu River section (Plotitsyn et al., 2014).

The most detailed biostratigraphic study of the Kozhim River key section was carried out on the basis of conodonts. It should be mentioned that conodonts from the Kozhim River key section have been studied for more than 40 years (Chermnykh et al., 1980, 1988; Deulin, 2006; Nemirovskaya et al., 1992; Sobolev et al., 2000, 2002; Zhuravlev, 2012). The latest revision work on the biostratigraphy of the Late Famennian-Middle Tournaisian conodonts of the section was carried out recently (Plotitsyn and Vevel, 2019). Currently, in the Upper Famennian-Middle Tournaisian part of the section, the following conodont zones are recognized: expansa, sulcata, Lower duplicata, Upper duplicata-sandbergi (undivided), quadruplicata, and isosticha. Terminal Famennian is provisionally correlated with the Lower and partly with the Middle praesulcata zones. The absence of the upper part of the Middle and Upper praesulcata zones is explained by a gap at the Devonian and Carboniferous boundary. The conodont associations of the terminal Famennian are extremely scarce and mainly represented by the *Palmatolepis gracilis* group, *Bispathodus stabilis* (Branson et Mehl), *Bispathodus aculeatus* group, and some surviving *Branmehla*, *Mehlina*, and *Polygnathus*. The Lower Tournaisian corresponds to the sulcata-sandbergi conodont zones. Finds of the index species *Siphonodella sulcata* (Huddle), *Siphonodella duplicata* (Branson et Mehl), and *Siphonodella wilberti* (Bardasheva, Bardashev, Weddige, Ziegler) allow us to distinguish the lower boundaries of the sulcata (= D/C boundary), Lower duplicata, and Upper duplicata-sandbergi conodont zones. The Middle Tournaisian quadruplicata and isosticha conodont zones are distinguished by the appearance of the species Siphonodella quadruplicata and by the extinction of all representatives of the genus *Siphonodella* (with the exception of *Siphonodella isosticha*), respectively. The conodont associations in the Lower-Middle Tournaisian part of the section are mainly composed of numerous representatives of *Siphonodella*, *Polyganthus*, *Pseudopolygnathus*, and neopolygnatids (for more information, see Plotitsyn and Vevel, 2019). The upper part of the Tournaisian is characterized by sporadic findings of the conodonts *Gnathodus typicus* Cooper and *Scaliognathus anchoralis* Branson et Mehl, which mark the typicus and anchoralis conodont zones. The Tournaisian-Visean boundary is marked by the FAD of *Mestognathus beckmanni* Bischoff (Sobolev et al., 2002).

2.2.2 Izyayol River and Vangyr River sections (inner slope of the shelf depression)

The Tournaisian Izyayol and Vangyr River sections are located approximately 70 to 75 km ENE of the town of Pechora, along the eastern tributaries of Kosyu and Vangyr rivers, respectively (Izyayol River section, N65° 21′53.4″ E58° 41′26.6″; Vangyr River section, N65° 07′15″ E58° 43′31″) (Figs. 7, 8). The sections are geographically separated (the distance between the sections is more than 20 km) and belong to different local tectonic structures. However, in the Tournaisian they were formed in the same facies belt, and, as a result, have identical lithological composition and structure.

The terminal Famennian (upper part of the expansa Zone—Upper praesulcata Zone) is represented by an alternation of dark gray, thin-bedded, clayey limestones and massive, gray, micritic limestones. In the Middle praesulcata to Late praesulcata conodont zones, an increase in the portion of allochthonous carbonate material is noted. As a result, there was an accumulation of lithoclastic and bioclastic, poorly sorted limestones of a turbiditic nature. Pelagic deposits are represented by thin alternations of dark gray, very clayey and siliceous limestones and black argillites. Earlier, traces of the transgressive and regressive phases of the Hangenberg event (HBS and HSS events) were detected here (Plotitsyn and Gruzdev, 2016). A decrease in the size of the debris material in the calciturbidites is noted in the lower Tournaisian part of the sections (sulcata—Lower duplicata zones). At the base of the Tournaisian, coarser and slightly sorted clastic material forms limestones and sandstones, which are overlain by thin, indistinct gradation cyclites, represented by elements of d and e in the Bouma sequence. In the Upper duplicata and quadruplicata zones, regressions are marked by beds of calciturbidites and fluxoturbidites, which are composed of clastic limestones or carbonate gravelites and conglobreccias. The transgressive phases are marked by finer clastic gradation cyclites. In the Upper Tournaisian (presumably from the isosticha Zone up to the base of the Visean?), the role of clayey material in sedimentation significantly increased. This part of the section is represented by argillites with lenses and thin layers of clayey limestones. The lower and upper parts of the section contain concretions of gray cherts and scattered silicification. Evidence of the recrystallization of micritic components and pyrite micro-concretions are common in the succession.

In the Upper Famennian—Middle Tournaisian part of the Vangyr and Izyayol sections, the sequence of conodont zones is recognized from the expansa Zone through to the quadruplicata Zone. The most detailed version of the biostratigraphy is obtained in the Izyayol River section. The lower boundaries of the expansa-Middle praesulcata, Upper praesulcata, sulcata,

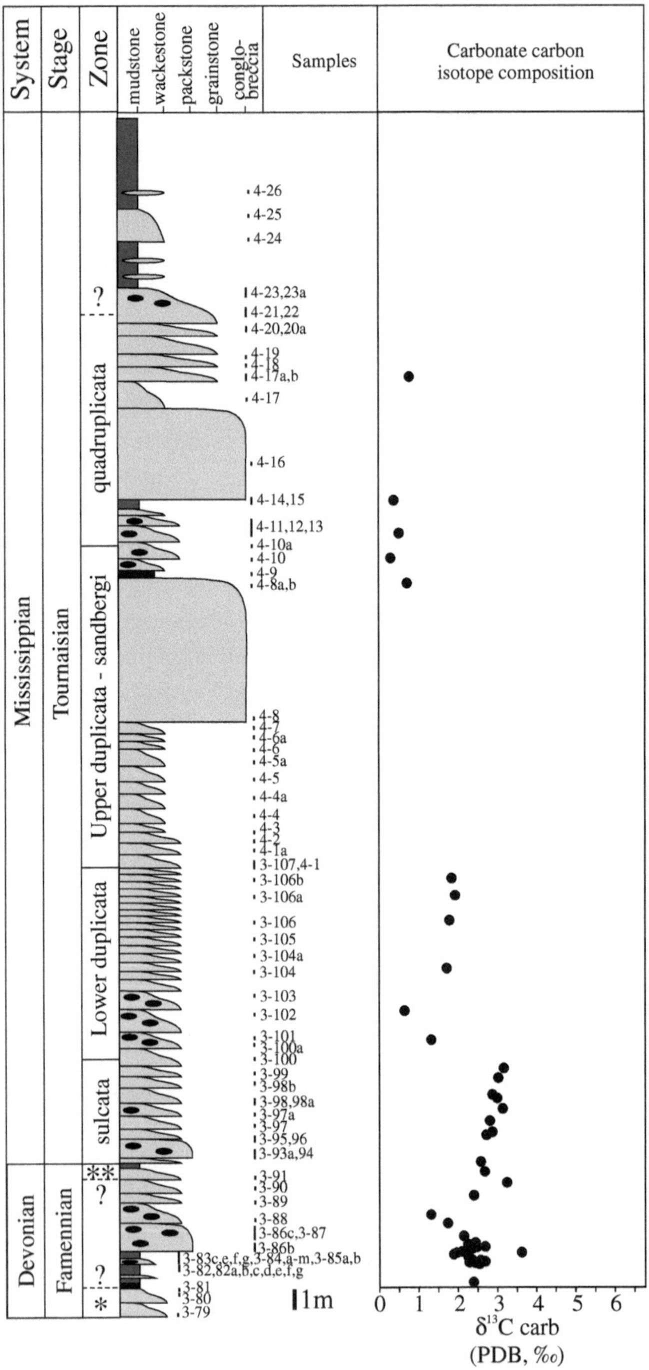

Fig. 7 Log and screened carbon isotope data of the Izyayol River section. See Fig. 4 for legend.

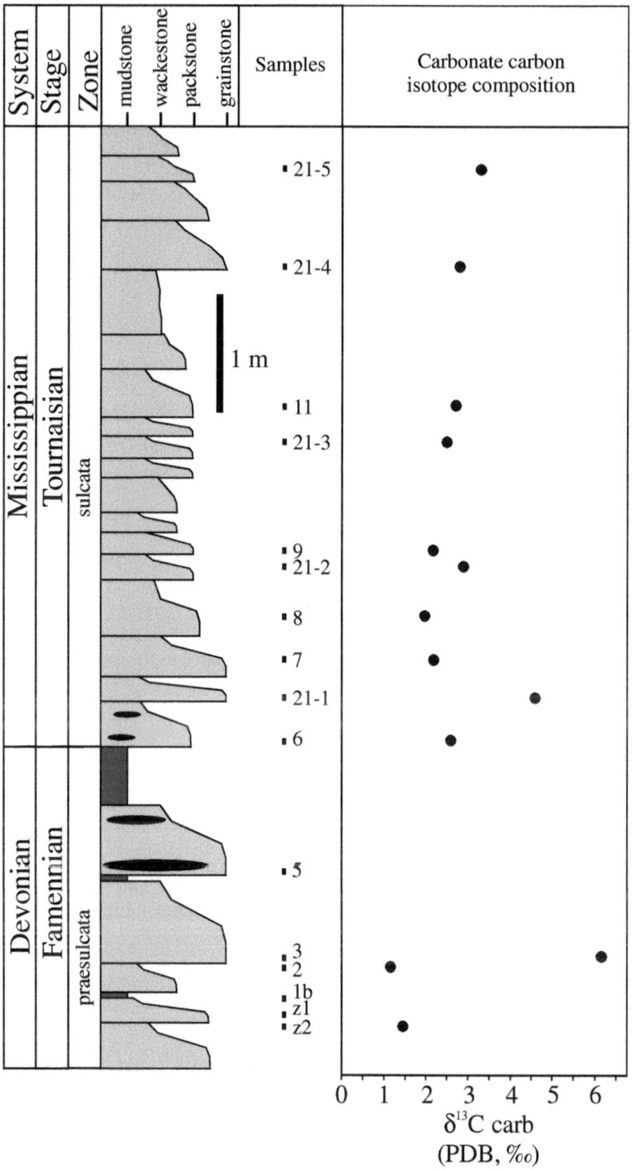

Fig. 8 Log and screened carbon isotope data of the Vangyr River section. See Fig. 4 for legend.

Lower duplicata, Upper duplicata-sandbergi, and quadruplicata zones are characterized by conodonts. The terminal Famennian yields rich conodont associations, composed of *Palmatolepis*, *Bispathodus*, *Neopolyganthus*, and *Protognathodus* (including *Prot. meischneri* Ziegler, *Prot. collinsoni* Ziegler, and *Prot. kockeli* Bischoff). The lower part of the Tournaisian (sulcata and Lower duplicata zones) is characterized by rich conodont associations too. These associations are represented by numerous early siphonodellids (*S. sulcata* (Huddle), *S. bransoni* Ji, *S. duplicata* (Branson et Mehl)), neopolygnatids, and bispathodids (*Bispthodus aculeatus aculeatus* (Branson et Mehl), *B. stabilis* (Branson et Mehl) M1, *B. stabilis* (Branson et Mehl) M2). The boundaries of the conodont zones are justified by FADs of the index species. The upper part of the Lower Tournaisian (Upper duplicata-sandbergi zones) and the Middle-Upper Tournaisian (quadruplicata Zone and the interval without conodonts) contain sporadic findings of conodonts. The findings only made it possible to establish the bases of the Upper duplicata Zone (FAD of *S. wilberti*) and quadruplicata Zone (FADs of *S. quadruplicata* (Branson et Mehl) and *S. lanei* Zhuravlev et Plotitsyn). For more detailed information on the conodont biostratigraphy of the Izyayol and Vangyr sections, see Plotitsyn and Gruzdev (2016), Sobolev and Zhuravlev (1994) and Zhuravlev and Sobolev (2019).

2.2.3 Malaya Usa River section (gentle slope of the shelf depression)

The Malaya Usa River section is located approximately 50 km NE of the town of Vorkuta (loc. mu2, N67° 39′42″ E65° 11′14″; mu3 N67° 39′47″ E65° 10′59″) (Fig. 9). The Famennian-Tournaisian interval crops out in scattered outcrops, which have been crushed into a series of tectonic folds. The Tournaisian part of the section comprises Upper duplicata-quadruplicata zones. Most of the deposits have been strongly recrystallized and are represented by heterocrystalline dolomites. Unaltered deposits are predominantly represented by mudstones, packstones and wackestones with a massive or gently undulating fabric. In the middle part of the Middle Tournaisian (approximately at the level of the FAD of *Siphonodella crenulata* (Cooper)), a transgressive member of dark gray, clayey limestones and dark gray to black, calcareous argillites is noted. This member is interpreted as representing traces of the transgressive phase of the Lower Alum shale event (Plotitsyn et al., 2018). The deposits cropping out in the Malaya Usa sections demonstrate signs of tectonic reworking, recrystallization of the carbonate rocks, and dolomitization.

The lower part of the section (lower part of the loc. mu2) is characterized by sporadic conodonts of a wide stratigraphic range (*Siphonodella sulcata* (Huddle), *Neopolyganthus dentatus* (Druce), *N. carinus* (Hass), *Hindeodus crassidentatus* (Branson et Mehl), and *Bispathodus stabilis* (Branson et Mehl)). These conodonts make it possible to approximately correlate this interval

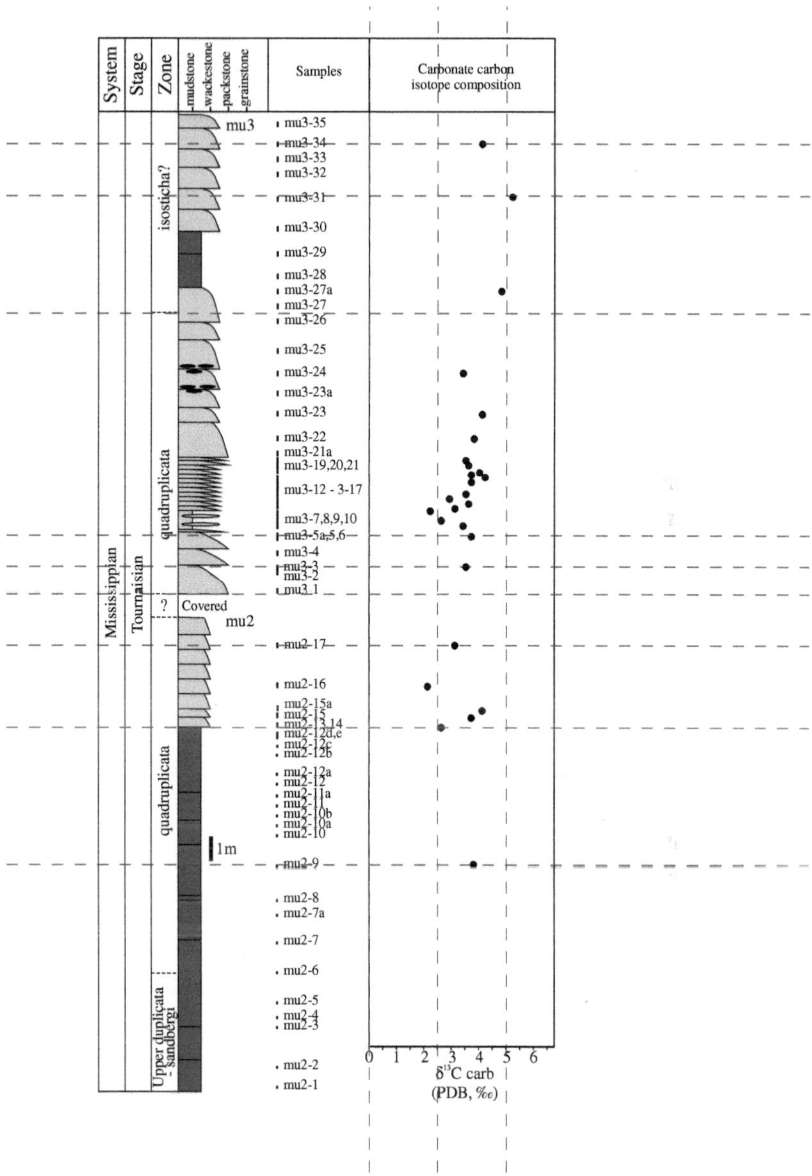

Fig. 9 Logs and screened carbon isotope data of the Malaya Usa River sections. See Fig. 4 for legend.

of the section with the Upper duplicata-sandbergi conodont zones. Most parts of the section (upper part of the loc. mu2 and loc. mu3) contain conodont associations, which are characteristic of the quadruplicata Zone. These associations contain *Siphonodella quadruplicata* (Branson et Mehl),

S. nandongensis Li, *S. gladia* Zhuravlev et Plotitsyn, *S. crenulata* (Cooper), and *Hindeodus scitulus* (Hinde), among others.

3. Methods

The bulk-rock samples were collected from fresh limestones with a stratigraphical spacing of decimeters to meters (Figs. 4–9). A total of 412 bulk-rock samples were analyzed for carbon and oxygen-stable isotopes (Appendix).

The carbonate powder for isotope analysis was extracted from fresh surfaces of rock samples with a steel microdrill, mainly from the micritic or fine-detritic matrix.

The carbon and oxygen isotope composition of the carbonates was studied with a DELTA V Advantage mass spectrometer with sample preparation on a Gas Bench II line by standard methods. $\delta^{13}C_{carb}$ values were reported relative to the PDB standard and $\delta^{18}O_{carb}$ values were reported relative to the SMOW standard. The precision of the $\delta^{13}C_{carb}$ value is ±0.04‰ and precision of the $\delta^{18}O_{carb}$ value is ±0.06‰. Isotope analysis was performed at the CKP "Geonauka" of the N.P. Yushkin Institute of Geology Komi SC UrB RAS (Syktyvkar, Russia).

A number of screening tests were used for evaluating the reliability of the isotope record (Brand et al., 2011). Screening tests for carbonate bulk-rock data included the following:

1. Visual examination of the samples. Fresh surfaces of samples were used to drilling out the carbonate powder for analyses.
2. Optical examination in thin and polished sections. Samples demonstrating significant recrystallization of the micritic component of the rock, as well as those having strong silicification were used with caution.
3. Elemental distribution. Some samples were analyzed for Sr content and the micromapping of elemental distributions of S, Ca, Mg, and Fe was carried out using a VEGA TESCAN microprobe (N.P. Yushkin Institute of Geology Komi SC UrB RAS). High contents of Mg, S, and Fe, and a low content of Sr are considered to be signs of diagenetic and/or epigenetic changes (Brand et al., 2011; Mii et al., 1999).
4. Distribution of carbon and oxygen-stable isotopes. A composite screening diagram, based on the works of Lohmann (1988), Immenhauser et al. (2003), Qie et al. (2011), Chen et al. (2016), and Huck et al. (2017), was used (Fig. 10). Samples located in the doubtful area of the diagram were excluded from the following analyses.

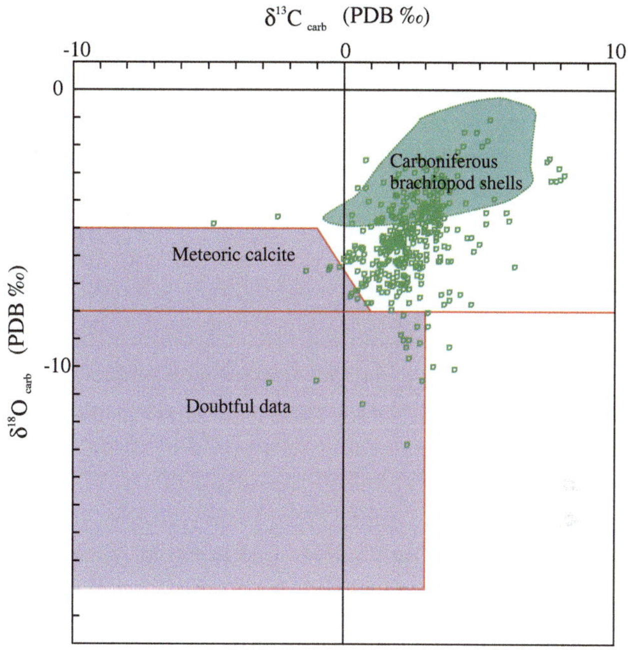

Fig. 10 Bivariate plots of the bulk carbonate $\delta^{13}C_{carb}$ and $\delta^{18}O_{carb}$ in the studied sections. The isotope data are in PDB notation.

5. Organic matter content. A high organic carbon content in carbonates (>1.7%, corresponding to a C_{org}/C_{carb} ratio less than 1:7) may lead to the incorporation of carbonates of ^{12}C derived from oxidized organic matter (Scholle and Arthur, 1980). This process potentially decreases the $\delta^{13}C_{carb}$ value. Samples containing TOC of more than 1.7% were used with caution.

If independent evidence indicated that postdepositional effects had been severe enough to alter the isotopic compositions, then a sample is considered as being doubtful. These samples, marked in Appendix, were excluded from the following analyses of the carbon isotope data.

4. Results

All 412 of the samples studied passed the first screening test, based on visual examination. The samples were distributed as follows: Kamenka River sections—115; Kozhva River sections—31; Izyayol River section—75; Vangyr River section—15; Malaya Usa River sections—62; and Kozhim River section—106. Optical examination in thin and polished sections allowed the exclusion of 141 samples, due to strong silicification, dolomitization, and/or recrystallization of the micritic component. Most of the samples failing this test were from the Kozhim River section and Malaya Usa River sections (50 out of 106 and 35 out of 62, respectively). The study of Sr content and distribution of S, Ca, Mg, and Fe in selected samples suggested weak geochemical impacts on the composition of the carbonates studied. A mean Sr content of 660 and 840 ppm (in the Kamenka River section and Vangyr River section, respectively) is quite high, and Mg, S, and Fe were very scattered in the samples (Fig. 11). These data support the suppositions regarding the near-primary composition of the studied samples. The distribution of carbon and oxygen stable isotopes on the screening diagram (Fig. 10) shows that most of the samples (386 out of 412) fall into the area corresponding to unaltered marine carbonates. The highest proportion of samples failing this test came from the Malaya Usa River sections (12 out of 62 samples from these sections). In other sections, doubtful samples comprise only 1%–6% of the total number of samples. Therefore, most samples passed this screening test. Organic matter content

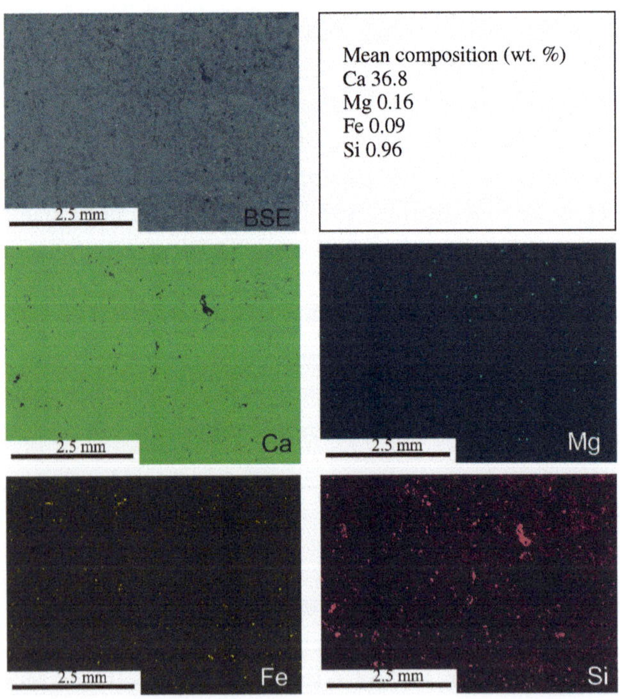

Fig. 11 Results of the element mapping of samples 111-7/19, S. carinata Zone, Kamenka River section, loc. 111.

in the studied samples was less than 1.5%, as a rule. Therefore, there is only a low probability of carbonates of ^{12}C, derived from oxidized organic matter, becoming incorporated. Several samples having higher TOC contents were excluded from the dataset. In total, out of 412 samples studied, 268 passed all of the screening tests. These samples were used for analyses of the carbonate carbon isotope record.

4.1 Shallow-water successions

Mean $\delta^{13}C_{carb}$ values in the shallow-water, carbonate platform successions (Kamenka River sections, Kozhva River section, Izyayu River section, and Kochmes-1 borehole) are approximately 2.8‰. A gradual decrease in $\delta^{13}C_{carb}$ can be observed in the interval from the bella Zone through the lower part of the semichatovae Zone (equivalent to the praesulcata-sulcata zones), with the lowest values being 1.5–2.0‰ in the middle of the semichatovae Zone in Section 121 (Kamenka River) (Fig. 4). The negative trend is interrupted by a weakly positive shift (up to 3.4‰) in the lower part of the semichatovae Zone (=lower part of sulcata Zone) (Fig. 4).

The interval comprising the upper part of the semichatovae Zone, through the ludmilae Zone (=L. duplicata – sandbergi zones) demonstrates high and variable $\delta^{13}C_{carb}$ values (Fig. 4). A threefold positive excursion had maximum values of about 3.6‰, with the last acme being found at the ludmilae/carinata zonal boundary (=sandbergi/Lower crenulata boundary or the boundary between the Lower and Middle Tournaisian). A subsequent, gradual decrease in $\delta^{13}C_{carb}$ can be observed in the lower part of the carinata Zone (=Lower crenulata Zone). The upper part of the carinata Zone was not studied in the shallow-water facies.

A prominent twofold positive shift in $\delta^{13}C_{carb}$ comprises the upper part of the Tournaisian carbonate succession in the shallow-water facies, this approximately corresponds to the isosticha-typicus zones. This shift, detected in the Kozhva River section (Fig. 5), Izyayu River section (Saltzman et al., 2004), and the Kochmes-1 borehole (Zhuravlev and Vevel, 2018), demonstrates $\delta^{13}C_{carb}$ values up to 8‰ with the highest values being found in the Kochmes-1 borehole. The double-peaked pattern of this shift can be observed in the Kozhva River section and Izyayu River section.

4.2 Deep-water shelf successions

Deep-water successions, corresponding to slope and depression paleoenvironments, with strong diagenetic overprinting and higher TOC content (0.1%–5.6%), demonstrate a mean $\delta^{13}C_{carb}$ value of 2.1‰

(2.3‰ based on the screened data). The analyzed carbonate material came mainly from calciturbidites, therefore a certain amount of mixing of isotope signals from the shallow-water allochthonous and the deep-water suballochthonous carbonate is presumed. The raw carbonate carbon isotope records in these sequences are characterized by a "trembling" appearance with stochastic variations of about 1.5‰ in amplitude (e.g., Sobolev, 2011). The screened $\delta^{13}C_{carb}$ values demonstrate smoother variations (Figs. 6–9).

The most complete isotope record came from the Kozhim River (Fig. 6) and Izyayol River (Fig. 7) sections, corresponding to a depression and a slope facies belt, respectively. A very weak positive shift in the lower part of the praesulcata Zone is followed by the interval of low values, increasing in the uppermost part of the zone near the Devonian/Carboniferous boundary. The Devonian/Carboniferous boundary interval studied in the Vangyr River section demonstrates a similar carbon isotope pattern in the praesulcata—sulcata conodont zones, excluding a positive anomaly (a single sample) in the praesulcata Zone (Fig. 8).

In the Tournaisian a twofolded positive excursion comprises the stratigraphic interval from the Upper duplicata Zone through to the lowermost part of the Lower crenulata Zone in the Kozhim River section (Fig. 6). A negative trend in the Lower crenulata Zone (approximately corresponding to the quadruplicata Zone) is followed by the positive trend in the upper part of the quadruplicata Zone and the isosticha Zone. The trends are distinguishable in the Kozhim River section (Fig. 6) and Malaya Usa River section (Fig. 9). $\delta^{13}C_{carb}$ values increase from 0.3‰ to 0.5‰ up to 5.2‰ in this interval.

4.3 General trends in carbonate carbon isotope record

General trends in the Tournaisian carbonate carbon isotope record in the Pechora Plate and North of Urals were evaluated by calculating the mean values and standard deviations for the biostratigraphic units, which are traceable in the study sections (Fig. 12). The mean values were calculated separately for different facies belts (Fig. 13).

Due to variations in the biostratigraphic subdivisions of different sections, the following biostratigraphic units that were traceable in all of the study sections were used (Fig. 12): praesulcata Zone (=bella Zone and

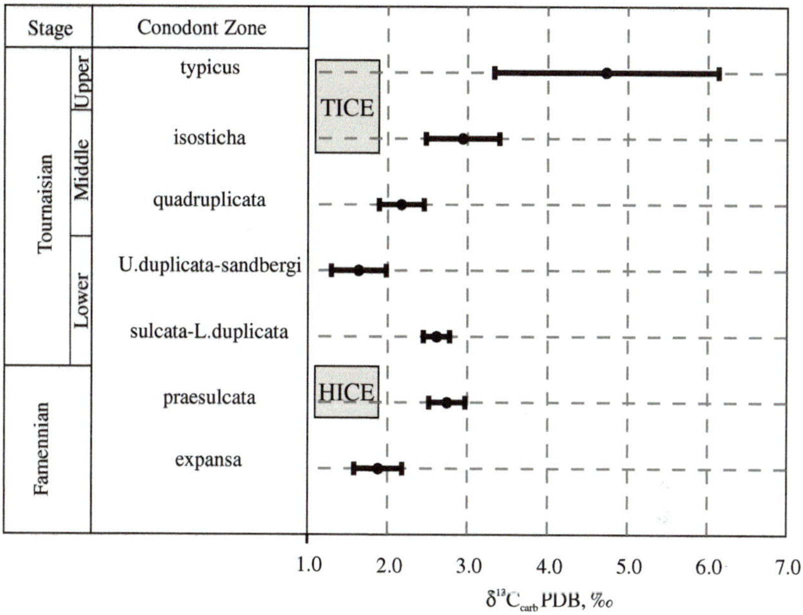

Fig. 12 Average values of $\delta^{13}C_{carb}$ in the studied sections. Whiskers mark 95% confidence intervals. Shaded areas mark the global isotope events.

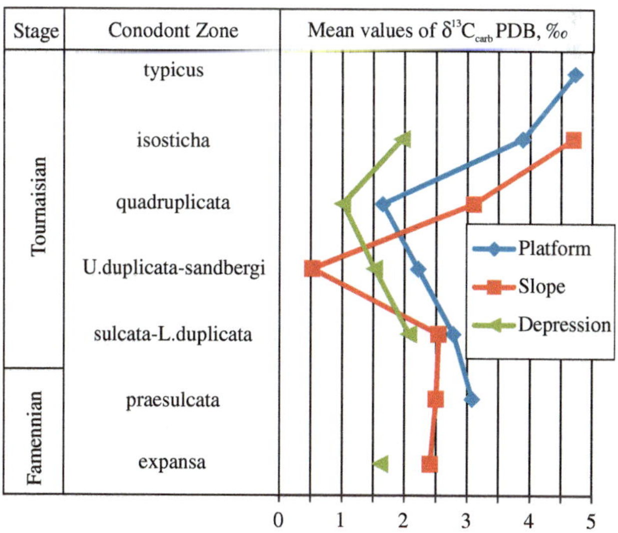

Fig. 13 Average values of $\delta^{13}C_{carb}$ in the studied sections separated by facies belts.

quasinuda Zone), sulcata-Lower duplicata zones (=semichatovae Zone), Upper duplicata-sandbergi zones (=ludmilae zone), quadruplicata Zone (=carinata Zone), isosticha Zone, and typicus Zone.

The mean $\delta^{13}C_{carb}$ value for the praesulcata Zone is 2.7‰ with a standard deviation of 0.78 ($N=45$). The value decreases in the sulcata-Lower duplicata and Upper duplicata-sandbergi zones from 2.6‰ to 1.6‰, and the standard deviation slightly increases from 0.71 up to 0.75. This decrease is followed by a significant rise in the mean $\delta^{13}C_{carb}$ (from 2.2‰ to 4.7‰) in the quadruplicata-typicus interval, associated with high standard deviations (1.17–2.86).

The same trends were observed in different facies belts (shallow-water carbonate platform, platform slope, and shelf depression) with greater mean values recorded from shallower settings, and lighter values recorded from shelf depression settings (Fig. 13). However, mean $\delta^{13}C_{carb}$ values in slope facies are extremely low in the Upper duplicata-sandbergi interval and high in quadruplicata-isosticha zones (Fig. 13). A low mean $\delta^{13}C_{carb}$ value in the Upper duplicata-sandbergi interval is of low reliability because of the small amount of samples (2) coming from the Izyayol River section only (Fig. 7). The positive deviation in the quadruplicata-isosticha interval may be caused by localized environments in the Malaya Usa River section, where an extremely heavy carbon isotope composition is detected in these zones (Fig. 9).

5. Discussion

In general, a lighter carbon isotope composition is known in offshore facies in the Paleozoic isotope record. For example, a similarly negative seaward spatial trend in $\delta^{13}C_{carb}$ was reported from the Silurian (Wenlock–Ludlow boundary interval) of the Midland Platform (Blain et al., 2016). However, the observed trend in $\delta^{13}C_{carb}$ in the facies in the studied sections may also be caused by increasing diagenetic processes, including silicification and recrystallization in the slope and depression successions. Also, $\delta^{13}C_{carb}$ decreases in the organic-rich, deep-water facies and might be related to the remineralization of organic matter within the carbonate rocks. These assumptions are supported by higher organic matter content and, generally, higher standard deviations of $\delta^{13}C_{carb}$ values in the slope and depression successions, in relation to those in carbonate platform successions. High $\delta^{13}C_{carb}$ scatter is considered to be possible evidence of local, diagenetic alteration (Metzger and Fike, 2013).

The two acmes in mean $\delta^{13}C_{carb}$ correspond to the terminal Famennian (praesulcata Zone) and Middle-Upper Tournaisian (isosticha-typicus zones) and resemble well-known HICE and TICE global isotope events, respectively (Qie et al., 2011, 2016) (Fig. 12). Minor variations and shifts in $\delta^{13}C_{carb}$ are not reflected in the mean values.

Considering the carbon isotope patterns in the studied successions allows the recognition of several positive shifts, which are traceable globally (Figs. 14, 15). The twofold pattern of the HICE event is distinguishable in the praesulcata Zone in the Izyayol River section and, approximately, in the Kamenka River section. The acme of the $\delta^{13}C_{carb}$ values is about 3.5‰, which is comparable with those reported from sections in South China, Carnic Alps, Belgium, and North America (Kaiser et al., 2006; Qie et al., 2016). A minor positive shift in the sulcata Zone, noted by Qie et al. as being P1 (2016), appears in the Kozhim River section and Vangyr River section, however, it is weakly represented in the lower part of the S. semichatovae Zone in the Kamenka River section and in the sulcata Zone in the Izyayol River section. The next two-peaked positive excursion, comprising the Upper duplicata-sandbergi interval, is distinguishable in the Kozhim River section and the Kamenka River section, where it corresponds to the upper part of the S. semichatovae and lowermost part of the *S. carinata* zones. The two positive peaks are separated by a negative excursion that resembles the negative shift reported by Kaiser et al. (2008) in the succession of the Rhenish Massif and Carnic Alps in the topmost Upper duplicata Zone. The traces of this twofold, positive excursion are muted in the Izyayol River section, where the duplicata-quadruplicata interval is characterized by a light carbon isotope composition. This excursion probably corresponds to the second minor positive shift (P2), mentioned by Qie et al. (2016), in the *S. sinensis* Zone, in South China, and in the duplicata-sandbergi interval, in Belgium.

High variability of $\delta^{13}C_{carb}$ values in the S. semichatovae—S. ludmilae zones in the Kamenka River section were probably caused by specific facies. This interval is represented by the alternation of semi-restricted lagoon, shoal, and open marine facies (Vevel et al., 2012). These facies changes and associated changes in water mass restriction probably caused local

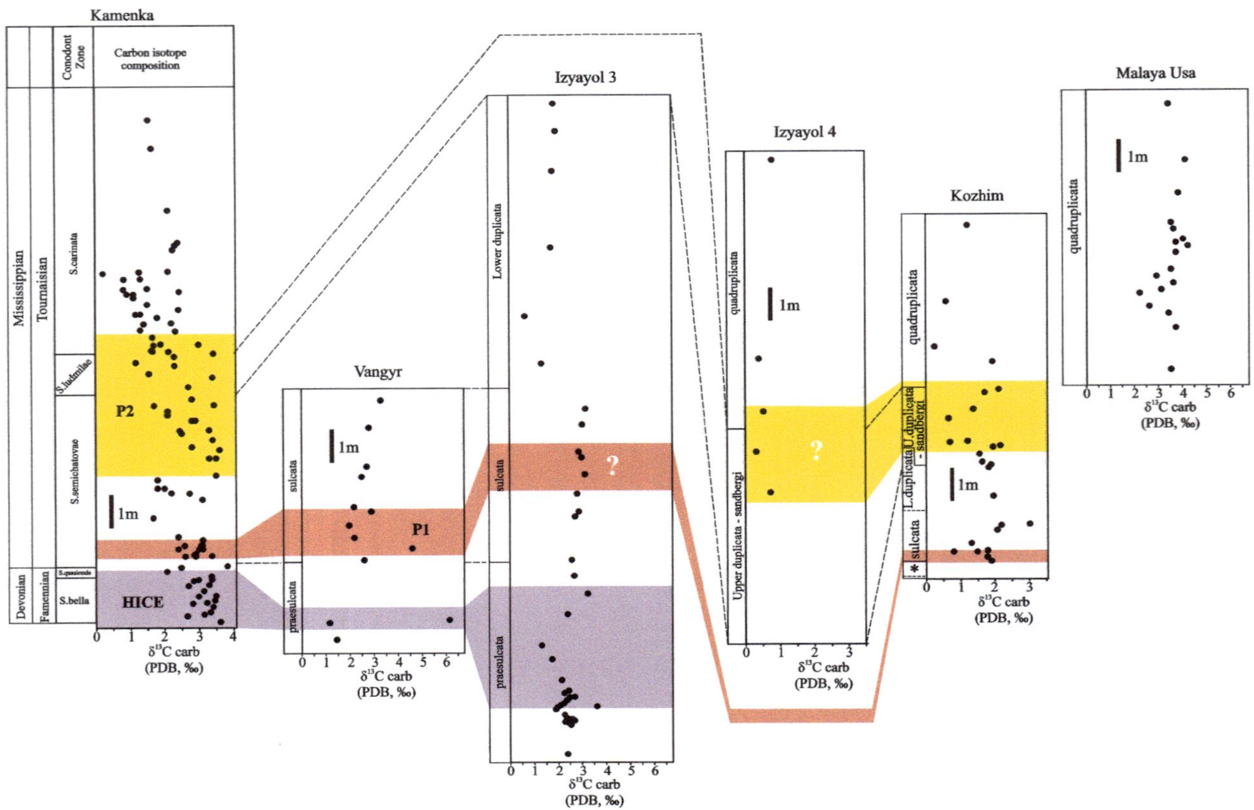

Fig. 14 Correlation of the carbonate carbon isotope variations in the terminal Famennian—Middle Tournaisian.

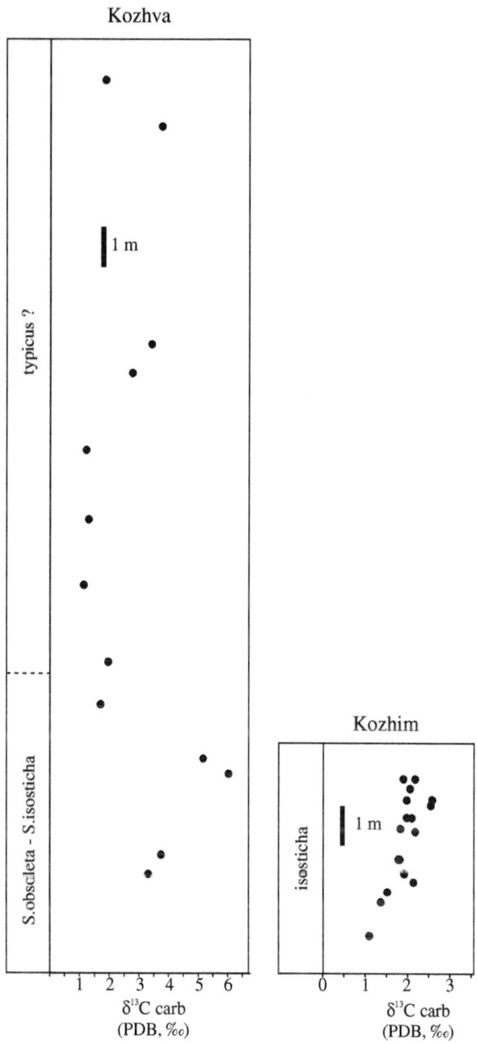

Fig. 15 Correlation of the carbonate carbon isotope variations in the Middle-Upper Tournaisian.

variations of the carbon isotope composition, which obscured weak global shifts in $\delta^{13}C_{carb}$. The P2 shift probably corresponds to the upper part of the S. semichatovae Zone and S. ludmilae Zone in the Kamenka River section (Fig. 14).

The prominent $\delta^{13}C_{carb}$ positive excursion, comprising the interval from the upper part of the quadruplicata Zone through the typicus Zone, is distinguished in a number of sections, characterizing various facies in the region under consideration, namely: Kozhva River section, Izyayu River section (Saltzman et al., 2004), Kochmes-1 borehole (Zhuravlev and Vevel, 2018), and Kozhim River section (Fig. 15). In the most complete sections, this excursion demonstrates a clear, twofold pattern, similar to those reported in the sections of South China (Qie et al., 2011, 2016). The lower peak approximately corresponds to the uppermost quadruplicata-isosticha interval but the upper peak comprises the lower part of the typicus Zone.

The positive shift of high amplitude (about 4‰) observed in the quadruplicata Zone in the Malaya Usa River section does not fit the global trend. It can be explained by local peculiarities of partly restricted water masses in the epicontinental basin (Immenhauser et al., 2003; Saltzman and Thomas, 2012).

6. Concluding remarks

The established regional $\delta^{13}C$ data record the distinct terminal Famennian (HICE) and mid-Tournaisian (TICE) carbon isotope events and two minor positive shifts in the sulcata Zone and Upper duplicata—lowermost quadruplicata interval, which are known in Chinese and European successions. The age of the isotope shifts is underpinned by conodont biostratigraphy. These isotope events and shifts can be traced through the facies gradient from shallow-water carbonate platforms to deep-water shelf depressions. In shallow-water facies, the carbon isotope record is obscured by variations, which are probably caused by local peculiarities in partly restricted water masses. The diagenetic processes, including recrystallization and silicification, mute the carbon isotope record, mainly in the deep-water facies. Generally, $\delta^{13}C$ values in the Tournaisian decrease seaward, and this might be related to the re-mineralization of organic matter within the carbonate rocks, as well as recrystallization of carbonate minerals.

The biostratigraphically-constrained carbonate carbon isotope record of the Tournaisian of the Pechora Plate and North of Urals matches the isotope variations known in different regions of the world (Europe, South China, and North America) and, therefore, reflects global isotope perturbations. These variations can be used as reliable chemostratigraphic markers for regional and global correlation.

Acknowledgment

The authors would like to thank A. Shuyskiy for help in SEM and microprobe investigations.

Appendix. Carbonate carbon and oxygen isotope data

Section	Sample	$d^{13}C$ PDB	$d^{18}O$ SMOW	Conodont zone	Facies belt	Screening test (C-O isotope diagram)	Screening test (lithology, thin sections)
Vangyr	w22-z1	1.7	25.4	praesulcata	Slope	Passed	Failed
Vangyr	w22-1b	2.2	23.8	praesulcata	Slope	Passed	Failed
Vangyr	w22-2A	1.2	23.6	praesulcata	Slope	Passed	Passed
Vangyr	w22-3	6.2	24.4	praesulcata	Slope	Passed	Passed
Vangyr	w22-5	1.7	24	praesulcata	Slope	Passed	Failed
Vangyr	w22-6	2.4	25.5	sulcata	Slope	Passed	Passed
Vangyr	w22-7	2.2	25	sulcata	Slope	Passed	Passed
Vangyr	w22-8	2	26.5	sulcata	Slope	Passed	Passed
Vangyr	w22-9	2.2	24.7	sulcata	Slope	Passed	Passed
Vangyr	w22-11	2.8	24.4	sulcata	Slope	Passed	Passed
Vangyr	21-26-1	4.6	25.5	sulcata	Slope	Passed	Passed
Vangyr	21-26-3	2.9	25.4	sulcata	Slope	Passed	Passed

Continued

—cont'd

Section	Sample	d^{13}C PDB	d^{18}O SMOW	Conodont zone	Facies belt	Screening test (C-O isotope diagram)	Screening test (lithology, thin sections)
Vangyr	21-26-5	2.5	27.9	sulcata	Slope	Passed	Passed
Vangyr	21-26-7	2.8	27.6	sulcata	Slope	Passed	Passed
Vangyr	21-26-9	3.3	25.4	sulcata	Slope	Passed	Passed
Izyayol	GER3-72	1.82	25.55	expansa	Slope	Passed	Passed
Izyayol	GER3-73	2.86	28.14	expansa	Slope	Passed	Passed
Izyayol	GER3-73a	2.93	28.00	expansa	Slope	Passed	Failed
Izyayol	GER3-74	2.64	28.39	expansa	Slope	Passed	Passed
Izyayol	GER3-75	2.00	26.09	expansa	Slope	Passed	Passed
Izyayol	GER3-76	1.55	25.33	expansa	Slope	Passed	Failed
Izyayol	GER3-77	2.05	25.24	expansa	Slope	Passed	Failed
Izyayol	GER3-78	2.94	28.28	expansa	Slope	Passed	Failed
Izyayol	GER3-79	2.80	26.69	expansa	Slope	Passed	Failed
Izyayol	GER3-80	2.71	24.55	expansa	Slope	Passed	Passed
Izyayol	GER3-81	2.42	27.08	praesulcata	Slope	Passed	Passed
Izyayol	GER3-82	2.57	24.92	praesulcata	Slope	Passed	Passed
Izyayol	GER3-82a	2.46	24.84	praesulcata	Slope	Passed	Passed
Izyayol	GER3-82b	2.31	24.92	praesulcata	Slope	Passed	Passed

Izyayol	GER3-82c	2.72	24.83	praesulcata	Slope	Passed	Passed
Izyayol	GER3-82d	2.60	24.70	praesulcata	Slope	Passed	Passed
Izyayol	GER3-82e	2.38	24.56	praesulcata	Slope	Passed	Passed
Izyayol	GER3-82g	2.29	24.83	praesulcata	Slope	Passed	Passed
Izyayol	GER3-83c	1.85	24.97	praesulcata	Slope	Passed	Failed
Izyayol	GER3-83e	1.91	24.65	praesulcata	Slope	Passed	Passed
Izyayol	GER3-83f	1.98	24.74	praesulcata	Slope	Passed	Passed
Izyayol	GER3-83g	3.66	27.60	praesulcata	Slope	Passed	Passed
Izyayol	GER3-84	1.30	26.10	praesulcata	Slope	Failed	Failed
Izyayol	GER3-84a	2.10	24.42	praesulcata	Slope	Passed	Failed
Izyayol	GER3-84b	2.13	24.03	praesulcata	Slope	Passed	Passed
Izyayol	GER3-84d	2.23	23.87	praesulcata	Slope	Passed	Passed
Izyayol	GER3-84g	2.39	26.32	praesulcata	Slope	Passed	Passed
Izyayol	GER3-84k	1.51	23.51	praesulcata	Slope	Passed	Failed
Izyayol	GER3-84h	2.46	25.30	praesulcata	Slope	Passed	Passed
Izyayol	GER3-84l	2.72	23.26	praesulcata	Slope	Passed	Passed
Izyayol	GER3-84m	1.67	23.06	praesulcata	Slope	Passed	Failed
Izyayol	GER3-85a	2.28	23.05	praesulcata	Slope	Passed	Passed
Izyayol	GER3-85b	2.23	24.00	praesulcata	Slope	Passed	Failed

Continued

—cont'd

Section	Sample	d^{13}C PDB	d^{18}O SMOW	Conodont zone	Facies belt	Screening test (C-O isotope diagram)	Screening test (lithology, thin sections)
Izyayol	GER3-86b	2.47	26.02	praesulcata	Slope	Passed	Passed
Izyayol	GER3-86c	2.17	26.22	praesulcata	Slope	Passed	Passed
Izyayol	GER3-87	1.76	25.27	praesulcata	Slope	Passed	Passed
Izyayol	GER3-88	1.33	25.54	praesulcata	Slope	Passed	Passed
Izyayol	GER3-89	2.43	24.85	praesulcata	Slope	Passed	Passed
Izyayol	GER3-90	3.28	25.09	praesulcata	Slope	Passed	Passed
Izyayol	GER3-91	2.71	25.42	praesulcata	Slope	Passed	Passed
Izyayol	GER3-93	2.61	24.49	sulcata	Slope	Passed	Passed
Izyayol	GER3-93a	3.18	27.78	sulcata	Slope	Passed	Failed
Izyayol	GER3-94	3.42	28.46	sulcata	Slope	Passed	Failed
Izyayol	GER3-95	2.75	26.42	sulcata	Slope	Passed	Passed
Izyayol	GER3-96	2.91	27.53	sulcata	Slope	Passed	Passed
Izyayol	GER3-97	2.84	25.81	sulcata	Slope	Passed	Passed
Izyayol	GER3-97a	3.17	26.79	sulcata	Slope	Passed	Passed
Izyayol	GER3-98	3.03	26.29	sulcata	Slope	Passed	Passed
Izyayol	GER3-98a	2.91	26.76	sulcata	Slope	Passed	Passed
Izyayol	GER3-98b	3.06	27.08	sulcata	Slope	Passed	passed
Izyayol	GER3-99	3.20	27.65	sulcata	Slope	Passed	Passed

Izyayol	GER3-100	2.48	27.51	L. duplicata	Slope	Passed	Failed
Izyayol	GER3-100a	1.34	27.56	L. duplicata	Slope	Passed	Passed
Izyayol	GER3-101	1.86	25.72	L. duplicata	Slope	Passed	Failed
Izyayol	GER3-102	0.65	24.87	L. duplicata	Slope	Passed	Passed
Izyayol	GER3-103	1.97	26.28	L. duplicata	Slope	Passed	Failed
Izyayol	GER3-104	1.74	27.62	L. duplicata	Slope	Passed	Passed
Izyayol	GER3-104a	1.78	27.67	L. duplicata	Slope	Passed	Failed
Izyayol	GER3-105	1.91	27.87	L. duplicata	Slope	Passed	Failed
Izyayol	GER3-106	1.81	27.10	L. duplicata	Slope	Passed	Passed
Izyayol	GER3-106a	1.96	26.25	L. duplicata	Slope	Passed	Passed
Izyayol	GER3-106b	1.87	27.03	L. duplicata	Slope	Passed	Passed
Izyayol	GER3-107	1.70	26.26	L. duplicata	Slope	Passed	Failed
Izyayol	GER4-7	1.54	26.55	U. duplicata-sandbergi	Slope	Passed	Failed
Izyayol	GER4-8b	0.72	28.41	U. duplicata-sandbergi	Slope	Passed	Passed

Continued

—cont'd

Section	Sample	d^{13}C PDB	d^{18}O SMOW	Conodont zone	Facies belt	Screening test (C-O isotope diagram)	Screening test (lithology, thin sections)
Izyayol	GER4-9	0.09	24.94	U. duplicata-sandbergi	Slope	Passed	Failed
Izyayol	GER4-10	0.31	25.20	U. duplicata-sandbergi	Slope	Passed	Passed
Izyayol	GER4-10a	2.13	28.08	quadruplicata	Slope	Passed	Failed
Izyayol	GER4-11	0.52	24.74	quadruplicata	Slope	Passed	Passed
Izyayol	GER4-12	0.68	27.35	quadruplicata	Slope	Passed	Failed
Izyayol	GER4-13	0.97	24.51	quadruplicata	Slope	Passed	Failed
Izyayol	GER4-14	0.39	27.37	quadruplicata	Slope	Passed	Passed
Izyayol	GER4-17	1.87	24.56	quadruplicata	Slope	Passed	Failed
Izyayol	GER4-17a	0.78	26.82	quadruplicata	Slope	Passed	Passed
Izyayol	GER4-17b	1.25	26.71	quadruplicata	Slope	Passed	Failed
Kamenka 121	7	2.6	25.7	sulcata-L. duplicata	Platform	Passed	Passed
Kamenka 121	8	2.9	25.7	sulcata-L. duplicata	Platform	Passed	Passed
Kamenka 121	9	2.9	25.9	sulcata-L. duplicata	Platform	Passed	Passed
Kamenka 121	10	3	26.1	sulcata-L. duplicata	Platform	Passed	Passed

Kamenka 121	11	2.4	25.7	sulcata-L. duplicata	Platform	Passed	Passed
Kamenka 121	12	3.1	26.4	sulcata-L. duplicata	Platform	Passed	Passed
Kamenka 121	15	2.4	25.8	sulcata-L. duplicata	Platform	Passed	Passed
Kamenka 121	18	3.1	26.1	sulcata-L. duplicata	Platform	Passed	Passed
Kamenka 121	19	2.2	24.7	sulcata-L. duplicata	Platform	Passed	Passed
Kamenka 121	20	2	25.4	sulcata-L. duplicata	Platform	Passed	Passed
Kamenka 121	21	1.8	25.6	sulcata-L. duplicata	Platform	Passed	Passed
Kamenka 121	22	1.8	25.9	sulcata-L. duplicata	Platform	Passed	Passed
Kamenka 121	23	3.5	26.3	sulcata-L. duplicata	Platform	Passed	Passed
Kamenka 121	26	3.5	26.6	sulcata-L. duplicata	Platform	Passed	Passed
Kamenka 121	27	3.3	26.4	sulcata-L. duplicata	Platform	Passed	Passed

Continued

—cont'd

Section	Sample	$d^{13}C$ PDB	$d^{18}O$ SMOW	Conodont zone	Facies belt	Screening test (C-O isotope diagram)	Screening test (lithology, thin sections)
Kamenka 121	28	3.6	26.2	sulcata-L. duplicata	Platform	Passed	Passed
Kamenka 121	30	2.8	26.5	sulcata-L. duplicata	Platform	Passed	Passed
Kamenka 121	31	3.4	26.2	sulcata-L. duplicata	Platform	Passed	Passed
Kamenka 121	32	2.5	26.4	sulcata-L. duplicata	Platform	Passed	Passed
Kamenka 121	33	3.3	26.3	sulcata-L. duplicata	Platform	Passed	Passed
Kamenka 121	34	2.8	26.4	sulcata-L. duplicata	Platform	Passed	Passed
Kamenka 121	35	2.9	26.5	sulcata-L. duplicata	Platform	Passed	Passed
Kamenka 121	36	2.1	25.3	sulcata-L. duplicata	Platform	Passed	Passed
Kamenka 121	37	1.7	23.7	sulcata-L. duplicata	Platform	Passed	Passed
Kamenka 121	38	2.8	26	sulcata-L. duplicata	Platform	Passed	Passed
Kamenka 121	39	2.7	26.6	U. duplicata-sandbergi	Platform	Passed	Passed

Kamenka 121	40	3.4	26.3	U. duplicata-sandbergi	Platform	Passed	Passed
Kamenka 121	43	3	26.1	quadruplicata	Platform	Passed	Passed
Kamenka 121	44	1.9	24.5	quadruplicata	Platform	Passed	Passed
Kamenka 121	45	1.3	25.1	quadruplicata	Platform	Passed	Passed
Kamenka 121	46	1.4	24.5	quadruplicata	Platform	Passed	Passed
Kamenka 121	47	1.8	26.2	quadruplicata	Platform	Passed	Passed
Kamenka 121	48	1.3	26.2	quadruplicata	Platform	Passed	Passed
Kamenka 121	49	1.5	24.6	quadruplicata	Platform	Passed	Passed
Kamenka 121	50	1.1	26.1	quadruplicata	Platform	Passed	Passed
Kamenka 121	51	1.5	23.8	quadruplicata	Platform	Passed	Passed
Kamenka 121	52	0.9	23.1	quadruplicata	Platform	Failed	Passed
Kamenka 121	53	1.3	25.4	quadruplicata	Platform	Passed	Passed
Kamenka 121	60	2.4	25.5	quadruplicata	Platform	Passed	Passed
Kamenka 121	64	2.1	25.2	quadruplicata	Platform	Passed	Passed
Kamenka 121	121-1	3.62	26.69	praesulcata	Platform	Passed	Passed
Kamenka 121	121-2	2.79	26.37	praesulcata	Platform	Passed	Failed
Kamenka 121	121-3	3.56	26.67	praesulcata	Platform	Passed	Failed
Kamenka 121	121-4	2.66	26.09	praesulcata	Platform	Passed	Passed

Continued

—cont'd

Section	Sample	$d^{13}C$ PDB	$d^{18}O$ SMOW	Conodont zone	Facies belt	Screening test (C-O isotope diagram)	Screening test (lithology, thin sections)
Kamenka 121	121-5	3.15	26.24	praesulcata	Platform	Passed	Passed
Kamenka 121	121-6	3.33	26.52	praesulcata	Platform	Passed	Passed
Kamenka 121	121-7	3.41	26.46	praesulcata	Platform	Passed	Passed
Kamenka 121	121-8	2.82	26.02	praesulcata	Platform	Passed	Passed
Kamenka 121	121-9	3.46	25.94	praesulcata	Platform	Passed	Passed
Kamenka 121	121-10	3.49	23.4	praesulcata	Platform	Passed	Passed
Kamenka 121	121-11	3.14	26.79	praesulcata	Platform	Passed	Passed
Kamenka 121	121-12	2.69	26.15	praesulcata	Platform	Passed	Passed
Kamenka 121	121-13a	3.29	26.64	praesulcata	Platform	Passed	Passed
Kamenka 121	121-14a	2.98	26.66	praesulcata	Platform	Passed	Passed
Kamenka 121	121-14в	3.37	27.1	praesulcata	Platform	Passed	Passed
Kamenka 121	121-15	3.36	26.96	praesulcata	Platform	Passed	Passed
Kamenka 121	121-17	2.48	27.14	praesulcata	Platform	Passed	Passed
Kamenka 121	121-18	2.86	26.79	sulcata-L. duplicata	Platform	Passed	Passed
Kamenka 121	121-19	2.58	26.26	sulcata-L. duplicata	Platform	Passed	Passed
Kamenka 121	121-19в	3.06	26.56	sulcata-L. duplicata	Platform	Passed	Passed

Kamenka 121	121-20	3.11	26.67	sulcata-L. duplicata	Platform	Passed	Passed
Kamenka 121	121-21	1.67	26.51	sulcata-L. duplicata	Platform	Passed	Passed
Kamenka 121	121-22	2.73	26.28	sulcata-L. duplicata	Platform	Passed	Passed
Kamenka 121	121-27	2.45	26.94	sulcata-L. duplicata	Platform	Passed	Passed
Kamenka 121	121-28	2.1	27.21	sulcata-L. duplicata	Platform	Passed	Passed
Kamenka 121	121-42	0.91	25.92	quadruplicata	Platform	Passed	Passed
Kamenka 121	121-42	1.1	24.6	quadruplicata	Platform	Passed	Passed
Kamenka 121	121-42-concretion	1.24	24.26	quadruplicata	Platform	Passed	Failed
Kamenka 121	121-36	3.43	25.58	quadruplicata	Platform	Passed	Passed
Kamenka 121	121-40	1.17	26.06	quadruplicata	Platform	Passed	Passed
Kamenka 121	121-1/18	3.37	26.77	sulcata-L. duplicata	Platform	Passed	Passed
Kamenka 121	121-2/18	1.55	25.09	U. duplicata-sandbergi	Platform	Passed	Passed
Kamenka 121	121-4/18	2.3	25.86	U. duplicata-sandbergi	Platform	Passed	Passed

Continued

—cont'd

Section	Sample	$d^{13}C$ PDB	$d^{18}O$ SMOW	Conodont zone	Facies belt	Screening test (C-O isotope diagram)	Screening test (lithology, thin sections)
Kamenka 121	121-5/18	1.17	26.17	U. duplicata-sandbergi	Platform	Passed	Passed
Kamenka 121	121-7/18	2.59	26.85	quadruplicata	Platform	Passed	Failed
Kamenka 121	121-8(1)/18	2.13	25.92	quadruplicata	Platform	Passed	Passed
Kamenka 121	121-8(2)/18	1.67	26.33	quadruplicata	Platform	Passed	Passed
Kamenka 121	121-9/18	1.62	25.39	quadruplicata	Platform	Passed	Passed
Kamenka 121	121-10/18	1.69	26.66	quadruplicata	Platform	Passed	Passed
Kamenka 121	121-11/18	1.66	25.07	quadruplicata	Platform	Passed	Passed
Kamenka 121	121-12/18	2.33	25.81	quadruplicata	Platform	Passed	Passed
Kamenka 121	121-13/18	2.21	25.89	quadruplicata	Platform	Passed	passed
Kamenka 121	121-14/18	2.42	25.5	quadruplicata	Platform	Passed	Passed
Kamenka 121	121-16/18	2.44	25.85	quadruplicata	Platform	Passed	Passed
Kamenka 122	122-1/16	0.81	23.95	quadruplicata	Platform	Passed	Passed
Kamenka 122	122-2/16	0.22	25.99	quadruplicata	Platform	Passed	Passed
Kamenka 122	122-2/16	0.21	26.31	quadruplicata	Platform	Passed	Passed
Kamenka 122	122-3/16	2.08	26.52	quadruplicata	Platform	Passed	Passed

Kamenka 122	122-3/16	1.25	26.14	quadruplicata	Platform	Passed	Passed
Kamenka 122	122-6/16	2.21	26.05	quadruplicata	Platform	Passed	Passed
Kamenka 122	122-8/16	1.32	24.52	quadruplicata	Platform	Passed	Failed
Kamenka 122	122-12/16	1.21	24.52	quadruplicata	Platform	Passed	Failed
Kamenka 111A	111A-1/19	3.24	26.81	praesulcata	Platform	Passed	Passed
Kamenka 111A	111A-2/19	3	26.5	praesulcata	Platform	Passed	Passed
Kamenka 111A	111A-3/19	2.56	27.47	praesulcata	Platform	Passed	Failed
Kamenka 111A	111A-4/19	2.86	25.72	praesulcata	Platform	Passed	Passed
Kamenka 111A	111A-5/19	2.08	26.82	praesulcata	Platform	Passed	Passed
Kamenka 111A	111A-6/19	2.96	26.58	sulcata-L. duplicata	Platform	Passed	Failed
Kamenka 111A	111A-6/19	3.84	24.63	sulcata-L. duplicata	Platform	Passed	Passed
Kamenka 111A	111A-7/19	3.13	26.83	sulcata-L. duplicata	Platform	Passed	Passed
Kamenka 111A	111A-9/19	3.25	25.81	sulcata-L. duplicata	Platform	Passed	Failed

Continued

—cont'd

Section	Sample	$d^{13}C$ PDB	$d^{18}O$ SMOW	Conodont zone	Facies belt	Screening test (C-O isotope diagram)	Screening test (lithology, thin sections)
Kamenka 111	111-1a/19	3.38	26.02	sulcata-L. duplicata	Platform	Passed	Passed
Kamenka 111	111-1b/19	1.63	26.57	sulcata-L. duplicata	Platform	Passed	Failed
Kamenka 111	111-5/19	0.37	23.55	quadruplicata	Platform	Failed	Passed
Kamenka 111	111-8/19	1.63	22.8	quadruplicata	Platform	Passed	Failed
Kamenka 111	111-11/19	1.48	25.05	quadruplicata	Platform	Passed	Passed
Kamenka 111	111-12/19	2.23	25.1	U. duplicata-sandbergi	Platform	Passed	Passed
Kamenka 111	111-13/19	3.18	26.14	quadruplicata	Platform	Passed	Failed
Kamenka 111	111-14/19	2.93	25.18	quadruplicata	Platform	Passed	Failed
Kamenka 111	111-2/19	2.19	25.7	quadruplicata	Platform	Passed	Failed
Kamenka 111	111-3/19	0.75	23.07	quadruplicata	Platform	Passed	Passed
Kamenka 111	111-6/19	2.25	17.83	quadruplicata	Platform	Failed	Passed
Kamenka 111	111-7/19	1.6	25.31	quadruplicata	Platform	Passed	Failed
Kamenka 111	111-9/19	0.13	23.34	quadruplicata	Platform	Failed	Failed
Kamenka 111	111-10/19	1.57	24.98	quadruplicata	Platform	Passed	Passed
Kozhva 4	KV4-1-1	4.9	25.25	isosticha	Platform	Passed	Passed
Kozhva 4	KV4-1-2	3.26	26.06	isosticha	Platform	Passed	Failed

Kozhva 4	KV4-1-3	3.46	24.77	isosticha	Platform	Passed	Passed
Kozhva 4	KV4-1-4	3.24	26.45	isosticha	Platform	Passed	Passed
Kozhva 4	KV4-1-5	3.3	25.64	isosticha	Platform	Passed	Failed
Kozhva 4	KV4-1-6	3.04	27.46	isosticha	Platform	Passed	Passed
Kozhva 4	KV4-9-1	3.62	27.39	isosticha	Platform	Passed	Passed
Kozhva 4	KV4-9-2	3.34	25.72	isosticha	Platform	Passed	Passed
Kozhva 4	KV4-9-3	−1.5	24.28	isosticha	Platform	Failed	Failed
Kozhva 4	KV4-9-4	4.72	24.96	isosticha	Platform	Passed	Failed
Kozhva 4	KV4-9-5	5.45	26.94	isosticha	Platform	Passed	Failed
Kozhva 4	KV4-9-6	5.13	25.41	isosticha	Platform	Passed	Passed
Kozhva 3	KV3/1	2.09	21.7	isosticha	Platform	Failed	Failed
Kozhva 3	KV3/2	3.3	25.52	isosticha	Platform	Passed	Passed
Kozhva 3	KV3/3	3.73	25.12	isosticha	Platform	Passed	Passed
Kozhva 3	KV3/5	4.47	25.52	isosticha	Platform	Passed	Failed
Kozhva 3	KV3/6	6.01	26.12	isosticha	Platform	Passed	Passed
Kozhva 3	KV3/7	5.15	26	isosticha	Platform	Passed	Passed
Kozhva 3	KV3/8	1.7	25.21	isosticha	Platform	Passed	Passed
Kozhva 3	KV3/9	1.95	24.52	typicus	Platform	Passed	Passed

Continued

—cont'd

Section	Sample	$d^{13}C$ PDB	$d^{18}O$ SMOW	Conodont zone	Facies belt	Screening test (C-O isotope diagram)	Screening test (lithology, thin sections)
Kozhva 3	KV3/10	1.13	24.19	typicus	Platform	Passed	Passed
Kozhva 3	KV3/11	2.95	24.88	typicus	Platform	Passed	Failed
Kozhva 3	KV3/12	1.29	23.48	typicus	Platform	Passed	Passed
Kozhva 3	KV3/13	1.21	23.64	typicus	Platform	Passed	Passed
Kozhva 3	KV3/14	2.75	24.05	typicus	Platform	Passed	Passed
Kozhva 3	KV3/15	3.4	24.63	typicus	Platform	Passed	Passed
Kozhva 3	KV3/16	2.97	24.68	typicus	Platform	Passed	Failed
Kozhva 3	KV3/17	2.77	23.79	typicus	Platform	Passed	Failed
Kozhva 3	KV3/18	3.44	24.62	typicus	Platform	Passed	Failed
Kozhva 3	KV3/19	3.75	24.64	typicus	Platform	Passed	Passed
Kozhva 3	KV3/20	1.85	23.31	typicus	Platform	Passed	Passed
Kozhim	Kj54a	1.61	25.39	expansa	Depression	Passed	Failed
Kozhim	Kj54b	1.69	25.39	expansa	Depression	Passed	Failed
Kozhim	Kj54c	1.34	24.98	expansa	Depression	Passed	Passed
Kozhim	Kj54d	1.46	24.86	expansa	Depression	Passed	Failed
Kozhim	Kj54e	1.35	25.33	expansa	Depression	Passed	Failed
Kozhim	Kj54f	1.12	25.08	expansa	Depression	Passed	Passed

Kozhim	Kj54g	1.93	25.45	expansa	Depression	Passed	Failed
Kozhim	Kj54h	2.48	25.5	expansa	Depression	Passed	Failed
Kozhim	Kj55a	1.42	25.29	expansa	Depression	Passed	Failed
Kozhim	Kj55b	1.6	25.05	expansa	Depression	Passed	Failed
Kozhim	54-1/89	0.6	19.3	expansa	Depression	Passed	Passed
Kozhim	Kj55c	2.02	25.51	expansa	Depression	Passed	Passed
Kozhim	Kj55d	1.8	25.37	expansa	Depression	Passed	Passed
Kozhim	Kj55e	1.92	25.3	expansa	Depression	Passed	Failed
Kozhim	1-9/91	-0.6	24.4	expansa	Depression	Passed	Failed
Kozhim	Kj56a	1.68	25.39	expansa	Depression	Passed	Passed
Kozhim	Kj56b	2.26	25.66	expansa	Depression	Passed	Failed
Kozhim	Kj56c	2.04	25.68	expansa	Depression	Passed	Passed
Kozhim	Kj56d	1.92	25.51	expansa	Depression	Passed	Failed
Kozhim	Kj56e	2.1	25.43	expansa	Depression	Passed	Passed
Kozhim	Kj56f	1.69	25.3	expansa	Depression	Passed	Passed
Kozhim	Kj56g	1.82	25.38	expansa	Depression	Passed	Passed
Kozhim	Kj56h	1.52	25.48	expansa	Depression	Passed	Failed
Kozhim	Kj56k	1.79	25.23	expansa	Depression	Passed	Failed
Kozhim	Kj57b	1.92	25.54	praesulcata	Depression	Passed	Failed

Continued

—cont'd

Section	Sample	$d^{13}C$ PDB	$d^{18}O$ SMOW	Conodont zone	Facies belt	Screening test (C-O isotope diagram)	Screening test (lithology, thin sections)
Kozhim	Kj58a	1.94	25.3	sulcata	Depression	Passed	Passed
Kozhim	Kj58b	1.79	24.61	sulcata	Depression	Passed	Passed
Kozhim	Kj58c	1.49	25.27	sulcata	Depression	Passed	Passed
Kozhim	1-14/91	0.8	23.1	sulcata	Depression	Passed	Passed
Kozhim	Kj58d	1.8	25.59	sulcata	Depression	Passed	Passed
Kozhim	Kj58e	2.01	25.99	sulcata	Depression	Passed	Failed
Kozhim	Kj58f	1.32	25.5	sulcata	Depression	Passed	Passed
Kozhim	Kj59	1.24	24.93	sulcata	Depression	Passed	Failed
Kozhim	Kj60a	2.08	25.26	sulcata	Depression	Passed	Passed
Kozhim	1-16/91	2.2	26.1	sulcata	Depression	Passed	Passed
Kozhim	Kj61	3.28	25.71	sulcata	Depression	Passed	Passed
Kozhim	Kj62b	2.99	25.63	L. duplicata	Depression	Passed	Passed
Kozhim	Kj62d	0.61	25.1	L. duplicata	Depression	Passed	Failed
Kozhim	Kj64a	3.23	25.79	L. duplicata	Depression	Passed	Passed
Kozhim	Kj66a	0.42	23.25	L. duplicata	Depression	Failed	Failed
Kozhim	Kj66b	-2.83	20.11	L. duplicata	Depression	Failed	Failed
Kozhim	Kj67a	1.97	25.64	L. duplicata	Depression	Passed	Passed
Kozhim	Kj67b	1.7	25.6	L. duplicata	Depression	Passed	Failed
Kozhim	Kj67c	1.58	25.37	L. duplicata	Depression	Passed	Failed

Kozhim	Kj69b	1.83	25.53	u. duplicata-sandbergi	Depression	Passed	Passed
Kozhim	Kj69c	1.9	25.43	u. duplicata-sandbergi	Depression	Passed	Passed
Kozhim	Kj69d	1.64	25.08	u. duplicata-sandbergi	Depression	Passed	Passed
Kozhim	Kj69e	1.6	25.75	u. duplicata-sandbergi	Depression	Passed	Failed
Kozhim	Kj69f	1.56	25.18	u. duplicata-sandbergi	Depression	Passed	Passed
Kozhim	Kj71b	1.96	25.18	u. duplicata-sandbergi	Depression	Passed	Passed
Kozhim	Kj71d	2.17	24.5	u. duplicata-sandbergi	Depression	Passed	Passed
Kozhim	Kj71g	0.21	23.45	u. duplicata-sandbergi	Depression	Failed	Failed
Kozhim	Kj71h	0.7	23.96	u. duplicata-sandbergi	Depression	Passed	Passed
Kozhim	Kj71l	1.21	23.93	u. duplicata-sandbergi	Depression	Passed	Passed
Kozhim	Kj71s	0.6	23.67	u. duplicata-sandbergi	Depression	Passed	Failed

Continued

—cont'd

Section	Sample	$d^{13}C$ PDB	$d^{18}O$ SMOW	Conodont zone	Facies belt	Screening test (C-O isotope diagram)	Screening test (lithology, thin sections)
Kozhim	Kj71u	2.7	24.38	u. duplicata-sandbergi	Depression	Passed	Failed
Kozhim	Kj72a	0.66	24.18	u. duplicata-sandbergi	Depression	Passed	Passed
Kozhim	Kj72c	1.38	25.22	u. duplicata-sandbergi	Depression	Passed	Passed
Kozhim	Kj74d	0.14	23.32	u. duplicata-sandbergi	Depression	Failed	Failed
Kozhim	Kj74h	1.72	23.87	u. duplicata-sandbergi	Depression	Passed	Passed
Kozhim	Kj75-1	2.13	24.99	quadruplicata	Depression	Passed	Passed
Kozhim	Kj75-2	0.97	24.17	quadruplicata	Depression	Passed	Failed
Kozhim	Kj75-3	0.57	23.81	quadruplicata	Depression	Passed	Failed
Kozhim	Kj76b	0.72	23.76	quadruplicata	Depression	Passed	Failed
Kozhim	Kj76c	0.38	23.54	quadruplicata	Depression	Failed	Failed
Kozhim	Kj76d	1.59	23.74	quadruplicata	Depression	Passed	Failed
Kozhim	Kj77a	1.95	24.47	quadruplicata	Depression	Passed	Passed
Kozhim	Kj77a2	1.55	24.66	quadruplicata	Depression	Passed	Failed
Kozhim	Kj77b	0.26	24.87	quadruplicata	Depression	Passed	Passed

Kozhim	Kj79b	-0.64	24.32	quadruplicata	Depression	Failed	Failed
Kozhim	Kj80o	0.59	25.73	quadruplicata	Depression	Passed	Passed
Kozhim	Kj81a	-4.9	25.99	quadruplicata	Depression	Passed	Failed
Kozhim	Kj81c	-0.27	24.4	quadruplicata	Depression	Failed	Failed
Kozhim	Kj83a	-2.57	26.28	quadruplicata	Depression	Passed	Failed
Kozhim	Kj83b	-0.08	24.74	quadruplicata	Depression	Passed	Passed
Kozhim	Kj85b	0.3	24.71	quadruplicata	Depression	Passed	Failed
Kozhim	Kj85c	0.01	24.63	quadruplicata	Depression	Passed	Failed
Kozhim	Kj85d	1.21	24.49	quadruplicata	Depression	Passed	Passed
Kozhim	Kj85e	0.06	24.85	quadruplicata	Depression	Passed	Failed
Kozhim	Kj85f	0.56	25.13	quadruplicata	Depression	Passed	Failed
Kozhim	Kj85g	0.18	24.69	quadruplicata	Depression	Passed	Failed
Kozhim	Kj85h	1.16	26.19	quadruplicata	Depression	Passed	Failed
Kozhim	Kj85l	0.34	25.11	quadruplicata	Depression	Passed	Failed
Kozhim	Kj85m	0.17	24.64	quadruplicata	Depression	Passed	Failed
Kozhim	Kj85n	0.6	24.62	quadruplicata	Depression	Passed	Failed
Kozhim	Kj85o	0.39	24.28	quadruplicata	Depression	Passed	Failed
Kozhim	Kj88v	1.12	24.02	isosticha	Depression	Passed	Passed
Kozhim	Kj89a	1.39	24.16	isosticha	Depression	Passed	Passed

Continued

—cont'd

Section	Sample	$d^{13}C$ PDB	$d^{18}O$ SMOW	Conodont zone	Facies belt	Screening test (C-O isotope diagram)	Screening test (lithology, thin sections)
Kozhim	Kj90d	1.54	24.11	isosticha	Depression	Passed	Passed
Kozhim	Kj90e	2.16	24.59	isosticha	Depression	Passed	Passed
Kozhim	Kj90f1	1.5	23.49	isosticha	Depression	Passed	Failed
Kozhim	Kj90f2	1.94	23.51	isosticha	Depression	Passed	Passed
Kozhim	K-89T	1.5		isosticha	Depression	Passed	Failed
Kozhim	Kj93a	1.84	24.08	isosticha	Depression	Passed	Passed
Kozhim	93-1/87	1.8		isosticha	Depression	Passed	Passed
Kozhim	Kj93v	2.2	24.77	isosticha	Depression	Passed	Passed
Kozhim	Kj94a	1.85	24.21	isosticha	Depression	Passed	Passed
Kozhim	Kj94v	2.12	24.09	isosticha	Depression	Passed	Passed
Kozhim	T-94	2		isosticha	Depression	Passed	Passed
Kozhim	Kj94g	2.57	24.43	isosticha	Depression	Passed	Passed
Kozhim	Kj94h	2	23.72	isosticha	Depression	Passed	Passed
Kozhim	95-1/87	2.6		isosticha	Depression	Passed	Passed
Kozhim	Kj95v1	2.08	23.57	isosticha	Depression	Passed	Passed
Kozhim	Kj95v2	1.28	23.47	isosticha	Depression	Passed	Failed
Kozhim	Kj95c	1.92	23.82	isosticha	Depression	Passed	Passed
Kozhim	96-1/87	2.2		isosticha	Depression	Passed	Passed

Malaya Usa 2	mu2-1	2.1	22.6	duplicata-sandbergi	Slope	Failed	Failed
Malaya Usa 2	mu2-2	3.6	23	duplicata-sandbergi	Slope	Passed	Failed
Malaya Usa 2	mu2-3	3.8	23.2	duplicata-sandbergi	Slope	Passed	Failed
Malaya Usa 2	mu2-4	2	25.2	duplicata-sandbergi	Slope	Passed	Failed
Malaya Usa 2	mu2-5	3.6	25.2	duplicata-sandbergi	Slope	Passed	Failed
Malaya Usa 2	mu2-6	2.3	23.6	quadruplicata	Slope	Passed	Failed
Malaya Usa 2	mu2-7	2	21.9	quadruplicata	Slope	Failed	Failed
Malaya Usa 2	mu2-7a	-1.1	20.2	quadruplicata	Slope	Failed	Failed
Malaya Usa 2	mu2-8	3	22.2	quadruplicata	Slope	Passed	Failed
Malaya Usa 2	mu2-9	3.8	23.5	quadruplicata	Slope	passed	Passed
Malaya Usa 2	mu2-10	2.6	22.2	quadruplicata	Slope	Failed	Failed
Malaya Usa 2	mu2-10a	2.2	21.4	quadruplicata	Slope	Failed	Failed
Malaya Usa 2	mu2-10b	4.2	23.4	quadruplicata	Slope	Passed	Failed
Malaya Usa 2	mu2-11	3	22.7	quadruplicata	Slope	Passed	Failed
Malaya Usa 2	mu2-11a	2.3	21.7	quadruplicata	Slope	Failed	Failed

Continued

—cont'd

Section	Sample	$d^{13}C$ PDB	$d^{18}O$ SMOW	Conodont zone	Facies belt	Screening test (C-O isotope diagram)	Screening test (lithology, thin sections)
Malaya Usa 2	mu2-12	2.7	21.6	quadruplicata	Slope	Failed	Failed
Malaya Usa 2	mu2-12a	4.6	23	quadruplicata	Slope	Passed	Failed
Malaya Usa 2	mu2-12b	3.8	21.4	quadruplicata	Slope	Failed	Failed
Malaya Usa 2	mu2-12c	2.3	21	quadruplicata	Slope	Failed	Failed
Malaya Usa 2	mu2-12d	2.8	20.2	quadruplicata	Slope	Failed	Failed
Malaya Usa 2	mu2-12e	3.2	20.7	quadruplicata	Slope	Failed	Failed
Malaya Usa 2	mu2-13	4	20.6	quadruplicata	Slope	Failed	Failed
Malaya Usa 2	mu2-14	2.6	24.1	quadruplicata	Slope	Passed	Passed
Malaya Usa 2	mu2-15	3.7	25.8	quadruplicata	Slope	Passed	Passed
Malaya Usa 2	mu2-15a	4.1	25.8	quadruplicata	Slope	Passed	Passed
Malaya Usa 2	mu2-16	2.1	23.8	quadruplicata	Slope	Passed	Passed
Malaya Usa 2	mu2-17	3.1	25.7	quadruplicata	Slope	Passed	Passed
Malaya Usa 3	mu3-1	4	28.4	quadruplicata	Slope	Passed	Failed
Malaya Usa 3	mu3-2	3.5	27.2	quadruplicata	Slope	Passed	Failed
Malaya Usa 3	mu3-3	3.5	27.8	quadruplicata	Slope	Passed	Passed
Malaya Usa 3	mu3-4	3.3	28	quadruplicata	Slope	Passed	Failed
Malaya Usa 3	mu3-5	3.7	27.4	quadruplicata	Slope	Passed	Failed

Malaya Usa 3	mu3-6	3.7	27.6	quadruplicata	Slope	Passed	Passed
Malaya Usa 3	mu3-8	3.4	27.3	quadruplicata	Slope	Passed	Passed
Malaya Usa 3	mu3-9	2.6	27.2	quadruplicata	Slope	Passed	Passed
Malaya Usa 3	mu3-10	2.2	27	quadruplicata	Slope	Passed	Passed
Malaya Usa 3	mu3-11	3.1	27.1	quadruplicata	Slope	Passed	Passed
Malaya Usa 3	mu3-12	3.6	27.4	quadruplicata	Slope	Passed	Passed
Malaya Usa 3	mu3-13	2.9	27.2	quadruplicata	Slope	Passed	Passed
Malaya Usa 3	mu3-14	3.5	27.7	quadruplicata	Slope	Passed	Passed
Malaya Usa 3	mu3-15	2.2	26.7	quadruplicata	Slope	Passed	Failed
Malaya Usa 3	mu3-16	3.7	27.4	quadruplicata	Slope	Passed	Passed
Malaya Usa 3	mu3-17	4.2	28.4	quadruplicata	Slope	Passed	Passed
Malaya Usa 3	mu3-18	3.7	27.4	quadruplicata	Slope	Passed	Passed
Malaya Usa 3	mu3-19	4	27.4	quadruplicata	Slope	Passed	Passed
Malaya Usa 3	mu3-20	3.6	27.7	quadruplicata	Slope	Passed	Passed
Malaya Usa 3	mu3-21	3.5	27.6	quadruplicata	Slope	Passed	Passed
Malaya Usa 3	mu3-22	3.8	27.2	quadruplicata	Slope	Passed	Passed
Malaya Usa 3	mu3-23	4.1	27.8	quadruplicata	Slope	Passed	Passed
Malaya Usa 3	mu3-24	3.4	26.3	quadruplicata	Slope	Passed	passed
Malaya Usa 3	mu3-25	4.4	26.8	quadruplicata	Slope	Passed	Failed
Malaya Usa 3	mu3-26	3.8	26.8	quadruplicata	Slope	Passed	Failed

Continued

—cont'd

Section	Sample	$d^{13}C$ PDB	$d^{18}O$ SMOW	Conodont zone	Facies belt	Screening test (C-O isotope diagram)	Screening test (lithology, thin sections)
Malaya Usa 3	mu3-27	3.2	27.2	isosticha	Slope	Passed	Failed
Malaya Usa 3	mu3-27a	4.8	29.4	isosticha	Slope	Passed	Passed
Malaya Usa 3	mu3-28	5	27.5	isosticha	Slope	Passed	Failed
Malaya Usa 3	mu3-29	3.4	27.7	isosticha	Slope	Passed	Failed
Malaya Usa 3	mu3-30	4.3	28.9	isosticha	Slope	Passed	Failed
Malaya Usa 3	mu3-31	5.2	29.1	isosticha	Slope	Passed	Passed
Malaya Usa 3	mu3-32	5.3	29.9	isosticha	Slope	Passed	Failed
Malaya Usa 3	mu3-33	5	28.9	isosticha	Slope	Passed	Failed
Malaya Usa 3	mu3-34	4.1	28.7	isosticha	Slope	Passed	Passed
Malaya Usa 3	mu3-35	5.9	26.4	isosticha	Slope	Passed	Failed
Kochmes 1	2919.5m	4.35	29.39	typicus	Platform	Passed	Passed
Kochmes 1	2945.4m	7.39	28.33	typicus	Platform	Passed	Passed
Kochmes 1	2962m	8.04	27.84	typicus	Platform	Passed	Passed
Kochmes 1	2977m	7.89	27.65	typicus	Platform	Passed	Passed
Kochmes 1	2992.3m	7.85	28.06	typicus	Platform	Passed	Passed
Kochmes 1	2993m	7.5	28.42	typicus	Platform	Passed	Passed
Kochmes 1	3006m	7.77	27.62	typicus	Platform	Passed	Passed
Kochmes 1	3024m	7.57	27.62	typicus	Platform	Passed	Passed

References

Barskov, I.S., Kononova, L.I., Migdisova, A.V., 1984. Konodonty' nizhneturnejskikh otlozhenij podmoskovnogo bassejna [Conodonts of Lower Tournaisian deposits of the Moscow Region basin]. In: Menner, V.V. (Ed.), Paleontologicheskaya kharakteristika stratotipicheskikh i opornykh razrezov karbona Moskovskoj sineklizy. MGU Publications, Moskow, pp. 3–31.

Blain, J.A., Ray, D.C., Wheeley, J.R., 2016. Carbon isotope (13Ccarb) and facies variability at the Wenlock–Ludlow boundary (Silurian) of the Midland Platform, UK. Can. J. Earth Sci. 53, 1–6. https://doi.org/10.1139/cjes-2015-0194.

Brand, U., Logan, A., Bitner, M.A., Griesshaber, E., Azmy, K., Buhl, D., 2011. What is the ideal proxy for Palaeozoic seawater chemistry? Mem. Assoc. Australas. Paleontol. 0810-8889. 41, 9–24.

Bruckschen, P., Veizer, J., 1997. Oxygen and carbon isotopic composition of Dinantian brachiopods: paleoenvironmental implications for the Lower Carboniferous of western Europe. Palaeogeogr. Palaeoclimatol. Palaeoecol. 132 (1–4), 243–264. https://doi.org/10.1016/s0031-0182(97)00066-7.

Buggisch, W., Joachimski, M.M., 2006. Carbon isotope stratigraphy of the Devonian of Central and Western Europe. Palaeogeogr. Palaeoclimatol. Palaeoecol. 240, 68–88.

Caputo, M.V., Melo, J.H.G., Streel, M., Isbell, J.L., 2008. Late devonian and early carboniferous glacial records of South America. In: Fielding, C.R., Frank, T.D., Isbell, J.L. (Eds.), Resolving the Late Paleozoic Ice Age in Time and Space: Geological Society of America Special Paper 441, pp. 161–173. https://doi.org/10.1130/2008.2441(11).

Chen, J., Montañez, I.P., Qi, Y., Wang, X., Wang, Q., Lin, W., 2016. Coupled sedimentary and $\delta^{13}C$ records of late Mississippian platform-to-slope successions from South China: insight into $\delta^{13}C$ chemostratigraphy. Palaeogeogr. Palaeoclimatol. Palaeoecol. 448, 162–178.

Cheng, K., Elrick, M., Romaniello, S.J., 2020. Early Mississippian ocean anoxia triggered organic carbon burial and late Paleozoic cooling: evidence from uranium isotopes recorded in marine limestone. Geology 48. https://doi.org/10.1130/G46950.1.

Chermnykh, V.A., Kononova, L.I., Pazukhin, V.N., Cheusova, G.N., 1980. Tri tipa razrezov otlozhenij, perekhodnykh mezhdu devonom i karbonom na severe Ural'skoj skladchatoj oblasti [Three types of the Devonian—Carboniferous boundary sections in the north of the Ural folded region]. In: Biostratigrafiya pogranichny'kh otlozhenij devona i karbona SSSR. Magadan, pp. 1–18.

Chermnykh, V.A., Kochetkova, N.M., Pazukhin, V.N., Lipina, O.A., 1988. Pogranichnye devonsko-kamennougol'nye otlozheniya Severnogo i Pripolyarnogo Urala [Devonian—Carboniferous boundary deposits of the Northern and Subpolar Urals]. In: Granicza devona i karbona na territorii SSSR. Minsk, pp. 145–151.

Cole, D., Myrowa, P.M., Fike, D.A., Hakima, A., Gehrels, G.E., 2015. Uppermost Devonian (Famennian) to Lower Mississippian events of the western U.S.: stratigraphy, sedimentology, chemostratigraphy, and detrital zircon geochronology. Palaeogeogr. Palaeoclimatol. Palaeoecol. 427, 1–19. https://doi.org/10.1016/j.palaeo.2015.03.014.

Corradini, C., Spaletta, C., Mossoni, A., Matyja, H., Over, D.J., 2016. Conodonts across the Devonian/Carboniferous boundary: a review and implication for the redefinition of the boundary and a proposal for an updated conodont zonation. Geol. Mag. (1), 1–15.

Cramer, B., Saltzman, M., Day, J.E., Witzke, B.J., 2008. Record of the Late Devonian Hangenberg global positive carbon-isotope excursion in an epeiric sea setting: carbonate production, organic-carbon burial and paleoceanography during the Late Famennian. In: Holmden, C., Pratt, B.R. (Eds.), Dynamics of Epeiric Seas: Sedimentological, Paleontological and Geochemical Perspectives. Geological Association of Canada Special Publication 48, pp. 103–118.

Deulin, Y.v., 2006. Konodonty i korrelyacziya produktivnykh na neft' verkhnedevonskikh otlozhenij severa Timano-Pechorskoj provinczii [Conodonts and correlation of the Upper Devonian oil productive deposits of the north of the Timan-Pechora province]. Pravda Severa, Arkhangel'sk, p. 253 (in Russian).

Erofeevskiy, A.V., Zhuravlev, A.V., 2019. Perspektivy ispol'zovaniya izotopno-stratigraficheskogo metoda (δ13Ckarb) dlya korrelyatsii verkhnedevonsko-kamennougol'nogo intervala vostoka Timano-Pechorskoy provintsii [Perspectives of application of isotope-stratigraphy (δ13Ccarb) for correlation of the Upper Devonian—Carboniferous interval of the east of Timan-Pechora Province]. In: Neftegazovaya Geologiya. Teoriya I Praktika. vol. 14. no. 1. Available at http://www.ngtp.ru/rub/2019/9_2019.html. (in Russian).

Huck, S., Wohlwend, S., Coimbra, R., Christ, N., Weissert, H., 2017. Disentangling shallow water bulk carbonate carbon isotope archives with evidence for multi stage diagenesis: an in depth component specific petrographic and geochemical study from Oman (mid Cretaceous). Deposit. Rec. 3 (2), 233–257. https://doi.org/10.1002/dep2.35.

Immenhauser, A., Della Porta, G., Kenter, J.A.M., Bahamonde, J.R., 2003. An alternative model for positive shifts in shallow marine carbonate $\delta^{13}C$ and $\delta^{18}O$. Sedimentology 50, 953–959. https://doi.org/10.1046/j.1365-3091.2003.00590.x.

Kaiser, S.I., Steuber, T., Becker, R.T., Joachimski, M.M., 2006. Geochemical evidence for major environmental change at the Devonian–Carboniferous boundary in the Carnic Alps and the Rhenish Massif. Palaeogeogr. Palaeoclimatol. Palaeoecol. 240, 146–160.

Kaiser, S.I., Steuber, T., Becker, R.T., 2008. Environmental change during the Late Famennian and Early Tournaisian (Late Devonian—Early Carboniferous)—implications from stable isotopes and conodont biofacies in southern Europe. In: Aretz, M., Herbig, H.G., Somerville, I.D. (Eds.), Carboniferous Platforms and Basins—Geological Journal Special Issue. vol. 43, pp. 241–260. 2–3.

Kumpan, T., Bábek, O., Kalvoda, J., Matys Grygar, T., Frýda, J., 2014. Sea-level and environmental changes around the Devonian–Carboniferous boundary in the Namur–Dinant Basin (S Belgium, NE France): a multi-proxy stratigraphic analysis of carbonate ramp archives and its use in regional and interregional correlations. Sediment. Geol. 311, 43–59. https://doi.org/10.1016/j.sedgeo.2014.06.007.

Lakin, J.A., Marshall, J.E.A., Troth, I., Harding, I.C., 2016. Greenhouse to icehouse: a biostratigraphic review of latest Devonian–Mississippian glaciations and their global effects. Geol. Soc. Lond. Spec. Publ. 423, 439–464. https://doi.org/10.1144/SP423.12.

Liu, J., Algeo, T.J., Qie, W., Saltzman, M.R., 2018. Intensified oceanic circulation during Early Carboniferous cooling events: Evidence from carbon and nitrogen isotopes. Palaeogeogr. Palaeoclimatol. Palaeoecol. 531, 108962. https://doi.org/10.1016/j.palaeo.2018.10.021.

Lohmann, K.C., 1988. Geochemical patterns of meteoric diagenetic systems and their application to studies of paleokarst. In: James, N.P., Choquette, P.W. (Eds.), Paleokarst. Springer, Berlin, pp. 58–80.

Metzger, J.G., Fike, D.A., 2013. Techniques for assessing spatial heterogeneity of carbonate δ13C values: implications for craton-wide isotope gradients. Sedimentology 60, 1405–1431. https://doi.org/10.1111/sed.12033.

Mii, H., Grossman, E.L., Yancey, T.E., 1999. Carboniferous isotope stratigraphies of North America: implications for Carboniferous paleoceanography and Mississippian glaciation. Geol. Soc. Am. Bull. 111 (7), 960–973. https://doi.org/10.1130/0016-7606(1999)111<0960:cisona>2.3.co;2.

Nemirovskaya, T.I., Chermnykh, V.A., Kononova, L.I., Pazukhin, V.N., 1992. Conodonts of the Devonian-Carboniferous boundary section, Kozhim, Polar Urals, Russia. Ann. Soc. Geol. Belg. 115, 629–647.

Plotitsyn, A.N., Gruzdev, D.A., 2016. Granicza devonskoj i kamennougol'noj sistem na r. Izyayol (yug gryady Chernysheva) [Devonian and Carboniferous boundary in the South of the Chernishev Swell (Izyael river section)]. In: Geologiya, geoekologiya i resursnyj potenczial Urala i sopredel'nykh territorij: Materials of VI All-Russian. youth. conf. Ufa, pp. 102–107. (in Russian).
Plotitsyn, A.N., Vevel, Y.A., 2019. Biostratigrafiya verkhnedevonsko-nizhnekamennougol' nykh otlozheniy opornogo razreza na r. Kozhim (Pripolyarnyy Ural) po konodontam [Conodonts of the Devonian-Carboniferous boundary deposits of the Kozhim River key section (Subpolar Urals)]. In: Neftegazovaya Geologiya. Teoriya I Praktika. vol. 14. no. 3, Available at: http://www.ngtp.ru/rub/2019/25_2019.html. https://doi.org/10.17353/2070-5379/25_2019. (in Russian).
Plotitsyn, A.N., Sobolev, D.B., Gruzdev, D.A., 2014. Novye dannye o pogranichnykh otlozheniyakh devona i karbona na Pripolyarnom Urale (r. Syv'yu) [New data on Devonian and Carboniferous boundary deposits in the Subpolar Urals (R. Syvyu)]. Vestnik IG Komi SC UBr RAS 6, 13–20. (in Russian).
Plotitsyn, A.N., Ponomarenko, E.S., Vevel, Y.A., 2018. Turnejskie glubokovodnye otlozheniya bassejna r. Usa (Polyarnyj Ural) [Tournaisisan deep-water deposits of the Usa river basin (Polar Urals)]. In: Neftegazovaya Geologiya. Teoriya I Praktika. vol. 13. no. 3, Available at: http://www.ngtp.ru/rub/2/32_2018.pdf. (in Russian).
Qie, W.K., Zhang, X.H., Du, Y.S., Zhang, Y., 2011. Lower Carboniferous carbon isotope stratigraphy in South China: Implications for the Late Paleozoic glaciation. Sci. China Earth Sci. 54, 84–92. https://doi.org/10.1007/s11430-010-4062-4.
Qie, W., Wang, X.-D., Zhang, X., Ji, W., Grossman, E.L., Huang, X., Liu, J., Luo, G., 2016. Latest Devonian to earliest Carboniferous conodont and carbon isotope stratigraphy of a shallow-water sequence in South China. Geol. J. 51, 915–935.
Qie, W., Algeo, T.J., Luo, G., Herrmann, A., 2019. Global events of the Late Paleozoic (Early Devonian to Middle Permian): a review. Palaeogeogr. Palaeoclimatol. Palaeoecol. 531. https://doi.org/10.1016/j.palaeo.2019.109259.
Saltzman, M.R., 2002. Carbon and oxygen isotope stratigraphy of the Lower Mississippian (Kinderhookian–lower Osagean), western United States: implications for seawater chemistry and glaciation. Geol. Soc. Am. Bull. 114, 96–108.
Saltzman, M.R., Thomas, E., 2012. Carbon isotope stratigraphy. In: The Geologic Time Scale. Elsevier, pp. 207–232. https://doi.org/10.1016/B978-0-444-59425-9.00011-1.
Saltzman, M.R., Groessens, E., Zhuravlev, A., 2004. Carbon cycle models based on extreme changes in δ13C: an example from the Lower Mississippian. Palaeogeogr. Palaeoclimatol. Palaeoecol. 213, 359–377. https://doi.org/10.1016/S0031-0182(04)00389-X.
Sandberg, C.A., Ziegler, W., Leuteritz, K., Brill, S.M., 1978. Phylogeny, speciation, and zonation of Siphonodella (Conodonta, Upper Devonian and Lower Carboniferous). Newsl. Stratigr. 7, 102–120.
Scholle, P.A., Arthur, M.A., 1980. Carbon isotope fluctuations in Cretaceous pelagic limestones: Potential stratigraphic and petroleum exploration tool. Am. Assoc. Pet. Geol. Bull. 64, 67–87.
Sedaeva, K.M., Ryabinkina, N.N., Kuleshov, V.N., Valyaeva, O.V., 2010. Otrazhenie Hangenbergskogo global'nogo geologicheskogo sobytiya rubezha devona i karbona v razrezakh zapadnogo sklona Pripolyarnogo (r. Kozhim) i Yuzhnogo (r. Sikaza) Urala [The Hangenberg global geological event on Devon-Carbon boundary strata on the western slope of Prepolar (the Kozhim River) and Southern (the Sikaza River) Urals]. In: Litosfera, pp. 25–37. no. 6. [in Russian].
Sobolev, D.B., 2011. Evolutsya i bioraznoobrazie rannekamennougolnykh ostrakod na sobytiynykh urovnyakh [Evolution and biodiversity of early Carboniferous ostracodes at event levels]. In: Tsyganko, V.S., Beznosova, T.M., Ponomarev, D.V. (Eds.),

Biogeologiya, evolutsya organizmov i bioraznoobraziya v geologicheskoy istorii Timano-Severouralskogo regiona, modelirovanie paleoekosistem, paleontologicheskie i stratigraficheskie geokorrelaytsyi, pp. 102–108. Syktyvkar, Geoprint. [in Russian].

Sobolev, D.B., Zhuravlev, A.V., 1994. Konodonty i ostrakody pogranichnykh otlozhenij devona i karbona na r. Vangyr [Conodonts and ostracods of the Devonian - Carboniferous boundary deposits of the Vangyr River section]. Obosnovanie granicz stratigraficheskikh podrazdelenij. Syktyvkar. pp. 35–44. (*Proceedings of the Institute of Geology of Komi Science Center*, Ural Branch of the Russian Academy of Sciences, issue 82). [in Russian].

Sobolev, D.B., Zhuravlev, A.V., Tsyganko, V.S., 2000. Stop 8. Upper Devonian-Lower Carboniferous succession on the Kozhym River. In: Pan-Arctic Palaeozoic Tectonics, Evolution of Basins and Faunas. Subpolar Urals Field Trip Guidebook. Syktyvkar, pp. 101–111 (IGCP406 Special Publication 6).

Sobolev, D.B., Zhuravlev, A.V., Tsyganko, V.S., 2002. Stop 3. Upper Devonian-Lower Carboniferous succession on the Kozhym River. In: Geology of the Devonian System. Subpolar Urals Field Trip Guidebook. Institute of Geology, Syktyvkar, pp. 21–40.

Sostoyanie izuchennosti, 2008. Sostoyanie izuchennosti stratigrafii dokembriya i fanerozoya Rossii. Zadachi dal'nejshikh issledovanij: Postanovleniya Mezhvedomstvennogo stratigraficheskogo komiteta i ego postoyanny'kh komissij [The state of knowledge of the Precambrian and Phanerozoic stratigraphy of Russia. Tasks for further research: resolutions of the Interagency Stratigraphic Committee and its standing commissions]. vol. 38 SPb: VSEGEI Publications, p. 131 (in Russian).

Spalletta, C., Perri, M.C., Over, D.J., Corradini, C., 2017. Famennian (Upper Devonian) conodont zonation: revised global standard. Bull. Geosci. 92, 31–57.

Tchernov, G.A., 1960. Nizhnij karbon Pechorskoj gryady [Lower Carboniferous of the Pechora Swell]. In: Sbornik trudov po geologii i paleontologii, Syktyvkar. Institute of Geology Publ, pp. 56–93 (in Russian).

Vevel, Y.A., Zhuravlev, A.V., Popov, V.V., 2012. Deposits of the Devonian and Carboniferous boundary in the Kamenka River section (Pechora-Kozhvinsky megaswell, Timan-Pechora province). In: Neftegazovaya geologiya. vol. 7. Teoriya i praktika (RUS). (1), Available at: http://www.ngtp.ru/rub/2/6_2012.pdf. (in Russian, with English abstract).

Yao, L., Qie, W., Luo, G., Liu, J., Algeo, T.J., Bai, X., Yang, B., Wang, X., 2015. The TICE event: Perturbation of carbon-nitrogen cycles during the mid-Tournaisian (Early Carboniferous) greenhouse-icehouse transition. Chem. Geol. 401, 1–14. https://doi.org/10.1016/j.chemgeo.2015.02.021.

Zhuravlev, A.V., 2012. Sedimentation model of the transitional zone from the Elets to Lemva Formation Belts (Devonian-Carboniferous boundary, Cis-Polar Ural). In: Neftegasovaa geologia. Teoria i practica. 7(4). http://www.ngtp.ru/rub/2/59_2012.pdf. (in Russian).

Zhuravlev, A.V., 2017. Shallow-water siphonodellids and definition of the Devonian-Carboniferous boundary. Newslett. Carbonifer. Stratigr. 33, 21–26.

Zhuravlev, A.V., Sobolev, D.B., 2019. Devonian-Carboniferous boundary in the North of the Urals (Kamenka River and Vangyr River sections). Vestnik IG Komi SC UB RAS 10, 16–22. https://doi.org/10.19110/2221-1381-2019-10-16-22.

Zhuravlev, A.V., Vevel, Y.A., 2018. Pozdneturneyskoe sobytie v razrezakh Kos'yu-Rogovskoy vpadiny i ee obramleniya (Timano-Pechorskaya provintsiya) [The late Tournaisian Event in the successions of the Kosyu-Rogovaya Depression and adjacent areas (Timan-Pechora Province)]. In: Neftegazovaya Geologiya. Teoriya I Praktika. vol. 13. no. 1, Available at http://www.ngtp.ru/rub/2/3_2018.pdf. https://doi.org/10.17353/2070-5379/3_2018. (in Russian).

Zhuravlev, A.V., Vevel, Y.A., Plotitsyn, A.N., Erofeevsky, A.V., Lukin, V.Y., 2019. Pozdneturneyskie melkovodnye otlozheniya vostoka Timano-Pechorskoy provintsii (r. Kozhva) [Late Tournaisian shallow-water sequenses of the eastern part of Timan-Pechora Province (Kozhva River)]. In: Neftegazovaya Geologiya. Teoriya i Praktika. 14(4), Available at http://www.ngtp.ru/rub/2019/37_2019.html. (in Russian).

CHAPTER TEN

The isotopic composition of pedogenic carbonate as a record of pCO_2: Insights from the Late Paleozoic of tropical western Pangea

Lawrence H. Tanner[a,]* and Spencer G. Lucas[b]
[a]Department of Biological Sciences, Le Moyne College, Syracuse, NY, United States
[b]New Mexico Museum of Natural History & Science, Albuquerque, NM, United States
*Corresponding author: e-mail address: tannerlh@lemoyne.edu

Contents

Abstract

The use of the carbon isotope composition ($\delta^{13}C$) of pedogenic calcite as a proxy for pCO_2 was established over three decades ago and remains a primary means of investigating climate history. Over time, an increased understanding of the systematics of pedogenic carbonate precipitation has improved the constraints on interpreting paleoatmospheric composition from the isotopic composition of calcrete. We review the development of the use of pedogenic $\delta^{13}C$ as a proxy for pCO_2 over geologic time,

Stratigraphy & Timescales, Volume 5
ISSN 2468-5178
https://doi.org/10.1016/bs.sats.2020.08.009

with focus on the late Paleozoic. We illustrate the discussion with specific examples from the late Carboniferous through early Permian icehouse climate interval and present new as well as previously published data from strata of Virgilian (Late Pennsylvanian) through Wolfcampian (early Permian) age collected in northern and south-central New Mexico, which was located in tropical western Pangea during the late Paleozoic. Our success in fitting pCO_2 values we generate through new analyses to previously published paleo-pCO_2 curves strengthens the rationale for use of these data as proxies for paleoclimate and suggests the potential for their use as a correlation tool.

1. Introduction

The sedimentary record of the upper Carboniferous through lower Permian of tropical Pangea has long been recognized as an invaluable archive of paleoclimatic data for its potential to inform our understanding of the timing and mechanisms of late Paleozoic global change. This interval of the late Paleozoic witnessed global environmental changes that included the northward drift of the Pangean continent (Fig. 1), potential changes

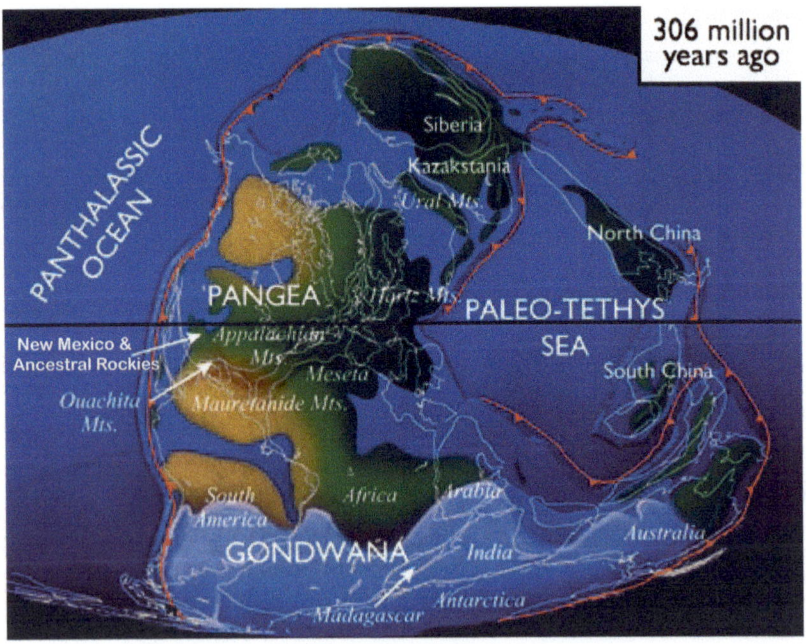

Fig. 1 Paleogeographic reconstruction of the Earth at the start of the late Pennsylvanian with various land areas and tectonic features labeled for reference.

in atmospheric circulation, substantial changes in paleoclimate coincident with changes in atmospheric composition and the end of Gondwanan glaciation, followed by the interpreted late Paleozoic aridification of the Pangean interior (e.g., DiMichele et al., 2006; Dinterman et al., 2000; Giles et al., 2013; Kessler et al., 2001; Mack et al., 2003; Montañez et al., 2007, 2016; Mountney, 2006; Mujal et al., 2018; Parrish, 1993; Soreghan et al., 2002; Tabor and Montañez, 2002, 2004; Tabor et al., 2008; Tanner and Lucas, 2017, 2018; Zhu and Tabor, 2014). Of particular interest is the interval that encompasses the Pennsylvanian-Permian transition, a time for which some computer models have indicated stable atmospheric circulation that followed a largely zonal configuration (Kutzbach and Gallimore, 1989; Kutzbach and Ziegler, 1993). As interpreted by multiple authors, a humid to subhumid-seasonal climate that prevailed at tropical latitudes during the latest Carboniferous (Virgilian) was replaced by a more arid climate during the earliest Permian (Wolfcampian, equivalent to the Asselian through early Artinskian); these conditions spread across central Pangea and at least as far north as the subtropical Appalachian basin (Cecil, 2013; DiMichele et al., 2006; Tabor and Montañez, 2002, 2004; Tabor et al., 2008). Some have interpreted this climate change as having been driven by a major reorganization of the climate system following the termination of Gondwanan glaciation, or to the reorganization of atmospheric circulation with the onset of the Pangean megamonsoon (Gibbs et al., 2002; Kutzbach and Gallimore, 1989; Kutzbach and Ziegler, 1993; Parrish, 1993; Peyser and Poulsen, 2008; Tabor and Poulsen, 2008).

In an effort to improve understanding of the environmental complexities of this time interval, several authors have examined the potential coupling of climate, paleo-pCO_2, (the past partial pressure of CO_2 in the atmosphere), Gondwanan glaciation and sea level (Cecil et al., 2003; Chen et al., 2018; Montañez et al., 2007, 2016; Peyser and Poulsen, 2008). Indeed, our understanding of the profound relationship between the cyclical variations in ice volume and pCO_2 during the Pleistocene (e.g., Lüthi et al., 2008) compels us to examine the late Paleozoic record for analogs. Relevant to these studies are the pedogenic features of the strata from this time of transition, with numerous studies citing specific pedologic morphologies and/or the stable isotopic composition of pedogenic minerals (e.g., DiMichele et al., 2006; Giles et al., 2013; Kessler et al., 2001; Mack et al., 2003; Montañez et al., 2007, 2016; Tabor and Montañez, 2002, 2004; Tabor et al., 2008; Tanner and Lucas, 2017, 2018; Zhu and Tabor, 2014). In an attempt to improve our understanding of this critical time interval, we examine previously published studies of paleo-pCO_2 as measured from the isotopic analysis of pedogenic carbonates and contribute to this discussion new data from the upper Carboniferous-lower Permian of New Mexico (USA), then located in tropical western Pangea.

2. Pedogenic carbonate

2.1 Basics

The identification of pedogenic carbonate in paleosols is widely applied as an indicator of arid to semi-arid paleoclimate due to the generally acknowledged propensity for carbonate precipitation to occur in the B horizons of soils in regions with seasonally or perennially dry climates (Schaetzl and Thompson, 2015). The mechanism of precipitation of calcium carbonate in soils involves the same carbonate/bicarbonate equilibria as in aqueous environments,

$$Ca^{2+} + 2HCO^{3-} \leftrightarrow CaCO_3 + H_2CO_3$$

where a decrease in either H_2O, as soil moisture, or CO_2 from root respiration or aerobic microbial decay, drives the reaction to the right (precipitation of $CaCO_3$). Conversely, more water or increased biotic activity (respiration) favors dissolution. Hence, the formation of soil carbonate horizons is heavily dependent on the availability of soil moisture. In modern soils, pedogenic carbonate accumulations are common in regions where the climate is warm and seasonally to mostly dry (i.e., torric, ustic and xeric soil-moisture regimes). Seasonally dry climates with mean annual precipitation of 100–500 mm have been cited as most conducive to the formation of calcareous soil horizons (Birkeland, 1999; Goudie, 1983; Reeves, 1976), although Retallack (2005) pointed out that the isohyet boundary between calcareous and noncalcareous soils may vary regionally from ca. 350 mm (or less) to >750 mm, depending on the magnitude of seasonality and temperature. Importantly, soil-forming processes operate over extended intervals of up to 10^4 years, 10^5 years under conditions of exceptional landscape stability, and the formation of calcite-cemented soil horizons, or calcretes, is no exception, as demonstrated by Machette (1985). Hence, paleoclimatic interpretations based on measurements derived from pedogenic carbonate must take the duration of paleosol formation into consideration to consider the potential that varying climate conditions have been "averaged."

Commonly, the simple presence or absence of calcareous horizons in paleosols is presented as evidence for aridity. As a case in point, Miller et al. (1996) interpreted the cyclical alternation between vertic and calcic paleosols in stacked sequences from the lower Permian (Wolfcampian) of the North American Midcontinent as indicating cyclical climate variations that correlate with the proposed orbitally forced glacial/interglacial cycles of this time interval. Tandon and Gibling (1997) similarly ascribed the presence of nodular calcretes in the coal-bearing cyclothems of the upper Carboniferous in the Canadian Maritimes to the alternation of arid climate conditions during glacio-eustatic lowstands and humid climates during highstands, at which time the peat accumulated that eventually became coal.

Clearly, a single pedogenic feature is not capable of recording the range of paleoclimate variations that may occur on time scales of the longer orbital cycles. Pedogenic carbonates will occur only in those portions of a stratigraphic section for which the threshold conditions were met for carbonate precipitation during soil formation. Therefore, the investigator must distinguish between the alternatives for the lack of pedogenic carbonate—does the absence of carbonate reflect a climate too humid for precipitation during soil formation, or was the landscape too actively accumulating sediment for soils to form?

2.2 Pedogenic Carbonate Isotope Systematics

Perhaps no single development has had a greater effect on the utility of pedogenic carbonates in paleoclimatic studies than the routine application of stable-isotope analysis. The analysis of carbon and oxygen isotopes, which can be measured relative to the standards PDB (Peedee Belemnite) or its equivalent VPDB (Vienna Peedee Belemnite), or SMOW (Standard Mean Ocean Water), has now become standard practice, in large part for the information it may provide about the paleoenvironment at the time of soil formation. The most commonly applied carbon isotope measurement is $\delta^{13}C_{carb}$, the ratio of ^{13}C to ^{12}C from pedogenic carbonate measured as:

$$\delta^{13}C_{carb} = \frac{\left({}^{13}C/{}^{12}C_{sample} - {}^{13}C/{}^{12}C_{standard}\right)}{{}^{13}C/{}^{12}C_{standard}} \times 1000$$

Cerling (1991) pioneered the use of $\delta^{13}C$ from pedogenic carbonate to estimate the concentration of atmospheric CO_2 in Earth's past, developing the methodology based on the simple principle that the isotopic composition of carbonate precipitating in the soil is determined by the relative contributions from the $\delta^{13}C$ of CO_2 respired in the soil and the $\delta^{13}C$ of CO_2 from the atmosphere. In general terms, the $\delta^{13}C$ of soil carbonate is controlled by multiple factors, including: the ratio of C_4 to C_3 vegetation (not relevant before the middle Cenozoic); the depth of carbonate accumulation below the soil surface, which controls the extent of atmospheric mixing; the temperature of carbonate precipitation, which determines fractionation; the isotopic composition of atmospheric carbon, a value well-constrained by analysis of marine carbonates; the rate of production of CO_2 within the soil; and the isotopic composition of the respired CO_2 within the soil (Cerling, 1991, 1999; Ekart et al., 1999). Recognition of these controlling factors allowed development of the diffusion-reaction model for calculation of the concentration of CO_2 in the atmosphere at the time carbonate accumulation in the past, the pCO_2 paleobarometer (Cerling, 1991, 1999; Ekart et al., 1999):

$$C_a = S(z)\frac{\delta^{13}C_s - 1.0044\delta^{13}C_\Phi - 4.4}{\delta^{13}C_a - \delta^{13}C_s}$$

where C_a is atmospheric pCO_2, which is often assumed proportional to the isotopic composition of marine carbonate by the conversion, $C_a = \delta^{13}C_{mc} - 8‰$; S(z) is pCO_2 (ppmV) contributed by soil respiration, a factor dependent on plant productivity, and often considered proportional to mean annual precipitation, although some authors assign S(z) values based on paleosol types or measurements of depth to carbonate (cf. Montañez, 2013); $\delta^{13}C_s$ is the isotopic composition of soil CO_2, calculated from the $\delta^{13}C_{carb}$ of pedogenic carbonate and the estimated soil temperature (Romanek et al., 1992); and $\delta^{13}C_\Phi$ is the isotopic composition of soil-respired CO_2, generally assumed to be equivalent or nearly so to the bulk isotopic composition of the soil organic matter.

Breecker (2013) developed a computer model (PBUQ) for calculating pCO_2 that allows researchers to evaluate the levels of uncertainty associated with the different parameters through Monte Carlo simulations. This approach allows a very nuanced and more statistically robust methodology for studying the history of paleo-pCO_2 variation than was possible only a decade earlier.

2.3 Limitations on use of pedogenic carbonate as a paleobarometer

There are significant restrictions on the proper application of the diffusion-reaction model. As noted by Ekart et al. (1999), not all pedogenic carbonate is suitable for this analysis. Specifically, the carbonate must be precipitated at least 50 cm below the ground surface to avoid direct atmospheric interaction. This can be problematic in many instances; due to erosional truncation of the soil epipedon, which typically is less resistant than the calcite-cemented calcrete, it is commonly impossible to determine the depth below the surface of carbonate accumulation. Further, the carbonate should be from mature paleosols, preferably Calcisols or calcic Argillisols rather than Protosols or Vertisols, the latter because the influence of isotopically-heavy atmospheric carbon can extend much deeper in soils with vertic fracturing (Driese and Mora, 1993).

Some authors (Breecker, 2013; Montañez, 2013) have noted that the greatest source of uncertainty in the application of the diffusion-reaction model paleobarometer is assigning an appropriate value to the volume of soil-respired CO_2, S(z). In particular, Montañez (2013) documented that there is a mismatch between S(z) and mean annual precipitation, which is commonly used as the basis for assigning a value of S(z). In most calcretes, the carbonate precipitation occurs only during the season with a soil-moisture deficit, during which S(z) will be considerably lower. Hence, values of S(z) should be chosen to reflect dry season values. Montañez (2013) specifically noted, from measurements in various orders of carbonate-bearing modern soils, that S(z) is better constrained to low values (<5000 ppmV) in Aridisols (the modern equivalent of Calcisols) and Alfisols (equivalent to Argillisols), but subject to much higher variability in Vertisols.

Importantly, the calculation requires knowledge of the isotopic composition of the soil-respired CO_2, which is assumed to be nearly identical to the composition of the soil organic matter (SOM). However, the identification and isotopic analysis of organic matter can present a challenge in strongly oxidized paleosols, including most Calcisols, in which recoverable SOM may be scarce to nonexistent. Some authors have compiled tables of published fossil SOM analyses across a wide range of ages, e.g., Ekart et al. (1999), who noted that $\delta^{13}C_{om}$ can be approximated from the relationship $\delta^{13}C_{om} = \delta^{13}C_a - 18‰$. This is certainly a very approximate relationship, and its application should be considered a last resort given that the isotopic composition of plants varies across groups and between ecological environments. Other authors have noted that not all SOM within a single soil profile is the same. Bowen and Beerling (2004), for example, presented data that demonstrated that soil organic matter from deeper in a soil profile has more likely undergone increased microbial processing that alters its isotopic signature, potentially biasing the pCO_2 calculation.

Furthermore, Myers et al. (2016) determined that occluded organic matter, that is, organic matter preserved within carbonate nodules, provides a more faithful record of the original organic matter composition and thus results in a more accurate calculation of pCO_2, whereas calculations based on nonoccluded SOM typically overestimate pCO_2.

In addition, the calcrete must be demonstrably free of evidence of diagenetic recrystallization, which potentially can alter the isotopic composition (Bera et al., 2010; Ekart et al., 1999; Kraus, 1999), although it has been noted that diagenesis seems to affect the isotopic composition of carbon substantially much less severely than it does oxygen (Bera et al., 2010; Cerling, 1991). Primary pedogenic carbonate generally precipitates as micrite, so identification of sparry textures on petrographic examination allows discrimination of primary and secondary carbonate (Alonso-Zarza, 2003; Alonso-Zarza and Wright, 2010; Tanner, 2010). Cathodoluminescence petrography also is used to distinguish generations of carbonate precipitation, identified by zonation due to differences in Mn^{+2} or Fe^{+2} substitution for Ca^{+2} (Hiatt and Pufahl, 2014). Because marine carbonate is generally isotopically heavy compared to typical pedogenic carbonate, it is essential that the source of the carbonate is free of marine influence. For a similar reason, the carbonate should be free of the influence of groundwater, which could import carbonate ions with a different isotopic signature. Fortunately, the features of groundwater calcretes are well known and distinctive (Alonso-Zarza and Wright, 2010).

Finally, the context of calcrete within the stratigraphic record is essential to consider. Because calcretes form within a limited range of climatic conditions, soils formed outside those conditions, e.g., tropical to temperate humid conditions, may not be represented by sampling of pedogenic carbonate. Thus, application of the carbonate paleobarometer to a stratigraphic succesion across which climate varied during deposition may fail to record variations in pCO_2 that may have driven the climate variations. Furthermore, all stratigraphic records—paleontologic, paleomagnetic, geochemical—are subject to the limitations of completeness. Hypothetically, major changes in base level driven by changing climate (e.g., glacial-interglacial cycles) could create unconformities that remove the sedimentary record of one climate regime, leaving an incomplete and thus biased stratigraphy.

2.4 Comparisons to other proxies

Values of paleo-pCO_2 derived from the diffusion reaction model are broadly similar to values from other sources, such as geochemical models

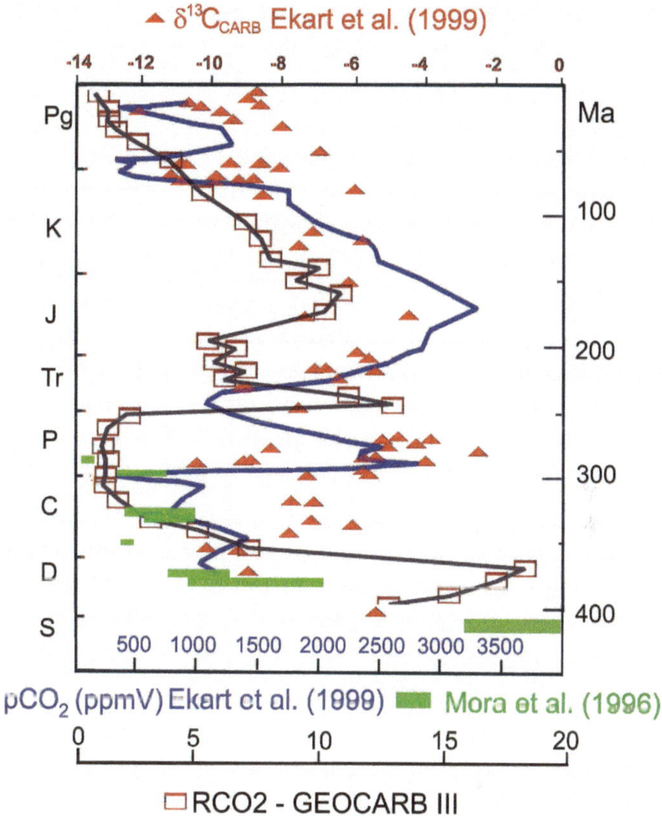

Fig. 2 Composite plot of pCO_2 as calculated from the GEOCARB III model of Berner and Kothavala (2001), pCO_2 ranges (based on range of S(z) values) calculated for specific formations by Mora et al. (1996), $\delta^{13}C_{carb}$ data compiled by Ekart et al. (1999) and pCO_2 curve calculated by Ekart et al. (1999) using S(z) = 5000 ppmV. *Adapted from Mora, C.I., Driese, S.G., Colarusso, L.A., 1996. Middle to Late Paleozoic atmospheric CO2 levels from soil carbonate and organic matter. Science 271, 1105–1107. doi:10.1126/science.271.5252.1105, Ekart, D.D., Cerling, T.E., Montanez, I.P., Tabor, N.J., 1999. A 400 million-year carbon isotope record of pedogenic carbonate: implications for paleoatmospheric carbon dioxide. Am. J. Sci. 299, 805–827 and Berner, R.A., Kothavala, Z., 2001. GEOCARB III: a revised model of atmospheric CO2 over Phanerozoic time. Am. J. Sci. 301, 182–204.*

(Berner, 1994, 2006; Berner and Kothavala, 2001) that use elemental fluxes between the atmosphere and geosphere over geologic time to estimate changes in pCO_2 (GEOCARB I through III and GEOCARBSULF; Fig. 2). More recently, Park and Royer (2011) made some adjustments to the GEOCARBSULF model of Berner (2006, 2008) that improve the model match to proxies, particularly for icehouse intervals (late Paleozoic, 340 to 260 Ma; late Cenozoic, 40 Ma to 0). Although not well resolved temporally, these models have provided and continue to provide useful context and constraints for examining qualitative measures of paleoclimate and paleoecology and for evaluating the consistency of other quantitative pCO_2 proxies. In particular, most paleosol-derived pCO_2 calculations match well geochemical models that indicate very high pCO_2 (10–15 times modern levels) during the Late Silurian (the earliest period for which pedogenic carbonates are available) through early Carboniferous, falling to near modern levels in the late Carboniferous through early Permian, before rising to levels that were 5–10 times higher than modern during the late Permian through Mesozoic (Berner, 2006; Berner and Kothavala, 2001; Cerling, 1991, 1999; Ekart et al., 1999; Mora et al., 1996; Royer et al., 2004; Schaller et al., 2015).

Another proxy commonly applied to the problem of determining paleo-pCO_2 is the use of stomatal indices measured from fossil plants. This proxy derives from observations by Woodward (1987) that in some modern plants stomatal density (the number of stomata per unit leaf area) and stomatal index (the number of stomata relative to the number of epidermal cells) have a demonstrated inverse correlation with atmospheric pCO_2. In principle, therefore, fossil leaves on which stomatal counts are possible can provide a record of paleo-pCO_2. Many studies have worked with the members of the order Gingkoales, which has a fossil record dating back to the Permian and is represented today by the single species *Gingko biloba*. Beerling et al. (1998) found that specimens of *G. biloba* grown in growth chambers for 3 years at elevated pCO_2 (560 ppm) exhibited a demonstrable decrease in both stomatal density and index compared to *G. biloba* grown at modern CO_2 levels. These effects are interpreted mainly as phenotypic adaptations that improve water-use efficiency in plants by limiting stomatal conductance, and therefore transpiration, when pCO_2 is high. Chen et al. (2001) analyzed fossils of *Gingko* spp. of various geologic ages and described an approximate inverse correlation between stomatal index and the pCO_2 as determined by geochemical models (Berner and Kothavala, 2001). Retallack (2001) utilized the fossil record of stomatal indices of Gingkoales and seed ferns and found evidence of transient increases in pCO_2 levels at multiple key paleontologic boundaries, including the Permian-Triassic, the Triassic-Jurassic, and the Cretaceous-Paleogene, although Haworth et al. (2005) severely critiqued the statistical validity of these data.

This paleofloral proxy is not without its own potential sources of error or uncertainty. The underlying assumption in the application of stomatal indices as a paleo-barometer is that variations in pCO_2 are the primary driver in the observed variations of the stomatal indices. In fact, numerous studies have shown that the stomatal indices used to calculate paleo-pCO_2 are also subject to variations forced by other confounding factors, including water availability, light availability, temperature and humidity extremes and nutrient limitations (Beerling and Chaloner, 1993; Fraser et al., 2009; Hill et al., 2019; Nejad and Van Meeteren, 2005; Onwueme and Johnston, 2000; Peñuelas and Matamala, 1990). Another key assumption of the method, particularly in earlier studies, has been that the fossil plant groups subjected to study produced stomatal responses that were quantitatively similar to their modern relatives and that these responses are scalable across much larger ranges of pCO_2 than used in experiments. This assumption also has been challenged (Boucot and Gray, 2001; Haworth et al., 2005; Royer et al., 2001; Tanner, 2002). Not surprisingly, results from earlier studies were often inconsistent and incongruent with the results of geochemical modeling or isotopic analysis of pedogenic carbonate (cf. McElwain, 2018; McElwain et al., 2015).

Most recently, the calibration of fossil data to experimental results has improved, and methodologies for collecting and treating the data have become more standardized, yielding results more consistent with other proxies (Zhou et al., 2020). The improvements have resulted from a combination of new or updated approaches to utilizing the paleofloral record. Empirical studies, the traditional methodology, measures the response of plant stomatal indices and/or $\delta^{13}C$ to changes in pCO_2 during long-term growth chamber experiments (e.g., Barclay and Wing, 2016). Schubert and Jahren (2012, 2015) developed a model based on the experimentally measured relationship between the $\delta^{13}C$ of the atmosphere and the $\delta^{13}C$ of the fixed carbon in the plant tissues. Mechanistic models are based on theoretical derivations from principles of gas exchange and CO_2 diffusion and carbon fractionation during photosynthesis combined with considerations of stomatal anatomy. For example, Franks et al. (2014) presented an equation for the calculation of pCO_2 from measurements of leaf $\delta^{13}C$ and stomatal dimensions. They demonstrated the accuracy of their methodology by testing against known pCO_2 for the late Pleistocene through the modern. Their calculations of paleo-pCO_2 match well with the levels predicted by GEOCARBSULF (Berner, 2006) for most of the Paleozoic, but are well below the high Mesozoic pCO_2 predicted by the geochemical models. Porter et al. (2019) compared the empirical method of Schubert and Jahren (2012, 2015) to the mechanistic method of Franks et al. (2014) and found both useful. However, Porter et al. (2019) suggested that the empirical method tended to overestimate pCO_2 and that the mechanistic method was most useful for using data from multiple plant species. In summary, several decades of improvements to the methodologies of pedogenic and paleofloral proxies have resulted in increased congruence of the methods of paleo-pCO_2 calculations.

3. Pedogenic carbonate $\delta^{13}C$ and pCO_2

3.1 Paleozoic $\delta^{13}C$ and pCO_2

As early as 1991, Mora et al. attempted the examination of long-term trends in pCO_2 by sampling calcrete-bearing formations of the Appalachian Basin in Pennsylvania ranging in age from Late Silurian to Late Mississippian. However, the methodology and constraints on sampling were not yet well established, and the results were inconsistent with later studies, yielding paleo-pCO_2 levels of 4000 to over 10,000 ppmV for the Late Silurian (Bloomsburg Formation), ca. 2000 ppmV for the Late Devonian (Catskill Formation) and 5000–8000 ppmV for the Mississippian (Mauch Chunk Formation) (Mora et al., 1991). The authors admitted that the Silurian results could be invalid due to the lack of true root systems in the primitive vascular plants then extant and the potential for contamination of the soil by relict marine carbonate.

In 1996, Mora and colleagues presented an expanded and updated version of the earlier work and sampled calcrete from a wider area of the Appalachian Basin in strata ranging in age from Late Silurian (Bloomsburg Formation) to early Permian (Dunkard Formation). The measured $\delta^{13}C_{carb}$ values ranged from a high of −5.3‰ (Bloomsburg Formation) to a low of −9.8‰ for the Late Devonian Duncannon Member of the Catskill Formation (Mora et al., 1996). Calculated paleo-pCO_2 values (Fig. 2) were at a maximum for the Late Silurian (3200–5200 ppmV), falling during the Devonian (700–1250 ppmV for the Duncannon Member), falling further still into the Early Mississippian (450–500 for the Maccrady Formation), holding fairly steady through the remainder of the Mississippian and Pennsylvanian (450–1000 ppmV for the Mauch Chunk, Hinton, Pennington and Conemaugh formations) and falling abruptly in the early Permian (150–200 ppmV for the Dunkard Formation). The authors calculated these paleo-pCO_2 levels using values of $\delta^{13}C_a$ ($\delta^{13}C$ of atmospheric carbon) calculated from published values of $\delta^{13}C$ for marine carbonate of similar age and assuming that $\delta^{13}C_a$ was lower by 7‰. In addition, the authors assumed S(z) of 3000–7000 ppmV for all samples.

Ekart et al. (1999) attempted a similar project, but on a more ambitious scale, compiling an inventory of 758 measurements of $\delta^{13}C$, both new and previously published (including those of Mora et al., 1996), ranging in age from Late Silurian to late Pliocene. The values they reported for individual formations represented averages for all samples from the formation. The Paleozoic $\delta^{13}C_{carb}$ values range widely from a low of −10.1‰ (for the earliest Permian Archer City Formation in Texas) to a high of −2.6‰ (from the early Permian Cedar Mesa Formation of Utah). The general trend demonstrates profound variation; from a high of −5.3‰ in the Late Silurian, values fall abruptly to ∼−9.0‰ for the most of the Devonian, rising and stabilizing at ∼−7.0‰ throughout the Carboniferous, before plunging briefly to −9.0‰ or lower in the earliest Permian, recovering to values of −5.0 to −4.5‰ or higher through much of the Permian, and finally falling to −7.4‰ at the very end of the Permian (Fig. 2).

For the purposes of calculating the pCO_2 associated with the calcrete $\delta^{13}C_{carb}$ measurements, they made the simplifying assumption that the $\delta^{13}C_a$ = marine carbonate $\delta^{13}C$—8‰ and that $\delta^{13}C$ of soil organic matter, $\delta^{13}C\varphi = \delta^{13}C_a - 18‰$. Furthermore, because most modern soils containing calcretes formed below 50 cm depth have S(z) between 4000 and 7000 ppmV, they set a constant S(z) = 5000 ppmV. The overall trends in paleo-pCO_2 as calculated by Ekart et al. (1999) are not terribly dissimilar to those of Mora et al. (1996) (Fig. 2), but with greater temporal resolution as a consequence of the more robust data set. The earliest pedogenic carbonate, from the Late Silurian, again suggests very high early Paleozoic pCO_2, at ca. 3000 ppmV, although the authors share the concerns of Mora et al. (1996) that calcretes formed so early in the history of plant colonization may not provide meaningful results. Samples from the Devonian suggest pCO_2 in the range of 1000 ppmV, rising to near 2000 ppmV in the early Carboniferous before falling back to near 1000 ppmV in the late Carboniferous. The earliest Permian recorded a plunge to levels of less than 200 ppmV followed by a precipitous rise to 2000–3000 ppmV through the remainder of the early Permian. There is a data gap until the very end of the Permian where a value of 1000 ppmV is calculated.

Breecker et al. (2009) concluded that the use of 4000 –10,000 ppmV for the S(z) values in earlier studies (cf. Driese et al., 2000; Ekart et al., 1999; Mora et al., 1996; Royer et al., 2001) was incorrect because the actual productivity of plants during the season in which carbonate precipitation actually occurs (under conditions of soil moisture deficit) is not represented by mean annual precipitation estimates. Hence, the authors recalculated paleo-pCO_2 from the earlier pedogenic $\delta^{13}C_{carb}$ data, including those of Ekart et al. (1999) and Driese et al. (2000), using the diffusion-reaction equation with S(z) = 2500 ppmV. Because the calculated pCO_2 varies directly with S(z), the resulting values are one-half those calculated by Ekart et al. using the same data. Hence, the resulting curve of pCO_2 through time (with a 10 million-year running average) has a shape similar to the original, but with lower amplitude of variation. The general trend for the Paleozoic continues to show initial high levels for the Late Silurian, falling abruptly in the earliest Devonian to <1000 ppmV, before rising again during the Early Devonian to ~2000 ppmV, falling again to ~1000 ppmV near the Devonian-Carboniferous boundary and continuing to fall during the early Carboniferous to levels ≪1000 ppmV. The late Carboniferous saw a recovery to near 1000 ppmV by the Carboniferous-Permian boundary. The Permian pCO_2 record suggests an abrupt plunge in the earliest Permian to levels possibly as low as ~100 ppmV, followed by a strong recovery to ~1500 ppmV and falling again to as low as ~200 ppmV during the early late Permian before recovering to around 500 ppmV in the latest Permian. The strength of the approach by Breecker et al. (2009) is that the curve generated by these data matches well the data from the stomatal proxy (Fig. 2B, p. 578).

3.2 Detailed late Paleozoic record

The work of Chen et al. (2018) provides a useful context for a detailed examination of the atmospheric and climatic changes during the latest Carboniferous through earliest Permian. Proxy records, including the pedogenic carbonate isotope data of Mora et al. (1996), Ekart et al. (1999) and Montañez et al. (2007, 2016), are consistent with the Sr-isotope data (Chen et al., 2018) in indicating cooling and decreasing paleo-pCO_2 from well over 1000 ppmV in the early Carboniferous (Viséan) to <600 ppmV in the middle late Carboniferous (Moscovian).

The most ambitious attempts at creating a high-resolution record of paleo-pCO_2 for a portion of the late Paleozoic have been produced by Montañez et al. (2007, 2016). The former study focused on the very latest Carboniferous through early Permian, while the latter work, which utilized the computer program (PBUQ) of Breecker (2013) to calculate pCO_2, mainly examined the late Middle through Late Pennsylvanian. While most climate reconstructions are based on the averaging of proxies compiled over intervals of 5 or 10 My, the study by Montañez et al. (2016) was notable for having assembled a data set collected at a 100 ky-scale resolution (Fig. 3). The authors

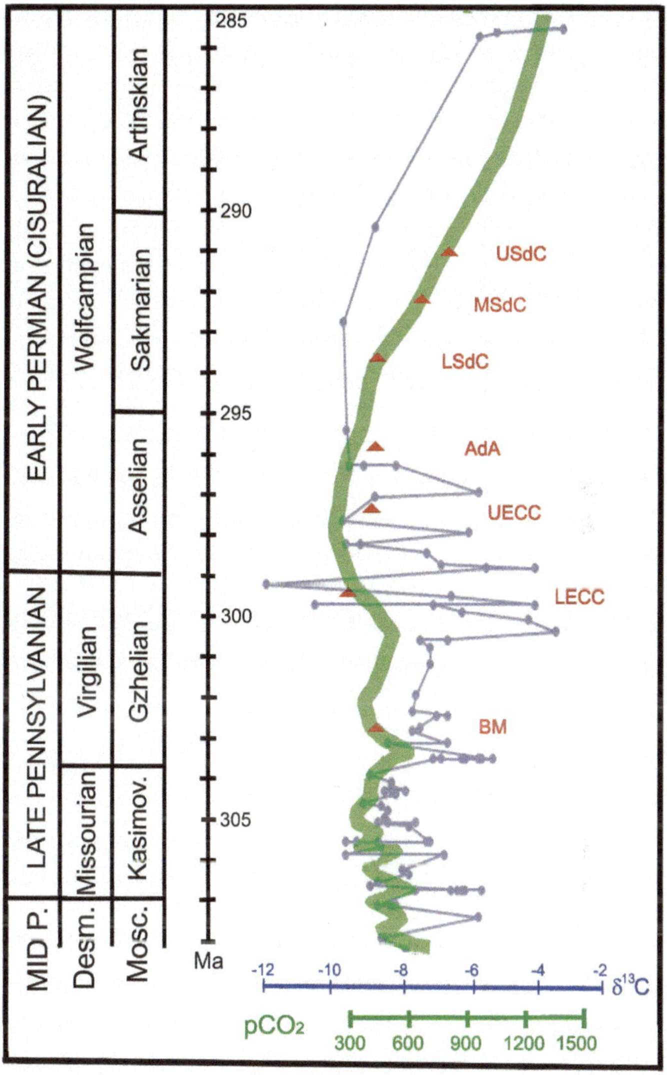

Fig. 3 Compilation of $\delta^{13}C_{carb}$ data from Montañez et al. (2016) (308–296 Ma) and Montañez et al. (2007) (296–285 Ma), the pCO_2 curve of Chen et al. (2018) calculated from the Montañez et al. data, and calculated pCO_2 values for the formations described in this chapter: BM = Burrego Member (of the Atrasado Fm.), LECC = lower El Cobre Canyon Fm., UECC = upper El Cobre Canyon Fm., AdA = Arroyo del Agua Fm., LSdC = lower Sangre de Cristo Fm., MSdC = middle Sangre de Cristo Fm., USdC = upper Sangre de Cristo Fm. *Adapted from Chen, J., Montañez, I.P., Qi, Y., Shen, S., Wang, X., 2018. Strontium and carbon isotopic evidence for decoupling of* pCO_2 *from continental weathering at the apex of the late Paleozoic glaciation. Geology 46, 395–398 and data of Montañez, I.P., Tabor, N.J., Niemeier, D., DiMichele, W.A., Frank, T.D., Fielding, C.R., et al., 2007.* CO_2*-forced climate and vegetation instability during late Paleozoic deglaciation. Science 315, 87–91, Montañez, I.P., McElwain, J.C., Poulsen, C.J., White, J.D., DiMichele, W.A., Wilson, J.P., et al., 2016. Climate* pCO_2 *and terrestrial carbon cycle linkages during Late Palaeozoic glacial-interglacial cycles. Nat. Geosci. 9, 824–831.*

collected over 300 samples of pedogenic carbonate, including calcrete nodules and rhizoliths, from Vertisols, Protosols and Calcisols in cyclothems in the Illinois, Appalachian and Donets (Ukraine) basins ranging in age from Desmoinesian to Asselian. The isotopic composition of organic matter in the soils was determined from analysis of fossil plant material, coal vitrinite macerals and occluded organic matter in the sampled paleosols. Some of the sampled formations include units previously reported by Mora et al. (1996) and Ekart et al. (1999), with $\delta^{13}C_{carb}$ values generally similar, but not always so. For example, the Dunkard Group (Asselian) samples were reported earlier (Mora et al., 1996) with a mean $\delta^{13}C_{carb} = -8.7‰$, well within the more recently reported range (by Montañez et al., 2016) of $\delta^{13}C = -8.0‰$ to $-11.7‰$. Similarly, the Nakoma Formation (Asselian) value $\delta^{13}C = -8.9‰$ reported earlier (Ekart et al., 1999) accords well with the recently reported range of $\delta^{13}C = -8.6‰$ to $-9.5‰$. In contrast, samples from the Conemaugh Group (Gzhelian) were reported earlier with a mean value $\delta^{13}C_{carb} = -7.2‰$ (Mora et al., 1996), whereas the later study (Montañez et al., 2016) reported a range of $\delta^{13}C_{carb} = -5.2‰$ to $-7.0‰$.

The data set of Montañez et al. (2016) does not suggest any discernible long-term trends of pCO_2 stability (Fig. 3). On the contrary, the data display great short-term (100 ky-scale) variability. The $\delta^{13}C_{carb}$ for individual formations varies from a low of $-11.7‰$ (Washington Formation of the Dunkard Group; Wolfcampian) to a high of $-3.4‰$ (Honaker Trail Formation; Virgilian/Gzhelian). Mean $\delta^{13}C_{carb}$ for individual stratigraphic units differ by as much 5‰ from one unit to the next, particularly around the Gzhelian-Asselian (Carboniferous-Permian) boundary. The individual unit $\delta^{13}C_{carb}$ measurements vary from a low of $-9.5‰$ to about $-4.0‰$ from late Desmoinesian (Moscovian) through mid-Virgilian time. Mid-Virgilian into late Asselian time is marked by more pronounced swings, from as low as ca. $-12‰$ to as high as $-3.5‰$.

The resultant time-averaged paleo-pCO_2 curve (Figs. 1 and 2, p. 825–826), which includes plant-based data from both empirical (McElwain and Chaloner, 1995) and mechanistic methods (Franks et al., 2014) in addition to the pedogenic carbonate data, has a shape similar to that of the raw data. It displays levels cycling between lows of ca. 150 ppmV and highs attaining a maximum of ca. 900 ppmV over a mean period of ca. 700 ky. However, from the Desmoinesian through mid-Virgilian the cycle period is very nearly 400 ky.

This short-term "cyclicity" disappears on locally weighted scatterplot smoothing (LOESS) with high-smoothing input. The general trend with high smoothing produces a trend of pCO_2 falling from ca. 600 ppmV in the late Desmoinesian to a low of ca. 320 ppmV in mid-Missourian (Kasimovian) time, rising to ca. 500 ppmV in the early Virgilian, then falling continuously to <200 ppmV by the middle Asselian, followed by a steady rise that continues through the late Asselian, and continuing through the Sakmarian. Chen et al. (2018) reproduced the data from Montañez et al. (2016), supplemented by the data from Mora et al. (1996), Ekart et al. (1999) and Montañez et al. (2007), to produce a similar paleo-pCO_2 curve (Fig. 3). The smoothed curves from both studies indicate most of the Late Pennsylvanian and earliest Permian was dominated by paleo-pCO_2 averaging around 200–300 ppmV with pronounced excursions to ca. 600 ppmV near the Missourian-Virgilian boundary and to as high as 900 ppmV during the middle Virgilian. However, the smoothing functions greatly reduce the amplitude of the 100 ky-scale variation noted in the raw data (Fig. 3).

4. Case studies from late Paleozoic of New Mexico

Given that atmospheric composition is approximately uniform at the global scale, the consistency of the pedogenic carbonate pCO_2-paleobarometer can be tested by comparing pCO_2 calculated from contemporaneous paleosols from disparate locations. Ideally, calculated paleo-pCO_2 can even be used as a correlation tool. Here, we compare previously published carbonate $\delta^{13}C$ data and calculated paleo-pCO_2 values (Chen et al., 2018; Ekart et al., 1999; Montañez et al., 2016) with data collected from four formations of late Paleozoic age (Late Pennsylvanian to early Permian) that crop out in New Mexico.

4.1 Regional setting

Paleogeographic reconstructions place New Mexico on the western edge of Pangea at tropical latitudes during the Late Pennsylvanian through early Permian (Fig. 1; Blakey, 2007; Scotese, 2014). Reconstructions by Scotese (2014) for the PALEOMAP Project place New Mexico evenly straddling the equator in Kasimovian time (305.3 Ma), but drifting northward very slowly so that by the start of Sakmarian time (289.5 Ma) almost the entirety of New Mexico is north of the equator. New Mexico was largely emergent during the late Pennsylvanian and Permian as uplifts shed arkosic sediments that filled the intervening basins (Tabor et al., 2008). Sedimentary rocks of late Paleozoic age in New Mexico are synorogenic deposits of the ancestral Rocky Mountain (ARM) orogeny. Driven by the Gondwana-Laurussia collision that amalgamated late Paleozoic-Mesozoic Pangea (Dickinson and Lawton, 2003; Kluth, 1986; Kluth and Coney, 1981), the ARM orogeny in New Mexico produced a series of basement-cored uplifts surrounded by sedimentary basins during Middle Pennsylvanian (Atokan)-early Permian (Wolfcampian) time (e.g., Brotherton et al., 2020; Kues and Giles, 2004). The result was an archipelago of islands (the uplifts) surrounded by shallow marine sedimentary basins during the Middle-Late Pennsylvanian followed by mostly nonmarine deposition in the sedimentary basins during the early Permian culmination of the ARM orogeny. Around the fringes and on the flanks of the uplifts, nonmarine fluvial deposition took place in some locations during the Pennsylvanian, and by early Permian time fluvial deposits (largely siliciclastic red beds) filled the formerly marine basins.

The late Paleozoic calcareous paleosols we studied in New Mexico are from these fluvial deposits. The oldest paleosols are from the early Virgilian Burrego Member of the Atrasado Formation, a mixed marine-nonmarine unit exposed across much of central New Mexico in the northern part of the ARM Orogrande basin, west of the ARM Pedernal uplift and around the ARM Joyita uplift (Lucas et al., 2014). Somewhat younger late Virgilian-Wolfcampian paleosols of the Cutler Group (El Cobre Canyon and Arroyo del Agua formations) in northern New Mexico are part of the siliciclastic red-bed succession deposited in the eastern part of the ARM Paradox basin and just south of (essentially on the flank of) the ARM Uncompahgre uplift (Eberth and Miall, 1991; Krainer and Lucas, 2010). The youngest calcareous paleosols we studied are from the Wolfcampian Sangre de Cristo Formation, siliciclastic red beds deposited in the ARM Rowe-Mora basin (also called the Taos trough) south of the ARM San Luis uplift (Fig. 4; Lucas et al., 2015).

4.2 Formations sampled

We sampled calcretes for isotopic analysis from four formations of Late Pennsylvanian to early Permian age (ranging in age from Virgilian to Wolfcampian) at multiple locations in northern and south-central New Mexico. In stratigraphic order, these formations are: the early Virgilian-age (Upper Pennsylvanian) Burrego Member of the Atrasado Formation; the Cutler Group formations—the Virgilian-age El Cobre Canyon Formation and Wolfcampian-age Arroyo del Agua Formation; and the

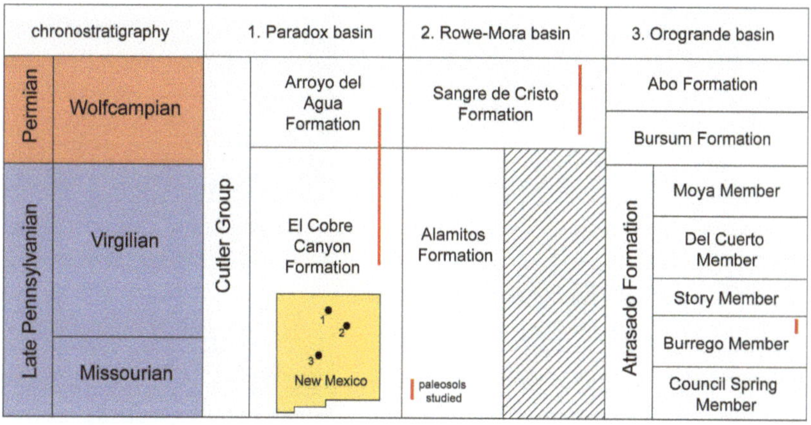

Fig. 4 Stratigraphy of the three basins in which formations were sampled for this study, with interbasinal correlations.

Sangre de Cristo Formation of Wolfcampian age. Where the paleosols in the section display sufficiently distinctive pedogenic features to allow classification, we apply the terminology of Mack et al. (1993) in assigning names of paleosol orders, modified by adjectives describing the most prominent subordinate characteristic.

4.2.1 Burrego Member (Atrasado Formation)

During the early Late Pennsylvanian (Virgilian), the site of deposition was located near the equator in western Pangea, near the eastern Panthalassa margin. The ca. 30-m section of the Burrego Member sampled for this study is located east of Socorro in south-central New Mexico (Socorro County). The section comprises calcareous paleosols hosted by red-bed mudstones that are interbedded with limestones that record brief marine incursions during the early Virgilian (Fig. 5). The section we studied includes meters-thick profiles of simple and cumulate paleosols, including calcic Argillisols, Calcisols and calcic Vertisols. Most samples were selected from Calcisols with well-developed (Stage III to Stage IV sensu Gile et al., 1966; Machette, 1985) calcrete horizons hosted in argillic red beds. Samples generally were selected from below the coalesced calcrete horizon to maximize depth below the original soil surface. Five paleosols were sampled for isotopic analysis. In ascending order these are: (1) a thin Calcisol with Stage III calcrete developed in a blocky very-fine grained sandstone containing calcareous rhizoliths; (2) a nodular calcrete formed on a marine limestone; (3) a calcic Vertisol exposure, 2.9-m thick, notable for the abundance of calcite-cemented rhizoliths up

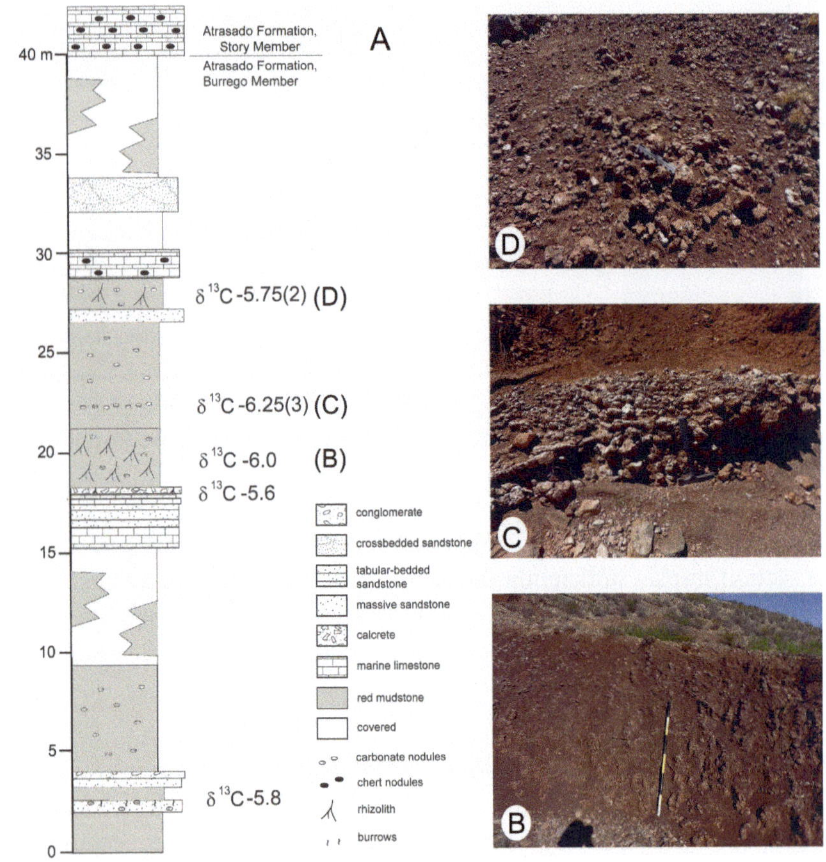

Fig. 5 Features of the Burrego Member (Atrasado Fm.). (A) Lithostratigraphy of the measured section with the isotopic composition of calcretes in the section. Parentheses indicate multiple samples comprising mean composition. (B), (C), (D) Specific paleosol horizons sampled for this study and described in the text at locations noted on the lithostratigraphic section. (B) Cumulate calcic Vertisol with abundant large rhizoliths. (C) Prominent nodular Bk horizon of a Calcisol. (D) Calcic argillisol with abundant calcrete nodules.

to 1 m in vertical length in a mudstone matrix with blocky ped fabric (Fig. 5B); (4) a Calcisol with a prominent Stage III nodular horizon (Fig. 5C); and (5) a calcic Argillisol containing two nodular horizons of Stage II and Stage III, likely representing stages in formation of a cumulate soil (Fig. 5D). Overall morphology of the paleosols in the Burrego Member suggests deposition and pedogenesis under subhumid, seasonally dry conditions.

4.2.2 Cutler Group

The Upper Pennsylvanian (upper Virgilian) to lowermost Permian (lower Wolfcampian) El Cobre Canyon and (lower Wolfcampian) Arroyo del Agua formations are well exposed in Cañon del Cobre, a large box canyon in southeastern Rio Arriba County, in the Chama Basin of northern New Mexico (Lucas et al., 2010a,b). The exposed El Cobre Canyon Formation is about 230m thick and is mostly siltstone and sandstone deposited in an ephemeral braided stream environment across an interval of perhaps as much as 3 million years (based on sedimentation rate: Brotherton et al., 2020) during the late Virgilian and the start of the Wolfcampian (Eberth and Miall, 1991; Krainer and Lucas, 2010; Lucas and Krainer, 2005a,b) (Fig. 4). The El Cobre Canyon Formation is considered mostly upper Virgilian, although the Wolfcampian base apparently occurs within the uppermost portion of the formation (Lucas, 2018; Lucas et al., 2010a,b).

The overlying Wolfcampian Arroyo del Agua Formation of the Cutler Group is as much as 160m thick. It also consists mostly of siltstone and sandstone, but was deposited on laterally extensive floodplains that developed between relatively stable channels during the first few million years of the Asselian (Eberth and Miall, 1991; Krainer and Lucas, 2010). Paleosols are prominent throughout both formations, including Protosols, calcic Protosols, Argillisols, calcic Argillisols, argillic Calcicols and Calcisols. All sampling was from Stage II to Stage IV calcretes in calcic Argillisols, argillic Calcisols and Calcisols (Fig. 6).

The studied El Cobre Canyon section contains multiple intervals of mudstone with Protosols and calcic Protosols that were not sampled. The seven paleosols sampled in this formation are Stage II to III calcretes in calcic Argillisol and argillic Calcisols (Fig. 6B) and more mature calcrete beds (Stage III to IV) in argillic Calcisols and Calcisols that form laterally continuous calcrete horizons. These are less common but do occur at multiple levels throughout the section (Tanner and Lucas, 2018). A pedogenic origin for these features is indisputable given the common association of roots and calcareous nodules in these strata.

The Wolfcampian Arroyo del Agua Formation contains a higher proportion of mudstone than the underlying El Cobre Canyon Formation, and many of the mudstone units host paleosols. Most abundant in the measured section of this formation are Argillisols and calcic Argillisols. Only three Calcisols and argillic Calcisols were found. These included one compound Calcisols with multiple Bk horizons (Fig. 6C), and one well-developed composite argillic Calcisol (Fig. 6D; Tanner and Lucas, 2018).

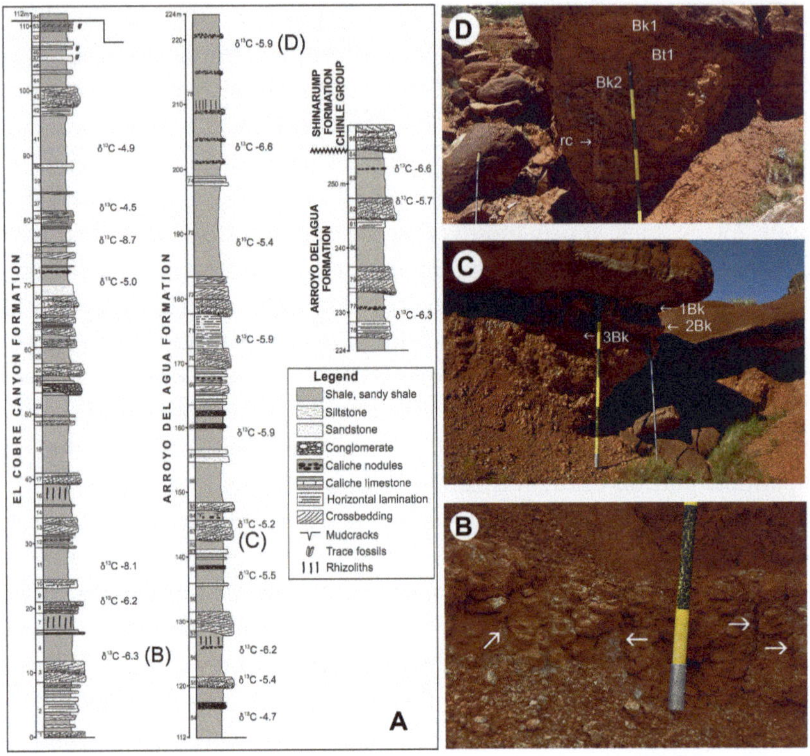

Fig. 6 Features of the Cutler Group (El Cobre Canyon and Arroyo del Agua formations). (A) Lithostratigraphy of the measured section with the isotopic composition of calcretes in the section. (B) Calcisol in El Cobre Canyon Fm. with Bk horizon illustrating crude platy-prismatic fabric and penetration by root traces (arrows). (C) Compound argillic Calcisol in the Arroyo del Agua Fm. with three distinct Bk horizons. (D) Composite argillic Calcisol in which the two Bk horizons are separated by a Bt horizon. An extended rhizocretion (rc) extends downward from the base of Bk2. *Adapted from multiple figures in Tanner, L.H., Lucas, S.G., 2018. Pedogenic record of climate change across the Pennsylvanian- Permian boundary in red-bed strata of the Cutler Group, northern New Mexico, USA. Sediment. Geol. 373, 98–10.*

We conducted isotopic analysis on 13 samples of Stage II to Stage III nodules from the Calcisols, argillic Calcisols, and calcic Argillisols from the formation, limiting sampling to below the Bk horizon, where present. Although we observed no distinct difference in paleosol type between the two formations, we noted that the calcretes in the Arroyo del Agua Formation are generally thicker than those in the underlying El Cobre Canyon Formation (Tanner and Lucas, 2018).

4.2.3 *Sangre de Cristo Formation*

The Sangre de Cristo Formation is an alluvial red-bed succession in the Rowe-Mora basin (or Taos trough) of San Miguel County, northern New Mexico. For the most part, the Sangre de Cristo Formation is of Wolfcampian age (Berman et al., 2013). The section we measured and sampled is from the upper part of the formation in southwestern San Miguel County, where the formation is overlain by Yeso Group strata of Leonardian Kungurian age (Fig. 7). Based on the estimated sedimentation rate, the measured section represents the last interval of no more than 2 million years near the end of the Wolfcampian (Brotherton et al., 2020). Details of the lithostratigraphy, sedimentology and paleontology are well-presented in Lucas et al. (2015).

Most of the paleosols in the ca. 85-m measured section of the uppermost part of the formation are Protosols that contain immature calcretes (Stage I to II). The lower and middle portions of the measured section contain mature calcrete beds (Stage III), with stacked nodules forming coalescing upper boundaries (Fig. 7B, C), and in a few places Stage IV calcrete; four of these were sampled (Fig. 7). The mature calcretes occur in calcic Argillisols (mainly) and Calcisols (Tanner and Lucas, 2017). Notably, a 1-m thick mudstone to very-fine-grained sandstone bed exposes the growth-position molds of the stems of 165 coniferophytic gymnosperms (Rinehart et al., 2015). The stem molds are partially to completely filled with nodular calcite. Eight samples from these stem molds and the calcareous nodules that occur in the host mudstone bed also were analyzed.

4.3 Stable isotope analysis

4.3.1 *Methodology*

To confirm the pedogenic origin of the carbonate in all calcrete samples, standard 30 μm petrographic thin sections were examined for pedogenic fabrics and textures and for evidence of diagenetic effects. Moreover, petrography revealed the presence of typical alpha-type calcrete fabrics in the nodules, such as grain coronas and corroded floating grains that indicate that these nodules represent the accumulation of carbonate in the B horizon of paleosols (Alonso-Zarza and Wright, 2010). Thus, we are confident in our interpretation that these carbonates represent calcretes formed within soils developed on alluvial muds. Carbonate samples were collected from the calcretes for isotopic analysis by selectively drilling micritic calcite in lapped slabs corresponding to prepared thin sections with an ultrafine engraving tool while viewing under a binocular microscope. The samples

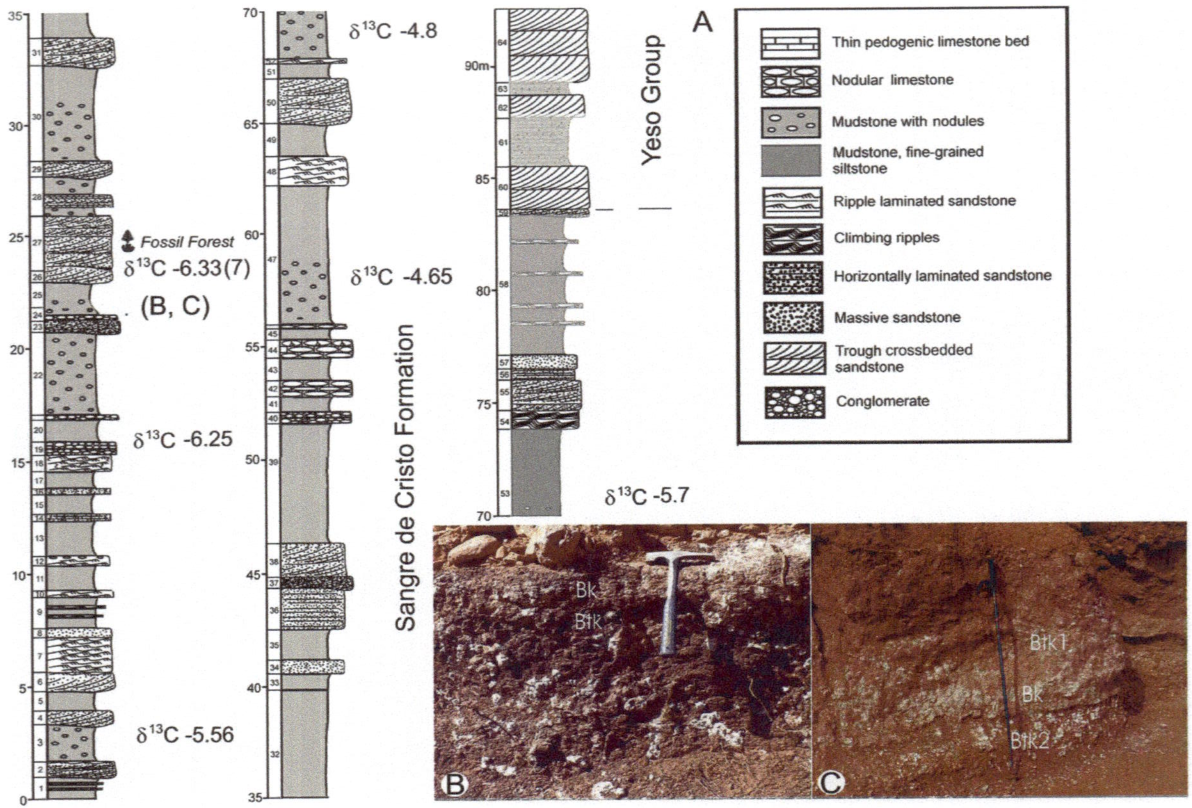

Fig. 7 Features of the Sangre de Cristo Fm. (A) Lithostratigraphy of the uppermost ca. 85 m of the formation with carbon isotope data. "Fossil forest" denotes stratigraphic level of layer containing numerous tree casts in growth position described by Rinehart et al. (2015). (B) Argillic Calcisol with distinct Bk and BtK horizons. (C) Compound Calcisol with multiple Bk and Btk horizons. *Adapted from multiple figures in Tanner, L.H., Lucas, S.G., 2017. Paleosols of the upper paleozoic sangre de cristo formation, north-central New Mexico: record of early Permian palaeoclimate in tropical Pangea. J. Palaeogeogr. 6(2), 144–162.*

were analyzed for $\delta^{13}C$ by the Duke Environmental Stable Isotope Laboratory (Nicholas School of the Environment, Duke University, Durham, North Carolina) using standard operating procedures (see Tanner and Lucas, 2018 for details).

We calculated pCO_2 for the samples from the diffusion-reaction equation, but as we did not sample for $\delta^{13}C_{om}$ we applied values from the literature for formations of similar age and geographic proximity (within two degrees latitude), primarily from the published data supplements to Montañez et al. (2007, 2016).

4.3.2 Results

Isotopic analysis of eight calcrete samples from the Burrego Member (Atrasado Formation) yielded tightly clustered $\delta^{13}C_{carb}$ values ranging from −6.4‰ to −5.6‰ with a mean of -6.0 ± 0.1‰ (VPDB) (Fig. 5). The composition of two samples from rhizoliths (−6.0‰, −5.8‰ (VPDB)) is indistinguishable from the remaining nodular calcrete samples, suggesting micritization of the root casts in the deep (>50 cm depth) soil environment.

Results of isotopic analyses from the 20 samples of the Cutler Group calcretes present $\delta^{13}C_{carb}$ values ranging from −8.4‰ to −4.5‰ (VPDB) (Fig. 6), with a mean for the entire section of $\delta^{13}C_{carb} = -5.9 \pm 0.2$‰. The means for the seven samples from the El Cobre Canyon Formation and the 13 Arroyo del Agua Formation samples are −6.2‰ and −5.8‰, respectively. However, much of the interval from 25 to 125 m, including most of the El Cobre Canyon Formation and the basal Arroyo del Agua Formation (Fig. 6), is notable for a lack of $\delta^{13}C_{carb}$ consistency that contrasts with the basal El Cobre Canyon and most of the Arroyo del Agua section. The two lightest values of $\delta^{13}C_{carb}$ (<−8.0‰) were both found in the El Cobre Canyon Formation, as is the heaviest value, $\delta^{13}C = -4.5$‰. The upper 130 m of the Arroyo del Ague Formation yield more consistent values of $\delta^{13}C_{carb}$, from a low of −6.6‰ to a maximum of −5.2‰, with a mean $\delta^{13}C_{carb} = -5.8$‰.

The isotopic analyses of the carbonate from the Sangre de Cristo calcrete nodules from five paleosol units demonstrates some variability. Values of $\delta^{13}C_{carb}$ from near the base and top range from −6.25 to −5.7, but an interval of enriched carbonate, $\delta^{13}C_{carb} = -4.8$ to −4.65, occurs in the middle of the section (Fig. 7A). The composition of the carbonate from the tree stem casts is tightly clustered, with $\delta^{13}C_{carb}$ ranging from −7.07 to −5.83 with mean = −6.1‰ (VPDB). The similarity of the $\delta^{13}C_{carb}$ of the calcrete from the stem casts to adjacent nodular calcrete suggests that the stem casts were micritized by calcite relatively deep in the soil environment.

4.4 Comparison to previous work

4.4.1 Borrego Member (Atrasado Formation)

Conodont and fusulinid biostratigraphy indicates an early Virgilian age for the Borrego calcretes (Barrick et al., 2013; Lucas et al., 2009), which yielded a mean $\delta^{13}C_{carb} = -6.0$. Similar-age samples from the Illinois Basin yielded $\delta^{13}C_{carb}$ values ranging from −7.5 to −6.5 for 12 samples, but these samples are too distant from the location of the Borrego Formation to provide analogous $\delta^{13}C_{om}$, T° and S(z) values. Notwithstanding, we estimated values of $\delta^{13}C_{om} = -21.0$, $C_a = -2.5$, $T = 25\,°C$ and $S(z) = 3000$ ppmV to obtain a paleo-pCO_2 for the early Virgilian of 411 ppmV. This accords well with values presented by Montañez et al. (2016) for the early-middle Virgilian across the extended interval from 303 to 301 Ma (Fig. 3).

4.4.2 Cutler Group

The same authors published analyses for samples from the Bursum Formation from south-central New Mexico to which they assigned an age of 299.7 Ma (late Virgilian). This location is near the field site where we collected the calcrete samples from the Burrego Member of the Atrasado Formation, which underlies the Bursum Formation. We estimate the El Cobre Formation of the Cutler Group as overlapping in age with the Bursum Formation, from late Virgilian to early Wolfcampian, and the mean $\delta^{13}C_{carb}$ of −5.9‰ is statistically identical to the mean value of −6.1‰ reported for the five samples from the Bursum Formation. Here, we note that Montañez et al. (2016) reported a mean $\delta^{13}C_{carb}$ of −4.2 (for two samples) from calcretes in the Hermit Shale, assigned an age of 299.9 Ma, and −4.0 (for 16 samples) from the upper Halgaito Formation, assigned an age of 299.5 Ma (both late Virgilian and both formations from Arizona). Thus, we suggest that the extended interval of high variability in the $\delta^{13}C_{carb}$ of the El Cobre Canyon Formation and basal Arroyo del Agua Formation described above may correlate with the data from the multiple formations reported by Montañez et al. For the lower El Cobre Formation (mean $\delta^{13}C_{carb} = -6.2‰$) we applied values of $\delta^{13}C_{om} = -20.5$, $C_a = -2.5$, $T = 25\,°C$ and $S(z) = 3000$ ppmV (by analogy to the Hermit Shale and Halgaito Formation) to obtain a pCO_2 of 239 ppmV. The middle and upper portion of the formation, in which the pedogenic carbonate is consistently heavier (mean $\delta^{13}C_{carb} = -4.9‰$), yields a pCO_2 of 530 ppmV with $\delta^{13}C_{om} = -20.5$, $C_a = -2.5$, $T = 25\,°C$ and $S(z) = 2500$ ppmV. This trend matches well the paleo-pCO_2 curve presented by Chen et al. (2018) that indicates declining pCO_2 through the late Virgilian into the earliest Wolfcampian followed by rising pCO_2 from the early Asselian through Sakmarian time (Fig. 3).

The Arroyo del Agua Formation, which overlies the El Cobre Canyon Formation, is Asselian (Wolfcampian) and overlaps in age with two other sets of Hermit Shale samples, $\delta^{13}C_{carb} = -5.4$ (three samples at 298.6 Ma) and −6.7 (two samples at 298.5 Ma), and the Abo Formation, $\delta^{13}C_{carb} = -5.6$ (11 samples at 297.6 Ma) reported by Montañez et al. (2016). The age of these samples in the early through middle Asselian suggests that the parameters estimated for pCO_2 calculation for these samples can be applied to the Arroyo del Agua Formation, for which the mean $\delta^{13}C_{carb} = -5.9$. Applying values of $\delta^{13}C_{om} = -20.5$, $C_a = -2.5$, $T = 25\,°C$ and $S(z) = 2500$ ppmV, we obtained a pCO_2 of 476–572 ppmV with $S(z) = 3000$ ppmV. The increased pCO_2 indicated by the El Cobre Canyon and Arroyo del Agua formations thus accords with the increase in the Asselian indicated by Chen et al. (2018).

4.4.3 Sangre de Cristo Formation

Montañez et al. (2007) reported $\delta^{13}C_{carb} = -5.6$ (for 11 samples) for a portion of the Abo Formation, to which they assigned a late Sakmarian (=Wolfcampian) age (285.2 Ma). The Abo and Sangre de Cristo formations are correlative units from southern and northern New Mexico (Berman et al., 2013; Lucas et al., 2015), respectively, but the deposition of both units effectively occupied the entirety of the Wolfcampian (Asselian through middle Artinskian). As our sampling was from the upper one-third of the Sangre de Cristo Formation, we compare our $\delta^{13}C_{carb}$ values to the correlative Abo Formation values of Montañez et al. and find they accord well. For the lower Sangre de Cristo samples, with a mean $\delta^{13}C_{carb} = -5.9‰$, we model $pCO_2 = 452$ ppmV based on assumptions of $\delta^{13}C_{om} = -21.0$, $C_a = -2.5$, $T = 28\,°C$ and $S(z) = 2500$ ppmV (derived from the Hermit Shale). The samples from the middle portion of our measured section yielded values with a mean $\delta^{13}C_{carb} = -4.7‰$. By applying estimates of $\delta^{13}C_{om} = -21.0$, $C_a = -2.5$, $T = 25\,°C$ and $S(z) = 2500$ ppmV we produced a calculated pCO_2 of 614 ppmV. The uppermost Sangre de Cristo samples have a mean $\delta^{13}C_{carb} = -5.7$. Applying estimates from contemporaneous formations (in particular, the Abo) of $\delta^{13}C_{om} = -21.0$, $C_a = -3.0$, $T = 30\,°C$ and $S(z) = 3000$ ppmV results in a pCO_2 of 707 ppmV. This tendency of rising pCO_2 in the middle through late Sakmarian also matches well the trend modeled in numerous reconstructions (Fig. 3; Montañez et al., 2007, 2016; Chen et al., 2018).

5. pCO_2 and late Paleozoic climate

The onset of late Paleozoic glaciation was driven by long-term trends in paleo-pCO_2, which, in turn, were driven by a tectonically-controlled long-term drawdown by silicate weathering (Goddéris et al., 2017). Modeling studies have suggested that the configuration of the Pangean land mass and oceans at this time made Gondwana susceptible to glaciation at a pCO_2 threshold value of 560 ppmV (Lowry et al., 2014). Gondwanan glaciation expanded greatly during the latest Mississippian through Middle Pennsylvanian, with peak glaciation occurring around the Mississippian-Pennsylvanian boundary, marked by a significant eustatic drop in sea level, and coinciding with the spread of lycopsid forests in the tropics (Montañez and Poulsen, 2013), which promoted substantial carbon sequestration (DiMichele, 2014; Nelsen et al., 2016).

Ice retreat during the late Early Pennsylvanian (Bashkirian) was followed by another major advance during the Middle Pennsylvanian (Moscovian) that peaked in the Virgilian. High-amplitude glacio-eustatic cycles that are superimposed on the longer-term trends at the eccentricity (100-ky scale) orbital scale are particularly pronounced during the Moscovian (Montañez and Poulsen, 2013). The Middle Pennsylvanian-Late Pennsylvanian (Desmoinesian-Missourian) boundary is marked by overall falling pCO_2, with superimposed higher frequency fluctuations, followed by rising pCO_2 from the middle Missourian through middle Virgilian, coinciding with rising sea level (Montañez et al., 2016). The detailed sampling by Montañez et al. (2016) of paleosols, supplemented by paleofloral data (stomatal indices and $\delta^{13}C_{plant}$) across (presumably) 415 ky cyclothems, allowed the authors to derive that pCO_2 varied from ca. 200 to 700 ppmV with a cyclicity at the 100-ky scale that coincides with demonstrated eustasy in the Donets Basin. The authors also noted that the rate of CO_2 flux from the atmosphere across these cycles is not dissimilar to that of the Pleistocene glacial-interglacial cycles, meaning that the scale of pCO_2 change indicated by pedogenic carbonate proxy is entirely feasible. Importantly, the Middle to Late Pennsylvanian transition is marked by a pronounced floral turnover during which coverage by moist tropical forest declined significantly and dry woodland and scrub biomes expanded (Montañez and Poulsen, 2013; Montañez et al., 2016; Pfefferkorn et al., 2008). Orbital forcing of climate at the eccentricity scale, superimposed on the longer-term trend of falling pCO_2, triggered vegetation feedbacks by increasing or decreasing carbon sequestration and pCO_2 (Montañez et al., 2016).

The latest Pennsylvanian through earliest Permian was characterized by increased shifts from moist to xerophytic floral assemblages, in addition to other evidence, including oxygen isotopes of pedogenic carbonate and phyllosilicates and shifting clay mineral assemblages (from kaolinitic to smectitic). These shifts suggest that across tropical Pangea a humid to subhumid-seasonal climate that prevailed at tropical latitudes during the latest Carboniferous was replaced during the earliest Permian by a generally more arid climate (Cecil, 2013; DiMichele et al., 2006; Tabor and Montañez, 2002, 2004; Tabor et al., 2002, 2008). The trend of aridification across the Carboniferous-Permian boundary has been interpreted by some authors as suggesting an increase in pCO_2, rather than a decrease (Michel et al., 2015; Tabor and Poulsen, 2008). For example, the isotopic compositions of phyllosilicates and hematite formed in paleosols in west-central Texas (eastern shelf of the Midland Basin) were cited by Tabor et al. (2002, 2013), Tabor and Montañez (2005) and Tabor (2007) as evidence of significant warming. Indeed, without the recent paleo-pCO_2 data, the pronounced shift in floristic assemblages from a wetter *Sigillaria* and pteridosperm-dominated biome in the Pennsylvanian, to the seasonally dry-adapted assemblage dominated by conifers, callipterids and other seed plants in the Permian (DiMichele et al., 2009; Tabor et al., 2013) could appear to have been temperature driven. Moreover, modeling of Carboniferous-Permian climate by GCM (Peyser and Poulsen, 2008) suggested widespread drying and warming consistent with increased pCO_2.

However, Montañez et al. (2007) cited brachiopod $\delta^{18}O$ data that suggest a cooling trend across the Carboniferous-Permian boundary. Moreover, the latest Pennsylvanian-earliest Permian is marked by a major eustatic sea-level fall as demonstrated by a basin-ward facies (or a hiatus) that marks the onset of a major glacial advance that initiated during the late Virgilian and continued into the early Permian (Rygel et al., 2008). This is now demonstrated as coincident with a major decline in pCO_2 that started in the middle Virgilian and continued through the Asselian before terminating in the Sakmarian (Chen et al., 2018; Montañez and Poulsen, 2013; Montañez et al., 2007, 2016). The causes of the environmental changes across the Pennsylvanian-Permian boundary are not well understood (DiMichele et al., 2009). Chen et al. (2018) noted that terrestrial sequestration of carbon during this interval is unlikely, given the trend toward increasing aridification since the Middle Pennsylvanian-Late Pennsylvanian boundary. The authors hypothesize that the aforementioned floral turnover caused decreased forest canopy cover, thereby allowing increased runoff and nutrient delivery to the oceans. Thus, they propose that the locus of carbon sequestration shifted from the terrestrial to the marine environment during the end Pennsylvanian and earliest Permian. This major stage of late Paleozoic glaciation terminated in the middle Sakmarian (Isbell et al., 2003; Montañez and Poulsen, 2013; Montañez et al., 2007), although the final termination of the Paleozoic Ice Age did not occur until the late early Permian (Leonardian), likely a consequence of both pCO_2-forced warming and the northward drift of Pangea (Montañez and Poulsen, 2013; Montañez et al., 2007).

6. Conclusions

The primary finding of our study is the continued validation of the use of the isotopic composition of pedogenic carbonate as a pCO_2 paleobarometer. The ability to match the pCO_2 values calculated from the samples we collected in New Mexico to curves generated from data generated through studies in such disparate locations as the Illinois, Appalachian, Midland and Donets basins (Montañez et al., 2007, 2016) demonstrates the ability of the technique to provide meaningful information regarding the trends of pCO_2 through most of the Phanerozoic. Admittedly, the precision of the results remains constrained by the accuracy of the various parameters required for application of the diffusion-reaction equation, but the understanding of how to produce appropriate values for these has improved greatly in recent decades as demonstrated by the convergence of pCO_2 trends generated by multiple proxies. The gradual development of a robust, relatively high-resolution paleo-pCO_2 curve for the late Paleozoic now provides a potential tool for correlation. Measurements of $\delta^{13}C_{carb}$ from pedogenic carbonate vary locally due to differences in hydrology and vegetation, but (ideally) the calculated pCO_2 from coeval formations will match (ideally) within the limits of measurement error.

References

Alonso-Zarza, A.M., 2003. Palaeoenvironmental significance of palustrine carbonates and calcretes in the geological record. Earth Sci. Rev. 60, 261–298.

Alonso-Zarza, A.M., Wright, V.P., 2010. Calcretes. In: Alonso-Zarza, A.M., Tanner, L.H. (Eds.), Carbonates in Continental Environments: Processes, Facies and Applications. Developments in Sedimentology, vol. 61. Elsevier, Amsterdam, pp. 226–267.

Barclay, R.S., Wing, S.L., 2016. Improving the Gingko CO_2 barometer: implications for early Cenozoic atmosphere. Earth Planet. Sci. Lett. 439, 158–171.

Barrick, J., Lucas, S.G., Krainer, K., 2013. Conodonts of the Atrasado Formation (uppermost Middle to Upper Pennsylvanian), Cerros de Amado region, Central New Mexico, USA. New Mex. Mus. Nat. Hist. Sci. Bull. 59, 239–252.

Beerling, D.J., Chaloner, W.G., 1993. Evolutionary responses of stomatal density to global CO_2 change. Biol. J. Linn. Soc. 48, 343–353.

Beerling, D.J., McElwain, J.C., Osborne, C.P., 1998. Stomatal responses of the "living fossil" *Gingko biloba* L. to changes in atmospheric CO_2 concentrations. J. Exp. Bot. 49, 1603–1607.

Bera, M.K., Sarkar, A., Tandon, S.K., Samanta, A., Sanyal, P., 2010. Does burial diagenesis reset pristine isotopic compositions in paleosol carbonates? Earth Planet. Sci. Lett. 300, 85–100.

Berman, D.S., Henrici, A.C., Lucas, S.G., 2013. *Ophiacodon* (Synapsida, Ophiacodontidae) from the lower Permian Sangre de Cristo Formation of New Mexico. New Mex. Mus. Nat. Hist. Sci. Bull. 60, 36–41.

Berner, R.A., 1994. GEOCARB II: a revised model of atmospheric CO_2 over Phanerozoic time. Am. J. Sci. 294, 56–91.

Berner, R.A., 2006. Inclusion of the weathering of volcanic rocks in the GEOCARBSULF model. Am. J. Sci. 306, 295–302.

Berner, R.A., 2008. Addendum to "Inclusion of the Weathering of Volcanic Rocks in the GEOCARBSULF Model" (R. A. Berner, 2006, V. 306, P. 295–302). Am. J. Sci. 308, 100–103.

Berner, R.A., Kothavala, Z., 2001. GEOCARB III: a revised model of atmospheric CO_2 over Phanerozoic time. Am. J. Sci. 301, 182–204.

Birkeland, P.W., 1999. Soils and Geomorphology, third ed. Oxford, New York, p. 536.

Blakey, R.C., 2007. Carboniferous-Permian paleogeography of the assembly of Pangaea. In: Wong, T.E. (Ed.), Proceedings of the XVth International Congress on Carboniferous and Permian Stratigraphy, Utrecht, the Netherlands, pp. 443–456.

Boucot, A.J., Gray, J., 2001. A critique of Phanerozoic climate models involving changes in the CO_2 content of the atmosphere. Earth Sci. Rev. 56, 1–159.

Bowen, G.J., Beerling, D.J., 2004. An integrated model for soil organic carbon and CO_2: implications for paleosol carbonate pCO2 paleobarometry. Glob. Biogeochem. Cycles 18. https://doi.org/10.1029/2003GB002117.

Breecker, D.O., 2013. Quantifying and understanding the uncertainty of atmospheric CO_2 concentrations determined from calcic paleosols. Geochem. Geophys. Geosyst. 14, 3210–3220.

Breecker, D., Sharp, Z.D., McFadden, L., 2009. Seasonal bias in the formation and stable isotope composition of pedogenic carbonate in modern soils from Central New Mexico, USA. Geol. Soc. Am. Bull. 121, 630–640.

Brotherton, J.L., Chowdhury, N.U.M.K., Sweet, D.E., 2020. Synthesis of late paleozoic sedimentation in central and eastern New Mexico: implications for timing of ancestral Rocky Mountains deformation. In: Late Paleozoic and Early Mesozoic Tectonostratigraphy and Biostratigraphy of Western Pangea. SEPM Spec. Publ. 113, https://doi.org/10.102110/sepmsp.113.03.

Cecil, C.C., 2013. An overview and interpretation of autocyclic and allocyclic processes and the accumulation of strata during the Pennsylvanian–Permian transition in the central Appalachian Basin, USA. Int. J. Coal Geol. 119, 21–31.

Cecil, C.B., Dulong, F.T., West, R.R., Stamm, R., Wardlaw, B.A., Edgar, N.T., 2003. Climate controls on the stratigraphy of a Middle Pennsylvanian cyclothem in North America. In: Blaine, C.C., Terence, E.N. (Eds.), Climate Controls on Stratigraphy. SEPM Spec. Publ. 77, Tulsa, pp. 151–182.

Cerling, T.E., 1991. Carbon dioxide in the atmosphere: evidence from Cenozoic and Mesozoic paleosols. Am. J. Sci. 291, 377–400.

Cerling, T.E., 1999. Stable carbon isotopes in palaeosol carbonates. In: Thiry, M., Simon-Coinçon, R. (Eds.), Palaeoweathering, Palaeosurfaces and Related Continental Deposits. Internat. Assoc. Sedimentol. Spec. Publ. 27, Blackwell, Oxford, pp. 43–60.

Chen, L.Q., Li, C.S., Chaloner, W.G., Beerling, D.J., Sun, Q.G., Collinson, M.E., Mitchell, P.L., 2001. Assessing the potential for the stomatal characters of extant and fossil Ginkgo leaves to signal atmospheric CO_2 change. Am. J. Bot. 88, 1309–1315.

Chen, J., Montañez, I.P., Qi, Y., Shen, S., Wang, X., 2018. Strontium and carbon isotopic evidence for decoupling of pCO2 from continental weathering at the apex of the late Paleozoic glaciation. Geology 46, 395–398.

Dickinson, W.R., Lawton, T.F., 2003. Sequential intercontinental suturing as the ultimate control of Pennsylvanian ancestral Rocky Mountains deformation. Geology 31, 609–612.

DiMichele, W.A., 2014. Wetland-dryland vegetational dynamics in the Pennsylvanian ice age tropics. Int. J. Plant Sci. 175, 123–164.

DiMichele, W.A., Tabor, N.J., Chaney, D.S., Nelson, W.J., 2006. From wetlands to wet spots: environmental tracking and the fate of Carboniferous elements in Early permian tropical floras. In: Greb, S.F., DiMichele, W.A. (Eds.), Wetlands Through Time. Geol. Soc. Am. Spec. Pap. 399, pp. 223–248.

DiMichele, W.A., Montañez, I.P., Poulsen, C.J., Tabor, N.J., 2009. Vegetation-climate feedbacks and regime shifts in the Late Paleozoic ice age earth. Geobiology 7, 200–226.

Dinterman, P.A., Mack, G.H., Leeder, M., Perez-Arlucea, M., 2000. Paleoclimatic control on Early Permian fluvial to eolian transition in New Mexico. Abstr. Prog. Geol. Soc. Am. 32 (7), 254.

Driese, S.G., Mora, C.K., 1993. Physico-chemical environment of pedogenic carbonate formation in Devonian vertic paleosols, central appalachians, USA. Sedimentology 40, 199–216.

Driese, S.G., Mora, C.I., Stiles, C.A., Joeckel, R.M., Nordt, L.C., 2000. Mass-balance reconstruction of a modern Vertisol: implications for interpreting the geochemistry and burial alteration of paleo-Vertisols. Geoderma 95, 179–204.

Eberth, D.A., Miall, A.D., 1991. Stratigraphy, sedimentology and evolution of a vertebrate-bearing braided to anastomosed fluvial system, Cutler Formation (Permian-Pennsylvanian), north-Central New Mexico. Sediment. Geol. 72, 225–252.

Ekart, D.D., Cerling, T.E., Montanez, I.P., Tabor, N.J., 1999. A 400 million-year carbon isotope record of pedogenic carbonate: implications for paleoatmospheric carbon dioxide. Am. J. Sci. 299, 805–827.

Franks, P.J., Royer, D.L., Beerling, D.J., Van de Water, P.K., Cantrill, D.J., Barbour, M.M., Berry, J.A., 2014. New constraints on atmospheric CO_2 concentration for the Phanerozoic. Geophys. Res. Lett. 41, 46854694.

Fraser, L.H., Greenhall, A., Carlyle, C., Turkington, R., Friedman, C.R., 2009. Adaptive phenotypic plasticity of *Pseudoroegneria spicata*: response of stomatal density, leaf area and biomass to changes in water supply and increased temperature. Ann. Rev. Bot. 103, 769–775.

Gibbs, M.T., Rees, P.M., Kutzbach, J.E., Ziegler, A.M., Behling, P.J., Rowlęy, D.B., 2002. Simulations of Permian climate and comparisons with climate-sensitive sediments. J. Geol. 110, 33–55.

Gile, L.H., Peterson, F.F., Grossman, R.B., 1966. Morphological and genetic sequences of accumulation in desert soils. Soil Sci. 100, 347–360.

Giles, J.M., Soreghan, M.J., Benison, K.C., Soreghan, G.S., Hasiotis, S.T., 2013. Lakes, loess, and paleosols in the Permian Wellington Formation of Oklahoma, U.S.A.: implications for paleoclimate and paleogeography of the Midcontinent. J. Sediment. Res. 83, 825–846.

Goddéris, Y., Donnadieu, Y., Carretier, S., Aretz, M., Dera, G., Macouin, M., Regard, V., 2017. Onset and ending of the late Palaeozoic ice age triggered by tectonically paced rock weathering. Nat. Geosci. 10, 382–386. https://doi.org/10.1038/ngeo2931.

Goudie, A.S., 1983. Calcrete. In: Goudie, A.S., Pye, K. (Eds.), Chemical Sediments and Geomorphology. Academic Press, New York, pp. 93–131.

Haworth, M., Hesselbo, S.P., McElwain, J.C., Robinson, S.A., Brunt, J.W., 2005. Mid-Cretaceous pCO_2 based on stomata of the extinct conifer Pseudofrenelopsis (Cheirolepidiaceae). Geology 33, 749–752.

Hiatt, E.E., Pufahl, P.K., 2014. Cathodoluminescence petrography of carbonate rocks: a review of applications for understanding diagenesis, reservoir quality and pre system evolution. In: Coulson, I. (Ed.), Cathodoluminescence and Its Application to Geoscience. Min. Assoc. Can. Short Course Series, vol. 45, pp. 75–96.

Hill, K.E., Barr, C., Tibby, J., Hill, R.S., Watling, J.R., 2019. A comparison of stomatal traits between contemporary and fossil leaves *Melaleuca quinquenervia*: do they reflect climate variation? Rev. Palaeobot. Palynol. 271. https://doi.org/10.1016/j.revpalbo.2019.104109.

Isbell, J.L., Miller, M.F., Wolfe, K.I., Lenaker, P.A., 2003. Timing of late Paleozoic glaciation in Gondwana: was glaciation responsible for the development of northern hemisphere cyclothems? In: Chan, M.A., Archer, A.W. (Eds.), Extreme Depositional Environments: Mega end Members in Geologic Time. Geol. Soc. Am. Spec. Pap. 370, pp. 5–24.

Kessler, J.L.P., Soreghan, G.S., Wacker, H.J., 2001. Equatorial aridity in western Pangea: lower Permian loessite and dolomitic paleosols in northeastern New Mexico, USA. J. Sediment. Res. 71, 817–832.

Kluth, C.F., 1986. Plate Tectonics of the Ancestral Rocky Mountains. Am. Assoc. Petrol. Geol. Mem. 41, 353–369.

Kluth, C.F., Coney, P.J., 1981. Plate tectonics of the ancestral rocky mountains. Geology 9, 10–15.

Krainer, K., Lucas, S.G., 2010. Sedimentology of the Pennsylvanian-Permian Cutler Group and Lower Permian Abo Formation, Northern New Mexico. New Mex. Mus. Nat. Hist. Sci. Bull. 49, 25–36.

Kraus, M.J., 1999. Paleosols in clastic sedimentary rocks: their geologic applications. Earth Sci. Rev. 47, 41–70.

Kues, B.S., Giles, K.A., 2004. The late Paleozoic Ancestral Rocky Mountains system in New Mexico. In: Mack, G.H., Giles, K.A. (Eds.), The Geology of New Mexico, A Geologic History. Spec. Publ. 11, New Mexico Geological Society, Socorro, pp. 95–136.

Kutzbach, J.E., Gallimore, R.G., 1989. Pangean climates: megamonsoons of the megacontinent. J. Geophys. Res. 94, 3341–3357.

Kutzbach, J.E., Ziegler, A.M., 1993. Simulation of Late Permian climate and biomes with an atmosphere/ocean model: comparisons with observations. Philos. Trans. R. Soc. Lond. B 341, 327–340.

Lowry, D.P., Poulsen, C.J., Horton, D.E., Torsvik, T.H., Pollard, D., 2014. Controls on Paleozoic ice sheet initiation. Geology 42, 627–630.

Lucas, S.G., 2018. Permian tetrapod biochronology, correlation and evolutionary events. In: Lucas, S.G., Shen, S.Z. (Eds.), The Permian Timescale. Geol. Soc. (London) Spec. Publ. 450, pp. 405–444.

Lucas, S.G., Krainer, K., 2005a. Cutler Group (Permo-Carboniferous) stratigraphy, Chama Basin, New Mexico. New Mex. Mus. Nat. Hist. Sci. Bull. 31, 90–100.

Lucas, S.G., Krainer, K., 2005b. Stratigraphy and correlation of the Permo-Carboniferous Cutler Group, Chama Basin, New Mexico. New Mex. Mus. Nat. Hist. Sci. Bull. 56, 145–159.

Lucas, S.G., Krainer, K., Barrick, J.E., 2009. Pennsylvanian stratigraphy and conodont biostratigraphy in the Cerros de Amado, Socorro County, New Mexico. New Mex. Geol. Soc. Guideb. 60, 183–212.

Lucas, S.G., Spielmann, J.A., Krainer, K., 2010a. Summary of geology of Canon del Cobre, Rio Arriba County, New Mexico. New Mex. Mus. Nat. Hist. Sci. Bull. 49, 15–24.

Lucas, S.G., Harris, S.K., Spielmann, J.A., Rinehart, L.F., Berman, D.S., Henrici, A.C., Krainer, K., 2010b. Vertebrate paleontology, biostratigraphy and biochronology of the Pennsylvanian-Permian Cutler Group, Cañon del Cobre, northern New Mexico. New Mex. Mus. Nat. Hist. Sci. Bull. 49, 115–123.

Lucas, S.G., Krainer, K., Allen, B.D., Vachard, D., 2014. The Pennsylvanian section at Cedro Peak: a reference section in the Manzanita Mountains, Central New Mexico. New Mex. Geol. 36, 3–24.

Lucas, S.G., Krainer, K., DiMichele, W.A., Voigt, S., Berman, D.S., Henrici, A.C., Tanner, L.H., Chaney, D.S., Elrick, S.D., Nelson, W.J., Rinehart, L.F., 2015. Lithostratigraphy, biostratigraphy and sedimentology of the Upper Paleozoic Sangre de Cristo Formation, southwestern San Miguel County, New Mexico. New Mex. Geol. Soc. Guideb. 66, 211–228.

Lüthi, D., Le Floch, M., Bereiter, B., Blunier, T., Barnola, J.-M., Siegenthaler, U., Raynaud, D., Jouzel, J., Fischer, H., Kawamura, K., Stocker, T.F., 2008. High-resolution carbon dioxide concentration record 650,000–800,000 years before present. Nature 453, 379–382.

Machette, M.N., 1985. Calcic soils of the southwestern United States. In: Weide, D.L. (Ed.), Soils and Quaternary Geology of the Southwest United States. Geol. Soc. Am. Spec. Pap. 203, pp. 1–21.

Mack, G.H., James, W.C., Monger, H.C., 1993. Classification of paleosols. Geol. Soc. Am. Bull. 105, 129–136.

Mack, G.H., Leeder, M., Perez-Arlucca, M., Bailey, B., 2003. Early Permian silt-bed fluvial sedimentation in the Orogrande basin of the Ancestral Rocky Mountains, New Mexico, USA. Sediment. Geol. 160, 159–178.

McElwain, J.C., 2018. Paleobotany and global change: important lessons for species to biomes from vegetation responses to past global change. Annu. Rev. Plant Biol. 69, 761–787.

McElwain, J.C., Chaloner, W.G., 1995. Stomatal density and index of fossil plants track atmospheric carbon dioxide in the Palaeozoic. Ann. Bot. 76, 389–395.

McElwain, J.C., Yiotis, C., Lawson, T., 2015. Using modern plant trait relationships between observed and theoretical maximum stomatal conductance and vein density to examine patterns of plant macroevolution. New Phytol. 209 (1), 94–103.

Michel, L.A., Tabor, N.J., Montañez, I.P., Schmitz, M.D., Davydov, V.I., 2015. Chronostratigraphy and paleoclimatology of the Lodève Basin, France: evidence for a pan-tropical aridification event across the Carboniferous-Permian boundary. Palaeogeogr. Palaeoclimatol. Palaeoecol. 430, 118–131.

Miller, K.B., McCahon, T.J., West, R.R., 1996. Lower Permian-Wolfcampian.paleosol-Bearing cycles of the U.S. Midcontinent; evidence of climatic cyclicity. J. Sediment. Res. 66A, 71–84.

Montañez, I.P., 2013. Modern soil system constraints on reconstructing deep time atmospheric CO_2. Geochim. Cosmochim. Acta 101, 57–75.

Montañez, I.P., Poulsen, C.J., 2013. The Late Paleozoic Ice Age: an evolving paradigm. Annu. Rev. Earth Planet. Sci. 41, 629–656.

Montañez, I.P., Tabor, N.J., Niemeier, D., DiMichele, W.A., Frank, T.D., Fielding, C.R., Isbell, J.L., 2007. CO_2-forced climate and vegetation instability during late Paleozoic deglaciation. Science 315, 87–91.

Montañez, I.P., McElwain, J.C., Poulsen, C.J., White, J.D., DiMichele, W.A., Wilson, J.P., Griggs, G., Hren, M.T., 2016. Climate pCO_2 and terrestrial carbon cycle linkages during Late Palaeozoic glacial-interglacial cycles. Nat. Geosci. 9, 824–831.

Mora, C.I., Driese, S.G., Seager, P.G., 1991. Carbon dioxide in the Paleozoic atmosphere: evidence from carbon-isotope compositions of pedogenic carbonate. Geology 19, 1017–1020.

Mora, C.I., Driese, S.G., Colarusso, L.A., 1996. Middle to Late Paleozoic atmospheric CO_2 levels from soil carbonate and organic matter. Science 271, 1105–1107. https://doi.org/10.1126/science.271.5252.1105.

Mountney, N.P., 2006. Periodic accumulation and destruction of aeolian sequences in the Permian Cedar Mesa Sandstone, White Canyon, southern Utah, USA. Sedimentol 53, 789–823.

Mujal, E., Fortuny, J., Marmi, J., Dinarès-Turell, J., Bolet, A., Oms, O., 2018. Aridification across the Carboniferous–Permian transition in central equatorial Pangea: the Catalan Pyrenean succession (NE Iberian Peninsula). Sed. Geol. 363, 48–68.

Myers, T.S., Tabor, N.J., Jacobs, L.L., Bussert, R., 2016. Effects of different organic- matter sources on estimates of atmospheric and soil pCO_2 using pedogenic carbonate. J. Sediment. Res. 86, 800–812.

Nejad, A.R., Van Meeteren, U., 2005. Stomatal response characteristics of *Tradescantia virginiana* grown at high relative air humidity. Physiol. Plant. 125, 324–332.

Nelsen, M.P., DiMichele, W.A., Peters, S.E., Boyce, C.K., 2016. Delayed fungal evolution did not cause the Paleozoic peak in coal production. Proc. Natl. Acad. Sci. U. S. A. 113, 2442–2447. https://doi.org/10.1073/pnas.1517943113.

Onwueme, I., Johnston, M., 2000. Influence of shade on stomatal density, leaf size and other leaf characteristics in the major tropical root crops, tannia, sweet potato, yam, cassava and taro. Exp. Agric. 36, 509–516.

Park, J., Royer, D.L., 2011. Geologic constraints on the glacial amplification of Phanerozoic climate sensitivity. Am. J. Sci. 311, 1–26.

Parrish, J.T., 1993. Climate of the supercontinent Pangea. J. Geol. 101, 215–253.

Peñuelas, J., Matamala, R., 1990. Changes in N and S content, stomatal density and specific leaf area of 14 plant species during the last three centuries of CO_2 increase. J. Exp. Bot. 41, 1119–1124.

Peyser, C.E., Poulsen, C.J., 2008. Controls on Permo-Carboniferous precipitation over tropical Pangea: a GCM sensitivity study. Palaeogeogr. Palaeoclimatol. Palaeoecol. 268, 181–192.

Pfefferkorn, H.W., Gastaldo, R.A., DiMichele, W.A., Phillips, T.L., 2008. Pennsylvanian tropical floras from the United States as a record of changing climate. Geol. Soc. Am. Spec. Pap. 441, 305–316.

Porter, A.S., Gerald, C.E., Yiotis, C., Montañez, I.P., MeElwain, J.C., 2019. Testing the accuracy of new paleoatmospheric CO_2 proxies based on plant stable carbon isotopic composition and stomatal traits in a range of simulated paleoatmospheric O_2: CO_2 ratios. Cosmochim. Acta. 259, 69–90.

Reeves Jr., C.C., 1976. Caliche; Origin, Classification, Morphology and Uses. Estacado Books, Lubbock, Texas, p. 233.

Retallack, G.J., 2001. A 300-million-year record of atmospheric carbon dioxide from fossil plant cuticles. Nature 411, 287–290. https://doi.org/10.1038/35077041.

Retallack, G.J., 2005. Pedogenic carbonate proxies for amount and seasonality of precipitation in paleosols. Geology 33, 333–336.

Rinehart, L.F., Lucas, S.G., Tanner, L.H., Nelson, W.J., Elrick, S.D., Chaney, D.S., DiMichele, W.A., 2015. Plant architecture and spatial structure of an early Permian woodland buried by flood waters, Sangre de Cristo Formation, New Mexico. Palaeogeogr. Palaeoclimatol. Palaeoecol. 424, 91–110.

Romanek, C.S., Grossman, E.L., Morse, J.W., 1992. Carbon isotopic fractionation in synthetic aragonite and calcite—effect of temperature and precipitation rate. Geochim. Cosmochim. Acta 56 (1), 419–430.

Royer, D.L., Berner, R.A., Beerling, D.J., 2001. Phanerozoic CO_2 change: evaluating geochemical and paleobiological approaches. Earth Sci. Rev. 54, 349–392.

Royer, D.L., Berner, R.A., Montañez, I.P., Tabor, N.J., Beerling, D.J., 2004. CO_2 as a primary driver of Phanerozoic climate. GSA Today 14 (3), 4–10.

Rygel, M.C., Fielding, C.R., Frank, T.D., Birgenheier, L.P., 2008. The magnitude of late Paleozoic glacioeustatic fluctuations: a synthesis. J. Sediment. Res. 78, 500–511.

Schaetzl, R.J., Thompson, M.L., 2015. Soils: Genesis and Geomorphology, second ed. 777 Cambridge, New York.

Schaller, M.F., Wright, J.D., Kent, D.V., 2015. A 30 Myr record of Late Triassic atmospheric pCO_2 variation reflects a fundamental control of the carbon cycle by changes in continental weathering. Geol. Soc. Am. Bull. https://doi.org/10.1130/B31107.1.

Schubert, B.A., Jahren, A.H., 2012. The effect of atmospheric CO_2 concentration on carbon isotope fractionation in C3 land plants. Geochim. Cosmochim. Acta 96, 29–43.

Schubert, B.A., Jahren, A.H., 2015. Global increase in plant carbon isotope fractionation following the Last Glacial Maximum caused by increase in atmospheric pCO_2. Geology 43 (5), 435–438.

Scotese, C.R., 2014. Atlas of Permo-Carboniferous Paleogeographic Maps (Mollweide Projection), Maps 53–64. In: The Late Paleozoic, PALEOMAP Atlas for ArcGIS, PALEOMAP Project, Evanston, IL. vol. 4.

Soreghan, G., Elmore, R., Lewchuck, M., 2002. Sedimentologic–magnetic record of western Pangean climate in upper Paleozoic loessite (lower Cutler beds, Utah). Geol. Soc. Am. Bull. 114, 1019–1035.

Tabor, N.J., 2007. Permo-Pennsylvanian palaeotemperatures from Fe-oxide and phyllosilicate δ18O values. Earth Planet. Sci. Lett. 253, 159–171.

Tabor, N.J., Montañez, I.P., 2002. Shifts in late Paleozoic atmospheric circulation over western equatorial Pangea: insights from pedogenic mineral δ18O compositions. Geology 30 (12), 1127–1130.

Tabor, N.J., Montañez, I.P., 2004. Permo-Pennsylvanian alluvial paleosols (north-central Texas): high-resolution proxy records of the evolution of early Pangean paleoclimate. Sedimentology 51, 851–884.

Tabor, N.J., Montañez, I.P., 2005. Oxygen and hydrogen isotope compositions of pedogenic phyllosilicates: development of modern surface domain arrays and implications for paleotemperature reconstructions. Palaeogeogr. Palaeoclimatol. Palaeoecol. 223, 127–146.

Tabor, N.J., Poulsen, C.J., 2008. Palaeoclimate across the Late Pennsylvanian–Early Permian tropical palaeolatitudes: a review of climate indicators, their distribution, and relation to palaeophysiographic climate factors. Palaeogeogr. Palaeoclimatol. Palaeoecol 268, 293–310.

Tabor, N.J., Montañez, I.P., Southard, R.J., 2002. Mineralogical and stable isotopic analysis of pedogenic proxies in Permo-Pennsylvanian paleosols: implications for paleoclimate and paleoatmospheric circulation. Geochim. Cosmochim. Acta 66, 3093–3107.

Tabor, N.J., Montañez, I.P., Scotese, C.R., Poulsen, C.J., Mack, G.H., 2008. Paleosol archives of environmental and climatic history in paleotropical western Pangea during the latest Pennsylvanian through early Permian. Geol. Soc. Am. Spec. Pap. 441, 291–303.

Tabor, N.J., DiMichele, W.A., Montañez, I.P., Chaney, D.S., 2013. Late Paleozoic continental warming of a cold tropical basin and floristic change in western Pangea. Internat. J. Coal Geol. 119, 177–186.

Tandon, S.K., Gibling, M.R., 1997. Calcretes at sequence boundaries in Upper Carboniferous cyclothems of the Sydney basin, Atlantic Canada. Sediment. Geol. 112, 43–67.

Tanner, L.H., 2002. Mesozoic atmospheric CO_2 spike: comment and reply. Nature 415, 388.

Tanner, L.H., 2010. Terrestrial carbonates as indicators of palaeoclimate. In: Alonsa-Zarza, A.M., Tanner, L.H. (Eds.), Carbonates in Continental Settings: Geochemistry, Diagenesis and Applications. Developments in Sedimentology, vol. 62. Elsevier, pp. 179–214.

Tanner, L.H., Lucas, S.G., 2017. Paleosols of the upper paleozoic sangre de cristo formation, north-central New Mexico: record of early Permian palaeoclimate in tropical Pangea. J. Palaeogeogr. 6 (2), 144–162.

Tanner, L.H., Lucas, S.G., 2018. Pedogenic record of climate change across the Pennsylvanian- Permian boundary in red-bed strata of the Cutler Group, northern New Mexico, USA. Sediment. Geol. 373, 98–110.

Woodward, F.I., 1987. Stomatal numbers are sensitive to increases in CO_2 from pre-industrial levels. Nature 327, 617–618.

Zhou, N., Wang, Y., Ya, L., Porter, A., Kürschner, W.M., Li, L., Lu, N., McElwain, J.C., 2020. An inter-comparison study of three stomatal-proxy methods for CO_2 reconstruction applied to early Jurassic Ginkgoales plants. Palaeogeogr. Palaeoclimat. Palaeoecol. 542. https://doi.org/10.1016/j.palaeo.2019.109547.

Zhu, L., Tabor, N.J., 2014. Lower Permian paleosol morphologies and paleoatmospheric pCO_2 estimates from north-central Texas, USA. Abstr. Progr. Geol. Soc. Am. 46 (6), 131.

CHAPTER ELEVEN

Globally enhanced Hg concentration and Hg and C isotopes in Permian–Triassic boundary successions: Possible linkage to volcanism

Alcides Nóbrega Sial[a,*], Jiubin Chen[b], Cristoph Korte[c], Manoj Kumar Pandit[d], Jorge Spangenberg[e], Juan Carlos Silva-Tamayo[f], Luis Drude de Lacerda[g], Valderez Pinto Ferreira[a], José Antônio Barbosa[a], Claudio Gaucher[h], Natan Silva Pereira[i], and Paulo Ricardo Riedel[a]

[a]NEG–LABISE, Department of Geology, Federal University of Pernambuco, Recife, Brazil
[b]Institute of Surface-Earth System Science, Tianjin University, Tianjin, China
[c]Department of Geosciences and Natural Resource Management, University of Copenhagen, ØsterVoldgade10,1350 and Nordic Center for Earth Evolution (NordCEE), Copenhagen, Denmark
[d]Department of Geology, University of Rajasthan, Jaipur, India
[e]Institute of Earth Surface Dynamics (IDYST), University of Lausanne, Lausanne, Switzerland
[f]Faculty of Sciences, Antonio Nariño University, Bogota, Colombia
[g]LABOMAR, Institute of Marine Sciences, Federal University of Ceará, Fortaleza, Brazil
[h]Facultad de Ciencias, Universidad de La República, Montevideo, Uruguay
[i]Department of Biology, State University of Bahia, Paulo Afonso, Brazil
*Corresponding author: e-mail address: alcides.sial@ufpe.br

Contents

Stratigraphy & Timescales, Volume 5
ISSN 2468-5178
https://doi.org/10.1016/bs.sats.2020.08.011

Abstract

This study presents C and Hg (Hg/TOC) chemostratigraphy data on identified Permian–Triassic successions across the globe with an aim to develop a better understanding on the "mass extinction" event during this time interval. Besides, it would also contribute to the global database on C and Hg stratigraphy and reinforce the application of the latter as a reliable proxy in global correlation, especially in cases of sedimentary sequences deposited coevally with LIP volcanism. In this endeavor, the $\delta^{13}C$ chemostratigraphic curves from thousands of kilometers apart from nine classical marine sedimentary sections from various paleolatitudinal zones as well as Hg/TOC stratigraphic pathways of eight of these sections, were examined in a transect across the Pangea supercontinent (~252 Ma). A strong negative $\delta^{13}C_{carb}$ shift at the Late Permian Mass Extinction (LPME), in Meishan D, Guryul Ravine and several other (Permian Triassic Boundary) PTB sections worldwide, is probably due to voluminous CO_2 degassing during volcanism in the Siberian Traps Large Igneous Province (STLIP). The C-isotope shift at the LPME is much stronger than in the PTB or Early Triassic Mass Extinction (ETME).

The Hg/TOC peaks are observed at LPME and ETME horizons in the Meishan D section, whereas the peaks were recorded at LPME and PTB horizons in Hovea-3, Ursula Creek, Idrijca and Rizvanuša sections. The Rizvanuša section displays a single peak at the ETME horizon, Zal and Abadeh successions at the LPME and ETME horizons, while the Misci section shows Hg/TOC enrichment at the LPME horizon. The largest Hg/TOC peaks at the LPME, PTB and ETME can be associated with the beginning of Stage 2 (extrusive hiatus) of STLIP. The origin of a Hg/TOC spike in the LPME–PTB interval in seven of these sections is unclear as well as the Hg/TOC spike below the LPME at the Misci and Zal sections. The latter may, possibly be, linked to the STLIP (Stage 1) or some local volcanism.

In the $\delta^{202}Hg$ (MDF) vs $\Delta^{201}Hg$ (MIF) cross plot, the majority of samples from the extinction interval plot within the "volcanic-emission box" and a few samples, in the "sediment/soil/peat box." Hg isotope signatures underline mixing of volcanic and terrestrial Hg, generating four trends with negligible variation in $\Delta^{201}Hg$. Lesser terrigenous-Hg influx was noticed in sections closer to the STLIP (Rizvanuša, Idrijca and Seres/Misci, all in Europe), recording close to zero $\Delta^{201}Hg$ (MIF) values. Marked influence is observed in sections far away from the STLIP (Meishan, Ursula Creek and Hovea-3) that exhibit negative a $\Delta^{201}Hg$ (MIF). The Zal and Abadeh sections, at intermediate distance from the STLIP, exhibit the highest, positive $\Delta^{201}Hg$ values (Abadeh) and the lowest, negative $\Delta^{201}Hg$ values (Zal). The $\Delta^{199}Hg$ (MIF) vs Hg ($n\,ng^{-1}$) cross plot testifies that volcanic Hg has been contaminated by terrigenously sourced-Hg influx, inducing the two major curved trends.

The negative C-isotope excursions and Hg/TOC enrichments in these sections are nearly coeval, substantiating a synchronism between the P–T transition biotic crises (LPME and ETME) and the start of the stage 2 of the Siberian Traps (sill-intrusions).

1. Introduction

There is a consensus that the most severe mass extinction of eukaryotic biota in the in the Earth's history took place at 251.9 Ma, i.e., at the end of the Permian (e.g., Benton, 2003; Burgess et al., 2014; Burgess and Bowring, 2015; Chen and Benton, 2012; Erwin et al., 2002; Renne et al., 2015) when over 90% of all the marine and about 70% of all the terrestrial species perished (e.g., Erwin, 2006). This mass extinction also caused profound degradation of marine bottom communities, resulting in an extreme reduction in bioturbation of the seafloor that led to eradication of the sedimentary mixed layer at an interregional scale (Hofmann et al., 2015). Likely the poor sediment oxygenation that resulted from the extinction of bioturbators has retarded the recovery from this mass extinction, during the Early Triassic.

This greatest decline in biodiversity is known as the P–Tr extinction, the P–T extinction, End-Permian Mass Extinction (EPME), Latest Permian Mass Extinction (LPME), and informally, as the "Great Dying." Several have been the causes invoked to explain this mass extinction and, among them, shallow oceanic water anoxia, bolide impact and volcanism seem to be the main ones.

2. Causes of the Permian–Triassic mass extinction

2.1 Photic-zone anoxia in the oceanic water column

Sedimentary deposits that document the anoxic conditions have been widely recognized in sections across the Permian–Triassic Boundary (PTB), except for some shallow-water locations in the earliest Triassic. Framboidal pyrite has been proposed as evidence of ocean anoxia/euxinia in PTB sections in eastern Greenland and Sosio Valley (Wignall and Twitchett, 2002), southwestern Japan (Isozaki, 1997), Italy and Slovenia (Wignall and Twitchett, 1996), Kashmir in northern India (Huang et al., 2019; Wignall et al., 2005) and British Columbia, Canada (Bond and Wignall, 2010; Wignall and Newton, 2003). This widespread episode of oceanic photic-zone anoxia, perhaps, lasted for about 10 million years, named as a "superanoxic event" (e.g., Grice et al., 2005; Isozaki, 1997) and, for this reason, anoxia has been regarded as one prime cause of the Permian–Triassic marine biotic crisis (e.g., Huang et al., 2019; Isozaki, 1997; Kaiho et al., 2016; Krystyn et al., 2003; Kumar et al., 2017; Lehrmann et al., 2003; Shen et al., 2007; Song et al., 2014; Wignall et al., 2005;

Wignall and Hallam, 1992). A detailed study of the size and distribution of framboidal pyrite across the Meishan Section (Shen et al., 2007) confirmed a worldwide dysoxic marine event during the P–T transition, reinforcing the possibility of a direct relationship between oxygen availability and the marine LPME.

Upwelling of toxic deepwaters has been invoked as an active mechanism during the marine LPME as documented by Algeo et al. (2008) in a marine PTB section at Nhi Tao, Vietnam. At this section, a series of pyritic horizons is characterized by patterns of concurrent decreases in pyrite S- (δ^{34}Spy) and carbonate C-isotopic compositions ($\delta^{13}C_{carb}$). The first pyritic horizon was precipitated in an event coincident with the LPME and this seems to strongly support upwelling of toxic deepwaters process at the P–T transition as a major.

A reconstruction at low latitudes of the ocean redox structure reveals that a gradual increase in temperature spanning the LPME–ETME interval have induced a gradual change from a mixed oxic to a stratified euxinic ocean prior to the LPME (Kaiho et al., 2016). This was followed by anoxia in near-shore surface waters and anoxia/euxinia in shallow-intermediate waters about 60 kyr later, at the ETME, almost for 1 million years.

Wignall and Newton (2003) recognized Late Permian–Early Triassic marine record in sections from the western margin of Pangea (Ursula Creek, British Columbia) and Neotethyan Ocean (Selong, South Tibet). The Ursula Creek section shows a decline in the seafloor oxygen availability in the Changxingian Stage (absence of bioturbation and presence of framboidal pyrite), followed by euxinic conditions in the latest Changxingian and Early Triassic, an event coeval to the disappearance of the siliceous sponge fauna. The Selong section recorded regression and erosion in the mid-Changxingian, followed by deepening in the late Changxingian, with the P–T transition, with a marked increase in the faunal diversity (appearance of foraminifera, brachiopods, and sponges). This suggests warming conditions without extinction in marine waters. Diverse Permian remnants perished only in the late Griesbachian Stage, at a level showing evidence for dysoxia (thinly-bedded, pyrite-rich limestone). Thus, the LPME, according to Wignall and Newton (2003), is diachronous by half a million years or more, with the late Changxingian extinction in Panathalassa, coinciding with the diversity increase that led to the migration of warm-water taxa into the high southerly paleolatitudinal regions of Neotethys.

The possibility of diachronous LPME was also recognized by Algeo et al. (2012) in sedimentary rocks of the West Blind Fiord, Ellesmere Island in the Canadian Arctic. They observed a regional extinction of siliceous sponges during the so-called Late Permian "Arctic extinction event" in this region. The horizon, equivalent to the LPME of Tethyan shallow-marine sections, is about 5.6m higher and ~100k.y. younger than the LPME horizon of this "Arctic extinction event," evidencing a diachronous nature of the marine biotic and environmental Late Permian crisis on a global scale, associated with intensification in anoxia.

A paleolatitude zonation with severe anoxia persisted in the higher paleolatitudes during the P–T transition while less intense and short-lived one was recorded at the equatorial settings (Bond and Wignall, 2010). These authors have evaluated the intensity and duration of anoxia on epicontinental PTB sections by analyzing the size of framboidal pyrite. They have inferred that: (a) intense anoxia with euxinia developed throughout the P–T interval from the Boreal (e.g., Spitsbergen, Greenland) and the Neotethys oceans (e.g., Hovea-3 section, Western Australia), and dysoxia in relatively shallow-water settings above the storm wave base; (b) dysoxic with rare euxinia conditions, in contrast, widely developed in a range of water depths (including shallow ones), at equatorial paleolatitudes; and (c) complex and unstable redox history was recorded in western and eastern Tethyan locations, with the replacement of anoxic (*Hindeodus praeparvus* Zone) by oxygenated conditions, one during the P–T interval (*H. changxingensis* to *H. parvus* zones), returning later to oxygen-poor deposition (*Isarcicella isarcica* Zone). However, little has been published on the extent and timing of anoxia during the P–T transition in Gondwana (Huang et al., 2019).

2.2 Extraterrestrial cause (bolide impact)

Studies on some of the PTB sites suggest that the LPME occurred very abruptly and as in the case of mass extinction recorded in the Cretaceous–Paleogene (K/Pg) boundary, one or multiple asteroid or a comet impact was proposed as a potential trigger (e.g., Retallack et al., 1998). This hypothesis was preferentially sought on the Tethyan realm and was based on limited occurrences of (a) iridium and microspherule near the PTB (Xu and Yan, 1993), (b) shocked quartz (Retallack et al., 1998), (c) gigantic release of sulfur from the mantle by a bolide impact at the end-Permian (Kaiho et al., 2001, 2006), (d) chondritic meteorite detritus

(Basu et al., 2003), (e) fullerenes with trapped helium and argon with isotope ratios of carbonaceous chondrites (Becker et al., 2001) and (f) a supposed 250 Ma impact crater (Becker et al., 2004). These pieces of evidence, however, have been questioned and discredited to a certain extent (e.g., Farley et al., 2005; Glikson, 2004; Koeberl et al., 2004; Müller et al., 2005).

2.3 Siberian traps LIP volcanism (STLIP)

The currently accepted hypothesis envisages that extensive volcanism of the Siberian Traps Large Igneous Province (ST LIP) has been the main trigger for this mass extinction event (e.g., Bowring et al., 1998; Burger et al., 2019; Campbell et al., 1992; Courtillot et al., 1999; Grard et al., 2005; Kamo et al., 2003; Korte et al., 2010a, b; Lo et al., 2002; Reichow et al., 2009; Renne et al., 1995; Yin et al., 1992, 2007). It is believed that the toxic effect of volatile-rich, mafic-ultramafic and sill-type intrusions depends on the size of the igneous province, geological setting of the host region, magma plumbing system, and eruption dynamics that control the magnitude and composition of the thermogenic outgassing (Racki, 2020). Furthermore, LIP eruptions can drive anoxia via global warming (Bond et al., 2020) and, therefore, the global PTB biodiversity crisis may have been mainly driven by volcanism-linked anoxia.

Climatic and environmental perturbations, including multiple oscillations in oceanic anoxia, that followed the LPME, further retarded the pace of the marine ecosystem recovery (~5 million years; Chen et al., 2020; Grasby et al., 2020; Lau et al., 2016, Zhang et al., 2018), and post-extinction recovery of lacustrine ecosystems was delayed for about 10 million years (Zhao et al., 2020). It is known that chemical weathering can supply nutrients to the ocean, influencing the C and the Ca cycles and biological recovery (e.g., Grasby et al., 2020; Silva-Tamayo et al., 2018). Chen et al. (2020) used Mg isotopic compositions of siliciclastic components of shallow-marine carbonates from well-known PTB sections (Iran and China), to track changes in the chemical weathering intensity after the LPME (Early Triassic). They have demonstrated the potential of the δ^{26}Mg patterns of the silicate fraction in marine carbonates as a proxy for chemical weathering intensity and proposed that the marine ecosystem recovery during the Early Triassic could be linked to recurrent episodes of intense chemical weathering, caused by CO_2 degassing and climate warming. It has been also suggested that in the Early Triassic high ocean temperatures depressed the marine nutricline, leading to nutrient limitations and stressing primary production, contributing to prolonged marine recovery (Grasby et al., 2020).

The lack of conodont specimens from the LPME horizon does not allow estimating the seawater temperature that prevailed during the biotic extinction event (Chen et al., 2014). However, an increase of 8 °C in tropical sea-surface temperature across the LPME horizon in the Meishan section (Bed 24e to Bed 27a) has been documented (Joachimski et al., 2012), and such elevated seawater temperatures (~30–40 °C) have potentially slowed down the marine biotic recovery of the Early Triassic, as proposed by Sun et al. (2012).

3. Aims of this study and geological overview

3.1 Aims of the study

Several studies have attempted to constrain the paleo-sedimentary geochemistry of the prolonged disturbance in the biological productivity of both marine and terrestrial environments during the LPME (e.g., Algeo et al., 2007; Burger et al., 2019; Fio et al., 2010; Holser et al., 1989; Shen et al., 2013; Yoshida et al., 2014; Zhou and Kyte, 1988). However, only two mid- to high-latitude PTB sections in north India (Kashmir valley), and Australia (Perth Basin) in the southern hemisphere Neothetys ocean, have been studied (e.g., Algeo et al., 2007; Ghosh et al., 2016; Thomas et al., 2004).

Sections in northern India contain good-quality marine sedimentary and fossil records and thus provide a unique opportunity to study the redox conditions associated with the biotic crisis in the Gondwana region. In an endeavor to fill this gap, Huang et al. (2019) have provided a high-resolution biotic and redox data from P–T sections in Kashmir. To evaluate redox conditions through an independent approach using the C and Hg isotope records, in this contribution we present high-resolution $\delta^{13}C$ stratigraphic data from the stratigraphically-expanded, siliciclastic-dominated, Guryul Ravine succession (Neotethys Ocean) from Kashmir Himalaya in north India and discuss the implications of the PTB extinction. These pathways are contrasted with the $\delta^{13}C$ and Hg stratigraphic patterns of the Meishan GSSP limestone-dominated condensed section (Paleotethys Ocean). Besides, Hg isotopes are used as a proxy to unravel the source of the Hg preserved in the Meishan section. Furthermore, the Hg/TOC stratigraphic patterns from Meishan section is contrasted with those from seven widely separated, classical PTB sections, reported by Sial et al. (2019b) that form a transect diagonal to the Pangea supercontinent (252 Ma ago): Ursula Creek (Canada), Misci (Italy), Rizvanuša (Croatia), Zal and Abadeh (Iran), Idrijca (Slovenia) and Hovea-3 (Australia). The global distribution of these nine PTB sections is found in Fig. 1 and location with coordinates

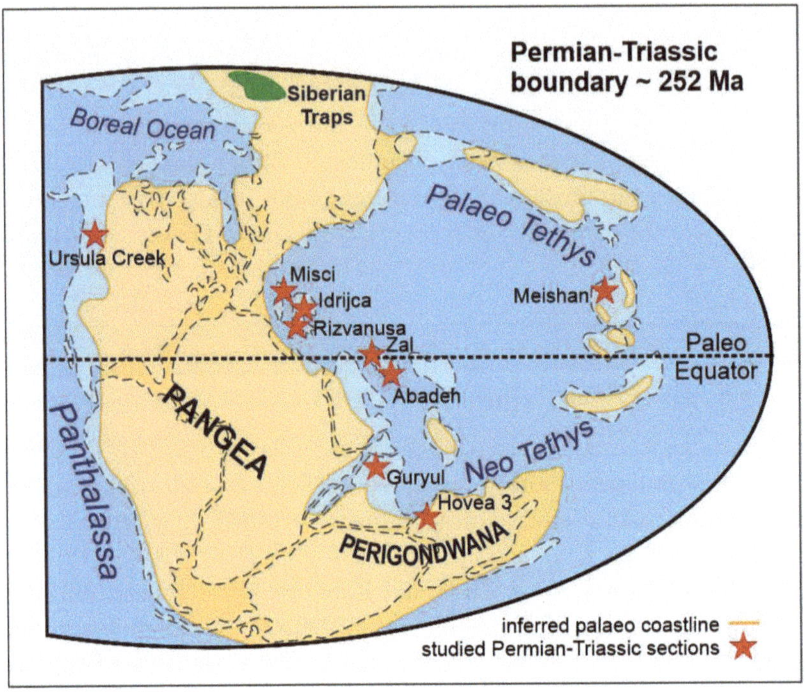

Fig. 1 Global palaeogeographic reconstruction during the Permian–Triassic transition and global geographical distribution of the Permian–Triassic sections in this study: Meishan GSSP (China), Guryul Ravine (India), Ursula Creek (Canada), Misci (Italy), Rizvanuša (Croatia), Zal and Abadeh (Iran), Idrijca (Slovenia), and Hovea-3 (Australia). *Modified from Wignall, P.B., Twitchett, R.J., 2002. Extent, duration, and nature of the Permian–Triassic superanoxic event. In: Koebel, C., MacLeod, K.G. (Eds.), Catastrophic Events and Mass Extinctions: Impacts and Beyond: Boulder, Colorado. Geological Society of America Special Paper 356, 395–413; Thomas, B.M., Willink, R.J., Grice, K., Twitchett, R.J., Purcell, R.R., Archbold, N.W., George, A.D., Tye, S., Alexander, R., Foster, C.B., Barber, C.J., 2004. Unique marine Permian–Triassic boundary section from Western Australia. Aust. J. Earth Sci. 51, 423–430; Sial, A.N., Chen, J., Lacerda, L.D., Korte, C., Spangenberg, J.E., Silva-Tamayo, J.C., Gaucher, C., Ferreira, V.P., Barbosa, J.A., Pereira, N.S., Benigno, A.P., 2019b. Globally enhanced Hg deposition and Hg isotopes in sections straddling the Permian–Triassic boundary: link to volcanism. Palaeogeogr. Palaeoclimatol. Palaeoecol., 540, https://doi.org/10.1016/j.palaeo.2019.109537.*

in Figs. 2 and 3. Besides, this study aims to contribute to an improved understanding of the relationship between the C and Hg chemostratigraphies, reinforcing the use of the latter as a reliable proxy in global correlation, especially in cases of sedimentary sequences deposited coevally with LIP volcanism.

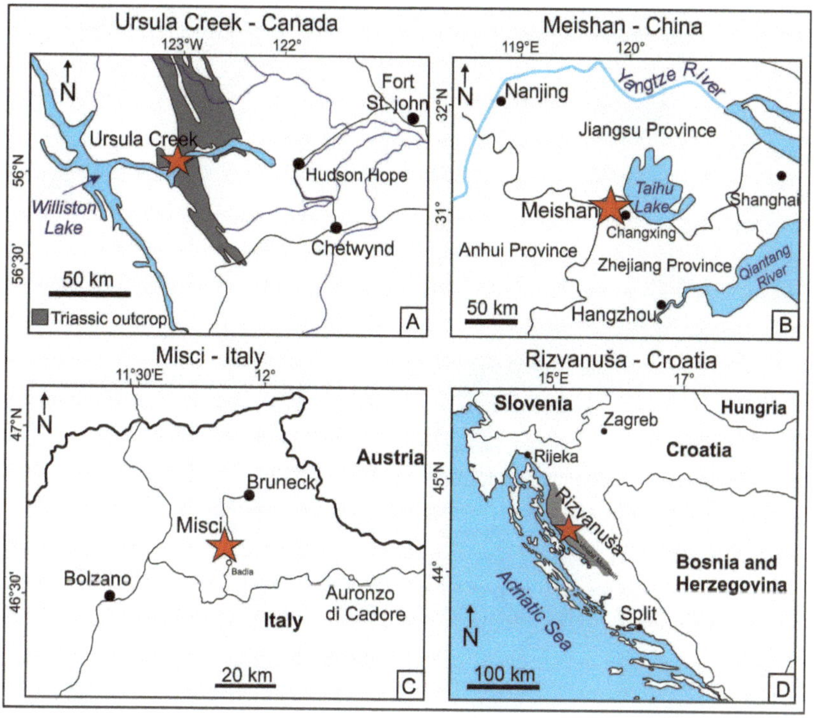

Fig. 2 Geographical sketch maps showing localities of four sections in this study: (A) Ursula Creek, Canada (after Wignall and Newton, 2003); (B) Meishan, China (based on Li and Jones, 2017); (C) Seres/Misci, southern Alps, modified from Kraus et al. (2013); (D) Rizvanuša, Croatia, modified from Fio et al. (2010). *Modified from Sial, A.N., Chen, J., Lacerda, L.D., Korte, C., Spangenberg, J.E., Silva-Tamayo, J.C., Gaucher, C., Ferreira, V.P., Barbosa, J.A., Pereira, N.S., Benigno, A.P., 2019b. Globally enhanced Hg deposition and Hg isotopes in sections straddling the Permian–Triassic boundary: link to volcanism. Palaeogeogr. Palaeoclimatol. Palaeoecol., 540, https://doi.org/10.1016/j.palaeo.2019.109537.*

3.2 Geological setting

3.2.1 Meishan succession (GSSP), China

All the nine stages of the Permian Period are defined using the first appearance datum (FAD) of conodonts, fossilized tooth-like phosphatic elements—usually less than 1 mm long, wormlike chordates, which lived in the marine realm from Cambrian to Triassic times. Kozur et al. (1995) summarized the Permian conodonts biostratigraphy, established by several earlier studies (e.g., Behnken, 1975; Clark and Behnken, 1971; Igo, 1981; Kozur, 1978; Movshovich et al., 1979; Perlmutter, 1975; Wang and Wang, 1981;

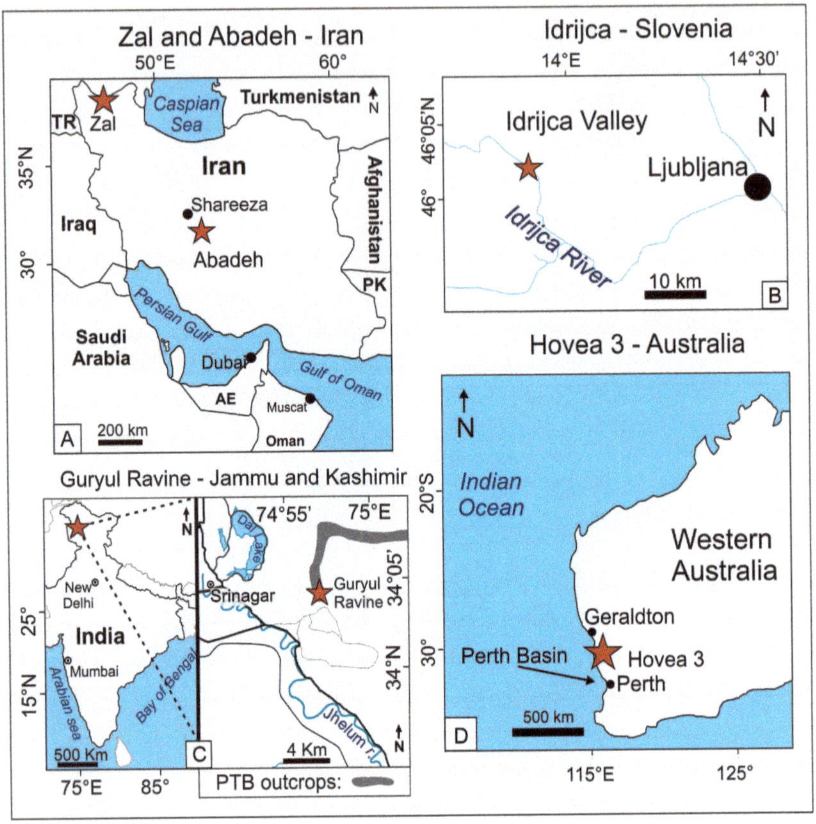

Fig. 3 Geographical localities of the other five sections: (A) Zal and Abadeh (central Iran, Middle East) sections, modified from Richoz et al. (2010); (B) Idrijca, western Slovenia, based on Schwab and Spangenberg (2004); (C) Guryul Ravine, Kashmir, and (D) Hovea-3, western Australia (based on Thomas et al., 2004).

Youngquist et al., 1951). The Upper Permian is "zone-based" on the partial range lineage interval of species of Clarkina (Kozur, 1990). These are all typical warm-water equatorial taxa and are not found in peri-Gondwana or in northern cool-water provinces where long-ranging species of Mesogondolella continued to evolve (Wardlaw and Collinson, 1979). The index fossil for the PTB strata is the *Hindeodus parvus*, which is of world-wide occurrence, not facies-related or latitude-restricted, found in both shallow-water and pelagic deposits (Kozur, 1996; Lai, 1998). It was first described by Rexroad and Furnish (1964) and proposed as an index fossil by Paull and Paull (1994). The Meishan succession has been subdivided into 115 beds and the Global Stratotype Section and Point (GSSP) of the PTB has been defined at the base of *Hindeodus parvus* horizon, i.e., the base of Bed 27c of the Meishan section D (Fig. 4), Changxing County, Zhejiang Province, South China (Yin et al., 2001), which has an inferred radiometric age of 251.902 ± 0.024 Ma (Burgess et al., 2014).

As the GSSP was defined within a condensed Tethyan carbonate-dominated section, it has some limitations, as already pointed out by Baresel et al. (2017) and Zuchuat et al. (2020). For example, the extinction interval is of only 0.36 m, corresponding to Beds 25–28 (Fig. 4) and this precludes some high-resolution proxy study rendering the correlation between PTB sections more difficult.

The demise of marine fauna during the LPME is well known from the detailed fossil record of the Meishan section (e.g., Erwin, 2006; Jin et al., 2000; Yin et al., 2001). A significant loss of most fusulinids and ammonoids and many brachiopods has been recorded below the Bed 25 (Jin et al., 2000; Yin et al., 2001, 2007). The onset of biodiversity reduction, most likely, started during the deposition of Bed 22 (Jin et al., 2000; Yin et al., 2001, 2007). This onset coincides with a negative shift in the conodonts $\delta^{44/40}$ Ca values at the same sections, which has been regarded as reflecting a major disturbation of the marine Ca cycle caused by ocean acidification (Hinojosa et al., 2012). The negative shift in the conodont $\delta^{44/40}$ Ca values displayed by the Meishan section is also displayed by marine carbonates from several P–T sections around the world (Payne et al., 2010; Silva-Tamayo et al., 2018). The global negative Ca isotope signal displayed by conodonts and marine carbonates has been regarded as reflecting a perturbation of the ocean water Ca isotope composition, and as a consequence of the massive emission of volcanic CO_2 to the P–T atmosphere by the P–T LIPs. The negative global shift in conodont $\delta^{44/40}$ Ca values continued until the LPME event, which is preserved in Beds 25–26. A second mass extinction event occurred during the deposition of Bed 28, in the earliest Triassic (ETME) (Song et al., 2013a, b; Xie et al., 2005). This mass extinction event coincides with an additional negative shift in the $\delta^{44/40}$ Ca values of globally distributed marine carbonates and thus, with a second ocean acidification event that postdated the LPME by 60 k.y. (Silva-Tamayo et al., 2018). This event also coincides with a major perturbation of the marine U isotope cycle (Lau et al., 2016), which suggests a clear link between volcanism, ocean acidification and ocean anoxia.

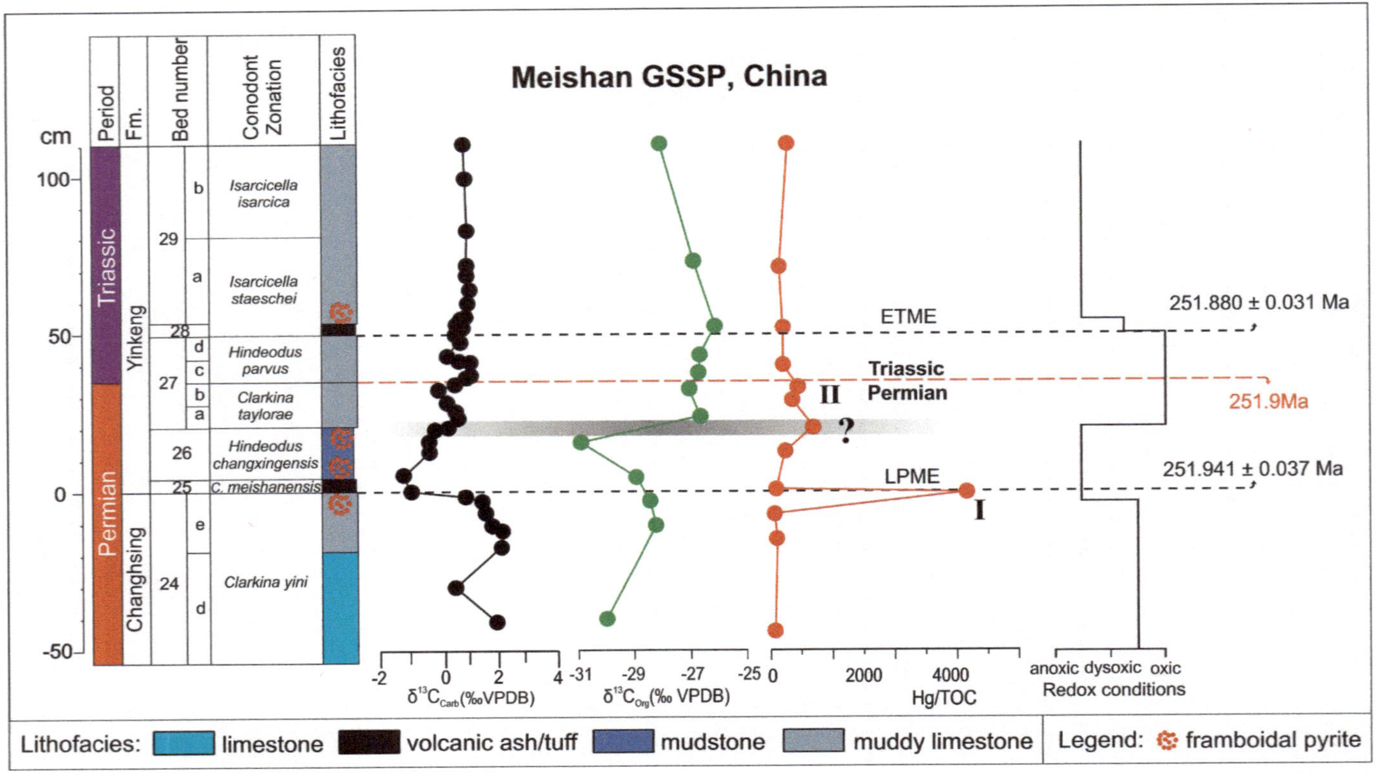

Fig. 4 Lithological succession for the Meishan GSSP section, according to Chen et al. (2015); $\delta^{13}C_{carb}$ variation pattern is from Chen et al. (2005); $\delta^{13}C_{org}$, Hg, TOC and Hg/TOC chemostratigraphic pathways are from Sial et al. (2019b); Hg/S, Hg/Al variation patterns are from this study. Variation of redox conditions according to Shen et al. (2007), based on size distribution of framboidal pyrite.

According to Burgess and Bowring (2015), this extinction interval corresponds to a maximum of 61 ± 48 k.y. and may contain several depositional hiatuses at or below the PTB (Cao and Zheng, 2007, 2009; Shen et al., 2018; Zheng et al., 2013). The U-Pb age data on the Siberian Traps LIP (STLIP) lava flows, sills and explosively erupted rocks, carried out by Burgess and Bowring (2015), coupled with fossil evidence, points to a single, brief catastrophic LPME event supporting previous assumptions by Jin et al. (2000), Rampino et al. (2000), Shen et al. (2011a, b), and Wang et al. (2014). These findings reveal that: (a) approximately two-thirds of the total lava/pyroclastic volume was erupted before and concurrent with the LPME for around 300 k.y., (b) after mass extinction ceased, lava eruptions continued for at least 500 k.y., and (c) there was a voluminous emplacement of subvolcanic intrusions (mainly sills) into the shallow crust for at least 500 k.y., concurrent with the LPME. It was perhaps this network of shallow magmatic intrusions of the Siberian Traps (stage 2, extrusive hiatus) into carbon-rich sediments, which induced the most important killing mechanism in the LPME (Burgess et al., 2017).

Although Jiang et al. (2006) determined sulfur content and isotopic composition near the PTB in the Meishan section, Shen et al. (2007) were the first ones to report abundant pyrite framboids from the upper part of Bed 24, and from Beds 25, 26 and 29 (were absent in Bed 27). Based on this fact, they suggested that the LPME was probably associated with a dramatic decline of benthic oxygen levels. According to these authors, framboids at this locality exhibit a narrow size distribution with average diameters from 4.6 to 8.7 μm, a range usually seen in pyrite framboids formed under anoxic/dysoxic conditions. Abundance and mean diameters of these framboids point to a change in redox conditions of deposition from upper dysoxia (Bed 24e, uppermost Changhsing Formation) and lower dysoxia (pyrite lamina, Beds 25 and 26, Yinkeng Formation) to oxygenation (Bed 27), and again to upper dysoxia (Bed 28) and lower dysoxia (Bed 29) (Shen et al., 2007).

Secular trends in aromatic hydrocarbons at Meishan, diagnostic for anoxygenic green sulfur bacteria, identified periods when euxinic conditions extended into the photic zone during the entire Changhsingian stage (Cao et al., 2009). Analyses of S isotopes in pyrites in sedimentary rocks from the Meishan section were reported by Shen et al. (2011a, b) who concluded that negative $\delta^{34}S$ with negative $\Delta^{33}S$ may have resulted from shortage in sulfate supply, related to near bioturbation shutdown during shoaling of anoxic water, in a prolonged deterioration of oceanic environments during the LPME.

3.2.2 Guryul Ravine succession, India

Kashmir was part of Gondwana during the late Paleozoic, located at the southern side of the Paleotethys Ocean and adjacent to Oman (Brookfield et al., 2013), when the separation of tectonic blocks from the northern margin of the Gondwana took place, the Neotethys Ocean opened with rapid subsidence of the Gondwanan margin (e.g., Brookfield, 1993; Brookfield et al., 2013, 2019; Garzanti et al., 1996) and widespread marine transgression in northern India (e.g., Garzanti et al., 1998; Kapoor, 1992). The P–T section at Guryul Ravine near Srinagar in the Kashmir (Fig. 3), was a candidate to be the GSSP, before the selection of Meishan D (Kapoor, 1996; Yin et al., 2001). Paleontologic, sedimentologic, stratigraphic aspects and record of redox conditions of the Guryul Ravine section have been the focus of several studies, including some more recent ones (e.g., Algeo et al., 2007; Brookfield et al., 2003, 2013, 2019; Huang et al., 2019; Kumar et al., 2017; Matsuda, 1981, 1982, 1983, 1984; Murata, 1981; Tewari et al., 2015; Wignall et al., 2005). This section and the one at the Spiti Valley, further east (Ghosh et al., 2016) are probably the two better geologically documented PTB sections in India. Although important advances were made toward the sedimentologic and biostratigraphic knowledge of these two sections, the International Commission on Stratigraphy (ICS), decided to point Meishan to be the GSSP of the PTB (Yin et al., 2001).

The Guryul Ravine succession (Fig. 5) was deposited in an outer-shelf or deep-ramp setting in the Neotethys (Algeo et al., 2007; Baud and Bhat, 2014; Brookfield et al., 2003; Wignall et al., 2005), at moderate sedimentation rates ($\sim$10–20 m Myr^{-1}) (Algeo et al., 2007). Flood basalts of the 289 Ma-old Panjal Traps (Shellnutt et al., 2011, 2014) are covered by 15 m thick siliceous shales of the Gangamopteris beds which, in turn, are overlain by the 100 m thick Zewan Formation. Nakazawa et al. (1975) subdivided this formation into four members, named A to D, and the Khunamuh Formation on top of it, into six members, named E to J, based on the variation in carbonate content. Similar to several other PT sections in the Neotethys, the Guryul Ravine one shows continuous sedimentation and gradual faunal changes across this boundary (e.g., Baud and Bhat, 2014; Brookfield et al., 2013, 2019; Nakazawa et al., 1970; Nakazawa and Kapoor, 1981; Sheng et al., 1984; Teichert et al., 1970). Much of the Tethyan sedimentary succession in Kashmir has experienced greenschist facies metamorphism (Algeo et al., 2007; Dèzes, 1999; Herren, 1987). The conodont biostratigraphy of this section was summarized by Murata (1981),

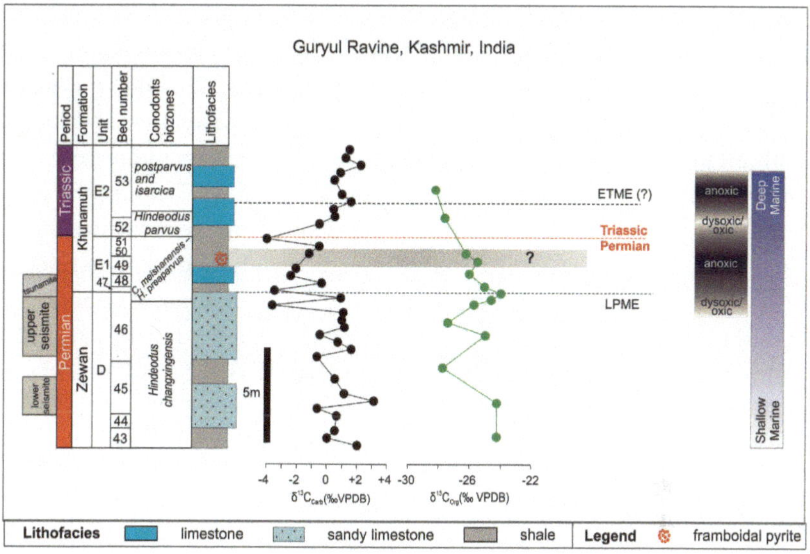

Fig. 5 Lithological succession for the Guryul Ravine section according to Brookfield et al. (2019); $\delta^{13}C_{org}$ curve from Algeo et al. (2007); $\delta^{13}C_{carb}$, Hg, TOC and Hg/TOC, Hg/S, Hg/Al chemostratigraphic pathways are from this study; variation of the redox conditions, based on Huang et al. (2019), two-marine anoxia based on distribution of framboidal pyrite. Relative depth of deposition (shallow, deep) is from Singh et al. (2015).

Matsuda (1981, 1982, 1983, 1984), Wignall et al. (1996), and Kozur (2003). The palynological features were described by Tewari et al. (2015) who observed a reduction in the latest Permian spore-pollen assemblages in the uppermost Zewan Formation and a rich palyno-assemblage in the basal Khunamuh Formation.

The Guruyl Ravine section displays a stratigraphically-expanded P–T transition when compared to the GSSP Meishan section (Yin et al., 2001). In the former, the LPME horizon marks the top of the Zewan Formation (Bed 46) and the base of the Khunamuh Formation (Bed 47), while the PTB marks the top of the Bed 51 and the base of Bed 52, within the Khunamuh Formation (Algeo et al., 2007). The LPME and the PTB horizons are separated by 0.36 m at Meishan (Yin et al., 2001), while at Guryul Ravine these two horizons are separated either by ~2.60 m (Algeo et al., 2007), or 3.10 m (Brookfield et al., 2013; Shen et al., 2011a, b), which helps the investigation of distinct events within this interval.

According to Chen et al. (2005), the Guryul Ravine beds D43 to D46–1 correlate with the Bed 24 at the Meishan section, and Guryul Ravine beds D46–3 to E51 are equivalent to beds 25–27 at Meishan. Likewise in the Meishan D GSSP, Guryul Ravine displays a stepwise extinction pattern, the largest peak between the LPME and the PTB and a less intense one, just above the PTB (Algeo et al., 2007), in the earliest Triassic (ETME). According to Shen et al. (2006), the main extinction event in the Guryul Ravine section is in the Unit E1 (Bed 47), although secondary extinction events (ETME) have been detected at the base of Unit E2 (Bed 52). A continuous sea-level rise has been recorded in the end-Permian in this section, from the base of Bed 46 up to the base of Bed 52 (PTB horizon) (Algeo et al., 2007). An anoxic event is widely believed to characterize the zenith of this transgression (Shen et al., 2006), which has been ascribed to a eustatic sea-level rise or the active vertical tectonism at the northern Gondwanan margin (Hallam and Wignall, 1999).

Deposition of seismite and tsunamite beds in the uppermost 5 m of the Zewan Formation, below the LPME horizon, perhaps represent the record of large earthquakes and tsunamis just below the PTB (Brookfield et al., 2013). Earthquake-induced soft-sediment deformation (i.e., seismites) from the Early Triassic at Guryul Ravine was also reported by Leu et al. (2014). Although a decent cross-correlation between biostratigraphic patterns of the Guryul and Meishan sections has been established (e.g., Algeo et al., 2007; Brookfield et al., 2013; Chen et al., 2005), seismites (two beds of sandstone to sandy limestone) and tsunamites (three beds of lenticular, graded, hummocky cross-stratified bioclastic grainstone) present at the Guryul Ravine seem to be absent from the Meishan GSSP. To our knowledge, except for Mandakpal, the occurrence of seismite/tsunamite beds has not been so far confirmed from PTB sections near Guryul Ravine in India (e.g., Spiti district, Barus Spur, Surghar Range, Salt Range, Ladakh, Malta Johar and Niti, Son-Damodar basins, Warda coalfield) and those described from Guryul Ravine have been rejected as such by Krystyn et al. (2014).

Several investigations on the Guryul Ravine PT section have suggested marine anoxia with increasing clay content during the LPME (e.g., Algeo et al., 2007; Brookfield et al., 2013). Geochemical proxies suggested that the environment in this section was oxic or suboxic across the PTB, but anoxia probably developed during the LPME event. This contention was based on a change from non-laminated, pyrite-poor mudstones to laminated shales, rich in syngenetic fine pyrite framboids of the Khunamuh (Bed E49), at about 1 m above the LPME horizon (Wignall et al., 2005). In a detailed geochemical study of the Guryul Ravine section, Kumar et al. (2017) examined the oceanic redox conditions across the PTB. They observed that C_{org}: P is <10:1 in the late Permian whereas >10:1 in the early Triassic, suggesting that the oceanic conditions changed from suboxic to anoxic, and this was one of the causes of oceanic redox regime at the PTB.

Based on pyrite framboid size and morphology, Huang et al. (2019) noted that there were two pronounced stages of oceanic oxygen deficiency at the Guryul Ravine PTB section: (a) in the latest Permian *Hindeodus praeparvus–Clarkina meishanensis* biozone and (b) in the earliest Triassic *Isarcicella staeschei* biozone. Besides, these authors observed three sharp species abundance declines in the marine invertebrate fossil records (a) one biotic decline within uppermost Permian storm beds and (b) two other biotic declines, coincident with two marine anoxic events, documented by pyrite framboid size distribution. Huang et al. (2019) concluded that these two anoxic events correlate with anoxic events in the Meishan GSSP, suggesting that this sequence of anoxic events might have been, perhaps, a global characteristic of the Permian–Triassic transition under some specific geological settings.

In the nearby Spiti Valley section, further east of Guryul, another remnant of the peri-Gondwanan shelf in India, the PTB is located at the top of a marly ferruginous limonitic layer of the Late Permian Gungri Formation (Ghosh et al., 2016). This uppermost layer of the Late Permian is underlain by organic-rich shale and overlain, by a sharp contact with biomicrite of the Early Triassic Mikin Formation at whose base the *Hindeodus parvus* has been described by Krystyn et al. (2004). Framboidal pyrites and phosphatic nodules, fossil assemblage and laminated lithology of the Late Permian organic-rich shales of the Gungri Formation suggest a limited circulation and anoxic to euxinic bottom water conditions (Ghosh et al., 2016; Singh, 2012). The Spiti section experienced a rapid shallowing, rendering the continental shelf under subaqueous conditions (Ghosh et al., 2016).

The connection between the LPME and deep-water anoxia in Guryul Ravine is in contrast with other peri-Gondwanan sections in which the onset of shallow-water anoxia and extinction took place during the earliest Triassic mass extinction (ETME), as pointed out by Wignall et al. (2005). As the only deep-water anoxia record of the P–T transition in this region, besides the Spiti Valley section, these two sections may represent the important key for unraveling the redox history of the Neothethys Ocean across the PTB.

4. Carbon and mercury (Hg) chemostratigraphy

The carbon isotopes are often used in chemostratigraphy as they can be conveniently analyzed and can reliably tell whether their original depositional signatures have been preserved or modified. Even so, Bagherpour et al. (2020) cautioned that the use of C-isotope stratigraphy in global correlation, should be used to: (a) make a biostratigraphical calibration of the obtained $\delta^{13}C_{carb}$ or $\delta^{13}C_{org}$ curves, (b) investigate the local bathymetry, (c) estimate the amount of buried terrestrial organic matter and (d) spot fluctuations due to diagenetic overprint thereof.

It is more or less agreed that the co-variation between $\delta^{13}C_{carb}$ and $\delta^{13}C_{org}$ can help to establish whether variations in the $\delta^{13}C_{carb}$ records reflect changes in the isotopic composition of the ancient dissolved inorganic carbon (DIC) pool (e.g., Oehlert and Swart, 2014, and references therein). It is also kept that covariant $\delta^{13}C_{carb}$ and $\delta^{13}C_{org}$ records are most probably indication that carbonate and organic matter were produced at the surface of the ocean water and that their original $\delta^{13}C$ values have been preserved (e.g., Korte and Kozur, 2010; Meyer et al., 2013) since it is unlikely that secondary processes would cause a shift in $\delta^{13}C_{carb}$ and $\delta^{13}C_{org}$ in the same direction at the same rate (Knoll et al., 1986). In turn, decoupled $\delta^{13}C_{carb}$ and $\delta^{13}C_{org}$ records suggest a diagenetic alteration effect (e.g., Grotzinger et al., 2011; Meyer et al., 2013) or that noise in the $\delta^{13}C_{org}$ record has resulted from local *syn*-sedimentary processes (Maloof et al., 2010).

It is well known that sedimentary successions across the PTB are marked by a prominent negative C-isotope excursion, observed in brachiopod fossils and whole-rock carbonate worldwide, and regarded as related to a global carbon cycle perturbation. The evaluation of potential diagenetic overprints on primary marine isotopic signals in PTB successions of the NW Iran led to the conclusion that Mn—and Sr concentrations and covariance of C–O isotopes are imperfect tracers of diagenesis (Schobben et al., 2017). According to these authors, this may be because Sr/Ca and Mn/Ca ratios can vary as a with whole-rock mineralogy and $\delta^{13}C$ vs $\delta^{18}O$ covariance depends largely on environmental factors rather than diagenesis. The PTB interval, marked by a low-carbonate "boundary clay" in some sections, is particularly prone to diagenetic alteration (e.g., late-stage dolomitization/recrystallization) (Schobben et al., 2017). However, because of the well-calibrated conodont biostratigraphy (e.g., Kozur, 2004, 2005, 2007; Mei et al., 1998; Wang and Wang, 1981), it is possible to have some control on the C-isotope trend across the PTB, in many cases (e.g., Cao et al., 2002; Korte et al., 2004b).

The C-isotope ratio of the ocean/atmosphere is mainly controlled by the burial and re-oxidation of ^{13}C-depleted organic matter in the oceans or on continents (Kump and Arthur, 1999; Scholle and Arthur, 1980). The negative $\delta^{13}C$ excursion across the PTB, however, seems to call for additional causes (e.g., Grasby et al., 2011; Korte and Kozur, 2010) such as: (a) release of ^{13}C-depleted CO_2, (b) destabilization of methane from permafrost soils and/or peat and coal beds, (c) release of methane from (pre) Siberian Traps coal beds, (d) thermal maturation and metamorphism of pre-Trap sedimentary organic matter, and (e) dispersion of coal fly ash into the ocean. One would expect that Hg as a fingerprint of volcanism would display a positive Hg/TOC spike if volcanism played an important role in shaping the carbon-isotope pathway, leading to a significant negative shift in $\delta^{13}C$ values.

As to the use of Hg, there is a growing agreement that Hg enrichment in sedimentary successions potentially indicates massive volcanism effects, allowing to examine the relationship between LIP activity and periods of extreme environmental turnover (e.g., Adatte et al., 2015; Burger et al., 2019; Charbonnier et al., 2017; Font et al., 2016; Gong et al., 2017; Grasby et al., 2016, 2017, 2019; Jones et al., 2017; Meyer et al., 2019; Nascimento-Silva et al., 2011, 2013; Percival et al., 2015, 2017, 2018; Sabatino et al., 2018; Sanei et al., 2012; Shen et al., 2019a, b, c; Sial et al., 2010, 2013, 2014, 2016, 2019a, b; Them et al., 2019; Thibodeau et al., 2016, Thibodeau and Bergquist, 2017). More recent studies on the entire Phanerozoic succession have pointed that Hg enrichment across some chronological boundaries can be the result of volcanogenic Hg loading onto the environment, recorded by an increase in the Hg/TOC ratio when organic matter is the dominant host phase of Hg (e.g., Bond and Grasby, 2017; Burger et al., 2019; Faggetter et al., 2019; Gong et al., 2017; Grasby et al., 2013, 2016; Jones et al., 2017; Korte et al., 2019; Kwon et al., 2019; Percival et al., 2015, 2017, 2018; Sanei et al., 2012; Shen et al., 2019a, b, c; Sial et al., 2016, 2019a, b). Hg isotopes, regarded as one possible key to screen the Hg source, have been receiving growing attention (e.g., Grasby et al., 2017, 2019; Gong et al., 2017; Shen et al., 2019a, b, c; Sial et al., 2014, 2016, 2019a, b; Thibodeau et al., 2016, Thibodeau and Bergquist, 2017; Them et al., 2019; Wang et al., 2018).

5. Analytical methods

One hundred and nine samples from the Guryul Ravine section, collected during this study were analyzed for carbon isotopes and results are provided in Table 1. Analytical data, including remaining eight sections, are shown in Tables 1 and 2, extracted from Sial et al. (2019b) where one can find the analytical details for the Hg concentration, and carbon and Hg isotope measurements.

6. Results and data interpretation

6.1 C-isotope chemostratigraphy

6.1.1 Meishan section

The $\delta^{13}C_{carb}$ and $\delta^{13}C_{org}$ variation patterns for the Meishan section are shown in Fig. 4. In constructing this $\delta^{13}C_{carb}$ curve, 40 samples (Table 1) from a previous work (Korte et al., 2010a, b) have been used. Some data points from the $\delta^{13}C_{carb}$ curve published by Chen et al. (2005), in the LPME–ETME interval, have been added to help producing the high-resolution $\delta^{13}C_{carb}$ curve in Fig. 4 which spans only an interval of 1.5 m, straddling the PTB. As the main target of this study is the extinction interval, just a fraction of this dataset (37 samples) is shown in this figure.

In the interval from Bed 24 through Bed 29, positive $\delta^{13}C_{carb}$ values predominate and a total variation from −1.2 to +4.7‰ is observed. All the analyzed samples from the Changsing Formation have positive $\delta^{13}C_{carb}$ values, except the topmost Bed 24 and Bed 25 (LPME) with value of −1.1‰. In Bed 27a (*Clarkina taylorae* Zone), a >1‰ positive shift coincides with a paleoredox-state change from anoxic to dysoxic/oxic. In the PTB horizon, the $\delta^{13}C_{carb}$ value is +1.1‰ (Bed 27c) while in the ETME horizon (base of Bed 28) +0.9‰. From Bed 29 upwards, the $\delta^{13}C_{carb}$ values are positive and gradually increase (Beds 30–40; not shown in Fig. 4).

Nineteen samples of organic carbon have been analyzed but only 12 samples are shown in the $\delta^{13}C_{org}$ variation curve (Fig. 4), which is from Sial et al. (2019b). The total variation is from −30.5 to −25.7‰ and the $\delta^{13}C_{org}$ curve show some co-variation with the $\delta^{13}C_{carb}$ curve. In the Bed 27a, a discrete positive shift of the $\delta^{13}C_{carb}$ coincides with a more marked positive shift of the $\delta^{13}C_{org}$ curve and with a change in the redox conditions from anoxic to dysoxic/oxic state.

Li and Jones (2017) have focused on diagenetic overprint on negative $\delta^{13}C$ excursions across the PTB with special emphasis on the Meishan Section D. They have carefully examined chemical and mineralogical composition, and cathodoluminescent features of Beds 24–62 and concluded that the $\delta^{13}C$ signals recorded in the GSSP spanning the PTB are a mixture

Table 1 TOC (%), Hg (ng g^{-1}) and Hg/TOC compositions of representative samples from Permian–Triassic sections at Ursula Creek (Canada), Misci (Seres, Alps), Idrijca (Slovenia), Rizvanuša (Croatia), Zal and Abadeh (Iran), Meishan (China) and Hovea-3 well, onshore Perth Basin, Western Australia.

(A) Ursula Creek, Canada

Sample	Height (m)	$\delta^{13}C_{carb}$(‰) vs V-PDB	$\delta^{13}C_{org}$(‰) vs V-PDB	TOC (%)	Hg (ng g^{-1})	Hg/TOC
UC-1	0.5			0.08	31.68	396.0
UC-3	3.7			0.10	27.99	279.9
UC-6	8			0.07	14.27	203.9
UC-9	10			0.06	22.34	372.3
UC-12	12.9			0.05	26.56	531.2
UC-13	12.5			0.12	29.76	248.0
UC-14	14.2			0.15	54.27	361.8
UC-17	15.6			0.12	67.53	562.8
UC-18	15.9			0.32	24.48	76.5
UC-19	16		−29.70	0.31	55.65	179.5
UC-20	16.1			0.50	68.97	137.9
UC-22	16.4		−27.40	0.45	23.34	51.9
UC-25	17.3		−28.46	0.80	308.35	385.4
UC-28	18.25		−32.57	0.50	43.21	86.4
UC-31	19.9		−32.15	0.80	249.38	311.7
UC-33	22		−29.90	0.90	460.35	511.5
UC-34	24			0.54	167.51	310.2
UC-35	26			0.67	154.42	230.5
UC-36	29			0.76	138.78	182.6
UC-37	32			1.10	132.39	120.4
UC-38	36			1.30	123.17	94.7
UC-40	39			1.60	31.66	19.8
UC-41	42			0.89	42.02	47.2
UC-41.5	44			1.10	80.57	73.2
UC-42	46			0.80	65.93	82.4
UC-43	48			0.30	48.68	162.3
UC-44	50			0.32	30.23	94.5

Continued

Table 1 TOC (%), Hg (ng g^{-1}) and Hg/TOC compositions of representative samples from Permian–Triassic sections at Ursula Creek (Canada), Misci (Seres, Alps), Idrijca (Slovenia), Rizvanuša (Croatia), Zal and Abadeh (Iran), Meishan (China) and Hovea-3 well, onshore Perth Basin, Western Australia.—cont'd

(B) Misci (Seres, Southern Alps)

Sample	Height (m)	$\delta^{13}C_{carb}$(‰) vs V-PDB	$\delta^{13}C_{org}$(‰) vs V-PDB	TOC (%)	Hg (ng g^{-1})	Hg/TOC
Ser-1	0.15	2.42	−25.07	0.07		
Ser-2	1.15	2.89	−26.39	0.02		
Ser-3	3.65	2.48	−26.3	0.07		
Ser-4	4.35	3.17	−26.3	0.04	13.3	333
Ser-5	5.55	3.21	−24.93	0.04		
Ser-6	6.75	3.53	−24.67	0.05	10.5	210
Ser-7	7.55	3.57	−25.67	0.04		
Ser-8	7.95	3.86	−25.96	0.02	11.2	560
Ser-9	8.45	2.64	−28.31	0.08	9.5	119
Ser-10	8.95	3.13	−29.91	0.2	15	75
Ser-101 (event layer)	9.05	2.98	−24.9	0.12	29	242
Ser-11	9.22	2.73	−29.75	0.06	12.8	213
Ser-12	9.3	1.42	−28.47	0.02	24.8	1240
Ser-13	9.48	0.74	−29.07	0.02		
Ser-14	9.57	1.16	−27.3	0.02		
Ser-15	9.75	1.02	−27.68	0.02		
Ser-16	9.91	1.02	−26.23	0.01	11.4	1140
Ser-17	10	0.79	−28.07	0.12	7.8	65
Ser-18	10.29	1.01	−26.51	0		
Ser-19	10.38	1.03	−28.8	0.01		
Ser-20	10.565		−26.34	0.03		
Ser-21	10.813	1.19	−28.91	0.02	11.3	565
Ser-22	10.975	0.88	−28.66	0.02		
Ser-23	11.095	0.94	−28.57	0.02		
Ser-24	11.335	0.37	−30.05	0.02		
Ser-25	11.435	0.47	−29.27	0.01		

Table 1 TOC (%), Hg (ng g^{-1}) and Hg/TOC compositions of representative samples from Permian–Triassic sections at Ursula Creek (Canada), Misci (Seres, Alps), Idrijca (Slovenia), Rizvanuša (Croatia), Zal and Abadeh (Iran), Meishan (China) and Hovea-3 well, onshore Perth Basin, Western Australia.—cont'd

(B) Misci (Seres, Southern Alps)

Sample	Height (m)	$\delta^{13}C_{carb}$(‰) vs V-PDB	$\delta^{13}C_{org}$(‰) vs V-PDB	TOC (%)	Hg (ng g^{-1})	Hg/TOC
Ser-26	11.8	0.2	−29.6	0.01	10.9	1090
Ser-27	12.055	0.46	−30.03	0.02		
Ser-28	12.6	0.39	−27.67	0.01		
Ser-29	13.195	−0.09	−29.58	0.01		
Ser-30	13.57	0.09	−29.33	0.01		
Ser-31	13.8	−0.25	−28.93	0.02	8.7	435
Ser-32	14.32	−0.3	−28.89	0.02		
Ser-33	14.51	−0.54	−29.12	0.02	3.3	165
Ser-34	14.88	−0.32	−29.48	0.01		
Ser-35	19.15	−2.33	−29.76	0.04		
Ser-36	19.65	−2.22	−30.85	0.05		
Ser-37	21.69	−3.74	−32.56	0.11		
Ser-38	21.99	−1.87	−30.79	0.04		
Ser-39	23.39	1.61	−30.84	0.03		

(C) Idrijca section (Idrjca Valley), Slovenia

Sample	Height (m)	$\delta^{13}C_{carb}$(‰) vs V-PDB	$\delta^{13}C_{org}$(‰) vs V-PDB	TOC (%)	Hg (ng g^{-1})	Hg/TOC
J-4b (VS-12)	−1.2	3	−28.4	0.28	930.18	3322
J-3a (VS-11)	−0.9	2.4	−27.9	0.24	907.26	3780
J-3b (VS-10)	−0.7	2.2	−27.6	0.2	831.89	4159
J-2b (VS-9)	−0.6	2.2	−26.7	0.17	937.44	5514
Job/a (VS-8a)	−0.2	1.2	−27.4	0.13	910.14	7001
Job/b (VS-8b)	−0.15	1	−26.7	0.11	490.79	4462

Continued

Table 1 TOC (%), Hg (ng g^{-1}) and Hg/TOC compositions of representative samples from Permian–Triassic sections at Ursula Creek (Canada), Misci (Seres, Alps), Idrijca (Slovenia), Rizvanuša (Croatia), Zal and Abadeh (Iran), Meishan (China) and Hovea-3 well, onshore Perth Basin, Western Australia.—cont'd

(C) Idrijca section (Idrjca Valley), Slovenia

Sample	Height (m)	$\delta^{13}C_{carb}$(‰) vs V-PDB	$\delta^{13}C_{org}$(‰) vs V-PDB	TOC (%)	Hg (ng g^{-1})	Hg/TOC
Jo (VS-8)	−0.1	1.2	−26	0.14	458.4	3274
IA-1/2 (VS-7)	0	1.1	−26.1	0.17	598.9	3523
IA-3 (VS-6)	0.1	1.3	−25.9	0.16	479.45	2997
IA-4/5 (VS-5)	0.3	0.8	−25.4	0.1	544.16	5442
IA-7 (VS-4)	0.7	0.7	−25.7	0.11	784.46	7131
IA-9 (VS-3)	1.2	0.5	−27.1	0.1	1092.39	10,924

(D) Rizvanuša, Velebit Mountain, Croatia

Sample	Height (m)	$\delta^{13}C_{carb}$(‰) vs V-PDB	$\delta^{13}C_{org}$(‰) vs V-PDB	TOC (%)	Hg (ng g^{-1})	Hg/TOC
PTR-35	−2.70	1.33	−26.3	0.006	30.89	5370
PTR-44	−0.90	1.40	−28.6	0.014	10.18	713
PTR-45	−0.70	1.70	−28.6	0.008	180.09	21,893
PTR-46	−0.50	1.89	−28.6	0.028	7.21	254
PTR-47	−0.30	1.02	−28.6	0.008	181.58	23,670
PTR-48	−0.10	1.48	−28.4	0.013	33.64	2509
PTR-49	0.10	−0.03	−29.2	0.030	284.80	9606
PTR-50	0.30	0.71	−28.8	0.010	42.90	4469
PTR-50/51	0.40	0.45	−29.1	0.011	70.51	6235
PTR-51	0.50	0.01	−29.6	0.024	299.29	12,722
PTR-52	0.70	0.04	−29.3	0.022	189.52	8682
PTR-53	0.90	0.03	−29.1	0.020	42.47	2131
PTR-59	2.10	0.48	−32	0.034	21.79	644

Table 1 TOC (%), Hg ($ng\,g^{-1}$) and Hg/TOC compositions of representative samples from Permian–Triassic sections at Ursula Creek (Canada), Misci (Seres, Alps), Idrijca (Slovenia), Rizvanuša (Croatia), Zal and Abadeh (Iran), Meishan (China) and Hovea-3 well, onshore Perth Basin, Western Australia.—cont'd

(E) Zal, Iran

Sample	Height (m)	$\delta^{13}C_{carb}$(‰) vs V-PDB	$\delta^{13}C_{org}$(‰) vs V-PDB	TOC (%)	Hg ($ng\,g^{-1}$)	Hg/TOC
Zal-19	0	3	–	0.03	19.6	653
Zal-18	0.70	2.8	−26.67	0.01	19.9	1990
Zal-17	2.30	2.6	−26.67	0.02	21.6	1080
Zal-16	2.40	2.4	−26.50	0.03	38.1	1270
Zal-15	2.43	2.4	−27.71	0.01	19.9	1990
Zal-14	2.46	2.2	−27.73	0.02	58.6	2930
Zal-13	2.48	2.1	−25.59	0.04	62.7	1567
Zal-11	2.61	1.8	−26.10	0.01	27	2700
Zal-10	2.64	1.5	−27.27	0.02	51.6	2580
Zal 9/10 (event layer)	2.66	1.3	−26.45	0.03	88.96	2953
Zal-9	2.67	1.3	−27.75	0.02	60.9	3045
Zal-8	2.77	1.1	−27.08	0.02	45.6	2280
Zal-7	3.07	0.8	−26.29	0.01	32.4	3240
Zal-6	3.19	1.2	−28.34	0.02	49.5	2475
Zal-5	3.31	0.6	−27.09	0.02	86	4330
Zal-4	3.86	0.4	−29.10	0.01	10.95	1095
Zal-3	4.11	−0.5	−31.11	0.01	17.84	1784
Zal-2	4.91	0	−29.22	0.02	12.48	624
Zal-1	5.28	−0.1	−30.77	0.03	23.12	770

(F) Abadeh section, Iran

Sample	Height (m)	$\delta^{13}C_{carb}$(‰) vs V-PDB	$\delta^{13}C_{org}$(‰) vs V-PDB	TOC (%)	Hg ($ng\,g^{-1}$)	Hg/TOC
Aba-47	47.80	3.26	−27.82	0.005	2.0	400
Aba-56	50.38	2.02	−25.80	0.007	2.0	285
Aba-61	50.93	0.62	−24.60	0.008	2.4	300
Aba-66	51.38	0.99	–	0.008	9.6	1200

Continued

Table 1 TOC (%), Hg (ng g^{-1}) and Hg/TOC compositions of representative samples from Permian–Triassic sections at Ursula Creek (Canada), Misci (Seres, Alps), Idrijca (Slovenia), Rizvanuša (Croatia), Zal and Abadeh (Iran), Meishan (China) and Hovea-3 well, onshore Perth Basin, Western Australia.—cont'd

(F) Abadeh section, Iran

Sample	Height (m)	$\delta^{13}C_{carb}$(‰) vs V-PDB	$\delta^{13}C_{org}$(‰) vs V-PDB	TOC (%)	Hg (ng g^{-1})	Hg/TOC
Aba-66-2					9.52	
Aba-68a	51.81	1.12	−27.22	0.012	2.1	175
Aba-68d	52.31	−0.12	−28.57	0.012	4.6	383
Aba-70	52.70	−0.08	−26.71	0.015	3.0	200
Aba-72	53.40	−0.32	−25.68	0.009	10.8	1200
Aba-73	53.60	−0.30	−25.68	0.009	2.6	289
Aba-76	55.00	–	−25.94	0.005	19.9	3980

(G) Meishan section, China

Sample	Height (m)	$\delta^{13}C_{carb}$(‰) vs V-PDB	$\delta^{13}C_{org}$(‰) vs V-PDB	TOC (%)	Hg (ng g^{-1})	Hg/TOC
Meishan-1	−273	4.7	−26.95	0.10	6.1	61
Meishan-2	−238	4.0				
Meishan-3	−198	3.6	−26.66	0.15	5.0	33.33
Meishan-4	−179	4.0				
Meishan-5	−168	4.2				
Meishan-6	−113	4.1	−28.84	0.53	14.7	27.73
Meishan-7	−106.5	3.7				
Meishan-8	−88	2.7	−30.05	1.87	26.6	12.24
Meishan-9	−73	2.5				
Meishan-10	−68	2.6				
Meishan-11	−68	2.6	−27.26	0.06	1.5	25
Meishan-12	−57	2.3				
Meishan-13	−56.5	2.5				
Meishan-14	−54.5	2.6				
Meishan-15	−22.5	2.1	−29.59	0.20	8.8	44
Meishan-16	−15	2.0				

Table 1 TOC (%), Hg ($ng\,g^{-1}$) and Hg/TOC compositions of representative samples from Permian–Triassic sections at Ursula Creek (Canada), Misci (Seres, Alps), Idrijca (Slovenia), Rizvanuša (Croatia), Zal and Abadeh (Iran), Meishan (China) and Hovea-3 well, onshore Perth Basin, Western Australia.—cont'd

(G) Meishan section, China

Sample	Height (m)	$\delta^{13}C_{carb}$(‰) vs V-PDB	$\delta^{13}C_{org}$(‰) vs V-PDB	TOC (%)	Hg ($ng\,g^{-1}$)	Hg/TOC
Meishan-17	−15	2.1				
Meishan-18	−9	1.9				
Meishan-19	−7.5	2.2				
Meishan-20	−7.5	2.2	−27.82	0.20	15.1	75.5
Meishan-21	−5.0	2.0				
Meishan-22	−3.5	1.8				
Meishan-23	−3.5	1.9	−28.03	0.09	3.6	40
Meishan-50 (24E top)	−0.1			0.11	346.4	3149.09
Meishan-51A	+0.8		−28.52	0.49	32.9	67.14
Meishan-51B	+1.9		−26.26	0.05	21.0	420
Meishan-24	+6.5	−1.0	−30.49	0.57	122.3	214.56
Meishan-25	+6.5	−1.1				
Meishan-26	+10.5	0.0	−26.23	0.07	46.9	670
Meishan-27	+10.5	0.0				
Meishan-28	+12.5	0.2				
Meishan-29	+15.0	0.1	−26.64	0.04	13.0	325
Meishan-30	+15.0	0.0				
Meishan-31	+18.0	0.3				
Meishan-32	+18.0	0.2				
Meishan-33	+20.5	0.7	−26.22	0.05	9.3	186
Meishan-34	+22.5	0.8				
Meishan-35	+22.5	0.5				
Meishan-36	+25.5	0.1	−25.68	0.05	9.3	186
Meishan-37	32.5	1.2				
Meishan-38	+36.5	1.2	−26.50	0.05	6.5	130

Continued

Table 1 TOC (%), Hg (ng g^{-1}) and Hg/TOC compositions of representative samples from Permian–Triassic sections at Ursula Creek (Canada), Misci (Seres, Alps), Idrijca (Slovenia), Rizvanuša (Croatia), Zal and Abadeh (Iran), Meishan (China) and Hovea-3 well, onshore Perth Basin, Western Australia.—cont'd

(G) Meishan section, China

Sample	Height (m)	$\delta^{13}C_{carb}$(‰) vs V-PDB	$\delta^{13}C_{org}$(‰) vs V-PDB	TOC (%)	Hg (ng g^{-1})	Hg/TOC
Meishan-39	+45.0	0.9				
Meishan-40	+56.5	0.9	−27.64	0.12	32.3	269.16
Meishan-41	+68.7		−24.97	0.05	26.8	536

(H) Guryul Ravine section, Kashmir, India

Sample	Height (m)	$\delta^{13}C_{carb}$(‰) vs V-PDB	$\delta^{13}C_{org}$(‰) vs V-PDB	TOC (%)	Hg (ng g^{-1})	Hg/TOC
4N	29.25	0.1				
4P	29.7	0.5				
4Q	30.6	0.7				
4R	31.05	−0.6				
4S	31.5	3.1				
4T	31.95	1.2				
4U	32.4	0.9				
4X	32.85	0.5				
4Y	34.2	−0.7				
4Z	34.65	1.6				
5A	35.1	0.7				
5B	35.55	−0.5				
5C	36	1.2				
5D	36.45	1.0				
5E	36.9	1.1				
5F	37.35	−3.6				
5G	37.8	0.9				
5H (LPME)	38.25	−3.5				
5I	38.7	−0.4				
5J	39.15	−2.4				

Table 1 TOC (%), Hg ($ng\,g^{-1}$) and Hg/TOC compositions of representative samples from Permian–Triassic sections at Ursula Creek (Canada), Misci (Seres, Alps), Idrijca (Slovenia), Rizvanuša (Croatia), Zal and Abadeh (Iran), Meishan (China) and Hovea-3 well, onshore Perth Basin, Western Australia.—cont'd

(H) Guryul Ravine section, Kashmir, India

Sample	Height (m)	$\delta^{13}C_{carb}$(‰) vs V-PDB	$\delta^{13}C_{org}$(‰) vs V-PDB	TOC (%)	Hg ($ng\,g^{-1}$)	Hg/TOC
5L	39.6	−2.1				
5M	40.5	−1.2				
5N	40.95	−0.5				
5P (PTB)	41.4	−4.0				
5Q	42.3	−0.5				
5R	42.75	0.5				
5S	43.2	0.4				
5T	43.65	1.6				
5U	44.1	1.0				
5V	44.55	0.8				
5W (ETME?)	45	0.5				
5X	45.45	0.9				
5Y	45.9	2.2				
5Z	46.35	1.2				
6A	46.8	1.5				
6B	47.25	1.2				
6C	47.7	1.3				
6D	48.15	0.9				

(I) Hovea-3 well, onshore Perth Basin, Western Australia

Black shale samples

Sample name	Core depth (m)	$\delta^{13}C_{org}$(‰) vs V-PDB	TOC	Hg ($ng\,g^{-1}$)	Hg/TOC
Hov-1	1969.1	−33.8	2.3	39.25	17.1
Hov-2	1973.29	−35.1	1.5	55.11	36.7
Hov-3	1977.19	−35.9	1.2	111.67	93.1
Hov-4	1979.13	−35.7	2.5	207.33	82.9

Continued

Table 1 TOC (%), Hg (ng g^{-1}) and Hg/TOC compositions of representative samples from Permian–Triassic sections at Ursula Creek (Canada), Misci (Seres, Alps), Idrijca (Slovenia), Rizvanuša (Croatia), Zal and Abadeh (Iran), Meishan (China) and Hovea-3 well, onshore Perth Basin, Western Australia.—cont'd

(I) Hovea-3 well, onshore Perth Basin, Western Australia

Black shale samples

Sample name	Core depth (m)	$\delta^{13}C_{org}$(‰) vs V-PDB	TOC	Hg (ng g^{-1})	Hg/TOC
Hov-5	1980	−35.7	1.9	77.98	41.0
Hov-6	1980.94	−36.3	1.5	158.74	105.8
Hov-7	1980.95	−36.2	1.6	108.1	67.6
Hov-8	1980.96	−36.3	1.7	273.36	160.8
Hov-9	1980.98	−35.9	1.7	123.77	72.8
Hov-10	1980.99	−35.6	1.8	96.73	53.7
Hov-11	1981.1	−34.1	3.9	77.21	19.8
Hov-12	1981.23	−34.9	2.3	97.43	42.4
Hov-13	1982.1	−34.8	2.2	78.58	35.7
Hov-14	1982.4	−31.9	0.8	92.01	115.0
Hov-15	1982.73	−31.3	1.2	77.3	64.4
Hov-16	1982.94	−31.1	1.5	75.86	50.6
Hov-17	1983.48	−30.4	0.8	437.05	546.3
Hov-18	1984.7	−27.3	0.9	62.28	69.2
Hov-19	1984.95	−27.6		116.07	

of primary and diagenetic signals. They cautioned that in dolomite-bearing whole-rock samples, diagenesis may have contributed with about 1.7‰ to the negative shift in the $\delta^{13}C$.

Kaiho et al. (2009) ascribed a −1‰ shift of $\delta^{13}C$ within around 13 ka before the LPME in the Meishan section to possible thermal dissociation of methane hydrate, regarding that subsequent mixing of anoxic stratified ocean led to another −1‰ $\delta^{13}C$ shift within around 5 ka during the extinction. The −1% shift around 13 ka after the LPME probably resulted from massive denudation of soils and sediments caused by the collapse of forests and possible dissociation of methane hydrate at the end of the Permian.

Table 2 Hg isotopes (‰ relative to NIST SRM 3133) in samples from PTB sections at Meishan (GSSP section, China), Rizvanuša (Croatia), Misci (Seres, Southern Alps), Ursula Creek (Canada); Idrijca (Slovenia) and Zal (Iran).

Samples	$\delta^{199}Hg$	$\delta^{200}Hg$	$\delta^{201}Hg$	$\delta^{202}Hg$	$\Delta^{199}Hg$	$\Delta^{200}Hg$	$\Delta^{201}Hg$
	Mass-dependent fractionation (MDF)				Mass-independent fractionation (MIF)		
Permian–Triassic Transition							
(A) Meishan, China (PTB, GSSP section)							
Layer-24	−0.35	−0.56	−0.93	−1.10	−0.08	0.00	−0.10
Layer-26	−0.29	−0.50	−0.90	−1.07	−0.03	0.04	−0.07
Layer-40	−0.34	−0.65	−1.12	−1.39	0.00	0.04	−0.09
Layer-50	−0.47	−0.85	−1.40	−1.79	−0.02	0.05	−0.06
Layer-51-A	−0.41	−0.67	−1.13	−1.38	−0.06	0.02	−0.10
(B) Rizvanuša (Velebit Mountain), Croatia							
PTR-47	−0.49	−0.93	−1.4	−1.82	−0.03	−0.02	−0.03
PTR-49	−0.50	−0.97	−1.5	−1.98	0.00	0.02	−0.01
PTR-51	−0.43	−0.80	−1.26	−1.62	−0.02	0.01	−0.04
PTR-52 (n=3)	−0.72	−1.46	−2.22	−2.96	0.03	0.03	0.01
(C) Ursula Creek, Canada							
UC-12	−1.07	0.36	−2.95	−3.79	−0.11	2.23	−0.1
UC-25	−0.10	−0.41	−0.57	−0.85	0.12	0.02	0.07
UC-33	−0.22	−0.72	−1.0	−1.44	0.14	0.00	0.08
(D) Idrijca, Slovenia							
VS-3 (n=3)	−0.58	−1.26	−1.89	−2.53	0.06	0.01	0.01
VS-4 (n=3)	−0.33	−0.70	−1.08	−1.43	0.03	0.02	0.00
VS-8b	−0.56	−1.26	−1.87	−2.52	0.08	0.00	0.03
(E) Zal, Iran							
Zal-2	−0.44	−1.12	−1.57	−2.29	0.14	0.03	0.16
Zal-6	−0.71	−1.50	−2.14	−3.04	0.05	0.03	0.14
Zal-9/10 (event layer)	0.08	−0.38	−0.57	−0.86	0.08	0.06	0.08
Zal-14	−0.37	−0.82	−1.11	−1.65	0.05	0.01	0.13

Continued

Table 2 Hg isotopes (‰ relative to NIST SRM 3133) in samples from PTB sections at Meishan (GSSP section, China), Rizvanuša (Croatia), Misci (Seres, Southern Alps), Ursula Creek (Canada); Idrijca (Slovenia) and Zal (Iran).—cont'd

Samples	$\delta^{199}Hg$	$\delta^{200}Hg$	$\delta^{201}Hg$	$\delta^{202}Hg$	$\Delta^{199}Hg$	$\Delta^{200}Hg$	$\Delta^{201}Hg$
	Mass-dependent fractionation (MDF)				Mass-independent fractionation (MIF)		
Permian–Triassic Transition							
(F) Abadeh, Iran							
ABA-72	−0.25	−0.58	−0.84	−1.23	0.06	0.04	0.09
ABA-66	−0.41	−1.00	−1.46	−2.08	0.12	0.04	0.11
ABA-66-2	−0.44	−0.99	−1.51	−2.14	0.10	0.08	0.10
(G) Seres/Misci, Southern Alps							
Misci (Seres-101)	0.10	−0.81	−0.97	−1.76	0.05	0.08	−0.09
(H) Hovea-3							
1980–94 (Hov-6)	−0.71	−0.38	−2.13	−2.84	0.00	0.03	0.00
1984–95 (Hov-19)	−0.65	−1.19	−1.88	−2.40	−0.04	0.02	−0.07
Cretaceous–Paleogene Transition							
	Mass-dependent fractionation (MDF)				Mass-independent fractionation (MIF)		
(I) K/Pg boundary layer at Højerup (Fiskeler Member), Denmark							
FCA	−0.51	−1.06	−1.76	−2.34	0.08	0.11	−0.01
FCB	−0.28	−0.74	−1.18	−1.62	0.13	0.07	0.04
FCC	−0.29	−0.55	−0.86	−1.06	−0.02	−0.02	−0.06
FCD	−0.43	−0.92	−1.44	−1.93	0.06	0.05	0.01
(J) K/Pg boundary layer, Bottaccione, Italy							
GP-15	−0.26	−0.59	−0.93	−1.28	0.06	0.05	0.03
(K) K/Pg boundary layer at Um Sohryngkew, Meghalaya, India							
KT-3	−0.35	−0.78	−1.16	−1.61	0.06	0.03	0.05
KT-bulk	−0.64	−0.94	−1.58	−1.89	−0.16	0.01	−0.16
(L) Poty Quarry drill hole, Brazil							
PO-104	−0.17	−0.97	−1.29	−1.87	0.30	−0.03	0.12
PO-105	−0.03	−0.62	−0.89	−1.36	0.31	0.07	0.13
PO-119	−0.31	−1.15	−1.59	−2.45	0.30	0.08	0.25

Table 2 Hg isotopes (‰ relative to NIST SRM 3133) in samples from PTB sections at Meishan (GSSP section, China), Rizvanuša (Croatia), Misci (Seres, Southern Alps), Ursula Creek (Canada); Idrijca (Slovenia) and Zal (Iran).—cont'd

Cretaceous–Paleogene Transition

	Mass-dependent fractionation (MDF)					Mass-independent fractionation (MIF)	
(M) Bidart, France							
K/Pg boundary layer	−0.04	−0.34	−0.42	−0.74	−1.12	−0.15	0.03
BI-9.28.5	−0.11	−0.57	−0.66	−1.44	−1.56	0.25	0.15
BI-9.34.2	0.05	−0.11	−0.13	−0.31	−0.73	0.12	0.04
BI-9.36.2	0.08	−0.10	−0.07	−0.25	−0.61	0.15	0.03
BI-10.35.1	−0.33	−1.28	−1.68	−2.66	−4.30	0.33	0.06
BI-11.7.3	0.13	−0.16	−0.04	−0.66	−0.73	0.30	0.17

K/Pg sections with Hg enrichments: Um Sohryngkew, Poty; Højerup, Bottaccione. All analyses are from Sial et al. (2019a, b). Analyses from Bidart (K/Pg) were added for comparison.

According to Kaiho et al. (2009), low $\Delta^{13}C$ ($=\delta^{13}C_{carb}-\delta^{13}C_{org}$) value recorded in Beds 27–29 indicates that phototrophic sulfur bacteria may have flowered and lasted more than 150 ka after the LPME in the low-latitude Paleo-Tethys.

6.1.2 Guryul Ravine section

The first carbon-isotope curve for the Guryul Ravine section was published by Baud and Magaritz (1988), further complemented by Baud et al. (1989, 1996) who, besides a $\delta^{13}C_{carb}$ variation curve for this section, provided an overview on the main Permian–Triassic sedimentary evolution. Korte et al. (2010a, b) provided another carbon-isotope curve for Guryul Ravine, when examining the role of the massive volcanism at the PTB and its impact on the isotopic composition of the ocean and atmosphere.

In this study, 109 samples were collected in a traverse perpendicular to the strata of the Zewan Formation, starting at about 71 m above the basalts of the Panjal Traps in the southeast side of the Guryul Ravine (coordinates of the start point: 34°04′24.13″N and 74°56′43.42″E; and of the end point: 34°04′6.66″N and 74°56′48.20″E). The obtained high-resolution $\delta^{13}C_{carb}$ variation curve for Bed 25 (the topmost bed of the Member B, Zewan Formation) through Bed 53 (Member E3, Khunamuh Formation) shows an overall similarity with the $\delta^{13}C_{carb}$ pathway, published in Korte et al. (2010a, b). Since the extinction interval is the main target of this study, the $\delta^{13}C_{carb}$ values of samples from beds 43 through 46 (Member D of the Zewan Formation) and beds 47 through 53 (Members E1 and part of Member E2 of the Khunamuh Formation) only, are listed in Table 1 and displayed in Fig. 5.

The Member D of the Zewan Formation is characterized by near zero to moderately positive $\delta^{13}C$ values (+0.1 to +3.1‰) wherein LPME and PTB horizons are marked by strong negative excursions (~ −4‰). A positive shift in $\delta^{13}C_{carb}$ values is observed at the transition from Bed 50 to 51 and another one at the base of Bed 53 (*postparvus and isarcica* Zone) that, according to Huang et al. (2019), was deposited during a stage of pronounced oceanic oxygen deficiency. We speculate that this latter shift, perhaps, records the ETME event in this section. In Bed 53 and up section, positive $\delta^{13}C_{carb}$ values predominate and progressively rise toward values around +3‰ (Member E3, not shown in Fig. 5).

The $\delta^{13}C_{org}$ curve is shown in Fig. 5 that contradicts with the $\delta^{13}C_{carb}$ variation curve. In this curve, the Zewan Formation rocks display values between −28 and −24‰, consistent with organic matter of both marine and terrestrial sources. A positive $\delta^{13}C_{org}$ shift is observed at the LPME horizon, in contrast with the negative shift in the $\delta^{13}C_{carb}$ values (base of Bed 47). There is a small positive $\delta^{13}C_{org}$ shift in the transition of Beds 49–50 (within Member E1), between the LPME and the PTB horizons (*C. meishanensis–H. praeparvus* Zone). The occurrence of framboidal pyrite in this horizon was initially mentioned by Brookfield et al. (2013). The $\delta^{18}O$ values are quite low (not shown in Table 1 or Fig. 5) and range from −13.9 to −9.1‰ (V-PDB) with almost no variation across the PTB, also pointed out by Baud and Bhat (2014).

A thorough evaluation of diagenetic effects, similar to Meishan GSSP (Li and Jones, 2017), is beyond the scope of this study. However, a comparison of $\delta^{13}C$ variation curves from geographically far apart but coeval sections may allow testing the authenticity of major $\delta^{13}C$ shifts, yet diagenetic signals may be globally synchronous (Fantle et al., 2020). In any case, the absolute value of a major $\delta^{13}C$ shift needs to be looked at with caution.

The C-isotope variation curves from seven well known, near-complete, marine PTB sections from thousands of kilometers apart paleolatitudinal zones (Ursula Creek, Misci/Seres, Idrijca, Rizvanuša, Zal, Abadeh and Hovea-3) discussed in Sial et al. (2019b) have been selected for comparison with the C-isotope data from the Meishan and Guryul Ravine sections (Figs. 6–8). These sections represent different margins of paleo-ocean—Panthalassa, western margin of Paleotethys, eastern margin of

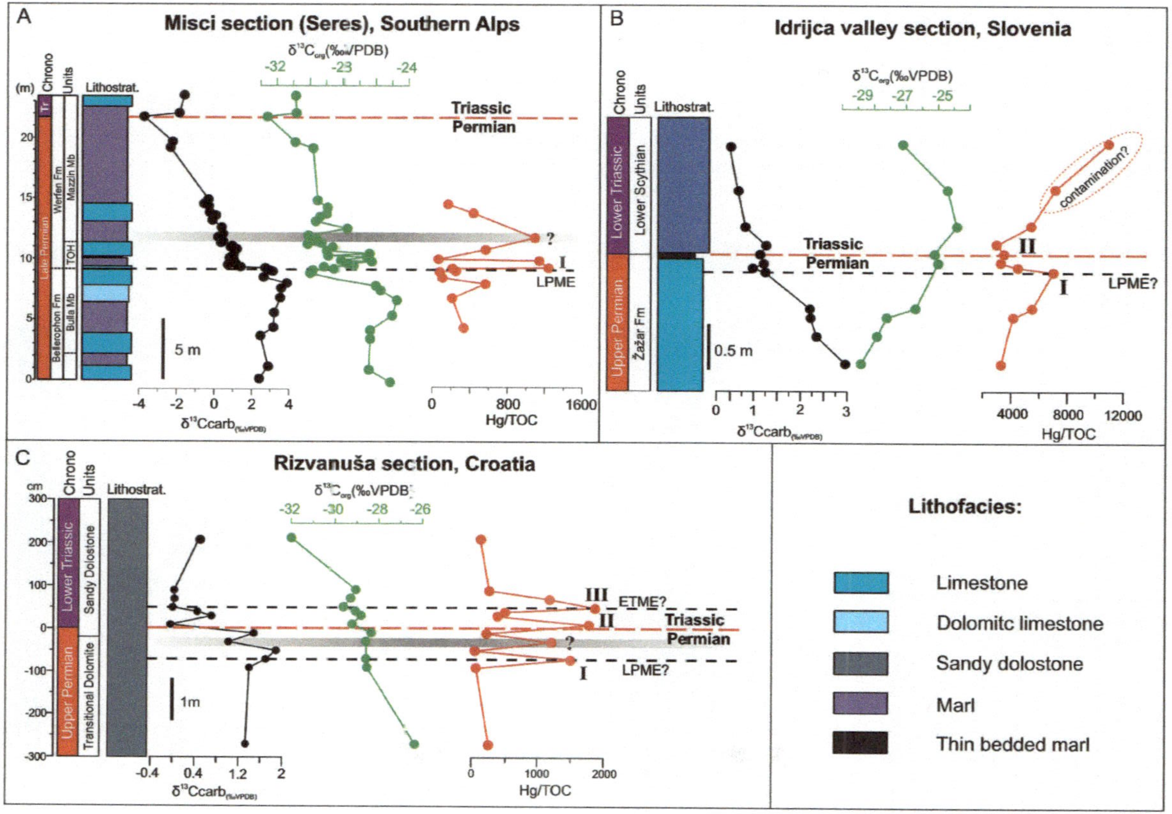

Fig. 6 Stratigraphy, lithology, carbon-isotope ratios and Hg/TOC for (A) Seres/Misci section (Southern Alps), (B) Idrijca Valley section (Slovenia), and (C) Rizvanuša section, Velebit Mountains, Croatia. $\delta^{13}C_{carb}$ and $\delta^{13}C_{org}$ for the Misci/Seres are from Kraus et al. (2013), for Idrijca are from Schwab and Spangenberg (2004) and for Rizvanuša, from Fio et al. (2010). Hg/TOC variation patterns, in all the three cases, are from Sial et al. (2019b).

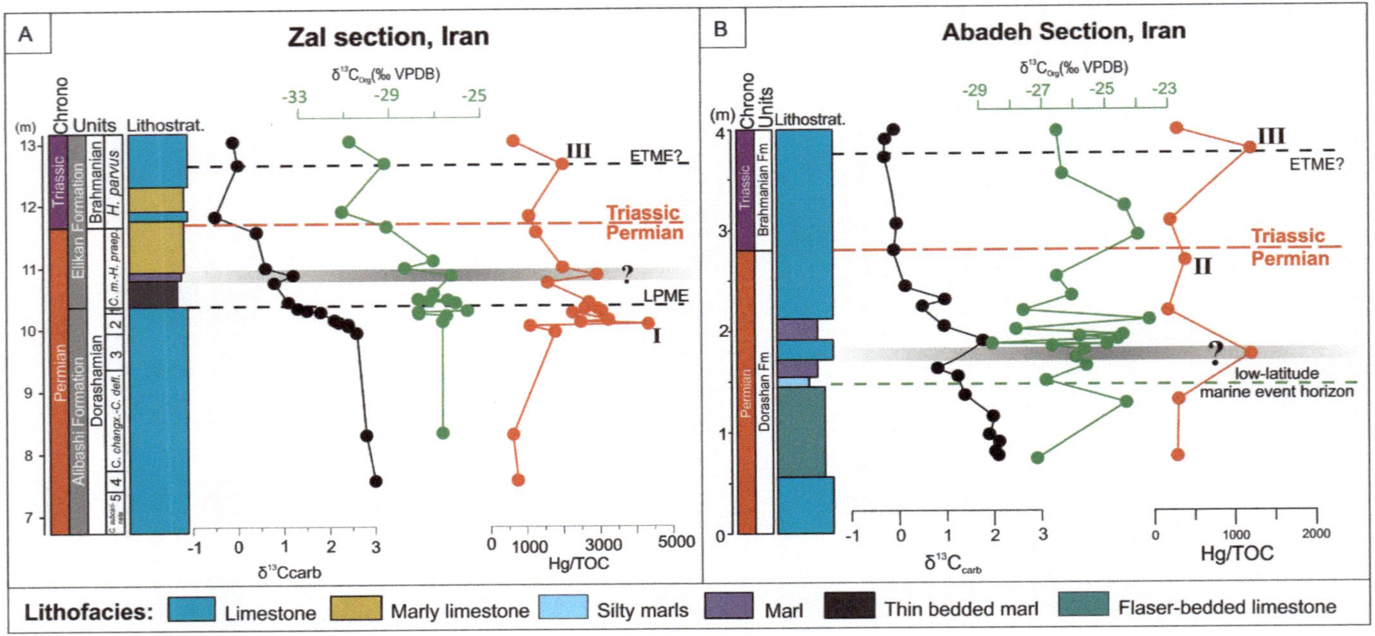

Fig. 7 Lithologic succession, biostratigraphy and chemostratigraphy for Zal and Abadeh sections, Iran. $\delta^{13}C_{carb}$ chemostratigraphy for Zal is from Korte et al. (2004a) and for Abadeh, from Korte et al. (2004c). Hg/TOC variation curves for both sections are from Sial et al. (2019b).

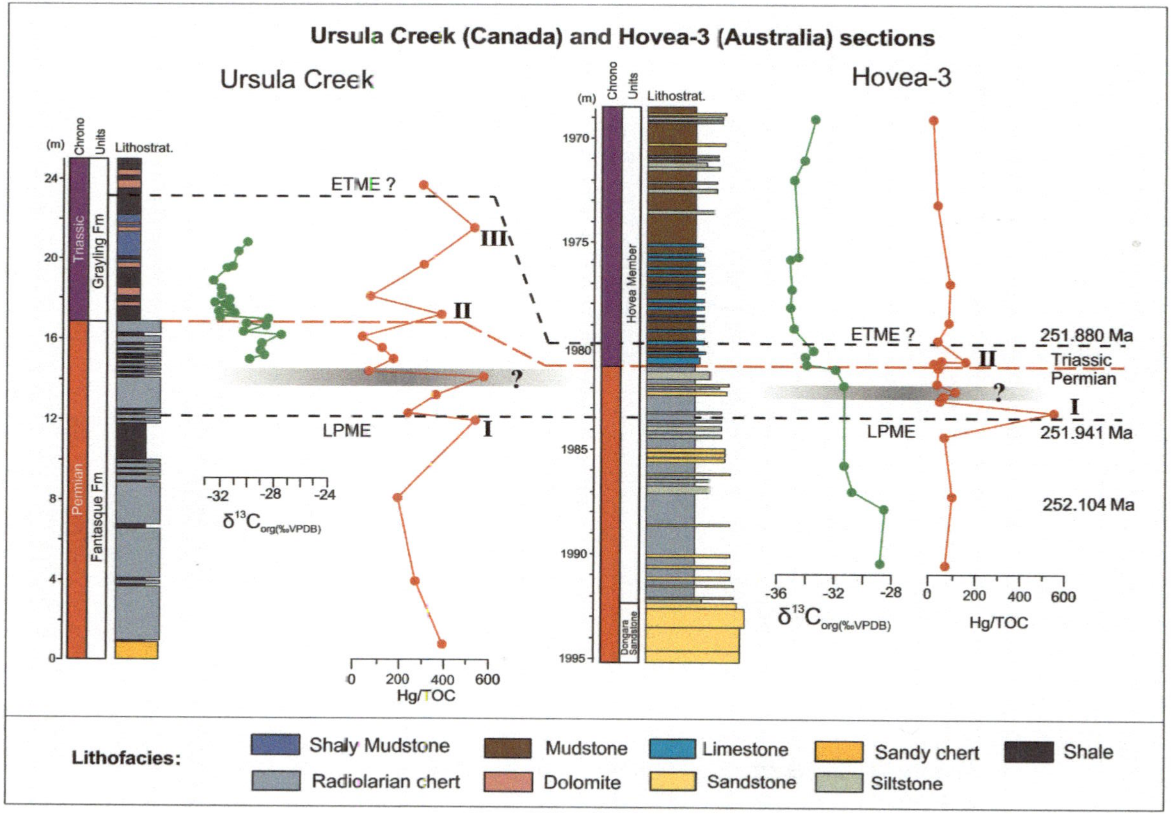

Fig. 8 Lithological and $\delta^{13}C_{org}$ variation patterns for the Ursula Creek section are, respectively, from Wignall and Newton (2003) and Wang et al. (1994) while for the Hovea-3 section, lithological column and $\delta^{13}C$ are from Thomas et al. (2004). Hg/TOC variation patterns for both sections are from Sial et al. (2019b).

Paleotethys and Neotethys margins; the variation raises the possibility of screening local from the global fingerprints. As a result, a NW–SE $\delta^{13}C$ pattern transect, diagonally across the Pangaea supercontinent (~252 Ma) is shown in Fig. 9. The locations of LPME and ETME horizons have been kept, as shown in Figs. 4–8 and the PTB has been used as the datum. The $\delta^{13}C_{org}$ variation curve is available for all the nine sections but Ursula Creek and Hovea-3 lack the $\delta^{13}C_{carb}$ trends.

The C-isotope data across the Permian–Triassic transition in this study, highlights the following features (Fig. 9): (a) it becomes clear that the prominent negative $\delta^{13}C$ shift is observed in the LPME horizon and is more pronounced in the Meishan, Guryul Ravine, Misci, Idrijca and Zal sections, (b) another negative $\delta^{13}C$ shift in the PTB horizon of Meishan, Guryul Ravine, Misci, Rizvanuša and Zal sections, (c) a discrete positive $\delta^{13}C_{carb}$ shift of unclear meaning is seen in all but Rizvanuša section, in the LPME–PTB interval, and a corresponding $\delta^{13}C_{carb}$ shift is also observed in the Meishan, Guryul Ravine, Zal, Abadeh and Misci/Seres sections, (d) the behavior of $\delta^{13}C_{carb}$ in Guryul Ravive section differs from the rest of the examined sections, shifting to positive values above the PTB, (e) the response of C isotopes on the environmental effects of the ETME event is less intense than the LPME resulting in a sharp increase in $\delta^{13}C_{carb}$ values only much above the ETME (e.g., Song et al., 2013a, b) in the Dinerian-Smithian boundary (251.17 Ma; not shown in Fig. 9) and was, perhaps, driven by upwelling of nutrient-rich deep waters, and (f) a strong variation of sedimentation rates among these sections is evident from Fig. 9.

Alternatively, it could have resulted from metamorphism of organic-rich shales during the emplacement of STLP sills in Stage 2 of STLIP magmatic activities (e.g., Burgess et al., 2017; Korte et al., 2010a, b; Racki, 2020; Svensen et al., 2009).

6.2 Hg (Hg/TOC) chemostratigraphy

$\delta^{13}C$ excursions across the P–T boundary may resulted from the influence of the STLIP or any other contemporaneous volcanism by: (a) exhalation of ^{13}C-depleted CO_2, (b) destabilization of methane from permafrost soils and/or pre-Trap coal beds, (c) thermal maturation and metamorphism of pre-Trap organic-rich sediments (Korte and Kozur, 2010), or (d) coal fly ash disposal in the ocean (Grasby et al., 2011), besides the impact on atmospheric pCO_2 by burial and post-burial oxidation of organic matter in the oceans or on continents (e.g., Kump and Arthur, 1999; Scholle and

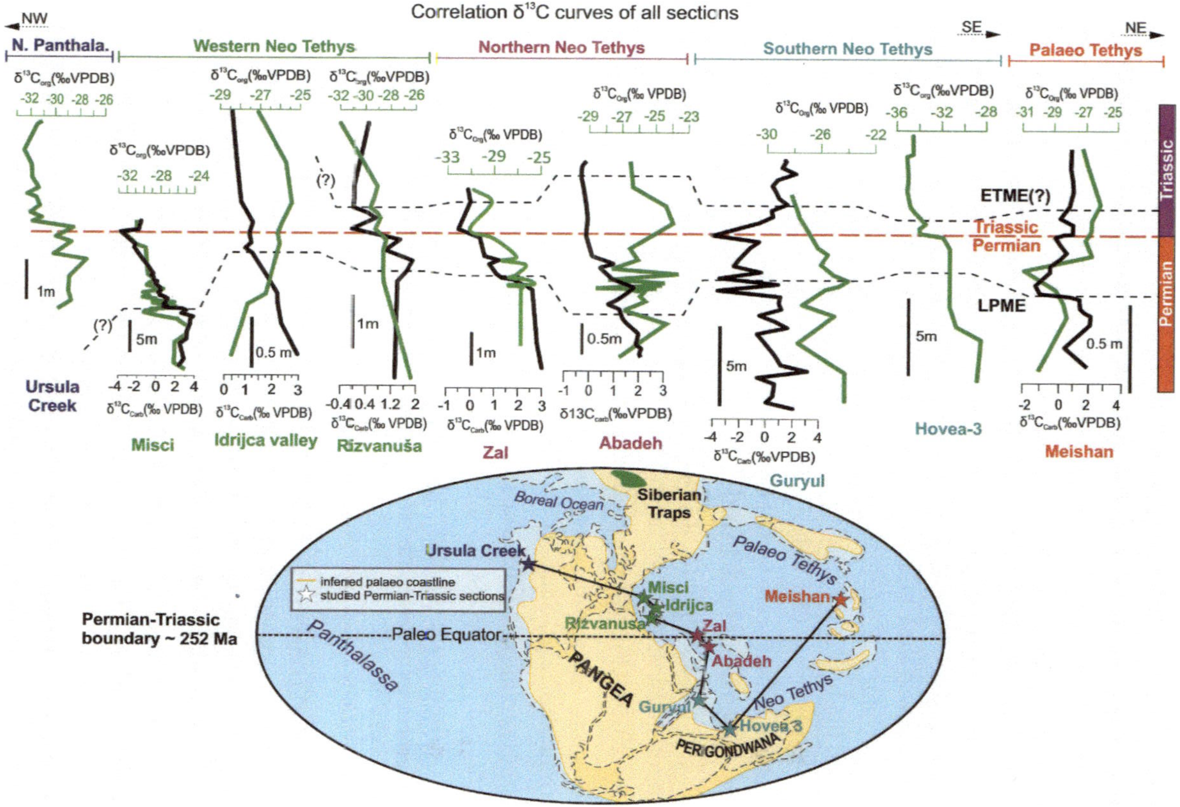

Fig. 9 A NW–SE $\delta^{13}C$ and $\delta^{13}C_{org}$ transect diagonal to the Pangea supercontinent (252 Ma ago) passing through eight sections (from the Ursula Creek through the Hovea-3). In building this transect, the position of the LPME and ETME horizons have been kept as shown in their original Figs. 5–8 and the PTB datum was used.

Arthur, 1980). If volcanism has played a direct role in shaping the $\delta^{13}C$ pathway, Hg concentrations, regarded as proxy for volcanism, should be elevated at the base of Neoproterozoic cap carbonate successions (Sial et al., 2010), K/Pg boundary horizon (Nascimento-Silva et al., 2011) or at the LPME and PTB horizons (e.g., Sanei et al., 2012), and this spike should stand normalization to total organic carbon, total sulfur and aluminum (e.g., Bond et al., 2020).

Typically, Hg concentrations are normalized to TOC because there is a propensity of Hg to be trapped in organic matter during sedimentation. Before assessing the normalized Hg values, however, it seems convenient to consider a few points raised by Grasby et al. (2019) in relation to the use of TOC values as normalizing factor. In their view, TOC values <0.2 wt% are likely unreliable, even when determined by the most accurate methods. They propose that only TOC values equal or >0.5 wt% should be used, but reminded that very high TOC values may be also a problem as they might not represent normal marine conditions as compared to modern values. Due to these problems, it has been suggested to normalize also Hg concentrations against total sulfur and aluminum (e.g., Bond et al., 2020; Shen et al., 2019b, 2020). In normalizing Hg to total sulfur, one evaluates whether the Hg spike was driven by sulfide drawdown under increased anoxic conditions (e.g., Sanei et al., 2012). Hg accumulates in sediments as organic compounds, sulfides and/or oxides depending upon pH, Hg oxidation state, presence of humic substances (e.g., Bond et al., 2020; Sanei et al., 2012). Aluminum, instead, is not actively involved in these reactions and is, therefore, an appropriate chemical species for normalizing Hg, especially when Hg has been adsorbed onto clays (Sanei et al., 2012; Sial et al., 2013).

Mercury (Hg) concentrations and TOC from eight of the nine sections under consideration in this review are found in Table 1 and curve of Hg/TOC, alongside $\delta^{13}C_{carb}$ and/or $\delta^{13}C_{org}$ pathways are seen in Fig. 4 (Meishan), Fig. 6 (Misci, Idrijca and Rizvanuša), Fig. 7 (Zal and Abadeh) and Fig. 8 (Ursula Creek and Hovea-3). The Hg content, TOC and Hg isotope data from the Guyul Ravine section can be found in another contribution which is under way (Sial et al., in preparation).

Only 30% of the Meishan samples exhibit TOC values >0.20 wt%, about 70% of the samples from the Ursula Creek section yielded TOC values >0.20 wt%, but all the Hovea-3 section display all samples with TOC values >0.20 wt%. Seres/Misci, Idrijca, Rizvanuša, Zal, Abadeh sections display TOC values are <20 wt%, with few exceptions. In principle, this picture would discourage the use of TOC to normalize the Hg values as it would potentially generate questionable Hg/TOC anomalies. As we do not avail total S (TS) and Al analyses for all these eight sections, we have not investigated the behavior of the Hg/TS and Hg/Al ratios. It seems, however, that the organic matter is the dominant Hg host phase in these sections as one observes a good correlation of Hg with TOC. Besides, the normalization of Hg against TOC in these sections has consistently led to Hg/TOC anomalies at the LPME and PTB, suggesting that these are true Hg loading to the environment, although some of them probably represent exaggerated Hg/TOC ratios due to very low TOC values.

Except for Abadeh, all sections display a short-lived Hg/TOC spike in the latest Permian mass extinction (LPME) horizon, labeled as spike I (in the top of the Bed # 24e, at Meishan). All sections display a less pronounced Hg/TOC spike at the PTB, labeled as spike II, with the exception of Seres/Misci and Zal sections. Zal, Abadeh, Ursula Creek and Meishan sections exhibit a discrete Hg/TOC spike and Rizvanuša a large one, at the Early Triassic mass extinction (ETME) horizon, labeled as spike III.

In addition, with the exception of Idrijca, seven of these sections display a Hg/TOC spike between the LPME and the PTB (at Meishan is in the top of Bed # 26 and corresponds to a discrete negative $\delta^{13}C_{org}$ excursion and a change from anoxic to the dysoxic/oxic redox state in that section). Similar anomaly has already been reported elsewhere from a shallow-marine section at Xiakou, China, and a deeper-marine section at Akkamori, Japan (Shen et al., 2019a). Curiously, a Hg/TOC spike at about 10 cm below the LPME (Figs. 6 and 7) is present only at the Misci and Zal sections.

The Hg/TOC spike in the interval between the LPME and the PTB horizons in all but at Idrijca is, perhaps, a primary feature and it is of unclear meaning. The spike III (ETME?), less prominent than spikes I and II except in the Rizvanuša section, is absent from three among the eight studied sections, perhaps due to depositional hiatus, intense chemical weathering/erosion that characterized the Early Triassic (e.g., Chen et al., 2020), or local secondary processes.

The Idrijca section displays a Hg/TOC pathway in Fig. 6 that departs from the rest of the other examined sections, increasing steadily in the Early Triassic. The Permo-Carboniferous to Middle Triassic hydrothermal Hg mineralizing activity in this region (Foucher et al., 2009; Lavric and Spangenberg, 2003) is, perhaps, the explanation for this behavior.

The spike I at the Meishan section (LPME horizon) coincides with the beginning of a sharp decrease in $\delta^{44/40}Ca$ values in conodonts, whose nadir is at the Bed 27 (Hinojosa et al., 2012) and supports a relationship between the LPME event and ocean acidification caused by LIP massive CO_2 emissions (Payne et al., 2010). The Meishan section also displays a pre-LPME negative shift of conodont $\delta^{44/40}Ca$ values from the base of Bed 23 (Hinojosa et al., 2012). This negative shift is also observed in carbonate successions elsewhere, perhaps related to a pre-LPME acidification episode (ca. 50 k.y.) and caused by massive volcanic CO_2 emission to the atmosphere (Silva-Tamayo et al., 2018) and also a possible explanation for the Hg/TOC spike I observed at the Zal section. Possible link between volcanism and ocean acidification is further reinforced by the Hg/TOC spike III, which coincides with a global post-LPME perturbation (~60 k.y.) of the marine calcium cycle during the ETME (Silva-Tamayo et al., 2018).

A NW–SE Hg/TOC transect disposed diagonally to the Pangaea supercontinent (at 252 Ma ago) passing through these eight studied sections (from Ursula Creek in Canada to Hovea-3 in Australia, including the Meishan section although it lies in the opposite side of the Palaeo Tethys) is found in Fig. 10. In this transect, the PTB has been used as the datum and the position of the LPME and ETME horizons have been kept as shown in their original Figs. 4, 6, 7, and 8. A global picture of the Hg/TOC behavior across the PTB transition in eight of these nine sections point to similar patterns and to a strong variation of sedimentation rates.

6.3 Mercury (Hg) isotopes

Mercury (Hg) isotopes undergo large mass-dependent (MDF) and mass-independent (MIF) fractionations in nature and their signatures can potentially help tracing Hg sources (e.g., Blum et al., 2014) as well as interpreting pathways of Hg and Hg/TOC fluctuations (e.g., Gong et al., 2017; Grasby et al., 2017; Thibodeau et al., 2016; Wang et al., 2018). Physical, chemical or biological reactions can trigger and modify the Hg mass-dependent fractionation and, therefore, should be taken into account when attempting to trace the source for Hg in the environment. Hg-MIF signatures are not affected by post-depositional processes and, therefore, they are more useful for source identification (Thibodeau et al., 2016).

Hg isotopes have been analyzed in a total of 25 samples from 8 of the PTB sections in this study and results are found in Table 2. Hg isotopes for the Guryul Ravine section were not available at the time of submission of this manuscript and they will be the scope of another contribution, under way. The majority of these samples are from the LPME–ETME interval

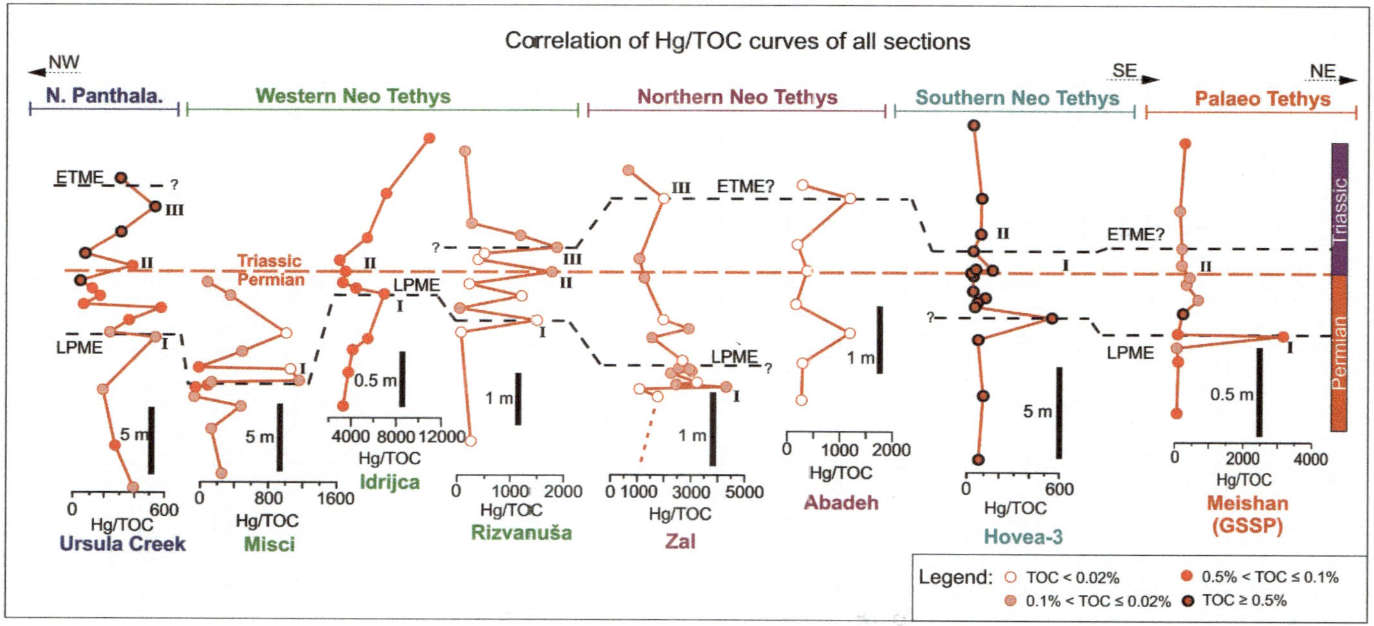

Fig. 10 A NW–SE Hg/TOC transect diagonal to the Pangea supercontinent (252 Ma ago) passing through the same sections as in Fig. 9. The position of the LPME and ETME horizons have been kept as shown in their original Figs. 5–8 and the PTB datum was used.

recorded in these sections, with high Hg/TOC values. Mass-dependent fractionation (MDF) reported here as δ^{202}Hg values are all negative, 16 of which are within the −2.00 to −0.85‰ range, 8 from −2.20 to −3.00‰ and 1 sample yielded a much lower value of −3.79‰ (Table 2).

Almost 60% of the analyzed samples plot within the "volcanic-emission box" (Fig. 11A) in a δ^{202}Hg vs Δ^{201}Hg (MIF) diagram and four horizontal trends are evident from this figure. The *trend I*, with positive Δ^{201}Hg (MIF) values, is defined by three samples from the Abadeh section plus the sample ZAL 9/10 (event layer at Zal section) while the *trend II*, exhibiting Δ^{201}Hg values around zero, is defined by samples from the three sections in Europe (Rizvanuša, Seres/Misci and Idrijca) plus one sample from Hovea-3 section. The *trend III* is defined by five samples from Meishan (Beds # 24, 26, 40, 50, 51A) with very consistent, slightly negative Δ^{201}Hg values, that cluster within the "volcanic-emission box," plus two samples from Ursula Creek and one from Hovea-3.

Likely Hg in the samples from Meishan in trend III as well as Rizvanuša samples in trend II originated from one same source each, with negligible or none Δ^{201}Hg (MIF) fractionation during atmospheric transport. The total range of measured Δ^{199}Hg values (−0.10 to 0.00) within the LPME–ETME interval in Rizvanuša, Meishan, Seres/Misci, Idrijca and Zal sections fall within experimental error of zero, suggesting a significant influx of volcanic Hg in this time interval. Long-term atmospheric transport and Hg loading to the environment by the STLIP magmatism is favored by small positive Δ^{201}Hg. The negative values displayed by δ^{202}Hg MDF(−1.07 to −1.79) and Δ^{199}Hg (0.00 to −008) in samples from Meishan in *trend III* can be explained by presence of a significant proportion of Hg from terrestrial sources as suggested elsewhere (Grasby et al., 2017; Wang et al., 2018). Three samples from the Zal section which exhibit the lowest negative Δ^{201}Hg (MIF) values in this study define the *trend IV*. These samples are in contrast with the Zal 9/10 sample (event layer) which exhibits the highest positive Δ^{201}Hg (MIF) value among the studied samples. These two sections from Iran display the highest (*trend I*) and the lowest (*trend IV*) Δ^{201}Hg (MIF) values in this study and require special attention.

Seven samples plot in the "sediment, soil and peat box" in the δ^{202}Hg vs Δ^{201}Hg diagram (Zal, Ursula Creek, Idrijca and Abadeh; Fig. 11A). Hg (MDF) isotopes seem to have suffered further fractionation in these samples, pushing δ^{202}Hg values toward more negative values due to terrestrial Hg influx. Alternatively, a coal source intruded by contemporary sill complex, during the stage 2 of the Siberian Trap magmatic activities (Burgess et al., 2017), could explain this feature as well.

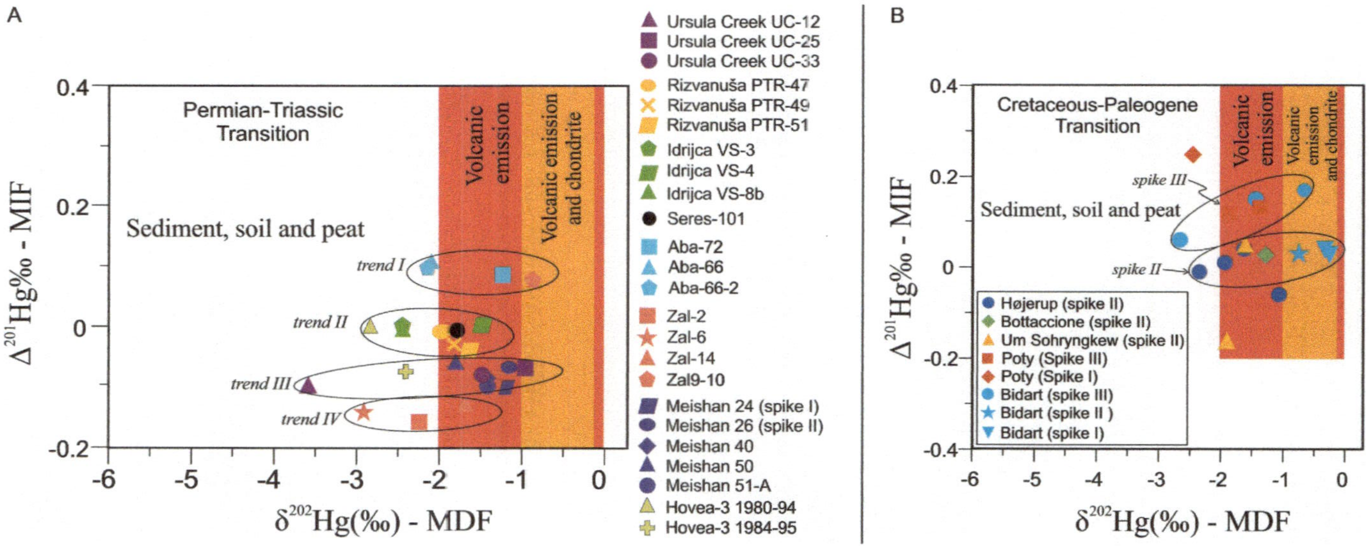

Fig. 11 (A) In a δ^{202}Hg‰ (MDF)–Δ^{201}Hg (MIF) diagram, most of the analyzed samples plot within the volcanogenic Hg "emission box" and display four trends: *trend I* is formed by Abadeh samples and the event layer of the Zal section; *trend II* is formed by Rizvanuša, Seres/Misci and Idrijca samples; *trend III* by Meishan and Ursula Creek samples and *trend IV*, by Zal samples. Four samples plotted within the "sediment, soil and peat box." (B) A δ^{202}Hg ‰ (MDF)–Δ^{201}Hg‰ plot for the Cretaceous–Paleogene transition (modified from Sial et al., 2019a) is added for comparison. While Δ^{201}Hg‰ (MIF) values tend to be negative in the PTB transition, they tend to be slightly positive to zero in the K/Pg transition. Two trends are displayed by the K/Pg samples: Hg/TOC spike II trend (K/Pg layer) and Hg/TOC spike III (slightly above the K/Pg). Ranges for (i) volcanogenic, (ii) chondritic Hg, and (iii) sediment, soil and peat are based on Bergquist and Blum (2009) and are shown as vertical bars. *Modified from Sial, A.N., Chen, J., Lacerda, L.D., Korte, C., Spangenberg, J.E., Silva-Tamayo, J.C., Gaucher, C., Ferreira, V.P., Barbosa, J.A., Pereira, N.S., Benigno, A.P., 2019b. Globally enhanced Hg deposition and Hg isotopes in sections straddling the Permian–Triassic boundary: link to volcanism. Palaeogeogr. Palaeoclimatol. Palaeoecol., 540, https://doi.org/10.1016/j.palaeo.2019.109537.*

To allow contrasting the Hg isotope behavior during mass extinction events in the P–T and K/Pg transitions, a δ^{202}Hg vs Δ^{201}Hg diagram for four K/Pg sections is shown in Fig. 11b (modified from Sial et al., 2019a). Two roughly parallel trends for the K/Pg transition can be depicted from this figure: (a) *spike II trend* defined by Hg/TOC analyses from the K/Pg boundary clay layer from K/Pg sections at Højerup (Stevns Klint, Denmark), Bottaccione (Gubbio, Italy), Sohryngkew (Meghalaya, India) and Bidart (Basque Basin, France) and (b) *spike III trend* defined by samples from the P1a planktic foraminiferal subzone from the Bidart, France, and the Poty Quarry (Brazil) sections. Samples in the *spike II trend* display Δ^{201}Hg values around zero and Hg was loaded by the Deccan volcanism or it is from the asteroid that collided with the Earth's surface at Chicxulub (Sial et al., 2019a). Just one sample in this trend (from Højerup) plotted inside the "sediment, soil and peat box," suggesting that samples have not undergone Hg-MIF fractionation during atmospheric transport, except for one sample that may have undergone Hg-MDF change by terrestrial influx. It is unlikely that post-depositional processes change Hg-MIF signatures, therefore the *spike II trend* is probably a primary one, delineated by Hg from the Deccan volcanism.

Two main differences between the behavior of Hg isotopes in the P–T and K/Pg transitions arise from this study: (a) samples from the K/Pg boundary clay layer tend to plot within the "chondrite/volcanic-emission box," on the contrary of PTB samples; (b) Hg in K/Pg clay layer samples tend to be aligned in the *spike II trend*, and apparently have undergone little or none Hg-MIF fractionation during transport from the Deccan volcanic site or further local change during deposition. Hg in samples in the PTB *trend II* are likely STLIP-sourced with some degree of terrigenous-Hg influx in the local of deposition.

The δ^{202}Hg vs Δ^{201}Hg crossplot suggests in a first approach that in the K/Pg transition, volcanic Hg isotopes received little influence from terrigenous-Hg influx. During the P–T transition, in contrast, volcanic Hg has received variable influence from terrigenous-source influx, more marked in samples in *trends III and IV* (Meishan, Ursula Creek, Zal and Hovea-3 sections, sites far distant from the STLIP), and less marked in *trend II* samples (Seres/Misci, Rizvanuša and Idrijca sections, in Europe, sites relatively closer to the STLIP).

Shen et al. (2019a) have documented that pre- and post-event beds of two continental P–T transition sections in the South China Craton are characterized by low Hg concentrations and more positive values of MIF odd Hg isotopes. The LPME– ETME interval, instead, is characterized by higher Hg concentrations and lower MIF odd Hg isotope values (usually negative values). From a crossplot of Δ^{199}Hg vs Hg concentrations, these authors concluded that samples tend to fall along a mixing trend defined by a background endmember with low Hg and high MIF and a volcanic endmember with high Hg and low MIF.

Two curved trends are observed in a Δ^{199}Hg‰ (MIF) and Hg ($ng\,g^{-1}$) crossplot in Fig. 12A for samples from eight of the nine sections in this study, probably resulting from two-component mixing between terrigenous and volcanic-source Hg. The *trend I* (three samples from Abadeh) exhibit positive Δ^{199}Hg values while two samples from Meishan and one from Ursula Creek display negative values. In *trend II*, three samples from Zal and one from Misci display positive values, while three samples from Meishan, two from Hovea-3, two from Ursula Creek and three from Rizvanuša sections display Δ^{199}Hg (MIF) around zero or with negative values. The sample Zal 9/10 (event layer) departs from this trend more than any other sample in this trend. Perhaps, this is a consequence of input from rapid erosion of Hg from soils as invoked to explain a similar situation in China (Shen et al., 2019b).

In *trend III*, three samples from Idrijca display positive Δ^{199}Hg (MIF) values and the highest Hg contents among the studied sections. However, these samples may have been strongly affected by hydrothermal Hg contamination from the Idrijca mining region (Foucher et al., 2009; Lavric and Spangenberg, 2003). Two samples from Ursula Creek plot in the vicinity of the *trend III*, a behavior for which we do not have so far a plausible explanation.

Using a limited data available in Shen et al. (2019a) a Δ^{199}Hg‰ (MIF) vs Hg ($ng\,g^{-1}$) crossplot for the K/Pg transition is shown in Fig. 12B. Except for two samples from the Poty Quarry (Brazil), collected about 1 m above the K/Pg boundary (*spike III trend*), and one from Um Sohryngkew (India), all samples from the K/Pg layer (*spike II trend*) consistently display Δ^{199}Hg‰ (MIF) values around zero. It is depicted from Fig. 12b that odd Hg-MIF isotopes in *spike II* have undergone negligible to none fractionation from the Deccan volcanic center to the local of deposition. Two-component mixing between volcanic and terrigenous-Hg influx at the *spike II* seems to have been more limited than in PTB layers.

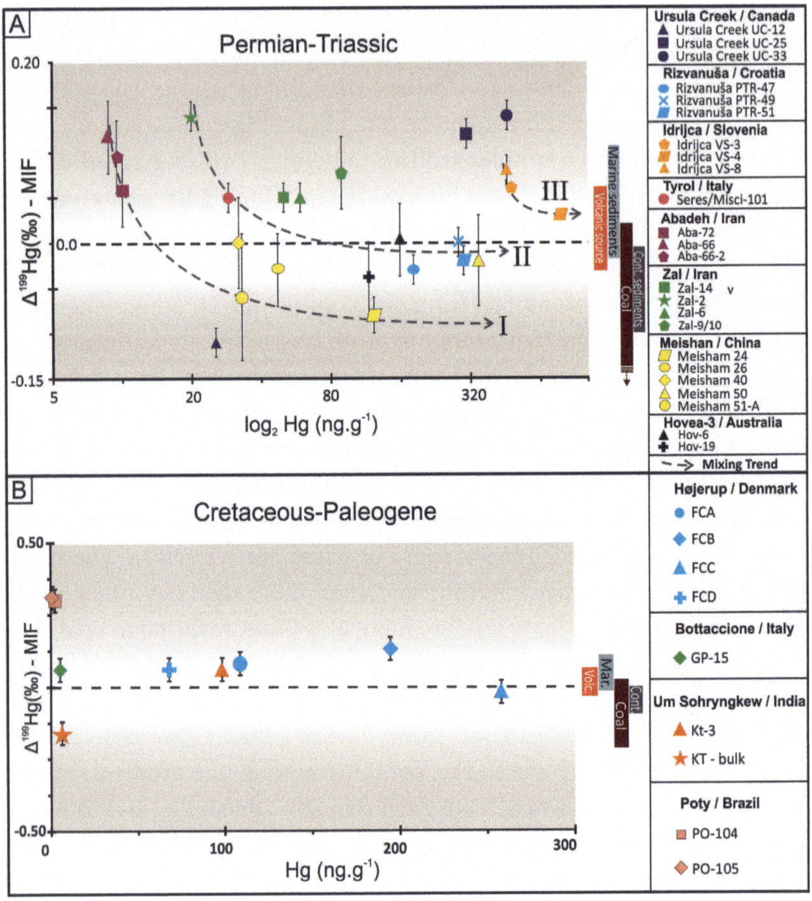

Fig. 12 (A) Crossplot between $\Delta^{199}Hg‰$ (MIF) and $\log_2 Hg$ (ng g^{-1}) for sections in this study. The four curved trends probably resulted from two-component mixing between terrestrial and volcanic-source Hg. Trends I (Meishan section) and II (Rizvanuša and Seres/Misci) are similar and close to each other. The *trend III* (Zal section) may have received an input of erosional concentration of Hg, and *trend IV* (Idrijca section) may have been strongly affected by a terrestrial Hg contamination from the Idrijca mining region; (B) A $\Delta^{199}Hg‰$ (MIF) vs Hg (ng g^{-1}) cross plot for the K/Pg transition using the Hg data available in Sial et al. (2019a) is shown in Fig. 12B. *Modified from Sial, A.N., Chen, J., Lacerda, L.D., Korte, C., Spangenberg, J.E., Silva-Tamayo, J.C., Gaucher, C., Ferreira, V.P., Barbosa, J.A., Pereira, N.S., Benigno, A.P., 2019b. Globally enhanced Hg deposition and Hg isotopes in sections straddling the Permian–Triassic boundary: link to volcanism. Palaeogeogr. Palaeoclimatol. Palaeoecol., 540, https://doi.org/10.1016/j.palaeo.2019.109537.*

7. Conclusions

Among the main conclusions drawn from this study one can list: (a) carbon-isotope and Hg chemostratigraphy seem to point to more drastic environmental changes in the LPME than at the ETME event horizons; (b) a negative $\delta^{13}C_{carb}$ shift coincide with a Hg/TOC spike at the LPME, synchronous with the beginning of the STLIP intrusive stage of magmatism (stage 2); (c) almost no sample in the LPME–ETME extinction interval lies in the "chondrite box" in a δ^{202}Hg vs Δ^{201}Hg crossplot while most samples from the K/Pg transition, on the contrary, plot within the "volcanic box" I this crossplot; (d) terrigenous-Hg influx in the PTB transition seem to be minor in those sections geographically located closer to the STLIP (Δ^{201}Hg around zero); higher terrigenous-Hg influx happened in PTB sections geographically located far distant from the STLIP (negative Δ^{201}Hg) and the highest terrigenous-Hg influx probably happened in sections geographically located in intermediate distance from the STLIP (highest positive Δ^{201}Hg values or lowest negative Δ^{201}Hg values); (e) in the K/Pg transition sections, Hg isotopes point to little terrigenous-Hg influx: (i) in the *spike II trend* (K/Pg clay layer) the Δ^{201}Hg is about zero while the *spike III trend* (P1a foraminiferal subzone) is characterized by positive Δ^{201}Hg values.

Among address-pending issues, the following seem to demand special attention: (1) whether the Hg/TOC spike in the LPME–PTB interval in all but Idrijca section is a global feature or not; (2) timing and duration of Hg/TOC spikes; (3) Hg concentrations and Hg stratigraphy in the STLIP and Deccan volcanic centers; (4) is Hg linked to basic or acidic lavas in the STLIP/Deccan volcanic centers?; (5) whether Hg is organic matter tied, sulfide-hosted or adsorbed onto clay to allow a suitable normalization of Hg concentration to TOC, total S or total Al; (6) the Hg/TOC spike at about 10 cm below the LPME in the Misci and Zal sections in this study represents a global feature, linked to the STLIP stage 1 volcanism, or whether it is simply coeval to nearby volcanic source.

Acknowledgments

Grants to A.N.S. (CNPq 471036/2012-9, 407171/2018-5; FACEPE APQ-1059-9.05/12, 1073-1.07/15), to L.D.L. (CNPq INCT-TMCOcean 573.601/2008-9, CNPq576.601/2009-1), to V.P.F. (CNPq 471034/2012-6; FACEPE APQ1738-1.07/12) and NSFC grants to J.B.C. (41625012, U1612442, 4196114028, 41830647) provided financial support to chemical and/or isotope analyses. M.K.P. thanks Ahsan Wani for the support and assistance during fieldwork and sampling at the Guryul Ravines. C.K. the Danish Natural

Science Research Council for providing financial support for sampling in China and to QuanFeng Zheng for leading him in the field in Meishan (grant number 11-103378). This is the NEG–LABISE scientific contribution n. 296.

References

Adatte, T., Keller, G., Schoene, B., Samperton, K.M., Font, E., Sial, A.N., Lacerda, L.D., Punekar, J., Fantasia, A., Khadri, S., 2015. Paleoenvironmental influence of Deccan volcanism relative to the KT extinction. Baltimore. Geol. Soc. Am. Abstr. Programs 47 (7), 210.

Algeo, T.J., Hanningan, R., Rowe, H., Brookfield, M., Baud, A., Krystyn, L., Ellwood, B.B., 2007. Sequencing events across the Permian–Triassic boundary, Guryul Ravine (Kashmir, India). Palaeogeogr. Palaeoclimatol. Palaeoecol. 252, 328–346.

Algeo, T.J., Shen, Y., Zhang, T., Lyons, T., Bates, S., Rowe, H., Nguyen, T.K.T., 2008. Association of ^{34}S-depleted pyrite layers with negative carbonate δ^{13}C excursions at the Permian–Triassic boundary: evidence for upwelling of sulfidic deep-ocean water masses. Geochem. Geophys. Geosyst 9, 2008. https://doi.org/10.1029/2007GC001823.

Algeo, T., Henderson, C.M., Ellwood, B., Rowe, H., Elswick, E., Bates, S., Lyons, T., Hower, J.C., Smith, C., Maynard, B., Hays, L.E., Summons, R.E., Fulton, J., Freeman, K.H., 2012. Evidence for a diachronous Late Permian marine crisis from the Canadian Arctic region. Geol. Soc. Am. Bull. 124, 1424–1448. https://doi.org/10.1130/B30505.1.

Bagherpour, B., Bucher, H., Vennemann, T., Schneebeli-Hermann, E., Yuan, D.-X., Leu, M., Zhang, C., Shen, S.-Z., 2020. Are Late Permian carbon isotope excursions of local or of global significance? Geol. Soc. Am. Bull. 132, 521–544.

Baresel, B., Bucher, H., Brosse, M., Cordey, F., Guodun, K., Schaltegger, U., 2017. Precise age for the Permian–Triassic boundary in South China from high-precision U-Pb geochronology and Bayesian age–depth modeling. Solid Earth 8, 361–378.

Basu, A.R., Petaev, M.I., Poreda, R.J., Jacobsen, S.B., Becker, L., 2003. Chondritic meteorite fragments associated with the Permian–Triassic boundary in Antarctica. Science 302, 1388–1392.

Baud, A., Bhat, G., 2014. IGCP 630 field guide book 1—the Permian–Triassic transition in the Kashmir Valley. In: IGCP 630 Field Workshop in Kashmir, p. 36.

Baud, A., Magaritz, M., 1988. Carbon isotope profile in the Permian–Triassic of the Central Tethys: The Kashmir Sections (India). In: Berichte der Geologisch. Bundesanst, Band 1, p. 2. Abstact.

Baud, A., Holser, W.T., Magaritz, M., 1989. Permian–Triassic of the Tethys: carbon isotope studies. Geogr. Rundsch. 78 (2), 649–677.

Baud, A., Atudorei, V., Sharp, Z., 1996. Late Permian and Early Triassic evolution of the Northern Indian margin: carbon isotope and sequence stratigraphy. Geodin. Acta 9, 57–77.

Becker, L., Poreda, R.J., Hunt, A.G., Bunch, T.E., Rampino, M., 2001. Impact event at the Permian–Triassic boundary: evidence from extraterrestrial noble gases in fullerenes. Science 291, 1530–1533.

Becker, L., Poreda, R., Basu, A., Pope, K.O., Harrison, T.M., Nicholson, C., Lasky, R., 2004. Bedout: a possible end-Permian impact crater offshore of Northwestern Australia. Science 304 (5676), 1469–1476. https://doi.org/10.1126/science.1093925.

Behnken, F.H., 1975. Leonardian and Guadalupian (Permian) conodont biostratigraphy in western and southwestern United States. J. Paleo. 49, 284–315.

Benton, M.J., 2003. When Life Nearly Died—The Greatest Mass Extinction of All Time. Thames & Hudson, London, p. 352.

Bergquist, B.A., Blum, J.D., 2009. The odds and evens of mercury isotopes: applications of mass-dependent and mass-independent isotope fractionation. Elements 5, 353–357.
Blum, J.D., Sherman, L.S., Johnson, M.W., 2014. Mercury isotopes in earth and environmental sciences. Annu. Rev. Earth Planet. Sci. 42, 249–269.
Bond, D.P.G., Grasby, S.E., 2017. On the causes of mass extinctions. Palaeogeogr. Palaeoclimatol. Palaeoecol. 478, 3–29.
Bond, D.P.G., Wignall, P.B., 2010. Pyrite framboid study of marine Permian-Triassic boundary sections: a complex anoxic event and its relationship to contemporaneous mass extinction. Geol. Soc. Am. Bull. 122, 1265–1279.
Bond, D.P.G., Wignal, P.B., Grasby, S.E., 2020. The capitanian (Guadalupian, Middle Permian) mass extinction in NW Pangea (Borup Fiord, Arctic Canada): a global crisis driven by volcanism and anoxia. Geol. Soc. Am. Bull. 132, 931–942. https://doi.org/10.1130/B35281.1.
Bowring, S.A., Erwin, D.H., Jin, Y.G., Martin, M.W., Davidek, K., Wang, W., 1998. U/Pb zircon geochronology and tempo of the end-Permian mass extinction. Science 280 (5366), 1039–1045.
Brookfield, M.E., 1993. The Himalayan passive margin from Precambrian to Cretaceous times. Sediment. Geol. 84, 1–35.
Brookfield, M.E., Twitchett, R.J., Goodings, C., 2003. Palaeoenvironments of the Permian–Triassic transition sections in Kashmir, India. Palaeogeogr. Palaeoclimatol. Palaeoecol. 198, 353–371.
Brookfield, M.E., Algeo, T.J., Hannigan, R., Williams, J., Bhat, G.M., 2013. Shaken and stirred: seismites and tsunamites at the Permian–Triassic boundary, Guryul Ravine, Kashmir, India. Palaios 28, 568–582.
Brookfield, M.E., Stebbins, A.G., Williams, J.C., Wolbach, W.S., Hannigan, R., Bhat, G.M., 2019. Palaeoenvironments and elemental geochemistry across the marine Permo-Triassic boundary section, Guryul Ravine (Kashmir, India) and a comparison with other North Indian passive margin sections. Depos. Rec. 6, 75–116. https://doi.org/10.1002/dep2.96.
Burger, B.J., Estrada, M.V., Gustin, M.S., 2019. What caused Earth's largest mass extinction event? New evidence from the Permian–Triassic boundary in northeastern Utah. Glob. Planet. Change 177, 81–100.
Burgess, S.D., Bowring, S.A., 2015. High-precision geochronology confirms voluminous magmatism before, during, and after Earth's most severe extinction. Sci. Adv. 1 (7). https://doi.org/10.1126/sciadv.1500470.
Burgess, S.D., Bowring, S.A., Shen, S.Z., 2014. High-precision timeline for Earth's most severe extinction. Proc. Natl Acad. Sci. USA 111, 3316–3321.
Burgess, S.D., Muirhead, J.D., Bowring, S.A., 2017. Initial pulse of Siberian Traps sills as the trigger of the end-Permian mass extinction. Nat. Commun. 8, 164.
Campbell, I.H., Czamanske, G.K., Fedorenko, V.A., Hill, R.I., Stepanov, V., 1992. Synchronism of the Siberian traps and the Permian–Triassic boundary. Science 258, 1760–1763.
Cao, C.Q., Zheng, Q.F., 2007. High-resolution lithostratigraphy of the Changhsingian stage in Meishan section D, Zhejiang. J. Stratigr. 31, 14–22.
Cao, C.Q., Zheng, Q.F., 2009. Geological event sequences of the Permian–Triassic transition recorded in the microfacies in Meishan section. Sci China Ser D Earth Sci 52 (10), 1529–1536.
Cao, C., Wang, W., Jin, Y., 2002. Carbon isotope excursions across the Permian–Triassic boundary in the Meishan section, Zhejiang Province, China. Chin. Sci. Bull. 47, 1125–1129.
Cao, C., Love, G.D., Hays, L.E., Wang, W., Shen, S., Summons, R.E., 2009. Biogeochemical evidence for euxinic oceans and ecological disturbance presaging the end-Permian mass extinction event. Earth Planet. Sci. Lett. 28, 188–201.

Charbonnier, G., Morales, C., Duchamp-Alphonse, S., Westermann, S., Adatte, T., Föllmi, K.B., 2017. Mercury enrichment indicates volcanic triggering of Valanginian environmental change. Sci. Rep. 7. https://doi.org/10.1038/srep40808.

Chen, Z.-Q., Benton, M.J., 2012. The timing and pattern of biotic recovery following the end-Permian mass extinction. Nat. Geosci. 5, 375–383.

Chen, Z.Q., Kaiho, K., George, A.D., 2005. Survival strategies of brachiopod faunas from the end-Permian mass extinction. Palaeogeogr. Palaeoclimatol. Palaeoecol. 224, 232–269.

Chen, Z.Q., Algeo, T.J., Bottjer, D.J., 2014. Global review of the Permian–Triassic mass extinction and subsequent recovery: part I. Earth Sci. Rev. 137, 1–5.

Chen, Z.Q., Yang, H., Luo, M., Benton, J.J., Kaiho, K., Zhao, L., Huang, Y., Zhang, K., Fang, Y., Jiang, H., Qiu, H., Li, Y., Tu, C., Shi, L., Zhang, L., Feng, X., Chen, L., 2015. Complete biotic and sedimentary records of the Permian–Triassic transition from Meishan section, South China: ecologically assessing mass extinction and its aftermath. Earth Sci. Rev. 149, 67–107.

Chen, X.-Y., Teng, F.-Z., Huang, K.-J., Algeo, T.J., 2020. Intensified chemical weathering during Early Triassic revealed by magnesium isotopes. Geochim. Cosmochim. Acta. doi.org/10.1016/j.gca.2020.02.035.

Clark, D.L., Behnken, F.H., 1971. Conodonts and biostratigraphyof the Permian. In: Stweet, W.C., Bergstrom, S.M. (Eds.), Symposium on Conodont Biostratigraphy. vol. 127. Geological Society of America, pp. 415–439. Memoirs.

Courtillot, V., Jaupart, C., Manighetti, I., Tapponnier, P., Besse, J., 1999. On causal links between flood basalts and continental breakup. Earth Planet. Sci. Lett. 166, 177–195.

Dèzes, P., 1999. Tectonic and metamorphic evolution of the Central Himalayan domain in southeast Zanskar (Kashmir India). Mém Géol. 32, 1–149.

Erwin, D.H., 2006. Extinction: How Life on Earth Nearly Ended 250 Million Years Ago. Princeton University Press, New Jersey, p. 296.

Erwin, D.H., Bowring, S.A., Yugan, J., 2002. End-Permian mass extinctions: a review. In: Koeberl, C., MacLeod, K.G. (Eds.), Catastrophic Events and Mass Extinctions: Impacts and Beyond. Geological Society of America Special Paper 356, pp. 363–383.

Faggetter, L.E., Wignall, P.B., Pruss, S.B., Jones, D.S., Grasby, S., Widdowson, M., Newton, R.J., 2019. Mercury chemostratigraphy across the Cambrian Series 2–Series 3 boundary: evidence for increased volcanic activity coincident with extinction? Chem. Geol. 510, 188–199.

Fantle, M.S., Barnes, B.D., Lau, K.V., 2020. The role of diagenesis in shaping the geochemistry of the marine carbonate record. Annu. Rev. Earth Planet. Sci. 48, 549–583.

Farley, K.A., Garrison, G., Mukhopadhyay, S., Ward, P., 2005. Absence of extraterrestrial ^{3}He in Permian-Triassic age sedimentary rocks. Earth Planet. Sci. Lett. 240, 265–275.

Fio, K., Spangenberg, J.E., Vlahović, I., Sremac, J., Velić, I., Mrinjek, E., 2010. Stable isotope and trace element stratigraphy across the Permian–Triassic transition: a redefinition of the boundary in the Velebit Mountain, Croatia. Chem. Geol. 278, 38–57.

Font, F., Adatte, T., Sial, A.N., Lacerda, L.D., Keller, G., Punekar, J., 2016. Mercury anomaly, Deccan volcanism and the end-Cretaceous mass extinction. Geology 44, 171–174.

Foucher, D., Ogrinc, N., Hintelmann, H., 2009. Tracing mercury contamination from the Idrija Mining Region (Slovenia) to the Gulf of Trieste using Hg isotope ratio measurements. Environ. Sci. Technol. 43, 33–39.

Garzanti, E., Angiolini, L., Sciunnach, D., 1996. The Permian Kuling Group (Spiti, Lahaul and Zanskar; NW Himalaya): sedimentary evolution during rift/drift transition and initial opening of Neo-Tethys. Riv. Ital. Paleontol. Stratigr. 102, 175–200.

Garzanti, E., Nicora, A., Rettori, R., 1998. Permo–Triassic boundary and lower to middle Triassic in South Tibet. J. Asian Earth Sci. 16, 143–157.

Ghosh, N., Basu, A.R., Bhargava, O.N., Shukla, U.K., Ghatak, A., Garzione, C.N., Ahluwalia, A.D., 2016. Catastrophic environmental transition at the Permian-Triassic Neo-Tethyan margin of Gondwanaland: geochemical, isotopic and sedimentological evidence in the Spiti Valley, India. Gondw. Res. 34, 324–345.

Glikson, A., 2004. Bedout: a possible end-Permian impact crater offshore of Northwestern Australia. Science 306, 613. https://doi.org/10.1126/science.1100404.

Gong, Q., Wang, X., Zhao, L., Grasby, S.E., Chen, Z.-Q., Zhang, L., Li, Y., Cao, L., Li, Z., 2017. Mercury spikes suggest volcanic driver of the Ordovician–Silurian mass extinction. Sci. Rep. 7, 5304.

Grard, A., Francois, L.M., Dessert, C., Dupre, B., Godderis, Y., 2005. Basaltic volcanism and mass extinction at the Permo-Triassic boundary: environmental impact and modeling of the global carbon cycle. Earth Planet. Sci. Lett. 234, 207–221.

Grasby, S.E., Sanei, H., Beauchamp, B., 2011. Catastrophic dispersion of coal fly ash into oceans during the latest Permian extinction. Nat. Geosci. 4, 104–107.

Grasby, S.E., Sanei, H., Beauchamp, B., Chen, Z., 2013. Mercury deposition through the Permo–Triassic biotic crisis. Chem. Geol. 351, 209–216.

Grasby, S.E., Beauchamp, B., Bond, D.P.G., Wignall, P.B., Sanei, 2016. Mercury anomalies associated with three extinction events (Capitanian Crisis, Latest Permian Extinction and the Smithian/Spathian Extinction) in NW Pangea. Geol. Mag. 153 (2), 285–297. https://doi.org/10.1017/S0016756815000436.

Grasby, S.E., Shen, W., Yin, R., Gleason, J.D., Blum, J.D., Lepak, R.F., Hurley, J.P., Beauchamp, B., 2017. Isotopic signatures of mercury contamination in latest Permian oceans. Geology 45, 55–58.

Grasby, S.E., Them II, T.R., Chen, Z., Yin, R., Ardakani, O.H., 2019. Mercury as a proxy for volcanic emissions in the geologic record. Earth Sci. Rev. 196. https://doi.org/10.1016/j.earscirev.2019.102880.

Grasby, S.E., Knies, J., Beauchamp, B., Bond, D.F.P.G., Wignall, P., Sun, Y., 2020. Global warming leads to Early Triassic nutrient stress across northern Pangea. Geol. Soc. Am. Bull. 132, 943–954. https://doi.org/10.1130/B32036.1.

Grice, K., Cao, C., Love, G.D., Bottcher, M.E., Twitchett, R.J., Grosjean, E., Summons, R.E., Turgeon, S.C., Dunning, W., Jin, W., 2005. Photic zone euxinia during the Permian Triassic superanoxic event. Science 307, 706–709.

Grotzinger, J.P., Fike, D.A., Fischer, W.W., 2011. Enigmatic origin of the largest known carbon isotope excursion in Earth's history. Nat. Geosci. 4, 285–292.

Hallam, A., Wignall, P.B., 1999. Mass extinctions and sea-level changes. Earth Sci. Rev. 48, 217–250.

Herren, E, 1987. Zanskar shear zone: northeast–southwest extension within the Higher Himalayas (Ladakh, India). Geology 15, 409–413.

Hinojosa, J.L., Brown, S.T., DePaolo, D.J., Paytan, A., Shen, S., Chen, J., Payne, J.L., 2012. Evidence for end-Permian ocean acidification from calcium isotopes in biogenic apatite. Geology 40, 743–746. https://doi.org/10.1130/G33048.1.

Hofmann, R., Buatois, L.A., MacNaughton, R.B., Mángano, M.G., 2015. Loss of the sedimentary mixed layer as a result of the end-Permian extinction. Palaeogeogr. Palaeoclimatol. Palaeoecol. 428, 1–11.

Holser, W.T., Schonlaub, H.-P., Attrep Jr., M., Boeckelmann, K., Klein, P., Magaritz, M., Orth, C.J., Fenninger, A., Jenny, C., Kralik, M., Mauritsch, H., Pak, E., Schramm, J.-M., Stattegger, K., Schmoller, R., 1989. A unique geochemical record at the Permian/Triassic boundary. Nature 337, 39–44.

Huang, Y., Chen, Z.-Q., Algeo, T.J., Zhao, L., Baud, A., Bhat, G.M., Zhang, L., Guo, Z., 2019. Two-stage marine anoxia and biotic response during the Permian–Triassic transition in Kashmir, northern India: pyrite framboid evidence. Glob. Planet. Change 172, 124–139. https://doi.org/10.1016/j.gloplacha.2018.10.002.

Igo, H., 1981. Permian conodont biostratigraphy of Japan. In: Palaeontological Society of Japan Special Papers, p. 24.
Isozaki, Y., 1997. Permo–Triassic boundary superanoxia and stratified superocean: records from lost deep sea. Science 276, 235–238.
Jiang, Y., Tang, Y., Dai, S., Zou, X., Qian, H., Zhou, G., 2006. Pyrites and sulfur isotopic composition near the Permian–Triassic boundary in Meishan Zhejiang. Acta Geol. Sin. 80 (8), 1202–1207.
Jin, Y.G., Wang, Y., Wang, W., Shang, Q.H., Cao, C.Q., Erwin, D.H., 2000. Pattern of marine mass extinction near the Permian–Triassic boundary in South China. Science 289, 432–436.
Joachimski, M.M., Lai, X., Shen, S., Jiang, H., Luo, G., Chen, B., Chen, J., Sun, Y., 2012. Climate warming in the latest Permian and the Permian–Triassic mass extinction. Geology 40, 195–198.
Jones, D.S., Martini, A.M., Fike, A., Kaiho, K., 2017. A volcanic trigger for the Late Ordovician mass extinction? Mercury data from South China and Laurentia. Geology 45, 631–634. https://doi.org/10.1130/G38940.1.
Kaiho, K., Kajiwara, Y., Nakano, T., Miura, Y., Kawahata, H., Tazaki, K., Ueshima, M., Chen, Z.Q., Shi, G.R., 2001. End-Permian catastrophe by a bolide impact: evidence of a gigantic release of sulfur from the mantle. Geology 29, 815–818.
Kaiho, K., Kajiwara, Y., Chen, Z.-Q., Gorjan, P., 2006. A sulfur isotope event at the end of the Permian. Chem. Geol. 235, 33–47.
Kaiho, K., Chen, Z.Q., Sawada, K., 2009. Possible causes for a negative shift in the stable carbon isotope ratio before, during and after the end-Permian mass extinction in Meishan, South China. Aust. J. Earth Sci. 56, 799–808.
Kaiho, K., Saito, R., Ito, K., Miyaji, T., Biswas, R., Tian, L., Sano, H., Shi, Z., Takahashi, S., Tong, J., Liang, L., Oba, M., Nara, F.W., Tsuchiya, N., Chen, Z.-Q., 2016. Effects of soil erosion and anoxic–euxinic ocean in the Permian–Triassic marine crisis. Heliyon 2. https://doi.org/10.1016/j.heliyon.2016.e00137.
Kamo, S.L., Czamanske, G.K., Amelin, Y., Fedorenko, V.A., Davis, D.W., Trofimov, V.R., 2003. Rapid eruption of Siberian flood-volcanic rocks and evidence for coincidence with the Permian–Triassic boundary and mass extinction at 251 Ma. Earth Planet. Sci. Lett. 214, 75–91.
Kapoor, H.M., 1992. Permo-Triassic boundary of the Indian subcontinent and its intercontinental correlation. In: Sweet, W.C., et al. (Eds.), Permo-Triassic Events in the Eastern Tethys. Cambridge University Press, pp. 21–36.
Kapoor, H.M., 1996. The Guryul ravine section, candidate of the global stratotype and point (GSSP) of the Permian–Triassic boundary (PTB). In: Yin, H. (Ed.), The Paleozoic–Mesozoic Boundary: Candidates of the Global Stratotype Section and Point of the Permian–Triassic Boundary. China University Geoscience Press, Wuhan, pp. 99–110.
Knoll, A.H., Hayes, J.M., Kaufman, A.J., Swett, K., Lambert, I.B., 1986. Secular variation in carbon isotope ratios from upper Proterozoic successions of Svalbard and East Greenland. Nature 321, 832–838.
Koeberl, C., Farley, K.A., Peucker-Ehrenbrink, B., Sephton, M.A., 2004. Geochemistry of the end-Permian extinction event in Austria and Italy: no evidence for an extraterrestrial component. Geology 32, 1053–1056.
Korte, C., Kozur, H.W., 2010. Carbon-isotope stratigraphy across the Permian–Triassic boundary: a review. J. Asian Earth Sci. 39, 215–235.
Korte, C., Kozur, H.W., Partoazar, H., 2004a. Negative carbon isotope excursion at the Permian/Triassic boundary section at Zal, NW-Iran. Hallesches Jahrbuch Geowissenschaften, Reihe B, Beihefte 18, 69–71.
Korte, C., Kozur, H.W., Mohtat-Aghai, P., 2004b. Dzhulfian to lowermost Triassic $\delta^{13}C$ record at the Permian/Triassic boundary section at Shahreza, Central Iran. Hallesches Jahrbuch fur Geowissenschaften B Beiheft 18, 73–78.

Korte, C., Kozur, H.W., Joachimski, M.M., Strauss, H., Veizer, J.N., Schwark, L., 2004c. Carbon, sulfur, oxygen and strontium isotope records, organic geochemistry and biostratigraphy across the Permian/Triassic boundary in Abadeh, Iran. Int. J. Earth Sci. 93, 565–581.

Korte, C., Pande, P., Kalia, P., Kozur, H.W., Joachimski, M.M., Oberhänsli, H., 2010a. Massive volcanism at the Permian–Triassic boundary and its impact on the isotopic composition of the ocean and atmosphere. J. Asian Earth Sci. 37, 293–311.

Korte, C., Pande, P., Kalia, P., Kozur, H.W., Joachimski, M.M., Oberhänsli, H., 2010b. Massive volcanism at the Permian–Triassic boundary and its impact on the isotope composition of the ocean and atmosphere. J. Asian Earth Sci. 37, 293–311.

Korte, C., Ruh, M., Palfy, J., Ullmann, C.V., Hesselbo, S.P., 2019. Chemostratigraphy across the Triassic–Jurassic boundary. In: Sial, A.N., Gaucher, C., Ramkumar, M., Ferreira, V.P. (Eds.), Chemostratigraphy Across Major Chronological Boundaries. AGU/Geophysical Monograph 240, John Wiley & Sons, Inc, pp. 185–210.

Kozur, H.W., 1978. Beitraege zur Stratigraphie des Perms; Teil II, Die Conodontenchronologie des Perms. Freiberger Forschungshefte, Reihe C: Geowissenschaften, Mineralogie Geochemie 334, 85–161.

Kozur, H.W., 1990. The taxonomy of the gondolellid conodonts in the Permian and Triassic. In: Courier Forschungsinstitut Institut Senckenberg. 117, pp. 409–469.

Kozur, H., 1996. The conodonts Hindeodus, Isarcicella and Sweetohindeodus in the Uppermost Permian and Lowermost Triassic. Geologia Croatia 49, 81–115.

Kozur, H.W., 2003. Integrated ammonoid, conodont and radiolarian zonation of the Triassic and some remarks to stage/substage subdivision and the numeric age of the Triassic stages. Albertiana 28, 57–84.

Kozur, H.W., 2004. Pelagic uppermost Permian and the Permian–Triassic boundary conodonts of Iran, part I: taxonomy. Hallesches Jahrbuch fur Geowissenschaften B Beiheft 18, 39–68.

Kozur, H.W., 2005. Pelagic uppermost Permian and the Permian–Triassic boundary conodonts of Iran. Part II: investigated sections and evaluation of the conodont faunas. Hallesches Jahrb. Geowiss. B Beiheft 19, 49–86.

Kozur, H.W., 2007. Biostratigraphy and event stratigraphy in Iran around the Permian–Triassic boundary (PTB): implications for the causes of the PTB biotic crisis. Glob. Planet. Change 55, 155–176.

Kozur, H., Brandner, R., Resch, W., Mostler, H., 1995. Permian conodont zonation and its importance for the Permian stratigraphic standard scale. Mitt. Geol. Paläontol. Innsbruck 20, 165–205.

Kraus, S.H., Brandner, R., Heubeck, C., Kozur, H.W., Struck, U., Korte, C., 2013. Carbon isotope signatures of latest Permian marine successions of the Southern Alps suggest a continental runoff pulse enriched in land plant material. Fossil Rec. 16, 97–109.

Krystyn, L., Richoz, S., Baud, A., Twitchett, R.J., 2003. A unique Permian–Triassic boundary section from the Neotethyan Hawasina Basin, Central Oman Mountains. Palaeogeogr. Palaeoclimatol. Palaeoecol. 191, 329–344.

Krystyn, L., Balini, M., Nicora, A., 2004. Lower and Middle Triassic stage and substage boundaries in Spiti. Albertiana 30, 40–53.

Krystyn, L., Horacek, M., Brandner, R., Suraj, P., 2014. Late Permian Tsunamites in Guryul Ravine (Kashmir, India)—revisited and rejected. In: Geophysical Research Abstracts, 16, EGU 2014, no. 15312.

Kumar, K., Tewari, R., Agnihotri, D., Sharma, A., Pandita, S.K., Pillai, S.S.K., Singh, V., Bhat, G.D., 2017. Geochemistry of the Permian–Triassic sequences of the Guryul Ravine section, Jammu and Kashmir, India: implications for oceanic redox conditions. GeoResJ 13, 114–125.

Kump, L.R., Arthur, M.A., 1999. Interpreting carbon-isotope excursions: carbonates and organic matter. Chem. Geol. 161, 181–198.

Kwon, H., Kim, G.M., Yong, I.L.L., 2019. Mercury evidence from the Sino-Korean 1 block for Emeishan volcanism during the Capitanian mass extinction. Geol. Mag. 156, 1105–1110.

Lai, X., 1998. A discussion on Permian-Triassic conodont studies. Albertiana 20, 25–30.

Lau, K.V., Maher, K., Altiner, D., Kelley, B.M., Kump, L.R., Lehrmann, D.J., Silva-Tamayo, J.C., Weaver, K.L., Yu, M., Payne, J.L., 2016. Marine anoxia and delayed Earth system recovery after the end-Permian extinction. Proc. Natl. Acad. Sci. U. S. A. 113 (9), 2360–2365. https://doi.org/10.1073/pnas.1515080113.

Lavric, J.V., Spangenberg, J.E., 2003. Stable isotope (C, O, S) systematics of the mercury mineralization at Idrija, Slovenia: constraints on fluid source and alteration processes. Miner. Deposita 38, 886–899. https://doi.org/10.1007/s00126-003-0377-9.

Lehrmann, D.J., Payne, J.L., Felix, S.V., Dillet, P.M., Wang, H.-M., Yu, Y.-Y., Wei, J.Y., 2003. Permian–Triassic boundary sections from shallow-marine carbonate platforms of the Nanpanjiang Basin, South China: implications for oceanic conditions associated with the end-Permian extinction and its aftermath. Palaios 18, 138–152.

Leu, M., Baud, A., Brosse, M., Meier, M., Bhat, G., Bucher, H., Goudemand, N., 2014. Seismites. In: 19th International Sedimentological Congress 2014, Geneva, Switzerland, p. 396. Abstract book.

Li, R., Jones, B., 2017. Diagenetic overprint on negative $\delta^{13}C$ excursions across the Permian/Triassic boundary: a case study from Meishan section, China. Palaeogeogr. Palaeoclimatol. Palaeoecol. 468, 18–33.

Lo, C.-H., Chung, S.-L., Lee, T.-Y., Wu, G., 2002. Age of the Emeishan flood magmatism and relations to Permian–Triassic boundary events. Earth Planet. Sci. Lett. 198, 449–458.

Maloof, A.C., Porter, S.M., Moore, J.H., Dudás, F.O., Bowring, S.A., Higgins, J.A., Fike, D.A., Michael, E., P., 2010. The earliest Cambrian record of animals and ocean geochemical change. Geol. Soc. Am. Bull. 122, 1731–1774.

Matsuda, T., 1981. Early Triassic conodonts from Kashmir, India; part 1, Hindeodus and Isarcicella. J. Geosci. Osaka City Univ. 23, 75–100.

Matsuda, T., 1982. Early Triassic conodonts from Kashmir, India; part 2, neospathodus. J. Geosci. Osaka City Univ. 25, 87–102.

Matsuda, T., 1983. Early Triassic conodonts from Kashmir, India. Part 3: neospathodus 2. J. Geosci. Osaka City Univ. 26, 87–110.

Matsuda, T., 1984. Early Triassic conodonts from Kashmir, India, part 4: gondolella and platyvillosus. J. Geosci. Osaka City Univ. 27, 119–141.

Mei, S., Zhang, K., Wardlaw, B.R., 1998. A refined succession of Changhsingian and Griesbachian neogondolellid conodonts from the Meishan section, candidate of the global stratotype section and point of the Permian–Triassic boundary. Palaeogeogr. Palaeoclimatol. Palaeoecol. 143, 213–226.

Meyer, K.M., Yu, M., Lehrmann, D., van de Schootbrugge, B., Payne, J.L., 2013. Constraints on early Triassic carbon cycle dynamics from paired organic and inorganic carbon isotope records. Earth Planet. Sci. Lett. 361, 429–435.

Meyer, K.W., Petersen, S.V., Lohmann, K.C., Blum, J.D., Washburn, S.J., Johnson, M.W., Gleason, J.D., Kurz, A.Y., Winkelstern, I.Z., 2019. End-Cretaceous mercury and climate records reveal influence of Deccan Traps. Nat. Commun. https://doi.org/10.1038/S41467-019-13366-0. in press.

Movshovich, E.V., Kozur, H., Pavlov, A.M., Pnev, V.P., Polozova, A.N., Chuvashov, B.N., Bogoslovskaya, M.R., 1979. Complexes of conodonts from the lower Permian of the pre-Urals and problems of correlation of lower Permian deposits. In: Papular, G.N., Puchkov, V.N. (Eds.), Conodonts of the Urals and Their Stratigraphic Significance. Trudy Institute of Geology and Geochemistry, vol. 145. Urals Science Centre, Akademii Nauk SSSR, pp. 94–133.

Müller, R.D., Goncharov, A., Kritski, A., 2005. Geophysical evaluation of the enigmatic Bedout basement high, offshore northwestern Australia. Earth Planet. Sci. Lett. 237, 264–284.

Murata, M., 1981. Late Permian and Early Triassic conodonts from Upper Permian and Lower Triassic Faunas of Kashmir. In: Palaeontologica Indica New Series. vol. 46. Geological Survey of India, pp. 179–186.

Nakazawa, K., Kapoor, H.M., 1981. The Upper Permian and Lower Triassic Faunas of Kashmir. In: Palaeontologica Indica New Series. vol. 46. Geological Survey of India, pp. 1–191.

Nakazawa, K., Kapoor, H.M., Ishii, Y., Bando, T., Maegoya, D., Shimizu, D., Nogami, T., Tokuoka, T., Nohda, S., 1970. Preliminary Report on the Permo–Trias of Kashmir, India: Memoirs of the Faculty of Science, Kyoto University, Series of Geology and Mineralogy. vol. 37, pp. 163–172.

Nakazawa, K., Kapoor, H.M., Ishii, K., Bando, Y., Okimura, Y., Tokuoka, T., 1975. The Upper Permian and Lower Triassic in Kashmir, India: Memoirs of the Faculty of Science, Kyoto University. Series of Geology and Mineralogy. vol. XLII, pp. 1–106.

Nascimento-Silva, V.M., Sial, A.N., Ferreira, V.P., Neumann, V.H., Barbosa, J.A., Pimentel, M.M., Lacerda, L.D., 2011. Cretaceous–Paleogene transition at the Paraíba Basin, northeastern, Brazil: carbon-isotope and mercury subsurface stratigraphies. J. South Am. Earth Sci. 32, 379–392.

Nascimento-Silva, M.V., Sial, A.N., Ferreira, V.P., Barbosa, J.A., Neumann, V.H., Pimentel, M.M., Lacerda, L.D., 2013. Carbon isotopes, rare-earth elements and mercury behavior of Maastrichtian–Danian carbonate succession of the Paraíba Basin, Northeastern Brazil. In: Bojar, A.V., Melinte-Dobrinescu, M.C., Smit, J. (Eds.), Isotopic Studies in Cretaceous Research. Geological Society, London, Special Publications, 382, pp. 85–104.

Oehlert, A.M., Swart, P.K., 2014. Interpreting carbonate and organic carbon isotope covariance in the sedimentary record. Nat. Commun. 5, 4672. https://doi.org/10.1038/ncomms5672.

Paull, R.K., Paull, R.A., *1994. Hindeodus parvus*—proposed index fossil for the Permian–Triassic boundary. Lethaia 27, 271–272.

Payne, J.L., Turchyn, A.V., Paytan, A., DePaolo, D.P., Lehrmann, D.J., Yu, Y., Wei, J., 2010. Calcium isotope constraints on the end-Permian mass extinction. Proc. Natl. Acad. Sci. USA 107, 8543–8548. https://doi.org/10.1073/pnas.0914065107.

Percival, L.M.E., Witt, M.L.I., Mather, T.A., Hermoso, M., Jenkyns, H.C., Hesselbo, S.P., Al-Suwaidi, A.H., Storm, M.S., Xu, W., Ruhl, M., 2015. Globally enhanced mercury deposition during the end-Pliensbachian extinction and Toarcian OAE: a link to the Karoo–Ferrar large Igneous Province. Earth Planet. Sci. Lett. 428, 267–280.

Percival, L.M.E., Ruhl, M., Hesselbo, S.P., Jenkyns, H.C., Mather, T.A., Whiteside, J.H., 2017. Mercury evidence for pulsed volcanism during the end-Triassic mass extinction. Proc. Natl. Acad. Sci. 114 (30), 7929–7934.

Percival, L.M.E., Jenkyns, H.C., Mather, T.A., Dickson, A.J., Batenburg, S.J., Ruhl, M., Hesselbo, S.P., Barclay, R., Jarvis, I., Robinson, S.A., Woelders, L., 2018. Does large igneous province volcanism always perturb the mercury cycle? Comparing the records of Oceanic Anoxic Event 2 and the end-Cretaceous to other Mesozoic events. Am. J. Sci. 318, 799–860.

Perlmutter, B., 1975. Conodonts from the uppermost Wabaunsee Group (Pennsylvanian) and the Admire and Council Grove groups (Permian) in Kansas. Geol. Palaeontol. 9, 95–115.

Racki, G., 2020. Volcanism as a prime cause of mass extinctions: retrospectives and perspectives. In: Adatte, T., Bond, D.P.G., Keller, G. (Eds.), Mass Extinctions, Volcanism, and Impacts: New Developments: Geological Society of America Special Paper 544, pp. 1–34. doi.org/10.1130/2020.2544(01.

Rampino, M.R., Prokoph, A., Adler, A., 2000. Tempo of the end-Permian event: high-resolution cyclostratigraphy at the Permian–Triassic boundary. Geology 28, 643–646.

Reichow, M.K., Pringle, M.S., Al'Mukhamedov, A.I., Allen, M.B., Andreichev, V.L., Buslov, M.M., Davies, C.E., Fedoseev, G.S., Fitton, J.G., Inger, S., Medvedev, A.Y., Mitchell, C., Puchkov, V.N., Safanova, I.Y., Scott, R.A., Saunders, A.D., 2009. The timing and extent of the eruption of the Siberian traps large igneous province: implications for the end-Permian environmental crisis. Earth Planet. Sci. Lett. 277, 9–20.

Renne, P.R., Black, M.T., Zichao, Z., Richards, M.A., Basu, A.R., 1995. Synchrony and causal relations between Permian–Triassic boundary crises and Siberian flood volcanism. Science 269, 1413–1416.

Renne, P., Sprain, C.J., Richards, M.A., Self, S., Vanderkluysen, L., Pande, K., 2015. State shift in Deccan volcanism at the Cretaceous–Paleogene boundary, possibly induced by impact. Science 350, 76–78.

Retallack, G.J., Seyedolali, A., Holser, W.T., Krinsley, D., Krull, E.S., 1998. Shocked quartz at the Permian–Triassic boundary in Australia and Antarctica. Geology 26, 979–982.

Rexroad, C.B., Furnish, W.M., 1964. Conodonts from the Pella Formation (Mississippian), South-Central Iowa. J. Paleo. 38, 667–676.

Richoz, S., Krystyn, L., Baud, A., Brandner, R., Horacek, M., Mohtat-Aghai, P., 2010. Permian– Triassic boundary interval in the Middle East (Iran and N. Oman): progressive environmental change from detailed carbonate carbon isotope marine curve and sedimentary evolution. J. Asian Earth Sci. 39, 238–253.

Sabatino, N., Ferraro, S., Coccioni, R., Bonsignore, M., Del Core, M., Tancredi, V., Sprovieri, M., 2018. Mercury anomalies in upper Aptian-lower Albian sediments from the Tethys realm. Palaeogeogr. Palaeoclimatol. Palaeoecol. 495, 163–170.

Sanei, H., Grasby, S.E., Beauchamp, B., 2012. Latest Permian mercury anomalies. Geology 40, 63–66.

Schwab, V., Spangenberg, J.E., 2004. Organic geochemistry across the Permian–Triassic transition at the Idrijca Valley, Western Slovenia. Applied Geochemistry 19, 55–72.

Schobben, Ullmann, C.V., Leda, L., Korn, D., Struck, U., Reimold, W.U., Ghaderi, A., Algeo, T.J., Korte, C., 2017. Discerning primary versus diagenetic signals in carbonate carbon and oxygen isotope records: an example from the Permian–Triassic boundary of Iran. Chem. Geol. 422, 94–107.

Scholle, P.A., Arthur, M.A., 1980. Carbon isotope fluctuations in Cretaceous pelagic limestones: potential stratigraphic and petroleum exploration tool. Am. Assoc. Pet. Geol. Bull. 64, 67–87.

Shellnutt, J.G., Bhat, G.M., Brookfield, M.E., Jahn, B.-M., 2011. No link between the Panjal Traps (Kashmir) and the late Permian mass extinction. Geophys. Res. Lett. 38 (19), L19308. https://doi.org/10.1029/2011GL049032.

Shellnutt, J.G., Bhat, G., Wang, K.L., Brookfield, M.E., Jahn, B.M., Dostal, J., 2014. Petrogenesis of the flood basalts from the Early Permian Panjal Traps, Kashmir, India: geochemical evidence for shallow melting of the mantle. Lithos 204, 159–171.

Shen, S.Z., Cao, C.Q., Henderson, C.M., Wang, X.D., Shi, G.R., Wang, Y., Wang, W., 2006. End-Permian mass extinction pattern in the northern peri-Gondwanan region. Palaeoworld 15, 3–30.

Shen, W., Lin, Y., Xu, L., Li, J.J., Wu, Y., Sun, Y., 2007. Pyrite framboids in the Permian–Triassic boundary section at Meishan, China: evidence for dysoxic deposition. Palaeogeogr. Palaeoclimatol. Palaeoecol. 253, 323–331.

Shen, S.Z., Crowley, J.L., Wang, Y., Bowring, S.A., Erwin, D.H., Sadler, P.M., Cao, C.Q., Rothman, D.H., Henderson, C.M., Ramezani, J., Zhang, H., Shen, Y., Wang, X.D., Wang, W., Mu, L., Li, W.Z., Tang, Y.G., Liu, X.L., Liu, L.J., Zeng, Y., Jiang, Y.F., Jin, Y.G., 2011a. Calibrating the end-Permian mass extinction. Science 334, 1367–1372.

Shen, Y., Farquhar, J., Zhang, H., Masterson, A., Zhang, T., Wing, B.A., 2011b. Multiple S-isotopic evidence for episodic shoaling of anoxic water during Late Permian mass extinction. Nat. Commun. 2, 210. https://doi.org/10.1038/ncomms1217.

Shen, J., Algeo, T.J., Hu, Q., Xu, G., Zhou, L., Feng, Q., 2013. Volcanism in South China during the Late Permian and its relationship to marine ecosystem and environmental changes. Glob. Planet. Change 105, 121–134.

Shen, S.-Z., Ramezani, J., Chen, J., Cao, C.Q., Erwin, D.H., Zhang, H., Xiang, L., Schoepfer, S.D., Henderson, C.M., Zheng, Q.-F., Bowring, S.A., Wang, Y., Li, X.-H., Wang, X.-D., Yuan, D.-X., Zhang, Y.-X., Mu, L., Wang, J., Wu, Y.-S., 2018. A sudden end-Permian mass extinction in South China. Geol. Soc. Am. Bull. 131, 205–223. https://doi.org/10.1130/B31909.1.

Shen, J., Chen, J., Algeo, T.A., Yuan, S., Feng, Q., Yu, J., Zhou, L., O'Connell, O., Planavsky, N.J., 2019a. Evidence for a prolonged Permian–Triassic extinction interval from global marine mercury records. Nat. Commun. 10, 1563. https://doi.org/10.1038/s41467-019-09620-0.

Shen, J., Yu, J., Chen, J.-B., Algeo, T.J., Xu, G., Feng, Q., Shi, X., Planavsky, N.H., Shu, W., Xie, S., 2019b. Mercury evidence of intense volcanic effects on land during the Permian–Triassic transition. Geology 47, 1117–1121. https://doi.org/10.1130/G46679.1.

Shen, J., Algeo, T.J., Planavsky, N.J., Yu, J., Feng, Q., Song, H., Song, H., Rowe, H., Zhou, L., Chen, J., 2019c. Mercury enrichments provide evidence of Early Triassic volcanism following the end-Permian mass extinction. Earth Sci. Rev. 195, 191–212.

Shen, J., Feng, Q., Algeo, T.J., Liu, J., Zhou, C., Wei, W., Liu, J., Them, T.R., Gill, B.C., Chen, J.-B., 2020. Sedimentary host phases of mercury (Hg) and implications for use of Hg as a volcanic proxy. Earth Planet. Sci. Lett. 543 (2020), 116333.

Sheng, J., Chen, C., Wang, Y., Rui, L., Liao, Z., Dando, Y., Ishii, K., Nakazawa, K., Nakamura, K., 1984. Permian–Triassic boundary in middle and eastern Tethys. J. Fac. Sci. Hokkaido Univ. 21, 133–181. Ser. 4.

Sial, A.N., Gaucher, C., Silva Filho, M.A., Ferreira, V.P., Pimentel, M.M., Lacerda, L.D., Silva Filho, E.V., Cezario, W., 2010. C-, Sr-isotope and Hg stratigraphies of Neoproterozoic cap carbonates of the Sergipano Belt, Northeastern Brazil. Precambrian Res. 182, 351–372.

Sial, A.N., Lacerda, L.D., Ferreira, V.P., Frei, R., Marquillas, R.A., Barbosa, J.A., Gaucher, C., Windmöller, C.C., Pereira, N.S., 2013. Mercury as a proxy for volcanic activity during extreme environmental turnover: the Cretaceous–Paleogene transition. Palaeogeogr. Palaeoclimatol. Palaeoecol. 387, 153–164.

Sial, A.N., Chen, J.-B., Lacerda, L.D., Peralta, S., Gaucher, C., Frei, R., Cirilli, S., Ferreira, V.P., Marquillas, R.A., Barbosa, J.A., Pereira, N.S., Belmino, I.K.C., 2014. High-resolution Hg chemostratigraphy: a contribution to the distinction of chemical fingerprints of the Deccan volcanism and Cretaceous – Paleogene boundary impact event. Palaeogeogr. Palaeoclimatol. Palaeoecol. 414, 98–115.

Sial, A.N., Chen, J., Lacerda, L.D., Frei, R., Tewari, V.C., Pandit, M.K., Gaucher, C., Ferreira, V.P., Cirilli, S., Peralta, S., Korte, C., Barbosa, J.A., Pereira, N.S., 2016. Mercury enrichment and mercury isotopes in cretaceous–Paleogene boundary successions: links to volcanism and palaeoenvironmental impacts. Cretac. Res. 66, 60–81. https://doi.org/10.1016/j.cretres.2016.05.006.

Sial, A.N., Chen, J., Lacerda, L.D., Frei, R., Higgins, J., Tewari, V.C., Gaucher, C., Ferreira, V.P., Cirilli, S., Peralta, S., Korte, C., Barbosa, J.A., Pereira, N.S., Ramos, D.S., 2019a. Chemostratigraphy across the Cretaceous–Paleogene (K/Pg) boundary: testing the impact and volcanism hypotheses. In: Sial, A.N., Gaucher, C., Ramkumar, M., Ferreira, V.P. (Eds.), Chemostratigraphy Across Major Chronological Boundaries. Geophysical Monograph Series 240, John Wiley & Sons, Inc., ISBN: 9781119382485, pp. 223–257.

Sial, A.N., Chen, J., Lacerda, L.D., Korte, C., Spangenberg, J.E., Silva-Tamayo, J.C., Gaucher, C., Ferreira, V.P., Barbosa, J.A., Pereira, N.S., Benigno, A.P., 2019b. Globally enhanced Hg deposition and Hg isotopes in sections straddling the Permian–Triassic boundary: link to volcanism. Palaeogeogr. Palaeoclimatol. Palaeoecol. 540. https://doi.org/10.1016/j.palaeo.2019.109537.

Sial, A.N., Chen, J-B., Korte, C., Pandit, M.K., Lacerda, L.D., Spangenberg, J., Silva-Tamayo, J.C., Gaucher, C., Ferreira, V.P., Barbosa, J., Pereira, N.S., Riedel, P.R. Hg chemostratigraphy and Hg isotopes of the Meishan and Guryul Ravine successions: proxies for volcanism across the Permian–Triassic Boundary. Front. Earth Sci. (in preparation).

Silva-Tamayo, J.C., Lau, K.V., Jost, A.B., Payne, J.L., Wignall, P.B., Newton, R.J., Eisenhauer, A., Depaolo, D.J., Brown, S., Mahler, K., Lehrmann, D.J., Altiner, D., Yu, M., Richoz, S., Paytan, A., 2018. Global perturbation of the marine calcium cycle during the Permian-Triassic transition. Bull. Geol. Soc. Am. 130, 1323–1338.

Singh, B.P., 2012. How deep was the early Himalayan foredeep? J. Asian Earth Sci. 56, 24–32.

Singh, V., Pandita, S.K., Tewari, R., Hengstum, P.J.V., Pillai, S.S.K., Agnihotri, D., Kumar, M.K., Bhat, G.D., 2015. Thecamoebians (testate amoebae) straddling the Permian–Triassic boundary in the Guryul ravine section, India: evolutionary and Palaeoecological implications. PLoS One 10 (8). https://doi.org/10.1371/journal.pone.0135593.

Song, H., Wignal, P.B., Tong, J., Yin, H., 2013a. Two pulses of extinction during the Permian–Triassic crisis. Nat. Geosci. 6, 52–56.

Song, H., Tong, J., Algeo, T.J., Horacek, M., Qiu, H., Song, H., Tian, L., Chen, Z.-Q., 2013b. Large vertical $\delta^{13}C_{DIC}$ gradients in Early Triassic seas of the South China craton: implications for oceanographic changes related to Siberian traps volcanism. Glob. Planet. Change 105, 7–20.

Song, H.Y., Tong, J.N., Tian, L., Song, H.J., Qiu, H.O., Zhu, Y.Y., Algeo, T., 2014. Paleo-redox conditions across the Permian–Triassic boundary in shallow carbonate platform of the Nanpanjiang Basin, South China. Sci. China Earth Sci. 57, 1030–1038. https://doi.org/10.1007/s11430-014-4843-2.

Sun, Y.D., Joachimski, M.M., Wignall, P.B., Yan, C.B., Chen, Y.L., Jiang, H.S., Wang, L., Lai, X.L., 2012. Lethally hot temperatures during the Early Triassic greenhouse. Science 338, 366–370.

Svensen, H, Planke, S, Polozov, A, Schmidbauer, N, Corfu, F, Podladchikov, Y, Jamtveit, B, 2009. Siberian gas venting and the end-Permian environmental crisis. Earth Planet. Sci. Lett. 277, 490–500.

Teichert, C., Kummel, B., Kapoor, H.M., 1970. Mixed Permian–Triassic fauna, Guryul Ravine, Kashmir. Science 167, 174–175.

Tewari, R., Awatar, R., Pandita, S.K., McLoughlin, S., Agnihotri, D., Pillai, S.S.K., Singh, V., Kumar, K., Bhat, G.D., 2015. The Permian–Triassic palynological transition in the Guryul Ravine section, Kashmir, India: implications for Tethyan–Gondwanan correlations. Earth Sci. Rev. 149, 53–66.

Them, T.R., Jagoe, C.H., Caruthers, A.H., Gill, B.C., Grasby, S.E., Gröcke, D.R., Yin, R., Owens, J.D., 2019. Terrestrial sources as the primary delivery mechanism of mercury to the oceans across the Toarcian Oceanic Anoxic Event (Early Jurassic). Earth Planet. Sci. Lett. 507, 62–72.

Thibodeau, A.M., Bergquist, B.A., 2017. Do mercury isotopes record the signature of massive volcanism in marine sedimentary records? Geology 45, 95–96.

Thibodeau, A.M., Ritterbush, K., Yager, J.A., West, A.J., Ibarra, Y., Bottjer, D.J., Berelson, W.M., Bergquist, B.A., Corsetti, F.A., 2016. Mercury anomalies and the timing of biotic recovery following the end-Triassic mass extinction. Nat. Commun. 7, 1–8.

Thomas, B.M., Willink, R.J., Grice, K., Twitchett, R.J., Purcell, R.R., Archbold, N.W., George, A.D., Tye, S., Alexander, R., Foster, C.B., Barber, C.J., 2004. Unique marine Permian–Triassic boundary section from Western Australia. Aust. J. Earth Sci. 51, 423–430.

Wang, C.Y., Wang, Z.H., 1981. Permian conodont biostratigraphy of China. In: Teichert, C., Liu, L., Chen, P.J. (Eds.), Paleontology in China, 1979. Geological Society of America, Special Papers 187, The Geological Society of America, pp. 227–236.

Wang, K., Geldsetzer, H.H.J., Krouse, H.R., 1994. Permian–Triassic extinction: organic $\delta^{13}C$ evidence from British Columbia, Canada. Geology 22, 580–584.

Wang, Y., Sadler, P.M., Shen, S.Z., Erwin, D.H., Zhang, Y.C., Wang, X.D., Wang, W., Crowley, J.L., Henderson, C.M., 2014. Quantifying the process and abruptness of the end-Permian mass extinction. Paleobiology 40, 113–129.

Wang, X., Cawood, P.A., Zhao, H., Zhao, L., Grasby, S.E., Chen, Z.-Q., Wignall, P.B., Lv, Z., Han, C., 2018. Mercury anomalies across the end Permian mass extinction in South China from shallow ad deep water depositional environments. Earth Planet. Sci. Lett. 496, 159–167.

Wardlaw, B.R., Collinson, J.W., 1979. Youngest Permian conodont faunas from the Great Basin and Rocky Mountain regions. Brigham Young University Geology Studies 26, 151–163.

Wignall, P.B., Hallam, A., 1992. Anoxia as a cause of the Permian–Triassic mass extinction: facies evidence from northern Italy and the western United States. Palaeogeogr. Palaeoclimatol. Palaeoecol. 93, 21–46.

Wignall, P.B., Newton, R., 2003. Contrasting deep-water records from the Upper Permian and lower Triassic of South Tibet and British Columbia: evidence for a diachronous mass extinction. Palaios 18, 153–167.

Wignall, P.B., Twitchett, R.J., 1996. Oceanic anoxia and the end Permian mass extinction. Science 272, 1155–1158.

Wignall, P.B., Twitchett, R.J., 2002. Extent, duration, and nature of the Permian–Triassic superanoxic event. In: Koebel, C., MacLeod, K.G. (Eds.), Catastrophic Events and Mass Extinctions: Impacts and Beyond: Boulder, Colorado. Geological Society of America Special Paper, vol. 356. The Geological Society of America, pp. 395–413.

Wignall, P.B., Kozur, H., Hallam, A., 1996. On the timing of palaeoenvironmental changes at the Permo–Triassic (P/Tr) boundary using conodont biostratigraphy. Hist. Biol. 12, 39–62.

Wignall, P.B., Newton, R., Brookfield, M.E., 2005. Pyrite framboid evidence for oxygen poor deposition during the Permian–Triassic crisis in Kashmir. Palaeogeogr. Palaeoclimatol. Palaeoecol. 216, 183–188.

Xie, S., Pancost, R.D., Yin, H., Wang, H., Evershed, R.P., 2005. Two episodes of microbial change coupled with Permo/Triassic faunal mass extinction. Nature 434, 494–497.

Xu, D.Y., Yan, Z., 1993. Carbon isotope and iridium event markers near the Permian/ Triassic boundary in the Meishan section. Palaeogeogr. Palaeoclimatol. Palaeoecol. 104, 171–176.

Yin, H., Huang, S., Zhang, K., Hansen, H.J., Yang, F., Ding, M., Bi, X., 1992. The effects of volcanism on the Permo-Triassic mass extinction in South China. In: Sweet, W.C., Yang, Z., Dickins, J.M., Yin, H. (Eds.), Permo-Triassic Events in the Eastern Tethys, Stratigraphy, Classification, and Relations with the Western Tethys. Cambridge University Press, pp. 146–157.

Yin, H., Zhang, K., Tong, J., Yang, Z., Wu, S., 2001. The global stratotype section and point (GSSP) of the Permian–Triassic boundary. Episodes 24, 102–114.

Yin, H., Feng, Q., Lai, X., Baud, A., Tong, J., 2007. The protracted Permo-Triassic crisis and multi-episode extinction around the Permian–Triassic boundary. Glob. Planet. Change 55, 1–20.

Yoshida, K., Kawamura, T., Suzuki, S., Regmi, A.D., Gyawali, B.R., Shiga, Y., Adachi, Y., Dhital, M.R., 2014. Continental weathering in the Early Triassic in Himalayan Tethys, Central Nepal: implications for abrupt environmental change on the northern margin of Gondwanaland. J. Asian Earth Sci. 79, 288–301.

Youngquist, W., Hawley, R.W., Miller, A.K., 1951. Phosphoria conodonts from southeastern Idaho. J. Paleontol. 25, 356–364.

Zhang, F., Romaniello, S.R., Algeo, T.J., Lau, K.V., Clapham, M.E., Richoz, S., Herrmann, A.D., Smith, H., Horacek, M., Anbar, A.D., 2018. Multiple episodes of extensive marine anoxia linked to global warming and continental weathering following the latest Permian mass extinction. Sci. Adv. 4, e1602921.

Zhao, X., Zheng, D., Xie, G., Jenkyns, H.C., Guan, C., Fang, Y., He, J., Yuan, X., Xue, N., Wang, H., Li, S., Jarzembowski, E.A., Zhang, H., Wang, B., 2020. Recovery of lacustrine ecosystems after the end-Permian mass extinction. Geology 48, 609–613. https://doi.org/10.1130/G47502.1.

Zheng, Q.F., Cao, C.Q., Zhang, M.Y., 2013. Sedimentary features of the Permian–Triassic boundary sequence of the Meishan section in Changxing County, Zhejiang Province. Sci. China Earth Sci. 56, 956–969.

Zhou, L., Kyte, F.T., 1988. The Permian-Triassic boundary event: a geochemical study of three Chinese sections. Earth Planet. Sci. Lett. 90, 411–421.

Zuchuat, V., Sleveland, A.R.N., Twitchett, R.J., Svensen, H.H., Turner, H., Augland, L.E., Jones, M.T., Hammer, Ø, Hauksson, B.T., Hafkidason, H., Midtkandal, I., Planke, S., 2020. A new high-resolution stratigraphic and palaeoenvironmental record spanning the End-Permian Mass Extinction and its aftermath in central Spitsbergen. Svalbard. Palaeogeogr. Palaeoclimatol. Palaeoecol. https://doi.org/10.1016/j.palaeo.2020.109732.

CHAPTER TWELVE

Early Toarcian glacio-eustatic unconformities and chemostratigraphic black holes

Wolfgang Ruebsam[a,*], Nicolas Thibault[b], and Moujahed Al-Husseini[c]

[a]Department of Organic and Isotope Geochemistry, University of Kiel, Kiel, Germany
[b]Department of Geosciences and Natural Resource Managment, University of Copenhagen, Copenhagen, Denmark
[c]Gulf PetroLink-GeoArabia, Manama, Bahrain
[*]Corresponding author: e-mail address: wolfgang.ruebsam@googlemail.com

Contents

Abstract

During Late Pliensbachian and Early Toarcian times (Early Jurassic Epoch, c. 185 to 180 Ma) major environmental instabilities and perturbations affected the evolution of the carbon cycle, climate, sea level and biota. The temporal and causal relationships

ISSN 2468-5178
https://doi.org/10.1016/bs.sats.2020.08.006

between these global-scale phenomena however remain poorly understood. This study starts by refining the accuracy of the ages of Lower Toarcian ammonite zones and subzones, and carbon-isotope excursions defined in the carbon-13 isotope ($\delta^{13}C$) reference curve of Ruebsam and Al-Husseini (2020). Based on an extensive review of stratigraphic data our study identified and calibrated by numerical age ammonite-dated transgressive-regressive (T-R) sequences, sequence boundaries (SB), maximum flooding intervals (MFI) and climate cycles. These calibrations placed in a common temporal framework the $\delta^{13}C$ reference curve, T-R sequences, climate cycles and the Karoo-Ferrar Large Igneous Province (K-F LIP).

The temporal framework revealed four intervals that are either fragmented, or completely missing in numerous Upper Pliensbachian and Lower Toarican $\delta^{13}C$ records (Stratigraphic Black Holes: SBH; Ruebsam and Al-Husseini, 2020). The identified SBHs closely correlated to erosional unconformities or sequence boundaries and global cooling. Unconformities occur in the *spinatum* zone (SBH 1), in the middle *tenuicostatum* zone (SBH 2) and in the middle *serpentinum* zone (SBHs 3 and 4). The older three SBHs further contain global sequence boundaries SBs JPl8, JTo1 and JTo2 of Haq (2018). The sea falls associated with the unconformities have amplitudes of many tens to possibly 75 m and exceed by more than an order of magnitude falls associated with thermo-eustasy or aquifer-eustasy (Davies et al., 2020). Sea level falls are therefore attributed to build-ups of polar ice caps and their ages and those of subsequent highstands were found to be consistent with the predictions of the Orbital Scale of glacio-eustasy (Matthews and Al-Husseini, 2010). We conclude tuning of insolation caused by eccentricity cycles with periods of c. 0.1, 0.405 and c. 2.4 myr drove sea-level fluctuations and also paced changes in the global carbon cycle and climate. Greenhouse gas emissions associated with volcanism in the Karoo-Ferrar Large Igneous Province affected the global carbon cycle and thereby amplified climate and sea level changes.

1. Introduction

Carbon-13 isotope ($\delta^{13}C$) data manifest orbitally tuned cycles, secular trends and carbon-isotope excursions (CIE) that are increasingly used to calibrate the durations of biozones and as global markers that facilitate correlations between sedimentary sections (Berger and Vincent, 1981; Kump and Arthur, 1999; Mercuzot et al., 2019; Salzman and Thomas, 2012; Scholle and Arthur, 1980; Storm et al., 2020). Despite being such a robust tool for global-scale stratigraphy, constraining and understanding the nature of changes in the carbon cycle and their relationship to eustatic fluctuations and glaciations remains challenging (Brenchley et al., 2003; Saltzman et al., 2004; see review in Al-Husseini and Ruebsam, 2020, this Volume). In this case study we investigate the relationships between chemostratigraphic data ($\delta^{18}O$, $\delta^{13}C$), climate and glacio-eustasy for the Late Pliensbachian and Early Toarcian times (Early Jurassic; c. 185 to 180 Ma). This interval is particularly relevant for such an investigation because it is characterized by marked eustatic fluctuations and a major lowstand associated with cooling in the Late Pliensbachian (Brandt, 1986; Dera et al., 2011; Guex et al., 2001; Korte and Hesselbo, 2011; Korte et al., 2015; Ruebsam et al., 2019; Suan et al., 2010), an earliest Toarcian eustatic fall associated with a "cold snap" (Krencker et al., 2019; Ullmann et al., 2020), followed by the global-scale Early Toarcian Toa-CIE, which was accompanied by a major eustatic sea level rise and a rise in global temperatures (e.g., Dera et al., 2011; Korte et al., 2015; Krencker et al., 2019; Ruebsam et al., 2019; Suan et al., 2010; Ullmann et al., 2020).

The present study starts by reviewing key aspects of the Early Toarcian carbon cycle and how Ruebsam and Al-Husseini (2020) recently constructed their $\delta^{13}C$ reference curve in a biostratigraphic framework calibrated in the Astronomical Time Scale (ATS). By comparing $\delta^{13}C$ data from numerous localities with this $\delta^{13}C$ reference curve, they documented four major discontinuities that consistently occur at specific biostratigraphic positions. Because these discontinuities distort time calibrations, Ruebsam and Al-Husseini (2020) referred to them as "Stratigraphic Black Holes" (SBH 1 to SBH 4). In this contribution, we test our ATS-calibrated $\delta^{13}C$ time-scale in the Yorkshire Section of the Cleveland Basin (UK), and build on this previous work by interpreting the four SBHs as unconformities/hiatuses. We consider how these SBHs relate to the global third-order sequence boundaries of Haq (2018) and stratigraphic discontinuities in local basins. We then reconstruct the eustatic and climate cycles of this interval from stratigraphic and proxy data (e.g., $\delta^{18}O$, paleofloral assemblages) according to the ATS and compare it to the predictions of the deterministic Orbital Scale of glacio-eustasy (Al-Husseini and Ruebsam, 2020, this Volume; Matthews and Al-Husseini, 2010). Our orbitally calibrated $\delta^{13}C$ reference curve is used here as a correlative tool for biostratigraphy, glacio-eustasy and sequence stratigraphy.

2. Early Toarcian carbon cycle

2.1 Carbon cycle perturbations

The Early Toarcian $\delta^{13}C$ record consists of a complex pattern of positive and negative CIEs, which document recurring perturbations of the global carbon cycle (Fig. 1). A long-lasting positive CIE started in the latest Pliensbachian and spanned the Early Toarcian (*tenuicostatum* and *serpentinum* ammonite zones, or coeval chronozones) (e.g., Hermoso et al., 2012; Hesselbo et al., 2000; Jenkyns et al., 2002; Storm et al., 2020). Superimposed onto the long-term positive excursion, negative CIEs have

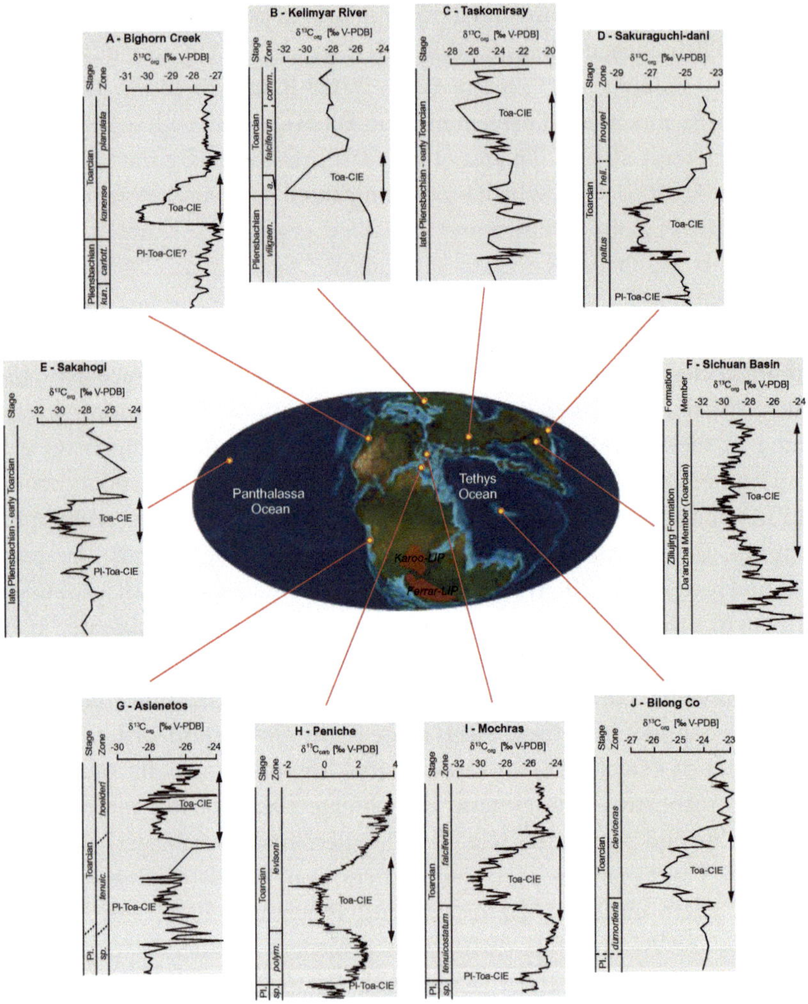

Fig. 1 Global records of Early Toarcian carbon cycle instabilities that are manifested in negative carbon isotope excursions at the Pliensbachian/Toarcian boundary (Pl-Toa-CIE) and in the Early Toarcian (Toa-CIE) (paleogeographic map from Blakey (2016); $\delta^{13}C$ data: (A) Bighorn Creek, Canada (Them et al., 2017a); (B) Kelimyar River, Siberia (Suan et al., 2011); (C) Taskomirsay, Kazakhstan (Schnyder et al., 2016); (D) Sakuraguchi-dani, Japan (Izumi et al., 2018); (E) Sakahogi, Japan (Ikeda et al., 2018); (F) Sichuan Basin, China (Xu et al., 2017); (G) Asienetos, Chile (Fantasia et al., 2018); (H) Peniche, Portugal (Hesselbo et al., 2007); (I) Mochras, UK (Xu et al., 2018a); (J) Bilong Co, Tibet (Fu et al., 2016). See Fig. 2 of Ruebsam and Al-Husseini (2020) for details of ammonite zones in different realms.

been globally documented at the Pliensbachian/Toarcian boundary (Pl-Toa-CIE) (e.g., Bodin et al., 2016; Fantasia et al., 2018; Hesselbo et al., 2007; Littler et al., 2010; Ruebsam et al., 2019; Xu et al., 2018b) and at the *tenuicostatum-serpentinum* zonal transition (Toa-CIE) (e.g., Al-Suwaidi et al., 2016; Bodin et al., 2016; Fantasia et al., 2018; Hesselbo et al., 2000, 2007; Izumi et al., 2018; Suan et al., 2011; Them et al., 2017a; Xu et al., 2017, 2018b). Both CIEs are documented in marine organic matter and carbonate, as well as in land plant-derived organic matter and attest of carbon cycle perturbations in various carbon reservoirs (e.g., Hesselbo et al., 2000, 2007; Kemp et al., 2019; Pienkowski et al., 2016; Ruebsam et al., 2020; Storm et al., 2020; Xu et al., 2017).

The Toa-CIE records a remarkable internal periodicity, expressed by recurring negative $\delta^{13}C$ shifts superimposed onto the negative excursion. This pattern suggests that changes in the carbon cycle were paced by astronomical parameters, namely Milankovitch cycles (Boulila and Hinnov, 2017; Boulila et al., 2014, 2019; Huang and Hesselbo, 2014; Kemp et al., 2005, 2011; Ruebsam et al., 2019). The Milankovitch frequencies preserved in the $\delta^{13}C$ records imply that the periodic release of ^{12}C-enriched carbon from climate-sensitive reservoirs was the main driver of the Early Toarcian carbon cycle perturbations. Such carbon reservoirs comprise marine gas hydrates (Hesselbo et al., 2000; Kemp et al., 2005), terrestrial biomass (Pienkowski et al., 2016), wetlands (Them et al., 2017a) and cryosphere-associated reservoirs (permafrost and permafrost/ice sheet-capped gas reservoirs) (Ruebsam et al., 2019).

2.2 Karoo-Ferrar Large Igneous Province

There is growing consensus that the release of volcanic CO_2 and/or thermogenic methane, related to the emplacement of the Karoo-Ferrar Large Igneous Province (K-F LIP), further impacted the global carbon cycle (Brazier et al., 2015; McElwain et al., 2005; Svensen et al., 2007). The role of the K-F LIP as an important trigger for environmental change is supported by the temporal and causal relationships between multiple phases of volcanic activity and environmental changes (e.g., Pálfy and Smith, 2000; Pálfy et al., 2002; Svensen et al., 2017; Xu et al., 2018a). The initial phase of volcanism in the Karoo Province correlates with the Pliensbachian/Toarcian boundary, while its second phase was coeval with the Toa-CIE. Moreover, the emplacement of the Ferrar Large Igneous Province correlates with the Toa-CIE (Ruebsam and Al-Husseini, 2020; Xu et al., 2018a).

Volcanism however fails in explaining the orbital-paced cyclicity in the carbon-isotope records. Moreover, due to the similar isotopic composition of the volcanic and atmospheric CO_2, huge amounts of carbon would have been required to produce a negative CIE with a magnitude of about -2 to -3‰ in the dissolved inorganic carbon (DIC) and about -4 to -5‰ in the atmosphere (Beerling and Brentnall, 2007; Ruebsam et al., 2020; Suan et al., 2015). This is however not supported by reconstructed changes in atmospheric CO_2 levels from concentrations of about 300–500 ppmv to a range of 1000–1500 ppmv (McElwain et al., 2005; Ruebsam et al., 2020). Nevertheless, emissions of volcanic CO_2 and thermogenic methane may have acted as important trigger mechanisms that led to the insolation-controlled destabilization of climate-sensitive reservoirs (Ruebsam et al., 2019; Them et al., 2017a).

2.3 Early Toarcian $\delta^{13}C$ reference curve

Major differences persist in regards to the age model of the Early Toarcian (e.g., Boulila and Hinnov, 2017; Boulila et al., 2014; Huang, 2018; Huang and Hesselbo, 2014; Ruebsam et al., 2014, 2015; Thibault et al., 2018), which led Ruebsam and Al-Husseini (2020) to propose an Early Toarcian $\delta^{13}C$ reference curve calibrated in the ATS (Figs. 2–4). They used visually evident, secular segments (rising and falling limbs, valleys and plateaus) from high-resolution, presumably biostratigraphically complete $\delta^{13}C$ records (bulk carbonate: $\delta^{13}C_{carb}$, bulk organic matter: $\delta^{13}C_{org}$) to construct the reference curve. Two key sections were used to construct the $\delta^{13}C$ reference curve: the Mochras Core (Wales, UK, Xu et al., 2018b) and the Peniche Section (Portugal) (Fantasia et al., 2019a; Hesselbo et al., 2007) that contains the Global Boundary Stratotype Section and Point (GSSP) for the lower Toarcian boundary (Rocha et al., 2016).

Ruebsam and Al-Husseini (2020) also used carbon-isotope data from sections with only minor stratigraphic gaps, like the records from the Yorkshire Section (Kemp et al., 2005; Littler et al., 2010; Thibault et al., 2018) and Core FR-210-078 from the NE Paris Basin (Ruebsam et al., 2019) (Fig. 2). The Sancerre-Couy Core record from the southern Paris Basin (France) was also included in this selection (Boulila et al., 2014). However, we consider this record as highly fragmented due to the occurrence of marine ravinement combined with an estimated hiatus of up to 750 kyr truncating basal Toarcian (Peti et al., 2017), as well as a condensation and/or multiple putative hiatuses in the Toa-CIE rising limb within the *serpentinum* zone (Thibault et al., 2018).

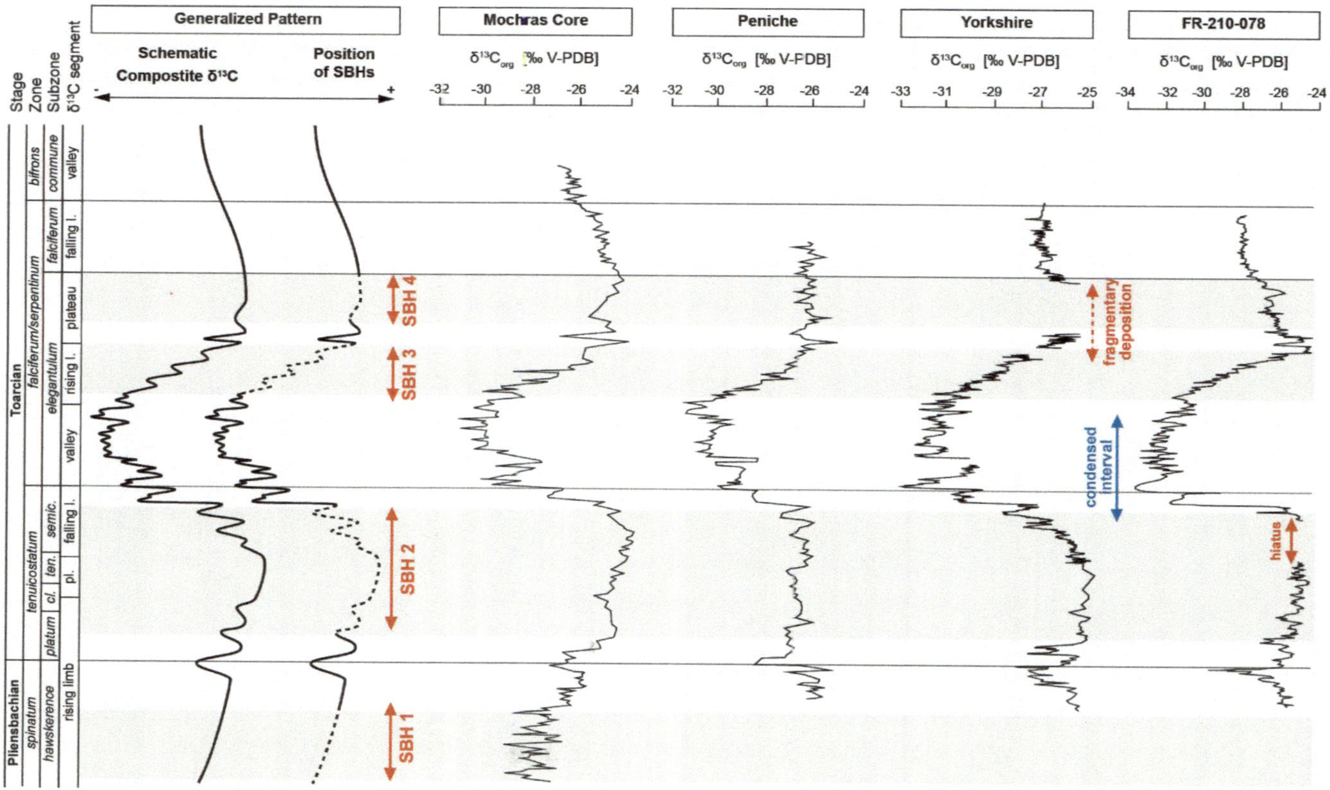

Fig. 2 Schematic $\delta^{13}C$ reference curve and positions of Stratigraphic Black Holes (SBH; Ruebsam and Al-Husseini, 2020), with examples of presumably complete and near-complete sections (pl.: plateau, falling l.: falling limb, rising l.: rising limb). The stratigraphic extent and thus the duration of SBHs can vary. Grey intervals only mark positions in which discontinuities are most frequently documented. Data for the Mochras Core (UK) are from Xu et al. (2018a,b), for Peniche (Portugal) from Fantasia et al. (2019a), for the Yorkshire Section (UK) from Kemp et al. (2005), Littler et al. (2010) and Thibault et al. (2018), and for Core FR-210-078 (Luxembourg) from Ruebsam et al. (2019).

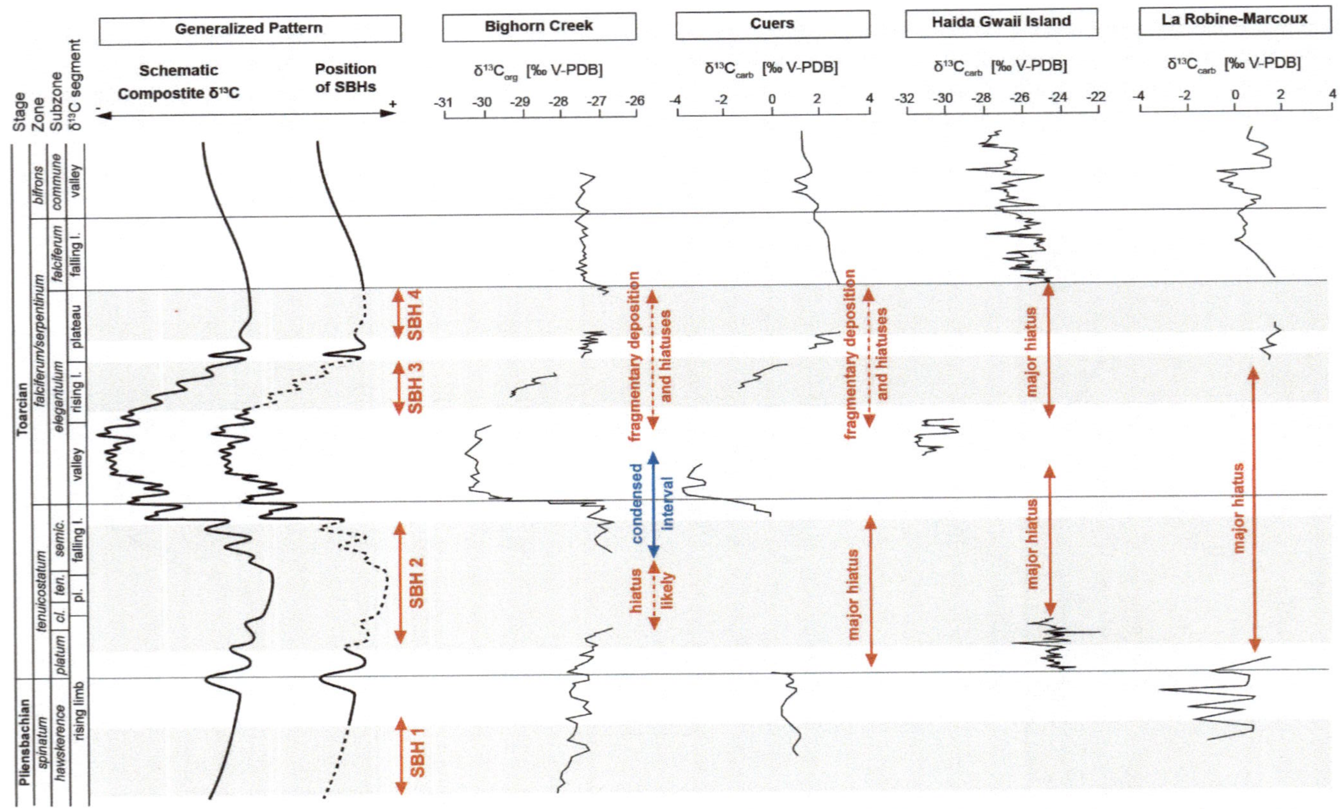

Fig. 3 Schematic $\delta^{13}C$ reference curve and positions of SBHs (Ruebsam and Al-Husseini, 2020), with examples of sections that contain major SBHs. The biostratigraphic extent and thus the duration of SBHs can vary. Grey intervals only mark positions in which discontinuities are most frequently documented. Data for the Bighorn Creek Section (Canada) are from Them et al. (2017a,b), for the Cures Section (France) from Léonide et al. (2012), for Haida Gwaii Island (Canada) from Caruthers et al. (2011) and for the La Robine-Marcoux Section (France) from Léonide et al. (2012).

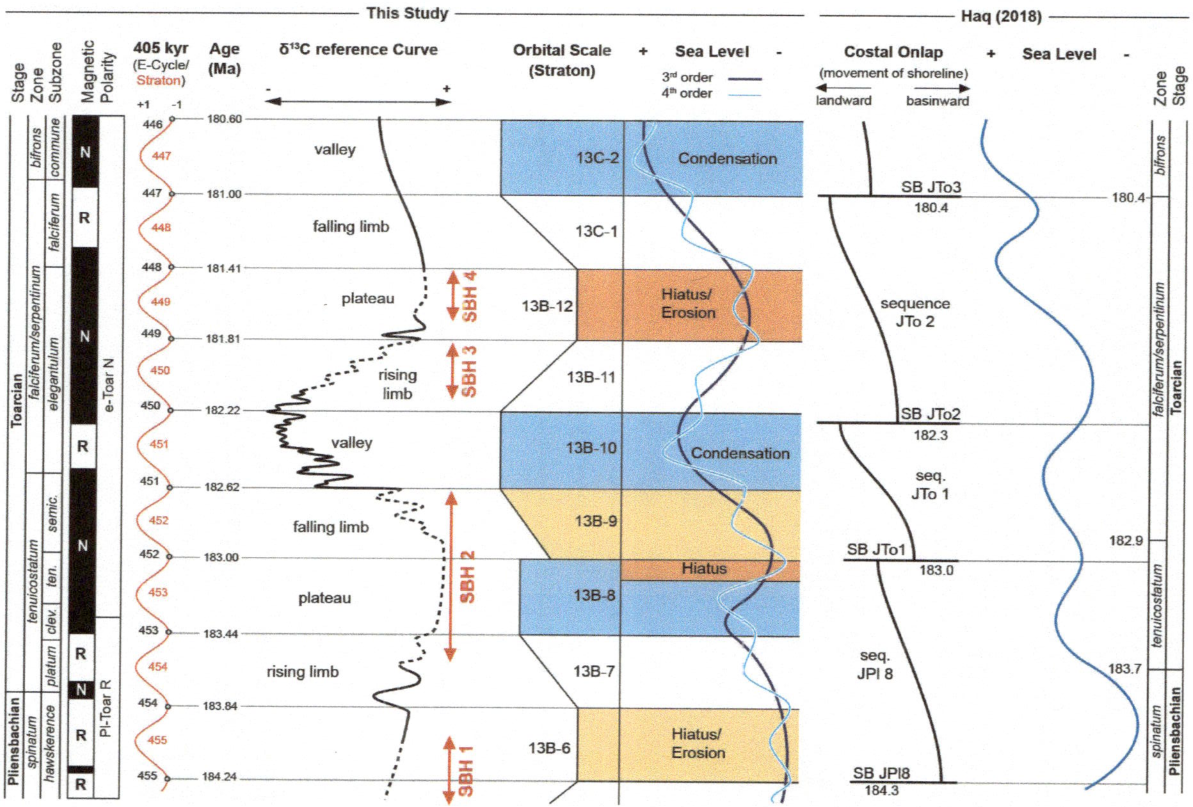

Fig. 4 Early Toarcian $\delta^{13}C$ reference curve with positions of SBH 1 to SBH 4 (Ruebsam and Al-Husseini, 2020) calibrated with 405-kyr eccentricity E-cycles in the Astronomical Time Scale (ATS, revised after Huang, 2018). Pl/Toa boundary assumed at 183.7 Ma is consistent with the 183.7 ± 0.5 Ma estimated by other authors (Huang, 2018; Ikeda and Hori, 2014; Ruhl et al., 2016; Storm et al., 2020; Xu et al., 2018b). Estimated ages of 3rd-order sequence boundaries (SB) based on ATS calibration of ammonite zones (Haq, 2018) are shown in terms of ammonite subzones and magnetic chrons. Model sea level dated with 405-kyr stratons in the Orbital Scale (Matthews and Al-Husseini, 2010) are correlated to 4th-order sea level cycles. Also shown coastal onlap and sea level cycles after Haq (2018). Note timescale used by Haq (2018) is based on Huang (2018) and Timescale Creater (v.2020/01; Ogg and Gradstein, 2020) that slightly differs from the timescale proposed by Ruebsam and Al-Husseini (2020).

The uppermost Pliensbachian *spinatum* zone and the lower Toarcian *tenuicostatum* zone (*paltum* and *clevelandicum* subzones) record a rising limb that marks the initial phase of the Early Toarcian positive CIE. A plateau was identified in the middle *tenuicostatum* zone. Transient negative carbon-isotope shifts occur superimposed onto the rising limb and mark the Pl-Toa-CIE. In the records from the Yorkshire Section (Littler et al., 2010) and Core FR-210-078 (Ruebsam et al., 2019) three negative shifts can be distinguished. A falling limb, starting in the upper *tenuicostatum* zone (*semicelatum* subzone), marks the onset of the Toa-CIE. The core of the Toa-CIE is marked by a $\delta^{13}C$ valley followed by a rising limb together spanning the lower part of the *serpentinum* zone (lower *elegantulum* subzone). The middle *serpentinum* zone (upper *elegantulum* subzone) records another $\delta^{13}C$ plateau, followed by a falling limb in the upper *serpentinum* zone (*falciferum* subzone) (Fig. 2). The falling limb, valley and rising limb of the Toa-CIE are recurring features of high-resolution reference sites that have also been described respectively as phases A, B and C by Thibault et al. (2018). In contrast, Boulila and Hinnov (2017) described a simpler stratigraphic model consisting only of descending and rising phases, which may lead to confusion for global-scale correlations.

Some of these three segments record a significant internal structure, expressed by transient $\delta^{13}C$ cycles. These cycles are most marked in the Toa-CIE falling limb and have been assigned to astronomical forcing (Kemp et al., 2005). However, there is ongoing debate as to whether they reflect short (100 kyr) eccentricity or obliquity (35–37 kyr) cycles (Boulila and Hinnov, 2017; Boulila et al., 2014, 2019; Huang and Hesselbo, 2014; Ruebsam et al., 2014, 2015, 2019). In Section 3.4 of the present study this debate is discussed in greater detail.

2.4 Stratigraphic black holes

Comparisons of $\delta^{13}C$ curves from numerous localities worldwide with the $\delta^{13}C$ reference curve reveal the presence of four major gaps that occur at specific biostratigraphic positions. They attest to the fragmentary nature of Late Pliensbachian and Early Toarcian sedimentary archives and are referred to "Stratigraphic Black Holes" (SBH 1 to SBH 4, Ruebsam and Al-Husseini, 2020).

The oldest discontinuity (SBH 1) in the study interval is a major unconformity in the Upper Pliensbachian *spinatum* zone. In the biostratigraphic record this major discontinuity is expressed by the partial or near-complete absence of Upper Pliensbachian ammonite subzones (*hawskerense* subzone) (e.g., Ansorge, 2007; Prauss, 1996). SBH 1 is however not clearly expressed in $\delta^{13}C$ records, mainly because high-resolution Late Pliensbachian $\delta^{13}C$ records are scarce and trends are less diagnostic, or not well defined (De Lena et al., 2019; Mercuzot et al., 2019; Peti et al., 2017; Storm et al., 2020). Most commonly SBH 1 occurs in the early rising limb and may thus impact its slope (see Fig. 3 in Storm et al., 2020).

SBH 2 occurs in the middle of the *tenuicostatum* zone, which in many regions is poorly represented or missing (Ansorge, 2007; Caswell and Coe, 2012; Fantasia et al., 2019b; Galbrun et al., 1988; Hermoso et al., 2014; Krencker et al., 2019; Léonide et al., 2012; Montero-Serrano et al., 2015; Pittet et al., 2014; Prauss, 1996; Ruebsam et al., 2019; Suan et al., 2011). In the $\delta^{13}C$ record SBH 2 is expressed by the absence of the plateau in the *tenuicostatum* zone and/or by the absence of $\delta^{13}C$ cycles at the Pliensbachian/Toarcian boundary marking the Pl-Toa-CIE. In some sections, SBH 2 is followed by a condensed interval that coincides with the Toa-CIE falling limb. The onset of the Toa-CIE is very sharp and abrupt, as expressed in the $\delta^{13}C$ records from the Cuers Section in France (Léonide et al., 2012), the Rietheim Core in Germany (Montero-Serrano et al., 2015), the Bighorn Creek Section in Canada (Them et al., 2017a), and from Core FR-210-078 from Luxembourg (Ruebsam et al., 2019) (Figs. 2 and 3).

SBH 3 occurs in the rising limb (recovery phase) of the Toa-CIE in middle *serpentinum* zone. This chemostratigraphic discontinuity has been documented from the Lusitanian Basin (Portugal), where it corresponds to the "D4" discontinuity (Pittet et al., 2014). SBH 3 is also evident in France's the Cuers Section (Léonide et al., 2012) and the Bighorn Creek Section in Canada (Them et al., 2017a), in which the Toa-CIE rising limb shows a remarkable stepped pattern. The youngest discontinuity in the studied interval (SBH 4) occurs in the upper *elegantulum* subzone (middle *serpentinum* zone). SBH 4 is not well recognized by biostratigraphy because it is an intra-zone discontinuity. It is however expressed in the $\delta^{13}C$ record and spans the post-Toa-CIE plateau (Ruebsam and Al-Husseini, 2020).

In some localities SBHs converge to form combined SBHs of variable durations. In sections, like for example Cueres (Léonide et al., 2012), or Gipf and Riniken (Switzerland) (Fantasia et al., 2019b), the Pl/Toa boundary and thus the Pl-Toa-CIE disappear in a combined SBH 1+2 (Fig. 3). In these cases, strata recording the Toa-CIE directly overlie Upper Pliensbachian strata. In localities where SBH 3 and 4 converge into a single discontinuity the entire recovery phase of the Toa-CIE (rising limb) is missing. Moreover, a box-like shape of the Toa-CIE, as documented from the East Midland Shelf (UK) (Caswell and Coe, 2012) and the Haida Gwaii

Island (Canada) (Caruthers et al., 2011), indicates that in addition to a combined SBH 1+2, SBH 3 and 4 converge into a single SBH that spans the middle *serpentinum* zone (Fig. 3). The most extreme SBH occurs in La Robine-Marcoux Section where sediments of the *falciferum* subzone (upper *serpentinum* zone) directly overlie Upper Pliensbachian sediments (Léonide et al., 2012). Accordingly, the uppermost *spinatum*, *tenuicostatum* and the lower and middle *serpentinum* zones are missing in a combined SBH 1-4 (Ruebsam and Al-Husseini, 2020) (Fig. 3).

3. Late Pliensbachian to early Middle Toarcian timescale and eustasy

3.1 Time scales

In the Astronomical Time Scale (ATS) a sine wave with a period of 405 kyr, oscillating between +1 and −1, schematically represents the long-eccentricity orbital signal (e.g., Hinnov, 2018; Huang, 2018; Laskar et al., 2011, 2004; Fig. 4). The 405-kyr time units of the ATS are numbered as integers between the maxima of the sine wave and abbreviated by "E." In Fig. 4 we show the ages of the sine wave at its minima (i.e., in mid E-cycles at -1) so as to calibrate when the Earth was colder and ice build-ups were more likely (Matthews and Frohlich, 2002). The eccentricity minima may therefore correlate to those of glacio-eustatic lowstands and associated maximum regression surfaces (MRS), third- and fourth-order sequence boundaries (SB) and unconformities. Ruebsam and Al-Husseini (2020) calibrated the $\delta^{13}C$ reference curve and SBHs with ammonite zones and subzones, magnetic chrons and E-cycles 455 to 446 in the ATS. Based on the analyses discussed in Section 3.4 the calibration of their $\delta^{13}C$ reference curve has been refined by approximately 100 kyr in Fig. 4.

The cycle between two consecutive eccentricity minima will be expressed as a c. 400-kyr stratigraphic package in the sedimentary record and generally referred to as a fourth-order sequence. To emphasize the astronomical origin of these glacio-eustatic sequences Matthews and Al-Husseini (2010) named them "stratons" and also numbered them as consecutive integers (shown in red in Fig. 4). E-cycles and stratons track the same sine wave, and therefore their estimates for the ages of stratigraphic units are identical; for example, the minimum of the sine wave in E-Cycle 450 in the ATS occurs at its mid-point and corresponds to the same minimum at the start of Straton 450 and both occur at 182.22 Ma.

3.2 Global eustasy

Haq (2018) depicted the Late Pliensbachian and Early Toarcian global sea level as a long-term rising trend and three third-order cycles, and designated their corresponding transgressive-regressive (T-R) sequences as JPl8, JTo1 and JTo2 by adding letter "J" to the nomenclature of Snedden and Liu (2010). Haq (2018) calibrated the four sequence boundaries (SB) that bound the T-R sequences according to biozones following Hardenbol et al. (1998) and the geological time scale of Ogg et al. (2016) (Fig. 4 and Table 1), and the same ages as in the ATS of Huang (2018). Haq (2018) also estimated the amplitude of sea-level falls associated with SB JTo1 and SB JTo2 as medium (25–75 m) and for SB JPl8 as major (>75 m). The supporting stratigraphic evidence for these sequences and eustatic cycles are not documented in the public domain, but nonetheless this framework is widely referred to in the literature. We will therefore review the supporting evidence for this framework and, where appropriate, revise or refine its calibration before we compare it to the $\delta^{13}C$ reference curve. Unlike Haq (2018) we will also introduce within each sequence its apparent fourth-order architecture.

Three types of global-scale temperature-driven cyclicity have been considered as possible causes of third- and fourth-order sea level fluctuations (National Research Council, NRC, 1990, and references therein): (1) thermal expansion and contraction of the volume of water in the oceans (thermo-eustasy), (2) transfer of ocean water to-and-from aquifers and other land-based reservoirs (e.g. lakes and rivers) during humid-arid cycles; and (3) transfer of ocean water to-and-from land-based ice sheets (glacio-eustasy). Davies et al. (2020) determined these types of sea level cycles are synchronous and estimated their relative contributions to sea level amplitudes. In regards to thermo-eustasy Davies et al. (2020, citing Sundquist, 1990) estimated in the

Table 1 Sequence boundaries (Haq, 2018).

Sequence boundary	Substage	Ammonite zone	Age (Ma in ATS)	Eustatic fall
SB JTo3	Lower-Middle Toarcian	*Serpentinum-bifrons* transition	180.4	Medium
SB JTo2	Lower Toarcian	*Serpentinum*	182.3	Medium
SB JTo1	Lower Toarcian	*Tenuicostatum*	183.0	Medium
SB JPl8	Upper Pliensbachian	*Spinatum*	184.3	Major

Medium: 25–75 m; major: >75 m.

extreme case of a c. 10°C rise of all the water in the oceans the eustatic rise would be c. 6 m. Therefore, thermo-eustatic fluctuations in even such an extreme and unrealistic scenario are much too small to cause eustatic cycles of many 10s of meters.

Davies et al. (2020) modeled the contribution of aquifer-eustasy to Cretaceous sea level fluctuations and specifically for the maximum sea-level falls of c. 50 m in the Maastrichtian, c. 30 m in the Turonian and c. 65 m in the Valanginian (Ray et al., 2019). In the most likely scenario aquifer-eustasy contributed falls that are in phase with temperature of 0.21 m in the Maastrichtian, 0.11 m in the Turonian and 0.10 m in the Valanginian, values that appear negligible compared to those attributed to glacio-eustasy. The results of this modeling approach tend to invalidate previous claims that aquifer-eustasy could be the driver of Cretaceous third- and fourth-order eustatic oscillations with an out-of-phase relationship to temperature (Sames et al., 2016; Wendler and Wendler, 2016; Wendler et al., 2015). Together with the discovery of new evidence that strongly support glacio-eustasy and the presence of continental ice in the Mesozoic (Alley et al., 2019; Krencker et al., 2019), the recent model of Davies et al. (2020) comes as a strong support for nearly continuous glacio-eustatic mechanisms throughout the Mesozoic. The 40 to 70 m amplitudes of the Late Pliensbachian and Early Toarcian sea level falls estimated by Krencker et al. (2019) and Haq (2018, Table 1) are of the same magnitude and duration as those in the Cretaceous and therefore, we see no plausible reason to assume that they would be anything else than glacio-eustatic in nature.

3.3 Orbital scale of glacio-eustasy

A second framework for relating the $\delta^{13}C$ reference curve to global eustasy is provided by the deterministic Orbital Scale of glacio-eustasy (Al-Husseini and Ruebsam, 2020, this Volume; Matthews and Al-Husseini, 2010). The Orbital Scale is based on the model of Matthews and Frohlich (2002) from which they concluded that so long as the Earth has a continental ice volume, orbital-forcing of glacio-eustasy will impose a 400-kyr periodicity upon fourth-order T-R sequences, and in turn, these sequences will build third-order T-R sequences (mainly 2.0 and 2.8 Myr).

Matthews and Al-Husseini (2010) used the results of the model to formulate the Orbital Scale entirely with stratons. In this scale the latest Pliensbachian to early Middle Toarcian between 184.24 and 180.60 Ma corresponds by numerical age to Stratons 455 to 447 in Orbiton 13 (Fig. 4).

These nine stratons are designated 13B-6 to 13B-12 in Dozon 13B and 13C-1 and 13C-2 in Dozon 13C. The Orbital Scale predicts glacio-eustatic lowstands will occur in Stratons 13B-6, 13B-9 and 13B-12, and highstands in 13B-8, 13B-10 and 13C-2 (see Al-Husseini and Ruebsam, 2020, this Volume). Sequence Boundary SB 13C separates the two dozons at base Straton 448 (base 13C-1 at 181.406 Ma) and is predicted to mark the initial eustatic rise following a major glacio-eustatic fall in Straton 13B-12 (Straton 449).

3.4 Testing the straton age-model on a reference section

Our straton age-model is directly in line with the tuning option of Huang and Hesselbo (2014) on the Peniche record who interpreted the well-expressed steps of the Toa-CIE falling limb as short-eccentricity cycles, contrasting with the interpretation of Boulila and Hinnov (2017) who reinterpreted the expression of these steps as 35-kyr cycles. This ongoing debate on the duration of the Early Toarcian is a fundamental aspect to solve as the distinct durations for the Toa-CIE have strong implications in the understanding of climatic controls and biogeochemical processes that took place across this transient environmental perturbation in Earth history. The 35-kyr cycle interpretation implies the maximum duration of the Toa-CIE of 500 kyr and a strong imprint of the obliquity on the carbon cycle. The 100-kyr cycle interpretation implies a total duration of the Toa-CIE of 1200 to 1300 kyr and a strong imprint of the short-eccentricity on the carbon cycle. For instance, the short duration option implies a much-enhanced chemical weathering susceptible to fundamentally change the slope of Sr-isotopic trends at a rate that is at odds with the long residence time of this element in the ocean (McArthur et al., 2016).

We tested our straton age-model across the Toa-CIE interval of Yorkshire as there seems to be a consensus in the literature that this record does not bear major discontinuities across the Toa-CIE and because this interval bears numerous excellent high-resolution datasets in various proxies (Kemp et al., 2011; Thibault et al., 2018) such as total organic carbon (TOC), total sulfur (TS), calcium carbonate content (%$CaCO_3$), bulk organic carbon isotopes ($\delta^{13}C_{org}$) and lithogenous proxies: aluminium (Al), zircon (Zr), silica (Si) and rubidium (Rb). The best proxy to tune the Yorkshire record across the Toa-CIE appears to be TS, which reveals high-amplitude cycles but also numerous significant frequency peaks (Fig. 5) besides the dominant 75 cm cycle highlighted in Kemp et al. (2005, 2011).

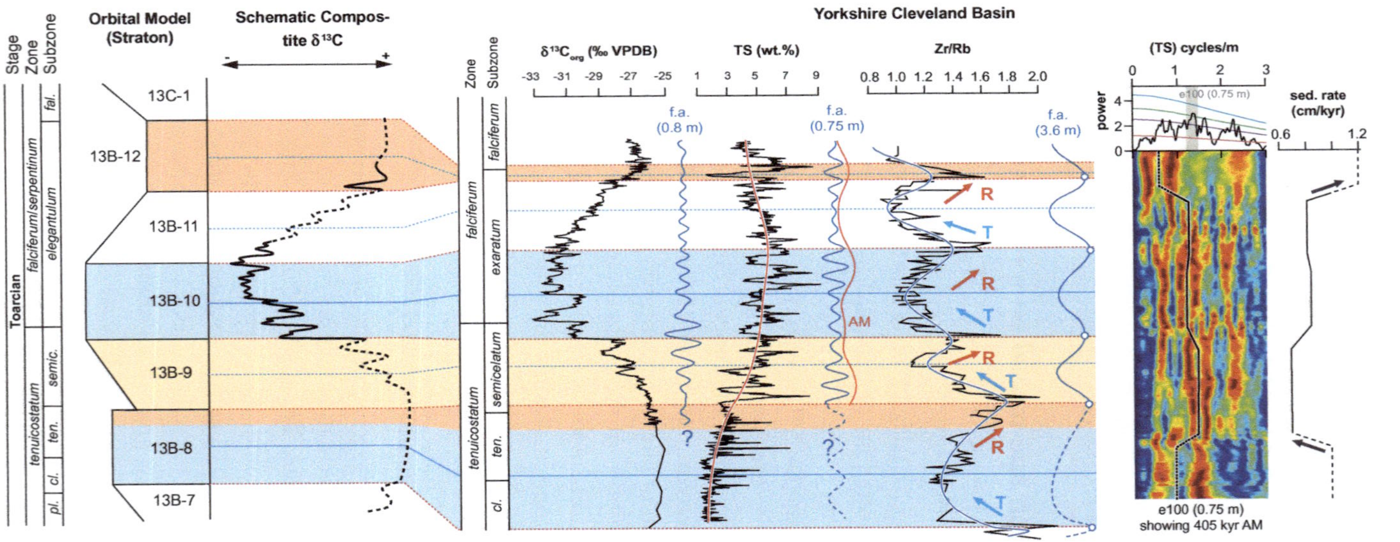

Fig. 5 Model-data comparason for the Early Toarcian interval of the Yorkshire Section. Dominant frequencies present in the geochemical records (here $\delta^{13}C$, TS: total sulfur, Zr/Rb) were extracted and correlated to the orbital model, using $\delta^{13}C_{org}$ trends as chemostratigraphic correlation tools. TS and $\delta^{13}C_{org}$ data show dominant frequencies with periods of 0.8 m and 0.75 m, respectively. Zr/Rb ratios are dominated by cycles with a period of 3.6 m. Based on the age model elaborated for the Early Toarcian (Ruebsam and Al-Husseini, 2020) and the frequency ratios, 0.75 m and 0.8 m cycles were attributed to the short eccentricity (100 kyr), while the 3.6 m cycles in the Zr/Rb record long eccentricity (405 kyr) periods (f.a., filter output; AM, amplitude modulation). Orbital cycles manifested in the Zr/Rb record precisely trace stratons. Sedimentation rates were calculated on the basis of the 100 kyr e-cycle from frequency stabilization. The evolutionary spectrogram indicates variable sedimentation rates, with significantly higher rates prior to and after the Toa-CIE interval (also see Huang and Hesselbo, 2014).

Our approach has thus been to re-examine the expression of TS in an evolutive Fast Fourier Transform (evo-FFT) which reveals potential amplitude modulations of the c. 75 cm cycle with a ratio of 1:4 suggesting modulation of the 100 kyr cycle by the 405 kyr (Fig. 6). Filtering of the c. 75 cm component from detrended TS and demodulation of the obtained filter output reveals indeed c. 5 amplitude modulation cycles which is in line with the 5.5 cycles extracted from a 360 cm filter output of Zr/Rb data (Thibault et al., 2018). The evo-FFT also suggests changes in sedimentation rate across the investigated interval from 0.63 cm/kyr across the Toa-CIE falling limb to between 1.2 and 1.4 cm/kyr across the Toa-CIE valley and the subsequent rising limb (Figs. 6 and 7). Our tuning was performed by picking the frequency on the evo-FFT in the center of amplitude modulation features of the main c. 75 cm cycle and calculating the corresponding sedimentation rate based on a 100 kyr short-eccentricity hypothesis.

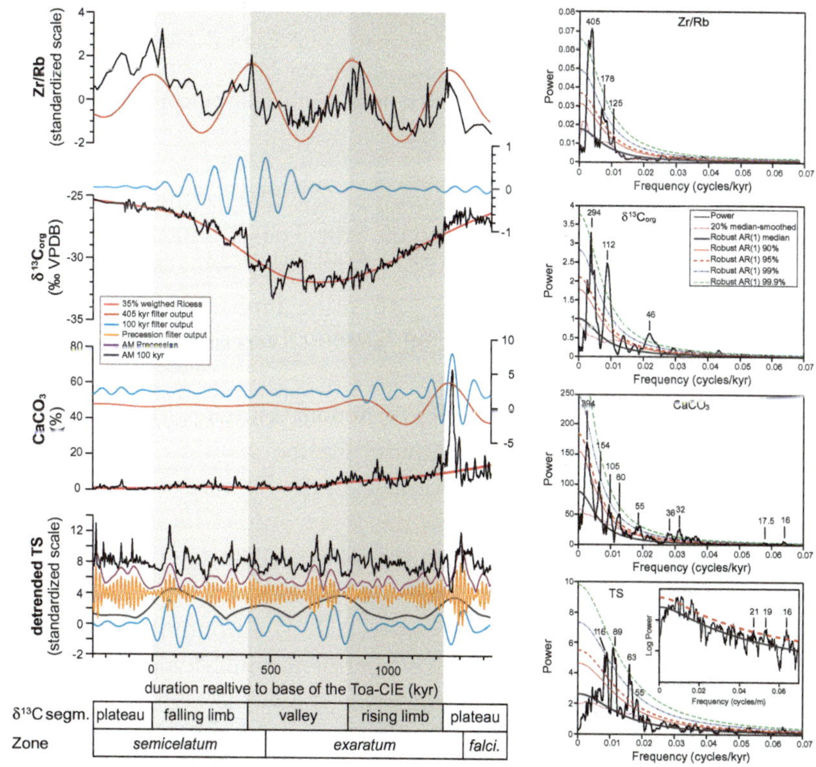

Fig. 6 Geochemical data from the Yorkshire Section shown in the time domain. Frequency stabilization tuning of the 100 kyr e-cycle (sed. rates in the range of 0.69 to 0.83 cm/kyr).

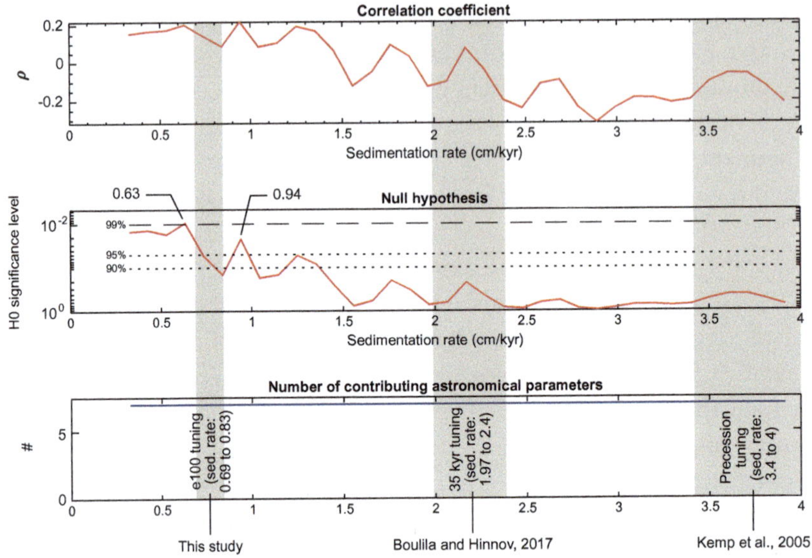

Fig. 7 COCO procedure on detrended TS for the Toa-CIE interval (5–21 m) of Yorkshire Section. Astronomical targets: La04 solution @ 182 Ma 405.0, 125.0, 95.0, 35.0, 21.5, 20.5, 17.7 kyr step: 0.1 cm/kyr (36 tested sed. rates) with Pearson correlation method, 2000 MC simulations.

Overall higher sedimentations rates must be assumed for the pre-Toa-CIE interval (lower *tenuicostatum* zone) and the post-Toa-CIE interval (*falciferum* subzone) (Huang and Hesselbo, 2014). Moreover, a major hiatus (SBH 4) occur directly at the top the Toa-CIE rising limb, which complicates cyclostratigraphic interpretaions of these intervals (Ruebsam and Al-Husseini, 2020).

We also calculated corresponding sedimentation rates for durations of 35 kyr (obliquity) and 21 kyr (precession) for the main c. 75 cm cyclicity. A COCO procedure was also performed on the TS record for a range of sedimentation rates that span across the three distinct interpretations for the c. 75 cm cycles (100 kyr, 35 kyr and 21 kyr). The COCO procedure rejects the 35 kyr and 21 kyr hypothesis that fall in a range of sedimentation rates where the correlation coefficient to La2004 astronomical solution is close to 0.0 and the H0 significance level of the null hypothesis is high (Fig. 7). In contrast, the COCO procedure shows the highest correlation coefficients ranging from 0.1 to 0.2 and H0 significance levels of the null hypothesis from close to 90% to above 95% across our calculated ranges of sedimentation rates (0.69–0.83 cm/kyr, Fig. 7). The highest significance level (99%) is obtained for a fixed sedimentation rate of 0.63 cm/kyr, which is rather close to our calculated minimum.

Tuning of the Yorkshire record by frequency stabilization of the c. 75 cm cycle interpreted as 100 kyr implies power spectra of various proxies showing frequencies that fall at expected orbital periods (Fig. 6). For instance, precession cycles ranging from 16 to 21 kyr are present in TS and % $CaCO_3$ data (Fig. 6), potential obliquity components at 32 and 36 kyr are present in %$CaCO_3$ data. A 55 kyr component, probably related to obliquity as well is present in both TS and %$CaCO_3$ data. Multiple short-eccentricity components are present in all datasets with a 112 kyr peak for $\delta^{13}C_{org}$, 116 and 89 kyr for TS, 105 and 80 for %$CaCO_3$ data and 125 kyr for Zr/Rb. The long-eccentricity component is expressed exactly at 405 kyr in Zr/Rb data (while our tuning was performed by frequency stabilization of the 100 kyr) and a 394 kyr peak is revealed in %$CaCO_3$ data. Other components are present in distinct proxies which cannot always be fully explained here such as the presence of a 294 kyr peak in $\delta^{13}C_{org}$, 63 kyr peak in TS, 154 kyr peak in %$CaCO_3$ and 178 kyr peak in Zr/Rb data. Unassigned frequency peaks are however explained by variations in the sedimention rate, related to environmental instabilities (e.g., climate, sea level).

Another argument in favor of our interpretation is a good match between amplitude modulations of the precession component and the filter output of the 100 kyr cycle both extracted from detrended TS data (Fig. 6). In the investigated 5 to 21 m tuned interval of the Yorkshire Section, a total of 21 short-eccentricity cycles are recorded in %$CaCO_3$ data while c. 23 short-eccentricity cycles are recorded from our 100 kyr filter output *versusc.* 25 to 27 cycles in amplitude modulations of the precession (Fig. 6). Two 405 kyr cycles extracted from %$CaCO_3$ data match the two uppermost 405 kyr cycles of the Zr/Rb record and the total investigated interval spans a little more than five complete 405 kyr cycles and a total calculated duration of c. 2,100 kyr obtained from our tuning. Hence, our orbital interpretation shows redundancy for various proxies, matching La2004 but also demonstrate a likelihood that is much higher than that of the 35 kyr hypothesis.

4. Late Pliensbachian to early Middle Toarcian sea level change: Evidence, timing and relation to $\delta^{13}C$ chemostratigraphy

In the following discussion we evaluate the interpretations of Sequences JPl8, JTo1 and JTo2 of Haq (2018) (Fig. 4 and Table 1) based on published sedimentary and geochemical evidence. Our evaluation includes the interpretation of clastic grain-size proxies expressed as trends in the ratios of quartz to quartz-plus-clay (Qz/(Qz + Clay)), Si/Al and Zr/Rb.

These proxy data are measured in four localities: Sancerre Core (Paris Basin, France, Hermoso et al., 2013, Fig. 8), Peniche Section (Lusitanian Basin, Portugal, Hesselbo et al., 2007; Pittet et al., 2014; Fantasia et al., 2019a, Fig. 9), Yorkshire Section (Cleveland Basin, UK, Kemp et al., 2005; Littler et al., 2010; Thibault et al., 2018, Fig. 10) and the Mochras Core (Cardigan Bay Basin, UK, Storm et al., 2020; Xu et al., 2018b, Fig. 11). Additionally, for the South German Basin (Dotternhausen Quarry, Frimmel et al., 2003; Röhl et al., 2001, Fig. 12), we evaluate trends in the Carbon Preference Índex (CPI) in terms of sea level evolution. The CPI assesses relative abundances of odd- *versus* even-number long-chain *n*-alkanes that vary as a function of organic matter source (aquatic *versus* land plant), organic matter preservation and thermal maturity of the organic matter (see Peters et al., 2005 and references therein).

In general, local maxima or increasing trends in the proxies (clastic and organic matter) reflect coastal progradation during sea level lowstands and erosional events. Conversely, local minima and decreasing trends in the proxies can be related to sea level rises and highstands. For an interpretation of clastic grain size proxies in terms of sea level change and retrograding/prograding coastlines we refer to the works by Williams et al. (2001), Coe (2003), Govin et al. (2012), Hermoso et al. (2013) and Thibault et al. (2018). For an interpretation of the CPI as a sea level indicator the reader is referred to Frimmel et al. (2003) and Zhang et al. (2010).

In Fig. 13 we show the biostratigraphic positions of minima in the clastic proxies and the CPI as grey and brownish diamonds for the Sancerre Core, the Yorkshire Section and for sections from the South German Basin, respectively. Also Fig. 13 shows the positions of condensed intervals and plaster build-ups of remains of pelagic organisms (e.g., fish and ammonite plasters) in the Yorkshire Section and the South German Basins (Caswell and Coe, 2014; Röhl and Schmid-Röhl, 2005; Suan et al., 2015). These intervals, especially when they coincide with minima in the clastic proxies and the CPI, highlight regional maximum flooding intervals (MFI) and support the interpretation of specific stratons. In the following discussion we will frequently refer to the sections in Figs. 4 and 8–12 without repeating the names of basins and countries.

4.1 T-R Sequence JPl8

4.1.1 Lower boundary SB JPl8

Haq (2018) interpreted a major sea level fall in the Late Pliensbachian *spinatum* zone and positioned SB JPl8 near the base of the lowstand at

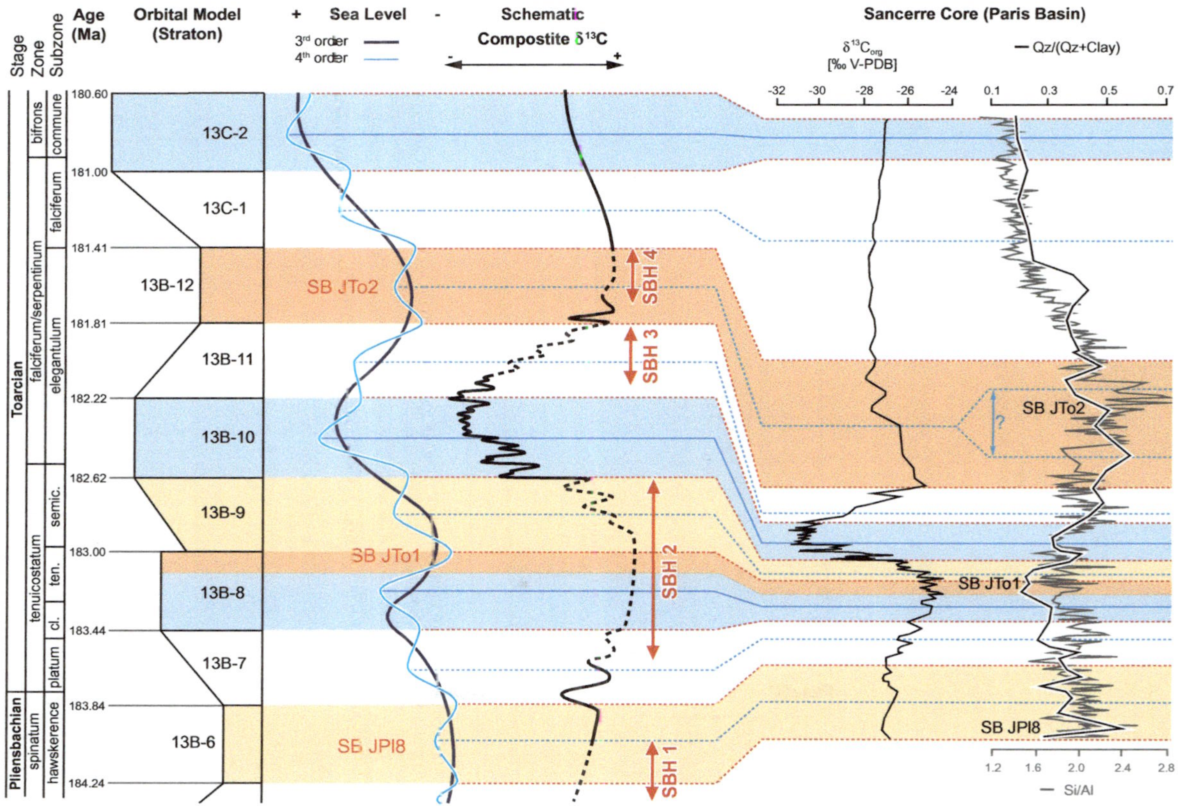

Fig. 8 Late Pliensbachian to early Middle Toarcian stratons, model sea level and the schematic $\delta^{13}C$ curve compared with trends in $\delta^{13}C_{org}$ and clastic proxies in the Sancerre Core (Hermoso et al., 2013; Thibault et al., 2018).

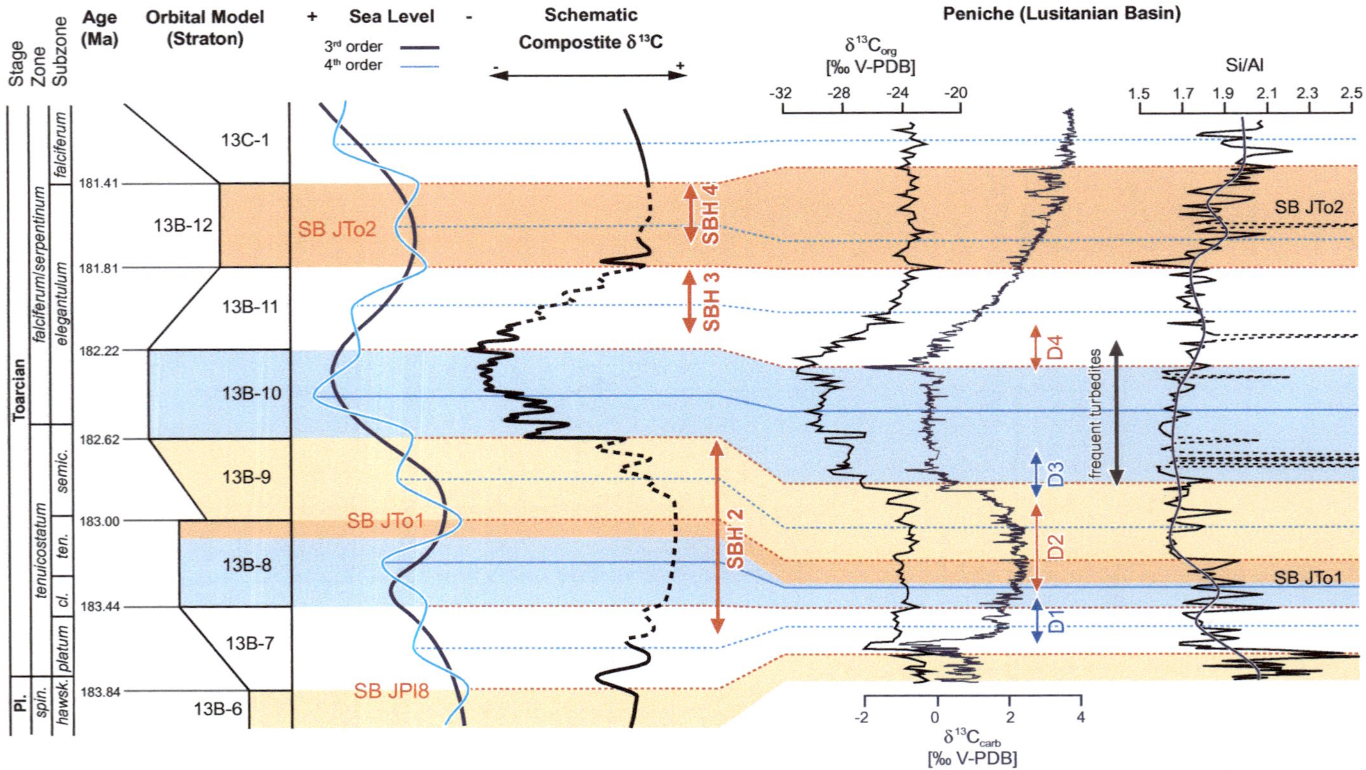

Fig. 9 Late Pliensbachian to Early Toarcian stratons, model sea level and the schematic $\delta^{13}C$ curve compared with trends in $\delta^{13}C_{org}$ and clastic proxies in the Peniche Section (Fantasia et al., 2019a; Hesselbo et al., 2007). Positions and extents of discontinuities D1 to D4 are from Pittet et al. (2014).

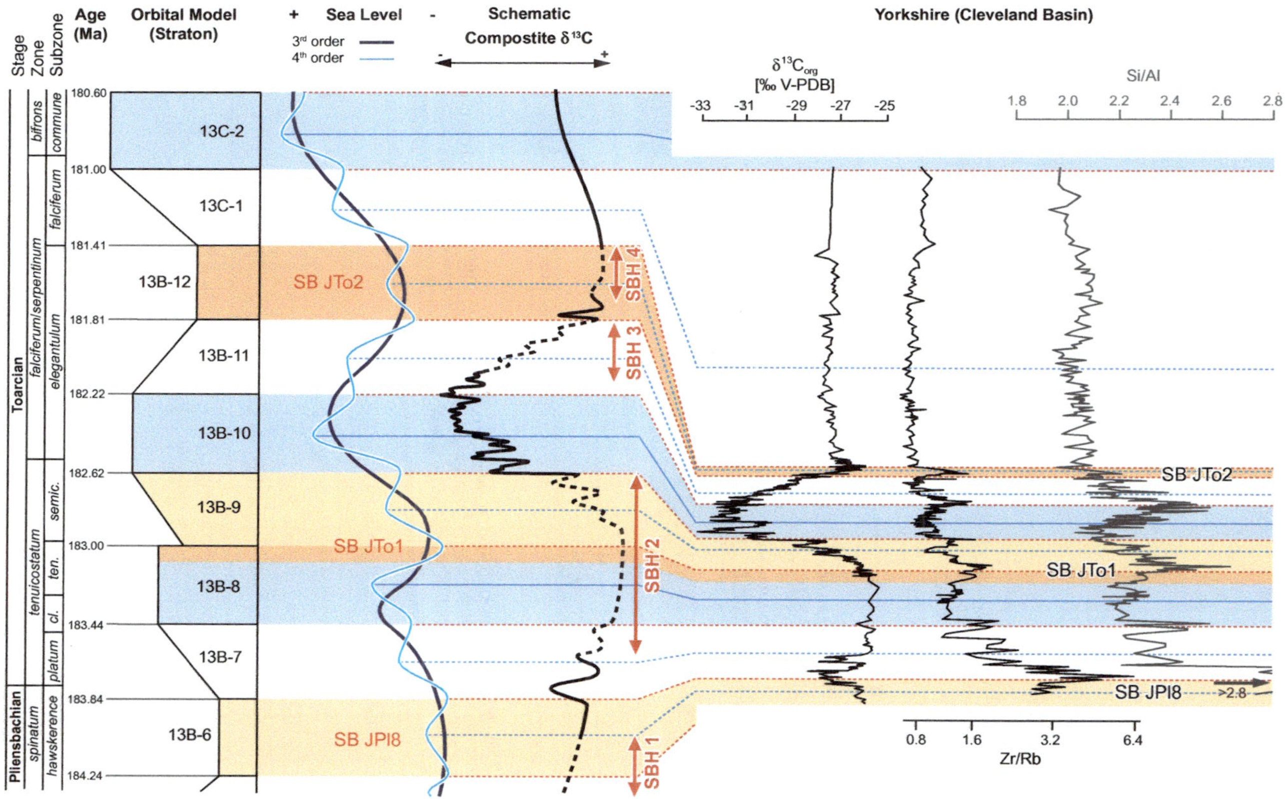

Fig. 10 Late Pliensbachian to early Middle Toarcian stratons, model sea level and the schematic $\delta^{13}C$ curve evolution compared with trends in $\delta^{13}C_{org}$ and clastic proxies in the Yorkshire Section (Kemp et al., 2005; Littler et al., 2010; Thibault et al., 2018).

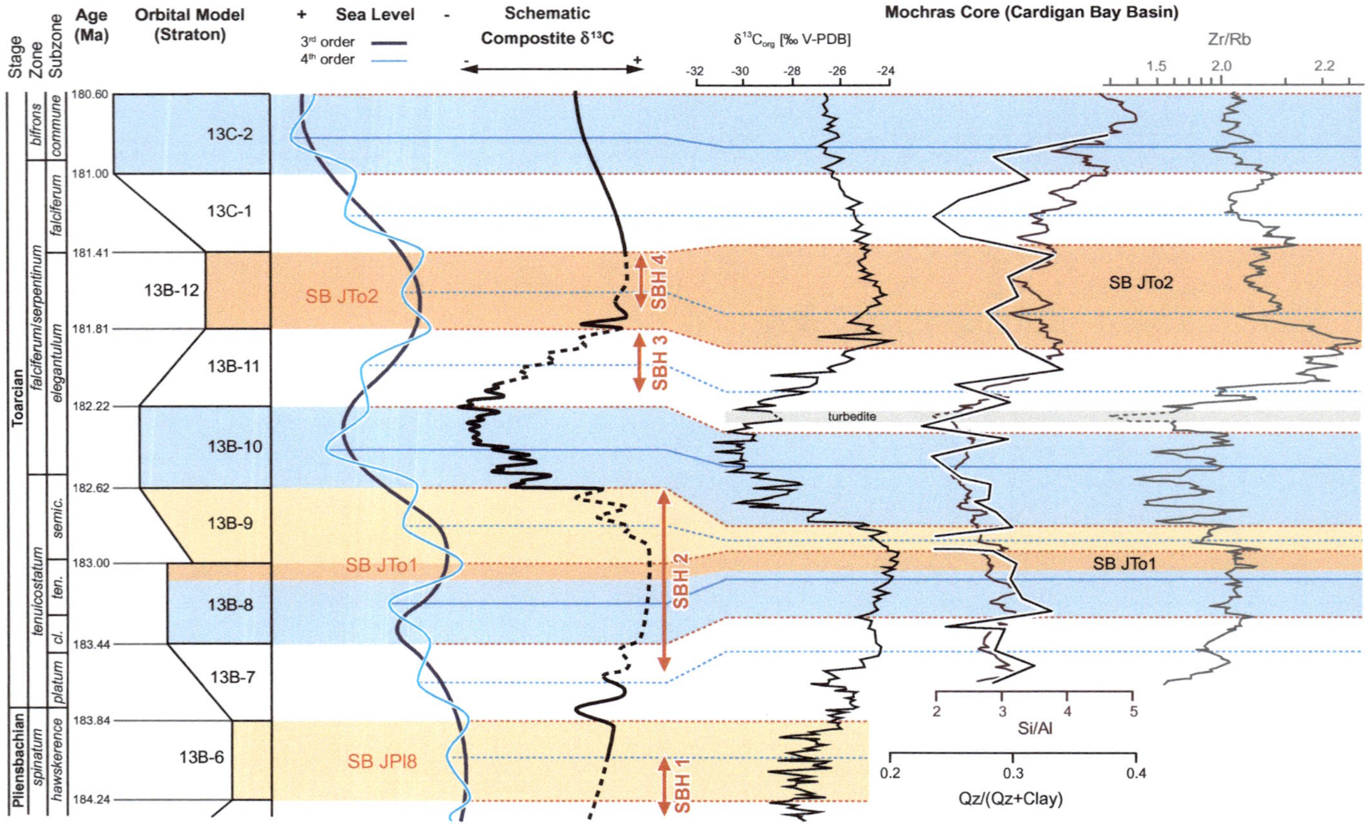

Fig. 11 Late Pliensbachian to early Middle Toarcian stratons, model sea level and the schematic $\delta^{13}C$ curve compared with trends in $\delta^{13}C_{org}$ and clastic proxies in the Mochras Core (Storm et al., 2020; Xu et al., 2018a).

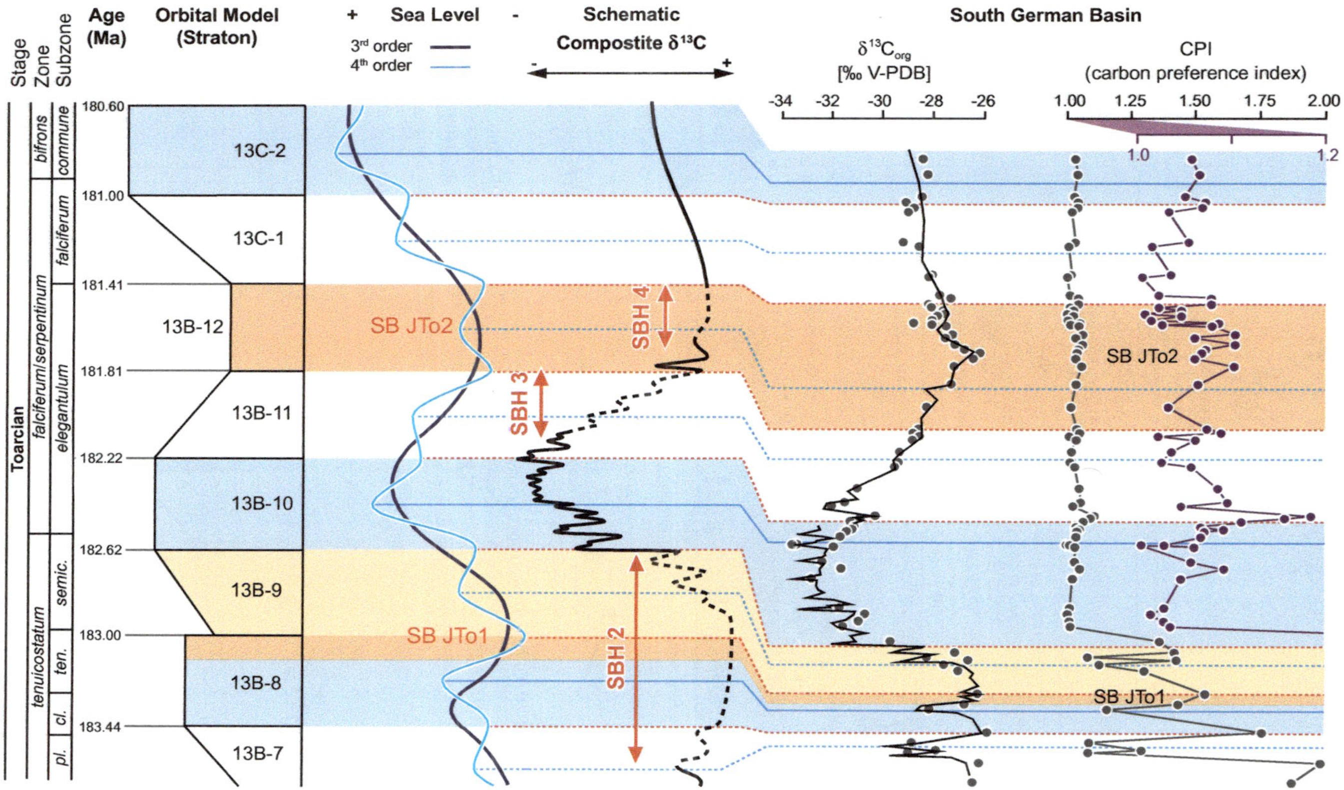

Fig. 12 Late Pliensbachian to early Middle Toarcian stratons, model sea level and the schematic $\delta^{13}C$ curve compared with trends in $\delta^{13}C_{org}$ and CPI data in the South German Basin. The continuous $\delta^{13}C_{org}$ trend is from the Denkingen Core (Suan et al., 2015), discrete data points are from the Dotternhausen Quarry (Frimmel et al., 2003).

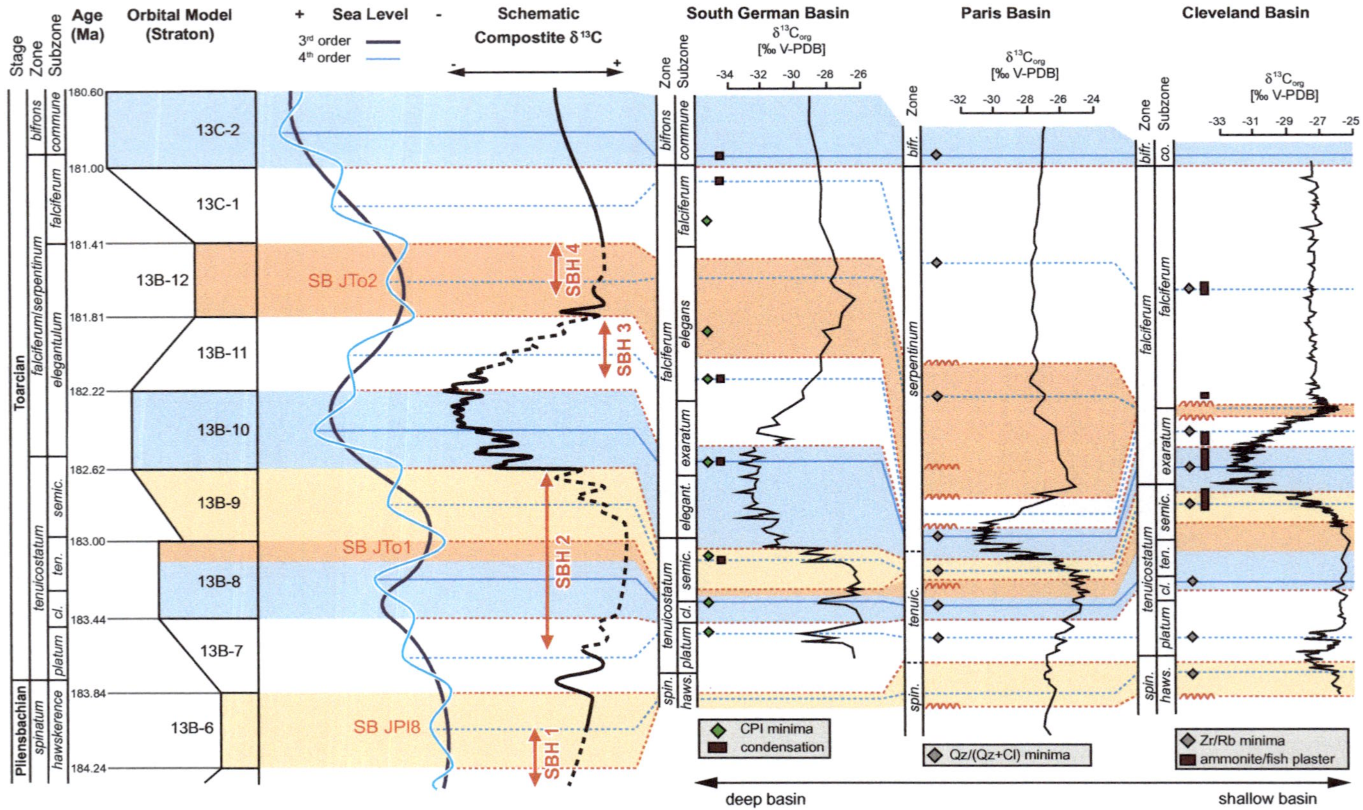

Fig. 13 See legend on opposite page.

184.3 Ma (Fig. 4 and Table 1). SB JPl8 is equivalent to MRS Pli 2 (Barth et al., 2018; Zimmermann et al., 2015) and is manifested in the sedimentary record by subaerial exposure surfaces, major erosional valleys, progradational stratal architectures, coarsening-upwards sequences and karstification of carbonate platforms (Ahokas et al., 2014; Barth et al., 2018; Guex et al., 2001; Léonide et al., 2012; Marjanac and Steel, 1997; Sabatino et al., 2009; Shemin et al., 2019; Suan et al., 2011; Zimmermann et al., 2015). Clastic proxies that cover the late *spinatum* zone are only available from the Sancerre Core (Hermoso et al., 2013, Fig. 8), where maxima in Qz/(Qz+Clay) and Si/Al ratios mark this lowstand.

The *spinatum*-aged eustatic fall may not have been as great as c. 90 m as depicted by Haq (2018), but more likely in the range of 50–70 m based on the depths of *spinatum*-aged incised valleys in offshore North Sea and Greenland paralic marine sediments (Ahokas et al., 2014; Marjanac and Steel, 1997). The 184.3 Ma estimate for SB JPl8 is consistent with older-than the Pliensbachian/Toarcian boundary estimated at 183.7±0.5 Ma (Ikeda and Hori, 2014; Ruebsam and Al-Husseini, 2020; Ruhl et al., 2016; Storm et al., 2020; Xu et al., 2018a).

4.1.2 Upper boundary SB JTo1

According to Haq (2018) SB JTo1 caps Sequence JPl8 in the Early Toarcian *tenuicostatum* zone at c. 183.0 Ma (Fig. 4). Evidence for this eustatic fall is manifested by major unconformities and the very fragmentary preservation

Fig. 13 Model sea level calibrated with 405-kyr stratons and schematic $\delta^{13}C$ curve compared to sedimentological and geochemical records from several basin of the West Tethys shelf. In the South German Basin highstand stratons generally correlate to condensed sediment intervals. In the Paris and Cleveland Basins, highstand stratons correlate with minima in the Qz/(Qz+Cl) and Zr/Rb ratios, respectively. In contrast, lowstand stratons correlate to erosional intervals and the maxima in Qz/(Qz+Cl) and Zr/Rb ratios. In the Yorkshire Section lowstand stratons further correlate with ammonite and fish plaster (Caswell and Coe, 2014). *Data from Röhl, H.J., Schmid-Röhl, A., Oschmann, W., Frimmel, A., Schwark, L., 2001. The Posidonia Shale (Lower Toarcian) of SW-Germany: an oxygen-depleted ecosystem controlled by sea level and palaeoclimate. Palaeogeogr. Palaeoclimatol. Palaeoecol. 165, 27–52; Hermoso, M., Minoletti, F., Pellenard, P., 2013. Black shale deposition during Toarcian super-greenhouse driven by sea level. Clim. Past 9, 2703–2712; Suan, G., van de Schootbrugge, B., Adatte, T., Fiebig, J., Oschmann, W., 2015. Calibrating and magnitude of the Toarcian carbon cycle perturbation. Paleoceanography 30, PA2758; Thibault, N., Ruhl, M., Ullmann, C.V., Korte, C., Kemp, D.B., Gröcke, D.R., Hesselbo, S.P., 2018. The wider context of the Lower Jurassic Toarcian oceanic anoxic event in Yorkshire coastal outcrops, UK. Proc. Geol. Assoc. 129, 372–391.*

of *tenuicostatum*-aged strata in NW Tethys Basins (e.g. France, Germany and UK) and the Arctic shelf (e.g., Pittet et al., 2014; Ruebsam et al., 2019; Suan et al., 2011). In the Peniche Section, SB JTo1 passes to within the "D2" discontinuity interval (Fig. 9, Pittet et al., 2014). Moreover, Krencker et al. (2019) showed that SB JTo1 passes to an erosional unconformity documented by major *tenuicoststaum*-aged erosional valleys in northern Gondwana paleomargin (Dades Valley, Central High Atlas Basin) and is expressed as a shallowing trend in coeval strata from eastern Greenland. In the Iberian Range in Spain, SB JTo1 is expressed as erosional discontinuities that occur in major regressive units in several sections (Comas-Rengifo et al., 1988; also see Danise et al., 2019). The sea level fall of Sequence JTo1 is inferred in all locations by the increasing trend in the clastic and organic matter proxies in the lower to middle *tenuicostatum* zone (*platum* and *clevelandicum* subzones) (Figs. 8–12). The fall may have been as great as 50–75 m based on the depths of incised valleys in the Dades Valley Section in Morocco (Krencker et al., 2019).

4.1.3 Architecture of Sequence JPl8

Following the lowstand in the *spinatum* zone Haq (2018) depicted the eustatic cycle of Sequence JPl8 as a rise across the Pliensbachian-Toarcian transitional interval and a fall in the Early Toarcian *tenuicostatum* zone. A similar eustatic cycle is interpreted in sections in the Lusitanian Basin (Pittet et al., 2014) and the Iberian Range (Comas-Rengifo et al., 1988; also see Danise et al., 2019). Sections in the Central High Atlas Basin record a significant SE retreat of carbonate facies in favor of olistolites and calciturbidites and the occurrence of numerous slumps and debris flows in the upper half of the *spinatum* zone at the Amellago-Ikis Section supporting the lowstand, and a return to a thick package of marls and claystone close to the top Pliensbachian supporting the following rise in sea-level (Mercuzot et al., 2019). In the three localities (Sancerre, Peniche, Yorkshire) where proxy data are measured across the Pliensbachian/Toarcian boundary they indicate a declining trend implying retreating coastlines and transgressions (Figs. 8–11).

4.1.4 Orbital interpretation and calibration

Sequence JPl8 has a duration of c. 1.3 myr between c. 184.3 and 183.0 Ma (Haq, 2018) and closely correlates to the 1.2-myr interval formed by Stratons 13B-6 to 13B-8 between 184.24 and 183.03 Ma (Fig. 4). We correlate the *spinatum*-aged lowstand to the lowstand predicted in Straton 13B-6 and the subsequent sea level cycle of Sequence JPl8 to Stratons 13B-7 and 13B-8.

These correlations imply Straton 13B-7 represents the transgresive systems tract (TST) of Sequence JPl8 and contains the stage boundary at c. 183.7 Ma. Straton 13B-8 is predicted as a sea level high and to contain the sea level peak in the middle *tenuicostatun* zone at c. 183.2 Ma. In sequence stratigraphic nomenclature Straton 13B-8 is the highstand systems tract (HST), however, Matthews and Al-Husseini (2010) suggested the term HST is confusing and instead favored "regressive systems tract" (RST) (see also discussion in Catuneanu et al., 2009). Their reasoning is that during the sea level highstand accommodation space decreases and forces the coastline to advance basinwards. Accordingly, sea-level highstands are generally characterized by regressive deposits as implied by the term RST.

4.1.5 Chemostratigraphic correlation

The major lowstand that hosts SB JPl8 (Straton 13B-6) in the Late Pliensbachian *spinatum* zone correlates to a $\delta^{13}C$ rising limb and SBH 1 (Fig. 4). The major sea level rise spanning the Pliensbachian-Toarcian transition (transgressive Straton 13B-7) records the negative Pl-Toa-CIE and interrupts the Early Toarcian $\delta^{13}C$ rising limb (Fig. 4). The highstand in the lower *tenuicostatum* zone (highstand Straton 13B-8) correlates to a $\delta^{13}C$ plateau (Figs. 4 and 8–13). The upper part of Straton 13B-8 represent the sea level fall culminating in SB JTo1 and closely correlates to SBH 2 (Figs. 3 and 4).

4.2 T-R Sequence JTo1

4.2.1 Sequence boundaries of T-R Sequence JTo1

The lower boundary SB JTo1 of Sequence JTo1 is discussed above in Section 4.1.2. Its upper boundary SB JTo2 is interpreted as a sea-level fall of about 30 to 40 m in the *serpentinum* zone at c. 182.3 Ma (Haq, 2018, Fig. 4). SB JTo2 is evident as erosional discontinuities in the lower to middle *serpentinum* zone in Spain's Iberian Range (Comas-Rengifo et al., 1988; Danise et al., 2019), in Portugal's Lusitanian Basin ("D4" discontinuity of Pittet et al., 2014), and in the East Midland Shelf (UK, Caswell and Coe, 2012). In France, in the northern Paris Basin, SB JTo2 is expressed by discontinuities in Core ANDRA HTM-102 (van Breugel et al., 2006) and Core ANDRA EST 433 (Lézin et al., 2012), and in the southern Paris Basin in the Sancerre Core (Hermoso et al., 2013; Thibault et al., 2018). In western France erosional channels occur in the middle *serpentinum* zone in the Airvault Section (Galbrun et al., 1988). In South America, erosional channels and unconformities are also documented in Toarcian sediments that post-date the *tenuicostatum* zone in the Arroyo Lapa South Section in Argentina (Al-Suwaidi et al., 2016), and the Asientos Section in Chile (Fantasia et al., 2018).

In the Sancerre Core (Fig. 8), the Peniche Section (Fig. 9) and the Mochras Core (Fig. 11) SB JTo2 correlates to marked maxima in the clastic proxies in the middle to upper *elegantulum* subzone (middle *serpentinum* zone). The maxima in clastic proxies further correlate to an interval with increased CPI values in the South German Basin (Fig. 12), implying that the middle *serpentinum* zone is prone to develop discontinuities consistent with global-scale SB JTo2 of Haq (2018) (Figs. 4–9). In the Yorkshire Section such a maxima is not clearly evident (Fig. 10), possibly because the lowstand sediments disappear in SBH 2. This interpretation supports the presences of a major discontinuity in the Yorkshire Section, as suggested by Jenkyns et al. (2002) and as evident in the carbon-13 chemostratigraphy (also see Section 4.2.4).

4.2.2 Architecture of T-R Sequence JTo1

Haq (2018) interpreted Sequence JTo1 as a transgression starting in uppermost *tenuicostatum* zone and a highstand in lower *serpentinum* zone (Fig. 4). The transgression in the *semicelatum* subzone (upper *tenuicostatum* zone) is evident from the decline in the clastic proxies and the drop in the CPI, thus documenting retrograding coastlines during a sea level rise (Figs. 9–12). The highstand of Sequence JTo1 in the lower *serpentinum* zone (lower *elegantulum* subzone) is indicated by a local minimum in the clastic proxies and CPI data followed by increasing trends (Figs. 9–12).

Some positive spikes in the clastic proxies (Si/Al) occur in the highstand in the Peniche Section (Fig. 9). The spikes are related to turbiditic levels and do not represent a eustatic signal (Fantasia et al., 2019a). Similar deposits also occur in other localities and have been related to changes in the weathering intensity and clastic supply during global warming (see Krencker et al., 2015; Xu et al., 2018b and references therein). Accordingly, caution must be taken when interpreting clastic proxies in terms of sea level variations. Nevertheless, we here document comparable trends from other basins, such as the CPI values from the South German Basin (Fig. 12). Our interpretation of the clastic and organic matter trends as sea level proxies is further supported by T-R sequences spanning the *tenuicostatum-serpentinum* zonal transition in the Lusitanian Basin (Pittet et al., 2014) and Iberian Range (Comas-Rengifo et al., 1988; Danise et al., 2019).

4.2.3 Orbital interpretation and calibration

Sequence JTo1 has an estimated duration of c. 700 kyr between c. 183.0 and 182.3 Ma (Haq, 2018) and closely correlates to the 800-kyr interval formed by Stratons 13B-9 and 13B-10 between 183.03 and 182.22 Ma (Fig. 4). This correlation implies the maximum sea level low occurred at the base of lowstand Straton 13B-9 at SB JTo1 and recovered during this straton. Straton 13B-9 represents a transgression in the upper *tenuicostatum* zone (approximately in the *semicelatum* subzone) and highstand Straton 13B-10 represents the RST in the lower *serpentinum* zone, which is in good agreement with sequence stratigraphic interpretations from the Iberian Range (Comas-Rengifo et al., 1988; Danise et al., 2019).

In the South German and Cleveland Basins several condensed horizons that are strongly enriched with the remains of pelagic organisms (e.g., fish and ammonite plaster) occur in the uppermost *tenuicostatum* and lower *serpentinum* zones (Caswell and Coe, 2014; Röhl and Schmid-Röhl, 2005; Suan et al., 2015) (Fig. 10). We interpret them as the MFIs of Stratons 13B-9 and 13B-10 that were paced by fourth-order sea level variations superimposed on the third-order sea level cycles.

4.2.4 Chemostratigraphic correlation

The transgression in upper *tenuicostatum* zone (*semicelatum* subzone) corresponds to Straton 13B-9 and correlates to the Toa-CIE falling limb, suggesting that the onset of the Early Toarcian carbon cycle perturbation occurred concomitant to a major sea level rise (Figs. 4–13). The transgression was apparently rapid and, in some locations like for example in France, Switzerland, and Canada, it is represented by strong stratigraphic condensation of the Toa-CIE falling limb (van Breugel et al., 2006; Léonide et al., 2012; Montero-Serrano et al., 2015; Pittet et al., 2014; Ruebsam et al., 2014, 2019; Them et al., 2017a). In other locations the Toa-CIE falling limb can partly disappear in the upper part of SBH 2 (Figs. 2 and 3). The subsequent sea level highstand in the lower *serpentinum* zone (lower *elegantulum* subzone, highstand Straton 13B-10) correlates to the $\delta^{13}C$ valley that marks the core of the Toa-CIE (Figs. 4–13).

4.3 T-R Sequence JTo2

4.3.1 Boundaries of T-R Sequence JTo2

The calibration by Haq (2018) of SB JTo2 in lower *serpentinum* zone at 182.3 Ma is consistent with estimates in the ATS and the Orbital Scale as discussed in Section 4.2.1. Haq (2018) positioned SB JTo3 corresponding to the upper boundary of Sequence JTo2 at the *serpentinum-bifrons* zonal boundary and estimated it at 180.4 Ma (Fig. 4 and Table 1). In the age model of Ruebsam and Al-Husseini (2020) this zonal boundary is estimated at c. 181.0 Ma, but due to difficulties in determining the exact duration of the *falciferum* subzone a slightly younger age is possible.

The interpretation of SB JTo3 by Haq (2018) is unclear and considered misleading in the present study. In his figure 1 Haq (2018) shows the positions of SB JPl8, SB JTo1 and SB JTo2 in sea level falls, however this relationship is reversed for SB JTo3 shown in a sea level rise (Fig. 4). Placing a third-order SB in a sea level rise is unjustified because this position marks the MFI. The MFI at the *serpentinum-bifrons* zonal boundary is evident in the Denkingen Core and Dotternhausen Section; it consists of condensed horizons, highly enriched in the remains from pelagic organisms (e.g. fish scales, ammonites) that occur below and above the *falciferum-bifrons* zonal boundary (Röhl and Schmid-Röhl, 2005; Suan et al., 2015; Fig. 13). Moreover, a minima in clastic proxies at the *falciferum-bifrons* zonal boundary is clearly evident in the Sancerre Core, the Yorkshire Section and the Mochras Core and is interpreted to reflect a MFI (Figs. 8, 10, and 11).

4.3.2 Architecture of T-R Sequence JTo2

Haq (2018) depicted two sea level lowstands in the *serpentinum* zone: a major one in mid-*serpentinum* and a lesser one in latest *serpentinum* (Fig. 4). The second lowstand is followed by the rise leading to the MFI at the *serpentinum-bifrons* zonal boundary, incorrectly interpreted as SB JTo3 by Haq (2018). As explained in Section 4.3.1 SB JTo3 should be positioned in upper *serpentinum* zone in a sea level fall rather than a rise. The rise is evident in the *falciferum* subzone (upper *serpentinum* zone) by decreasing trends in the clastic proxies and declining CPI in the Sancerre and Mochras Cores (Figs. 8 and 11).

4.3.3 Orbital interpretation and calibration

The Orbital Scale predicts a falling sea level in Straton 13B-11 between 182.22 (SB JTo2) and 181.81 Ma consistent with the age of the mid-*serpentinum* fall. It predicts a major glacio-eustatic lowstand in Straton 13B-12 between 181.81 Ma and 181.41 Ma wherein one or more unconformities may occur. In sequence stratigraphic nomenclature the top of Straton 13B-12 at 181.41 Ma is referred to as the maximum regressive surface (MRS). Following the lowstand in Straton 13B-12, Straton 13C-1 represents a transgression between 181.41 and 181.0 Ma. Its base at 181.41 Ma is often referred at as the transgressive surface or wave ravinement surface. Transgressive Straton 13C-1 closely spans the *falciferum* subzone according to the age model of Ruebsam and Al-Husseini (2020). The *serpentinum-bifrons* zonal transition represents a third-order MFI with a minor fourth-order SB separating Stratons 13C-1 and 13C-2 (Fig. 4).

4.3.4 Chemostratigraphic correlation

The sea level fall and the subsequent lowstand in middle *serpentinum* zone (middle to upper *elegantulum* subzone), corresponding to the regressive Straton 13B-11 and the lowstand Straton 13B-12, correlate to the c. 400-kyr Toa-CIE rising limb and the c. 400-kyr post-Toa-CIE plateau, respectively (Figs. 4–13). SBH 3 occurs in the Toa-CIE rising limb and SBH 4 in the post-Toa-CIE plateau implying these $\delta^{13}C$ patterns represent a regression and lowstand, respectively. This interpretation implies SBH 3 and 4 were paced by fourth-order sea level cycles superimposed on the third-order sea level fall. In some sections both SBHs can merge into a single chemostratigraphic unconformity in the middle *serpentinum* zone (Fig. 3, Ruebsam and Al-Husseini, 2020), suggesting that the post Toa-CIE sea level fall must have been particularly significant. The late *serpentinum* zone (*falciferum* subzone) sea level rise predicted by transgressive Straton 13C-1 is correlated to the falling limb. The MFI at the *serpentinum-bifrons* zonal boundary and subsequent highstand in the bifrons zone are assigned to Straton 13C-2 characterized by the gentle $\delta^{13}C$ valley (Figs. 4–13).

5. Early Toarcian climate swings

In Section 4 we documented sea level falls in Late Pliensbachian *spinatum* zone, and Early Toarcian *tenuicostatim* and *serpentinum* zones, and estimated their numerical ages according to the ATS (Ruebsam and Al-Husseini, 2020) and the deterministic Orbital Scale of glacio-eustasy (Matthews and Al-Husseini, 2010). The remarkable correlation between the model and data and the great amplitude of the falls (both in the range 45–75 m) imply the SBHs are glacio-eustatic unconformities. In this section we seek further evidence for these glaciations by examining $\delta^{18}O$ data and other climate proxies (Figs. 14 and 15).

5.1 Glacio-Eustatic Cycle JPl8

Ruebsam et al. (2019) compiled sedimentological evidence indicating the Late Pliensbachian (*spinatum* zone) eustatic fall associated with SB JPl8

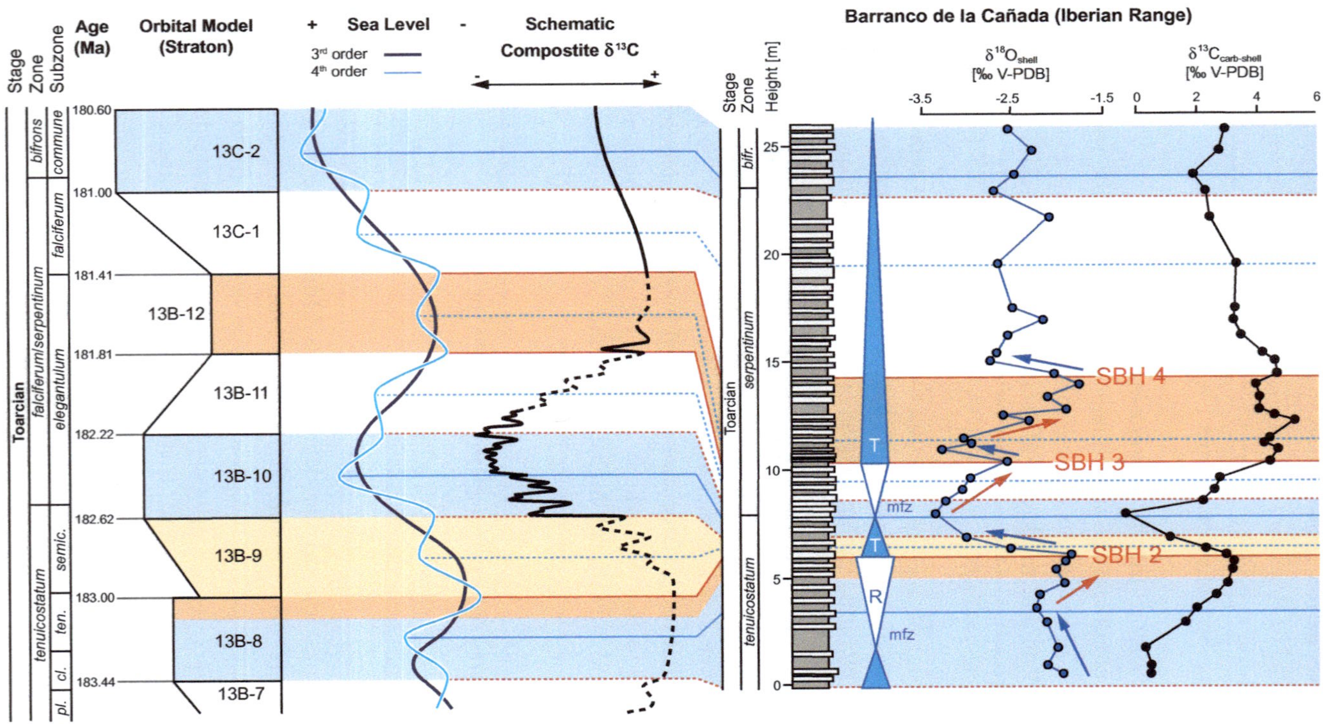

Fig. 14 Early Toarcian to early Middle Toarcian stratons, model sea level and the schematic $\delta^{13}C$ curve compared with trends in $\delta^{18}O$, $\delta^{13}C$ and T-R cycles for early Toarcian successions from the Iberian Range (Ullmann et al., 2020).

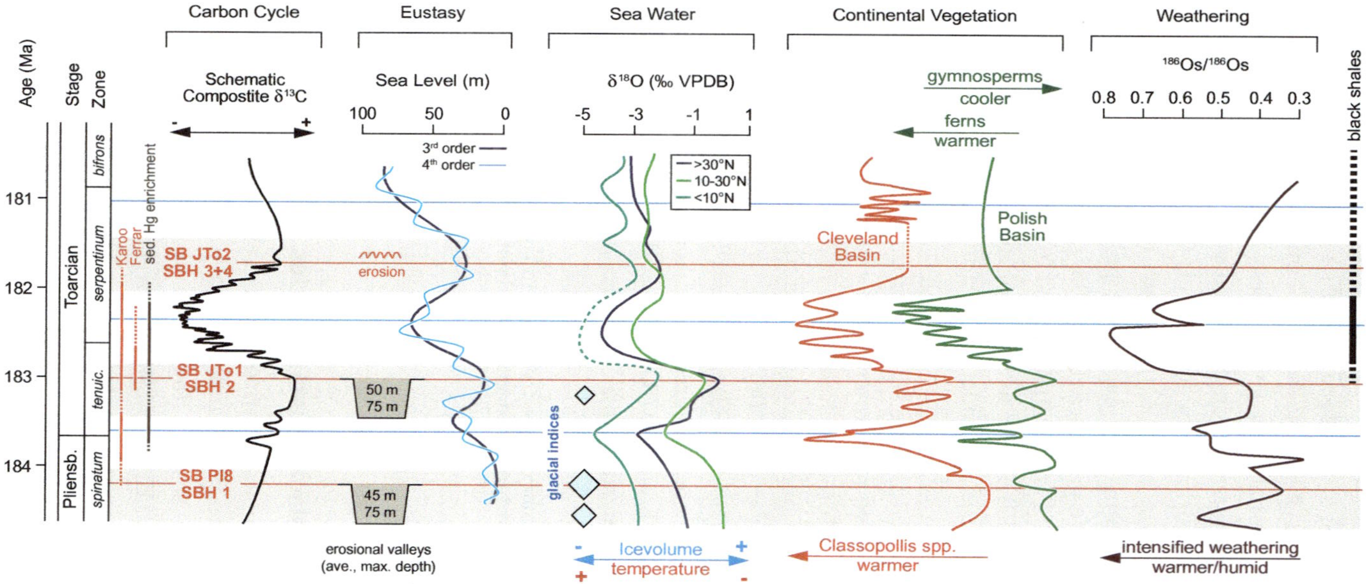

Fig. 15 Late Pliensbachian to early Middle Toarcian chronostratigraphic framework showing Karoo and Ferrar large igneous provinces (Ruebsam et al., 2019), and the parallel evolution of the global carbon cycle ($\delta^{13}C$ curve, this study), fourth- and third-order sea level cycles (this study), climate trends ($\delta^{18}O$, Baghli et al., 2020), continental vegetation (Pienkowski et al., 2016; Slater et al., 2019), continental weathering intensity (Percival et al., 2016). Environmental changes at the *tenuicostatum-serpentinum* zonal transition favored the widespread deposition of black shales over vast shelf areas (Jenkyns, 1988, 2010).

was caused by the expansion of one or both polar ice caps. Beside this major sea level lowstand Brandt (1986) documented a marked cyclicity in the Late Pliensbachian sediments from the South German Basin and interpreted their origin as glacio-eustatic sea level cycles. The fall in *spinatum* zone was accompanied by an increase in $\delta^{18}O$ values in marine fossilized shell material (Danise et al., 2019; Dera et al., 2011; Gómez et al., 2016; Korte and Hesselbo, 2011; Suan et al., 2010), which reflects a depletion in ^{16}O of the seawater, consistent with expanding polar ice caps and declining seawater temperatures (e.g., Shackelton, 1987), and with a decrease of continental runoff in the Paris Basin suggested from clay mineralogical data (Bougeault et al., 2017). The analyses of fossilized land vegetation indicate preferentially cold to moderately warm climates during the Late Pliensbachian in present-day northern Europe (Pienkowski et al., 2016; Slater et al., 2019) (Fig. 15). The lack of evidence at high paleo-latitudes for vegetation accustomed to warmer climates supports cold polar climates (Vakhrameev, 1991). Moreover, some areas in the present-day southwest Siberia appear to have been unvegetated during certain periods in the Early Jurassic (Le Heron et al., 2008).

Following the *spinatum*-aged glaciation, sea level rose as the ice sheets melted as evident from the transgression of Sequence JPl8 spanning the Pliensbachian/Toarcian boundary (Ruebsam et al., 2019). The transgression was accompanied by a decline in $\delta^{18}O$ values implying a rise in seawater temperatures and a decline in continental ice volume (Korte and Hesselbo, 2011; Korte et al., 2015; Suan et al., 2010). This short-lived global warming episode is also interpreted from floral data (Pienkowski et al., 2016; Slater et al., 2019) and an increase in chemical weathering as inferred from Osmium-isotope and sedimentological data (Kemp et al., 2020; Percival et al., 2016; Xu et al., 2018b).

5.2 Early Toarcian glacio-eustatic cycles

One of the best examples illustrating the relationship between Early Toarcian T-R Sequences JTo1 and JTo2, $\delta^{18}O$ and $\delta^{13}C$ occurs in the Barranco de la Cañada Section (Iberian Range) (Ullmann et al., 2020) (Fig. 14). In this section SBH 2 to SBH 4 correlate to positive $\delta^{18}O$ excursions implying they represent cold phases. These cold phases are also evident from continental climate trends inferred from the paleofloral characteristics (Pienkowski et al., 2016; Slater et al., 2019; Fig. 15). However sedimentological evidence for intra-*serpentinum* glaciations is limited to strata in mid-latitudinal Bulgaria where matrix-supported and unsorted conglomerates, and cold-adapted floral remains attest to the presence of Alpine glaciers during cool climates (Tchoumatchenco et al., 2008).

The late *semicelatum* sea level rise correlates to a major decline in $\delta^{18}O$ values commonly linked to major global warming and a dramatic rise in seawater temperatures (e.g., Danise et al., 2019; Dera et al., 2011; Goméz et al., 2016; Korte et al., 2015; Ullmann et al., 2020). The magnitude of the $\delta^{18}O$ decline suggests a significant change in the isotopic composition of seawater oxygen, consistent with post-glacial warming and melting of continental ice sheets (Ruebsam et al., 2019; Shackleton, 1987; Suan et al., 2010). A major rise in global temperatures has also been interpreted based on changes in floral assemblages (Pienkowski et al., 2016; Slater et al., 2019). Osmium isotopes further attest to drastically increased weathering rates due to an acceleration of the hydrological cycle during warm and humid climates (Cohen et al., 2004; Kemp et al., 2020; Percival et al., 2016; Them et al., 2017b).

A shift towards more positive $\delta^{18}O$ values in the middle *serpentinum* zone is documented from various basin (e.g. Danise et al., 2019; Goméz et al., 2016; Suan et al., 2010; also see Baghli et al., 2020 and references therein) and can be correlated to SBH 3 and SBH 4 (Fig. 15). In the Barranco de la Cañada Section (Iberian Range) (Ullmann et al., 2020) SBH 3 and SBH 4 are further distinguished by local $\delta^{18}O$ maxima (Fig. 14). Finally, the sea level rise in the *falciferum* subzone was accompanied by a decline in $\delta^{18}O$ values that are clearly expressed in Barranco de la Cañada Section (Ullmann et al., 2020), and also evident from other localities (e.g., Baghli et al., 2020) (Fig. 15).

6. Discussion

The 405-kyr straton is a robust time-rock unit that is commonly recognized in sequence stratigraphic interpretations as the *fourth-order sequence.* Stratons track the long-eccentricity 405-kyr component of the Earth's orbit around the Sun and their sequence boundaries occurred when the orbit attained a minimum every c. 405 kyr on average. At these minima the orbit approached a circle, the climate became colder, polar ice caps may have formed or expanded causing sea level to fall, and breaks were likely to have been marked in the stratigraphic record. Therefore, stratons are the most natural units for tying eccentricity-forcing of climate and glacio-eustasy to fourth-order sequences. Stratons can be used to build T-R sequences consisting of multiples of 405 kyr and according to the

Orbital Scale their contribution to sea level can be predicted (Matthews and Al-Husseini, 2010; see Al-Husseini and Ruebsam, 2020, this Volume).

Applying the predictions of the Orbital Scale to the Late Pliensbachian to Early Toarican times (c. 185 to 180 Ma) resulted in a remarkable similarity between the model and proxy data for sea level changes. In the study interval sea level fell several times with amplitudes as great as 50–75 m. Each fall was accompanied by decreasing sea-surface temperature (SST), as well as cooler and more arid climates. In contrast, sea level rises were accompanied by rising SST, global warming and a more humid climate (e.g., Cohen et al., 2004; Kemp et al., 2020; Montero-Serrano et al., 2015; Percival et al., 2016; Pienkowski et al., 2016; Them et al., 2017b; Xu et al., 2018b). Our findings agree with the analyses and conclusions of Davies et al. (2020) that such great falls recurring approximately at one million year intervals cannot be caused by the warming and cooling of oceans (thermo-eustasy), nor by the transfer of water back and forth between oceans and land reservoirs during arid-humid cycles (aquifer-eustasy). Only the waxing and waning of continental ice sheets can cause sea-level fluctuations of many 10s of meters.

The oldest cycle in the study interval started with a glacio-eustatic fall in the range of 50 and 70 m at c. 184.22 Ma consistent with the estimate of 184.3 Ma for SB JPl8 in the Late Pliensbachian *spinatum* zone (Haq, 2018). The fall is interpreted to have been followed by a 400-kyr sea level low between c. 184.22 and 183.84 Ma characterized by a $\delta^{13}C$ plateau (Straton 13B-6). In many sections the plateau is not evident and is apparently lost in Stratigraphic Black Hole SBH 1.

The lowstand was followed by a glacio-eustatic cycle between 183.84 and 183.0 Ma containing the Pliensbachian/Toarcian Boundary marked by Pl-Toa CIE at c. 183.7 Ma. The upper limit of this cycle is interpreted as a fall of between 50 and 75 m centered at 183.0 Ma, consistent with the position of SB JTo1 in Lower Toarcian *tenuicostatum* zone similarly dated at 183.0 Ma by Haq (2018). The interval between SB JPl8 and SB JTo1 forms a c. 1.2-myr glacio-eustatic cycle corresponding to T-R Sequence JPl8 of Haq (2018) and Stratons 13B-6 to 13B-8.

Following the maximum lowstand at SB JTo1 at 183.0 Ma, an Early Toarcian sea level rise and high, each lasting c. 400 kyr, are interpreted to have spanned the interval between c. 183.0 and 182.22 Ma. The rise occurred in the *semiculatum* subzone (*tenuicostatum* zone) characterized by the Toa-CIE falling limb. The sea-level high straddled the uppermost *tenuicostatum* zone and the lower *elegantulum* subzone (*serpentinum* zone) and is characterized by the Toa-CIE valley. At c. 182.22 Ma sea level is interpreted to have started falling as consistent with the position of SB JTo2 in lower *serpentinum* zone, estimated at 182.3 Ma (Haq, 2018). The interval between SB To1 and SB JTo2 corresponds to the c. 0.80-myr T-R Sequence JTo1 of Haq (2018) and Stratons 13B-9 and 13B-10.

In Straton 13B-11, between 182.22 and 181.81 Ma, sea level is interpreted to have been falling; this interval occurs in middle *elegantulum* subzone (*serpentinum* zone) and is characterized by the Toa-CIE rising limb. The amplitude of the fall is estimated by Haq (2018) as c. 30 m but could not be independently estimated in our study. In this interval, numerous sections contain unconformities corresponding to SBH 3. Between c. 181.81 and 181.41 Ma, sea level is interpreted as a low in Straton 13B-12 and characterized by the post-Toa-CIE plateau in the upper *elegantulum* subzone. This lowstand is also an unconformity-prone interval that is hidden in many sections by SBH 4. Its glacio-eustatic nature is supported by the colder climate proxy data.

Haq (2018) depicted cycle JTo2 as a broad sea level low in middle *serpentinum* zone capped by SB JTo3 near base *bifrons* zone. We modify this interpretation and use Stratons 13B-11 to 13C-2 to describe sea level in the *serpentinum* to *bifrons* interval. As explained above we interpret Straton 13B-11 as a sea level fall and Straton 13B-12 as a lowstand. These two stratons apparently correspond to the broad mid-*serpentinum* lowstand of cycle JTo2 of Haq (2018). The entire 800-kyr interval represented by Stratons 13B-11 and 13B-12 is unconformity-prone and may contain SBH 3 and SBH 4 in many sections, which in some sections can merge into a super SBH 3+4. Assigning a precise position and age to a single SB in this 800-kyr interval is therefore ambiguous and misleading. In the straton interpretation the post-glacial transgression starts near the *elegantulum-falciferuum* subzone boundary marked by the start of the falling limb at c. 181.41 Ma (base Straton 13C-1) and the *bifrons* MFS/MFI occurs near or within the lower part of Straton 13C-2 at c. 181.0 or 180.9 Ma.

We here identify orbital forcing as the major process pacing eustatic sea level changes, climate swings and changes in the global carbon cycle. The impact of orbital forcing on Early Toarcian carbon cycle dynamics has been demonstrated previously (e.g., Boulila et al., 2019; Huang and Hesselbo, 2014; Kemp et al., 2005, 2011; Ruebsam et al., 2019; Thibault et al., 2018). Furthermore, the direct coupling of carbon cycle and climate-induced changes in the weathering intensity has been confirmed by Kemp et al. (2020), which substantiates the role of orbital forcing as the major driver of Early Toarcian climate and environmental changes.

Greenhouse gas emissions associated with the K-F-LIP volcanism has been considered an important factor contributing to the Early Toarcian climate instabilities, which is supported by the temporal and causal relation between volcanism and environmental perturbations (e.g., Pálfy and Smith, 2000; Percival et al., 2015; Ruebsam and Al-Husseini, 2020; Xu et al., 2018a). However, volcanism and related climate feedacks rather acted as an amplifier of orbitally paced climate cycles, but were not the sole drivers of Early Toarcian environmental changes.

7. Conclusions

In this case study we have built a temporal bridge between biostratigraphy, chemostratigraphy, sequence stratigraphy, cyclostratigraphy and geochronology in the interval spanning the Late Pliensbachian and Early Toarcian (185–180 Ma). In this relatively brief 5 myr interval we found evidence for three major sea-level falls with amplitudes of many tens of meters. In this chronostratigraphic framework sea level falls closely correlate to intervals that are either fragmented, or completely missing in various $\delta^{13}C$ records, suggesting that Stratigraphic Black Holes (Ruebsam and Al-Husseini, 2020) are directly related to unconformities or sequence boundaries. Our age model and interpretations of sea level and T-R sequences are very similar to those of Haq (2018) except at the *serpentinum-bifrons* transition: we interpret this transitional zone as older by c. 500 kyr and as a maximum flooding interval rather than a third-order sequence boundary.

Based on the great amplitudes and the orbital forcing of the sea level and climate cycles we conclude large continental ice sheets waxed and waned during the study interval. Our conclusions extend into the Jurassic the conclusions of Davies et al. (2020) for Cretacoues glacio-eustasy: namely, aquifer- and/or thermo-eustasy cannot account for sea level fluctuations of many tens of meters. Our conclusion is further strengthened by how closely our sea level model resembles the predictions of the Orbital Scale of glacio-eustasy. Our study indicates the age of the Karoo-Ferrar Large Igneous Province coincides with the Early Toarcian but its contribution to sea level and climate cycles requires further evaluations. Finally we find no reason for dividing the Phanerozoic (and older times) into ice-free greenhouse and icehouse intervals, when the evidence indicates glacio-eustasy prevailed throughout these times.

Acknowledgments

The authors would like to thank Micha Ruhl for sharing his insights on the Early Toarcian carbon isotope stratigraphy, and Michael Montenari for inviting this contribution.

References

Ruhl, M., Hesselbo, S.P., Hinnov, L., Jenkyns, H.C., Xu, W., Storm, M.S., Riding, J.B., Minisini, D., Ullmann, C.V., Leng, M.J., 2016. Astronomical constraints on the duration of the Early Jurassic Pliensbachian stage and global climatic fluctuations. Earth Planet. Sci. Lett. 455, 149–165.

Sabatino, N., Neri, R., Bellanca, A., Jenkyns, H.C., Baudin, F., Prisi, G., Masertti, D., 2009. Carbon-isotope records of the Early Jurassic (Toarcian) oceanic anoxic event from the Valdorbia (Umbria–Marche Apennines) and Monte Mangart (Julian Alps) sections: palaeoceanographic and stratigraphic implications. Sedimentology 56, 1307–1328.

Ahokas, J.M., Nystuen, J.P., Martinius, A.W., 2014. Stratigraphic signatures of punctuated rise in relative sea level in an estuary-dominated heterolithic succession: incised valley fills of the Toarcian Ostreaelv Formation, Neill Klinter Group (Jameson Land, East Greenland). Mar. Petrol. Geol. 50, 103–129.

Al-Husseini, M., Ruebsam, W., 2020. Interpreting phanerozoic $\delta^{13}C$ patterns as periodic glacio-eustatic sequences. In: Montenari, M. (Ed.), Stratigraphy & Timescales vol. 5. Elsevier, pp. 41–105.

Al-Suwaidi, A.H., Hesselbo, S.P., Damborenea, S.E., Manceñido, M.O., Jenkyns, H.C., Riccardi, A.C., Angelozzi, G.N., Baudin, F., 2016. The Toarcian Oceanic Anoxic Event (Early Jurassic) in the Neuquén Basin, Argentina: a reassessment of age and carbon isotope stratigraphy. J. Geol. 124, 171–193.

Alley, N.F., Hore, S.B., Frakes, L.A., 2019. Glaciationsat high-latitude Southern Australia during the Early Cretaceous. Aust. J. Earth Sci. https://doi.org/10.1080/08120099.2019.1590457.

Ansorge, J., 2007. Lower Jurassic clay pit of Klein Lehmhagen near Grimmen. In: Dobracki, R., Schütze, K. (Eds.), Niedermeyer. Geo-Pomerania–Excursion Guide, Biuletyn Państwowego.

Baghli, H., Mattioli, E., Spangenberg, J.E., Bensalah, M., Arnaud-Godet, F., Pittet, B., Suan, G., 2020. Early Jurassic climate trends in the southern Tethys margin. Gondwana Res. 77, 67–81.

Barth, G., Pienkowski, G., Zimmermann, J., Franz, M., Kuhlmann, G., 2018. Palaeogeographical evolution of the Lower Jurassic: high-resolution biostratigraphy and sequence stratigraphy in the Central European Basin. In: Kilhams, B., Kukla, P.A., Mazur, S., Mckie, T., Mijnlieff, H.F., Van Ojik, K. (Eds.), Mesozoic Resource Potential in the Southern Permian Basin. Geological Society, London, Special Publications 469, pp. 341–369.

Beerling, D.J., Brentnall, S.J., 2007. Numerical evaluation of mechanisms driving early Jurassic changes in global carbon cycling. Geology 35, 247–250.

Berger, W.H., Vincent, E., 1981. Chemostratigraphy and biostratigraphic correlation: exercises in systemic stratigraphy. Oceanol. Acta 1981, 115–127.

Blakey, R.C., 2016. Global Jurassic Paleogeographic Map (180MaBP): Molleweide. Global Paleogeography and Tectonics in Deep Time © 2016. Colorado Plateau Geosystems Inc. https://deeptimemaps.com.

Bodin, S., Krencker, F.N., Kothe, T., Hoffmann, R., MAttioli, E., Heimhofer, U., Kabiri, L., 2016. Perturbation of the carbon cycle during the late Pliensbachian e early Toarcian: new insight from high-resolution carbon isotope records in Morocco. J. Afr. Earth Sci. 116, 89–104.

Bougeault, C., Pellenard, P., Deconninck, J.F., Hesselbo, S.P., Dommergues, J.L., Bruneau, L., Cocquerez, T., Laffont, R., Huret, E., Thibault, N., 2017. Climatic and palaeoceanographic changes during the Pliensbachian (Early Jurassic) inferred from clay mineralogy and stable isotope (C-O) geochemistry (NW Europe). Global Planet. Change 149, 139–152.

Boulila, S., Galbrun, B., Huret, E., Hinnov, L.A., Rouget, I., Gardin, S., Bartolini, A., 2014. Astronomical calibration of the toarcian stage: implications for sequence stratigraphy and duration of the early Toarcian OAE. Earth Planet. Sci. Lett. 386, 98–111.

Boulila, S., Galbrun, B., Sadki, D., Gardin, S., Bartolini, A., 2019. Constraints on the duration of the early Toarcian T-OAE and evidence for carbon-reservoir change from the High Atlas (Morocco). Global Planet. Change 175, 113–128.

Boulila, S., Hinnov, L.A., 2017. A review of tempo and scale of the early Jurassic Toarcian OAE: implications for carbon cycle and sea level variation. Newslett. Stratigr. 50, 363–389.

Brandt, K., 1986. Glacioeustatic cycles in the early Jurassic? Neues Jahrbuch Geol. Paläontol. 5, 257–274.

Brazier, J.M., Suan, G., Tacail, T., Simon, L., Martin, J.E., Mattioli, E., Blater, V., 2015. Calcium isotope evidence for dramatic increase of continental weathering during the Toarcian oceanic anoxic event (Early Jurassic). Earth Planet. Sci. Lett. 41, 1164–1176.

Brenchley, P.J., Carden, G.A., Hints, L., Kaljo, D., Marshall, J.D., Martma, T., Meldha, T., Nõlvak, J., 2003. High-resolution stable isotope stratigraphy of upper ordovician sequences: constraints on the timing of bioevents and environmental changes associated with mass extinctions and glaciation. GSA Bullet. 115, 89–104.

Caruthers, A.H., Gröcke, D.R., Smith, P.L., 2011. The significance of an Early Jurassic (Toarcian) carbon-isotope excursion in Haida Gwaii (Queen Charlotte Islands), British Columbia, Canada. Earth Planet. Sci. Lett. 307, 19–26.

Caswell, B.A., Coe, A.L., 2012. A high-resolution shallow marine record of the Toarcian (Early Jurassic) Oceanic Anoxic event from the East Midlands Shelf, UK. Palaeogeogr. Palaeoclimatol. Palaeoecol. 365, 124–135.

Caswell, B.A., Coe, A.L., 2014. The impact of anoxia on pelagic macrofauna during the Toarcian Oceanic anoxic event (Early Jurassic). Proc. Geol. Assoc. 125, 383–391.

Catuneanu, O., Abreu, V., Bhattacharya, J.P., Blum, M.D., Dalrymple, R.W., Eriksson, P.G., Fielding, C.R., Fisher, W.L., Galloway, W.E., Gibling, M.R., Giles, K.A., Holbrook, J.M., Jordan, R., Kendall, C.G.S.C., Macurda, B., Martinsen, O.J., Miall, A.D., Neal, J.E., Nummedal, D., Pomar, L., Posamentier, H.W., Pratt, B.R., Sarg, J.F., Shanley, K.W., Steel, R.J., Strasser, A., Tucker, M.E., Winker, C., 2009. Towards the standardization of sequence stratigraphy. Earth-Sci. Rev. 92, 1–33.

Coe, A.L., 2003. The Sedimentary Record of Se-Level Change. Cambridge University Press, p. 288.

Cohen, A.S., Coe, A.L., Harding, S.M., Schwark, L., 2004. Osmium isotope evidence for the regulation of atmospheric CO_2 by continental weathering. Geology 32, 157–160.

Comas-Rengifo, M.J., Goy, A., Yébenes, A., 1988. El Lias en el Sector Sur occidental de la Sierra de la Demanda (Castrovido, Burgos). Cienc. Tierra Geol. 11, 119–141.

Danise, S., Clémence, M.E., Price, G.D., Murphy, D.P., Gómez, J.J., Twitchett, R.J., 2019. Stratigraphic and environmental control on marine benthic community change through the early Toarcian extinction event (Iberian Range, Spain). Palaeogeogr. Palaeoclimatol. Palaeoecol. 524, 183–200.

Davies, A., Gréselle, B., Hunter, S., Baines, G., Robson, C., Haywood, A., Ray, D.C., Simmons, M.D., van Buchem, F.S.P., 2020. Assessing the impact of aquifer-eustasy on short-term cretaceous sea level. Cretaceous Res. 112, 104445.

De Lena, L.F., Taylor, D., Guex, J., Bartolini, A., Adatte, T., van Acken, D., Spangenberg, J.E., Samankassou, E., Vennemann, T., Schaltegger, U., 2019. The driving mechanisms of the carbon cycle perturbations in the late Pliensbachian (Early Jurassic). Sci. Rep. 9, 18430.

Dera, G., Brigaud, B., Monna, F., Laffont, R., Pucéat, E., Deconinck, J.-P., Pellenard, P., Joachimski, M.M., Durlet, C., 2011. Climatic ups and downs in a disturbed Jurassic world. Geology 39, 215–218.

Fantasia, A., Adatte, T., Spangenberg, J.E., Font, E., Duarte, L.V., Föllmi, K.B., 2019a. Global versus local processes during the Pliensbachian–Toarcian transition at the Peniche GSSP, Portugal: a multi-proxy record. Earth-Sci. Rev. 198, 102932.

Fantasia, A., Föllmi, K.B., Adatte, T., Spangenberg, J.E., Mattioli, E., 2019b. Expression of the Toarcian Oceanic anoxic event: new insights from a Swiss transect. Sedimentology 66, 262–284.

Fantasia, A., Föllmi, K.B., Adette, T., Bernárdez, E., Spangenberg, J.E., Mattioli, E., 2018. The Toarcian Oceanic anoxic event in southwestern Gondwana: an example from the Andean Basin, northern Chile. J. Geol. Soc. 175, 883–902.

Frimmel, A., Oschmann, W., Schwark, L., 2003. Chemostratigraphy of the Posidonia Black Shale, SW Germany: I. Influence of sea-level variation on organic facies evolution. Chem. Geol. 206, 199–230.

Fu, X., Wang, J., Feng, X., Wang, D., Chen, W., Song, C., Zeng, S., 2016. Early Jurassic carbon-isotope excursion in the Qiangtang Basin (Tibet), the eastern Tethys: implications for the Toarcian Oceanic anoxic event. Chem. Geol. 442, 62–72.

Galbrun, B., Gabilly, J., Rasplus, L., 1988. Magnetostratigraphy of the Toarcian stratotype sections at Thouars and Airvault (Deux-Sevres, France). Earth Planet. Sci. Lett. 87, 453–462.

Gómez, J.J., Comas-Rengifo, M.J., Goy, A., 2016. Palaeoclimatic oscillations in the Pliensbachian (Early Jurassic) of the Asturian Basin (Northern Spain). Clim. Past 12, 1199–1214.

Govin, A., Holzwarth, U., Heslop, D., Ford Keeling, L., Zabel, M., Mulitza, S., Collins, J.A., Chiessi, C.M., 2012. Distribution of major elements in Atlantic surface sediments (36°N-49°S): Imprint of terrigenous input and continental weathering. Geochem. Geophys. Geosyst. 13, 1–23.

Guex, J., Morard, A., Bartolini, A., Morettini, E., 2001. Découverte d'une importante lacune stratigraphique à la limite Domérien-Toarcien: implications paléo-océanographiques. Bull. Soc. vaudoise Sci. Natur. 87, 277–284.

Haq, B.U., 2018. Jurassic sea level variations: a reappraisal. GSA Today (Geol. Soc.) 28, 4–10.

Hardenbol, J., Thierry, J., Farley, M.B., Jacquin, T., de Graciansky, P.C., Vail, P.R., 1998. Mesozoic and cenozoic sequence chronostratigraphic framework of European basins. In: de Graciansky, P.C., Hardenbol, J., Jacquin, T., Vail, P.R. (Eds.), Mesozoic and Cenozoic Sequence Stratigraphy of European Basins. SEPM Special Publication, 60. Tulsa, OK, pp. 3–13.

Hermoso, M., Delsate, D., Baudin, F., Le Callonnec, L., Minoletti, F., Renard, M., Faber, A., 2014. Record of early Toarcian carbon cycle perturbations in a nearshore environment: bascharage section (easternmost Paris Basin). Solid Earth Discuss. 6, 1073–1100.

Hermoso, M., Minoletti, F., Pellenard, P., 2013. Black shale deposition during Toarcian super-greenhouse driven by sea level. Clim. Past 9, 2703–2712.

Hermoso, M., Minoletti, F., Rickaby, R.E.M., Hesselbo, S.P., Baudin, F., Jenkyns, H.C., 2012. Dynamics of a stepped carbon-isotope excursion: ultra high-resolution study of Early Toarcian environmental change. Earth Planet. Sci. Lett. 319–320, 45–54.

Hesselbo, S.P., Gröcke, D.R., Jenkyns, H.C., Bjerrum, C.J., Farrimod, P., Morgens Bell, H.S., Green, O.R., 2000. Massive dissociation of gas hydrate during the Jurassic oceanic anoxic event. Nature 406, 392–395.

Hesselbo, S.P., Jenkyns, H.C., Duarte, L.V., Oliveira, L.C.V., 2007. Carbon-isotope record of the early jurassic (Toarcian) oceanic anoxic event from fossil wood and marine carbonate (Lusitanian Basin, Portugal). Earth Planet. Sci. Lett. 253, 455–470.

Hinnov, L., 2018. Cyclostratigraphy and astrochronology in 2018. In: Montenari, M. (Ed.), Stratigraphy and Timescales, vol. 3. Elsevier Inc, pp. 4–80.

Huang, C., 2018. Astronomical time scale for the mesozoic. In: Montenari, M. (Ed.), Stratigraphyand timescales, vol. 3. Elsevier Inc, pp. 81–150.

Huang, C., Hesselbo, S.P., 2014. Pacing of the toarcian oceanic anoxic event (early Jurassic) from astronomical correlation of marine sections. Gondwana Res. 25, 1348–1356.

Ikeda, M., Hori, R.S., 2014. Effects of Karoo–Ferrar volcanism and astronomical cycles on the Toarcian Oceanic anoxic events (early Jurassic). Palaeogeogr. Palaeoclimatol. Palaeoecol. 410, 134–142.

Ikeda, M., Hori, R.S., Ikehara, M., Miyashita, R., Chino, M., Yamada, K., 2018. Carbon cycle dynamics linked with Karoo-Ferrar volcanism and astronomical cycles during Pliensbachian-Toarcian (Early Jurassic). Global Planet. Change 170, 163–171.

Izumi, K., Kemp, D.B., Itamiya, S., Inui, M., 2018. Sedimentary evidence for enhanced hydrological cycling in response to rapid carbon release during the early Toarcian oceanic anoxic event. Earth Planet. Sci. Lett. 481, 162–170.

Jenkyns, H.C., 1988. The early Toarcian (Jurassic) anoxic event—stratigraphic, sedimentary, and geochemical evidence. Am. J. Sci. 288, 101–151.

Jenkyns, H.C., 2010. Geochemistry of oceanic anoxic events. Geochem. Geophys. Geosyst. 11, Q03004. https://doi.org/10.1029/2009GC002788.

Jenkyns, H.C., Jones, C.E., Gröcke, D.R., Hesselbo, S.P., Parkinson, D.N., 2002. Chemostratigraphy of the Jurassic system: applications, limitations and implications for palaeoceanography. J. Geol. Soc. 159, 351–378.

Kemp, D.B., Baranyi, V., Izumi, K., Burgess, R.D., 2019. Organic matter variations and links to climate across the early Toarcian oceanic anoxic event (T-OAE) in Toyora area, southwest Japan. Palaeogeogr. Palaeoclimatol. Palaeoecol. 530, 90–102.

Kemp, D.B., Coe, A.L., Cohen, A.S., Schwark, L., 2005. Astronomical pacing of methane release in the Early Jurassic period. Nature 437, 396–399.

Kemp, D.B., Coe, A.L., Cohen, A.S., Weedon, G.P., 2011. Astronomical forcing and chronology of the early Toarcian (Early Jurassic) oceanic anoxic event in Yorkshire, UK. Paleoceanography 26. https://doi.org/10.1029/2011PA002122.

Kemp, D.B., Selby, D., Izumi, K., 2020. Direct coupling between carbon release and weathering during the Toarcian oceanic anoxic event. Geology 48. https://doi.org/10.1130/G47509.1.

Korte, C., Hesselbo, S.P., 2011. Shallow marine carbon and oxygen isotope and elemental records indicate icehouse-greenhouse cycles during the early Jurassic. Paleoceanography 26, 1–18.

Korte, C., Hesselbo, S.P., Ullmann, C.V., Dietl, G., Ruhl, M., Schweigert, G., Thibault, N., 2015. Jurassic climate mode governed by ocean gateway. Nat. Commun. 6, 10015.

Krencker, F.-N., Bodin, S., Suan, G., Heimhofer, U., Kabiri, L., Immenhauser, A., 2015. Toarcian extreme warmth led to tropical cyclone intensification. Earth Planet. Sci. Lett. 425, 120–130.

Krencker, F.N., Lindström, S., Bodin, S., 2019. A major sea-level drop briefly precedes the Toarcian oceanic anoxic event: implication for early Jurassic climate and carbon cycle. Sci. Rep. 9, 12518.

Kump, L.R., Arthur, M.A., 1999. Interpreting carbon-isotope excursions: carbonates and organic matter. Chem. Geol. 161, 181–198.

Laskar, J., Fienga, A., Gastineau, M., Manche, H., 2011. La2010: a new orbital solution for the long-term motion of the earth. Astron. Astrophys. 532, A89, 15 p.

Laskar, J., Robutal, P., Joutel, F., Gastineau, M., Correia, A., Levrard, B., 2004. A long-term numerical solution for the insolation quantities of the earth. Astron. Astrophys. 428, 261–286.

Le Heron, D.P., Buslov, M.M., Davies, C., Richards, K., Safonova, I., 2008. Evolution of mesozoic fluvial systems along the SE flank of the West Siberian Basin, Russia. Sediment. Geol. 208, 45–60.

Léonide, P., Floquet, M., Christophe, D., Baudin, F., Pittet, B., Lécuyer, C., 2012. Drowning of a carbonate platform as a precursor stage of the Early Toarcian global anoxic event (Southern Provence Sub-Basin, South-east France). Sedimentology 59, 156–184.

Lézin, C., Andreu, B., Pellenard, P., Bouchez, J.L., Emmanuel, L., Fauré, P., Landrain, P., 2012. Geochemical disturbance and paleoenvironmental changes during the Early Toarcian in NW Europe. Chem. Geol. 341, 1–15.

Littler, K., Hesselbo, S.P., Jenkyns, H.C., 2010. A carbon-isotope perturbation at the Pliensbachian–Toarcian boundary: evidence from the Lias Group, NE England. Geol. Mag. 147, 181–192.

Marjanac, T., Steel, R.J., 1997. Dunlin Group sequence stratigraphy in the Northern North Sea: a model for cook sandstone deposition. AAPG Bull. 81, 276–292.

Matthews, R.K., Al-Husseini, M.I., 2010. Orbital-forcing glacio-eustasy: a sequence stratigraphic time scale. GeoArabia 15 (3), 155–167.

Matthews, R.K., Frohlich, C., 2002. Maximum flooding surface and sequence boundaries: comparisons between observation and orbital forcing in the Cretaceous and Jurassic (65-190 Ma). GeoArabia 7 (3), 503–538.

McArthur, J.M., Steuber, T., Page, K., Landman, N.H., 2016. Sr-isotope stratigraphy: assigning time in the campanian, pliensbachian, toarcian, and valanginian. J. Geol. 124, 569–586.

McElwain, J.C., Wade-Murphy, J., Hesselbo, S.P., 2005. Changes in carbon dioxide during an oceanic anoxic event linked to intrusion into Gondwana coals. Nature 435, 479–482.

Mercuzot, M., Pellenard, P., Durlet, C., Bougeault, C., Meister, C., Dommergues, J.L., Thibault, N., Baudin, F., Mathieu, O., Bruneau, L., Huret, E., El Hmidi, K., 2019. Carbon-isotope events during the Pliensbachian (Lower Jurassic) on the African and European margins of the NW Tethyan realm. Newslett. Stratigr. 53, 41–69.

Montero-Serrano, J.C., Föllmi, K.B., Adatte, T., Spangenberg, J.E., Tribovillard, N., Fantasia, A., Suan, G., 2015. Continental weathering and redox conditions during the early Toarcian oceanic anoxic event in the northwestern Tethys: insight from the Posidonia Shale section in the Swiss Jura mountains. Palaeogeogr. Palaeoclimatol. Palaeoecol. 429, 83–99.

National Research Council, 1990. Sea-Level Change. The National Academies Press, Washington, DC. https://doi.org/10.17226/1345.

Ogg, J.G., Gradstein, F., 2020. TSCreator. In: Geological TimeScale Foundation. https://engineering.purdue.deu.

Ogg, J., Ogg, G., Gradstein, F., 2016. A Concise Geological Timesclae 2016. Elsevier, 240 p.

Pálfy, J., Smith, P.L., 2000. Synchrony between Early Jurassic extinction, oceanic anoxic event, and the Karoo-Ferrar Flood basalt volcanism. Geology 28, 740–747.

Pálfy, J., Smith, P.L., Mortensen, J.K., 2002. Dating the end-Triassic and early Jurassic mass extinctions, correlative large igneous provinces, and isotopic events. In: Koeberl, C., MacLeod, K.G. (Eds.), Catastrophic Events and Mass Extinctions: Impacts and Beyond. Geological Society of America Special Paper 356, pp. 523–532.

Percival, L.M.E., Cohen, A.S., Davies, M.K., Hesselbo, S.P., Jenkyns, H.C., Leng, M.J., Mather, T.A., Storm, M.S., Xu, W., 2016. Osmium isotope evidence for two pulses of increased continental weathering linked to Early Jurassic volcanism and climate change. Geology 44, 759–762.

Percival, L.M.E., Witt, M.L.I., Mather, T.A., Hermoso, M., Jenkyns, H.C., Hesselbo, S.P., Al-Suwaidi, A.H., Storm, M.S., Xu, W., Ruhl, M., 2015. Globally enhanced mercury deposition during the end-Pliensbachian and Toarcian OAE: a link to the Karoo–Ferrar Large Igneous Province. Earth Planet. Sci. Lett. 428, 267–280.
Peters, K.E., Walters, C.C., Moldowan, J.M., 2005. The Biomarker Guide: Volume 2, Biomarkers and Isotopes in Petroleum Systems and Earth. Cambridge University Press, 1132 p.
Peti, L., Thibault, N., Clémence, M.-E., Korte, C., Dommergues, J.-L., Bougeault, C., Pellenard, P., Jelby, M.E., Ullmann, C.V., 2017. Sinemurian-Pliensbachian calcareous nannofossil biostratigraphy and organic carbon isotope stratigraphy in the Paris Basin: Calibration to the ammonite biozonation of NW Europe. Palaeogeogr. Palaeoclimatol. Palaeoecol. 468, 142–161.
Pienkowski, G., Hodbod, M., Ullmann, C.V., 2016. Fungal decomposition of terrestrial organic matter accelerated Early Jurassic climate warming. Nat. Sci. Rep. 6, 31930.
Pittet, B., Suan, G., Lenoir, F., Duarte, L.V., Mattioli, E., 2014. Carbon isotope evidence for sedimentary discontinuities in the lower Toarcian of the Lusitanian Basin (Portugal): sea-level change at the onset of the oceanic anoxic event. Sediment. Geol. 303, 1–14.
Prauss, M., 1996. The lower Toarcian Posidonia Shale of Grimmen, Northeast Germany—Implications from the palynological analysis of a near-shore section. N. Jb. Geol. Paläont. Abh. 200, 107–132.
Ray, D.C., van Buchem, F.S.P., Baines, G., Davies, A., Greselle, B., Simmons, M.D., Robson, C., 2019. The magnitude and cause of short-term eustatic Cretaceous sea-level change: a synthesis. Earth-Sci. Rev. 197, 1–20.
Rocha, R.B., Mattioli, E., Duarte, L.V., Pittet, B., Elmi, S., Mouterde, R., Cabral, M.C., Comas-Rengifo, M.J., Gómez, J.J., Goy, A., Hesselbo, S.P., Jenkyns, H.C., Littler, K., Mailliot, S., de Oliveira, C.V., Osete, M.L., Perilli, N., Pinto, S., Ruget, C., Suan, G., 2016. Base of the Toarcian stage of the Lower Jurassic defined by the Global Boundary Stratotype Section and Point (GSSP) at the Peniche Section (Portugal). Episodes 39 (3), 460–481.
Röhl, H.J., Schmid-Röhl, A., 2005. Lower Toarcian (Upper Liassic) black shales of the Central European Epicontinental Basin: a sequence stratigraphic case study from the SW German Posidonia Shale. SEPM Special Publ. 82, 165–189.
Röhl, H.J., Schmid-Röhl, A., Oschmann, W., Frimmel, A., Schwark, L., 2001. The Posidonia Shale (Lower Toarcian) of SW-Germany: an oxygen-depleted ecosystem controlled by sea level and palaeoclimate. Palaeogeogr. Palaeoclimatol. Palaeoecol. 165, 27–52.
Ruebsam, W., Al-Husseini, M., 2020. Calibrating the Early Toarcian (Early Jurassic) with stratigraphic black holes (SBH). Gondwana Res. 82, 317–336.
Ruebsam, W., Mayer, B., Schwark, L., 2019. Cryosphere carbon dynamics control early Toarcian global warming and sea level evolution. Global Planet. Change 172, 440–453.
Ruebsam, W., Münzberger, P., Schwark, L., 2014. Chronology of the early Toarcian environmental crisis in the Lorraine Sub-Basin (NE Paris Basin). Earth Planet. Sci. Lett. 404, 273–282.
Ruebsam, W., Münzberger, P., Schwark, L., 2015. Reply to the comment by Boulila and Hinnov towards "Chronology of the early Toarcian environmental crisis in the Lorraine Sub-Basin (NE Paris Basin)" by W. Ruebsam, P. Munzberger, L. Schwark [Earth Planetary Science Letters 404 (2014) 273–282]. Earth Planet. Sci. Lett. 416, 147–150.
Ruebsam, W., Reolid, M., Schwark, L., 2020. $\delta^{13}C$ of terrestrial vegetation records Toarcian CO_2 and climate gradients. Nat. Sci. Rep. 10, 117.
Salzman, M.R., Thomas, E., 2012. Chapter 11—Carbon isotope stratigraphy. In: Gradstein, F.M., Ogg, J., Schmitz, M., Ogg, G. (Eds.), The Geological Timescale 2012. Elsevier Inc, pp. 207–232.

Same, B., Wagenreich, M., et al., 2016. Review: Short-term sea-level changes in a greenhouse world—A view from the Cretaceous. Palaeogeogr. Palaeoclimatol. Palaeoecol. 441, 393–411.

Saltzman, M.R., Runkel, A.C., Cowan, C.A., Runnegar, B., Stewart, M.C., Palmer, A.R., 2004. The upper Cambrian SPICE (d13C) event and the Sauk II-Sauk III regression: New evidence from Laurentian basins in Utah, Iowa and Newfoundland. J. Sediment. Petrol. 74, 366–377.

Schnyder, J., Pons, D., Yans, J., Tramoy, R., Abdulanova, S., 2016. Integrated stratigraphy of a continental Pliensbachian–Toarcian Boundary (Lower Jurassic) section at Taskomirsay, Leontiev Graben, southwest Kazakhstan. In: Brunet, M.F., McCann, T., Sobel, E.R. (Eds.), Geological Evolution of Central Asian Basins and the Western Tien Shan Range. London, Special Publications, Geological Society, p. 427. https://doi.org/10.1144/SP427.15.

Scholle, P.A., Arthur, M.A., 1980. Carbon isotope fluctuations in Cretaceous pelagic limestones: Potential stratigraphic and petroleum exploration tool. Am. Assoc. Petrol. Geol. Bull. 64, 67–87.

Shackleton, N.J., 1987. Oxygen isotopes, ice volume and sea level. Quat. Sci. Rev. 6, 183–190.

Shemin, G., Deev, E., Vernikovsk, V.A., Drachev, S.S., Moskvin, V., Vakulenko, L., Pervukhina, N., Sapyanik, V., 2019. Jurassic paleogeography and sedimentation in the northern West Siberia and South Kara Sea, Russian Arctic and Subarctic. Mar. Petrol. Geol. 104, 286–312.

Slater, S.A., Twitchett, R.J., Danise, S., Vajda, V., 2019. Substantial vegetation response to Early Jurassic global warming with impacts on oceanic anoxia. Nat. Geosci. 12, 462–467.

Snedden, J.W., Liu, C., 2010. A compilation of Phanerozoic sea level change, coastal onlaps, and recommended sequence designations. AAPG Search Discov. Art., 40594.

Storm, M.S., Hesselbo, S.P., Jenkyns, H.C., Ruhl, M., Ullmann, C.V., Xu, W., Leng, M.J., Riding, J.B., Gorbanenko, O., 2020. Orbital pacing and secular evolution of the Early Jurassic carbon cycle. Proc. Natl. Acad. Sci. U. S. A. 117, 3974–3982.

Suan, G., Mattioli, E., Pittet, B., Lécuyer, C., Suchéras-Marx, B., Duarte, L.V., Philippe, M., Reggiani, L., Martineau, F., 2010. Secular environmental precursors to Early Toarcian (Jurassic) extreme climate changes. Earth Planet. Sci. Lett. 290, 448–458.

Suan, G., Nikitenko, B.L., Rogov, M.A., Baudin, F., Spangenberg, J.E., Knyazev, V.G., Glinskikh, L.A., Goryacheva, A.A., Adatte, T., Riding, J.B., Föllmi, K.B., Pittet, B., Mattioli, E., Lécuyer, C., 2011. Polar record of early Jurassic massive carbon injection. Earth Planet. Sci. Lett. 312, 102–113.

Suan, G., van de Schootbrugge, B., Adatte, T., Fiebig, J., Oschmann, W., 2015. Calibrating and magnitude of the Toarcian carbon cycle perturbation. Paleoceanography 30, PA2758.

Sundquist, E.T., 1990. Long-term aspects of future atmospheric CO_2 and sea-level changes. In: National Research Council Staff (Ed.), Sea-Level Change. Academies Press, pp. 193–207.

Svensen, H.H., Planke, S., Chevallier, L., Malthe-Sørenssen, A., Corfu, F., Jamtveit, B., 2007. Hydrothermal venting of greenhouse gases triggering Early Jurassic global warming. Earth Planet. Sci. Lett. 256, 554–566.

Svensen, H.H., Torsvik, T.H., Callegaro, S., Augland, L., Heimdal, T.H., Jerram, D.A., Planke, S., Pereira, E., 2017. Gondwana Large Igneous Provinces: plate reconstructions, volcanic basins and sill volumes. Geol. Soc. Lond. Spec. Publ. 463, 17–40.

Tchoumatchenco, P., et al., 2008. Did glaciation occur during the Toarcian (Early Jurassic) in the east Strata Planina Mts. (East Bulgaria)? Comptes rendus de l'Académie bulgare des Sciences 61, 1301–1308.

Them, T.R., Gill, B.C., Caruthers, A.H., Gröcke, D.R., Tulsky, E.T., Martindale, R.C., Poulton, T.P., Smith, P.L., 2017a. High-resolution carbon isotope records of the Toarcian Oceanic Anoxic Event (Early Jurassic) from North America and implications for the global drivers of the Toarcian carbon cycle. Earth Planet. Sci. Lett. 459, 118–126.

Them, T.R., Gill, B.C., Selby, D., Gröcke, D.R., Friedman, R.M., Owens, J.D., 2017b. Evidence for rapid weathering response to climatic warming during the Toarcian Oceanic anoxic event. Nat. Sci. Rep. 7, 5003.

Thibault, N., Ruhl, M., Ullmann, C.V., Korte, C., Kemp, D.B., Gröcke, D.R., Hesselbo, S.P., 2018. The wider context of the Lower Jurassic Toarcian oceanic anoxic event in Yorkshire coastal outcrops, UK. Proc. Geol. Assoc. 129, 372–391.

Ullmann, C., Boyle, R., Duarte, L., Hesselbo, S.P., Kasemann, S.A., Klein, T.M., Lenton, T.M., Piazza, V., Aberhan, M., 2020. Warm afterglow from the Toarcian Oceanic anoxic event drives the success of deep-adapted brachiopods. Sci. Rep. 10, 6549.

Vakhrameev, V.A., 1991. Jurassic and Cretaceous Floras and Climates of the Earth. Cambridge University Press, 318 p.

van Breugel, Y., Baas, M., Schouten, S., Mattioli, E., Sinninghe Damsté, J.S., 2006. Isorenieratane record in black shales from the Paris Basin, France: constraints on recycling of respired CO_2 as a mechanism for negative carbon isotope shifts during the Toarcian oceanic anoxic event. Paleoceanography 21, PA4220.

Wendler, J.E., Wendler, I., 2016. What drove sea-levelfluctuations during the mid-Cretaceous greenhouse climate? Palaeogeogr. Palaeoclimatol. Palaeoecol. 441 (2016), 412–419.

Wendler, I., Wendler, J.E., Clarke, L.J., 2015. Sea-levelreconstruction for Turonian sediments fromTanzaniabased on integration of sedimentology, microfacies,geochemistry and micropaleontology. Palaeogeogr. Palaeoclimatol. Palaeoecol. 441, 528–564.

Williams, C.J., Hesselbo, S.P., Jenkyns, H.C., Morgans-Bell, H.S., 2001. Quartz silt in mudrocks as a key to sequence stratigraphy (Kimmeridge Clay Formation, Late Jurassic, Wessex Basin, UK). Terra Nova 13, 449–455.

Xu, W., Ruhl, M., Jenkyns, H.C., Hesselbo, S.P., Riding, J.B., Selby, D., Naafs, B.D.A., Weijers, J.W.H., Pancost, R.D., Telelaar, E.W., Idez, E.F., 2017. Carbon sequestration in an expanded lake system during the Toarcian oceanic anoxic event. Nat. Geosci. 10, 129–134.

Xu, W., Niocaill, C.M., Ruhl, M., Jenkyns, H.C., Riding, J.B., Hesselbo, S.P., 2018a. Magnetostratigraphy of the Toarcian Stage (Lower Jurassic) of the Llanbedr (Mochras Farm) Borehole, Wales: basis for a global standard and implications for volcanic forcing of palaeoenvironmental change. J. Geol. Soc. 175 (4), 594.

Xu, W., Ruhl, M., Jenkyns, H.C., Leng, M.J., Huggett, J.M., Minisini, D., Ullmann, C.V., Riding, J.B., Weijers, J.W.H., Storm, M.S., Percival, L.M.E., Tosca, N.J., Idiz, E.F., Tegelaar, E.W., Hesselbo, S.P., 2018b. Evolution of the Toarcian (Early Jurassic) carbon cycle and global climatic controls on local sedimentary processes (Cardigan Bay Basin, UK). Earth Planet. Sci. Lett. 484, 396–411.

Zhang, J., Yu, H., Jia, G., Chen, F., Lui, Z., 2010. Terrestrial *n*-alkane signatures in the middle Okinawa Trough during the post-glacial transgression: control by sea level and paleovegetation confounded by offshore transport. Geo-Mar. Lett. 30, 143–150.

Zimmermann, J., Franz, M., Heunisch, C., Luppold, F.W., Mönnig, E., Wolfgramm, M., 2015. Sequence stratigraphic framework of the lower and middle Jurassic in the North German Basin: epicontinental sequences controlled by Boreal cycles. Palaeogeogr. Palaeoclimatol. Palaeoecol. 440, 395–416.

CHAPTER THIRTEEN

Assessing shale gas reservoir potential using multi-scaled SEM pore network characterizations and quantifications: The Ciñera-Matallana pull-apart basin, NW Spain

Ethan J. Richardson[a,b] **and Michael Montenari**[a,*]
[a]Geography, Geology and the Environment, Keele University, Newcastle under Lyme, United Kingdom
[b]35 Thoresby Road, York, United Kingdom
[*]Corresponding author: e-mail address: m.montenari@keele.ac.uk

Contents

Abstract

In recent years, unconventional hydrocarbon reservoirs (UHRs), specifically shale gas plays, have become a vast source of economically viable hydrocarbons, in response

Stratigraphy & Timescales, Volume 5
ISSN 2468-5178
https://doi.org/10.1016/bs.sats.2020.07.001

to better drilling and stimulation techniques. To further improve recovery, more detailed pore network modeling needs to be conducted by means of detailed pore network characterizations, based upon predefined pore size and pore type classification schemes. Through the quantification of the nano/microporous properties of the pore network, a better understanding on how hydrocarbon storage and migration phenomena operate within UHRs and how well they might respond to stimulation can be developed. Kasimovian-Gzhelian (Late Carboniferous) lacustrine sediments from the Upper Pastora Formation of the Ciñera-Matallana coalfield, NW Spain, were investigated to identify the potential for shale gas, by conducting a detailed pore network analysis using SEM-based porosity data, alongside traditional methods of formation evaluation. Here we show that the Upper Pastora Formation demonstrates shale gas potential and that the nanoporous network is a complex multifaceted and multi-scaled system composed of various pore types and sizes, each with specific dynamic properties and varying contributions toward total porosity. The present contribution provides a full pore network characterization and quantification, based on pore type and size, followed by a quantification of the porosity percentage held within each pore type and size. This work will provide an insight into the need for pore network characterizations alongside pore property quantifications with regard to computational fluid flow models and as such the need to carry out this work in conjunction with more traditional methods of shale gas exploration.

1. Introduction

Unconventional hydrocarbon reservoirs (UHRs) are reservoirs characterized by low permeability but with great hydrocarbon potential, most notable of which are shale gas plays, tights gas sands and coalbed methane (Chengzao, 2017). One of the most productive UHRs which has currently been reshaping the global energy industry, are shale gas plays, characterized by high organic matter (OM) content and low–ultra low permeability (0.0001 mD). These reservoirs hold a much greater volume of hydrocarbons than their conventional counterparts (Zendehboudi and Bahadori, 2016). Acting as both source and seal, large quantities of gas are held in very small nano/μm scale pores within dissolved, adsorbed and free gas states, throughout a multi-scaled system, spread over several orders of magnitude. With shales representing the most abundant type of sedimentary rock on Earth (Boyer et al., 2013) and unproven, but technically recoverable shale gas reserves estimated at 7576.6 trillion cubic feet (Tcf), from 46 of the world's most shale rich countries, this UHR is a highly valuable resource (EIA, 2013). However, traditional methods of recovery cannot be used due to poor flow properties within the pore network, requiring hydraulic fracturing to increase recovery (Guo et al., 2018).

Extending over 19,000 km^2, the Cantabrian Massif has been subject to a limited number of hydrocarbon exploration campaigns, with only two hydrocarbon wells spanning the entire area (González et al., 2017). Despite the Central Carboniferous Basin being identified as offering source rock potential and the presence of mine gas, predominately consisting of methane and traces of ethane, having long been recognized, no published shale gas/oil reports for the Cantabrian Massif exist (European Commission, 2016). The Ciñera-Matallana Basin, located in the Southern Cantabrian Zone, was identified by Lemos de Sousa and Pinheiro (1996) as the most important coalfield with real natural gas prospects in the Cantabrian Massif. Further work by Cienfuegos and Loredo (2010) within the Asturian Central Coal Basin located further to the north, estimated that coal bed methane (CBM) resources contain a minimum of 25,100 Mm^3 of gas. This is clear evidence of the great potential in the Cantabrian Massif for CBM, but currently, no methane has been economically extracted in the region (Álvarez et al., 2018). As with most studies into UHRs within the Cantabrian Massif, CBM has been the dominant focus, while shale gas potential has remained untested and is at a very preliminary stage of understanding and exploration (Mackenzie, 2018). A preliminary study by Álvarez et al. (2018), aimed at assessing UHR source rock potential of 12 Silurian and Carboniferous lithostratigraphic units across the Cantabrian Zone, found that according to TOC results, the Naranco, San Emiliano, Fito Ereñes, and Barcaliente Formations offer a moderately good potential for hydrocarbon generation. However, thermal maturity data based on vitrinite reflectance indicated that all Silurian/Carboniferous dated strata were too immature for the development of thermogenic gas as they clearly plotted within the oil generation window.

While the search for shale gas potentials within the Cantabrian Massif is still in an initial exploration phase, around the world the potential of shale gas has already been realized as a cleaner alternative to coal which can be used as a "bridging fuel" during the transition to a CO_2 neutral future. Furthermore, shale gas has proved to be a cost efficient source of energy which reduces expenditure on imports and increases energy security while bolstering domestic economic prosperity through the generation of regional job opportunities (Bazilian et al., 2014). However, our understanding of the physicochemical properties and characteristics of micro- and nanometric pores, which largely control hydrocarbon migration and storage in shale gas reservoirs, still remains largely insufficient in both predicting and maximizing well productivity. It appears that the rapidly increasing demand and the subsequent production of shale gas was not accompanied by an adequate investment into the detailed research of the very fundamentals of shale gas generation and exploitation. The knowledge of how these fundamental properties within the pore network operate in response to recovery, is still in an early if not rudimentary stage, especially with regard to the physical properties that control the flow regimes (King et al., 2015).

One of the major issues facing the industry is the sustainability of the so-called shale gas revolution with the significant decline in production after the initial 12-month flow-back period. Production data from the Late Devonian to Early Carboniferous Bakken Shale and the early Late Cretaceous (Cenomanian to Turonian) Eagleford Shale gas plays in the U.S. have shown a 60% decrease in production after the first 12 months, while gas wells in 5 of the most productive shale gas plays across the U.S. have reported that after 3 years, wells only achieve between 5% and 20% of their initial production rate (Hu et al., 2015). Further production data from wells located in the Early Carboniferous Barnett Shale (Texas) also show that over a 10-year period high performing wells produce 33% of their gas within the first 12 months and low performing wells 22% (Middleton et al., 2017). Such production data has given rise to the assumption that there are two fundamentally different physical mechanisms controlling shale gas production, with regard to initially high, followed by later sustained but significantly lower production rates seen in tail-end production. This significantly reduced production rate is generally attributed to desorption processes. The majority of shale gas production, however, occurs in this latter stage but over a prolonged period, which highlights the need for research to maximize the recoverability during tail-end production. It is estimated that by enhanced desorption alone, wells could yield up to 20–30% more gas over a 5–10-year period (Middleton et al., 2015). This phenomenon, which is becoming more and more apparent in mature wells across U.S. shale gas plays, has led to the conclusion that future decline rates of shale gas have been underestimated and as a result well longevity and estimated ultimate recovery (EUR) severely overestimated (McGlade et al., 2013). To overcome this, new wells are constantly being drilled to make up for the rapid decline of mature wells in order to maintain shale gas production (Wang, 2018). However, another way in which these problems could be addressed is through developing a deeper understanding of the multi-scaled distribution of different pore types throughout the pore network, using an integrated analytical approach aimed at providing a full, visual, quantitative analysis of the pore network and associated properties. Upon completion, associated changes in specific pore type abundances, permeability and flow-back data will provide a deeper understanding of pore type controls on gas storage and migration mechanisms. Further research into hydrocarbon migration and storage in fine clastic materials such as shales is vital for evaluating the long term behavior of shale gas reservoirs, the economic viability of recovery operations and the reduction of exploration risks (Ross and Bustin, 2009). However, up to the present date there remains only a moderate understanding of the full characteristics of pore systems in shales, with only limited work conducted on the systematic quantification of pore networks, and with detailed imaging being rare (Curtis et al., 2012; Ma et al., 2018).

In order to better understand important productivity controlling factors, the pore network must be systematically measured and quantitatively described (Anovitz and Cole, 2015). In this paper, we approach this problem by (1) identifying and quantifying the petrographic loci of the porosity; (2) establishing a pore network characterization, followed by (3) identifying which pore types have a similar mode of occurrence and as such, display similar properties, which in turn are considered to be different from other pore types found throughout the network. This is an area which has previously been understudied, with most published research focusing on understanding the factors controlling pore type genesis and rarely conducting an in-depth analysis into pore type contributions toward total porosity (Chen et al., 2019). As the majority of hydrocarbons within shale gas plays are thought to be stored within nanopores, which have a very low permeability (Scotchman, 2016), this characterization will quantify both the nanoporous and microporous systems as separate entities. As decreasing pore radii result in multiple flow regimes operating within one and the same pore network, Darcy's law can no longer be utilized to accurately describe gas flow within extremely small nanopore systems. Instead, a wide variety of non-Darcy flow regimes controlled by increases in intermolecular and molecule-wall collisions (such as viscous flow, slip flow, transition flow, and Knudsen flow) are being realized within these systems (Shen et al., 2017; Zhang et al., 2015). A greater understanding of the dimensions and characteristics of pores and the resulting pore networks within shale gas plays is critical for any further development and exploration (Nie et al., 2018), especially with regard to pore types and sizes as these factors strongly affect the reservoir properties (Goral et al., 2015).

2. Geological setting

The Late Carboniferous Pastora Formation is developed in the western part of the Ciñera-Matallana Basin (northern Castilla and Leon, NW Spain), which is home to one of Europe's largest opencast coalmines. The coalmine contains medium to high-rank coals ranging from Bituminous C to Anthracite A. Located 42 km north of Leon and 2.5 km east from the mining village of Santa Lucia (Fig. 1A, B), this small pull-apart basin was formed during the Variscan Orogenesis and takes the form of a complex half-graben system. It extends over 12 km east-west and is considered to have formed as a result of multiple transtensional/transpressional movements along the Sabero-Gordon strike-slip fault (Ayllon et al., 2008). The mine displays in total a 1500 m succession of Late Carboniferous (Kasimovian-Gzhelian; Fig. 1C) strata (Colmenero et al., 2008), which rest unconformably on the Asturian discordance (Frings et al., 2004).

Siliciclastic in nature, these sediments are interpreted to be largely swamp-, lacustrine-, fluvial and alluvial fan deposits (Heward, 1978). The Pastora Formation itself is an 85–200 m thick formation consisting of lacustrine and swamp deposits and is one of two stratigraphic formations out of the seven that occupy the mine, which demonstrates the greatest potential for natural gas prospecting (Lemos de Sousa and Pinheiro, 1996). The numerous coal seams which are present within the formations, most notably within the Pastora Formation, which consists of up to 30 separate seams ranging from 0.5 to 15 m in thickness, initiated the research into the natural gas potential. The largest of these seams can be found in the Mid-Upper Pastora Formation where coal seams are most abundant. The regional-scale facies distribution of this largest coal seam horizon, the so-called "Pastora Coal Seam," provides evidence for an NW-SE striking lake, with large coal deposits forming in past swamp environments located at the lake margins (Wagner, 1971). Toward the top of the formation, sedimentation appears to have formed in a solely lacustrine environment, characterized by interbedded organic matter (OM)-rich, finely laminated shale successions, very fine-grained siltstones, and intercalated up to meter thick coarse-grained sandstones. These lacustrine successions formed the target of the present study.

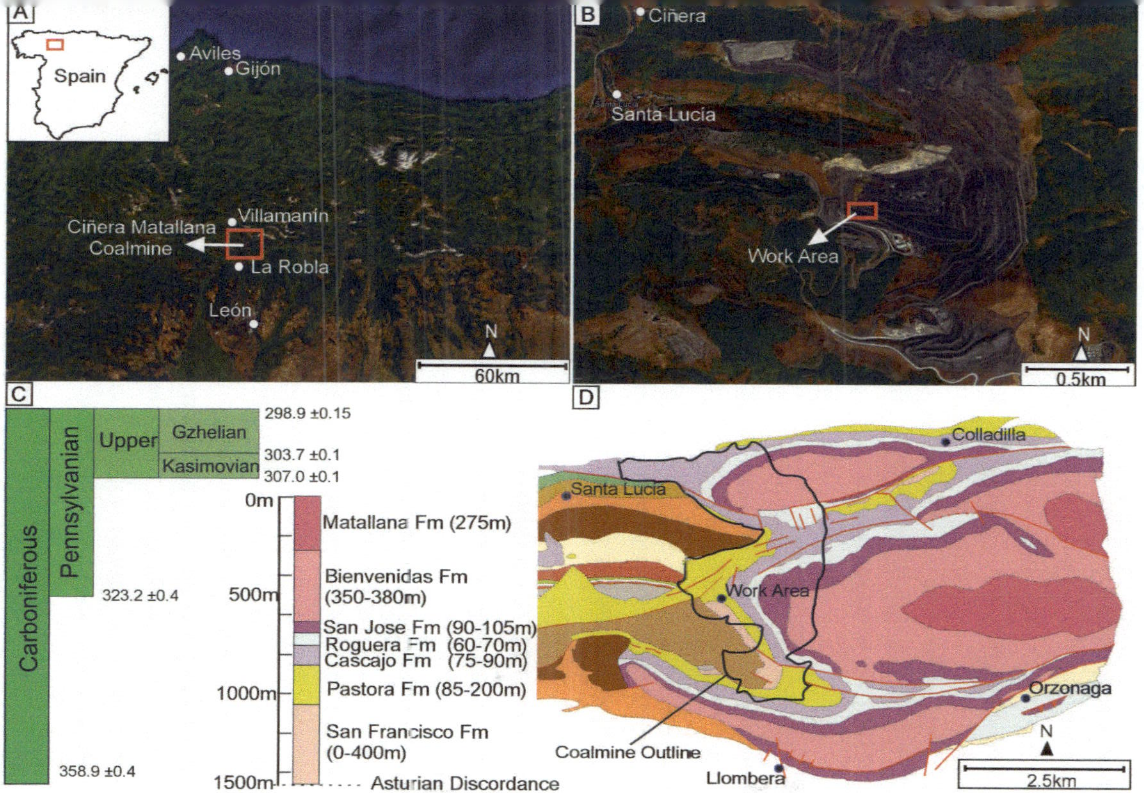

Fig. 1 Location map of the Ciñera-Matallana Coalmine. (A) Aerial view of Cantabria NW Spain. (B) Aerial view of the Ciñera-Matallana coalmine. (C) Late Carboniferous timescale (D) Geological map of the area surrounding the Ciñera-Matallana coalmine, with the coalmine's outline highlighted in black. *(A) and (B) Adapted from Google Maps, (C) adapted from Cohen, K.M., Finney, S.C., Gibbard, P.L., Fan, J.X., 2013. The ICS International Chronostratigraphic Chart. Episodes 36 (3), 199–204, (D) adapted from IGME, 1989. MAGNA 50—Sheet Number 103 (LA POLA DE GORDÓN) Scale 1:50.000 [image]. Available at: http://info.igme.es/cartografiadigital/geologica/Magna50.aspx [Accessed 15 October 2019].*

3. Material and methods

3.1 Material

For the present study, a total of 90 samples were extracted in stratigraphic order (bed-by-bed) from the Upper Pastora Formation. Sampling locality was the southern part of the Santa Lucia coal mine (which forms part of the Ciñera-Matallana coal field complex) as it provided the best exposure and accessibility. The samples were taken in conjunction with a high-resolution sedimentary log of a 12 m section of the Upper Pastora Formation (Figs. 2 and 3) which consists of interbedded sandstone/siltstone and shale beds (Fig. 4A–C). A spectral gamma ray log to complement the sedimentary log was generated using a handheld RS-230-BGO gamma ray spectrometer (Radiation Solutions Inc., Canada). Spectral gamma ray data were obtained as a "gliding average" equidistantly every 5 cm with an analytical time of 360 s per analytical point. The concentrations of the major element potassium (K) are expressed as weight percentages (wt%), whereas the concentrations of the trace elements thorium (Th) and uranium (U) are given as parts per million (ppm).

3.2 Facies and microfacies analysis

Samples were subject to a detailed facies- and microfacies analysis based on field observations, hand specimens and thin sections in order to identify the hydrocarbon resource potential and to categorize any changes in micro/nanometric porosity and pore-based properties, in association with petrological/mineralogical changes that might impact on hydrocarbon migration and accumulation. Microfacies analysis was conducted using a ZEISS petrological microscope, a HITACHI-TM3000 scanning electron microscope (SEM) equipped with an energy dispersive X-ray analyzer (EDAX) and a NITON-XL3 XRF analyzer. This facies analysis enabled the identification of key mineralogical variations between the shale/siltstone/sandstone beds and the mineralogical/geochemical subdivision based on geochemical composition, OM observations and sedimentary/biogenic structures.

3.3 Pore network characterization

To characterize the pore network and identify changes in porosity and permeability, binarized images were described and pore types quantified, followed by a series of detailed image-based analyses of the pore space.

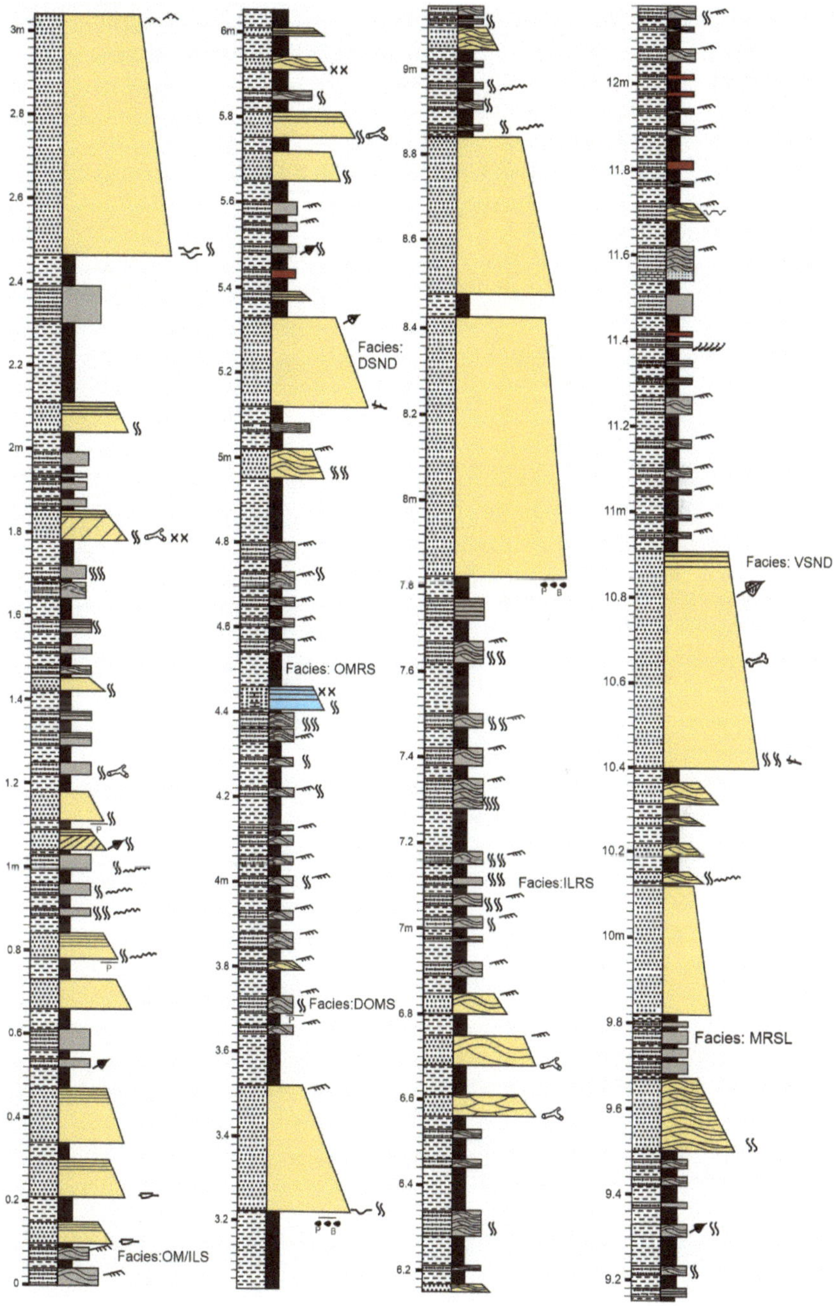

Fig. 2 High-resolution 12-m sedimentary log of the Upper Pastora Formation with sampled microfacies locations.

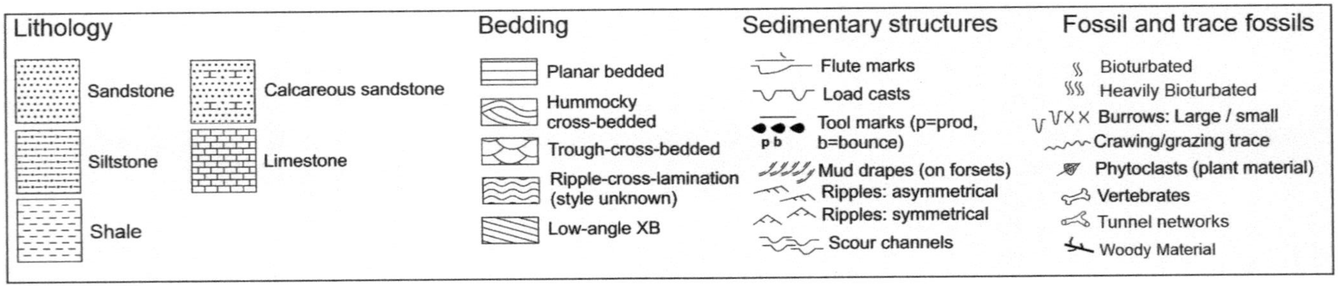

Fig. 3 Key for sedimentary structures.

Fig. 4 Outcrop of the Upper Pastora Formation. (A–C) Interbedded sequence of argillaceous sediment of the Upper Pastora Formation with 5-m staff for scale. The formation is clearly marked with large intermittent sandstone beds which interrupt the more finely bedded siltstone/shale sequence. (D) Sandstone bed with weak hummocky cross-stratification. (E) Very finely interbedded shales/siltstones between two large dolomitized sandstone beds. (F) Siltstone displaying quasi-planar laminations at the top of the bed with a carbonate base. (G) Basal bioturbation (thalassinoides) in a larger sandstone bed. (H) Calamite fossil. (I) Clustered group of fossilized plant material.

This procedure was executed by generating backscattered electron (BSE) data using a HITACHI-TM3000 SEM. The resulting BSE images were digitally binarized using ImageJ and Corel Draw software packages to generate clear black and white images, with the black parts representing the pores and white parts representing the non-porous background. BSE images were taken systematically at a magnification of 500 times and 1200 times. The 1200 times magnification was identified as the most appropriate scale to analyze nanometric variability whereas the 500 times magnification proved to be effective in analyzing the micrometric variability. The choice of such magnification ranges and scales ran in accordance with the pore size definitions by Loucks et al. (2012; see Fig. 5) and with the limits of SEM resolution. Although the acquisition of representative images of such heterogeneous samples is an inherent problem in pore network modeling (Pak et al., 2016), the scales chosen in this study proved to provide an effective basis for the integration of different scale observations and also allowed for hierarchal data separation and integration.

The acquired BSE images were saved in TIF format; these images were then adjusted appropriately using the software components ImageJ Fiji to acquire a gray range, which contained black, white and intermediate gray levels (Jiang, 2013). The binarized image was then exported to Corel Draw 2017 and a high-resolution trace of the binarized image was generated. Once generated, the white background was digitally removed and replaced with the original BSE image which allowed for the manual selection of individual and groups of pores in preparation for the pore characterization. Using the BSE image as a base image to guide the pore identification, pore types were identified based on the pore type classification scheme adapted from Loucks et al. (2012) and subsequently color coded, to provide a multi-colored binarized BSE image (Figs. 5–7). The pore size- and type classification schemes of Loucks et al. (2012), who analyzed and described a broad and varied spectrum of different pore types within mudrocks, enabled the identification of the most frequent nano- and micropore types found within the various microfacies. Once all pore types were characterized and color coded, specific pore types were digitally extracted from the image and placed on a separate BSE image, thus representing and illustrating a single pore type within a given pore network. These images were transformed into a grayscale (8-bit) format and returned to the software package ImageJ and its components for re-binarization and the subsequent analysis and quantification (Fig. 7).

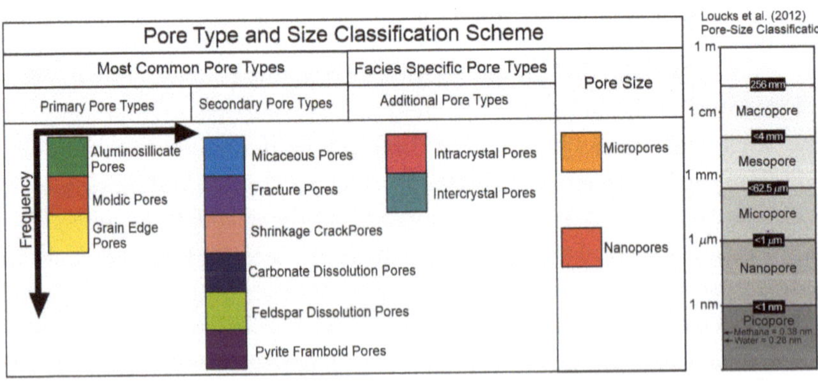

Fig. 5 Pore type and size classification key listing the most frequent pore types found throughout all microfacies.

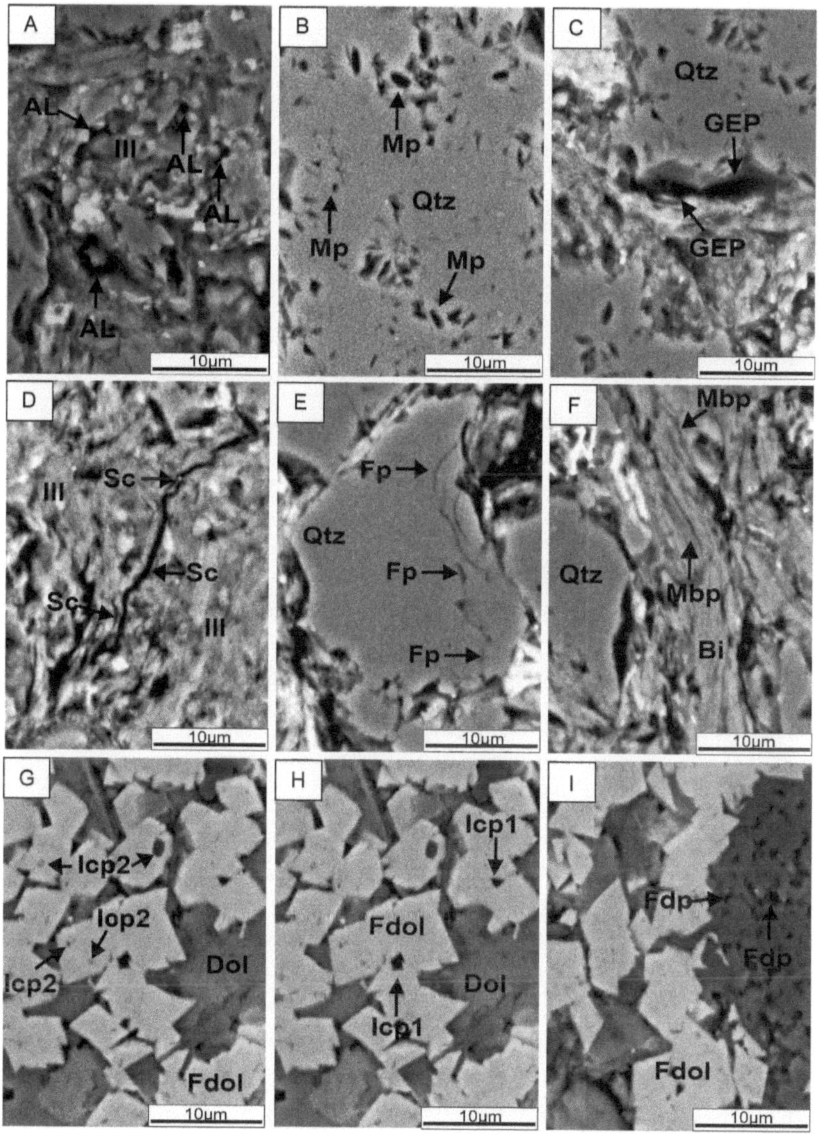

Fig. 6 Scanning electron microscope (backscatter electron, BSE) images depicting the spectrum of pore types used for the characterization of microfacies pore networks. (A) Aluminosilicate pores (AL) found between grains. (B) Moldic pores (MP) isolated within mineral phases, showing elongate pore morphologies. (C) Grain edge pore (GEP) located adjacent to a rigid grain. (D) Shrinkage crack pore (SCP) located within the illite matrix. (E) Facture pores (FP) concentrated around a small area within a quartz grain. (F) Micaceous pores (MBP) found closely grouped within a biotite grain. (G) Intracrystal pores (ICP1) found isolated within ferroan dolomite crystals. (H) Intercrystal pores (ICP2) found between closely packed ferroan dolomite crystals. (I) Feldspar dissolution pores (FDP) found within a highly weathered feldspar grain displaying a high pore density. There are also carbonate dissolution pores which are referred to as (CDP).

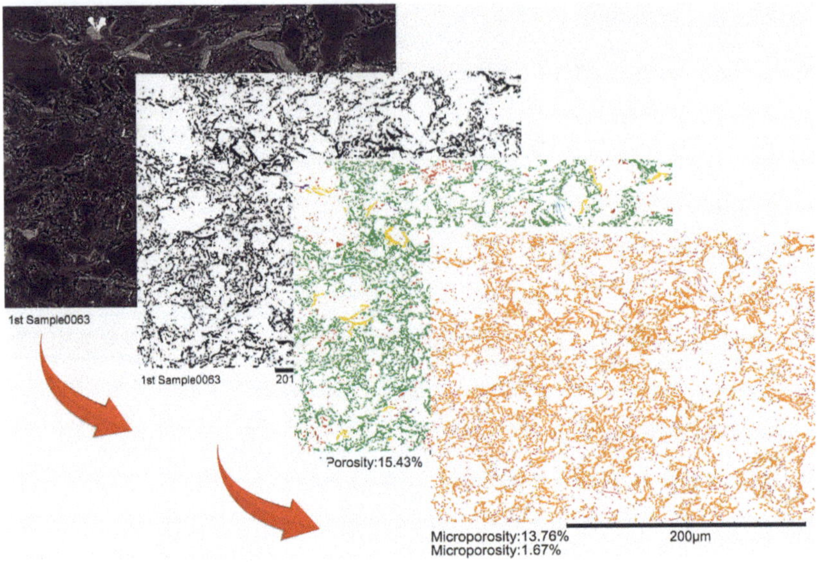

Fig. 7 The process of image binarization and subsequent pore type and pore size characterization from a back scattered electron image (example: OMRS microfacies, for explanations see text).

4. Results

4.1 Lithological description

The stratigraphically youngest parts of the Late Carboniferous Pastora Formation are characterized by a continuous succession of interbedded thinly layered fissile shales, very fine-grained siltstones, and sandstones, which generally display weak quasi-planar (QP)/hummocky cross-stratification (HCS) with vitrified bases (Fig. 4A, F). These dominant lithologies also occur with less frequent limestone and calcareous/iron-rich siltstone intercalations (Fig. 4F). Commonly, this background sedimentation pattern is interrupted by 10–60 cm thick, massive bedded, coarse-grained sandstones (Fig. 4B). These larger sandstone beds typically show the widest variety of textural, bioclastic and sedimentological features, including basal erosional features, (Fig. 4H), bone fragments, palynomorphs, and phytoclastic material (Fig. 4G–I). The largest sandstone bodies (up to 60 cm in thickness) found in this predominantly argillaceous sedimentary succession, are also characterized by a dolomitic diagenetic cement, which greatly reduces porosity, acting as an intermittent seal above the more nano- and microporous shale and siltstone facies. Most shales and siltstone facies show evidence of only minor dolomitization (varying between 1% and 5%). However, it can be demonstrated that this varying degrees of dolomitization do have an impact on hydrocarbon migration by reducing the overall porosity and more importantly the pore network connectivity. Generally, all facies types show a strong degree of bioturbation with more diverse ichnofabrics found in the fine-grained sandstones such as the decapod dwelling structure *Thalassinoides* (Fig. 4G) and traces probably produced by annelids within the fine-grained mudstones (Fig. 4C). Organic matter (OM) is ubiquitous and present in all samples as either bed-parallel laminae or pervasive matrix-based OM. Strong enrichments in OM are typically found within the Organic Matter/Illite Rich Siltstone facies (OM/ILS), the Organic Matter Rich/Dolomitized Siltstone facies (DOMS), the Organic Matter Rich Shale facies (OMRS), and the Varved Sandstone facies (VSND, see Fig. 8A, B, F, G). Randomly distributed

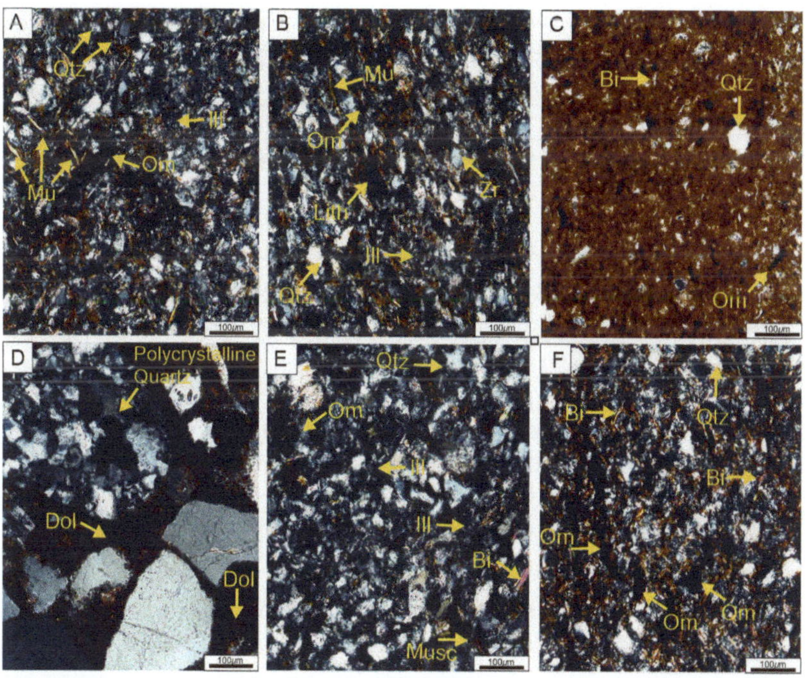

Fig. 8 Annotated thin section (100 μm-scale) of the seven most abundant microfacies found along the 12-m logged section. (A) OM/ILS. (B) DOMS. (C) OMRS. (D) DSND. (E) ILRS. (F) MRSL (all images with crossed polarizers). Bi = Biotite, Dol = Dolomite, Ill = Illite, Mu and Musc = Muscovite, OM = Organic matter, Qtz = Quartz, Zr = Zircon.

OM shows a variety of structured morphologies that have been interpreted as woody tissue and other phytoclast fragments. Structures interpreted here as varves are typically formed by bundles of OM rich- and siltstone laminae. These varves most commonly occur in the DSND and VSND facies with up to 60 varves in just 5 cm. As such, this section of the Upper Pastora Formation can be described and interpreted as a multi-layered sealing sequence (sensu Bertoni et al., 2018), in which the paleoenvironment is interpreted to be a hydrologically open lake, located close to a river mouth distributing coarser grained sediment.

4.2 Mineralogy

Mineralogically, all mudstones and siltstones show similar compositions but in varying quantities. The intergranular space is dominated by a mixed clay content, which is predominantly Illite (20–33%) with localized dolomitization (<6%). The most common grains are detrital quartz grains (40–60%), which show varying textures and structures. Albite (8–14%), biotite (2–8%), muscovite (2–13%) which are prominent within the shale facies and lithic fragments (1–11%) which are most prominent in sandstones. Accessory minerals include pyrite (2–3%), rutile (<3%) and zircon (<1%) (Figs. 8–10).

4.3 Facies types

The overall petrological characteristics of the 90 samples were described and quantified using SEM-EDS, XRF and conventional light microscopy and allowed for the differentiation and establishment of 7 different facies types:

1. Organic Matter Rich Shale (OMRS)
2. Organic Matter/Illite Rich Siltstone (OM/ILS)
3. Organic Matter Rich/Dolomitised Siltstone (DOMS)
4. Illite Rich Siltstone (ILRS)
5. Mica Rich Siltstone (MRSL)
6. Dolomitised Sandstone (DSND)
7. Varved Sandstone (VSND)

These facies types occur in a repetitive sealing succession, whereby smaller interchanging shale and sandstone/siltstones units are repeatedly capped by much more prominent sandstone bodies. This rhythmic sedimentation pattern has created a dual self-sealing system whereby shales, which already act as a self-sealing and self-sourcing hydrocarbon reservoirs, are locked within a larger sealing system capped by very low porosity sandstones (VSND), confining any escaped gas to these layers (Fig. 2).

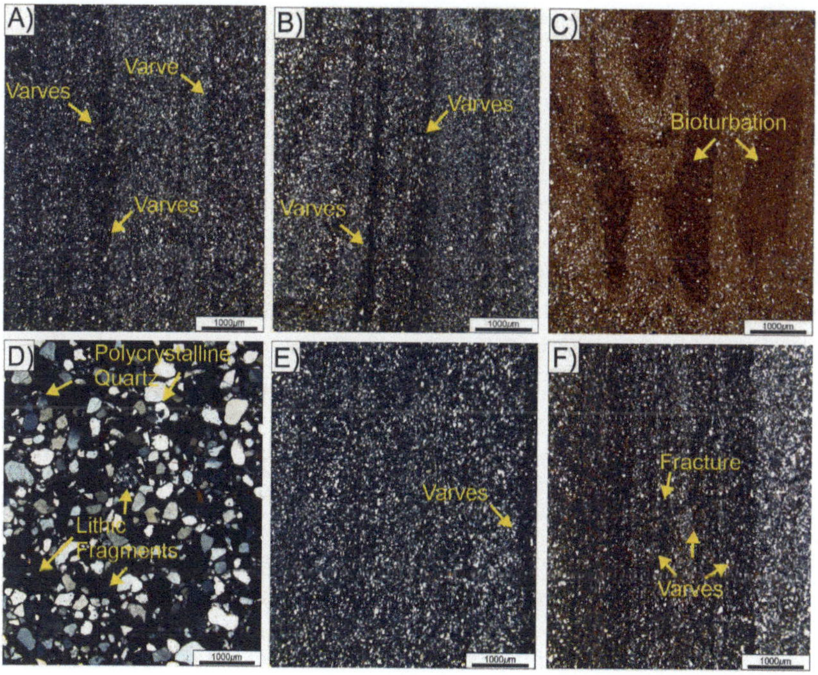

Fig. 9 Annotated thin section (1000 μm-scale) of the seven most abundant microfacies found along the 12-m logged section. (A) OM/ILS. (B) DOMS. (C) OMRS. (D) DSND. (E) ILRS. (F) MRSL (all images with crossed polarizers).

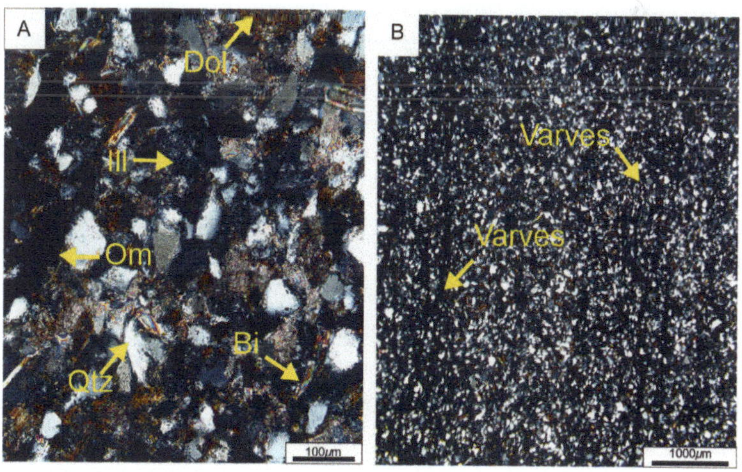

Fig. 10 Annotated thin section of VSND facies. (A) 100 μm-scale and (B) 1000 μm-scale. Bi = Biotite, Dol = Dolomite, Ill = Illite, Om = Organic matter, Qtz = Quartz (all images with crossed polarizers).

4.4 Organic matter content

4.4.1 *Pervasive organic matter*

Pervasive organic matter (OM) found within the matrix and inter-varve regions have a variety of distinct structured morphologies indicating a diverse range of OM sources. Macerals were identified by variations in morphology which correspond to observations made by the International Committee for Coal and Organic Petrology (ICCP) and other published studies on maceral identification (Cardott and Curtis, 2018; Mastalerz et al., 2010; Pickel et al., 2017). Major maceral groups identified were vitrinite and liptinite, derived from woody tissue, leaf cuticles and cell walls of plant tissues. Some of the maceral subgroups identified include secretinite with cell lumens filled with clay minerals, cutinite with serrated edges on one side and smooth on the other, and finally various vitrinite subgroups related to woody tissue, characterized by thin elongate and straight edge structures (Fig. 11A–I). Various palynomorph and botryococcoid green algal cysts have also been identified, providing further evidence for a diverse range in secondary sourced OM but also evidence from primary production of OM within the lake.

4.4.2 *Organic matter rich varves*

Varves, unlike other dispersed organic matter, consist of lateral concentrations of organic matter in a laminar and occasionally, undulatory fashion. There are two distinctive types of varves present throughout the sequence. The most abundant type shows excellent lateral continuity and connectivity, which can display a sinuous flow pattern and, in some cases, large concentrations of pyrite framboids. These varves contain a high and concentrated OM content which appears at lower magnifications to be amorphous. However, at higher resolution the OM of these varves does show a bubble-like nanomorphology, indicating that the OM within these varves is algal in nature, as they have a similar morphology to unicellular algal colonies found in the Wufeng-Longmaxi Formation in the Sichuan Basin, SW China (Nie et al., 2018). The bundle like nature of these varves could well be a photosynthetic response to solar variations, providing an insight into seasonal variability in lacustrine primary paleoproductivity. More importantly, they provide evidence for a continued and prolonged source of type I kerogen with high H/C values. The second varve type present is more common in the upper reaches of coarser-grained sandstone bodies and consists of quasi-laminar concentrations of structured organic matter, with a

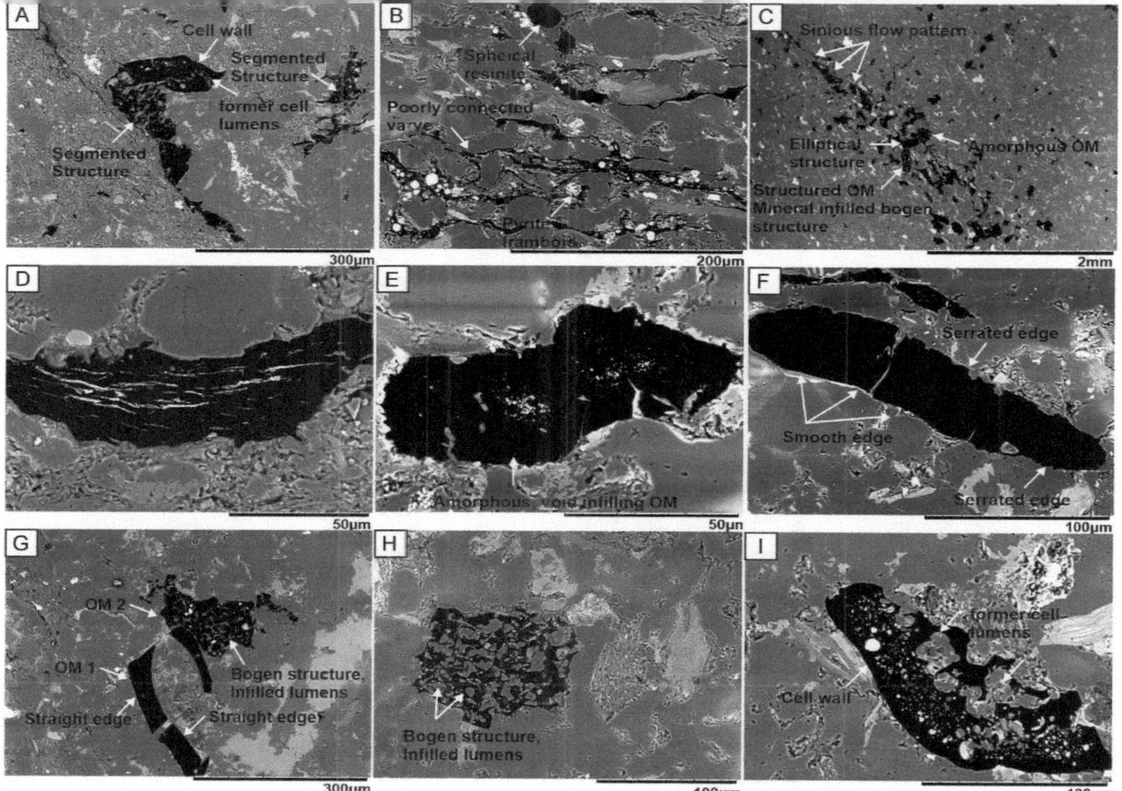

Fig. 11 Scanning electron microscope (BSE) images depicting organic matter morphologies with interpretation of maceral type. (A) Segmented structures representing remnant woody cell wall structures (vitrinite). (B) Poorly connected varve showing sinuous flow patterns presumed to be of algal/phytoplanktonic origin. (C) Phytoclastic varve showing various OM structures but poor connectivity with discrete sinuous flow paths. (D) Inertodetrinite, structured morphology with a mineral infilled bogen structure. (E) Unidentifiable OM, amorphous with low relief (void-filling). (F) Cutinite, serrated pattern more dominate on one side. (G) Two types of OM, Om1 with straight edge morphology (Telovitrinite), Om2 bogen structure and infilled cell lumens (fusinite). (H) Fusinite, well pronounced bogen structure. (I) Unidentifiable organic maceral, cell wall and infilled lumens are well pronounced.

variety of morphologies similar to the pervasive OM structures found in the matrix. These varves lack continuity, displaying poor connectivity, and represent the settling of phytoclastic material located in the catchment basin of the paleolake, with the rivers transporting this material together with coarser sediments.

4.4.3 Thermal maturation

The most fundamental parameters in controlling the thermal maturation of kerogen and expulsion of hydrocarbons during pyrolysis are temperature, time and pressure. Pressure plays only a subordinate role in the thermal maturation of kerogen, though it can alter chemical reactions involved in the process of thermal maturation. Temperature is more important, and the amount of time kerogen is exposed to higher temperatures. The reconstruction of the paleo-geothermal gradient in the Ciñera-Matallana Basin provides evidence for conditions conducive to the formation of thermogenic gas. The rapid subsidence of the Pastora Formation during the Late Carboniferous and Permian—estimated to be 0.23 m/1000 years by Frings et al. (2004) and 0.21–0.36 m/1000 years by Heward (1978), has provided a window of around 58 Myrs for the generation of thermogenic gas. Erosion rates increased during the Triassic, which continued to drive the Pastora Formation into the upper reaches of the oil window during a long period of non-deposition, until subsidence resumed associated with the diachronous sedimentation of transgressive sandstones and conglomerates during the Albian-Santonian time interval.

4.5 Gamma ray log

4.5.1 Overview

A 3.4 m section from the base of the investigated section was analyzed using gamma ray spectrometry to identify important paleoenvironmental parameters that control and influence hydrocarbon formation, storage, and migration (Fig. 12). Chemically reducing and especially anoxic environments are a highly important factor in determining organic matter preservation. Without anoxic water, organic matter is broken down by the activity of aerobic bacteria, resulting in the loss of OM, which would have provided a source for hydrocarbons. Quantifying the changes in reducing and oxidizing conditions within the lake at the time of deposition, will be used to identify organic matter preservation potential through Th/U ratio analysis and whether authigenic U values, which are considered to be associated

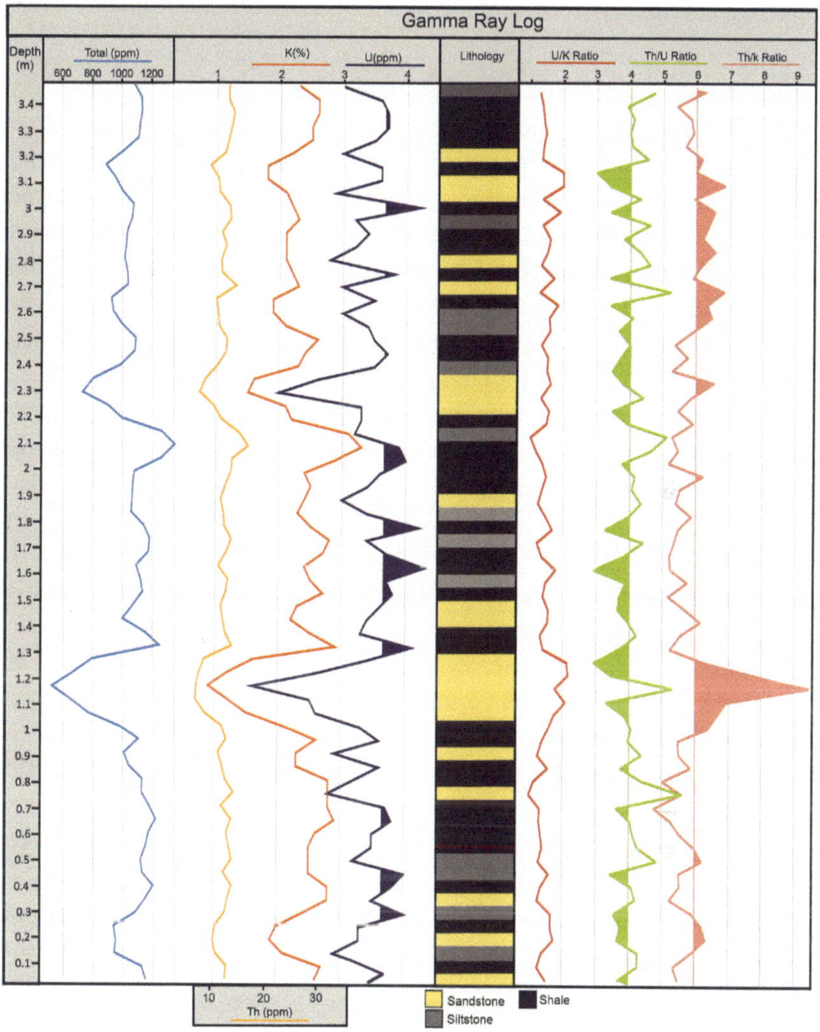

Fig. 12 3.4-m high-resolution gamma ray log depicting variations in the total signal, thorium (Th), potassium (K), and uranium (U) concentrations along the sedimentary log as well as Th/U, Th/K and U/K ratios.

with total organic carbon (TOC), remain constant or fluctuate and to what degree uranium is enriched within the shales. Finally, Th/K ratios were used to aid facies identification and grouping as well as an indicator for potential changes in provenance (Rider, 1996).

4.5.2 TOC-U values

The systematic association between TOC and uranium values is well documented and was pioneered by Adams and Weaver (1958) in marine sediments. Because organic matter can extract uranium from solution very efficiently via red/ox-reactions, organic matter rich shales are often enriched in chemically reduced authigenic uranium (U^{IV}). The association between TOC and authigenic uranium within the lacustrine environment is far more complicated and often there is no covariation between TOC and U due to the relative low concentration of uranium within freshwater compared to seawater (Lüning and Kolonic, 2003). Uranium values within shales of the Upper Pastora Formation range from 3.2 to 4.3 ppm, which is similar to the "average" uranium value in shale of 3.7 ± 5 ppm reported by Adams and Weaver (1958). When compared with highly productive shale gas plays such as the Late Devonian to Early Carboniferous Woodford Shale in Oklahoma (U.S.), which shows a strong correlation between U and TOC and can have U values >100 ppm, the U-TOC systematics of the Pastora mudstones would present little hope for source rock potential. However, the U values presented here are also very similar to those recorded by Felder et al. (2001) in oil shales from the Eocene Prinz von Hessen Lake in Germany, where U values consistently remained between 3 and 4 ppm, which would indicate a poor TOC content. However, a revisiting of the lake sediments by Hofmann et al. (2005) found that organic matter rich laminae which had shown no correlation between observable high TOC content and uranium values, actually contained a TOC content of between 30% and 45%. It is important to stress that authigenic uranium values alone—even though widely used as red/ox indicators within the published literature—can be misleading and should always be combined with further independent lines of evidence.

4.5.3 Th/U ratios

Th/U ratios throughout the logged section seem to indicate that the Upper Pastora Formation was subject to subtle fluctuations in red/ox conditions. Th/U ratios between 3 and 5.6 and typically higher uranium values associated with a higher TOC content can be correlated with lower Th/U ratios, indicating a more reducing chemical environment. Th/U ratios in average shales vary from 3.26 to 4.73 ppm which, following Adams and Weaver (1958), places these shales between a reducing and oxidizing environment. However, their work was conducted on marine shales and the applicability of such values to the ratios derived from lacustrine successions has remained controversial. According to Jones and Manning (1994), anoxic-oxic conditions in the lacustrine environment are defined as Th/U ratios ranging from anoxic <1.5, dysoxic 1.5–3 to oxic <3. This would place the OMRS shales in an oxic environment, which shows fluctuations toward a more dysoxic environment. Combined with observations made regarding sedimentary structures which show frequent alternations between hummocky cross-stratification and parallel lamination, the environment of deposition likely underwent oxygen fluctuations in response to eolian interactions with the water-sediment interface, indicating that these sediments are shallow, as opposed to deep lacustrine sediments. The interpreted relative proximity to a river, as indicated by the large sandstone bodies, also provides a source for oxygenation. It is observable on the log that the most continuous Th/U ratios can be found within shale bundles which remain uninterrupted from coarser sedimentary influxes.

4.6 Pore network characterization (pore type abundance)

4.6.1 Pore network overview

Pore networks within shales are usually heterogeneous and contain a variety of pore types, geometries and sizes spread over several orders of magnitude (Clarkson et al., 2013). To better understand hydrocarbon migration within UHRs the heterogeneous pore network has to be characterized first based on universally accepted pore type classification schemes followed by a quantification of the porosity associated with each pore type. Subsequently, a detailed analysis and quantification of the hydrocarbon storage and migration factors associated with the specific pore types can be carried out. This enables the identification of the potential relationship between changes in the fluid flow activity, the geomechanical behavior during both pre- and post-convective flux and the principal pore type. Finally, systematic changes in the petrological features are linked to the development of distinctive pore type associations and their abundances. This enables the determination of the most relevant pore-scale factors that exert control over both the multiphase fluid flows and the desorption processes. Both of these processes have been documented to be paramount for the assessment and evaluation of well productivity and longevity. Therefore, the pore networks present within the Pastora Formation have been characterized using the pore type and pore size classification schemes established by Loucks et al. (2012). Subsequently, each pore type's total abundance has been quantified together with each pore type's contribution toward the total porosity of the pore network. The same procedure was then applied to the various pre-defined pore sizes, as the size of pores plays a fundamental role in hydrocarbon migration and gives rise to the establishment of non-Darcy flow regimes and multiphase fluid flow (Javadpour, 2009; Shen et al., 2017; Singh and Myong, 2018; Swami et al., 2012; Yang and Bao, 2017; Zhang et al., 2019). The systematic multi-scaled characterization and quantification provides vital information on the multifaceted nature of the different pore networks, on both the microporous and nanoporous scale. In effect this permits the detailed analysis of two seemingly separate systems which, despite being present in one semi-interconnected network, store and migrate hydrocarbons in fundamentally different ways. The wider implications of such an analysis is that more accurate multiphase fluid flow models, even on a reservoir scale, can and have to be generated.

4.6.2 Pore types

The most frequent pore types developed within the nanoporous and microporous systems of the Pastora Formation are aluminosilicate pores (AL), moldic pores (MP) and more variably, pores at the edges of rigid grains (GEP). Among these are the aluminosilicate pores found within the illite matrix by far the most common, followed in abundance by the moldic pores within isolated mineral phases and finally the grain edge pores that are surrounding single grains. These three dominant pore types are referred to as the *primary pore types*, as they make the most significant contributions toward the pore network in both frequency and porosity (Fig. 16). *Secondary pore types* which contribute significantly less to the overall porosity and are also significantly less frequent within the pore networks are the fracture pores (FP), the shrinkage crack pores (SCP), the mica bearing pores (MBP), the dissolution pores (DP) and framboidal pores (FHP). In addition to these generally developed pore types, subordinate pore types that only occur within distinct lithological units such as the dolomitized sandstones have been documented. These pores are classed within the present study as *intra*crystalline pores (ICP1) and *inter*crystalline pores (ICP2) and discussed as "additional pore types" in Fig. 6.

4.6.3 Primary pore type abundance

4.6.3.1 Aluminosilicate pores (AL)

Aluminosilicate pores are by far the most abundant pore type found throughout all the different facies- and microfacies types that are associated with an aluminosilicate matrix and are most frequent in the OM/ILS and OMRS. At the two separately analyzed and discussed scales, aluminosilicate pores represent the following percentage of all pore types found throughout the pore network:

- 50 μm Scale: 46.96–73.77%
- 200 μm Scale: 60.46–80.09%

4.6.3.1.1 Nanoporous system (50 μm) Aluminosilicate nanopores are the most frequent pore type found in the nanopore system, typically developed between detrital grains within the illite matrix. Aluminosilicate nanopores make up 69.08–94.69% of all aluminosilicate pore types present at this scale, apart from the DSND microfacies, which has undergone diagenetic dolomitization of the intergranular pore space and is characterized by dissolution-based porosity. The microfacies with the largest quantities of aluminosilicate nanopores are the OMRS and MRSL in which 85.9% and 94.69% of pores are nanoporous. Aluminosilicate nanopores show a great variability in morphology at the 50 μm scale, with larger nanopores becoming increasingly irregular in shape, while the smallest nanopores are diametrically constrained but elongate, showing a fibrous morphology as observed by Desbois et al. (2010).

4.6.3.1.2 Microporous system (200 μm) Aluminosilicate pores within the microporous system are still the most abundant pore type, however, only 34.54–47.25% of all aluminosilicate pores present are indeed microporous in size. This indicates that even as the scale increases nanopores are still by far the more frequent pore size. Aluminosilicate micropores are most abundant in the ILRS and VSND microfacies representing 46.21% and 47.25% of the pore network. At this scale, the morphological complexity of aluminosilicate pore types dramatically increases as larger pore structures become apparent. These were potentially formed from pore amalgamations during diagenesis.

4.6.3.2 Moldic pores (MP)

Moldic pores are the second most abundant pore type found throughout all microfacies types, except the DSND. Moldic pore frequency can vary significantly, as does the area occupied by individual pores. Found isolated within mineral phases, predominately quartz grains, their abundance seems to be closely related to quartz grain provenance. At the two separate scales moldic pores represent the following percentage of all pore types found throughout the pore network:

- 50 μm Scale: 1.19–44.78%
- 200 μm Scale: 14.48–36.86%

4.6.3.2.1 Nanoporous system (50 μm) Because moldic pores are predominately nanoporous in size, they account for 87.5–97.73% of all moldic pores in the nanopore system, a clear evidence that specific pore types are dominated by a certain size. Moldic pores are most abundant in the siltstone and sandstone facies (OM/ILS, DOMS, ILRS, VSND), with a reduced presence in the OMRS and MRSL facies and nearly absent in the DSND facies, accounting there for only 1.19% of all pore types present. Moldic pores display highly variable morphologies which also accounts for their variable contributions toward the total porosity. This morphometric variability is most prominently documented by variations in area and frequency within the OM/ILS and DOMS microfacies. Despite a similar frequency of nanopores, the area occupied in the OM/ILS microfacies is significantly greater due to the presence of elongated pores (Figs. 14–16). The abundance of these pores shows a strong correlation with the presence of detrital quartz grains from a presumed secondary source. Larger sandstone bodies with quartz grains exhibiting polycrystallinity and a large lithic fragment content have the lowest quantity of moldic pores, while all other microfacies lacking these quartz grains have much higher moldic porosities (Fig. 13A–G). Observations under SEM provide evidence for quartz grain dissolution. This dissolution process is forming quartz grain remnants with a lower percentage of moldic pores that can be found adjacent to grains with a high abundance of moldic and fracture pores. As to whether the increase in these pore types is a result of selective quartz framework dissolution, a greater degree of weathering of the secondary sourced material prior to sedimentation, or a combination of both is unknown. What is known is that these detrital quartz grains generate another pore type located at the grain's edge, the GEP pores.

4.6.3.2.2 Microporous system (200 μm) Moldic pores once again are the second most abundant pore type within the microporous system but moldic micropores constitute just 10.6–34.34% of all moldic pores found at the 200 μm scale, with the highest content of micropores found in the ILRS (34.34%), OM/ILS (25.55%) and OMRS (21.23%) microfacies indicating that this primary pore type is dominated by nanopores.

4.6.3.3 Grain edge pores (GEP)

Despite being classified as a primary pore type in the existing literature, GEPs are not always the third most abundant pore type developed. Sometimes they form only the fourth or even fifth most frequent pore type

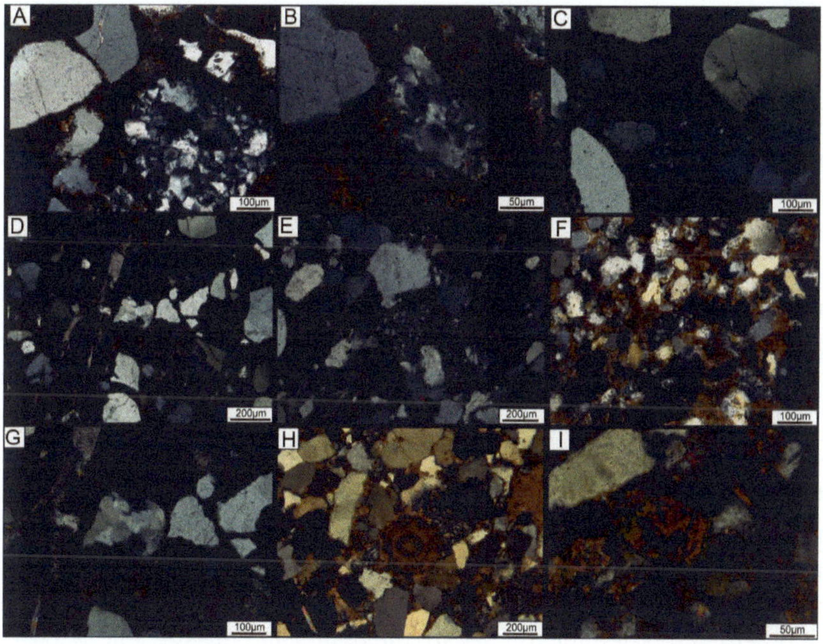

Fig. 13 (A–G) Polycrystalline quartz grains found in coarse grained sandstones showing varying degrees of polycrystallinity and variable microfabrics (all images with crossed polarizers). (H) Bone fragment. (I) Palynomorph.

within the entire pore network. However, their contributions toward total porosity due to their size is so great that they are considered a primary pore type nonetheless. Found throughout all the different microfacies types the GEPs are always located at the edges of single grains. Their size varies greatly and can therefore also significantly increase the overall porosity even if they are only present in low frequencies. Often displaying highly complex but continuous structures, they are predominantly found alongside grains that have undergone selective quartz framework dissolution. At the two separate scales GEPs represent the following percentages of all pore types found throughout the pore network:

- 50 μm Scale: 1.53–8.78%
- 200 μm Scale: 1.84–3.63% (DSND = 10.4%)

4.6.3.3.1 Nanoporous system (50 μm) Nanoporous GEPs represent between 29.41% and 100% of all GEPs present within the nanoporous system. Each microfacies type shows a similar quantity of nanoporous

GEPs except for those containing more of the aforementioned angular grains, such as the OM/ILS, OMRS and VSND microfacies. The frequency may not vary significantly but the area occupied does. This is due to GEPs being predominantly microporous in size and characterized by a large diameter and an even larger elongate morphology which wraps around detrital grains. The reason for such a high abundance of nanometric GEPs is the discontinuity of larger pores, resulting in the rise of multiple smaller pores wrapping around the same grain. However, although nanoporous GEPs are few in frequency and occupy a smaller area than their microporous counterparts, within the nanoporous system they are still comparatively large in relation to other nanopore types due to their broad and elongate morphology.

4.6.3.3.2 Microporous system (200 μm) GEPs form a significantly smaller portion of the micropore system, representing only 1.84–3.63% of all pore types present, with the exception of the DSND facies (which displays a total of GEPs of 10.4%, of which 23.8% are microporous). In all other microfacies, micropores are the dominant pore size among GEPs and represent between 59.38% and 95.24% of which most microfacies micropore content is >60%. GEPs are most abundant in the OM/ILS microfacies and OMRS, with fewest found in the MRSL microfacies.

4.6.4 *Secondary pore type abundance*

Pore types that make up significantly less of the pore network yet continue to be present nonetheless, are referred to as secondary pore types and possess unique pore-based properties and features. Secondary pore types consist of fracture, shrinkage, framboidal, micaceous and dissolution pores, and are present in similar abundances throughout all microfacies except the DSND facies. A trend observable across all secondary pore types is an increase in abundance within the nanoporous system. This is a prime example of why downscaling and higher resolution SEM imaging is necessary when analyzing smaller, less frequent pore types. At the two separate scales secondary pores represent the following percentage of all pore types found throughout the pore network:

- 50 μm Scale: 4.82–9.35% (VSND = 0.5%)
- 200 μm Scale: 1.4–5.58% (VSND = 0.4%)

4.6.4.1 Fracture pores (FP)

Fracture pores are found throughout all microfacies at both scales in low frequencies, making them one of the most consistently found secondary pore type. At the two separate scales fracture pores represent the following percentage of all pore types found within the entire pore network:

- 50 μm Scale: 1.13–5.87%
- 200 μm Scale: 0.21–2.35%

4.6.4.1.1 Nanoporous system (50 μm) Nanoporous fracture pores are typically localized phenomena, small in width but pronounced elongate. In the OMRS and MRSL microfacies, where they are most abundant, they display a lack of continuous connectivity across mineral phases. At the 50 μm scale, fracture pores are 100% nanometric in size. Although they only occupy a small area they show the potential to act as conduits for nanometric hydrocarbon migration from adjacent pore types, such as moldic pores, where discontinuous fracture pores cross mineral phases and can transport hydrocarbons isolated within moldic pores. However, fracture pores cannot contribute significant porosity and have a limited potential to serve as an efficient migratory pathway for hydrocarbons due to their low frequency and lack of continuity.

4.6.4.1.2 Microporous system (200 μm) Fracture pores contribute only a very small proportion to the overall pore network, typically between 0.21% and 2.35%, of which micropores represent <40% of total fracture pores present. However, this data is somewhat unreliable given that so few fracture pores are present within a single area. The largest amount of fracture pores are found within the OMRS microfacies in which facture pores are present, but only at a negligible quantity when considering the proportions of aluminosilicate and moldic pores present within a given network.

4.6.4.2 Shrinkage cracks pores (SCP)

Shrinkage crack pores are microfacies specific and are found isolated within certain areas of the aluminosilicate matrix. At the two separate scales shrinkage crack pores represent the following percentage of all pore types found throughout the pore network:

- 50 μm Scale: 0.4–0.52%
- 200 μm Scale: 0.17–0.19%

4.6.4.2.1 Nanoporous system (50 μm) Shrinkage crack pores, unlike their fracture pore counterparts, are generally more elongate and larger in diameter, and more continuous. These pore types are not abundant throughout all microfacies and are only developed within the MRSL and VSND microfacies. Representing 0.4–0.52% of the nanoporous network of which 80–100% is nanoporous, this pore type is most notable in VSND microfacies where selective diagenetic processes appear to have played a role in their formation. The aluminosilicate matrix, where the shrinkage cracks can be found, is isolated, unlike other cements. This isolation of the cement may be due to the alteration of large feldspar grains, which resulted in the formation of illite, which may serve as an explanation as to why so many shrinkage cracks both large and small are found in confined and isolated spaces. The morphology of the space occupied would also indicate that the illite present is a product of feldspar alteration.

4.6.4.2.2 Microporous system (200 μm) Shrinkage pores are equally uncommon, only present in the OM/ILS and VSND, representing 0.17–0.19% of all pore types present, of which only 16.6–20% is microporous. With similar properties as fracture pore types, shrinkage cracks are uncommon and where present, predominantly nanoporous, despite decreases in scale.

4.6.4.3 Micaceous pores (MBP)

Mica-based pore types are a small contributor to the nanoporous network, their frequency is marginally variable given that their presence is a function of the quantity of biotite and muscovite present within each microfacies, with most microfacies displaying relatively similar quantities of micas. At the two separate scales mica-based pores represent the following percentages of all pore types found throughout the entire pore network:

- 50 μm Scale: 2.54–7.04%
- 200 μm Scale: 0.18–1.87%

4.6.4.3.1 Nanoporous system (50 μm) Mica-based pores represent 0.6–7% of the nanoporous system, of which 92.31–100% of micaceous pores are nanometric. Micaceous nanopores are most abundant in the mica rich MRSL microfacies followed by the OM/ILS and OMRS microfacies. Typically characterized by an elongate, but constrained pore morphology, these pores demonstrate the potential to contribute to nanometric permeability, but their limitations in width will likely result in a complex mode of hydrocarbon migration, due to small pore throat diameters.

4.6.4.3.2 Microporous system (200 μm) Representing between 0.18% and 1.87% of the microporous system, mica-based pores contribute significantly less to the overall pore network than mica-based pores in the nanoporous system. Microporous mica-based pores represent between 6.81% and 10% (within the DOMS facies up to 50%) of all mica-based pores indicating that this pore type is predominately nanoporous.

4.6.4.4 Framboidal pores (FHP)

Framboidal pore abundance and size is a function of red/ox conditions, with pyrite framboids varying in size from 5 to 11 μm, affecting framboidal pore size. At the two separate scales framboidal pores represent the following percentages of all pore types found throughout the entire pore network:

- 50 μm Scale: 1.65% (OM/ILS)
- 200 μm Scale: 0.06–0.3%

4.6.4.4.1 Nanoporous system (50 μm) Framboidal pores are highly localized and largely confined to varves and large pieces of amorphous OM and are rarely seen residing in the matrix. Where present, Framboidal pores are 100% nanometric, but that is a function of pyrite framboid size, which across all microfacies ranges from 5 to 11 μm with larger pyrite framboids found in varves, which are consequently more microporous. Representing 1.65% of the pore network in the OM/ILS microfacies, framboidal pores are 100% nanometric.

4.6.4.4.2 Microporous system (200 μm) Framboid pores, as within the nanoporous system, are not only microfacies specific, but the wide and variable distribution within the matrix and preferential concentration in OM varves provide varied results. Most prominent in the ILRS and MRSL microfacies, microporous framboid pores within the microporous system make up a very small portion of the pore network of which 42.86–100% are microporous.

4.6.4.5 Dissolution pores (DP)

Dissolution pores are microfacies specific and are more abundant in microfacies where dolomitization of the intergranular pore space is more pervasive. At the two separate scales dissolution pores represent the following percentages of all pore types found throughout the entire pore network:

- 50 μm Scale: <2.55%
- 200 μm Scale: 0.86–3.73%

4.6.4.5.1 Nanoporous system (50 μm) The largest quantity of dolomitic dissolution pores can be found within the DSND facies and occur sporadically throughout most other microfacies, but never in significant quantities. Dolomitic dissolution pores represent <2.55% of the nanopore system of which 75–100% of dolomitic dissolution pores are nanometric. With the exception of the DSND microfacies where dolomitic dissolution pores represent 56.17% of the nanopore system, of which 97.62% is nanometric. Pores associated with dolomitization are dominated by nanopores, not micropores.

4.6.4.5.2 Microporous system (200 μm) Present in only the DOMS and ILRS microfacies at the 200 μm scale, dissolution pores constitute between 0.86% and 3.37% of the pore network of which just 20–23.86% is microporous, indicating that dolomitic dissolution pores are predominately nanoporous. In the DSND microfacies, where the intergranular space is dominated by dolomite, dissolution pores represent of 62.77% of the pore network, of which 16.52–21.31% are micrometric, depending on the *type* of dissolution.

4.7 Pore network characterization (pore type porosity)

4.7.1 Pore network overview

Porosity is heavily controlled by specific pore types, with the vast majority of porosity found within primary pore types (AL, MP, GEP). However, the latter two contribute significantly less porosity to the overall pore network. Combined nano/microporosity across all microfacies ranges from 0.91% to 24.69% (Fig. 16).

Shale microfacies

1. OMRS: Total Porosity (503 μm) = 17.08%. Total Porosity (200 μm) = 15.43%

Siltstone microfacies

1. OM/ILS: Total porosity (50 μm) = 19.35%. Total porosity (200 μm) = 24.47%
2. ILRS: Total porosity (50 μm) = 10.56%. Total porosity (200 μm) = 7.7%
3. MRSL: Total porosity (50 μm) = 6.94%. Total porosity (200 μm) = 6.34%
4. DOMS: Total porosity (50 μm) = 4.86%. Total porosity (200 μm) = 5.26%

Sandstone microfacies

1. VSND: Total porosity (508 μm) = 8.53%. Total porosity (200 μm) = 6.97%
2. DSND: Total porosity (50 μm) = 2.8%. Total porosity (200 μm) = 0.91%

4.7.2 Primary pore type porosity

4.7.2.1 Aluminosilicate pores (AL)

Aluminosilicate pores are by far the largest contributor toward total porosity in all microfacies and in both scale systems. The largest aluminosilicate porosity can be found in the OM/ILS and OMRS microfacies with aluminosilicate porosities of 13.36% and 12.04%, respectively, with the lowest aluminosilicate porosity located in the DOMS, MRSL, and VSND microfacies types with porosities of 3.23%, 5.24%, and 5.28%, respectively. The reason for a lack of cement-based porosity in these sandstone/siltstone microfacies, which characteristically should exhibit high cement-based porosity, is due to more closely spaced grains and diagenetic dolomite destroying the intergranular aluminosilicate porosity. Proportionally, aluminosilicate contributions toward total porosity across all microfacies at the two separate scales are as follows:

- 50 μm Scale: 61.9–75.5%
- 200 μm Scale: 65.6–88.52%

4.7.2.1.1 Nanopore system (50 μm) Aluminosilicate pores are the largest contributor toward total porosity. They are contributing between 3.23% and 13.36% toward the total porosity, of which nanopores contribute 1.25–3.68%, representing 17.31–44.58% of the total aluminosilicate porosity. As Figs. 14 and 15 show, despite being a smaller contributor toward porosity than larger aluminosilicate micropores, nanopore contributions toward porosity cannot be understated or overlooked, with the data highlighting the importance of the matrix being a primary source for hydrocarbon storage in the free gas state. The OMRS microfacies has the greatest amount of nanoporosity held within the cement phase, contributing 3.68% toward total porosity with a limited dolomitic overprint and a smaller grain size which has greatly increased nanoporosity. In total, aluminosilicate nanoporosity equals:

1. *Siltstone microfacies*: (ILRS, OM/ILS, MRSL, and DOMS) contribute 1.25–2.32% toward total porosity, representing 11.99–33.64% of pore network porosity.

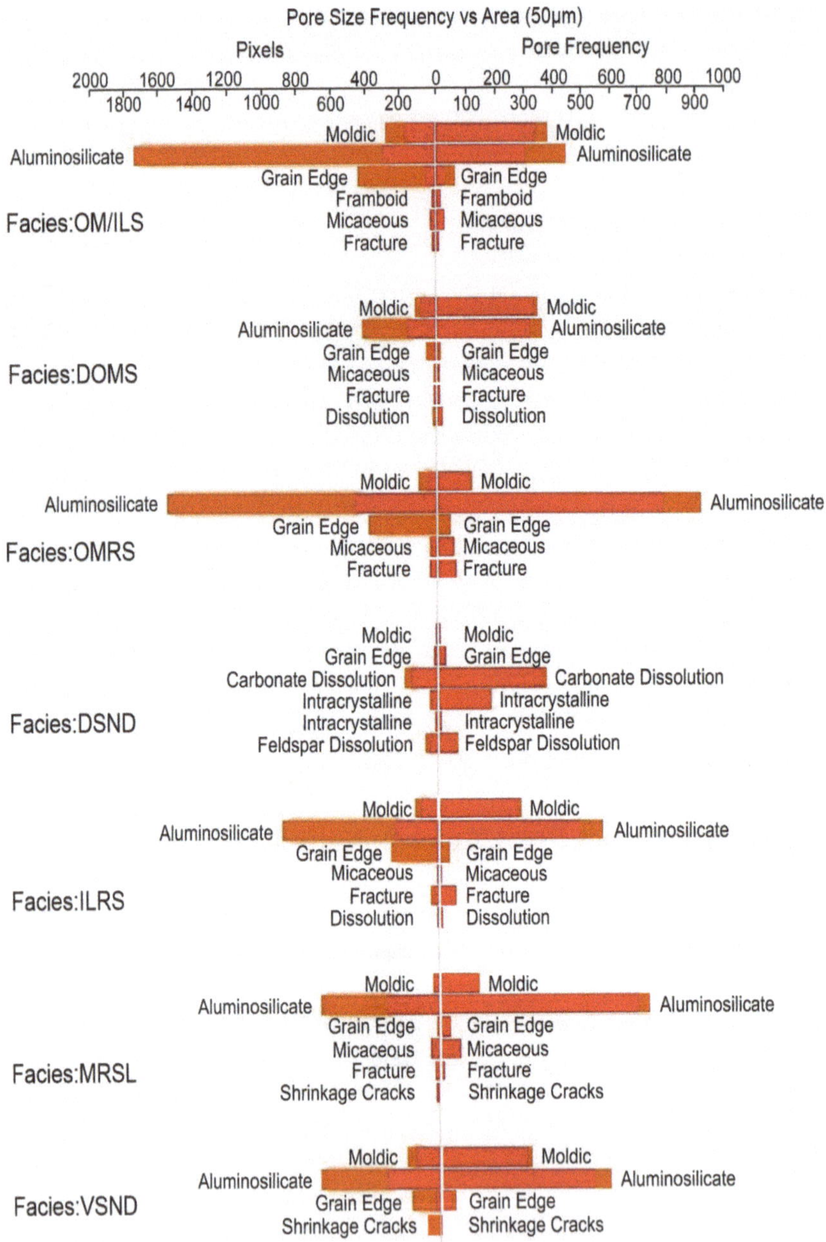

Fig. 14 Pore size frequency and area changes in relation to pore types within different microfacies, highlighting the disproportionate relationship between nanopore frequency and area occupied within the pore network (50 μm-scale).

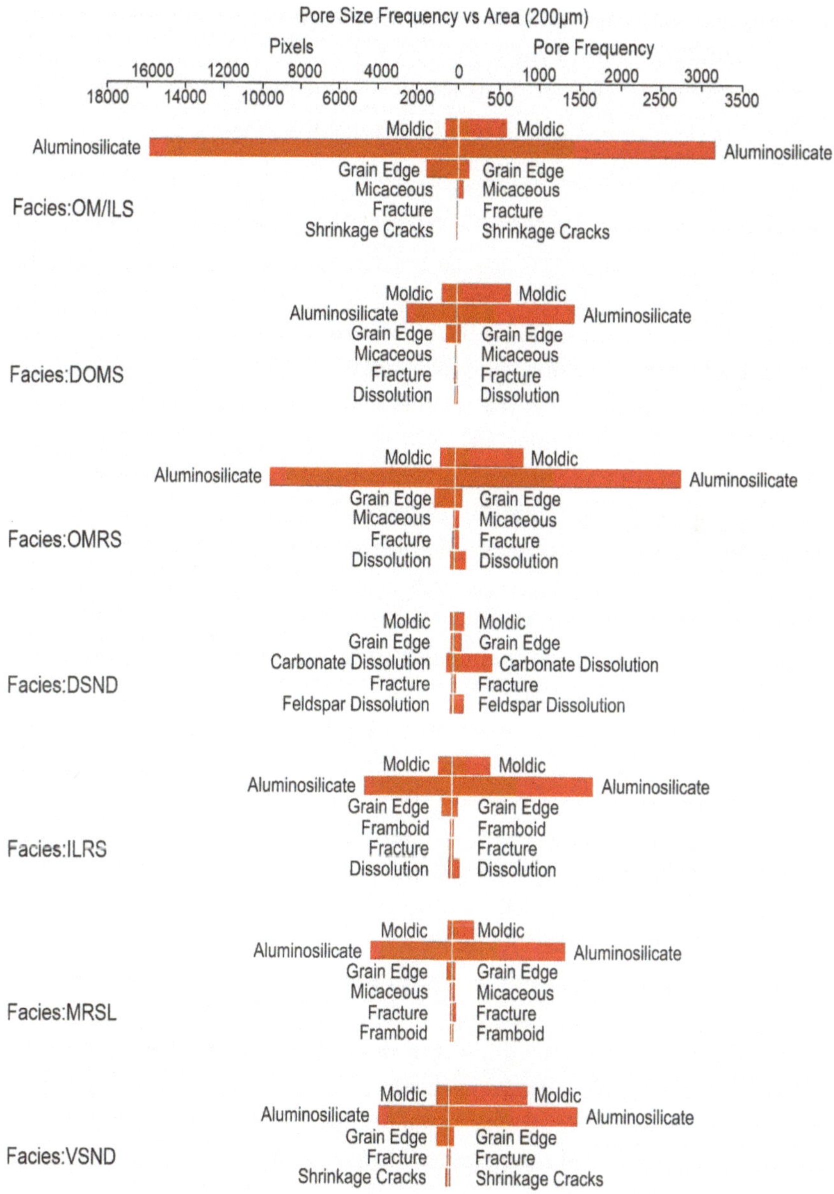

Fig. 15 Pore size frequency and area changes in relation to pore types within different microfacies, highlighting the disproportionate relationship between nanopore frequency and area occupied within the pore network. (200 µm - scale).

2. *Sandstone microfacies*: (VSND) contributes 2.34% toward total porosity, representing 27.7% of the pore network.
3. *Shale microfacies* (OMRS) nanopores contribute 3.68% toward total porosity, representing 21.54% of the pore network.

4.7.2.1.2 Micropore system (200 μm) Aluminosilicate pores are also within the micropore system the largest contributor toward total porosity, contributing between 3.44% and 21.38%, of which micropores contribute 2.86–20.2% representing 83.05–94.56% of the total aluminosilicate porosity. This close relationship between total aluminosilicate porosity and microporosity demonstrates that at this scale we cannot quantify the role nanopores play, due to their size and the resolution of the BSE image, but at the 50 μm scale we have already identified their increase in frequency and contributions toward porosity. As demonstrated by the nanopore system analysis, the ability of nanopores to contribute significant pore space for free gas, stems from their widespread abundance, thus on the 200 μm scale, where this is not apparent, their total contribution decreases, while micropore contributions increase, as large pore structures become observable. The microfacies with the greatest amount of aluminosilicate porosity, contributing 21.36% toward total porosity of which 94.56% is microporous, is the OM/ILS siltstone microfacies, with no other microfacies being able to contribute similar levels of porosity. The microfacies with the second most abundant aluminosilicate content, and second most porous by far, is the shale OMRS microfacies with aluminosilicate pores contributing 12.75% toward total porosity, of which 91.15% is microporous. The large discrepancy now apparent between the OM/ILS and OMRS microfacies which was not present at the 50 μm scale, is due to the presence of larger micropores in the OM/ILS siltstone microfacies where the OMRS shale microfacies is predominantly more nanoporous. Other microfacies such as DOMS, ILRS, MRSL and VSND have aluminosilicate porosities, which only contribute between 3.44% and 5.96% toward total porosity which is significantly lower than the aforementioned microfacies. However, all microfacies continue to show that the aluminosilicate matrix is dominated, almost exclusively, by micropores at this scale with micropores making up 83.05–90.15% of aluminosilicate pore types. Unlike the nanopore system, the role of moldic pores as the second most dominant contributor toward porosity is less apparent, as it often interchanges with pores located at the edge of rigid grains. In total aluminosilicate microporosity equals:

1. *Siltstone microfacies* (ILRS, OM/ILS, MRSL, and DOMS) contribute 2.86–20.2% toward total porosity and represent 54.5–82.54% of the pore network.
2. *Sandstone microfacies* (VSND) contributing 4.23% toward total porosity, representing 60.64% of the pore network.
3. *Shale microfacies* (OMRS) contribute 11.62% toward total porosity, representing 75.33% of the pore network.

4.7.2.2 Moldic pores (MP)

Despite being the dominant second most abundant pore type present across all microfacies, moldic pores are in some microfacies the third largest contributor toward total porosity or have similar porosity contributions as GEPs due to moldic pores being predominately nanometric. Proportionally, moldic pore contributions toward total porosity across all microfacies at the two separate scales are as follows:

- 50 μm Scale: 4.57–18.54%
- 200 μm Scale: 3.54–22.54%

4.7.2.2.1 Nanopore system (50 μm) Moldic pores are the second largest contributor toward total porosity, contributing between 0.03% and 2.17%. Unlike the aluminosilicate matrix, moldic pores are dominated by nanopores, while only contributing between 0.02% and 1.34% toward total porosity, nanopores represent 45.04–74.67% of total moldic porosity. The smallest and most frequent moldic nanopores display a pseudospherical morphology, creating a pitted texture, and in other microfacies display a more elongated pore structure which greatly increases pore size. By far the greatest contributors to overall moldic nanoporosity can be found in the OM/ILS and VSND microfacies, where moldic nanopores contribute 1.34% and 1.08% to total porosity. Other microfacies with high moldic porosity include DOMS, OMRS, and ILRS with the smallest amount located in MRSL and are essentially non-existent in DSND. Quartz grains with a large number of GEP are closely associated with the presence of larger moldic pores which could imply that the elongated moldic pores form in response to selective diagenetic dissolution or that these quartz grains are the product of a highly weathered secondary sourced material. In addition to the elongated pore structure, there is a second type of moldic pores which is morphometrically dissimilar, forming a diamond shape whereby pore elongation is coupled with an equal increase in diameter. This is creating a much larger and deep-set pore which, although far less abundant, can contribute a large amount toward total porosity. Often these larger moldic pores begin to coalesce and detach segments of the quartz grains to form a separate and more importantly rugged independent grain, which promotes the formation of GEPs. Evidence of this selective quartz framework dissolution is very similar to observations made by Leneuf (1973), who observed deeper dissolution features, triangular etch pits, and interconnected networks of grooves and crevasses in association with increased quartz weathering. In total, moldic nanoporosity equals:

1. *Siltstone microfacies* (ILRS, OM/ILS, MRSL, and DOMS) has contributed 0.36–1.34% toward total porosity, representing 5–13.1% of the pore network.
2. *Sandstone microfacies* (DSND and VSND) contributes 0.02–1.08% toward total porosity, representing 0.1–12.8% of the pore network.
3. *Shale microfacies* (OMRS) Contributes 0.35% toward total porosity, representing 2.1% of the pore network.

4.7.2.2.2 Micropore system (200 μm) Moldic pores contribute between 0.21% and 1.24% toward total porosity of which micropores contribute 0.14–0.79%, representing 56.58–78.57% of moldic porosity. The five main microfacies which show very similar results are all siltstones and shales, OM/ILS, DOMS, OMRS, ILRS, and VSND and have moldic porosities between 0.87% and 1.24%, while the DSND sandstones exhibit a low porosity of 0.21%. However, the percentage of micropores present within moldic pore porosity is variable across the two groups ranging from 8.44% to 55.85%. The lower quantity of micropores present with the moldic pore phase indicates that even at a higher resolution moldic porosity still contains significant quantities of nanopores. In total moldic microporosity equals:

1. *Siltstone microfacies* (ILRS, OM/ILS, MRSL, and DOMS) contribute 0.17–0.7% toward total porosity and represent 2.46–13.4% of the pore network.
2. *Sandstone microfacies* (VSND and DSND) contributing 0.14–0.79% toward total porosity representing 11.37–16.26% of the pore network.
3. *Shale microfacies* (OMRS) contribute 0.58% toward total porosity representing 3.74% of the pore network.

4.7.2.3 Grain edge pores (GEP)

Stressing the disproportionality between pore type abundance and contributions toward total microfacies porosity, predominantly in the OM/ILS and OMRS microfacies, GEPs are the second largest contributor toward total porosity due to their size, not their abundance. Proportionally, GEP contributions toward total porosity across all microfacies are similar and sometimes much greater than moldic pore contributions. At the two separate scales proportional contribution toward total porosity is as follows:

- 50 μm Scale: 4.78–21.59%
- 200 μm Scale: 4.54–13.13%

4.7.2.3.1 Nanopore system (50 μm) The third-largest contributor toward total porosity are GEPs, contributing 0.13–3.69% toward total microfacies porosity. Due to these pores being characteristically microporous in size, GEP nanopores only contribute 0.08–0.65% toward total microfacies porosity representing 3.73–45.28% of the total GEP porosity. This, however, is with the exception of the DSND microfacies in which GEP pores are 100% nanoporous. Characteristically elongate and microporous, their nanoporous counterparts contribute very little toward total porosity. Similar to moldic pores these pore types are also grain controlled and are abundant in immature sediments with angular quartz grains. In total grain edge nanoporosity equals:

1. *Siltstone microfacies* (ILRS, OM/ILS, MRSL and DOMS) contributed 0.1–0.43% toward total porosity, representing 0.54–3.46% of the total pore network.
2. *Sandstones microfacies* (DSND and VSND) contribute 0.13–0.19% toward total porosity, representing 2.23–4.81% of the pore network.
3. *Shale microfacies* (OMRS) contribute 0.65% toward total porosity, representing 3.8% of the pore network.

4.7.2.3.2 Micropore system (200 μm) As previously noted, GEPs are predominantly microporous in nature as they follow the length of rigid grains. GEPs contribute between 0.12% and 2.13% toward the total microfacies porosity, of which micropores contribute 0.28–2.11%. Representing 95.52–99.83% of total GEP porosity. Apart from the DSND microfacies, where microporous GEP pores represent 63.61% of total GEP porosity. The mechanism of their formation has ensured that they remain large and elongate and thus are ideal pore types for facilitating hydrocarbon migration around the matrix and drastically increase gas mobility and freedom to flow. In total grain edge microporosity equals:

1. *Siltstone microfacies* (ILRS, OM/ILS, MRSL, and DOMS) contribute 0.28–2.13% toward total porosity and represent 4.33–12.5% of the pore network.

2. *Sandstone microfacies* (VSND and DSND) contributing 0.76% and 0.12% toward total porosity, representing 8.3% and 10.8% of their respective pore networks.
3. *Shale microfacies* (OMRS) contribute 1.35% toward total porosity, representing 8.7% of the pore network.

4.7.3 Secondary pore type porosity

Secondary porosity increases as scale increases, highlighting the importance of factoring in the contributions of secondary pore types in pore network modeling. Proportionally, secondary pore typecontributions toward total porosity across all microfacies at the two separate scales are as follows:

- 50 μm Scale: 0.7–9.22%
- 200 μm Scale: 0.16–4.13%

4.7.3.1 Nanopore system (50 μm)

Unlike primary pore types, secondary pore types are predominantly nanoporous in size and consist of fractures, shrinkage cracks, framboidal pores, dissolution pores and mica-based pore types. All five types only contributing between 0.06% and 0.64%. Although some of these pore type may be plentiful within the pore network, as confirmed by earlier results in pore abundance, they occupy a very small area (Figs. 14–16), with shrinkage cracks and fractures displaying elongate but diametrically constrained morphologies. Nanoporous mica-based pores contribute between 0.02% and 0.35% toward total porosity. As these pores are very narrow, 65–100% of mica pores are nanoporous in all microfacies. Their extremely narrow but elongate morphology prohibits larger contributions toward overall porosity, despite large quantities of micas being present. Fracture nanopores contribute between 0.07% and 0.32% toward total porosity, where present, fractures are predominately nanoporous. Shrinkage crack nanopores are not a common pore type, in microfacies MRSL and VSND they only contribute 0.11% and 0.06% toward total porosity. Shrinkage cracks in microfacies MRSL are completely nanoporous. However, in the VSND larger microporous shrinkage cracks are present. Framboidal pores are less abundant in the matrix and are primarily concentrated in the varves; where present in the matrix, pores are exclusively nanoporous and contribute only 0.11% toward total porosity. At this scale, combined dissolution-based porosity can only be demonstrated from the DOMS and the DSND microfacies with porosities of 0.13% and 2.12%, of which nanopores contribute 0.05% and 1.59%, respectively. Within the DSND microfacies, which is

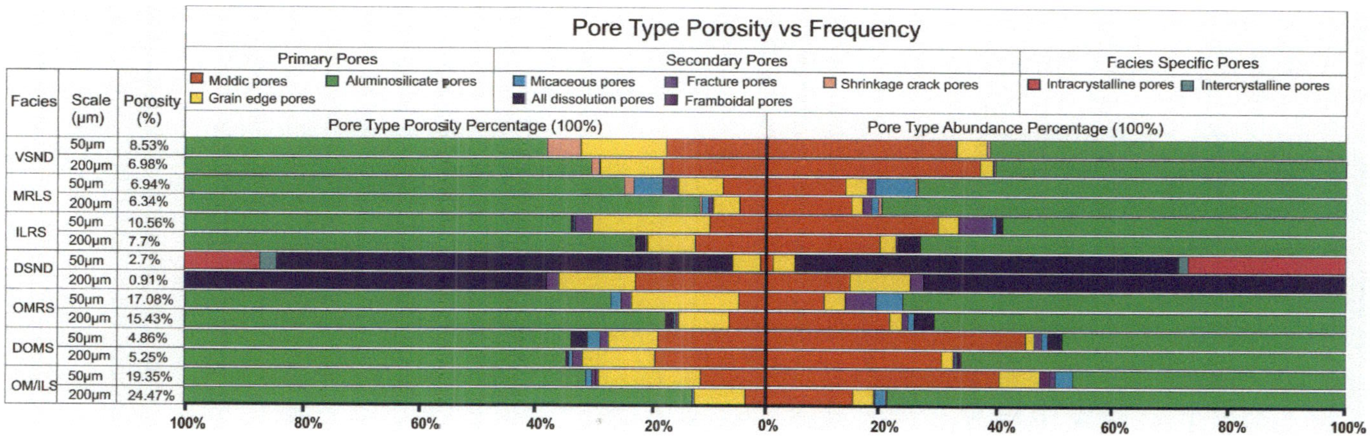

Fig. 16 Pore type abundance in relation to porosity across all facies types within the nanopore system (50 µm) and micropore system (200 µm). The chart highlights the disparity between the abundance of specific pore types throughout the pore network and their disproportionate contribution toward porosity, best exemplified by GEP and MP pores.

dominated by intergranular dolomitic cement, a significant amount of dissolution-based nanoporosity is present. This nanoporosity contributes 1.22% toward total porosity and represents 45.19% of the total pore network. In total combined secondary pore type nanoporosity equals:

1. *Siltstone microfacies* (ILRS, OM/ILS, MRSL, and DOMS) contribute 0.14–0.64% toward total porosity, representing 2.23–9.22% of the pore network.
2. *Sandstone microfacies* (VSND and DSND, excluding the dolomitic dissolution pores in the DSND), contribute 0.06% and 0.78% toward total porosity, representing 0.7% and 21.1% of pore network porosity.
3. *Shale microfacies* (OMRS) contributes 0.51% toward total porosity, representing 2.99% of total network porosity.

4.7.3.2 Micropore system (200 μm)

The reduction in resolution has reduced the number of observable nanopores and as such the observable porosity that they can contribute. Mica-based porosity is negligible, contributing between 0.04% and 0.07% representing between 0.25% and 1.13% of the total network. Of this percentage, micropores only constitute 0.01–0.02%. Fracture pores contribute even less toward total porosity contributing just 0.01–0.12% of which microporous only contribute between 0.006% and 0.03%. Shrinkage cracks also produce negligible quantities of porosity and microporosity where present. Shrinkage crack porosity ranges from 0.04% to 0.1% of which 0.02–0.09% is microporous. Dissolution-based porosity within argillaceous sediments ranges from 0.12% to 0.57%, of which 0.02% to 0.17% is microporous. The DSND microfacies, which has a dolomite dominated intergranular pore space, has a dissolution-based porosity of 0.43% of which microporosity contributes 0.22%, representing 51.1% of dolomite dissolution-based porosity (Figs. 17–44). In total combined secondary pore type microporosity equals:

1. *Siltstone microfacies* (ILRS, OM/ILS, MRSL, and DOMS) contribute 0.04–0.21% toward total porosity, representing 0.07–1.56% of the pore network.
2. *Sandstone microfacies* (VSND and DSND, excluding the dissolution pores in the DSND), contribute 0.1% and 0.08% toward total porosity, representing 1.38% and 8.79% of pore network porosity.
3. *Shale microfacies* (OMRS) contributes 0.21% toward total porosity, representing 1.36% of total network porosity.

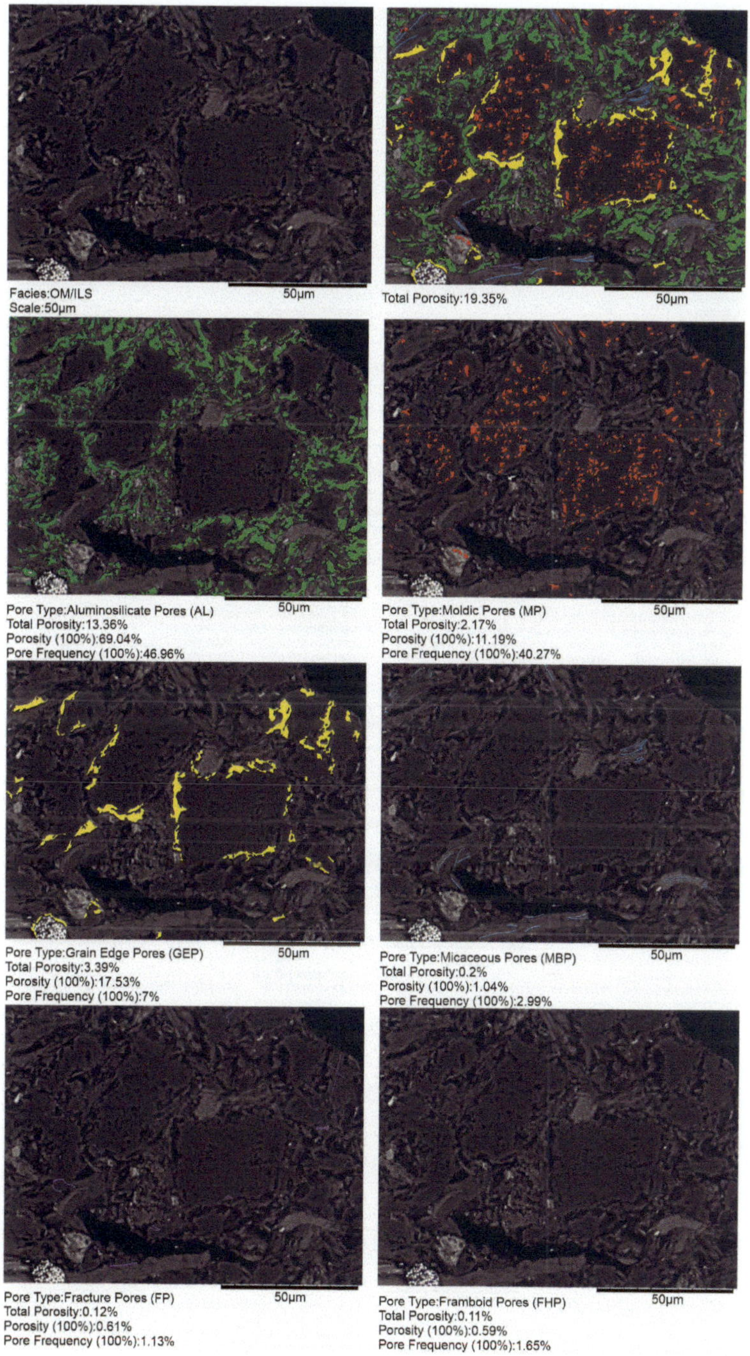

Fig. 17 Facies OMILS pore type characterization (50 µm Scale).

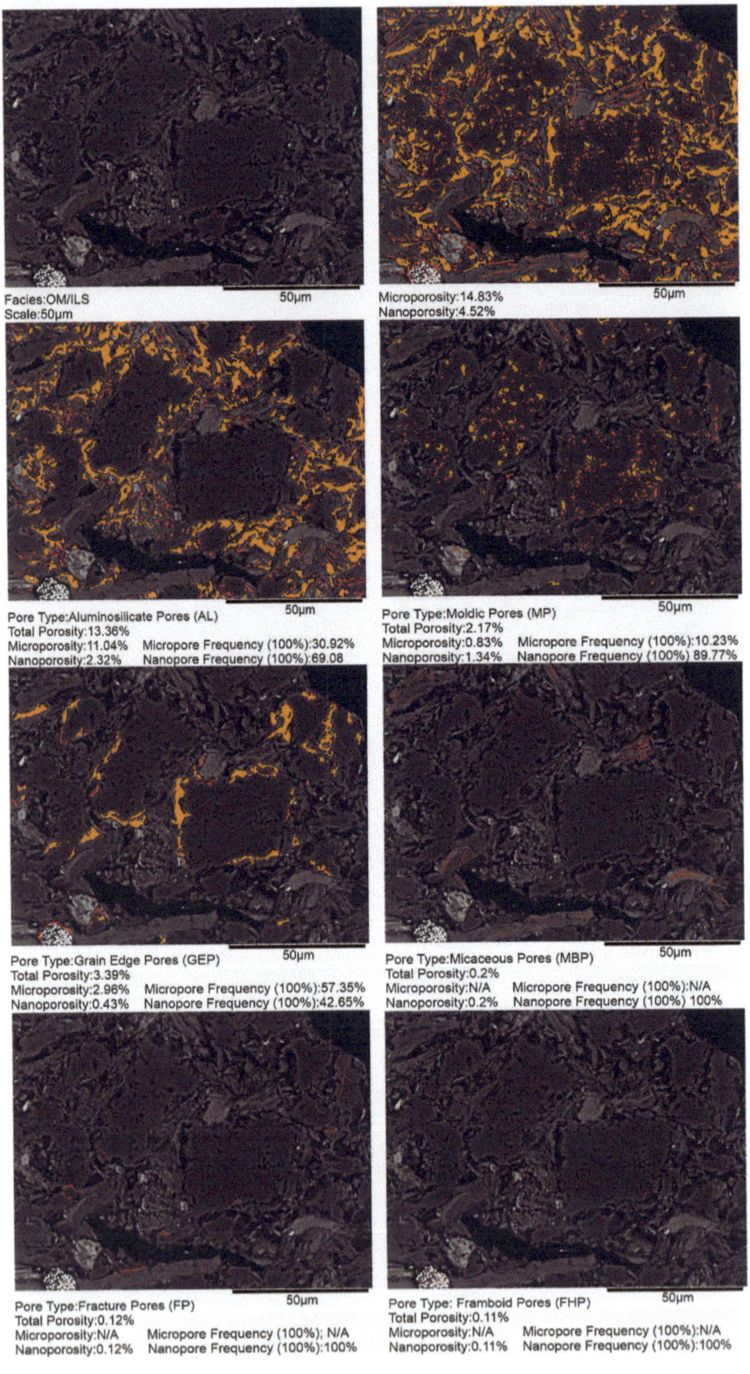

Fig. 18 Facies OMILS pore size characterization (50 μm Scale).

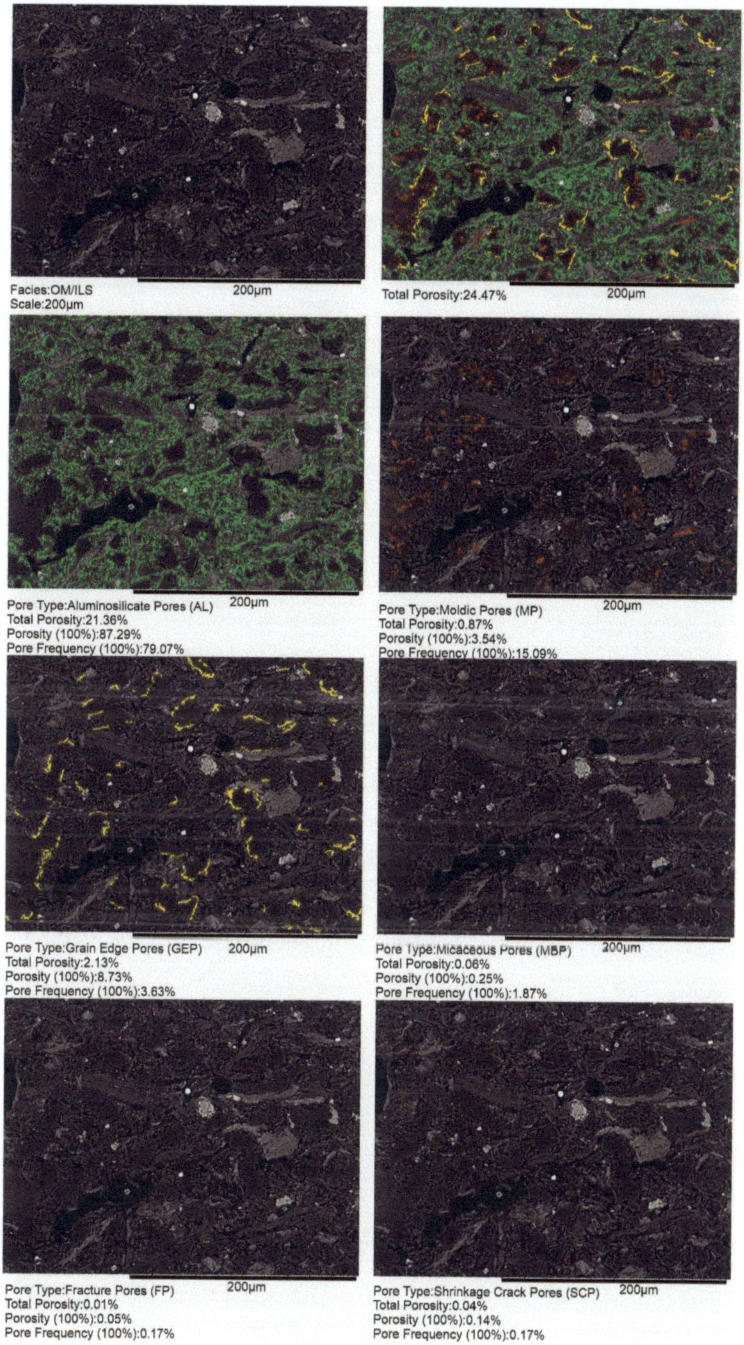

Fig. 19 Facies OMILS pore type characterization (200 μm Scale).

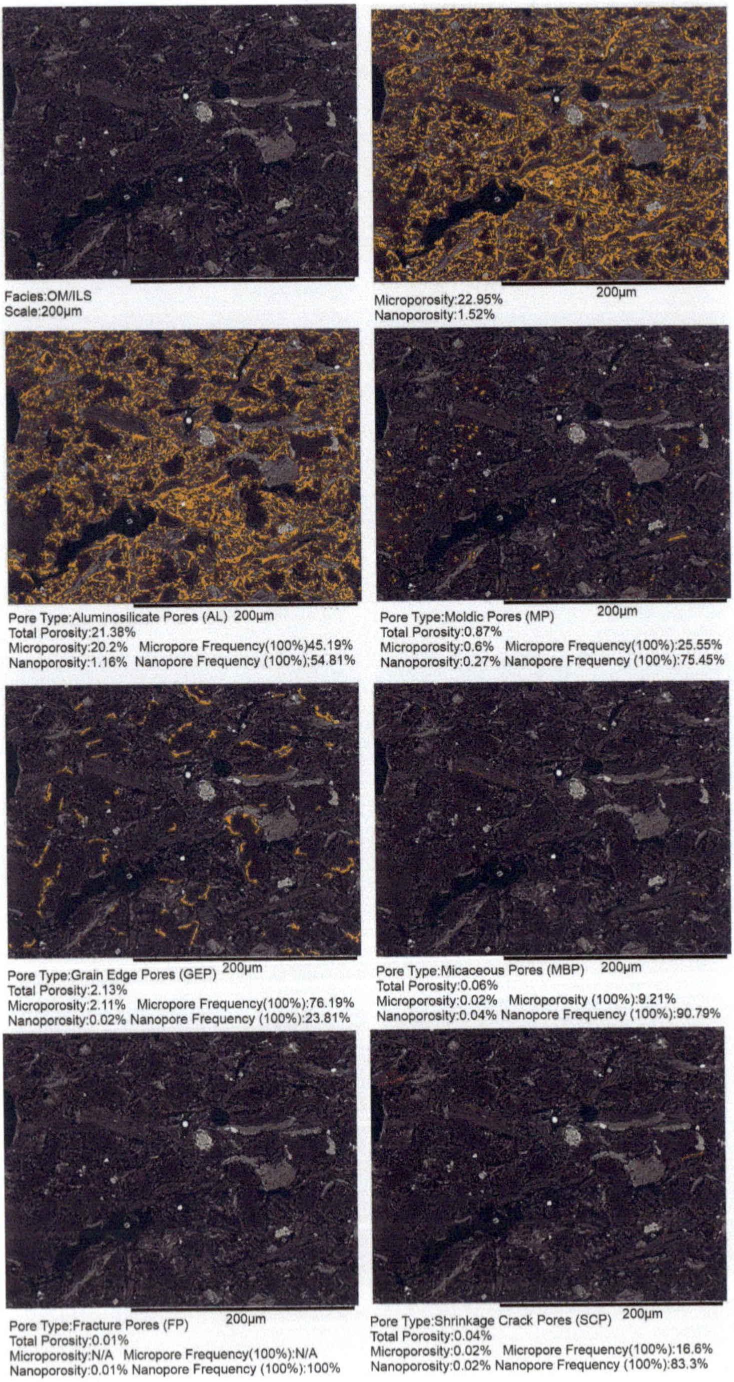

Fig. 20 Facies OMILS pore size characterization (200 μm Scale).

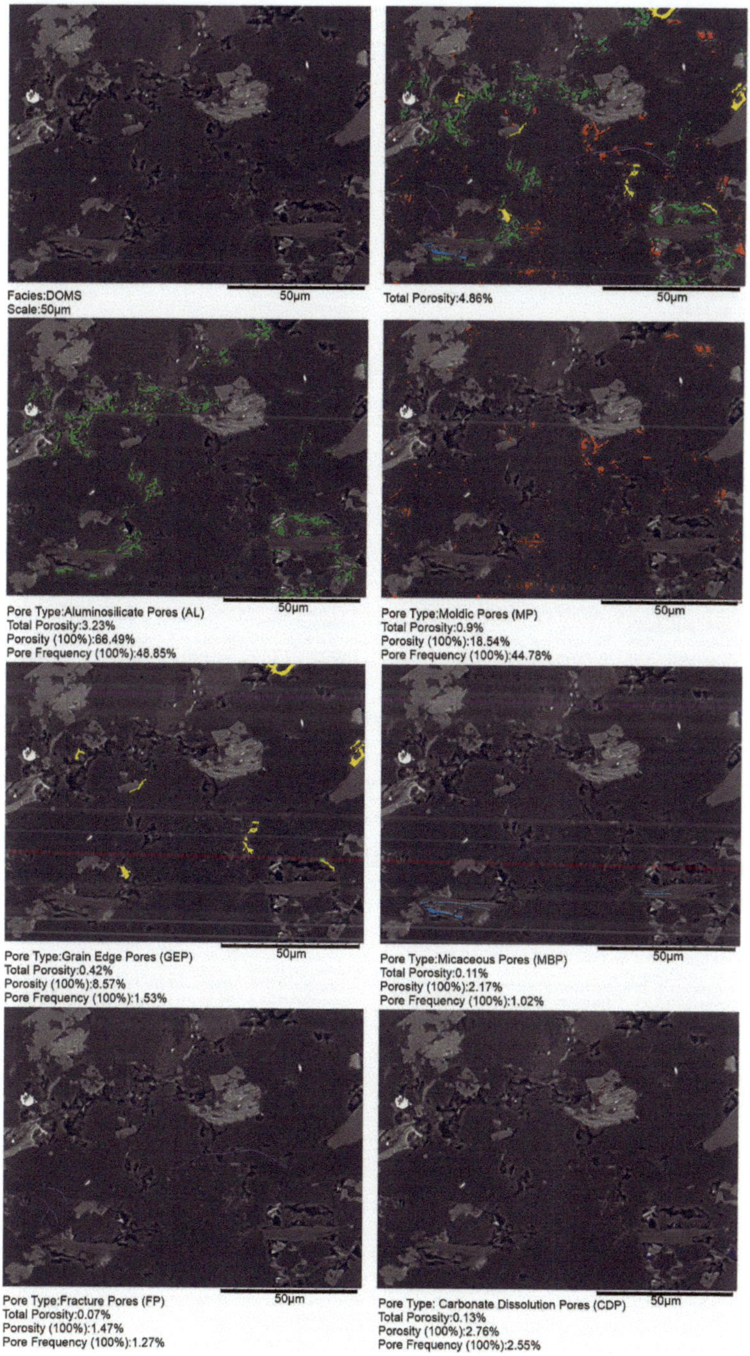

Fig. 21 Facies DOMS pore type characterization (50 μm Scale).

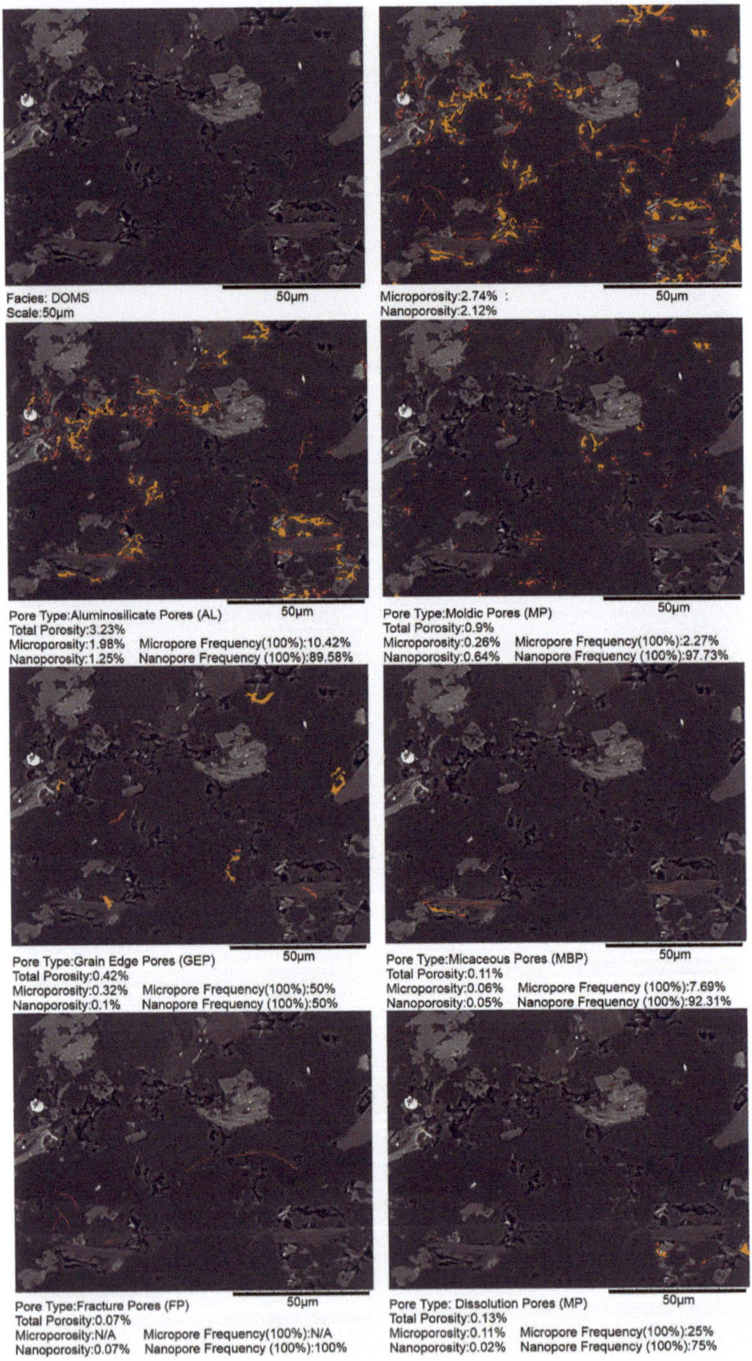

Fig. 22 Facies DOMS pore size characterization (50 μm Scale).

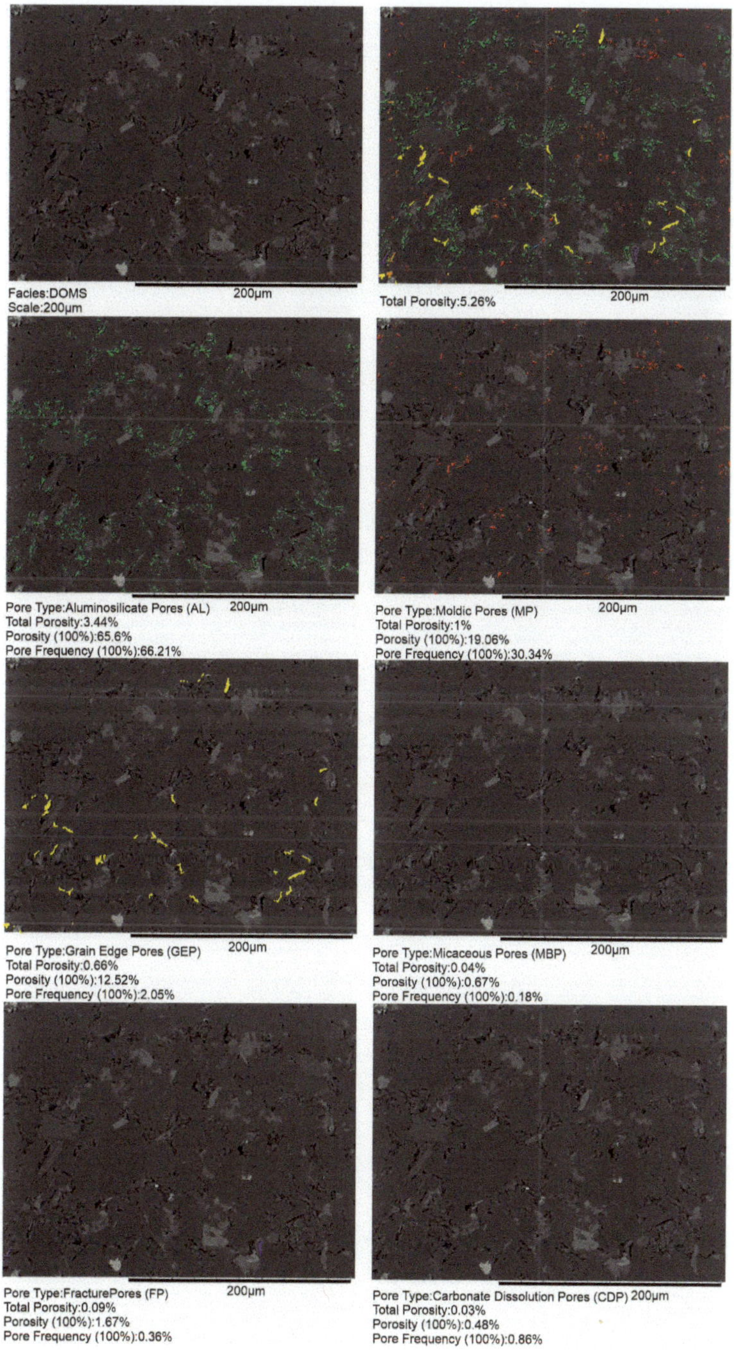

Fig. 23 Facies DOMS pore type characterization (200 μm Scale).

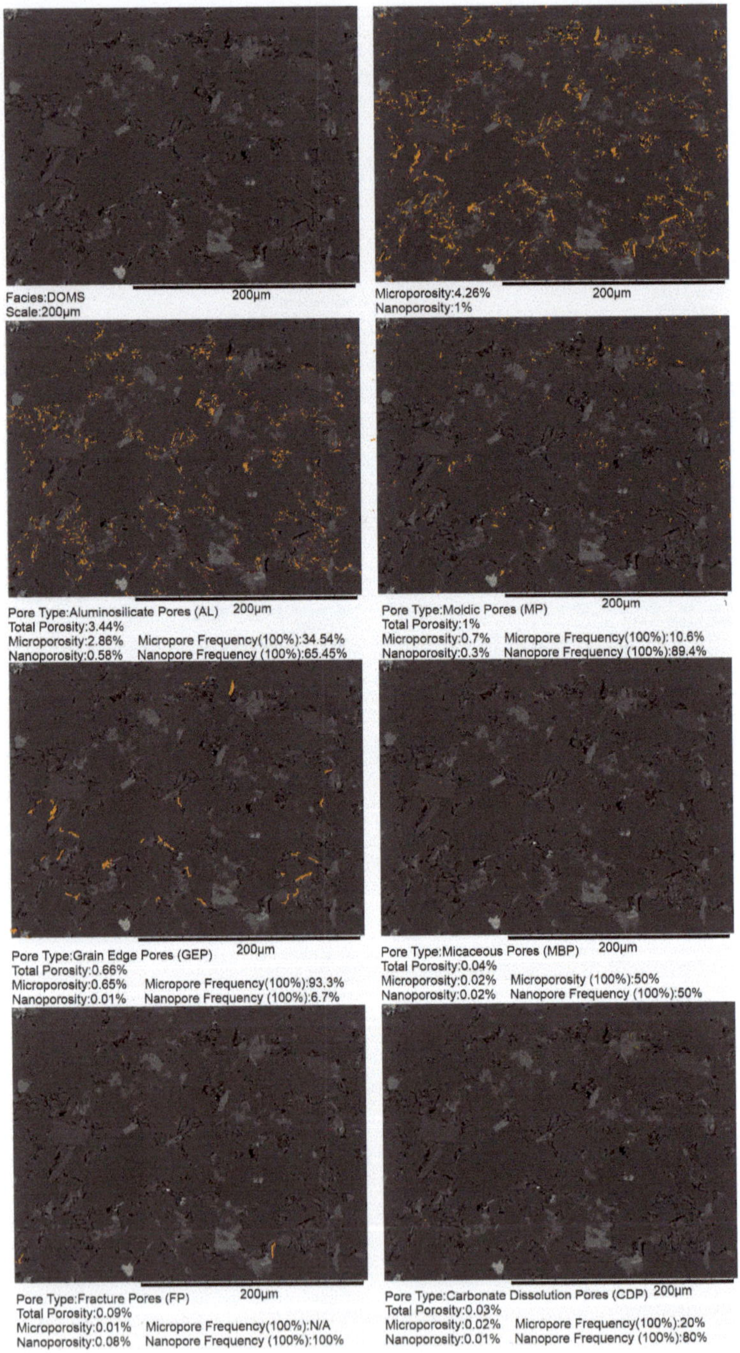

Fig. 24 Facies DOMS pore size characterization (200 μm Scale).

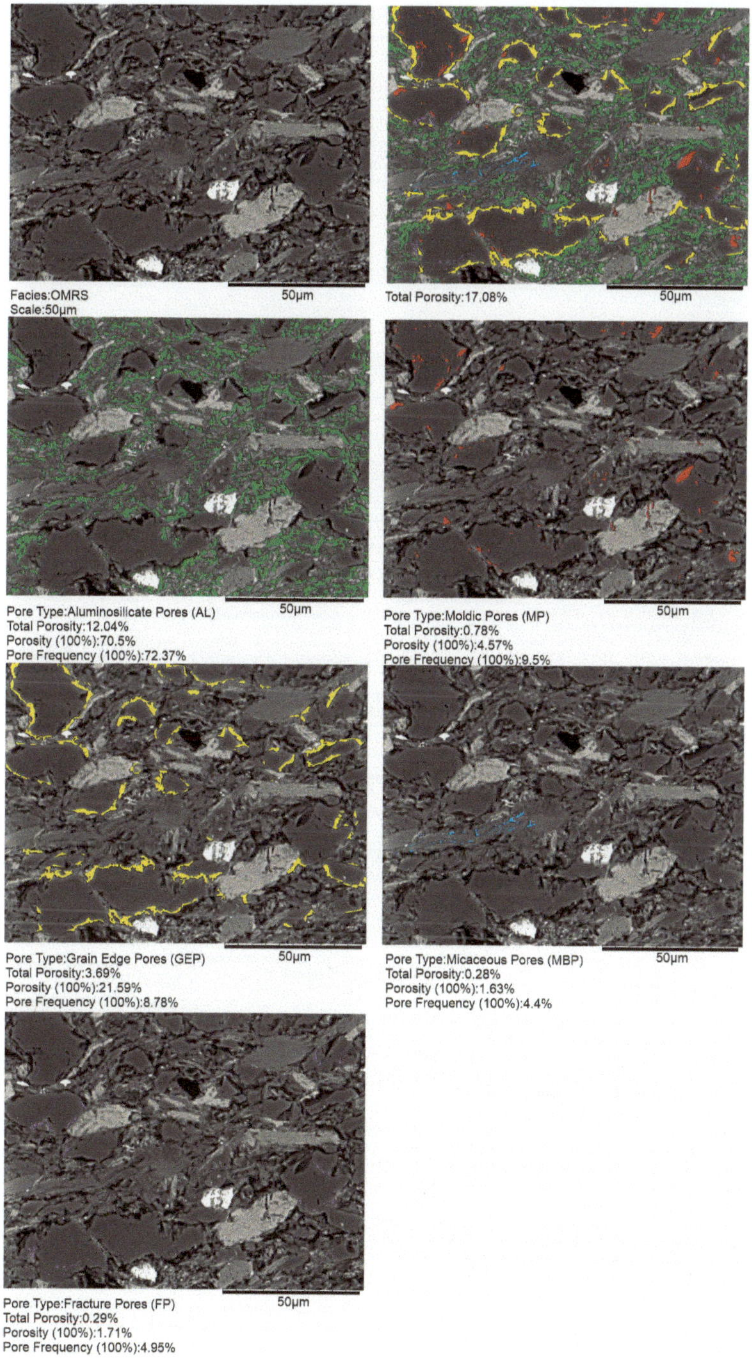

Fig. 25 Facies OMRS pore type characterization (50 μm Scale).

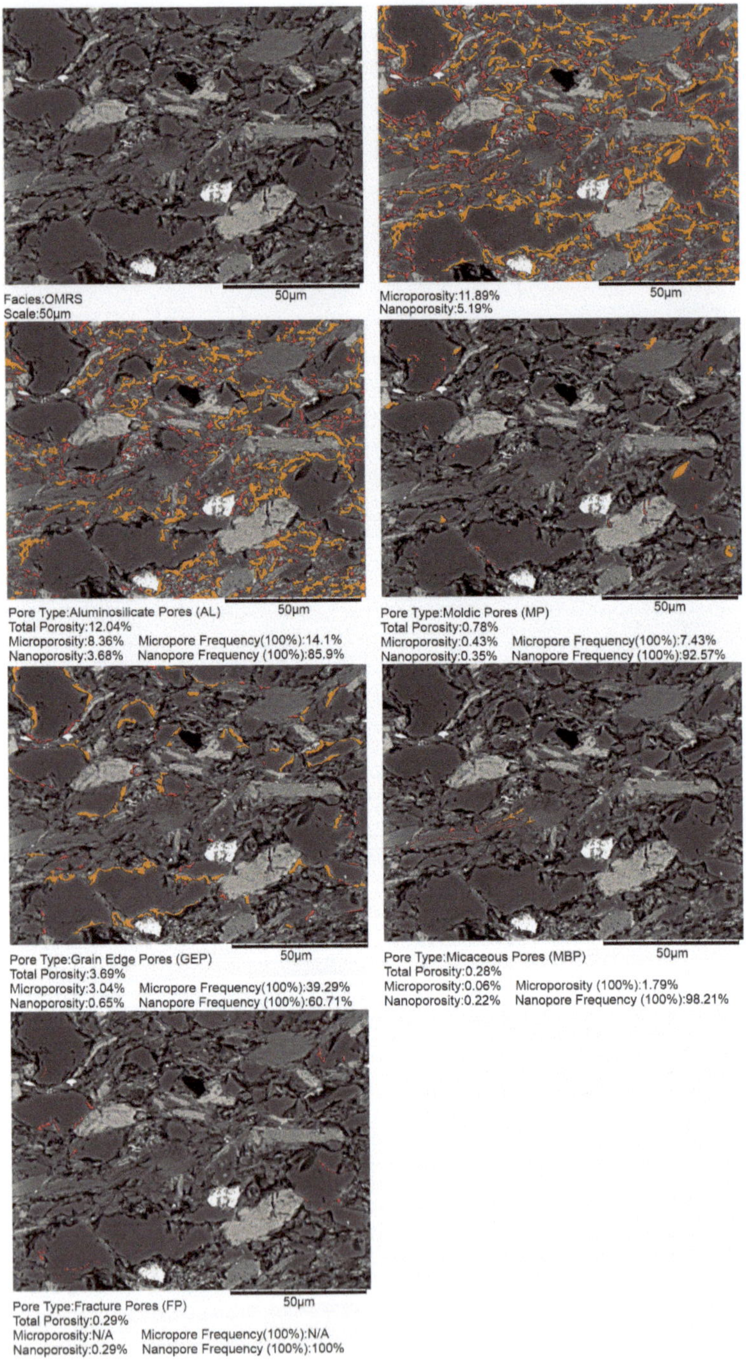

Fig. 26 Facies OMRS pore size characterization (50 μm Scale).

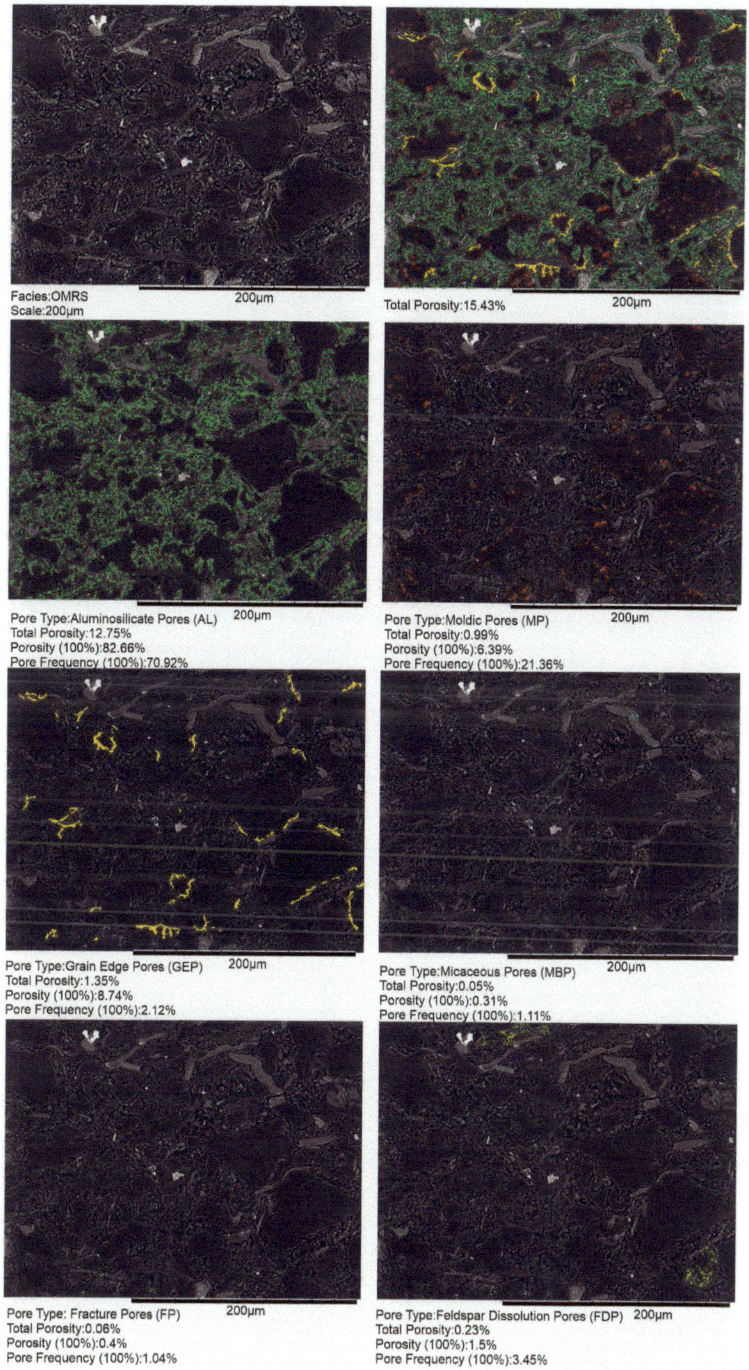

Fig. 27 Facies OMRS pore type characterization (200 μm Scale).

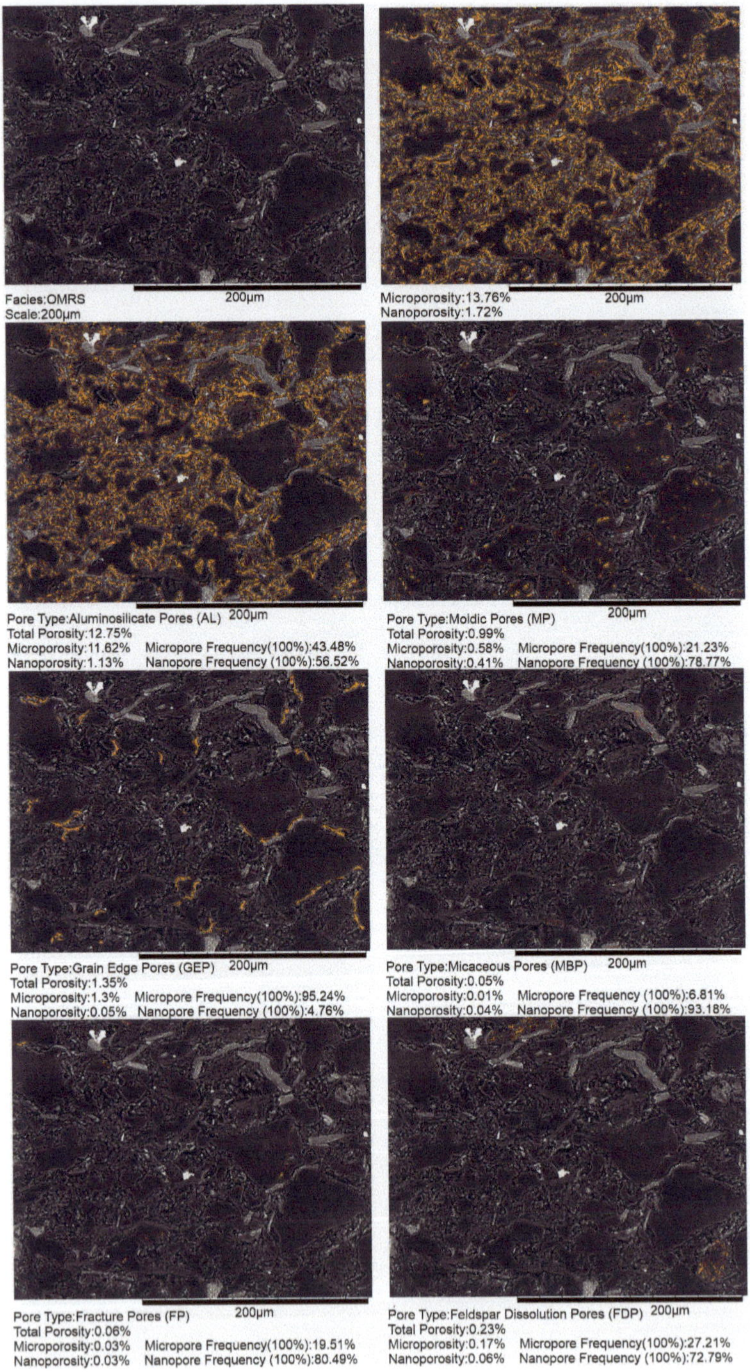

Fig. 28 Facies OMRS pore size characterization (200 μm Scale).

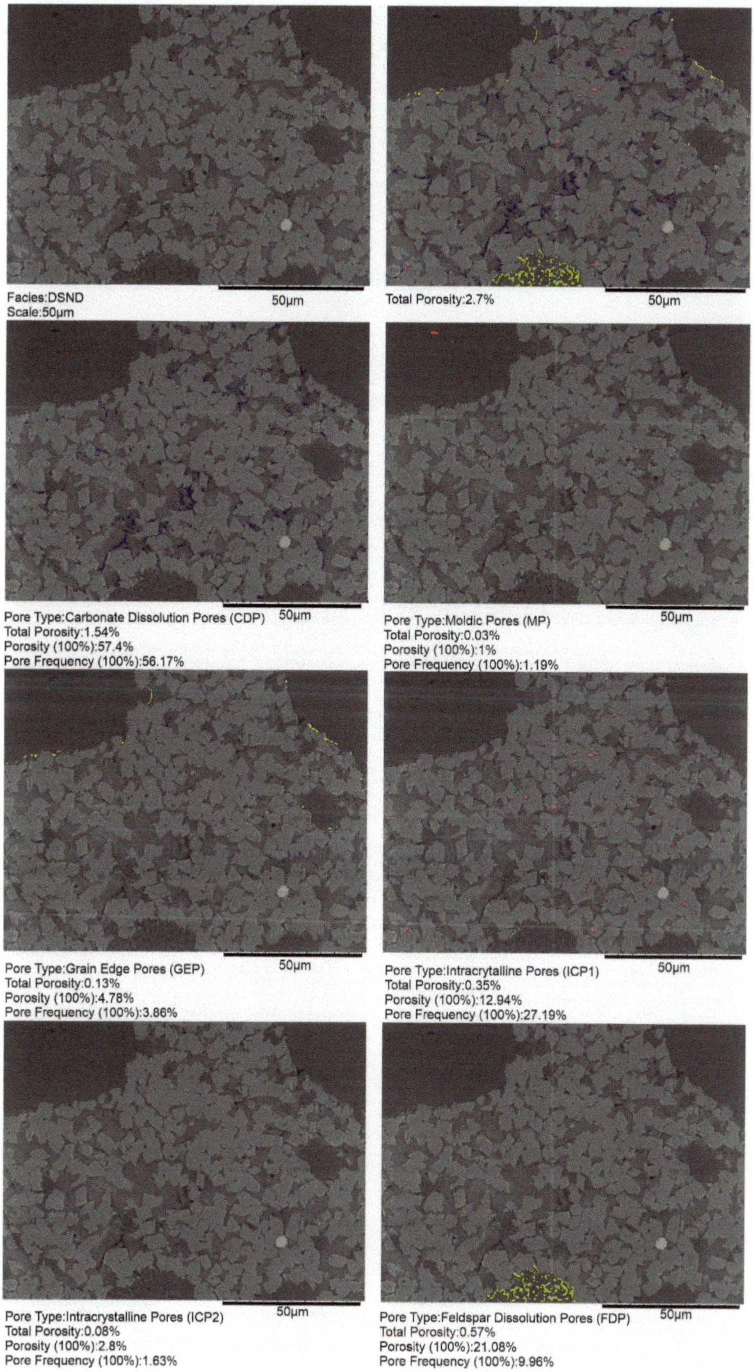

Fig. 29 Facies DSND pore type characterization (50 μm Scale).

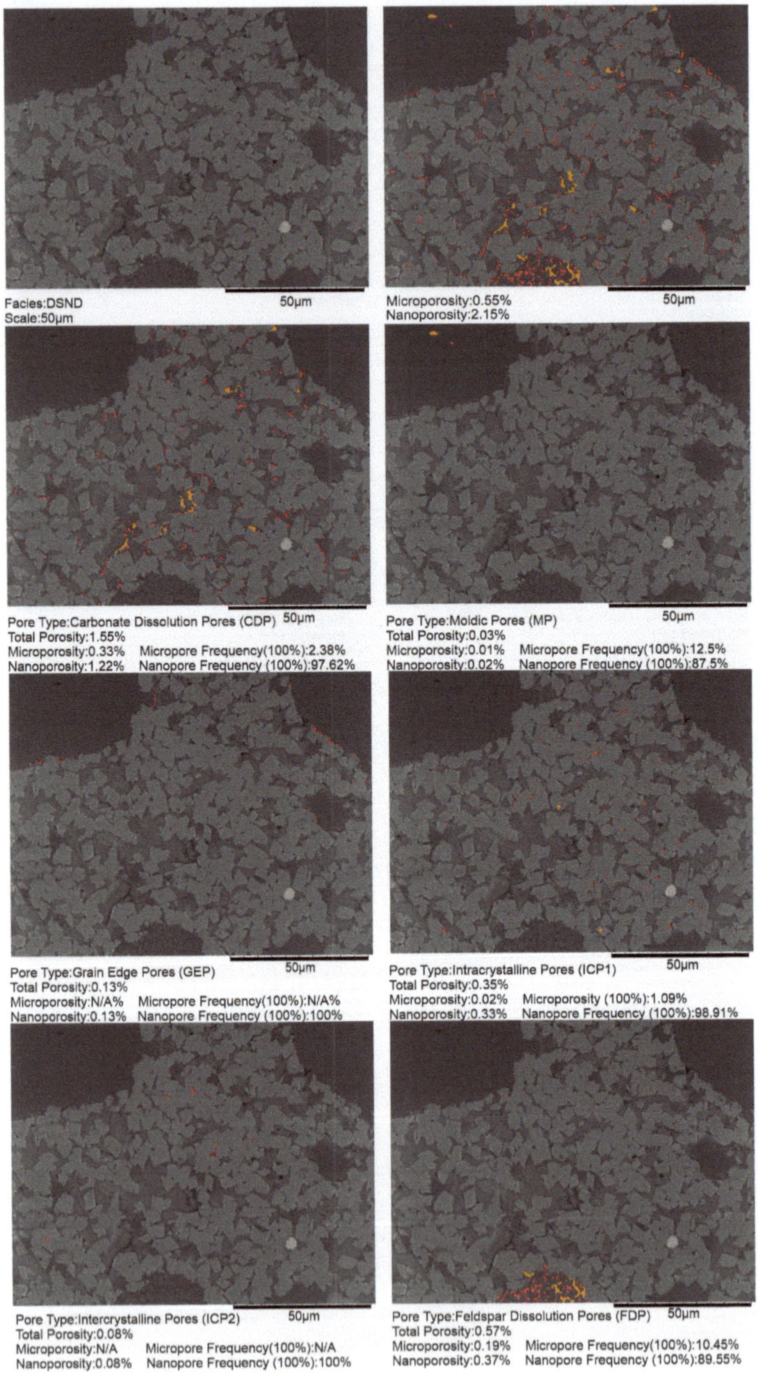

Fig. 30 Facies DSND pore size characterization (50 μm Scale).

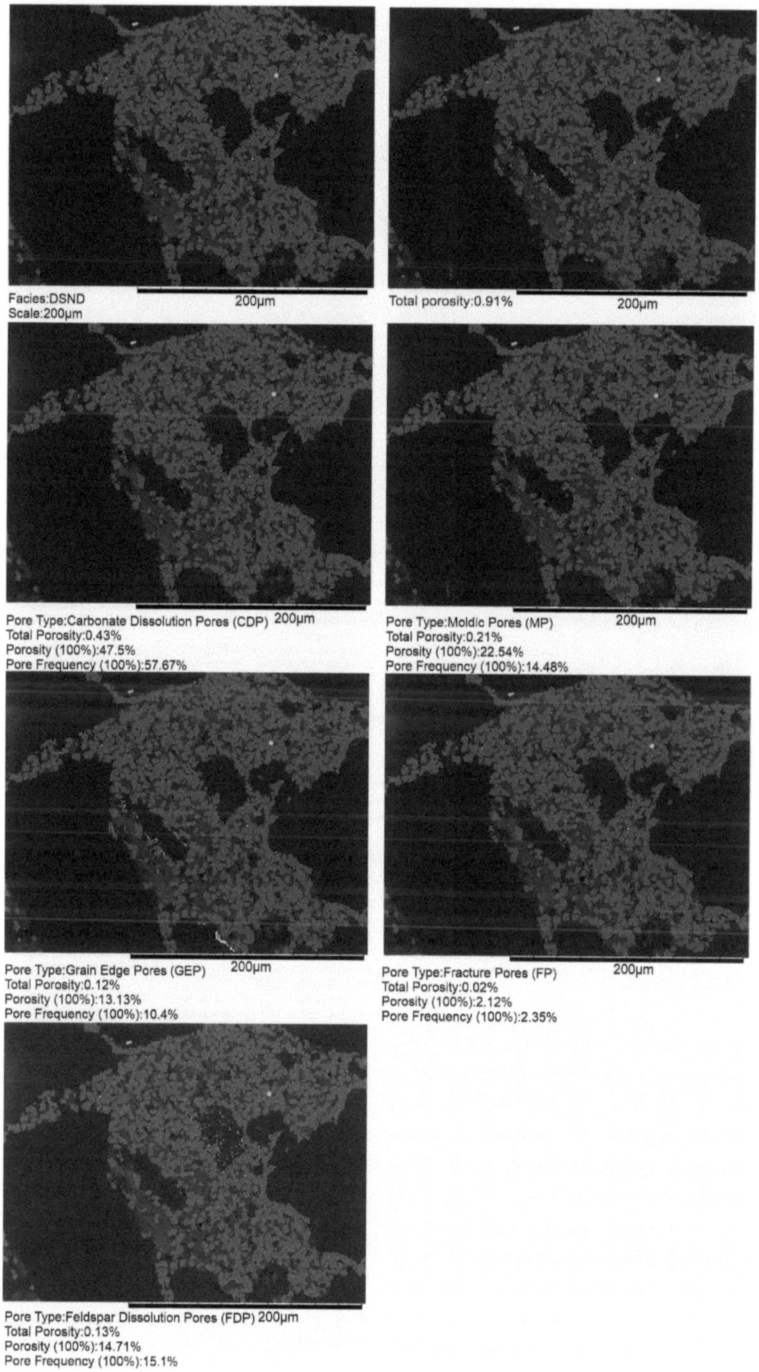

Fig. 31 Facies DSND pore type characterization (200 μm Scale).

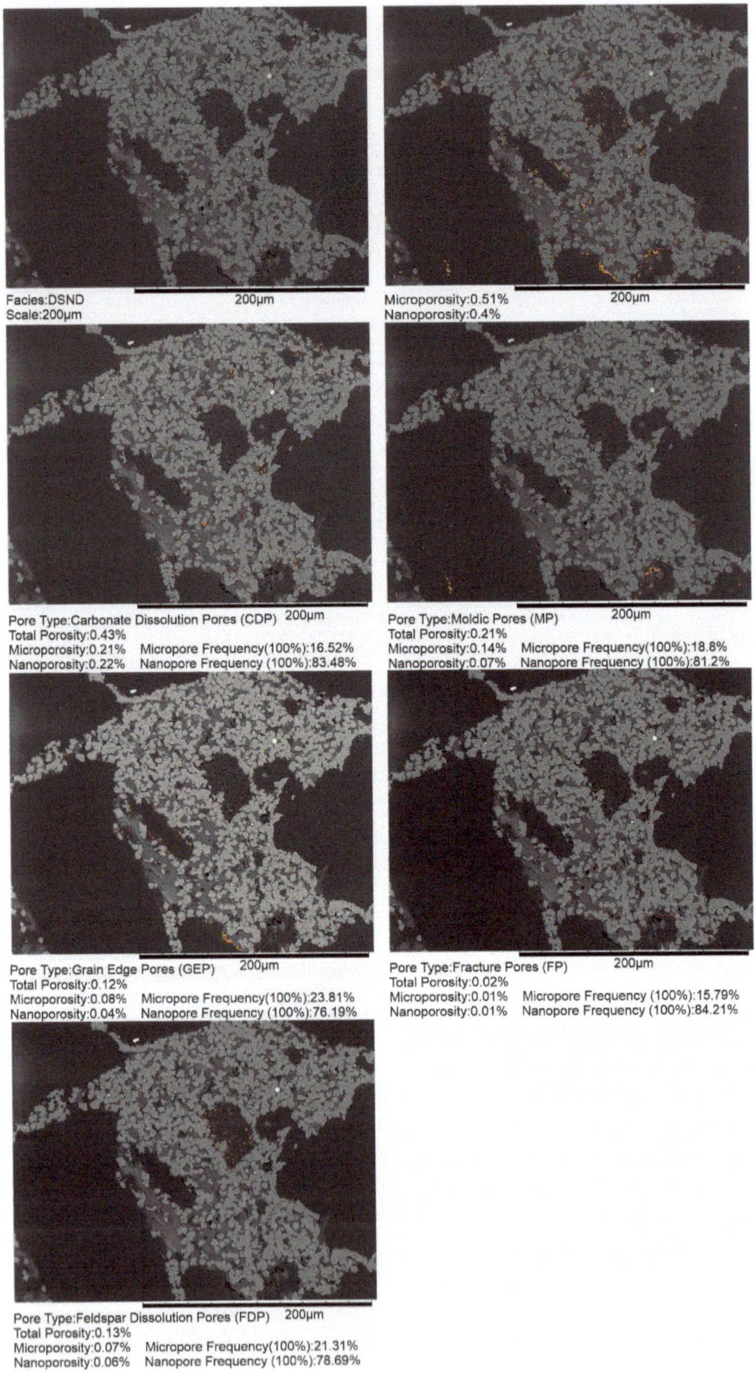

Fig. 32 Facies DSND pore size characterization (200 μm Scale).

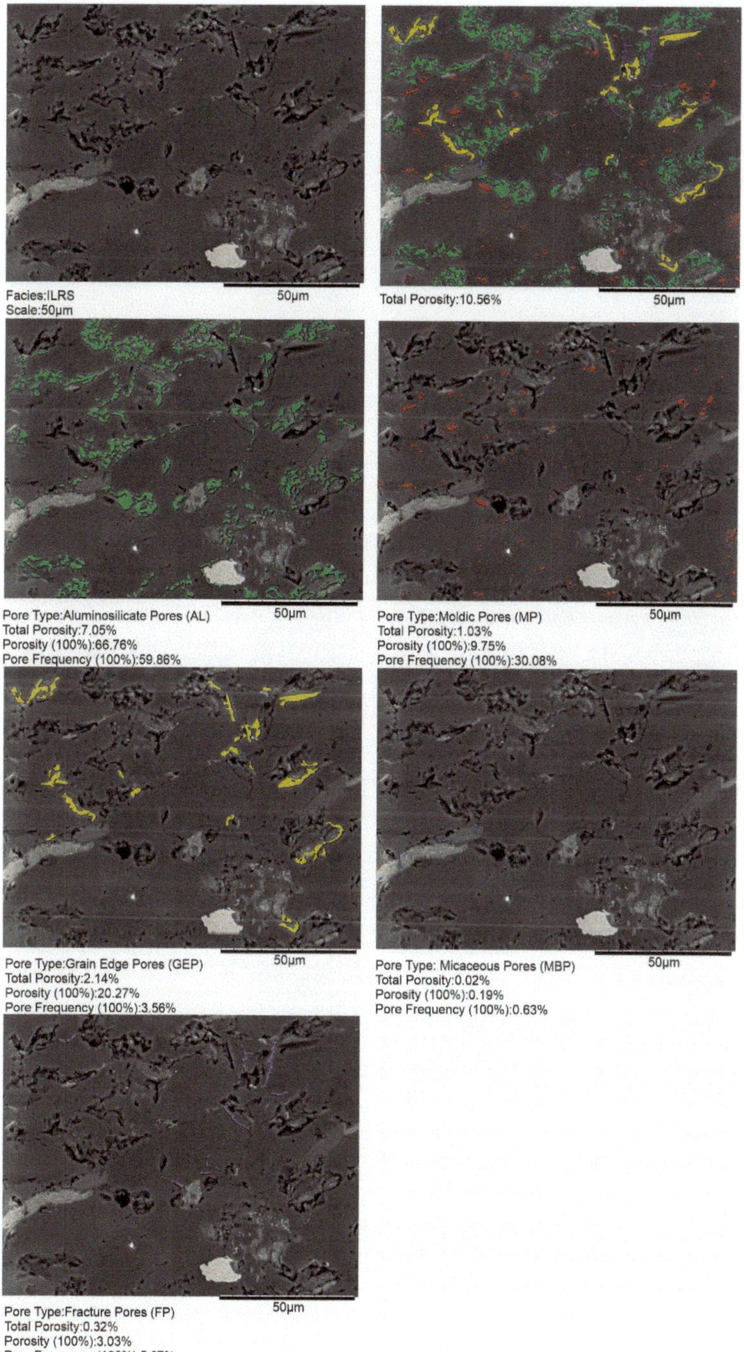

Fig. 33 Facies ILRS pore type characterization (50 µm Scale).

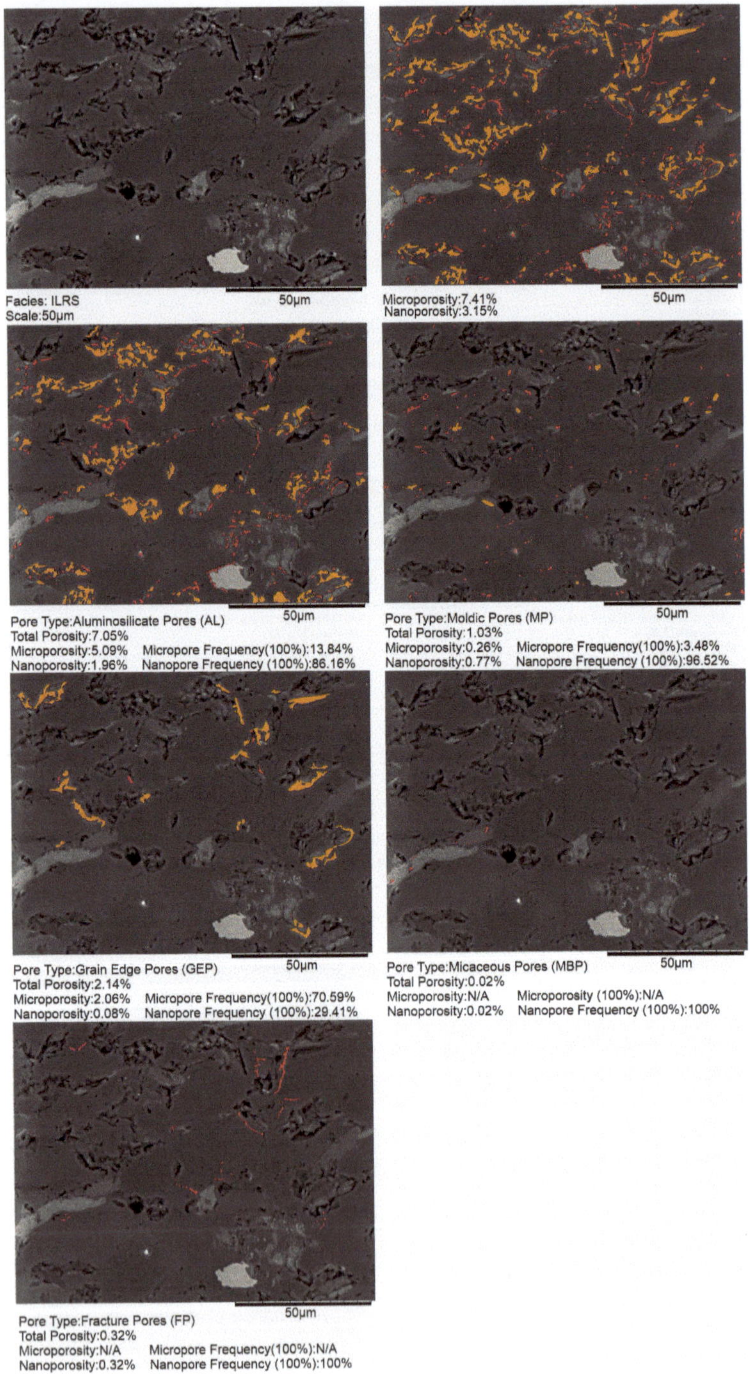

Fig. 34 Facies ILRS pore size characterization (50 μm Scale).

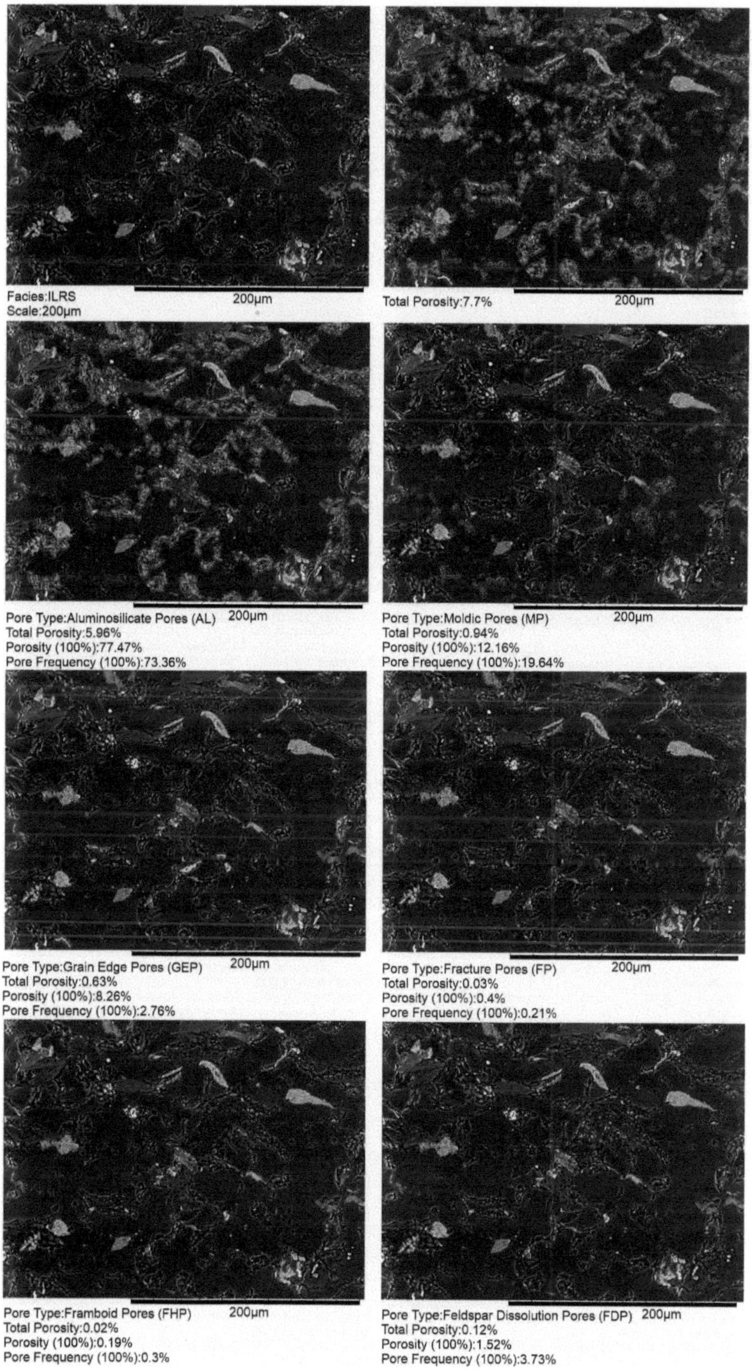

Fig. 35 Facies ILRS pore type characterization (200 μm Scale).

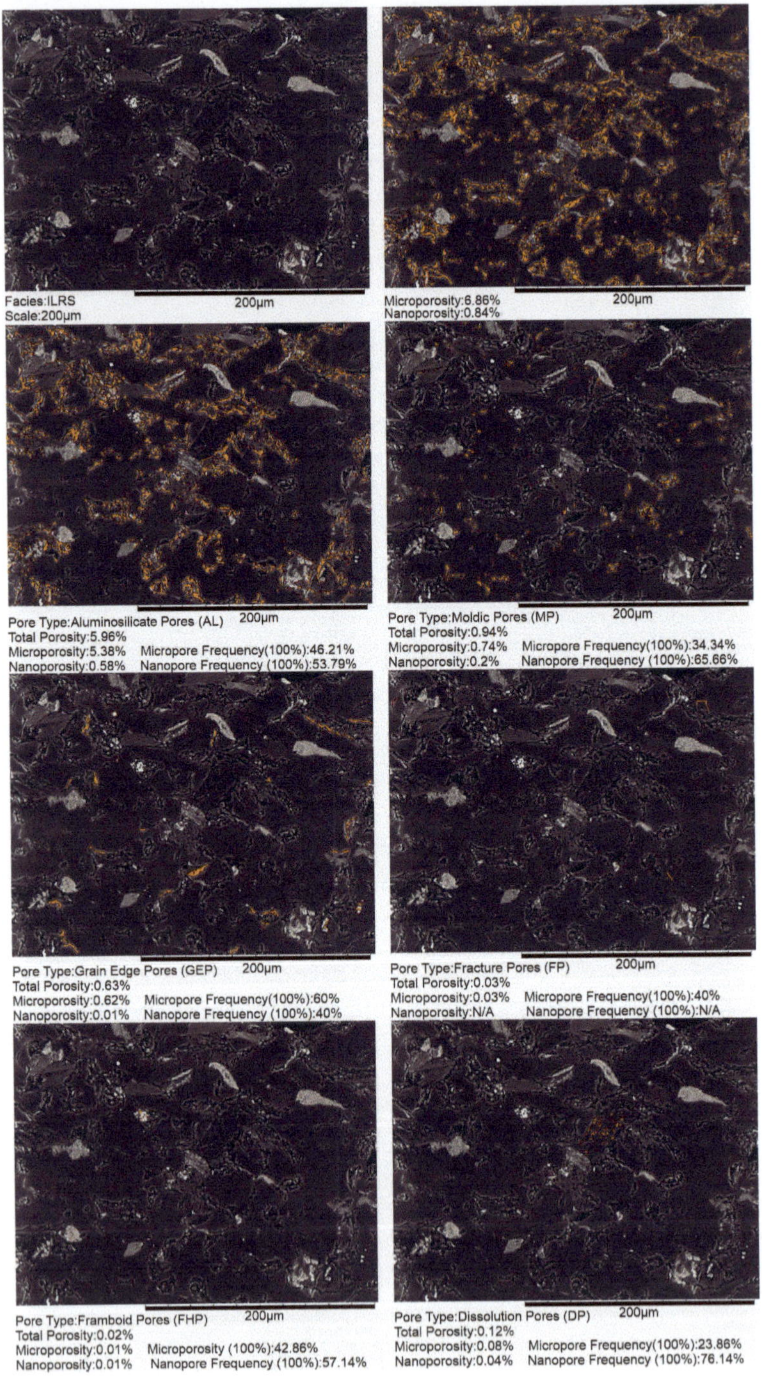

Fig. 36 Facies ILRS pore size characterization (200 μm Scale).

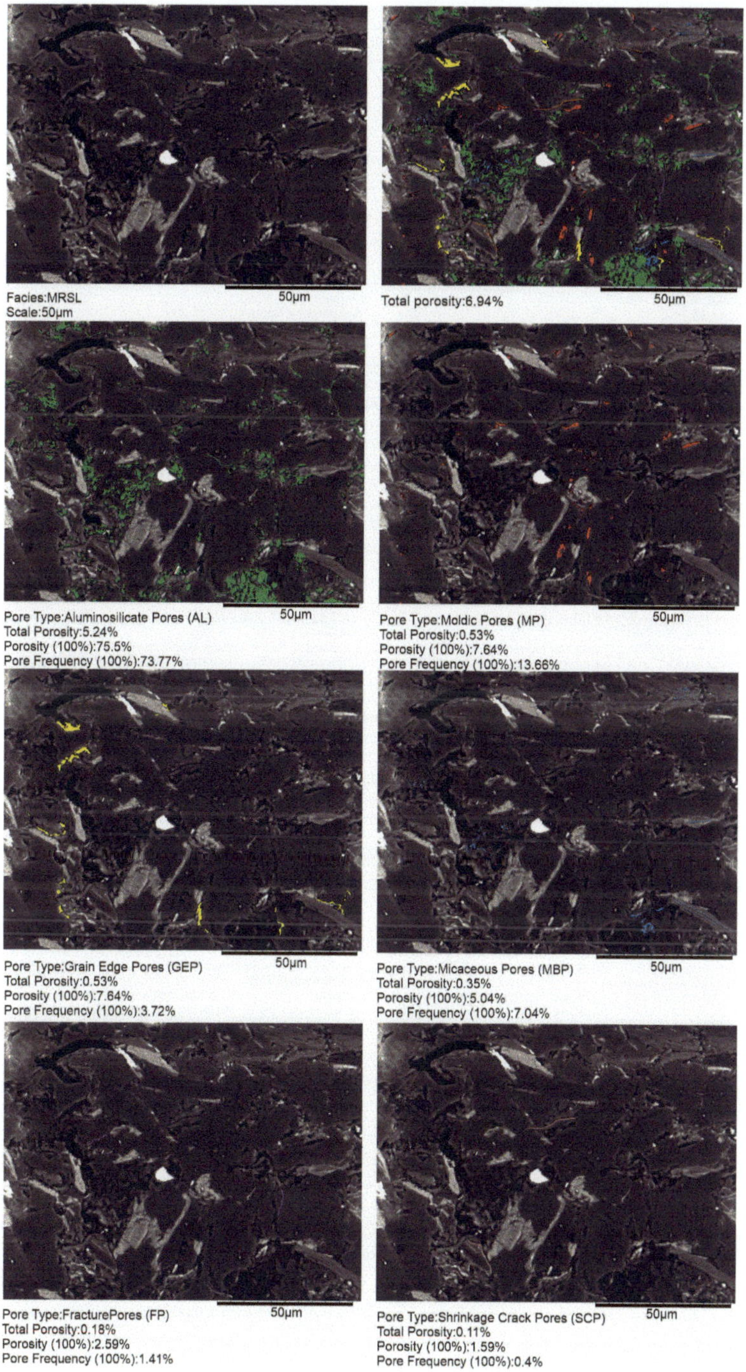

Fig. 37 Facies MRSL pore type characterization (50 μm Scale).

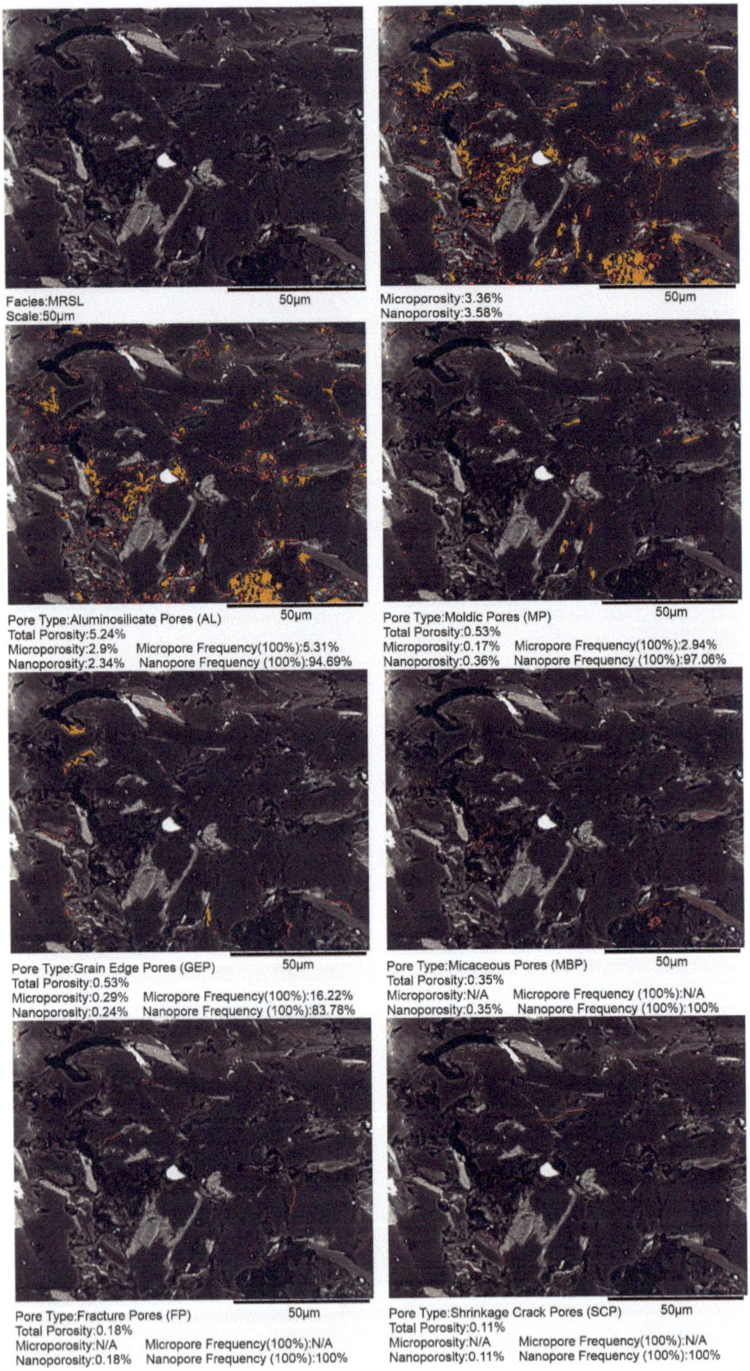

Fig. 38 Facies MRSL pore size characterization (50 μm Scale).

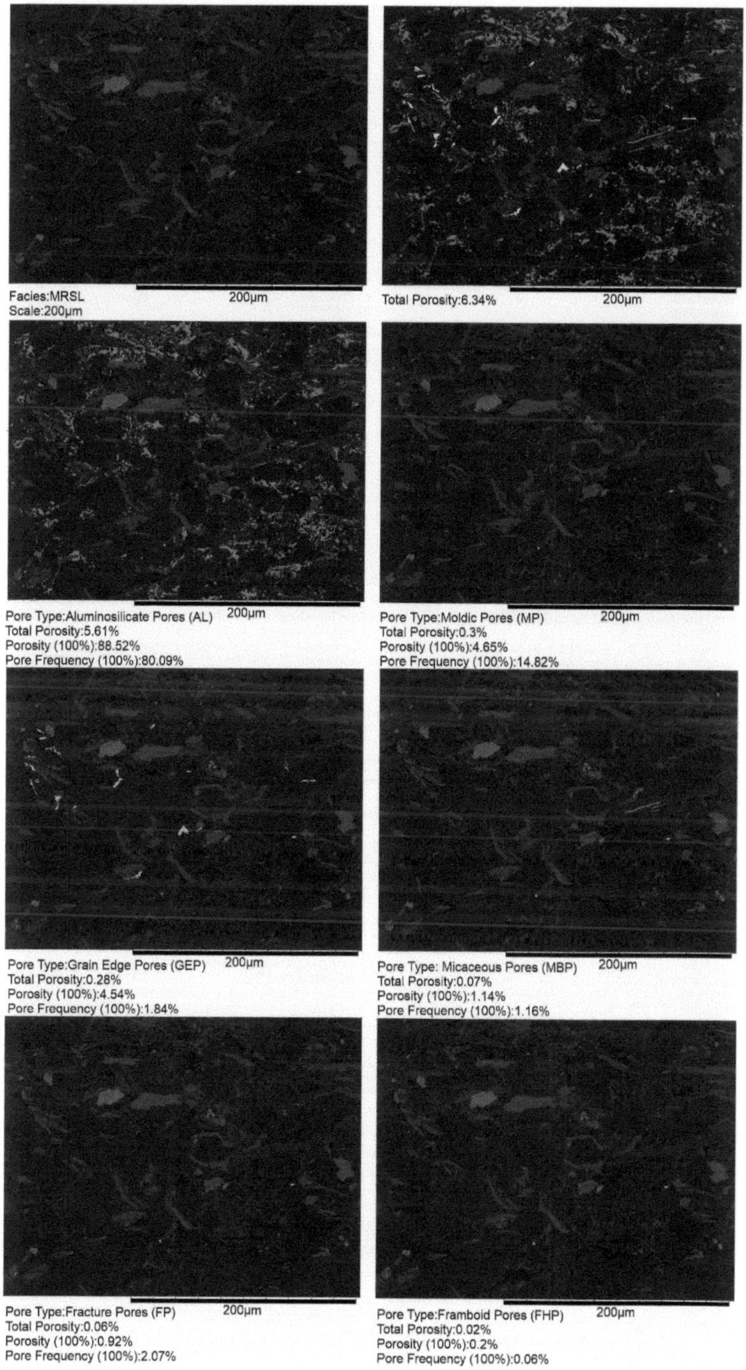

Fig. 39 Facies MRSL pore type characterization (200 μm Scale).

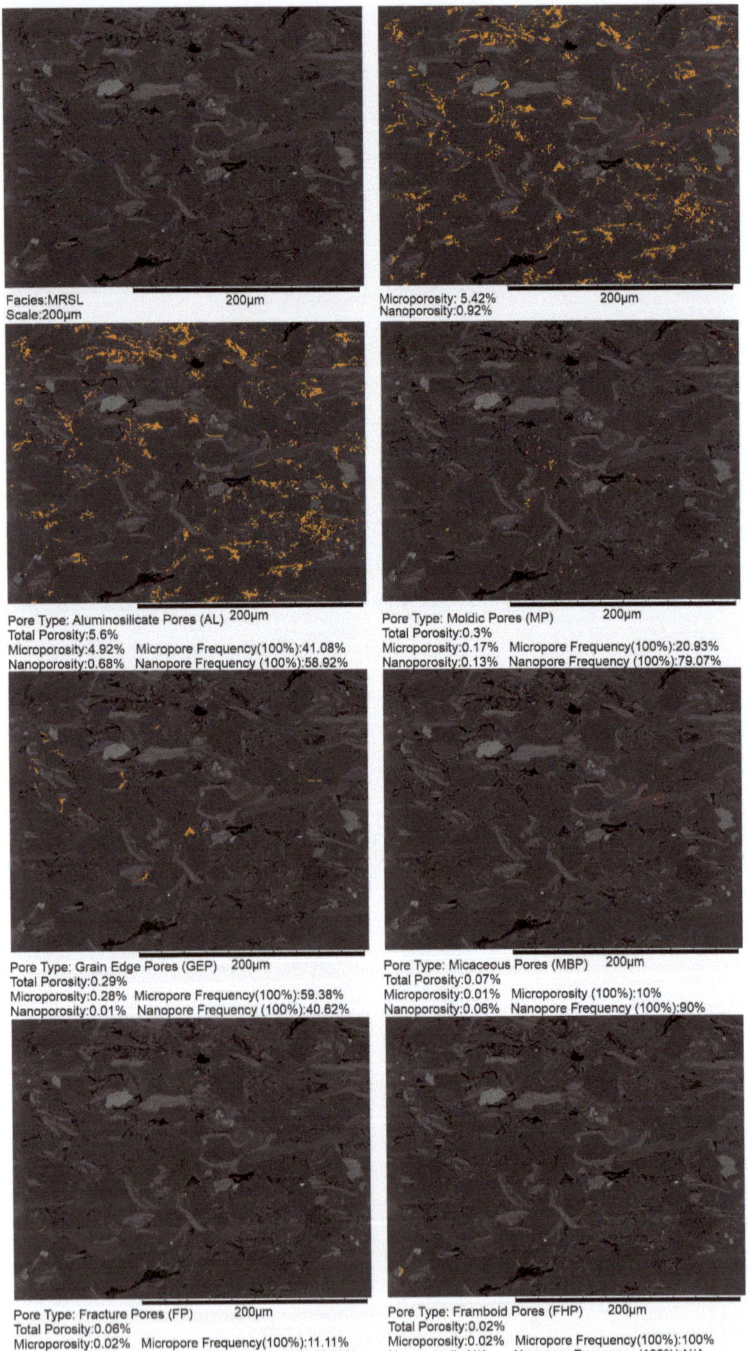

Fig. 40 Facies MRSL pore size characterization (200 μm Scale).

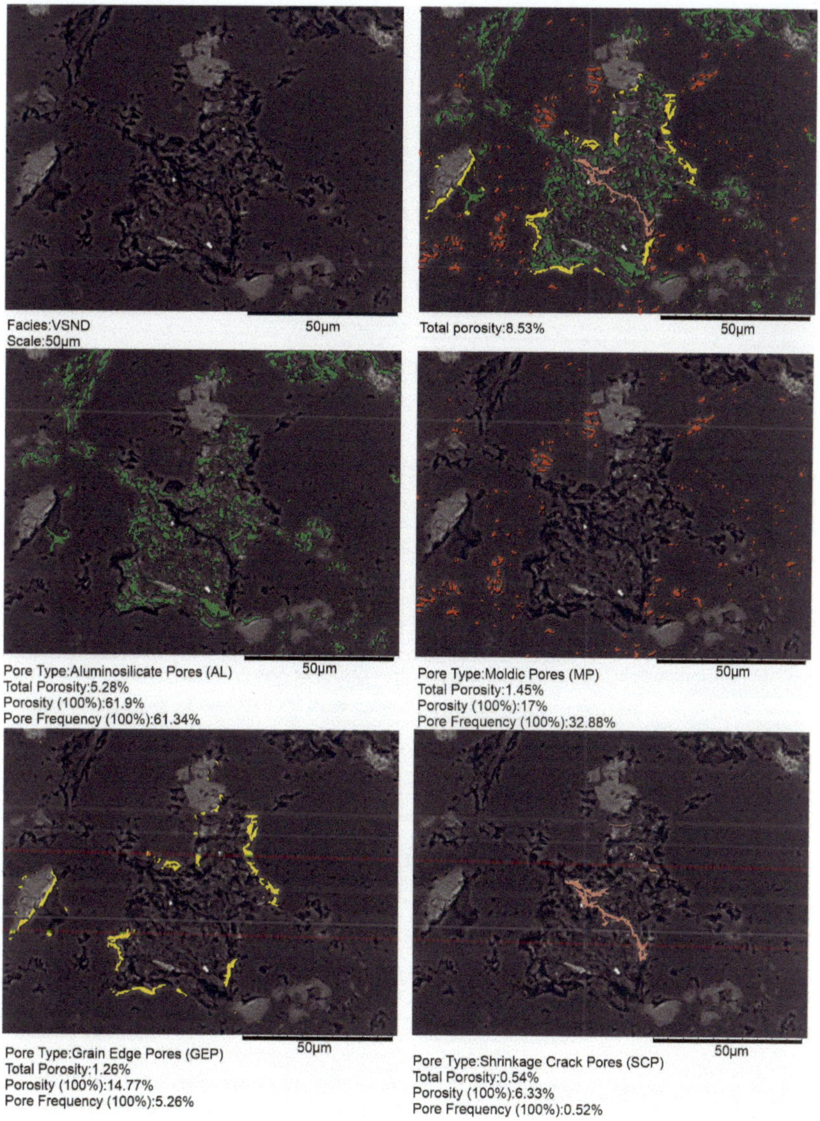

Fig. 41 Facies VSND pore type characterization (50 µm Scale).

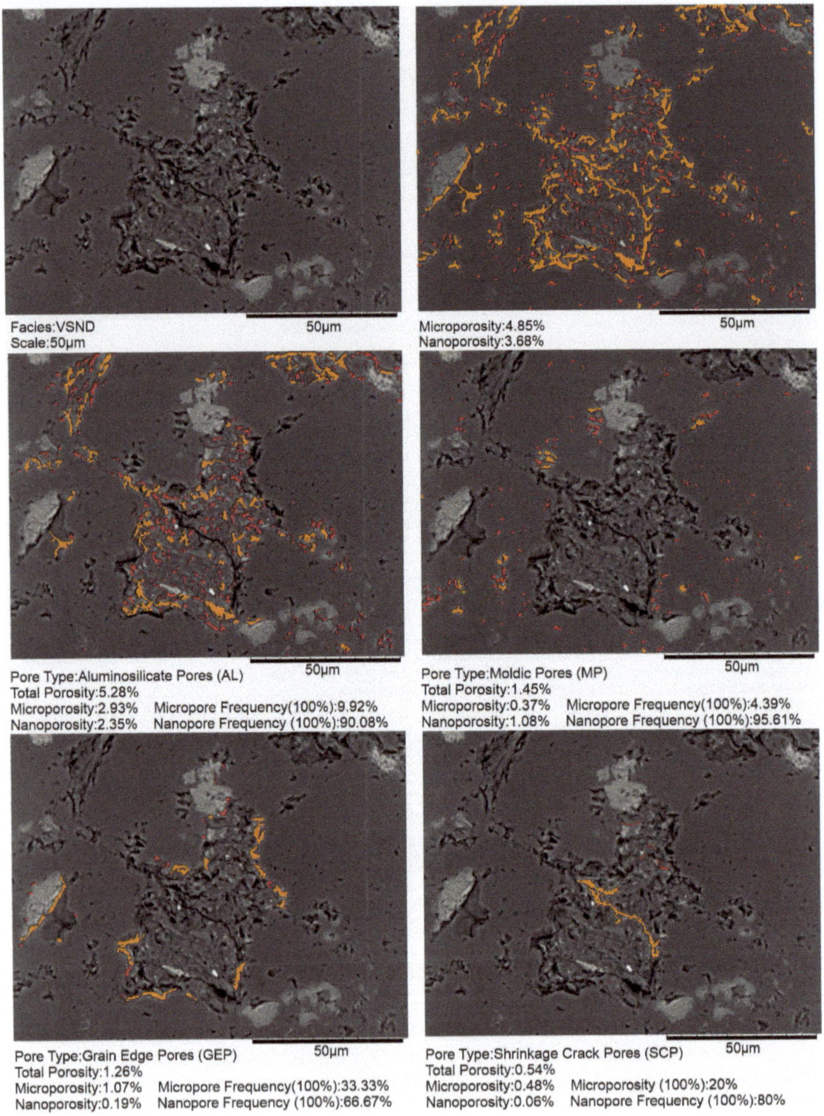

Fig. 42 Facies VSND pore size characterization (50 μm Scale).

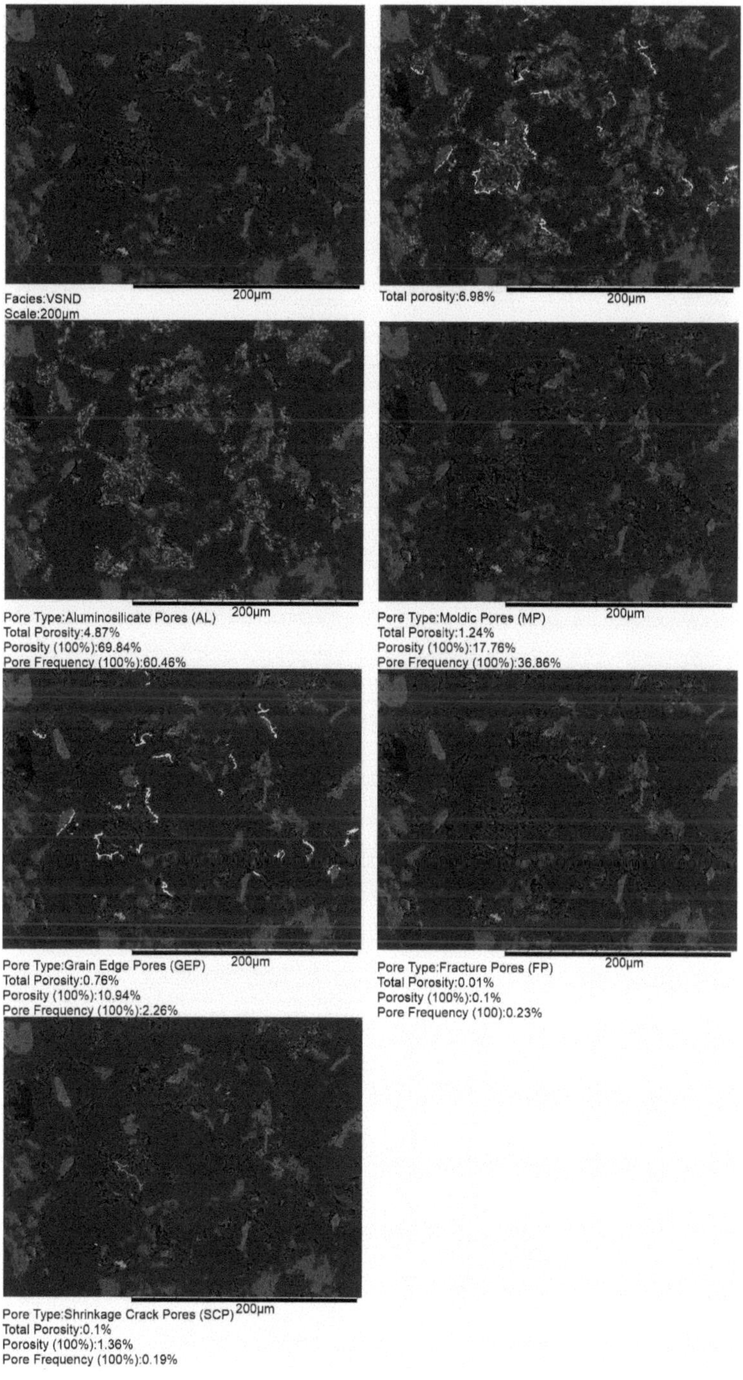

Fig. 43 Facies VSND pore type characterization (200 µm Scale).

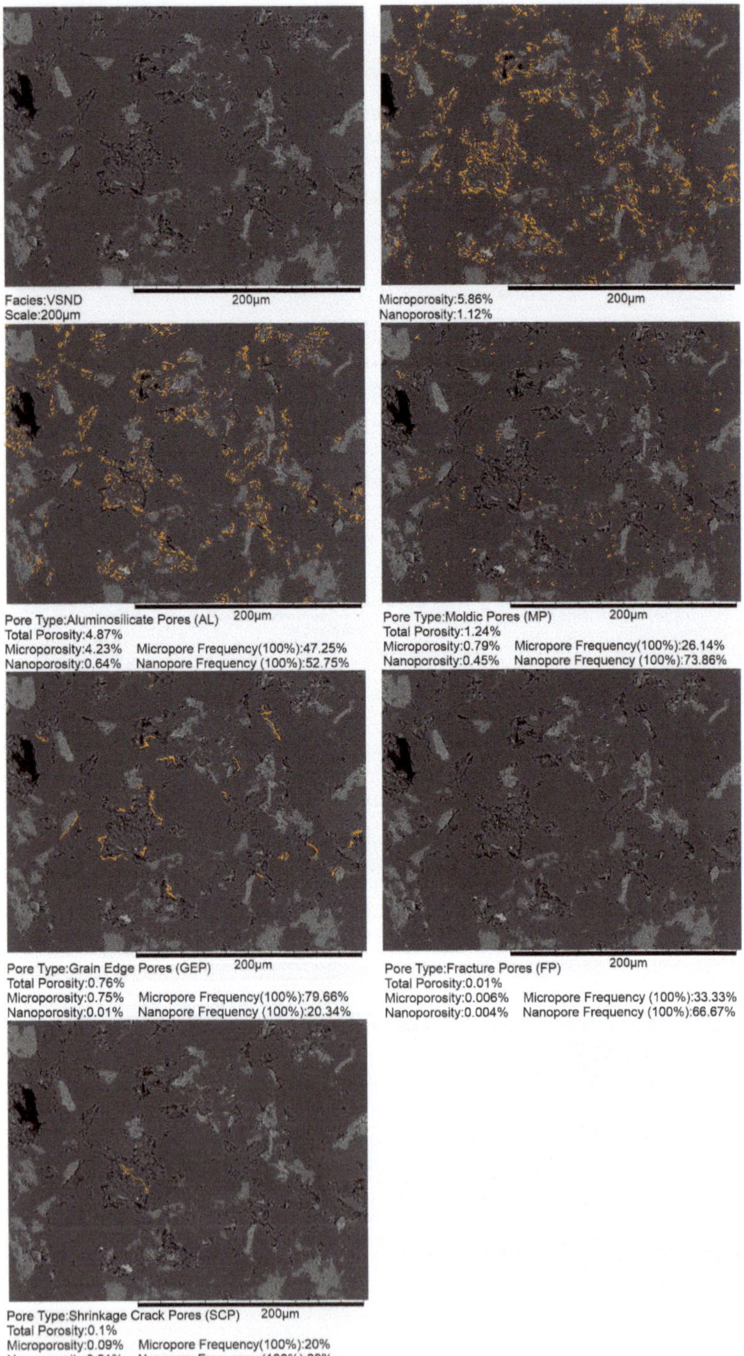

Fig. 44 Facies VSND pore size characterization (200 μm Scale).

5. Discussion

5.1 Resource assessment

5.1.1 Upper Pastora Formation (formation overview)

- The Upper Pastora Formation is characterized by an interbedded succession of organic matter rich shales and siltstones which are intermittently interrupted by massive dolomitized sandstone beds. The negligible porosity of these sandstone units (DSND) and isolation of dissolution based pores within their pore networks, has created a multi-layered sealing sequence. Within this multi-layered sealing sequence, shale gas is confined to shales as a self-sealing, self-sourcing reservoir as well as any escaping gas will be trapped between larger, regularly occurring DSND sandstone units, acting as a seal and trapping any gas within the more porous fine-grained siltstones. The presence of interbedded siltstone and sandstones layers within shale gas plays can improve the permeable capacity of shale reservoirs (Zou, 2013), however, the porosities demonstrated here show that most interbedded siltstone are of low porosity.

Pervasive OM content and porosity is highest in the OM/ILS siltstone and OMRS shale facies, with other facies showing more negligible quantities of both.

- Shale microfacies: Demonstrate porosities of 17.08% at the 50 μm scale and 15.43% at the 200 μm scale with an increase in porosity associated with a greater nanopore frequency.
- Siltstone microfacies: Exhibit porosities of 4.86–10.6% at the 50 μm scale and 5.26–7.7% at the 200 μm scale, except for the high degree of porosity found in the OM/ILS siltstone with porosities of 19.35% at the 50 μm scale and 24.47% at the 200 μm scale.
- Sandstone microfacies: DSND microfacies have a porosity of 2.7% at the 50 μm scale and 0.91% at the 200 μm while the VSND microfacies have a porosity of 8.52% at the 50 μm and 6.97% at the 200 μm in the varved sandstone, which is an argillaceous and dolomitic hybrid. This indicates that the microporous system is responsible for the majority of porosity.
- Gamma ray data combined with SEM analysis of organic matter indicate uranium values between 3.2 and 4.3 ppm, and OM identification under the SEM could be an indication of a highly TOC enriched shale which consists of limited type I, and enriched type II and III kerogen.

- Shales were deposited under an environment which underwent oxygenation fluctuations with Th/U ratios between 3.26 and 4.77, which places the Pastora paleolake in an oxic-dysoxic environment.

Should exploration occur, these shale/siltstone bundles should be targeted, specifically shale bundles in which OMRS microfacies are located directly above OM/ILS microfacies. Not only the shale self-generating and self-sealing sources but any gas which escapes these shale beds, is likely to accumulate in the more porous OM/ILS siltstone microfacies, which demonstrate the best pore network characteristics for hydrocarbon migration, as mentioned through the nanoporous and more importantly microporous sections. The intermittent occurrence of the DSND microfacies through the sealing sequence, act as gas seals and keep methane trapped between the coarser sandstone beds, while all other siltstone microfacies (DOMS, ILRS, MRSL) play a significantly less important role in trapping and storing gas, as they have lower porosities, lower effective permeabilities and limited sealing capacity, due to only partial dolimitization. There is currently sufficient evidence for the presence of methane gas migration, as gas between the various lithotypes has already been established by proxy. In 2013, seven miners died in a methane gas blow out in Pozo Emilio de Valle, located in the southeast of the mine. This is an area where a large section of the Upper Pastora Formation crops out and it may very well be that the methane blowout was a result of a penetration into the large dolomitized sandstones beds (DSND). The initial high porosity prior to diagenetic dolomitization of the intergranular space, may have caused the development of a geopressurized zone. As shale and very fine-grained siltstones, with a high water and gas bearing capacity within the illite matrix, were expelled during lithostatic compaction, they likely migrated into the large porous sandstone beds and were further locked in by later dolomitization. This presents both problems and solutions, in that there is clear evidence of methane generation and migration within the multi-layered sealing sequence and a potential target microfacies has been identified for exploration. We also have a good understanding of the degree of porosity and permeability and which pores can be found. However, it has become clear that rapid sedimentation rates in the Late Carboniferous/Permian, during the primary phase of thermogenic methane generation, may have generated a series of geopressurized sequences marked by gas charged sand bodies. Although it remains a prominent source of blowouts, which complicates extraction, it has been said that geopressurized UHRs contain the greatest quantities of natural gas (Speight, 2019).

5.1.2 Upper Pastora Formation (pore network overview)

As the pore type frequency and porosity data has shown, there is a highly disproportional relationship between these two factors, the most important of which are GEP and moldic pores. Despite moldic pores being the second most abundant pore type, these pore types are predominately nanometric and as BSE imaging shows, isolated from the pore network, indicating that they have an inherent inability to migrate hydrocarbons throughout the pore network. GEP pores however, are lower in frequency but contribute much greater amounts of porosity and are located adjacent to the aluminosilicate matrix, offering great migration potential for gas seeking to escape the matrix. In terms of pore size distribution, we see that as we downscale from the 200 to 50 μm scale there is a 25% average increase in nanopore frequency associated with a 27% increase in nanoporosity throughout the pore network (Fig. 45). This results in an increase in total rock porosity of between 1.12% and 3.47% of which the greatest increase in nanoporosity can be found in the organic matter rich OMILS and OMRS siltstone sand shale facies with an increase of 3% and 3.47%, respectively. This highlights the importance of nanopore contributions toward porosity and the need for multi-scaled imaging, but also the close association between increase in TOC content and nanoporosity.

The two most important microfacies types in the Upper Pastora Formation are the OM/ILS and OMRS. Both show great hydrocarbon potential in response to recovery operations, given the high quantity of porosity found within the aluminosilicate matrix, and the hydrocarbon migration potential offered from the elongate structure of grain edge pores. These two facies offer the greatest potential for shale gas extraction due to the following characteristics:

1. The most abundant pore types are: aluminosilicate pores, moldic pores and grain edge pores. Despite moldic pores being the second most abundant, they are much smaller than GEP pores, contributing less to total porosity and are more isolated from the pore network. The inverse can be said for GEP pores, stressing the disproportionality between moldic and GEP pore type abundance and porosity (Fig. 16).
2. All microfacies demonstrate a low effective porosity, with the greatest potential for effective hydrocarbon migration found in the OM/ILS and OMRS microfacies, due to higher quantities of microporous aluminosilicate and GEP pores. In addition to this, these facies have the highest frequency of aluminosilicate nanopores, which despite contributing proportionally smaller quantities of porosities, can offer further adsorbed gas potential due to a larger specific surface area.

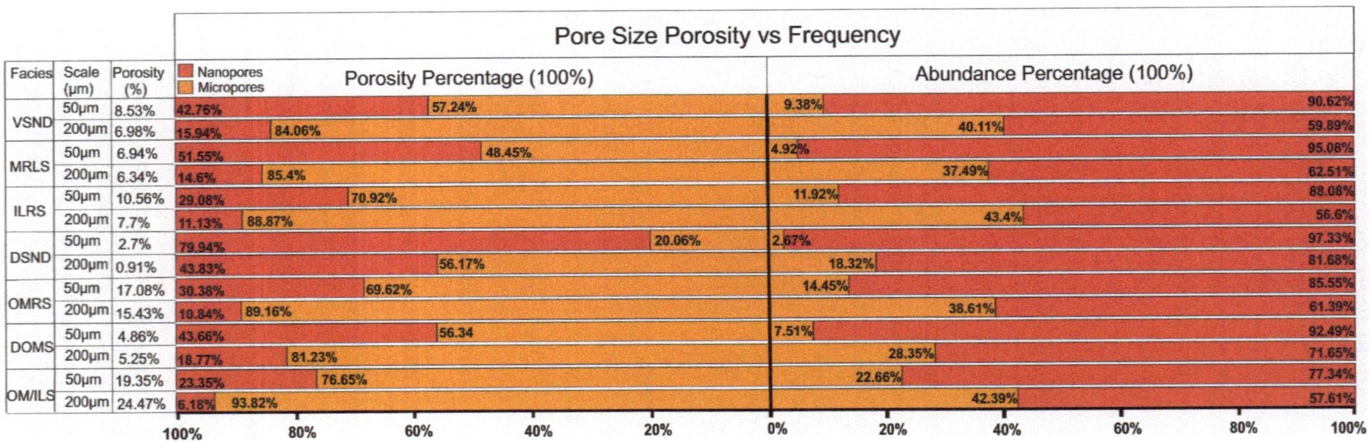

Fig. 45 Pore size abundance and porosity distribution across all facies types within the nanopore system (50 μm) and micropore system (200 μm). The chart shows the abundance of nanopores throughout all facies at all scales is always higher than their microporous counterparts, yet micropores still contribute the majority of porosity except for the DSND (50 μm) and MRSL (50 μm) facies.

3. Illite-based aluminosilicate porosity, at the 50 μm scale, is the greatest contributor in the two most porous microfacies types OM/ILS-13.36%, OMRS-12.04%) with nanoporosity more dominant in the shale microfacies (OM/ILS-2.32%, OMRS-3.68%). At the 200 μm scale, Illite-based aluminosilicate porosity is still the biggest contributor toward total microfacies porosity in the OM/ILS (21.36%) and OMRS microfacies (12.75%).
4. In the/OM/ILS and OMRS as we transition from the micrometric system to the nanometric system we see that nanoporosity in primary pore types can increase by up to 2.5%.

As confirmed by the characterized BSE images, specific pore types visually demonstrate varying structures and contributions toward total porosity. Given the level of discontinuity and lack of pore connectivity within the pore network of shales and siltstone both here, and recorded around the world, we must begin to view the pore network differently. Rather than a collective group of unconnected void spaces, it is vital to identify the importance of their genesis, the morphological variability associated with specific pore types and identify their potential to form effective hydrocarbon migration pathways. This can only be done by first characterizing pore types. Once completed, we can begin to identify pore network properties prior to recovery operations, which indicate conditions most conducive to fluid flow and enhance our ability to predict well longevity and productivity. Furthermore, research and development (R&D) regarding increased hydrocarbon recovery could greatly benefit from identifying pore type characteristics which are most conducive to specific types of recovery operations such as acidization. Although in its preliminary stages, combining matrix acidization research in shales, with characterized pore network and geochemical data, could yield interesting results. However, first a deeper understanding of shale mineralogy, permeability, porosity and fluid dynamics is required (Gomez, 2019), something which characterized pore networks from SEM-based binary imagery can provide.

6. Conclusion

In this paper with we provide evidence for a multi-layered sealing sequence within the Upper Pastora Formation, consisting of interbedded OM rich shales/siltstones and dolomitic sandstones, which were deposited in a shallow-intermediate lacustrine setting and contain mostly type II and III kerogen sources, with kerogen type I confined to rhythmic varves.

The intermittent presence of larger, low porosity sandstone beds within this sequence has provided a dual sealing sequence which has charged the overlying sandstone bodies with gas. Porosity throughout the sequence is variable, ranging from 2.7% to 19.35% (50 μm) and 0.91% to 24.47% (200 μm), and >3% within dolomitized sandstones, with the greatest potential for gas in the very fine-grained siltstone microfacies OM/ILS and shale microfacies OMRS. The porosity in these microfacies is dominated by primary pores types, predominantly within the aluminosilicate matrix and then by GEP pores, with moldic pores being isolated from the pore network showing poor migration potential. The structure of these larger microporous GEP pores provides excellent hydrocarbon migration pathways for free gas locked within the aluminosilicate matrix and also adsorbed gas, which shows strong potential due to the large frequency of nanopores within the illite matrix. This characterization of the pore network and quantification of both pore frequency and porosity has highlighted the individual importance of each pore type within the network. Further statistical analysis of the factors associated with pore-scale fluid flow activity such as pore geometry, pore size distribution, spatial distribution and other important morphological parameters, can now be quantified and associated with each individual pore type. With recent work on pore type genesis having greatly improved our understanding of the formation of specific pore types within shales, we can now begin to correlate parameters surrounding pore type genesis with changes in the structure of the pore network. This subsequently allows us to identify conditions which are likely to generate pore types which are known to be more conducive to hydrocarbon migration, and contribute the most toward pore network porosity. Although this requires a large amount of data synthesis from a wide range of sources and adoption of a universal characterization scheme, this method can provide valuable data for generating more accurate multiphase fluid flow models and provide insights into pore-scale geomechanics at both the nanometric and micrometric scale.

Acknowledgments

We thank the technical members of staff at GGE (Keele University, UK) for their support and the staff of the Pensión Rabocán in Beberino (Castilla y Leon, Spain) for their hospitality. The comments and suggestions of two anonymous reviewers helped to improve the original manuscript. E.J.R. acknowledges the financial support through the GGE-MSc program of Keele University (UK) and M.M. acknowledges support from Total SA (CSTJF—Centre Scientifique et Technique Jean Féger, Pau, France). The authors declare no conflict of interests.

References

Adams, J., Weaver, C., 1958. Thorium-to-uranium ratios as indicators of sedimentary processes: example of concept of geochemical facies. AAPG Bulletin 42, 387–430.

Álvarez, R., Ordóñez, A., Canteli, P., De Miguel, E., 2018. Unconventional gas resources in the Cantabrian zone (NW Spain): a comprehensive preliminary assessment. Geol. J. 54 (4), 2608–2620.

Anovitz, L., Cole, D., 2015. Characterization and analysis of porosity and pore structures. Rev. Mineral. Geochem. 80 (1), 61–164.

Ayllon, F., Bakker, R., Warr, L., 2008. Re-equilibration of fluid inclusions in diagenetic anchi zonal rocks of the Ciñera-Matallana coal basin (NW Spain). Geofluids 3 (1), 49–68.

Bazilian, M., Brandt, A., Billman, L., Heath, G., Logan, J., Mann, M., Melaina, M., Statwick, P., Arent, D., Benson, S., 2014. Ensuring benefits from north American shale gas development: towards a research agenda. J. Unconv. Oil Gas Res. 7, 71–74.

Bertoni, C., Cartwright, J.A., Foschi, M., Martin, J., 2018. Spectrum of gas migration phenomena across multi-layered sealing sequences. AAPG Bulletin 102 (6), 1011–1034.

Boyer, C., Clark, B., Jochen, V., Lewis, R., 2013. Shale Gas: A Global Resource. Schlumberger, pp. 28–39.

Cardott, B., Curtis, M., 2018. Identification and nanoporosity of macerals in coal by scanning electron microscopy. Int. J. Coal Geol. 190, 205–217.

Chen, L., Jiang, Z., Liu, Q., Jiang, S., Liu, K., Tan, J., Gao, F., 2019. Mechanism of shale gas occurrence: insights from comparative study on pore structures of marine and lacustrine shales. Mar. Pet. Geol. 104, 200–216.

Chengzao, J., 2017. Breakthrough and significance of unconventional oil and gas to classical petroleum geology theory. Pet. Explor. Dev. 44 (1), 1–10.

Cienfuegos, P., Loredo, J., 2010. Coalbed methane resources assessment in Asturias (Spain). Int. J. Coal Geol. 83 (4), 366–376.

Clarkson, C., Solano, N., Bustin, R., Bustin, A., Chalmers, G., He, L., Melnichenko, Y., Radliński, A., Blach, T., 2013. Pore structure characterization of north American shale gas reservoirs using USANS/SANS, gas adsorption, and mercury intrusion. Fuel 103, 606–616.

Colmenero, J., Suárez-Ruiz, I., Fernández-Suárez, J., Barba, P., Llorens, T., 2008. Genesis and rank distribution of upper carboniferous coal basins in the Cantabrian Mountains, northern Spain. Int. J. Coal Geol. 76 (3), 187–204.

Curtis, M., Sondergeld, C., Ambrose, R., Rai, C., 2012. Microstructural investigation of gas shales in two and three dimensions using nanometer-scale resolution imaging. AAPG Bulletin 96 (4), 665–677.

Desbois, G., Enzmann, F., Urai, J., Baerle, C., Kukla, P., Konstanty, J., 2010. Imaging pore space in tight gas sandstone reservoir: insights from broad ion beam cross-sectioning. EPJ Web Conf. 6, 22022.

EIA, 2013. World Shale Resource Assessments. [online] U. S Energy Infomation Administration. Available at: https://www.eia.gov/analysis/studies/worldshalegas/>. (Accessed 1 Jun 2020).

European Commission, 2016. Overview of the current status and development of shale gas and shale oil in Europe. In: European Unconventional Oil and Gas Assessment (EUOGA), pp. 90–94.

Felder, M., Harms, H.-J., Liebig, V., Hottenrott, M., Rolf, C., Wonik, T., 2001. Lithologische Beschreibungen der Forschungsbohrungen Groß-Zimmern, Prinz von Hessen und Offenthal sowie zweier Lagerstättenbohrungen bei Eppertshausen (Sprendlinger Horst, Eozän, Messel-Formation, Süd Hessen). Geol. Jb. Hess. 128, 29–82.

Frings, K., Lutz, R., de Wall, H., Warr, L., 2004. Coalification history of the Stephanian Ciñera-Matallana pull-apart basin, NW Spain: combining anisotropy of vitrinite reflectance and thermal modelling. Int. J. Earth Sci. 93 (1), 92–106.

Gomez, J., 2019. Design, Set-up, and Testing of Matrix Apparatus. Msc, Texas A&M University.
González, A., Molinero, R., Crespo, J., Arribas, J., Nita, R., 2017. Proyecto de exploracion de hidrocarburos no convencionales en Espana (HCNC). Ministerio de Economia, Industria Y Competitividad.
Goral, J., Miskovic, I., Gelb, J., 2015. Pore network investigation in Marcellus shale rock matrix. In: society of petroleum engineers. In: SPE Asia Pacific Unconventional Resources Conference and Exhibitio: Society of Petroleum Engineers, pp. 1–8.
Guo, C., Wei, M., Liu, H., 2018. Study of gas production from shale reservoirs with multistage hydraulic fracturing horizontal well considering multiple transport mechanisms. PLoS One 13 (1). https://doi.org/10.1371/journal.pone.0188480.
Heward, A., 1978. Alluvial fan and lacustrine sediments from the Stephanian a and B (La Magdalena, Ciñera-Matallana and Sabero) coalfields, northern Spain. Sedimentology 25 (4), 451–488.
Hofmann, P., Duckensell, M., Chpitsglous, A., Schwark, L., 2005. Geochemical and organic petrological characterization of the organic matter of lacustrine Eocene oil shales (Prinz von Hessen, Germany): reconstruction of the depositional environment. J. Paleolimnol. 33 (2), 155–168.
Hu, Q., Ewing, R., Rowe, H., 2015. Low nanopore connectivity limits gas production in Barnett formation. J. Geophys. Res. Solid Earth 120 (12), 8073–8087.
Javadpour, F., 2009. Nanopores and apparent permeability of gas flow in mudrocks (shales and siltstone). J. Can. Pet. Technol. 48 (08), 16–21.
Jiang, P., 2013. Pore Morphometrics and Thermal Evolution of Organic-Matter Microporosity, Colorado Group, Western Canada Foreland Basin. Msc, Western University.
Jones, B., Manning, D., 1994. Comparison of geochemical indices used for the interpretation of palaeoredox conditions in ancient mudstones. Chem. Geol. 111 (14), 111–129.
King, H., Eberle, A., Walters, C., Kliewer, C., Ertas, D., Huynh, C., 2015. Pore architecture and connectivity in gas shale. Energy Fuel 29 (3), 1375–1390.
Lemos de Sousa, M., Pinheiro, H., 1996. Prospecting for Coalbed Methane: Preliminary Investigation into the Possibilities in Spain. 90 (1), Real Academia de Ciencias Exactas, Físicas y Naturales, pp. 63–74.
Leneuf, N., 1973. Observations stereoscopiques sur les figures de corrosion du quartz dans certaines formations superficielles. Cah. ORSTOM, Serie Pid. 11 (1), 43–51.
Loucks, R., Reed, R., Ruppel, S., Hammes, U., 2012. Spectrum of pore types and networks in mudrocks and a descriptive classification for matrix-related mudrock pores. AAPG Bulletin 96 (6), 1071–1098.
Lüning, S., Kolonic, S., 2003. Uranium spectral gamma-ray response as a proxy for organic richness in black shales: applicability and limitations. J. Pet. Geol. 26 (2), 153–174.
Ma, L., Slater, T., Dowey, P., Yue, S., Rutter, E., Taylor, K., Lee, P., 2018. Hierarchical integration of porosity in shales. Sci. Rep. 8 (1), 11683. https://doi.org/10.1038/s41598-018-30153-x.
Mackenzie, W., 2018. Cantabrian Basin Shale Gas Unconventional Concept Play. Wood Mackenzie, Edinburgh.
Mastalerz, M., Drobniak, A., Hower, J., O'Keefe, J., 2010. Coal and Peat Fires: A Global Perspective, first ed. Elsevier, pp. 47–62.
McGlade, C., Speirs, J., Sorrell, S., 2013. Methods of estimating shale gas resources – comparison, evaluation and implications. Energy 59, 116–125.
Middleton, R., Carey, J., Currier, R., Hyman, J., Kang, Q., Karra, S., Jiménez-Martínez, J., Porter, M., Viswanathan, H., 2015. Shale gas and non-aqueous fracturing fluids: opportunities and challenges for supercritical CO2. Appl. Energy 147, 500–509.

Middleton, R., Gupta, R., Hyman, J., Viswanathan, H., 2017. The shale gas revolution: barriers, sustainability, and emerging opportunities. Appl. Energy 199, 88–95.

Nie, H., Jin, Z., Zhang, J., 2018. Characteristics of three organic matter pore types in the Wufeng-Longmaxi shale of the Sichuan Basin, Southwest China. Sci. Rep. 8 (1), 7014. https://doi.org/10.1038/s41598-018-25104-5.

Pak, T., Butler, I., Geiger, S., van Dijke, M., Jiang, Z., Surmas, R., 2016. Multiscale pore network representation of heterogeneous carbonate rocks. Water Resour. Res. 52 (7), 5433–5441.

Pickel, W., Kus, J., Flores, D., Kalaitzidis, S., Christanis, K., Cardott, B., Misz-Kennan, M., Rodrigues, S., Hentschel, A., Hamor-Vido, M., Crosdale, P., Wagner, N., 2017. Classification of liptinite - ICCP system 1994. Int. J. Coal Geol. 169, 40–61.

Rider, M., 1996. The Geological Interpretation of Well Logs, second ed. Rider-French consulting LTD, Sutherland, pp. 67–86.

Ross, D., Bustin, R., 2009. The importance of shale composition and pore structure upon gas storage potential of shale gas reservoirs. Mar. Pet. Geol. 26 (6), 916–927.

Scotchman, I., 2016. Shale gas and fracking: exploration for unconventional hydrocarbons. Proc. Geol. Assoc. 127 (5), 535–551.

Shen, W., Li, X., Xu, Y., Sun, Y., Huang, W., 2017. Gas flow behavior of nanoscale pores in shale gas reservoirs. Energies 10 (6), 751.

Singh, H., Myong, R., 2018. Critical review of fluid flow physics at micro- to nano-scale porous media applications in the energy sector. Adv. Mater. Sci. Eng. 2018, 1–31.

Speight, J., 2019. Natural gas, second ed. Gulf Professional Publishing, Oxford, pp. 59–98.

Swami, V., Clarkson, C.R., Settari, A., 2012. Non-Darcy flow in shale nanopores: do we have a final answer? In: SPE Canadian Unconventional Resources Conference. https://doi.org/10.2118/162665-MS.

Wagner, R., 1971. The stratigraphy and structure of the Ciñera-Matallana coalfield (Prov Leon N W Spain). In: Wagner, R. (Ed.), Trabajos de Geologia, fourth ed. Universidad de Oviedo, Oviedo, pp. 385–429.

Wang, S., 2018. Shale gas exploitation: status, problems and prospect. Nat. Gas Ind. B 5 (1), 60–74.

Yang, Y., Bao, F., 2017. Characteristics of shale nanopore system and its internal gas flow: a case study of the lower Silurian Longmaxi formation shale from Sichuan Basin, China. J. Nat. Gas Geosci. 2 (5–6), 303–311.

Zendehboudi, S., Bahadori, A., 2016. Shale Oil and Gas Handbook. Gulf Professional Publishing, Oxford, pp. 1–26.

Zhang, P., Hu, L., Meegoda, J., Gao, S., 2015. Micro/nano-pore network analysis of gas flow in shale matrix. Sci. Rep. 5 (1), 13501. https://doi.org/10.1038/srep13501.

Zhang, L., Chen, Z., Zhao, Y., 2019. Well Production Performance Analysis for Shale Gas Reservoirs. Elsevier Science.

Zou, C., 2013. Unconventional Petroleum Geology, first ed. Elsevier, pp. 149–190.

FSC
www.fsc.org
MIX
Papier | Fördert
gute Waldnutzung
FSC® C083411

Kolibri 360 GmbH
Ferdinand-Jühlke-Straße 7
99095 Erfurt, Deutschland
produktsicherheit@kolibri360.de